Gene Transfer

DELIVERY AND EXPRESSION OF DNA AND RNA

A LABORATORY MANUAL

ALSO FROM COLD SPRING HARBOR LABORATORY PRESS

RELATED LABORATORY MANUALS

Molecular Cloning: A Laboratory Manual

Live Cell Imaging: A Laboratory Manual

Manipulating the Mouse Embryo: A Laboratory Manual, Third Edition

Imaging in Neuroscience and Development: A Laboratory Manual

RNAi: A Guide to Gene Silencing

OTHER RELATED TITLES

Lab Math: A Handbook of Measurements, Calculations, and Other Quantitative Skills for Use at the Bench

Lab Ref Volume 2: A Handbook of Recipes, Reagents, and Other Reference Tools for Use at the Bench

Lab Ref: A Handbook of Recipes, Reagents, and Other Reference Tools for Use at the Bench

Lab Dynamics: Management Skills for Scientists

At the Bench: A Laboratory Navigator

At the Helm: A Laboratory Navigator

Gene Transfer

DELIVERY AND EXPRESSION OF DNA AND RNA

A LABORATORY MANUAL

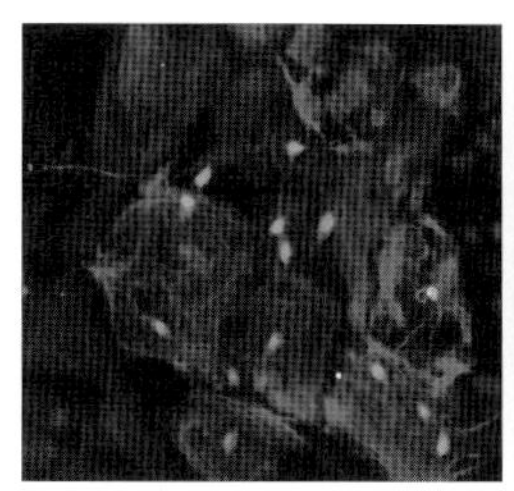 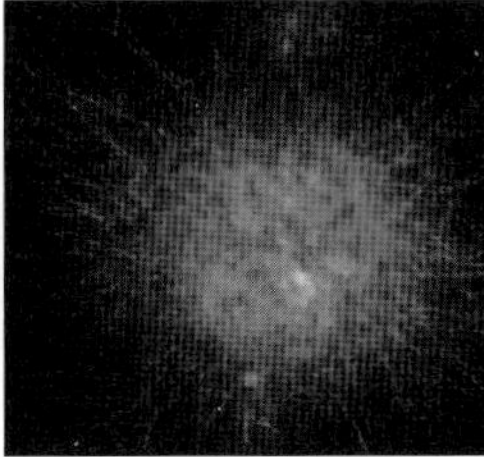 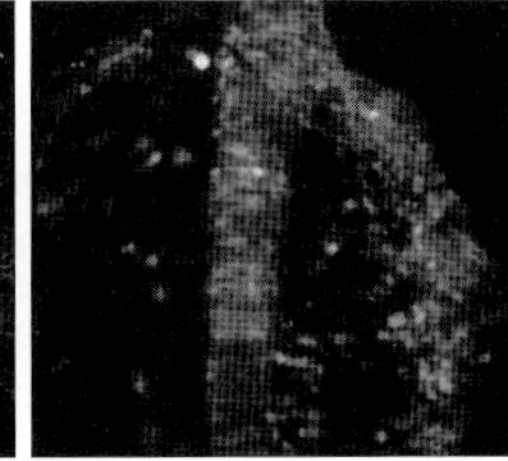 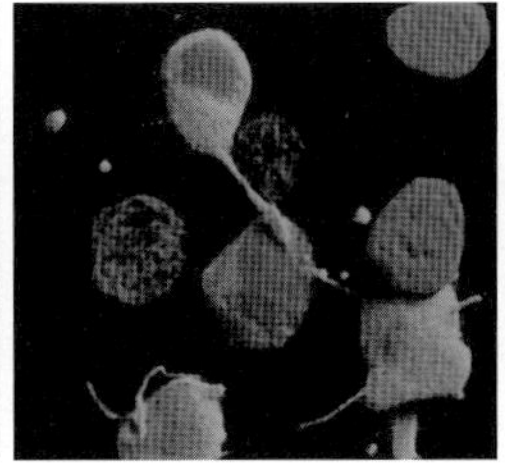 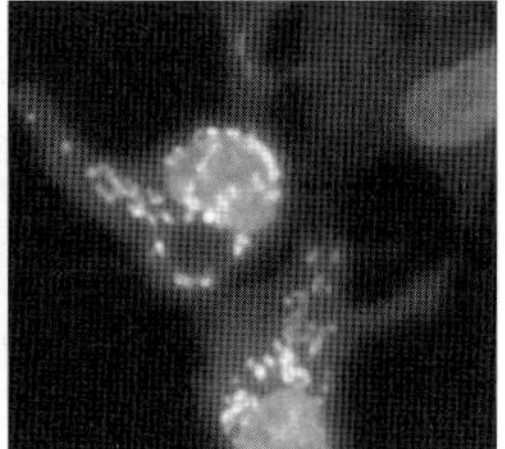

EDITED BY

Theodore Friedmann

University of California, San Diego
School of Medicine

John Rossi

Beckman Research Institute
City of Hope National Medical Center

http://www.cshprotocols.org

COLD SPRING HARBOR LABORATORY PRESS
Cold Spring Harbor, New York

Gene Transfer: Delivery and Expression of DNA and RNA

A Laboratory Manual

Printed in the United States of America

Publisher	John Inglis
Acquisition Editor	David Crotty
Development Director	Jan Argentine
Managing Editor	Kaaren Janssen
Developmental Editors	Tracy Kuhlman, Irene Pech, and Martin Winer
Project Coordinator	Inez Sialiano
Production Editor	Rena Steuer
Copy Editor	Dotty Brown
Desktop Editor	Lauren Heller
Production Manager	Denise Weiss
Cover Design	Ed Atkeson

Front cover artwork: (*Main image and box 2 from top*) Immunostaining of the neurospheres formed by rat primary embryonic spinal cord cultures transduced with HIV1-based vector expressing EGFP from neuron-specific synapsin promoter (HIV1-Syn-EGFP); (*box 5*) immunostaining of rat primary embryonic spinal cord culture transduced with HIV1-Syn-EGFP vector. Photographs by Drs. Atsushi Miyanohara and Martin Marsala, University of California, San Diego. (*Boxes 1, 3,* and *4*) Multicolor labeled cells within the living chick embryo, shown here by multispectral confocal imaging, were fluorescently labeled by injection and electroporation of a cocktail of constructs coding for fluorescent proteins targeted to distinct regions of the cell. The technique allows for more accurate identification of individual cells. Images collected by Jessica Teddy and Paul Kulesa, Stowers Institute for Medical Research. Fluorescent protein constructs were generated in collaboration with Rusty Lansford at the California Institute of Technology.

Library of Congress Cataloging-in-Publication Data

Gene transfer : delivery and expression of DNA and RNA / edited by Theodore Friedmann, John Rossi.
p. ; cm.
Includes bibliographical references and index.
ISBN-13: 978-0-87969-764-8 (cloth : alk. paper)
ISBN-13: 978-0-87969-765-5 (pbk. : alk. paper)
1. Genetic transformation--Laboratory manuals. 2. Genetic vectors--Laboratory manuals. 3. Genetic engineering--Laboratory manuals. I. Friedmann, Theodore, 1935- II. Rossi, John J.
[DNLM: 1. Gene Transfer Techniques. 2. Gene Expression Regulation. 3. Genetic Vectors. 4. Viruses--genetics. QU 450 G326 2006]
QH448.4.G435 2006
572.8'619--dc22

2006020456

10 9 8 7 6 5 4 3 2 1

All Cold Spring Harbor Laboratory Press publications may be ordered directly from Cold Spring Harbor Laboratory Press, 500 Sunnyside Blvd., Woodbury, New York 11797-2924. Phone: 1-800-843-4388 in Continental U.S. and Canada. All other locations: (516) 422-4100. FAX: (516) 422-4095. E-mail: cshpress@cshl.edu/. For a complete catalog of all Cold Spring Harbor Laboratory Press publications, visit our World Wide Web site http://www.cshlpress.com/.

To those who have been most influential in leading me into and along the path of gene transfer: my wife Ingrid for her love and patience; Eric, Carl, Tai, and Oscar for proving the principle of gene transfer; and my colleague Renato Dulbecco for friendship and inspiration.

—TF

To my wife Mary Jane and son Daniel for their love, patience, and understanding, and to the members of my laboratory and staff for their dedication and enthusiasm for science.

—JR

Contents

NONVIRAL TECHNIQUES AND VECTORS

Preface

IN 1890, THE GERMAN BACTERIOLOGIST AND PHYSICIAN ROBERT KOCH identified four criteria, formally known as Koch's Postulates, that should be satisfied in order to prove the infectious etiology of a disease. We now recognize certain exceptions, yet these criteria continue to define the essential components of a rigorous proof that connects a microorganism with a human disease. Koch argued that a suspected organism must be identified in every case of a disease and must be cultured in pure form from all affected patients. The disease must be recapitulated when the purified organism is introduced into an uninfected organism and, finally, the suspected organism must be recovered from the infected host organism. Until the mid 20th century, genetics was largely a descriptive science. Koch's approach has converted the study of human disease from the merely descriptive to a much more rigorous experimental science.

With the explosive growth of medical and molecular genetics during the past century, a similar set of criteria has evolved for confirming the connection between a genetic mechanism and a normal or pathological human trait. In his famous 1908 *Croonian Lectures to the Royal Society in England*, Archibald Garrod introduced the concept of chemical pathology through his demonstration that errors in genetic information were responsible for human genetic diseases (called "inborn errors of metabolism"). And, beginning in 1910, Thomas Hunt Morgan, with his parade of remarkable colleagues and associates in the famous "fly room" at Columbia University, established the concepts of genetic linkage and the relationship between genetic factors and the phenotypic features of *Drosophila*. These systematic studies revealed just how genetic information defines the properties and development of the traits of living organisms. It was not until the 1940s, however, that specific molecular underpinnings of the genetic inheritance of normal and pathological traits came to be identified. In 1941, George Beadle and Edward Tatum showed that radiation-induced mutations in *Neurospora crassa* were expressed as specific metabolic aberrations. The findings that genes express their normal and mutated functions through the production of enzymes conferred a formal mechanistic understanding to Garrod's intuitive concept of the inborn errors of metabolism. In a series of remarkable experiments in 1944, Oswald Avery, Maclyn McCarty, and Colin McLeod showed that DNA, transferred from one strain of *Pneumococci* into another strain, transformed the properties of the recipient strain into those of the donor strain. This observation proved that it was indeed DNA that determines all of the heritable properties of a microorganism. Further proof that DNA represents the repository of genetic information came from the series of bacteriophage-mediated gene transfer experiments, conducted in 1952 by Alfred Hershey and Martha Chase, showing that the genetic determinant of the phage was also DNA.

The discovery by James Watson and Francis Crick in 1953 of the tertiary structure of the DNA molecule suddenly clarified just how DNA codes for and stores genetic information, and suggested the mechanism for its replication. Other discoveries followed quickly. Within the next two decades, Frederick Sanger demonstrated that proteins have specific amino acid sequences, and Vernon Ingram and his colleagues identified the genetic mistakes responsible for naturally occurring human disease, revealing how these changes wrought havoc through the production of proteins with abnormal amino acid sequences. The ensuing insights into the flow of genetic information during the "golden era" of molecular biology in the 1950s through the 1970s was succeeded over the following decades by the surprising and remarkable success of the "genome projects." These discoveries have, collectively, illuminated the genetic mechanisms responsible for normal and abnormal functions of all living systems.

In the spirit of Koch's postulates, the concept underlying current modern molecular genetics is that the presumed mechanisms of normal and disease-related gene expression require formal functional proof. In the final analysis, rigorous proof of a suspected genetic mechanism requires an experimental recapitulation of the connection between genotype and phenotype. In many cases, unequivocal proof can be provided by the introduction of specific genetic changes into an organism or by blocking functional expression of genetic information, followed by phenotypic analyses. These studies rely on the development of methods for introducing exogenous genetic information into cells or even into whole organisms. The power to modify the genotype of an organism, either transiently or permanently, has opened vast new opportunities to expand our understanding of immensely complex genetic mechanisms and, in a more pragmatic sense, even to imagine the correction of genetic aberrations in the context of disease therapy—"gene therapy."

This manual presents the features of many methods for introducing functional foreign genetic information into mammalian cells and even into whole organisms. The chapters provide background information as well as detailed protocols that describe a variety of gene delivery vectors and techniques, both viral and nonviral. The protocols are presented in a stepwise, "cookbook" style, thus making it feasible for any investigator to take advantage of the gene delivery/expression system. In addition to gene delivery, there are also protocols for regulated gene expression of the genetic information under study. For any given experimental model or gene therapy application there exist multiple methods for introduction and expression of the genetic material. To guide the reader in selecting the appropriate system, we have included in each chapter descriptions of the advantages and disadvantages of the particular approach.

Obviously, in such a rapidly evolving field, new methods will continue to emerge to augment, and even to replace, some existing approaches. Nevertheless, the techniques included in this volume constitute much of the current methodology for gene transfer into mammalian cells and tissues.

We thank those at Cold Spring Harbor Laboratory Press who made this project possible—John Inglis, Executive Director of the Press, David Crotty, Commissioning Editor, and Jan Argentine, Editorial Development Manager. We appreciate the devoted efforts of the editorial staff at the Press—Kaaren Janssen, for originating and overseeing the project, Inez Sialiano, and Cher Mattes, as well as developmental editors Irene Pech, Martin Winer, and Tracy Kuhlman. We also thank Denise Weiss, Rena Steuer, Dotty Brown, and Lauren Heller for their expertise in the production of this work.

We hope that this manual will serve as a useful and convenient standard laboratory manual for any investigator wishing to use gene delivery as an investigative or therapeutic tool.

THEODORE FRIEDMANN
JOHN ROSSI

1 Introduction

Theodore Friedmann* and John Rossi†

*University of California, San Diego; †Beckman Research Institute of the City of Hope, Duarte, California 92093-0634

The techniques for introducing foreign genetic material into mammalian cells can conveniently be thought of as virus-based and nonviral methods, and it is this general classification that we use in this manual. Both approaches have advantages and shortcomings and find important uses in the laboratory and even now in the clinic. No single vector type is suitable for all gene delivery applications, and despite the fact that favorite vector systems continually gain and lose popularity, most of them have useful and important roles in either broad or niche applications. Of course, nature beat scientists to the punch and designed and evolved agents, i.e., viruses, to deliver foreign genetic information into prokaryotic and eukaryotic cells. It is the job description of viruses to introduce their genomes into cells in the most efficient manner possible, generally with the selfish evolutionary intent of propagating themselves but with the frequent secondary effect of causing damage to their hosts.

Although the phenomenon of viral disease has long been known, it could not be understood at a mechanistic level until the birth of molecular genetics in the latter half of the twentieth century. Nevertheless, even without such detailed knowledge, it was possible to intuit the role of such agents in human disease and to use this primitive understanding to develop approaches to control of viral disease in humans. When the English physician Edward Jenner deliberately infected a boy with cowpox material in 1796 to protect against smallpox, he was acting on his observation that milkmaids who were infected with cowpox seldom contracted smallpox, although of course he had no idea of the involvement of viruses or the nature of the protection. Even so, we now understand that this was the result of gene transfer from the virus to the milkmaids.

A further molecular understanding of the mechanisms and effects of virus-based gene transfer came in the modern era from the demonstration by Renato Dulbecco and his colleagues in the mid 1960s (Sambrook et al. 1968) that the papovaviruses SV40 and polyomavirus induce permanent and even tumorigenic changes in infected cells by integrating copies of some of their genes into the genomes of infected cells. If native viruses have the capacity to act as vectors for the transfer of their own genes, might it be possible to design engineered viruses that could transfer other genes into mammalian cells without causing viral cytotoxicity? After the birth of recombinant DNA technology in 1973, that possibility became a reality in the early 1980s with the construction of the initial retroviral vectors (Shimotohno and Temin 1981; Wei et al. 1981; Tabin et al. 1982).

Viral vectors are certainly efficient, but their frequent cytotoxic, immunogenic and biohazardous properties represent serious potential flaws in some gene-transfer applications. This underscores the need to develop chemical or synthetic complexes that approach the same levels of gene-transfer efficiency as displayed by viral vectors but that avoid their disadvantages. For many decades, investigators have struggled with exceedingly inefficient methods for introducing foreign genes into mammalian cells, using cations or polyamines to facilitate the passage of negatively charged nucleic acids through the negatively charged cell surface. The birth of reasonably efficient nonviral gene-transfer methods received a major boost in 1973 when Frank Graham and Alexander van der Eb described the markedly enhanced cellular uptake and infectivity of adenoviral DNA as a coprecipitate with calcium phosphate (Graham and van der Eb 1973).

Despite the more recent development of the many other nonviral methods reviewed in this volume, the calcium phosphate method remains one of the bedrock methods for mammalian cell gene transfer. Overall, however, even the better nonviral methods suffer from a markedly lower efficiency of gene transfer and reduced stability of transgene expression compared with the virus-based methods. This is due at least in part to the fact that viruses are able to wrap the incoming genes into a tight package that recognizes and attaches to specific sites on the surface of the target cell, permitting entry of the vector and its genetic payload into cells through pathways that protect the incoming material from cellular degradative processes while directing the transgenes to the appropriate cellular site.

Many of the methods described in this volume, particularly the plethora of nonviral techniques, are very early in their development and, in the long run, may have only limited and niche applications. To become truly useful, future gene-transfer methods should provide important advantages over existing methods. Such advantages, especially for in vivo gene transfer, should eventually include an ability to select a cell-specific or tissue-specific target for the gene transfer, a capability that is still sorely lacking. Furthermore, in the case of integrating vectors, important impediments to safe and truly effective in vivo gene delivery would be eased by the development of methods to specify the genomic site of integration, thus avoiding the cellular damage that would inevitably emerge as a result of insertional mutagenesis. In addition, methods to produce a specific sequence change in a target sequence instead of a promiscuous addition of genetic information would allow a hit-and-run method for gene modification that produces only the desired genetic change without footprints of the vectors themselves. The beginnings of such a technology have recently emerged through the development of vectors containing zinc finger elements that define the sites of interaction of a transgene with the target genome (Porteus and Carroll 2005). Finally, methods for producing genetic changes in mammalian cells cannot be fully faithful to the native state unless they contain the full range of the regulatory elements that define the native expression of the target locus. This often requires insertion of long stretches of genetic material containing regulatory information. Such a requirement alone will often dictate which vector system is most well suited for the delivery of this genetic information, since vectors vary widely in their carrying capacity for genetic information.

The collection of protocols in this volume offers the widest possible range of methods for gene delivery into mammalian cells. There are also protocols for regulated gene expression that accompany the gene delivery methods. It is our intent to provide the scientific community with a user-friendly guide to gene delivery and expression that includes both the tried-and-true methods and some exciting new technologies. We believe these protocols will stand the test of time in their usefulness.

REFERENCES

Graham F.L. and van der Eb A. 1973. A new technique for the assay of infectivity of human adenovirus 5 DNA. *Virology* **52:** 456–467.

Porteus M.H. and Carroll D. 2005. Gene targeting using zinc finger nucleases. *Nat. Biotechnol.* **23:** 967–973.

Sambrook J., Westphal H., Srinivasan P.R., and Dulbecco R. 1968. The integrated state of viral DNA in SV40-transformed cells. *Proc. Natl. Acad. Sci.* **60:** 1288–1295.

Shimotohno K. and Temin H.M. 1981. Formation of infectious progeny virus after insertion of he herpes simplex thymidine kinase gene into DNA of an avian retrovirus. *Cell* **26:** 67–77.

Tabin C.J., Hoffmann J.W., Goff S.P., and Weinberg R.A. 1982. Adaptation of a retrovirus as a eukaryotic vector transmitting the herpes simplex thymidine kinase gene. *Mol. Cell. Biol.* **2:** 426–436.

Wei C., Gibson M., Spear P.G., and Scolnick E.M. 1981. Construction and isolation of a transmissible retrovirus containing the src gene from Harvey murine sarcoma virus and the thymidine kinase gene from herpes simplex virus type 1. *J. Virol.* **39:** 935–944.

2 Retroviral Vectors

Kenneth Cornetta,* Karen E. Pollok,[†] and A. Dusty Miller[‡]

*Department of Medical and Molecular Genetics, Indiana University School of Medicine, Indianapolis, Indiana 46202; [†]Department of Pediatrics, Section of Pediatric Hematology/Oncology, Indiana University School of Medicine, Indianapolis, Indiana 46202; [‡]Divisions of Human Biology and Basic Sciences, Fred Hutchinson Cancer Research Center, Seattle, Washington 98109

ABSTRACT

This chapter focuses on retroviral vectors derived from the γ-retrovirus genus. These were the first retroviral vectors to be developed, and they have been called oncoretroviral vectors or simple retroviral vectors because of their derivation from oncogenic retroviruses having a simple *gag-pol-env* genome structure. Later additions to the retroviral vector family include the lentiviral and foamy viral vectors derived from more complex retroviruses that contain multiple accessory genes in addition to the standard *gag-pol-env* genes.

Retroviral vectors have a number of advantages for laboratory research and for clinical gene therapy applications. Unlike many viruses, retroviruses efficiently integrate into the genomes of infected cells. Transgenes carried by retroviral vectors also integrate into target cells such that the gene is copied to all progeny of the cell, making these vectors ideal for altering stem cells, progenitor cells, or other cells that are expected to expand in great number in vivo (e.g., T cells responding to an immune response). The efficiency of retroviral gene transfer is significantly greater than that of nonviral gene transfer. Furthermore, vector production is easily performed in most research laboratories and is amenable to large-scale production, facilitating its use in clinical settings.

Retroviral vectors do have several disadvantages for gene therapy applications. First, they are generally not useful for systemic administration because of their inactivation by protein and cellular components of human blood and typically have been used to transduce cells ex vivo. Second, they require cell division for efficient integration. Third, integration has been associated with oncogene activation after transduction of hematopoietic cells, a rare and complex process that must be considered when calculating the risk/benefit ratio for this method of gene transfer. Despite these limitations, retroviral vectors are a well-defined system with many novel reagents resulting from the extensive experience with them during the past 20 years. They remain an attractive system for transducing target cells where integration of transgene sequences is required.

INTRODUCTION

Retroviruses as Gene Delivery Vehicles

Retroviruses are attractive gene delivery vehicles due to certain unique aspects of their life cycle as well as their ability to efficiently integrate into target cell DNA. The viral genome is flanked by two regulatory regions, called long terminal repeats (LTRs), that contain promoter and enhancer functions and are required for integration. A packaging (ψ) sequence greatly facilitates the uptake of viral RNA into virions. There are three viral gene regions: *gag* which encodes the viral structural proteins; *pol* which encodes enzymatic proteins, most notably reverse transcriptase and integrase; and *env* which generates an envelope glycoprotein that spans the lipid coat of the virus and mediates infection by targeting specific receptors on target cells. Within the viral capsid are two copies of the viral RNA genome along with reverse transcriptase and integrase. Carrying the enzymatic genes allows the virus to infect a cell, make DNA copies from the RNA template, and integrate the DNA without expressing any of the viral genes. This allows for the deletion of the viral protein-coding genes (*gag, pol,* and *env*) and substitution of this region with exogenous genes of interest (Fig. 1). This renders the vector replication-defective and presents a technical problem in packaging vector RNA into virions. This challenge can be met using transient transfection of plasmids expressing the viral genes along with a plasmid

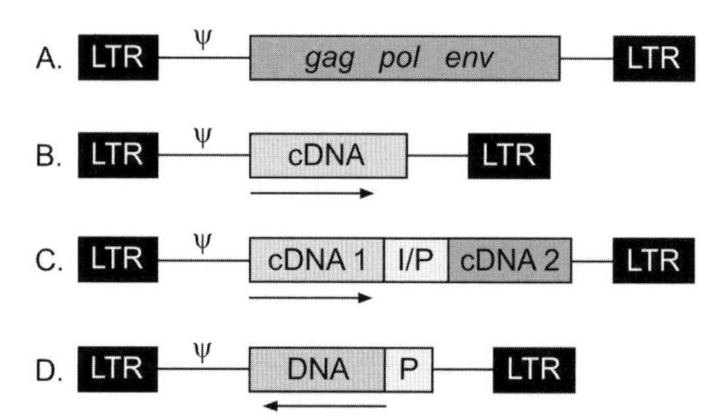

FIGURE 1. Retroviral vector design. (*A*) Schematic representation of the Moloney murine leukemia. The three viral gene regions, *gag, pol,* and *env* are flanked by the LTRs, which contain promoter and enhancer functions. The psi (ψ) sequence is required for efficient packaging of viral RNA into virions. (*B*) A vector construct retains the LTRs and ψ sequence with deletion of the majority of the viral gene region. The promoter in the 5′ LTR is used to express the cDNA of interest. (*C*) More than one gene product or sequence can be expressed by use of an internal ribosome entry site (IRES) sequence (I) or introduction of a second promoter (P). (*D*) When the sequence to be expressed contains introns or other sequences that may interfere with production of a full-length vector transcript, a gene can be created and inserted in the opposite orientation. This allows the full-length transcript to be incorporated into virions, and the gene of interest will be expressed after integration into target cells. Arrows indicate the orientation of the gene with respect to the 5′ LTR.

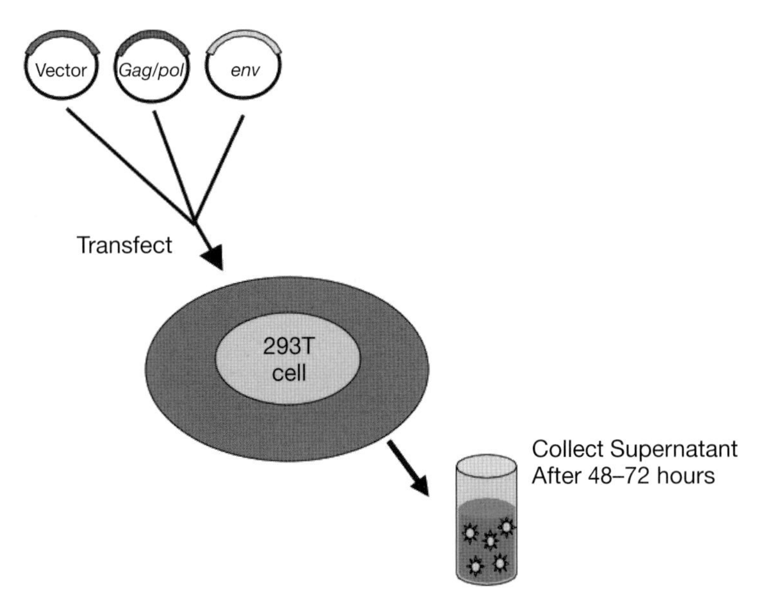

FIGURE 2. Vector production by transient transfection. Plasmids expressing the vector, the viral *gag* and *pol* genes, and the viral envelope are introduced into a cell that transfects with high efficiency (e.g., HEK 293T cells) using calcium phosphate, lipofection, or electroporation. Vector supernatant is harvested after 48–72 hours and can be used immediately or frozen at –70°C for later use.

expressing the vector genome (Fig. 2) or, more commonly, through the use of retroviral packaging cell lines as discussed in detail below.

Considerations in Retroviral Vector Design

A number of important issues must be considered when constructing retroviral vectors. Simplistically, the vector can be thought of as containing two major components: the vector backbone and the transgene cassette. As depicted in Figure 1, the vector backbone contains those sequences derived from the parent retroviruses. The majority of initial studies utilized vectors derived from the Moloney murine leukemia virus (Mo-MLV). The promoter and enhancer regions within the Mo-MLV LTR drive expression in most cell lines and in differentiated primary cells. A limitation of this LTR is poor expression in a variety of primitive cells such as preimplantation embryo cells, embryonic stem cells, and primitive hematopoietic progenitor cells (Jahner et al. 1982; Challita and Kohn 1994). It has subsequently been shown that the Mo-MLV LTR and primer binding site contains at least four silencer elements. The mechanisms by which vectors are silenced is complex and not completely understood but has prompted a variety of investigators to develop novel vectors using components of other viruses or mutation of silencer sequences (for review, see Pannell and Ellis 2001). During the past 15 years, a significant number of vector backbones have been generated with LTRs derived from alternative retroviruses or with engineered LTRs that have been shown to improve expression in specific cell types (e.g., hematopoietic cells of myeloid lineage) and may be less likely to undergo in vivo silencing due to methylation or other cellular mechanisms (for review, see Hawley 2001). Investigators seeking to express vectors in primitive cell types now have a variety of novel backbones that decrease, but do not completely eliminate, retroviral vector silencing.

Once a vector backbone has been selected, the transgene cassette must be inserted. The most simplistic design utilizes the LTR promoter to express the transgene. Generally, the transgene sequence lacks introns to prevent splicing during RNA processing. In situations where intron sequences are important for transgene expression, or where tissue-specific promoters are preferred to the nonspecific expression associated with the viral LTR, the transgene cassette can be placed in reverse orientation (Fig. 1).

Most recently, the documented ability of retroviral vectors to cause malignancy by insertional mutagenesis has led to further considerations of vector design. Insertional mutagenesis occurs when retroviral regulatory sequences (most commonly the enhancer) integrate near susceptible oncogenes, leading to overexpression of the oncogene. Insertional mutagenesis is believed to require alterations in multiple oncogenes and/or tumor suppressor genes. T-cell lymphomas that arise from infection with the Mo-MLV have multiple viral integrations per cell. The ability of replication-competent γ-retroviruses to cause malignancy has been shown in nonhuman primates (Donahue et al. 1992; Cornetta et al. 1993), but the risk of insertional mutagenesis with a single integration (as typically occurs with retroviral vectors) was believed to be very low (Cornetta 1992; Li et al. 2002). Clinical trials using retroviral vectors had not reported insertional mutagenesis until recently, when a single vector integration near the *LMO2* gene was associated with leukemia in at least 2 of 11 subjects participating in a gene therapy trial for X-linked severe combined immunodeficiency disease (SCID) (Cavazzana-Calvo et al. 2000; Hacein-Bey-Abina et al. 2002, 2003a,b). The reasons these children have developed leukemia is complex, but preliminary evidence suggests that the transgene in this study (the common cytokine receptor γ-chain) is also acting as an oncogene (Berns 2004; Dave et al. 2004). Emerging data suggest that different LTRs may have different potentials for causing malignancy, and new vectors are being developed that eliminate the enhancer sequence using self-inactivating vector design (Kraunus et al. 2004). In theory, these vectors should provide a higher safety profile, but they do require other regulatory regions to drive transgene expression. These concerns have also led to evaluation of insulator sequences, matrix attachment regions, and locus control regions. Such sequences have the potential to prevent undesired activation of surrounding genetic sequences and may also protect the transgene cassette from silencing due to positional effects related to the site of integration.

Considerations in Retroviral Packaging Cell Line Choice

Many retroviral packaging cell lines have been made since the first such cell lines were described (Mann et al. 1983; Watanabe and Temin 1983). The key considerations are (1) the range of cell types that can be transduced, which is primarily determined by the Env protein produced by the cells; (2) the propensity of the cells to generate replication-competent virus (also called helper virus), which was a problem with early packaging cell lines but has been largely resolved with newer designs; (3) the susceptibility of vector produced by the packaging cells to inactivation by serum from humans, which is sometimes important for gene therapy applications; and (4) copackaging of endogenous retroviral sequences into virions, especially from packaging cells derived from mouse cells, which is of concern for gene therapy applications.

Table 1 provides a list of some commonly available packaging cells. For standard laboratory usage, a typical choice is a packaging line that produces vectors which can transduce a broad range of mammalian and avian cell lines, such as the PT67 cells. Alternatively, for transfer of oncogenes that represent a potential hazard, the investigator might choose a cell line that produces vectors capable of transducing only rodent cells but not human cells, such as the GP+E-86 cells. For genetic studies, the PG13 cell line has the useful property that vectors produced from these cells cannot reinfect the packaging cells, unlike many other packaging cells that undergo reinfection with time of cultivation.

For generation of stable vector-producing packaging cells, the best approach is to transfect one packaging cell line and use the virus from these cells to transduce a second packaging cell line. The second packaging cell line is then selected for the presence of the vector, and clonal isolates can be analyzed for the presence of an intact, single copy of the vector. Vector stocks generated from such cells are likely to be as genetically homogeneous as possible, since the genomic RNA in the vector virions all originates from a single stable integrated provirus. In contrast, vector produced from stably or transiently transfected packaging cells is more heterogeneous because vector RNA can

TABLE 1. Common Retrovirus Packaging Cell Lines

Env protein source	Cell line name	Parental cell line	Marker gene(s) expressed	Source(s)	Reference
Ecotropic MLV	PE501	NIH-3T3 mouse	*tk*	Authors	Miller and Rosman (1989)
	GP+E-86	NIH-3T3 mouse	*gpt*	ATCC CRL-9642	Markowitz et al. (1988)
Amphotropic MLV	PA317	NIH-3T3 mouse	*tk*	ATCC CRL-9078	Miller and Buttimore (1986)
	ProPak-A	293 human	*hpt, pac*	ATCC CRL-12479	Forestell et al. (1997)
	FLYA	HT-1080 human	*bsr, ble*	Authors	Cosset et al. (1995)
10A1 MLV	PT67	NIH-3T3 mouse	*tk, dhfr**	BD Biosciences, ATCC CRL-12284	Miller and Chen (1996)
GALV	PG13	NIH-3T3 mouse	*tk, dhfr**	ATCC CRL-10686	Miller et al. (1991)
RD114	FLYRD	HT-1080 human	*bsr, ble*	Authors	Cosset et al. (1995)

Marker genes used during construction of packaging cells include herpes simplex virus thymidine kinase (*tk*), a mutant methotrexate-resistant dihydrofolate reductase (*dhfr**), xanthine-guanine phosphoribosyltransferase (*gpt*), hygromycin phosphotransferase (*hpt*), puromycin-*N*-acetyl-transferase (*pac*), blasticidin S (*bsr*), and bleomycin and phleomycin resistance (*ble*). These genes are indicated because selection for vectors carrying a particular selectable marker cannot be performed in packaging cells that already express the marker

arise from multiple vector copies, some of which can be rearranged. In the protocol described below for generation of stable vector-producing cells, the example uses the PE501 packaging cells for transient transfection, followed by transduction of the PT67 packaging cells. It is important to use a packaging cell line for transfection that will result in virus capable of transducing the second packaging cells. The Env protein made in packaging cells can bind to and block the receptor(s) used by this Env for cell entry, but it will not block other cell surface receptors used by other Env proteins. In this example, the 10A1 Env protein made by PT67 cells binds to Pit1 and Pit2 receptors, but it does not bind to the Cat1 receptor used by virus produced by the GP+E-86 packaging cells. For a broader discussion of these virus classes, see Overbaugh et al. (2001).

Protocol 1

Vector Production by Transient Transfection

This protocol describes vector production by transient transfection. The production of retroviral vectors requires a full-length copy of the vector RNA to be incorporated into virions. This is accomplished by coexpressing vector RNA and the viral proteins required for virion formation from expression plasmids. To avoid generation of replication-competent virus, the viral genes are carried by separate plasmids. Generally, the *gag* and *pol* genes are on one plasmid and the viral envelope gene is on a second plasmid. The viral protein-coding regions can be expressed using various promoters to decrease homology and thereby decrease recombination. As these plasmids do not contain the ψ sequence, the viral genes are unlikely to be incorporated into virions.

MATERIALS

CAUTION: See Appendix for appropriate handling of materials marked with <!>.

Reagents

$CaCl_2$ (2.0 M) <!>
D-10 medium: Dulbecco's modified Eagle's medium with 10% fetal bovine serum (DMEM-10), 2 mM L-glutamine, 100 units/ml penicillin, and 100 μg/ml streptomycin <!>
HEK 293T cells (ATCC, CRL-11268)
HEPES-buffered saline (HBS). Mix 100 μl of 500 mM HEPES-NaOH (pH 7.1), 125 μl of 2.0 M NaCl, 10 μl of 150 mM Na_2HPO_4-NaH_2PO_4 (pH 7.0), and H_2O to make 1 ml. Prepare fresh.
Phosphate-buffered saline (PBS)
Plasmids
Packaging plasmid(s) containing retroviral *gag, pol,* and *env* genes
Vector plasmid
Prepare endotoxin-free plasmid stocks (e.g., by using QIAGEN Endotoxin-free Purification Kit) and determine plasmid DNA concentration.

Equipment

Biosafety cabinet
CO_2 incubator
Pipettes (sterile disposable)
Syringe filter (0.45 μm, pore size)
Tissue-culture centrifuge
Tissue-culture flasks (75 cm^2)

METHOD

1. Prepare a single-cell suspension of HEK 293T cells and seed 5 x 10^6 cells in a 75-cm^2 tissue-culture flask. Incubate overnight.

2. Remove flasks from the incubator and aspirate the medium. Add 12 ml of fresh D-10 medium to each flask and return flasks to the incubator.
3. Sterilize all reagents before use by filtration through 0.22-μm pore-size sterile filters. Prepare DNA for transfection by diluting plasmids in H_2O to a total volume of 876 μl. Add 124 μl of 2.0 M $CaCl_2$. Mix gently. Add DNA mix (1 ml) to 1 ml of HBS dropwise. A faint cloudiness should form. Let stand at room temperature for 30 minutes. Mix gently and add 1.5 ml of DNA/HBS suspension to the flask. Incubate overnight.
4. Aspirate and discard the medium from the flasks. Add 5 ml of PBS and then aspirate. Add 12 ml of medium to each flask and return them to the incubator for 20–24 hours.
5. Remove medium containing vector from the flasks. Filter through a 0.45-μm syringe filter to remove cells. Freeze in aliquots based on anticipated needs and store at or below –70°C.

Protocol 2

Generation of Stable Vector-producing Cells

This procedure describes the generation of clonal vector-producing cells that will provide an unlimited amount of unrearranged retroviral vector. The procedure (Fig. 3) involves transfection of one packaging cell line to generate a vector that is used to transduce a second packaging cell line. The resultant vector-producing clones generally contain a single integrated copy of the retroviral vector, and virus produced from this integrated vector is as genetically homogeneous as possible. This is in contrast to transient (Protocol 1) or stable (not shown) transfection techniques that produce mixed populations of unrearranged and rearranged vector genomes (e.g., see Bender et al. 1988; Miller et al. 1988; Lynch et al. 1993). Suitable combinations of packaging cell lines are shown for mouse (Table 2) and human (Table 3) packaging cell lines. Differences in these tables relate to the inability of ecotropic vectors to transduce human cells and the inability of gibbon ape leukemia virus (GALV) and RD114 vectors to transduce mouse cells. Although the vector produced by a given packaging cell line can sometimes be used to transduce the same cell line, the transduction rate is typically low due to receptor blockage by the Env protein made by the target packaging cells. Indeed, this procedure will select for target cells that express low Env protein levels, and thus are less resistant to transduction, but at the same time will ultimately produce less vector because of low Env production. Therefore, to obtain the highest vector titers, it is important to use pairs of packaging cells such that receptor blockage is not an issue. In this example we use PE501 ecotropic packaging cells for transfection and broad-host-range PT67 packaging cells to make stable vector-producing cells, as outlined in Figure 3.

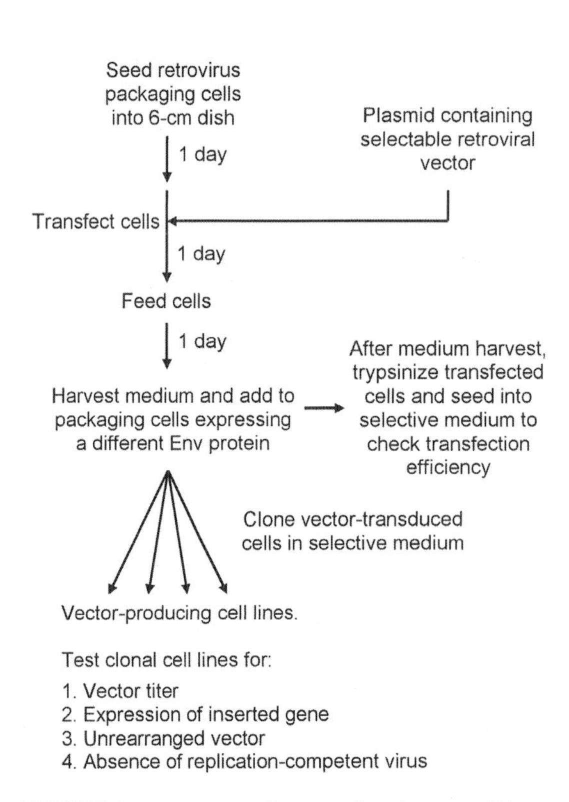

FIGURE 3. Generation of retroviral packaging cell lines.

TABLE 2. Susceptibility of Mouse Packaging Cell Lines to Transduction with Vectors Produced by Other Packaging Cell Lines

Vector-producing packaging cells	Virus receptor	Susceptibility of target mouse packaging cells				
		ecotropic	amphotropic	10A1	GALV	RD114
Ecotropic	CAT-1	–	+	+	+	+
Amphotropic	Pit2	+	–	–	+	+
10A1	Pit1 or Pit2	+	+	–	+	+
GALV	Pit1	–	–	–	–	–
RD114	RDR	–	–	–	–	–

MATERIALS

CAUTION: See Appendix for appropriate handling of materials marked with <!>.

Reagents

$CaCl_2$ (2.0 M) <!>
Dulbecco's modified Eagle's medium (DMEM) with 10% fetal bovine serum (FBS)
G418 <!>, histidinol, or hygromycin B <!>, depending on the selectable marker in the vector
Packaging cells (see Tables 2 and 3)
Polybrene in phosphate-buffered saline (PBS) (4 mg/ml;1000x, sterile-filtered) (Sigma-Aldrich)
Precipitation buffer

Mix 100 µl of 500 mM HEPES-NaOH (pH 7.1), 125 µl of 2.0 M NaCl, 10 µl of 150 mM Na_2HPO_4-NaH_2PO_4 <!> (pH 7.0), and H_2O to make 1 ml total. Prepare fresh.

Recipient cells (e.g., NIH-3T3 or HeLa)
Tris-Cl (10 mM, pH 7.5) <!>
Trypsin-EDTA
Vector plasmid (must carry selectable marker)

Include a vector that carries only the selectable marker gene and not the desired cDNA insert as a positive control to monitor success of the various steps of the procedure. Use a plasmid that does not carry a retroviral vector as a negative control.

Prepare endotoxin-free vector plasmid stocks (e.g., by using the QIAGEN Endotoxin-free Purification Kit) and determine plasmid DNA concentration.

Equipment

Biosafety cabinet
Cloning rings
CO_2 incubator
Felt-tip pen

TABLE 3. Susceptibility of Human Packaging Cell Lines to Transduction with Vectors Produced by Other Packaging Cell Lines

Vector-producing packaging cells	Virus receptor	Susceptibility of target human packaging cells				
		ecotropic	amphotropic	10A1	GALV	RD114
Ecotropic	CAT-1	–	–	–	–	–
Amphotropic	Pit2	+	–	–	+	+
10A1	Pit1 or Pit2	+	+	–	+	+
GALV	Pit1	+	+	–	–	+
RD114	RDR	+	+	+	+	–

Filters (0.22 μm, sterile)
Petri dish (10-cm glass)
Pipettes (sterile, disposable)
Silicone grease (Dow Corning high-vacuum grease or equivalent)
Tissue-culture dishes (6 and 10 cm)
Tube (clear polystyrene, 12 x 75 mm) (Falcon 2054 or equivalent)
Tweezers

METHODS

1. Day 1: Seed PE501 ecotropic retroviral packaging cells at 5×10^5 cells per 6-cm dish and incubate overnight.
2. Day 2: Replace the culture medium on the PE501 cells with 4 ml of fresh medium and transfect the cells with vector plasmid DNA using the calcium phosphate precipitation procedure as follows.

   Sterilize all reagents before use by filtration through 0.22-μm, pore-size sterile filters.

   a. For each plasmid sample, prepare a DNA-$CaCl_2$ solution by mixing the following together.

      25 μl of 2.0 M $CaCl_2$

      10 μg of plasmid DNA (in 10 mM Tris-Cl at pH 7.5)

      H_2O to make 200 μl total

   b. Prepare fresh precipitation buffer. In a clear 12 x 75-mm polystyrene tube, add 200 μl of DNA-$CaCl_2$ solution dropwise to 200 μl of precipitation buffer with constant agitation.

      A faint cloudiness in the solution should be immediately apparent. If the mixture remains clear or a precipitate consisting of large clumps develops, something is wrong.

   c. After 30 minutes at room temperature, add the resultant fine precipitate to a dish of cells and swirl the dish to distribute the precipitate. Incubate overnight.
3. Day 3: Aspirate the medium from the transfected PE501 cells and add 4 ml of fresh medium. Seed PT67 packaging cells at 10^5 cells per 6-cm dish, two dishes for each dish of transfected PE501 cells. Incubate overnight.
4. Day 4: Replace the medium on the PT67 cells with medium containing 4 μg/ml Polybrene. Remove 3 ml of virus-containing medium from each dish of transfected PE501 cells (retain 1 ml to keep the cells from drying out until they are trypsinized). Centrifuge the medium at 3000*g* for 5 minutes to remove cells and debris. From each dish of transfected PE501 cells, use 1 ml of virus-containing medium to infect one dish of PT67 cells, and 10 μl to infect another dish of PT67 cells. Return PT67 cells to the incubator.
5. Trypsinize and seed the PE501 cells at a 1:20 dilution into 6-cm dishes containing medium with 0.75 mg/ml G418 (active concentration), 4 mM histidinol, or 0.4 mg/ml hygromycin B, depending on the selectable marker in the vector. Returns cells to the incubator.
6. Day 5: Trypsinize the infected PT67 cells and seed the cells at 9:10 and 1:10 dilutions into 10-cm dishes containing 10 ml of medium plus the appropriate drug for selection (see Step 5).

   The 9:10 and 1:10 dilutions of PT67 cells infected with 1 ml or 10 μl of virus result in a 4 log range of dilutions, some of which should yield appropriate numbers of colonies for isolation of clonal cell lines.

7. Day 9: Stain the PE501 cells and evaluate for colony formation after 5 days of selection as a measure of the efficiency of DNA transfection. A transfection efficiency of about 1000 colonies per microgram of plasmid DNA is typical.
8. Days 9–14: Passage the transduced PT67 cells for 5–10 days, until drug-resistant colonies form and are clearly visible.
9. Use cloning rings to isolate clones from dishes containing small numbers of colonies.
   a. To prepare the cloning rings for use, spread a thin coating of silicone grease on the bottom plate of a 10-cm glass Petri dish. Place the rings in the dish so that the grease coats one open end and autoclave the dish to sterilize.
   b. To isolate clones, locate colonies and draw a circle around each colony on the bottom of the dish with a felt-tip pen.

      Colonies can be most easily visualized by holding the dish up to the light, taking care not to spill the medium. We find it useful to turn off the airflow in the laminar airflow hood to avoid desiccation of the colonies during placement of cloning rings.
   c. Aspirate the medium, place cloning rings over colonies to be isolated, and press down with tweezers.
10. Add a drop of trypsin-EDTA to each cylinder and monitor the extent of trypsinization microscopically.
11. When the cells have rounded up, add medium to each ring (one at a time) and, fairly vigorously, force the medium in and out of a pipette to dislodge the cells.

    We typically isolate approximately ten colonies for analysis.
12. After expansion, assay the clonal lines for an intact vector structure by Southern analysis, for the production of high vector titer, for the presence of helper virus (see Marker Rescue Assay on following page), and for expression of the inserted gene.

Virus Harvest and Assay

1. To prepare virus, replace the medium on confluent cultures of vector-producing cells. Collect the medium 12–24 hours later and centrifuge the medium at 3000*g* for 5 minutes to remove cells and debris.

   This process can be repeated three or four times at 12-hour intervals from the same dish of cells. The virus-containing medium can be used immediately to infect recipient cells or frozen at –70°C for later use.
2. Determine vector titer as follows.
   a. Seed recipient cells (e.g., NIH-3T3 or HeLa) at 5×10^5 per 6-cm dish and grow overnight.
   b. Change the medium to medium containing 4 μg/ml Polybrene and add various dilutions of test virus. Incubate overnight.
   c. Trypsinize and dilute the cells 1:20 into medium containing 0.75 mg/ml G418 (active concentration) for vectors carrying the *neo* gene, 4 mM histidinol for vectors carrying the *hisD* gene, and 0.4 mg/ml hygromycin B for vectors carrying the *hph* gene. Adjust these concentrations depending on the cell line. Return to incubator.
   d. After 5–7 days, stain and count colonies. Calculate virus titer in cfu/ml by dividing the number of colonies by the volume (in milliliters) of virus used for infection and multiplying by 20 to correct for the 1:20 cell dilution.

Marker Rescue Assay for Helper Virus

This assay measures the ability of a viral sample to rescue or mobilize a retroviral vector from cells that contain a vector but do not produce a vector. The ability of this assay to detect a given helper virus depends on whether the helper virus can infect the cells used in the assay. For example, ecotropic helper virus cannot be detected by using human cells. Thus, the assay cells should be chosen to match the expected helper viruses (see ref. 22 for an example). This assay is somewhat tedious and slow, but it is very sensitive and measures the most important property of helper viruses in the context of retroviral vector design, the ability to mobilize vectors. An alternative is described in the boxed text below Step 8.

1. To make cells that harbor a vector but do not release a vector, infect NIH-3T3 or HeLa cells with a helper-free vector carrying a selectable marker (we use the LN vector that carries the *neo* gene) and select the cells for the presence of the selectable gene (G418 for *neo*).

   This virus can be obtained from a packaging line that produces any vector at high titer.

2. Passage the cells for 2 weeks to allow potential helper virus (which should not be present) to spread. Assay the cells for vector production by using NIH-3T3 or HeLa cells as indicator cells for virus production, respectively.
3. Preserve cells that do not produce the vector (nonproducer cells) for use in the marker rescue assay described beginning with Step 4.
4. Seed nonproducer cells containing a *neo* vector (NIH-3T3 or HeLa cells) identified in Step 3 at 5×10^5 per 6-cm dish and grow overnight.
5. Infect nonproducer cells by adding 1 ml of test virus (centrifuged at 3000*g* for 5 minutes to remove cells and debris), 3 ml of regular medium, and 4 μg/ml Polybrene. Infect control positive dishes with a small amount of amphotropic helper virus plasmid (e.g., 1 μl or less of virus produced by NIH-3T3 cells transfected with pAM-MLV [Miller and Buttimore 1986] and passaged for 2 weeks to allow complete infection of the cells) or other helper virus capable of replicating in the nonproducer cells.
6. Passage cells for 2 weeks to allow helper virus spread (3 weeks if attempting to qualify clinical grade material; see Wilson et al. 1997). Take care not to cross-contaminate the cultures, some of which may begin to make helper virus at very high titer. Trypsinize the cells two to three times a week and replate the cells at 1:10 to 1:40 dilutions. Keep the cells at relatively high density to facilitate viral spread.
7. After 13 or 20 days of passaging nonproducer cells (see Step 20), plate naive NIH-3T3 or HeLa cells (same cell type as nonproducer cell line used) at 10^5 per 6-cm dish. Grow overnight. Feed confluent dishes of nonproducer cells (which now may be "producing" virus).

EXTENDED S^+/L^- ASSAY

An alternative to the marker rescue assay is the extended S^+/L^- assay (Reeves et al. 2002). In this assay, virus is amplified in a manner identical to that of marker rescue except that the amplification cells do not contain vector. Instead, virus is detected using indicator cell lines referred to as S^+/L^- cell lines. For example, the cat cell line PG-4 contains the murine sarcoma virus (MSV) genome (S^+) but lacks the MLV genome (L^-). Cells that express the MSV induce a transformed phenotype but only in cells coexpressing an MLV. If medium collected from the amplification phase cells contains replication-competent retrovirus, the PG-4 cells will be transformed and can be detected as foci within the PG-4 cell culture. The S^+/L^- assay can also be performed without the amplification phase (a direct S^+/L^- assay). This is useful for determining the replication-competent retrovirus titer. As with the marker rescue assay, the selection of the amplification phase cell line and the indicator cell assay will depend on the vector pseudotype (Wilson et al. 1997; Chen et al. 2001; Duffy et al. 2003).

8. Harvest medium from the nonproducer cells and use 1-ml samples to infect the naive NIH-3T3 or HeLa cells in the presence of 4 μg/ml Polybrene. Centrifuge the medium at 3000*g* for 5 minutes to remove cells and debris. Any live cells that are transferred along with the medium will be drug-resistant and could give a false-positive result. Return the newly infected cells to the incubator and grow overnight.
9. Replace the medium on the newly infected cells with medium containing G418 (0.75 mg/ml active concentration for NIH-3T3 or 1.0 mg/ml active concentration for HeLa).
10. After 5 days, stain and count colonies.

    The presence of colonies indicates that the *neo* vector was rescued by helper virus in the test sample. Usually this is very obvious, and positive dishes are covered with drug-resistant colonies.

Protocol 3

Transduction of Cell Lines

This protocol is suitable for transduction of many adherent cell lines. The number of target cells transduced can be varied as needed by maintaining the ratio of surface area to volume and using plates/flasks of various sizes. The protocol can also easily be adapted for nonadherent cells using similar vector-to-cell ratios.

MATERIALS

CAUTION: See Appendix for appropriate handling of materials marked with <!>.

Reagents

Cell culture medium (normal growth medium for the cells being transduced)
Protamine sulfate or Polybrene
Retroviral vector stock
Target cells

Equipment

Biosafety cabinet
Incubator, preset to 37°C; 5% CO_2
Pipettes (sterile, disposable)
Tissue-culture centrifuge
Tissue-culture flasks (75 cm^2)
Water bath, preset to 37°C

METHOD

1. Prepare a single-cell suspension of target cells and calculate the number of cells/ml. Prepare cultures for transduction as well as a mock transduction control. If drug selection of the transduced population is anticipated, prepare a second mock-transduced control that will serve as a drug selection control to document drug potency. For adherent cells, plate 5×10^2 cells in each 75-cm^2 tissue-culture flask with 12 ml of appropriate medium. Incubate overnight in a 37°C, 5% CO_2 incubator.
2. Rapidly thaw the vector in a 37°C water bath and prepare dilutions, if appropriate. Minimize the time between thawing and cell exposure.
3. Remove medium from the cells by aspiration and discard.
4. Perform transduction as follows.
   a. Vector transduction: Add 4 ml of vector-containing medium supplemented with 8 μg/ml Polybrene or 10 μg/ml protamine sulfate to each 75-cm^2 flask.

b. Mock transduction: Add 4 ml of fresh medium supplemented with 4 μg/ml Polybrene or 10 μg/ml protamine sulfate to the appropriate control cultures.

Polycations such as Polybrene or protamine sulfate are added to increase the interaction of the negatively charged cells and vector particles and can increase transduction tenfold or more. A stock solution of 1000x polycation can be made for convenience. Polycations can be toxic to primary cells and some cell lines, in which case, test a variety of lower concentrations to identify the dose with minimal toxicity and optimal gene transfer.

c. Return cells to 37°C, 5% CO_2 incubator. Expose cells to vector for 4 hours.

With cell lines, incubation over 4 hours generally does not increase transduction efficiency significantly. To increase gene transfer, it is preferable to repeat the transduction procedure on the subsequent day(s), presumably when additional cells have entered into a favorable part of the cell cycle for transduction.

d. After the incubation, aspirate the medium from the flasks, add fresh medium, and return the cells to incubator.

5. Optional drug selection or cell sorting: Vectors are frequently generated that express the specific transgene of interest along with a "marker" gene that allows for enrichment of transduced populations. If this option is chosen, add drug selection markers 24–48 hours after transduction. Controls should include mock transduced cultures with and without drug selection. In cases where surface expression molecules or fluorescent proteins are expressed that facilitate sorting of vector expressing cells, wait 72 hours to allow adequate time for vector expression in the majority of transduced cells.

6. Stable expression of integrated transgene is seen approximately 72 hours after transduction of most cell lines. Cells can be used for analysis or expanded for subsequent use.

TROUBLESHOOTING

Problem (Step 4): Gene-transfer efficiency is low.

Solution: Three factors should be considered: multiplicity of infection (moi), vector titer, and cell density at the time of transduction. The optimal moi varies with each cell line. For most immortalized cell lines, an moi of 10–20 infectious vector particles per each cell is adequate, especially if selection of transduced cells is planned. Primary cells usually require higher moi (when possible), colocalization (as described below), or multiple rounds of transduction. Above the optimal moi, the transduction efficiency tends to plateau, although the number of integrations per cell can increase. The vector titer (or concentration) is also a factor, and generally, the more concentrated the material, the higher the level of gene transfer (up to the plateau level).

Protocol 4

Transduction of Primary Hematopoietic Cells

This protocol describes transduction of primary hematopoietic cells. Noncycling cells are relatively resistant to transduction with retroviral vectors. Since most immortalized cell lines are actively proliferating, this is not an issue. However, for many primary cells, especially quiescent populations such as primitive progenitor and stem cells, the gene-transfer rate can be particularly low. Two interventions are now combined to maximize gene transfer in hematopoietic progenitor cells. First, investigators have used cytokines and other growth factors to stimulate cell cycling in hematopoietic cells (Bodine et al. 1989, 1991; Nolta and Kohn 1990; Luskey et al. 1992). Second, matrix proteins such as fibronectin, which mediates colocalization of target cells (via VLA-4- and VLA-5-binding sites) and vector (via heparin binding sites), have been utilized (Moritz et al. 1994). The highest gene transfer is obtained with a recombinant protein (CH-296, Retronectin) which brings the target cell and vector into close proximity (Hanenberg et al. 1996).

MATERIALS

CAUTION: See Appendix for appropriate handling of materials marked with <!>.

Reagents

Cell culture growth medium (complete or serum-free)

Iscove's modified Dulbecco's medium (IMDM) containing 10% fetal bovine serum (FBS), 1% glutamine, and 20 μg/ml gentamycin

or

X-Vivo 10 medium (Cambrex) containing 1% human serum albumin and 20 μg/ml gentamycin

Cell dissociation buffer (Invitrogen)

Cells

For mouse stem and progenitor cells, bone marrow is harvested from 5-fluorouracil-treated mice or is enriched for lineage negative cells using VarioMACS (Miltenyi Biotec). For human stem and progenitor cells, cells derived from bone marrow, peripheral blood, or cord blood are isolated by VarioMACS (Miltenyi Biotec). Fluorescent-activated cell sorting (FACS) is used for enrichment of the most primitive hematopoietic cells (mouse-Lin^{neg} c-Kit^{+}, Sca-1^{+}; human-$CD34^{+}CD38^{-}$).

Cytokine cocktails (commonly used for stimulation of cells)

Flt3-ligand + thrombopoietin (TPO) + stem cell factor (SCF) (100 ng/ml each)

Granulocyte colony-stimulating factor (G-CSF) + Flt3-ligand + TPO + SCF (50–100 ng/ml each)

G-CSF + TPO + SCF (100 ng/ml each)

Interleukin-6 (IL-6) (100 units/ml) + SCF (100 ng/ml)

Phosphate-buffered saline (PBS)

Retronectin

Retronectin is available as lyophilized powder or on precoated 35-mm dishes from Takara (www.takaramirusbio.com) or Cambrex (www.cambrex.com). Prepare Retronectin stock (1 mg/ml) in endotoxin-free, sterile distilled H_2O.

Retrovirus vector stock

Equipment

Centrifuge
Fluorescent-activated cell sorter (FACS)
Microscope
Petri dishes or multiwell plates, disposable plastic (nontissue-culture treated; BD Biosciences)
Polypropylene tubes
VarioMACS (Miltenyi Biotec) for enrichment of the most primitive hematopoietic cells

METHOD

1. About 12–48 hours prior to transduction, seed cells to be transduced at 2×10^5 to 4×10^5 per milliliter in cytokine-containing medium. Determine the length of the prestimulation period and the number of transduction cycles in a pilot experiment (Hanenberg et al. 1997).
2. On the day of transduction, coat plates with Retronectin as follows.
   a. Dilute Retronectin stock solution into PBS in polypropylene tubes and then coat nontissue-culture plates at 2–5 μg/cm² (see table below for coating at 2 μg/cm²). For example, when coating a 24-well plate, add 4 μl of Retronectin to 0.5 ml of PBS for each well coated. Mix by pipetting and add to wells.

      Use nontissue-culture plates because the hydrophobic surface of the nontissue-culture plates promotes optimal binding of Retronectin.

      Dish Size and Retronectin Coating Concentration

Dish type	Working surface area (cm²)	Retronectin at 2 μg/cm² (μg/well)	Retronectin (μl/well)	PBS (ml/well)
96 wells	0.3	(1.0)	1.0	0.1
24 wells	2.0	4.0	4.0	0.5
12 wells	3.3	7.0	7.0	1.0
6 wells	9.6	19	19	2.0
10 cm	58.1	116	116	5.0

   b. Incubate at room temperature for 2 hours or wrap in plastic wrap and incubate overnight at 4°C.
   c. Before use, aspirate off Retronectin/PBS solution.
3. Determine approximate number of cells to be transduced. Cells are transduced on Retronectin at 1×10^5 to 2×10^5 cells/cm² (see table below).

   Cell Number per Retronectin-coated Well or Plate

Dish type	Working surface area (cm²)	Cell number/ well (10⁵/cm²)	Retrovirus vector and medium (ml/well)
96 wells	0.3	3.2×10^4	0.2
24 wells	2.0	2.0×10^5	1.5
12 wells	3.3	3.3×10^5	3.0
6 wells	9.6	9.6×10^5	5.0
10 cm	58.1	5.8×10^6	10.0

   This protocol can also be used for gene transfer into adherent and nonadherent cell lines. For adherent cell lines, it may be necessary to begin the transduction with fewer cells per well than stated in the table above. This should be determined in a pilot experiment.
4. Following the prestimulation, harvest and count cells. Transduce cells at a 1:1 ratio of medium : retrovirus vector supernatant and supplement with the cytokine cocktail.

5. After 4 hours, harvest nonadherent cells and centrifuge. Add fresh medium supplemented with cytokines. Following centrifugation, add back nonadherent cells and place in incubator overnight. Perform this for each transduction cycle.

   Cells are typically transduced for two to four cycles on consecutive days. The cells can remain on the same Retronectin-coated plates as long as they do not become too crowded. If the cells become crowded, harvest and place on fresh Retronectin-coated plates.

6. After the last transduction cycle, incubate overnight.
7. Resuspend cells to detach from the plate and place in a tube. Wash each plate with PBS (5 ml per 10-cm dish; 3 ml per well of a six-well plate) and add washing to tube. Add dissociation buffer (same volumes as PBS) and place in a chemical fume hood for 2–3 minutes. Gently tap the side of the plate and harvest remaining cells. Check under a microscope that no cells remain.

TROUBLESHOOTING

Problem (Step 4): Long-term engraftment is compromised.

Solution: It is possible that exposure of hematopoietic cells to retrovirus-containing supernatants can promote differentiation of these cells and compromise their long-term engraftment. This is not true for all supernatants and should be determined empirically. Spinoculation can be used to preload the viral particles on the Retronectin and thus prevent the cells from being exposed to the potentially harmful effects of the retroviral supernatant. For spinoculation, add the retroviral supernatant to the Retronectin-coated wells and centrifuge at 2000*g* for 2 hours at 22°C. The volume of retroviral supernatant that results in optimal viral particle attachment in 1 well of a 24-well plate is reported to be 1.0 ml (Tonks et al. 2005). In this protocol, the cells are incubated in fresh medium containing cytokines. The addition of a 5-minute centrifugation at 2000*g* will also facilitate attachment of the cells to the Retronectin-coated plates.

Problem (Step 7): Insufficient long-term engraftment.

Solution: A 2-day resting period following transduction may increase long-term engraftment properties of hematopoietic cells (Takatoku et al. 2001). In this procedure, cells remain on the Retronectin-coated plates. On the day following the last transduction, harvest nonadherent cells and centrifuge. Remove the old medium and add fresh medium containing stem cell factor (100 ng/ml). Following centrifugation, add back the nonadherent cells (Takatoku et al. 2001).

REFERENCES

Bender M.A., Miller A.D., and Gelinas R.E. 1988. Expression of the human beta-globin gene after retroviral transfer into murine erythroleukemia cells and human BFU-E cells. *Mol. Cell. Biol.* **8:** 1725–1735.

Berns A. 2004. Good news for gene therapy. *N. Engl. J. Med.* **350:** 1679–1680.

Bodine D.M., Karlsson S., and Nienhuis A.W. 1989. Combination of interleukins 3 and 6 preserves stem cell function in culture and enhances retrovirus-mediated gene transfer into hematopoietic stem cells. *Proc. Natl. Acad. Sci.* **86:** 8897– 8901.

Bodine D.M., McDonagh K.T., Seidel N.E., and Nienhuis A.W. 1991. Survival and retrovirus infection of murine hematopoietic stem cells in vitro: Effects of 5-FU and method of infection. *Exp. Hematol.* **19:** 206–212.

Cavazzana-Calvo M., Hacein-Bey S., de Saint Basile G., Gross F., Yvon E., Nusbaum P., Selz F., Hue C., Certain S., Casanova J.L., Bousso P., Deist F.L., and Fischer A. 2000. Gene therapy of human severe combined immunodeficiency (SCID)-X1 disease. *Science* **288:** 669–672.

Challita P.M. and Kohn D.B. 1994. Lack of expression from a retroviral vector after transduction of murine hematopoietic stem cells is associated with methylation in vivo. *Proc. Natl. Acad. Sci.* **91:** 2567–2571.

Chen J., Reeves L., and Cornetta K. 2001. Safety testing for replication-competent retrovirus associated with gibbon ape leukemia virus-pseudotyped retroviral vectors. *Hum. Gene Ther.* **12:** 61–70.

Cornetta K. 1992. Safety aspects of gene therapy. *Br. J. Haematol.* **80:** 421–426.

Cornetta K., Nguyen N., Morgan R.A., Muenchau D.D., Hartley J.W., Blaese R.M., and Anderson W.F. 1993. Infection of human cells with murine amphotropic replication-competent retroviruses. *Hum. Gene Ther.* **4:** 579–588.

Cosset F.L., Takeuchi Y., Battini J.L., Weiss R.A., and Collins M.K. 1995. High-titer packaging cells producing recombinant retroviruses resistant to human serum. *J. Virol.* **69:** 7430– 7436.

Dave U.P., Jenkins N.A., and Copeland N.G. 2004. Gene therapy insertional mutagenesis insights. *Science* **303:** 333.

Donahue R.E., Kessler S.W., Bodine D., McDonagh K., Dunbar C., Goodman S., Agricola B., Byrne E., Raffeld M., Moen R., et al. 1992. Helper virus induced T cell lymphoma in nonhuman primates after retroviral mediated gene transfer. *J. Exp. Med.* **176:** 1125–1135.

Duffy L., Koop S., Fyffe J., and Cornetta K. 2003. Extended S^+/L^- assay for detecting replication competent retroviruses pseudotyped with the RD114 viral envelope. *Preclinica* **1:** 53–59.

Forestell S.P., Dando J.S., Chen J., de Vries P., Bohnlein E., and Rigg R.J. 1997. Novel retroviral packaging cell lines: Complementary tropisms and improved vector production for efficient gene transfer. *Gene Ther.* **4:** 600–610.

Hacein-Bey-Abina S., Le Deist F., Carlier F., Bouneaud C., Hue C., De Villartay J.P., Thrasher A.J., Wulffraat N., Sorensen R., Dupuis-Girod S., et al. 2002. Sustained correction of X-linked severe combined immunodeficiency by ex vivo gene therapy. *N. Engl. J. Med.* **346:** 1185–1193.

Hacein-Bey-Abina S., von Kalle C., Schmidt M., Le Deist F., Wulffraat N., McIntyre E., Radford I., Villeval J.L., Fraser C.C., Cavazzana-Calvo M., and Fischer A. 2003a. A serious adverse event after successful gene therapy for X-linked severe combined immunodeficiency. *N. Engl. J. Med.* **348:** 255–256.

Hacein-Bey-Abina S., Von Kalle C., Schmidt M., McCormack M.P., Wulffraat N., Leboulch P., Lim A., Osborne C.S., Pawliuk R., Morillon E., et al. 2003b. LMO2-associated clonal T cell proliferation in two patients after gene therapy for SCID-X1. *Science* **302:** 415–419.

Hanenberg H., Hashino K., Konishi H., Hock R.A., Kato I., and Williams D.A. 1997. Optimization of fibronectin-assisted retroviral gene transfer into human CD34+ hematopoietic cells. *Hum. Gene Ther.* **8:** 2193–2206.

Hanenberg H., Xiao X.L., Dilloo D., Hashino K., Kato I., and Williams D.A. 1996. Colocalization of retrovirus and target cells on specific fibronectin fragments increases genetic transduction of mammalian cells. *Nat. Med.* **2:** 876–882.

Hawley R.G. 2001. Progress toward vector design for hematopoietic stem cell gene therapy. *Curr Gene Ther* **1:** 1–17.

Jahner D., Stuhlmann H., Stewart C.L., Harbers K., Lohler J., Simon I., and Jaenisch R. 1982. De novo methylation and expression of retroviral genomes during mouse embryogenesis. *Nature* **298:** 623–628.

Kaleko M., Garcia J.V., Osborne W.R., and Miller A.D. 1990. Expression of human adenosine deaminase in mice after transplantation of genetically-modified bone marrow. *Blood* **75:** 1733–1741.

Kraunus J., Schaumann D.H., Meyer J., Modlich U., Fehse B., Brandenburg G., von Laer D., Klump H., Schambach A., Bohne J., and Baum C. 2004. Self-inactivating retroviral vectors with improved RNA processing. *Gene Ther.* **11:** 1568– 1578.

Li Z., Dullmann J., Schiedlmeier B., Schmidt M., von Kalle C., Meyer J., Forster M., Stocking C., Wahlers A., Frank O., Ostertag W., Kuhlcke K., Eckert H.G., Fehse B., and Baum C. 2002. Murine leukemia induced by retroviral gene marking. *Science* **296:** 497.

Luskey B.D., Rosenblatt M., Zsebo K., and Williams D.A. 1992. Stem cell factor, interleukin-3, and interleukin-6 promote retroviral-mediated gene transfer into murine hematopoietic stem cells. *Blood* **80:** 396–402.

Lynch C.M., Israel D.I., Kaufman R.J., and Miller A.D. 1993. Sequences in the coding region of clotting factor VIII act as dominant inhibitors of RNA accumulation and protein production. *Hum. Gene Ther.* **4:** 259–272.

Mann R., Mulligan R.C., and Baltimore D. 1983. Construction of a retrovirus packaging mutant and its use to produce helper-free defective retrovirus. *Cell* **33:** 153–159.

Markowitz D., Goff S., and Bank A. 1988. A safe packaging line for gene transfer: Separating viral genes on two different plasmids. *J. Virol.* **62:** 1120–1124.

Miller A.D. and Buttimore C. 1986. Redesign of retrovirus packaging cell lines to avoid recombination leading to helper virus production. *Mol. Cell. Biol.* **6:** 2895–2902.

Miller A.D. and Chen F. 1996. Retrovirus packaging cells based on 10A1 murine leukemia virus for production of vectors that use multiple receptors for cell entry. *J. Virol.* **70:** 5564–5571.

Miller A.D. and Rosman G.J. 1989. Improved retroviral vectors for gene transfer and expression. *Biotechniques* **7:** 980–990.

Miller A.D., Bender M.A., Harris E.A., Kaleko M., and Gelinas R.E. 1988. Design of retrovirus vectors for transfer and expression of the human beta-globin gene. *J. Virol.* **62:** 4337–4345.

Miller A.D., Garcia J.V., von Suhr N., Lynch C.M., Wilson C., and Eiden M.V. 1991. Construction and properties of retrovirus packaging cells based on gibbon ape leukemia virus. *J. Virol.* **65:** 2220–2224.

Moritz T., Patel V.P., and Williams D.A. 1994. Bone marrow extracellular matrix molecules improve gene transfer into human hematopoietic cells via retroviral vectors. *J. Clin. Invest.* **93:** 1451–1457.

Nolta J.A. and Kohn D.B. 1990. Comparison of the effects of growth factors on retroviral vector-mediated gene transfer and the proliferative status of human hematopoietic progenitor cells. *Hum. Gene Ther.* **1:** 257–268.

Overbaugh J., Miller A.D., and Eiden M.V. 2001. Receptors and entry cofactors for retroviruses include single and multiple transmembrane-spanning proteins as well as newly described glycosylphosphatidylinositol-anchored and secreted proteins. *Microbiol. Mol. Biol. Rev.* **65:** 371–389.

Pannell D. and Ellis J. 2001. Silencing of gene expression: Implications for design of retrovirus vectors. *Rev Med Virol* **11:** 205–217.

Reeves L., Duffy L., Koop S., Fyffe J., and Cornetta K. 2002. Detection of ecotropic replication-competent retroviruses: Comparison of s(+)/l(−) and marker rescue assays. *Hum. Gene Ther.* **13:** 1783–1790.

Takatoku M., Sellers S., Agricola B.A., Metzger M.E., Kato I., Donahue R.E., and Dunbar C.E. 2001. Avoidance of stimulation improves engraftment of cultured and retrovirally transduced hematopoietic cells in primates. *J. Clin. Invest.* **108:** 447–455.

Tonks A., Tonks A.J., Pearn L., Mohamad Z., Burnett A.K., and Darley R.L. 2005. Optimized retroviral transduction protocol which preserves the primitive subpopulation of human hematopoietic cells. *Biotechnol. Prog.* **21:** 953–958.

Watanabe S. and Temin H.M. 1983. Construction of a helper cell line for avian reticuloendotheliosis virus cloning vectors. *Mol. Cell. Biol.* **3:** 2241–2249.

Wilson C.A., Ng T.H., and Miller A.E. 1997. Evaluation of recommendations for replication-competent retrovirus testing associated with use of retroviral vectors. *Hum. Gene Ther.* **8:** 869–874.

3 Development of Lentiviral Vectors Expressing siRNA

Gustavo Tiscornia, Oded Singer, and Inder M. Verma

The Salk Institute for Biological Studies, Laboratory of Genetics, La Jolla, California 92037

ABSTRACT

This chapter describes the use of lentiviral vectors to deliver small interfering RNA (siRNA)-mediated silencing cassettes. The combination of these two technologies allows for the development of a powerful tool to achieve long-term down-regulation of specific target genes both in vitro and in vivo. It combines the specificity of RNA interference with the versatility of lentiviral vectors to stably transduce a wide range of cell types.

INTRODUCTION

RNA interference (RNAi) has recently emerged as a novel pathway that allows modulation of gene expression. The pathway has been under intense study, and details of its basic biological mechanism are described elsewhere (Denli and Hannon 2003). Briefly, long double-stranded RNA molecules are processed by the endonuclease Dicer into short 21–23-nucleotide siRNAs that are then incorporated into RNA-induced silencing complex (RISC), a multicomponent nuclease complex that selects and degrades mRNAs that are homologous to the double-stranded RNA initially delivered (Fjose et al. 2001; Hannon 2002). In mammalian systems, siRNAs can be delivered exogenously or expressed endogenously from polymerase III (pol III) promoters, resulting in sustained and specific down-regulation of target mRNAs (Elbashir et al. 2001; Brummelkamp et al. 2002; Miyagishi and Taira 2002; Oliveira and Goodell 2003). To exploit this technique, efficient siRNA delivery methods must be developed. In this chapter, we describe the design and preparation of lentiviral vectors expressing siRNA for down-regulation of specific target genes.

Overview of Lentiviral Vectors

During the past decade, gene delivery vehicles based on human immunodeficiency virus type 1 (HIV-1), the best characterized of the lentiviruses, have been developed. Lentiviral vectors derived from HIV-1 are capable of transducing a wide variety of dividing and nondividing cells, integrate stably into the host genome, and result in long-term expression of the transgene. The HIV-1 genome contains nine open reading frames (ORFs) encoding at least 15 distinct proteins involved in the infectious cycle, including structural and regulatory proteins. In addition, there are a number of *cis*-acting elements required at various stages of the viral life cycle (for review, see Trono 2002). The general strategy used to produce vector particles has been to eliminate all dispensable genes from the HIV-1 genome and separate the *cis*-acting sequences from those *trans*-acting factors that are absolutely required for viral particle production, infection, and integration.

The widely used third generation of lentiviral vectors consists of four plasmids (Fig. 1A). The transfer vector contains the transgene to be delivered in a lentiviral backbone containing all of the *cis*-acting sequences required for genomic RNA production and packaging. The packaging system involves three additional plasmids (pMDL, pRev, and pEnv) that provide the required *trans*-acting factors, namely, Gag-Pol, Rev, and an envelope protein, respectively. Gag-Pol codes for integrase, reverse transcriptase, and structural proteins. The structural proteins are required for particle production, whereas integrase and reverse transcriptase molecules are packaged into the viral particle and are required upon subsequent infection. Rev interacts with the Rev-responsive element (RRE), a sequence contained in the transfer vector, enhancing the nuclear export of unspliced viral genomic RNA and thus increasing viral titer.

Viral particles can be pseudotyped with a variety of envelope proteins. One commonly used envelope protein is the vesicular stomatitis virus protein G (VSV-G), which is incorporated into the viral membrane and confers the ability to transduce a broad range of cell types, including primary cells, stem cells, and early embryos. The transfer vector also contains the woodchuck hepatitis virus regulatory element (WPRE) that enhances expression of the transgene (Zufferey et al. 1999) and a central polypurine tract (cPPT) purported to increase efficiency of nuclear import of the preintegration complex (Zennou et al. 2000). In addition, an important safety feature is provided by a deletion in the 3′ LTR (long terminal repeat) that results in replication-defective particles. During reverse transcription, the proviral 5′ LTR is copied from the 3′ LTR, thus transferring the deletion to the 5′ LTR. The deleted 5′ LTR is transcriptionally inactive, preventing viral genomic RNA production from the integrated provirus (Fig. 1B) (Miyoshi et al. 1998). When these four plasmids are transfected into 293T human embryonic kidney cells, viral

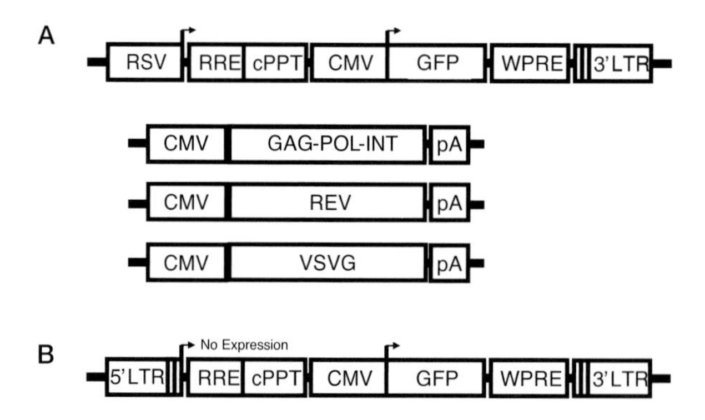

FIGURE 1. (*A*) Third-generation lentiviral vector system. The transfer vector contains all *cis*-acting elements required for replication and packaging of transfer vector RNA into viral particles. Carrying capacity is about 8 kb. The hatched box represents the self-inactivating deletion. Three helper plasmids provide all required *trans*-acting factors (see text for details). (*B*) Structure of an integrated provirus.

particles accumulate in the supernatant and high-titer viral preparations can be obtained by ultracentrifugation.

Design of Lentivectors Expressing shRNAs

A crucial breakthrough occurred with the report that siRNAs could be expressed as short hairpin RNA (shRNA) from pol III promoters cloned into plasmids (Brummelkamp et al. 2002; Miyagishi and Taira 2002). The two pol III promoters most commonly used are H1 and U6 (both human and mouse). Pol III promoters are characterized by their compact size (less than 400 bp) (Myslinski et al. 2001) and by the fact that all sequences required for promoter function are upstream of the +1 transcriptional start site. Pol III promoters have ubiquitous expression and efficiently express short RNAs (shRNAs). Thus, they are ideally suited to express shRNAs consisting of a 21–23-nucleotide sense sequence that is identical to the target sequence in the mRNA to be down-regulated, followed by a 9-bp loop and an antisense 21–23-nucleotide sequence. A stretch of five Ts provides a pol III transcriptional termination signal. The total length of the silencing cassette is about 350 bp. Thus, when this construct is expressed, a short 21–23-bp hairpin is formed; the loop is digested by Dicer and the resulting siRNA triggers degradation of the mRNA target.

Ideally, a silencing lentiviral vector would contain both a marker gene such as EGFP (enhanced green fluorescent protein) or an antibiotic resistance gene and the shRNA silencing cassette. We have designed two different versions of lentiviral silencing vectors that differ both in the position of the silencing cassette and in the cloning strategy required to construct them.

Version A (Protocol 1) involves a lentiviral vector carrying GFP as a marker and cloning the human H1-driven silencing cassette into a unique restriction site in the 3′ LTR (Tiscornia et al. 2003). As during integration, the 5′ LTR of the provirus is copied from the 3′ LTR, cloning the H1-driven shRNA into the 3′ LTR, which results in duplication of the silencing cassette. Although this strategy maximizes the silencing power of the lentiviral vector, in our experience, the main parameter determining level of silencing is the multiplicity of infection (moi). The moi required to silence a given target will depend on the levels of expression of the target mRNA, siRNA efficiency, and the transducibility of the cell type involved.

One undesirable consequence of the version-A design is that the siRNA target sequence is also present in the mRNA expressing the marker gene, resulting in somewhat lower expression of the marker. In version-B silencing lentivectors, the position of the silencing cassette is upstream of the marker expression cassette, thus avoiding down-regulation of the marker. Because the silencing cassette is not in the 3′ LTR, only one copy of the silencing cassette is delivered per viral particle. This design has been adapted to Gateway cloning technology, allowing an alternative and efficient cloning method (Protocol 2).

Design and Cloning of Lentiviral Silencing Vectors

Development and validation of an efficient lentiviral silencing vector involves the following steps: (1) selection of siRNA target sequences and design of shRNAs, (2) cloning and validation of shRNAs' effect on the target gene, and (3) cloning and testing the lentiviral silencing vector. The effectiveness of a particular siRNA is largely unpredictable and presumably reflects both mechanistic constraints of the RNAi pathway and accessibility of the target sequence within the tertiary structure of the target mRNA. A number of algorithms have been developed to predict effective siRNA sequences (Reynolds et al. 2004), and many of them are available online for free or commercial use (for example, see www.ambion.com or http://sfold.wadsworth.org). In general, the target sequence should be 21–23 bases long, but lengths of up to 28 bases have been reported (Paddison et al. 2002). Longer targets should be avoided, because longer double-stranded RNA molecules can trigger a PKR (protein kinase activated by double-stranded RNA) response

(Clemens and Elia 1997). A database search is recommended to filter out candidate targets that are present in other genes to avoid silencing of these loci. GC content should be between 40% and 55%. shRNAs to be driven by the H1 promoter can begin with any base, but the U6 promoter requires a G as its first base. shRNAs can be directed to 5′-untranslated region (5′ UTR), ORF, or 3′ UTR of the target mRNA. Strings of identical bases should be avoided. As a loop, we generally use the 9-bp sequence (TTC AAG AGA) (Brummelkamp et al. 2002).

Typically, several shRNAs must be generated and tested for every target gene. The candidate shRNAs are cloned into a simple plasmid containing only the silencing cassette and tested. Efficient silencing cassettes are then transferred to the lentiviral vector. Initial screening is best achieved by cotransfection of an shRNA-expressing plasmid and a vector-expressing tagged (myc, FLAG, etc.) cDNA of the target into 293T followed by western blot against the tag. Alternatively, provided the candidate plasmids can be efficiently transfected into a cell type of interest, the effectiveness of target down-regulation can be followed by analysis of target mRNA with quantitative reverse transcriptase–polymerase chain reaction (RT-PCR), northern blot, or analysis of target protein levels by western blot against the endogenous target.

When one or more efficient shRNA candidates have been identified, the silencing cassettes must be cloned into the lentiviral vector. High-titer viral preparations are made and should be tested by transduction of a cell line expressing the target protein, followed by measurement of target expression. The final validation of the lentiviral silencing vector against the endogenous target is crucial and should include use of a lentiviral vector lacking a silencing cassette or carrying a silencing cassette against a different target as a specificity control, because overexpression of any siRNA can cause some nonspecific down-regulation of gene expression. Several mois should be tested. Precise determination of efficiency of down-regulation of the target will require testing homogeneously transduced cell populations, which can be obtained by fluorescence-activated cell sorting (FACS) for GFP-positive cells or applying selection if an antibiotic resistance gene is used as a marker.

Protocol 1

Design and Cloning of an shRNA into a Lentiviral Silencing Vector: Version A

This protocol describes version A for the design and cloning of an shRNA for a given target. We use a plasmid containing the poll III promoter as a template and a single round of PCR to amplify the silencing cassette. We employ a 5′ forward primer upstream of the pol III promoter and a 3′ reverse primer that includes the entire shRNA sequences (sense, loop, and antisense sequences followed by five Ts), followed by 22 bases complementary to the last 22 bp upstream of the +1 transcriptional start site of the pol III promoter. An NheI compatible restriction site is included at the 5′ end of both forward and reverse primers (such as XbaI). PCR amplification will result in a DNA fragment containing an shRNA expression cassette that can be cloned into a simple cloning vector, tested, and then transferred to the lentiviral vector, or cloned into the lentiviral vector directly (Fig. 2).

MATERIALS

CAUTION: See Appendix for appropriate handling of materials marked with <!>.

Reagents

Advantage GC-2 polymerase mix (BD)

2x BES-buffered saline (BBS) solution (50 mM BES, 280 mM NaCl, 1.5 mM Na_2HPO_4 <!>)

Mix 16.36 g of NaCl, 10.65 g of BES (Calbiochem 391334), and 0.21 g of Na_2HPO_4. Add double-distilled H_2O up to 900 ml. Dissolve, titrate to pH 6.95 with 1 M NaOH, and bring volume to 1 liter. Filter-sterilize and store 14-ml aliquots at 4°C.

$CaCl_2$ (2.5 M) stock solution <!>

Mix 36.75 g of $CaCl_2$ in 70 ml of double-distilled H_2O. Adjust to a final volume of 100 ml. Aliquot into 1.5-ml microcentrifuge tubes and store at –20°C.

Cells: 293T human embryonic kidney (Invitrogen)

Dimethylsulfoxide (DMSO; 7%) <!>

Dulbecco's modified Eagle's medium (DMEM)

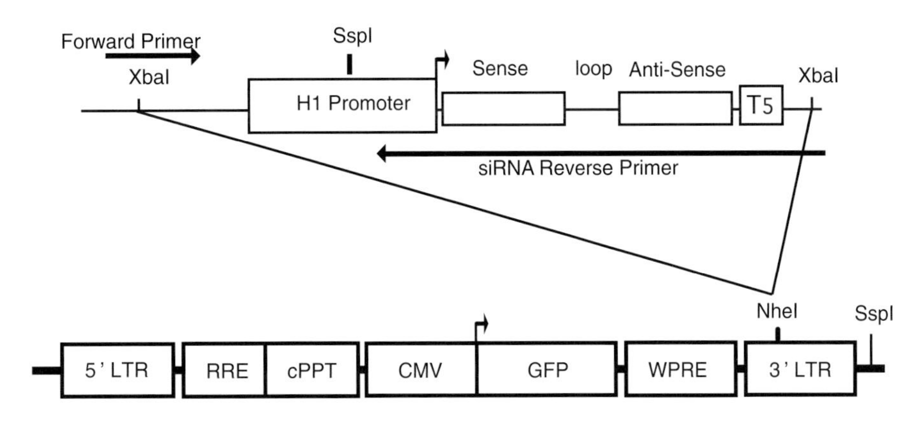

FIGURE 2. Cloning scheme for version A (see text for details).

ELISA (enzyme-linked immunosorbent assay) Kit, p24 (New England Nuclear Life Science products NEK050B) (*Optional:* See Step 15)
Fetal bovine serum (FBS)
Hank's balanced salt solution (HBSS; Invitrogen)
H1 promoter cloned into pGEM-T (Promega)
Plasmids
pMDL (Gag-Pol)
pREV
pVSV-G
For plasmid preparation, use QIAGEN plasmid maxipreps at 1 μg/μl.
Poly-L-lysine (Sigma-Aldrich P 4832)
Use 0.001% in phosphate-buffered saline (PBS). Filter-sterilize and store at –20°C.
Restriction endonucleases: NheI, SspI, XbaI
Sucrose (20% in HBSS)

Equipment

Beckman tubes (358126 and 326819)
Filters (0.22 or 0.45 μm)
Incubators, preset to 37°C (3% and 10% CO_2)
Microcentrifuge tubes
PCR machine
SW 28 and SW 55 rotors (Beckman)
Tissue-culture dishes (15 cm and six well)

METHODS

Design and Cloning of shRNAs

1. Select a target within the gene to be silenced, e.g., for GFP: GCAAGCTGACCCTGAAGTTC (Tiscornia et al. 2003).
2. Design primers to amplify the silencing cassette. As template, we use an H1 promoter cloned into pGEM-T. The 5′ forward primer must contain an XbaI site. The 3′ reverse primer contains 22 nucleotides from the 3′ end of the pol III promoter and a 5′ tail including the entire shRNA, loop, transcriptional stop signal (T5), and XbaI site sequences. For the target suggested above, design the following 3′ reverse primer: 5′CTGTCTAGACAAAAAGCAAGCTG ACCCTGAAGTTC**TCTCTTGAA**GAACTTCAGGGTCAGCTTGc***GGGGATCTGTGGTCTC ATACA***3′, where the H1 sequence is in italic bold, XbaI (NheI compatible) is underlined, the loop is bold underlined, and nucleotide +1 is in small caps.
3. With the 5′ forward primer and the 3′ reverse primer (final primer concentration is 10 μm) described above, use 10 ng of plasmid containing the H1 promoter as template to amplify the silencing cassette by PCR. We use Advantage GC-2 polymerase mix and use the GC-melt additive as 10x. It is essential to add 7% DMSO or a similar agent to a regular *Taq* polymerase reaction to prevent hairpin formation. Amplify using the parameters listed below:

Cycle number	Denaturation	Annealing	Polymerization
First cycle	3 min at 94°C		
30 cycles	30 sec at 94°C	30 sec at 55°C	40 sec at 72°C
Last cycle			10 min at 72°C

4. The result of the PCR is an amplified fragment of approximately 400 bp that can be cloned in an A/T vector for sequencing or directly cloned in the lentivector plasmid. Digest the insert with the XbaI and gel-purify. Digest the lentivector plasmid with NheI, gel-purify, and then dephosphorylate.

   Typically, 50 ng of vector is ligated to 100 ng of insert and transformed into competent bacteria. Plasmid DNA from the resulting colonies can be screened by digestion with SspI. The parental vector should have only one SspI site, whereas the vector containing the insert will acquire an additional SspI site located in the H1 promoter. It is important to verify the integrity of the hairpin by sequencing using the following H1-F primer 5′-TGGCAGGAAGATGGCT-GTGA-3′, because mutations in the hairpin can significantly reduce the efficiency because down-regulation.

5. Validate the cloned shRNA cassettes by transfecting or transducing (as lentiviral particles) to a cell line that expresses the target gene. Alternatively, coexpress a tagged cDNA of the target gene together with shRNA silencing cassettes in an easily transfected cell line (e.g., 293T).

   This is very useful when target mRNA is restricted to certain cell types or a specific antibody against the target is unavailable. Typically, we transfect 200 ng of target cDNA plasmid plus 500–1000 ng of the plasmid containing the silencing cassette per well (six-well cluster) and harvest the cells for immunoblot analysis 48–72 hours after transfection.

Preparation of Lentiviral Vectors

6. For a 12 x 15-cm–dish lentiviral preparation: Twenty-four hours before transfection, prepare plates and cells.

   a. To increase cell adherence, precoat 12 15-cm dishes with 10 ml of poly-L-lysine, incubate 15 minutes at room temperature, and aspirate off the liquid.

   b. Immediately seed 293T cells from two confluent 15-cm plates to the 12 15-cm plates in DMEM + 10% FBS. Addition of 1% antibiotic-antimycotic solution does not interfere with transfection. Cells should be of low-passage number and should not be used after passage 20 or if growth is slow. Certain brands of FBS do not support efficient transfection and can result in low viral titers.

   c. Grow the cells overnight.

7. Make sure that the cells are 70–80% confluent and evenly distributed at the time of transfection to optimize viral titer. Transfect the plasmid mix into the cells using the $CaPO_4$ precipitation method as follows:

   a. Aliquot the four plasmids into a 50-ml tube. For a 12 x 15-cm dish, use:

      270 μg of lentivector

      176 μg of pMDL (Gag-Pol)

      95 μg of pVSV-G

      68 μg of pREV

   b. Prepare a working solution of $CaCl_2$ (13.5 ml of 0.25 M $CaCl_2$) and add to the plasmid mix. Add 13.5 ml of 2x BBS solution. Mix gently by inversion and incubate for 15 minutes at room temperature.

   c. Add the transfection mixture (spreading in drops) to each plate (2.25 ml/plate). Swirl the plates gently and incubate overnight at 37°C in a 3% CO_2 atmosphere.

8. Approximately 16–20 hours after transfection, remove media. Add 15 ml of fresh DMEM + 2% FBS to each plate and incubate overnight at 37°C in a 10% CO_2 atmosphere.

9. Collect the supernatant from the plates and filter through 0.22- or 0.45-μm filters. Add 15 ml of fresh medium to each plate and incubate overnight.

   Filtered supernatants can be stored for several days at 4°C.

10. Collect media and filter as in Step 9.
11. Pool collected supernatants from Steps 9 and 10. Transfer to Beckman tubes (358126), using 25–29 ml per tube. Concentrate viral particles by centrifuging in an SW 28 rotor at 19,400 rpm for 2 hours at 20°C.
12. Resuspend all pellets in a total of 1 ml of HBSS. Wash tubes a second time with 1 ml of HBSS.
13. Increase the combined volume from 2 to 3 ml with HBSS and layer the resuspended pellets on 1.5 ml of a 20% sucrose (in HBSS) cushion in Beckman tubes (326819). Centrifuge using an SW 55 rotor at 21,000 rpm for 1.5 hours at 20°C.
14. Resuspend the pellet in 100 μl of HBSS and wash the tube with an additional 100 μl of HBSS. Shake the resuspended viral preparation on a low-speed vortexer for 15–30 minutes. Centrifuge for 10 seconds to remove debris. Aliquot the cleared viral solution and store at –80°C. It can be stored for many months. Avoid repeated freeze-thaw cycles.
15. Titrate the viral preparations by quantitating levels of the capsid protein p24 using a p24 ELISA Kit or by biological titration if an adequate marker is contained in the lentivector.

    Titers normally range between 10^9 and 10^{10} viral particles/ml but can be lower if transfection efficiency of packaging plasmids is suboptimal.

Protocol 2

Cloning an shRNA into a Lentiviral Silencing Vector: Version B

This protocol describes version B of the lentivector silencing system. It has been adapted to Gateway cloning technology (Invitrogen), allowing for a fast and convenient cloning procedure (see Fig. 3). Initially, an shRNA is cloned into an entry vector (pENTR/U6, Invitrogen) immediately downstream from an hU6 promoter. The silencing cassette is flanked by recombination sites from bacterophage λ (*attL1* and *attL2*). Once an effective shRNA is obtained, it can be transferred to the destination vector. The destination vector is a lentiviral vector carrying a marker (GFP or a selection marker) with a destination cassette cloned upstream of the marker (*attR1* and *attR2* flanking a *ccdB* toxic gene). Thus, the silencing cassette can be transferred from the entry vector to the destination vector in a simple LR cloning reaction.

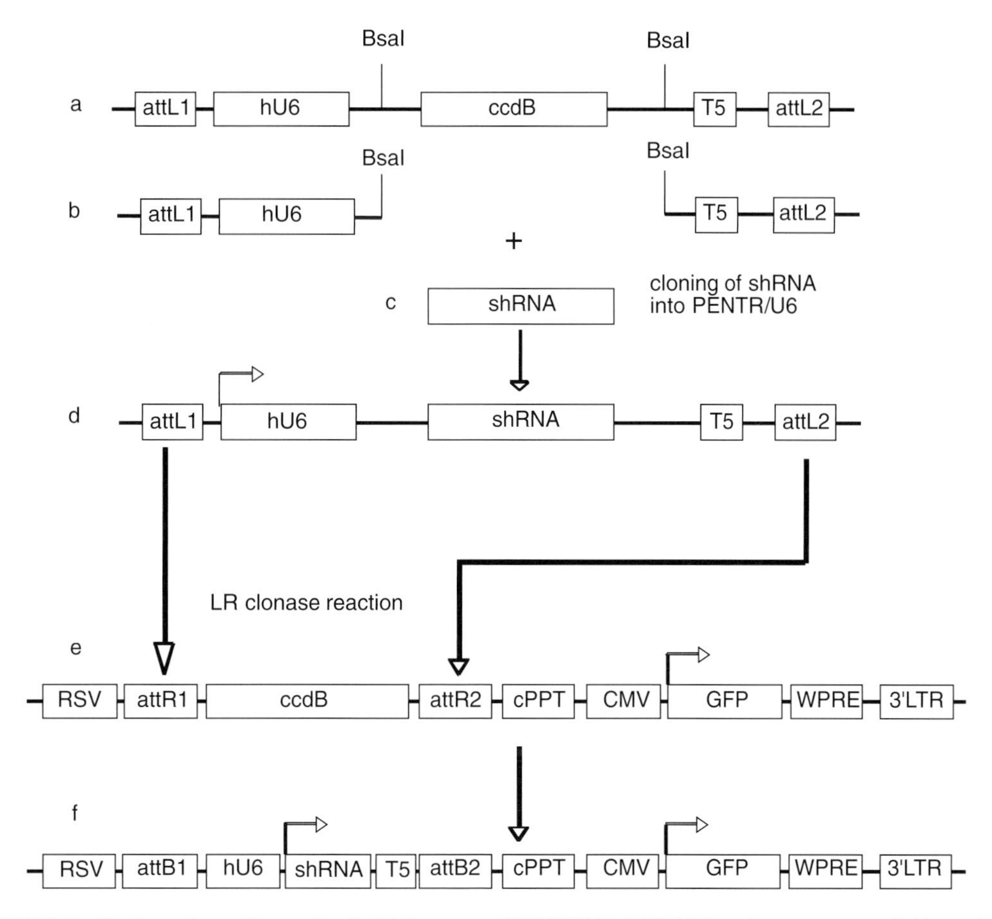

FIGURE 3. Cloning scheme for version B: (*a*) Structure of PENTR/U6. (*b*) BsaI digestion creates termini into which an shRNA duplex (*c*) can be cloned to obtain the silencing cassette (*d*). In turn, bacteriophage λ *att* recombination sites allow transfer of the silencing cassette to a lentiviral destination vector (*e*) to obtain the lentiviral transfer vector containing both a silencing cassette and a GFP marker (*f*).

MATERIALS

CAUTION: See Appendix for appropriate handling of materials marked with <!>.

Reagents

Agarose gel (4%)
10x Annealing buffer
  100 mM Tris-HCl (pH 8) <!>
  10 mm EDTA
  1 M NaCl
Escherichia coli DB3.1 (Invitrogen)
Kanamycin <!> and ampicillin <!> plates
5x LR buffer (Invitrogen)
Oligonucleotides; order as appropriate (200-µm concentration)
pENTR/U6 (Invitrogen)
Proteinase K <!>
Restriction endonucleases: AvaI, BsaI, ClaI, NdeI, XbaI
STBL3 (Invitrogen or equivalent recombination-deficient bacterial strain)

Equipment

PCR machine

METHOD

1. Select a target within the gene to be silenced; e.g., for GFP: GCAAGCTGACCCTGAAGTTC (Tiscornia et al. 2003).
2. Design an shRNA and clone into pENTR/U6. For a diagram of pENTR/U6, see Figure 3a. A detailed map can be downloaded from the Invitrogen Web site. pENTR/U6 has a *ccdB* toxic gene and must therefore be propagated in tolerant strains such as *E. coli* DB3.1. When this plasmid is digested with BsaI, the vector is left with the following termini:

   XXXXX 3′   5′ TTTTTXXXXXXX

   XXXXXGTGG 5′   3′ AXXXXXXX

   To clone the shRNA, design two complementary oligonucleotides with termini compatible to those in the BsaI-digested pENTR/U6, followed by annealing and ligation as described below (Fig. 3b,c,d). For the GFP target sequence given above, the required oligonucleotides, after annealing, would look as follows:

   5′ CACC *GCAAGCTGACCCTGAAGTTC* **TTCAAGAGA** . . .

   3′ *CGTTCGACTGGGACTTCAAG* **AAGTTCTCT** . . .

   . . . GAACTTCAGGGTCAGCTTGC 3′

   . . . CTTGAAGTCCCAGTCGAACG AAAA 5′

   where the sense strand of the hairpin is in italics, the loop is in bold, and the antisense strand of the hairpin is underlined.

3. Anneal and clone according to the following procedure:

   a. Make annealing mixture:

      5 μl of P1 (50 μm final concentration)
      5 μl of P2 (50 μm final concentration)
      2 μl of 10x annealing buffer
      8 μl of H_2O

   b. Heat to 94°C for 5 minutes. Use a PCR machine to cool to 25°C at 0.1°C per second.

      Even at 50-μm final concentration, the efficiency of the reaction is only about 50%. Efficiency of annealing can be ascertained by running an aliquot of the annealed oligonucleotides (5 μl of a 1/100 dilution of the annealing mix) on a 4% agarose gel. Single-stranded oligonucleotides will run as a hairpin of approximately 30 bases, whereas the annealed product should run at its predicted size (~55 bp).

   c. Dilute annealing mix 1/1000 in H_2O and clone into the pENTR/U6 vector. Typically, 50 ng of vector are ligated to 2–3 μl of a 1/1000 dilution of the annealed product. Transform competent bacteria and select on kanamycin plates. Resulting plasmids will have the structure shown in Figure 3d.

   d. Digest plasmid DNA from colonies with NdeI-XbaI and run on a 4% agarose gel. Positive clones will contain an approximately 127-bp insert compared to 76 bp for colonies without an insert. Sequence the insert with hU6-F (GGACTATCATATGCTTACCG) and M13-R primers to check hairpin integrity.

4. Silencing cassettes can be tested as in Step 5 of Protocol 1. Once a suitable candidate is found, transfer the silencing cassette from the entry vector to the destination vector by performing an LR recombination reaction. For a typical destination vector, see Figure 3e.

   a. For LR recombination reaction, mix:

      100–300 ng of entry vector
      150 ng of destination vector
      4 μl of 5x LR buffer
      H_2O to 20 μl final volume

   b. Incubate overnight at room temperature.

   c. Add 2 μl of proteinase K. Incubate for 10 minutes at 37°C.

   d. Transform STBL3 competent bacteria (or equivalent recombination-deficient strain) with 2 μl of LR reaction mix. Plate on ampicillin plates.

      The resulting constructs will have the structure depicted in Figure 3f. The use of recombination-deficient bacteria is highly recommended, because use of other strains can result in unwanted recombination events within the plasmid.

   e. Digest plasmid DNA with ClaI-AvaI. Clones containing the silencing cassette should have an approximately 700-bp insert, whereas the unrecombined parental destination vector will show a 1.9-kb insert. Sequence positive clones with hU6-F (5′-GGA CTA TCA TAT GCT TAC CG-3′).

5. For preparation of the vector, follow Steps 6–15 in Protocol 1.

REFERENCES

Brummelkamp T.R., Bernards R., and Agami R. 2002. A system for stable expression of short interfering RNAs in mammalian cells. *Science* **296:** 550–553.

Clemens M.J. and Elia A. 1997. The double-stranded RNA-dependent protein kinase PKR: Structure and function. *J. Interferon Cytokine Res.* **17:** 503–524.

Denli A.M. and Hannon G.J. 2003. RNAi: An ever-growing puzzle. *Trends Biochem. Sci.* **28:** 196–201.

Elbashir S.M., Harborth J., Lendeckel W., Yalcin A., Weber K., and Tuschl T. 2001. Duplexes of 21-nucleotide RNAs mediate RNA interference in cultured mammalian cells. *Nature* **411:** 494–498.

Fjose A., Ellingsen S., Wargelius A., and Seo H.C. 2001. RNA interference: Mechanisms and applications. *Biotechnol. Annu. Rev.* **7:** 31–57.

Hannon G.J. 2002. RNA interference. *Nature* **418:** 244–251.

Miyagishi M. and Taira K. 2002. U6 promoter-driven siRNAs with four uridine 3′ overhangs efficiently suppress targeted gene expression in mammalian cells. *Nat. Biotechnol.* **20:** 497–500.

Miyoshi H., Blomer U., Takahashi M., Gage F.H., and Verma I.M. 1998. Development of a self-inactivating lentivirus vector. *J. Virol.* **72:** 8150–8157.

Myslinski E., Ame J.C., Krol A., and Carbon P. 2001. An unusually compact external promoter for RNA polymerase III transcription of the human H1RNA gene. *Nucleic Acids Res.* **29:** 2502–2509.

Oliveira D.M. and Goodell M.A. 2003. Transient RNA interference in hematopoietic progenitors with functional consequences. *Genesis* **36:** 203–208.

Paddison P.J., Caudy A.A., Bernstein E., Hannon G.J., and Conklin D.S. 2002. Short hairpin RNAs (shRNAs) induce sequence-specific silencing in mammalian cells. *Genes Dev.* **16:** 948–958.

Reynolds A., Leake D., Boese Q., Scaringe S., Marshall W.S., and Khvorova A. 2004. Rational siRNA design for RNA interference. *Nat. Biotechnol.* **22:** 326–330.

Tiscornia G., Singer O., Ikawa M., and Verma I.M. 2003. A general method for gene knockdown in mice by using lentiviral vectors expressing small interfering RNA. *Proc. Natl. Acad. Sci.* **100:** 1844–1848.

Trono D. 2002. *Lentiviral vectors.* Springer-Verlag, Berlin-Heidelberg.

Zennou V., Petit C., Guetard D., Nerhbass U., Montagnier L., and Charneau P. 2000. HIV-1 genome nuclear import is mediated by a central DNA flap. *Cell* **101:** 173–185.

Zufferey R., Donello J.E., Trono D., and Hope T.J. 1999. Woodchuck hepatitis virus posttranscriptional regulatory element enhances expression of transgenes delivered by retroviral vectors. *J. Virol.* **73:** 2886–2892.

4 HIV-2 Vectors in Human Gene Therapy: Design, Construction, and Therapeutic Potential

Kevin V. Morris* and Flossie Wong-Staal†

*Division of Rheumatology, Department of Molecular and Experimental Medicine, The Scripps Research Institute, La Jolla, California 92037; †Immusol, Inc., San Diego, California 92121

ABSTRACT

Viruses in the subfamily Lentivirinae from the family Retroviridae are unique in their ability to stably infect both dividing and nondividing cells. In addition, they preferentially integrate into gene-coding regions. For these reasons, they provide a useful tool as gene-transfer vectors for virtually any target cell. Indeed, there has been much development and testing of various lentivirus-based vector systems, including systems based on human immunodeficiency viruses (HIV-1 and HIV-2), simian immunodeficiency virus (SIV), and feline immunodeficiency virus (FIV). Lentiviral vectors can be designed to be either replication-incompetent (SIN) or conditionally replicating (mobilizable) vectors. HIV-2 vectors may be particularly useful for anti-HIV-based gene therapy, because they are less pathogenic in humans than their HIV-1 counterparts and are less likely to recombine with HIV-1 in an infected cell. This chapter discusses the design, use, and testing of HIV-2 vectors as vehicles for delivering genes of interest, with an emphasis on anti-HIV-based gene therapies.

INTRODUCTION

Basic Lentiviral Biology (HIV-1, HIV-2, SIV, and FIV)

The lentiviruses (*lenti*, Latin for slow) are nononcogenic retroviruses that cause persistent infections associated with a wide range of immunological and inflammatory disorders. HIV-1 and HIV-2 specifically deplete the immune regulatory cells ($CD4^+$ and macrophages). HIV-1 is the major causative agent of acquired immunodeficiency syndrome (AIDS) worldwide (Barre-Sinoussi et al. 1983; Gallo et al. 1984). AIDS is a global pandemic with about 95% of the infected individuals living in developing countries (Temesgen 1999). HIV-2 was discovered in West Africa in 1986 (Clavel et al. 1986a,b) and was shown to be closely related to the simian immunodeficiency virus found in sooty mangabe (SIVsm). The relatively close homology of SIVsm and HIV-2 supports the hypothesis that HIV-2 originated from cross-species transmission of SIVsm from sooty mangabeys to humans (Gao et al. 1994). On the basis of both clinical and in vitro data, HIV-2 is overall less pathogenic than HIV-1, although it is also associated with AIDS (Azevedo-Pereira et al. 2005).

Like all retroviruses, lentiviruses are enveloped, positive-strand RNA viruses that use reverse transcriptase to convert their RNA genome into a DNA "provirus." The provirus is then integrated into the cellular genome. Lentiviruses have a relatively large RNA genome (~10 kb). The prototype virus, HIV-1, encodes approximately 15 proteins, including the three coding regions *gag, pol,* and *env*, which are common to all retroviruses. *gag* encodes the capsid proteins, *pol* encodes the viral enzymes (reverse transcriptase, protease, and integrase) that are necessary for replication, and *env* encodes the external glycoprotein responsible for the infectivity of the viral particle via its attachment to specific cellular receptors (Fig. 1A) (Coffin 1998). In addition, the HIV-1 genome encodes six genes (*tat, rev, vif, vpr, vpu,* and *nef*) with different roles for viral infection and replication (Fig. 1A). The HIV-2 genome is very similar to that of HIV-1, with the exception of the presence of the *vpx* gene and the absence of the *vpu* gene (Fig. 1B). Only *tat* and *rev*, however, are essential for viral replication in vitro. The HIV provirus contains a long terminal repeat (LTR) motif at both ends consisting of U3/R/U5 sequences (Fig. 1C). The single promoter for transcription is located within the U3 region, and the AAUAAA polyadenylation signal is within the R region of the LTRs. Viral transcription commences at the 5′ U3/R junction and finishes at the 3′ R/U5 junction, thus making R terminally redundant on the viral RNA genome (Fig. 1C). The production of the many viral proteins from a single primary transcript is through a complex mechanism that involves (1) generation and proteolytic processing of precursor polyproteins by protease (Pettit et al. 2003); (2) ribosomal frameshifting, which suppresses translational termination; and (3) alternative splicing of the primary transcript (Reinhart et al. 1996) or the production of bicistronic mRNAs capable of essentially producing two proteins (Coffin 1998).

Lentiviral Vectors in Gene Therapy

The discovery and knowledge gained from working with lentiviruses, especially HIV-1, led to the use of these viruses to develop gene-transfer vectors with unique properties. Lentiviral vectors are engineered to be replication-defective and to incorporate gene(s) of interest (e.g., therapeutic genes) for delivery to target cells. Their ability to stably integrate the gene(s) into both dividing and nondividing cells (Naldini et al. 1996) makes them remarkably useful.

Lentiviral vectors have been widely used in various experimental models. For example, they have been used to deliver small interfering RNAs (siRNAs) for targeted gene silencing, cDNA libraries for functional gene discovery, and specific genes for the generation of transgenic animals (for review, see Wiznerowicz and Trono 2005). Lentiviral vectors are also used in preclinical development of gene therapy. Of particular interest is the potential to use lentiviral vectors (including HIV-1, HIV-2/SIV, and FIV) to deliver anti-HIV-1 therapeutic genes to viral target cells (Buchschacher and Wong-Staal 2001; Mautino and Morgan 2002). Many of the candidate vector systems have demonstrated promising in vitro anti-HIV-1 efficacy, but the HIV-2-based vector system appears to offer the most advantages for treating HIV-1 infection. It shares many of the advantages of HIV-1 vectors, e.g., it can be mobilized by wild-type HIV-1 in vivo (D'Costa et al. 2001) and can spread to additional cells targeted by the virus; it can also compete for viral proteins such as Gag (required for virus packaging), Tat, and Rev (required for viral replication), providing additional antiviral effects in the HIV-1-infected target cells. Additionally, the HIV-2-based

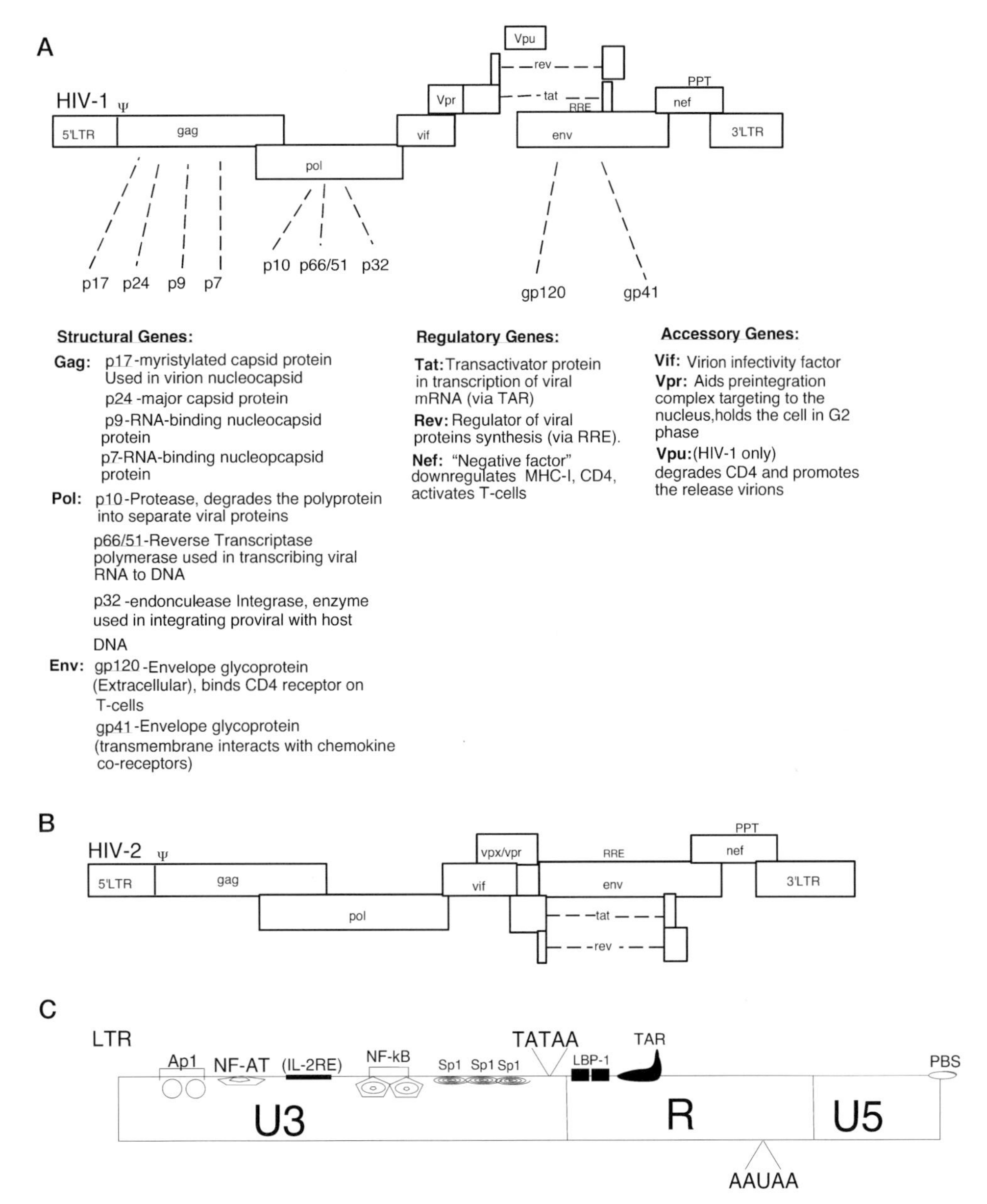

FIGURE 1. (*A*) Genomic organization of HIV-1. Dashed lines depict the splice reactions necessary to produce Tat and Rev, as well as the genetic locations of accessory and structural proteins. The structural genes and their respective gene products are listed on the left, and the regulatory genes and their respective gene products are listed on the right. The polypurine tract (PPT) and Rev response element (RRE) sites are also shown. (*B*) Genomic organization of HIV-2 with dashed lines showing the splice reactions necessary to produce the early expressed Tat and Rev. (*C*) The LTR of HIV-2 shown with transcription-factor-binding sites, the TATAA box, Tat-activating region (TAR), the AAUAA polyadenylation site, and the tRNA-binding site (PBS) are shown.

vector should be safer than an HIV-1 vector because of its lower pathogenicity and the reduced probability of generating recombinants with HIV-1 due to their overall sequence divergence.

HIV-2 lentiviral vector systems are basically a three-plasmid system composed of the HIV-2 vector, the packaging plasmid, and an envelope-expressing plasmid. The packaging plasmid provides the structural proteins encoded by *gag*, the replicative enzymes encoded in *pol*, and the essential regulatory proteins of *tat* and *rev*. The accessory genes *vif*, *vpr*, *vpx*, and *nef* are usually not incorporated into the vector system. The three plasmids are cotransfected into producer cells, usually 293 fibroblasts, and the vectors harvested from the supernatants are used to transduce the chosen target cells as depicted in Figure 2. More detailed information on the three plasmid constructs are given below.

The HIV-2 Vector

An efficacious HIV-2 vector must produce an RNA that can be packaged into infectious particles capable of completing the viral entry, reverse transcription, and proviral integration steps in the target cell as well as expressing the transgene(s) of interest upon integration. To accommodate these requirements, the appropriate HIV-2 vector should contain a few key factors as depicted in Figure 3 and described in detail below.

The vector and transgene promoters

The basic HIV-2 vector contains two LTRs. The 5′ LTR functions essentially as the promoter for vector RNA (Fig. 1C). Expression from the LTR requires the Tat protein, which can be provided in *trans* by the packaging plasmid in the producer cells (see Fig. 3). Thus, Tat is expressed in the producer cells but not in the target transduced cells unless the latter are infected with HIV-1. Interestingly, the Tat-response element (TAR) in the HIV-2 LTR can use HIV-1 Tat for transcriptional activation and can therefore compete with HIV-1 TAR in infected cells for Tat (for review, see Morris and Rossi 2004). The goal of gene therapy in this case, however, is to introduce the

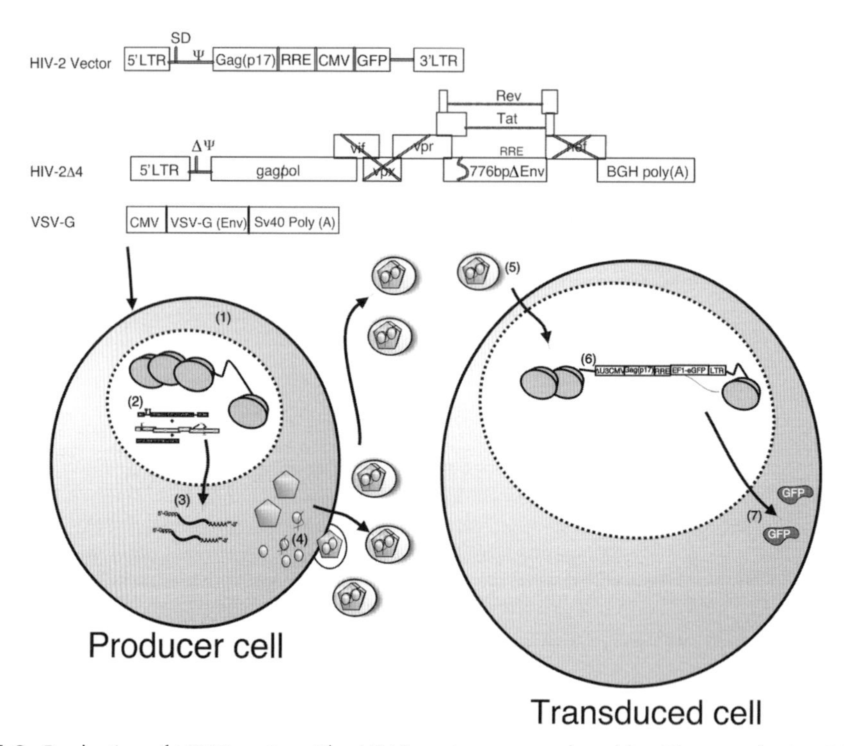

FIGURE 2. Production of HIV-2 vectors. The HIV-2 vectors are produced by (*1*) cotransfecting 293T producer cells with the HIV-2 vector, the packaging plasmid (HIV-2Δ4), and the envelope plasmid (VSV-G). (*2*) Both the vector RNA and mRNA encoding the viral regulatory and structural proteins are transcribed in the nucleus. (*3*) Upon the expression of functional Tat and Rev proteins, transcription is enhanced from the LTR/promoters, and the vector RNA and mRNA for *gag-pol* are exported from the nucleus to the cytoplasm. (*4*) All of the necessary components of the vector are expressed and (*5*) vector RNA is packaged into the budding particles. (*6*) Approximately 48–72 hours later, the culture supernatants are collected, and vector concentration is determined by titering on target cells as measured by GFP expression (*7*).

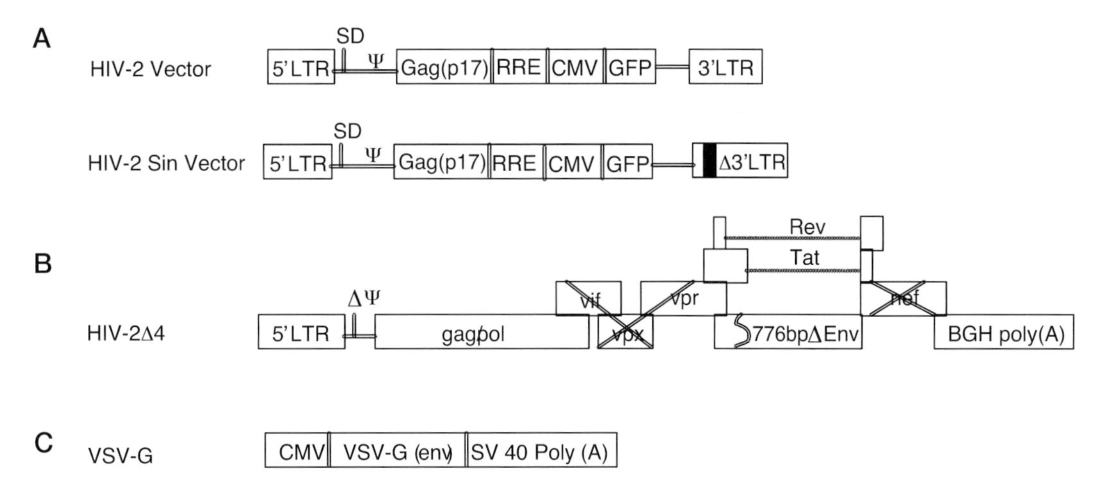

FIGURE 3. HIV-2 vector systems. (*A*) HIV-2 vectors can be designed to be mobilization-competent (HIV-2 vector), i.e., containing a functional 3′ LTR, or self-inactivated (HIV-2 Sin vector), i.e., containing a deletion in the U3 of the 3′ LTR. Vectors are shown with the GFP marker gene expressed from the CMV promoter. However, antibiotic resistance or other marker genes could be used in place of GFP. Antiviral modalities such as siRNAs or ribozymes can be cloned upstream of the CMV-GFP marker gene. (*B*) The HIV-2 packaging plasmid containing deletions of *vif, vpr, vpx,* and *nef* as well as an approximately 776-bp deletion of the HIV-2 envelope is shown. The RRE site is also shown in the packaging plasmid to ensure efficient export of the packaging mRNA from the nucleus to the cytoplasm. (*C*) The VSV-G envelope used to enhance cell tropism as well as function in place of the Δ776-bp HIV-2 envelope.

antiviral gene(s) before the cells become infected; therefore, a Tat-independent internal promoter should be used to drive transgene expression. This promoter can be a strong, broadly expressing promoter such as cytomegalovirus (CMV) (Fig. 3A) or one with cell-type specificity. To minimize expression of any HIV-2 sequences in the target cells, and to avoid upstream promoter interference with the downstream transgene promoter, the 3′ LTR can be mutated to generate self-inactivating (SIN) vectors. SIN vectors contain a deletion of the promoter and enhancer elements in the U3 region of the 3′ LTR (Figs. 1C and 3A). The deletion will be maintained in the U3 region of the 3′ LTR in the vector RNA, and upon reverse transcription in the target cells, it will be duplicated in both LTRs of the provirus. The mutation in the U3 region of the 5′ LTR of the integrated vector DNA will abolish any promoter function that drives the expression of vector RNA. SIN vectors offer an added layer of safety, because they reduce the likelihood that the integrated vector can express viral sequences and subsequently, in the context of an HIV-1-infected cell, be packaged or recombine with HIV-1. Careful analysis of SIN vectors revealed that the optimal deletions include the downstream binding factor 1 (DBF1) and SP1 sites in U3 (Figs. 1C and 3A) (Logan et al. 2004).

Primer-binding sequence and polypurine tract

The LTRs and adjacent sequences at both ends of the vector genome are essential for the reverse transcription process. An 18-nucleotide sequence immediately downstream from U5 toward the 5′ end constitutes the primer-binding site where reverse transcription is initiated to generate minus (–) strand strong-stop DNA. The primer used in this reaction is tRNAlys[3], which anneals to the primer-binding site and is incorporated into the virion in the producer cell. Positive-strand DNA synthesis is initiated discontinuously in two halves, with one half beginning at a polypurine tract located toward the 3′ end (3′ PPT) and another half initiating from a centrally located polypurine tract (cPPT). As a consequence, a short overlap in the positive strand is generated during reverse transcription, resulting in a triple-strand structure referred to as the central DNA flap. It has been reported that this central DNA flap may have a critical role in nuclear import of the preintegration complex in nondividing cells (Sirven et al. 2000; Zennou et al. 2000). However, other studies also showed that cPPT viral mutants are fully capable of infecting nondividing cells (Dvorin et al. 2002; Limon et al. 2002).

Rev response element

The HIV-2 vector RNA expressed in the producer cells needs to be exported from the nucleus to the cytoplasm as unspliced RNA. This requires the presence of the Rev response element (RRE) in *cis* and the Rev protein provided in *trans* (Malim and Cullen 1993). Rev binds to the RRE through its arginine-lysine-rich domain, whereas its leucine-rich nuclear export sequence (NES) binds to the nuclear export protein CRM-1. These interactions result in trafficking of the Rev/RRE-containing vector RNA complex to the cytoplasm (for review, see Vaishnav and Wong-Staal 1991; Tang et al. 1999). For these reasons, the RRE site must be incorporated into the vector RNA (see Fig. 3A).

Packaging signal (ψ)

After export to the cytoplasm, the vector RNA needs to be encapsulated into infectious particles to be used as gene-transfer vectors. The encapsidation of viral and vector RNAs requires a *cis*-acting element known as ψ or packaging signal (Arya 1988; Garzino-Demo et al. 1995; Poeschla et al. 1998; D'Costa et al. 2001). ψ has been mapped by deletion analysis in HIV-2 to a region upstream of the splice donor site (McBride et al. 1997). Other investigators, however, have shown that regions located downstream from the major splice donor function in packaging (Poeschla et al. 1998) and as a negative regulator (Garzino-Demo et al. 1995). Thus, for efficient packaging of the HIV-2 vectors, ψ is required as well as a short stretch of *gag* (~250 bp; Fig. 3A).

Packaging Vectors

The HIV-2 vector RNA contains all of the *cis*-acting elements for vector production as well as the transgene expression cassette. All of the *trans*-acting viral structural and replicative proteins, with the exception of the envelope proteins, are provided by the packaging plasmid (Fig. 3B). The HIV-2 packaging plasmid contains sequences that encode the Gag and Pol proteins as well as Tat and Rev. This complement of proteins enables transcription, export, and encapsidation of the vector RNA. At the same time, the packaging plasmid contains a deletion of ψ to remove the possibility of its incorporation into the vector particle(s). Also deleted from the HIV-2 packaging plasmid are the *vif, vpr, vpx,* and *nef* genes, as well as an approximately 700-bp fragment of envelope. Interestingly, HIV-2 vectors can be cross-packaged by HIV-1 and FIV packaging systems (Browning et al. 2001; Morris et al. 2004). The protocol for cross-packaging HIV-2 vectors with either HIV-1 or FIV follows the same protocol for homologous packaging of HIV-2 vectors, with the HIV-2 packaging plasmid being exchanged for either the HIV-1 or FIV packaging plasmids (Fig. 2). The use of cross-packaged vectors could provide an added layer of safety when using HIV-2 vectors and FIV proteins in a therapeutic setting.

Envelope Plasmids

The envelope of HIV-2, which has a narrow tropism for $CD4^+$ cells, is deleted, and a heterologous envelope gene is provided to either broaden the tropism or confer one of a different selectivity. Usually, a vesicular stomatitis virus G (VSV-G) glycoprotein is used because of its broad tropism and high efficiency of infection of most target cells (see Fig. 3C). However, other viral envelope proteins, e.g., xenotropic or amphotropic murine retroviruses, can also be used.

Therapeutic Potential of HIV-2 Vectors

Both HIV and Moloney murine leukemia virus (Mo-MLV) vectors can stably transduce cell populations. However, the lentiviral HIV-based vectors offer significant advantages. For instance, Mo-MLV-based vectors have been plagued by reduced transgene expression and a predisposition for the vector to integrate into active promoter regions and induce oncogenesis (Wu et al. 2003). Lentiviruses, on the other hand, tend to preferentially integrate downstream from active promoters within the active transcriptional unit, potentially limiting their overall oncogenicity (Wu et al. 2003). Moreover, lentivirus-based vectors are capable of transducing nondividing cells, as their preintegration complex is able to traverse the intact nuclear membrane (Buchschacher and Wong-Staal 2000; Greber and Fassati 2003). This feature is valuable for gene-therapy targeting cells such as neuronal cells, which are not actively dividing. A final advantage to the use of lentiviral vectors, such as HIV-1 or HIV-2, is specific to their use in anti-HIV-1 therapy. These vectors have intrinsic antiviral effects due to their ability to compete with the wild-type virus for packaging and *trans* elements and their ability to be mobilized and spread to the same target cells infected by the virus.

Lentiviral vectors such as those based on HIV-1, HIV-2/SIV, or FIV are generally produced as depicted in Figure 2 and are capable of stably transducing many cell types, including $CD34^+$ hematopoietic precursor cells (Gervaix et al. 1997), and integrating and expressing the desired transgenes (Poeschla et al. 1996; Price et al. 2002; Quinonez and Sutton 2002; Yam et al. 2002). The current experimental protocol for using lentiviral vectors to treat HIV-1 is to transduce the primary T lymphocytes or the relatively quiescent hematopoietic stem cells (HSC) ex vivo with lentiviral vectors containing anti-HIV-1 modalities such as siRNAs or ribozymes (for review, see Morris and Rossi 2004) and then reinfuse the transduced cultures into the autologous patients after myeloablation. A similar approach has been carried out experimentally in baboons and shown to require the addition of multiple growth factors to the transduced, repopulating HSC (Horn et al. 2002). For reasons that are not yet entirely clear, HIV-1- or HIV-2-based lentiviral vectors offer better delivery and expression of foreign transgenes in HSC and human T lymphocytes than do FIV vectors (Price et al. 2002; Yam 2002) and as such are good candidate vector systems to employ in gene-therapy-based strategies to treat HIV-1 infection. Although the FIV vectors can transduce target genes into HSCs, their expression appears to be severely limited (Price et al. 2002).

Recently, lentiviruses such as HIV-1 and HIV-2 have been shown to be capable of cross-packaging each other (White et al. 1999; Browning et al. 2001; Goujon et al. 2003), and FIV has been shown to cross-package HIV-1- and HIV-2-based lentiviral vectors (Morris et al. 2004). The FIV-packaged HIV-2 vectors containing four ribozymes targeting HIV-1 can stably transduce and protect human primary blood mononuclear cells from HIV-1 infection (Morris et al. 2004). These relatively new observations of lentiviral cross-packaging offer alternative methods that combine the positive features of different vector systems. For instance, FIV-packaged HIV-1 or HIV-2 vectors reduce the likelihood of immune recognition, or seroconversion, due to exposure to HIV structural proteins. At the same time, in particular cell types such as stem cells, HIV-based vectors are markedly enhanced in marker gene expression relative to FIV vectors (Price et al. 2002). Thus, FIV-packaged HIV-2 vectors could be used to both reduce seroconversion or immune recognition while simultaneously maintaining the highest level of transgene expression within the transduced HSC. Finally, lentiviral vectors can be pseudotyped (Kobinger et al. 2001; Sandrin et al. 2003) or designed with a receptor-ligand bridge (Boerger et al. 1999) to target specific cell types.

Protocol

Production of HIV-2 Vector Particles for Gene Delivery

The production of the HIV-2 vector and packaging can be initiated from HIV-2 proviral clones, some of which are available through the AIDS Research and Reference Reagent Program at the U.S. National Institute of Allergy and Infectious Diseases (http://www.aidsreagent.org/). Many groups have now constructed various versions of the respective HIV-2 vectors and packaging plasmids, so it is advisable to contact the investigators directly for these materials (Arya et al. 1998; Morris et al. 2004). The CMV-expressed VSV-G envelope plasmids are readily available from many commercial sources. The three plasmids are cotransfected into producer cells and the resultant vector is collected over time.

MATERIALS

CAUTION: See Appendix for appropriate handling of materials marked with <!>.

Reagents

$CaCl_2$ (1 M) <!>
Cell lines for transfection and transduction: 293T or 293FT human embryonic kidney cells
DNA for producing HIV-2 vector particles
  HIV-2 vector DNA
  packaging vector (HIV-1, HIV-2, or FIV)
  envelope plasmid (VSV-G)
Dulbecco's modified Eagle's medium (DMEM) supplemented with 10% fetal bovine serum (FBS), amino acids, and an antibiotic-antimycotic mixture, if desired
2x HEPES-buffered saline (HBS)
  1 g of HEPES acid (Sigma-Aldrich H3375)
  1.6 g of NaCl
  0.72 ml of 0.25 M Na_2HPO_4 <!>
  1 ml of 1 M KCl <!>
    Bring to 100 ml with double distilled H_2O. Adjust the pH to exactly 7.12 using NaOH <!> (5 M and 1 M).
Paraformaldehyde (2%) <!> in PBS
Phosphate-buffered saline (PBS)
Polybrene stock solution (hexadimethrine bromide; 8 mg/ml in PBS)
Trypsin <!>

Equipment

Flow cytometry equipment
Fluorescence microscope
Rotors
  fixed-angle 45 Ti (Beckman-Coulter)
  swinging bucket SW 41 Ti (Beckman-Coulter)

Standard tissue-culture equipment, including 100-mm tissue-culture dishes and 12-well plates
Syringe filter (0.45-μm pore size)
Tabletop centrifuge (Beckman TJ-6)

METHODS

HIV-2 Vector Production (Calcium Phosphate Method)

1. One day before transfection, plate approximately 4.0 x 10^6 293T or 293FT human embryonic kidney cells in a 100-mm tissue-culture dish, generally in 8–10 ml of DMEM with 10% FBS, amino acids, and antibiotic if desired. Incubate overnight at 37°C in a humidified incubator with an atmosphere of 5% CO_2.
2. Change the medium 1–2 hours before transfection.
3. Prepare the calcium phosphate–DNA coprecipitate by mixing together 15 μg of HIV-2 vector DNA, 10 μg of packaging vector (HIV-1, HIV-2, or FIV), and 5 μg of the VSV-G envelope plasmid. Add 374 μl of sterile H_2O and 126 μl of 1 M $CaCl_2$ to the plasmid mixture.
4. Transfer the plated 293T producer cells from the incubator to a tissue-culture hood.
5. While vortexing the plasmid mixture from Step 3, add 500 μl of 2x HBS dropwise with a sterile 1-ml pipette. Use the same pipette to transfer the calcium phosphate–DNA suspension onto the producer cells in a dropwise manner. Transfer the culture back into the incubator.
6. Approximately 6–8 hours after transfection (Step 5), gently remove the medium from the dish and replace with approximately 10 ml of fresh medium.

Collection of Vector Particles

7. On day 2 after transfection, carefully transfer the culture supernatant to a 50-ml conical tube and replace with fresh medium. Store the removed supernatant at 4°C.
8. On day 3 after transfection, carefully collect the supernatant again and add it directly to the sample that has been saved in conical tubes at 4°C from the previous day. If the cells remain viable, add fresh medium and repeat this step 24 hours later.
9. Centrifuge the collected supernatants at 2800 rpm in a tabletop centrifuge (e.g., Beckman TJ-6) for 15 minutes to remove unwanted cellular debris. Transfer the supernatant to a new tube and sterilize using a 30-ml syringe and a 0.45-μm pore size filter. Save the filtered supernatant in 2-ml aliquots at –80°C or concentrate as described in Steps 10–12.

   Store the filtered supernatant as aliquots to avoid thawing the entire stock of vector when determining titers.

Vector Concentration

10. Centrifuge the filtered supernatant containing the vector at about 42,000*g* (19,000 rpm using a fixed-angle 45 Ti rotor) for 2 hours.
11. Discard the supernatant, invert the tube on sterile gauze, and allow it to dry for 2 minutes.
12. Wipe the walls of the tube with sterile gauze and resuspend the pellet in 20–50 μl of 1x PBS. Combine all aliquots and freeze at –80°C until needed.

    The vector can be stored for at least 1 year. Longer storage periods may reduce the titer of the vector.

Vector Titer Determination

Vector titering can be done on virtually any cell type. This protocol assumes that titering is performed on 293T cells and is based on green fluorescent protein (GFP) transgene expression (Fig. 2).

13. The day before the titer experiment, plate 293T cells at a density of 2×10^4 to 4×10^4 cells/well in a 12-well plate. Incubate overnight at 37°C in a humidified incubator with an atmosphere of 5% CO_2.
14. The day of the titer experiment, make serial 1.0-ml dilutions (five- or tenfold intervals) of the vector from approximately 10^2 particles/ml to 10^6 particles/ml in medium containing 8 µg/ml Polybrene. Count the cells in one well of the 12-well plate in triplicate to obtain a cell count for the day of transduction. Transfer the diluted vectors to each respective well and incubate overnight.

    The expected titer of the vector is about 10^6 transduction units (TU)/ml.
15. The next day, aspirate the medium and replace with fresh medium. Culture the cells for a further 2–4 days.
16. Remove the medium and add approximately 500 µl of trypsin to the cell monolayer. Incubate for about 1–2 minutes at room temperature. Remove the trypsin, knock the cells loose, and collect the contents of each well in approximately 1 ml of PBS. Fix the samples and the negative control (nontransduced) cells with 2% paraformaldehyde in PBS. Measure GFP expression of the samples by flow cytometry.
17. Determine the dilution in which 50% of the cells are GFP-positive by plotting the number of vector particles per well versus the percentage of GFP-positive cells. Compute the number of TU/ml of vector stock from the resultant dilution.

    Another formula for calculating the titer of the vector is TU/ml = (% GFP-positive cells/100) × (number of cells at transduction) × (dilution factor).

Transduction of Adherent Cells

18. The day before transduction, plate adherent cells at a density of approximately 1×10^5 cells/well in a six-well plate or 4×10^6 cells in a 100-mm dish. Incubate overnight at 37°C in a humidified incubator with an atmosphere of 5% CO_2.
19. The day of transduction, remove the medium. Add 2 ml of new medium without FBS, 1 ml of freshly thawed frozen vector stock, and Polybrene to a final concentration of 8 µg/ml. Incubate for 2 hours at 37°C in a humidified incubator with an atmosphere of 5% CO_2.
20. Aspirate the supernatants and add fresh culture medium to the cells. Incubate for 2–4 days at 37°C in the humidified incubator with an atmosphere of 5% CO_2. Assay for GFP expression by flow cytometry or microscopy.

Transduction of Suspension Cells

21. Collect the suspension cells, count, and pellet approximately 3×10^6 to 5×10^6 cells by centrifuging in a 15-ml polypropylene tube at 500*g* for 5 minutes.
22. Resuspend the pellet in 1 ml of medium without FBS.
23. Add 1 ml of freshly thawed vector (from the frozen stock) and mix gently.
24. Add Polybrene to a final concentration of 8 µg/ml.
25. Centrifuge the cell/vector/Polybrene solution at 1000*g* for 2 hours at room temperature.

26. Aspirate the supernatants and resuspend the cell pellet in culture medium. Incubate for 2–4 days at 37°C in a humidified incubator with an atmosphere of 5% CO_2. Assay GFP expression by flow cytometry or microscopy.

ACKNOWLEDGMENTS

The authors thank Dr. David J. Looney for his insightful discussions on HIV-2 vector design and application.

REFERENCES

Arya S.K. 1988. Human and simian immunodeficiency retroviruses: Activation and differential transactivation of gene expression. *AIDS Res. Hum. Retrovir.* **4:** 175–186.

Arya S.K., Zamani M., and Kundra P. 1998. Human immunodeficiency virus type 2 lentivirus vectors for gene transfer: Expression and potential for helper virus-free packaging. *Hum. Gene Ther.* **9:** 1371–1380.

Azevedo-Pereira J.M., Santos-Costa Q., and Moniz-Pereira J. 2005. HIV-2 infection and chemokine receptors usage—Clues to reduced virulence of HIV-2. *Curr. HIV Res.* **3:** 3–16.

Barre-Sinoussi F., Chermann J.-C., Rey R., Nugeyre M.T., Chamaret S., Gruest J., Dauguet C., Axler-Blin C., Vezinet-Brun F., Rouzioux C., Rozenbaum W., and Montagnier L. 1983. Isolation of a T-lymphotropic retrovirus from a patient at risk for acquired immune deficiency syndrome (AIDS). *Science* **220:** 868–871.

Boerger A.L., Snitkovsky S., and Young J.A.T. 1999. Retroviral vectors preloaded with a viral receptor-ligand bridge protein are targeted to specific cell types. *Proc. Natl. Acad. Sci.* **96:** 9867–9872.

Browning M.T., Schmidt R.D., Lew K.A., and Rizvi T.A. 2001. Primate and feline lentivirus vector RNA packaging and propagation by heterologous lentivirus virions. *J. Virol.* **75:** 5129–5140.

Buchschacher G.L., Jr. and Wong-Staal F. 2000. Development of lentiviral vectors for gene therapy for human diseases. *Blood* **95:** 2499–2504.

———. 2001. Approaches to gene therapy for human immunodeficiency virus infection. *Hum. Gene Ther.* **12:** 1013–1019.

Clavel F., Brun-Vezinet F., Guetard D., Chamaret S., Laurent A., Rouzioux C., Rey M., Katlama C., Rey F., Champelinaud J.L., et al. 1986a. LAV type II: A second retrovirus associated with AIDS in West Africa (in French). *C.R. Acad. Sci. III* **302:** 485–488.

Clavel F., Guetard D., Brun-Vezinet F., Chamaret S., Rey M.A., Santos-Ferreira M.O., Laurent A.G., Dauguet C., Katlama C., Rouzioux C., et al. 1986b. Isolation of a new human retrovirus from West African patients with AIDS. *Science* **233:** 343–346.

Coffin J.M. 1996. Retroviridae: The viruses and their replication. In *Fields virology* (ed. B.N. Fields, et al.), pp. 1767–1847. Raven Press, New York.

D'Costa J., Brown H., Kundra P., Davis-Warren A., and Arya S. 2001. Human immunodeficiency virus type 2 lentiviral vectors: Packaging signal and splice donor in expression and encapsidation. *J. Gen. Virol.* **82:** 425–434.

Dvorin J.D., Bell P., Maul G.G., Yamashita M., Emerman M., and Malim M.H. 2002. Reassessment of the roles of integrase and the central DNA flap in human immunodeficiency virus type 1 nuclear import. *J. Virol.* **76:** 12087–12096.

Gallo R.C., Salahuddin S.Z., Popovic M., Shearer G.M., Kaplan M., Haynes B.F., Palker T.J., Redfield R., Oleske J., Safai B., et al. 1984. Frequent detection and isolation of cytopathic retrovirus (HTLV-III) from patients with AIDS and at risk for AIDS. *Science* **224:** 500–503.

Gao F., Yue L., Robertson D.L., Hill S.C., Hui H., Biggar R.J., Neequaye A.E., Whelan T.M., Ho D.D., Shaw G.M., et al. 1994. Genetic diversity of human immunodeficiency virus type 2: Evidence for distinct sequence subtypes with differences in virus biology. *J. Virol.* **68:** 7433–7447.

Garzino-Demo A., Gallo R.C., and Arya S.K. 1995. Human immunodeficiency virus type 2 (HIV-2): Packaging signal and associated negative regulatory element. *Hum. Gene Ther.* **6:** 177–184.

Gervaix A., Schwarz L., Law P., Ho A.D., Looney D., and Wong-Staal F. 1997. Gene therapy targeting peripheral blood CD34+ hematopoietic stem cells of HIV-infected individuals. *Hum. Gene Ther.* **8:** 2229–2238.

Goujon C., Jarrosson-Wuilleme L., Bernaud J., Rigal D., Darlix J., and Cimarelli A. 2003. Heterologous human immunodeficiency virus type 1 lentiviral vectors packaging a simian immunodeficiency virus-derived genome display a specific postentry transduction defect in dendritic cells. *J. Virol.* **787:** 9295–9304.

Greber U.F. and Fassati A. 2003. Nuclear import of viral DNA genomes. *Traffic* **4:** 136–143.

Horn P.A., Morris J.C., Bukovsky A.A., Andrews R.G., Naldini L., Kurre P., and Kiem H.-P. 2002. Lentivirus-mediated gene transfer into hematopoietic repopulating cells in baboons. *Gene Ther.* **9:** 1464–1471.

Kobinger G.P., Weiner D.J., Yu Q., and Wilson J.M. 2001. Filovirus-pseudotyped lentiviral vector can efficiently and stably transduce airway epithelia in vivo. *Nat. Biotechnol.* **19:** 225–230.

Limon A., Nakajima N., Lu R., Ghory H.Z., and Engelman A. 2002. Wild-type levels of nuclear localization and human immunodeficiency virus type 1 replication in the absence of the central DNA flap. *J. Virol.* **76:** 12078–12086.

Logan A.C., Haas D.L., Kafri T., and Kohn D.B. 2004. Integrated self-inactivating lentiviral vectors produce full-length genomic transcripts competent for encapsidation and integration. *J. Virol.* **78:** 8421–8436.

Malim M.H. and Cullen B.R. 1993. Rev and the fate of pre-mRNA in the nucleus: Implications for the regulation of

RNA processing in eukaryotes. *Mol. Cell. Biol.* **13:** 6180–6189.

Mautino M.R. and Morgan R.A. 2002. Gene therapy of HIV-1 infection using lentiviral vectors expressing anti-HIV-1 genes. *AIDS Patient Care STDs* **16:** 11–26.

McBride M.S., Schwartz M.D., and Panganiban A.T. 1997. Efficient encapsidation of human immunodeficiency virus type 1 vectors and further characterization of *cis* elements required for encapsidation. *J. Virol.* **71:** 4544–4554.

Morris K.V. and Rossi J.J. 2004. Anti-HIV-1 gene expressing lentiviral vectors as an adjunctive therapy for HIV-1 infection. *Curr. HIV Res.* **2:** 185–191.

Morris K.V., Gilbert J., Wong-Staal F., Gasmi M., and Looney D.J. 2004. Transduction of cell lines and primary cells by FIV-packaged HIV vectors. *Mol. Ther.* **10:** 181–190.

Naldini L., Blomer U., Gallay P., Ory D., Mulligan R., Gage F.H., Verma I.M., and Trono D. 1996. In vivo gene delivery and stable transduction of nondividing cells by a lentiviral vector. *Science* **272:** 263–267.

Pettit S.C., Gulnik S., Everitt L., and Kaplan A.H. 2003. The dimer interfaces of protease and extra-protease domains influence the activation of protease and the specificity of GagPol cleavage. *J. Virol.* **77:** 366–374.

Poeschla E., Corbeau P., and Wong-Staal F. 1996. Development of HIV vectors for anti-HIV gene therapy. *Proc. Natl. Acad. Sci.* **93:** 11395–11399.

Poeschla E., Gilbert J., Li X., Huang S., Ho A., and Wong-Staal F. 1998. Identification of a human immunodeficiency virus type 2 (HIV-2) encapsidation determinant and transduction of nondividing human cells by HIV-2 based lentivirus vectors. *J. Virol.* **72:** 6527–6536.

Price M.A., Case S.S., Carbonaro D.A., Yu X.J., Petersen D., Sabo K.M., Curran M.A., Engel B.C., Margarian H., Abkowitz J.L., Nolan G.P., and Kohn D.B. 2002. Expression from second-generation feline immunodeficiency virus vectors is impaired in human hematopoietic cells. *Mol. Ther.* **6:** 645–652.

Quinonez R. and Sutton R.E. 2002. Lentiviral vectors for gene delivery into cells. *DNA Cell Biol.* **12:** 937–951.

Reinhart T.A., Rogan M.J., and Haase AT. 1996. RNA splice site utilization by simian immunodeficiency viruses derived from sooty mangabey monkeys. *Virology* **224:** 338–344.

Sandrin V., Russell S.J., and Cosset F.L. 2003. Targeting retroviral and lentiviral vectors. *Curr. Top. Microbiol. Immunol.* **281:** 137–178.

Sirven A., Pflumio F., Zennou V., Titeux M., Vainchenker W., Coulombel L., Dubart-Kupperschmitt A., and Charneau P. 2000. The human immunodeficiency virus type-1 central DNA flap is a crucial determinant for lentiviral vector nuclear import and gene transduction of human hematopoietic stem cells. *Blood* **96:** 4103–4110.

Tang H., Kuhen K.L., and Wong-Staal F. 1999. Lentivirus replication and regulation. *Annu. Rev. Genet.* **33:** 133–170.

Temesgen Z. 1999. Overview of HIV infection. *Ann. Allergy Asthma Immunol.* **83:** 1–5.

Vaishnav Y.N. and Wong-Staal F. 1991. The biochemistry of AIDS. *Annu. Rev. Biochem.* **60:** 577–630.

White S.M., Renda M., Nam N.Y., Klimatcheva E., Zhu Y., Fisk J., Halterman M., Rimel B.J., Federoff H., Pandya S., Rosenblatt J.R., and Planelles V. 1999. Lentivirus vectors using human and simian immunodeficiency virus elements. *J. Virol.* **73:** 2832–2840.

Wiznerowicz M. and Trono D. 2005. Harnessing HIV for therapy, basic research and biotechnology. *Trends Biotechnol.* **23:** 42–47.

Wu X., Li Y., Crise B., and Burgess S.M. 2003. Transcription start regions in the human genome are favored targets for mlv integration. *Science* **300:** 1749–1751.

Yam P.Y., Li S., Wu J.U., Hu J., Zaia J.A., and Yee J. 2002. Design of HIV vectors for efficient gene delivery into human hematopoietic cells. *Mol. Ther.* **5:** 479–484.

Zennou V., Petit C., Guetard D., Nerhbass U., Montagnier L., and Charneau P. 2000. HIV-1 genome nuclear import is mediated by a central DNA flap. *Cell* **101:** 173–185.

WWW RESOURCE

http://www.aidsreagent.org/ NIH AIDS Research and Reference Reagent Program.

5 SIV Vectors as Vehicles for DNA Delivery

Els Verhoeyen, François-Loïc Cosset, and Didier Nègre

INSERM, U758, Lyon, F-69007 France; Ecole Normale Supérieure de Lyon, Lyon, F-69007 France; IFR128 BioSciences Lyon-Gerland, Lyon, F-69007 France

ABSTRACT

The acquisition of vectors suitable for in vivo gene delivery has been a recurrent theme in gene therapy research. Several challenging hurdles must be overcome to reach this goal. First, the vectors must be prepared at high titers. Second, to avoid inactivation, the gene-transfer vectors should not be recognized by the host immune system. Third, the vectors should be able to replicate and to express a transgene in cells that are nonproliferating or slowly proliferating, a predominant situation in vivo.

Lentiviral vectors derived from simian immunodeficiency virus (SIV) have now been generated in several laboratories (Nègre et al. 2002). Characterization of these vectors has indicated that they are similar to those derived from human immunodeficiency virus (HIV-1 or HIV-2) in that they can insert transgenes in nonproliferating cells. It is becoming clear, however, that SIV vectors perform better than HIV-1 vectors in simian cells (Nègre et al. 2000) and thus may be a valid alternative to HIV-1-based vectors in the early phases of the clinical testing of lentiviral vectors in nonhuman primate models. This chapter presents methods that use minimal SIV packaging systems and SIV gene-transfer vectors to generate high titers and safe lentiviral vectors derived from SIV.

INTRODUCTION

SIV Proviral Genome: Key Features for the Design of SIV-derived Vectors

The genetic organization of SIVs is similar to that of HIV-1 and HIV-2, although there are some exceptions (Gardner et al. 1993; Clements and Zink 1996). SIV proviral genomes (Fig. 1A) are composed of two identical long terminal repeats (LTRs) flanking the coding regions for the structural genes, *gag, pol,* and *env*, which form the viral particles. Like the human lentiviruses, SIVs harbor two regulatory proteins, Tat and Rev, that govern the expression of the viral gene at both transcriptional and posttranscriptional levels. Most SIV isolates also code for virulence factors or accessory proteins, Vpr, Vpx, Vif, and Nef, that are necessary for pathogenicity but not for replication of the viruses in cell culture. The *vpx* gene is found also in HIV-2 but not in HIV-1 isolates, whereas the *vpu* gene, found in HIV-1 but not in HIV-2, is absent from most SIV isolates. Expression of some of these accessory genes in vector producer cells has a strong impact on the performance of the vectors (Mangeot et al. 2002). Like other retroviruses, the SIV LTR is divided into the U3, R, and U5 regions. Signal sequences in U3 regulate the level and the start site of transcription. The R region contains a *cis*-acting element, the *trans*-acting responsive (TAR) sequence, which is an essential regulatory motif mediating *trans*-activation by the virus-encoded Tat protein. The TAR element forms a stable secondary structure that binds Tat as a complex with cellular factors, allowing elongation of the viral transcript. The R region also contains the polyadenylation site for SIV RNAs. An essential determinant of the viral RNA, the packaging sequence, drives the packaging of two copies of the genomic RNA into the viral particles (Guan et al. 2000); additional sequences located in R-U5 and in the beginning of the *gag* gene are also likely to be part of this critical motif (Das et al. 1997; Kaye and Lever 1998; Dorman and Lever 2000).

Reverse transcription requires the presence of several *cis*-acting elements that are responsible for the initiation of DNA synthesis. They include (Fig. 1A) (1) the primer-binding site (PBS), located at the beginning of the leader region between the 5′ LTR and the *gag* gene, which binds a tRNA that primes synthesis of the negative strand of proviral DNA; (2) the central polypurine track (cPPT) and the 3′ PPT regions that are both used to prime the synthesis of the positive strand of the viral DNA; and (3) a conserved uridine-rich sequence located immediately upstream of the PPT in various lentiviral strains (Ilyinskii and Desrosiers 1998). During the reverse transcription process, the 3′ U3 region is used as a template for formation of the U3 regions of both the 5′ and 3′ LTRs. Interaction of integrase with both ends of the LTRs leads to insertion of the viral DNA into the host-cell chromatin.

Optimizing SIV Transfer Vectors

Replication-competent vectors derived from SIV have been generated by replacing the *nef* gene with heterologous sequences (Giavedoni and Yilma 1996). These vectors were shown to be capable of replicating not only their own genome, but also the transgenes inserted into their genomes. Although they represent attenuated and probably not pathogenic forms of the initial viruses due to the deletion of the *nef* virulence gene, such vectors are nevertheless unsuitable for most gene therapy applications and have in fact been designed as live attenuated vaccine candidates against AIDS (Ruprecht 1999).

In general, retroviral vectors can be rendered replication-defective by deleting the critical genes involved in viral replication, i.e., the *gag-pol* and *env* genes. The transgenes and some regulatory elements (promoters, enhancers, and intron sequences) are inserted into the deleted virus, referred to as the transfer vector (Fig. 1B), which still retains most of the viral *cis*-acting elements such as LTRs, packaging sequences, and regions involved in reverse transcription (PBS and PPT). Some indispensable *cis*-acting sequences in the vector overlap with *trans*-acting regions in all retroviral genomes (Fig. 1A). Consequently, most transfer vectors still contain residual sequences derived from genes of the parental virus and whose coding capacities have been inactivated. This, for example, is true of the packaging sequence that extends into the *gag* gene, the Rev-responsive element (RRE) region located in the *env*-coding sequence, the cPPT/CTS (central termination sequence) sequences positioned in the *pol* gene, and the PPT sequence that extends into the *nef* gene (Fig. 1A). Such redundancies between packaging and vector genomes should be avoided whenever possible, or at least minimized, to reduce the possibility of recombination with the

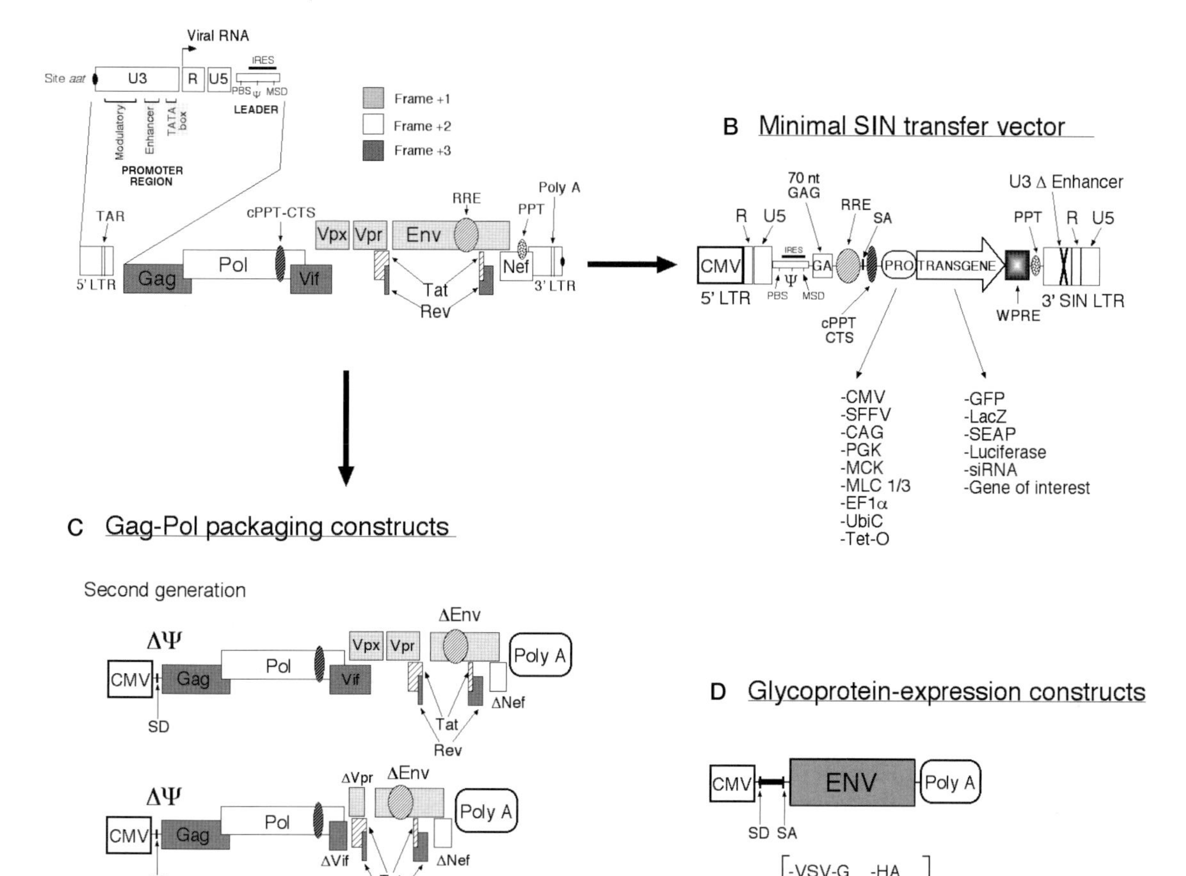

C Gag-Pol packaging constructs

Second generation

ΔΨ CMV SD Gag Pol Vif Vpx Vpr ΔEnv Poly A Tat Rev ΔNef

ΔΨ CMV SD Gag Pol ΔVif ΔVpr ΔEnv Poly A Tat Rev ΔNef

Third generation

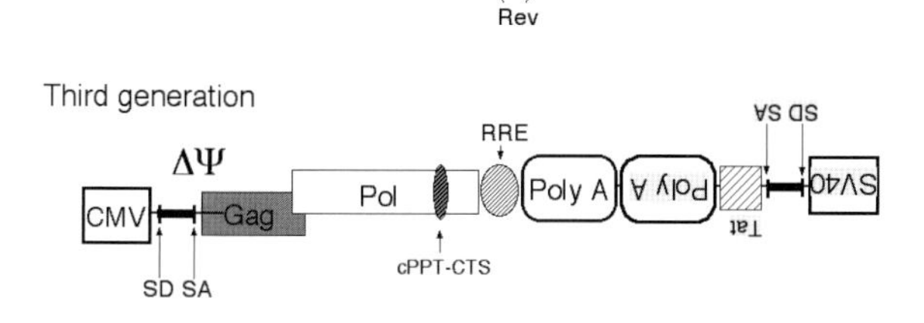

D Glycoprotein-expression constructs

CMV SD SA ENV Poly A

-VSV-G -RD114 -EBO -MLV-A -MLV-E -HA -LCMV -GALV -10A1 -HCV

E Accessory gene construct

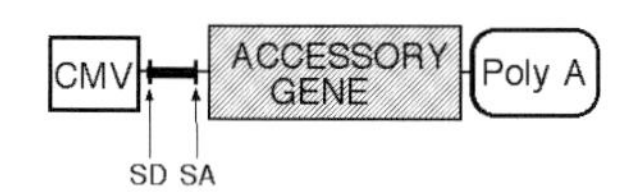

FIGURE 1. Generation of vectors derived from SIV. The genomes of infectious molecular clones of SIV (*A*) were dismantled to derive constructs carrying the transfer vector (*B*) and constructs encoding the packaging functions (*C*). Different versions of the Gag-Pol packaging constructs were generated to progressively eliminate the viral sequences unnecessary for formation of vector particles. A similar approach was used to optimize efficient and safe transfer vectors. Only a minimal transfer vector is shown here. (*Open boxes*) *cis*-acting sequences; (*closed boxes*) viral genes. (LTR) Long terminal repeat; (CMV) human cytomegalovirus early promoter; (PBS) primer binding site; (MSD) major splice donor site; (Ψ) packaging sequence; (cPPT/CTS) central polypurine track and central termination sequence; (RRE) Rev-responsive element; (Poly [A]) polyadenylation site; (SD) splice donor site; (SA) splice acceptor site; (SV40) simian virus 40 early promoter; (IRES) internal ribosomal entry signal; (Pro) internal promoter. *Upside-down text* indicates that the SV40-Tat-Poly(A) expression cassette is in antisense orientation as compared to the CMV-gagpol-Poly(A) cassette. Expression constructs that express the viral glycoproteins (ENV; *D*) and/or the Rev accessory protein (*E*) were also designed. Vector particles are produced by cotransfection of plasmids harboring the packaging constructs (Gag-Pol/Tat/Env/Rev) and the transfer vector into 293 cells. The supernatants of transfected cells are collected during transient expression and used for transduction of target cells.

packaging genome that carries the *trans*-acting functions. Indeed, despite the deletion of most of the packaging sequence in the latter type of genome, low levels of packaging may still occur (Patience et al. 1998).

On the basis of these considerations, several transfer vector and packaging genomes have recently been derived from different types of SIVs, SIVmac251 (Mangeot et al. 2000; Nègre et al. 2000; Schnell et al. 2000; Wagner et al. 2000), and SIVagm (Nakajima et al. 2000; Stitz et al. 2001). The minimal configuration required to efficiently propagate these different vectors is shown in Figure 1B. The genomic RNA of the transfer vector can be expressed using wild-type or hybrid 5′ LTRs and required Tat expression in producer cells in *trans* for optimal transcription (Mangeot et al. 2000; Nègre et al. 2000). Vectors may also retain the RRE sequence, although its presence was found dispensable in the vector backbone (Schneider et al. 1997; Mangeot et al. 2000; Schnell et al. 2000). Because the packaging sequence is thought to extend into the beginning of the *gag* gene, the most efficient SIV-based vectors still contain 50 residual nucleotides derived from this gene (Mangeot et al. 2000). Further deletions of the *gag* gene and of the 3′ end of the leader region have a strong negative influence on vector titers (Mangeot et al. 2000). Following vector integration in target cells, the transgenes can be efficiently expressed from internal transcription units that do not require Tat/TAR and Rev/RRE sequences for optimal expression. This has allowed the construction of vectors that carry large deletions in the 3′ LTR U3 regions, encompassing the binding sites for several transcription factors and the TATA box (Fig. 1B), that dramatically reduce transcription capacities from the 5′ LTR after a round of reverse transcription and integration in target cells (Mangeot et al. 2000; Nakajima et al. 2000; Schnell et al. 2000). These self-inactivating (SIN) vectors offer several advantages such as (1) the prevention of possible vector mobilization upon coinfection with a replication-competent retrovirus (RCR), (2) the elimination of the Tat requirement for transgene expression in target cells, and (3) the reduction of interferences of the SIV LTR with the internal promoter, e.g., tissue-specific or regulatable internal promoters inserted in the vector backbone (Mangeot et al. 2002). Other *cis*-acting elements, including sequences that increase RNA export or other posttranscriptional regulation (Belshan et al. 2000) and transgene expression (Zufferey et al. 1999), have been inserted in SIV-derived vectors and shown to improve vector efficiency in a target-cell-type-dependent manner (Mangeot et al. 2002). Depending on the particular configurations of the vectors, infectious titers higher than 10^7 infectious particles (i.p.)/ml can be obtained.

Minimal Packaging Genomes

To propagate the transfer vector as replication-defective viral particles, it is necessary to derive *trans*-complementing genomes that provide in *trans* the packaging proteins whose genes have been deleted from the transfer vector (Fig. 1C). The *trans*-complementing genomes, also referred to as the packaging vectors, are designed such that they cannot replicate by themselves. This is usually achieved by removing essential *cis*-acting sequences that pertain to the initial retrovirus, such as the packaging sequence, the PBS and PPT regions, and both LTRs. These modifications of the packaging genomes are in practice sufficient to prevent transfer of SIV genes into transduced cells (Mangeot et al. 2000) and formation of RCRs (Nègre et al. 2000). Additionally, the packaging proteins are usually encoded by two separate and complementary genomes, one for the Gag-Pol proteins and another for the viral glycoproteins, which does not necessarily originate from the parental lentivirus (Fig. 1D). This physical separation ensures both high levels of biosafety, by minimizing the possibility of recombination, and flexibility, by easily allowing the exchange of glycoproteins of different viral origins. Widely used second- and third-generation packaging vectors are shown in Figure 1C. Because second-generation packaging constructs still contain most of the accessory genes and bear high resemblance to the parental SIV genome, they are not considered safe. Packaging constructs of the third generation contain only the genes encoding the structural proteins (Nègre et al. 2000; Wagner et al. 2000).

Pseudotyping SIV-derived Vectors with Heterologous Envelope Glycoproteins

Protein incorporation on retroviruses is not specific to the homologous viral glycoproteins. Additionally, many heterologous viral glycoproteins can be incorporated into retroviral particles and mediate infectivity (Swanstrom and Wills 1997). This process, known as pseudotyping, allows retroviral vectors to transduce a broader range of cells and tissues. Expression of glycoproteins can be achieved from packaging-deficient expression constructs driven by strong constitutive promoters (Fig. 1D). No sequence homology is found between these constructs, the Gag-Pol packaging vector, and the transfer vector constructs, thus minimizing the possibility of generating recombinant viruses that have integrated the glycoprotein-coding sequences.

There is considerable interest in exploring the properties of lentiviral vectors pseudotyped with alternative viral glycoproteins (Mochizuki et al. 1998; Salmon et al. 2000; Stitz et al. 2000; Christodoulopoulos and Cannon 2001; Desmaris et al. 2001; Kobinger et al. 2001; Lewis et al. 2001; Sandrin et al. 2002). This parameter is likely to modulate the physicochemical properties of the vectors, their interaction with the host immune system, and their host range. Several studies have shown that the transduction efficiency of target cells is dependent on the type of glycoprotein used to coat retroviral vectors (Porter et al. 1996; Marandin et al. 1998; Movassagh et al. 1998; Kelly et al. 2000; Gatlin et al. 2001; Goerner et al. 2001; Sandrin et al. 2002). Additionally, some in vivo gene-transfer applications will require vectors that are targeted for specific cell entry and/or gene expression after systemic administration (Peng et al. 2001). Lentiviral vectors have been mostly generated with vesicular stomatitis virus (VSV)-G, as this glycoprotein simplifies the recovery and concentration of pseudotyped vectors (Burns et al. 1993; Naldini et al. 1996). These vectors, however, have drawbacks that might limit their use in systemic gene delivery. Because of the wide distribution of the VSV-G receptor, a lipid component of the plasma membrane (Seganti et al. 1986), VSV-G pseudotypes may bind to the surfaces of all cells encountered after inoculation before reaching the target cells. Moreover, VSV-G-pseudotyped vectors are rapidly inactivated by human serum (DePolo et al. 2000; Sandrin et al. 2002). For these reasons, formation of SIV vectors pseudotyped with several glycoproteins has been recently explored (Nègre et al. 2000; Sandrin et al. 2002).

Summary of the Assembly of SIV Vector Particles

Lentiviral vectors are usually prepared from transient expression of the vector components in highly transfectable cells (e.g., 293T cells), although alternative and novel methods to generate SIV vectors rely on the use of an inducible packaging cell line (Kuate et al. 2002).

The expression of both the transfer vector and the packaging vector(s) into cotransfected cells allows the release in the culture supernatant of viral particles that have packaged the genome of the transfer vector and are replication-defective. They can usually be produced in large quantities, concentrated, and purified before being used to transduce the target cells. The recombinant virus retains the ability to enter into the cell via a specific receptor that is recognized by the viral glycoprotein and to permanently integrate its genetic material into the host genome, thus ensuring efficient and long-term gene delivery. Because no replication-competent virus is used to complement the defective vector, retrovirus-mediated gene transfer is a single-round operation. This results in the insertion of one to a few copies of the transgene into the host-cell genome.

Protocol

Production of SIV Vectors for Gene Delivery

This protocol describes the production of SIV vector particles and the assay used to measure transduction efficiency. In addition, methods for detecting RCRs and testing the putative mobilization of the green fluorescent protein (GFP)-containing SIV vector are included. These safety assays make use of sMAGI target cells that are permissive to SIVmac251 replication and harbor a stably integrated *lacZ* transcription unit driven by the SIV/HIV *tat*-inducible HIV-1 LTR (Chackerian et al. 1995).

MATERIALS

CAUTION: See Appendix for appropriate handling of materials marked with <!>.

Reagents

Bovine serum albumin (BSA Fraction V)

Use 1% BSA in PBS as freezing medium and store at 4°C.

Calphos Mammalian Transfection Kit for calcium-phosphate-mediated transfection (Clontech)

Cell lines for transfection and transduction

293T cells

TE671 cells (human rhabdomyosarcoma, ATCC CRL-8805)

sMAGI macaque cells

These cells should be permissive to SIVmac251 replication and harboring a stably integrated *lacZ* transcription unit driven by the SIV/HIV *tat*-inducible HIV-1 LTR (Chackerian et al. 1995). sMAGI cells are grown in DMEM/10% FCS/hygromycin B (50 μg/ml) / G418 (200 μg/ml).

Dulbecco's modified Eagle's medium (DMEM) with 0.11 g/liter sodium pyridoxine and pyridoxine

Supplement DMEM with 10% FCS, 100 μg/ml streptomycin <!>, and 100 units/ml penicillin. Store at 4°C.

Fetal calf serum (FCS), sterile

G418 (100 mg/ml in H_2O, Geneticin; Invitrogen) <!>

Sterilize through a 0.22-μm filter and store at –20°C.

Glutaraldehyde solution (0.5% in PBS, store at 4°C) <!>

Hygromycin B in PBS (50 mg/ml, store at 4°C; Invitrogen) <!>

Nucleic acids

Lentiviral vector DNA encoding an SIV-derived self-inactivating vector with the internal promoter of choice driving the reporter gene GFP

The choice of promoter used in the internal transcription cassette of the vector should be based on the promoter's expected properties in the target cells of interest.

Envelope-glycoprotein-expressing plasmids (e.g., VSV-G)

Viral structural protein (Gag-Pol)-expressing plasmid

Paraformaldehyde solution (4% in PBS, store at room temperature) <!>

Phosphate-buffered saline (PBS) without calcium and magnesium, without sodium bicarbonate, sterile

Polybrene stock solution (hexadimethrine bromide, 8 mg/ml in H_2O; Sigma-Aldrich)

Sterilize through a 0.22-µm filter, aliquot, and store at –20°C. Use at a final concentration of 6 µg/ml during infection.

Trypsinization reagents

Trypsin-ethylenediaminetetraacetic acid (EDTA) in 1x Hank's balanced salt solution (HBSS) without calcium and magnesium, sterile

Versene 1:5000 (Invitrogen)

X-gal buffer

43.5 mM sodium deoxycholic acid (monohydrate; Sigma-Aldrich) <!>

2 mM $MgCl_2 \cdot 6H_2O$ (Sigma-Aldrich) <!>

5 mM potassium ferricyanide (Sigma-Aldrich) <!>

5 mM potassium ferrocyanide (Sigma-Aldrich) <!>

0.002% Nonidet P-40 (NP-40) (Roche Diagnostics)

Dilute in PBS. For sMAGI cells, adjust the pH to 9.0 with NaOH <!>.

X-gal (5-bromo-4-chloro-3-indolyl-β-D-galactopyranoside) (Euromedix) <!>

Create a 60x stock by diluting in dimethyl formamide (DMF; Sigma-Aldrich) <!> to a final concentration of 40 mg/ml. Store at –20°C. Use X-gal that has been freshly diluted in X-gal buffer before staining cells.

Equipment

FACSCalibur flow cytometry system (Becton Dickinson)

FACS tubes (Becton Dickinson)

Fluorescence microscope

Membrane filters (0.45 µm)

Reverse transcriptase assay (Roche Diagnostics)

Standard tissue-culture equipment, including 100-mm and six-well tissue-culture dishes

Ultracentrifugation tubes (26 ml, 25 x 89 mm, polyallomer; Beckman)

METHODS

Production, Concentration, and Storage of SIV Vector Particles

1. Day –1: The day before transfection, seed 2.4 x 10^6 293T cells in 100-mm tissue-culture dishes in a final volume of 12 ml/dish.
2. Day 0: Cotransfect 8.1 µg of SIV packaging constructs with 8.1 µg of the transfer vector constructs and 2.5 µg of the envelope-expressing construct using calcium-phosphate-mediated transfection.
3. Day 1: 15 hours after transfection, replace the medium in each dish with 8 ml of fresh medium.
4. Day 2: 36 hours after transfection, harvest the virus: Centrifuge the medium from each plate at low speed to remove cellular debris and filter through 0.45-µm membranes.

   The filtered supernatants can be stored at –80°C.
5. Transfer the supernatants to 26-ml ultracentrifugation tubes. Pellet the virions by centrifugation at 110,000*g* (32,000 rpm) in a 70Ti Beckman rotor for 2 hours at 4°C.
6. Remove the supernatants and rinse the walls of the tubes with 5 ml of PBS. Invert the tubes on absorbant papers for 5 minutes. Allow to dry for 5 minutes and wipe.
7. Resuspend the viral pellets by adding a volume of ice-cold 1% BSA in PBS equal to 1/100 of the viral supernatant's initial volume and incubating for 2 hours on ice.

8. Aliquot the virions in microfuge tubes and store at –80°C or in liquid nitrogen.

   Samples can be kept for at least 6 months at –80°C without impact on the infectious titers.

Transduction Assay

9. Day –1: Seed TE671 or sMAGI target cells at a density of 4×10^5 cells/well in six-well plates in a final volume of 2 ml/well.
10. Day 0: Create serial dilutions of vector preparations in 1 ml of DMEM/10% FCS. Add to the cells in the presence of 6 µg/ml polybrene.
11. Incubate the cultures for ≥4 hours at 37°C. Replace the vector-containing medium with 2 ml of fresh culture medium and incubate the cells for 72 hours at 37°C.
12. Day 3: Trypsinize the cells, fix in 4% formaldehyde in PBS for 30 minutes, and transfer to FACS tubes. Analyze the cells by flow cytometry.

    In this assay, **transduction efficiency** is usually determined by calculating the percentage of GFP-positive cells after transduction of 4×10^5 target cells with 1 ml of viral supernatant.

    Infectious titers are expressed as transducing units (TU)/ml and can be calculated by using the formula:

    Titer = $\%inf \times (4 \times 10^5/100) \times d$

    where d is the dilution factor of the viral supernatant and $\%inf$ is the percentage of GFP-positive cells as determined by FACS (fluorescence-activated cell sorting) analysis using dilutions of the viral supernatant that transduce between 5% and 10% of GFP-positive cells.

    Multiplicity of infection (**moi**) is the ratio of infectious particles to target cells.

Safety Testing of Vector Preparations: Detection of RCRs

13. Day –1: Seed sMAGI target cells at a density of 4×10^5 cells per well in six-well plates in a final volume of 2 ml/well.
14. Day 0: Perform a primary infection of the sMAGI cells with SIV vector particles. Create serial dilutions of the vector preparations in 1 ml of DMEM/10% FCS. Add the dilutions to the cells in the presence of 6 µg/ml polybrene. Incubate the cultures for ≥4 hours at 37°C. Replace the vector-containing medium with 2 ml of normal culture medium. Incubate the cells for 72 hours at 37°C.
15. Day 3: Trypsinize the transduced cells.

    a. Determine the percentage of GFP-positive cells: Fix the first third of the cell suspension in 4% formaldehyde in PBS for 30 minutes and analyze by FACS.

    b. Check β-galactosidase expression: Reseed the second third of the cells in 24-well plates in a final volume of 1 ml/well. When the cells are confluent, fix them with 1 ml of PBS/0.5% glutaraldhehyde for 10 minutes, rinse with 1 ml of PBS, and incubate with X-gal solution (pH 9.0) for 2 hours at 32°C.

    c. Allow the spread of an eventual RCR: Reseed the last third of the cells in six-well plates in a final volume of 2 ml/well. Maintain the cells for 10–15 days in culture.
16. Day ~14: The day before the secondary infection, seed sMAGI target cells at a density of 4×10^5 cells/well in six-well plates in a final volume of 2 ml/well.
17. Day ~15: Use the supernatants of the infected primary target cells (from the 13–18-day cultures) to infect sMAGI cells as secondary target cells. Incubate 1 ml of each supernatant on the target cells for 4 hours at 37°C. Replace the medium with 2 ml of normal culture medium. Incubate the cells for 48–72 hours postinfection at 37°C. Evaluate reverse transcriptase activity and GFP and β-galactosidase expression.

To demonstrate the absence of mobilization of the GFP-containing SIV vector and to show that the stocks of SIV vectors are devoid of both RCRs and *tat*-recombinant retroviruses, GFP and β-galactosidase expression must remain negative in all experiments 48–72 hours postinfection. Additionally, reverse transcriptase activity must remain negative in the supernatants of both the primary and secondary infected cells after several passages. As a positive control in these experiments, assess the capacity of recombinant retroviruses to undergo replication and amplification within this time frame in these cells by measuring reverse transcriptase activity and *tat* gene expression in sMAGI cells infected7 by wild-type SIVmac251.

REFERENCES

Belshan M., Park G.S., Park G.S., Bilodeau P., Stoltzfus C.M., and Carpenter S. 2000. Binding of equine infectious anemia virus rev to an exon splicing enhancer mediates alternative splicing and nuclear export of viral mRNAs. *Mol. Cell. Biol.* **20:** 3550–3557.

Burns J.C., Friedmann T., Driever W., Burrascano M., and Yee J.K. 1993. Vesicular stomatitis virus G glycoprotein pseudotyped retroviral vectors: Concentration to very high titer and efficient gene transfer into mammalian and nonmammalian cells. *Proc. Natl. Acad. Sci.* **90:** 8033–8037.

Chackerian B., Haigwood N.L., and Overbaugh J. 1995. Characterization of a CD4-expressing macaque cell line that can detect virus after a single replication cycle and can be infected by diverse simian immunodeficiency virus isolates. *Virology* **213:** 386–394.

Christodoulopoulos I. and Cannon P. 2001. Sequences in the cytoplasmic tail of the gibbon ape leukemia virus envelope protein that prevent its incorporation into lentivirus vectors. *J. Virol.* **75:** 4129–4138.

Clements J. and Zink M. 1996. Molecular biology and pathogenesis of animal lentivirus infections. *Clin. Microbiol. Rev.* **9:** 100–117.

Das A., Klaver B., Klasens B.I., van Wamel J.L., and Berkhout B. 1997. A conserved hairpin motif in the R-U5 region of the human immunodeficiency virus type 1 RNA genome is essential for replication. *J. Virol.* **71:** 2346–2356.

DePolo N.J., Reed J.D., Sheridan P.L., Townsend K., Sauter S.L., Jolly D.J., and Dubensky T.W., Jr. 2000. VSV-G pseudotyped lentiviral vector particles produced in human cells are inactivated by human serum. *Mol. Ther.* **2:** 218–222.

Desmaris N., Bosch A., Salaun C., Petit C., Prevost M.C., Tordo N., Perrin P., Schwartz O., de Rocquigny H., and Heard J.M. 2001. Production and neurotropism of lentivirus vectors pseudotyped with lyssavirus envelope glycoproteins. *Mol. Ther.* **4:** 149–156.

Dorman N. and Lever A. 2000. Comparison of viral genomic RNA sorting mechanisms in human immunodeficiency virus type 1 (HIV-1), HIV-2, and moloney murine leukemia virus. *J. Virol.* **74:** 11413–11417.

Gardner M., Endres M., and Barry P. 1993. The simian retroviruses SIV and SRV. In *The retroviridae* (ed. J. Levy), vol. 3, pp. 133–276. Plenum Press, New York.

Gatlin J., Melkus M.W., Padgett A., Kelly P.F., and Garcia J.V. 2001. Engraftment of NOD/SCID mice with human CD34$^+$ cells transduced by concentrated oncoretroviral vector particles pseudotyped with the feline endogenous retrovirus (RD114) envelope protein. *J. Virol.* **75:** 9995–9999.

Giavedoni L. and Yilma T. 1996. Construction and characterization of replication-competent simian immunodeficiency virus vectors that express gamma interferon. *J. Virol.* **70:** 2247–2251.

Goerner M., Horn P.A., Peterson L., Kurre P., Storb R., Rasko J.E.J., and Kiem H.P. 2001. Sustained multilineage gene persistence and expression in dogs transplanted with CD34$^+$ marrow cells transduced by RD114-pseudotype oncoretrovirus vectors. *Blood* **98:** 2065–2070.

Guan Y., Whitney J., Diallo K., and Wainberg M.A. 2000. Leader sequences downstream of the primer binding site are important for efficient replication of simian immunodeficiency virus. *J. Virol.* **74:** 8854–8860.

Ilyinskii P. and Desrosiers R. 1998. Identification of a sequence element immediately upstream of the polypurine tract that is essential for replication of simian immunodeficiency virus. *EMBO J.* **17:** 3766–3774.

Kaye J.F. and Lever A.M. 1998. Nonreciprocal packaging of human immunodeficiency virus type 1 and type 2 RNA: A possible role for the p2 domain of Gag in RNA encapsidation. *J. Virol.* **72:** 5877–5885.

Kelly P., Vandergriff J., Nathwani A., Nienhuis A.W., and Vanin E.F. 2000. Highly efficient gene transfer into cord blood nonobese diabetic/severe combined immunodeficiency repopulating cells by oncoretroviral vector particles pseudotyped with the feline endogenous retrovirus (RD114) envelope protein. *Blood* **96:** 1206–1214.

Kobinger G.P., Weiner D.J., Yu Q.C., and Wilson J.M. 2001. Filovirus-pseudotyped lentiviral vector can efficiently and stably transduce airway epithelia in vivo. *Nat. Biotechnol.* **19:** 225–230.

Kuate S., Wagner R., and Uberla K. 2002. Development and characterization of a minimal inducible packaging cell line for simian immunodeficiency virus-based lentiviral vectors. *J. Gene Med.* **4:** 347–355.

Lewis B.C., Chinnasamy N., Morgan R.A., and Varmus H.E. 2001. Development of an avian leukosis-sarcoma virus subgroup A pseudotyped lentiviral vector. *J. Virol.* **75:** 9339–9344.

Mangeot P.E., Duperrier K., Nègre D., Boson B., Rigal D., Cosset F.L., and Darlix J.L. 2002. High levels of transduction of human dendritic cells with optimized SIV vectors. *Mol. Ther.* **5:** 283–290.

Mangeot P.E., Nègre D., Dubois B., Winter A.J., Leissner P., Mehtali M., Kaiserlian D., Cosset F.L., and Darlix J.L. 2000. Development of minimal lentiviral vectors derived from simian immunodeficiency virus (SIVmac251) and their use for the gene transfer in human dendritic cells. *J. Virol.* **74:** 8307–8315.

Marandin A., Dubart A., Pflumio F., Cosset F.L., Cordette V., Chapel-Fernandes S., Coulombel L., Vainchenker W., and Louache F. 1998. Retroviral-mediated gene transfer into human CD34$^+$/38$^-$ primitive cells capable of reconstituting long-term cultures in vitro and in nonobese diabetic-severe combined immunodeficiency mice in vivo. *Hum. Gene Ther.* **9:** 1497–1511.

Mochizuki H., Schwartz J.P., Tanaka K., Brady R.O., and Reiser J. 1998. High-titer human immunodeficiency virus type 1-

based vector systems for gene delivery into nondividing cells. *J. Virol.* **72:** 8873–8883.

Movassagh M., Desmyter C., Baillou C., Chapel-Fernandes S., Guigon M., Klatzmann D., and Lemoine F.M. 1998. High-level gene transfer to cord blood progenitors using gibbon ape leukemia virus pseudotyped retroviral vectors and an improved clinically applicable protocol. *Hum. Gene Ther.* **9:** 225–234.

Nakajima T., Nakamaru K., Ido E., Terao K., Hayami M., and Hasegawa M. 2000. Development of novel simian immunodeficiency virus vectors carrying a dual gene expression system. *Hum. Gene Ther.* **11:** 1863–1874.

Naldini L., Blömer U., Gallay P., Ory D., Mulligan R., Gage F.H., Verma I.M., and Trono D. 1996. In vivo gene delivery and stable transduction of nondividing cells by a lentiviral vector. *Science* **272:** 263–267.

Nègre D., Duisit G., Mangeot P.E., Moullier P., Darlix J.L., and Cosset F.L. 2002. Lentiviral vectors derived from simian immunodeficiency virus (SIV). *Curr. Top. Microbiol. Immunol.* **261:** 53–74.

Nègre D., Mangeot P., Duisit G., Blanchard S., Vidalain P.O., Leissner P., Winter A.J., Rabourdin-Combe C., Mehtali M., Moullier P., Darlix J.L., and Cosset F.L. 2000. Characterization of novel safe lentiviral vectors derived from simian immunodeficiency virus (SIVmac251) that efficiently transduce mature human dendritic cells. *Gene Ther.* **7:** 1613–1623.

Patience C., Takeuchi Y., Cosset F.L., and Weiss R.A. 1998. Packaging of endogenous retroviral sequences in retroviral vectors produced by murine and human packaging cells. *J. Virol.* **72:** 2671–2676.

Peng K.W., Pham L., Ye H., Zufferey R., Trono D., Cosset F.L., and Russell S.J. 2001. Organ distribution of gene expression after intravenous infusion of targeted and untargeted lentiviral vectors. *Gene Ther.* **8:** 1456–1463.

Porter C.D., Collins M.K.L., Tailor C.S., Parkar M.H., Cosset F.L., Weiss R.A., and Takeuchi Y. 1996. Comparison of efficiency of infection of human gene therapy target cells via four different retroviral receptors. *Hum. Gene Ther.* **7:** 913–919.

Ruprecht R. 1999. Live attenuated AIDS viruses as vaccines: Promise or peril? *Immunol. Rev.* **170:** 135–149.

Salmon P., Nègre D., Trono D., and Cosset F.-L. 2000. A chimeric GALV-derived envelope glycoprotein harboring the cytoplasmic tail of MLV envelope efficiently pseudotypes HIV-1 vectors. *J. Gene Med.* (suppl.) **2:** 23.

Sandrin V., Boson B., Salmon P., Gay W., Nègre D., Le Grand R., Trono D., and Cosset F.L. 2002. Lentiviral vectors pseudotyped with a modified RD114 envelope glycoprotein show increased stability in sera and augmented transduction of primary lymphocytes and $CD34^+$ cells derived from human and non-human primates. *Blood* **100:** 823–832.

Schneider R., Campbell M., Nasioulas G., Felber B.K., and Pavlakis G.N. 1997. Inactivation of the human immunodeficiency virus type 1 inhibitory elements allows Rev-independent expression of Gag and Gag/protease and particle formation. *J. Virol.* **71:** 4892–4903.

Schnell T., Foley P., Wirth M., Munch J., and Uberla K. 2000. Development of a self-inactivating, minimal lentivirus vector based on simian immunodeficiency virus. *Hum. Gene Ther.* **11:** 439–447.

Seganti L., Superti F., Girmenia C., Melucci L., and Orsi N. 1986. Study of receptors for vesicular stomatitis virus in vertebrate and invertebrate cells. *Microbiologica* **9:** 259–267.

Stitz J., Buchholz C., Engelstadter M., Uckert W., Bloemer U., Schmitt I., and Cichutek K. 2000. Lentiviral vectors pseudotyped with envelope glycoproteins derived from gibbon ape leukemia virus and murine leukemia virus 10A1. *Virology* **273:** 16–20.

Stitz J., Muhlebach M.D., Blömer U., Scherr M., Selbert M., Wehner P., Steidl S., Schmitt I., Konig R., Schweizer M., Cichutek K. 2001. A novel lentivirus vector derived from apathogenic simian immunodeficiency virus. *Virology* **291:** 191–197.

Swanstrom R. and Wills J.W. 1997. Synthesis, assembly, and processing of viral proteins. *Retroviruses* (ed. J.M. Coffin et al.), pp. 263–334. Cold Spring Harbor Laboratory Press, Cold Spring Harbor, New York.

Wagner R., Graf M., Bieler K., Wolf H., Grunwald T., Foley P., and Uberla K. 2000. Rev-independent expression of synthetic *gag-pol* genes of human immunodeficiency virus type 1 and simian immunodeficiency virus: Implications for the safety of lentiviral vectors. *Hum. Gene Ther.* **11:** 2403–2413.

Zufferey R., Donello J.E., Trono D., and Hope T.J. 1999. Woodchuck hepatitis virus posttranscriptional regulatory element enhances expression of transgenes delivered by retroviral vectors. *J. Virol.* **73:** 2886–2892.

6 Production and Use of Feline Immunodeficiency Virus-based Lentiviral Vectors

Dyana T. Saenz, Román Barraza, Nils Loewen, Wulin Teo, and Eric M. Poeschla

Molecular Medicine Program, Mayo Clinic College of Medicine, Rochester, Minnesota 55905

ABSTRACT

Feline immunodeficiency virus (FIV)-based lentiviral vectors are useful for introducing integrated transgenes into nondividing human cells. This chapter describes methods for production and use of advanced generation FIV vectors. Key properties are discussed in comparison to other lentiviral vectors and detailed protocols for both large- and small-scale vector production are provided. Additional topics include the practical implications of recently identified species-specific retroviral restriction factors and the production of nonintegrating FIV vectors.

INTRODUCTION

Lentiviral vectors were first derived from the pathogen human immunodeficiency virus type 1 (HIV-1) (Parolin and Sodroski 1995). Exploitation of new envelope-pseudotyping options (Burns et al. 1993) with improved recombinant HIV-1 packaging constructs and transfer vectors led to decisive demonstration that HIV-1-based vectors could transduce nondividing cells in vitro and in vivo at levels that were promising for gene therapy (Naldini et al. 1996). Nonprimate lentivirus-based vector systems were devised soon thereafter, and feline immunodeficiency virus (FIV)-based vectors were shown to transduce nondividing human cells such as primary human neurons and macrophages (Poeschla et al. 1998). FIV vectors effectively transduce cells in the brain, eye, airway, hematopoietic system, liver, muscle, and pancreas (Poeschla et al. 1998; Johnston et al. 1999; Wang et al. 1999; Alisky et al. 2000; Curran et al. 2000, 2002; Loewen et al. 2001, 2002, 2003a,b; Stein et al. 2001; Brooks et al. 2002; Curran and Nolan 2002a,b; Derksen et al. 2002; Hughes et al. 2002; Kang et al. 2002; Lotery et al. 2002; Price et al. 2002; Stein and Davidson 2002; Haskell et al. 2003; Sinn et al. 2003; Saenz and Poeschla 2004). They have also been used with good results in explanted human organs (Wang et al. 1999; Loewen et al. 2001, 2002).

It is possible that FIV and other nonprimate lentivirus-based vectors will provide biosafety advantages in human gene therapy, as FIV does not propagate in humans, cause human disease, or even cross-react with HIV serologically. One incentive for adapting FIV has been the extensive natural epidemiology of human exposure, unique among lentiviruses. FIV is pandemic in domestic cat populations, where it causes an immunodeficiency similar to that caused by HIV-1 in humans. Human populations have had extensive and presumably efficient exposure to FIV, through the predominant means operative in interfeline transmission, biting (Pedersen 1993). Nevertheless, no disease, productive replication, or seroreactivity has been observed in humans. A second consideration, for both patients and personnel, is that lentiviral virions contain large amounts of capsid and other viral proteins, and they are often pseudotyped to expand both infectivity and host range at the cell entry (fusion) step. In this regard, FIV does not cross-react immunologically with HIV-1 (Pedersen 1993; Elder and Phillips 1995). Thus, HIV-1-specific immune responses cannot be generated by therapeutic use or inadvertent personnel exposure. Although HIV-1 vectors are demonstrably safe in the research setting, it has been our experience that in research laboratories not oriented toward virology or gene therapy, some personnel remain reluctant to work with even replication-defective HIV-1 systems and find nonprimate systems more acceptable. Finally, in HIV-1-infected recipients, rescue of the FIV transfer vector genome by replicating HIV-1 will not occur.

Postentry Restriction: Potential Advantages and Disadvantages for Nonprimate Lentiviral Vectors

Using nonhuman lentiviruses for gene therapy naturally raises practical questions about vector tropisms and efficiencies in human tissues. An important question is the relative per particle efficacy compared to HIV vectors. Although as cited above, good practical results have been achieved in human targets with FIV vectors, until recently, little firm scientific data were available in the way of direct comparisons of different lentiviral vectors on a particle-normalized basis. The recent energy brought to bear on species-specific postentry restriction patterns such as Ref1 and Lv1 has led to greater clarity on this issue for both equine infectious anemia virus (EIAV) and FIV. Most recently, these studies culminated in the discovery and characterization of the tripartite motif protein TRIM5α as the protein accounting for the Lv1 restriction activity (Stremlau et al. 2004). Ref1, the activity previously known to restrict EIAV and N-tropic murine leukemia virus (N-MLV) in human cells, is mediated by human TRIM5α. Like these vectors, FIV vectors were also shown recently to display a Ref1-restricted phenotype in some human cell lines, i.e., to be restricted in a saturable manner at a step prior to or concurrent with reverse transcription (Saenz et al. 2005). The degree of restriction is in general equivalent to that of EIAV vectors, and it appears to be largely accounted for by primate TRIM5α proteins (Saenz et al. 2005).

It should be emphasized that only some human target cells display practically significant postentry restrictions for nonprimate lentivirus-derived vectors compared to primate lentivirus-derived vectors and these must be empirically determined. In addition, postentry restrictions of this type have the potential to be either desirable or undesirable depending on the particular gene-therapy setting. For characterizations of the susceptibility of FIV to primate restriction factors and a discussion of the implications for gene therapy, see Saenz et al. (2005). Briefly, Ref1/human TRIM5α-mediated postentry restriction is significant only when particle input is quite low. When cells are infected with substantial amounts of virus or vector, or coinfected with other restricted retroviruses or virus-like particles (VLPs), the cellular TRIM5α pool can be saturated and restriction is neutralized. This input-dependent titration of the restriction requires only that the coinfecting particles be entry-competent and contain mature capsids of any retrovirus that is also restricted; they need not contain RNA genomes or enter via the same receptor pathway (Besnier et al. 2002, 2003; Towers et al. 2002; Hatziioannou et al. 2003; Berthoux et al. 2004; Saenz et al. 2005). In many gene therapy applications, particle inputs without VLPs result in local concentrations substantially greater than that needed to saturate Ref1/TRIM5α activity. In this situation, or if a target tissue expresses little TRIM5α, Ref1 restriction might provide a net advantage: a strong innate barrier to systemic propagation of potential replication-competent retrovirus (RCR). Restriction may also be surmountable by adding VLPs, either from FIV or other restricted retroviruses, e.g., N-MLV. Whether the FIV capsid can be engineered to evade human TRIM5α interaction remains to be determined, but the cost of such a maneuver would be to lose the evolution-tested systemic RCR barrier.

Once infection has proceeded past cytoplasmic stages where postentry restriction occurs, FIV preintegration complexes integrate efficiently in human cells (Saenz et al. 2005). After proviral integration has occurred, one other block to *wild-type* viral function in human cells becomes evident, this one more severe than postentry restriction. FIV long terminal repeat (LTR)-driven transcription is virtually nil in human cells (Poeschla and Looney 1998; Poeschla et al. 1998). However, in this case, the block in transcriptional inactivity is desirable from the standpoint of gene therapy, since it mitigates potential insertional mutagenesis effects and interference with internal promoters, which are now standard for lentiviral vector transgenes. The block was an initial obstacle to transfer vector production in desirable human producer cells such as 293T cells. Substitution of the 5′ U3 with a heterologous promoter was found to permit effective vector production (Poeschla and Looney 1998; Poeschla et al. 1998).

Optimization of Vector Components

We have substantially revised and optimized the proof-of-principle FIV vector system (for review, see Loewen et al. 2003a). Minimal overlap now exists between vectors and packaging elements. For the packaging construct, additional *cis*-acting noncoding and protein-coding sequences were deleted (Fig. 1). We also systematically mapped encapsidation determinants in the FIV genomic mRNA (Kemler et al. 2002, 2004) and, except for a short segment of the *gag* open reading frame (ORF), then removed them from packaging constructs. Viral leader sequences are absent from packaging construct pFP93 (replaced with a 9-nucleotide canonical splice donor sequence), as are *vif* and U3 sequences. Less residual *env* sequence flanks the Rev response element (Fig. 1). Viral sequences terminate at the 3′ end with the *rev* stop codon. The central DNA flap, consisting of the central polypurine tract (cPPT) and central termination sequence (CTS) (Whitwam et al. 2001), has been incorporated into transfer vectors, as has the woodchuck hepatitis virus posttranscriptional regulatory element (WPRE) (Donello et al. 1998; Zufferey et al. 1999). An internal ribosome entry site (IRES) allows a marker gene (*neoR, gfp*) to be coordinately expressed with the gene of interest (Fig. 1). Fluorescent markers can be used to enrich for transduced cells by live-cell fluorescence-activated cell sorting (FACS). Selectable markers (e.g., puromycin, hygromycin, and neomycin) can also be used to select for transduced cells. Both types of genes can be used for titering vectors.

The above-described transcriptional silence of the FIV LTR in most human cells also provides a first-level self-inactivating (SIN) property for FIV vectors. We have also derived a conventional SIN vector to eradicate the possibility of LTR-mediated insertional mutagenesis effects by deleting the central 167 nucleotides in the 3′ U3 element; 25- and 24-nucleotide flanking segments remain,

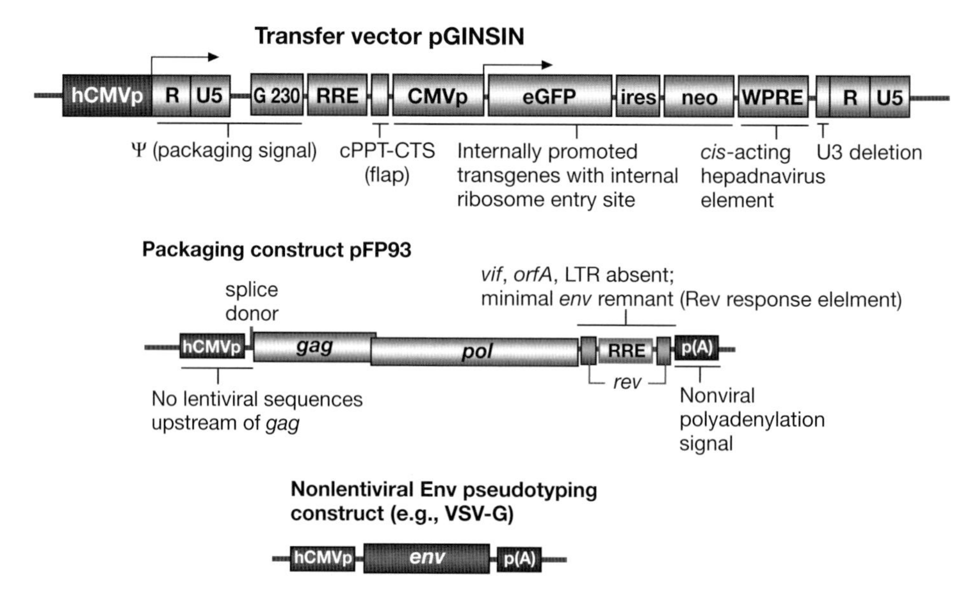

FIGURE 1. Diagram of the three-component FIV-based lentiviral vector system. FIV transfer vector pGINSIN: (ψ) Encapsidation signal; (G230) only the 5′ 230 bp of *gag* are present. The WPRE facilitates mRNA stability. U3 has a central 167-nucleotide deletion. (pFP93) Packaging construct. A variety of viral envelope proteins can be used for pseudotyping.

but the main transcription-factor-binding sites and the TAT box are eliminated. Plasmid names in this chapter refer to commonly used vector construct derivatives. pFP93 is a minimal FIV packaging plasmid encoding Gag, Gag/Pol precursor, and Rev protein. pGINSIN is a SIN vector with a U3 deletion. It encodes enhanced green fluorescent protein (GFP), which is linked monocistronically via an IRES to the gene for neomycin phosphotransferase (*neoR*). Alternatives include the β-galactosidase vector pCT26. pMD.G is an envelope-pseudotyping construct that expresses the envelope glycoprotein of vesicular stomatitis virus (VSV-G), a rhabdovirus. Other viral envelope glycoproteins are possible, e.g., amphotropic Env, gibbon ape leukemia virus Env, or filovirus glycoproteins. Nonrhabdoviral envelopes may be less stable and benefit from concentration through a sucrose cushion or by other methods. Further iterations or incrementally modified versions of these plasmids, particularly of the transfer vector, are in use and can be substituted as appropriate. We have not yet pursued codon optimization of FIV *gag/pol*, since it is not clear whether freeing *gag/pol* mRNA export and stability from the requirement for Rev in this way, although eliminating another viral gene, yields a net vector safety advantage or disadvantage (Hope 2002).

General Information for Vector Production

Safety considerations that apply to murine oncoretroviral vectors are operative. FIV vectors can generally be used with biosafety level 2 (BL2) practices. Transgene identity is pertinent, since transduction of dominant oncogenes by any retroviral vector requires stringent precautions (generally BL3 level; for guidelines, see Richmond and McKinney 1999). Since optimal practices will depend on the specific transgene and uses, all vector work requires preapproval of the local Institutional Biosafety Committee. The unlikely possibility of spontaneously arising RCR should be remembered, although none have been reported for lentiviral vector systems to date. Routine assays for their presence in preclinical basic scientific experimental material is not considered as necessary as in validation of clinical-grade material.

The following material uses a now preferred nomenclature that reserves "transfection" for physical introduction of plasmid DNA, "transduction" for gene transfer by a replication-defective viral vector, and "infection" for infection by a replication-competent virus. Calcium phosphate coprecipitation-based transfection of plasmid DNAs into 293T cells is an efficient method for producing FIV vectors. These cells are highly transfectable by this method. Nevertheless, fastidious technique and optimization of specific reagents, particularly the pH of the 2x HBS, are required to produce high-quality lentiviral vector stocks. For example, vector titer does not scale arithmetically with percent transfection of 293T producer cells. Rather, log scale variability in titers can result from a fall in producer cell transfection efficiency. Transfection problems can be traced to relatively small variations in reagents or technique or even the particular plasticware. Scale-up of a well-working transfection protocol to larger volumes may require reoptimization; 10-cm dishes or 75-cm^2 flasks, with 10–15 ml of vector supernatant per flask, can suffice for smaller-scale production. The following protocol also describes scaled-up vector production in multichamber Cell Factories (Nunc) with 1000-ml (CF10) or 200-ml (CF2) volumes. These devices significantly facilitate larger-volume production. One CF10 provides a surface area equivalent to more than 80 75-cm^2 flasks. Centrifugation in six 220-ml buckets will yield 10–400-fold concentration. A second centrifugation step can be used if needed.

Some lentiviral vector production protocols specify relatively large quantities of high-quality plasmid DNA (35–70 μg per 75 cm^2 of transfected monolayer). We found that less input DNA produced equivalent results. (DNAs were produced with QIAGEN Maxi Kits.) Generally, approximately 7 μg of total DNA per 75 cm^2 produces the same results as higher amounts. Personnel time and material costs are thereby considerably reduced. 293 human embryonic kidney (HEK), the precursor cell line to 293T cells, appear to have originated from rare cells of neural progenitor lineage that were transformed when human embryonic kidney cultures were transfected with sheared adenovirus type 5 DNA (Shaw et al. 2002). 293T cells are a 293 derivative engineered to express the simian virus 40 (SV40) large T antigen (DuBridge et al. 1987), which was designed to allow T-antigen-driven plasmid amplification. Generally, transfection is done in the late afternoon because medium must be changed 6–18 hours later, with 12–16 hours being most suitable for vectors pseudotyped with VSV-G (Burns et al. 1993). The protocols in this chapter are for small- and large-scale production for basic science laboratory use and in vivo experimentation. They do not meet standards for clinical-grade (GMP) production.

Protocol 1

Vector Production with Cells Grown in CF10 or CF2 Devices

This protocol describes the production of FIV-based lentiviral vectors using cells grown in CF10 or CF2 devices.

MATERIALS

CAUTION: See Appendix for appropriate handling of materials marked with <!>.

Reagents

Cells for transfection: 293T

Early-passage 293T cells are preferred. They should be continuously and uniformly dividing prior to seeding for transfection.

Dulbecco's modified Eagle's medium with 10% fetal calf serum (DMEM-10), 100 units/ml penicillin G <!>, 100 μg/ml streptomycin <!>, and 2 mM L-glutamine

Plasmid DNAs: pMD.G, pFP93, pGINSIN

For in vivo applications, plasmid DNA should be sterile and endotoxin-free. Commercially available kits, e.g., the Endotoxin-free Plasmid Maxi Kit (QIAGEN 2362), are reliable. Cesium-chloride-based purification is an alternative.

Transfection reagents

Tris-HCl <!> (10 mM, pH 8.0) for diluting plasmid DNA prior to $CaCl_2$ addition

$CaCl_2$ (2.5 M) <!>

Autoclave and store at –20°C.

2x HEPES-buffered saline (HBS)

Stock solution—dibasic Na_2HPO_4 <!>: 52.5 g of Na_2HPO_4, 5000 ml of H_2O. 2x HBS: 80 g of NaCl, 65 g of HEPES (sodium salt), 100 ml of Na_2HPO_4 stock solution. Careful pH optimization is important. Prepare several batches and adjust pH with 1 N NaOH to several set points within a narrow range, e.g., pH 6.95, 7, and 7.05. Bring to a final volume of 5000 ml. Optimal pH is generally 6.95 in our hands, but final preferred stocks will depend on empirical determination of transfection efficiency. Store frozen (–20°C). pH shifts with time, as may percent transfection obtained. Prepare fresh every 3–6 months. Optimal transfection reagents will yield 95% transfection of 293T cells as assessed by transfection of GFP-encoding plasmids. Store transfection reagents cold or frozen, but it is essential to use each of the reagents at room temperature at the time of transfection.

Trypsin-EDTA (tissue culture grade) <!>

Equipment

Aspirator bottle (2-liter Kimax) (Kimble Glass 14607–2000) or similar aspirator

Beakers (1 liter and same volume as culture media used)

Bottle (sterile plastic; CF2: 250 ml, CF10:1000 ml)

Cell Factory loading bottle (sterile) with silicon tube and connector

Funnel (sterile)

Incubator, preset to 37°C, 5% CO_2 (shelves must be precisely level for Cell Factories)

Microscope

Cells in CF2 factories can be examined with most tissue-culture microscopes. For CF10 factories, use Nikon Eclipse TE300 or similar.

Nunc Cell Factory (1-layer CF1, 2-layer CF2, 10-layer CF10, or 40-layer CF40)

Nunc Cell Factory Start-up Kit (170769) with HDPE connectors (171838), cover caps (171897), blue sealing caps (167652), and Gelman 4210 bacterial air vent filter

Tissue-culture flasks (175 cm^2; T175) with gas-permeable cap

METHOD

1. Prepare the cells for transfection.

   For CF10 device

   a. Three days before transfection, seed six T175 flasks with 6×10^6 293T cells per flask. Incubate the cells for 48 hours.

   b. One day before transfection, trypsinize the T175 flasks and seed 2.5×10^8 of these 293T cells into the CF10.

   For CF2 device

   a. Three days before transfection, seed two to three T175 flasks with 1×10^7 293T cells per flask. Incubate the cells for 48 hours.

   b. One day before transfection, trypsinize the T175 flasks and seed 5×10^7 of these 293T cells into the CF2.

2. Adjust three plasmid DNAs (pMD.G, pFP93, pGINSIN) to a ratio of 1/3/3.

   For CF10: Mix 84.5/253.5/253.5 μg in a 250-ml sterile plastic bottle. Adjust volume to 60.5 ml with 10 mM Tris-HCl (pH 8), add 6.5 ml of 2.5 M $CaCl_2$, and swirl vigorously in a closed bottle.

   For CF2: Mix 16.9/50.7/50.7 μg in a 250-ml sterile plastic bottle. Adjust volume to 12.1 ml with 10 mM Tris-HCl (pH 8), add 1.3 ml of 2.5 M $CaCl_2$, and swirl vigorously in a closed bottle.

3. Pool the contents of the bottle by tilting it and then add 67 ml (for CF10) or 13.4 ml (for CF2) of 2x HBS.

4. Immediately mix bottle by inverting gently four to six times (preferable to vortexing). Precipitate DNAs for 3–5 minutes.

   Adjust this interval empirically for maximal transfection efficiency for a particular batch of transfection reagents.

5. Stop precipitation by decanting the DNA-Ca_2PO_4 mix.

   For CF10: Decant into 800 ml of fresh medium in a sterile plastic 1000-ml bottle with a cap. The bottle that comes with the filter unit can be used.

   For CF2: Decant into 185 ml of fresh medium in a sterile 250-ml plastic bottle with a cap.

6. Change the culture medium as follows:

   a. Pour Cell Factory medium into a waste beaker and discard.

   b. Connect sterile silicon tube with funnel to the Cell Factory.

   c. Place the Cell Factory on its side such that the open vent spout is closest to the work surface.

d. Elevate the funnel and slowly pour the transfection mix with fresh medium into the funnel. Ensure equal distribution of medium throughout all chambers.

e. Lift the factory at its connector/spout end such that the medium flows away from the spout opening end. This will prevent medium in upper chambers from leaking to lower chambers.

f. Place the Cell Factory back on the incubator shelf in an exactly horizontal orientation (cells will be barely covered by the transfection mix).

7. Incubate the cells for 16–18 hours and then observe the tissue-culture surface under the microscope. A fine DNA-hydroxyapatite precipitate (fine black specks much smaller than cells) should be evident on the surface of the cells.

8. Change the culture medium as follows:

   a. Pour the Cell Factory medium into a waste beaker and discard.

   b. Connect sterile silicon tube with funnel to the Cell Factory and place the Cell Factory on its side.

   c. Elevate the funnel and slowly pour 1000 ml (for CF10) or 200 ml (for CF2) of fresh, prewarmed medium into the funnel. Ensure equal distribution of medium throughout all chambers.

   d. Lift the Cell Factory at its connector/spout end such that the medium flows away from the spout opening end as in Step 6.

   e. Place the Cell Factory back in a horizontal orientation on the incubator shelf.

Protocol 2

Vector Harvest and Concentration for Cells Grown in CF10 or CF2 Devices

This protocol describes the harvesting and concentration of vector from cells grown in CF10 or CF2 devices.

MATERIALS

CAUTION: See Appendix for appropriate handling of materials marked with <!>.

Reagents

Bleach (10%) <!>
Ethanol (70% in tissue-culture-grade distilled H_2O; molecular grade)
Phosphate-buffered saline (PBS; tissue-culture grade)
Sucrose (20%), 20 mM Tris <!> (pH 7.4), 100 mM NaCl solution
Filter through a 0.45- or 0.2-μm filter.

Equipment

Autoclave
Bottle (250-ml plastic with screw cap) (e.g., Corning polystyrene storage bottle, 430281)
Cell strainer (e.g., BD Falcon, 70 μm Nylon, 352350)
Cryovials with screw cap (1.0–1.8 ml, sterile; Nunc)
Filtration devices (1000 ml, 0.22-μm pore size) (e.g., Nalgene MF75 Series)
Microcentrifuge tubes (1.5 and 0.5 ml, sterile)
Micropipettor
Scale to balance centrifuge tubes
Ultracentrifuge (Sorvall Discovery 100SE, Beckman L8–80M, or equivalent device)
Ultracentrifuge bottles (250 ml): polyallomer, Oakridge (Sorvall 54477) with fluorocarbon caps (Sorvall 54421) and A612 rotor (Sorvall 11997) or equivalent
Ultracentrifuge tubes (36 ml): disposable polyallomer (Sorvall 03141) for SureSpin 630 rotor (Sorvall 79367) or SW 28 rotor (Beckman 342207); 12-ml disposable polyallomer (Sorvall 03699) for TH641 rotor (Sorvall 08224) or SW 41 Ti rotor (Beckman 331302)

METHOD

1. Forty-eight hours after replacement of transfected cell medium, collect the supernatant in a large sterile container. Centrifuge at low speed for a few minutes to clear large-cell debris.
   Alternatively, allow detached cells to settle to bottom for 3–5 minutes. The Cell Factories can be washed twice with autoclaved distilled H_2O in the tissue-culture hood and then sealed and stored at 4°C for repeat use.
2. Pour the supernatant through a 50-μm cell strainer to clarify. Then filter through a 0.22-μm filter device. One Nalgene 1000-ml device suffices for 600–800 ml of vector. One Nalgene 500-

ml device suffices for 300–400 ml of vector supernatant before clogging. Aliquot a small portion (~2 ml) of the filtered vector supernatant into small cryovials for later titration and recovery comparison. Store at –80°C.

VSV-G-pseudotyped FIV vectors are quite stable in the short term at higher temperatures, with half-lives of 24 hours at 37°C and 72 hours at 4°C.

3. Wash the appropriate rotor with 70% ethanol.
4. Centrifuge the vector supernatant as follows:

For CF10

a. Wash the interior of 250-ml polyallomer Oakridge ultracentrifuge bottles and fluorocarbon caps with 70% ethanol. Aspirate dry.

The A621 rotor uses reusable ultracentrifuge bottles with white caps. These bottles cannot be autoclaved; use 10% bleach.

b. Add 200 ml of filtered vector supernatant to each 250-ml ultracentrifuge bottle. After filtration, the total volume of vector supernatant will be less than 1 liter. Adjust the volume of the last bottle to 200 ml with PBS. Close lids tightly.

Partial collapse of tubes or bottle may occur during ultracentrifugation if they are not full. Balance tubes on arm balance together with lids.

c. Centrifuge at 19,000 rpm in an A612 rotor in a Sorvall Discovery 100SE ultracentrifuge (67,000g_{rmax}), or equivalent rotor and device, for 6 hours at 4°C.

Shorter times may be possible with a different ultracentrifuge or rotor capable of higher *g* forces. A pellet will be visible at the outer bottom surface after decanting.

d. Decant supernatant and place 250-ml bottles on ice at a 45° angle with pellet oriented upward. Aspirate the pooled liquid from the face opposite the pellet. Then, rotate the pellet downward and add 6–7 ml of PBS if two rounds of concentration are desired. Otherwise, resuspend in an appropriate smaller volume.

e. Begin resuspension by pipetting with PBS along the pellet side. Direct the wash over the full length of the tube at first, with the stream directed toward the pellet. This step will take approximately 5 minutes per bottle. Pipette all material into one 250-ml bottle for a second round of ultracentrifugation.

About 35–45 ml of resuspended vector supernatant will be pooled from the six bottles. Some first-round concentrated vector may also be aliquoted (e.g., 800 μl) and frozen at –80°C in small tubes for later titration and recovery comparison. The first-round concentration step results in recoveries of 50–80% of transducing units (TUs).

f. For second-round concentration, wash the insides of disposable 12-ml polyallomer ultracentrifuge tubes with nanopure H_2O to remove cardboard fibers from the packaging, which can otherwise persist and copurify with vector particles. Autoclave tubes and allow them to cool to room temperature before use.

Tubes are not reusable. Ultracentrifugation weakens the plastic and the tubes leak if used a second time.

g. Wash buckets and lids of SW 41 Ti rotor with 70% ethanol. Aspirate dry. Fill 12-ml disposable polyallomer ultracentrifuge tubes with 9 ml of resuspended vector from the first round and 1 ml of 20% sucrose solution. Add PBS if necessary to fill tubes to maximum volume (10 ml), and to balance. Close tightly.

It is recommended for this step that 5 ml of resuspended vector be added to the ultracentrifuge tube, followed by 1 ml of sucrose solution with the pipette tip directed to the bottom of the tube, followed by the remaining 4 ml of resuspended vector added to the top of the total volume. This technique ensures an optimal interphase between the recovered vector and sucrose solutions.

h. Centrifuge at 24,000 rpm (67,000g_{rmax} at the bottom of tube and 31,000g_{rmin} at the top of tube) for 1.5 hours at 4°C in an SW 41 Ti rotor in a Sorvall Discovery 100SE ultracentrifuge, or equivalent. A pellet will again be visible.

i. Resuspend in 100 µl of PBS by washing the entire bottom of the tube for 5 minutes. Optimally, use a micropipettor set at 100-µl volume. Keep the tip submerged to prevent foaming. The resuspended liquid will be milky and slightly viscous. Pipette the resuspended vector into a 1.5-ml microcentrifuge tube and centrifuge at 3000*g* for 3 minutes to remove unresuspended material.

    The second ultracentrifugation and resuspension results in 30–70% recovery of TUs compared to the original unconcentrated supernatant. Some recoveries may exceed 100%, which may reflect dispersal of aggregated TUs present in the raw supernatant.

j. Aliquot the vector in appropriate volumes (e.g., 50 or 100 µl) and store at –80°C.

k. Clean the A621 rotor with 70% ethanol to remove any vector supernatant that may have leaked. Store rotor at 4°C.

For CF2

a. Wash the interior of new 36-ml disposable polyallomer ultracentrifuge tubes with nanopure H_2O to remove cardboard fibers. Any debris that is not washed away pellets out with the vector particles. Autoclave tubes and allow them to cool to room temperature before use.

    Tubes are not reusable. Ultracentrifugation weakens the plastic and the tubes leak if used a second time.

b. Add 25 ml of vector supernatant to each 36-ml tube. Slowly pipette 5 ml of 20% sucrose solution below that. Fill to maximum by adding 2.5–3 ml of vector supernatant.

    Partial collapse of tubes may occur during ultracentrifugation if they are not full. Balance tubes on arm balance together with SW 28 buckets and screw caps.

c. Centrifuge at 19,000 rpm for 1.5 hours without sucrose cushion or 25,000 rpm for 2 hours with a sucrose cushion in an SW 28 rotor in a Sorvall Discovery 100SE ultracentrifuge (67,000g_{rmax}) or equivalent rotor and device at 4°C. A pellet is not visible.

d. Aspirate all but a small volume of supernatant from the tubes. Resuspend the vector in PBS or medium as desired. Add 100 µl of PBS or medium to each tube. Use a micropipettor to wash the walls of the tubes. Begin resuspension by pipetting along the bottom of the tube. This step will take about 5 minutes per tube. Pool the resuspended vector supernatant in one tube. Aim for a total of 1–2 ml of resuspended vector. Aliquot in small volumes (e.g., 50 µl) and store at –80°C.

    The first-round concentration step results in recoveries of 50–80% of TUs.

e. Clean the buckets with 70% ethanol to remove any vector supernatant that may have leaked. Store rotor and buckets at 4°C.

Protocol 3

Vector Production and Harvesting for Cells Grown in T75 Tissue-culture Flasks

This protocol describes the production and harvesting of vector from cells grown in T75 tissue culture flasks.

MATERIALS

CAUTION: See Appendix for appropriate handling of materials marked with <!>.

Reagents

Cells for transfection: 293T

Early-passage 293T cells are preferred. They should be continuously and uniformly dividing prior to seeding for transfection.

Dulbecco's modified Eagle's medium with 10% fetal calf serum (DMEM-10), 100 units/ml penicillin G <!>, 100 μg/ml streptomycin <!>, and 2 mM L-glutamine

Plasmid DNAs: pMD.G, pFP93, pGINSIN

Transfection reagents

Tris-HCl <!> (10 mM, pH 8.0) for diluting plasmid DNA before $CaCl_2$ addition

$CaCl_2$ (2.5 M) <!>

Autoclave and store at –20°C.

2x HEPES-buffered saline (HBS)

See Protocol 1 for recipe.

Equipment

Filtration device (0.22 μm, e.g., 50 ml; Nalgene)

Tissue-culture flasks with gas-permeable cap (75 cm^2, T75)

Tubes (5-ml clear polystyrene; e.g., Falcon 352058)

Vortex mixer for mixing transfection mixtures in Falcon tubes

METHOD

1. Two days before transfection, split the cell culture to be transfected. One day before transfection, seed 3×10^6 cells per flask and grow overnight.
2. Four hours before transfection, replace the culture medium with 10 ml of fresh medium.
3. Adjust three plasmid DNAs (pMD.G, pFP93, pGINSIN) to a ratio of 1/3/3 by combining 1 μg of pMD.G, 3 μg of pFP93, and 3 μg of pGINSIN in a sterile 5-ml Falcon tube. Adjust volume to 800 μl with 10 mM Tris-HCl (pH 8).
4. Add 800 μl of 2.5 M $CaCl_2$ while vortexing at moderate speed and then allow the $CaCl_2$ to precipitate for 3 minutes. A faint milky color will appear.

5. Pipette mixture directly into the medium at the bottom of a tilted T75 flask, such that the cell monolayer is not detached. Swirl gently to distribute the precipitate evenly over the monolayer without dislodging cells.
6. Incubate the cells for 16–18 hours. Then, to change the culture medium, tip the flask, remove the medium by gentle aspiration, and replace it with 10–15 ml of fresh medium. Take care not to disturb the monolayer.
7. Incubate the cells for a further 48 hours. Collect the supernatant by centrifugation and filter it through a 0.22-μm filter.
8. Aliquot the supernatant into cryovials and freeze at –80°C.

Protocol 4

Titration

This protocol describes methods for measuring and calculating vector titers in TU/ml.

MATERIALS

CAUTION: See Appendix for appropriate handling of materials marked with <!>.

Reagents

Cell line or lines for titration: HT1080, HeLa, RD, Crandell feline kidney (CrFK) cells, 293T, canine D17, or NIH-3T3

Suspension cells can also be used for GFP vectors.

Culture medium appropriate for cells

Formalin (1%) (*optional,* see Step 5) <!>

Glutaraldehyde (1%) <!>

X-gal staining solution <!>

Prepare 10x buffer stock: 50 mM potassium ferricyanide <!>, 50 mM potassium ferrocyanide <!>, 20 mM $MgCl_2$ <!> in H_2O.

Prepare 40x X-gal stock: 40 mg/ml X-gal <!> in DMSO <!>.

Final solution is 1x buffer and 1 mg/ml X-gal diluted in PBS. Store at –80°C in small aliquots.

Phosphate-buffered saline (PBS; tissue-culture grade)

Vector stock

Equipment

Fluorescence-activated cell sorter (FACS)

Microscope

Tissue-culture plates (6 and 24 well)

Transparent counting grid

METHOD

1. Seed 2.5×10^5 cells (see Reagents list) into each well of a six-well plate. Incubate overnight.
2. Thaw vector stock and make fourfold to tenfold serial dilutions depending on the vector concentration. Use higher-fold dilutions for higher-titer preparations. These can be made in 24-well plates. Seed 4×10^4 to 6×10^4 cells/well depending on the cell line. For six-well plates, seed 2.5×10^5 cells/well. Mix thoroughly and change pipette tip at each step to avoid carryover.

   For unconcentrated vector, use dilutions over a 10^1-fold to 10^3-fold range. For singly concentrated vector, use dilutions over a 10^2-fold to 10^5-fold range, and for doubly concentrated vector, used dilutions over a 10^3-fold to 10^6-fold range. Carry out in duplicate or triplicate.

3. Aspirate medium from the cells and add each mixture to the wells, either changing the tip each time or moving from most dilute to least dilute. Leave at least one well untransduced as a control.

   Polybrene (8 μg/ml) may increase the transduction efficiency up to tenfold, but it is often toxic in vitro and is unsuitable for in vivo use. We avoid its use in titration. Qualify any titer operationally by the exact set of conditions used for titering (cell line, density, Polybrene, etc.).

4. After 6 hours, replace supernatant with fresh medium. Return to incubator.
5. Determine titer.

   For GFP vectors

   a. After 48 hours, collect cells (trypsinize adherent cells), wash with PBS, and resuspend.
   b. Fix cells in 1% formalin in PBS to stabilize fluorescence for later scoring, or store at 4°C without fixation and analyze within 24 hours.
   c. Determine percent transduction by flow cytometry. Choose wells with 5–30% enhanced GFP-positive cells for most reliable computation of titer.

      Wells with more than 30% transduced cells contain a significant fraction of multiply transduced cells. The number of cells present at the time of transduction rather than at the time of seeding is most properly used to calculate the correct titer. We estimate one round of cell division over 18–24 hours for CrFK cells grown in DMEM with 10% FCS. For a transduction volume of 1 ml/well, calculate the titer as cells present at time of transduction × percentage of FACS-positive cells × dilution factor = TU/ml. Determine mean and standard deviation of duplicate or triplicate wells.

      The human CMV promoter yields abundant gene expression in many cell lines by 48 hours. Alternative promoters (e.g., PGK and EF1α) may be required for better GFP expression in certain cell lines such as T cells. Different promoter-cell combinations may also require more time for transgene products to accumulate to peak levels.

   For β-galactosidase vectors

   a. After 48 hours, fix cells with 1% glutaraldehyde for 3 minutes at room temperature and wash with PBS.
   b. Incubate with X-gal-staining solution overnight at 37°C. Wash once with PBS and either replace with PBS or dry the plate.
   c. Count positive foci in ten random squares at 100× magnification and use this to determine the number of positive colonies per well.

      A transparent counting grid facilitates counting. For a volume of 1 ml, calculate the titer as total number of β-galactosidase-positive colonies per well × dilution factor = TU/ml. Determine mean and standard deviation of duplicate or triplicate wells.

      Titrations can also employ selectable markers such as *neoR* (G418). Count colonies after selection for the required time in the antibiotic. Immunofluorescence can be used to detect foci for other transgene products. Alternatively, use direct linkage of the protein of interest to a marker gene via an IRES. Normalization by reverse transcriptase (RT) activity or capsid protein ELISAs (enzyme-linked immunosorbent assays) does not accurately establish a surrogate for titration since infectivity only approximately correlates with these parameters.

ALTERNATIVE PROTOCOL: INTEGRASE MUTANT VECTORS

Mutations in certain residues in the catalytic center of retroviral integrase (IN) proteins confer selective abrogation of integrase catalytic activity (D64, D116, and E152 in the case of HIV-1 IN). Such "class I mutations" were also recently validated for FIV IN (Saenz et al. 2004). Other IN mutations display "class II" phenotypes, i.e., additional effects are apparent. The distinction is important, because IN arises by protease-mediated cleavage from the Gag/Pol precursor. Often, subtle defects in various other parts of the life cycle are produced by class II mutations. Deletions in IN, for example, and many single-amino-acid mutations, produce pleiotropic effects on the virus. Class I IN mutants were originally used as controls to establish that transgene expression in vivo occurs from integrated vector DNA (and not from pseudotransduction, for example) (Naldini et al. 1996). Using an FIV packaging construct with a single (D66V) or double (D66V, D116A) class I mutation permits elegantly controlled vector comparisons, in which viral particles are identical to the active vector except for one or two amino acids in one internal, proportionally low molar constituent, the integrase (Naldini et al. 1996; Loewen et al. 2003b; Saenz et al. 2004). This allows precise functional study and direct examination of alternative endpoints to integration, e.g., one- and two-LTR circles.

There is growing interest, however, in the potential utility of class I IN mutant vectors for gene therapy (Lu et al. 2004; Saenz et al. 2004; Vargas et al. 2004). One potential use is in noncycling cells, where unintegrated DNA, both linear and circular forms, may persist and be transcribed (Wu and Marsh 2001; Saenz et al. 2004). In tissue culture, expression of class I mutant lentiviral vectors can be as abundant as that from wild-type vectors, provided an internal promoter is used and the cells are growth-arrested (Saenz et al. 2004). The second situation where such vectors might be useful is where nondividing cells are the targets, but only transient expression of a transgene is desirable. Gene repair with custom zinc finger nucleases is one such example of a transiently required protein (Urnov et al. 2005). To prepare such vectors, all other aspects of the preparative protocols described here remain the same, except that pFP93.D66V or pFP93.D66V + D116A is used instead of the usual packaging plasmid pFP93.

REFERENCES

Alisky J.M., Hughes S.M., Sauter S.L., Jolly D., Dubensky T.W., Jr., Staber P.D., Chiorini J.A., and Davidson B.L. 2000. Transduction of murine cerebellar neurons with recombinant FIV and AAV5 vectors. *Neuroreport* **11:** 2669–2673.

Berthoux L., Sebastian S., Sokolskaja E., and Luban J. 2004. Lv1 inhibition of human immunodeficiency virus type 1 is counteracted by factors that stimulate synthesis or nuclear translocation of viral cDNA. *J. Virol.* **78:** 11739–11750.

Besnier C., Takeuchi Y., and Towers G. 2002. Restriction of lentivirus in monkeys. *Proc. Natl. Acad. Sci.* **99:** 11920– 11925.

Besnier C., Ylinen L., Strange B., Lister A., Takeuchi Y., Goff S.P., and Towers G.J. 2003. Characterization of murine leukemia virus restriction in mammals. *J. Virol.* **77:** 13403–13406.

Brooks A.I., Stein C.S., Hughes S.M., Heth J., McCray P.M., Jr., Sauter S.L., Johnston J.C., Cory-Slechta D.A., Federoff H.J., and Davidson B.L. 2002. Functional correction of established central nervous system deficits in an animal model of lysosomal storage disease with feline immunodeficiency virus-based vectors. *Proc. Natl. Acad. Sci.* **99:** 6216–6221.

Burns J.C., Friedmann T., Driever W., Burrascano M., and Yee J.K. 1993. Vesicular stomatitis virus G glycoprotein pseudotyped retroviral vectors: Concentration to very high titer and efficient gene transfer into mammalian and nonmammalian cells. *Proc. Natl. Acad. Sci.* **90:** 8033–8037.

Curran M.A. and Nolan G.P. 2002a. Nonprimate lentiviral vectors. *Curr. Top. Microbiol. Immunol.* **261:** 75–105.

———. 2002b. Recombinant feline immunodeficiency virus vectors. Preparation and use. *Methods Mol. Med.* **69:** 335–350.

Curran M.A., Kaiser S.M., Achacoso P.L., and Nolan G.P. 2000. Efficient transduction of nondividing cells by optimized feline immunodeficiency virus vectors. *Mol. Ther.* **1:** 31–38.

Curran M.A., Ochoa M.S., Molano R.D., Pileggi A., Inverardi L., Kenyon N.S., Nolan G.P., Ricordi C., and Fenjves E.S. 2002. Efficient transduction of pancreatic islets by feline immunodeficiency virus vectors. *Transplantation* **74:** 299–306.

Derksen T.A., Sauter S.L., and Davidson B.L. 2002. Feline immunodeficiency virus vectors. Gene transfer to mouse retina following intravitreal injection. *J. Gene Med.* **4:** 463–469.

Donello J.E., Loeb J.E., and Hope T.J. 1998. Woodchuck hepatitis virus contains a tripartite posttranscriptional regulatory element. *J. Virol.* **72:** 5085–5092.

DuBridge R.B., Tang P., Hsia H.C., Leong P.M., Miller J.H., and Calos M.P. 1987. Analysis of mutation in human cells by using an Epstein-Barr virus shuttle system. *Mol. Cell. Biol.* **7:** 379–387.

Elder J.H. and Phillips T.R. 1995. Feline immunodeficiency virus as a model for development of molecular approaches to intervention strategies against lentivirus infections. *Adv. Virus Res.* **45:** 225–247.

Haskell R.E., Hughes S.M., Chiorini J.A., Alisky J.M., and Davidson B.L. 2003. Viral-mediated delivery of the late-infantile neuronal ceroid lipofuscinosis gene, TPP-I to the mouse central nervous system. *Gene Ther.* **10:** 34–42.

Hatziioannou T., Cowan S., Goff S.P., Bieniasz P.D., and Towers G.J. 2003. Restriction of multiple divergent retroviruses by Lv1 and Ref1. *EMBO J.* **22:** 385–394.

Hope T. 2002. Improving the post-transcriptional aspects of lentiviral vectors. *Curr. Top. Microbiol. Immunol.* **261:** 179–189.

Hughes S.M., Moussavi-Harami F., Sauter S.L., and Davidson B.L. 2002. Viral-mediated gene transfer to mouse primary neural progenitor cells. *Mol. Ther.* **5:** 16–24.

Johnston J.C., Gasmi M., Lim L.E., Elder J.H., Yee J.K., Jolly D.J., Campbell K.P., Davidson B.L., and Sauter S.L. 1999. Minimum requirements for efficient transduction of dividing and nondividing cells by feline immunodeficiency virus vectors. *J. Virol.* **73:** 4991–5000.

Kang Y., Stein C.S., Heth J.A., Sinn P.L., Penisten A.K., Staber P.D., Ratliff K.L., Shen H., Barker C.K., Martins I., Sharkey C.M., Sanders D.A., McCray P.B., Jr., and Davidson B.L. 2002. In vivo gene transfer using a nonprimate lentiviral vector pseudotyped with Ross River Virus glycoproteins. *J. Virol.* **76:** 9378–9388.

Kemler I., Azmi I., and Poeschla E.M. 2004. The critical role of proximal *gag* sequences in feline immunodeficiency virus genome encapsidation. *Virology* **327:** 111–120.

Kemler I., Barraza R., and Poeschla E.M. 2002. Mapping of the encapsidation determinants of feline immunodeficiency virus. *J. Virol.* **76:** 11889–11903.

Loewen N., Barraza R., Whitwam T., Saenz D., Kemler I., and Poeschla E. 2003a. FIV vectors. *Methods Mol. Biol.* **229:** 251–271.

Loewen N., Fautsch M., Peretz M., Bahler C., Cameron J.D., Johnson D.H., and Poeschla E.M. 2001. Genetic modification of human trabecular meshwork with lentiviral vectors. *Hum. Gene Ther.* **12:** 2109–2119.

Loewen N., Leske D., Chen Y., Teo W., Saenz D., Peretz M., Holmes J., and Poeschla E.M. 2003b. Comparison of wild-type and class I integrase mutant-FIV vectors in retina demonstrates sustained expression of integrated transgenes in retinal pigment epithelium. *J. Gene Med.* **5:** 1009–1017.

Loewen N., Bahler C., Teo W., Whitwam T., Peretz M., Xu R., Fautsch M., Johnson D.H., and Poeschla E.M. 2002. Preservation of aqueous outflow facility after second-generation FIV vector-mediated expression of marker genes in anterior segments of human eyes. *Investig. Ophthalmol. Vis. Sci.* **43:** 3686–3690.

Lotery A.J., Derksen T.A., Russell S.R., Mullins R.F., Sauter S., Affatigato L.M., Stone E.M., and Davidson B.L. 2002. Gene transfer to the nonhuman primate retina with recombinant feline immunodeficiency virus vectors. *Hum. Gene Ther.* **13:** 689–696.

Lu R., Nakajima N., Hofmann W., Benkirane M., Jeang K.T., Sodroski J., and Engelman A. 2004. Simian virus 40-based replication of catalytically inactive human immunodeficiency virus type 1 integrase mutants in nonpermissive T cells and monocyte-derived macrophages. *J. Virol.* **78:** 658–668.

Naldini L., Bloemer U., Gallay P., Ory D., Mulligan R., Gage F.H., Verma I.M., and Trono D. 1996. In vivo gene delivery and stable transduction of nondividing cells by a lentiviral vector. *Science* **272:** 263–267.

Parolin C. and Sodroski J. 1995. A defective HIV-1 vector for gene transfer to human lymphocytes. *J. Mol. Med.* **73:** 279–288.

Pedersen N.C. 1993. The feline immunodeficiency virus. In *The Retroviridae* (ed. J.A. Levy), pp. 181–228. Plenum Press, New York.

Poeschla E. and Looney D. 1998. CXCR4 is required by a non-primate lentivirus: Heterologous expression of feline immunodeficiency virus in human, rodent and feline cells. *J. Virol.* **72:** 6858–6866.

Poeschla E., Wong-Staal F., and Looney D. 1998. Efficient transduction of nondividing cells by feline immunodeficiency virus lentiviral vectors. *Nat. Med.* **4:** 354–357.

Price M.A., Case S.S., Carbonaro D.A., Yu X.J., Petersen D., Sabo K.M., Curran M.A., Engel B.C., Margarian H., Abkowitz J.L., Nolan G.P., Kohn D.B., and Crooks G.M. 2002. Expression from second-generation feline immunodeficiency virus vectors is impaired in human hematopoietic cells. *Mol. Ther.* **6:** 645–652.

Richmond Y.S. and McKinney R.W., eds. 1999. *Biosafety in microbiological and biomedical laboratories (BMCL)*, 4th edition. U.S. Department of Health and Human Services, Public Health Service & Centers for Disease Control and Prevention, National Institutes of Health. U.S. Government Printing Office, Washington, D.C.

Saenz D.T. and Poeschla E.M. 2004. FIV: From lentivirus to lentivector. *J. Gene Med.* (suppl. 1) **16:** S95–104.

Saenz D., Teo I., Olsen J.C., and Poeschla E. 2005. Restriction of feline immunodeficiency virus by Ref1, LV1 and primate TRIM5α proteins. *J. Virol.* **79:** 15175–15188.

Saenz D., Loewen N., Peretz M., Whitwam T., Barraza R., Howell K., Holmes J.H., Good M., and Poeschla E.M. 2004. Unintegrated lentiviral DNA persistence and accessibility to expression in nondividing cells: Analysis with class I integrase mutants. *J. Virol.* **78:** 2906–2920.

Shaw G., Morse S., Ararat M., and Graham F.L. 2002. Preferential transformation of human neuronal cells by human adenoviruses and the origin of HEK 293 cells. *FASEB J.* **16:** 869–871.

Sinn P.L., Hickey M.A., Staber P.D., Dylla D.E., Jeffers S.A., Davidson B.L., Sanders D.A., and McCray P.B., Jr. 2003. Lentivirus vectors pseudotyped with filoviral envelope glycoproteins transduce airway epithelia from the apical surface independently of folate receptor alpha. *J. Virol.* **77:** 5902–5910.

Stein C.S. and Davidson B.L. 2002. Gene transfer to the brain using feline immunodeficiency virus-based lentivirus vectors. *Methods Enzymol.* **346:** 433–454.

Stein C.S., Kang Y., Sauter S.L., Townsend K., Staber P., Derksen T.A., Martins I., Qian J., Davidson B.L., and McCray P.B., Jr. 2001. In vivo treatment of hemophilia A and mucopolysaccharidosis type VII using nonprimate lentiviral vectors. *Mol. Ther.* **3:** 850–856.

Stremlau M., Owens C.M., Perron M.J., Kiessling M., Autissier P., and Sodroski J. 2004. The cytoplasmic body component TRIM5α restricts HIV-1 infection in Old World monkeys. *Nature* **427:** 848–853.

Towers G., Collins M., and Takeuchi Y. 2002. Abrogation of Ref1 retrovirus restriction in human cells. *J. Virol.* **76:** 2548–2550.

Urnov F.D., Miller J.C., Lee Y.L., Beausejour C.M., Rock J.M., Augustus S., Jamieson A.C., Porteus M.H., Gregory P.D., and Holmes M.C. 2005. Highly efficient endogenous human gene correction using designed zinc-finger nucleases. *Nature* **435:** 646–651.

Vargas J., Jr., Gusella G.L., Najfeld V., Klotman M.E., and Cara A. 2004. Novel integrase-defective lentiviral episomal vectors for gene transfer. *Hum. Gene Ther.* **15:** 361–72.

Wang G., Slepushkin V., Zabner J., Keshavjee S., Johnston J.C., Sauter S.L., Jolly D.J., Dubensky T.W., Jr., Davidson B.L., and McCray P.B., Jr. 1999. Feline immunodeficiency virus vectors persistently transduce nondividing airway epithelia and correct the cystic fibrosis defect (see comments). *J. Clin. Invest.* **104:** R55–62.

Whitwam T., Peretz M., and Poeschla E.M. 2001. Identification of a central DNA flap in feline immunodeficiency virus. *J. Virol.* **75:** 9407–9414.

Wu Y. and Marsh J.W. 2001. Selective transcription and modulation of resting T cell activity by preintegrated HIV DNA. *Science* **293:** 1503–1506.

Zufferey R., Donello J.E., Trono D., and Hope T.J. 1999. Woodchuck hepatitis virus posttranscriptional regulatory element enhances expression of transgenes delivered by retroviral vectors. *J. Virol.* **73:** 2886–2892.

7 Lentivirus Transduction of Hematopoietic Cells

Ming-Jie Li and John J. Rossi

Division of Molecular Biology, Beckman Research Institute of the City of Hope, Duarte, California 91010

ABSTRACT

Efficient transfer and sustained expression of transgenes are among the most important issues in gene delivery technologies. The majority of hematopoietic cells, including hematopoietic stem/progenitor cells and terminally differentiated cells, such as primary T cells and macrophages, are nondividing or slowly self-renewing. These cell types are refractory to most nonviral or retroviral delivery methods. Lentiviral vectors are capable of transducing nondividing cells and maintaining long-term and sustained expression of the transgenes. Many hematopoietic cell types have been successfully transduced with lentiviral vectors carrying a variety of genes. Lentiviral vectors are becoming useful for many delivery protocols, such as long-term expression of short hairpin RNA (shRNA) and functional genetics. They may also have great potential in gene therapy.

INTRODUCTION

One of the major problems in gene transfer is the need for a delivery system that ensures efficient gene transduction and expression. Retroviral vectors derived from murine leukemia virus (MLV) have long been favored in gene delivery and human gene therapy protocols for their efficient integration into the genome of the target cells (Miller et al. 1990; Verma 1994; Leiden 1995). However, these vectors require cell division for efficient gene transfer, which limits their application in nonproliferating cells, such as hematopoietic stem cells (Miller et al. 1990). In contrast, lentiviral vectors derived from human immunodeficiency virus type 1 (HIV-1) are able to transduce a wide array of quiescent cell types with sustained long-term expression of the transgenes (Naldini et al. 1996a,b; Blomer et al. 1997; Kafri et al. 1997). We have reported that a single lentiviral vector can efficiently deliver and express up to three polymerase III (pol III) expression cassettes along with a pol-II-driven reporter gene in hematopoietic stem cells (Li et al. 2005).

Using pHIV7-GFP (green fluorescent protein) as a prototype, a typical third generation self-inactivating (SIN) vector and its production system are depicted in Figure 1. The plasmid pCgp contains *gag/pol* genes, which provide structural proteins and reverse transcriptase. pCMV-*rev* encodes Rev, which binds to the Rev-responsive element (RRE) for efficient RNA export from the nucleus. pCMV-G encodes the vesicular stomatitis virus glycoprotein (VSV-G) that replaces HIV-1 Env. VSV-G expands the tropism of the vectors and allows concentration via ultracentrifugation. In the transfer vector, pHIV7-GFP, the U3 region in the 5′ LTR (long terminal repeat), is replaced with the cytomegalovirus (CMV) promoter and enhancer sequences. The packaging signal (ψ) is essential for encapsidation and the RRE is required for producing high-titer vectors. The flap sequence is important for nuclear import of the vector DNA, a feature required for transducing nondividing cells (Sirven et al. 2000). The enhanced GFP (EGFP) reporter gene is useful for vector titration, measurements of transduction efficiency, and selection of transduced cells. The woodchuck hepatitis virus posttranscription regulation element (WPRE) (Zufferey et al. 1999) following the EGFP sequence improves the expression of the reporter gene. In the 3′ LTR, the *cis*-regulatory sequences were completely removed from the U3 region.

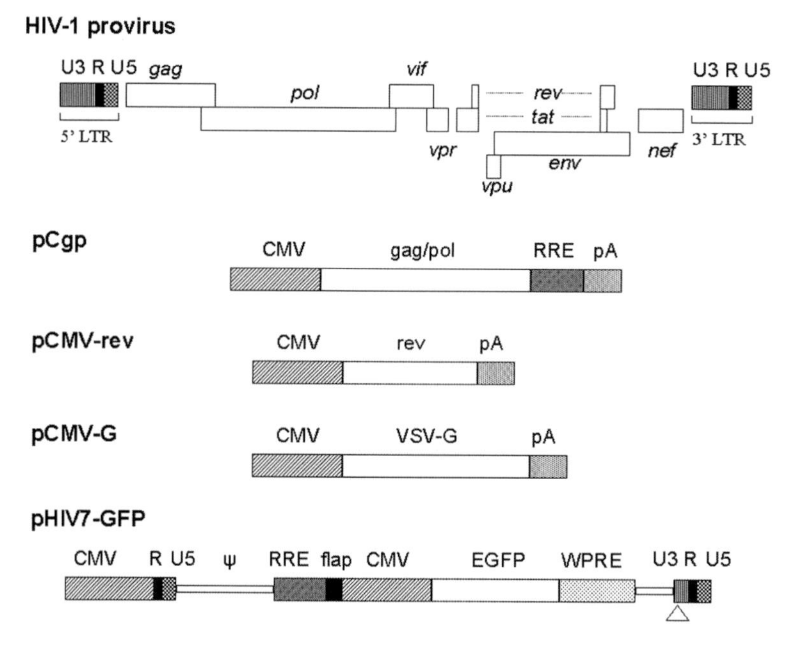

FIGURE 1. Schematic representation of HIV-1 provirus, the HIV-based lentiviral vector, and the packaging plasmids. The elements for vector production are separated into four different plasmids. pCgp contains the *gag* and *pol* genes and the RRE sequence from HIV-1 under the control of the CMV promoter. pCMV-*rev* contains the coding sequence of *rev* driven by the CMV promoter. pCMV-G contains the VSV-G protein gene under the control of the CMV promoter. The transfer vector, pHIV7-GFP, contains a hybrid 5′ LTR in which the U3 region is replaced by the CMV promoter, packaging signal (ψ), RRE sequence, flap sequence, EGFP gene driven by the CMV promoter, WPRE, and the 3′ LTR in which the *cis*-regulatory sequences are completely removed from the U3 region. A gene of interest can be cloned to the polylinker located directly upstream of the CMV promoter of EGFP in the transfer vector. pA indicates the polyadenylation signal from the human β-globin gene.

To package the transfer vector into transducing particles, a cotransfection of these four plasmids into the suitable producer cells is required. The calcium phosphate transfection method is inexpensive and highly efficient for delivery of plasmids to 293T cells. Inclusion of sodium butyrate in the medium during vector production can enhance the titer of lentiviral vectors more than tenfold (Gasmi et al. 1999). The titer can be further increased by ultracentrifugation concentration.

Polybrene has been widely used for gene-transfer protocols. Low-speed centrifugation (spinoculation) is also reported to enhance the transduction efficiency in some cell types (Zielske and Gerson 2002). RetroNectin, a recombinant human fibronectin fragment CH-296, is commonly used for $CD34^+$ hematopoietic cell transduction (Hanenberg et al. 1997). Even though lentiviral vectors do not require cell division to enter the cell nucleus, prestimulation with cytokines can remarkably enhance transduction efficiency in hematopoietic stem cells (Zielske and Gerson 2003).

Protocol

Lentivirus Transduction of Hematopoietic Cells

This protocol describes lentivirus-vector-based delivery of foreign genes to hematopoietic cells. The method is applicable to various cell types in experiments that require long-term transgene expression.

MATERIALS

CAUTION: See Appendix for appropriate handling of materials marked with <!>.

Reagents

Anti-CD34 antibody-coupled magnetic beads (if transducing primary hematopoietic stem cells; Miltenyi Biotec Inc.)

Bovine serum albumin (BSA; 2% in PBS)

Dissolve 2 g of BSA (Fraction V) in 100 ml of PBS. Sterilize by passage through a 0.2-μm filter. Store at 4°C.

$CaCl_2$ (2 M) <!>

Dissolve 29.4 g of $CaCl_2$ (J.T. Baker) in 70 ml of H_2O. Adjust the volume to 100 ml. Sterilize by passage through a 0.2-μm filter.

Cells

293T (human embryonic kidney cells containing SV40 large T antigen)

HT1080 (a human fibrosarcoma cell line)

Cells to be transduced (target cells)

Formaldehyde (3.7% in PBS) <!>

Prepare a 1:10 dilution of 37% formaldehyde solution (Sigma-Aldrich) with PBS.

2x HEPES-buffered saline (HBS; 0.05 M HEPES, 0.28 M NaCl, and 1.5 mM Na_2HPO_4 at pH 7.12)

Dissolve 2.38 g of HEPES, 3.28 g of NaCl, and 42.6 mg of Na_2HPO_4 in 200 ml of H_2O. Sterilize by passage through a 0.2-μm filter.

Medium for 293T cells and HT1080 cells

Use Dulbecco's modified Eagle's medium (DMEM) with high glucose (4500 mg/liter) supplemented with 10% fetal bovine serum (FBS), 100 units/ml penicillin, and 100 μg/ml streptomycin <!>.

Medium for target cells

Choose appropriate media, cytokines, and other supplements depending on cell types and the experimental design. For the example in this protocol, use $CD34^+$ transduction medium: Iscove's modified Dulbecco's medium (IMDM) supplemented with 20% BIT9500 (Stem Cell Technology, Vancouver, B.C., Canada), 100 ng/ml of stem cell factor (SCF), 100 ng/ml of Flt3-ligand, and 10 ng/ml of thrombopoietin (TPO).

Phosphate-buffered saline (PBS, pH 7.4)

137 mM NaCl

2.7 mM KCl

8.1 mM Na_2HPO_4 <!>

1.5 mM KH_2PO_4 <!>

Plasmid DNA

The plasmid DNAs in this protocol include pHIV7-GFP containing the desired transgenes, pCgp, pCMV-*rev*, and pCMV-G. Purify all the plasmid DNAs using the QIAGEN Plasmid Maxi Kit according to the manufacturer's instructions.

Polybrene (4 mg/ml)

Dissolve 40 mg of polybrene (hexadimethrine bromide) in 10 ml of H_2O. Sterilize by passage through a 0.2-µm filter. Store at 4°C.

RetroNectin (Takara Mirus Bio Inc., Madison, Wisconsin)

Prepare 25 µg/ml RetroNectin solution according to the manufacturer's instructions. Store at 4°C.

Sodium butyrate (0.6 M)

Dissolve 3.3 g of sodium butyrate in 50 ml of H_2O. Sterilize by passage through a 0.2-µm filter. Store at 4°C.

TE 79/10

Prepare a 1:10 dilution of TE79 (10 mM Tris-HCl and 1 mM EDTA at pH 7.9) with H_2O. Sterilize by passage through a 0.2-µm filter.

Equipment

Centrifuge tubes (1 x 3.5-inch polyallomer [Beckman])

Autoclave (15 psi, 15–20 min) before use.

Fluorescence-activated cell sorter (FACS)

Incubator (37°C)

Nontissue-culture-treated plates (48 and 24 well)

Plate (12 well)

Syringe (30 ml)

Syringe filter (0.2 µm)

Tissue culture dishes (100 mm)

Ultracentrifuge and SW 28 rotor (Beckman)

METHODS

Production of Lentiviral Vectors: Packaging of the Vectors

1. Maintain 293T cells in complete culture medium in a 37°C incubator with 5% CO_2. Twenty-four hours before transfection, plate exponentially growing 293T cells in 100-mm tissue culture dishes at 4 x 10^6 cells/plate.

   Cell density should be approximately 80% confluent for transfection.

2. Prepare 1 ml of calcium phosphate–DNA suspension for each 100-mm plate of cells as follows:

   a. Set up two sterile tubes for transfection of one plate. Label the tubes 1 and 2.

   b. Add 0.5 ml of 2x HBS to Tube 1.

   c. Add TE 79/10 to Tube 2. The volume of TE 79/10 = 440 µl minus the volume of the DNA solution.

   d. Add 15 µg of the transfer vector containing the transgene, 15 µg of pCgp, 5 µg of pCMV-*rev*, and 5 µg of pCMV-G to Tube 2 and mix.

   e. Add 60 µl of 2 M $CaCl_2$ solution to Tube 2 and mix gently.

f. Transfer the contents from Tube 2 to Tube 1 dropwise with gentle mixing.

g. Allow the suspension to sit for 30 minutes at room temperature.

3. Mix the precipitate well by pipetting or vortexing.
4. Add 1 ml of the suspension to a 100-mm plate containing cells. Add the suspension *slowly*, dropwise while gently swirling the medium in the plate. Return the plates to the 37°C incubator and leave the precipitate for 4 hours.
5. Replace the old medium with 6 ml of fresh culture medium. Add 60 µl of 0.6 M sodium butyrate. Return to the incubator.
6. After 48 hours of culture, collect the supernatant and freeze it at –80°C or proceed to the concentration step.

Concentration of the Vectors

7. Centrifuge the supernatant (freshly collected or thawed from the freezer) at 900*g* for 10 minutes to remove any cell debris in the supernatant.
8. Filter the supernatant through a 0.2-µm syringe filter.
9. Transfer the supernatant to autoclaved polyallomer tubes. Concentrate the supernatant by ultracentrifugation for 1.5 hours at 4°C in a Beckman SW 28 swinging bucket rotor at 24,500 rpm.
10. Remove the supernatant and resuspend the pellet in an appropriate amount of culture medium, e.g., 300 µl for 30 ml of original supernatant if a 100-fold concentration is desired.
11. Divide the concentrated vector into 10–50-µl aliquots and store at –80°C until use.

Titration of the Vectors

12. Seed 5×10^4 HT1080 cells/well in a 12-well plate in complete medium and culture overnight in a 37°C incubator with 5% CO_2.
13. Add serial diluted vector stock and 4 µl/ml polybrene to the cultured cells. Continue culture for 48 hours.
14. Trypsinize the cells. Following centrifugation, remove the supernatant and resuspend the pellet in 300 µl of 3.7% formaldehyde in PBS.
15. Determine the percentage of EGFP-positive cells by FACS analysis.

    The titer will be represented as transduction units (TUs) per milliliter concentrated vector (TU/ml).

$$\text{Titer} = \frac{\text{cell number} \times \text{percentage of EGFP}^{+}\text{ cells} \times \text{dilution}}{\text{vector volume (ml)} \times 100}$$

Transduction of Lentiviral Vectors to Target Cells

For transducing cultured hematopoietic cells

16. Seed exponentially growing cells at 2×10^5 cells/well in 1 ml of culture medium into a 24-well plate.
17. Add various amounts of concentrated vector stock depending on cell type. For the K562 cell line (CML leukemia cell line), a multiplicity of infection (moi) of 10 can achieve virtually 100% transduction.

18. Add 4 μg/ml polybrene. Return the cells to a 37°C incubator.
19. After overnight incubation, centrifuge the cells, discard the supernatant, and resuspend the cells with fresh culture medium. Return the cells to culture.
20. Determine transduction efficiency by FACS analysis 48 hours after transduction (see Troubleshooting).

For transducing primary hematopoietic stem cells

21. Purify $CD34^+$ hematopoietic stem cells from umbilical cord blood or bone marrow using anti-CD34 antibody-coupled magnetic beads following the manufacturer's protocol.
22. Forty-eight hours before transduction, culture the $CD34^+$ cells in $CD34^+$ transduction medium.
23. Coat a 48-well nontissue-culture-treated plate with 0.2 ml of 25 μg/ml RetroNectin (~5 $\mu g/cm^2$) for 2 hours at room temperature.
24. Remove RetroNectin and then add 0.2 ml of 2% BSA in PBS for blocking. Store the plate for 30 minutes at room temperature.
25. After washing the wells with PBS, adjust the lentiviral vector stock to the appropriate moi (range 5–40) with plain IMDM medium to 200-μl volume and load it into the well of the coated plate.
26. After incubation for 2 hours at 37°C, remove the vector supernatant and then wash the well with PBS.
27. Add the prestimulated $CD34^+$ cells to the well at 1×10^5 cells/well in 0.2 ml of growth medium and return the cells to the 37°C incubator.
28. After overnight culture, centrifuge the cells, resuspend the cell pellet in 1 ml of culture medium, and transfer the cells to a 24-well plate. Return the cells to the incubator.
29. Determine transduction efficiency by FACS analysis 6 days after transduction.

TROUBLESHOOTING

Problem (Step 20): Some cell types, such as CEM (a human T cell line), are difficult to transduce by the standard protocol.

Solution: For those cell types that resist transduction by normal means, centrifugation can substantially enhance the transduction efficiency. Treat these cells as follows. Combine 2×10^5 cells in 1 ml of culture medium in a 15-ml centrifuge tube with vector and 4 μg/ml polybrene. Centrifuge at 900*g* for 30 minutes at 20°C. Then, without removing the supernatant, use a pipette to resuspend the cell pellet and transfer the cells to the culture plate. Incubate the cells overnight, replace the medium, and continue the culture.

ACKNOWLEDGMENTS

We thank Dr. Jiing-Kaun Yee for providing pHIV7-GFP and the packaging plasmids. The authors were supported by National Institutes of Health grants AI29329, AI42552, and HL074704.

REFERENCES

Blomer U., Naldini L., Kafri T., Trono D., Verma I.M., and Gage F.H. 1997. Highly efficient and sustained gene transfer in adult neurons with a lentivirus vector. *J. Virol.* **71:** 6641–6649.

Gasmi M., Glynn J., Jin M.J., Jolly D.J., Yee J.K., and Chen S.T. 1999. Requirements for efficient production and transduction of human immunodeficiency virus type 1-based vectors. *J. Virol.* **73:** 1828–1834.

Hanenberg H., Hashino K., Konishi H., Hock R.A., Kato I., and Williams D.A. 1997. Optimization of fibronectin-assisted retroviral gene transfer into human $CD34^+$ hematopoietic cells. *Hum. Gene Ther.* **8:** 2193–2206.

Kafri T., Blomer U., Peterson D.A., Gage F.H., and Verma I.M. 1997. Sustained expression of genes delivered directly into liver and muscle by lentiviral vectors. *Nat. Genet.* **17:** 314–317.

Leiden J.M. 1995. Gene therapy—Promise, pitfalls, and prognosis. *N. Engl. J. Med.* **333:** 871–873.

Li M.-J., Kim J., Li S., Zaia J., Yee J.-K., Anderson J., Akkina R., and Rossi J.J. 2005. Long term inhibition of HIV-1 infection in primary hematopoietic cells by lentiviral vector delivery of a triple combination of anti-HIV shRNA, anti-CCR5 ribozyme and a nucleolar localizing TAR decoy. *Mol. Ther.* **12:** 900–909.

Miller D.G., Adam M.A., and Miller A.D. 1990. Gene transfer by retrovirus vectors occurs only in cells that are actively replicating at the time of infection. *Mol. Cell. Biol.* **10:** 4239–4242.

Naldini L., Blomer U., Gage F.H., Trono D., and Verma I.M. 1996a. Efficient transfer, integration, and sustained long-term expression of the transgene in adult rat brains injected with a lentiviral vector. *Proc. Natl. Acad. Sci.* **93:** 11382–11388.

Naldini L., Blomer U., Gallay P., Ory D., Mulligan R., Gage F.H., Verma I.M., and Trono D. 1996b. In vivo gene delivery and stable transduction of nondividing cells by a lentiviral vector. *Science* **272:** 263–267.

Sirven A., Pflumio F., Zennou V., Titeux M., Vainchenker W., Coulombel L., Dubart-Kupperschmitt A., and Charneau P. 2000. The human immunodeficiency virus type-1 central DNA flap is a crucial determinant for lentiviral vector nuclear import and gene transduction of human hematopoietic stem cells. *Blood* **96:** 4103–4110.

Verma I.M. 1994. Gene therapy: Hopes, hypes, and hurdles. *Mol. Med.* **1:** 2–3.

Zielske S.P. and Gerson S.L. 2002. Lentiviral transduction of P140K MGMT into human $CD34^+$ hematopoietic progenitors at low multiplicity of infection confers significant resistance to BG/BCNU and allows selection in vitro. *Mol. Ther.* **5:** 381–387.

———. 2003. Cytokines, including stem cell factor alone, enhance lentiviral transduction in nondividing human LTCIC and NOD/SCID repopulating cells. *Mol. Ther.* **7:** 325–333.

Zufferey R., Donello J.E., Trono D., and Hope T.J. 1999. Woodchuck hepatitis virus posttranscriptional regulatory element enhances expression of transgenes delivered by retroviral vectors. *J. Virol.* **73:** 2886–2892.

8 Spleen Necrosis Virus-based Vectors

Zahida Parveen,* Muhammad Mukhtar,* and Roger J. Pomerantz†

*The Dorrance H. Hamilton Laboratories, Division of Infectious Diseases, Department of Medicine, Jefferson Medical College, Thomas Jefferson University, Philadelphia, Pennsylvania 19107;
†Tibotec Inc., Yardley, Pennsylvania 19067

ABSTRACT

Genetically engineered retroviruses are widely used for gene delivery into human cells. A number of investigators have studied spleen necrosis virus (SNV) as a vehicle for gene delivery. We have used vectors developed from SNV and its closely associated avian reticuloendotheliosis virus strain A (REV-A) for gene transfer into a variety of cells, including primary hematopoietic cells and human brain and postmitotic neuronal cells that are difficult to transduce with various other vector systems. SNV-based vector systems have the advantage of being quite safe, because wild-type SNV is unable to infect human cells and has less preference for integration into transcriptionally active sites or genes. The generation of retroviral vectors requires cotransfection of more than one plasmid into a packaging cell line, which is a tedious process. The development of stable packaging cell lines expressing the structural proteins Gag-Pol and envelope (Env) proteins will enhance mass production of retroviral vectors for future gene therapy experiments both in vitro and in vivo.

INTRODUCTION

Genetically engineered retroviruses are the most popular tools for the delivery of genes into human cells (Dornburg 1995; Dornburg and Pomerantz 2000; Kim et al. 2000; Kurian et al. 2000; Palu et al. 2000). In the past decade, a variety of retroviral vectors for gene delivery have been developed and used in clinical trials. The pros and cons of each vector system in transferring genes into human cells is still an active area of investigation. SNV-based vectors have shown promise for targeted and efficient gene delivery into human cells (Dougherty and Temin 1987; Dornburg 1995, 2003; White and Gilmore 1996; Martinez and Dornburg 1995b; Parveen et al. 2000, 2003; Acheampong et al. 2005; Marusich et al. 2005). This vector system is quite safe, due to the inability of wild-type SNV to infect human cells and less preference for integration into transcriptionally active sites or genes (Gautier et al. 2000; Mitchell et al. 2004).

Currently, retroviral vectors are produced by monolayer culture of packaging cell lines in tissue-culture dishes or flasks. They are generated as particles with particular half-lives and responses toward various stimuli. Retroviral vectors provide an opportunity to deliver genes into a variety of dividing cells as well as quiescent cells.

The protocol described in this chapter is based on our experience with vectors developed from SNV and the closely associated REV-A (Parveen et al. 2000). We have used both of these viral vectors for gene transfer into a variety of cells, including primary isolates from hematopoietic cells and human brain and postmitotic neuronal cells (Parveen et al. 2000). As in other retroviral vectors, the three major elements of the SNV-based vector system are (1) the packaging signal sequence (Ψ) that ensures the encapsidation of vector RNA into retroviral particles, (2) reverse transcription machinery, and (3) the sequence necessary for integration of the proviral vector into the host (Kurian et al. 2000; Dornburg 2003). To harness the therapeutic potential of viral particles, the viral life cycle is genetically altered such that, upon infection, these particles can only transfer a therapeutic gene without any further replication. This has been accomplished by expressing various elements necessary for viral morphogenesis from separate plasmids (Naldini et al. 1996; Naldini 1999; Naldini and Verma 2000; Ailles and Naldini 2002). Thus, cotransfection of more than one plasmid into a packaging cell line is necessary for the generation of retrovirus particles, as shown in Figure 1. Cotransfection of multiple plasmids is a tedious process involving optimization of the plasmids' DNA ratio (Mukhtar et al. 2000). To overcome this hurdle, efforts have been directed toward the generation of stable packaging cell lines expressing the structural proteins Gag-Pol and envelope (Env) proteins (Martinez and Dornburg 1995a). This will enhance mass production of retroviral vectors for gene therapy experiments both in vitro and in vivo. A detailed list of commonly used SNV-based vectors and packaging cell lines is given in Table 1.

As producer cells, most laboratories use the cell lines 293T and D17 (dog osteosarcoma-derived cell line) (Riggs et al. 1974; Watanabe and Temin 1983), a highly permissive cell line for SNV replication and infection. The D17 cell line is preferred for packaging SNV particles with the wild-type envelope. Several transfection protocols have been described. The calcium phosphate

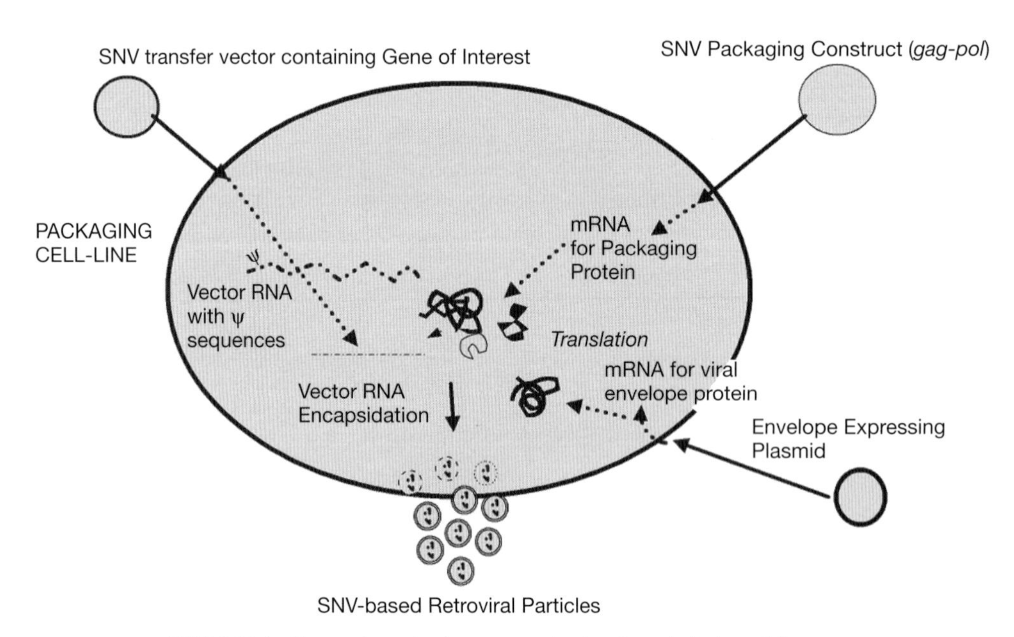

FIGURE 1. Generation of spleen necrosis virus-based viral particles.

TABLE 1. Commonly Used SNV-based Vectors and Packaging Cell Lines

Vector	Description	Reference
(a) SNV-based Avian Retroviral Vectors		
pJD214	SNV-based splicing vector containing ~1 kb of virus sequence	Dougherty and Temin (1987)
pMH105	Bicistronic SNV retroviral vector containing v-*rel*-IRES-*neo*	White and Gilmore (1996)
pJD220SVHy	SNV-based vector containing an internal SV40 promoter expressing the hygromycin phosphotransferase B gene (*hygro*)	Dougherty and Temin (1987)
pIM29	Expresses SNV wild-type SU region of envelope	Martinez and Dornburg (1995a,b)
pRD134	SNV envelope plasmid expressing wild-type proteins	Martinez and Dornburg (1995a,b)
pRD136	SNV packaging construct expressing wild-type *gag-pol*	Martinez and Dornburg (1995a,b)
pZP32	SNV packaging construct with mutation in matrix region	Parveen et al. (2000, 2003)
pZP36	SNV-based retroviral vector used for generation of viral particles expressing LacZ	Parveen et al. (2003)
pEM12 and 13	Bicistronic vectors expressing anti-HIV-1 genes	Marusich et al. (2005)
(b) Packaging Cell Lines for SNV-based Vectors		
D17	Canine osteosarcoma cell line, permissive to infection by SNV, REV-A, and murine leukemia virus[a]	Riggs et al. (1974)
C3A2	A derivative of D17 cell line expressing REV-A *gag-pol* and *env*	Watanabe and Temin (1983)
DSH134G	SNV-based packaging cell-line	Martinez and Dornburg (1995a,b)
DSH-cxl	SNV-based packaging cell line with β-gal as marker	

[a]REV-A and SNV are more than 90% homologous. REV-A proteins have also been previously reported to package SNV-derived vectors with high efficiency (Yin and Hu 1997).

($CaPO_4$) transfection method is a very efficient means of introducing DNA into packaging cell lines, particularly for retroviral vector generation. Mixing of DNA with calcium chloride leads to formation of $CaPO_4$-DNA complexes, which transfer DNA into the packaging cells by endocytosis. Although numerous transfection protocols are quite sensitive to the amount of input plasmid, in $CaPO_4$ transfection, the total amount of transfected DNA can be scaled up to 30 μg for a 100-mm plate. This is particularly helpful for the production of retroviral vectors involving cotransfection of three plasmids.

The G protein of vesicular stomatitis virus (VSV) is often used to replace the envelope protein of retroviruses including SNV (Lee et al. 2001). REV-A and SNV are naturally not infectious in human cells (Gautier et al. 2000). However, these viral vectors efficiently infect human cells when pseudotyped with VSV-G (Parveen et al. 2003) or envelope proteins expressing targeted ligands specific for human cell surface receptors (Parveen et al. 2003). One of the major advantages of VSV-G protein is that the retroviral particles can be concentrated by ultracentrifugation to attain high titer compared to the wild-type envelope of SNV. Postmitotically differentiated cells are difficult to transduce with any viral vector system. Ease of pseudotyping SNV-based particles has made it possible to selectively transduce postmitotically differentiated cells such as neurons both in vitro and in vivo (Parveen et al. 2003). This will be extremely useful for the development of gene therapy protocols for various neurological disorders.

The methods in this chapter are mainly for in vitro studies. Regarding clinical application of retroviral vectors, several vector production programs exist and more are undergoing standardization. The United States DNA Recombinant Advisory Board of the National Institutes of Health (NIH) oversees each protocol individually for clinical trials involving gene transfer in the United States. The minimum safety level for handling retroviral vectors is Biosafety Level 2. In the United States, the National Institutes of Health and Centers for Disease Control require review of the handling of retroviral vectors by the Institutional Review Board (IRB) and Institutional Biosafety Committee of each institution. However, universal safety measures should always be exercised while working with retroviral vectors. The titer of viral particles generated through conventional methods is insufficient for clinical application. For the future, it will be essential to develop optimal strategies for viral vector production and to focus on the stability of viral particles for their use in gene therapy protocols.

Protocol

Generation of Retroviral Particles for the SNV-based Vector System and Their Use in Transduction of Various Cell Types

This protocol describes the generation of retroviral particles for the SNV-based vector system. These particles are then used for transduction of various cell types.

MATERIALS

CAUTION: See Appendix for appropriate handling of materials marked with <!>.

Reagents

Calcium chloride ($CaCl_2$) (2 M) <!>

Dissolve 14.7 g of $CaCl_2$ in 100 ml of H_2O. Filter-sterilize or autoclave.

Cell lines for transfection: 293T and D17 (ATCC)

Glycerol (10–20%) (optional)

Hanks's balanced salt solution (HBSS)

2x HEPES-buffered saline (HBS: 280 mM NaCl, 10 mM KCl <!>, 1.5 mM Na_2HPO_4 <!>, 12 mM dextrose, and 50 mM HEPES at pH 7.5)

LacZ staining reagents (Chemicon)

Media

- Complete medium: Dulbecco's modified Eagle's medium (DMEM) containing 10% fetal calf serum (FCS) (for 293T cells) or fetal bovine serum (FBS) (for D17 cells), 1% penicillin-streptomycin <!>, and 1% glutamine
- Serum-free medium with 1% penicillin-streptomycin and 1% glutamine
- Phosphate-buffered saline (PBS) (optional; required if glycerol shock is being performed)

Nuclease-free H_2O

Plasmids used to generate retroviral particles

- Plasmid encoding viral core proteins
- Gene-transfer vector containing gene of interest
- Env protein encoding plasmid

SNV- and REV-A-based viruses are pseudotyped with VSV-G and other Env proteins displaying targeted ligands specific for human cell surface receptors.

The quality of plasmid DNA used for transfection is critical for the generation of high-titer retroviral vectors. It should be ultrapure, i.e., free from any chemical contaminants, RNA, or protein. We suggest column-purified DNA with an A260:A280 ratio above 1.8 for the generation of retroviral particles.

Polybrene

Transfectam reagent (Promega) (optional)

Equipment

Cell culture plates (100 mm [or 12- or 24-well plates for multiple transfections] and six well)

Centrifuge tubes (15 ml)

Filters: cellulose acetate or polysulfonic (low protein binding) and 0.45-μm membrane

Incubator, preset to 37°C
Microcentrifuge tubes (1.5 ml, sterile)
Microscope
Pipette (1.0 ml or pasteur)
Rocker platform
Ultracentrifuge (Beckman) with SW 41 rotor

METHODS

Preparation of Cells

1. Twenty-four hours before transfection, seed actively growing 293T or D17 cells in 100-mm plates in complete medium and grow the cells overnight.

   Plating density is dependent on cell viability. Typically, 0.5×10^6 to 0.8×10^6 cells are seeded in a 100-mm plate.

2. Observe the cell density on the plates. The cells should be 75–80% confluent. Replace the medium with 10 ml of fresh complete medium. Allow the cells to incubate for 4–6 hours.

   Cell confluency has a major role in achieving optimal viral titer.

Transfection

This protocol is optimized for 100-mm cell culture plates. The DNA input can be modified with respect to the size of the plates being used for transfection. Reagents ($CaCl_2$, 2x HEPES buffered saline, nuclease-free H_2O) can be prepared or purchased (e.g., ProFection Mammalian Transfection System–$CaPO_4$ from Promega).

3. For each 100-mm dish, aliquot 0.5 ml of 2x HEPES buffered saline into a sterile 1.5-ml microcentrifuge tube and label it Tube 1.

   For multiple transfections, it is easier to use 12- or 24-well plates to prepare the transfection mix.

4. To another tube labeled Tube 2, add 10 μg of Gag-Pol encoding plasmid, 4–5 μg of Env (VSVG or SNV wild type) encoding plasmid, and 10 μg of transfer vector. The total amount of plasmid DNA should not exceed 30 μg. Add 62 μl of 2 M $CaCl_2$ and enough nuclease-free H_2O to bring the total volume to 0.5 ml.
5. Use a 1-ml pipette (with the eject button depressed) or pasteur pipette to bubble the mixture in Tube 1 vigorously while adding the contents of Tube 2 dropwise.

   Alternatively, vortex Tube 1 while adding the contents of Tube 2. The resulting solution may appear milky with the formation of precipitates.

6. Allow the mixture to stand for 20–30 minutes at room temperature.
7. Mix the precipitate well by pipetting up and down or vortexing.
8. Add 1 ml of suspension mix to the 100-mm plate containing 293T or D17 cells and 5 ml of serum-free medium. Slowly add the suspension mix dropwise, with gentle swirling.
9. Incubate the plate for 4–6 hours at 37°C. Replace the medium with fresh medium containing 10% FCS. Incubate the plate for 36–40 hours to allow for recombinant virus production.

Glycerol Shock (Optional)

Some laboratories incorporate a glycerol shock at this stage to enhance nuclear uptake of the introduced DNA. In our experience with SNV-based retroviral particle generation, there is no major difference in cells treated or not treated with glycerol shock.

10. If glycerol shock is chosen, perform it as follows.

    a. Aspirate the medium and wash each plate containing the cells twice with 5 ml of PBS or medium. To ensure that most of the precipitate has been removed, swirl the dish while washing and then check the plate under the microscope. Repeat the wash if much of the precipitate remains.

    b. Add 2 ml of a predetermined concentration of glycerol to the cells. Allow them to sit for 2–4 minutes.

       The percentage of glycerol to be used varies from 10% to 20% and should be determined empirically. Generally, use the higher percentage of glycerol if the cells can tolerate this without any loss in viability.

    c. Aspirate the glycerol, wash the cells with PBS, and then add 10 ml of fresh medium.

Harvesting of Retroviral Vector Supernatant

11. For SNV-based vector-generated virus, collect the supernatant from transfected 293T or D17 cells within 36–40 hours after change of medium. Place the supernatant in 15-ml centrifuge tubes and centrifuge at 1500 rpm for 5 minutes to remove cell debris.

12. Remove the supernatant. Use it immediately to determine the titer or freeze it at –80°C until needed for infection. Determine virus production by biological titering for SNV- and REV-A-based vector systems. Avoid multiple freeze-thaw cycles of the supernatant, because this causes significant reduction in retroviral titer.

Concentrating the Retroviral Particles

13. Filter the harvested retroviral vector supernatant through a 0.45-μm membrane.

    Never use a nitrocellulose filter, because the proteins bind to the filter and can destroy the membrane of virus.

14. Centrifuge the supernatant in an SW 41 rotor in a Beckman ultracentrifuge at 25,000 rpm for 2 hours at 4°C.

15. Carefully remove the supernatant without disturbing the pellet. Resuspend the pellet, which contains the retroviral particles, in 50 μl of HBSS.

16. Determine the titer of the concentrated retroviral particles as described in the next section or freeze at –80°C until further use.

    Because freezing and thawing significantly reduce the titer, especially for SNV, it is best to use SNV particles fresh.

Determining Transduction Efficiency of Retroviral Particles

This procedure requires that *lacZ* or another reporter gene has been incorporated into the transfer vector.

17. Determine the titer of the retroviral particles as follows.

    For adherent cells

    a. Seed the target cells, D17 (for determination of the titer of SNV wild-type envelope pseudotypes) or 293T, in complete medium at a density of less than 1×10^6 cells per 100-mm dish. Allow the cells to grow for 24 hours.

    b. Aspirate the medium and add 1 ml of viral vector supernatant containing the retroviral particles and polybrene at a concentration of 10 μg/ml.

Alternatively, use Transfectam reagent at a concentration of 2 μg/ml. It has been observed that Transfectam reagent is less cytotoxic than polybrene.

c. After 3–5 hours (depending on the cell type), replace the virus-containing medium with 10 ml of fresh medium. Grow the cells for an additional 48 hours.

Semimaximal infection occurs 5–6 hours after exposure of cells to virus. Maximize infection by exposing cells to the virus overnight.

d. If a marker gene such as *lacZ* is used, determine the titer as colony forming units (cfu)/ml by staining with LacZ reagents and counting the blue cells.

If another reporter gene is used, methods such as immunocytochemistry or real-time polymerase chain reaction (PCR) can be used to determine gene expression and thus the titer accurately.

For transduction of nonadherent cells (in suspension)

a. Seed the target cells 12–18 hours before the infection in a six-well plate with a density of 1.5×10^6 cells in 1 ml of medium.

b. Centrifuge the cells in suspension. Resuspend them at a density of 1.5×10^6 cells/ml in viral vector supernatant in the presence of either polybrene (8 μg/ml) or Transfectam (2 μg/ml). As described earlier, Transfectam is preferable due to less cytotoxicity.

c. If polybrene is used as a transfecting reagent, it is essential to observe its effects on cell viability and growth, because suspension cells are more sensitive to cytotoxic responses than are adherent cells. Swirl the plate on a rocker platform.

The concentration of polybrene may be titrated 8–30 μg/ml to optimize the efficiency of infection for SNV.

d. Remove the viral particles by centrifugation after 4–5 hours and incubate the cells in complete medium for 48 hours as described for the adherent cells.

e. To further increase efficiency, perform multiple rounds of infection within 12–24 hours after the initial infection.

f. Fix and count the infected cells to determine the cfu/ml.

For transduction of postmitotic neurons with SNV-based retroviral particles

This procedure is for the transfer of a marker *lacZ* gene into postmitotically differentiated neurons.

a. Prepare postmitotically differentiated neurons according to a procedure previously described by Mukhtar et al. (2000).

b. Overlay the cells with the retroviral particle suspension mixed with Transfectam reagent (2 μg/ml). Allow the suspension to sit for 2 hours.

c. Repeat the overlay procedure with retroviral particles three times, at 24-hour intervals.

d. Confirm expression of the marker or therapeutic gene in these postmitotic cells via complementary techniques.

ACKNOWLEDGMENTS

This work was supported in part by grants from the National Institutes of Health: MH074359 (Z.P.) and MH074375 (M.H.). The authors thank Dr. Edward Acheampong, Rita M. Victor, and Brenda O. Gordon for their assistance in the preparation of the manuscript.

REFERENCES

Acheampong E.A., Parveen Z., Muthoga L.W., Kalayeh M., Mukhtar M., and Pomerantz R.J. 2005. Human immunodeficiency virus type 1 Nef potently induces apoptosis in primary human brain microvascular endothelial cells via the activation of caspases. *J. Virol.* **79:** 4257–4269.

Ailles L.E. and Naldini L. 2002. HIV-1-derived lentiviral vectors. *Curr. Top. Microbiol. Immunol.* **261:** 31–52.

Dornburg R. 1995. Reticuloendotheliosis viruses and derived vectors. *Gene Ther.* **2:** 301–310.

———. 2003. The history and principles of retroviral vectors. *Front. Biosci.* **8:** d818–835.

Dornburg R. and Pomerantz R.J. 2000. HIV-1 gene therapy: Promise for the future. *Adv. Pharmacol.* **49:** 229–261.

Dougherty J.P. and Temin H.M. 1987. A promoterless retroviral vector indicates that there are sequences in U3 required for 3′ RNA processing. *Proc. Natl. Acad. Sci.* **84:** 1197–1201.

Gautier R., Jiang A., Rousseau V., Dornburg R., and Jaffredo T. 2000. Avian reticuloendotheliosis virus strain A and spleen necrosis virus do not infect human cells. *J. Virol.* **74:** 518–522.

Kim S.H., Kim S., and Robbins P.D. 2000. Retroviral vectors. *Adv. Virus Res.* **55:** 545–563.

Kurian K.M., Watson C.J., and Wyllie A.H. 2000. Retroviral vectors. *Mol. Pathol.* **53:** 173–176.

Lee H., Song J.J., Kim E., Yun C.O., Choi J., Lee B., Kim J., Chang J.W., and Kim J.H. 2001. Efficient gene transfer of VSV-G pseudotyped retroviral vector to human brain tumor. *Gene Ther.* **8:** 268–273.

Martinez I. and Dornburg R. 1995a. Improved retroviral packaging lines derived from spleen necrosis virus. *Virology* **208:** 234–241.

———. 1995b. Mapping of receptor binding domains in the envelope protein of spleen necrosis virus. *J. Virol.* **69:** 4339–4346.

Marusich E.I., Parveen Z., Strayer D., Mukhtar M., Dornburg R.C., and Pomerantz R.J. 2005. Spleen necrosis virus-based vector delivery of anti-HIV-1 genes potently protects human hematopoietic cells from HIV-1 infection. *Virology* **332:** 258–271.

Mitchell R.S., Beitzel B.F., Schroder A.R., Shinn P., Chen H., Berry C.C., Ecker J.R., and Bushman F.D. 2004. Retroviral DNA integration: ASLV, HIV, and MLV show distinct target site preferences. *PLoS Biol.* **2:** E234.

Mukhtar M., Duke H., BouHamdan M., and Pomerantz R.J. 2000. Anti-human immunodeficiency virus type 1 gene therapy in human central nervous system-based cells: An initial approach against a potential viral reservoir. *Hum. Gene Ther.* **11:** 347–359.

Naldini L. 1999. In vivo gene delivery by lentiviral vectors. *Thromb. Haemostasis* **82:** 552–554.

Naldini L. and Verma I.M. 2000. Lentiviral vectors. *Adv. Virus Res.* **55:** 599–609.

Naldini L., Blomer U., Gallay P., Ory D., Mulligan R., Gage F.H., Verma I.M., and Trono D. 1996. In vivo gene delivery and stable transduction of nondividing cells by a lentiviral vector. *Science* **272:** 263–267.

Palu G., Parolin C., Takeuchi Y., and Pizzato M. 2000. Progress with retroviral gene vectors. *Rev. Med. Virol.* **10:** 185–202.

Parveen Z., Krupetsky A., Engelstadter M., Cichutek K., Pomerantz R.J., and Dornburg R. 2000. Spleen necrosis virus-derived C-type retroviral vectors for gene transfer to quiescent cells. *Nat. Biotechnol.* **18:** 623–629.

Parveen Z., Mukhtar M., Rafi M., Wenger D.A., Siddiqui K.M., Siler C.A., Dietzschold B., Pomerantz R.J., Schnell M.J., and Dornburg R. 2003. Cell-type-specific gene delivery into neuronal cells in vitro and in vivo. *Virology* **314:** 74–83.

Riggs J.L., McAllister R.M., and Lennette E.H. 1974. Immunofluorescent studies of RD-114 virus replication in cell culture. *J. Gen. Virol.* **25:** 21–29.

Watanabe S. and Temin H.M. 1983. Construction of a helper cell line for avian reticuloendotheliosis virus cloning vectors. *Mol. Cell. Biol.* **3:** 2241–2249.

White D.W. and Gilmore T.D. 1996. Bcl-2 and CrmA have different effects on transformation, apoptosis and the stability of I kappa B-alpha in chicken spleen cells transformed by temperature-sensitive v-Rel oncoproteins. *Oncogene* **13:** 891–899.

Yin P.D. and Hu W.S. 1997. RNAs from genetically distinct retroviruses can copackage and exchange genetic information in vivo. *J. Virol.* 6237–6242.

9 Foamy Virus Vector Production and Transduction of Hematopoietic Cells

Neil C. Josephson*† and David W. Russell†

*Puget Sound Blood Center, Seattle, Washington 98104; †Division of Hematology, Department of Medicine, University of Washington, Seattle, Washington 98112

ABSTRACT

Foamy viruses (FVs), or spumaviruses, are nonpathogenic retroviruses that have been developed as integrating viral vectors. Several potential advantages of FV vectors include a broad host range, large packaging capacity, and stable preintegration complex containing a double-stranded DNA genome. FV vectors are especially effective at transducing stem cells, including hematopoietic stem cells from several species.

INTRODUCTION

FVs are integrating viruses of the spumavirus family, which is a class of retroviruses distinct from oncoviruses such as murine leukemia virus and lentiviruses such as human immunodeficiency virus. Wild-type FV genomes contain *gag*, *pol*, and *env* genes, as do other retroviral genomes (Fig. 1), and additional *bel* genes between *env* and the viral long terminal repeats (LTRs). The *bel1* or *tas* gene, transcribed from an internal promoter in *env* (Lochelt et al. 1993), encodes a transcriptional activator required for expression from the LTR promoter (Keller et al. 1991; Rethwilm et al. 1991; Venkatesh et al. 1991). Our laboratory (Hirata et al. 1996; Russell and Miller 1996; Trobridge and Russell 1998; Trobridge et al. 2002b) and others (Erlwein et al. 1998; Heinkelein et al. 1998; Wu et al. 1998; Schwantes et al. 2002) have developed retroviral vectors based on FVs, which have steadily improved over the years. Early vectors were either replication-competent (Schmidt and Rethwilm 1995) or frequently contaminated with replication-competent FV (Russell and Miller 1996). Subsequent improvements eliminated replication-competent FV by removing the essential *bel1* (*tas*) gene (Trobridge and Russell 1998) and deleting nonessential sequences from the vector backbone (Wu et al. 1998; Heinkelein et al. 2002; Trobridge et al. 2002b). Potential advantages of FV vectors include a lack of pathogenicity of the wild-type virus, a wide host range, stable virions

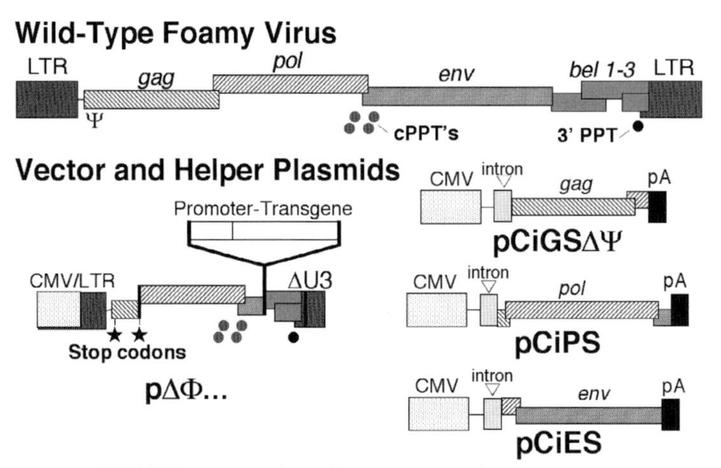

FIGURE 1. Four-plasmid FV vector production system.

that can be concentrated by centrifugation, a double-stranded DNA genome that is reverse-transcribed in the vector-producing cells, and the largest packaging capacity of any retrovirus.

Our current FV vector production protocol is a four-plasmid transient transfection system (Fig. 1) (Trobridge et al. 2002b). The pΔΦ (deleted foamy) vector plasmid retains only essential *cis*-acting sequences, including the packaging signal Ψ, a *cis*-acting region at the 3′ end of *pol* that includes central polypurine tract (cPPT) sequences, and the 3′ PPT sequence. A hybrid cytomegalovirus (CMV)-LTR fusion promoter drives transcription of the vector genome. Stop codons are introduced into the remaining 5′ portion of *gag* so that the vector does not encode any viral gene products. There are separate helper plasmids for *gag*, *pol*, and *env*. The vector transgene is driven by an internal promoter. FV vector stocks are produced by transfecting 293 cells, harvesting and filtering the culture medium, and concentrating vector virions by ultracentrifugation (Trobridge et al. 2002a). The resulting stocks are free of replication-competent helper virus based on a sensitive marker rescue assay (Trobridge and Russell 1998). A typical stock made from 23 10-cm dishes has a final volume of 2 ml with a titer of 10^7 to 10^8 transducing units/ml.

FV vectors are especially useful for transducing hematopoietic cells (Table 1). Because hematopoietic stem cells have the ability to self-renew, proliferate, and repopulate the bone marrow after transplantation, efficient transduction of these cells offers the promise to cure many inherited and acquired diseases. Hematopoietic stem cells can be obtained from marrow, peripheral blood, and umbilical cord blood. However, if kept in culture for more than 24 hours, they rapidly lose their repopulating ability (Peters et al. 1996; Gothot et al. 1998). Importantly, studies with human stem cells (Josephson et al. 2002) demonstrate efficient FV vector gene transduction with an ex vivo culture time of only 10 hours.

TABLE 1. Published Reports of Hematopoietic Cell Transduction by FV Vectors

FV vector strain	Hematopoietic Cells	Reporter gene(s)	Transduction frequency (%)	Reference
PFV	murine marrow	AP	43	Hirata et al. (1996)
	baboon $CD34^+$ marrow	AP	11	
	human $CD34^+$ marrow	AP	27–40	
PFV	murine, 5FU-treated marrow	AP, GFP	5–55	Vassilopoulos et al. (2001)
PFV	human, $CD34^+$ cord blood	GFP	12–80	Josephson et al. (2002)
SFV-1	human, $CD34^+$ cord blood	LacZ	4–27	Zucali et al. (2002)
PFV	murine, lineage-depleted marrow	GFP	44	Vassilopoulos et al. (2003)
PFV	human, $CD34^+$ cord blood	GFP	62–84	Leurs et al. (2003)
PFV	human, $CD34^+$ cord blood	GFP	65–75	Josephson et al. (2004)
	human $CD34^+$ peripheral blood	GFP	35–42	

(PFV) Prototype foamy virus (Achong et al. 1971); (SFV-1) simian foamy virus type 1; (AP) alkaline phosphatase; (GFP) green fluorescent protein; (LacZ) β-galactosidase.

Protocol

Production of FV Vector and Transduction of Hematopoietic Cells

This protocol presents methods for producing high-titer FV vector stocks, free of contaminating replication competent retrovirus, to be used for transducing hematopoietic stem cells.

MATERIALS

CAUTION: See Appendix for appropriate handling of materials marked with <!>.

Reagents

Use distilled, deionized H_2O, and sterile technique.

Bovine serum albumin (BSA) Fraction V

$CaCl_2$ (2.0 M) <!>

Dissolve 29.4 g of $CaCl_2$ (dihydrate) in 100 ml of H_2O. Filter-sterilize. Store in aliquots at 4°C.

Cells

Many different subclones of 293 cells (Graham et al. 1977) are available. Use a subclone known to have high transfection efficiencies. Freeze multiple vials to ensure reproducibility. When cells are expanded before transfection, do not allow them to overgrow. A single frozen vial can be thawed and kept in growing culture for 4–6 weeks for use in transfections. After this time, thaw a new vial.

Chloroquine (100 mM) <!>

Dissolve 0.52 g of chloroquine diphosphate salt in 10 ml of H_2O. Filter-sterilize. Store in aliquots at 4°C.

Cytokines (species-specific to the target cells)
- stem cell factor (SCF)
- Flt3-ligand (FL)
- thrombopoietin (TPO)

D10

D10 is DMEM supplemented with 10% heat-inactivated FBS, 100 units/ml penicillin G, and 100 μg/ml streptomycin <!>.

Dimethylsulfoxide (DMSO, 5%) <!>

Dulbecco's modified Eagle's medium (DMEM)

Fetal bovine serum (FBS), heat-inactivated (30 minutes at 56°C)

Hematopoietic cells

Human $CD34^+$ cells from marrow, peripheral blood, or umbilical cord blood can be isolated with magnetic beads by using commercially available reagents and protocols (Miltenyi Biotec, Auburn, California). Murine hematopoietic cells can be obtained from bone marrow and lineage-depleted with the Miltenyi system. Pretreatment of donor mice with 5-fluorouracil (5FU) can be used to enrich for stem cells (Vassilopoulos et al. 2001). This protocol is optimized for transduction of hematopoietic repopulating cells in nontissue-culture-treated six-well plates. If dishes or wells with a different area are used, scale the cell density and reagent volumes accordingly.

2x HEPES saline

Dissolve 8.18 g of NaCl and 5.96 g of HEPES in 400 ml of H_2O. Adjust pH to 7.10 with 0.5 M NaOH, bring up to 500 ml with H_2O, and filter-sterilize. The final solution is 280 mM NaCl and 50 mM HEPES. Store in aliquots at 4°C.

Human fibronectin fragment CH-296 (RetroNectin, Takara Shuzo, Otsu, Japan)

Dissolve lyophilized RetroNectin in sterile H_2O to a concentration of 1 mg/ml. Sterilize by passing through a 0.22-μm filter. Store at –20°C.

Phosphate-buffered saline (PBS), calcium- and magnesium-free

Phosphate mix

Combine 4.95 ml of 1.0 M NaH_2PO_4 <!> (12 g of monobasic salt in 100 ml of H_2O), 10.05 ml of 1.0 M Na_2HPO_4 <!> (26.8 g of dibasic salt in 100 ml of H_2O), and 85 ml of H_2O. Filter-sterilize. Store in aliquots at 4°C.

Plasmid DNA

ΔΦ Vector plasmid (containing desired transgene)

Helper plasmids: pCiGSΔΨ, pCiPS, pCiES

Plasmid DNAs can be purified with QIAGEN kits or banded on CsCl gradients, extracted with phenol and chloroform, precipitated with ethanol, and resuspended in TE (pH 8.0). Plasmids can be heat-treated for 30 minutes at 68°C to destroy most bacteria that might contaminate the preparation. For details of the four-plasmid FV vector production system, see Figure 1.

Sodium butyrate (500 mM)

Dissolve 5.5 g of sodium butyrate in 100 ml of DMEM. Filter-sterilize. Store in aliquots at –20°C.

Transduction medium

Concentrated vector stock in DMEM supplemented with 20% FBS and 100 ng/ml of each the following cytokines (species-specific to the target cells): stem cell factor (SCF), Flt3-ligand (FL), and thrombopoietin (TPO).

TE: 1x (recipe below) and 0.1x solutions

10 mM Tris-HCl (pH 8.0)

1 mM EDTA

Equipment

Centrifuge tubes (two 250-ml conical or several 50-ml tubes)

Centrifuge (Beckman SW 28 rotor and tubes)

Millipore filters (0.45 and 0.22 μm; 0.45-μm Stericup/Steritop, Durapore PVDF, SCHVU02RE)

Plates (six well, nontissue-culture treated)

Slide-A-Lyzer cassette (10K m.w. cutoff; Pierce Biotechnology, 66450)

Tissue-culture dishes (10 cm)

Waterbath, preset to 37°C

METHODS

FV Vector Stock Production

1. Twenty-four hours before transfection, seed 23 tissue-culture dishes (10 cm) with 3.25×10^6 293 cells each in 10 ml of D10 and grow the cells overnight.
2. Transfect the cells using a calcium phosphate–DNA precipitate.
   a. Prepare 9.2 ml of DNA solution by combining the following:

      1.15 ml of 2.0 M $CaCl_2$

      272 μg of ΔΦ vector plasmid (containing desired transgene)

      272 μg of pCiGSΔΨ

      34.5 μg of pCiPS

      16.9 μg of pCiES

      8.05 ml of sterile H_2O

b. Prepare 9.2 ml HEPES solution by combining the following:

   9.1 ml of 2x HEPES saline

   92 μl of phosphate mix

c. Add DNA solution dropwise to HEPES solution while vortexing gently. Once mixed, immediately add 64 μl of 100 mM chloroquine (final concentration of 25 μM).

d. Incubate the precipitate for 10 minutes and add 800 μl directly to each 10-cm dish. Avoid dislodging the cells. Gently rock the plate to spread the precipitation.

e. Incubate the cells for 4–6 hours to allow them to take up the DNA and then add 200 μl of 500 mM sodium butyrate to each dish. Continue incubating overnight. (See Troubleshooting.)

3. In the morning, change the medium with fresh, warm D10 (avoid dislodging the cells) and continue incubating for 48 hours.

4. Collect the culture medium into two 250-ml conical centrifuge tubes (or several 50-ml tubes) and centrifuge at about 300*g* for 5 minutes. Remove the supernatant and filter through a 250-ml, 0.45-μm Millipore Stericup/Steritop filter. Transfer the supernatant to Beckman SW 28 rotor tubes (36 ml/tube) and centrifuge at approximately 50,000*g* for 2 hours at 20°C.

5. Carefully aspirate the supernatant (avoiding the bottom of the tube). Add to each SW 28 tube 250 μl of the tissue-culture medium to be used in subsequent transductions (serum-containing is fine) before the pellets dry out. Solubilize the vector pellet by repeated pipetting while minimizing the generation of foam. Transfer the resolubilized pellet from one tube to the next tube and chase with 500 μl of tissue-culture medium. The pooled pellets should have a final volume of 2.0 ml.

   The stock can be used fresh or frozen in aliquots at –80°C in the presence of 5% DMSO. Minimize direct contact of pure DMSO with stocks by resuspending pellets in media containing 5% DMSO or by dripping in pure DMSO while agitating.

Transduction of Hematopoietic Cells

6. Coat nontissue-culture-treated six-well plates with CH-296 (RetroNectin) by adding 1 ml of stock solution, diluted to a final concentration of 50 μg/ml in PBS. Incubate the plates for 2 hours at room temperature and then remove the RetroNectin solution and add 2 ml of PBS with 2% BSA. Incubate the plates a further 30 minutes and then wash the wells with 2 ml of PBS. The coated dishes are now ready to use.

7. If a frozen FV vector stock is used, thaw it quickly in a 37°C waterbath and reduce the final DMSO concentration to ≤1%, either by dilution or by sterile dialysis in a Slide-A-Lyzer cassette in 500 ml of DMEM for 2 hours at room temperature.

8. Prepare the transduction medium using a dialyzed, diluted, or freshly concentrated vector stock. Adjust the stock volume and cell number (see Steps 9 and 10) to achieve a multiplicity of infection of 1–20 transducing particles/cell.

9. Resuspend hematopoietic cells in transduction medium at a concentration of 0.5×10^6 to 1.0×10^6 cells/ml. If cryopreserved cells are used, thaw them in a 37°C waterbath just before transduction. Wash the cells in 10 ml of DMEM supplemented with 20% heat-inactivated FBS, centrifuge at 220*g* for 5 minutes, and then resuspend the cells in vector-containing transduction media.

10. Add 1.5 ml of transduction medium with cells to the CH-296 (RetroNectin)-coated wells prepared in Step 6 (0.75×10^6 to 1.5×10^6 cells/well) and place the plates in a tissue-culture incubator for 10–24 hours.

11. Following transduction, remove the weakly attached cells from the well by pipetting medium on the plate surface. Repeat the collection with 2.0 ml of PBS in each well and combine with the initial collection. Confirm that all cells were collected by examining the wells under the microscope.

    The transduced cells can be grown in liquid culture, plated in progenitor colony assays, or used in animal transplantation experiments.

TROUBLESHOOTING

Problem (Step 2): Low transfection efficiency.

Solution: The pH of the HEPES saline solution can be critical in determining transfection efficiency. If titers are low, prepare a series of solutions covering a range of pH values from 7.0 to 7.2 and test them by transfecting a simple reporter gene plasmid to identify the best solution pH.

REFERENCES

Achong B.G., Mansell P.W., Epstein M.A., and Clifford P. 1971. An unusual virus in cultures from a human nasopharyngeal carcinoma. *J. Natl. Cancer Inst.* **46:** 299–307.

Erlwein O., Bieniasz P.D., and McClure M.O. 1998. Sequences in pol are required for transfer of human foamy virus-based vectors. *J. Virol.* **72:** 5510–5516.

Gothot A., van der Loo J.C., Clapp D.W., and Srour E.F. 1998. Cell cycle-related changes in repopulating capacity of human mobilized peripheral blood CD34+ cells in non-obese diabetic/severe combined immune-deficient mice. *Blood* **92:** 2641–2649.

Graham F.L., Smiley J., Russell W.C., and Nairn R. 1977. Characteristics of a human cell line transformed by DNA from human adenovirus type 5. *J. Gen. Virol.* **36:** 59–74.

Heinkelein M., Schmidt M., Fischer N., Moebes A., Lindemann D., Enssle J., and Rethwilm A. 1998. Characterization of a *cis*-acting sequence in the pol region required to transfer human foamy virus vectors. *J. Virol.* **72:** 6307–6314.

Heinkelein M., Dressler M., Jarmy G., Rammling M., Imrich H., Thurow J., Lindemann D., and Rethwilm A. 2002. Improved primate foamy virus vectors and packaging constructs. *J. Virol.* **76:** 3774–3783.

Hirata R.K., Miller A.D., Andrews R.G., and Russell D.W. 1996. Transduction of hematopoietic cells by foamy virus vectors. *Blood* **88:** 3654–3661.

Josephson N.C., Trobridge G., and Russell D.W. 2004. Transduction of long-term and mobilized peripheral blood-derived NOD/SCID repopulating cells by foamy virus vectors. *Hum. Gene Ther.* **15:** 87–92.

Josephson N.C., Vassilopoulos G., Trobridge G.D., Priestley G.V., Wood B.L., Papayannopoulou T., and Russell D.W. 2002. Transduction of human NOD/SCID-repopulating cells with both lymphoid and myeloid potential by foamy virus vectors. *Proc. Natl. Acad. Sci.* **99:** 8295–8300.

Keller A., Partin K.M., Lochelt M., Bannert H., Flugel R.M., and Cullen B.R. 1991. Characterization of the transcriptional *trans*-activator of human foamy retrovirus. *J. Virol.* **65:** 2589–2594.

Leurs C., Jansen M., Pollok K.E., Heinkelein M., Schmidt M., Wissler M., Lindemann D., von Kalle C., Rethwilm A., Williams D.A., and Hanenberg H. 2003. Comparison of three retroviral vector systems for transduction of nonobese diabetic/severe combined immunodeficiency mice repopulating human CD34+ cord blood cells. *Hum. Gene Ther.* **14:** 509–519.

Lochelt M., Muranyi W., and Flugel R.M. 1993. Human foamy virus genome possesses an internal, Bel-1-dependent and functional promoter. *Proc. Natl. Acad. Sci.* **90:** 7317–7321.

Peters S.O., Kittler E.L., Ramshaw H.S., and Quesenberry P.J. 1996. Ex vivo expansion of murine marrow cells with interleukin-3 (IL-3), IL-6, IL-11, and stem cell factor leads to impaired engraftment in irradiated hosts. *Blood* **87:** 30–37.

Rethwilm A., Erlwein O., Baunach G., Maurer B., and ter Meulen V. 1991. The transcriptional transactivator of human foamy virus maps to the bel 1 genomic region. *Proc. Natl. Acad. Sci.* **88:** 941–945.

Russell D.W. and Miller A.D. 1996. Foamy virus vectors. *J. Virol.* **70:** 217–222.

Schmidt M. and Rethwilm A. 1995. Replicating foamy virus-based vectors directing high level expression of foreign genes. *Virology* **210:** 167–178.

Schwantes A., Ortlepp I., and Lochelt M. 2002. Construction and functional characterization of feline foamy virus-based retroviral vectors. *Virology* **301:** 53–63.

Trobridge G.D. and Russell D.W. 1998. Helper-free foamy virus vectors. *Hum. Gene Ther.* **9:** 2517–2525.

Trobridge G., Vassilopoulos G., Josephson N., and Russell D.W. 2002a. Gene transfer with foamy virus vectors. *Methods Enzymol.* **346:** 628–648.

Trobridge G., Josephson N., Vassilopoulos G., Mac J., and Russell D.W. 2002b. Improved foamy virus vectors with minimal viral sequences. *Mol. Ther.* **6:** 321–328.

Vassilopoulos G., Wang P.R., and Russell D.W. 2003. Transplanted bone marrow regenerates liver by cell fusion. *Nature* **422:** 901–904.

Vassilopoulos G., Trobridge G., Josephson N.C., and Russell D.W. 2001. Gene transfer into murine hematopoietic stem cells with helper-free foamy virus vectors. *Blood* **98:** 604–609.

Venkatesh L.K., Theodorakis P.A., and Chinnadurai G. 1991. Distinct *cis*-acting regions in U3 regulate trans-activation of the human spumaretrovirus long terminal repeat by the viral bel1 gene product. *Nucleic Acids Res.* **19:** 3661–3666.

Wu M., Chari S., Yanchis T., and Mergia A. 1998. *cis*-Acting sequences required for simian foamy virus type 1 vectors. *J. Virol.* **72:** 3451–3454.

Zucali J.R., Ciccarone T., Kelley V., Park J., Johnson C.M., and Mergia A. 2002. Transduction of umbilical cord blood $CD34^+$ NOD/SCID-repopulating cells by simian foamy virus type 1 (SFV-1) vector. *Virology* **302:** 229–235.

10 Simian Foamy Virus Type-1 Vectors

Jeonghae Park and Ayalew Mergia

Department of Pathobiology, College of Veterinary Medicine, University of Florida, Gainesville, Florida 32610

ABSTRACT

Foamy viruses (FVs) are nonpathogenic retroviruses that offer opportunities for efficient and safe gene transfer in various cell types from different species. These viruses have unique replication mechanisms that are distinct from other retroviruses, which may give an advantage to FV-mediated gene transfer. Reverse transcription takes place late in the virus life cycle, and at least 20% of the virions have an infectious double-stranded DNA genome (Yu et al. 1996, 1999; Enssle et al. 1997). This may provide an advantage for FV vectors over other retroviruses because reverse transcription would have already been completed before infection of the target cells. The FV vector can efficiently deliver genes into human CD34$^+$ hematopoietic stem cells (Josephson et al. 2002; Zucali et al. 2002). A comparative study of transduction efficiency in CD34$^+$ cells revealed that FV vectors are superior to human immunodeficiency virus (HIV) vectors (Leurs et al. 2003). Efficient transduction of quiescent cells by FV vector has been reported (Patton et al. 2004; Trobridge and Russell 2004). However, cell division is required for transgene expression, suggesting that mitosis is necessary for transport to the nucleus and/or integration of the provirus genome (Patton et al. 2004; Trobridge and Russell 2004). Thus far, attempts to establish efficient packaging cell lines have not been successful (Wu and Mergia 1999). Therefore, vector production relies mainly on cells transiently transfected with vector and packaging plasmids.

INTRODUCTION

FVs are a group of retroviruses that belong to the Spumaretrovirinae subfamily. These viruses are found in many mammalian species and are nonpathogenic in naturally infected and experimentally infected animals. FVs can be propagated efficiently in various cell types of several

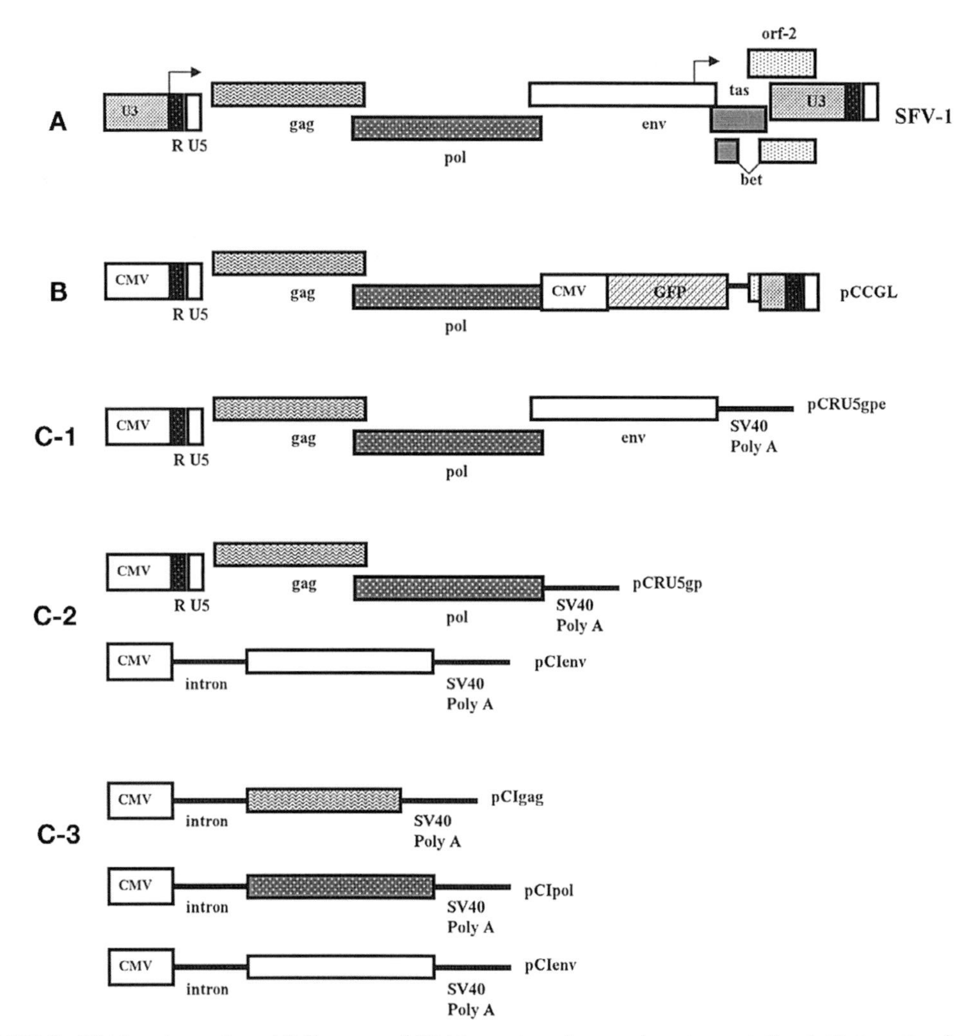

FIGURE 1. SFV-1 vector system: (*A*) Genome of SFV-1 provirus. Arrows show transcription initiation sites for the SFV-1 RNA genome and for early gene products under the control of an internal promoter located at the end of *env*. (*B*) SFV-1 vector containing a GFP reporter gene. (*C*) Packaging constructs where *gag-pol-env* is expressed as one transcriptional unit (*C-1*); *gag-pol* and *env* as two independent transcriptional units (*C-2*); and *gag*, *pol*, and *env* as three transcriptional units (*C-3*). CMV is the human cytomegalovirus immediate early gene promoter.

species (Mergia et al. 1996; Hill et al. 1999). Humans do not harbor FVs except for accidental occupational infections of animal handlers (Schweizer et al. 1995, 1997; Heneine et al. 1998; Callahan et al. 1999; Wolfe et al. 2004). We have molecularly characterized simian FV type 1 (SFV-1) originally isolated from a macaque monkey (Fig. 1). SFV-1 has a complex genome that encodes the virion structural genes, *gag*, *pol*, and *env*, as well two large open reading frames (ORFs) located at the 3′ end of *env* (Mergia et al. 1991; Mergia and Wu 1998). Other strains of FVs also have two ORFs in the same region (Flugel et al. 1987; Maurer and Flugel 1988; Renne et al. 1993; Herchenroder et al. 1994; Renshaw and Casey 1994; Winkler et al. 1997; Tobaly-Tapiero et al. 2000). Viral gene expression in FV is temporally regulated and involves differential gene expression controlled by two promoters. An internal promoter at the 3′ end of *env* regulates the expression of the early gene products Tas and Bet (Lochelt et al. 1993, 1994; Campbell et al. 1994; Mergia 1994). Tas is encoded by the first ORF located between *env* and the long terminal repeat (LTR), overlapping the 3′ end of *env*. It is a transcriptional *trans*-activator that strongly augments gene expression directed by viral promoters (Keller et al. 1991; Mergia et al. 1991; Rethwilm et al. 1991; Venkatesh et al. 1991; Renne et al. 1993; Renshaw and Casey 1994; Herchenroder et al. 1995). Bet is a product of a spliced message containing the first 88 amino acids of Tas fused to the last 390 amino acids of Orf-2 (Muranyi and Flugel 1991; Hahn et al. 1994; Mergia 1994). Mutational analysis in FVs has revealed that the *tas* gene is essential for viral replication (Lochelt et al. 1991; Baunach et al. 1993; Mergia and Wu 1998), whereas the *orf-2* region is dispensable (Lochelt et al. 1991; Baunach et al. 1993; Mergia and Wu 1998). Different functions have been reported for Bet (Meiering and Linial 2002; Lochelt 2003), suggesting multiple roles of this protein in FV replication. However, because the *orf-2* region is not critical, Bet is also dispensable for virus replication in a cell culture system.

FV vectors that are replication-defective can easily be constructed by deleting the *tas* gene and placing the genomes of the vector and the helper under the control of a heterologous promoter (Trobridge and Russel 1998; Park and Mergia 2002). The lack of virus replication in the absence of *tas* despite potential recombination eliminates the concern of generating replication-competent FV vector. We have constructed several SFV-1 vectors (Mergia and Wu 1998; Wu et al. 1998; Wu and Mergia 1999; Park et al. 2002). The FV vector can accommodate a 9-kb size heterologous DNA that is larger than most other retrovirus vectors can tolerate (Park et al. 2002; Trobridge et al. 2002). An SFV-1 vector with a 9-kb heterologous DNA insert was constructed by removing the *env-tas* and regions in *gag* and *pol*, as well as the *orf-2* sequence upstream of the polypurine tract (PPT) and deleting 1000 nucleotides in the 3′ U3 region of the LTR (Park et al. 2002). The 5′-untranslated region and the 5′-*gag*-, and the 3′-*pol*-coding sequences are important *cis* sequences required for FV-vector-mediated gene transfer. Because the *pro-pol* gene product is translated from a subgenomic message and lacks the *gag* sequences (Bodem et al. 1996; Yu et al. 1996), the 3′ *pol* region may have a role in Pol incorporation into virus particles (Heinkelein et al. 2002). Although the SFV-1 vector can accommodate 9-kb inserts, a higher titer of vector production (10–100-fold higher) is attained when an intact *gag-pol* region is kept in the vector (pCCGL, Fig. 1B). Insert sizes of up to 6.5 kb can be placed into such a vector. In pCCGL, the vector genome is constitutively expressed from the cytomegalovirus (CMV) promoter. Helper plasmids containing structural genes can be supplied either as one transcriptional (*gag-pol-env*), two transcriptional (*gag-pol* and *env*), or three transcriptional (*gag*, *pol*, and *env*) units (Fig. 1C). These helper plasmids are also expressed under the control of a CMV promoter. The levels of vector production with these packaging constructs are similar. None of these approaches result in wild-type virus production because the *tas* gene is missing. We routinely use the packaging plasmids where the *gag*, *pol*, and *env* expression cassettes are in independent plasmids as previously described to limit recombination events (Trobridge et al. 2002).

SFV-1 vector particles are generated by cotransfection of vector and packaging plasmids. The transfection procedure provided here is slightly modified from the calcium phosphate precipitation method described previously (Jordan et al. 1996). No core particles are released from infected cells in the absence of the envelope protein, suggesting that FV particle assembly is unique among retroviruses (Baldwin and Linial 1998; Fischer et al. 1998). In fact, unlike other retroviruses, the FV vectors cannot be pseudotyped with other envelopes such as vesicular stomatitis virus G (VSV-G) envelope protein, confirming the unique maturation process of FVs (Wu and Mergia 1999). Fortunately, FV can be concentrated without significant loss of infectivity, indicating that the FV envelope is not as fragile as that of other retroviruses (Hill et al. 1999).

Protocol

Preparation of SFV-1 Vectors

This protocol describes a method for SFV-1 vector preparation and concentration (Fig. 2). A transient transfection of vector and packaging constructs allows generation of the SFV-1 vector with titers of 10^7/ml. The vectors can be further concentrated by 100–200-fold without significant loss of vector titer.

MATERIALS

CAUTION: See Appendix for appropriate handling of materials marked with <!>.

Reagents

$CaCl_2$ (2 M) <!>
Chloroquine (50 mM) <!>
Complete Dulbecco's modified Eagle's medium (DMEM) with 10% FBS, 2 mM L-glutamine, 100 units/ml penicillin G, and 100 μg/ml streptomycin <!>
Fetal bovine serum (FBS)
2x HEPES-buffered saline (HBS; 50 mM HEPES, 0.78 mM Na_2HPO_4 <!>, and 273 mM NaCl)
Hank's balanced salt solution (HBSS)
Human embryonic kidney fibroblast-derived 293T cells

Passage cells by splitting 1:5 every 3–4 days. 293T cells can be obtained from American Type Culture Collection, Rockville, Maryland (ATCC: CRL 1573).

Plasmids
Packaging plasmids pCI*gag*, pCI*pol*, and pCI*env*, containing the structural gene *gag*, *pol*, and *env*, respectively (Fig. 1C-3)
SFV-1 vector pCCGL (Fig. 1B) containing a green fluorescent protein (GFP) expression cassette for monitoring transduction efficiency

Purify all plasmids used for transfection using the QIAGEN plasmid kit.

Sodium butyrate (1 M)
Trypsin-EDTA
0.05% trypsin
0.53 mM EDTA in HBSS

Equipment

Apollo Centrifugal Spin Concentrators (70 kD; Orbital Biosciences, Topsfield, Massachusetts)
Culture flasks (25 and 75 cm^2)
Filters (0.45 μm, sterile-packed, low protein binding; Millex-HV, Millipore)
Microscope, inverted fluorescence
Syringe (10 ml, disposable)
Tissue-culture plates (6 and 24 well)
Tubes (conical, 15 and 50 ml)
Tubes (1.5 ml, screw-cap; Sarstedt)

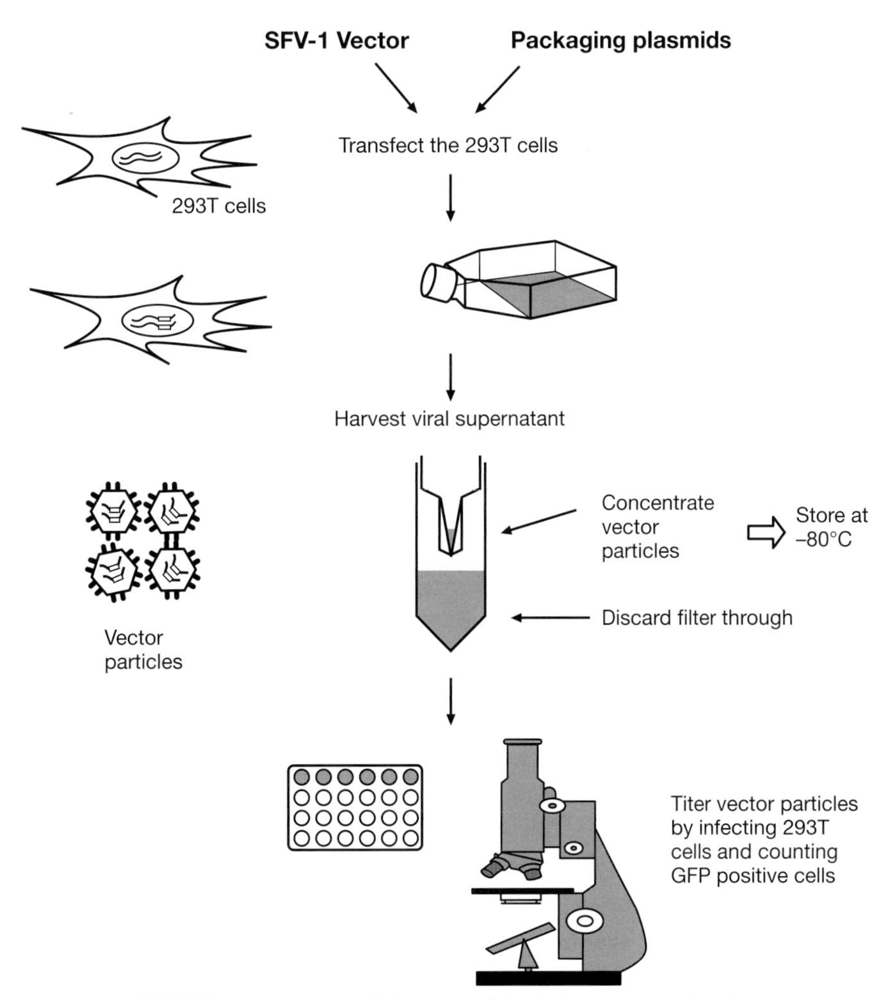

FIGURE 2. Overview of the protocol for SFV-1 vector production.

METHODS

DNA Transfection Using Calcium Phosphate Precipitation

1. One day before transfection, seed cells at a density of 5×10^6 per 75-cm^2 flask.

   On the day of transfection, cell density should be at least 80% confluent.

2. To prepare the DNA mixture, dilute 12 µg of SFV-1 vector, 12 µg of pCI*gag*, 6 µg of pCI*pol*, and 2 µg of pCI*env* in 2.634 ml (2634 µl) of H_2O.

   The H_2O must be ultrapure (autoclaved, Millipore).

3. Add 366 µl of 2 M $CaCl_2$ to the DNA mixture and mix well. Add 3 ml of 2x HBS to the DNA-$CaCl_2$ mixture and let stand for 10 minutes at room temperature.

4. Prepare a 25 µM chloroquine solution by adding 4 µl of 50 mM chloroquine to 6 ml of complete DMEM. Replace the medium with 6 ml of this chloroquine solution.

5. Mix the DNA-$CaCl_2$-HBS mixture with a pipette while creating bubbles for 10 seconds and then add it to the cells immediately.

6. After 6–8 hours of incubation, remove the medium and replace with complete medium containing 10 mM sodium butyrate.

   The sodium butyrate treatment should not exceed 24 hours. Sodium butyrate has been shown to enhance vector production (Tanaka et al. 1991).

Viral Vector Harvesting and Concentration

7. The next morning, replace the medium with 6.5 ml of fresh complete medium containing 3 mM sodium butyrate.

   At least 80% of the cells should express GFP, which can be monitored directly under an inverted fluorescent microscope. Supernatant is usually harvested 4–5 days after transfection.

8. To determine vector titer, infect fresh 293T cells and count GFP-positive cells.

   Routinely, 10^7 /ml SFV-1 vector particles can be produced. Vector stock can be kept at 4°C without significant loss for 2 weeks. One cycle of freeze/thaw results in a tenfold loss of vector concentration.

9. To enhance transduction efficiency, concentrate the SFV-1 supernatant 100-fold by using a spin column. Centrifuge the viral supernatant at 4000*g* for 20 minutes to remove cell debris and filter through a 0.45-μm filter.

10. Load the filtered supernatant into 70-kD Apollo Centrifugal Spin Concentrator filter tubes and centrifuge for 30 minutes at room temperature. Aliquot viral stock. Keep at 4°C for short-term storage or at –80°C for long-term storage.

ACKNOWLEDGMENTS

The research to develop an SFV-1 vector system was supported by a grant from the National Institutes of Health (AI39126) to A.M.

REFERENCES

Baldwin D.N. and Linial M.L. 1998. The roles of Pol and Env in the assembly pathway of human FV. *J. Virol.* **72:** 3658–3665.

Baunach G., Maurer B., Hahn H., Kranz M., and Rethwilm A. 1993. Functional analysis of human FV accessary reading frames. *J. Virol.* **67:** 5411–5418.

Bodem J., Lochelt M., Winkler I., Flower R.P., Delius H., and Flugel R.M. 1996. Characterization of the spliced *pol* transcript of feline foamy virus: The splice acceptor site of the *pol* transcript is located in *gag* of foamy viruses. *J. Virol.* **70:** 9024–9027.

Callahan M.E., Switzer W.M., Matthews A.L., Roberts B.D., Heneine W., Folks T.M., and Sandstrom P.A. 1999. Persistent zoonotic infection of a human with simian foamy virus in the absence of an intact *orf-2* accessory gene. *J. Virol.* **73:** 9619–9624.

Campbell M., Renshaw-Gegg L., Renne R., and Luciw P.A. 1994. Characterization of the internal promoter of simian foamy viruses. *J. Virol.* **68:** 4811–4820.

Enssle J., Fischer N., Moebes A., Mauer B., Smola U., and Rethwilm A. 1997. Carboxy-terminal cleavage of the human foamy virus Gag precursor molecule is an essential step in the viral life cycle. *J. Virol.* **71:** 7312–7317.

Fischer N., Heinkelein M., Lindemann D., Enssle J., Baum C., Werder E., Zentgraf H., Muller J.G., and Rethwilm A. 1998. Foamy virus particle formation. *J. Virol.* **72:** 1610–1615.

Flugel R.M., Rethwilm A., Maurer B., and Darai G. 1987. Nucleotide sequence analysis of the *env* gene and its flanking regions of the human spumaretrovirus reveals two novel genes. *EMBO J.* **6:** 2077–2084.

Hahn H., Gerald B., Brautigam S., Mergia A., Neumann-Haefelin D., Daniel M.D., McClure M.O., and Rethwilm A. 1994. Reactivity of primate sera to foamy virus Gag and Bet proteins. *J. Gen. Virol.* **75:** 2635–2644.

Heinkelein M., Leurs C., Rammling M., Peters K., Hanenberg H., and Rethwilm A.A. 2002. Pregenomic RNA is required for efficient incorporation of pol polyprotein into foamy virus capsids. *J. Virol.* **76:** 10069–10073.

Heneine W., Switzer W.M., Sandstrom P., Brown J., Vedapuri S., Schable C.A., Khan A.S., Lerche N.W., Schweizer M., Neumann-Haefelin D., Chapman L.E., and Folks T.M. 1998. Identification of a human population infected with simian foamy viruses. *Nat. Med.* **4:** 403–407.

Herchenroder O., Turek R., Neumann-Haefelin D., Rethwilm A., and Schneider J. 1995. Infectious proviral clones of chimpanzee foamy virus (SFVcpz) generated by long PCR reveal close functional relatedness to human foamy virus. *Virology* **214:** 685–689.

Herchenroder O., Renne R., Loncar D., Cobb E.K., Murthy K.K., Schneider J., Mergia A., and Luciw P.A. 1994. Isolation, cloning, and sequencing of simian foamy viruses from chim-

panzees (SFVcpz): High homology to human foamy virus (HFV). *Virology* **201:** 187–199.
Hill C.L., Bieniasz P.D., and McClure M.O. 1999. Properties of human foamy virus relevant to its development as a vector for gene therapy. *J. Gen. Virol.* **80:** 2003–2009.
Jordan M., Schalhorn A., and Wurm F.W. 1996. Transfecting mammalian cells: Optimization of critical parameters affecting calcium-phosphate precipitate formation. *Nucleic Acids Res.* **24:** 596–601.
Josephson N.C., Vassilopoulos G., Trobridge G.D., Priestley G.V., Wood B.L., Papayannopoulou T., and Russell D.W. 2002. Transduction of human NOD/SCID-repopulating cells with both lymphoid and myeloid potential by foamy virus vectors. *Proc. Natl. Acad. Sci.* **99:** 8295–8300.
Keller A., Partin K.M., Lochelt M., Bannert H., Flugel R.M., and Cullen B.R. 1991. Characterization of the transcriptional *trans*-activator of human foamy virus. *J. Virol.* **65:** 2589–2594.
Leurs C., Jansen M., Pollok K.E., Heinkelein M., Schmidt M., Wissler M., Lindemann D., Kalle C.V., Rethwilm A., Williams D.A., and Hanenberg H. 2003. Comparison of three retroviral vector systems for transduction of nonobese diabetic/severe combined immunodeficiency mice repopulating human CD34$^+$ cord blood cells. *Hum. Gene Ther.* **10:** 509–519.
Lochelt M. 2003. Foamy virus transactivation and gene expression. *Curr. Top. Microbiol. Immunol.* **277:** 27–61.
Lochelt M., Flugel R.M., and Aboud M. 1994. The human foamy virus internal promoter directs the expression of the functional Bel 1 and Bet protein early after infection. *J. Virol.* **68:** 638–645.
Lochelt M., Muranyi W., and Flugel R.M. 1993. Human foamy virus genome possesses an internal, Bel-1-dependent and functional promoter. *Proc. Natl. Acad. Sci.* **90:** 7317–7321.
Lochelt M., Zentgraf H., and Flugel R.M. 1991. Construction of an infectious DNA clone of the full-length human spumaretrovirus genome and mutagenesis of the *bel1* gene. *Virology* **184:** 43–54.
Maurer B. and Flugel R.M. 1988. Genomic organization of the human spumaretrovirus and its relatedness to AIDS and other retroviruses. *AIDS Res. Hum. Retrovir.* **4:** 467–473.
Meiering C.D. and Linial M.L. 2002. Reactivation of a complex retrovirus is controlled by a molecular switch and is inhibited by a viral protein. *Proc. Natl. Acad. Sci.* **99:** 15130–15135.
Mergia A. 1994. Simian foamy virus type 1 contains a second promoter located at the 3′ end of the *env* gene. *Virology* **199:** 219–222.
Mergia A. and Wu M. 1998. Characterization of provirus clones of simian foamy virus type 1 (SFV-1). *J. Virol.* **72:** 817–822.
Mergia A., Leung N.J., and Blackwell J. 1996. Cell tropism of the simian foamy virus type 1 (SFV-1). *J. Med. Primatol.* **25:** 2–7.
Mergia A., Shaw K.E.S., Pratt-Lowe E., Barry P.A., and Luciw P.A. 1991. Identification of the simian foamy virus transcriptional transactivator gene (*taf*). *J. Virol.* **65:** 2903–2909.
Muranyi W. and Flugel R.M. 1991. Analysis of splicing patterns of human spumaretrovirus by polymerase chain reaction reveals complex RNA structures. *J. Virol.* **65:** 727–735.
Park J. and Mergia A. 2002. Simian foamy virus vectors: Preparation and use. In *Gene therapy protocols*, 2nd edition (ed. J.R. Morgan), pp. 319–333. Humana Press, Totowa, New Jersey.
Park J., Nadeau P.E., and Mergia A. 2002. A minimal genome simian foamy virus type 1 (SFV-1) vector system with efficient gene transfer. *Virology* **302:** 336–344.
Patton G.S., Erlwein O., and McClure M.O. 2004. Cell-cycle dependence of foamy virus vectors. *J. Gen. Virol.* **85:** 2925–2930.
Renne R., Mergia A., Renshaw-Gegg L.W., Neumanm-Haefelin D., and Luciw P.A. 1993. Regulatory elements in the long terminal repeat (LTR) of simian foamy virus type 3. *Virology* **192:** 365–369.
Renshaw R.W. and Casey J.W. 1994. Transcriptional mapping of the 3′ end of the bovine syncytial virus genome. *J. Virol.* **68:** 1021–1028.
Rethwilm A., Otto E., Baunach G., Maurer B., and Meulen V. 1991. The transcriptional transactivater of human foamy virus maps to the *bel1* genomic region. *Proc. Natl. Acad. Sci.* **88:** 941–945.
Schweizer M., Falcone V., Gange J., Turek R., and Neumann-Haefelin D. 1997. Simian foamy virus isolated from an accidentally infected human individual. *J. Virol.* **71:** 4821–4824.
Schweizer M., Turek R., Hahn H., Schliephake A., Netzer K.O., Eder G., Reinhardt M., Rethwilm A., and Neumann-Haefelin D. 1995. Markers of foamy virus infections in monkeys, apes, and accidentally infected humans: Appropriate testing fails to confirm suspected foamy virus prevalence in humans. *AIDS Res. Hum. Retrovir.* **11:** 161–170.
Tanaka J., Sadanari H., Sato H., and Fukuda S. 1991. Sodium butyrate-inducible replication of human cytomegalovirus in a human epithelial cell line. *Virology* **185:** 271–280.
Tobaly-Tapiero J., Bittoun P., Neves M., Guillemin M.C., Lecellier C.H., Puvion-Dutilleul F., Gicquel B., Zientara S., Giron M.L., de The H., and Saib A. 2000. Isolation and characterization of an equine foamy virus. *J. Virol.* **74:** 4064–4073.
Trobridge G., Josephson N., Vassilopoulos G., Mac J., and Russell D.W. 2002. Improved foamy virus vectors with minimal viral sequences. *Mol. Ther.* **6:** 321–328.
Trobridge G.D. and Russel D.W. 1998. Helper-free foamy virus vectors. *Hum. Gene Ther.* **9:** 2517–2525.
———. 2004. Cell cycle reqirements for transduction by foamy virus vectors compared to those of oncovirus and lentivirus vectors. *J. Virol.* **78:** 2327–2335.
Venkatesh L.K., Theodorakis P.A., and Chinnadurai G. 1991. Distinct *cis*-acting regions in the U3 regulate *trans*-activation of the human spumaretrovirus long terminal repeat by the viral *bel1* gene product. *Nucleic Acids. Res.* **19:** 3661–3666.
Winkler I., Bodem J., Haas L., Zemba M., Delius H., Flower R., Flugel R.M., and Lochelt M. 1997. Characterization of the genome of feline foamy virus and its proteins shows distinct features different from those of primate spumaviruses. *J. Virol.* **71:** 6727–6741.
Wolfe N.D., Switzer W.M., Carr J.K., Bhullar V.B., Shanmugam V., Tamoufe U., Prosser A.T., Torimiro J.N., Wright A., Mpoudi-Ngole E., McCutchan F.E., Birx D.L., Folks T.M., Burke D.S., and Heneine W. 2004. Naturally acquired simian retrovirus infections in central African hunters. *Lancet* **363:** 932–937.
Wu M. and Mergia A. 1999. Packaging cell lines for simian foam virus type 1 (SFV-1) vectors. *J. Virol.* **73:** 4498–4501.
Wu M., Chari S., Yanchis T., and Mergia A. 1998. *cis*-Acting sequences required for simian foamy virus type 1 (SFV-1) vectors. *J. Virol.* **72:** 3451–3454.
Yu S.F., Sullivan M.D., and Linial M.L. 1999. Evidence that the human foamy virus genome is DNA. *J. Virol.* **73:** 1565–1572.
Yu S.F., Baldwin D.N., Gwynn S.R., Yendapalli S., and Linial M.L. 1996. Human foamy virus replication: A pathway distinct from that of retroviruses and hepadnaviruses. *Science* **271:** 1579–1582.
Zucali J.R., Ciccarone T., Kelley V., Park J., Johnson C.M., and Mergia A. 2002. Transduction of umbilical cord blood CD34$^+$ NOD/SCID-repopulating cells by simian foamy virus type 1 (SFV-1) vector. *Virology* **302:** 229–335.

11 Generation of VSV-G-pseudotyped Retroviral Vectors

Jiing-Kuan Yee

Division of Virology, Beckman Research Institute, City of Hope National Medical Center, Duarte, California 91010

ABSTRACT

Retroviral vectors are used increasingly as gene delivery vehicles for basic research and for potential treatment of human diseases. The viral envelope protein that recognizes cell surface receptors to allow cell entry generally dictates the cell tropism of the vector system. However, gene delivery into specific cell types with this vector system is not always efficient due to either the complete absence of receptor expression or low receptor concentrations on the cell surface. To improve transduction, a strategy termed "pseudotype formation" is adopted to incorporate other virus-encoded envelope proteins into retroviral particles. As these pseudotyped vector particles use different surface receptors for cell entry, they may deliver transgenes into mammalian cells more efficiently than the same vector containing the native retroviral envelope protein. Depending on the envelope protein, they may also confer tissue-specific gene delivery if the receptor has a restricted tissue distribution. Here, we use the glycoprotein (G) of vesicular stomatitis virus (VSV) to demonstrate how the pseudotyped vector can be generated within a few days. A similar strategy has also been used to generate pseudotyped lentiviral vectors. Besides VSV-G, various other virus-encoded envelope proteins can also be incorporated into either retrovirus- or lentivirus-based vectors using an approach similar to the one described here. These pseudotyped vectors exhibit various degrees of infectivity among different cell types that allow preferential delivery of a transgene into different tissues.

INTRODUCTION

Retroviral vectors have been used widely as gene delivery vehicles for the study of gene functions and for therapeutic treatment of human diseases. However, there are limitations for the application of this vector system, including its inability to transduce nonproliferating cells, poor infectivity in some cell types, and low vector titers. Because the host range of a retrovirus is generally determined by the interaction of a virus-encoded envelope protein and its cell surface receptor, poor infectivity could be attributed to low receptor concentration in a particular cell type. Poor infectivity demands the production of a higher vector titer to improve gene delivery efficiency. In addition, direct gene transfer in vivo also requires high-titer vector preparation because in vivo application is frequently limited by the volume of the vector-containing solution that can be injected. Because of these problems, it is often necessary to apply physical concentration steps such as centrifugation to increase the vector titer. However, traditional retroviral vectors containing either the ecotropic or amphotropic envelope protein cannot sustain centrifugation force and lose infectivity after such procedures (Burns et al. 1993).

"Pseudotype formation" is used to overcome the limitation of the traditional retroviral vector (Emi et al. 1991). Pseudotype formation is based on the observation that mixed infection of a cell by VSV, a member of the rhabdovirus family, and retroviruses such as murine leukemia virus (MLV) resulted in the generation of progeny viruses containing the genome of one virus packaged with the envelope protein of the other (Zavada 1972; Huang et al. 1973; Love and Weiss 1974). The exact mechanism for pseudotype formation remains unclear at present. However, the pseudotyped viral particle has an altered host range, depending on the envelope protein in the viral particle. MLV pseudotyped with the VSV-G protein has an expanded host range similar to that of VSV. Although CHO cells are susceptible to VSV infection, they are refractory to MLV infection. However, VSV-G-pseudotyped MLV can infect CHO cells with high efficiency (Emi et al. 1991). The infection was attributed to the presence of the VSV-G protein in the MLV particle, as antibody against VSV-G abolished the infectivity. The host range of the VSV-G-pseudotyped retroviral vector was also shown to extend to nonmammalian species including cells derived from insects, *Xenopus*, and fish (Lin et al. 1994; Burns et al. 1996; Matsubara et al. 1996). This suggests that a major block for cross-species infection by a mammalian retrovirus is the availability of cell surface receptor. The identity of the VSV-G receptor remains unknown, but it is believed that the receptor is widely distributed throughout different species because VSV has an extremely wide host range. Besides the expanded host range, an MLV vector pseudotyped with VSV-G also transduced some mammalian cells more efficiently than the same vector containing the amphotropic envelope protein (Emi et al. 1991). VSV can sustain centrifugation force and get concentrated to high titers (Burns et al. 1993). Similar to VSV, the VSV-G-pseudotyped MLV vector can be concentrated more than 1000-fold. This allows the production of vector titers reaching more than 10^9 infectious particles/ml, making direct in vivo gene delivery into animal models feasible. Overall, VSV-G pseudotypes increase the infectivity of MLV vectors in mammalian cells and allow the vectors to be concentrated to high titers for in vivo application.

Production of the VSV-G-pseudotyped MLV vector is a four-step process: (1) construction of an MLV vector containing the gene of interest, (2) production and vector harvest from transiently transfected 293T cells, (3) partial vector purification and concentration, and (4) titer determination. An MLV vector containing the gene encoding green fluorescent protein (GFP) pseudotyped with VSV-G will be used as an example here to illustrate this process. MLV pseudotyped with other envelope proteins can be generated by a similar approach. The whole process relies on the extremely high transfection efficiency of human kidney 293T cells that express high-level adenovirus E1A, E1B, and simian virus 40 (SV40) large T antigen (Pear et al. 1993). The transient transfection efficiency of this cell line under optimal conditions can reach more than 90% even with the calcium phosphate coprecipitation method. Use of other more efficient transfection methods such as liposome-mediated transfection is therefore not necessary with this cell line. For vector construction, MLV vector constructs with modification in the 5′ long terminal repeat (LTR) should be used to maximize vector production. This modification is based on the observation that the enhancer activity of the MLV LTR is suppressed by the adenovirus E1A protein, whereas the enhancer activity of the cytomegalovirus (CMV) immediate early (IE) gene is stimulated by E1A (Naviaux et al. 1996). The MLV 5′ LTR enhancer is removed, and the TATA-box-containing promoter is then fused with the CMV enhancer (Fig. 1, pPY-1). Because the structure of the MLV 3′ LTR remains intact, this modification does not affect the overall genomic structure of the vector produced.

A number of retroviral vector constructs suitable for gene insertion are available from different laboratories. The most important feature of these vectors is the fusion between the CMV IE enhancer and the basal promoter of the MLV 5′ LTR to allow efficient production of the pseudotyped vectors from 293T cells. The pCL vector system provides four such MLV vectors containing the *neo* gene (Naviaux et al. 1996). The transgene can be placed under the transcriptional control of either 5′ LTR or one of the three chosen internal promoters (Fig. 2a). The replication origin of SV40 was inserted into the packaging plasmid, the envelope expression plasmid, and the vector construct in an attempt to amplify these DNAs in the transfected cells. Coexpression of SV40 T antigen in 293 cells significantly increased the copy number of the transfected plasmid DNAs. However, no concomitant increase in vector titer was observed, suggesting that cell capacity to produce an infectious vector was saturated (Naviaux et al. 1996). The GFP-containing pPY-1 retroviral vector in Figure 1 is derived from pCMV-LL-SA-2 containing the extended *gag* region and the splicing signals from the MLV (Fig. 2b) (Peng et al. 2001). The extended *gag* region allows more efficient packaging of the vector genome, and the MLV splicing signals led to enhanced translation of the transgene. Similar to the pCL system, the MLV enhancer in the 5′ LTR of pCMV-LL-SA-2 was replaced with the CMV enhancer to enhance vector production from 293T cells.

For vector production, the vector construct is cotransfected with pCMV-GP, the packaging plasmid, and pCMV-G, a VSV-G expression plasmid, into 293T cells (Fig. 1). The *gag* and *pol* genes in pCMV-GP and the VSV-G gene in pCMV-G are under the transcriptional control of the strong CMV IE promoter to maximize gene expression in 293T cells. Separating the vector production system into three different plasmids minimizes the risk of producing replication-competent retroviruses (RCRs) through DNA recombination. Alternatively, the vector construct and pCMV-G can be directly transfected into 293-based packaging cell lines that stably express high levels of the MLV *gag* and *pol* proteins. One such cell line, GP2-293, is commercially available from BD Biosciences, Mountain View, California. Overexpression of VSV-G in 293T cells induces cell detachment and cytotoxicity (Burns et al. 1993). Therefore, infectious vectors should be harvested before complete cell detachment occurs. The harvested vectors are subjected to a polyethylene glycol (PEG) purification step to remove potential toxic substances in the cell lysate. This step also

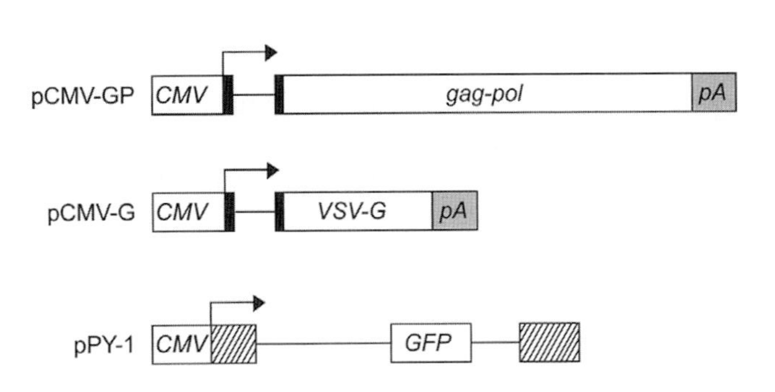

FIGURE 1. Packaging system for VSV-G-pseudotyped vectors. *Black boxes* and the *horizontal line* connecting the two boxes in pCMV-GP and pCMV-G represent the β-globin exons (*black boxes*) and intron (*horizontal line*) sequences. Arrows indicate the approximate transcription initiation site and the direction of transcription. *Hatched boxes* in pPY-1 indicate the MLV LTRs. CMV in pCMV-GP and pCMV-G denotes the CMV IE promoter and enhancer, whereas CMV in pPY-1 denotes the CMV IE enhancer only. pA denotes the polyadenylation signal from the β-globin gene.

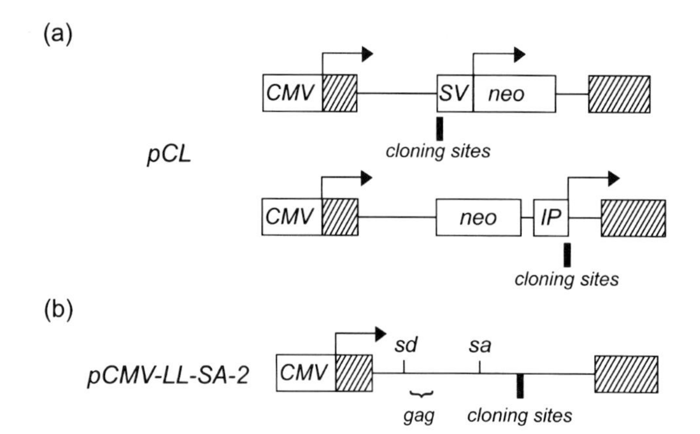

FIGURE 2. Structures of the MLV vectors. *Arrows* indicate the approximate transcription initiation site and the direction of transcription. The position of the multicloning site for the transgene in each vector is indicated. CMV denotes the CMV IE enhancer. (*a*) Vectors of the pCL system. SV denotes the SV40 early promoter. IP denotes the internal promoter including the CMV IE promoter, Rous sarcoma virus LTR, or dihydrofolate reductase promoter. (*b*) The structure of pCMV-LL-SA-2. The extended gag region is indicated. *sd* and *sa* denote the splicing donor and acceptor sites of MLV.

partially concentrates the vector titer 10–30-fold because a smaller volume of solution is used to resuspend the PEG-precipitated vector. This purification step is relatively quick and substantially reduces cytotoxic substances present in vector preparation. Ultracentrifugation is then applied to concentrate the VSV-G-pseudotyped MLV vector to high titer for direct in vivo application. Titer determination is usually performed in human fibrosarcoma line HT1080 cells or 293T cells. These two cell lines proliferate rapidly in cell culture and can be transduced efficiently with MLV. Cells transduced by the GFP gene-containing vector can be scored by fluorescence-activated cell sorting (FACS) analysis between 3 and 7 days after transduction. "Pseudotransduction" caused by the GFP protein copackaged into the pseudotyped particle can give rise to false-positive cells (Liu et al. 1996), prolonging the culturing time (7 days instead of 3 days after transduction) for the transduced cells before FACS analysis allows more accurate titer estimation. This problem with GFP is not an issue if the vector contains a selectable marker such as the gene encoding neomycin phosphotransferase (*neo*). In this case, transduced cells are subjected to G418 selection for at least 2 weeks, and the titer is scored by colony formation.

Pseudotype formation with VSV-G does not bypass the problem that the MLV vector cannot transduce nonproliferating cells. In contrast, lentiviral vectors can transduce quiescent cells, and VSV-G can be incorporated efficiently into lentiviral particles (Naldini et al. 1996). The VSV-G-pseudotyped lentiviral vector can be generated, purified, and concentrated by the same procedures described here.

Pseudotype formation with retroviral vectors is not limited to the VSV-G protein. MLV vectors can be pseudotyped with the envelope protein of gibbon ape leukemia virus (GALV) (Miller et al. 1991). Such a pseudotype delivered transgenes into a baboon's marrow, repopulating cells more efficiently than a similar vector pseudotyped with the amphotropic envelope protein (Kiem et al. 1997). This increase in infectivity is consistent with higher GALV receptor expression in hematopoietic stem cells. MLV vectors pseudotyped with the envelope protein derived from feline endogenous virus (FEV) RD114, the MLV 10A1, or lymphocytic choriomeningitis virus (LCMV) have also been reported (Takeuchi et al. 1992; Miller and Chen 1996; Miletic et al. 1999). Because these viruses use different cell surface receptors for cell entry, cell susceptibility to these pseudotypes is expected to be different. MLV vectors pseudotyped with the RD114 envelope protein transduced human cord blood $CD34^+$ cells more efficiently than a similar vector pseudotyped with the amphotropic envelope protein (Kelly et al. 2000). However, there was little difference in the transduction efficiency between the MLV vector pseudotyped with the RD114 or GALV envelope protein when dog CD34-enriched cells were used as the target (Goerner et al. 2001). MLV vectors pseudotyped with the MLV 10A1 envelope protein transduced human $CD4^+$ and $CD8^+$ lymphocytes more efficiently than the same vectors pseudotyped with the GALV, RD114, or amphotropic envelope protein (Gladow et al. 2000; Uckert et al. 2000). In this case, the longer half-life of the 10A1 pseudotype in culture medium may partially account for its increased transduction efficiency. The ability of these pseudotypes to sustain centrifugation force also varied. MLV vectors pseudotyped with the RD114 or LCMV envelope protein were concentrated by ultracentrifugation, whereas vectors pseudotyped with the amphotropic or GALV envelope protein lost significant infectivity at the concentration step.

Pseudotype formation resulting in MLV vectors with restricted tissue tropism has also been reported. The Sendai virus fusion (SV-F) protein is capable of binding specifically to the hepatic asialoglycoprotein receptor, mediating the fusion of the viral envelope with the cell membrane. SV-F can be incorporated into MLV particles and the resulting pseudotype transduced hepatocytes specifically (Spiegel et al. 1998). These studies clearly demonstrate that MLV pseudotype formation can occur with the envelope protein from a wide spectrum of viruses. However, the infectious titers of these pseudotypes may vary, depending on the efficiency of envelope protein incorporation into the MLV particle. The infectivity of each pseudotype depends on the availability of the cell surface receptor, the receptor concentration, the affinity of the envelope protein with its receptor, and the stability of the pseudotype in solution.

Because overexpression of VSV-G by transient transfection induces severe cytotoxicity, it is difficult to establish a stable packaging cell line for mass production of the VSV-G-pseudotyped MLV vector. Using the tetracycline inducible system for conditional VSV-G expression, two 293-based MLV packaging cell lines were established (Chen et al. 1996; Ory et al. 1996). Without induction, only basal levels of VSV-G expression can be detected, and cells continue to proliferate in culture. Any MLV vector can be introduced into these cell lines under noninducible conditions, and stable vector producer cell lines can be established. These producer cells can generate MLV vectors upon induction with titers between 10^5 and 10^6 infectious particles/ml. Nevertheless, transient transfection to produce infectious pseudotyped vectors is still preferred for cell culture work and animal model studies because proliferation of these inducible packaging cell lines can be retarded due to residue VSV-G expression, and the establishment of stable producer cells is tedious and time-consuming. This is a lesser problem with vectors pseudotyped with other envelope proteins derived from GALV or RD114 because expression of these proteins is generally nontoxic. Stable producer cell lines that continue to release such pseudotyped particles can readily be established.

Protocol

Generation of VSV-G-pseudotyped Retroviral Vectors

This protocol describes the use of VSV-G to demonstrate how the pseudotyped vector can be generated within a few days. The resulting vector has an expanded host range similar to that of VSV. Incorporation of VSV-G into retroviral particles also increases vector stability, allowing some simple schemes to be used for vector purification and concentration that lead to high-titer vector preparation.

MATERIALS

Reagents

$CaCl_2$ (2 M)

Dissolve 29.41 g of $CaCl_2 \cdot 2H_2O$ in H_2O to make 100 ml. Filter-sterilize.

Cells

293T (ATCC: CRL-11268)

HT1080 (ATCC: CRL-121)

Dulbecco's modified Eagle's medium (DMEM) containing 4.5 g/liter glucose, 3.7 g/liter $NaHCO_3$, 2 mM L-glutamine, and 100 mg/liter gentamycin

Ethanol (100%)

Fetal bovine serum (FBS)

2x HBS (HEPES-buffered saline)

Dissolve 2.38 g of HEPES, 3.28 g of NaCl, and 42.4 mg of Na_2HPO_4 <!> in 180 ml of H_2O and adjust pH to 7.12. Add H_2O to 200 ml and filter-sterilize.

Lactose solution (4%)

Dissolve 4 g α-lactose monohydrate (Sigma-Aldrich, cell-culture-tested) in 100 ml of PBS. Filter-sterilize.

PEG 8000 (40%; Sigma-Aldrich)

Dissolve 40 g of PEG in PBS to make 100 ml of stock solution. Stir overnight at room temperature and then autoclave. Store the stock solution at 4°C.

Phosphate-buffered saline (PBS)

Dissolve 8 g of NaCl, 0.2 g of KCl <!>, 1.44 g of Na_2HPO_4 <!>, and 0.24 g of KH_2PO_4 <!> in 800 ml of H_2O. Adjust pH to 7.4. Add H_2O to 1 liter and autoclave.

Plasmids

Packaging plasmid: pCMV-GP

VSV-G expression plasmid: pCMV-G

Polybrene (4 mg/ml in PBS; Sigma-Aldrich)

Retroviral vectors containing the GFP gene or a drug-selectable marker

TE79/10

1 mM Tris (pH 7.9)

0.1 mM EDTA

Filter-sterilize.

Equipment

Acrodisc syringe filters

Use 0.45-μm low-protein-binding filters (Pall Gelman laboratory, Ann Arbor, Michigan).

Centrifuge (Beckman Allegra 6 tabletop) with a PTS-2000 rotor

Culture dishes (100 mm)

Eppendorf thermomixer R mixer (Eppendorf, Westbury, New York)

Fluorescence-activated cell sorter (FACS)

Syringe (10 ml)

Tube (12 x 75-mm polystyrene) for DNA transfection

Tube (ultracentrifuge, 25 x 89 mm; Ultra-Clear, Beckman Coulter)

Ultracentrifuge (Beckman Coulter) with an SW 28 rotor

METHODS

Vector Production

1. Plate 293T cells at 2×10^6 cells per 100-mm dish in complete DMEM/10% FBS. Grow the cells overnight.

   The generation time for 293T cells is approximately 20 hours. This cell line can be maintained for up to 2 months and still retain its ability to be transfected with high efficiency. However, cells grown in culture for more than 3 months can suddenly lose the ability to produce vector. Thus, do not grow the cells for more than 3 months in culture.

2. Replace the culture medium with 10 ml of fresh culture medium and continue to culture the cells for a further 1–2 hours in a CO_2 incubator (see Troubleshooting).

   293T cells proliferate rather quickly in culture. Changing the medium on the day of transfection assures optimal pH of the culture medium for efficient transfection. Cell density at the time of transfection should reach at least 70% confluence. Lower cell density will result in reduced vector titers. Cells with a density of up to 80% confluence have been transfected and can produce high-titer vectors. Higher cell density not only increases vector production, but also prolongs the period for vector harvest because cells in high density tend to better resist VSV-G-induced cell detachment.

3. Prepare fresh transfection solution as follows:

   a. Mix 15 μg of pPY-1, 15 μg of pCMV-GP, and 2 μg of pCMV-G in a 12 x 75-mm polystyrene tube.

   b. Add TE79/10 solution to make a total volume of 437 μl.

   c. Add 63 μl of 2 M $CaCl_2$ and mix well.

   d. Add 500 μl of 2x HBS solution dropwise with constant agitation.

   e. Incubate for 30 minutes at room temperature to allow the calcium phosphate DNA precipitate to form.

   The calcium phosphate precipitate should form and turn the solution turbid after the 30-minute incubation. If this is not observed, repeat the precipitation procedure with fresh transfection solutions. The pH of the 2x HBS solution is the most critical factor for successful transfection. If the pH of this solution changes during storage, prepare a new solution.

   VSV-G is toxic to mammalian cells. Increasing the amount of pCMV-G in the transfection can lead to decreased vector titers. This could be caused by premature death of the transfected 293T cells due to VSV-G overexpression. Thus, keep the amount of pCMV-G to a minimum for vector production.

4. Add the DNA precipitate to the 293T cells prepared in Steps 1 and 2. Mix well. Incubate the dish for another 5–6 hours. Replace the medium with 6 ml of fresh medium and incubate overnight.

   Use 6 ml of culture medium (instead of 10 ml) to reduce the volume of vector harvest for subsequent vector purification and concentration steps.

5. Collect the culture medium from the same culture dish containing the vector at 24, 40, 48, 64, and 72 hours after initial DNA transfection. With each collection, replace the medium with 6 ml of fresh culture medium for the next vector collection.

   Fresh culture medium must be carefully added from the edge of the culture dish because cells become easily detachable due to transfection and toxicity induced by VSV-G expression.

   Cytotoxicity induced by VSV-G overexpression will become increasingly apparent. Most of the transfected cells should detach from the culture dish by 72 hours after transfection. If no cytotoxicity such as syncytia formation or cell detachment is observed at this point, the transfection may have failed and low vector titer would be expected.

6. Centrifuge the culture medium in a PTS-2000 rotor in a Beckman Allegra 6 tabletop centrifuge at 2500 rpm for 5 minutes at room temperature. Collect the supernatant and filter it through a 0.45-μm syringe filter. Store the vector immediately at –80°C until the purification step.

Vector Purification

7. Thaw crude vector preparations and mix with 40% PEG stock solution to make a final PEG concentration of 10%.

8. Place the PEG-vector mixture on ice for at least 4 hours or in a refrigerator overnight.

   Vectors in PEG are very stable: Storing the vector in PEG for up to 1 week at 4°C results in little loss of the infectious titer.

   Harvested vectors can be concentrated directly by ultracentrifugation without PEG purification. However, the resulting vector induces cytotoxicity more frequently in different cell types than the PEG-purified vector. The PEG step is therefore highly recommended.

9. Centrifuge the vector in the PTS-2000 rotor in the Beckman Allegra 6 tabletop centrifuge at 3000 rpm for 30 minutes.

10. Use a pipette to gently resuspend the vector pellet in PBS containing 4% lactose or in DMEM/10% FBS. Use 1/10 to 1/30 of the original harvest volume to resuspend the vector.

    The partially purified vector can be either used directly for transduction or continue to be processed by ultracentrifugation for high-titer vector preparation. The presence of lactose or FBS in the resuspension solution stabilizes the MLV vector during storage. The solution containing the resuspended vector is generally turbid due to the presence of residual PEG that fails to redissolve. Increasing the volume of resuspension solution can reduce the turbidity. However, cell transduction with such preparations is not affected by the presence of residual PEG.

    Residual PEG also seems to stabilize the vector in storage. Loss of vector infectivity due to prolonged storage at –80°C occurs less frequently and multiple freeze-thaw cycles affect the vector titer less dramatically than the non-PEG-treated vector. PEG precipitation removes a majority of contaminating cellular proteins in the vector preparation as determined by gel electrophoresis and silver stain of the vector preparation before and after PEG purification. Judging from the band intensity, more than 95% of FBS is removed by this procedure. Based on titer, the efficiency for vector recovery with this procedure is high (between 80% and 100%).

Vector Concentration

11. Sterilize an Ultra-Clear ultracentrifuge tube (25 x 89 mm), the adapter, and the screw cap of an SW 28 rotor with 100% ethanol and air-dry in a laminar flow hood.
12. Transfer the PEG-purified vector preparation into the sterile ultracentrifuge tube and add either PBS or DMEM to a volume of 37 ml.
13. Centrifuge the vector in an SW 28 rotor at 25,000 rpm for 90 minutes in a Beckman Coulter ultracentrifuge at 4°C.
14. Discard the supernatant and invert the tube in a laminar flow hood for a few minutes to remove any trace of liquid in the centrifuge tube. Add 0.5–1 ml of 4% lactose solution or DMEM to resuspend the vector pellet.
15. Vortex the pellet in lactose-containing solution or DMEM for 3–4 hours at 4°C in an Eppendorf thermomixer R mixer with mixing speed set at 800 rpm (see Troubleshooting).

    The solution containing the resuspended vector should be turbid due to the presence of residual PEG. If the vector is resuspended in DMEM, add FBS to 10% after vector resuspension is completed. Aliquot the concentrated vector preparation and store at –80°C.

Titer Determination

16. Plate HT1080 cells at 2 x 10^6 cells per 100-mm dish in complete DMEM/10% FBS. Prepare one extra plate for cell count the following day. Grow the cells overnight.
17. Trypsinize the cells in one dish and determine the cell count. Replace the culture medium in other dishes with fresh medium containing 4 μg/ml polybrene. Dilute the concentrated vector with DMEM and add to the HT1080 dishes. Incubate overnight.
18. Replace the medium with fresh medium without polybrene and continue the incubation overnight.
19. Trypsinize the cells, centrifuge, and resuspend them in 1 ml of DMEM/10% FBS for FACS analysis.
20. Calculate the vector titer (infectious particles/ml) as follows: (Total cell number on the day of transduction) x (Percentage of GFP^+ cells as determined by FACS) x (Vector dilution factor) (see Troubleshooting).

    If a drug selectable marker is used for titer determination, initially plate only 2 x 10^5 HT1080 cells per 100-mm dish, because antibiotics used for selection are most effective when cells are actively proliferating. Extended selection is required to eliminate untransduced background if the cells in a dish become confluent.

TROUBLESHOOTING

Problem (Step 2): 293T cells tend to detach from culture dish immediately after culture medium change.
Solution: DNA transfection immediately following medium change is therefore not recommended. Continue to grow the cells for an extra 1–2 hours after the change of medium to allow firm attachment of the cells to the dish.

Problem (Step 15): DMEM in the presence of FBS foams during vortexing, leading to significant loss of infectious vector titer.
Solution: Add FBS after the vector is resuspended.

Problem (Step 15): Attempting to clear the turbid vector-containing solution by centrifugation results in significant loss of the infectious titer.

Solution: In most cases, the turbidity does not interfere with infectivity and no obvious cytotoxicity is observed when cells are transduced with such a vector preparation.

Problem (Step 20): With the GFP marker, pseudotransduction can occur due to copackaging of the GFP protein into viral particles during vector production from 293T cells. GFP protein released into the transduced HT1080 cells can generate false-positive cells. Treating cells with AZT abolishes "true" expression from the vector-delivered transgene but does not inhibit pseudotransduction.

Solution: To avoid overestimation of the GFP titer, delay FACS analysis until day 7 or 8 after transduction to minimize the effect of pseudotransduction. For HT1080 cell selection, neomycin and hygromycin are generally used in concentrations of 400–600 μg/ml (active components).

ACKNOWLEDGMENTS

We thank Dr. David Hsu for his advice on vector purification and concentration.

REFERENCES

Burns J.C., Friedmann T., Driever W., Burrascano M., and Yee J.K. 1993. Vesicular stomatitis virus G glycoprotein pseudotyped retroviral vectors: Concentration to very high titer and efficient gene transfer into mammalian and nonmammalian cells. *Proc. Natl. Acad. Sci.* **90:** 8033–8037.

Burns J.C., McNeill L., Shimizu C., Matsubara T., Yee J.K., Friedmann T., Kurdi-Haidar B., Maliwat E., and Holt C.E. 1996. Retrovirol gene transfer in *Xenopus* cell lines and embryos. *In Vitro Cell Dev. Biol. Anim.* **32:** 78–84.

Chen S.T., Iida A., Guo L., Friedmann T., and Yee J.K. 1996. Generation of packaging cell lines for pseudotyped retroviral vectors of the G protein of vesicular stomatitis virus by using a modified tetracycline inducible system. *Proc. Natl. Acad. Sci.* **93:** 10057–10062.

Emi N., Friedmann T., and Yee J.K. 1991. Pseudotype formation of murine leukemia virus with the G protein of vesicular stomatitis virus. *J. Virol.* **65:** 1202–1207.

Gladow M., Becker C., Blankenstein T., and Uckert W. 2000. MLV-10A1 retrovirus pseudotype efficiently transduces primary human CD4$^+$ T lymphocytes. *J. Gene Med.* **2:** 409–415.

Goerner M., Horn P.A., Peterson L., Kurre P., Storb R., Rasko J.E., and Kiem H.P. 2001. Sustained multilineage gene persistence and expression in dogs transplanted with CD34$^+$ marrow cells transduced by RD114-pseudotype oncoretrovirus vectors. *Blood* **98:** 2065–2070.

Huang A.S., Besmer P., Chu L., and Baltimore D. 1973. Growth of pseudotypes of vesicular stomatitis virus with N-tropic murine leukemia virus coats in cells resistant to N-tropic viruses. *J. Virol.* **12:** 659–662.

Kelly P.F., Vandergriff J., Nathwani A., Nienhuis A.W., and Vanin E.F. 2000. Highly efficient gene transfer into cord blood nonobese diabetic/severe combined immunodeficiency repopulating cells by oncoretroviral vector particles pseudotyped with the feline endogenous retrovirus (RD114) envelope protein. *Blood* **96:** 1206–1214.

Kiem H.P., Heyward S., Winkler A., Potter J., Allen J.M., Miller A.D., and Andrews R.G. 1997. Gene transfer into marrow repopulating cells: Comparison between amphotropic and gibbon ape leukemia virus pseudotyped retroviral vectors in a competitive repopulation assay in baboons. *Blood* **90:** 4638–4645.

Lin S., Gaiano N., Culp P., Burns J.C., Friedmann T., Yee J.K., and Hopkins N. 1994. Integration and germ-line transmission of a pseudotyped retroviral vector in zebrafish. *Science* **265:** 666–669.

Liu M.L., Winther B.L., and Kay M.A. 1996. Pseudotransduction of hepatocytes by using concentrated pseudotyped vesicular stomatitis virus G glycoprotein (VSV-G)–Moloney murine leukemia virus-derived retrovirus vectors: Comparison of VSV-G and amphotropic vectors for hepatic gene transfer. *J. Virol.* **70:** 2497–2502.

Love D.N. and Weiss R.A. 1974. Pseudotypes of vesicular stomatitis virus determined by exogenous and endogenous avian RNA tumor viruses. *Virology* **57:** 271–278.

Matsubara T., Beeman R.W., Shike H., Besansky N.J., Mukabayire O., Higgs S., James A.A., and Burns J.C. 1996. Pantropic retroviral vectors integrate and express in cells of the malaria mosquito, *Anopheles gambiae*. *Proc. Natl. Acad. Sci.* **93:** 6181–6185.

Miletic H., Bruns M., Tsiakas K., Vogt B., Rezai R., Baum C., Kuhlke K., Cosset F.L., Ostertag W., Lother H., and von Laer D. 1999. Retroviral vectors pseudotyped with lymphocytic choriomeningitis virus. *J. Virol.* **73:** 6114–6116.

Miller A.D. and Chen F. 1996. Retrovirus packaging cells based on 10A1 murine leukemia virus for production of vectors that use multiple receptors for cell entry. *J. Virol.* **70:** 5564–5571.

Miller A.D., Garcia J.V., von Suhr N., Lynch C.M., Wilson C., and Eiden M.V. 1991. Construction and properties of retrovirus packaging cells based on gibbon ape leukemia virus. *J. Virol.* **65:** 2220–2224.

Naldini L., Blomer U., Gallay P., Ory D., Mulligan R., Gage F.H., Verma I.M., and Trono D. 1996. In vivo gene delivery and stable transduction of nondividing cells by a lentiviral vector. *Science* **272:** 263–267.

Naviaux R.K., Costanzi E., Haas M., and Verma I.M. 1996. The pCL vector system: Rapid production of helper-free, high-titer, recombinant retroviruses. *J. Virol.* **70:** 5701–5705.

Ory D.S., Neugeboren B.A., and Mulligan R.C. 1996. A stable human-derived packaging cell line for production of high titer retrovirus/vesicular stomatitis virus G pseudotypes. *Proc. Natl. Acad. Sci.* **93:** 11400–11406.

Pear W.S., Nolan G.P., Scott M.L., and Baltimore D. 1993. Production of high-titer helper-free retroviruses by transient transfection. *Proc. Natl. Acad. Sci.* **90:** 8392–8396.

Peng H., Chen S.T., Wergedal J.E., Polo J.M., Yee J.K., Lau K.H., and Baylink D.J. 2001. Development of an MFG-based retroviral vector system for secretion of high levels of functionally active human BMP4. *Mol. Ther.* **4:** 95–104.

Spiegel M., Bitzer M., Schenk A., Rossmann H., Neubert W.J., Seidler U., Gregor M., and Lauer U. 1998. Pseudotype formation of Moloney murine leukemia virus with Sendai virus glycoprotein F. *J. Virol.* **72:** 5296–5302.

Takeuchi Y., Simpson G., Vile R.G., Weiss R.A., and Collins M.K. 1992. Retroviral pseudotypes produced by rescue of a Moloney murine leukemia virus vector by C-type, but not D-type, retroviruses. *Virology* **186:** 792–794.

Uckert W., Becker C., Gladow M., Klein D., Kammertoens T., Pedersen L., and Blankenstein T. 2000. Efficient gene transfer into primary human $CD8^+$ T lymphocytes by MuLV-10A1 retrovirus pseudotype. *Hum. Gene Ther.* **11:** 1005–1014.

Zavada J. 1972. Pseudotypes of vesicular stomatitis virus with the coat of murine leukaemia and of avian myeloblastosis viruses. *J. Gen. Virol.* **15:** 183–191.

12 Targeted Gene Transfer with Surface-engineered Lentiviral Vectors

Els Verhoeyen and François-Loïc Cosset

INSERM, U412, Lyon, F-69007 France; Ecole Normale Supérieure de Lyon, Lyon, F-69007 France; IFR128 BioSciences Lyon-Gerland, Lyon, F-69007 France

ABSTRACT

Vectors derived from retroviruses such as lentiviruses and oncoretroviruses are especially suitable tools for long-term gene transfer, because they allow stable integration of a transgene and its propagation in daughter cells. Lentiviral vectors are preferred over vectors derived from oncoretroviruses such as murine leukemia virus (MLV) vectors, because they can transduce nonproliferating target cells. Moreover, lentiviral vectors that can target tissues specifically will be valuable for various gene-transfer approaches in vivo. To achieve targeted gene transfer, two types of surface modifications have been made to lentiviral vectors: (1) Heterologous viral glycoproteins have been incorporated to exploit the tropism of other viruses (this is called pseudotyping), and (2) heterologous polypeptides have been fused to viral glycoproteins to retarget the lentiviral particles to a cell of interest. This chapter provides an overview of innovative approaches to upgrade lentiviral vectors for tissue targeting, as well as an example protocol for targeted transduction of hematopoietic stem cells (HSCs).

INTRODUCTION

Host-range Modification Using Natural Tropism of Glycoproteins

Retrovirus-derived vectors are particularly useful for gene-transfer applications because of the many possible combinations of viral surface glycoproteins, which determine cell tropism, with viral cores (Nègre et al. 2002). The envelope glycoprotein of the amphotropic strain of MLV was used in some early experiments with human immunodeficiency virus (HIV)-derived vectors (Naldini et al. 1996), but its receptor, Pit-2, is present only at very low levels on HSCs, which are an important target for gene therapy. In contrast, the association of the vaccinia stomatitis virus (VSV)-G glycoprotein with lentivirus-derived cores results in vector pseudotypes that have broad tropism and can integrate into nonproliferating target cells including HSCs (Naldini et al. 1996). Moreover, these pseudotypes are resistant to freeze-thaw cycles and ultracentrifugation, two important qualities for gene delivery vectors. These vectors do have drawbacks. For instance, some in vivo gene-transfer applications require vectors that are target-specific after systemic administration (Peng et al. 2001). Because of the wide distribution of the VSV-G receptor, a lipid component of the plasma membrane (Seganti et al. 1986), VSV-G pseudotypes may bind to the surfaces of all cells encountered after inoculation before reaching the target cells. Moreover, VSV-G-pseudotyped vectors are rapidly inactivated by human serum (DePolo et al. 2000). These properties limit the use of VSV-G as a glycoprotein in pseudotype vectors for in vivo gene delivery.

More selective tropisms have been achieved by taking advantage of the natural tropisms of glycoproteins from other membrane-enveloped viruses. For instance, the use of surface glycoproteins derived from viruses that cause lung infection and infect via the airway epithelia (e.g., ebola virus or influenza virus) may prove useful for gene therapy of the human airway (Kobinger et al. 2001). Exclusive transduction of retinal pigmented epithelium has been obtained following subretinal inoculations of some vector pseudotypes in rat eyes (Duisit et al. 2002). Likewise, screening of a large panel of pseudotyped vectors has established the superiority of the gibbon ape leukemia virus (GALV) and the cat endogenous retroviral glycoproteins (RD114) for transduction of progenitor and differentiated hematopoietic cells (Porter et al. 1996; Marandin et al. 1998; Movassagh et al. 1998; Kelly et al. 2000; Hanawa et al. 2002; Sandrin et al. 2002).

Coexpression of a given glycoprotein with a heterologous viral core will not necessarily give rise to highly infectious viral particles. This is the case for GALV and RD114 lentiviral pseudotypes (Takeuchi et al. 1992; Lindemann et al. 1997; Mammano et al. 1997; Schnierle et al. 1997; Stitz et al. 2000; Christodoulopoulos and Cannon 2001; Sandrin et al. 2002).

The lack of infectivity of simian immunodeficiency virus (SIV) vectors generated with the GALV and RD114 glycoproteins is caused by defective incorporation of glycoprotein on the lentiviral cores. Indeed, replacement of the cytoplasmic tail of the RD114 and GALV glycoproteins with that of the MLV-A glycoprotein strongly increases incorporation (Stitz et al. 2000; Christodoulopoulos and Cannon 2001; Sandrin et al. 2002). These chimeric GALV and RD114 glycoproteins (named GALV/TR and RD114/TR) preserve the host range of the initial glycoproteins and confer 25-fold increased titers in comparison with SIV vectors pseudotyped with the wild-type glycoproteins. Lentiviral vectors pseudotyped with GALV/TR or RD114/TR are able to transduce $CD34^+$ cells after a short exposure of the cells to virus along with cytokine treatment that does not allow MLV vectors to transduce these cells. Moreover, RD114/TR-pseudotyped SIV vectors very efficiently transduce human and macaque T lymphocytes, as opposed to vectors pseudotyped with either VSV-G or MLV-A glycoprotein. This discrepancy may be caused by a difference in expression of the receptors for these glycoproteins on the target cells. Several reports, however, have shown that transduction efficiency does not correlate with the level of receptor expression (Uckert et al. 1998; Goerner et al. 2001) but instead depends on postbinding events such as receptor clustering, membrane fusion, the site of fusion, viral particle uncoating, and migration of the viral particle from the site of uncoating to the nucleus (Rodrigues and Heard 1999; Lavillette et al. 2001).

ENGINEERING RETROVIRAL GLYCOPROTEINS FOR RETARGETING

Retroviral envelope glycoproteins can tolerate a variety of genetically encoded modifications. Initial attempts to engineer retroviral tropism have consisted of the insertion of various ligand types, i.e., growth factors, hormones, peptides, or single-chain antibodies, in several locations on the viral surface glycoproteins. These modifications changed the viral particles' host range by allowing them to bind human cell surface molecules different from the parent virus receptor. In some cases (Battini et al. 1998; Lorimer and Lavictoire 2000), the chosen insertion was rationalized by the structural information available for the MLV receptor-binding domain (Linder et al. 1992, 1994; Fass et al. 1997). Such modifications included (1) domain replacement (Kasahara et al. 1994; Benedict et al. 1999; Barnett et al. 2001), (2) peptide insertion in prefolded domains (Valsesia-Wittmann et al. 1994; Battini et al. 1998; Wu et al. 1998, 2000; Gollan and Green 2002), and (3) display of polypeptides as additional folded domains (Russell et al. 1993; Cosset et al. 1995; Valsesia-Wittmann et al. 1996, 1997; Buchholz et al. 1998; Fielding et al. 1998, 2000; Chadwick et al. 1999; Kayman et al. 1999; Maurice et al. 1999, 2002; Martin et al. 2002). Many of these chimeric glycoproteins fold correctly, are stably incorporated on virions, and allow efficient, retargeted virion binding to the expected cell surface molecules. Most amino-terminally substituted chimeric envelope glycoproteins, however, show low or undetected levels of viral incorporation, possibly as a consequence of suboptimal folding, assembly, or transport to the cell surface. Coexpression of such chimeras with unmodified glycoproteins appears to enhance their incorporation (Chu et al. 1994; Kasahara et al. 1994; Matano et al. 1995), suggesting that they need to be chaperoned as hetero-oligomeric complexes for correct processing, transport to the cell surface, and incorporation into vector particles.

Direct Targeting through Specific Ligand–Receptor Binding

Direct targeting strategies rely on the simple idea that fusion activation of the chimeric envelope should be triggered by the interaction of the displayed ligand with its specific receptor on the target cell. In practice, this concept has met with limited success. Although most ligand-displaying envelope glycoproteins specifically and efficiently bind their targeted cells, the infectivity of the corresponding viral particles is generally very poor. The poor infectivity of these vectors is due to two general properties of retargeted retroviral glycoproteins and their receptors: (1) their inability to induce membrane fusion and subsequent penetration of the viral core in the cytosol, and (2) the sequestration of the targeted receptor-bound retroviral particles by some types of cell surface molecules.

The effect of the displayed polypeptides on the basic functions of the backbone envelope glycoprotein cannot always be predicted. Even after optimization of the binding and fusion domains of the chimeric glycoprotein, by using interdomain spacers, for example (Ager et al. 1996; Valsesia-Wittmann et al. 1996, 1997), such pseudotyped vectors still give very low titers with cells that express only the targeted cell surface molecules.

Indirect Targeting

Indirect targeting strategies restrict the vectors' host-range properties. Host-range restrictions are achieved by modifications to the envelope glycoproteins that destroy the vectors' ability to infect nontarget cells. An important advantage of these strategies is that the use of natural viral entry pathways makes fusion activation of the chimeric glycoproteins easier. Two different targeting strategies have been developed: inverse targeting and protease targeting.

Inverse targeting is an indirect targeting strategy that exploits receptor-mediated virus neutralization (Fielding et al. 1998). Several ligands displayed at the amino terminus of C-type retroviral envelope glycoproteins have been shown to inhibit infectivity on cells expressing the targeted receptor (Cosset et al. 1995; Nilson et al. 1996; Chadwick et al. 1999; Fielding et al. 1998, 2000). The original observation was that amphotropic vectors displaying epidermal growth factor (EGF) could efficiently bind to EGF receptor-positive target cells but could not infect them (Cosset et al. 1995). EGF-displaying vectors could efficiently infect EGF receptor-negative hematopoietic cells but were noninfectious for EGF receptor-positive epithelial carcinoma cells. Conversely, stem cell factor (SCF)-displaying vectors failed to infect c-*kit*-positive hematopoietic cells, but efficiently infected c-*kit*-negative epithelial carcinoma cells. Thus, retroviral vectors displaying EGF or SCF were shown to efficiently discriminate between hematopoietic and nonhematopoietic cell populations in tissue culture. Inverse targeting has potential utility, therefore, for the selective transduction of hematopoietic cells with therapeutic transgenes that confer resistance to cytotoxic drug therapy. The value of inverse targeting in this situation is that it can be used to minimize the risk of inadvertent transduction of contaminating cancer cells that express abundant EGF receptors.

Protease targeting takes advantage of the sequestration properties of some cell surface receptors expressed on target cells (Peng et al. 1997, 1998, 1999; Chadwick et al. 1999; Martin et al. 1999, 2002). This strategy depends on the use of a molecular device that allows the vectors to escape the sequestering receptor. This can be achieved by inserting, between the displayed virion-sequestration ligand and the viral glycoprotein, peptide substrates cleaved by cell-surface-specific proteases. Infection proceeds in two steps. First, the vector attaches to the cells via the displayed binding domain. Second, the linker that anchors the binding domain to the viral glycoprotein is cleaved by the protease. The underlying glycoprotein is then exposed and can subsequently interact with the natural viral receptor on the target cell and promote viral entry (Nilson et al. 1996).

Three different strategies have exploited protease-cleavable linkers to target retroviral vector entry into specific cells. In one, the receptor-binding domain EGF is displayed on a retroviral vector as a cleavable extension of the MLV envelope glycoprotein. Vector pseudotyped with this chimeric glycoprotein is blocked for gene delivery in EGF receptor-positive target cells, presumably because the virus is sequestered onto cell surface receptors that are unable to support subsequent steps in viral entry (Nilson et al. 1996; Buchholz et al. 1998; Peng et al. 1999; Schneider et al. 2003; Hartl et al. 2005). EGF receptor-positive cells secreting matrix metalloproteases (MMPs) activate this MMP-sensitive vector. This strategy is limited, however, because it requires the presence of the EGF receptor on the target cells and infectivity is not impaired on receptor-negative cells, which reduces specific targeting.

In a second strategy, trimeric polypeptides (e.g., CD40 ligand) are fused to the amino terminus of the MLV glycoprotein, which prevents binding to the MLV receptor due to steric hindrance and blocks infectivity on cells in general, irrespective of their receptor phenotype (Morling et al. 1997). Protease sites positioned between the polypeptides and the envelope glycoproteins are cleaved by target cell proteases, which renders the vector infectious. This strategy severely impairs gene-transfer vector binding, which might lead to nonspecific targeting of neighbor cells after activation by proteases.

In contrast to these strategies that rely on the display of heterologous blocking domains, a recently reported alternative takes advantage of the fact that the uncleaved forms of viral surface glycoproteins derived from some enveloped viruses with broad tropisms (e.g., the hemagglutinin [HA] of influenza virus) naturally block infection at a postbinding stage. The cell entry properties of HA are reversibly abolished by replacing the cleavage site located between its HA1 surface and HA2 transmembrane subunits with substrates for heterologous proteases (J. Szécsi et al., in press). Vector particles harboring these engineered HA glycoproteins are unable to induce gene delivery in cells that do not express the corresponding proteases unless they are activated by HA1/HA2 digestion. Such vectors have been shown to readily infect tumor cells that express cell-surface-associated MMPs, leading to as much as 200-fold increased gene transfer in these cells as compared to MMP-poor cells. In mixed cell populations containing MMP-rich and MMP-poor cells, these vectors resulted in 50-fold preferential transduction of MMP-rich cells. These results open new possibilities for targeting tumor cells with specific protease activity. Whereas the other targeting approaches rely on heterologous polypeptides to conditionally block infection, the ubiquitously expressed HA receptor, a sialic acid, ensures that vector particles harboring the modified HA glycoproteins are efficiently captured by all target cells, and cell entry is completely abrogated on nontarget cells.

These results have opened the possibility that cell surface proteases, rather than receptors, could be used to target gene delivery, because of the specific expression of many proteases at or close to the surface of target cells. In principle, any protease that does not degrade the envelope glycoprotein can be targeted using this strategy if a specific and sensitive cleavage signal has been identified.

Selective Targeting of Tissues

Escorting viral entry

As discussed above, direct targeting strategies are only useful if the binding and activation of the envelope fusion machinery are recoupled. Alternatively, coexpression of the wild-type glycoprotein with a second "escorting" glycoprotein also permits the tethering of virion binding to tissues that abundantly express the target molecules. This second glycoprotein carries cell-specific binding determinents and is usually defective for fusion. These strategies are called "selective" targeting strategies and lead to preferential (but not exclusive) infection of target cells. In essence, such an approach cannot give rise to highly specific targeting, because the broad distribution of the auxiliary viral receptor will lead to transduction of cells lacking the targeted cell surface molecule.

Vector-mediated target cell activation

Another indirect targeting strategy relies on the interaction of a ligand displayed on the vector's surface with its specific receptor to induce signaling and stimulation of the target cells. The specific stimulation significantly enhances gene transfer into the target cell. This concept was first shown with amphotropic MLV retroviral vectors whose glycoproteins were engineered to display hepatocyte growth factor (HGF) or interleukin 2 (IL-2). Such vectors can infect primary hepatocytes (Nguyen et al. 1998) or IL-2-dependent cells (Maurice et al. 1999), respectively, that are refractory to gene transfer by vectors carrying unmodified envelope glycoproteins. Because MLV vectors require cell division to integrate, it is clear that the increased transduction levels of the target cells are induced by activation of the cell cycle following interaction of the virions with these cells (Maurice et al. 1999).

Although lentiviral vectors are capable of transducing nonproliferating cells both in vitro and in vivo, some cell types that are important gene therapy targets resist gene transfer from these vectors. Cell types of particular interest that fall into this category include early progenitor HSCs in G_0 (Sutton et al. 1999), monocytes (Kootstra et al. 2000; Neil et al. 2001), and resting T lymphocytes (Dardalhon et al. 2001). To overcome the inability to transduce nonactivated T cells, HIV-1 vector particles that display a T-cell-activating polypeptide have been created. These particles target and stimulate the T cells at the time of infection. A single-chain antibody variable fragment (scFv) derived from the anti-CD3 OKT3 monoclonal antibody, which recognizes and activates the T-cell receptor, is fused to the amino terminus of the surface (SU) subunit of the MLV envelope glycoprotein. This chimeric, CD3-targeted MLV glycoprotein demonstrates reduced infectivity, so coexpression of an "escorting" wild-type VSV-G envelope protein is necessary to render the lentiviral particles fully infectious. Stimulation by this vector increases gene transfer in lymphocytes to 48%, 100-fold higher than the performance of unmodified lentiviral vectors in nonactivated T cells (Maurice et al. 2002). It has been shown, however, that the phenotype of the transduced naive T cells was modified to memory cells. Therefore, to transduce resting T cells while conserving their phenotype (Soares et al. 1998; Webb et al. 1999), the human IL-7 gene was fused to the amino terminus of the MLV envelope glycoprotein. IL-7-displaying lentiviral vectors allowed efficient transduction, up to 40%, of naive $CD4^+$ T cells as well as memory $CD4^+$ T cells, and the functional characteristics of the naive T cells were maintained (Verhoeyen et al. 2003).

Another major limitation of current lentiviral vectors is their inability to govern efficient gene transfer into human $CD34^+$ cells that reside in the G_0 phase of the cell cycle and include a large population of HSCs. To overcome this restriction, novel lentiviral vectors have been designed that display early-acting cytokines on their surface. Display of thrombopoietin (TPO), SCF, or both cytokines on the lentiviral surface allows efficient gene delivery into quiescent cord blood $CD34^+$ cells (Verhoeyen et al. 2005). These surface-engineered lentiviral vectors preferentially transduce and promote survival of resting $CD34^+$ cells, rather than cycling cells. Importantly, they also allow superior gene transfer in the most immature $CD34^+$ cells as compared to conventional lentiviral vectors, even when the latter vectors are used to transduce cells in the presence of recombinant cytokines. This was demonstrated by their ability to promote selective transduction of $CD34^+$ cells in long-term culture-initiating cell colonies (LTC-ICs) derived in vitro and of long-term nonobese diabetic/severe combined immunodeficiency (NOD/SCID) repopulating cells (SRCs) in vivo.

Protocol

HSC Targeting with Surface-engineered Lentiviral Vectors

In the targeting strategy presented here, HSCs are specifically transduced with a vector displaying the HSC-activating polypeptides, SCF and TPO. Targeted HSC transduction is evaluated in the NOD/SCID mouse model.

MATERIALS

CAUTION: See Appendix for appropriate handling of materials marked with <!>.

Reagents

293T cells
Anti-CD34–phycoerythrin (PE) antibody (BD Pharmingen)
Calphos Mammalian Transfection Kit for calcium-phosphate-mediated transfections (Clontech)
Cellgro medium (serum-free medium; Cellgenix)
Dulbecco's modified Eagle's medium (DMEM) with 0.11 g/liter sodium pyridoxine and pyridoxine

Supplement DMEM with 10% FCS, 100 μg/ml streptomycin <!>, and 100 units/ml penicillin. Store at 4°C.

Fetal calf serum (FCS; sterile)
Ficoll-Paque Plus (sterile; Amersham)
HeLa cells
Hematopoietic stem cells (HSCs; purified from fresh neonatal cord blood as described below)

Many studies suggest that HSCs reside in a cell population expressing $CD34^+$ antigen. Approximately 1% of mononuclear cells from cord blood are $CD34^+$.

NOD/SCID mice
Nucleic acids
- Lentiviral vector DNA encoding an HIV-1-derived self-inactivating vector with the internal EF1-α promoter driving the reporter gene green fluorescent protein (GFP)
- Envelope glycoprotein-expressing plasmids
  - fusion glycoprotein: VSV-G
  - activating and targeting glycoproteins for hematopoietic stem cells: (1) TPOHA (TPO fused to HA envelope glycoprotein) and (2) SCFSUx (SCF fused to MLV envelope glycoprotein)
- Virus structural protein (Gag-Pol)-expressing plasmid (pCMV8.91)
- Neuraminidase-expressing plasmid pCMV-NA

Phosphate-buffered saline (PBS) without calcium and magnesium, without sodium bicarbonate, sterile
Trypsinization reagents
- Trypsin <!>/EDTA
- 1x Hank's balanced salt solution (HBSS) without calcium and magnesium, sterile

Equipment

CD34+ cell separation kit from Miltenyi containing anti-human CD34 MicroBeads and Blocking Reagent
Filter membranes (0.45 μm)
Flow cytometry system capable of cell sorting and three-color analysis
Flow cytometry tubes
MACS cell separation columns (Miltenyi Biotec)
MACS Separator (Miltenyi Biotec)
Standard tissue-culture reagents, including 100-mm, 6-well, and 48-well tissue-culture plates

METHODS

Production of Lentiviral Vectors Displaying Activating Polypeptides

1. Day 0: The day before transfection, seed 293T cells in DMEM at a density of 2.5×10^6 cells per 100-mm tissue-culture dish in a final volume of 10 ml/dish.
2. Day 1: Use calcium phosphate transfection to cotransfect 8.6 μg of the HIV packaging construct with 8.6 μg of the lentiviral gene-transfer vector and the two glycoproteins: VSV-G (1.5 μg) and TPOHA (1.5 μg).

   Cotransfect TPOHA with the plasmid encoding neuraminidase (pCMV-NA) to allow efficient release of virus from the producer cell. Otherwise, the vector particles bind to the producer cells via an interaction between the HA envelope and the receptor sialic acid.
3. Day 2: 15 hours after transfection, replace the medium with 6 ml of fresh Cellgro medium.

   This medium is adapted for culture of $CD34^+$ cells and allows the cytokines displayed on the vector's surface to maintain their functionality.
4. Day 3: 36 hours after transfection, harvest the vectors, filter through a 0.45-μm membrane, and store at –80°C.

   Vectors can be stored for 2–3 months.

Immunoselection of Human $CD34^+$ Cells

5. Dilute cord blood (CB) 1:1 with PBS. Gently layer 35 ml of the diluted blood on 15 ml of Ficoll-Paque Plus in a 50-ml tube.
6. Centrifuge the cells at 850*g* for 30 minutes at 20°C with no brake. Collect the layer that contains the mononuclear cells.

   The mononuclear cells are found in a white band at the top of the Ficoll layer.
7. Wash the collected mononuclear cells in PBS/2% FCS and centrifuge at 850*g* for 10 minutes at 20°C.
8. Resuspend the cells to a concentration of 1×10^8 to 2×10^8 cells/ml in PBS/2% FCS. To magnetically label the cells, add anti-h$CD34^+$ MicroBeads according to the manufacturer's instructions. Incubate while rocking for 30 minutes at 4°C.
9. Wash the cells to remove unbound antibody and resuspend in PBS/2% FCS.
10. Wash a MACS separation column with 200 μl of PBS/2% FCS. Add the labeled cells to the column and place it in the MACS separator. Allow the unlabeled cells to pass through the column. Wash once with PBS/2% FCS. Remove the column from the separator to elute the $CD34^+$ cells. Repeat this procedure once.

    The purity of the $CD34^+$ cells is routinely 90–95%.

Titer Determination

11. Day –1: Seed HeLa cells in DMEM at a density of 2×10^5 cells/well in six-well plates in a final volume of 2 ml/well. Incubate overnight.
12. Day 0: Add serial dilutions of vector preparations to the HeLa cells and incubate overnight.
13. Day 1: Replace the medium on the cells with 2 ml of fresh DMEM and incubate for 72 hours.
14. Day 3: Trypsinize the cells and transfer to flow cytometry tubes. Determine the percentage of GFP-positive cells by flow cytometry.

    Infectious titer is expressed as transducing units (TU)/ml and is calculated by the following formula:

    Titer = % GFP-positive cells (number of cells at the time of infection/100) x dilution

    Multiplicity of infection (moi) is the ratio of infectious particles to target cells.

Transduction of Human CD34$^+$ Cells

15. Seed CD34$^+$ CB cells in CellGro medium at a density of 5×10^4 cells/well in 48-well plates. Transduce with fresh lentiviral vector supernatant at an moi of 20 or 4. Incubate for 24 hours.
16. Wash the transduced cells and resuspend in CellGro medium. Incubate for 48 hours. Determine transduction efficiency by analyzing enhanced GFP expression by flow cytometry after immunolabeling the cells with an anti-CD34-PE antibody.

NOD/SCID Repopulating Assay

17. After a 24-hour (moi of 4) transduction with TPOHA-, SCFSUx-displaying lentiviral vectors, inject CD34$^+$ CB cells into the tail veins of sublethally irradiated (3.5 Gy) NOD/SCID mice without in vivo administration of cytokines.
18. Harvest the bone marrow from femurs 6–8 weeks after transplantation.
19. Use three-color flow cytometry to detect GFP$^+$ human cells of various lineages in the NOD/SCID bone marrow using anti-hCD45-CyChrome and anti-hCD19-PE, anti-hCD14-PE, anti-hCD13-PE, and anti-hCD34-PE antibodies. In all cases, corresponding PE-conjugated mouse IgG controls should be used to evaluate specific labeling.

REFERENCES

Ager S., Nilson B.H.K., Morling F.J., Peng K.W., Cosset F.-L., and Russell S.J. 1996. Retroviral display of antibody fragments; interdomain spacing strongly influences vector infectivity. *Hum. Gene Ther.* **7:** 2157–2164.

Barnett A.L., Davey R.A., and Cunningham J.M. 2001. Modular organization of the Friend murine leukemia virus envelope protein underlies the mechanism of infection. *Proc. Natl. Acad. Sci.* **98:** 4113–4118.

Battini J.L., Danos O., and Heard J.M. 1998. Definition of a 14-amino-acid peptide essential for the interaction between the murine leukemia virus amphotropic envelope glycoprotein and its receptor. *J. Virol.* **72:** 428–435.

Benedict C.A., Tun R.Y., Rubinstein D.B., Guillaume T., Cannon P.M., and Anderson W.F. 1999. Targeting retroviral vectors to CD34-expressing cells: Binding to CD34 does not catalyze virus-cell fusion. *Hum. Gene Ther.* **10:** 545–557.

Buchholz C.J., Peng K.-W., Morling F.J., Zhang J., Cosset F.-L., and Russell S.J. 1998. In vivo selection of protease cleavage sites from retrovirus display libraries. *Nat. Biotechnol.* **16:** 951–954.

Chadwick M.P., Morling F.J., Cosset F.-L., and Russell S.J. 1999. Modification of retroviral tropism by display of IGF-I. *J. Mol. Biol.* **285:** 485–494.

Christodoulopoulos I. and Cannon P. 2001. Sequences in the cytoplasmic tail of the gibbon ape leukemia virus envelope protein that prevent its incorporation into lentivirus vectors. *J. Virol.* **75:** 4129–4138.

Chu T.H., Martinez I., Sheay W.C., and Dornburg R. 1994. Cell targeting with retroviral vector particles containing antibody-envelope fusion proteins. *Gene Ther.* **1:** 292–299.

Cosset F.-L., Morling F.J., Takeuchi Y., Weiss R.A., Collins M.K., and Russell S.J. 1995. Retroviral retargeting by envelopes expressing an N-terminal binding domain. *J. Virol.* **69:** 6314–6322.

Dardalhon V., Herpers B., Noraz N., Pflumio F., Guetard D., Leveau C., Dubart-Kupperschmitt A., Charneau P., and Taylor N. 2001. Lentivirus-mediated gene transfer in primary T cells is enhanced by a central DNA flap. *Gene Ther.* **8:** 190–198.

DePolo N.J., Reed J.D., Sheridan P.L., Townsend K., Sauter S.L., Jolly D.J., and Dubensky T.W. 2000. VSV-G pseudotyped lentiviral vector particles produced in human cells are inactivated by human serum. *Mol. Ther.* **2:** 218–122.

Duisit G., Conrath H., Saleun S., Folliot S., Provost N., Cosset F.-L., Sandrin V., Moullier P., and Rolling F. 2002. Five recombinant simian immunodeficiency virus pseudotypes lead to exclusive transduction of retinal pigmented epithelium in rat. *Mol. Ther.* **6:** 446–454.

Fass D., Davey R.A., Hamson C.A., Kim P.S., Cunningham J.M., and Berger J.M. 1997. Structure of a murine leukemia virus receptor-binding glycoprotein at 2.0 angstrom resolution. *Science* **277:** 1662–1666.

Fielding A.K., Maurice M., Morling F.J., Cosset F.-L., and Russell S.J. 1998. Inverse targeting of retroviral vectors: Selective gene transfer in a mixed population of hematopoietic and nonhematopoietic cells. *Blood* **91:** 1802–1809.

Fielding A.K., Chapel-Fernandes S., Chadwick M., Bullough F., Cosset F.-L., and Russell S. 2000. A hyperfusogenic Gibbon ape leukemia virus envelope glycoprotein: Targeting of a cytotoxic gene by ligand display. *Hum. Gene Ther.* **11:** 817–826.

Goerner M., Horn P.A., Peterson L., Kurre P., Storb R., Rasko J.E., and Kiem H.P. 2001. Sustained multilineage gene persistence and expression in dogs transplanted with $CD34^+$ marrow cells transduced by RD114-pseudotype oncoretrovirus vectors. *Blood* **98:** 2065–2070.

Gollan T.J. and Green M.R. 2002. Redirecting retroviral tropism by insertion of short, nondisruptive peptide ligands into envelope. *J. Virol.* **76:** 3558–3563.

Hanawa H., Kelly P.F., Nathwani A.C., Persons D.A., Vandergriff J.A., Hargrove P., Vanin E.F., and Nienhuis A.W. 2002. Comparison of various envelope proteins for their ability to pseudotype lentiviral vectors and transduce primitive hematopoietic cells from human blood. *Mol. Ther.* **5:** 242–251.

Hartl I., Schneider R.M., Sun Y., Medvedovska J., Chadwick M.P., Russell S.J., Cichutek K., and Buchholz C.J. 2005. Library-based selection of retroviruses selectively spreading through matrix metalloprotease-positive cells. *Gene Ther.* **12:** 918–926.

Kasahara N., Dozy A.M., and Kan Y.W. 1994. Tissue-specific targeting of retroviral vectors through ligand-receptor interactions. *Science* **266:** 1373–1376.

Kayman S.C., Park H., Saxon M., and Pinter A. 1999. The hypervariable domain of the murine leukemia virus surface protein tolerates large insertions and deletions, enabling development of a retroviral particle display system. *J. Virol.* **73:** 1802–1808.

Kelly P., Vandergriff J., Nathwani A., Nienhuis A., and Vanin E. 2000. Highly efficient gene transfer into cord blood nonobese diabetic/severe combined immunodeficiency repopulating cells by oncoretroviral vector particles pseudotyped with the feline endogenous retrovirus (RD114) envelope protein. *Blood* **96:** 1206–1214.

Kobinger G.P., Weiner D.J., Yu Q.C., and Wilson J.M. 2001. Filovirus-pseudotyped lentiviral vector can efficiently and stably transduce airway epithelia in vivo. *Nat. Biotechnol.* **19:** 225–230.

Kootstra N.A., Zwart B.M., and Schuitemaker H. 2000. Diminished human immunodeficiency virus type 1 reverse transcription and nuclear transport in primary macrophages arrested in early G_1 phase of the cell cycle. *J. Virol.* **74:** 1712–1717.

Lavillette D., Boson B., Russell S., and Cosset F.-L. 2001. Membrane fusion by murine leukemia viruses is activated in *cis* or in *trans* by interactions of the receptor-binding domain with a conserved disulfide loop at the carboxy-terminus of the surface glycoproteins. *J. Virol.* **75:** 3685–3695.

Lindemann D., Bock M., Schweizer M., and Rethwilm A. 1997. Efficient pseudotyping of murine leukemia virus particles with chimeric human foamy virus envelope proteins. *J. Virol.* **71:** 4815–4820.

Linder M., Wenzel V., Linder D., and Stirm S. 1994. Structural elements in glycoprotein 70 from polytropic Friend mink cell focus-inducing virus and glycoprotein 71 from ecotropic Friend murine leukemia virus, as defined by disulfide-bonding pattern and limited proteolysis. *J. Virol.* **68:** 5133– 5141.

Linder M., Linder D., Hahnen J., Schott H.H., and Stirm S. 1992. Localization of the intrachain disulfide bonds of the envelope glycoprotein 71 from Friend murine leukemia virus. *Eur. J. Biochem.* **203:** 65–73.

Lorimer I.A. and Lavictoire S.J. 2000. Targeting retrovirus to cancer cells expressing a mutant EGF receptor by insertion of a single chain antibody variable domain in the envelope glycoprotein receptor binding lobe. *J. Immunol. Methods* **237:** 147–157.

Mammano F., Salvatori F., Indraccolo S., de Rossi A., Chieco-Bianchi L., and Göttlinger H.G. 1997. Truncation of the human immunodeficiency virus type 1 envelope glycoprotein allows efficient pseudotyping of moloney murine leukemia virus particles and gene transfer into $CD4^+$ cells. *J. Virol.* **71:** 3341–3345.

Marandin A., Dubart A., Pflumio F., Cosset F.-L., Cordette V., Chapel-Fernandes S., Coulombel L., Vainchenker W., and Louache F. 1998. Retroviral-mediated gene transfer into human $CD34^+/38^-$ primitive cells capable of reconstituting long-term cultures in vitro and in nonobese diabetic-severe combined immunodeficiency mice in vivo. *Hum. Gene Ther.* **9:** 1497–1511.

Martin F., Chowdhury S., Neil S., Phillipps N., and Collins M.K. 2002. Envelope-targeted retrovirus vectors transduce melanoma xenografts but not spleen or liver. *Mol. Ther.* **5:** 269–274.

Martin F., Neil S., Kupsch J., Maurice M., Cosset F.-L., and Collins M. 1999. Retrovirus targeting by tropism restriction to melanoma cells. *J. Virol.* **73:** 6923–6929.

Matano T., Odawara T., Iwamoto A., and Yoshikura H. 1995. Targeted infection of a retrovirus bearing a CD4-Env chimera into human cells expressing human immunodeficiency virus type 1. *J. Gen. Virol.* **76:** 3165–3169.

Maurice M., Verhoeyen E., Salmon P., Trono D., Russell S.J., and Cosset F.-L. 2002. Efficient gene transfer into human primary blood lymphocytes by surface-engineered lentiviral vectors that display a T cell-activating polypeptide. *Blood* **99:** 2342–2350.

Maurice M., Mazur S., Bullough F.J., Salvetti A., Collins M.K.L., Russell S.J., and Cosset F.-L. 1999. Efficient gene delivery to quiescent IL2-dependent cells by murine leukemia virus-derived vectors harboring IL2 chimeric envelopes glycoproteins. *Blood* **94:** 401–410.

Morling F.J., Peng K.W., Cosset F.L., and Russell S.J. 1997. Masking of retroviral envelope functions by oligomerizing polypeptide adaptors. *Virology* **234:** 51–61.

Movassagh M., Desmyter C., Baillou C., Chapel-Fernandes S., Guigon M., Klatzmann D., and Lemoine F.M. 1998. High-level gene transfer to cord blood progenitors using gibbon ape leukemia virus pseudotyped retroviral vectors and an improved clinically applicable protocol. *Hum. Gene Ther.* **9:** 225–234.

Naldini L., Blömer U., Gallay P., Ory D., Mulligan R., Gage F.H., Verma I.M., and Trono D. 1996. In vivo gene delivery and stable transduction of nondividing cells by a lentiviral vector. *Science* **272:** 263–267.

Nègre D., Duisit G., Mangeot P.-E., Moullier P., Darlix J.-L., and Cosset F.-L. 2002. Lentiviral vectors derived from simian immunodeficiency virus (SIV). *Curr. Top. Microbiol. Immunol.* **261:** 53–74.

Neil S., Martin F., Ikeda Y., and Collins M. 2001. Postentry restriction to human immunodeficiency virus-based vector transduction in human monocytes. *J. Virol.* **75:** 5448–5456.

Nguyen T., Pages J.-C., Farge D., Briand P., and Weber A. 1998. Amphotropic retroviral vectors displaying hepatocyte growth factor-envelope fusion proteins improve transduction efficiency of primary hepatocytes. *Hum. Gene Ther.* **9:** 2469–2479.

Nilson B.H.K., Morling F.J., Cosset F.-L., and Russell S.J. 1996. Targeting of retroviral vectors through protease-substrate interactions. *Gene Ther.* **3:** 280–286.

Peng K.W., Morling F.J., Cosset F.-L., and Russell S.J. 1998. A retroviral gene delivery system activatable by plasmin. *Tumor Targeting* **3:** 112–120.

Peng K.-W., Vile R.G., Cosset F.-L., and Russell S.J. 1999. Selective transduction of protease-rich tumors by matrix-metalloproteinase-targeted retroviral vectors. *Gene Ther.* **6:** 1552–1557.

Peng K.W., Morling F.J., Cosset F.-L., Murphy G., and Russell S.J. 1997. A gene delivery system activatable by disease-associated matrix metalloproteinases. *Hum. Gene Ther.* **8:** 729–738.

Peng K.W., Pham L., Ye H., Zufferey R., Trono D., Cosset F.-L., and Russell S.J. 2001. Organ distribution of gene expression after intravenous infusion of targeted and untargeted lentiviral vectors. *Gene Ther.* **8:** 1456–1463.

Porter C.D., Collins M.K.L., Tailor C.S., Parker M.H., Cosset F.-L., Weiss R.A., and Takeuchi Y. 1996. Comparison of efficiency of infection of human gene therapy target cells via four different retroviral receptors. *Hum. Gene Ther.* **7:** 913–919.

Rodrigues P. and Heard J.M. 1999. Modulation of phosphate uptake and amphotropic murine leukemia virus entry by posttranslational modifications of PIT-2. *J. Virol.* **73:** 3789–3799.

Russell S.J., Hawkins R.E., and Winter G. 1993. Retroviral vectors displaying functional antibody fragments. *Nucleic Acids Res.* **21:** 1081–1085.

Sandrin V., Boson B., Salmon P., Gay W., Nègre D., LeGrand R., Trono D., and Cosset F.-L. 2002. Lentiviral vectors pseudotyped with a modified RD114 envelope glycoprotein show increased stability in sera and augmented transduction of primary lymphocytes and $CD34^+$ cells derived from human and non-human primates. *Blood* **100:** 823–832.

Schneider R.M., Medvedovska Y., Hartl I., Voelker B., Chadwick M.P., Russell S.J., Cichutek K., and Buchholz C.J. 2003. Directed evolution of retroviruses activatable by tumour-associated matrix metalloproteases. *Gene Ther.* **10:** 1370–1380.

Schnierle B.S., Stitz J., Bosch V., Nocken F., Merget-Millitzer H., Engelstadter M., Kurth R., Groner B., and Cichutek K. 1997. Pseudotyping of murine leukemia virus with the envelope glycoproteins of HIV generates a retroviral vector with specificity of infection for CD4-expressing cells. *Proc. Natl. Acad. Sci.* **94:** 8640–8645.

Seganti L., Superti F., Girmenia C., Melucci L., and Orsi N. 1986. Study of receptors for vesicular stomatitis virus in vertebrate and invertebrate cells. *Microbiologica* **9:** 259–267.

Soares M.V., Borthwick N.J., Maini M.K., Janossy G., Salmon M., and Akbar A.N. 1998. IL-7-dependent extrathymic expansion of $CD45RA^+$ T cells enables preservation of a naive repertoire. *J. Immunol.* **161:** 5909–5917.

Stitz J., Buchholz C., Engelstadter M., Uckert W., Bloemer U., Schmitt I., and Cichutek K. 2000. Lentiviral vectors pseudotyped with envelope glycoproteins derived from gibbon ape leukemia virus and murine leukemia virus 10A1. *Virology* **273:** 16–20.

Sutton R.E., Reitsma M.J., Uchida N., and Brown P.O. 1999. Transduction of human progenitor hematopoietic stem cells by human immunodeficiency virus type 1-based vectors is cell cycle dependent. *J. Virol.* **73:** 3649–3660.

Szécsi J., Drury R., Josserand V., Grange M.P., Boson B., Hartl I., Schneider R., Buchholz C., Russell S.J., Cosset F.L., and Verhoeyen E. 2006. Targeted retroviral vectors displaying a cleavage site-engineered hemagglutinin (HA) through HA-protease interactions. *Mol. Ther.* (in press).

Takeuchi Y., Simpson G., Vile R., Weiss R., and Collins M. 1992. Retroviral pseudotypes produced by rescue of moloney murine leukemia virus vector by C-type, but not D-type, retroviruses. *Virology* **186:** 792–794.

Uckert W., Willimsky G., Pedersen F.S., Blankenstein T., and Pedersen L. 1998. RNA levels of human retrovirus receptors Pit1 and Pit2 do not correlate with infectibility by three retroviral vector pseudotypes. *Hum. Gene Ther.* **9:** 2619–2627.

Valsesia-Wittmann S., Morling F.J., Hatziioannou T., Russell S.J., and Cosset F.-L. 1997. Receptor co-operation in retrovirus entry: Recruitment of an auxilliary entry mechanism after retargeted binding. *EMBO J.* **16:** 1214–1223.

Valsesia-Wittmann S., Morling F.J., Nilson B.H.K., Takeuchi Y., Russell S.J., and Cosset F.-L. 1996. Improvement of retroviral retargeting by using amino acid spacers between an additional binding domain and the N terminus of Moloney murine leukemia virus SU. *J. Virol.* **70:** 2059–2064.

Valsesia-Wittmann S., Drynda A., Deleage G., Aumailley M., Heard J.-M., Danos O., Verdier G., and Cosset F.-L. 1994. Modifications in the binding domain of avian retrovirus envelope protein to redirect the host range of retroviral vectors. *J. Virol.* **68:** 4609–4619.

Verhoeyen E., Dardalhon V., Ducrey-Rundquist O., Trono D., Taylor N., and Cosset F.-L. 2003. IL-7 surface-engineered lentiviral vectors promote survival and efficient gene transfer in resting primary T-lymphocytes. *Blood* **101:** 2167–2174.

Verhoeyen E., Wiznerowicz M., Olivier D., Izac B., Trono D., Dubart-Kupperschmitt A., and Cosset F.L. 2005. Novel lentiviral vectors displaying "early acting cytokines" selectively promote survival and transduction of NOD/SCID repopulating human hematopoietic stem cells. *Blood* **106:** 3386–3395.

Vigna E. and Naldini L. 2000. Lentiviral vectors: Excellent tools for experimental gene transfer and promising candidates for gene therapy. *J. Gene Med.* **2:** 308–316.

Webb L.M., Foxwell B.M., and Feldmann M. 1999. Putative role for interleukin-7 in the maintenance of the recirculating naive $CD4^+$ T-cell pool. *Immunology* **98:** 400–405.

Wu B.W., Cannon P.M., Gordon E.M., Hall F.L., and Anderson W.F. 1998. Characterization of the proline-rich region of murine leukemia virus envelope protein. *J. Virol.* **72:** 5383–5391.

Wu B.W., Lu J., Gallaher T.K., Anderson W.F., and Cannon P.M. 2000. Identification of regions in the Moloney murine leukemia virus SU protein that tolerate the insertion of an integrin-binding peptide. *Virology* **269:** 7–17.

13 Preparation of Pseudotyped Lentiviral Vectors Resistant to Inactivation by Serum Complement

Ghiabe H. Guibinga and Theodore Friedmann

Department of Pediatrics, Center for Molecular Genetics, University of California, School of Medicine, San Diego, California 92093-0634

ABSTRACT

A major obstacle to in vivo delivery of lentivirus or other retroviral vectors is their lability in the presence of serum. In vivo, these viral particles are rapidly destroyed by nonspecific complement-mediated degradation mechanisms. The eventual effective use of retroviral vectors for in vivo gene delivery would be greatly facilitated by the development of methods to protect the viral particles from such degradation. This protocol describes methods for the production of complement-stabilized lentiviral vectors either by pseudotyping the viral particles with a fusion envelope protein containing the complement-regulatory protein CD55 (decay accelerating factor, DAF) or by coassembly with the native DAF protein.

INTRODUCTION

Lysis of retroviruses by human serum was first reported three decades ago (Welsh et al. 1975, 1976; Cooper et al. 1976). Early generations of retroviral vectors made predominantly in murine and other nonhuman packaging cell lines demonstrated high sensitivity to serum complement. Indeed, viral and/or cellular components can trigger a serum-complement-mediated cascade that leads to serum-mediated inactivation and destruction of the retroviral particles (Takeuchi et al. 1994; Pensiero et al. 1996). In contrast, retroviral vectors derived from human packaging cell lines are considerably more resistant to inactivation by primate serum complement (DePolo et al. 1999).

The development of methods to pseudotype lentiviral vectors of human, feline, or equine origin with a variety of surrogate viral envelope proteins such as vesicular stomatitis virus G protein (VSV-G), baculovirus gp64 protein, and others have made possible the production of vectors with high titer and broad tropism (Yee et al. 1994; Kumar et al. 2003). However, lentiviral vectors pseudotyped with VSV-G or gp64 are very sensitive to serum complement inactivation in vitro, regardless of the packaging cell lines used (DePolo et al. 2000; Schauber et al. 2004; Guibinga and Friedmann 2005). Furthermore, the stability of retroviral vectors in serum in vivo after intravenous administration is correlated with their lability to serum in vitro (DePolo et al. 2000; Croyle et al. 2004). This indicates that the retroviral vectors' lability in vivo is also influenced by the innate immune response and mediated by the complement pathway. This property has greatly limited the use of this class of vectors in large animal gene-transfer models. In contrast, other classes of viral vector particles pseudotyped with different surrogate envelope proteins (such as the RD114 protein of the feline endogenous virus) are highly resistant to serum-complement-mediated inactivation. However, these vector systems are not as useful or attractive as VSV-G-pseudotyped particles for many uses, mostly due to their narrower tropism and/or lower titers (Sandrin et al. 2002).

The mammalian complement system helps to protect the host organism against pathogens, but it can in some cases cause serious damage when directed against "self" tissues. A set of complement regulatory proteins (CRPs) exists in mammalian systems to help prevent autologous complement-mediated autoimmune damage. These CRPs have been extensively characterized in the human, including the decay accelerating factor (DAF) CD55 (Kirschfink 1997). This chapter describes a short protocol for preparing pseudotyped retroviral vectors resistant to serum complement inactivation. In this approach, a transgene encoding the complement-regulatory protein DAF is added during viral preparation. Alternatively, lentiviral pseudotyping can be performed with transgenes encoding fusion proteins of DAF with VSV-G- or gp64-pseudotyped lentivectors or simply carrying out the viral preparation in the presence of the complement-regulatory protein DAF (Guibinga and Friedmann 2005; Schauber-Plewa et al. 2005).

Protocol 1

Preparation of Lentiviral Vectors

Human immunodeficiency virus type-1 (HIV-1)-based lentiviral vectors pseudotyped with VSV-G or gp64 are produced by transient transfection with three plasmids: a packaging construct (pack), a transgene construct (trans), and an envelope-encoding construct (env). However, to prepare complement-stabilized lentiviral vectors, a fourth expression plasmid must be added to the standard transfection system. This fourth transgene encodes either a DAF protein or a fusion protein VSV-G–DAF (for VSV-G-pseudotyped vectors) or gp64-DAF (for gp64-pseudotyped vectors). Transient triplicate and quadruplicate transfections can be performed using any nonviral transfection reagents or the established calcium phosphate protocols as described below.

MATERIALS

CAUTION: See Appendix for appropriate handling of materials marked with <!>.

Reagents

$CaCl_2$ (2 M) <!>
Dulbecco's modified Eagle's medium (DMEM) containing 4.5 g/liter glucose (serum-free and complete) and 10% fetal calf serum (FCS)
2x HBS (HEPES-buffered saline)
  50 mM HEPES
  280 mM NaCl
  1.5 mM Na_2HPO_4 <!>
    Adjust pH to 7.12 with 5 M NaOH <!>.
HEK 293 T cells (ATCC, CRL 11268)
Phosphate-buffered saline (PBS)
Plasmids
  DAF-encoding transgene construct
  envelope-encoding construct (env)
  gp64-DAF fusion protein-encoding transgene construct
  packaging construct (pack)
  transgene construct (trans)
  VSV-G–DAF fusion protein-encoding transgene construct
0.1x TE
  10 mM Tris-HCl (pH 8.0) <!>
  1 mM EDTA (pH 8.0)
    Dilute 1:10 with double-distilled H_2O (ddH_2O).

Equipment

Beckman rotor, SW28 or equivalent
Filters (0.45 μm, 500 ml)
Incubator, 30°C or 37°C
SW 28 Ultra-Clear Beckman tubes
Tissue-culture plates (10 cm)

METHODS

Plasmid Transfection

1. One day before transfection, seed 293 T cells in DMEM at 10% confluency onto 10-cm plates.
   Cells should be 40–70% confluent on the day of transfection.
2. Two hours prior to transfection, replace the medium with fresh DMEM.
3. Combine env, pack, and trans plasmids with plasmid constructs encoding DAF or the fusion proteins gp64-DAF or VSV-G–DAF in the proportions indicated (Tables 1 and 2) as required by the experiment.
4. Add 125 μl of 2 M $CaCl_2$ and 20 μl of 0.1x TE to the plasmid DNA mix. Add ddH_2O to a total volume of 1 ml.
5. Add 1 ml of the DNA mix dropwise to 1 ml of 2x HBS while vortexing at maximum speed. Let stand 30 minutes.
6. Slowly add the precipitate dropwise onto the cells.
7. Incubate for 16 hours.
   Generation of DAF-modified gp64 or VSV-G-pseudotyped HIV-1-based lentiviral vectors by codisplay of gp64-DAF or VSV-G–DAF fusion protein is carried out at 30°C instead of standard 37°C.
8. Discard the medium. Replace with 10–12 ml of fresh complete medium.

Virus Collection, Filtration, and Concentration

9. Forty-eight hours after transfection, collect the conditioned culture medium. Replace the medium with 10–12 ml of fresh complete medium.
10. Filter the collected medium through a 0.45-μm filter.
    Store filtrate at –80°C.
11. Repeat Steps 8 and 9 daily (2–3 days after the first harvest for VSV-G-pseudotyped virus and for up to 4–6 days for gp64-pseudotyped virus).
    Because of the cytotoxicity of VSV-G, very few cells producing VSV-G-pseudotyped particles survive more than 4 days. In contrast, the reduced toxicity of gp64 allows cell survival and medium collection for up to 1 week after transfection with the gp64 constructs.
12. Pool conditioned media. Concentrate by centrifugation at 25,000 rpm for 2 hours at 4°C in an SW28 using Ultra-clear tubes (maximum 30 ml/tube).
13. Aspirate and discard the supernatant. Resuspend the pellet in complete DMEM. Store resuspended vector preparations at –80°C for further characterization and use.

TABLE 1. Ratios of DNA Plasmids Used for Preparation of VSV-G/DAF-modified Pseudotyped HIV-1-based Lentiviral Vectors

	VSV-G/DAF	VSV-G/ VSV-G–DAF
DNA plasmid ratio (Pack:Trans:Env:DAF)	1:1:0.5:0.1	1:1:0.5:05
DNA/10-cm plate (μg)	10:10:5:1	10:10:5:5

TABLE 2. Ratios of DNA Plasmids Used for Preparation of gp64/DAF-modified Pseudotyped HIV-1-based Lentiviral Vectors

	gp64/DAF	gp64/ gp64-DAF
DNA plasmid ratio (Pack:Trans:Env:DAF)	1:1:1:0.1	1:1:0.5:05
DNA/10-cm plate (μg)	10:10:10:1	10:10:5:5

Protocol 2

In Vitro Serum Inactivation Assay

Titration is carried out in presence of normal human serum complement in order to evaluate complement-stabilizing effect of the viral preparations (see Protocol 1). Viral titers are determined by a standard endpoint dilution method. To characterize the complement sensitivity of pseudotyped lentiviral vector preparations, we recommend the use of human serum complement with a CH_{50} value of 100 or above. The CH_{50} value measures the ability of a given serum sample to lyse 50% of a standardized suspension of sheep erythrocytes. Samples with CH_{50} values of 0 to 100 are considered to have low levels of complement, whereas values above 100 are suitable. Samples with high complement activity are optimal (>300).

On the basis of prior experience, lentiviral vector preparations pseudotyped with a DAF-displaying envelope will show no significant reduction in titer after incubation in normal human serum complement. In contrast, lentiviral vector preparations pseudotyped with native gp64 or VSV-G envelopes and incubated in normal serum complement should exhibit titer reductions of up to approximately 10% of their infectivity relative to control incubations in DMEM or heat-inactivated serum.

It is likely (although still unproven) that increased stability of pseudotyped lentiviral vectors to human serum complement in vitro is likely to lead to at least partial enhancement of stability in vivo. However, sensitivity to complement inactivation is not the sole pharmacokinetic determinant of effective vector stability and delivery in vivo. Many other factors such as opsonization and interaction of viral particles with serum proteins may have a role in retroviral vector stability and in vivo gene delivery.

MATERIALS

Reagents

Cells, exponentially growing cultures of 293 or HT1080 (human fibrosarcoma)

Dulbecco's modified Eagle's medium (DMEM; complete and serum-free) containing 10% fetal calf serum (FCS)

Human serum complement (CH_{50} >100; normal and heat-inactivated) (Quidel Corp., San Diego, California)

To inactivate complement, incubate for 1 hour at 56°C.

Polybrene

Pseudotyped HIV-1-based lentiviral vectors (>10^7 transduction units/ml)

Vectors should be constructed to encode the green fluorescent protein (GFP).

Equipment

Water bath (37°C and 56°C)

Vortex mixer

METHOD

1. Dilute 100 µl of the vector preparations with an equal volume of normal human serum complement. Use heat-inactivated human serum complement and/or complete DMEM as controls.
2. Incubate the vector-complement mixture for 1 hour at 37°C. Vortex for 2 seconds every 15 minutes.
3. Following incubation, dilute the reaction tenfold with serum-free DMEM.
4. Infect target HT1080 or 293 cells with serial dilutions of the GFP-encoding viral preparation in the presence of 8 µg/ml Polybrene.
5. Forty-eight hours after transduction, count the number of GFP-positive foci in the plate containing the highest dilution of virus that contains positive foci.
6. Calculate the viral titer (expressed as GFP-transduction units/ml; GFP-TU/ml) as follows:

   Virus titer = (number of GFP foci × dilution factor)/volume of vector used for transduction.

REFERENCES

Cooper N.R., Jensen F.C., Welsh R.M., Jr., and Oldstone M.B. 1976. Lysis of RNA tumor viruses by human serum: Direct antibody-independent triggering of the classical complement pathway. *J. Exp. Med.* **144:** 970–984.

Croyle M.A., Callahan S.M., Auricchio A., Schumer G., Linse K.D., Wilson J.M., Brunner L.J., and Kobinger G.P. 2004. PEGylation of a vesicular stomatitis virus G pseudotyped lentivirus vector prevents inactivation in serum. *J. Virol.* **78:** 912–921.

DePolo N.J., Reed J.D., Sheridan P.L., Townsend K., Sauter S.L., Jolly D.J., and Dubensky T.W., Jr. 2000. VSV-G pseudotyped lentiviral vector particles produced in human cells are inactivated by human serum. *Mol. Ther.* **2:** 218–222.

DePolo N.J., Harkleroad C.E., Bodner M., Watt A.T., Anderson C.G., Greengard J.S., Murthy K.K., Dubensky T.W., Jr., and Jolly D.J. 1999. The resistance of retroviral vectors produced from human cells to serum inactivation in vivo and in vitro is primate species dependent. *J. Virol.* **73:** 6708–6714.

Guibinga G.H. and Friedmann T. 2005. Baculovirus GP64-pseudotyped HIV-based lentivirus vectors are stabilized against complement inactivation by codisplay of decay accelerating factor (DAF) or of a GP64-DAF fusion protein. *Mol. Ther.* **11:** 645–651.

Kirschfink M. 1997. Controlling the complement system in inflammation. *Immunopharmacology* **38:** 51–62.

Kumar M., Bradow B.P., and Zimmerberg J. 2003. Large-scale production of pseudotyped lentiviral vectors using baculovirus GP64. *Hum. Gene Ther.* **14:** 67–77.

Pensiero M.N., Wysocki C.A., Nader K., and Kikuchi G.E. 1996. Development of amphotropic murine retrovirus vectors resistant to inactivation by human serum. *Hum. Gene Ther.* **7:** 1095-1101.

Sandrin V., Boson B., Salmon P., Gay W., Negre D., Le Grand R., Trono D., and Cosset F.L. 2002. Lentiviral vectors pseudotyped with a modified RD114 envelope glycoprotein show increased stability in sera and augmented transduction of primary lymphocytes and CD34$^+$ cells derived from human and nonhuman primates. *Blood* **100:** 823–832.

Schauber C.A., Tuerk M.J., Pacheco C.D., Escarpe P.A., and Veres G. 2004. Lentiviral vectors pseudotyped with baculovirus gp64 efficiently transduce mouse cells in vivo and show tropism restriction against hematopoietic cell types in vitro. *Gene Ther.* **11:** 266–275.

Schauber-Plewa C., Simmons A., Tuerk M.J., Pacheco C.D., and Veres G. 2005. Complement regulatory proteins are incorporated into lentiviral vectors and protect particles against complement inactivation. *Gene Ther.* **12:** 238–245.

Takeuchi Y., Cosset F.-L.C., Lachmann P.J., Okada H., Weiss R.A., and Collins M.K.L. 1994. Type C retrovirus inactivation by human complement is determined by both the viral genome and the producer cell. *J. Virol.* **68:** 8001–8007.

Welsh R.M., Jr., Cooper N.R., Jensen F.C., and Oldstone M.B. 1975. Human serum lyses RNA tumour viruses. *Nature* **257:** 612–614.

Welsh R.M., Jr., Jensen F.C., Cooper N.R., and Oldstone M.B. 1976. Inactivation of lysis of oncornaviruses by human serum. *Virology* **74:** 432–440.

Yee J.-K., Friedmann T., and Burns J.C. 1994. Generation of high-titer pseudotyped retroviral vectors with very broad host range. *Methods Cell Biol.* **43:** 99–112.

14 Generation of 2A Peptide-linked Multicistronic Vectors

Andrea L. Szymczak-Workman, Kate M. Vignali, and Dario A.A. Vignali

Department of Immunology, St. Jude Children's Research Hospital, Memphis, Tennessee 38105

ABSTRACT

The need for reliable, multicistronic vectors for multigene delivery is at the forefront of biomedical technology. This chapter describes a novel method for expressing multiple proteins from a single open reading frame (ORF) using 2A peptide-linked multicistronic vectors. These small sequences, when cloned between genes, allow for efficient, stoichiometric production of discrete protein products within a single vector through a novel "cleavage" event within the 2A peptide sequence.

Expression of more than two genes using conventional approaches has several limitations, most notably imbalanced protein expression and large size. The use of 2A peptide sequences alleviates these concerns. They are small (18–22 amino acids) and have divergent amino-terminal sequences, which minimizes the chance for homologous recombination and allows for multiple, different 2A peptide sequences to be used within a single vector. Importantly, separation of genes placed between 2A peptide sequences is nearly 100%, which allows for stoichiometric and concordant expression of the genes, regardless of the order of placement within the vector.

Multicistronic expression vectors incorporating 2A peptide sequences have been used in a variety of applications including biomedical research, gene therapy, and plant biotechnology. Potential applications for this technology are almost unlimited. The expression of multiple drug resistance genes, for example, could protect hematopoietic cells during cancer treatment, or multiple suicide, antiangiogenic, and/or immunoactive proteins could enhance treatments against tumors.

INTRODUCTION

During the past few years, researchers have searched for new ways to express multiple genes (de Felipe 2002). To date, the most exciting approach takes advantage of strategies that single-stranded RNA viruses use to mediate multigene expression. Picornaviruses use 2A peptides, which function as *cis*-acting hydrolase elements (CHYSELs), to mediate "cleavage" between two proteins (de Felipe 2004). These sequences were first discovered in the foot-and-mouth disease virus (FMDV) (Ryan and Drew 1994) which encodes a single, long ORF in which two of the gene products are separated by the short 2A sequence. A novel "cleavage" event takes place between the two products within a highly conserved consensus sequence in the carboxyl terminus of the 2A sequence between the glycine residue of the 2A sequence and the proline residue of 2B (Asp-Val/Ile-Glu-X-Asn-Pro-Gly-↓-Pro) (Fig. 1). The 2A peptide sequence appears to impair the formation of a normal peptide bond via a "ribosomal skip" mechanism without affecting translation of 2B (Donnelly et al. 2001b). Since their discovery, many 2A-like sequences have been identified in other viruses and some parasites, including insect viruses, rotaviruses, and *Trypanosoma* spp (for more details, see http://www.st-andrews.ac.uk/ryanlab/Index.htm; Szymczak and Vignali 2005). The most complex 2A-based multicistronic vector encoded the four CD3 chains of the T-cell receptor (TCR) complex using three different 2A peptide sequences (Szymczak et al. 2004b). This vector restored T-cell development in CD3-deficient mice, demonstrating that all four CD3 chains were efficiently cleaved and expressed on T cells.

Vector Design and Construction Considerations

Protein function and subcellular localization

The use of 2A peptide sequences is compatible with targeting proteins to a variety of cellular locations, including the cell surface, cytosolic and luminal compartments, the nucleus, and mitochondria (de Felipe and Ryan 2004; Lorens et al. 2004; Szymczak et al. 2004b). There is, however, one important consideration when targeting proteins to the lumen (e.g., for proteins that contain a signal sequence but not a transmembrane domain). In mammalian cells, but seemingly not in yeast, "slip-streaming" can occur for proteins that follow luminally targeted proteins (de Felipe and Ryan 2004). For instance, if a secreted protein is followed by green fluorescent protein (GFP), which would normally reside in the cytosol, the ribosome remains attached to the translocon, and therefore, GFP is also translocated into the lumen. This would clearly be a problem for proteins that require cytosolic, nuclear, or mitochondrial localization. Thus, if such a mixture of differentially targeted proteins is required and includes one that is targeted to the lumen, one should consider the order in which the proteins are included in 2A-linked vectors (e.g., place the secreted/luminal protein last).

Family	Example(s)	Abbr.	Sequences (2A 'Cleavage' Site ↓)
Picornaviridae	FMDV	F2A	VKQTLNFDLLKLA**GDVESNPG** ↓ **P**
	ERAV	E2A	QCTNYALLKLA**GDVESNPG P**
	PTV1	P2A	ATNFSLLKQA**GDVEENPG P**
Tetraviridae	TaV	T2A	EGRGSLLTC**GDVEENPG P**

FIGURE 1. 2A peptide sequences successfully used in vivo. These sequences were successfully tested in vitro and in vivo (Szymczak and Vignali 2005). (FMDV) Foot-and-mouth disease virus; (ERAV) equine rhinitis A virus; (PTV1) porcine teschovirus-1; (TaV) *Thosea asigna* virus. Conserved residues are bold with the site of cleavage noted. For a comprehensive list of publications using 2A peptide sequences, see www.st-andrews.ac.uk/ryanlab/Index.htm.

A 2A "tag" attached to the upstream protein

Cleavage occurs at the end of the 2A peptide sequence; therefore, most of the peptide remains attached to the carboxyl terminus of the upstream protein. Thus far, no adverse effects have been reported from the presence of this residual sequence. This may be a consideration, however, for proteins that might be affected by such additions (e.g., if their function is affected by addition of other tags such as FLAG, Myc, 6xHis, etc.). For secreted proteins, the insertion of a furin cleavage site between the protein and the 2A sequence has been shown to result in the removal of this "tag" (Fang et al. 2005). It is quite possible that alternative methods could be developed for the proteolytic removal of the 2A tags. It should be emphasized that this residual tag does have two advantages. First, the presence of the 2A sequence usually results in a shift in protein size which can be useful if mutant and endogenous proteins are coexpressed and need to be distinguished (Fig. 2A,B) (Szymczak et al. 2004b). Second, antibodies have been generated against the 2A peptide consensus motif that work effectively in western blot and immunohistochemistry (Fig. 2C) (de Felipe et al. 2006; Holst et al. 2006). Therefore, the 2A tag becomes a useful tool for protein identification. Finally, it should also be noted that all proteins generated after the first 2A peptide will have an amino-terminal proline (P) (Fig. 3). Thus far, we and other investigators have seen no deleterious effects of this addition (de Felipe 2002; Szymczak and Vignali 2005). Proteins with signal sequences are processed efficiently and have this proline removed. Proteins generated in the cytosol are stable as predicted by the N-end rule (Varshavsky 1992).

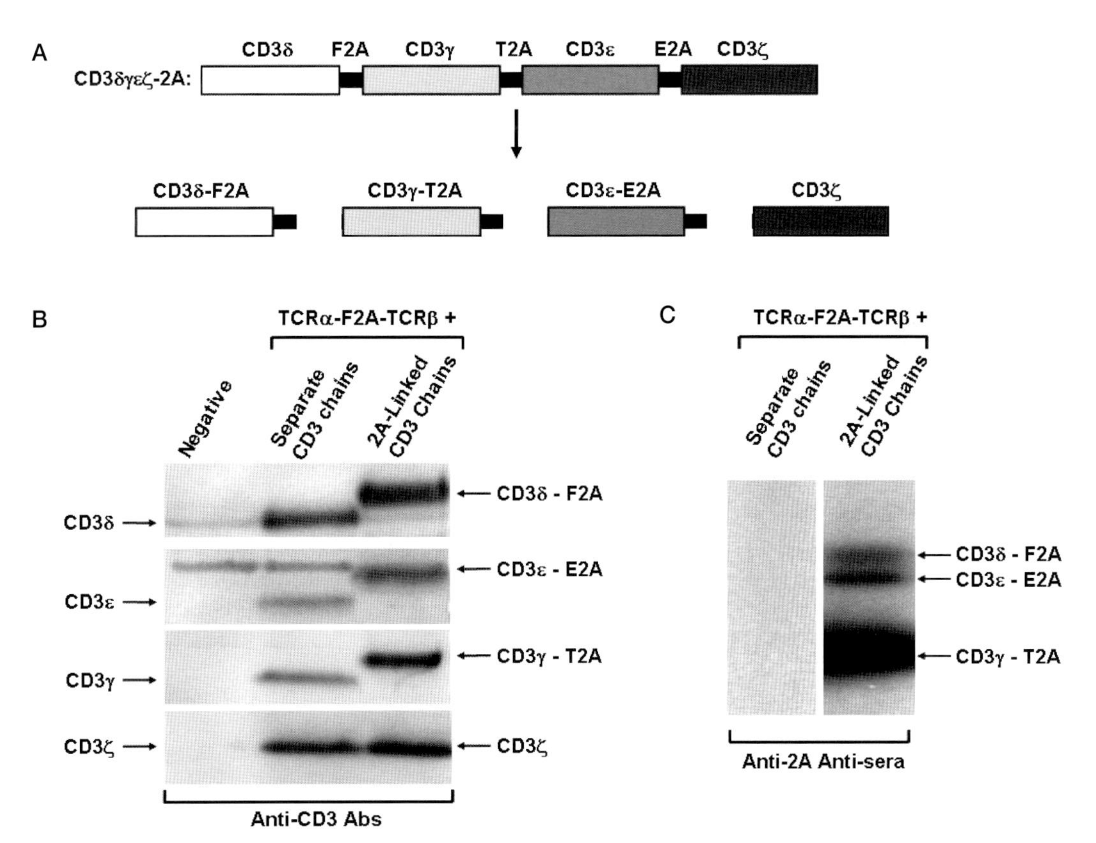

FIGURE 2. Biochemical analysis of 2A-peptide-mediated cleavage. 3T3 fibroblasts were transduced with the separate CD3 chains (CD3δ, CD3γ, CD3ε, CD3ζ) or 2A-peptide-linked CD3 chains along with 2A-peptide-linked TCRα chains. (*A*) Schematic of 2A-peptide-linked construct. (*B*) Cells were lysed and immunoprecipitated with anti-TCR antibodies and resolved by SDS-PAGE. Membranes were probed with antibodies to the individual CD3 chains. A shift in protein size is observed with the CD3 chains that have the 2A peptide still attached (all but CD3ζ). (*C*) Cells were lysed and proteins were resolved by SDS-PAGE. Membranes were probed with antisera against the T2A peptide [GDVEENPG]. Intensity of the bands is weaker with F2A- and E2A-linked proteins as this sequence contains a serine residue in place of the second glutamine [GDVESNPG] (J. Holst and D. Vignali, unpubl.).

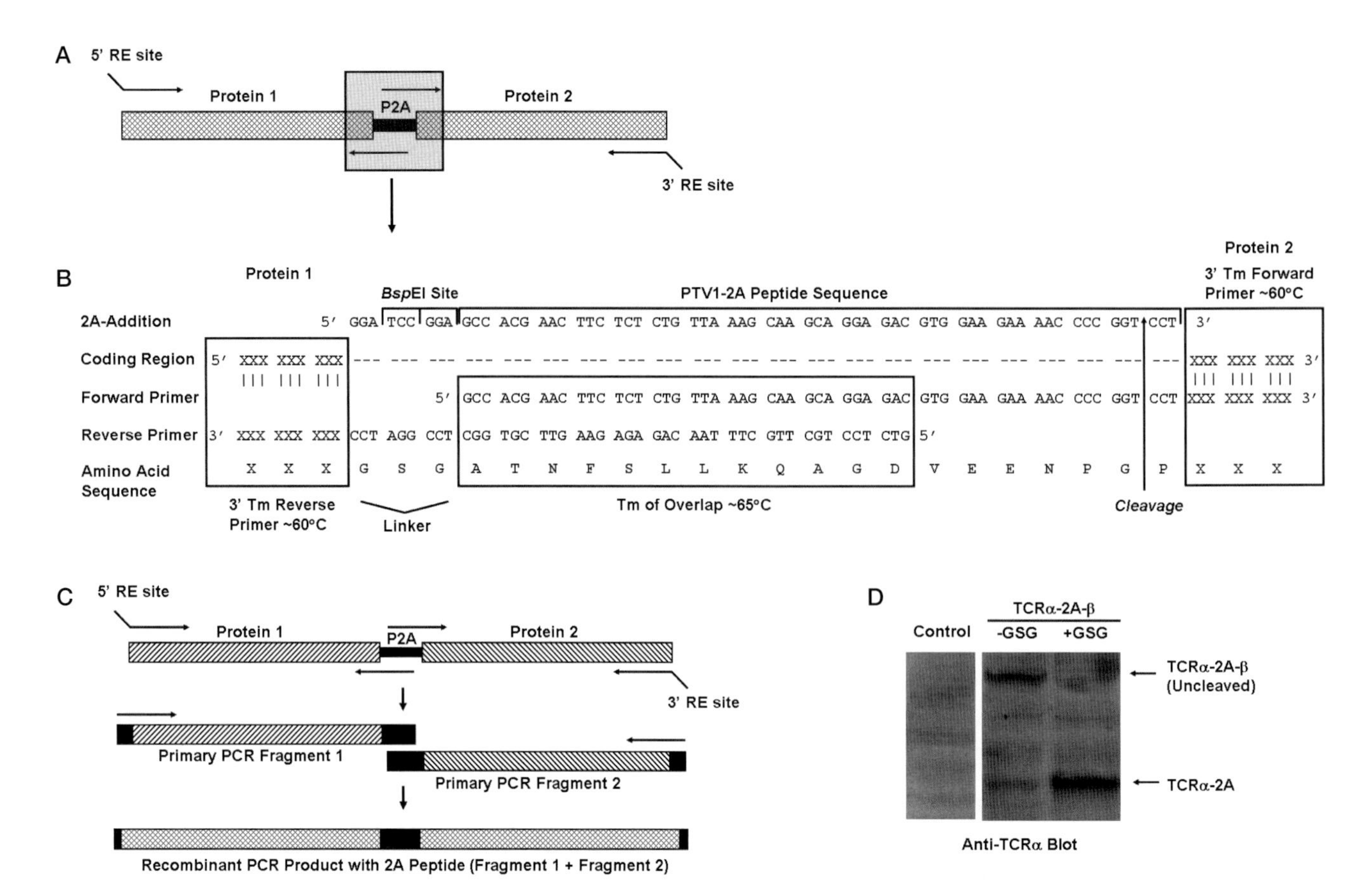

FIGURE 3. Primer design strategy for the addition of 2A peptides to proteins. (*A*) Schematic of two proteins linked together via the 2A peptide from porcine teschovirus-1 (PVT1; P2A). The 5′ and 3′ primers used to clone into the vector are shown with restriction enzyme sites (RE) noted. A BspEI site is incorporated within the GSG linker between protein 1 and the 2A sequence. The shaded area is shown in detail in *B*. (*B*) A schematic representation of the primers used to link two cistrons together via a 2A peptide. The 3′ regions for both the forward and reverse primers are boxed. The base pairs and amino acids for the hypothetical proteins are designated with an "X." The number of base pairs of the target cistrons that should be added is defined by the number required to reach a 3′ T_m of 60–65°C. The T_m of the overlap between the two primary fragments (*boxed*) is ideally around 65°C. A BspEI site that also incorporates the GSG linker is shown between Protein 1 and P2A. (*C*) Cloning strategy is shown for fragments generated during primary and recombinant PCR steps using the primers from *B*. (*D*) 293T cells were transfected with 2A-peptide-linked TCRα and TCRβ chains with or without a GSG linker along with the CD3 chains or an empty vector control. Cells were lysed, and proteins were resolved by SDS-PAGE and probed with anti-TCRα antibodies. Cleavage via the 2A peptide is enhanced in cells with the TCRαβ-GSG proteins as shown by the lack of uncleaved product compared to cells receiving TCRβ without the GSG linker (A. Burton and D. Vignali, unpubl.).

Expression in eukaryotic organisms

The 2A peptide system has thus far worked successfully in all eukaryotic systems tested, from mammalian cells, yeast, and plants to in vitro transcription/translation experiments using wheat-germ lysate. The only requirement for 2A-peptide-based cleavage appears to be translation by 80S ribosomes; thus, it is likely that the system will work in any eukaryotic system. Cleavage does not occur following translation with 70S ribosomes and thus does not work in prokaryotes.

The 2A peptide sequence and gene order

The 2A sequence from FMDV has been the most widely studied in the literature; additional sequences, however, have been used successfully, including those from the porcine teschovirus-1 (PTV1), equine rhinitis A virus (ERAV), and *Thosea asigna* virus (TAV) (Szymczak and Vignali

2005). Although all of these sequences work very well, our order of preference is P2A, T2A, F2A, E2A (see Fig. 1). Additional 2A peptides can be found at http://www.st-andrews.ac.uk/ryanlab/2A_2Alike.pdf. To minimize the risk of homologous recombination, it is important to use different 2A peptide sequences if more than two genes are being linked in a retroviral vector. In addition, codon-diversified 2A sequences can be constructed in which changes are made in the base pair composition of the sequence while leaving the amino acid sequence unaltered. The sequence encoding the cleavage consensus motif at the carboxyl terminus of the 2A peptide must remain the same (Fig. 1). Finally, the order in which the genes are expressed within the vector must be considered, keeping in mind potential problems with slipstreaming as well as issues associated with the presence of the residual 2A tag on all but the last protein.

Incorporation of a GSG linker

The cleavage efficiency of the 2A peptide motif is affected by the preceding peptide sequence and the upstream protein (Donnelly et al. 2001a; Szymczak et al. 2004a). Our own studies have clearly shown that inclusion of a Gly-Ser-Gly (GSG) linker between the upstream cistron and the 2A peptide nullified this variable (Fig. 3). We have obtained essentially 100% cleavage in all vector/protein combinations tested in which this linker was included. Presumably, this linker affords more flexibility and may enhance cleavage by relaxing unfavorable conformations.

Primer design

It is important to note that the final cassette will constitute a single ORF and thus must have only one start codon with an effective Kozak consensus sequence (Kozak 1987) and one stop codon. Our 5′ primers consist of the following: 6-bp GC-rich clamp (to prevent "breathing" and facilitate restriction enzyme [RE] cleavage)—RE site—Kozak consensus sequence with ATG (i.e., GxxAxxATGG)—annealing sequence (e.g., CGC TCT GAA TTC GCC AGC ATG . . . . [EcoRI site underlined]). Our 3′ primers consist of the following: annealing sequence—RE site—6-bp GC-rich clamp (e.g., ... TGA CTC GAG TGT CGG [XhoI site underlined]). The appropriate restriction sites for inserting the 2A-cistron cassette into the multicloning site (MCS) of the vector could be encoded within the 5′ and 3′ polymerase chain reaction (PCR) primers to facilitate cloning as shown. The starting methionine codon of all proteins can be retained if so desired, but all stop codons except for the last should be removed. The 2A-cistron cassette can also be engineered with unique restriction sites preceding (ideally) the 2A peptide sequences (Holst et al. 2006). This, in essence, creates a simple way to "shuttle" in mutants.

Optimal melting temperatures for the annealing sequences

Our own experience with recombinant PCR has shown that fewer mutations are introduced if the annealing temperature of the reaction is high. Consequently, the melting temperature (T_m) of the primers must be set appropriately. We have empirically determined that the 3′ T_m of the sequence that is homologous to the target cistron should be around 60–65°C (Fig. 3). Thus, the length of the annealing sequence is defined solely by the number of residues required to achieve this target T_m. Furthermore, we ensure (where possible) that the 3′ base is a C or G to reduce primer-template "breathing." We have also had greater success and a reduced number of mutations if we use staggered primers encoding the 2A sequence, rather than forward and reverse primers that are entirely overlapping (Fig. 3). In this instance, we adjust the length of the overlap to equal a T_m of approximately 65°C. This strategy results in primers that are about 80 bp long, but we frequently perform PCR amplifications using primers in excess of 120 bp without complication.

Protocol 1

Generation of 2A-linked Multicistronic Cassettes by Recombinant PCR

This protocol describes the use of recombinant PCR to connect multiple 2A-linked protein sequences. The final construct is subcloned into an expression vector.

MATERIALS

CAUTION: See Appendix for appropriate handling of materials marked with <!>.

Reagents

Agarose gel suitable for purifying DNA

Melt high-purity agarose in the appropriate volume of 1x TAE by boiling for 2–3 minutes. Place in a 55°C water bath until the temperature has equilibrated. Add 3–5 µl of ethidium bromide just before casting the gel. After the gel has solidified, fill the apparatus with 1x TAE and remove the combs. A 1% agarose gel is generally used for fragments of more than 500 bp. For smaller products, the amount of agarose can be increased up to 2%.

Cloning vector

There are many commercially available cloning vectors that encode various selection cassettes, antibiotic resistance genes, and fluorescent proteins. It is important to note that the activity of 2A-peptide-mediated "cleavage" has not been tested in all vector systems and should be considered when designing the cloning strategy.

DNA polymerase and polymerase buffer

Product length, GC content of the template DNA, and the melting temperature (T_m) of the primer influences enzyme selection. We typically use the Advantage-HF 2 PCR Kit (Clontech), which has consistently yielded products of up to 3 kb with low error frequency. Other high-fidelity enzymes (Advantage-GC, Clontech; Phusion High-Fidelity DNA Polymerase, Finnzymes; Expand Long Template PCR System, Roche Applied Science) have been successfully used in our lab, depending on the template used and length of fragment desired, and may be tested according to manufacturer's instructions.

DNA mass ladder

dNTPs

Prepare dNTPs (commercially available) as a 10 mM stock in sterile, PCR-grade deionized H_2O. Store small aliquots at –80°C for long-term use (>2 months) and –20°C for short-term use (<2 months).

Ethidium bromide (10 mg/ml) <!>

Add 1 g of ethidium bromide to 100 ml of deionized H_2O and stir for several hours until the dye has dissolved. Store in a dark bottle at room temperature.

10x Gel-loading buffer for agarose gels (e.g., 0.25% bromophenol blue, 0.25% xylene cyanol FF, 50% glycerol in H_2O)

H_2O, PCR grade

Oligonucleotide PCR primer preparations

Prepare working concentrations of 20 µM for each primer with sterile, PCR-grade H_2O. Store aliquots at –20°C.

Template DNA

Plasmid DNA or cDNA; use approximately 100 ng per PCR.

50X TAE (Tris/Acetate/EDTA) buffer

Dissolve 242 g of Tris base <!> with 57.1 ml of glacial acetic acid <!>, 100 ml of 0.5 M EDTA (186.1 g disodium EDTA·$2H_2O$ per liter of H_2O at pH. 8.0), and deionized H_2O to a volume of 1 liter. Dilute with deionized H_2O and use at 1X concentration.

Equipment

Horizontal gel electrophoresis apparatus for nucleic acid separation, including a power source and the appropriately sized combs for separation

Test digests can be run using combs with 3- to 5-mm-wide teeth. A larger reaction volume is used for gel extraction; therefore, a comb with teeth at least 7- to 10-mm wide is recommended.

PCR clean-up or gel extraction kit (available commercially)

Thermocycler

Thin-walled PCR tubes

UV light source

METHOD

1. To generate primary PCR products, prepare the following reaction.

template DNA (100 ng)	1.0 μl
polymerase buffer	5.0 μl
forward (5′) primer (20 μM stock)	1.25 μl
reverse (3′) primer (20 μM stock)	1.25 μl
dNTPs (10 mM stock)	2.0 μl
DNA polymerase (Advantage-HF 2)	1.0 μl

Add PCR-grade deionized H_2O to a final volume of 50 μl.

The Advantage-HF 2 system provides two different buffers: an Advantage-2 buffer and a high-fidelity (HF) buffer. The latter is generally used for most reactions, but longer fragments or more difficult templates may require manipulation of buffer volumes as described by the manufacturer or the addition of 5% DMSO <!> and/or 1 M betaine (Rees et al. 1993; Varadaraj and Skinner 1994; Baskaran et al. 1996; Henke et al. 1997).

2. Perform the PCR under the following conditions.

Number of Cycles	Denaturation	Annealing	Extension
1	94°C for 1 min		
2	94°C for 10–30 sec	58°C for 1 min	72°C for 2–4 min
33	94°C for 10–30 sec	68°C for 30 sec	72°C for 2–4 min
1			72°C for 5–10 min
Hold at 4–8°C			

Use a two-stage reaction protocol. In the first stage, only the 3′ sequence homologous to the target cistron will anneal. After two rounds of this reaction, the entire primer will be incorporated into the template and the effective T_m of the primer will increase significantly. The second stage, in which the annealing temperature is increased to 68°C, reduces the possibility of errors in the annealing sequence. It might also be possible to run a two-step reaction for the second stage. The annealing temperature for the first stage depends on the T_m of the PCR primers. We design primers with a 3′ T_m of 60–65°C. Generally, the initial annealing temperature is approximately 3°C lower than the 3′ T_m of the primer that will anneal to the denatured template during the first two cycles of the PCR.

It is best to use the shortest possible denaturation time during the second stage, because exposure of single-stranded DNA to high temperatures may cause some depurination. A loss of enzyme activity may also be seen. The extension time depends on the length of the PCR product. For the reaction conditions and reagents outlined above, use 1 minute per kilobase of PCR product. The optimum number of cycles depends on the starting concentration of DNA. To reduce nonspecific background products, use the minimal number of cycles necessary to generate the product. If the desired products are not generated, it may be necessary to alter the annealing temperatures. The addition of DMSO or betaine may also enhance the yield and specificity of the PCR, particularly for templates with a high GC content. Alternative reagents and parameters may yield similar success, but may require optimization.

3. Verify primary PCR product size and purity by loading 5 μl of the reaction on an agarose gel.

   Mix 5 μl of the reaction with the amount of 10x loading buffer and H_2O appropriate for the well volume. Include a DNA ladder for size verification. Run the gel (typically at 80–90 volts) until the loading dye has migrated a sufficient distance to allow separation of the DNA fragments. Place the gel on a UV light source to visualize the products.

4. Purify the primary PCR products to remove any remaining primers. Use a PCR clean-up kit or perform gel extraction.

   Use a gel extraction approach if background bands are generated in the recombinant PCR. Many PCR clean-up kits do not effectively remove long (>70 bp) PCR primers and/or result in the loss of small fragments (<500 bp).

5. Quantitate the primary PCR products and verify purity by running 2–5 μl of the reaction on an agarose gel with DNA mass markers.

   For the recombinant PCR, it is very important to use equimolar amounts of the primary PCR products as templates, i.e., the absolute number of fragments should be equal for each PCR product. Do not forget to compensate for differences in product length when determining how much of each product is needed. Use 100 ng of the largest primary PCR product and use the following equation to calculate the necessary amount of the smaller primary product(s): (100)(bp size of the smaller primary product)/(bp size of largest primary PCR product) = (ng of smaller primary product per 100 ng of the largest).

6. Assemble the recombinant PCR using the conditions described in Step 1 with the primary PCR products as the templates and the external primers, i.e., those containing the start and stop codons, rather than those containing the 2A sequences (see Fig. 3).

   It is not usually necessary to change the annealing temperature for the recombinant PCR. The extension time should be increased, however, as the PCR product will be larger.

   Recombinant PCR can be performed with more than two primary PCR products, providing the overlap between each fragment is sufficiently different and there are equimolar amounts of each fragment. If generating a recombinant product proves to be difficult, perform the recombinant PCR in several stages with two or three primary templates until the full product is generated. Be aware, however, that additional rounds of PCR may increase the possibility of introducing errors.

7. Verify PCR product size as in Step 3 and purify as in Step 4.

8. Subclone the final PCR product into the expression vector using the restriction sites that are encoded in the external primers.

9. Sequence the final construct using primers specific to the gene of interest or to the expression vector.

   It is important to verify the correct sequence of the 2A peptide, as any alteration in amino acid composition can affect the cleavage efficiency.

Protocol 2

Verification of 2A Peptide Cleavage

The easiest and most effective way to assess 2A cleavage is to perform transient transfection of 293T cells (human embryonic kidney cells) followed by western blot analysis, as described below. 293T cells are easy to grow and can be efficiently transfected with a variety of vectors. Cleavage can be assessed by detection with antibodies against the target proteins or anti-2A serum (not yet commercially available). Alternate methods may be necessary depending on the proteins or vector system under examination. We have also effectively used in vitro transcription/translation using cell-free systems to assess 2A cleavage (Szymczak et al. 2004b).

MATERIALS

Reagents

Antibodies directed against the protein of interest or 2A

Dulbecco's modified Eagle's medium (DMEM)

> Complete and serum-free media are used. Complete medium is supplemented with 10% fetal bovine serum, 2 mM L-glutamine, 1 mM sodium pyruvate, 100 μM MEM nonessential amino acids, 5 mM HEPES, 5.5×10^{-5} units of β-mercaptoethanol <!>, 100 units/ml penicillin, and 100 μg/ml streptomycin <!>. In addition, 20 μg/ml ciprofloxacin can be used to prevent mycoplasma contamination. All media and supplements are commercially available.

Lysis buffer (e.g., Nonidet-P40 or RIPA lysis buffers; see Harlow and Lane 1999)

Phosphate-buffered saline (PBS)

Plasmid DNA of interest

Transfection reagent

> Many transfection reagents are available, but we have found the highest transfection efficiency and cell viability with 293T cells using *Trans*IT-LT1, *Trans*IT-293T (Mirus, Madison, Wisconsin) or FuGENE 6 Transfection Reagent (Roche). If these reagents are not available, transfection using calcium phosphate also works well.

Trypsin/versene <!>

> 293T cells are adherent; treating them with trypsin/versene will facilitate removal from the plate and reduce cell clumping.

293T cells in exponential growth

Equipment

SDS-PAGE and western blot equipment

Standard tissue culture equipment, including 100-mm dishes

> This protocol is designed for 293T cells grown in 100-mm tissue-culture dishes. If tissue-culture wells or dishes of a different size are used, the cell density and reagent volumes used must be scaled accordingly.

METHOD

1. Harvest the 293T cells by trypsinization 24 hours before transfection.
   a. Aspirate the medium (DMEM) from the adherent cells, wash with PBS to remove all traces of medium, and add 3–4 ml of trypsin/versene.
   b. Incubate for 2–3 minutes at room temperature or until cells begin to slough off the bottom of the plate. Gently resuspend the cells and transfer to a conical tube. Wash the plate with 10 ml of complete medium; transfer the wash medium to the tube to neutralize the trypsin.
   c. Centrifuge at 1000 rpm in a bench-top centrifuge for 10 minutes. Remove the supernatant, resuspend the pellet in 5 ml of medium, and count the cells. Plate the cells in 100-mm tissue-culture dishes at a density of 2×10^6 cells/dish in 10 ml of medium. Allow the cells to adhere overnight at 37°C.
2. Transfect the cells with the reagents of choice according to the manufacturer's instructions. Use as much as 10 μg of vector per plate and incubate the cells for ≥48 hours at 37°C.
3. Harvest the cells by trypsinization as in Step 1.
4. To assess cleavage, lyse the cells, separate the proteins by SDS-PAGE, and perform a western blot. Block and probe the blot using standard procedures appropriate for the target proteins. If antibodies to the proteins under study are available and work in a western blot, they can be used to test cleavage efficiency. Antibodies to the 2A peptide sequences are not yet commercially available but they can be generated. Use them to identify the protein of interest and distinguish cleaved material from uncleaved material based on protein size.

   Expression of proteins generated from cells transfected with the multicistronic vector can be compared to that of cells transfected with those proteins in separate plasmids. Remember, a slight shift in protein size will be caused by the presence of the 2A tag.

REFERENCES

Baskaran N., Kandpal R.P., Bhargava A.K., Glynn M.W., Bale A., and Weissman S.M. 1996. Uniform amplification of a mixture of deoxyribonucleic acids with varying GC content. *Genome Res.* **6:** 633–638.

de Felipe F.P. 2002. Polycistronic viral vectors. *Curr. Gene Ther.* **2:** 355–378.

———. 2004. Skipping the co-expression problem: The new 2A "CHYSEL" technology. *Genet. Vaccines Ther.* **2:** 13.

de Felipe F.P. and Ryan M.D. 2004. Targeting of proteins derived from self-processing polyproteins containing multiple signal sequences. *Traffic* **5:** 616–626.

de Felipe P., Luke G.A., Hughes L.E., Gani D., Halpin C., and Ryan M.D. 2006. *E unum pluribus*: Multiple proteins from a self-processing polyprotein. *Trends Biotechnol.* **24:** 68–75.

Donnelly M.L., Hughes L.E., Luke G., Mendoza H., ten Dam E., Gani D., and Ryan M.D. 2001a. The "cleavage" activities of foot-and-mouth disease virus 2A site- directed mutants and naturally occurring "2A-like" sequences. *J. Gen. Virol.* **82:** 1027–1041.

Donnelly M.L., Luke G., Mehrotra A., Li X., Hughes L.E., Gani D., and Ryan M.D. 2001b. Analysis of the aphthovirus 2A/2B polyprotein "cleavage" mechanism indicates not a proteolytic reaction, but a novel translational effect: A putative ribosomal "skip". *J. Gen. Virol.* **82:** 1013–1025.

Fang J., Qian J.J., Yi S., Harding T.C., Tu G.H., VanRoey M., and Jooss K. 2005. Stable antibody expression at therapeutic levels using the 2A peptide. *Nat. Biotechnol.* **23:** 584–590.

Harlow E. and Lane D. 1999. *Using antibodies: A laboratory manual.* Cold Spring Harbor Laboratory Press, Cold Spring Harbor, New York.

Henke W., Herdel K., Jung K., Schnorr D., and Loening S.A. 1997. Betaine improves the PCR amplification of GC-rich DNA sequences. *Nucleic Acids Res.* **25:** 3957–3958.

Holst J., Vignali K.M., Burton A.R., and Vignali D.A. 2006. Rapid analysis of T-cell selection in vivo using T-cell receptor retrogenic mice. *Nat. Methods* **3:** 191–197.

Kozak M. 1987. At least six nucleotides preceding the AUG initiator codon enhance translation in mammalian cells. *J. Mol. Biol.* **196:** 947–950.

Lorens J.B., Pearsall D.M., Swift S.E., Peelle B., Armstrong R., Demo S.D., Ferrick D.A., Hitoshi Y., Payan D.G., and Anderson D. 2004. Stable, stoichiometric delivery of diverse protein functions. *J. Biochem. Biophys. Methods* **58:** 101–110.

Rees W.A., Yager T.D., Korte J., and von Hippel P.H. 1993. Betaine can eliminate the base pair composition dependence of DNA melting. *Biochemistry* **32:** 137–144.

Ryan M.D. and Drew J. 1994. Foot-and-mouth disease virus 2A oligopeptide mediated cleavage of an artificial polyprotein. *EMBO J.* **13:** 928–933.

Szymczak A.L. and Vignali D.A. 2005. Development of 2A peptide-based strategies in the design of multicistronic vectors. *Expert Opin. Biol. Ther.* **5:** 627–638.

Szymczak A.L., Workman C.J., Wang Y., Vignali K.M., Dilioglou S., Vanin E.F., and Vignali D.A. 2004a. Addendum: Correction of multi-gene deficiency in vivo using a single "self-cleaving" 2A peptide-based retroviral vector. *Nat. Biotechnol.* **22:** 1590.

———. 2004b. Correction of multi-gene deficiency in vivo using a single "self-cleaving" 2A peptide-based retroviral vector. *Nat. Biotechnol.* **22:** 589–594.

Varadaraj K. and Skinner D.M. 1994. Denaturants or cosolvents improve the specificity of PCR amplification of a G + C-rich DNA using genetically engineered DNA polymerases. *Gene* **140:** 1–5.

Varshavsky A. 1992. The N-end rule. *Cell* **69:** 725–735.

WWW RESOURCES

http://www.st-andrews.ac.uk/ryanlab/Index.htm Picornavirus Group Homepage, Ryan Lab, University of St. Andrews.

http://www.st-andrews.ac.uk/ryanlab/2A_2Alike.pdf Peptide Sequences, Ryan Lab, University of St. Andrews.

15 Construction of First-generation Adenoviral Vectors

P. Joel Ross[*†] and Robin J. Parks[*†‡]

*Molecular Medicine Program, Ottawa Health Research Institute, Ottawa, Ontario, Canada; †Department of Biochemistry, Microbiology, and Immunology, Centre for Neuromuscular Disease, and ‡Department of Medicine, University of Ottawa, Ottawa, Ontario, Canada

ABSTRACT

Genetically modified adenoviruses (Ads) make attractive vectors for the delivery of exogenous DNA to mammalian cells for basic science and gene therapy applications. Ad vector production consists of (1) cloning a trangene into an infectious plasmid by in vivo recombination in bacteria, (2) rescuing and propagating the vector in complementing cells, and (3) purifying the vector. All of this can be accomplished using commercially available reagents, plasmids, and cell lines. First-generation Ads have a large cloning capacity (5–14 kbp) and efficiently transduce a wide range of both quiescent and proliferating cell types. They are readily propagated to produce high-titer stocks (10^{11} to 10^{13} vector particles from a 3-liter culture). Furthermore, Ads rarely integrate into the host genome and are relatively safe. However, Ad vector production typically takes 4–6 weeks, and promiscuous host-cell transduction can occur in vivo. Furthermore, immune responses against viral proteins encoded by the vector backbone can occur, which limits the duration of transgene expression in vivo. Regardless of these limitations, Ad remains one of the more versatile and efficient gene delivery systems. Here, we discuss the current methods for the generation, propagation, purification, and characterization of first-generation Ad vectors.

INTRODUCTION

Genetically modified Ads make attractive vectors for the delivery of exogenous DNA to mammalian cells for basic science and medical applications. Ad vectors are relatively easy for nonspecialists to produce, because current methods for Ad vector generation require only basic laboratory skills in molecular biology and cell culture. These vectors have a relatively large cloning capacity and are easily propagated and purified to high titer. Furthermore, Ad vectors do not integrate, but they are capable of long-term gene expression. Although Ad vectors transduce a wide range of mammalian cell types in a cell-cycle-independent manner, this promiscuity makes delivery to specific tissues in vivo difficult.

Most Ad vectors are based on the well-characterized human Ad serotypes 2 or 5. The 36-kbp linear, double-stranded DNA genome encodes genes that are classified based on the timing of their expression (Fig. 1). Early genes (E1, E2, E3, and E4) are expressed before the onset of DNA replication. Proteins encoded by the early genes function to activate other Ad genes, replicate the viral DNA, interfere with immune recognition of infected cells, and modify the host-cell environment to make it more conducive to viral replication. The late genes (the major late transcription unit, pIX, and IVa2) are expressed after DNA replication and primarily encode proteins involved in capsid production and packaging of the Ad genome. The viral DNA also contains the origins of replication (the inverted terminal repeats [ITR], ~100 bp located at both the left and right end) and the packaging sequence (~150 bp located immediately adjacent to the left ITR).

The most commonly used type of Ad vector is the E1-deleted or first-generation Ad (fgAd) vector, which has a cloning capacity of approximately 5 kbp (Danthinne and Imperiale 2000). Typically, the transgene is inserted in place of the E1 region (Fig. 1). Since E1 is essential for virus replication, these vectors must be propagated in E1-complementing cell lines such as 293 (Graham et al. 1977), 293N3S (Graham 1987), 911 (Fallaux et al. 1996), or PER.C6 (Murakami et al. 2002). The E3 region is unnecessary for replication in vitro, and its removal increases vector cloning capacity to 8 kbp. Currently, the most efficient method for producing fgAd vectors is by construction of "infectious" Ad plasmids in bacteria (Chartier et al. 1996; He et al. 1998). This method uses recombination-proficient (*RecA*$^{+}$) bacteria to transfer a transgene cassette into an Ad genomic plasmid, generating a DNA molecule that is essentially identical to the final virus construct. Transfection of the infectious plasmid into an E1-complementing cell line results in recovery of the desired recombinant Ad vector at a very high frequency (He et al. 1998).

The commercially available AdEasy system is widely used for recombination-mediated construction of Ad vectors. This consists of several variations of two plasmids (Table 1 and Fig. 2) (He

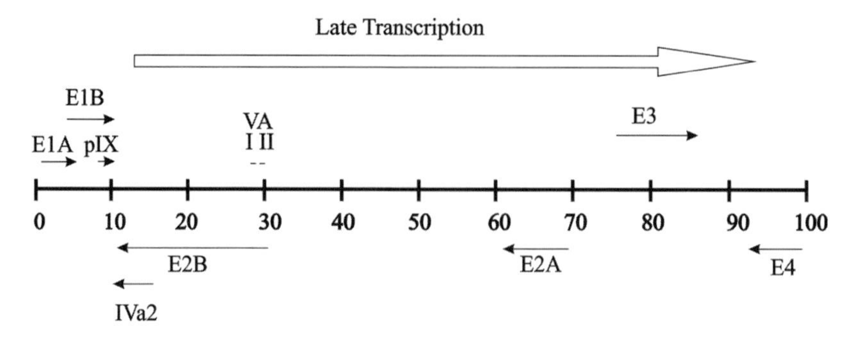

FIGURE 1. Simplified transcription map of the Ad5 genome. See text for details.

TABLE 1. Summary of the Suppliers and Contents of AdEasy Vector Preparation Kits

Supplier	Kit name	Catalog no.	Components
ATCC	AdEasy starter kit	JHU-23	AdEasier-1 cells (stably transformed with pAdEasy-1) pShuttle pShuttle-CMV pAdTrack pAdTrack-CMV
	AdEasy supplement kit	JHU-24	BJ5183 cells pAdEasy-1 pAdEasy-2 pAdEasy1-GFP + β-gal (amplification control) pAdEasy2-GFP + β-gal
Stratagene	AdEasy adenoviral vector kit	240009	pAdEasy-1 pShuttle pShuttle-CMV pShuttle-CMV-*lacZ* BJ5183 electrocompetent cells XL10-Gold ultracompetent cells pUC18 (control DNA)
	AdEasy XL adenoviral vector kit	240010	BJ5183-Ad1 cells (similar to AdEasier-1 cells) pShuttle pShuttle-CMV pShuttle-CMV-*lacZ* (amplification control) XL10-Gold ultracompetent cells pUC18 (control DNA) AD-293 cells (Stratagene 293 cell derivative)
Qbiogene	AdEasy basic kit	AES1001	BJ5183 cells DH5α cells pAdEasy-1 QBI-Infect+ (purified E1/E3-deleted Ad vector encoding *lacZ*) Reagents for calcium phosphate transfection
		AES1000	Basic kit components QBI-293A cells (Qbiogene 293 cell derivative)
		AES1000A	Basic kit components QBI-293A cells pShuttle pShuttle-CMV
		AES1001A	Basic kit components pShuttle pShuttle-CMV

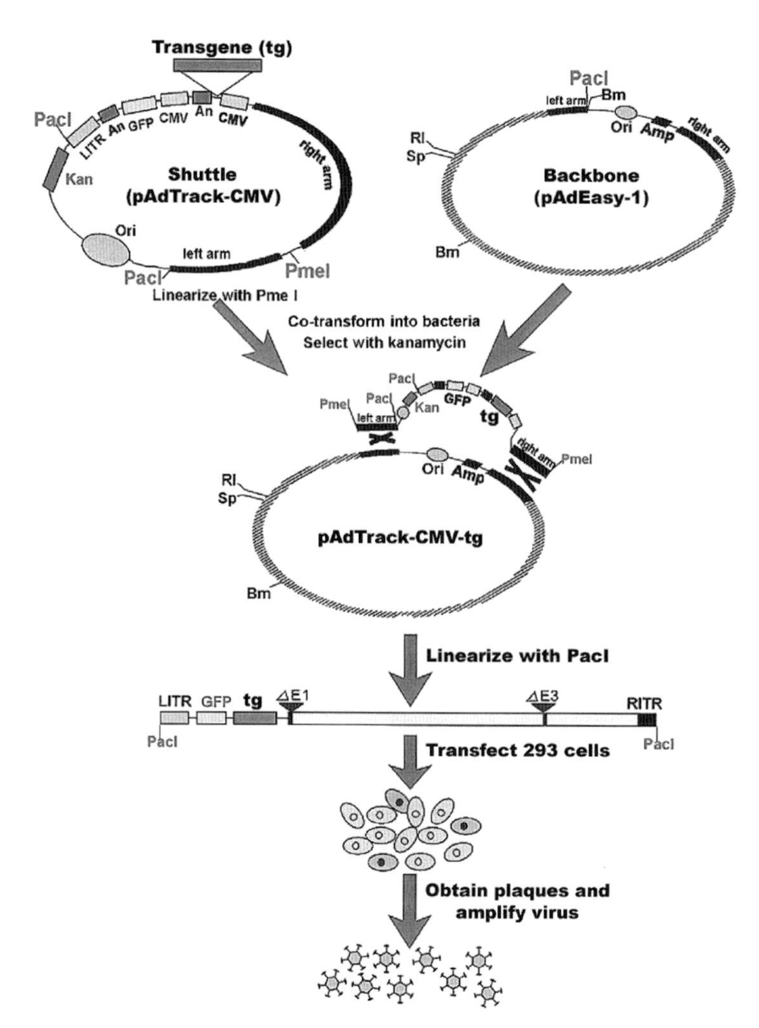

FIGURE 2. Schematic representation of the AdEasy system for Ad vector production. See text for details. (Modified, with permission, from He et al. 1998 [©National Academy of Sciences].)

et al. 1998; He 2001). The shuttle plasmid (Table 2) contains a kanamycin (*kan*) resistance cassette flanked by the left and right ends of an E1-deleted Ad vector genome, a multiple cloning site located in the E1 locus, and several kilobase pairs of Ad5 DNA downstream from the E1 region (called the right arm of homology). The backbone plasmid encodes the majority of the Ad5 genome and an ampicillin resistance cassette. In vivo recombination between homologous sequences contained in both plasmids transfers the *kan* resistance cassette, left ITR and packaging sequence, and the transgene from the shuttle into the backbone plasmid, generating an infectious recombinant Ad vector (Fig. 2). The choices of shuttle and vector depend on the needs of the investigator, with the primary considerations being the promoter to be used, the size of the insert, and whether vector tracking is desired (Table 2). Two variations of pAdEasy (the Ad genomic plasmid) are available: pAdEasy-1 has an intact E4 region, whereas pAdEasy-2 is deleted of E4. Deletion of E4 increases the cloning capacity of resultant vectors, but this requires the use of 911- or 293-based cell lines that express E4 (Amalfitano and Parks 2002).

TABLE 2. Shuttle Vectors Available for Use with the AdEasy System

Shuttle vector	Size (kbp)	Maximum insert size (kbp)[a]	Supplier[b]	GFP-tracer[c]	Notes[d]
pShuttle	6.6	1–7.5 2–10.2	A, S, Q	no	Maximal cloning capacity; entire expression cassette ligated into MCS
pShuttle-CMV	7.5	i–6.6 ii–9.1	A, S, Q	no	Transgene ligated into MCS between CMV promoter and poly(A)
pAdTrack	8.3	1–5.9 2–8.6	A	yes	GFP expression regulated by CMV promoter; entire expression cassette ligated into MCS
pAdTrack-CMV	9.2	1–5.0 2–7.7	A	yes	GFP and transgene each regulated by CMV promoters in adjacent (head-to-tail) expression cassettes; transgene ligated into MCS between CMV promoter and poly(A)
pShuttle-IRES-hrGFP-1	8.9	1–5.3 2–8.0	S	yes	One promoter regulates transgene and GFP expression in a dicistronic expression cassette; transgene cloned in-frame with carboxy-terminal FLAG epitope tag in MCS, between CMV promoter and poly(A).
pShuttle-IRES-hrGFP-2	8.9	1–5.3 2–8.0	S	yes	One promoter regulates transgene and GFP expression in dicistronic expression cassette; transgene cloned in-frame next to carboxy-terminal HA epitope tag in MCS, between CMV promoter and poly(A)

[a]The maximum size insert will differ depending on whether the shuttle vector is recombined with pAdEasy-1 (intact E4; i) or pAdEasy-2 (E4-deleted; ii), as indicated.

[b]Abbreviations: (A) American Type Culture Collection; (S) Stratagene; (Q) Qbiogene.

[c]GFP Tracer: Vector with a green fluorescent protein (GFP) expression cassette can be tracked visually during propagation, obviating the need for plaque purification.

[d]Abbreviations: (MCS) Multiple cloning site; (poly[A]) polyadenylation signal; (CMV) cytomegalovirus immediate-early promoter/enhancer.

Protocol

Construction and Characterization of Ad Vectors

The shuttle, containing the transgene of interest, is produced using standard cloning procedures. Once obtained, the shuttle vector is linearized by digestion with PmeI and cotransformed into BJ5183 cells with the AdEasy backbone vector. After small-scale purification of DNA from BJ5183 cells (which do not maintain high copy numbers of plasmids), the DNA is transformed into a general-purpose cloning strain, such as DH5α. Once the structure of the vector is verified by restriction endonuclease mapping, clones are subjected to large-scale purification in CsCl gradients, and the resulting purified Ad vectors are characterized. For Troubleshooting guide, see Table 3.

MATERIALS

CAUTION: See Appendix for appropriate handling of materials marked with <!>.

Reagents

Agarose (1% w/v)

Dissolve 1 g of agarose (UltraPure; Invitrogen, 15510-027) in 100 ml of sterile distilled H_2O. Sterilize by autoclaving for 20 minutes at 121°C. Store at room temperature and melt in a microwave oven before use.

Bacteria

BJ5183 cells (*RecA*-proficient)

These cells are supplied in a number of commercially available AdEasy kits. For details, see Table 1.

DH5α (*RecA*-deficient; Invitrogen, 18258-012)

Transformation-competent cells are prepared with rubidium chloride using a protocol similar to that of Sambrook et al. (1989) and stored at –80°C in 0.2-ml aliquots.

Buffer-saturated phenol <!>

Cell lines

Low-passage 293 cells (Microbix; Toronto, Canada, PD-02-01)

A549 cells (ATCC; Manassas, Virginia, CCL-185)

Grow 293 and A549 cells in 150-mm dishes in a 37°C, 5% CO_2 incubator and split 1 to 2 or 1 to 3 when they reach about 90% confluence.

293N3S cells (Microbix, PD-02-02)

Maintain 293N3S cells in 150-mm dishes as described above or in suspension. Grow the cells in suspension in maintenance medium in suspension flasks, agitated at 70 rpm. When they reach a density of 5×10^5 cells/ml, dilute 1 to 2 or 1 to 3 with maintenance medium.

Cesium chloride <!> gradient solutions (1.25 and 1.35 g/ml)

Dissolve 54.0 and 70.4 g of solid CsCl (Fisher Scientific, BP1595-1) in 146.0 and 129.6 ml of dialysis buffer, respectively. Sterilize by filtration through a 0.2-μm filter.

Chloroform <!> :isoamyl alcohol <!> (24:1 v/v)

1x Citric saline

135 mM KCl <!>

15 mM sodium citrate <!>

Sterilize by autoclaving for 45 minutes at 121°C.

2x Citric saline

270 mM KCl <!>

30 mM sodium citrate <!>

Sterilize by autoclaving for 45 minutes at 121°C.

Complete medium

Minimum essential medium (MEM; Sigma-Aldrich, M2279-500ml, M2279)

10% fetal bovine serum (FBS; Sigma-Aldrich, F2442-500ml)

2 mM GlutaMAX (Invitrogen, 35050-061)

1x antibiotic-antimycotic (Invitrogen, 15240-062)

Dialysis buffer (10 mM Tris-HCl <!> at pH 8.0)

Sterilize by filtration through a 0.2-μm filter.

DNase I (Sigma-Aldrich, D-5025)

Prepare 10 mg/ml of DNase I in 20 mM Tris-HCl (pH 7.6), 50 mM NaCl, 1 mM dithiothreitol <!>, and 50% (v/v) glycerol.

Ethidium bromide <!>

Isopropanol <!>

Kanamycin <!> (25 μg/ml)

Luria broth (LB) medium

$MgCl_2$ <!> (2 M)

Sterilize by autoclaving for 45 minutes at 121°C.

Maintenance medium

MEM (Sigma-Aldrich)

5% FBS (Sigma-Aldrich)

2 mM GlutaMAX (Invitrogen)

1x antibiotic-antimycotic (Invitrogen)

2x Maintenance medium

2x MEM (Invitrogen, 11935-046)

10% FBS (Sigma-Aldrich)

2x antibiotic/antimycotic (Invitrogen)

Phosphate-buffered saline (PBS)

137 mM NaCl

8.2 mM Na_2HPO_4 <!>

1.5 mM KH_2PO_4 <!>

2.7 mM KCl

Sterilize by autoclaving for 45 minutes at 121°C.

Plasmids: Shuttle and backbone plasmids (see Tables 1 and 2)

Restriction endonucleases

PacI restriction endonuclease (New England BioLabs, R0547L)

PmeI (New England BioLabs, R05605)

Other enzymes may be needed for cloning the gene of interest into the shuttle vector.

RNase A <!> (10 mg/ml; Sigma-Aldrich, R-4875)

NaCl (5 M)

Sterilize by autoclaving for 45 minutes at 121°C.

Sodium deoxycholate <!> (5% w/v)

Sterilize by filtration through a 0.2-μm filter.

SDS-proteinase K solution

10 mM Tris-HCl (pH 7.4)

10 mM EDTA

1% SDS <!> (w/v)
1 mg/ml proteinase K <!>

SDS (0.1% w/v)-TE

Dissolve 0.1 g of SDS in 100 ml of TE.

Sucrose (40% w/v)

Dissolve 40 g of sucrose in 100 ml of PBS. Sterilize by filtration through a 0.2-μm filter.

SuperFect transfection reagent (QIAGEN, 301305)

1x TE

10 mM Tris-HCl (pH 7.5)
1 mM EDTA

Sterilize by autoclaving for 45 minutes at 121°C. For 0.1x TE, dilute the 1x stock solution with H_2O and then sterilize by autoclaving for 45 minutes at 121°C.

1x Trypsin <!> -EDTA

Dilute a 10x solution (Invitrogen, 15400-054) with PBS.

Equipment

Centrifuge (Avanti J-25 I; Beckman) equipped with a JLA 10.500 rotor (Beckman)
Heat sealer (e.g., Beckman Model 7700 Cordless Tube Topper)
Hemacytometer (VWR International, 15170-172)
Incubator (37°C, 5% CO_2)
Laminar flow hood
Magnetic stirrer (five position; Bellco Glass, 7785-D2005)
Microcentrifuge tubes (1.5 ml)
Pasteur pipettes, sterile cotton-plugged
Petri dishes containing 1.5% agar (granulated agar; Fisher Scientific, BP1423-500) in LB medium supplemented with kanamycin (25 μg/ml)
Quick-Seal (16 x 76-mm ultracentrifuge tubes; Beckman, 344322)
Slide-A-Lyzer dialysis cassettes (10,000-kD m.w. cutoff, 0.5–3.0 ml; Pierce, Rockford, Illinois, 66425)
Polypropylene tubes (13 ml capped)
Polystyrene tubes (5 ml)
Spinner flasks with impeller assembly: 250 ml (Bellco Glass, 1965-61002) and 3000 ml (Bellco Glass, 1965-61030)
Syringe (3 cc) equipped with a 22-gauge needle
Ultracentrifuge (e.g., Beckman Optima XL-100K ultracentrifuge)
  70.1 Ti rotor (Beckman)
  SW 41 Ti swinging bucket rotor (Beckman)
Ultra-Clear (14 x 89 mm) ultracentrifuge tubes (Beckman, 344059)

METHODS

Ad manipulations must be performed in a laboratory operating at biosafety level 2. A laminar flow hood and incubator should be dedicated to Ad work. Care should be taken to use only sterile equipment, reagents, and technique. Decontaminate and dispose of liquid and solid waste and disinfect contaminated surfaces.

Generation of Infectious Ad Plasmids

1. Generate a shuttle vector containing the transgene of interest using standard cloning procedures.

2. Digest 2 μg of the shuttle vector with 1 μl of PmeI in a total volume of 20 μl overnight at 37°C.
3. Add 3 μl of PmeI-digested shuttle and 3.3 μl of 0.1 μg/μl supercoiled pAdEasy to a 13-ml capped polypropylene tube. Add 3 μl of PmeI-digested shuttle without pAdEasy to another tube as a negative control. Chill on ice.
4. Thaw two 0.2-ml aliquots of BJ5183 competent cells on ice.
5. Add 0.2 ml of BJ5183 cells to the DNA and incubate for 25 minutes on ice.
6. Incubate the tubes for 90 seconds at 42°C and then for 2 minutes on ice.
7. To each tube, add 1 ml of LB and incubate with shaking at 225 rpm for 25 minutes at 37°C.
8. Transfer the bacteria to 1.5-ml microcentrifuge tubes and centrifuge at 8500*g* for 1 minute at room temperature.
9. Aspirate 1 ml of LB and resuspend the pellet in a total volume of 0.2 ml.
10. Spread 0.1 ml of bacterial suspension on the surface of two petri dishes containing 1.5% solid agar in LB supplemented with 25 μg/ml kanamycin. Use one plate for the control suspension.

    No colonies should form on the negative control plate. However, if "background" colonies do appear, recombinant colonies will be noticeably smaller than those produced by bacteria transformed with undigested shuttle plasmid.
11. Choose four to eight small colonies and purify the plasmid DNA by small-scale alkali lysis. Suspend the DNA in a total volume of 25 μl of 0.1x TE.
12. Repeat Steps 5–10 with 5 μl of DNA and 100 μl of DH5α, rather than BJ5183. Pick two colonies from each plate.
13. Purify the DNA by small-scale alkali lysis.

    Digest the DNA with several different restriction enzymes to verify the structure of the plasmid. Correct clones should by subjected to large-scale purification by CsCl buoyant density centrifugation and then used for generation of a recombinant Ad vector (see Step 15).

Cell Preparation

14. Prepare the cultured cells for transfection.

    To prepare adherent 293 and A549 cells

    a. Remove medium from 150-mm dishes of 293 or A549 cells.
    b. Rinse monolayer twice with 5 ml of 1x citric saline (for 293 cells) or 2 ml of trypsin-EDTA (for A549 cells).
    c. For 293 cells, remove all but 0.5 ml of the citric saline after the second rinse and leave the dishes for 5–10 minutes at 22°C until cells begin to detach. For A549 cells, add 2 ml of trypsin-EDTA and leave the dishes for 5 minutes at 37°C until the cells begin to detach.
    d. Tap the sides of the dishes to detach all cells.
    e. Bring the cells to 12 ml with complete medium and distribute into new dishes.

       Incubate the plates so that the cells are about 90% confluent on the day of transfection.

    To prepare 293N3S cells in suspension

    a. Transfer six confluent 150-mm plates of 293N3S cells to a 3-liter spinner flask. Bring the total volume in the flask to 1 liter with maintenance medium.
    b. Every 1–2 days, remove 2 ml of suspension cells to a 15-ml polystyrene conical tube.

c. Add 2 ml of 2x citric saline and vortex vigorously for 10 seconds.

d. Incubate the cells for 15 minutes at 37°C.

e. Vortex vigorously for 10 seconds.

f. Count cells using a hemacytometer.

If cells are in clusters too large to count, continue to incubate at 37°C and vortex until clusters are broken up.

g. When the cell density is greater than 4×10^5 cells/ml, add 1 liter of maintenance medium until 3 liters of cells with a final density of 3×10^5 to 5×10^5 is obtained.

Transfection for Rescue of Ad Vectors

15. Digest 10 µg of the plasmid encoding the recombinant Ad genome (obtained from Step 13) with 2 µl of PacI in a total volume of 50 µl overnight at 37°C.
16. Mix 40 µl of PacI-digested DNA with 360 µl of MEM and 16 µl of SuperFect reagent in a round-bottom 5-ml polystyrene tube.

    Other transfection reagents may be used in place of the SuperFect reagent.
17. Vortex vigorously for 10 seconds. Allow complexes to form for 15 minutes at room temperature.
18. Add 2.4 ml of MEM to each tube containing the DNA-SuperFect complexes. Mix well by tituration.
19. Rinse four 35-mm plates seeded with 293 cells (as described above) twice with 2 ml of PBS.
20. Transfer 0.7 ml of the plasmid complex suspension to each 35-mm dish of 293 cells.
21. Incubate the cells for 3 hours at 37°C, 5% CO_2.
22. After approximately 2.5 hours, melt 1% (w/v) agarose solution (minimum 1.5 ml per 35-mm dish) in a microwave oven. Equilibrate to 42°C in a water bath.
23. Equilibrate 2x maintenance medium (1.5 ml per 35-mm dish) to 37°C.
24. After 3 hours, remove the transfectant from the cells. Rinse the monolayer with 1 ml of PBS.
25. Mix equal volumes of the agarose solution and 2x maintenance medium. Add 3 ml to each 35-mm dish of 293 cells.

    This must be done quickly to avoid solidification of the agarose, but gently to avoid disturbing the monolayer.
26. Allow the overlay to solidify (~15 minutes at 22°C) and return the cells to the incubator.
27. Incubate until plaques form (~7–12 days).
28. Choose several well-isolated plaques. Use a sterile cotton-plugged pasteur pipette to remove a plug of agarose over each plaque. Place each plug in a vial containing 1 ml of 4% sucrose PBS. Vortex briefly and store plaques at –80°C.

    Initial Ad plaque isolates should be plaque-purified a second time (Steps 22–28) to ensure that the resulting virus is from a single clone.

Analysis and Expansion of Plaque-isolated Ad Vector

29. For each sample of plaque-purified vector obtained above, add 100 µl to two 35-mm plates of 293 cells (~90% confluence). Return cells to the incubator.
30. Allow the virus to adsorb for 1 hour, rocking the dishes every 10–15 minutes.
31. After the adsorption period, add 2 ml of maintenance medium to each plate and return them to the incubator.

32. Examine plates for cytopathic effect (CPE).

    The cells should have a rounded morphology or be detached from the plate. Use one plate for analysis of the recombinant Ad structure. Reserve the contents of the other plate for vector expansion.

    To analyze the plaque-isolated Ad vector

    a. Once complete CPE is achieved, leave plates undisturbed for about 10 minutes in a laminar flow hood so that detached cells will come to rest on the bottom of the plate.
    b. Carefully remove the medium. Resuspend the remaining cells in 0.2 ml of SDS–proteinase K solution. Transfer the suspension to a microcentrifuge tube.
    c. Incubate the lysate overnight at 37°C.
    d. Add 0.3 ml of TE to the lysate. Extract with 0.5 ml of buffer-saturated phenol, followed by 0.5 ml of chloroform:isoamyl alcohol.
    e. Add 0.1 ml of 5 M NaCl and 0.5 ml of isopropanol.
    f. Pellet the DNA by centrifugation at 20,000*g* for 10 minutes at 4°C.
    g. Resuspend the DNA in 20–50 μl of TE and use 5–10 μl for digestion with appropriate restriction enzymes.
    h. Examine the resulting banding pattern by electrophoresis on a 0.8% agarose gel, followed by staining with ethidium bromide.

    To expand plaque-isolated Ad vector

    a. Once complete CPE is achieved (Step 32), scrape the cells from the dish into the medium and remove to a 4-ml cryovial.
    b. Add 40% sucrose PBS to a final concentration of 4%. Vortex briefly. Use immediately or store at –80°C until use.
    c. Infect a 150-mm dish of 293 cells with 1 ml of inoculum. Return cells to the incubator.
    d. Allow the virus to adsorb for 1 hour, rocking the dishes every 10–15 minutes.
    e. After the adsorption period, add 20 ml of maintenance medium to each plate and return them to the incubator.
    f. Examine plates for CPE. Once complete CPE is evident, remove the cells and medium to a 50-ml conical tube. Add 1/10 volume of 40% sucrose PBS to a final concentration of 4% sucrose and store inoculum at –80°C.

Large-scale Preparation of Ad Vectors

33. Thaw the inoculum in a 22°C water bath. If necessary, increase the volume of the inoculum to 30 ml using MEM.

    For large-scale preparation of Ad vectors using adherent 293 cells

    a. Remove medium from 30 150-mm dishes of 293 cells (~90% confluent at the time of infection), ten at a time, and replace with 1 ml of inoculum.
    b. Allow the virus to adsorb for 1 hour in a 37°C, 5% CO_2 incubator, rocking the plates every 10–15 minutes.
    c. Add approximately 20 ml of maintenance medium to each dish and return the cells to the incubator.
    d. Examine cells daily until complete CPE is evident (~2–3 days).

e. Remove medium and cells to two 500-ml polypropylene bottles.

   Most of the cells should be detached in the medium; however, any remaining cells are usually loosely attached and can be removed by tapping the sides of the dish.

f. Use the same pipette to rinse groups of ten dishes twice with 10 ml of PBS.

g. Centrifuge cells at 650*g* for 20 minutes at 4°C. Decant the medium and retain the cell pellets.

h. Resuspend the cells in 3 ml of 4% sucrose PBS and transfer to a 50-ml conical tube.

i. Use the same pipette to rinse the bottles once with 2 ml of 4% sucrose PBS and once with 4–5 ml of 4% sucrose PBS.

   The total volume of the cell pellet should be about 15 ml. The cells can be processed immediately for vector purification or they can be stored at –80°C.

For large-scale preparation of Ad vectors using suspension-adapted 293N3S cells

a. Distribute a 3-liter culture of 293N3S cells into eight 500-ml centrifuge bottles.

b. Centrifuge at 650*g* for 20 minutes at room temperature. Decant the medium into sterile 1-liter bottles.

   Retain 1 liter of spent medium and add it to the 3-liter flask. Return the flask to the incubator.

c. Use the spent medium to resuspend the cell pellets to a final volume of approximately 40 ml. Transfer the suspension to a 250-ml spinner flask. Rinse the bottles twice with 10 ml of spent medium and transfer to the spinner flask.

d. Add the thawed inoculum to the spinner flask.

   The total volume in the flask should be about 100 ml.

e. Transfer the flask to the incubator and agitate the cells at 70 rpm for 2 hours.

f. Transfer the cells to the 3-liter suspension flask containing 1 liter of spent medium. Rinse the 250-ml spinner flask twice with approximately 250 ml of fresh maintenance medium and transfer to the 3-liter spinner flask. Add 500 ml of fresh maintenance medium to a final volume of 2 liters.

g. Remove 2 ml of the cell suspension to a 35-mm plate and place in a 37°C, 5% CO_2 incubator.

   The cells should reattach to the plate.

h. Return the suspension flask to the incubator.

i. When complete CPE is evident on the 35-mm plate (~2–3 days), decant the suspension culture into 500-ml centrifuge bottles.

j. Centrifuge cells at 650*g* for 20 minutes at 4°C. Decant the medium and retain the cell pellets.

k. Resuspend the cells in 3 ml of 4% sucrose PBS and transfer to a 50-ml conical tube.

l. Use the same pipette to rinse the bottles once with 2 ml of 4% sucrose PBS and once with 4–5 ml of 4% sucrose PBS.

   The total volume of the cell pellet should be about 15 ml. The cells can be processed immediately for vector purification or they can be stored at –80°C.

Vector Purification

34. Thaw pellets obtained by either method for large-scale preparation of Ad vectors in a 37°C water bath.

    All volumes stated below are for a cell pellet with a total volume of 15 ml. Scale volumes accordingly.

35. Add 1.5 ml of 5% deoxycholate to the pellet. Incubate with frequent inversion for 30 minutes at 22°C.

    The lysate should have a thick, highly viscous consistency.

36. Add 0.3 ml of 2 M $MgCl_2$, 0.15 ml of 10 mg/ml RNase A, and 0.15 ml of 10 mg/ml DNase I. Incubate with occasional inversion for 30–60 minutes at 37°C.

37. Once the viscosity of the lysate is near that of water, centrifuge at 1000*g* for 10 minutes at 22°C.

38. Prepare CsCl step gradients in Ultra-Clear ultracentrifuge tubes (two tubes per virus).

    a. Add 2 ml of 1.35 g/ml CsCl to each tube.

    b. Carefully (i.e., with a steady stream, at a rate of ~30 sec/ml) overlay with 3 ml of 1.25 g/ml CsCl.

    c. Carefully add equal volumes of cleared lysate (~6.5–7 ml) to each tube.

    d. Balance the tubes and transfer to the buckets of a SW 41 rotor.

39. Use slow acceleration and deceleration profiles (500 rpm over ~5 minutes) to centrifuge the samples at 35,000 rpm for 1 hour at 10°C.

    The viral band is the lowest band visible on the gradient and will be found at the interface between the 1.25- and 1.35-g/ml layers of the gradient.

40. Use a 3-cc syringe and a 22-gauge needle to pierce the tube approximately 1 cm below the virus. Turn the bevel so that it is parallel to the band and slowly remove, lowering the needle as the band lowers.

    Virus from both step gradient tubes can be combined in a single Quick-Seal ultracentrifuge tube.

41. Fill the Quick-Seal tube containing the virus to the base of the neck with 1.35 g/ml CsCl. Use a heat sealer to seal the Quick-Seal tubes.

    A balance tube can be prepared by filling another Quick-Seal tube to the base of the neck with 1.35 g/ml CsCl.

42. Centrifuge at 35,000 rpm with maximal acceleration and deceleration in a 70.1 Ti rotor overnight at 10°C.

43. Pierce the top of the sealed 70.1 Ti tube to form an air inlet. Use a 3-cc syringe and a 22-gauge needle to pierce the tube about 1 cm below the virus. Turn the bevel so that it is parallel to the band and slowly remove, lowering the needle as the band lowers.

    Take care to minimize the volume extracted.

44. Inject the Ad into a prepared dialysis cassette. Remove the air bubble with the syringe.

45. Dialyze the Ad vector for 24 hours at 4°C against two 500-ml volumes of dialysis buffer.

46. Remove the vector from the dialysis cassette. Retain the syringe and 0.9 ml of dialysis buffer.

47. Add 40% sucrose PBS to the vector and the dialysis buffer to a final concentration of 4%. Store purified vector in small aliquots (~100–200 μl) at –80°C and the buffer at –20°C.

    Ad vector stocks are stable for years at –80°C.

Characterization of Purified Ad Vectors

48. Further characterize the Ad vectors by assessing the genetic structure, determining titer, and examining for contamination with replication-competent adenovirus (RCA).

To confirm the genomic structure of the purified Ad vector

a. Rinse the syringe used to remove the vector from the dialysis cassette with 0.2 ml of SDS-proteinase K. Transfer the liquid into a 1.5-ml microcentrifuge tube.

b. Incubate overnight at 37°C.

c. Add 0.3 ml of TE to the lysate. Extract with 0.5 ml of buffer-saturated phenol, followed by 0.5 ml of chloroform:isoamyl alcohol.

d. Add 0.1 ml of 5 M NaCl and 0.5 ml of isopropanol.

e. Pellet the DNA by centrifugation at 20,000*g* for 10 minutes at 4°C.

f. Resuspend the DNA in 20–50 μl of TE and use 5–10 μl for digestion with appropriate restriction enzymes.

g. Examine the resulting banding pattern by electrophoresis on a 0.8% agarose gel, followed by staining with ethidium bromide.

To determine titer in infectious units by plaque-forming unit (pfu) assay

a. Prepare serial dilutions (10^{-4} to 10^{-9}) of Ad vector in MEM.

b. Infect 293 cells (~90% confluent) in the wells of a six-well dish with 0.1-ml aliquots of each dilution. Return cells to the incubator.

c. Allow the virus to adsorb for 1 hour, rocking the dishes every 10–15 minutes.

d. After the adsorption period, overlay with agarose as described previously (Steps 22–26).

e. Incubate for 10–12 days. Count the number of plaques.

   Multiply the number of plaques in a well by the dilution factor to determine the vector titer in pfu/ml.

To determine vector titer in particles/milliliter

a. Dilute 20–50 μl of purified vector to a final volume of 1 ml in 0.1% SDS-TE.

   Use the dialysis buffer (from Step 46) as a blank control.

b. Incubate for 10 minutes at 56°C, vortex briefly, and centrifuge briefly.

c. Determine OD_{260}.

d. Calculate the number of particles/milliliter, based on the extinction coefficient of 1.1×10^{12} for wild-type Ad (Maizel et al. 1968):

$$(OD_{260})(\text{dilution factor})(1.1 \times 10^{12})$$

A typical Ad vector preparation examined as described should have a ratio of particle to pfu of about 10 (Mittereder et al. 1996).

To detect presence of RCA in purified vector preparations

Recombination between vector DNA and Ad5 DNA present in 293 or 911 cells can result in transfer of E1 to the vector (Lochmuller et al. 1994), generating RCA. A549 cells do not express E1 and cannot support efficient replication of E1-deleted vectors. Therefore, after infection with purified Ad, only contaminating RCA will induce CPE in A549 cells.

a. Infect one 60-mm dish of A549 cells (~90% confluence) with 10^6 pfu in 250 µl of MEM. Infect a second 60-mm dish with 10^7 pfu in 250 µl of MEM. Infect a 150-mm dish with 10^8 pfu in 1 ml of MEM. Return cells to the incubator.

b. Allow the virus to adsorb for 1 hour, rocking the dishes every 10–15 minutes.

c. After the adsorption period, add 5 or 20 ml of maintenance medium to each plate and return them to the incubator.

d. Once complete CPE is evident or 7 days pass, harvest the monolayer by scraping the cells into the medium, add 40% sucrose PBS to a final concentration of 4% sucrose, and store at –80°C.

e. Thaw the viruses obtained above and use 1 ml of each culture to infect an individual 150-mm dish of A549 cells. Add 1 ml of MEM to a fourth plate as a negative control.

f. Allow the virus to adsorb for 1 hour, rocking the dishes every 10–15 minutes.

g. After the adsorption period, add 20 ml of maintenance medium to each plate and return them to the incubator.

h. Compare the infected cells with the uninfected control daily for signs of CPE. Change the medium every 5 days, if necessary.

   If CPE is evident (usually apparent by ~14 days postinfection), RCA is present in the purified stock. The relative amount of RCA to pfu can be inferred by comparing CPE on the three infected dishes.

i. Extract DNA from dishes showing signs of CPE. Analyze by restriction enzyme digestion and agarose gel electrophoresis.

   The left end of the RCA genome will have a structure identical to that of wild-type Ad, due to the presence of E1.

ACKNOWLEDGMENTS

The authors thank Robert Lanthier for a critical reading of the manuscript. Research in the Parks laboratory is supported by grants from the Canadian Institutes of Health Research (CIHR) and the Jesse Davidson Foundation for Gene and Cell Therapy, a CIHR/Muscular Dystrophy Canada/Amyotrophic Lateral Sclerosis Society of Canada Partnership Grant, and the Premier's Research Excellence Award. R.J.P. is a CIHR New Investigator. P.J.R. is supported by a Canada Graduate Scholarship from the National Science and Engineering Research Council.

TABLE 3. Troubleshooting

Problem	Possible cause(s)	Suggestion(s)
Few or no colonies after cotransformation of BJ5183 cells	Cotransformation conditions suboptimal	Try another, more efficient, transformation method, such as electroporation
		Try using AdEasier-1 cells
	Incorrect antibiotic, or too much kan used	Plate the cells on 1.5% agar LB plates supplemented with 25 μg/ml kan
	Wrong strain of competent cells used	Ensure that BJ5183 cells are used
	Competence of BJ5183 cells too low	Check the competence of the cells and generate new competent cells if necessary
		Use a different transformation method (e.g., electroporation)
		Obtain competent cells from a commercial source
Too many colonies after cotransformation of BJ5183 cells	Incomplete digestion of shuttle by PmeI	Use less DNA or more PmeI (also, ensure that PmeI is active)
		Check the digestion efficiency by agarose gel electrophoresis
Failure to generate plaques on 293 cells after initial transfection	Incomplete digestion with PacI	Ensure that PacI is active
		Examine digestion efficiency by agarose gel electrophoresis
	Transfection efficiency too low	Optimize the transfection protocol by trying different amounts of DNA and transfection reagent
		Try another transfection reagent or method
	DNA preparation not appropriate	Prepare DNA by CsCl gradient centrifugation
		Verify DNA concentration
	293 cell passages too high	Thaw a new aliquot of 293 cells
	Defect in Ad vector backbone	Analyze the plasmid structure by digestion with several restriction enzymes; if a defect is detected, generate a new clone
	Insert size exceeds upper limit of Ad packaging	Consult Table 2
	Transgene product cytotoxic	Use a weaker or inducible promoter
No virus band visible on CsCl gradients	Density of CsCl solutions incorrect	Verify densities by weighing 1 ml of each solution
	1.25 and 1.35 g/ml CsCl solutions mixed in the step gradient	Overlay the 1.35 g/ml CsCl with the 1.25 g/ml CsCl very carefully; ensure that a continuous, slow stream of 1.25 g/ml CsCl is ejected and that the phases do not mix
No transgene expression detected	The transgene or promoter is mutated	Analyze purified capsid DNA by restriction analysis and sequencing
		If an error is detected, screen other plaque isolates for transgene expression
	The transgene not efficiently expressed	Verify that the promoter is active in the cell type being used
		Ensure that a Kozak sequence and polyadenylation sequence have been included in the construct
	Purified virus is RCA	Examine replication on noncomplementing cells, such as A549
		If virus is RCA, or if RCA levels are high, purify the vector again, starting from the plaque purification step

REFERENCES

Amalfitano A. and Parks R.J. 2002. Separating fact from fiction: Assessing the potential of modified adenovirus vectors for use in human gene therapy. *Curr. Gene Ther.* **2:** 111–133.

Chartier C., Degryse E., Gantzer M., Dieterle A., Pavirani A., and Mehtali M. 1996. Efficient generation of recombinant adenovirus vectors by homologous recombination in *Escherichia coli*. *J. Virol.* **70:** 4805–4810.

Danthinne X. and Imperiale M.J. 2000. Production of first generation adenovirus vectors: A review. *Gene Ther.* **7:** 1707–1714.

Fallaux F.J., Kranenburg O., Cramer S.J., Houweling A., van Ormondt H., Hoeben R.C., and van der Eb A.J. 1996. Characterization of 911: A new helper cell line for the titration and propagation of early region 1-deleted adenoviral vectors. *Hum. Gene Ther.* **7:** 215–222.

Graham F.L. 1987. Growth of 293 cells in suspension culture. *J. Gen. Virol.* **68:** 937–940.

Graham F.L., Smiley J., Russell W.C., and Nairn R. 1977. Characteristics of a human cell line transformed by DNA from human adenovirus type 5. *J. Gen. Virol.* **36:** 59–74.

He T.-C. 2001. Adenoviral vectors. In *Current protocols in human genetics* (ed. N.C. Dracopoli et al.), pp. 12.4.1–12.4.21. John Wiley & Sons, New York.

He T.-C., Zhou S., da Costa L.T., Yu J., Kinzler K.W., and Vogelstein B. 1998. A simplified system for generating recombinant adenoviruses. *Proc. Natl. Acad. Sci.* **95:** 2509–2514.

Lochmuller H., Jani A., Huard J., Prescott S., Simoneau M., Massie B., Karpati G., and Acsadi G. 1994. Emergence of early region 1-containing replication-competent adenovirus in stocks of replication-defective adenovirus recombinants (ΔE1 + ΔE3) during multiple passages in 293 cells. *Hum. Gene Ther.* **5:** 1485–1491.

Maizel J.V., Jr., White D.O., and Scharff M.D. 1968. The polypeptides of adenovirus. I. Evidence for multiple protein components in the virion and a comparison of types 2, 7A, and 12. *Virology* **36:** 115–125.

Mittereder N., March K.L., and Trapnell B.C. 1996. Evaluation of the concentration and bioactivity of adenovirus vectors for gene therapy. *J. Virol.* **70:** 7498–7509.

Murakami P., Pungor E., Files J., Do L., van Rijnsoever R., Vogels R., Bout A., and McCaman M. 2002. A single short stretch of homology between adenoviral vector and packaging cell line can give rise to cytopathic effect-inducing, helper-dependent E1-positive particles. *Hum. Gene Ther.* **13:** 909–920.

Sambrook J., Fritsch E.F., and Maniatis T. 1989. *Molecular cloning: A laboratory manual,* 2nd edition. Cold Spring Harbor Laboratory Press, Cold Spring Harbor, New York.

16 Production and Characterization of Helper-dependent Adenoviral Vectors

Donna J. Palmer and Philip Ng

Department of Molecular and Human Genetics, Baylor College of Medicine, Houston, Texas 77030

ABSTRACT

This chapter describes in detail the rescue, amplification, and large-scale production of helper-dependent adenoviral vectors (HDAds) for gene transfer and gene therapy. These techniques use suspension-culture-adapted 116 producer cells, AdNG163 helper virus, and HDAd based on the pΔ28E4 backbone. As of this writing, the improved methods and reagents described herein represent the most efficient available for large-quantity, high-quality HDAd production. Characterization of HDAd with respect to physical titer, helper virus contamination, and genomic structure using standard molecular biology techniques are also described.

INTRODUCTION

HDAds (also referred to as gutless, gutted, mini, fully deleted, high-capacity, Δ, pseudo) are deleted of all virus-coding sequences. HDAds retain the advantages of early-generation Ad vectors including high-efficiency in vivo transduction and high-level transgene expression. However, the absence of viral gene expression in transduced cells permits long-term transgene expression in the absence of chronic toxicity. Moreover, the deletion of the viral sequences permits a cloning capacity of approximately 37 kb. This allows for the delivery of whole genomic loci, multiple transgenes, and large *cis*-acting ele-

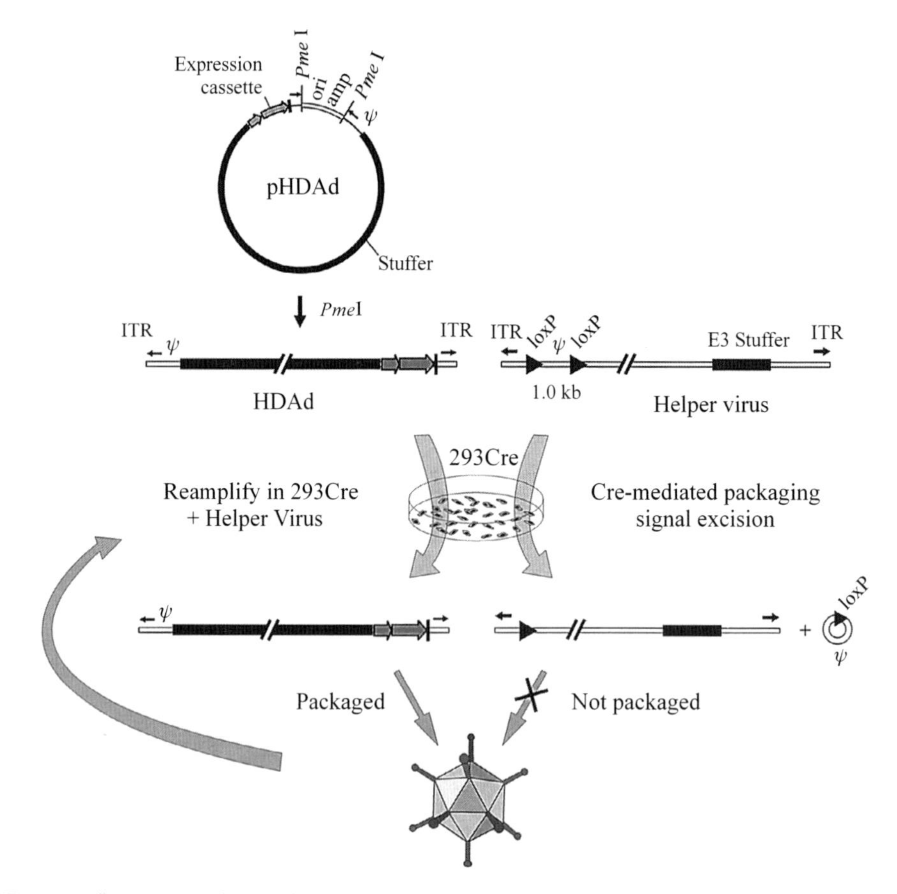

FIGURE 1. Cre/*loxP* system for producing HDAd. The HDAd contains only about 500 bp of *cis*-acting Ad sequences required for DNA replication (ITRs) and packaging (ψ). The remainder of the genome consists of the desired transgene and non-Ad "stuffer" sequences. The HDAd genome is constructed as a bacterial plasmid (pHDAd) and is liberated by restriction enzyme digestion (e.g., PmeI). The liberated genome is transfected into 293 cells expressing Cre and infected with a helper virus bearing a packaging signal (ψ) flanked by *loxP* sites. Cre-mediated excision of ψ renders the helper virus genome unpackagable, but still able to replicate and provide all of the necessary *trans*-acting factors for propagation of the HDAd. The titer of the HDAd is increased by serial coinfections (passages) of 293Cre cells with the HDAd and the helper virus.

ments to enhance, prolong, and regulate transgene expression. In addition, because the vector genome exists episomally in transduced cells, the risks of germ-line transmission and insertional mutagenesis leading to oncogenic transformation are negligible. There are a number of excellent reviews on HDAd with different emphases and perspectives (Kochanek 1999; Parks 2000; Parks and Amalfitano 2002; Ng and Graham 2004; Palmer and Ng 2005).

Because HDAds are devoid of virus-coding sequences, a helper virus is required for their propagation (Parks et al. 1996). The HDAd is first constructed as a bacterial plasmid. The plasmid form of the HDAd (pHDAd) is digested with a restriction endonuclease to release the HDAd genome and transfected into 293 producer cells (Graham et al. 1977) that express the bacteriophage P1 site-specific recombinase Cre. The transfected cells are then infected with a helper virus bearing a packaging signal (ψ) flanked by *loxP* sites. Cre-mediated recombination between the two *loxP* sites results in ψ excision, rendering the helper virus genome unpackagable but still able to replicate and *trans*-complement HDAd replication and encapsidation (Fig. 1). The HDAd is propagated by serial coinfections of 293Cre cells with the HDAd and helper virus and finally purified by CsCl ultracentrifugation.

A number of factors affect the production efficiency of HDAd (Ng et al. 2002a; Palmer and Ng 2005). Efficient transfection of the producer cell line with the pHDAd is critical. The higher the transfection efficiency, the lower the number of serial coinfections (passages) required to achieve maximum HDAd titer. The amount of helper virus added to each passage is also critical. To achieve the maximum yield of HDAd per cell, the optimal amount of helper virus should be the minimum amount required to simultaneously infect 100% of the producer cells without adversely affecting their ability to propagate the vector (Ng et al. 2001). The nature of the HDAd also has a critical role. HDAd with genome sizes outside the optimal packagable range (27.7 to 37.8 kb) will not amplify efficiently and may undergo rearrangements (Bett et al. 1993; Parks and Graham 1997). Likewise, repetitive sequences or unstable elements within the HDAd or homologous sequences between the HDAd and helper virus can cause rearrangement during vector amplification and should be avoided. Other factors related to the HDAd include the nature of the expression cassette, which may not be compatible with producer cell growth (e.g., expression of a toxic or otherwise interfering product), and the stuffer sequence (Parks et al. 1999). Unidentified open reading frames within the HDAd backbone may also contribute in this regard. This chapter focuses on the methods used to produce and characterize HDAd based on the pΔ28E4 backbone (Toietta et al. 2002) using the producer cell line 116 (Palmer and Ng 2003) and the helper virus AdNG163 (Palmer and Ng 2004). Results using different reagents may vary.

Protocol 1

Rescue, Amplification, and Large-scale Production of Helper-dependent Adenoviral Vectors

Production of HDAd can be divided into three parts (Fig. 2): (1) rescue, which involves converting pHDAd (plasmid form) to HDAd (viral form); (2) amplification, in which the amount of HDAd is increased by serial coinfections (passages) of the producer cells with the HDAd and the helper virus; and (3) large-scale production to generate large quantities of HDAd.

MATERIALS

CAUTION: See Appendix for appropriate handling of materials marked with <!>.

Reagents

116 cells

These cells are 293 derivatives that express high levels of Cre and can grow as monolayers or in suspension.

AdNG163 helper virus (HV; Fig. 3A)

Stocks of helper virus should always originate from a single plaque isolate to ensure homogeneity (i.e., no Cre-resistant mutants) and be titrated by plaque assay (Ng and Graham 2002). Accurate determination of the infectious titer of the helper virus is critical for the production of HDAd.

$CaCl_2$ (2.5 M) <!>

Sterilize by filtration through a 0.2-µm filter. Store aliquots in tightly sealed conical tubes at 4°C.

Cesium chloride gradient solutions (1.25 and 1.35 g/ml) <!>

For 1.25- and 1.35-g/ml solutions, dissolve 54.0 and 70.4 g of solid CsCl in 146.0 and 129.6 ml of dialysis buffer (see below), respectively. Sterilize by filtration through a 0.2-µm filter. Weigh

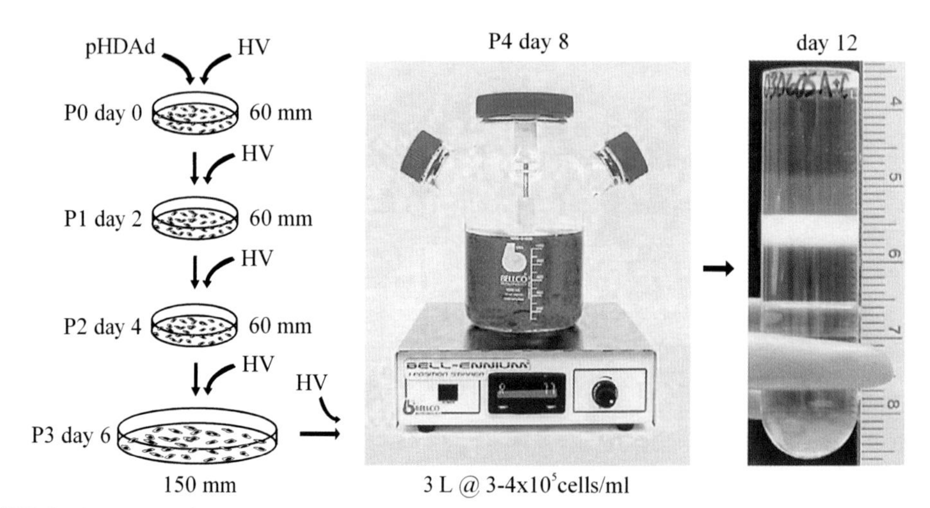

FIGURE 2. Overview of rescue (passage 0 or P0), amplification (P1–P3), and large-scale (P4) production of HDAd. Producer cells are transfected with pHDAd followed by infection with the helper virus (P0). Each subsequent passage involves coinfecting producer cells with the appropriate amounts of HDAd and helper virus. The entire process can take as little as 2 weeks with total yields of more than 10^{13} viral particles (vp)/3-liter culture and specific yields of more than 10,000 vp/cell. The amplified vector is purified by ultracentrifugation on CsCl gradients. (HV) Helper virus; (pHDAd) HDAd plasmid.

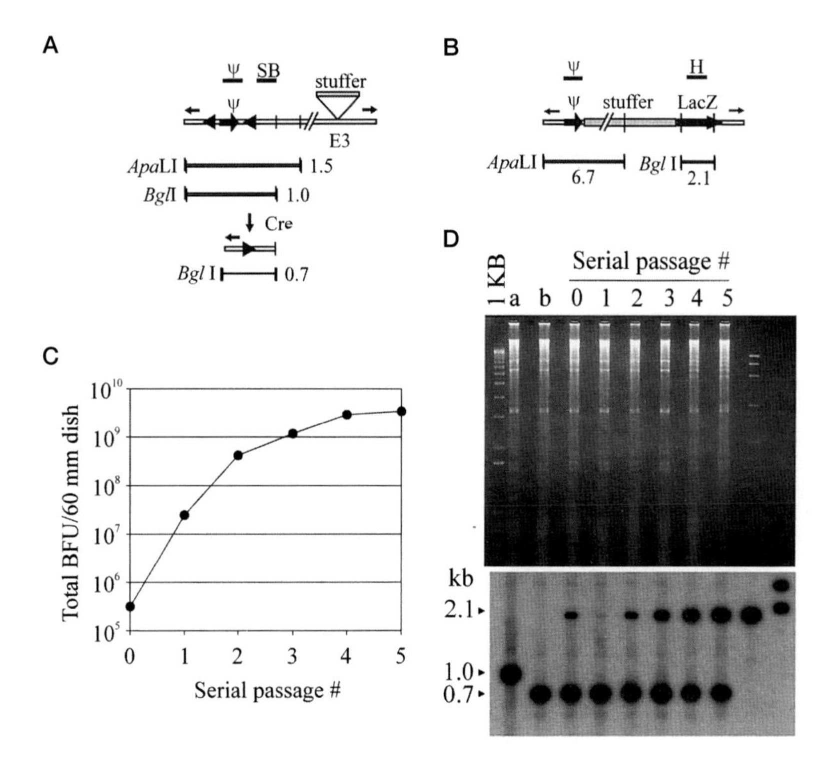

FIGURE 3. (*A*) Helper virus AdNG163 is a first-generation (E1-deleted) adenovirus that contains a ψ flanked by *loxP* sites and a DNA stuffer inserted into the wild-type E3 region to preclude replication-competent adenovirus formation. The 1-kb BglI fragment is reduced to 0.7 kb following Cre-mediated ψ excision. Shown are the sizes of the relevant restriction enzyme fragments and the positions of probe ψ and the helper-virus-specific probe SB used for Southern analysis. (*Small triangles*) *loxP* sites; (*small horizontal arrows*) viral ITRs. (*B*) HDAd HDΔ28E4LacZ contains an MCMV-*lacZ* expression cassette. Shown are the sizes of the relevant restriction enzyme fragments and the positions of probe ψ and HDAd-specific probe H used for Southern analysis. (*Small triangles*) *loxP* sites; (*small horizontal arrows*) viral ITRs. (C) Amplification of HDΔ28E4LacZ with 116 cells and AdNG163. The total yield of HDΔ28E4LacZ per 60-mm dish in blue-forming units (BFU) was determined at each serial passage by X-gal staining following titration on 293 cells (Ng et al. 2002a). (*D*) 116 cells were coinfected with HDΔ28E4LacZ and AdNG163. Total intracellular DNA was extracted from each serial passage shown (*lanes 0–5*) and digested with BglI. (*Lane a*) 293 cells infected with AdNG163; (*lane b*) 116 cells infected with AdNG163; (*lane c*) HDAd plasmid pΔ28E4LacZ digested with BglI and PmeI; (*lane d*) plasmid bearing probe SB and H fragments digested with AvaII. The 2.3-kb fragment encompasses probe SB and the 3.1-kb fragment encompasses probe H. (*Upper panel*) Ethidium bromide–stained agarose gel. The HDAd-specific bands are visible in *lanes 3–5*. (*Lower panel*) Southern blot hybridized simultaneously with probes SB and H. The helper virus genome bearing a packaging signal (1-kb BglI fragment) is undetectable, indicating minimal helper virus contamination. The intensity of the HDAd-specific bands increases with each subsequent passage. In this example, P3 was chosen to make inoculum for large-scale HDAd production but individual amplifications may vary.

1.00 ml to confirm density.

2x Citric saline

270 mM KCl <!>

30 mM sodium citrate <!>

Sterilize by autoclaving for 45 minutes at 121°C.

Dialysis buffer (10 mM Tris-HCl, pH 8.0) <!>

Sterilize by autoclaving for 45 minutes at 121°C.

DNase I

Prepare 100 mg of bovine pancreatic deoxyribonuclease I in 10 ml of 20 mM Tris-HCl (pH 7.4), 50 mM NaCl, 1 mM dithiothreitol <!>, 0.1 mg/ml bovine serum albumin, and 50% (v/v) glycerol. Store aliquots at –20°C.

Ethanol (70% and 95% v/v) <!>

Ethidium bromide <!>

Fetal bovine serum (FBS), heat-inactivated

L-glutamine

Glycerol

Sterilize by autoclaving for 45 minutes at 121°C.

HDΔ28E4LacZ

This HDAd contains an mCMV-*lacZ* expression cassette (Fig. 3B).

HEPES-buffered saline (HBS)

21 mM HEPES
137 mM NaCl
5 mM KCl
0.7 mM Na_2HPO_4 <!>
5.5 mM glucose

Adjust pH to 7.1 with NaOH <!> and sterilize by filtration though a 0.2-μm filter. Store aliquots in tightly sealed conical tubes at 4°C. The pH of HBS is critical for efficient transfection.

Hygromycin <!>

$MgCl_2$ (2 M) <!>

Sterilize by autoclaving for 45 minutes at 121°C.

Minimal essential medium (MEM; Invitrogen 61100)

PBS^{++} (phosphate-buffered saline)

Supplement PBS with 0.01 volume of 68 mM sterile $MgCl_2$ and 0.01 volume of 50 mM sterile $CaCl_2$.

Penicillin <!>/streptomycin <!>

pHDAd

Pronase solution

Prepare 20 mg/ml pronase in 10 mM Tris-HCl (pH 7.5). Preincubate for 15 minutes at 56°C and then for 1 hour by 37°C. Store aliquots at –20°C.

Pronase-SDS solution

Prepare 0.5 mg/ml pronase in 0.5% SDS <!>, 10 mM Tris-HCl (pH 7.4), and 10 mM EDTA (pH 8.0).

RNase A <!> (10 mg/ml)

Salmon sperm DNA (2 μg/μl in TE)

Sodium deoxycholate <!> (5% w/v)

Sterilize by filtration though a 0.2-μm filter.

Sucrose solution (40% w/v)

Sterilize by filtration through a 0.2-μm filter.

TE

10 mM Tris-HCl (pH 8.0)
1 mM EDTA (pH 8.0)

Sterilize by autoclaving for 45 minutes at 121°C.

Tris-HCl (100 mM, pH 8.0)

Sterilize by autoclaving for 45 minutes at 121°C.

Virion lysis buffer

Supplement TE with SDS to a final concentration of 0.1% (w/v).

Equipment

Beckman SW 40 Ti rotor and ultraclear tubes (342413)

Beckman SW 55 rotor and ultraclear tubes (344057)

Culture dishes (60 and 150 mm)
Magnetic stirrer (Bellco Glass 7785-D200)
Slide-A-Lyzer dialysis cassettes (Pierce 66381)
Southern blot hybridization materials and reagents
Spectrophotometer and cuvettes
Spinner flasks (Bellco Glass): 250 ml (1965-61002) and 3 liters (1965-61030)

METHODS

Growth of 116 Cells in Monolayers

116 cells are maintained in 150-mm dishes in MEM supplemented with 10% heat-inactivated FBS, 0.1 mg/ml hygromycin, 100 units/ml penicillin/streptomycin, and 2 mM L-glutamine. Split cells 1 to 2 (every 2 days) or 1 to 3 (every 3 days) when they reach about 90% confluency. Always prewarm media to 37°C before use.

1. Remove medium.
2. Tap the side of the dish to detach cells.
3. Resuspend the cells in an appropriate volume of fresh medium at 37°C and distribute to new dishes.

   In general, infect 116 cells 2 days after seeding to permit firm attachment of cells. One confluent 150-mm dish of 116 cells can be split into 20 60-mm dishes.

Rescue of HDAd

The pHDAd must first be constructed. We use conventional molecular biology to insert the expression cassette of interest into pΔ28E4. The size of the HDAd genome should be between 37.8 and 27.7 kb for efficient packaging (Bett et al. 1993; Parks and Graham 1997). The restriction endonuclease used to liberate the HDAd genome from the plasmid must not be present elsewhere within the HDAd genome. For other considerations in HDAd design, refer to the Introduction. High-quality plasmid DNA is critical for efficient vector rescue and we use CsCl ultracentrifugation for plasmid purification (Ng et al. 2002b). The steps below are for rescue and amplification of one HDAd. Scale up accordingly for more than one vector.

4. Seed 116 cells into 60-mm dishes (one dish per vector) to reach 70% confluency in 2 days for transfection. One hour before transfection, remove the medium from the dishes. Replace with 5 ml of fresh MEM supplemented with 10% FBS.
5. Digest 10 μg of pHDAd with the appropriate restriction enzyme in a total volume of 25–50 μl.
6. Incubate for 20 minutes at 65°C.
   a. Use the digested DNA immediately or store at 4°C for transfection the next day.
   b. Analyze 0.5 μl of digested DNA by agarose gel electrophoresis followed by ethidium bromide staining.

      Complete digestion should result in only two bands: the HDAd genome and the bacterial plasmid sequences.
7. Add 5 μg of salmon sperm DNA to 0.5 ml of HBS buffer in a polystyrene tube. Vortex for 1 minute at maximum setting.
8. Add the digested pHDAd. Mix thoroughly but gently.
9. Add 25 μl of 2.5 M $CaCl_2$ dropwise with mixing.

   The solution should appear slightly cloudy without large clumps.

10. Incubate the solution for 30 minutes at room temperature.
11. Apply 0.5 ml of the pHDAd solution dropwise to the 116 cell monolayer without removing the medium.
12. Rock the dish to distribute the precipitate evenly. Incubate overnight.
13. The next day, remove the medium from the transfected monolayer.
14. Wash twice with 1 ml of MEM containing 10% FBS.
15. Immediately infect the transfected cells with AdNG163 at a multiplicity of infection (moi) of 5 pfu/cell and PBS^{++} to a total volume of 0.1 ml.
16. Adsorb for 1 hour in the incubator, rocking the dishes every 10 minutes.
17. Following adsorption, add 2.5 ml of MEM supplemented with 5% FBS to the monolayer.
18. Forty-eight hours postinfection, examine the cells for complete cytopathic effect (CPE).

    More than 90% of the cells should be rounded up and detached from the dish. If not, see Troubleshooting at the end of this chapter.

19. Scrape the cells into the medium. Transfer the cell suspension into a vial. Add 0.1 volume of 40% sucrose and store at –80°C.

    Reserve 0.5 ml of the cell suspension for extraction of total DNA to monitor vector amplification (see Step 26).

Amplification of HDAd

20. Thaw the cell suspension at 37°C and allow it to equilibrate to 37°C. Mix well.
21. Coinfect a 60-mm dish of 90% confluent 116 cells with helper virus at an moi of 2 pfu/cell and 0.4 ml of the transfected cell suspension.
22. Adsorb for 1 hour in the incubator, rocking the dishes every 10 minutes.
23. Add 2.5 ml of MEM containing 5% FBS to the monolayer 1 hour postcoinfection.
24. Repeat Steps 18–23 (passages) until maximum HDAd titer is obtained.

    The number of passages required to achieve maximum HDAd titer may vary.

Monitoring HDAd Amplification

We always amplify an HDAd bearing a *lacZ* expression cassette (such as HDΔ28E4LacZ; see Fig. 3B) in parallel because the titer of this vector can be easily and quickly determined by X-gal staining following titration on 293 cells (Ng et al. 2002a). The titer of this control HDAd should increase tenfold to 100-fold with each passage until maximum yield is achieved and confirms that the system is performing as expected (Fig. 3C). To monitor amplification of most vectors that do not contain a reporter such as *lacZ*, total DNA is extracted from the coinfected cells from each passage and analyzed by ethidium bromide staining following agarose gel electrophoresis (Fig. 3D).

25. Determine the titer of the *lacZ*-containing vector by X-gal staining following titration on 293 cells (Ng et al. 2002a).

    The titer of this control HDAd should increase tenfold to 100-fold with each passage until maximum yield is achieved and confirms that the system is performing as expected (Fig. 3C).

26. Assess HDAd amplification for the vector of interest.

    a. Centrifuge the cell suspension (collected in Step 19) in a microcentrifuge at 750g for 1 minute. Discard the supernatant.

    b. Resuspend the cell pellet in 200 ml of pronase-SDS solution. Incubate overnight at 37°C.

    c. Add 0.5 ml of 95% ethanol. Mix by inverting the tube until a visible precipitate is formed.

d. Pellet the DNA by microcentrifugation at maximum speed for 1 minute.

e. Wash the DNA twice with 70% ethanol and dry.

f. Resuspend the DNA in an appropriate volume of TE.

g. Heat the DNA at 65°C and vortex until dissolved.

h. Digest an aliquot with an appropriate restriction enzyme.

i. Analyze by agarose gel electrophoresis followed by ethidium bromide staining.

The passage(s) with the maximum amount of HDAd is identified as the one in which both the HV- and HDAd-specific bands are visible (Fig. 3D). Note that the presence of HV-specific bands in the total cellular DNA is expected and does not represent HV contamination because these HV genomes do not contain a packaging signal; this can be verified by Southern analysis (Fig. 3D). If several passages meet this criterion, the earliest passage should be chosen because the chance of rearrangement of either the vector or helper may increase with higher passage numbers. If no passages meet this criterion, continue the serial passages until one is obtained.

27. Infect one 150-mm dish of 90% confluent 116 cells with helper virus at an moi of 2 pfu/cell and 0.5 ml of lysate from the passage containing the maximum HDAd titer.

28. Forty-eight hours postinfection, examine the cells to ensure complete CPE. Scrape the cells into the medium.

29. Centrifuge the cell suspension at 750*g* for 5 minutes. Discard the supernatant.

30. Resuspend the cell pellet in 1 ml of 100 mM Tris-HCl (pH 8.0) supplemented with 10% glycerol and freeze at –80°C.

Large-scale HDAd Production

Helper virus and high-titer HDAd are used to coinfect 116 cells grown in suspension.

31. Prepare 116 cells in suspension as follows:

a. Transfer spent media from eight confluent dishes of 116 cells into a 3-liter spinner flask.

b. Tap the dishes to detach 116 cells. Transfer them into the 3-liter spinner flask containing the spent media.

c. Add fresh MEM supplemented with 5% heat-inactivated FBS, 0.1 mg/ml hygromycin, 100 units/ml penicillin/streptomycin, and 2 mM L-glutamine at 37°C to a final volume of 1 liter. Tighten all lids.

d. Incubate overnight at 37°C on a magnetic stirrer set at 60 rpm.

A humid, CO_2 environment is not needed for cell growth in spinner flasks.

e. Add 0.5 liter the next day. Continue incubation overnight with stirring.

f. Add 0.5 liter the next day. Continue incubation overnight with stirring.

g. Add 1 liter the next day. Continue incubation overnight with stirring.

The final volume should be 3 liters. The cell density should increase each day and can be confirmed by daily counting before addition of medium.

32. Count 116 cells in suspension as follows:

Accurate determination of the cell density at the time of coinfection is critical because it will dictate the amount of helper virus added and thus the HDAd yield.

a. Transfer 2 ml of cells into a 15-ml conical tube.

b. Add 2 ml of 2x citric saline at 37°C. Vortex for 10 seconds at maximum setting.

c. Incubate for 5–10 minutes at 37°C.

d. Vortex for 10 seconds at maximum setting.

e. Use a hemocytometer to obtain two independent cell counts.

If the two independent cell counts are significantly different from each other, repeat the cell count. There should be no large numbers of uncountable cell clumps. The cell density should be 2×10^5 to 4×10^5 cells/ml for coinfection.

f. Transfer 0.1 ml of cell suspension from the 3-liter culture to one well of a 24-well dish containing 1 ml of fresh media. Incubate at 37°C.

This sample serves as a control for the health and appearance of uninfected cells, which should reattach and form a monolayer within a few hours.

33. Perform large-scale coinfection of 116 cells as follows:

a. Harvest cells from the 3-liter culture by centrifugation at 750*g* for 5 minutes at room temperature. Reserve 0.5 liter of the spent medium.

b. Resuspend the cell pellet in 100 ml of the spent medium. Transfer to a 250-ml spinner flask.

c. Coinfect cells by adding helper virus at an moi of 2 pfu/ml and the HDAd lysate (Step 30). Incubate at 60 rpm for 2 hours at 37°C.

d. Transfer the coinfected cells to a 3-liter spinner flask. Add 0.5 liter of spent medium and 1.5 liters of fresh MEM containing 5% FBS (2 liters total volume).

e. Transfer 0.1 ml of coinfected cells into one well of a 24-well dish that contains 1 ml of fresh MEM containing 5% FBS.

These cells initially reattach to the dish but should all round up and detach by 48 hours. If not, see Troubleshooting at the end of this chapter.

f. Incubate the spinner flask at 60 rpm for 48 hours at 37°C.

g. Harvest coinfected cells by centrifugation at 750*g* for 5 minutes at room temperature.

h. Resuspend cells in 15 ml of 100 mM Tris-HCl (pH 8.0). Transfer to a 50-ml conical tube and proceed with HDAd purification.

Coinfected cells can also be resuspended in 15 ml of 100 mM Tris-HCl (pH 8.0) supplemented with 10% glycerol and stored at –80°C for processing later.

HDAd Purification

34. Add 2.0 ml of 5% sodium deoxycholate to the coinfected cells.

The mixture should immediately become thick and gelatinous.

35. Incubate for 30 minutes at room temperature with frequent mixing.

36. Add 170 µl of 2 M $MgCl_2$, 150 µl of RNase A (10 mg/ml), and 150 µl of DNase I (10 mg/ml). Incubate for 1 hour at 37°C with frequent mixing.

The viscosity should be reduced significantly.

37. Centrifuge in a tabletop centrifuge at maximum speed for 10 minutes at room temperature. Remove and save supernatant.

38. Prepare CsCl step gradients in ultraclear ultracentrifuge tubes (two tubes per virus).

a. Add 2 ml of 1.35 g/ml CsCl to each tube.

b. Carefully overlay with 3 ml of 1.25 g/ml CsCl.

c. Carefully overlay with supernatant from Step 37 on top of the two CsCl steps.

If necessary, fill the remainder of the tubes evenly with the 100 mM Tris-HCl (pH 8.0).

39. Centrifuge in an ultracentrifuge with an SW 40 rotor at 151,000*g* (35,000 rpm) for 1 hour at 4°C.
40. Use an 18-gauge needle attached to a 3-cc syringe to pierce the tube below the vector band and slowly retrieve the vector from the tubes.

    The HDAd is the lowest band in the step gradient (Fig. 4A).
41. Transfer the vector from both tubes into one SW 55 Ultra-Clear tube. Fill the tube with 1.35 g/ml CsCl.

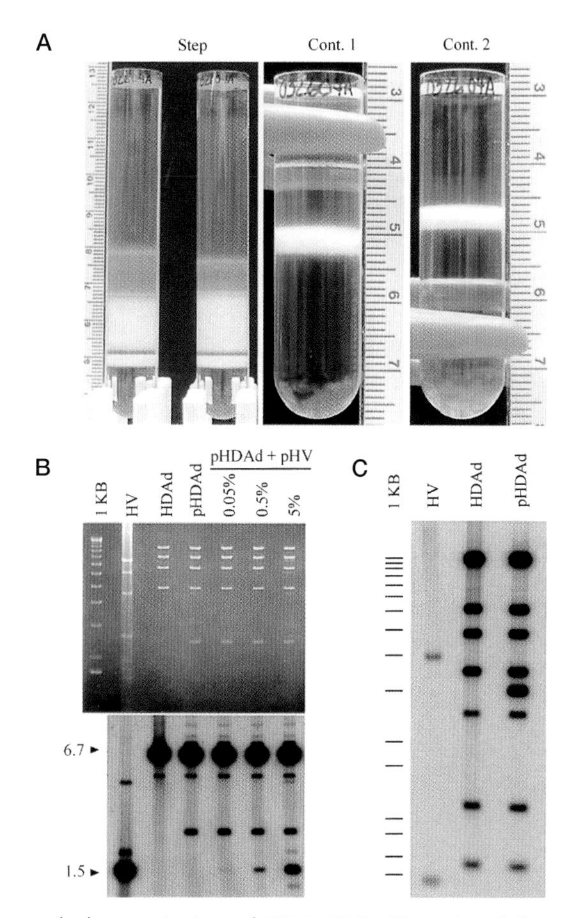

FIGURE 4. Purification and characterization of HDΔ28E4LacZ produced from 3 liters of 116 cells using AdNG163. (*A*) HDAd is purified by a single step (*Step*) and two continuous CsCl gradients (*Cont. 1*, *Cont. 2*). Only a single virus band is visible in all CsCl gradients. The lowest band in the CsCl step gradient is the virions. Total HDAd yield in this example was 2.6 x 10^{13} vp with a specific yield of 29,126 vp/cell. (*B*) Helper virus contamination analysis of HDAd. DNA was extracted from the purified virions and digested with ApaLI. (1 KB) 1KB PLUS standard (Life Technologies); (HV) AdNG163 DNA; (HDAd) HDΔ28E4LacZ DNA; (pHDAd) parental pΔ28E4LacZ plasmid DNA. Control lanes (pHDAd + pHV): ApaLI- and PmeI-digested pΔ28E4LacZ mixed with tenfold serial dilutions (0.05–5%) of ApaLI- and PacI-digested helper virus plasmid pNG163 (pHV). (*Upper panel*) Ethidium bromide–stained agarose gel; (*lower panel*) Southern blot hybridized with probe ψ. The structure of HDAd is indistinguishable from that of pHDAd, except for the expected absence of the 2.5-kb ApaLI-PmeI fragment bearing the bacterial plasmid sequences. Southern analysis with probe ψ revealed the expected 6.7-kb HDΔ28E4LacZ-specific band but no detectable AdNG163-specific 1.5-kb band. Extraneous bands in HDAd are also present in the pHDAd lane and thus represent nonspecific hybridization of the probe to other vector bands, rather than rearrangement products. Based on comparison to the controls, the helper virus contamination of this HDAd preparation was less than 0.05%. (*C*) Genomic structure of HDΔ28E4LacZ. DNA was extracted from purified HDAd, digested with HpaI, and analyzed by Southern blot hybridization using pΔ28E4LacZ as the probe. (1 KB) 1KB PLUS standard (Life Technologies); (HV) AdNG163; (HDAd) HDΔ28E4LacZ; (pHDAd) parental pΔ28E4LacZ plasmid. The genomic structure of HDAd is indistinguishable from that of pHDAd except for the expected absence of the 3.0-kb PmeI fragment bearing the bacterial plasmid sequences. The bands in HV represent terminal fragments that hybridize to the homologous ψ and ITR sequences in the pΔ28E4LacZ probe.

42. Centrifuge with an SW 55 rotor at 115,000*g* (35,000 rpm) overnight at 4°C.
43. Use a 22-gauge needle attached to a 1-cc syringe to pierce the tube below the vector band and retrieve the vector band from the CsCl gradient.
44. Transfer the vector into a new SW 55 ultracentrifuge tube. Fill the tube with 1.35 g/ml of CsCl.
45. Centrifuge with an SW 55 rotor at 115,000*g* (35,000 rpm) overnight at 4°C.
46. Use a 1-cc syringe and 22-gauge needle to retrieve the virus band from the CsCl gradient. Repeat Steps 44 and 45.

    A single band containing the HDAd should be present. See Troubleshooting at the end of the chapter if more than one band is present.
47. Use a 1-cc syringe and 22-gauge needle to retrieve the virus band from the CsCl gradient. Transfer into a Slide-A-Lyzer dialysis cassette presoaked with dialysis buffer at 4°C, as per manufacturer's instructions.
48. Dialyze overnight at 4°C with three 500-ml changes of dialysis buffer with slow stirring to remove CsCl.
49. Use a 22-gauge needle attached to a 1-cc syringe to retrieve the virus from the cassette. Record the volume of vector in the syringe and transfer into a vial. Add glycerol to a final concentration of 10% and mix well.

    Remove at least two small aliquots of the vector preparation and transfer them into microfuge tubes for DNA extraction and for absorbance at 260 nm. The volume of these aliquots depends on the amount of vector obtained but is typically 10–50 μl. Additional small aliquots can be taken if desired for other characterizations.
50. Aliquot vector and store at –80°C.

    Purified HDAd can be used to produce more of itself. Using purified HDAd instead of crude lysate greatly simplifies production and improves consistency. Follow the procedure described in Step 33, but use 200 vp/cell of the purified vector instead of the lysate from a 150-mm dish as indicated in Step 33c.

Protocol 2

Characterization of HDAd

At a minimum, purified vector should be assessed for its physical titer, degree of helper virus contamination, and analysis of the HDAd genomic structure. If desired and depending on the application, additional characterization (i.e., endotoxin, sterility, adventitious agents, etc.) should be performed according to protocols established for E1-deleted Ad vectors.

METHODS

Physical Titer

1. Assess the physical titer of the vector preparation:

   The physical titer is the concentration of viral particles in a vector preparation and is obtained by measuring the absorbance at 260 nm following virion lysis (Maizel et al. 1968) and correction for vector genome size (Ng et al. 2002a). This method essentially measures the amount of viral DNA in a vector preparation and is expressed as vp/ml.

   a. Add virion lysis buffer to the aliquot of purified vector (Step 49) to a total volume of 0.5 ml.

      Prepare a blank sample by adding virion lysis buffer to the same volume of vector vehicle (10 mM Tris-HCl [pH 8.0] supplemented with 10% glycerol).

   b. Vortex both samples briefly. Avoid bubbles.

   c. Incubate samples for 10 minutes at 56°C.

   d. Vortex samples for 1 minute.

   e. Use the blank sample to set references at 260 and 280 nm. Measure the absorbance of the vector sample at 260 and 280 nm.

      The 260/280 ratio, a measure of relative purity, should be approximately 1.3.

   f. Calculate vp/ml from absorbance at 260 nm using the formula below:
      vp/ml = (absorbance at 260 nm)(dilution factor)(1.1×10^{12})(36)/(size of vector in kb)

Helper Virus Contamination

HDAd is invariably contaminated with helper virus and thus accurate determination of this level is critical.

2. Determine helper virus contamination by Southern blot hybridization analysis (Fig. 4B)

   Alternatively, use real-time quantitative polymerase chain reaction (PCR) for this purpose (Palmer and Ng 2003). Determining helper virus contamination by plaque assay is not as reliable (Palmer and Ng 2005).

HDAd Genomic Structure

3. Determine the genomic structure of the vector.

   a. Extract DNA and digest a sample of the purified vector with an appropriate restriction endonuclease.

b. Verify the genomic structure of the HDAd by agarose gel electrophoresis and ethidium bromide staining.

Alternatively, Southern blot hybridization provides greater sensitivity (Fig. 4B,C).

Vector Infectivity

In addition to the physical titer (vp/ml), HDAd also possess an infectious titer defined as the proportion of viral particles capable of actually transducing target cells. The infectious titer is expressed as the number of infectious units (IU)/ml. The particle-to-infectious unit ratio (vp:IU) is the infectivity of the vector and represents the proportion of the total particles that are infectious. By definition, the higher the vp:IU ratio, the lower the infectivity. This is an important parameter for meaningful comparisons among different vectors (or among different preparations of the same vector) because erroneous conclusions can be drawn if they have different infectivities. Moreover, because the viral particle itself may mediate dose-dependent acute toxicity (Brunetti et al. 2004, 2005; Muruve et al. 2004), noninfectious particles (reflected in a high vp:IU ratio) may increase risk and decrease benefit. Therefore, the Food and Drug Administration has recommended that the infectivity of clinical grade Ad vector be less than 30:1 to address this concern (Simek et al. 2002). Because HDAd cannot propagate by themselves, methods based on cytopathic effect used for early-generation Ad vectors are not applicable, and the infectious titer of HDAd without reporter transgenes such as *lacZ* is difficult to determine. However, a method has recently been developed to determine the infectious titer of HDAd (Palmer and Ng 2004).

Handling and Storage

Adenoviruses including HDAd are relatively stable when handled and stored properly. Vectors are always stored at –70°C to –80°C. The number of freeze-thaws should be minimized by making multiple aliquots of the vector preparation immediately after purification (before the first freeze). For use, the vector should be thawed at room temperature and returned to –70°C to –80°C as soon as possible. Proper handling and storage are critical for preserving vector infectivity. Exposure of vector to dry ice may negatively impact infectivity due to the reduction in pH from the CO_2 (Nyberg-Hoffman and Aguilar-Cordova 1999). Therefore, for transportation of vector in dry ice, special precautions should be taken to ensure proper protection. These precautions also apply to the HV and indeed to all Ad-based vectors.

TROUBLESHOOTING

Problem: During the rescue of the HDAd or large-scale HDAd production, complete CPE of the 116 cells is not observed by 48 hours.
Solution: The amount of HV used is insufficient. Add more HV.

Problem: Agarose gels of samples from serial amplification passages do not exhibit both helper-virus- and HDAd-specific bands.
Solution: Continue serial amplification passages (Steps 21–25) until a sample exhibiting both bands is obtained.

Problem: More than one band is visible in the CsCl gradients.
Solution: Carefully collect each band separately and subject to DNA analysis to determine their identity. Two main causes for multiple bands are helper virus contamination (see below) and vector genome rearrangement. Rearrangements can occur if the genome size of the HDAd is outside the optimal size for encapsidation (Bett et al. 1993; Parks and Graham 1997) or if the HDAd genome contains repetitive sequences or other unstable elements. Modify the original pHDAd to address these problems.

Problem: High HV contamination level.

Solution: HDAd is invariably contaminated with HV. If HV contamination is high, two distinct bands will be visible in the continuous CsCl gradient following ultracentrifugation, provided that the sizes of the HDAd and HV genomes are sufficiently different. For this reason, HDAd is optimally engineered to have a smaller genome size than the HV. In this case, the HDAd is the upper band, whereas the HV is the lower band. Remove the lower HV band first from the gradient and then collect the upper HDAd. The chief reason for HV contamination is inefficient excision of the HV packaging signal due to low-level Cre expression in the producer cell line (Ng et al. 2002a; Palmer and Ng 2003). One solution is to use a producer cell line that expresses higher levels of Cre. Other sources of high HV contamination include mutation/rearrangement of the HV that prevents packaging signal excision, but these instances are generally quite rare (Ng et al. 2002a). Be sure that every HV stock is generated from a single, well-isolated plaque. Also, bear in mind that the probability of such events increases with increasing number of passages performed.

ACKNOWLEDGMENTS

P.N. is supported by grants from the National Institutes of Health (P50 HL59314, R01 DK067324, and P51 RR13986), the Texas Affiliate of the American Heart Association (0465102Y), and the Cystic Fibrosis Foundation (NG0530 and NG05G0).

REFERENCES

Bett A.J., Prevec L., and Graham F.L. 1993. Packaging capacity and stability of human adenovirus type 5 vectors. *J. Virol.* **67:** 5911–5921.

Brunetti-Pierri N., Palmer D.J., Beaudet A.L., Carey D., Finegold M., and Ng P. 2004. Acute toxicity following high-dose systemic injection of helper-dependent adenoviral vectors into non human primates. *Hum. Gene Ther.* **15:** 35–46.

Brunetti-Pierri N., Palmer D.J., Mane V., Finegold M., Beaudet A.L., and Ng P. 2005. Increased hepatic transduction with reduced systemic dissemination and proinflammatory cytokines following hydrodynamic injection of helper-dependent adenoviral vectors. *Mol. Ther.* **12:** 99–106.

Graham F.L., Smiley J., Russell W.C., and Nairn R. 1977. Characteristics of a human cell line transformed by DNA from human adenovirus type 5. *J. Gen. Virol.* **36:** 59–74.

Kochanek S. 1999. High-capacity adenoviral vectors for gene transfer and somatic gene therapy. *Hum. Gene Ther.* **10:** 2451–2459.

Maizel J.V., White D., and Scharff M.D. 1968. The polypeptides of adenovirus. I. Evidence of multiple protein components in the virion and a comparison of types 2, 7a, and 12. *Virology* **36:** 115–125.

Muruve D.A., Cotter M.J., Zaiss A.K., White L.R., Liu Q., Chan T., Clark S.A., Ross P.J., Meulenbroek R.A., Maelandsmo G.M., and Parks R.J. 2004. Helper-dependent adenovirus vectors elicit intact innate but attenuated adaptive host immune responses in vivo. *J. Virol.* **78:** 5966–5972.

Ng P. and Graham F.L. 2002. Construction of first generation adenoviral vectors. *Meth. Mol. Med.* **69:** 389–414.

———. 2004. Helper-dependent adenoviral vectors for gene therapy. In *Gene therapy: Therapeutic mechanisms and strategies*, 2nd edition (ed. N.S. Templeton), pp. 53–70. Marcel Dekker, New York.

Ng P., Parks R.J., and Graham F.L. 2002a. Preparation of helper-dependent adenoviral vectors. *Meth. Mol. Med.* **69:** 371–388.

Ng P., Evelegh C., Cummings D., and Graham F.L. 2002b. Cre levels limit packaging signal excision efficiency in the Cre/*loxP* helper-dependent adenoviral vector system. *J. Virol.* **76:** 4181–4189.

Ng P., Beauchamp C., Evelegh C., Parks R., and Graham F.L. 2001. Development of a FLP/frt system for generating helper-dependent adenoviral vectors. *Mol. Ther.* **3:** 809–815.

Nyberg-Hoffman C. and Aguilar-Cordova E. 1999. Instability of adenoviral vectors during transport and its implication for clinical studies. *Nat. Med.* **5:** 955–957.

Palmer D.J. and Ng P. 2003. Improved system for helper-dependent adenoviral vector production. *Mol. Ther.* **8:** 846–852.

———. 2004. Physical and infectious titers of helper-dependent adenoviral vectors: A method of direct comparison to the adenovirus reference material. *Mol. Ther.* **10:** 792–798.

———. 2005. Helper-dependent adenoviral vectors for gene therapy. *Hum. Gene Ther.* **16:** 1–16.

Parks R.J. 2000. Improvements in adenoviral vector technology: Overcoming barriers for gene therapy. *Clin. Genet.* **58:** 1–11.

Parks R.J. and Amalfitano A. 2002. Separating fact from fiction: Assessing the potential of modified adenovirus vector for use in human gene therapy. *Curr. Gene Ther.* **2:** 111–133.

Parks R.J. and Graham F.L. 1997. A helper-dependent system for adenovirus vector production helps define a lower limit for efficient DNA packaging. *J. Virol.* **71:** 3293–3298.

Parks R.J., Bramson J.L., Wan Y., Addison C.L., and Graham F.L. 1999. Effects of stuffer DNA on transgene expression from helper-dependent adenovirus vectors. *J. Virol.* **73:** 8027–8034.

Parks R.J., Chen L., Anton M., Sankar U., Rudnicki M.A., and Graham F.L. 1996. A helper-dependent adenovirus vector system: Removal of helper virus by Cre-mediated excision of the viral packaging signal. *Proc. Natl. Acad. Sci.* **93:** 13565–13570.

Simek S., Byrnes A., and Bauer S. 2002. FDA perspective on the use of the adenovirus reference material. *BioProcessing* **1:** 40–42.

Toietta G., Pastore L., Cerullo V., Finegold M., Beaudet A.L., and Lee B. 2002. Generation of helper-dependent adenoviral vectors by homologous recombination. *Mol. Ther.* **5:** 204–210.

17 Cell and Tissue Targeting

Yosuke Kawakami* and David T. Curiel*†

*Division of Human Gene Therapy, Departments of Medicine, Surgery, Pathology and
†Gene Therapy Center, University of Alabama, Birmingham, Alabama, 35294

ABSTRACT

The balance between target and nontarget cell toxicity determines the efficacy of any therapeutic agent. Therapeutic gene-transfer strategies aim to maximize gene transfer and expression in target cells. However, correlative laboratory studies have demonstrated that current gene delivery systems fail to preferentially transduce target cells in mixed cell populations. In addition, limitations in vector specificity can lead to transduction of nontarget cells, resulting in untoward toxicity, even with compartmental dosing. Thus, vector optimization is critical for the development of efficient genetic experiments. The life cycle and biology of adenovirus (Ad) infection in cells have been thoroughly characterized. Three distinct sequential steps are required for Ad infection and transgene expression: (1) binding of the virus to the surface receptor, (2) internalization, and (3) transfer of the viral genome to the target cell's nucleus. The protocols in this chapter describe the creation of Ad vectors that enable vector targeting at the level of Ad binding and entry in targeted cells through primary and/or secondary Ad receptors (transduction), and protein expression of the transgene in the targeted cells (transcription/translation).

Continued

INTRODUCTION

Adenovirus serotype 5 (Ad5) is commonly used for gene therapy because it can transduce a wide variety of both dividing and nondividing cell types. The genome, capsid protein function, replication cycle, and biology of Ad5 infection in cells have been thoroughly characterized (Varga et al. 1991). Briefly, three distinct sequential steps are required for Ad infection and transgene expression (Fig. 1). On the basis of these steps, two opportunities for targeting intervention are implied: (1) Ad binding and entry in targeted cells through primary and/or

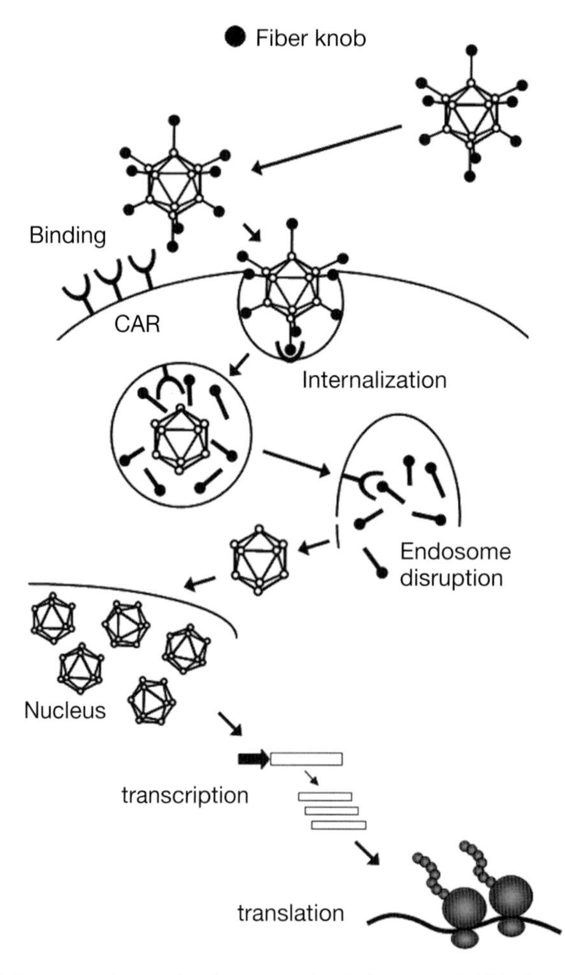

FIGURE 1. Pathway of Ad entry and strategies for retargeting Ad vectors. Ad binds to specific receptors (CAR) on the surface of the target cell via its fiber knob. The virus is internalized and digested in endosomes, and its genome is translocated to the nucleus where the transgene is expressed through transcription and translation. Retargeting can occur at the level of binding (by direct or indirect modification of the fiber knob's specificity) and/or at the level of transcription (by insertion of cell/tissue-specific promoters).

secondary Ad receptors and (2) protein expression of the transgene in the targeted cells through transcription and translation.

Ad infection is initiated by recognition of the native Ad5 receptor, coxsackievirus–adenovirus receptor (CAR), on target cells by the carboxy-terminal portion of the fiber protein, termed the knob (Bergelson et al. 1997). However, a number of tumor types are relatively refractory to Ad5 infection due to the paucity of CAR (Freeman and Zwiebel 1993; Culver and Blaese 1994; Schmidt-Wolf and Schmidt-Wolf 1994; Herrmann 1995). In addition, its broad tropism can result in the transduction of many nontarget CAR-expressing tissues. This precludes in vivo targeting and provokes immune responses that can prevent repeated administration and shorten the duration of therapeutic gene expression.

The development of genetically modified Ad vectors with transductional specificity for a single cell type requires both the ablation of endogenous tropism and the introduction of novel tropism determinants for target cells, a concept known as "retargeting." The advantage of this strategy is that the retargeted Ad vector functions not only with enhanced specificity, but also with enhanced transgene activity, which is often abrogated in most cases of transcriptional targeting. Retargeting of Ad vectors generally takes two different approaches, both of which are designed to attain CAR-independent transduction (Goldman et al. 1997; Wickham et al. 1997): direct genetic modification of the vector itself or the use of adapter targeting molecules.

Direct retargeting of Ad5 can be achieved via genetic alterations of the capsid, modifications of the fiber knob, or construction of chimeric fibers composed of a foreign fiber knob domain fused to the Ad5 fiber tail and shaft (Stevenson et al. 1995, 1997; Krasnykh et al. 1996). Certain targeting motifs may be incorporated in this manner to route the virus toward CAR-independent cellular entry pathways. For example, Ad vectors were constructed by inserting Arg-Gly-Asp (RGD) or polylysine at the carboxyl terminus of the fiber knob region. These redirected virus binding to either the α_v integrin or heparan sulfate cellular receptors, respectively (Wickham et al. 1993, 1997; Dmitriev et al. 1998). These vectors demonstrated a fivefold to 500-fold transduction increase in several cell types lacking sufficient levels of the primary Ad receptor, including macrophages, endothelial cells, smooth muscle cells, fibroblasts, and T cells (Davidoff et al. 1999; Von Seggern et al. 2000; Haviv et al. 2002; Kanerva et al. 2002; Kawakami et al. 2003; Volk et al. 2003).

Other transductional retargeting strategies use bifunctional adapter molecules (Dmitriev et al. 2000; Haisma et al. 2000; Hemminki et al. 2001; de Gruijl et al. 2002; Kashentseva et al. 2002; Hakkarainen et al. 2003; Breidenbach et al. 2004; Korn et al. 2004; Nettelbeck et al. 2004; Pereboev et al. 2004). One element of the bispecific adapter binds to the Ad fiber knob, blocking its interaction with CAR and, thus, its native tropism. This attachment can be accomplished with anti-Ad knob Fab, single-chain Fv (scFv) fragments, or the extracellular domain of CAR. The second component of the bispecific adapter introduces specificity for the target cells and is chemically or genetically conjugated to the knob-binding portion. Some of these molecules are antibodies (Fab, scFv) or ligands that bind to specific receptors expressed on target cells (Douglas et al. 1999). Recombinant fusion proteins offer a number of technological advantages over chemical conjugates, including simplified production and purification.

One goal of cell and tissue targeting is to minimize ectopic expression to prevent toxicity when the transduced gene is a cytocidal drug or a prodrug-activating enzyme. Indeed, adverse effects due to gene expression in nontarget cells, most commonly liver or bone marrow toxicity, have been reported in vivo following Ad gene transfer (Huard et al. 1995; Reynolds et al. 1999; Varnavski et al. 2005). Furthermore, ectopic gene expression in immune cells may cause an immune response to the transgene, thus limiting therapeutic efficacy (Huard et al. 1995; van der Eb et al. 1998; Reynolds et al. 1999; Bilbao et al. 2000; Kirn et al. 2001). Efficient gene therapy regimens therefore require transgene expression in target cells and absence of expression in nontarget cells. Such target-cell-specific gene expression can be accomplished using tissue-specific

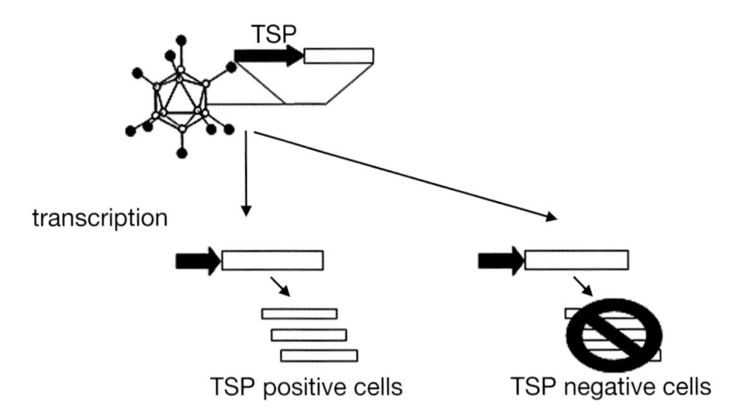

FIGURE 2. Schematic representation of the types of transcriptional targeting. Efficient gene therapy regimens require transgene expression in target cells (TSP-positive) resulting in the "on" state and absence of expression in nontarget (TSP-negative) cells producing an "off" state.

promoters (TSPs), DNA elements that restrict expression to specific cellular subsets (Fig. 2). The major drawback of transcriptional retargeting is that pathological change in target tissues or organs, such as degeneration or malignant transformation, can ectopically activate TSPs.

Protocol 1

Construction of Ad Vectors with RGD-modified Fiber for Transductional Targeting

A rescue plasmid with a mutant fiber is generated through homologous recombination between the pAdEasy-1 and the fiber shuttle plasmid (pFiber-dE3-RGD). The resultant fiber-modified rescue plasmid (pAdEasy-RGD) is isogenic to AdEasy-1 (except the fiber region) and can therefore be used for another round of homologous recombination with pShuttle-promoter-Luc to make a final Ad vector construct that expresses the luciferase gene and contains an RGD-modified fiber knob. (For detailed methods on homologous recombination, amplification, and purification of Ad vectors using the Ad-Easy system, refer to Ross and Parks, Chapter 15, this volume.)

MATERIALS

CAUTION: See Appendix for appropriate handling of materials marked with <!>.

Reagents

AdEasy system (Qbiogene)

This system consists of pAdEasy-1 (an E1/E3-deleted rescue plasmid) and pShuttle (a promoter-transgene expression cassette).

Agarose

Ammonium sulfate (cold-saturated) <!>

Bacteria: BJ5183 and DH5α

Kanamycin (25 μg/ml) <!>

Luria broth (LB) medium

Plasmid: pABS.4 (Microbix)

PCR primers

- Primer A (5′-AAG CTA GCC CTG CAA ACA TCA-3′)
- Primer B (5′-GCA GAA GCA GTC TCC TCG GCA GTC GCA ACT TGT GTC TCC TGT TTC CTG-3′)
- Primer C (5′-TGC GAC TGC CGA GGA GAC TGC TTC TGC CCA AGT GCA TAC TCT ATG TCA TTT-3′)
- Primer D (5′-TGC AAT TGA AAA ATA AAC ACG TTG AAA-3′)

Restriction enzymes: EcoRI, BstBI, BamHI, MfeI, NheI, PacI, SwaI

For additional materials required for homologous recombination, amplification, and purification of Ad vectors, see Ross and Parks, Chapter 15, this volume.

Equipment

Polymerase chain reaction (PCR) thermal cycler

For additional equipment required for homologous recombination, amplification, and purification of Ad vectors, see Ross and Parks, Chapter 15, this volume.

METHODS

Construction of the Shuttle Plasmid for Fiber Modification

1. Isolate the ampicillin-resistant fiber region from the rescue plasmid.
   a. Digest 10 μg of ampicillin-resistant pAdEasy-1 with EcoRI for 4 hours.
   b. Recover the 9.6-kb DNA fragment by 1% agarose gel extraction.
   c. Recirculize the plasmid by ligation.
      The resultant plasmid (pFiber-dE3-Amp) includes the whole fiber region.
2. Isolate the kanamycin resistance gene.
   a. Digest 10 μg of pABS.4 with EcoRI.
   b. Blunt the termini by a Klenow enzyme fill-in reaction.
   c. Recirculize the resulting plasmid by ligation (pABS.4.ΔEcoRI).
   d. Digest pABS.4.ΔEcoRI with BamHI.
   e. Blunt the BamHI site using a Klenow enzyme fill-in reaction.
   f. Digest pABS.4.ΔEcoRI with BstBI.
      The resulting 1.3-kb fragment includes the kanamycin resistance gene and two SwaI sites on both ends.
3. Insert the kanamycin resistance gene in the ampicillin-resistant fiber plasmid.
   a. Digest the pFiber-dE3-Amp with MfeI.
   b. Blunt the MfeI site of the digest using a Klenow enzyme fill-in reaction.
   c. Digest the pFiber-dE3-Amp with BstBI.
   d. Clone the 1.3-kb BstBI blunt fragment into the MfeI-BstBI site of pFiber-dE3-Amp.
      The resultant plasmid (pFiber-dE3) includes the fiber region as well as both the ampicillin and kanamycin resistance genes.

Construction of Mutant Fiber Knob (Fig. 3)

4. Generate complementary RGD-4C fragments by PCR.
   a. To generate NheI-(RGD-4C), add 30 μM primer A (forward primer) and primer B (reverse primer) to 500 ng of pFiber-dE3.
   b. To generate the (RGD-4C)-MfeI fragment, add 30 μM primer C (forward primer) and primer D (reverse primer) to 500 ng of pFiber-dE3.
   c. Perform PCR on both samples with an initial denaturation step of 94°C for 10 minutes, followed by 35 cycles of 94°C for 30 seconds, 50°C for 30 seconds, and 72°C for 60 seconds, and a final step of 72°C for 10 minutes.
   d. Extract the NheI-(RGD-4C) (1199 bp) and (RGD-4C)-MfeI (180 bp) PCR fragments from a 2% agarose gel.
5. Generate the mutant fiber knob sequence.
   a. Mix the purified NheI-(RGD-4C) and (RGD-4C)-MfeI fragments.
   b. Amplify 500 ng of the mixture without any primers, using an initial denaturation step of 94°C for 4 minutes and 30 seconds, and ten cycles of 94°C for 60 seconds, 52°C for 60 seconds, and 72°C for 4 minutes, and a final extension step of 72°C for 6 minutes.

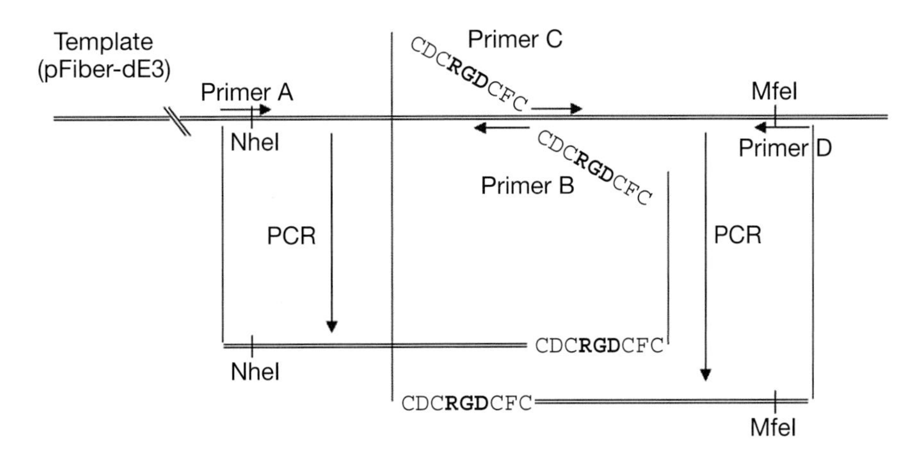

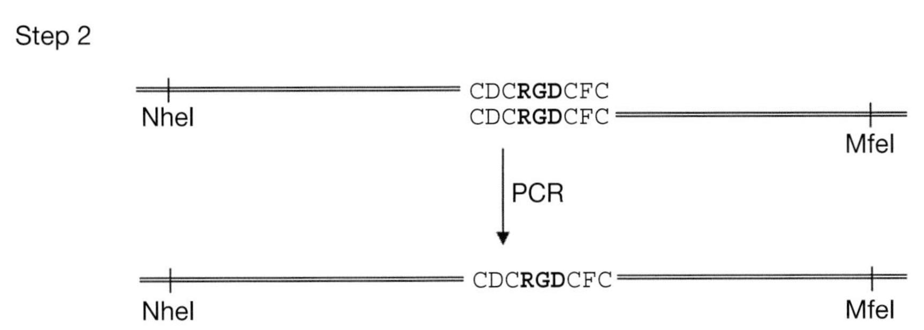

FIGURE 3. Strategy for RGD-4C insertion into the HI loop of the Ad5 knob. In the first round of PCR, complementary RGD-4C fragments are generated from the NheI and MfeI sites of the pFiber-dE3 plasmid. The fragments are mixed, and a second round of PCR is performed to produce a single NheI-MfeI fragment that includes the mutant fiber knob. This fragment can then be cloned into a suitable shuttle plasmid.

c. Extract the second PCR product (1352 bp) from a 1% agarose gel.

6. Insert the mutant fiber knob sequence in the shuttle plasmid.

   a. Digest 10 μg of the second PCR product with NheI and MfeI for 4 hours.

   b. Extract the fragment from a 1% agarose gel.

   c. Clone the second NheI- and MfeI-digested PCR product into the NheI-MfeI site of the pFiber-dE3 plasmid.

      The resultant fiber shuttle plasmid (pFiber-dE3-RGD) has an RGD-modified mutant fiber and is resistant to both ampicillin and kanamycin.

Construction of Fiber-Modified Rescue Plasmid

7. Digest 10 μg of pFiber-dE3-RGD with EcoRI for 4 hours and then with PacI for another 4 hours.

8. Recover the 7.2-kb DNA fragment from a 1% agarose gel.

   This fragment includes the whole fiber region and the kanamycin resistance gene.

9. Mix the 7.2-kb shuttle plasmid fragment and undigested pAdEasy-1 in a 10:1 molar ratio.

10. Transform BJ5183 cells with the plasmid mixture and plate on LB medium containing kanamycin (25 μg/ml). Incubate overnight at 30°C.

11. Prepare plasmid minipreps from individual colonies. Analyze for both size (35 kb) and presence of pAdEasy-RGD-A/K, which has both ampicillin- and kanamycin-resistance genes.
12. Transform DH5α cells with pAdEasy-RGD-A/K and subject to large-scale amplification.

Construction of Fiber-modified Ad Vector

13. Digest 3 μg of pAdEasy-RGD-A/K with SwaI and run on a 2% agarose gel for 4 hours.

    This process removes the 1.3-kb kanamycin fragment.

14. Recover the 35-kb Ad backbone from the gel.
15. Recirculize the plasmid by ligation.

    The resultant fiber-modified rescue plasmid (pAdEasy-RGD) is an E1/E3-deleted rescue plasmid that has a fiber modified with the RGD-4C insertion in the HI loop in the knob.

16. Generate and purify the Ad transgene vector by homologous recombination with pShuttle CMV-Luc (shuttle plasmid) and the pAdEasy-RGD (fiber-modified backbone plasmid).

    The constitutively active cytomegalovirus (CMV) promoter is used for samples requiring only transductional targeting. For experiments examining both transductional and transcriptional targeting, the CMV promoter can be replaced by a TSP (see Protocol 3, Step 1).

17. Test the resultant Ad vector for targeting efficacy against cells expressing α_v integrin cell surface receptors.

Protocol 2

Construction of Fusion Proteins for Transductional Targeting

A recombinant fusion protein is generated that ablates the vector's native tropism by blocking the CAR receptor and retargets the vector to cells expressing epidermal growth factor (EGF; Fig. 4).

MATERIALS

CAUTION: See Appendix for appropriate handling of materials marked with <!>.

Reagents

Agarose
Ammonium sulfate (cold-saturated) <!>
Cell lines: 293 cells
G418 (500 μg/ml) <!>
Phosphate-buffered saline (PBS)
Plasmids
  pBsF5slEGF
  pcDNA3.1 (Invitrogen)
  pQBI-AdCMV5 (Qbiogene)

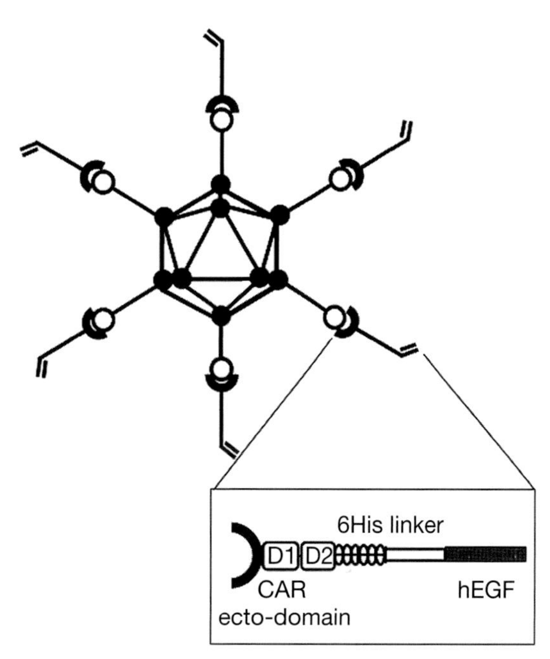

FIGURE 4. Targeting with the CAR ectodomain and an hEGF fusion protein. The amino terminus of the fusion protein consists of the extracellular portion of human CAR (D1 and D2) connected to human EGF (hEGF) via a short peptide linker (PSASASASAPGS) and preceded by a 6 histidine (6His) purification tag. The CAR domain binds to the Ad5 fiber knob, whereas the hEGF domain binds to EGFR present on the cells, thus facilitating Ad docking to EGFR-expressing cells.

PCR primers
  Primer E (5′-AAA CCG CCT ACC TGC AGC CG-3′)
  Primer F (5′-GAG CTT TAT TTG AAG GAG GGA CAA CG-3′)
  Primer G (5′-GAT CCC CCC GAT ATC ACC ATC ACC ATC ACT AAT AAA-3′)
  Primer H (5′-GGC CTT TAT TAG TGA TGG TGA TGG TGA TAT CGG GGG-3′)
  Primer I (5′-CCC ATT GGC CAT CAG CCT CCG CAT C-3′)
  Primer J (5′-GCC CCC GCT CGA GGT CGA CGG TAT C-3′)
Restriction enzymes: EcoRV, BamHI, MscI, NotI, PmeI, PvuI, SalI, XhoI

Equipment

Polymerase chain reaction (PCR) thermal cycler
TALON affinity resin, cobalt-immobilized (Clontech) <!>
Western blot hybridization materials and reagents
For additional equipment required for homologous recombination, amplification, and purification of Ad vectors, see Ross and Parks, Chapter 15, this volume.

METHODS

Generation of the shCAR-6His Fragment

1. Prepare the 6His purification tag.
   a. Synthesize the oligonucleotides 5′-GAT CCC CCC GAT ATC ACC ATC ACC ATC ACT AAT AAA-3′ and 5′-GAT CTT TAT TAG TGA TGG TGA TGG TGA TAT CGG GGG-3′.
      These form a DNA duplex encoding the His tag, two in-frame stop codons, and the BamHI and EcoRV restriction sites for fusing the CAR open reading frame with the 6His-coding sequence.
   b. Digest pQBI-AdCMV5 with BamHI.
   c. Clone the oligonucleotide duplex into BamHI-digested pQBI-AdCMV5.
   d. Sequence the resulting plasmid (pQBI-AdCMV5.6h) in the region of the insert.
      The correct orientation has a PmeI site upstream of the 6His-coding sequence.
2. Introduce the 6His purification tag into the carboxyl terminus of CAR.
   a. Digest pQBI-AdCMV5.6h with PmeI and EcoRV.
   b. Amplify the extracellular domain of human CAR (position 20-767) by PCR using primer E and primer F.
   c. Clone the PCR product into the PmeI/EcoRV digest.
      This plasmid (pQBIshCAR.6h) encodes the 236 amino-terminal amino acids of the extracellular domain of human CAR, including the signal sequence, fused with a carboxy-terminal 6His purification tag.
3. Insert the CAR-6His sequence into the pcDNA3.1 plasmid.
   a. Amplify by PCR the DNA sequence encoding the shCAR-6His from the pQBIshCAR.6h plasmid using primers G and H.
      The PCR fragment contains a unique 5′ BamHI site and 3′ NotI site to facilitate subsequent cloning.
   b. Digest both the PCR product and pcDNA3.1 with BamHI and NotI.
   c. Clone the PCR digest into BamHI- and NotI-digested pcDNA3.1.

This plasmid (pcDNAshCAR.6h) encodes the amino-terminal amino acids of the extracellular human CAR domain, including signal sequence, fused with a carboxy-terminal 6His purification tag.

Preparation of the Short Linker and Human EGF Fragment

4. Amplify by PCR the DNA sequence coding for a short flexible linker and human EGF (hEGF) from plasmid pBsF5slEGF using primers I and J.

   The PCR-derived DNA fragment contains a unique 5′ MscI site and 3′ SalI site to facilitate subsequent cloning.

5. Digest the PCR product with MscI and SalI.
6. Recover the purified 282-bp DNA fragment from an agarose gel.

Generation of the Recombinant CAR/hEGF Fusion Protein Sequence

7. Cleave pcDNAshCAR.6h with NotI.
8. Fill in the 3′ recessed ends with the Klenow fragment of *Escherichia coli* DNA polymerase I. Heat-inactivate the polymerase.
9. Cleave the plasmid with XhoI.
10. Ligate the 282-bp MscI- and SalI-digested PCR fragment (Step 6) into the digested pcDNAshCAR.6h.

    The resulting plasmid (pcDNAshCAR-EGF) encodes the recombinant fusion protein with bifunctional specificity for CAR and EGF receptors (EGFR).

Production of Stable CAR/hEGF Fusion Protein-expressing Cells

11. Linearize the pcDNAshCAR-EGF plasmid vector with PvuI.
12. Transfect 293 cells with the linearized vector.
13. The day after transfection, seed the cells in a 96-well plate and culture in medium supplemented with 500 μg/ml G418 until about two thirds of the well surface is covered with cells.
14. Expand the protein-expressing clones for future analyses.

Purification of the CAR/hEGF Fusion Protein

15. Collect the medium from stable CAR/hEGF fusion protein-expressing 293 cells.
16. Precipitate the proteins by adding an equal volume of cold-saturated ammonium sulfate.
17. Collect the precipitate by centrifugation. Dissolve the pellet in PBS to 1/20 of the original medium volume.
18. Dialyze against PBS.
19. Purify the recombinant protein via immobilized metal affinity chromatography using cobalt-immobilized TALON affinity resin.
20. Dialyze against PBS.
21. Verify protein expression by western blotting analysis.
22. Test the resulting adapter protein for targeting efficacy with an Ad vector.
    a. Complex Ad vectors with the adapter proteins.
    b. Assess the ability of the resulting human EGFR-targeted Ad to infect EGFR-expressing cells.

Protocol 3

Construction of Ad Vectors for Transcriptional Targeting

Recombinant Ad vectors that express the luciferase gene are constructed through homologous recombination in *E. coli* using the AdEasy system (see Ross and Parks, Chapter 15, this volume). TSPs are placed in front of the luciferase gene for selective expression. All vectors used in these experiments have the transgene cassettes in the E1-deleted region of the Ad vector backbone.

MATERIALS

CAUTION: See Appendix for appropriate handling of materials marked with <!>.

Reagents

AdEasy system (Qbiogene)
Agarose
Bacteria: BJ5183, DH5α
Cell lines: 293 cells
Cesium chloride <!>
Ethanol <!>
Glycerol
Kanamycin (25 μg/ml) <!>
Luria broth (LB) medium
Phosphate-buffered saline (PBS)
Plasmids: pGL3 Basic Vector (Promega)
Restriction enzymes: KpnI, PacI, PmeI, SalI
For additional materials required for homologous recombination, amplification, and purification of Ad vectors, see Ross and Parks, Chapter 15, this volume.

Equipment

For additional equipment required for homologous recombination, amplification, and purification of Ad vectors, see Ross and Parks, Chapter 15, this volume.

METHOD

1. Derive an appropriate TSP and clone it into the multicloning site of the pGL3 Basic Vector (Promega).

   Use the CMV promoter as a positive control.

2. Excise the "TSP-luciferase-poly(A)" region of the vector and clone it into pShuttle using the KpnI-SalI site.

   The resultant plasmid, pShuttle-promoter-Luc, has the TSP and luciferase gene in the ΔE1 region.

3. Perform homologous recombination between the shuttle (pShuttle-TSP-Luc) and rescue (pAdEasy-1) plasmids.
   a. Digest 10 μg of shuttle plasmid with PmeI for 4 hours and obtain a linear DNA fragment by 1% agarose gel extraction.
   b. Mix linear shuttle and undigested rescue plasmids in a 10:1 molar ratio.
   c. Transform BJ5183 cells with the plasmid mixture and plate on LB medium containing kanamycin (25 μg/ml). Incubate overnight at 30°C.
   d. Test individual colonies for both the size (35 kb) and the presence of the desired insert by plasmid minipreps.
      The correct plasmid is the Ad genome plasmid (pAd-TSP-Luc).
   e. Transform DH5α cells with pAd-TSP-Luc. Subject to large-scale amplification. Purify the DNA by small-scale alkali lysis.
4. Digest 3 μg of the plasmid encoding pAd-TSP-Luc with PacI for 4 hours. Collect the DNA by ethanol precipitation.
5. Transfect the linearized Ad genome DNA into a monolayer of 293 cells using a lipofection method.
   Plaque formation of 293 cells occurs at days 5–14 postinfection.
6. When the complete cytopathic effect is observed, collect the recombinant Ad virions from the cytoplasm by the freeze-thaw method.
7. Amplify and purify the Ad vector as follows:
   a. Amplify the vector by culture of 293 cells in suspension.
   b. Purify the vector by ultracentrifugation on step and continuous CsCl density gradients.
   c. Remove the CsCl by dialysis against PBS with 10% glycerol.
8. Determine virus titer by plaque assay in 293 cells.
9. Test the resultant Ad vector for targeting efficacy.
   Luciferase expression should only occur in the cell or tissue types targeted by the promoter of interest.

REFERENCES

Bergelson J.M., Cunningham J.A., Droguett G., Kurt-Jones E.A., Krithivas A., Hong J.S., Crowell R.L., and Finberg R.W. 1997. Isolation of a common receptor for coxsackie B viruses and adenoviruses 2 and 5. *Science* **275:** 1320–1323.

Bilbao R., Gerolami R., Bralet M.P., Qian C., Tran P.L., Tennant B., Prieto J., and Brechot C. 2000. Transduction efficacy, antitumoral effect, and toxicity of adenovirus-mediated herpes simplex virus thymidine kinase/ganciclovir therapy of hepatocellular carcinoma: The woodchuck animal model. *Cancer Gene Ther.* **7:** 657–662.

Breidenbach M., Rein D.T., Everts M., Glasgow J.N., Wang M., Passineau M.J., Alvarez R.D., Korokhov N., and Curiel D.T. 2004. Mesothelin-mediated targeting of adenoviral vectors for ovarian cancer gene therapy. *Gene Ther.* **12:** 187–193.

Culver K.W. and Blaese R.M. 1994. Gene therapy for cancer. *Trends Genet.* **10:** 174–178.

Davidoff A.M., Stevenson S.C., McClelland A., Shochat S.J., and Vanin E.F. 1999. Enhanced neuroblastoma transduction for an improved antitumor vaccine. *J. Surg. Res.* **83:** 95–99.

de Gruijl T.D., Luykx-De Bakker S.A., Tillman B.W., van den Eertwegh A.J., Buter J., Lougheed S.M., van der Bij G.J., Safer A.M., Haisma H.J., Curiel D.T., Scheper R.J., Pinedo H.M., and Gerritsen W.R. 2002. Prolonged maturation and enhanced transduction of dendritic cells migrated from human skin explants after in situ delivery of CD40-targeted adenoviral vectors. *J. Immunol.* **169:** 5322–5331.

Dmitriev I., Kashentseva E., Rogers B.E., Krasnykh V., and Curiel D.T. 2000. Ectodomain of coxsackievirus and adenovirus receptor genetically fused to epidermal growth factor mediates adenovirus targeting to epidermal growth factor receptor-positive cells. *J. Virol.* **74:** 6875–6884.

Dmitriev I., Krasnykh V., Miller C.R., Wang M., Kashentseva E., Mikheeva G., Belousova N., and Curiel D.T. 1998. An adenovirus vector with genetically modified fibers demonstrates expanded tropism via utilization of a coxsackievirus and adenovirus receptor-independent cell entry mechanism. *J. Virol.* **72:** 9706–9713.

Douglas J.T., Miller C.R., Kim M., Dmitriev I., Mikheeva G.,

Krasnykh V., and Curiel D.T. 1999. A system for the propagation of adenoviral vectors with genetically modified receptor specificities. *Nat. Biotechnol.* **17:** 470–475.

Freeman S.M. and Zwiebel J.A. 1993. Gene therapy of cancer. *Cancer Invest.* **11:** 676–688.

Goldman C.K., Rogers B.E., Douglas J.T., Sosnowski B.A., Ying W., Siegal G.P., Baird A., Campain J.A., and Curiel D.T. 1997. Targeted gene delivery to Kaposi's sarcoma cells via the fibroblast growth factor receptor. *Cancer Res.* **57:** 1447–1451.

Haisma H.J., Grill J., Curiel D.T., Hoogeland S., Van Beusechem V.W., Pinedo H.M., and Gerritsen W.R. 2000. Targeting of adenoviral vectors through a bispecific single-chain antibody. *Cancer Gene Ther.* **7:** 901–904.

Hakkarainen T., Hemminki A., Pereboev A.V., Barker S.D., Asiedu C.K., Strong T.V., Kanerva A., Wahlfors J., and Curiel D.T. 2003. CD40 is expressed on ovarian cancer cells and can be utilized for targeting adenoviruses. *Clin. Cancer Res.* **9:** 619–624.

Haviv Y.S., Blackwell J.L., Kanerva A., Nagi P., Krasnykh V., Dmitriev I., Wang M., Naito S., Lei X., Hemminki A., Carey D., and Curiel D.T. 2002. Adenoviral gene therapy for renal cancer requires retargeting to alternative cellular receptors. *Cancer Res.* **62:** 4213–4281.

Hemminki A., Dmitriev I., Liu B., Desmond R.A., Alemany R., and Curiel D.T. 2001. Targeting oncolytic adenoviral agents to the epidermal growth factor pathway with a secretory fusion molecule. *Cancer Res.* **61:** 6377–6381.

Herrmann F. 1995. Cancer gene therapy: Principles, problems, and perspectives. *J. Mol. Med.* **73:** 157–163.

Huard J., Lochmuller H., Acsadi G., Jani A., Massie B., and Karpati G. 1995. The route of administration is a major determinant of the transduction efficiency of rat tissues by adenoviral recombinants. *Gene Ther.* **2:** 107–115.

Kanerva A., Mikheeva G.V., Krasnykh V., Coolidge C.J., Lam J.T., Mahasreshti P.J., Barker S.D., Straughn M., Barnes M.N., Alvarez R.D., Hemminki A., and Curiel D.T. 2002. Targeting adenovirus to the serotype 3 receptor increases gene transfer efficiency to ovarian cancer cells. *Clin. Cancer Res.* **8:** 275–280.

Kashentseva E.A., Seki T., Curiel D.T., and Dmitriev I.P. 2002. Adenovirus targeting to c-*erb*B-2 oncoprotein by single-chain antibody fused to trimeric form of adenovirus receptor ectodomain. *Cancer Res.* **62:** 609–616.

Kawakami Y., Li H., Lam J.T., Krasnykh V., Curiel D.T., and Blackwell J.L. 2003. Substitution of the adenovirus serotype 5 knob with a serotype 3 knob enhances multiple steps in virus replication. *Cancer Res.* **63:** 1262–1269.

Kirn D., Martuza R.L., and Zwiebel J. 2001. Replication-selective virotherapy for cancer: Biological principles, risk management and future directions. *Nat. Med.* **7:** 781–787.

Korn T., Nettelbeck D.M., Volkel T., Muller R., and Kontermann R.E. 2004. Recombinant bispecific antibodies for the targeting of adenoviruses to CEA-expressing tumour cells: A comparative analysis of bacterially expressed single-chain diabody and tandem scFv. *J. Gene Med.* **6:** 642–651.

Krasnykh V.N., Mikheeva G.V., Douglas J.T., Curiel D.T. 1996. Generation of recombinant adenovirus vectors with modified fibers for altering viral tropism. *J. Virol.* **70:** 6839–6846.

Nettelbeck D.M., Rivera A.A., Kupsch J., Dieckmann D., Douglas J.T., Kontermann R.E., Alemany R., and Curiel D.T. 2004. Retargeting of adenoviral infection to melanoma: Combining genetic ablation of native tropism with a recombinant bispecific single-chain diabody (scDb) adapter that binds to fiber knob and HMWMAA. *Int. J. Cancer* **108:** 136–145.

Pereboev A.V., Nagle J.M., Shakhmatov M.A., Triozzi P.L., Matthews Q.L., Kawakami Y., Curiel D.T., and Blackwell J.L. 2004. Enhanced gene transfer to mouse dendritic cells using adenoviral vectors coated with a novel adapter molecule. *Mol. Ther.* **9:** 712–720.

Reynolds P.N., Dmitriev I., and Curiel D.T. 1999. Insertion of an RGD motif into the HI loop of adenovirus fiber protein alters the distribution of transgene expression of the systemically administered vector. *Gene Ther.* **6:** 1336–1339.

Schmidt-Wolf G. and Schmidt-Wolf I.G. 1994. Human cancer and gene therapy. *Ann. Hematol.* **69:** 273–279.

Stevenson S.C., Rollence M., Marshall-Neff J., and McClelland A. 1997. Selective targeting of human cells by a chimeric adenovirus vector containing a modified fiber protein. *J. Virol.* **71:** 4782–4790.

Stevenson S.C., Rollence M., White B., Weaver L., and McClelland A. 1995. Human adenovirus serotypes 3 and 5 bind to two different cellular receptors via the fiber head domain. *J. Virol.* **69:** 2850–2857.

van der Eb M.M., Cramer S.J., Vergouwe Y., Schagen F.H., van Krieken J.H., van der Eb A.J., Rinkes I.H., van de Velde C.J., and Hoeben R.C. 1998. Severe hepatic dysfunction after adenovirus-mediated transfer of the herpes simplex virus thymidine kinase gene and ganciclovir administration. *Gene Ther.* **5:** 451–458.

Varga M.J., Weibull C., and Everitt E. 1991. Infectious entry pathway of adenovirus type 2. *J. Virol.* **65:** 6061–6070.

Varnavski A.N., Calcedo R., Bove M., Gao G., and Wilson J.M. 2005. Evaluation of toxicity from high-dose systemic administration of recombinant adenovirus vector in vector-naive and pre-immunized mice. *Gene Ther.* **12:** 427–436.

Volk A.L., Rivera A.A., Kanerva A., Bauerschmitz G., Dmitriev I., Nettelbeck D.M., and Curiel D.T. 2003. Enhanced adenovirus infection of melanoma cells by fiber-modification: Incorporation of RGD peptide or Ad5/3 chimerism. *Cancer Biol. Ther.* **2:** 511–515.

Von Seggern D.J., Huang S., Fleck S.K., Stevenson S.C., and Nemerow G.R. 2000. Adenovirus vector pseudotyping in fiber-expressing cell lines: Improved transduction of Epstein-Barr virus-transformed B cells. *J. Virol.* **74:** 354–362.

Wickham T.J., Mathias P., Cheresh D.A., and Nemerow G.R. 1993. Integrins $\alpha_v\beta_3$ and $\alpha_v\beta_5$ promote adenovirus internalization but not virus attachment. *Cell* **73:** 309–319.

Wickham T.J., Tzeng E., Shears L.L 2nd, Roelvink P.W., Li Y., Lee G.M., Brough D.E., Lizonova A., and Kovesdi I. 1997. Increased in vitro and in vivo gene transfer by adenovirus vectors containing chimeric fiber proteins. *J. Virol.* **71:** 8221–8229.

18 Stable Producer Cell Lines for AAV Assembly

Gilliane Chadeuf and Anna Salvetti

Laboratoire de Thérapie Génique, INSERM U649, CHU Hôtel-Dieu, Bâtiment Jean Monnet, 44035 Nantes, France

ABSTRACT

Stable producer cell lines containing both the *rep* and *cap* genes and rAAV (recombinant adeno-associated virus) vectors provide a reliable and efficient procedure for the production of rAAV stocks, simply by infecting such producer cell clones with a helper virus. However, the development of such cell lines is time-consuming. Therefore, this method is recommended only for studies requiring the production of high amounts of rAAV, such as preclinical studies performed in large animals.

INTRODUCTION

In 1995, Clark et al. demonstrated that rAAV could be produced by stable cell lines. These authors generated HeLa cell clones with the AAV-2 *rep* and *cap* genes and an rAAV vector integrated into their genome. The clones efficiently assembled rAAV particles upon infection with wild-type adenovirus. Such stable cell lines significantly amplify the *rep-cap* genes upon adenovirus infection, despite the absence of viral inverted terminal repeats (Liu et al. 1999; Chadeuf et al. 2000; Liu et al. 2000; Tessier et al. 2001; Gao et al. 2002). This protocol describes a technique for obtaining stable rAAV producer cells. In a two-step process, stable cell clones containing the *rep* and *cap* genes (packaging cell clones) are generated. These are subsequently used to generate producer cell clones containing an integrated rAAV vector (Fig. 1).

Step 1: Isolation of stable packaging cell clones	
Transfection of HeLa or A549 cells with pRC and pPGK-neo^R	Day 2
Trypsinization and plating at different dilutions	Day 3
Geneticin selection	Day 4
Isolation of neo^R clones and amplification	3-4 weeks
PCR analysis to detect rep-cap containing clones	5-8 weeks
Screening by Southern blot of PCR-positive clones for rep-cap amplification	8-10 weeks
Screening of rep-cap clones for rAAV production	10-12 weeks

2: Isolation of stable producer cell clones	
Transfection of packaging cells with pAAV and pSV-hygro^R	Day 2
Trypsinization and plating at different dilutions	Day 3
Hygromycin B selection	Day 4
Isolation of hygro^R clones and amplification	3-4 weeks
PCR analysis to detect rAAV containing clones	5-8 weeks
Screening PCR-positive clones for rAAV production	8-10 weeks
Stability studies and large scale rAAV production	10-14 weeks

FIGURE 1. Selection procedure for the isolation of stable packaging and producer cell clones.

Protocol

Techniques for Obtaining Stable rAAV Producer Cells

MATERIALS

CAUTION: See Appendix for appropriate handling of materials marked with <!>.

Reagents

AAV plasmids

pAAV: Construct containing the rAAV vector cloned in a plasmid backbone

pRC: Plasmid containing the AAV *rep* and *cap* genes expressed under their native promoters with no viral inverted terminal repeats

The nature of these plasmids will vary according to the AAV serotype to be produced and with the rAAV vector used.

Culture medium

Dulbecco's modified Eagle's medium (Sigma-Aldrich) containing 10% fetal bovine serum (Research Grade, EU approved, Hyclone, Perbio), 1% penicillin (50 units/ml) <!>, and streptomycin (Cambrex, 50 μm/ml) <!>

Geneticin stock solution (100 mg/ml, w/v)

Dissolve 5 g of geneticin (Invitrogen) in 50 ml of sterile distilled H_2O. Filter the stock solution with a 0.2-μm filter and store aliquots at –20°C.

HeLa (ATCC CCL-2) or A549 cells (ATCC CCL-185)

Hygromycin B stock solution (10 mg/ml, w/v) <!>

Dissolve 1 g of hygromycin B (Sigma-Aldrich) in 100 ml of sterile distilled H_2O. Filter the stock solution with a 0.2-μm filter and store aliquots at –20°C.

Selection plasmids

pPGK-*neo*: Plasmid encoding for the neomycin resistance gene under the control of the mouse phosphoglycerate kinase-1 promoter

pSV-*hygro*: Plasmid encoding for the hygromycin resistance gene under the control of the viral SV40 promoter region

Any construct expressing a selection gene can substitute for these plasmids.

1x Trypsin <!> /EDTA solution

Dilute the 10x trypsin/EDTA stock solution (Sigma-Aldrich) with sterile 0.9% NaCl. Aliquots may be stored at –20°C.

Wild-type adenovirus 5 (Ad5) (ATCC VR-5)

METHODS

Generation of Stable Packaging Lines

1. Twenty-four hours before transfection, seed HeLa or A549 cells in 10-cm tissue-culture dishes at a density of 3 x 10^6.

   Adjust cell density for each cell line. Make sure that cells are 50–80% confluent before transfection.

2. Cotransfect the cells with 10 μg of pRC and 1.0 μg of pPGK-*neo* using the $CaPO_4$ procedure.
3. Twenty-four hours later, trypsinize cells, dilute suspensions serially (1/5, 1/10, 1/20, 1/40), and replate each dilution in duplicate in 10-cm plates. Incubate the cultures for 20–24 hours.
4. Remove the culture medium by aspiration and add fresh culture medium containing 1 mg/ml of geneticin.
5. Continue incubation for 3–4 weeks. Change the culture medium every 2 days with fresh medium containing 1 mg/ml of geneticin.

   Massive cell death usually occurs 7–14 days posttransfection. Individual clones appear around 3–4 weeks posttransfection. It is important to keep the cells continuously under selection with geneticin during this period.
6. Mark and isolate individual clones by viewing the plates from the bottom. Trypsinize clones using cloning rings or use a pipette tip to directly scrape clones from the plate.
7. Amplify clones by incubation with geneticin-free culture medium to obtain two confluent 10-cm plates.
8. Harvest cells on one plate and store aliquots (~5 x 10^6 cells /vial) at –80°C. With the second plate, prepare genomic DNA.

Screening of Clones for Presence of *rep* and/or *cap* DNA

9. Twenty-four hours before viral infection, thaw and amplify PCR (polymerase chain reaction)-positive clones and seed the cells in two wells in a six-well plate (approx. 5 x 10^5 cells/well).

   Screen clones for the presence of *rep* and/or *cap* DNA by PCR performed on genomic DNA using appropriate primers. Use the pRC plasmid as a positive control.
10. Infect the cells in one of the wells with wild-type Ad5 using an moi (multiplicity of infection) of 50 for HeLa and A549 cells.

    Adjust the moi to obtain at least 90% of infected cells. A cytopathic effect should be clearly visible in the infected well.
11. Harvest the cells 48 hours postinfection and prepare genomic DNA using standard methods. Analyze *rep-cap* amplification by Southern blot analysis using a probe hybridizing in the *rep* or the *cap* gene (Fig. 2).
12. Seed clones exhibiting adenovirus-induced *rep-cap* amplification in a six-well plate, transfect with the pAAV plasmid (2 μg/well), and infect with wild-type Ad5 (moi = 50).

    The number of rAAV particles produced can be directly measured in the cell lysate by dot blot analysis (Snyder et al. 1996). As a positive control, produce rAAV using 293 cells cotransfected with the pRC and pAAV plasmids and infected with wild-type Ad5 (moi = 5).

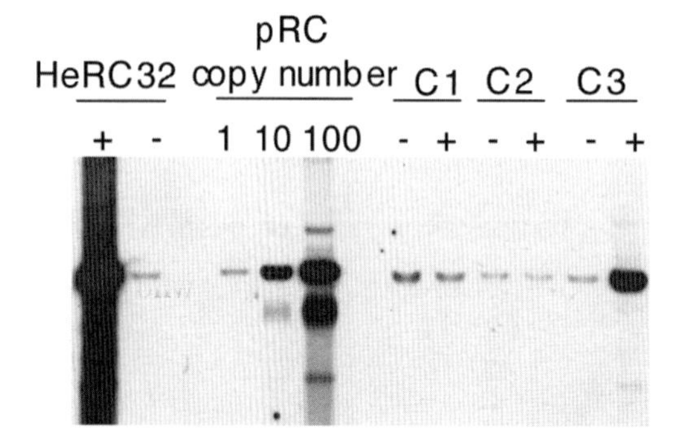

FIGURE 2. Detection of *rep* gene amplification by Southern blot hybridization. Genomic DNA was extracted from three different stable *rep-cap* cell clones (C1, C2, and C3) infected (+) or not (–) with wild-type Ad5, digested with the appropriate enzyme, and analyzed by Southern blot hybridization using a *rep* probe. Despite the presence of integrated *rep* sequences in the three clones analyzed, the adenovirus-induced amplification of these sequences is detected only in clone C3, as also observed with the control HeRC32 packaging cell line (see Chadeuf et al. 2000).

13. Amplify clones producing at least 10^4 rAAV particles/cell and store aliquots at –20°C.

    For further analyses of these cell clones, include a stability study to determine if rAAV is produced at a constant level during at least 30 passages. Also, analyze Rep and Cap protein synthesis by western blot hybridization after infection of the cells with wild-type Ad5.

Generation of Stable Producer Clones

14. Twenty-four hours before transfection, seed cloned packaging cells as prepared above in 10-cm tissue culture dishes at a density of 3 x 10^6 cells/well.
15. Cotransfect cells with 10 µg of pAAV and 1.0 µg of pSV-*hygro* using the $CaPO_4$ procedure.
16. Twenty-four hours later, trypsinize cells, dilute suspensions serially (1/5, 1/10, 1/20, 1/40), and replate each dilution in duplicate in 10-cm plates. Incubate cultures for 20–24 hours.
17. Remove the culture medium by aspiration and add fresh culture medium containing 0.3 mg/ml of hygromycin B.
18. Continue incubation for 3–4 weeks. Change the culture medium every 2 days with fresh medium containing 0.3 mg/ml of hygromycin B.
19. Mark and isolate individual clones by viewing the plates from the bottom. Trypsinize clones using cloning rings or use a pipette tip to directly scrape clones from the plate.
20. Amplify the clones by incubation with hygromycin-B-free culture medium to obtain two confluent 10-cm plates.
21. Harvest cells on one plate and store aliquots (~5 x 10^6 cells /vial) at –80°C. With the second plate, prepare genomic DNA.

    Screen clones for the presence of the rAAV vector by PCR analysis performed on genomic DNA using appropriate primers. Use the rAAV vector plasmid as a positive control.

22. Seed PCR-positive clones in a six-well plate and infect with wild-type Ad5 (moi = 50). Determine the efficiency of rAAV particle production by dot-blot analysis (Snyder et al. 1996).

    For further analyses on the positive clones, include Southern blot analysis to measure the number of rAAV vector copies per cell, a stability assay to measure the level of rAAV production achieved during at least 30 passages, and determination of large-scale rAAV production (5–20 15-cm dishes) (Salvetti et al. 1998).

TROUBLESHOOTING

Problem: Contaminating helper virus particles are generated.
Solution: The wild-type Ad5 can be substituted with a replication-defective helper virus (Toublanc et al. 2004).

ACKNOWLEDGMENTS

We thank Jacques Tessier, Véronique Blouin, and Estelle Toublanc who contributed to the establishment of this protocol. This work was funded by the French Muscular Dystrophy Association, the Association Nantaise pour la Thérapie Génique (ANTG), the Fondation pour la Thérapie Génique en Pays de la Loire, and the INSERM.

REFERENCES

Chadeuf G., Favre D., Tessier J., Provost N., Nony P., Kleinschmidt J., Moullier P., and Salvetti A. 2000. Efficient recombinant adeno-associated virus production by a stable *rep-cap* HeLa cell line correlates with adenovirus-induced amplification of the integrated *rep-cap* genome. *J. Gene Med.* **2:** 260–268.

Clark K.R., Voulgaropoulou F., Fraley D.M., and Johnson P.R. 1995. Cell lines for the production of recombinant adeno-associated virus. *Hum. Gene Ther.* **6:** 1329–1341.

Gao G.P., Lu F., Sanmiguel J.C., Tran P.T., Abbas Z., Lynd K.S., Marsh J., Spinner N.B., and Wilson J.M. 2002. *rep/cap* gene amplification and high-yield production of AAV in an A549 cell line expressing *rep/cap*. *Mol. Ther.* **5:** 644–649.

Liu L., Clark K.R., and Johnson P.R. 1999. Production of recombinant adeno-associated virus vectors using a packaging cell line and a hybrid recombinant adenovirus. *Gene Ther.* **6:** 293–299.

Liu X., Voulgaropoulou F., Chen R., Johnson P.R., and Clark K.R. 2000. Selective *rep-cap* gene amplification as a mechanism for high-titer recombinant AAV production from stable cell lines. *Mol. Ther.* **2:** 394–403.

Salvetti A., Orève S., Chadeuf G., Favre D., Cherel Y., Champion-Arnaud P., David-Ameline J., and Moullier P. 1998. Factors influencing recombinant adeno-associated virus production. *Hum. Gene Ther.* **9:** 695–706.

Snyder R., Xiao S., and Samulski R.J. 1996. Production of recombinant adeno-associated viral vectors. In *Current protocols in human genetics* (ed. N. Dracopoli et al.), pp. 12.1.1–12.1.23. John Wiley & Sons, New York.

Tessier J., Chadeuf G., Nony P., Avet-Loiseau H., Moullier P., and Salvetti A. 2001. Characterization of adenovirus-induced inverted terminal repeat-independent amplification of integrated adeno-associated virus *rep-cap* sequences. *J. Virol.* **75:** 375–383.

Toublanc E., Abdellatif B., Bonnin D., Blouin V., Brument N., Cartier N., Epstein A.L., Moullier P., and Salvetti A. 2004. Identification of a replication-defective herpes simplex virus for recombinant adeno-associated virus type 2 (rAAV2) particle assembly using stable producer cell lines. *J. Gene Med.* **6:** 555–564.

19 Strategies for the Design of Hybrid Adeno-associated Virus Vectors

Aravind Asokan and R. Jude Samulski

Gene Therapy Center, University of North Carolina, Chapel Hill, North Carolina 27599

ABSTRACT

Hybrid adeno-associated virus (AAV) vectors can be exploited to achieve tissue-specific gene expression or the transduction of refractory cell types. Conventionally, restriction of AAV transduction to specific tissue types has been achieved through a battery of biochemical techniques including chemical conjugation, insertion of peptide ligands, and generation of epitope-tagged capsid subunit fusion constructs. Recent trends in the field of retargeting AAV vectors include the application of combinatorial techniques such as AAV capsid peptide display libraries, error-prone PCR, and DNA shuffling. Expansion of AAV tropism to refractory tissue types can be achieved by exploiting the diverse tissue tropisms of AAV serotypes generated by natural mutagenesis and recombination. Alternatively, serotype homology can be exploited to engineering hybrid vectors from different serotypes by transcapsidation or marker rescue. Because AAV2 is the best-characterized serotype, this serotype often serves as a convenient starting template for generating hybrid AAV vectors. This chapter focuses on strategies to engineer hybrid AAV vectors with expanded tissue tropisms.

INTRODUCTION

Due to their relatively low immunogenicity and high transduction efficiency, recombinant AAV serotype 2 (AAV2) vectors have advanced to the forefront of human gene therapy. However, some studies suggest that the transduction efficiency of AAV2 vectors in certain tissue types falls short of requirements for adequate levels of transgene expression (Zabner et al. 2000; Thomas et al. 2004). There is an urgent need to enable gene expression in cell types refractory to AAV2, control vector dissemination, and reduce vector dose administered in vivo. To these ends, research has focused on the use of tissue-specific promoters and manipulation of AAV capsids to generate vectors with enhanced transduction and/or targeting efficiency. One approach is to exploit the diverse tissue tropisms of naturally occurring AAV serotypes (Table 1). To date, 11 primate AAV serotypes have been cloned, sequenced, and developed as gene therapy vectors. Due to the high efficiency of these serotypes, the vector load can be reduced. They also have an advantage over AAV2 vectors in their ability to evade preexisting neutralizing antibodies generated by a humoral immune response to natural infection or prior treatment with AAV vectors (Peden et al. 2004).

Tissue tropisms of AAV serotype vectors likely arise due to the cumulative effects of viral binding to specific cell surface receptors and intracellular processing (Ding et al. 2005). An understanding of crystal structure and functional correlates of AAV serotype capsids is therefore critical for elucidating the mechanism(s) of tissue tropism as well as the design of hybrid vectors. The crystal structure of AAV2 has been determined at a resolution of 3 Å using X-ray crystallography (Xie et al. 2002), whereas the structures of AAV4 (Padron et al. 2005) and AAV5 (Walters et al. 2004) have been elucidated using cryo-electron microscopy and image reconstruction at resolutions of 13 Å and 16 Å, respectively. Crystal structures of other key AAV serotype capsids, notably AAV1 and AAV8, are currently in progress.

Currently, a "cross-packaging" strategy is used to generate transcapsidated AAV vectors. Expression of AAV2 Rep proteins together with capsid proteins of a different serotype results in formation of viral particles that package AAV2 vector genomes (Rabinowitz et al. 2002). Alternatively, baculovirus expression vector systems for AAV vector production in insect *Sf9* cells have been developed (Urabe et al. 2002). Several groups have exploited the cross-packaging strategy to compare the transduction efficiencies of AAV serotype vectors in different tissues (Grimm et al. 2003; Burger et al. 2004; Wang et al. 2005).

TABLE 1. Relative Transduction Efficiencies of AAV Serotype Vectors in Major Tissues

Tissue type	Optimal serotype(s)	References
Liver	AAV8, AAV9[a]	Grimm et al. (2003)
Pancreas	AAV8	Wang et al. (2004); Loiler et al. (2005)
Lung	AAV1, AAV6, AAV9[a]	Halbert et al. (2001)
Heart	AAV8, AAV9[a]	Wang et al. (2005)
Skeletal muscle	AAV1, AAV6, AAV7, AAV8, AAV9[a]	Xiao et al. (1999); Blankinship et al. (2004); Louboutin et al. (2005); Wang et al. (2005)
Brain	AAV1, AAV5, AAV4	Davidson et al. (2000); Burger et al. (2004)
Kidney	AAV2	Takeda et al. (2004)
Eye	AAV4, AAV5	Lotery et al. (2003); Weber et al. (2003)

[a]Multiple abstracts from the American Society of Gene Therapy Annual Meeting (2005).

AAV VECTORS WITH SELECTIVE TROPISM

The AAV virion has a $T = 1$ icosahedral capsid consisting of 60 copies of three related proteins, VP1, VP2, and VP3 (Rose et al. 1971). The three proteins share a common carboxy-terminal region, but have different amino termini resulting from alternative start codon usage. The core of the protein is composed of a conserved eight-stranded antiparallel β-barrel motif. The majority of the variable surface structure consists of large loops inserted between strands of the β barrel. This variable region spans the center of the primary sequence (residues 440–600, AAV2 VP1 numbering), whereas residues located at the amino and carboxyl termini are conserved. Other regions that show the most variability among AAV serotypes are near residues 260 and 380 (AAV2 VP1 numbering). With rapidly advancing knowledge of AAV capsid structure, rational manipulation of the AAV capsid surface through site-directed/insertional mutagenesis presents a promising strategy for engineering novel AAV vectors. In this regard, restriction of AAV transduction to specific tissue types can be achieved through a battery of techniques that include chemical conjugation, insertion of peptide ligands, and epitope-tagged capsid subunit fusion constructs (for review, see Choi et al. 2005; Muzyczka and Warrington 2005). Other recent trends in the field of retargeting AAV vectors include the application of combinatorial techniques such as the generation of large random mutant AAV libraries by error-prone polymerase chain reaction (PCR) and DNA shuffling (see, e.g., Perabo et al. 2005; Maheshri et al. 2006). Future collaboration between rational design and combinatorial approaches will likely enable the directed evolution of novel AAV vectors tailored to fit individual disease and/or patient profiles.

Insertion of Peptide Ligands

AAV vectors can be retargeted to cell-type-specific receptors by inserting corresponding peptide ligands into the capsid sequence. This involves engineering short peptide sequences that encode specific receptor ligands into the capsid open reading frame. Care must be taken to ensure that the site of insertion does not affect the viability of the virion and permits proper surface display and presentation of the ligand to the target receptor. Thus far, all attempts to retarget AAV vectors have been limited to modifying the surface architecture of AAV2 (Nicklin and Baker 2002; Buning et al. 2003).

Before the availability of AAV2 crystal structure information (Xie et al. 2002), insertional (Rabinowitz et al. 1999; Shi et al. 2001) and extensive scanning (Wu et al. 2000) mutagenesis of the capsid open reading frame identified a number of sites that tolerated insertion of surface-displayed foreign epitopes. For example, hemagglutinin epitope tag insertions identified several regions on the capsid surface (Wu et al. 2000). These included insertions at amino acids 1, 34, 138, 266, 447, 591, and 664 (VP1 numbering) with positions located in the VP1 region or the putative loop regions of VP3. A related study established that the display of foreign peptide epitopes on the AAV capsid surface is dependent on the inclusion of appropriate scaffolding sequences (Shi et al. 2001). Optimal scaffolding sequences and five preferred sites for the insertion of targeting peptide epitopes were identified. These sites are located within each of the three AAV capsid proteins, and thus display inserted epitopes 3, 6, or 60 times per vector particle. Similarly, position 138, which represents the end of the VP1 unique region, can tolerate insertion of a peptide ligand from apolipoprotein E that targets the low-density lipoprotein receptor (Loiler et al. 2003).

An important consideration in designing retargeted AAV vectors is whether the peptide insertion should simultaneously disrupt endogenous receptor usage. Conservation of endogenous receptor-binding sites can result in vectors with expanded, rather than selective, tropism. On the other hand, insertion of the peptide ligand at a position that disrupts native receptor binding facilitates selective retargeting. For example, the broad tropism of AAV2 is likely due to ubiquitous cell surface expression of heparan sulfate proteoglycans (Summerford and Samulski 1998). Identification and mutation of residues critical for heparan sulfate binding (Kern et al. 2003; Opie et al. 2003) provide a "detargeted" platform for engineering heparan-sulfate-independent targeting mechanisms into the AAV capsid. In a particularly striking example, mutation of arginine residues at positions 484 and 585 (critical for heparin binding) to glutamate residues resulted in vectors that transduced the heart instead of the liver (Kern et al. 2003).

Of several positions tested for peptide insertion, residue 587 has received the most attention due to its ability to disrupt heparin binding by AAV2. Girod et al. (1999) exploited this surface loop position by insertion of the RGD motif recognized by various integrins. The modified vectors were capable of transducing the B16F10 cell line expressing integrin receptors that were normally refractory to recombinant AAV2 infection. Furthermore, Grifman et al. (2001) reported selective transduction of tumor vasculature by inserting an NGR-sequence-containing peptide known to bind the aminopeptidase or CD13 receptor. In similar studies, other cell-type-selective peptides were identified by phage display. The sequence SIGYPLP and MTP-containing peptide (which selectively target endothelial cells) and peptides containing the EYH core motif (targeting smooth muscle cells) have been inserted into position 587, resulting in successful cell-type-specific transduction (Nicklin et al. 2001; White et al. 2004; Work et al. 2004). At present, rigorous studies to determine the effect of substitution of different receptor-binding ligands on AAV entry/intracellular trafficking pathways have not been done. A thorough understanding of the intricate role of receptor-ligand interactions in AAV infection will require a combination of crystal structure information and biochemical/biological studies in vitro and in vivo.

Peptide Display Libraries

The principle of phage display libraries has been extended to AAV particles with randomized peptide libraries at position 587. Such a combinatorial approach enables biological panning of AAV vector mutants on a specific cell type in vitro (Muller et al. 2003; Perabo et al. 2003). Unlike insertion of peptides through rational structural analysis, these methods involve rescue, propagation, and sequence analysis of capsid genes selected for infection of a specific cell type. In principle, this approach ensures that the modified virus is optimized not only for cell surface receptor binding, but also for efficient transduction of target cells. Perabo et al. (2003) created an AAV library with seven random residues flanked by alanines. These were serially passaged in MO7e and Mec1 cells, which are known to be nonpermissive for AAV2 transduction. Vectors rescued from MO7e cells had a core RGD motif, whereas those rescued from Mec1 cells were also able to transduce primary B-CLL cells with high efficiency. In a similar study, Muller et al. (2003) selected for vectors with up to a 600-fold increase in transduction efficiency of coronary artery endothelial cells. In addition, one of these vectors tested in vivo displayed a tropism for heart tissue and lower transduction in the liver.

Capsid Subunit-Ligand Fusion Proteins

The isolated expression of individual AAV capsid proteins offers several advantages to tailoring such proteins, including the fusion of larger peptides or proteins to the AAV capsid subunits. Techniques used to synthesize such vectors are similar to the production of mosaic AAV vectors (Rabinowitz et al. 2004), wherein unmodified subunits preserve capsid integrity while subunit fusion constructs are incorporated at well-tolerated levels within the capsid during assembly. One of the first examples of retargeting AAV2 using a large targeting moiety utilized a single-chain antibody fragment recognizing CD34 fused to the amino terminus of the VP2 capsid protein (Yang et al. 1998). When produced in the presence of wild-type VP2, this construct formed mosaic particles containing both wild-type and modified VP2 proteins. To limit the fusion to specific capsid proteins, all three capsid proteins were expressed separately under the control of a strong cytomegalovirus (CMV) promoter. Whereas VP1 and VP3 could not tolerate this fusion, the single-chain antibody fused to the VP2 amino terminus generated vector particles with selective transduction of the CD34 myeloleukemic cell line, albeit at extremely low efficiency. However, this work has not been reproduced by other laboratories and it remains unclear whether stable virions were indeed being assembled or aggregated clusters of virion proteins were being formed.

A recent study established the amino terminus of the VP2 subunit as a novel site for fusion of proteins as large as 30 kD (Warrington et al. 2004). The VP2 fusion construct was transfected in *trans* at different ratios with unmodified VP1 and VP3 subunits. Using this technique, nearly wild-type levels of AAV particles with large ligands could be inserted after residue 138 in VP1 and VP2 or in VP2 exclusively, including the 8-kD chemokine-binding domain of rat fractalkine, the 18-kD human hormone leptin, or green fluorescent protein (GFP). Although insertions at residue 138 in VP1 significantly decreased infectivity, insertions at residue 138 that were exclusively in VP2 had a minimal effect on viral assembly or infectivity.

Chemical Conjugation Strategies

Several indirect methods to retarget AAV vectors that do not involve modification of the AAV capsid gene sequence have been reported. These methods have the advantage of allowing conjugation of targeting ligands of any size. For example, bispecific antibodies can mediate interaction between the AAV capsid and a specific cell surface receptor expressed on human megakaryocytes (Bartlett et al. 1999). Another approach involves chemical derivatization of the AAV with biotin moieties followed by the noncovalent adsorption of streptavidin-ligand fusion proteins to the capsid surface (Ponnazhagan et al. 2002). Although successful in vitro for targeting αIIbβ3 integrin, epidermal growth factor, and fibroblast growth factor receptors (Ponnazhagan et al. 2002), the stability of these conjugated vectors in vivo remains to be seen. Direct chemical conjugation of targeting ligands that risk inactivating key amino acids on the capsid surface has not been explored.

Another approach to conjugate targeting motifs to the AAV capsid surface has been proposed by Ried et al. (2002). The strategy involves insertion of a 34-residue fragment of the protein A (Z34C) IgG-binding domain at position 587. Subsequently, monoclonal antibodies to CD29, CD117, or CXCR4 are conjugated to the vector surface for selective targeting of respective receptors, albeit at low efficiency. Limiting the number of Z34C fragments per particle had no effect on the ability of monoclonal antibodies to be conjugated to the surface of the vectors and allowed specific retargeting of recombinant AAV2 to the CD29 and CD117 receptors with significantly higher transduction efficiency. Although insertion of the Z34C peptide resulted in low particle yields and decreased infectious titer, use of the mosaic capsid strategy can improve Z34C vector particle yields (Gigout et al. 2005).

AAV VECTORS WITH EXPANDED TROPISM

The following protocols describe strategies to engineer AAV vectors with expanded tissue tropisms. Several AAV serotypes that have evolved through natural mutagenesis and recombination provide alternative vectors for gene therapy (Grimm and Kay 2003). In addition, strategies that exploit serotype homology can be utilized to mix and match features from different serotypes. Thus, expansion of AAV tropism to refractory tissue types can be achieved by using AAV vectors derived from AAV serotypes (Protocol 1) and by engineering hybrid AAV serotype vectors through transcapsidation (Protocol 2) or marker rescue (Protocol 3).

Protocol 1

Production of AAV Serotype Vectors

Production of recombinant AAV serotype vectors requires transfection of the following plasmid components in host 293 cells: (1) vector genome containing the transgene expression cassette flanked by two inverted terminal repeats (ITRs), (2) an AAV helper plasmid in *trans* capable of expressing AAV2 Rep and AAV serotype-specific Cap proteins, and (3) adenovirus helper genes (Xiao et al. 1998).

MATERIALS

CAUTION: See Appendix for appropriate handling of materials marked with <!>.

Reagents

Benzonase (Sigma-Aldrich)
Cells: Human embryonic kidney (HEK) 293 cells (ATCC)
Cesium chloride (Sigma-Aldrich) <!>
Dulbecco's modified Eagle's medium (DMEM) with 10% fetal bovine serum and 1% penicillin/streptomycin (Sigma-Aldrich) <!>
Leupeptin (3 μg/ml) in double distilled (dd)H_2O (Sigma-Aldrich) <!>
Phosphate-buffered saline (PBS) without Ca^{++}/Mg^{++} (Sigma-Aldrich)
Plasmids
  AAV serotype helper plasmids (e.g., pXR1, pXR2, pXR3, pXR4, and pXR5; UNC Vector Core)
  Adenoviral helper plasmid (e.g., pXX6-80; UNC Vector Core)
  AAV transgene packaging cassette (e.g., pTR-CMV-GFP; UNC Vector Core)
Superfect (QIAGEN)
Trypsin-EDTA (Sigma-Aldrich) <!>

Equipment

Centrifuge and ultracentrifuge systems (Sorvall RT 6000B)
Centrifuge tubes (16 x 76 mm) (Beckman polyallomer Quick-Seal)
Dialysis cassettes (Pierce Slide-A-Lyzer) (10,000 kD m.w. cutoff)
Sonicator (3-mm probe)
Ultracentrifuge (Beckman Optima TLX) with TLN 100 rotor

METHOD

1. Trypsinize a 15-cm plate of confluent HEK 293 cells (2×10^7). Dilute with medium in a 1:3 ratio. Replate and allow the cells to recover overnight.
2. Transfect each plate (total 5–10) with 7 μg of pTR-CMV-GFP (transgene packaging cassette), 15 μg of pXX6-80 (adenoviral helper plasmid), and 10 μg of a given AAV serotype helper plasmid (pXR1-5) using the Superfect reagent according to the manufacturer's instructions.

3. Harvest the cells at 48–60 hours posttransfection by scraping.

   Perform all subsequent manipulations on ice unless otherwise noted.
4. Centrifuge the cells at 1000 rpm (Sorvall RT 6000B). Resuspend the pellet in leupeptin.
5. Sonicate the cell lysate with 25 bursts at 50% duty and an output control setting of 2.
6. Remove cell debris by centrifugation at 1000 rpm.
7. Add benzonase to a final concentration of 1.2 units/μl.
8. Incubate for 30–60 minutes at 37°C.
9. Add 0.6 g of cesium chloride to each milliliter of lysate. Transfer the suspension into Beckman polyallomer tubes.
10. Centrifuge the mixtures in a Beckman Optima TLX ultracentrifuge at 100,000 rpm for 4 hours using a TLN 100 rotor.
11. Determine peak fractions from the gradient by dot-blot hybridization (Rabinowitz et al. 1999) using the radiolabeled (GFP transgene) probe.
12. Dialyze the peak fractions using Pierce dialysis cassettes against PBS in a cold room overnight and again determine viral titer (number of vector genomes/ml).

Protocol 2

Generation of Mosaic Vectors through Transcapsidation

A mosaic virion can be defined as a capsid structure composed of a mixture of capsid subunits from different serotypes. Such particles can be generated using a mixture of helper constructs that encode capsid proteins from different serotypes, wild-type and mutant capsid proteins of the same serotype, or two different mutant capsid subunits of the same serotype (Fig. 1). In theory, the ratio of capsid subunits from different sources in the mosaic virion should reflect the input ratio of different AAV helper constructs, although this has not been proved experimentally. The unique advantage of this technique is the ability to combine selective features from different sources that synergistically enhance transgene expression. The following protocol, adapted from Rabinowitz et al. (2004), can generate mosaic vectors by mixing pairwise combinations of different AAV serotypes at several input ratios.

Materials and equipment for mosaic AAV production are the same as those described in Protocol 1.

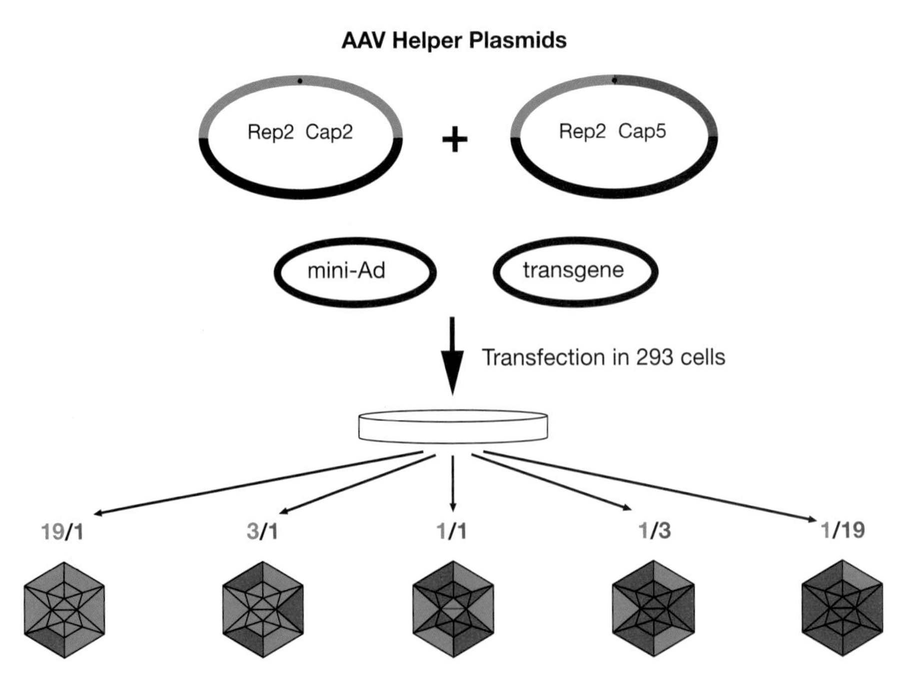

FIGURE 1. Plasmid mixing approach for generating mosaic AAV capsids (Protocol 2) (Rabinowitz et al. 2004). Helper plasmid DNA containing the capsid gene from any two AAV serotypes (e.g., Cap2 in *blue* and Cap5 in *red*) are cotransfected at different mass ratios (19:1, 3:1, 1:1, 1:3, and 1:19) using the triple plasmid transfection protocol (Xiao et al. 1998) for AAV vector production in 293 cells. The topology maps indicate mosaic virion shells produced using this technique and are colored to represent potential proportions of subunits that reflect the ratio of helper plasmids with each capsid gene in the transfection mixtures.

METHOD

1. Trypsinize a 15-cm plate of confluent HEK 293 cells (2×10^7). Dilute with medium in a 1:3 ratio. Replate and allow the cells to recover overnight.
2. Transfect each plate (total 5–10) with 7 μg of pTR-CMV-GFP (transgene packaging cassette), 15 μg of pXX6-80 (adenoviral helper plasmid), and 10 μg of a mixture of AAV serotype helper plasmids (pXR1-5) using the Superfect reagent according to the manufacturer's instructions.

   To generate AAV virions containing capsid components of different serotypes, the amount of each AAV serotype helper plasmid transfected must be varied depending on the ratio being assessed. For example, if AAV1 and AAV2 are to be mixed at a ratio of 19:1, then 9.5 μg of pXR1 and 0.5 μg of pXR2 must be added to the transfection mixture.
3. Subsequent production of mosaic AAV vectors is achieved by following Steps 3–12 described in Protocol 1.

Protocol 3

Generation of Chimeric AAV Vectors through Marker Rescue

Chimeric virions are vectors containing capsid proteins that have been modified by domain swapping between different serotypes. Strategies for the generation of chimeric virions primarily involve the marker rescue approach or mutagenesis of AAV virions to swap surface domains ranging from single to multiple amino acid residues. The marker rescue strategy (Bowles et al. 2003) exploits the sequence homology between AAV serotypes to serve as crossover points for recombination initiated by cellular proteins (Fig. 2). The recombination of sequences can result in the "rescue" of infectious or targeted phenotypes in mutant virions through directed selection of functional capsid subunits that assemble into viable virions. Such "forward engineering" could serve as a powerful tool for generating chimeric virions with unique properties governed by the criteria set in the screening process.

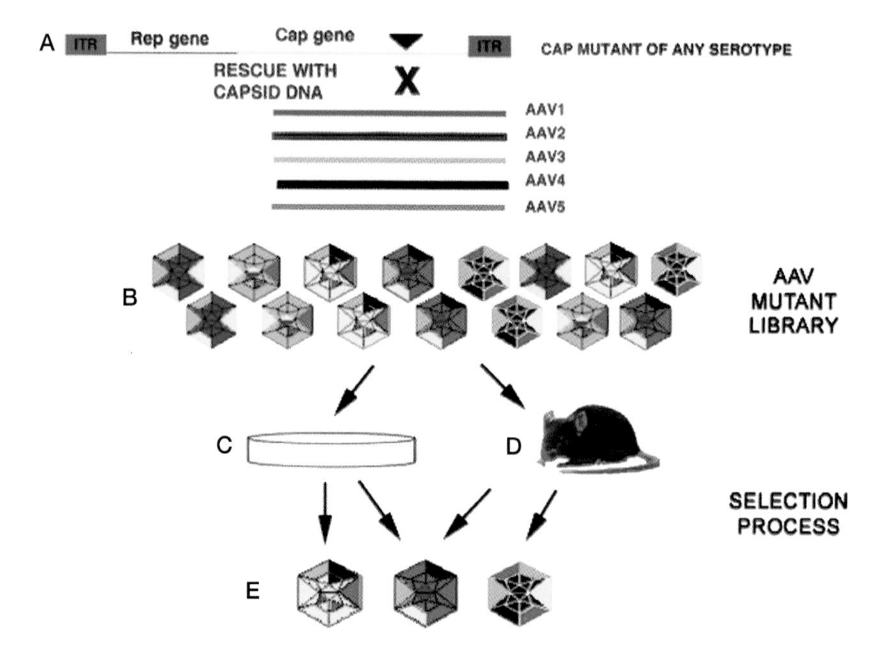

FIGURE 2. Marker rescue approach for generating chimeric AAV vectors (Protocol 3) (Bowles et al. 2003). (*A*) A capsid mutant of any serotype serves as the marker rescue plasmid template. The mutation can be rescued via cotransfection with capsid DNA from any other AAV serotype serving as template. (*B*) Generation of a library of viral genomes upon transfection of mutant template and capsid DNA results in a mixed population of viral genomes that can be amplified by PCR, cloned into a shuttle vector, and used to produce individual clones or a library of clones. These can then be (*C*) selected by repeated infectious cycling onto any cell line of interest or (*D*) selected in specific tissue types in vivo to enrich for a predominant chimeric AAV recombinant vector (*E*).

MATERIALS

CAUTION: See Appendix for appropriate handling of materials marked with <!>.

Reagents

Benzonase (Sigma-Aldrich)
Cells: HeLa (ATCC) or other target cell type for selection/cycling and HEK 293 cells (ATCC)
Cesium chloride (Sigma-Aldrich) <!>
Dulbecco's modified Eagle's medium (DMEM) with 10% fetal bovine serum and 1% penicillin/streptomycin (Sigma-Aldrich) <!>
Leupeptin (Sigma-Aldrich) <!>
Phosphate-buffered saline (PBS) (Sigma-Aldrich)
Plasmids
  AAV helper shuttle vector (e.g., pXR2AN)
    The pXR2AN shuttle vector used in this procedure contains a unique AflII site near the beginning of the capsid gene and a unique NotI site after the end of the capsid gene. This facilitates cloning of all PCR products directly into an AAV helper plasmid for sequencing and vector production (Bowles et al. 2003).
  AAV serotype helper plasmids (e.g., pXR1, pXR2, pXR3, pXR4, and pXR5; UNC Vector Core)
  Adenoviral helper plasmid (e.g., pXX6-80; UNC Vector Core)
  AAV transgene packaging cassette (e.g., pTR-CMV-GFP; UNC Vector Core)
  AAV serotype plasmids with Rep, Cap, and ITR regions (UNC Vector Core)
  Mutated AAV serotype plasmids as per user requirements (e.g., H/N3761 noninfectious AAV2 mutants)
Primers to amplify PCR fragments of AAV serotype Cap (e.g., AAV3b)
Superfect (QIAGEN)
Trypsin-EDTA (Sigma-Aldrich) <!>
Virus: Adenovirus *dl*309 (UNC Vector Core) <!>

Equipment

Equipment for chimeric AAV vector production are the same as those described in Protocol 1.

METHOD

1. Digest H/N3761 AAV2 mutant plasmids with PvuII.
2. Plate approximately 10^6 HEK 293 cells in a 6-cm dish.
3. Transfect the cells with 1 µg of the digested H/N3761 AAV2 mutant plasmid and 1 µg of AAV3b PCR fragments using the Superfect reagent according to the manufacturer's instructions.
4. Immediately after transfection, coinfect the 293 cells with adenovirus *dl*309 at a multiplicity of infection of four. Allow the virus to attach for 1 hour.
5. Replace the virus-containing medium with 5 ml of fresh medium.
6. Harvest the cells at 48–60 hours posttransfection by scraping.
   Perform all subsequent manipulations on ice unless otherwise noted.
7. Centrifuge the cells at 1000 rpm (Sorvall RT 6000B).
8. Suspend the cell pellet in PBS and subject to three cycles of freeze-thaw to release the viruses.
9. Incubate the lysate for 30 minutes at 56°C to inactivate the adenovirus.

10. Add 100 µl of lysate to HeLa cells with adenovirus *dl*309 at a multiplicity of infection of four.
    Depending on the requirements of the specific study, an alternate target cell type can be used.
11. Repeat Steps 5–10.
12. After each cycle, perform dot-blot analyses of viral cell lysates to confirm the generation of viable cell-type-specific chimeric virions.
13. Perform PCR analyses to determine the sequence of chimeric viral genomes generated by cycling.
14. To test for production of recombinant AAV, perform the triple-plasmid transfection technique described in Protocol 1 using plasmids pTR-CMV-GFP. and pXX6-80, and each individual chimeric AAV helper plasmid.

REFERENCES

Bartlett J.S., Kleinschmidt J., Boucher R.C., and Samulski R.J. 1999. Targeted adeno-associated virus vector transduction of nonpermissive cells mediated by a bispecific $F(ab'\gamma)_2$ antibody. *Nat. Biotechnol.* **17:** 181–186.

Blankinship M.J., Gregorevic P., Allen J.M., Harper S.Q., Harper H., Halbert C.L., Miller D.A., and Chamberlain J.S. 2004. Efficient transduction of skeletal muscle using vectors based on adeno-associated virus serotype 6. *Mol. Ther.* **10:** 671–678.

Bowles D.E., Rabinowitz J.E., and Samulski R.J. 2003. Marker rescue of adeno-associated virus (AAV) capsid mutants: A novel approach for chimeric AAV production. *J. Virol.* **77:** 423–432.

Buning H., Ried M.U., Perabo L., Gerner F.M., Huttner N.A., Enssle J., and Hallek M. 2003. Receptor targeting of adeno-associated virus vectors. *Gene Ther.* **10:** 1142–1151.

Burger C., Gorbatyuk O.S., Velardo M.J., Peden C.S., Williams P., Zolotukhin S., Reier P.J., Mandel R.J., and Muzyczka N. 2004. Recombinant AAV viral vectors pseudotyped with viral capsids from serotypes 1, 2, and 5 display differential efficiency and cell tropism after delivery to different regions of the central nervous system. *Mol. Ther.* **10:** 302–317.

Choi V.W., McCarty D.M., and Samulski R.J. 2005. AAV hybrid serotypes: Improved vectors for gene delivery. *Curr. Gene Ther.* **5:** 299–310.

Davidson B.L., Stein C.S., Heth J.A., Martins I., Kotin R.M., Derksen T.A., Zabner J., Ghodsi A., and Chiorini J.A. 2000. Recombinant adeno-associated virus type 2, 4, and 5 vectors: Transduction of variant cell types and regions in the mammalian central nervous system. *Proc. Natl. Acad. Sci.* **97:** 3428–3432.

Ding W., Zhang L., Yan Z., and Engelhardt J.F. 2005. Intracellular trafficking of adeno-associated viral vectors. *Gene Ther.* **12:** 873–880.

Gigout L., Rebollo P., Clement N., Warrington K.H., Jr., Muzyczka N., Linden R.M., and Weber T. 2005. Altering AAV tropism with mosaic viral capsids. *Mol. Ther.* **11:** 856–865.

Girod A., Ried M., Wobus C., Lahm H., Leike K., Kleinschmidt J., Deleage G., and Hallek M. 1999. Genetic capsid modifications allow efficient re-targeting of adeno-associated virus type 2 (erratum *Nat. Med.* [1999] **5:** 1438). *Nat. Med.* **5:** 1052–1056.

Grifman M., Trepel M., Speece P., Gilbert L.B., Arap W., Pasqualini R., and Weitzman M.D. 2001. Incorporation of tumor-targeting peptides into recombinant adeno-associated virus capsids. *Mol. Ther.* **3:** 964–975.

Grimm D. and Kay M.A. 2003. From virus evolution to vector revolution: Use of naturally occurring serotypes of adeno-associated virus (AAV) as novel vectors for human gene therapy. *Curr. Gene Ther.* **3:** 281–304.

Grimm D., Zhou S., Nakai H., Thomas C.E., Storm T.A., Fuess S., Matsushita T., Allen J., Surosky R., Lochrie M., Meuse L., McClelland A., Colosi P., and Kay M.A. 2003. Preclinical in vivo evaluation of pseudotyped adeno-associated virus vectors for liver gene therapy. *Blood* **102:** 2412–2419.

Halbert C.L., Allen J.M., and Miller A.D. 2001. Adeno-associated virus type 6 (AAV6) vectors mediate efficient transduction of airway epithelial cells in mouse lungs compared to that of AAV2 vectors. *J. Virol.* **75:** 6615–6624.

Kern A., Schmidt K., Leder C., Muller O.J., Wobus C.E., Bettinger K., Von der Lieth C.W., King J.A., and Kleinschmidt J.A. 2003. Identification of a heparin-binding motif on adeno-associated virus type 2 capsids. *J. Virol.* **77:** 11072–11081.

Loiler S.A., Tang Q., Clarke T., Campbell-Thompson M.L., Chiodo V., Hauswirth W., Cruz P., Perret-Gentil M., Atkinson M.A., Ramiya V.K., and Flotte T.R. 2005. Localized gene expression following administration of adeno-associated viral vectors via pancreatic ducts. *Mol. Ther.* **12:** 519–527.

Loiler S.A., Conlon T.J., Song S., Tang Q., Warrington K.H., Agarwal A., Kapturczak M., Li C., Ricordi C., Atkinson M.A., Muzyczka N., and Flotte T.R. 2003. Targeting recombinant adeno-associated virus vectors to enhance gene transfer to pancreatic islets and liver. *Gene Ther.* **10:** 1551–1158.

Lotery A.J., Yang G.S., Mullins R.F., Russell S.R., Schmidt M., Stone E.M., Lindbloom J.D., Chiorini J.A., Kotin R.M., and Davidson B.L. 2003. Adeno-associated virus type 5: Transduction efficiency and cell-type specificity in the primate retina. *Hum. Gene Ther.* **14:** 1663–1671.

Louboutin J.P., Wang L., and Wilson J.M. 2005. Gene transfer into skeletal muscle using novel AAV serotypes. *J. Gene Med.* **7:** 442–451.

Maheshri N., Koerber J.T., Kaspar B.K., and Schaffer D.V. 2006. Directed evolution of adenoassociated virus yields enhanced gene delivery vectors. *Nat. Biotechnol.* **24:** 198–204.

Muller O.J., Kaul F., Weitzman M.D., Pasqualini R., Arap W., Kleinschmidt J.A., and Trepel M. 2003. Random peptide libraries displayed on adeno-associated virus to select for targeted gene therapy vectors. *Nat. Biotechnol.* **21:** 1040–1046.

Muzyczka N. and Warrington K.H., Jr. 2005. Custom adeno-associated virus capsids: The next generation of recombinant vectors with novel tropism. *Hum. Gene Ther.* **16:** 408–416.

Nicklin S.A. and Baker A.H. 2002. Tropism-modified adenoviral and adenoassociated viral vectors for gene therapy. *Curr. Gene Ther.* **2:** 273–293.

Nicklin S.A., Buening H., Dishart K.L., de Alwis M., Girod A., Hacker U., Thrasher A.J., Ali R.R., Hallek M., and Baker A.H. 2001. Efficient and selective AAV2-mediated gene transfer directed to human vascular endothelial cells. *Mol. Ther.* **4:** 174–181.

Opie S.R., Warrington K.H., Jr., Agbandje-McKenna M., Zolotukhin S., and Muzyczka N. 2003. Identification of amino acid residues in the capsid proteins of adeno-associated virus type 2 that contribute to heparan sulfate proteoglycan binding. *J. Virol.* **77:** 6995–7006.

Padron E., Bowman V., Kaludov N., Govindasamy L., Levy H., Nick P., McKenna R., Muzyczka N., Chiorini J.A., Baker T.S., and Agbandje-McKenna M. 2005. Structure of adeno-associated virus type 4. *J. Virol.* **79:** 5047–5058.

Peden C.S., Berger C., Muzyczka N., and Mandel R.J. 2004. Circulating anti-wild-type adeno-associated virus type 2 (AAV2) antibodies inhibit recombinant AAV2 (rAAV2)-mediated, but not rAAV5-mediated, gene transfer in the brain. *J. Virol.* **78:** 6344–6359.

Perabo L., Endell J., King S., Lux K., Goldnau D., Hallek M., and Buning H. 2005. Combinatorial engineering of a gene therapy vector: Directed evolution of adeno-associated virus. *J. Gene Med.* **8:** 155–162.

Perabo L., Buning H., Kofler D.M., Ried M.U., Girod A., Wendtner C.M., Enssle J., and Hallek M. 2003. In vitro selection of viral vectors with modified tropism: The adeno-associated virus display. *Mol. Ther.* **8:** 151–157.

Ponnazhagan S., Mahendra G., Kumar S., Thompson J.A., and Castillas M., Jr. 2002. Conjugate-based targeting of recombinant adeno-associated virus type 2 vectors by using avidin-linked ligands. *J. Virol.* **76:** 12900–12907.

Rabinowitz J.E., Xiao W., and Samulski R.J. 1999. Insertional mutagenesis of AAV2 capsid and the production of recombinant virus. *Virology* **265:** 274–285.

Rabinowitz J.E., Bowles D.E., Faust S.M., Ledford J.G., Cunningham S.E., and Samulski R.J. 2004. Cross-dressing the virion: The transcapsidation of adeno-associated virus serotypes functionally defines subgroups. *J. Virol.* **78:** 4421–4432.

Rabinowitz J.E., Rolling F., Li C., Conrath H., Xiao W., Xiao X., and Samulski R.J. 2002. Cross-packaging of a single adeno-associated virus (AAV) type 2 vector genome into multiple AAV serotypes enables transduction with broad specificity. *J. Virol.* **76:** 791–801.

Ried M.U., Girod A., Leike K., Buning H., and Hallek M. 2002. Adeno-associated virus capsids displaying immunoglobulin-binding domains permit antibody-mediated vector retargeting to specific cell surface receptors. *J. Virol.* **76:** 4559–4566.

Rose J.A., Maizel J.V., Jr., Inman J.K., and Shatkin A.J. 1971. Structural proteins of adeno-associated viruses. *J. Virol.* **8:** 766–770.

Shi W., Arnold G.S., and Bartlett J.S. 2001. Insertional mutagenesis of the adeno-associated virus type 2 (AAV2) capsid gene and generation of AAV2 vectors targeted to alternative cell-surface receptors. *Hum. Gene Ther.* **12:** 1697–1711.

Summerford C. and Samulski R.J. 1998. Membrane-associated heparan sulfate proteoglycan is a receptor for adeno-associated virus type 2 virions. *J. Virol.* **72:** 1438–1445.

Takeda S., Takahashi M., Mizukami H., Kobayashi E., Takeuchi K., Hakamata Y., Kaneko T., Yamamoto H., Ito C., Ozawa K., Ishibashi K., Matsuzaki T., Takata K., Asano Y., and Kusano E. 2004. Successful gene transfer using adeno-associated virus vectors into the kidney: Comparison among adeno-associated virus serotype 1–5 vectors in vitro and in vivo. *Nephron Exp. Nephrol.* **96:** e119–e126.

Thomas C.E., Storm T.A., Huang Z., and Kay M.A. 2004. Rapid uncoating of vector genomes is the key to efficient liver transduction with pseudotyped adeno-associated virus vectors. *J. Virol.* **78:** 3110–3122.

Urabe M., Ding C., and Kotin R.M. 2002. Insect cells as a factory to produce adeno-associated virus type 2 vectors. *Hum. Gene Ther.* **13:** 1935–1943.

Walters R.W., Agbandje-McKenna M., Bowman V.D., Moninger T.O., Olson N.H., Seiler M., Chiorini J.A., Baker T.S., and Zabner J. 2004. Structure of adeno-associated virus serotype 5. *J. Virol.* **78:** 3361–3371.

Wang L., Peng P.D., Erhardt A., Storm T.A., and Kay M.A. 2004. Comparison of adenoviral and adeno-associated viral vectors for pancreatic gene delivery in vivo. *Hum. Gene Ther.* **15:** 405–413.

Wang Z., Zhu T., Qiao C., Zhou L., Wang B., Zhang J., Chen C., Li J., and Xiao X. 2005. Adeno-associated virus serotype 8 efficiently delivers genes to muscle and heart. *Nat. Biotechnol.* **23:** 321–328.

Warrington K.H., Jr., Gorbatyuk O.S., Harrison J.K., Opie S.R., Zolotukhin S., and Muzyczka N. 2004. Adeno-associated virus type 2 VP2 capsid protein is nonessential and can tolerate large peptide insertions at its N terminus. *J. Virol.* **78:** 6595–6609.

Weber M., Rabinowitz J., Provost N., Conrath H., Folliot S., Briot D., Cherel Y., Chenuaud P., Samulski R.J., Moullier P., and Rolling F. 2003. Recombinant adeno-associated virus serotype 4 mediates unique and exclusive long-term transduction of retinal pigmented epithelium in rat, dog, and nonhuman primate after subretinal delivery. *Mol. Ther.* **7:** 774–781.

White S.J., Nicklin S.A., Buning H., Brosnan M.J., Leike K., Papadakis E.D., Hallek M., and Baker A.H. 2004. Targeted gene delivery to vascular tissue in vivo by tropism-modified adeno-associated virus vectors. *Circulation* **109:** 513–519.

Work L.M., Nicklin S.A., Brain N.J., Dishart K.L., Von Seggern D.J., Hallek M., Buning H., and Baker A.H. 2004. Development of efficient viral vectors selective for vascular smooth muscle cells. *Mol. Ther.* **9:** 198–208.

Wu P., Xiao W., Conlon T., Hughes J., Agbandje-McKenna M., Ferkol, Flotte T., and Muzyczka N. 2000. Mutational analysis of the adeno-associated virus type 2 (AAV2) capsid gene and construction of AAV2 vectors with altered tropism. *J. Virol.* **74:** 8635–8647.

Xiao W., Chirmule N., Berta S.C., McCullough B., Gao G., and Wilson J.M. 1999. Gene therapy vectors based on adeno-associated virus type 1. *J. Virol.* **73:** 3994–4003.

Xiao X., Li J., and Samulski R.J. 1998. Production of high-titer recombinant adeno-associated virus vectors in the absence of helper adenovirus. *J. Virol.* **72:** 2224–2232.

Xie Q., Bu W., Bhatia S., Hare J., Somasundaram T., Azzi A., and Chapman M.S. 2002. The atomic structure of adeno-associated virus (AAV-2), a vector for human gene therapy. *Proc. Natl. Acad. Sci.* **99:** 10405–10410.

Yang Q., Mamounas M., Yu G., Kennedy S., Leaker B., Merson J., Wong-Staal F., Yu M., and Barber J.R. 1998. Development of novel cell surface CD34-targeted recombinant adenoassociated virus vectors for gene therapy. *Hum. Gene Ther.* **9:** 1929–1937.

Zabner J., Seiler M., Walters R., Kotin R.M., Fulgeras W., Davidson B.L., and Chiorini J.A. 2000. Adeno-associated virus type 5 (AAV5) but not AAV2 binds to the apical surfaces of airway epithelia and facilitates gene transfer. *J. Virol.* **74:** 3852–3858.

20 Recombinant Herpes Simplex Virus Vectors

William F. Goins, David M. Krisky, James B. Wechuck, Darren Wolfe, Shaohua Huang, and Joseph C. Glorioso

Department of Molecular Genetics and Biochemistry, University of Pittsburgh School of Medicine, E1240 Biomedical Sciences Tower, Pittsburgh, Pennsylvania 15261

ABSTRACT

Advances in the identification and characterization of gene products responsible for specific diseases have opened opportunities for novel therapies using gene-transfer vectors for gene replacement. Engineering effective vectors has been crucial to the efficient delivery and expression of therapeutic gene products in vivo. Viruses have been adapted for use as gene-transfer vectors, utilizing their natural mechanisms to efficiently and effectively deliver nucleic acids to cells. Among these, herpes simplex virus type 1 (HSV-1) represents an excellent candidate vector for delivery to the peripheral and central nervous systems. The natural biology of HSV-1 includes the establishment of a lifelong latent state in neurons in which the viral genome persists as an episomal molecule. Genomic HSV vectors can be produced that are totally replication-defective, nontoxic, and capable of long-term transgene expression. Replication-competent recombinant genomic HSV vectors that retain the ability to replicate selectively in specific cells (i.e., oncolytic) have also been engineered and used in Phase I and II human trials in patients with glioblastoma multiforme. Recent efforts have focused on the use of HSV vectors for treatment of non-nervous-system disorders. Scalable systems capable of manufacturing these vectors for preclinical efficacy and safety testing have also been developed. This chapter describes the construction of recombinant HSV vectors and their use for transduction of various cell types.

INTRODUCTION

HSV Gene-transfer Vectors

HSV-1 has many biological features that make it attractive for gene delivery to the nervous system (Wolfe et al. 1999; Burton et al. 2001; Fink et al. 2003; Goins et al. 2004). In natural infection, the virus establishes latency in neurons, a state in which viral genomes may persist for the life of the host as intranuclear episomal elements. The natural lifelong persistence of latent genomes in trigeminal ganglia without the development of sensory loss or histologic damage to the ganglion attests to the effectiveness of these natural latency mechanisms. Although the wild-type virus may be reactivated from latency under the influence of a variety of stresses, completely replication-defective viruses can be constructed that retain the ability to establish persistent quiescent genomes in neurons, but which are unable to replicate or reactivate in the nervous system. These persistent genomes are devoid of lytic gene expression, but they retain the ability to express latency-associated transcripts.

HSV Life Cycle

HSV-1 is a large neurotropic human virus containing 152 kb of linear double-stranded DNA encoding more than 80 gene products (Roizman and Knipe 2001). The viral genome consists of two segments (Fig. 1)—the unique long (U_L) segment and unique short (U_S) segment—each flanked by inverted repeats containing important immediate-early (IE) transcriptional regulatory

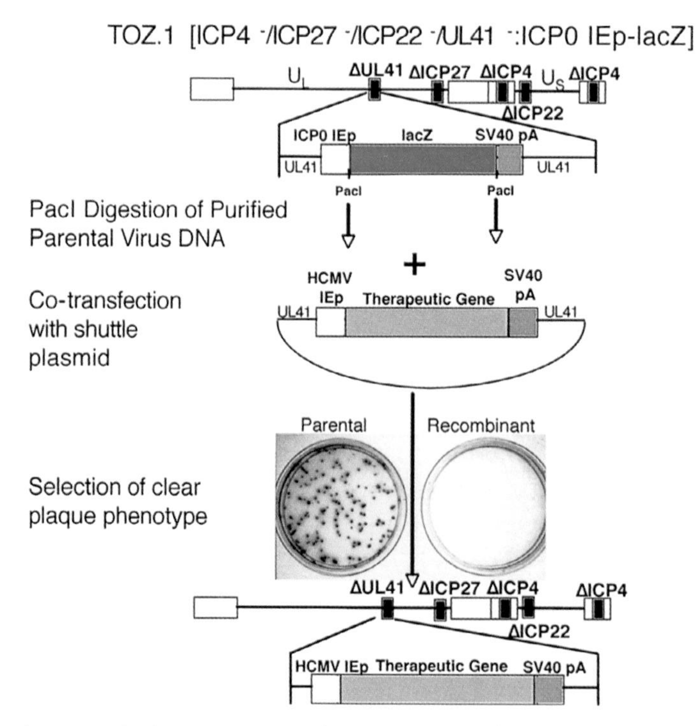

FIGURE 1. Recombination of a therapeutic gene of interest into a replication-defective HSV-1 vector. The replication-defective HSV-1 vector TOZ.1 is deleted for ICP4, ICP27, and ICP22 and contains an ICP0 promoter-driven *lacZ* expression cassette that possesses unique PacI restriction sites. To achieve homologous recombination of the therapeutic gene of interest into TOZ.1, digest purified TOZ.1 DNA with PacI and transfect this DNA into 7b cells along with the UL41 transfer plasmid containing the gene of interest. If homologous recombination occurs, the gene of interest replaces *lacZ*. Positive recombinants can be identified by the production of a clear-plaque phenotype, compared to the blue-plaque phenotype produced by the parental vector in limiting dilution assays.

genes and the latency-associated transcription unit. With few exceptions, the viral genes are almost entirely found as contiguous transcribable units, making their genetic manipulation relatively straightforward. The viral particle consists of a nucleocapsid surrounded by an envelope containing glycoproteins essential for viral attachment and penetration into cells. Between the capsid and the envelope is a protein matrix known as the tegument. The tegument contains a number of structural proteins, including VP16 (Mackem and Roizman 1982), that act in concert with cellular transcription factors to activate HSV IE gene promoters and the UL41 virion gene product that shuts off host protein synthesis (Oroskar and Read 1989).

The virus initially replicates in epithelial cells of skin or mucosa. Progeny virions are then taken up by sensory nerve terminals and carried by retrograde axonal transport to peripheral nerve cell bodies. Viral DNA is injected through a modified capsid penton into the nucleus. At this point, two alternative viral life cycles may ensue:

1. The virus may enter a lytic replication cycle, in which expression of viral IE genes serve to transactivate expression of early (E) genes. These products in turn are the principal components of the DNA replication complex. IE gene expression does not require the de novo expression of viral gene products but is augmented by the HSV tegument protein VP16. Expression of the E genes leads to the production of progeny viral genome concatemers. Following viral DNA synthesis in conjunction with IE gene product transcriptional and post-translational regulatory processes, the late (L) genes are transcribed. The L genes encode primarily the structural proteins required for viral particle assembly in the nucleus, particle budding through a modified portion of the nuclear membrane, transport to the cell surface, and egress from the cells.
2. Alternatively, the virus may enter a latent state, in which the gene products characteristic of lytic infection are either not transcribed or are silenced over time.

These properties of the virus hold true for both replication-competent (i.e., oncolytic) and replication-defective vectors, except that the replication-defective vectors have been rendered deficient through the deletion of one or more essential HSV gene products. This chapter describes the generation and use of replication-defective HSV vectors. Indeed, many of the techniques can be applied to both vector systems, except that the replication-competent vectors do not require a complementing cell line for their growth and propagation.

Replication-defective HSV Vectors

HSV genes are expressed in a sequential, interdependent order during lytic infection (Honess and Roizman 1974). Consequently, removal of the essential IE gene ICP4 prevents expression of later genes (DeLuca and Schaffer 1985), resulting in a defective vector that is incapable of producing viral particles or expressing E or L gene products. However, the remaining IE genes continue to be expressed and, with the exception of ICP47, are individually toxic to cells in culture (Johnson et al. 1992). Accordingly, only the combined elimination of multiple IE genes reduces the cytotoxicity of HSV-based vectors for tissue-culture cells (Johnson et al. 1994; Samaniego et al. 1995, 1997; Marconi et al. 1996; Wu et al. 1996; Krisky et al. 1998b). In contrast, the number of deleted viral genes does not affect vector-associated toxicity when infecting neurons in vivo. At multiplicities of infection (moi) below ten, the ICP4-deleted vector background (as well as further viral gene-deleted mutants) can be used for high-level expression of multiple transgenes in a variety of cells, particularly primary peripheral nervous system neurons and in vivo (Krisky et al. 1998b) without vector-related toxicity. These replication-defective mutants readily establish persistence in sensory neurons and in other cell types, making these vectors ideal backbones for the expression of therapeutic genes.

A variety of methods have been developed to engineer replication-defective HSV-based vectors, and studies have been performed to determine the efficiency of gene transfer, the ability of specific vectors to persist and express reporter or therapeutic gene products in the absence of vector-mediated toxicity. Methods exist to delete HSV-1 IE gene functions and subsequently introduce foreign genes into the HSV-1 genome using homologous recombination (Krisky et al. 1997). Because some of the HSV-1 IE genes (e.g., ICP4 and ICP27) are essential for growth of the virus in cell culture, complementing cell lines expressing these gene products in *trans* are required for isolation and propagation of replication-defective viral mutants. First-generation replication-defective recombinant viruses are deleted for a sole essential gene, either ICP4 or ICP27. The ICP4-deficient recombinants can be grown on a complementing cell line that expresses ICP4 in *trans* upon infection with an ICP4 deletion mutant virus (DeLuca et al. 1985). The introduction of an ICP0 IE gene promoter–β-galactosidase (*lacZ*) reporter cassette into the UL41 locus of the genome of this replication-defective mutant creates the vector SOZ (Krisky et al. 1997), which expresses the transgene in a variety of cell types in culture and in various tissues in vivo (Krisky et al. 1998a,b; Moriuchi et al. 2000).

Second-generation replication-defective vectors are deleted for multiple IE essential gene functions such as ICP4 and ICP27. To propagate such mutants, a cell line must be constructed to complement both ICP4 and ICP27. This is accomplished by transfecting Vero cells with a plasmid containing the HSV-1 sequences coding for ICP4 and ICP27, along with a neomycin expression cassette for the selection of individual clones (Marconi et al. 1996). The plasmid containing the coding regions for ICP4 and ICP27 was engineered to avoid overlap of these sequences with the deletions present within the virus to eliminate the chance of homologous recombination and rescue of the mutant viruses during propagation in the complementing line. The 7b cell line (Marconi et al. 1996) is able to complement the growth of the IE mutants deleted for either ICP4 (DeLuca et al. 1985) or ICP27 (McCarthy et al. 1989).

Because deletion of only two IE genes failed to dramatically reduce toxicity (Krisky et al. 1998b), third-generation recombinants deleted for ICP4, ICP27, and ICP22 were engineered. The recombinant TOZ.1 was engineered by genetic viral cross of the SOZ.1 vector with the T.1 recombinant deficient for ICP4, ICP27, and ICP22 (Krisky et al. 1997, 1998b). Positive recombinants ($lacZ^+$) were propagated and isolated on the 7b complementing cell line, compared to $lacZ^+$ viruses that would only plaque on the ICP4-expressing line (i.e., parental SOZ.1). New HSV vectors can be constructed using the TOZ.1 vector backbone by homologous recombination of a plasmid containing a cassette expressing the gene of interest inserted into the UL41 gene sequence (Krisky et al. 1997, 1998b). The TOZ.1 vector (Fig. 1) expresses a reporter gene (*lacZ*) in the UL41 locus, such that recombination of the transgenic cassette into the UL41 locus results in the loss of readily assayable reporter gene activity. The TOZ.1 vector also contains a unique PacI endonuclease site for digestion of parental viral DNA that substantially reduces nonrecombinant background. In most cases, the proportion of viral plaques that represent recombinants rises to 10–50% using this technique (Krisky et al. 1997). The new recombinants can be isolated following homologous recombination of the shuttle plasmid into the PacI-digested TOZ.1 genome by the identification of clear plaques (Fig. 1). Following three rounds of limiting dilution analysis, the structure of the recombinants can be confirmed by Southern blot or polymerase chain reaction (PCR) analysis.

Protocol

Generation of Replication-competent and -defective HSV Vectors

MATERIALS

CAUTION: See Appendix for appropriate handling of materials marked with <!>.

Reagents

Cells: 7b derivative of standard of Vero African green monkey kidney cells (Marconi et al. 1996), Vero cells (ATCC: CCL81), or other complementing cell lines

These cells are required to propagate HSV-1 replication-competent or -defective viruses.

Crystal Violet solution (1% in methanol:dH_2O, 50:50 v/v) <!>

DMEM/10% fetal bovine serum (FBS)
- Dulbecco's modified Eagle's essential medium supplemented with nonessential amino acids
- 100 units/ml penicillin G <!>
- 100 µg/ml streptomycin sulfate <!>
- 2 mM glutamine
- 10% FBS

Ethanol (70%) <!>

Glycerol

HEPES (1 M, pH 7.35)

Iodixanol (60%; Invitrogen)

Dilute 60% iodixanol to a final concentration of 20% with PBS before use.

Isopropanol <!>

Lipofectamine 2000 (Invitrogen)

Lysis buffer (10 mM Tris-HCl <!> at pH 8.0 and 10 mM EDTA)

Methylcellulose overlay (1.0%)

Add 25 g of methylcellulose to 100 ml of PBS at pH 7.5 in a 500-ml sterile bottle containing a stir bar and autoclave the bottle on liquid cycle (45 min). After the solution cools, add 350 ml of DMEM supplemented with nonessential amino acids, 100 units/ml of penicillin G, 100 µg/ml of streptomycin sulfate, and 2 mM glutamine. Mix well and place the bottle on a stir plate at 4°C overnight. Once the methylcellulose has entered solution, add 50 ml of FBS.

PacI restriction endonuclease (New England BioLabs)

PBS (pH 7.5)
- 135 mM NaCl
- 2.5 mM KCl <!>
- 1.5 mM KH_2PO_4 <!>
- 8.0 mM Na_2HPO_4 <!>

Plasmids
- Transfer plasmid pUL41 (Krisky et al. 1997)
- Plasmid containing gene of interest

Proteinase K (Boehringer Mannheim) <!>

Tris-buffered saline (TBS) (pH 7.5) (50 mM Tris-HCl at pH 7.5, 150 mM NaCl, and 1 mM EDTA)

TE (10 mM Tris-HCl at pH 8.0 and 1 mM EDTA)

TE-equilibrated phenol:chloroform:isoamyl alcohol (25:24:1 v/v) <!>

Virus: TOZ.1 virus (Krisky et al. 1997, 1998b)
X-gal staining solution
  300 µg of X-gal/ml TBS <!>
  14 mM $K_4Fe(CN)_6$ <!>
  14 mM $K_3Fe(CN)_6$ <!>

X-gal is highly insoluble and must be dissolved in dimethyl formamide <!> before adding to the staining solution.

Equipment

Cell scrapers
Centrifuge bottle (500-ml polypropylene)
CO_2 incubator
Cup-horn sonicator
Flasks (T-75 cm^2)
Multichannel pipettor
Nutator rocker platform
Phase Lock Gel Heavy Tube (Eppendorf)
Plates (6 well, 12 well, and 96 well flat-bottomed)
Preparative centrifuge with JLA 10.5 rotor (Beckman)
Roller bottles (850 cm^2)
Syringe with needle (3 ml)
Tubes:
  13-ml Beckman 17-mm path-length seal
  15-ml conical polypropylene
Ultracentrifuge (Beckman XL-90) with NVT65 rotor
Wide-bore pipette tips (Bio-Rad)

METHODS

Isolation of Viral DNA for Transfection

The quality of viral DNA determines the frequency at which the desired recombinant virus is generated. The optimized protocol below produces highly infectious viral DNA; 1 µg of purified viral DNA yields 100–1000 plaques. The DNA produced by this protocol contains a mixture of cellular and viral DNAs. The cellular DNA acts as carrier during precipitation, increasing the yield. Cellular DNA also acts as a carrier during transfections, increasing the overall efficiency of precipitate formation and thus the chance of obtaining the desired recombinant. If necessary, pure viral DNA can be prepared from virus harvested solely from the media of infected cells or from viral particles that have been gradient-purified. However, the yield of DNA obtained in these instances is significantly reduced.

1. Add 2.4×10^7 pfu of TOZ.1 virus to 8×10^6 complementing 7b cells (moi = 3). Incubate for 1 hour at 37°C with rocking.
2. Transfer the infected cells into a T-75 flask. Grow in a 37°C CO_2 incubator for approximately 18–24 hours.

   All cells should be infected with virus and remain adherent to the flask.
3. Dislodge the cells with a cell scraper. Transfer into a 15-ml conical polypropylene tube.
4. Centrifuge at 2060*g* for 10 minutes at 4°C. Remove supernatant.

5. Add 1 ml of lysis buffer plus 0.1 mg/ml proteinase K to the pellet.
6. Incubate the tube overnight at 37°C with continuous agitation on a Nutator rocker platform.
7. Transfer the solution into a Phase Lock Gel Heavy Tube.
8. Add 1 ml of phenol:chloroform:isoamyl alcohol (25:24:1). Mix gently for about 1–2 minutes.
9. Centrifuge the tube at 3020*g* for 5 minutes.
10. Transfer the aqueous phase to a new tube.
11. Add 2 volumes of isopropanol. Mix thoroughly.
12. Spool the precipitated DNA (white slurry) onto a glass pasteur pipette.
13. Transfer the spooled DNA to a new tube. Rinse once with 70% ethanol. Allow the DNA to air-dry.
14. Add 0.5 ml of TE buffer to the dry pellet. Incubate overnight at 25°C.
15. Resuspend the DNA by gently pipetting using a wide-bore pipette tip.

    The use of wide-bore Pipetman tips (Gilson) minimizes shearing of the viral DNA, thereby increasing the infectivity of the viral DNA preparation.
16. Measure DNA concentration spectrophotometrically (OD_{260nm}).

Construction of Recombinant Virus

The gene of interest to be inserted into the virus must first be cloned into a transfer plasmid (pUL41) containing at least 500–1000 bp of flanking HSV-1 sequences. The HSV-1 gene of interest (in this case, UL41) is deleted, while maintaining sufficient HSV-1 flanking sequences to allow efficient recombination into the viral genome. It is then possible to recombine these sequences into the virus at the UL41 locus. This is accomplished by transfecting both linearized plasmid and PacI-digested purified viral DNA into 7b permissive cells. To detect the recombinants containing the desired therapeutic gene of interest, the target parental virus backbone should possess a reporter gene cassette (*lacZ* for TOZ.1) at the desired site of recombination (UL41). Positive recombinants obtained from the transfection will produce clear plaques, compared to the blue-plaque phenotype of the parental virus (Fig. 1).

17. Clone the gene of interest into the UL41 shuttle plasmid.
18. One day before transfection, seed 5 x 10^5 7b cells in a six-well tissue-culture plate in DMEM/10% FBS.
19. Digest viral DNA (obtained above in Step 15) with PacI for 6 hours at 37°C according to the manufacturer's protocol.
20. Transfect the cells with both viral and plasmid DNA using Lipofectamine 2000, following the manufacturer's instructions.

    The plasmid construct should be linearized before transfection. This increases the recombination frequency relative to that obtained with uncut supercoiled plasmid. Digestion of the plasmid to release the insert followed by purification of the restriction fragment does not increase the recombination frequency. However, use of a purified fragment is preferred, because it eliminates the chances for insertion of plasmid vector sequences into the virus by semihomologous recombination.
21. Add fresh DMEM/10% FBS to the cells and incubate at 37°C.

    Depending on the virus, plaques develop in 3–5 days.
22. Once plaques have formed, harvest the media and cells.
23. Subject the cells to three cycles of freeze-thaw. Sonicate the cells.

24. Centrifuge at 2060*g* for 5 minutes at 4°C to remove cell debris. Combine this supernatant with the media from the original harvest (Step 22).

    Store the supernatant at –80°C for use as a stock.

25. Determine the titer of the stock of recombinant virus.

    a. Prepare a series of tenfold dilutions (10^{-2} to 10^{-10}) of the virus stock in serum-free media.

    b. Add 100 μl of each dilution to a well of a 12-well plate containing 4×10^5 7b cells/well.

    c. Incubate the plates in a CO_2 incubator for 1 hour at 37°C. Add 1 ml of DMEM/10% FBS and place in the incubator overnight.

    d. Within the next 24 hours, remove the media. Overlay the monolayer with 1 ml of 1% methylcellulose/10% FBS solution.

    e. Incubate the plates for 3–5 days until well-defined plaques appear.

    f. Aspirate the methylcellulose. Stain with Crystal Violet solution for 5 minutes.

    g. Count the plaques and calculate the number of pfu/1 ml of original stock.

26. Add 30 pfu of virus to 1 ml of 2×10^6 cells in suspension (in DMEM/10% FBS) in a 15-ml conical tube. Place the tube on a Nutator rocker platform for 1.5 hours at 37°C.

27. Add 9 ml of fresh medium. Plate 100 μl in each well of a 96-well flat-bottomed plate.

28. Incubate the plates in a CO_2 incubator for 3–5 days at 37°C, until plaques appear.

29. Stain wells with X-gal to determine which samples contain plaques derived from the desired recombinant virus.

    If recombination has occurred, the gene of interest will have replaced the *lacZ* reporter gene.

    a. Transfer the media from the wells of the 96-well plate to a new 96-well plate.

       Store this plate at –80°C for use as a stock for the next round of limiting dilution.

    b. Pipette 100 μl of X-gal staining solution in each well of a 96-well plate of infected cells.

    c. Incubate plates for 1–18 hours at 37°C, until blue color appears.

    d. Identify wells containing only single clear plaques.

       Clear plaques are formed from viral recombinants in which the gene of interest has replaced *lacZ* in TOZ.1.

30. Repeat two rounds of limiting dilution plaque isolation (Steps 26–29) using the stock of virus stored at –80°C (Step 29a).

31. Isolate viral DNA and confirm the presence of the insert as well as the absence of the deleted sequences by Southern blot analysis.

    At this point, the viral stock can be used to produce a midistock for the eventual preparation of a high-titer stock for general experimental use.

Viral Stock Preparation and Purification

The following procedure calls for preparing a viral stock from two roller bottle's worth of cells. It can be scaled either up or down, depending on specific needs.

32. Seed two 850-cm^2 roller bottles with 2×10^7 complementing cells per bottle in 100 ml of DMEM/10% FBS per bottle. Incubate at 37°C.

    Add 1.75 ml of HEPES (pH 7.35) per bottle to buffer the medium if the roller bottles are not incubated in a CO_2 incubator.

33. When the cell monolayer is approximately 75% confluent (~2–3 days), aspirate the medium.

Add virus at an moi of 0.02–0.05 in 10 ml of fresh serum-free DMEM. Incubate for 1.5 hours at 37°C.

34. At the end of the adsorption period, add 90 ml of DMEM/10% FBS plus 1 ml of HEPES. Incubate for 2–3 days at 37°C.

    Observe the bottles by light microscopy to determine the optimal time for harvesting virus. This time varies dramatically depending on the particular vector backbone. Generally, viral cytopathic effect should be obvious with the majority of the cells rounded up, yet not detached from the roller bottle surface.

35. Use a cell scraper to detach the cells into the media.
36. Harvest the medium containing the cell suspension. Centrifuge at 2060*g* for 10 minutes at 4°C.
37. Decant the supernatant into a 500-ml polypropylene centrifuge bottle. Store on ice.
38. Resuspend the cell pellet in a minimal volume (2–5 ml) of serum-free DMEM. Subject it to three rounds of sonication to disrupt the pellet and release cell-associated virus.
39. Centrifuge at 2060*g* for 10 minutes at 4°C. Decant the supernatant. Combine with the supernatant from Step 37. Store on ice.
40. Centrifuge the combined supernatants in a Beckman preparative centrifuge with a JLA 10.5 rotor at 18,600*g* for 1 hour.
41. Discard the supernatant. Resuspend by thoroughly vortexing the virus pellet in 500 µl of PBS per roller bottle of virus-infected cells.
42. Transfer the suspension to 13-ml Beckman 17-mm path-length seal tubes. Add 20% iodixanol to the tubes to a total volume of 13 ml. Seal the tubes.
43. Centrifuge in a Beckman XL-90 ultracentrifuge using a NVT65 rotor at 342,000*g* for 4.5 hours.
44. Use a 3-ml syringe to insert a needle and aspirate the 1- to 1.5-ml viral (upper) band (Fig. 2).

    Approximately 60% of the virus is present in the second band from the bottom of the gradient. Occasionally, with wild-type preparations, a higher band of defective particles is visible. If there are three bands, take the middle band. Alternatively, fractionate the tubes and determine the titer of each fraction.

45. Aliquot and titrate the virus. Store viral stocks at –80°C. Optiprep serves as a cryoprotectant.

    Low moi of 3–5 are usually sufficient to transduce 100% of the cells, unlike other viruses such as adenovirus, AAV, standard retroviruses, and lentiviruses. Virus can be delivered in vivo by direct injection of various doses.

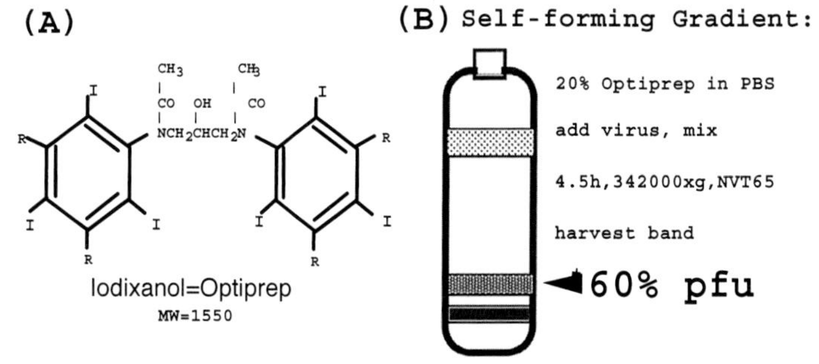

FIGURE 2. Optiprep self-forming gradients and HSV-1 vector purification. (*A*) Chemical structure of Optiprep (iodixanol). (*B*) Self-forming Optiprep gradient and the presence of HSV virions within the gradient at harvest.

TROUBLESHOOTING

Problem: Titer of the viral stock is too low.

Solution: The virus can be furthered concentrated by centrifugation in a microcentrifuge at 2060g for 10 minutes at 4°C. Following centrifugation, resuspend the viral pellet in a minimal volume of PBS containing a final concentration of 10% glycerol as a cryoprotectant.

Problem: Wild-type virus is rescued during the propagation of viruses carrying deletions of essential gene(s).

Solution: Viral stocks should be maintained at a low passage. Use one vial of a newly prepared stock as a stock for preparing all future stocks. Routinely prepare stocks from single plaque isolates.

REFERENCES

Burton E.A., Wechuck J.B., Wendell S.K., Goins W.F., Fink D.J., and Glorioso J.C. 2001. Multiple applications for replication-defective herpes simplex virus vectors. *Stem Cells* **19:** 358–377.

DeLuca N.A. and Schaffer P.A. 1985. Activation of immediate-early, early, and late promoters by temperature-sensitive and wild-type forms of herpes simplex virus type 1 protein ICP4. *Mol. Cell. Biol.* **5:** 1997–2008.

DeLuca N.A., McCarthy A.M., and Schaffer P.A. 1985. Isolation and characterization of deletion mutants of herpes simplex virus type 1 in the gene encoding immediate-early regulatory protein ICP4. *J. Virol.* **56:** 558–570.

Fink D.J., Glorioso J., and Mata M. 2003. Therapeutic gene transfer with herpes-based vectors: Studies in Parkinson's disease and motor nerve regeneration. *Exp. Neurol.* **184:** S19–24.

Goins W.F., Wolfe D., Krisky D.M., Bai Q., Burton E.A., Fink D.J., and Glorioso J.C. 2004. Delivery using herpes simplex virus: An overview. *Methods Mol. Biol.* **246:** 257–299.

Honess R. and Roizman B. 1974. Regulation of herpes simplex virus macromolecular synthesis. I. Cascade regulation of the synthesis of three groups of viral proteins. *J. Virol.* **14:** 8–19.

Johnson P., Wang M., and Friedmann T. 1994. Improved cell survival by the reduction of immediate-early gene expression in replication-defective mutants of herpes simplex virus type 1 but not by mutation of the virion host shutoff function. *J. Virol.* **68:** 6347–6362.

Johnson P., Miyanohara A., Levine F., Cahill T., and Friedmann T. 1992. Cytotoxicity of a replication-defective mutant herpes simplex virus type 1. *J. Virol.* **66:** 2952–2965.

Krisky D., Marconi P., Oligino T., Rouse R., Fink D., and Glorioso J. 1997. Rapid method for construction of recombinant HSV gene transfer vectors. *Gene Ther.* **4:** 1120–1125.

Krisky D.M., Marconi P.C., Oligino T.J., Rouse R.J., Fink D.J., Cohen J.B., Watkins S.C., and Glorioso J.C. 1998a. Development of herpes simplex virus replication-defective multigene vectors for combination gene therapy applications. *Gene Ther.* **5:** 1517–1530.

Krisky D.M., Wolfe D., Goins W.F., Marconi P.C., Ramakrishnan R., Mata M., Rouse R.J., Fink D.J., and Glorioso J.C. 1998b. Deletion of multiple immediate-early genes from herpes simplex virus reduces cytotoxicity and permits long-term gene expression in neurons. *Gene Ther.* **5:** 1593–1603.

Mackem S. and Roizman B. 1982. Structural features of the herpes simplex virus α gene 4, 0, and 27 promoter-regulatory sequences which confer α regulation on chimeric thymidine kinase. *J. Virol.* **44:** 939–949.

Marconi P., Krisky D., Oligino T., Poliani P.L., Ramakrishnan R., Goins W.F., Fink D.J., and Glorioso J.C. 1996. Replication-defective HSV vectors for gene transfer in vivo. *Proc. Natl. Acad. Sci.* **93:** 11319–11320.

McCarthy A.M., McMahan L., and Schaffer P.A. 1989. Herpes simplex virus type 1 ICP27 deletion mutants exhibit altered patterns of transcription and are DNA deficient. *J. Virol.* **63:** 18–27.

Moriuchi S., Krisky D.M., Marconi P.C., Tamura M., Shimizu K., Yoshimine T., Cohen J.B., and Glorioso J.C. 2000. HSV vector cytotoxicity is inversely correlated with effective TK/GCV suicide gene therapy of rat gliosarcoma. *Gene Ther.* **7:** 1483–1490.

Oroskar A. and Read G. 1989. Control of mRNA stability by the virion host shutoff function of herpes simplex virus. *J. Virol.* **63:** 1897–1906.

Roizman B. and Knipe D.M. 2001. Herpes simplex viruses and their replication. In *Fields virology*, 4th edition (ed. D.M. Knipe and P.M. Howley), pp. 2399–2459. Lippincott Williams and Wilkins, Philadelphia, Pennsylvania.

Samaniego L., Naxin W., and DeLuca N. 1997. The herpes simplex virus immediate-early protein ICP0 affects transcription from the viral genome and infected-cell survival in the absence of ICP4 and ICP27. *J. Virol.* **71:** 4614–4625.

Samaniego L., Webb A., and DeLuca N. 1995. Functional interaction between herpes simplex virus immediate-early proteins during infection: Gene expression as a consequence of ICP27 and different domains of ICP4. *J. Virol.* **69:** 5705–5715.

Wolfe D., Goins W.F., Yamada M., Moriuchi S., Krisky D.M., Oligino T.J., Marconi P.C., Fink D.J., and Glorioso J.C. 1999. Engineering herpes simplex virus vectors for CNS applications. *Exp. Neurol.* **159:** 34–46.

Wu N., Watkins S.C., Schaffer P.A., and DeLuca N.A. 1996. Prolonged gene expression and cell survival after infection by a herpes simplex virus mutant defective in the immediate-early genes encoding ICP4, ICP27, and ICP22. *J. Virol.* **70:** 6358–6368.

21 Herpes Simplex Virus Type-1-derived Amplicon Vectors

William J. Bowers*[‡] and Howard J. Federoff*[†‡]

Departments of Neurology and Microbiology and Immunology,[†] and the Center for Aging and Developmental Biology,[‡] University of Rochester School of Medicine and Dentistry, Rochester, New York 14642*

ABSTRACT

Until recently, herpes simplex virus (HSV) amplicon vectors were packaged into virions using a helper-virus-based methodology. This previously employed packaging technology has been supplanted by "helper-virus-free" methodologies that use bacterial artificial chromosomes (BACs) carrying an HSV genome devoid of its cognate cleavage/packaging sequences. Cotransfection of this DNA construct with an amplicon plasmid using an optimized packaging method results in packaging of only the amplicon component. Following concentration using sucrose-gradient-based ultracentrifugation, titers of helper-virus-free amplicon stocks using BAC-based methods can approach approximately 1×10^8 transducing units (TU)/ml without replication-competent revertants with specifically designed amplicons. The protocols presented here are designed to enable the successful generation of viable HSV amplicon vectors for in vitro and in vivo gene-transfer applications. They also describe a new titering method based on real-time quantitative polymerase chain reaction (PCR) that permits enumeration of transducing particles, leading to more accurate assessment of expression and gene-transfer efficacy differences in amplicon comparison experiments.

Continued

INTRODUCTION

The HSV-1-derived amplicon vector is a highly versatile platform due to its facility for molecular genetic manipulation and broad tropism (Oehmig et al. 2004). The amplicon is a eukaryotic expression plasmid modified by the incorporation of an HSV origin of DNA replication (ori) and cleavage/packaging sequence (Fig. 1, "a" sequence). Neither sequence encodes a viral gene product, but both act as *cis* elements to facilitate vector genome replication and packaging into HSV particles. Because each vector genome contains concatenated transcription units, a given transduction event results in robust expression of the delivered transgene. Heterologous transcription units of various sizes can be cloned into an amplicon plasmid where an upper limit of 150 kb has been experimentally verified (Wade-Martins et al. 1999, 2001, 2003).

Amplicon vectors have been used in a number of gene-transfer paradigms. The vectors have broad tropism, thus allowing highly efficient transduction into many cell types. The major cellular receptors for HSV (HveA and Nectin-1) have been identified, and their widespread expression accounts for broad amplicon tropism (Rajcani and Vojvodova 1998). Although natural HSV infections involve localized replication typically at an epithelial surface with subsequent infection of innervating sensory neurons, HSV-1 enveloped vectors can and do infect many cell types with great efficiency. Highly efficient transduction of numerous terminally differentiated cell types has been documented extensively (Freese and Geller 1991; Federoff et al. 1992; Geschwind et al. 1993, 1994, 1996; Xu et al. 1994; Fong et al. 1995; Geller et al. 1995; Lu and Federoff 1995; Jin et al. 1996; Fraefel et al. 1997; Lu et al. 1997; Carew et al. 1998; Coffin et al. 1998; Federoff 1998; Parry et al. 1998; Antonawich et al. 1999; D'Angelica et al. 1999; Halterman et al. 1999; Kutubuddin et al. 1999; Yamada et al. 1999; Chen et al. 2001; Tolba et al. 2001; Willis et al. 2001).

Until recently, HSV amplicon vectors were packaged into virions using a helper-virus-based methodology. Such an approach produces high-titer amplicon stocks that are unfortunately contaminated with helper virus particles. Although the copropagated helper viruses are replication defective, low-level viral gene expression occurs in transduced cells, ultimately resulting in cytotoxicity. This previously employed packaging technology has been supplanted by "helper-virus-free" methodologies that use BACs carrying an HSV genome devoid of its cognate cleavage/packaging sequences. Cotransfection of this DNA construct with an amplicon plasmid using an optimized packaging method results in packaging of only the amplicon component (Fig. 2). Following concentration using sucrose-gradient-based ultracentrifugation, titers of helper-virus-free amplicon stocks using BAC-based methods can approach approximately 1 x 10^8 TU/ml

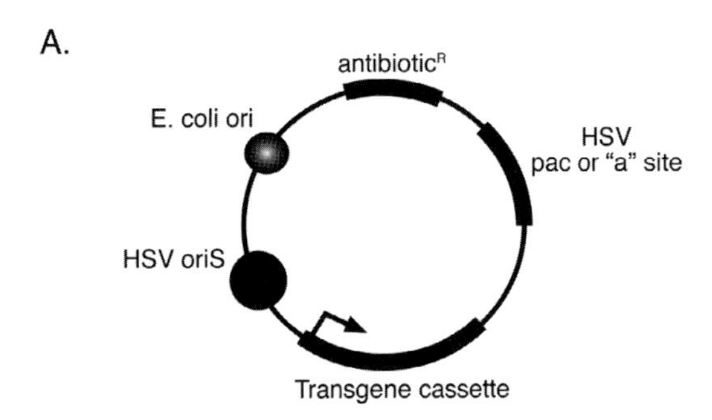

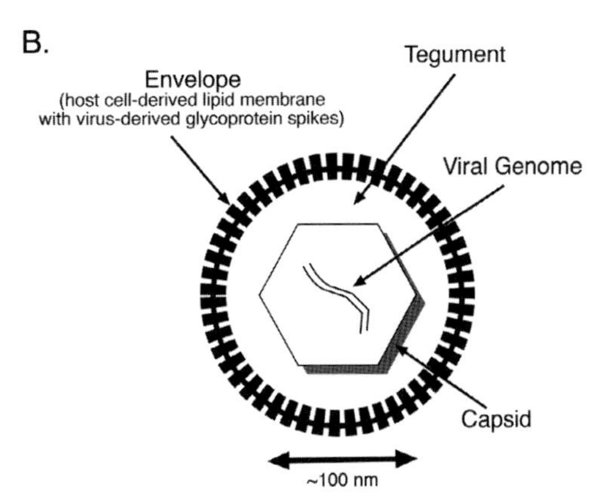

FIGURE 1. Schematic representations of an HSV-1 amplicon plasmid and viral particle. (*A*) In its simplest iteration, the HSV-1-derived amplicon vector plasmid harbors two nonencoding *cis* sequences from HSV (HSV ori and HSV "a" or pac signal), a transgene cassette, and bacterial replication and antibiotic resistance genes. (*B*) Once packaged into a viral particle, the HSV amplicon resembles a wild-type HSV-1 particle in that a membranous envelope, tegument, and nucleocapsid structures surround an approximately 150-kb genome that possesses head-to-tail concatemeric repeats of the unit-sized amplicon genome.

without replication-competent revertants with specifically designed amplicons (Bowers et al. 2001; Saeki et al. 2001, 2003). The following protocols are designed to enable the successful generation of viable HSV amplicon vectors for in vitro and in vivo gene-transfer applications.

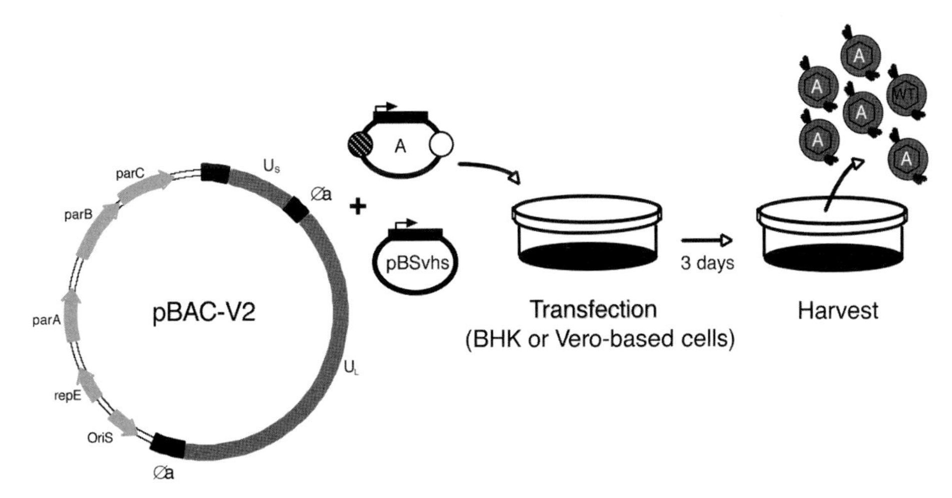

FIGURE 2. Illustration of helper-virus-free HSV-1 amplicon packaging. The packaging of helper-virus-free HSV amplicon vectors requires high-efficiency cotransfection of the amplicon plasmid, pBAC-V2 BAC, and pBSKS(vhs) plasmid into cells permissive to herpes replication and particle production. Three days posttransfection, cell monolayers are processed to release virus that is then purified for subsequent in vitro and in vivo gene-transfer applications.

Protocol 1

Plasmid DNA Preparation

Helper-virus-free HSV amplicon packaging requires the use of high-quality stocks of amplicon plasmid and the accessory pBSKS virion host shutoff (vhs) plasmid DNAs. Standard CsCl-based ultracentrifugation methodologies and commercially available plasmid DNA preparation kits are amenable for efficient amplicon packaging. Final plasmid preparations should be handled with care to minimize shearing and bacterial contamination because these constructs will be eventually used for eukaryotic cell transfection. Large-scale purification using the Bio-Rad Quantum Plasmid Maxiprep Kit consistently prepares high-quality plasmid DNA stocks for amplicon packaging and is described in detail.

MATERIALS

CAUTION: See Appendix for appropriate handling of materials marked with <!>.

Reagents

Agarose (electrophoresis grade; Invitrogen or equivalent)
Antibiotics
  ampicillin (100 μg/ml; Sigma-Aldrich or equivalent) <!>
  kanamycin (10 μg/ml; Sigma-Aldrich or equivalent) <!>
Escherichia coli glycerol stock transformed with plasmid DNA of interest
Endonucleases (New England BioLabs or equivalent)
Ethanol (100%) <!>
LB broth (EMD Chemicals or equivalent)
Plasmid purification kit: Quantum Plasmid Maxiprep Kit (Bio-Rad 732-6130)
Sodium chloride (5 M)
TB broth (250 ml; Invitrogen or equivalent)
Water, double distilled (ddH_2O)

Equipment

Centrifuge (Sorvall RC5B or equivalent) with SLC-1500 rotor (Sorvall or equivalent)
Cheesecloth (VWR or equivalent)
Clinical centrifuge (IEC GP8R Centra or equivalent)
Conical tubes (50 ml; Corning or equivalent)
Environmental shaking incubator (Lab-Line or equivalent)
Electrophoresis gel box (Owl or equivalent)
Erlenmeyer flask (2 liters)
Latex examination gloves (powder free)
Microfuge tubes (1.7 ml; Axygen or equivalent)
Oakridge bottles (36 and 250 ml; Thermo-Electron or equivalent)
Pipette-Aid (Drummond or equivalent)
Pipettors (adjustable volume; Rainin P-20, P-200, and P-1000 or equivalent)
Serological pipettes (5- and 10-ml capacity; Kimble or equivalent)

Spectrophotometer (Beckman DU 800 or equivalent)
Test tube with cap (13 x 100 mm; VWR or equivalent)

METHOD

1. Inoculate 3 ml of sterile LB plus antibiotic in a 13 x 100-mm test tube with the *E. coli* glycerol stock containing the plasmid of interest. Incubate in a shaking incubator for 8 hours at 37°C.
2. Inoculate 250 ml of sterile TB plus antibiotic in a 2-liter flask with 3 ml of cultured *E. coli* (from Step 1). Culture in a shaking incubator overnight (16–20 hours) at 37°C.
3. Pellet cells from the 250-ml culture using 250-ml centrifuge bottles in the Sorvall RC5B at 3795*g* for 10 minutes. Discard the supernatant.

Isolation of Plasmid DNA

This protocol uses the Bio-Rad Quantum Plasmid Maxiprep Kit with the following procedural modifications.

4. Resuspend the bacterial pellet in 15 ml of resuspension buffer.
5. Add 23 ml of lysis buffer. Mix gently. Incubate for 4 minutes at room temperature (~22°C).
6. Add 15 ml of neutralization buffer. Mix gently. Incubate for 3 minutes at room temperature.
   A white flocculant should be clearly visible.
7. Centrifuge in a Sorvall SLC-1500 rotor at 16,736*g* for 20 minutes at 4°C.
8. Decant the supernatant through a cheesecloth into a 50-ml conical tube to remove cellular debris.
9. Split the supernatant evenly into two 50-ml conical tubes. Add 5 ml of Quantum Matrix to each tube. Mix by inversion ten times.
10. Centrifuge in the clinical centrifuge at 1290*g* for 5 minutes at room temperature.
11. Decant the supernatant from both tubes. Resuspend each matrix pellet in 12.5 ml of wash buffer and combine into one 50-ml conical tube.
12. Centrifuge in the clinical centrifuge at 1290*g* for 5 minutes at room temperature. Discard the supernatant. Resuspend the matrix pellet in 15 ml of wash buffer.
13. Place a spin basket into one of the 50-ml conical tubes contained in the Maxiprep Kit. Transfer the suspension to the spin basket.
14. Centrifuge in the clinical centrifuge at 1290*g* for 5 minutes at room temperature. Discard the supernatant.
15. Remove the spin basket and decant the filtrate.
16. Add 10 ml of wash buffer to the spin basket. Centrifuge in the clinical centrifuge at 1290*g* for 5 minutes at room temperature.
17. Transfer the spin basket to a clean 50-ml conical tube (from the Maxiprep Kit). Add 5 ml of ddH_2O. Centrifuge in the clinical centrifuge at 1398*g* for 5 minutes at room temperature.
18. Add 2 volumes of chilled (–20°C) ethanol and 60 μl of 5 M NaCl per milliliter of eluate to 1 volume of eluate in a 36-ml Oakridge tube.
19. Centrifuge in the Sorvall RC5B at 20,846*g* for 30 minutes at 4°C.
    Note the orientation of the tubes when loading.

20. Carefully decant the supernatant and invert the tube. Allow the plasmid DNA pellet to air-dry for 10 minutes at room temperature.
21. Add 500 μl of ddH_2O to the tube. Recap the tube and rock overnight, pellet side down, on a flat rocker.
22. Centrifuge the tube in the Sorvall RC5B at 3619*g* for 5 minutes at 4°C. Transfer the resuspended plasmid DNA to a 1.7-ml microfuge tube.
23. Determine the A_{260}/A_{280} using a spectrophotometer. Calculate the DNA concentration.
24. Digest 2 μg of plasmid DNA using appropriate restriction endonuclease(s) and analyze the restriction digestion pattern on a 1% agarose gel.

Protocol 2

pBAC-V2 DNA Preparation

F-plasmid-based BACs have been designed to harbor packaging signal-deleted versions of the wild-type HSV-1 genome for use in helper-virus-free amplicon vector packaging protocols (Stavropoulos and Strathdee 1998; Bowers et al. 2001; Saeki et al. 2001, 2003). BACs are typically maintained as single-copy DNA entities within *E. coli* strains such as DH10b and can be purified using modified commercially available kits and protocols. Due to their large sizes, BAC DNAs must be handled with care to prevent shearing because failure to do so will drastically diminish the efficiency of HSV amplicon packaging. Therefore, the use of wide-bore pipette tips and a gentle pipetting technique are crucial. The following protocol uses the QIAGEN Endo-free Plasmid Mega Kit with several modifications demonstrated to increase the yield and purity of BAC DNA.

MATERIALS

CAUTION: See Appendix for appropriate handling of materials marked with <!>.

Reagents

Agarose (electrophoresis grade; Invitrogen or equivalent)
BamHI endonuclease (New England BioLabs or equivalent)
Chloramphenicol (Sigma-Aldrich or equivalent) <!>
E. coli glycerol stock transformed with pBAC-V2 construct
Ethanol (70% and 100%) <!>
Isopropanol (100%) <!>
LB broth (EMD Chemicals or equivalent)
Plasmid purification kit: Endo-free Plasmid Mega Kit (QIAGEN 12381)

Equipment

Centrifuge (Sorvall RC5B centrifuge or equivalent) with SLC-1500 and SLC-4000 rotors (Sorvall or equivalent)
Cheesecloth (VWR or equivalent)
Electrophoresis gel box (Owl or equivalent)
Erlenmeyer flasks (125 ml and 2 liters; Corning or equivalent)
Filter paper (S&S grade 588 or 606; Whatman 59728435)
Floor shaker (Lab-Line or equivalent)
Latex examination gloves (powder free)
Oakridge bottles (250 ml and 1 liter; Thermo-Electron or equivalent)
Pipette-Aid (Drummond or equivalent)
Pipettors with sterile barrier tips (adjustable volume; Rainin P-20, P-200, and P-1000 or equivalent)
Serological pipettes (5-, 10-, and 25-ml capacity, sterile; Corning or equivalent)
Spectrophotometer (Beckman DU 800 or equivalent)

METHODS

Amplification of *E. coli* Stock

1. Inoculate 25 ml of sterile LB plus 17 μg/ml chloramphenicol in a 125-ml Erlenmeyer flask with pBAC-V2 *E. coli* glycerol stock. Incubate in a shaking incubator for 6 hours at 37°C.
2. Prepare four 2-liter flasks, each containing 1 liter of LB plus 17 μg/ml chloramphenicol. Inoculate each with 5 ml of cultured pBAC-V2 (from Step 1). Culture in a shaking incubator overnight (20–24 hours) at room temperature.
3. Use 1-liter centrifuge bottles to pellet cells from the culture using the SLC-4000 rotor at 1700*g* for 20 minutes. Discard the supernatant.

Isolation of BAC DNA

This protocol uses the QIAGEN Endo-free Plasmid Mega Kit with the following procedural modifications.

4. Resuspend the bacterial pellets in 125 ml of P1 buffer (with RNase added) and pool the cell suspensions into one bottle.
5. Add 125 ml of P2 buffer. Mix gently. Incubate the suspension for 5 minutes at room temperature.
6. Add 125 ml of ice-chilled P3 buffer. Mix immediately and gently. Incubate for 30 minutes on ice.
7. Centrifuge in a Sorvall SLC-4000 rotor at 5200*g* for 10 minutes at 4°C. Promptly remove centrifuge bottles after centrifugation is complete.
8. Filter supernatant through two layers of cheesecloth dampened with ddH_2O to remove cellular debris.
9. Filter-clarify supernatant through wet filter paper.
10. Add 40 ml of ER buffer to the filtered supernatant. Mix by swirling the bottle ten times. Place on ice.
11. Equilibrate two QIAGEN-tip 2500 columns by applying 25 ml of QBT buffer.
12. Apply half of the filtered lysate to each column. Allow the lysate to enter the resin by gravity flow.
13. Wash each column with 220 ml of QC buffer.
14. Elute the DNA by washing each column five times with QN buffer at 56°C. Use 10 ml of warmed QN buffer for each wash. Pool the eluates into 250-ml Oakridge bottles.
15. Precipitate the pBAC-V2 DNA with 0.7 volume of room-temperature isopropanol.
16. Centrifuge immediately using a SLA-1500 rotor at 15,000*g* for 40 minutes at 4°C.
    Note orientation of bottles when loading.
17. Carefully remove the isopropanol and blot bottles.
18. Resuspend the pBAC-V2 DNA in each bottle using 2.5 ml of endotoxin-free H_2O. Incubate at room temperature with bottle loosely capped so that any residual isopropanol will evaporate.

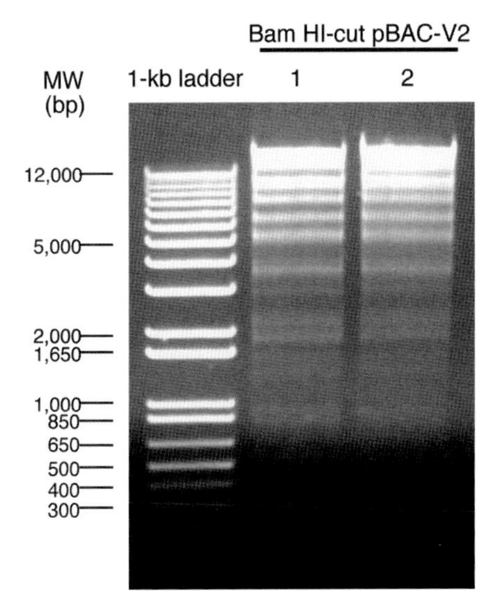

FIGURE 3. Representative DNA-banding pattern following BamHI-mediated digestion of the pBAC-V2 BAC construct. pBAC-V2 DNA (2 μg) was fully digested with 4 units of BamHI restriction endonuclease and analyzed on an ethidium bromide–stained 0.8% agarose gel alongside a 1-kb molecular-weight ladder standard.

19. Refrigerate overnight at 4°C, pellet side down.
20. Rock tubes for 2 hours, pellet side down, on flat rocker.
21. Determine the A_{260}/A_{280} using a spectrophotometer. Calculate DNA concentration.
22. Digest 2 μg of pBAC-V2 with BamHI. Analyze the restriction digest pattern on a 0.8% agarose gel (Fig. 3).

Protocol 3

Maintenance and Expansion of BHK Packaging Cell Line

The preparation of helper-virus-free HSV amplicon vectors requires the use of a packaging cell line. The desired characteristics of the packaging cell include permissiveness to HSV replication and propagation, high-density growth in cell culture, low-passage number in vitro, and high efficiency of transfection. The most commonly used amplicon packaging cell lines that meet such requirements are those derived from Vero (African green monkey) and baby hamster kidney (BHK-21) cells. It is imperative to (1) maintain these cell lines by passaging the cultures regularly (every 2–3 days) and (2) prevent contamination by using standard aseptic techniques within the clean environment of a laminar flow-based biosafety cabinet.

MATERIALS

CAUTION: See Appendix for appropriate handling of materials marked with <!>.

Reagents

BHK-21 cells (ATCC: CCL-10)
Complete growth medium
  Dulbecco's modified Eagle's medium (DMEM), sterile, high glucose with L-glutamine, without *S*-pyruvate (Invitrogen or equivalent)
  10% fetal bovine serum (FBS), sterile, mycoplasma-tested
  1% penicillin/streptomycin (10,000 IU/ml; Invitrogen or equivalent) <!>
Dulbecco's phosphate-buffered saline (DPBS), sterile, without Ca^{++}/Mg^{++}
TE (0.25% trypsin/0.02% EDTA; Invitrogen) <!>

Equipment

Bioguard biosafety cabinet (Baker Steriguard II or equivalent)
Centrifuge tubes (conical, 15 and 50 ml, sterile; Corning or equivalent)
Clinical tabletop centrifuge (IEC GP8R Centra or equivalent)
CO_2/H_2O-jacketed incubators (37 ± 1.0°C, 5% CO_2; Forma or equivalent)
Culture flasks (T-150, sterile; Corning or equivalent)
Freezers: standard (–20°C) and ultralow (–80°C)
Hemocytometer counting chamber
Inverted phase-contrast microscope (Zeiss Telaval31 or equivalent)
Latex examination gloves (powder free)
Microcentrifuge tubes (sterile, 0.5 and 1.7 ml)
Minivortex (VWR or equivalent)
Pipette-Aid (Drummond or equivalent)
Pipettors with sterile barrier tips (adjustable volume; Rainin P-20, P-200, and P-1000 or equivalent)
Serological pipettes (5-, 10-, and 25-ml capacity, sterile)
Water bath, 37 ± 1°C (VWR 1235 or equivalent)

METHOD

1. Prewarm TE and complete growth medium in a 37°C water bath.
2. Examine BHK cells to be subcultured microscopically for confluency and any signs of bacterial or fungal contamination.

   The cells should be 80–100% confluent.
3. Working in a biosafety cabinet, aseptically remove and discard the growth medium.
4. Wash cells with 5 ml of DPBS without Ca^{++}/Mg^{++}. Discard wash.
5. Dilute the prewarmed T/E 1:1 with DPBS without Ca^{++}/Mg^{++}. Add 2 ml of the diluted TE solution to each flask of cells. Rock gently to coat the cell monolayers evenly.
6. Incubate the flasks at room temperature or at 37°C until cells have detached.

   Observe status of detachment at approximately 5-minute intervals and gently rock flasks to dislodge cells.
7. Wash remaining attached cells off the bottom of the flask with 10 ml of prewarmed complete growth medium. Suspend the cells by gently pipetting up and down several times.
8. Remove 8–10 ml of suspended cells. Use this volume to suspend the cells from a second flask. Repeat until all the cells from up to five T-150 flasks have been suspended. Pool all suspended cells into one flask.
9. Remove 1 ml of resuspended cells to a 15-ml conical tube. Add 9 ml of complete growth medium. Mix well by gently pipetting up and down.
10. Load hemocytometer with approximately 10 μl of the diluted cells.
11. Count all the cells in four-square millimeters on each side of the hemocytometer (four corners of the grid). Determine the average number of cells per square millimeter. Multiply average by dilution factor. Multiply the result by 10^4.

    This represents the number of cells per milliliter of pooled cells.
12. Use the cell count to determine the seed volume as per the packaging protocol.

    The volume calculated should provide approximately 1.5×10^7 cells (see Protocol 4).
13. Subculture an aliquot of the remaining cells at approximately one-half the volume determined in Step 12 (ratio of 1:5 to 1:20). Passage BHK cells every 2–4 days.

Protocol 4

Helper-virus-free HSV-1 Amplicon Packaging

The packaging of helper-virus-free HSV amplicon vectors (Fig. 2) requires high-efficiency cotransfection of three DNA constructs into BHK-21 cells: the amplicon plasmid, pBAC-V2 BAC, and pBSKS(vhs) plasmid. Several transfection techniques have been tested in this paradigm with varying success. Invitrogen's Lipofectamine-Plus reagent has been found to perform consistently in helper-virus-free amplicon packaging. Other lipofection-based methodologies can be used, but their inherent efficiencies in producing viable amplicon stocks must be determined empirically. Moreover, the utmost care must be taken when handling the pBAC-V2 construct to prevent shearing, because this will drastically reduce the titers of resultant vector stocks. The following protocol typically yields 1×10^6 to 5×10^6 transducing particles per T-150 flask of BHK-21 monolayers used.

MATERIALS

CAUTION: See Appendix for appropriate handling of materials marked with <!>.

Reagents

BHK-21 cells (ATCC: CCL-10)

These cells are plated as specified in Protocol 3, Step 13. At least 3 days before the packaging procedure and when the cells reach 80–100% confluency, passage the BHK cells and plate half the number of T-150 flasks needed for the packaging experiment. Passage number should be <30 from time of in vitro establishment.

Bleach solution (10% v/v) <!>

Dulbecco's modified Eagle's medium (DMEM), sterile, high glucose with L-glutamine, without *S*-pyruvate (Invitrogen or equivalent)

Dulbecco's phosphate-buffered saline (DPBS), sterile, with and without Ca^{++}/Mg^{++}

Ethanol (70% v/v and 100%) <!>

Fetal bovine serum (FBS), sterile, mycoplasma-tested

Hexamethylene *bis*-acetamide (HMBA; 0.5 M, filter-sterilized) (Sigma-Aldrich)

H_2O, distilled (dH_2O), sterile

Lipofectamine (Invitrogen 18324-012)

Opti-MEM (Invitrogen 31985-070)

Penicillin/streptomycin (10,000 IU/ml; Invitrogen or equivalent) <!>

Plasmids

- Amplicon plasmid DNA (prepared according to Protocol 1)
- pBAC-V2 DNA (or equivalent, prepared according to Protocol 2)
- pBSKS(vhs) plasmid DNA (prepared according to Protocol 1)

PLUS Reagent (Invitrogen 11514-015)

Sodium chloride (5 M)

Equipment

Equipment for helper-virus-free HSV-1 amplicon packaging is the same as that required in Protocol 3, with the following additions:

Cell scraper (large; Nunc or equivalent)

CO_2/H_2O-jacketed incubators (34 ± 1.0°C, 5% CO_2; Forma or equivalent)

METHOD

1. Working aseptically in a biosafety cabinet, seed T-150 flasks with 1.5 x 10^7 BHK-21 cells per flask. Incubate overnight at 37°C.
2. Precipitate amplicon plasmid and pBSKS(vhs) DNAs with ethanol.
3. In separate microfuge tubes, add a sufficient volume (in microliters) of amplicon vector plasmid and pBSKS(vhs) vector plasmid to yield 7 μg of each plasmid per flask of virus to be packaged. Add sterile H_2O to bring the final volume of each tube to 100 μl.
4. To each 100 μl of diluted DNA, add 6 μl of 5 M NaCl and 200 μl of 100% ethanol and mix. Freeze at least 30 minutes (or overnight) at –20°C.
5. Centrifuge at 10,000*g* for 20 minutes to pellet DNA.
6. Working in a biosafety cabinet, decant the ethanol from each tube. Carefully rinse DNA pellets with 1.7 ml of 70% ethanol.
7. Decant ethanol and dry pellets inverted for 10 minutes in a biosafety cabinet (do not overdry).
8. Resuspend pellets in 50 μl of sterile H_2O. Refrigerate overnight at 4°C.

   If samples were frozen overnight in 100% ethanol (Step 4), omit the overnight refrigeration. Proceed immediately to the next step after ethanol precipitation (Steps 5–8).
9. Working in biosafety cabinet, prepare the lipofection mixture.

   Prepare a separate batch of the lipofection mixture for each flask of HSV-1 amplicon to be packaged.

   a. In a 15-ml tube, add 3 ml of Opti-MEM (warmed to 37°C), 25 μg of pBAC-V2 DNA, 7 μg of amplicon plasmid DNA, and 7 μg of pBSKS(vhs) DNA. Mix gently.

   b. Add 71 μl of PLUS Reagent in 1 ml of Opti-MEM to this tube.

      If DNA precipitates during the addition of the PLUS reagent, discard and repeat the above.

   c. To a second 15-ml tube, add 2 ml of Opti-MEM.

   d. Add 107 μl of Lipofectamine to the second tube.

   e. Incubate both tubes for 20 minutes at room temperature.

   f. Combine the contents of the first and second tubes very slowly (i.e., over a 30-second period). Allow the lipofection mixture to stand for 20 minutes at room temperature.

      If DNA precipitates during the combination of these solutions, discard the tubes and repeat Step 9.
10. While incubating the lipofection mixture, remove media from the BHK packaging cells. Add 12 ml of Opti-MEM to each flask.

11. Use a 10-ml pipette to add the lipofection mixture to the cells in the T-150 flasks. Gently distribute. Incubate the flasks for 5 hours at 37°C.
12. Add 16 ml of DMEM containing 20% FBS and 2 mM HMBA to each flask. Incubate the flasks overnight at 34°C.
13. Working in a biosafety cabinet, remove media from each flask. Replace with 25 ml of DMEM containing 10% FBS and 2 mM HMBA.
14. Incubate the flasks for 2–3 days at 34°C.
15. Working in a biosafety cabinet, remove media from each flask.
16. Scrape the cells into 10 ml of cold DPBS without Ca^{++}/Mg^{++}. Pool the scraped cells from each flask into a 50-ml tube. Place the tube on ice.
17. Wash any residual cells from the flasks with 10 ml of cold DPBS without Ca^{++}/Mg^{++}. Keep tubes on ice continuously.
18. Centrifuge the tubes at 250*g* for 10 minutes.
19. Decant the DPBS into a beaker containing 10% bleach solution to inactivate released virus.
20. Resuspend the pellet (which harbors cell-bound amplicon viral particles) in 1.5 ml/flask of DPBS with Ca^{++}/Mg^{++} by gentle pipetting.
21. Freeze the resuspended pellets for at least 1 hour at –80°C before amplicon viral particle purification (see Protocol 5).

Protocol 5

Ultracentrifugation-based Purification of Packaged HSV Amplicons

Once amplicon vector stocks have been propagated (Protocol 4), viral particles must be purified and concentrated for subsequent use in vitro and in vivo. Combined sonication, low-speed centrifugation, and sucrose-gradient-based ultracentrifugation reproducibly produce amplicon vector stock titers ranging from 1×10^7 to 5×10^7 transducing particles per milliliter.

MATERIALS

CAUTION: See Appendix for appropriate handling of materials marked with <!>.

Reagents

Dry ice <!>
Dulbecco's phosphate-buffered saline (DPBS) with Ca^{++}/Mg^{++}, sterile
Ethanol (70% v/v) <!>
Sucrose solution (30% w/v) in DPBS with Ca^{++}/Mg^{++}, filter-sterilized

Equipment

Bioguard biosafety cabinet (Baker Steriguard II or equivalent)
Centrifuge tubes (conical, 15 and 50 ml, sterile) (Corning or equivalent)
Clinical tabletop centrifuge (IEC GP8R Centra or equivalent)
Cup sonicator (Misonix XL2010 or equivalent)
Forceps
Freezer: ultralow temperature (–80°C)
Latex examination gloves (powder free)
Microcentrifuge (IEC Micromax or equivalent)
Microcentrifuge tubes (0.5 and 1.7 ml, sterile)
Minicentrifuge (National Labnet C-1200 or equivalent)
Minivortex (VWR or equivalent)
PA centrifuge tube (36-ml capacity; Sorvall 03141 or equivalent)
Pipette-Aid (Drummond or equivalent)
Pipettors with sterile barrier tips (adjustable volume; Rainin P-20, P-200, and P-1000 or equivalent)
Preparative ultracentrifuge (Sorvall Discovery 100S or equivalent)
  Swinging bucket rotor (Sorvall 630/17 Surespin or equivalent)
  Swinging buckets (prechilled; Sorvall 79338 or equivalent)
Q-tips (sterile)
Serological pipettes (5-, 10-, and 25-ml capacity, sterile)
Water bath, 37 ± 1°C (VWR 1235 or equivalent)

METHOD

1. Precool the ultracentrifuge buckets (in a refrigerator) and the rotor (in the ultracentrifuge).
2. Wash the 36-ml ultracentrifuge tubes with 70% ethanol twice. Remove any remaining alcohol with a sterile Q-tip. Invert in biosafety hood to air-dry.
3. Thaw the previously harvested virus (from Protocol 4, Step 21) in a 37°C water bath. Place on ice.
4. Use a cup sonicator (cooled on ice) to sonicate each sample three times at setting 8, 30-sec on/30-sec off.

   Sonicate two thawed samples at a time, adding fresh ice frequently.
5. Centrifuge viral stock at 1290*g* for 10 minutes.
6. Add 5–7 ml of cold DPBS with Ca^{++}/Mg^{++} to the ultracentrifuge tubes.
7. Underlay 7–10 ml of cold, sterile 30% sucrose in DPBS in each tube.
8. Transfer the viral supernatant (from Step 5) to a fresh 50-ml conical tube. Centrifuge again at 1290*g* for 10 minutes.
9. Following the second centrifugation, overlay the clarified viral supernatant drop-wise onto the sucrose gradient in ultracentrifuge tubes.
10. Add sufficient cold DPBS with Ca^{++}/Mg^{++} to each tube to within 0.25 cm of the top.
11. Use forceps sterilized with 70% ethanol to place the tubes in the prechilled buckets. Carefully replace and secure the caps on the buckets.
12. Centrifuge gradients in the ultracentrifuge at 78,000*g* for 1 hour at 4°C.
13. Decant the supernatant from each ultracentrifuge tube. Remove any residual sucrose with a sterile Q-tip.

    The virus appears at the bottom of the tube as a slightly yellow-colored pellet.
14. Add 100 μl of PBS with Ca^{++}/Mg^{++} per each T-150 pooled during harvest. Incubate for no more than 30 minutes on ice to soften pellet.
15. Resuspend the pellet carefully by pipetting up and down, breaking up clumps.

    Take care not to generate foamy bubbles.
16. Transfer the suspension to sterile 1.7-ml microfuge tubes. Quickly centrifuge (<3 seconds) to pellet any unsuspended protein debris.
17. Aliquot 25–100 μl of viral suspension to prelabeled, chilled, sterile 0.5-ml microfuge tubes. Freeze on dry ice immediately. Store frozen aliquots at –80°C in a labeled freezer box or bags.

Protocol 6

Titering of β-galactosidase-expressing Amplicon Vectors

Packaged HSV amplicon vector particles that express β-galactosidase as the resident reporter gene can be readily enumerated using a standard histochemical technique. This involves transducing monolayers of a titering cell line (e.g., NIH-3T3 fibroblasts) with serial volumes of purified viral stock and subsequently processing using X-gal histochemistry. Cells positively expressing β-galactosidase produce a blue-colored insoluble precipitate formed by cleavage of the X-gal substrate. These cells can be counted on an inverted phase-contrast microscope fitted with an eyepiece micrometer (reticle). Cell counts for each volume of original viral stock used in the transduction step can be used to calculate vector titers, expressed as TU/ml.

MATERIALS

CAUTION: See Appendix for appropriate handling of materials marked with <!>.

Reagents

Dulbecco's modified Eagle's medium (DMEM), sterile, high glucose with L-glutamine, without S-pyruvate (Invitrogen or equivalent)
Dulbecco's phosphate-buffered saline (DPBS), sterile, without Ca^{++}/Mg^{++}
Ethanol (70% v/v) <!>
Fetal bovine serum (FBS; sterile, mycoplasma-tested)
Glutaraldehyde (25%) <!>
Iron solution
- 165 mg of potassium ferricyanide <!>
- 211 mg of potassium ferrocyanide <!>
- 20 mg of NP-40 (Tergitol) <!>
- 10 mg of sodium deoxycholate <!>
- 19 mg of magnesium chloride <!>

Bring total volume to 100 ml in PBS. Wrap bottle in foil and store at 4°C.

NIH-3T3 cells (ATCC: CRL-1658)
TE (0.25% trypsin/0.02% EDTA; Invitrogen) <!>
Water, distilled (dH_2O), sterile
X-gal stock solution (5-bromo-4-chloro-3-indolyl-β-D-galactopyranoside; 20 mg/ml in dimethylsulfoxide) <!>

Equipment

Bioguard biosafety cabinet (Baker Steriguard II or equivalent)
Cell culture dish (24 well; Corning or equivalent)
Cell scraper (large; Nunc or equivalent)
Centrifuge tubes (conical, 15 and 50 ml, sterile; Corning or equivalent)
CO_2/H_2O-jacketed incubators (37 ± 1.0°C, 5% CO_2; Forma or equivalent)
Culture flasks (T-150, sterile; Corning or equivalent)
Freezers: standard (–20°C) and ultralow (–80°C)

Hemocytometer counting chamber
Inverted phase-contrast microscope (Zeiss Telaval31 or equivalent)
Latex examination gloves (powder free)
Microcentrifuge tubes (0.5 and 1.7 ml, sterile; Axygen or equivalent)
Minicentrifuge (National Labnet C-1200 or equivalent)
Minivortex (VWR or equivalent)
Pipette-Aid (Drummond or equivalent)
Pipettors with sterile barrier tips (adjustable volume; Rainin P-20, P-200, and P-1000 or equivalent)
Serological pipettes (5-, 10-, and 25-ml capacity, sterile)
Water bath, 37 ± 1°C (VWR 1235 or equivalent)

METHODS

1. Prewarm sufficient TE and DMEM containing 5% FBS to 37°C for the number of wells required for titering.
2. Working in a biosafety cabinet, aseptically remove complete growth medium from cells and discard.
3. Carefully rinse cells with 5 ml of DPBS without Ca^{++}/Mg^{++}. Discard wash.
4. Dilute TE 1:1 with DPBS without Ca^{++}/Mg^{++}. Add 2 ml of diluted TE to each flask of NIH-3T3 cells.
5. Incubate cells at room temperature or at 37°C until cells have completely detached.
   Observe at 5-minute intervals, rocking gently to dislodge cells.
6. Add approximately 10 ml of DMEM containing 5% FBS to detached cells. Carefully suspend the cells by pipetting.
7. Dilute 0.5 ml of suspended cells into 4.5 ml of growth medium in a 15-ml conical tube. Suspend well by pipetting.
8. Load hemocytometer counting chamber with approximately 10 µl of the suspended dilution.
9. Count all of the cells in four-square millimeters on each side of the hemocytometer (four corners of the grid). Determine the average number of cells per square millimeter. Multiply average by the dilution factor. Multiply this result by 10^4.
   This represents the number of cells/milliliter in the original cell suspension.
10. Based on the cell count, dilute the cells to a final concentration of 2×10^5 cells/ml using prewarmed growth medium. Mix well by pipetting.
11. Add 0.5 ml of DMEM containing 5% FBS to each well of a 24-well culture dish.
12. Use a 5-ml pipette to add 0.5 ml of the diluted cells to each well to obtain a final plating concentration of 1×10^5 cells/well.
13. Incubate the culture plate overnight at 37°C.

Transduction

14. Dilute 1- and 5-µl aliquots of the amplicon viral stock to be titered in 350 ml of prewarmed DMEM containing 5% FBS. Vortex each tube for about 1 second.
15. Working in a biosafety cabinet, aseptically remove growth medium from cells and discard. Replace with the prepared viral suspension.

16. Incubate the culture plate for 1 hour at 37°C. Gently rock the plate every 15 minutes to evenly disperse virus.
17. Working in a biosafety cabinet, aseptically add 400 μl of prewarmed DMEM containing 5% FBS to each well.
18. Incubate the culture plate for 20–48 hours (depending on the amplicon-harbored promoter driving the β-galactosidase transgene) at 37°C.

Fixation of Transduced Cells

19. Remove the culture plate from the incubator and aspirate growth medium.
    This step does not have to be aseptically performed.
20. Carefully rinse each well with approximately 1 ml of PBS.
21. Add 400 μl of 1% glutaraldehyde to each well.
22. Incubate for 10 minutes at room temperature.
23. Remove glutaraldehyde. Rinse each well with 1 ml of PBS.
    The fixed cell monolayers can be stored at 4°C in PBS until ready to perform histochemical stain for β-galactosidase activity.

Histochemical Staining of Transduced Cells

24. Prepare the X-gal stain solution (40 μl X-gal stock solution per 1 ml of iron solution).
    Prepare 500 μl of X-gal stain solution for each transduced cell monolayer to be processed.
25. Remove PBS rinse (from Step 23) from each well.
26. Add 500 μl of X-gal stain solution to each well.
27. Incubate overnight at 37°C.
28. Remove X-gal stain solution. Rinse monolayers with 500 μl of PBS.
29. Remove rinse and add another 500 μl of PBS.
    Cells can be stored at 4°C until analyzed.

Enumeration of β-Galactosidase-expressing Cells and Titer Calculations

30. Visualize β-galactosidase-expressing cells using the 20x objective and reticle on the inverted phase-contrast microscope (Fig. 4).
31. Count blue cells in 10–20 representative fields.

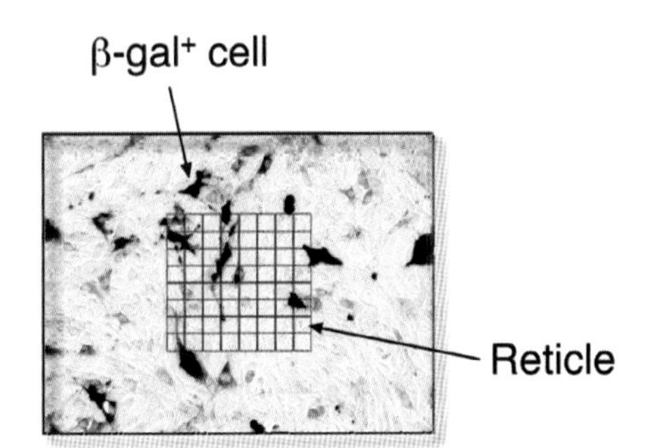

FIGURE 4. Expression-based titering analysis of NIH-3T3 cells transduced with an HSV amplicon expressing β-galactosidase (HSVlac). NIH-3T3 cells were transduced with 0.5 μl of purified HSVlac amplicon stock. Twenty-four hours later, the monolayer was fixed with 1% glutaraldehyde and processed by X-gal histochemistry. β-galactosidase-positive cells stain blue. A line illustration of the reticle used to enumerate positive cells is also depicted.

32. Average the number of stained cells per field.
33. Multiply the average by a factor of 752.
34. Divide by the number of microliters used to initially transduce cell monolayers.
35. Multiply by 1000 to obtain the titer of the concentrated viral stock, expressed as blue-forming units (bfu)/ml.

Protocol 7

Amplicon Transduction Titering

Accurate assessments regarding vector comparison and improvements in vector design require a standard for titering prepared viral stocks. Packaged amplicon vectors are routinely titered using reporter gene expression units to quantitate numbers of infectious amplicon virions (Protocol 6). The strength of the promoter, sensitivity of detection of the gene product, and choice of titering cell type can greatly influence the apparent numbers of infectious viral particles. This is especially evident when comparisons are made between two amplicon vectors that possess different promoters. To this end, a new titering method based on a real-time quantitative PCR technique allows for enumeration of transducing particles (Bowers et al. 2000). This approach ensures that amplicon comparison experiments are initiated with equivalent transduction units, thus allowing for a fair assessment of expression and gene-transfer efficacy differences.

MATERIALS

CAUTION: See Appendix for appropriate handling of materials marked with <!>.

Reagents

Bleach solution (10% v/v) <!>
Chloroform/isoamyl alcohol mix (24:1 v/v) <!>
2x Digestion buffer
- 0.2 M NaCl
- 0.02 M Tris-HCl (pH 8.0) <!>
- 50 mM EDTA (pH 8.0)
- 0.5% SDS <!>

Dithiothreitol (DTT; 1 M) <!>
Dulbecco's modified Eagle's medium (DMEM), high glucose with L-glutamine, without *S*-pyruvate, sterile (Invitrogen or equivalent)
Dulbecco's phosphate-buffered saline (DPBS), sterile, with and without Ca^{++}/Mg^{++}
Ethanol (70% v/v and 100%) <!>
FAM/TAMRA (5-carboxyfluorescein/*N,N,N′,N′*-tetramethyl-6-carboxyrhodamine)-labeled fluorogenic probe (Table 1)
- amplicon specific (Synthegen or equivalent)
- cell specific (Synthegen or equivalent)

Fetal bovine serum (FBS), sterile, mycoplasma-tested
Lysis buffer
- 100 mM potassium phosphate (pH 7.8) <!>
- 0.2% Triton X-100 <!>

NIH-3T3 cells (ATCC: CRL-1658)
PCR primers, sense and antisense (Table 1)
- amplicon specific (Synthegen or equivalent)
- cell specific (Synthegen or equivalent)

TABLE 1. TaqMan Primer/Probe Sets Used for Transduction Titering

Set name	Sense primer	Antisense primer	Probe
β-galactosidase	5′-GGGATCTGCCATTGTCAGACAT-3′	5′-TGGTGTGGGCCATAATTCAA-3′	5′-ACCCCGTACGTCTTCCCGAGCG-3′
Kanamycin	5′-CCGGCTACCTGCCCATTC-3′	5′-CCGGCTTCCATCCGAGTA-3′	5′-AACATCGCATCGAGCGAGCACG-3′
β-lactamase	5′-CTGGATGGAGGCGGATAAAGT-3′	5′-CGGCTCCAGATTTATCAGCAAT-3′	5′-CAGGACCACTTCTGCGCTCGGC-3′
18S RNA	5′-CGGCTACCACATCCAAGGAA-3′	5′-GCTGGAATTACCGCGGCT-3′	5′-TGCTGGCACCAGACTTGCCCTC-3′

PCR standards
- 18S standards: NIH-3T3 cell-derived genomic DNA (50, 5, 0.5, 0.05, and 0.005 ng/ml)
- amplicon standards: Amplicon plasmid DNA specific to the virus to be titered (1.0, 0.1, 0.01, 0.001, and 0.0001 ng/ml)

Phenol (saturated) <!>
Proteinase K (20 mg/ml) <!>
Sodium chloride (5 M)
TE (0.25% trypsin/0.02% EDTA; Invitrogen) <!>
2x Universal PCR Master Mix (Applied Biosystems or equivalent)
Water, distilled (dH_2O), sterile

Equipment

Bench paper (absorbent)
Bioguard biosafety cabinet (Baker Steriguard II or equivalent)
Cell culture dish (24 well; Corning or equivalent)
Cell scraper (large; Nunc or equivalent)
Centrifuge tubes (conical, 15 and 50 ml, sterile; Corning or equivalent)
CO_2/H_2O-jacketed incubators, 37 ± 1.0°C, 5% CO_2; Forma or equivalent)
Culture flasks (T-150; sterile; Corning or equivalent)
Freezers: standard (–20°C) and ultralow (–80°C)
Hemocytometer counting chamber
Ice bucket
Inverted, phase-contrast microscope (Zeiss Telaval31 or equivalent)
Latex examination gloves (powder free)
Microcentrifuge tubes (0.5 and 1.7 ml, sterile)
Minicentrifuge (National Labnet C-1200 or equivalent)
Minivortex (VWR or equivalent)
Parafilm
PCR tubes and caps (optical grade; Applied Biosystems or equivalent)
Pipette-Aid (Drummond or equivalent)
Pipettors with sterile barrier tips (adjustable volume; Rainin P-20, P-200, and P-1000 or equivalent), two sets

> To minimize the potential for nucleic acid contamination, designate one set of filter tips and one set of pipettors solely for quantitative PCR analysis.

Protective arm sleeves (disposable)
Sequence detector (Applied Biosystems Prism 7700 or equivalent)
Serological pipettes (5-, 10-, and 25-ml capacity, sterile)
Ultraviolet/visible spectrophotometer (Beckman Coulter DU800 or equivalent)
Water bath, 37 ± 1°C (VWR 1235 or equivalent)

METHODS

Plating of NIH-3T3 cells

1. Prewarm sufficient TE and DMEM containing 5% FBS to 37°C for the number of wells required for titering.
2. Working in a biosafety cabinet, aseptically remove growth medium from cells and discard.
3. Carefully rinse cells with 5 ml of DPBS without Ca^{++}/Mg^{++}. Discard wash.
4. Dilute TE 1:1 with DPBS without Ca^{++}/Mg^{++}. Add 2 ml of diluted TE to each flask of NIH-3T3 cells.
5. Incubate cells at room temperature or at 37°C until cells have completely detached.
   Observe at 5-minute intervals, rocking gently to dislodge cells.
6. Add approximately 10 ml of DMEM containing 5% FBS to detached cells. Carefully suspend cells by pipetting.
7. Dilute 0.5 ml of suspended cells into 4.5 ml of growth medium in a 15-ml conical tube. Suspend well by pipetting.
8. Load hemocytometer counting chamber with approximately 10 µl of the suspended dilution.
9. Count all the cells in four-square millimeters on each side of the hemocytometer (four corners of the grid). Determine the average number of cells per square millimeter. Multiply average by the dilution factor. Multiply this result by 10^4.
   This represents the number of cells/ml in the original cell suspension.
10. Based on the cell count, dilute the cells to a final concentration of 2×10^5 cells/ml using prewarmed growth medium. Mix well by pipetting.
11. Add 0.5 ml of DMEM containing 5% FBS to each well of a 24-well culture dish.
12. Use a 5-ml pipette to add 0.5 ml of the diluted cells to each well to obtain a final plating concentration of 1×10^5 cells/well.
13. Incubate the culture plate overnight at 37°C.

Transduction

14. Dilute 1- and 5-µl aliquots of the amplicon viral stock to be titered in 350 ml of prewarmed DMEM containing 5% FBS. Vortex each tube for about 1 second.
15. Working in a biosafety cabinet, aseptically remove growth medium from cells and discard. Replace with the prepared viral suspension.
16. Incubate the culture plate for 1 hour at 37°C. Gently rock the plate every 15 minutes to evenly disperse virus.
17. Working in a biosafety cabinet, aseptically add 400 µl of prewarmed DMEM containing 5% FBS to each well.
18. Incubate the culture plate for 20–24 hours at 37°C.

Lysis and Digestion of Transduced Cells

19. Remove the culture medium from cell monolayers.
20. Add 100 µl of lysis buffer containing 1 mM DTT to each well. Incubate for 15 minutes at room temperature.

21. Transfer lysates to 1.7-ml microfuge tubes.
22. Add 100 μl of 2x digestion buffer containing 0.2 mg/ml proteinase K to each tube. Incubate for 3 hours at 56°C.

Phenol/Chloroform Extraction

23. Dilute saturated phenol 1:1 with chloroform:isoamyl alcohol mix. Add 200 μl of phenol:chloroform:isoamyl alcohol mix to each tube. Mix by inverting the tube ten times.
24. Centrifuge at 13,000*g* for 5 minutes.
25. Carefully remove aqueous (top) layer and transfer to a fresh 1.7-ml microfuge tube.
26. Add 200 μl of chloroform:isoamyl alcohol mix to each tube. Mix by inverting the tube ten times.
27. Centrifuge at 13,000*g* for 5 minutes.
28. Carefully remove aqueous (top) layer and transfer to a fresh 1.7-ml microfuge tube.
29. Add 375 μl of 100% ethanol to each tube. Incubate samples for 15 minutes at –80°C or 1–2 hours at –20°C.
30. Centrifuge samples at 13,000*g* for 20 minutes.
31. Decant supernatant. Carefully rinse DNA pellet with 500 μl of 70% ethanol. Allow pellet to air-dry for 10 minutes.
32. Resuspend DNA pellet in 50 μl of dH_2O.
33. Determine the A_{260}/A_{280} using a spectrophotometer. Calculate the DNA concentration.
34. Dilute each sample to a final concentration of 5–20 ng/μl for PCR analysis.

    The optimal dilution is instrument-specific. Diluted stocks of 5 ng/μl are readily detectable and yield adequate amplification profiles on the Applied Biosystems Prism 7700 Sequence Detector. Other instruments may require similar or more concentrated template DNA stocks.

Real-time Quantitative PCR Analysis

35. Prepare a sample template (see Table 2, for example) using the software on the computer attached to the Sequence Detector.

    CAUTION: Wear labcoat, gloves, and protective arm sleeves during all steps. Do not handle DNA standards or titering DNA samples until Step 46.
36. Set up fresh bench paper at the chosen work area.
37. Wipe pipettors, tube racks, ice bucket, and lab benchtop with fresh 10% bleach solution. Allow 15 minutes to dry.
38. Based on the reaction setup information (Table 3), calculate the volumes of reagents needed for the entire quantitative analysis. Assume 1.15 volumes x the number of samples.
39. Change gloves and protective arm sleeves.
40. Place tubes into a PCR tube tray appropriate for the quantitative PCR instrument to be used in accordance with the template (Step 35). Cover tubes with a fresh piece of parafilm. Cut two extra sheets of parafilm.
41. Change gloves.
42. Remove Universal PCR Master Mix, primers, and fluorogenic probes from –20°C freezer. Thaw on ice.

TABLE 2. Example of Quantitative PCR Sample Setup

	1	2	3	4	5	6	7	8	9	10	11	12
A	NTC	NTC	STD Ampl	STD Ampl	STD Ampl	STD Ampl	STD Ampl	STD Ampl	STD Ampl	STD Ampl	STD Ampl	STD Ampl
B	UNK Ampl	UNK Ampl	UNK Ampl	UNK Ampl	UNK Ampl	UNK Ampl	UNK Ampl	UNK Ampl	UNK Ampl	UNK Ampl	UNK Ampl	UNK Ampl
C	UNK Ampl	UNK Ampl	UNK Ampl	UNK Ampl	UNK Ampl	UNK Ampl	UNK Ampl	UNK Ampl	UNK Ampl	UNK Ampl	UNK Ampl	UNK Ampl
D	UNK Ampl	UNK Ampl	UNK Ampl	UNK Ampl	UNK Ampl	NTC	NTC	18S STD	18S STD	18S STD	18S STD	18S STD
E	STD 18S	STD 18S	STD 18S	UNK 18S	UNK 18S	UNK 18S	UNK 18S	UNK 18S	UNK 18S	UNK 18S	UNK 18S	UNK 18S
F	UNK 18S	UNK 18S	UNK 18S	UNK 18S	UNK 18S	UNK 18S	UNK 18S	UNK 18S	UNK 18S	UNK 18S	UNK 18S	UNK 18S
G	UNK 18S	UNK 18S	UNK 18S	UNK 18S	UNK 18S	UNK 18S	UNK 18S	UNK 18S	UNK 18S	UNK 18S	UNK 18S	UNK 18S
H	UNK 18S	UNK 18S	UNK 18S	UNK 18S	UNK 18S	UNK 18S	UNK 18S	UNK 18S	UNK 18S	UNK 18S	NTC	NTC

(NTC) No template control (no DNA added); (STD Ampl) standard amplicon plasmid DNA control with FAM-labeled amplicon-specific fluorogenic probe; (UNK Ampl) unknown transduced DNA sample with FAM-labeled amplicon-specific fluorogenic probe; (STD 18S) standard NIH-3T3 cell DNA control with FAM-labeled 18S gene-specific fluorogenic probe; (UNK 18S) unknown transduced DNA sample with FAM-labeled 18S gene-specific fluorogenic probe.

43. Change gloves.
44. Use the PCR-dedicated pipettors and tips to prepare working master mix (Table 3) on ice. Return any unused stock reagents to –20°C.
45. Change labcoat, gloves, and protective arm sleeves.
46. Move tube tray and caps to a separate area to load DNA standards and unknown sample DNAs (from Step 34). Add DNA using only filter tips to avoid cross-contamination.
47. Cover tubes with fresh parafilm as each row of the PCR tube rack is loaded.

    Prepare no-template control (NTC) samples with the same reagents used for master mix.
48. Change gloves and protective arm sleeves.
49. Use a different set of pipettors to add working master mix (from Step 44) to the appropriate reaction tubes. Seal tubes with optical caps.

TABLE 3. Quantitative Real-time PCR Sample Setup

Order of reagent addition	Reagent	Stock concentration	Final concentration	Volume (μl)/ reaction
1	dH_2O	–	–	7.5
2	TaqMan Universal Mix	2x	1x	12.5
3	sense primer	20 mM	900 nM	1.125
4	antisense primer	20 mM	900 nM	1.125
5	fluorogenic probe	10 mM	50 nM	0.25
6	target DNA sample	5–20 ng/μl	–	2.5
Total volume				25

TABLE 4. Cell Line Conversion Factors Based on Nanograms of DNA Used in Quantitative PCR

	Mass of template DNA assayed by qPCR		
Titering cell line	50 ng	25 ng	12.5 ng
NIH-3T3	0.028	0.014	0.007

50. Transfer tube rack to quantitative PCR instrument. Initiate the following PCR profile:

    Stage 1: 50°C for 2 minutes.

    Stage 2: 95°C for 10 minutes.

    Stage 3: 40 cycles of 95°C for 15 seconds and 60°C for 1 minute.

51. Once run is complete, analyze and save data for calculation of transduction titer.

    To prevent contamination of the instrument and reaction setup area, discard reaction tubes in a location removed from the quantitative PCR instrument.

52. Calculate the number of transducing units (TU) per milliliter:

    TU = ng target sequence detected x (1/97,500) x ($6020/10^{-8}$) x (amplicon plasmid size [in kb]/150 kb)

    TU/ml = (TU [as calculated above]/ng of template DNA originally assayed) x ([1000/μl used to transduce cell monolayers]/cell line conversion factor [see Table 4])

REFERENCES

Antonawich F.J., Federoff H.J., and Davis J.N. 1999. BCL-2 transduction, using a herpes simplex virus amplicon, protects hippocampal neurons from transient global ischemia. *Exp. Neurol.* **156:** 130–137.

Bowers W.J., Howard D.F., and Federoff H.J. 2000. Discordance between expression and genome transfer titering of HSV amplicon vectors: Recommendation for standardized enumeration. *Mol. Ther.* **1:** 294–299.

Bowers W.J., Howard D.F., Brooks A.I., Halterman M.W., and Federoff H.J. 2001. Expression of vhs and VP16 during HSV-1 helper virus-free amplicon packaging enhances titers. *Gene Ther.* **8:** 111–120.

Carew J.F., Federoff H., Halterman M., Kraus D.H., Savage H., Sacks P.G., Schantz S.P., Shah J.P., and Fong Y. 1998. Efficient gene transfer to human squamous cell carcinomas by the herpes simplex virus type 1 amplicon vector. *Am. J. Surg.* **176:** 404–408.

Chen X., Frisina R.D., Bowers W.J., Frisina D.R., and Federoff H.J. 2001. HSV amplicon-mediated neurotrophin-3 expression protects murine spiral ganglion neurons from cisplatin-induced damage. *Mol. Ther.* **3:** 958–963.

Coffin R.S., Thomas S.K., Thomas D.P., and Latchman D.S. 1998. The herpes simplex virus 2 kb latency associated transcript (LAT) leader sequence allows efficient expression of downstream proteins which is enhanced in neuronal cells: Possible function of LAT ORFs. *J. Gen. Virol.* **79:** 3019–3026.

D'Angelica M., Karpoff H., Halterman M., Ellis J., Klimstra D., Edelstein D., Brownlee M., Federoff H., and Fong Y. 1999. In vivo interleukin-2 gene therapy of established tumors with herpes simplex amplicon vectors. *Cancer Immunol. Immunother.* **47:** 265–271.

Federoff H.J. 1998. Replication-defective herpesvirus amplicon vectors and their use for gene transfer. In *Cells: A laboratory manual*, vol. 2. *Light microscopy and cell structure* (ed. D.L. Spector et al.), pp. 91.1–91.10. Cold Spring Harbor Laboratory Press, Cold Spring Harbor, New York.

Federoff H.J., Geschwind M.D., Geller A.I., and Kessler J.A. 1992. Expression of nerve growth factor in vivo, from a defective HSV-1 vector prevents effects of axotomy on sympathetic ganglia. *Proc. Natl. Acad. Sci.* **89:** 1636–1640.

Fong Y., Federoff H.J., Brownlee M., Blumberg D., Blumgart L.H., and Brennan M.F. 1995. Rapid and efficient gene transfer in human hepatocytes by herpes viral vectors. *Hepatology* **22:** 723–729.

Fraefel C., Jacoby D.R., Lage C., Hilderbrand H., Chou J.Y., Alt F.W., Breakefield X.O., and Majzoub J.A. 1997. Gene transfer into hepatocytes mediated by helper virus-free HSV/AAV hybrid vectors. *Mol. Med.* **3:** 813–825.

Freese A. and Geller A. 1991. Infection of cultured striatal neurons with a defective HSV-1 vector: Implications for gene therapy. *Nucleic Acids Res.* **19:** 7219–7223.

Geller A.I., During M.J., Oh Y.J., Freese A., and O'Malley K. 1995. An HSV-1 vector expressing tyrosine hydroxylase causes production and release of L-DOPA from cultured rat striatal cells. *J. Neurochem.* **64:** 487–496.

Geschwind M.D., Amat J., and Federoff H.J. 1993. Expression of BDNF in cultured neurons and non-neuronal cells from a defective HSV-1 vector. *Soc. Neurosci. Abstr.* **19:** 255.

Geschwind M.D., Kessler J.A., Geller A.I., and Federoff H.J. 1994. Transfer of the nerve growth factor gene into cell lines and cultured neurons using a defective herpes simplex virus vector. Transfer of the NGF gene into cells by a HSV-1 vector.

Mol. Brain Res. **24:** 327–335.

Geschwind M.D., Hartnick C.J., Liu W., Amat J., Van De Water T.R., and Federoff H.J. 1996. Defective HSV-1 vector expressing BDNF in auditory ganglia elicits neurite outgrowth: Model for treatment of neuron loss following cochlear degeneration. *Hum. Gene Ther.* **7:** 173–182.

Halterman M.W., Miller C.C., and Federoff H.J. 1999. Hypoxia-inducible factor-1α mediates hypoxia-induced delayed neuronal death that involves p53. *J. Neurosci.* **19:** 6818–6824.

Jin B.K., Belloni M., Conti B., Federoff H.J., Starr R., Son J.H., Baker H., and Joh T.H. 1996. Prolonged *in vivo* gene expression driven by a tyrosine hydroxylase promoter in a defective herpes simplex virus amplicon vector. *Hum. Gene Ther.* **7:** 2015–2024.

Kutubuddin M., Federoff H.J., Challita-Eid P.M., Halterman M., Day B., Atkinson M., Planelles V., and Rosenblatt J.D. 1999. Eradication of pre-established lymphoma using herpes simplex virus amplicon vectors. *Blood* **93:** 643–654.

Lu B. and Federoff H.J. 1995. Herpes simplex virus type 1 amplicon vectors with glucocorticoid-inducible gene expression. *Hum. Gene Ther.* **6:** 421–430.

Lu B., Federoff H.J., Wang Y., Goldsmith L.A., and Scott G. 1997. Topical application of viral vectors for epidermal gene transfer. *J. Invest. Dermatol.* **108:** 803–808.

Oehmig A., Fraefel C., and Breakefield X.O. 2004. Update on herpesvirus amplicon vectors. *Mol. Ther.* **10:** 630–643.

Parry S., Holder J., Halterman M.W., Weitzman M.D., Davis A.R., Federoff H., and Strauss J.F., III. 1998. Transduction of human trophoblastic cells by replication-deficient recombinant viral vectors. Promoting cellular differentiation affects virus entry. *Am. J. Pathol.* **152:** 1521–1529.

Rajcani J. and Vojvodova A. 1998. The role of herpes simplex glycoproteins in the virus replication cycle. *Acta Virol.* **42:** 103–118.

Saeki Y., Breakefield X.O., and Chiocca E.A. 2003. Improved HSV-1 amplicon packaging system using ICP27-deleted, oversized HSV-1 BAC DNA. *Methods Mol. Med.* **76:** 51–60.

Saeki Y., Fraefel C., Ichikawa T., Breakefield X.O., and Chiocca E.A. 2001. Improved helper virus-free packaging system for HSV amplicon vectors using an ICP27-deleted, oversized HSV-1 DNA in a bacterial artificial chromosome. *Mol. Ther.* **3:** 591–601.

Stavropoulos T.A. and Strathdee C.A. 1998. An enhanced packaging system for helper-dependent herpes simplex virus vectors. *J. Virol.* **72:** 7137–7143.

Tolba K.A., Bowers W.J., Hilchey S.P., Halterman M.W., Howard D.F., Giuliano R.E., Federoff H.J., and Rosenblatt J.D. 2001. Development of herpes simplex virus-1 amplicon-based immunotherapy for chronic lymphocytic leukemia. *Blood* **98:** 287–295.

Wade-Martins R., Frampton J., and James M.R. 1999. Long-term stability of large insert genomic DNA episomal shuttle vectors in human cells. *Nucleic Acids Res.* **27:** 1674–1682.

Wade-Martins R., Saeki Y., and Chiocca E.A. 2003. Infectious delivery of a 135-kb LDLR genomic locus leads to regulated complementation of low-density lipoprotein receptor deficiency in human cells. *Mol. Ther.* **7:** 604–612.

Wade-Martins R., Smith E.R., Tyminski E., Chiocca E.A., and Saeki Y. 2001. An infectious transfer and expression system for genomic DNA loci in human and mouse cells. *Nat. Biotechnol.* **19:** 1067–1070.

Willis R.A., Bowers W.J., Turner M.J., Fisher T.L., Abdul-Alim C.S., Howard D.F., Federoff H.J., Lord E.M., and Frelinger J.G. 2001. Dendritic cells transduced with HSV-1 amplicons expressing prostate-specific antigen generate antitumor immunity in mice. *Hum. Gene Ther.* **12:** 1867–1879.

Xu H., Federoff H.J., Maragos J., Parada L.F., and Kessler J.A. 1994. Viral transduction of *trk A* into cultured nodose and spinal motor neurons conveys NGF responsiveness. *Dev. Biol.* **163:** 152–161.

Yamada M., Oligino T., Mata M., Goss J.R., Glorioso J.C., and Fink D.J. 1999. Herpes simplex virus vector-mediated expression of Bcl-2 prevents 6-hydroxydopamine-induced degeneration of neurons in the substantia nigra *in vivo*. *Proc. Natl. Acad. Sci.* **96:** 4078–4083.

22 γ-2 Herpesvirus Saimiri-based Vectors

Adrian Whitehouse

Institute of Molecular and Cellular Biology, Faculty of Biological Sciences, and Astbury Centre for Structural Molecular Biology, University of Leeds, LS2 9JT, United Kingdom

ABSTRACT

Herpesvirus saimiri (HVS) is capable of infecting a range of human cell types with high efficiency. The viral genome persists as high-copy-number, circular, nonintegrated episomes that segregate to progeny upon cell division. This allows HVS-based vectors to transduce stably a dividing cell population and provide sustained transgene expression for an extended period of time both in vitro and in vivo. Moreover, the insertion of a bacterial artificial chromosome cassette into the HVS genome simplifies the incorporation of large amounts of heterologous DNA for gene delivery. These properties offer characteristics similar to that of an artificial chromosome combined with an efficient delivery system.

INTRODUCTION

HVS is the prototype γ-2 herpesvirus, or rhadinovirus (Albrecht et al. 1992; Fickenscher and Fleckenstein 2001). It persistently infects its natural host, the squirrel monkey, without causing any obvious disease. Although HVS infection of other species of New World primates can result in lymphoproliferative diseases, this can be completely eliminated by deletion of the transforming genes, *stp* and *tip* (Fickenscher and Fleckenstein 2001). The vectors used herein possess these deletions and as such are incapable of transforming any cell type.

HVS offers the potential to incorporate large amounts of heterologous DNA and infect a broad range of human cell lines, including carcinoma and hematopoietic cells (Grassmann and Fleckenstein 1989; Simmer et al. 1991; Stevenson et al. 1999, 2000). On infection, the viral genome can persist by virtue of episomal maintenance and stably transfer heterologous gene expression (Grassmann and Fleckenstein 1989; Simmer et al. 1991; Stevenson et al. 1999, 2000; Hall et al. 2000; Calderwood et al. 2004). The long-term presence of the HVS episome is believed not to affect cell growth, suggesting that the maintenance of the HVS episome does not disrupt the growth machinery of human cells in vitro (Smith et al. 2001). Moreover, in ex vivo tumor xenograft experiments, the HVS-based vector remained latent in the xenograft without spreading to other organs. The long-term maintenance of the HVS genome as a nonintegrated circular episome provides efficient sustained expression of a heterologous transgene in vivo (Smith et al. 2001; Giles et al. 2003). In addition, HVS can efficiently infect solid-tumor xenografts derived from a variety of human carcinoma cells via direct intratumoral injections. On infection of both the tumor xenografts and spheroid cultures, HVS-based vectors can establish a persistent episomal infection within the tumor xenograft, allowing expression of a heterologous transgene (Smith et al. 2005a). The delivery profiles of the HVS vector have also been examined in vivo after systemic administration. Results demonstrate that an HVS-based vector can infect and establish a persistent episomal infection in a range of mouse tissues. Moreover, the HVS episome provided sustained expression of the heterologous transgene, particularly in the liver (Smith et al. 2005b).

The HVS genome has recently been cloned into a bacterial artificial chromosome (HVS-BAC) that allows easier manipulation and production of HVS-based vectors. The large packaging capacity offered by such vectors permits the incorporation of complete eukaryotic *cis*-regulatory elements to provide the correct level of tissue-specific expression of a therapeutic gene or even complete genomic loci. However, further development of HVS-based vectors is required; at present, the humoral and cellular immune responses to HVS-based vectors have not addressed. Although long-term maintenance and expression were obtained in immunologically incompetent female athymic BALB/c nude mice, there is no indication that these results will be similar in an immunocompetent animal. Moreover, therapeutic applications will require the clarification of a number of biosafety aspects. Safety mechanisms must be incorporated into these vectors, including replication-deficient and packaging cell-dependent viruses to eliminate any viral replication. Methods contained in this chapter allow production of an HVS-based vector expressing a heterologous transgene of choice as well as working and high-purification stocks of the recombinant vector. Furthermore, methods are discussed that allow the assessment of infectivity efficiencies of HVS-based vectors and the establishment of stably transduced cell lines where the HVS genome remains as a circular nonintegrated episome.

Protocol 1

Production of Recombinant HVS-based Vectors

To insert and express a heterologous gene in an HVS-based vector, a recombinant virus must be constructed. An HVS-BAC is used to simplify and enhance the production of recombinant viruses (White et al. 2003; Smith et al. 2005b). This requires a two-step process to insert the heterologous expression cassette first into the pHVS-Shuttle and then into the HVS-BAC (Fig. 1).

MATERIALS

CAUTION: See Appendix for appropriate handling of materials marked with <!>.

Reagents

Bovine serum albumin (BSA; 10%)
Cells: Electromax DH10 B cells (Invitrogen) and owl monkey kidney (OMK) cells
Chloramphenicol (50 μg/ml) <!>
DMEM (Dulbecco's modified Eagle's medium) supplemented with the following:
 2 mM glutamine
 0.1% (w/v) sodium bicarbonate
 200 units/ml penicillin <!>
 200 units/ml streptomycin <!>

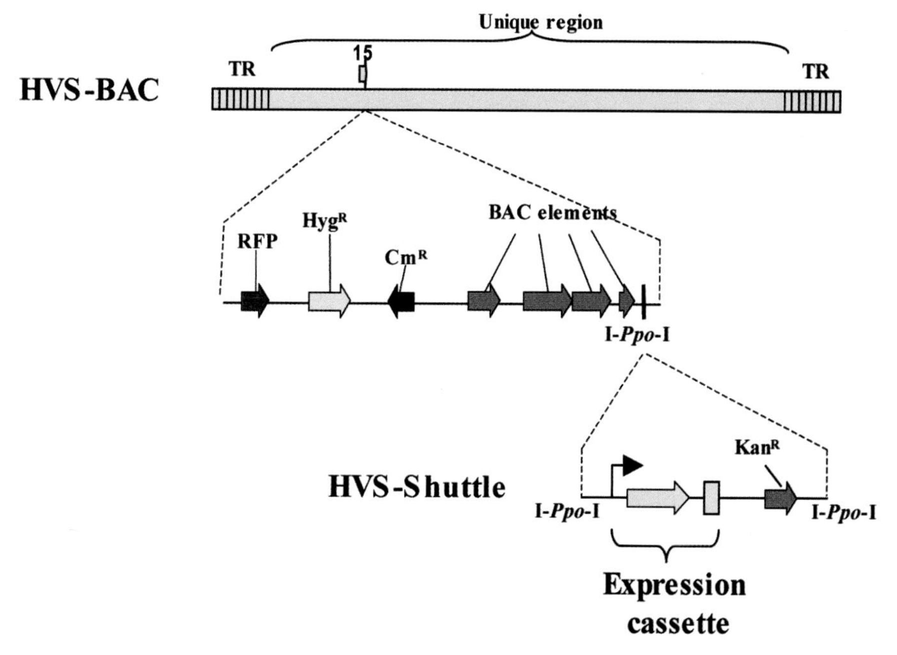

FIGURE 1. Schematic representation of the production of recombinant HVS-based vector using the BAC system. The heterologous expression construct (containing the appropriate promoter, gene, and polyadenylation signals) is inserted into pHVS-Shuttle. This heterologous expression cassette and a kanamycin resistance gene is then excised from the pHVS-Shuttle using I-PpoI and ligated directly into the HVS-BAC via its unique I-PpoI restriction site.

Fetal bovine serum (FBS)
Glucose (20% and 40% w/v) in PBS
Sterilize by filtration through a 0.2-μm filter.
I-PpoI restriction enzyme and restriction buffer (Promega)
Kanamycin (25 μg/ml) <!>
Lipofectamine 2000 (Invitrogen)
Luria broth (LB, pH 7.5)
1% (w/v) bactopeptone
0.5% (w/v) yeast extract
0.5% (w/v) NaCl
$MgCl_2$ (10 mM) <!>
Phosphate-buffered saline (PBS)
Plasmids: HVS-BAC and pHVS-Shuttle
QIAquick Gel Extract Column Kit (QIAGEN)
Virus resuspension buffer
100 mM Tris-HCl (pH 8.0) <!>
50 mM NaCl
10 mM EDTA

Equipment

Beckman SW 27 rotor
Flask (75 cm^3)
Gradient mixer
LB agar plates (14 cm)
Maxiprep column (QIAGEN)
Roller bottles

METHODS

1. Insert the heterologous expression construct (containing the appropriate promoter, gene, and polyadenylation signals) into the shuttle vector, pHVS-Shuttle.
2. Excise the heterologous expression cassette and a kanamycin resistance gene from pHVS-Shuttle using the I-PpoI restriction enzyme.

   I-PpoI restriction digests are performed in 1x I-PpoI restriction buffer supplemented with 1x BSA and 10 mM $MgCl_2$.
3. Purify the fragment using the QIAquick Gel Extract Column Kit, according to the manufacturer's instructions.
4. Linearize 5–10 μl of HVS-BAC DNA with I-PpoI.
5. Heat-inactivate both the shuttle plasmid and HVS-BAC digests for 20 minutes at 65°C.
6. Drop-dialyze the HVS-BAC digest against distilled H_2O (dH_2O) for 2 hours.
7. Ligate the linearized HVS-BAC and shuttle plasmid fragment together overnight at 16°C using standard conditions.

   A ratio of 3 μl of HVS-BAC to 1 μl of shuttle plasmid fragment in 10 μl of ligation mix often gives good results. This may vary depending on the quality and quantity of both DNA preparations.
8. Drop-dialyze the ligations against dH_2O for 2 hours.

9. Electroporate the salt-free ligation into 20 μl of Electromax DH10 B cells.
10. Add 400 μl of LB. Allow the cells to recover.
11. Incubate with shaking (100 rpm) for 1 hour at 37°C.
12. Transfer the electroporated cells to a 14-cm LB agar plate. Select overnight with chloramphenicol (50 μg/ml) and kanamycin (25 μg/ml).
13. Extract and purify recombinant HVS-BAC DNA from the colonies using a Maxiprep column according to the manufacturer's low-copy-plasmid protocol.
    Confirm insertion of heterologous DNA using restriction digests and Southern blot analysis.
14. Transfect 2 μg of purified HVS-BAC DNA into OMK cells using Lipofectamine 2000, as described by the manufacturer.
15. Incubate cells in DMEM for 4–5 days, until the cytopathic effect is evident.
16. Release the progeny virus into the extracellular medium.

Growth of Working Virus Stocks

The most successful cells for lytic infection and growth of HVS are OMK cells maintained in DMEM supplemented with 10% FBS.

17. Seed a 75-cm^3 flask with 1 x 10^7 OMK cells. Allow to grow to confluency.
18. Infect cultures at a low moi of about 0.1 pfu/cell in a total volume of 5 ml of DMEM containing 2% FBS.
19. Allow adsorption of the virus for 1 hour at 37°C. Remove medium.
20. Add 10 ml of DMEM supplemented with 5% FBS. Incubate for 4–5 days at 37°C.
21. When cytopathic effect is observed, release the progeny viruses into the extracellular medium.
    Working stocks of HVS can be stored long term at a final concentration of approximately 2 x 10^6 to 2 x 10^7 pfu/ml at –70°C. For medium-term storage, HVS is surprisingly stable at 4°C.

Growth and Purification of High-titer Virus

22. Grow OMK monolayers in roller bottles (~2 x 10^8 cells) in DMEM supplemented with 10% FBS.
23. Infect cultures using an moi of approximately 0.5 pfu/cell. Incubate for 4–5 days.
24. Detach the cells and collect the medium.
25. Centrifuge the suspension in a Beckman SW 27 rotor at 20,000 rpm for 90 minutes at 4°C.
26. Resuspend the pellet in 3 ml of virus resuspension buffer per two roller bottles. Incubate overnight at 4°C.
27. Prepare a 20–40% glucose gradient using a gradient mixer. Layer the disrupted pellet onto the glucose gradient.
28. Centrifuge in a Beckman SW 27 rotor at 20,000 rpm for 30 minutes at 4°C.
29. Harvest the central band, which contains 90% enveloped particles.
30. Centrifuge in a Beckman SW 27 rotor at 25,000 rpm for 1 hour at 4°C. Resuspend the pellet in PBS.

Protocol 2

Assessment of Infectivity Using an HVS Recombinant that Expresses HVS-GFP

Several studies have demonstrated that HVS can infect a wide range of human cells including human hematopoietic cells and human carcinoma cell lines (Simmer et al. 1991; Stevenson et al. 1999, 2000). To assess the infectivity of a specific cell line, an HVS recombinant virus expressing green fluorescent protein (GFP) (HVS-GFP) (Hall et al. 2000; White et al. 2003) can be used (Fig. 2a).

MATERIALS

CAUTION: See Appendix for appropriate handling of materials marked with <!>.

Reagents

Cells: Cell line of interest
DMEM (Dulbecco's modified Eagle's medium)
Fetal bovine serum (FBS)
Hygromycin (250 μg/ml; Invitrogen) <!>
Virus: HVS-GFP

Equipment

Dishes (35 mm)
Fluorescent microscope

METHODS

1. Infect the cell line of interest with the HVS-GFP.

   To infect adherent monolayers

   a. Seed a 35-mm dish with 5×10^5 cells of the cell line of interest. Grow them to 80% confluency.
   b. Infect cultures using an moi of one in a total volume of 0.5 ml of DMEM supplemented with 2% FBS.
   c. Allow adsorption of the virus for 1 hour at 37°C.
   d. Aspirate the medium and replace with fresh DMEM containing 10% serum. Incubate at 37°C.
   e. Twenty-four hours postinfection, assess infectivity rates using fluorescence microscopy.

   To infect suspension cultures

   a. Mix 2×10^5 of the cell line of interest and HVS-GFP at an moi of ten in a total volume of 0.5 ml of DMEM containing 2% FBS.

      High rates of infectivity of suspension cell cultures can be attained using spinoculation.

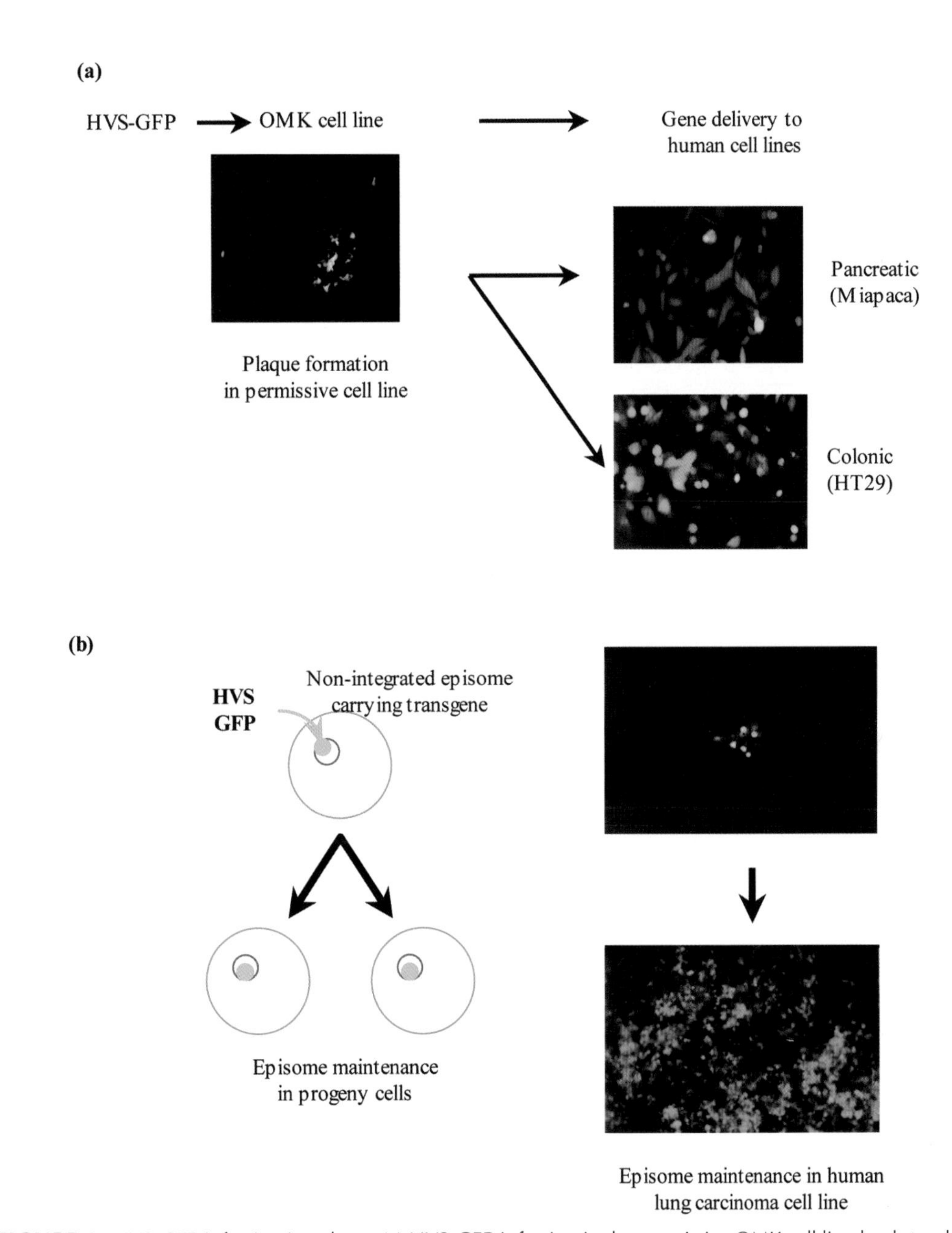

FIGURE 2. HVS-GFP infection in culture. (a) HVS-GFP infection in the permissive OMK cell line leads to plaque formation. OMK cells are a fully permissive cell line used to propagate the virus. In contrast, infection of human colorectal (HT29) and pancreatic (Miapaca) carcinoma cell lines leads to latent infection where the virus persists as a circular nonintegrated episome. (*b*) Establishment of a persistent infection of HVS-GFP in the human lung A549 carcinoma cell line. A549 cells were infected with HVS-GFP and selected in the presence of hygromycin. Shown is the development of an individual clone over a 6-week period.

b. Centrifuge the suspension at 1500 rpm for 90 minutes at 20°C to allow adsorption.

c. Wash the cells in DMEM containing 10% FBS.

d. Resuspend the pellet in 2 ml of DMEM. Incubate at 37°C.

e. Twenty-four hours postinfection, assess infectivity rates using fluorescence microscopy.

Production of HVS-infected Stably Transduced Cell Lines

A major advantage of HVS-based vectors is the ability to establish and persist as a nonintegrated episome in a dividing cell population (Hall et al. 2000; White et al. 2003).

2. Repeat monolayer or suspension infectivity protocols as described above to establish a long-term persistent infection.
3. Twenty-four hours postinfection, culture the cells in the presence of 250 µg/ml hygromycin for 2 weeks. Change the media and hygromycin every 4 days.

   After 2 weeks in culture, only cells that have been successfully transduced remain viable.
4. Assess the efficiency of stably transduced cell lines using fluorescence microscopy, where 100% of cells should exhibit the GFP-expressing phenotype (Fig. 2b).

Protocol 3

Gardella Gel Analysis

Gardella gel analysis is performed to determine whether the HVS genome is maintained in a non-integrated episomal form. This method allows the identification of chromosomal/integrated DNA and episomal and linear forms of viral DNA, which run at the top, middle, and bottom of the gel, respectively. To obtain the required sensitivity to detect HVS episomes within host tissues, a modified polymerase chain reaction (PCR)-based Gardella gel can also be utilized.

MATERIALS

CAUTION: See Appendix for appropriate handling of materials marked with <!>.

Reagents

Agarose gel
- 0.75% (w/v) agarose in TBE buffer
- 0.8% (w/v) low-melting-point agarose in TBE buffer containing 2% SDS <!> and 1 mg/ml self-digested pronase (Sigma-Aldrich)

PCR primers
- ORF73f (5′-CGCGGATCCATGGAAGCAGGACCAAGTACTCCA-3′)
- ORF73r (5′-CCGCTCGAGCCTTCTATAGGCAGGCTTTTGCT-3′)

Sample buffer, in PBS
- 15% Ficoll
- 0.01% bromophenol blue <!>

TBE buffer
- 89 mM Tris-base <!>
- 2 mM EDTA
- 80 mM boric acid <!>

Equipment

PCR thermal cycler
Southern blot hybridization materials and reagents

METHOD

1. Prepare the horizontal gels in two steps.
   a. Pour a 0.75% agarose gel in TBE buffer.
   b. Once solidified, remove 5 cm of the top of the gel.
   c. Replace with 0.8% agarose containing SDS and pronase.
2. Pellet 2 x 10^6 infected cells. Resuspend in 50 µl of sample buffer.
3. Load samples into the wells.

4. Electrophorese at 40 volts for 2 hours at 4°C. Continue electrophoresis at 160 volts for 18 hours.
5. Identify the episomal DNA using standard Southern blot analysis and a suitable probe (e.g., a ^{32}P-radiolabeled random-primed probe specific for the KpnE fragment of the HVS genome).

Modified PCR-based Gardella Gel

6. Process the samples as described in Steps 1–4 above.
7. After electrophoresis, cut the Gardella gel into horizontal slices.
8. Melt the agarose at 65°C.
9. Perform a single round of PCR on 5 µl of each gel slice to amplify ORF73 using primers ORF73f and ORF73r.

ACKNOWLEDGMENTS

The author thanks past and present members of the laboratory and collaborating groups for their scientific contributions. This work was supported in part by grants from the Biotechnology and Biological Sciences Research Council, Medical Research Council, Wellcome Trust, Candlelighters Trust, Yorkshire Cancer Research, and Association of International Cancer Research.

REFERENCES

Albrecht J., Nicholas J., Biller D., Cameron K., Biesinger B., Newman C., Wittmann S., Craxton M., Coleman H., and Fleckenstein B. 1992. Primary structure of the herpesvirus saimiri genome. *J. Virol.* **66:** 5047–5058.

Calderwood M.A., White R.E., and Whitehouse A. 2004. Development of herpesvirus-based episomally maintained gene delivery vectors. *Expert Opin. Biol. Ther.* **4:** 493–505.

Fickenscher H. and Fleckenstein B. 2001. Herpesvirus saimiri. *Philos. Trans. R. Soc. Lond. B Biol. Sci.* **356:** 545–567.

Giles M.S., Smith P.G., Coletta P.L., Hall K.T., and Whitehouse A. 2003. The herpesvirus saimiri ORF 73 regulatory region provides long-term transgene expression in human carcinoma cell lines. *Cancer Gene Ther.* **10:** 49–56.

Grassmann R. and Fleckenstein B. 1989. Selectable recombinant herpesvirus saimiri is capable of persisting in a human T-cell line. *J. Virol.* **63:** 1818–1821.

Hall K.T., Giles M.S., Goodwin D.J., Calderwood M.A., Carr I.M., Stevenson A.J., Markham A.F., and Whitehouse A. 2000. Analysis of gene expression in a human cell line stably transduced with herpesvirus saimiri. *J. Virol.* **74:** 7331–7337.

Simmer B., Alt M., Buckreus I., Berthold S., Fleckenstein B., Platzer E., and Grassmann R. 1991. Persistence of selectable herpesvirus saimiri in various human haematopoietic and epithelial cell lines. *J. Gen. Virol.* **72:** 1953–1958.

Smith P.G., Coletta P.L., Markham A.F., and Whitehouse A. 2001. In vivo episomal maintenance of a herpesvirus saimiri-based gene delivery vector. *Gene Ther.* **8:** 1762–1769.

Smith P.G., Burchill S.A., Brooke D., Coletta P.L., and Whitehouse A. 2005a. Efficient infection and persistence of a herpesvirus saimiri-based gene delivery vector into human tumour xenografts and multicellular spheroid cultures. *Cancer Gene Ther.* **12:** 248–256.

Smith P.G., Oakley F., Fernandez M., Mann D.A., Lemoine N.R., and Whitehouse A. 2005b. Herpesvirus saimiri-based vector biodistribution using non-invasive optical imaging. *Gene Ther.* **12:** 1465–1476.

Stevenson A.J., Clarke D., Meredith D.M., Kinsey S.E., Whitehouse A., and Bonifer C. 2000. Herpesvirus saimiri-based gene delivery vectors maintain heterologous expression throughout mouse embryonic stem cell differentiation in vitro. *Gene Ther.* **7:** 464–471.

Stevenson A.J., Cooper M., Griffiths J.C., Gibson P.C., Whitehouse A., Jones E.F., Markham A.F., Kinsey S.E., and Meredith D.M. 1999. Assessment of herpesvirus saimiri as a potential human gene therapy vector. *J. Med. Virol.* **57:** 269–277.

White R.E., Calderwood M.A., and Whitehouse A. 2003. Generation and precise modification of a herpesvirus saimiri bacterial artificial chromosome demonstrates that the terminal repeats are required for both virus production and episomal persistence. *J. Gen. Virol.* **84:** 3393–3403.

23 Gene Delivery Using HSV/AAV Hybrid Amplicon Vectors

Okay Saydam, Daniel L. Glauser, and Cornel Fraefel

Institute of Virology, University of Zurich, Switzerland

ABSTRACT

Herpes simplex virus type 1 (HSV-1)-based amplicon vectors conserve most of the properties of the parental virus: broad host range, the ability to transduce dividing and nondiving cells, and a large transgene capacity. This permits incorporation of genomic sequences as well as cDNA, large transcriptional regulatory sequences for cell-specific expression, multiple transgene cassettes, or genetic elements from other viruses. Hybrid vectors use elements from HSV-1 that allow replication and packaging of large-vector DNA into highly infectious particles, and elements from other viruses that confer genetic stability to vector DNA in the transduced cell. For example, adeno-associated virus (AAV) has the unique ability to integrate its genome into a specific site on human chromosome 19. The viral *rep* gene and the inverted terminal repeats (ITRs) that flank the AAV genome are sufficient for this process. However, AAV-based vectors have a very small transgene capacity and do not conventionally contain the *rep* gene to support site-specific genomic integration. HSV/AAV hybrid vectors contain both HSV-1 replication and packaging functions and the AAV *rep* gene and a transgene cassette flanked by the AAV ITRs. This combines the large transgene capacity of HSV-1 with the capability of site-specific genomic transgene integration and long-term transgene expression of AAV. This protocol describes the preparation of HSV/AAV hybrid vectors using a helper-virus-free packaging system. The advantages of such vectors over standard HSV-1 amplicon vectors, as well as their limitations, are also discussed.

INTRODUCTION

There are two fundamentally different types of HSV-1-based vector systems: recombinant and amplicon. Recombinant HSV-1 vectors are created by introducing the transgene cassette into the viral genome, often replacing one or several viral genes. HSV-1-based amplicon vectors are bacterial plasmids that contain only two HSV-1 *cis* elements: an origin of DNA replication (ori), and a DNA packaging/cleavage signal (pac), but no viral genes (Fig. 1a). These two elements are sufficient to support replication of the vector DNA and packaging of the concatemeric replication products into virions in the presence of HSV-1 helper functions. Helper functions can be provided by replication-conditional HSV-1 helper viruses (Lim et al. 1996), replication-competent/conditional, packaging-defective HSV-1 genomes cloned as sets of cosmids (Fraefel et al. 1996), or bacterial artificial chromosomes (BACs) (Saeki et al. 1998, 2001; Stavropoulos and Strathdee 1998). Amplicon vectors have a large transgene capacity (up to 150 kbp) and can transduce most cell types, both dividing and nondividing (Wade-Martins et al. 2001, 2003). However, transgene expression is generally transient (Oehmig et al. 2004), largely due to loss of the nonreplicating and nonintegrating vector DNA in the host cells.

The wild-type AAV genome is a linear, single-stranded DNA of 4680 nucleotides containing 145-nucleotide ITRs at both ends that flank two clusters of genes, *rep* and *cap* (Lusby et al. 1980; Srivastava et al. 1983). The ITRs contain the origin of DNA replication and the packaging signal. The *rep* gene encodes four overlapping proteins, Rep78, Rep68, Rep52, and Rep40, from two different promoters, p5 and p19. ITRs and either Rep78 or Rep68 are sufficient for replication of the AAV genome in the presence of a helper virus, such as adenovirus or a herpesvirus. In the absence of helper virus, ITRs and either Rep78 or Rep68 are sufficient to mediate the integration of the AAV genome into a specific site, termed AAVS1, on chromosome 19 of human cells (Kotin et al. 1990; Linden et al. 1996).

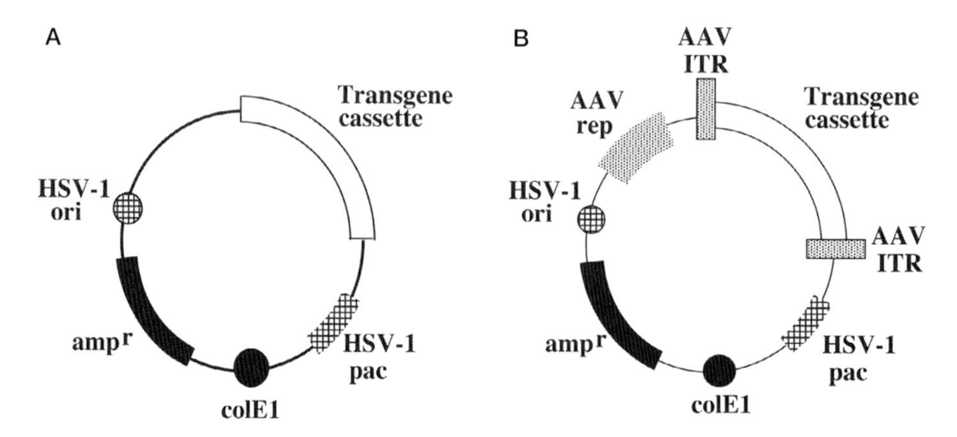

FIGURE 1. Schematic representation of HSV-1 amplicon (*a*) and HSV/AAV hybrid vector (*b*). The HSV-1 amplicon contains an antibiotic resistance gene (*amp*^{r}) and a bacterial origin of DNA replication (ColE1) to allow plasmid propagation in *E. coli*. It also features a transgene cassette and two HSV-1 *cis* elements: an origin of DNA replication (ori) and a DNA packaging/cleavage signal (pac) to support replication and packaging into HSV-1 particles in the presence of helper functions. In addition to these standard genetic elements of amplicons, HSV/AAV hybrid vectors contain the AAV *rep* gene and a transgene cassette flanked by AAV ITRs to support site-specific genomic integration of the transgene DNA in the transduced cell.

HSV/AAV hybrid vectors have been designed to combine the large transgene capacity of HSV-1 with the potential for site-specific integration and long-term gene expression of AAV in a single vector. In addition to HSV-1 ori and pac, HSV/AAV hybrid vectors contain the AAV *rep* gene and a transgene cassette flanked by AAV ITRs (Fig. 1b). These vectors can mediate site-specific integration at AAVS1 and thus provide stable transgene expression in human cells (Heister et al. 2002; Wang et al. 2002). However, the presence of the *rep* gene on the vector backbone inhibits HSV-1 replication and therefore markedly reduces the titers of hybrid vector stocks, compared to those of standard HSV-1 amplicons or HSV/AAV hybrid vectors that contain an AAV ITR-flanked transgene cassette but no *rep* gene (Heister et al. 2002; Wang et al. 2002). This protocol describes the preparation of HSV/AAV hybrid vector stocks using a replication-competent/conditional, packaging-defective HSV-1 genome cloned as a BAC to provide helper functions for vector replication and packaging. Strategies to overcome the problem of the inhibition of HSV-1 replication by AAV *rep* expression are being investigated.

Protocol

Construction and Packaging of HSV/AAV Hybrid Amplicon Vectors

This protocol uses the HSV-1 BAC deleted for both the essential immediate-early 2 gene (ICP27) and the pac signals (fHSVΔpacΔ27ΔKn), as described in Saeki et al. (2001). ICP27, which is essential for replication, is provided by a separate plasmid (pEBHICP27) in *trans*. An overview of the preparation and analysis of HSV/AAV hybrid vector stocks is shown in Table 1.

MATERIALS

CAUTION: See Appendix for appropriate handling of materials marked with <!>.

Reagents

Cells: Vero cells (clone 76; ECACC 85020205) and VERO 2-2 cells (Smith et al. 1992)
Dry ice <!>
Dulbecco's modified Eagle's medium (DMEM; Invitrogen)
Ethanol <!>
Fetal bovine serum (FBS; 2%, 6%, or 10%)
G418 (Geneticin; Invitrogen, 11811-031) <!>
Hank's buffered salt solution (HBSS; GIBCO-BRL)
Lipofectamine reagent (Invitrogen)
Plasmids
- HSV/AAV hybrid amplicon plasmid (pHSV/AAV; Heister et al. 2002)
- Packaging-defective HSV-1 helper DNA (Saeki et al. 2001)
  - fHSVΔpacΔ27ΔKn
  - pEBHICP27

Opti-MEM I (Invitrogen)
Phosphate-buffered saline (PBS)
Sucrose
Trypsin (0.25%) <!>/EDTA (0.02%) (Invitrogen)

Equipment

Disposable syringes (20 ml)
Hemacytometer
Incubator (humidified, 37°C, 5% CO_2)

TABLE 1. Packaging of HSV/AAV Hybrid Amplicons into HSV-1 Particles

Day	Procedure
0	Prepare cells for transfection
1	Cotransfect HSV-1 helper DNA and vector DNA
4	Harvest vector stocks, concentrate (optional), infect fresh cells for titration
5	Detect transgene expression, count positive cells, calculate titers

Probe sonicator (e.g., Sonifier 250; Branson)
Syringe-tip filters (0.45 mm, polyethersulfone membrane; Sarstedt)
Tissue-culture flasks (75 cm^2)
Tissue-culture plates (60-mm diameter)
Ultracentrifuge (Sorvall SS-34 rotor)
Ultra-Clear centrifuge tubes (30 ml, 25 x 89 mm; Beckman)

METHODS

Cotransfection of Packaging-defective HSV-1 Helper DNA and Vector DNA

HSV-1 amplicons and hybrid amplicons are based on bacterial plasmids and can be constructed and amplified in *Escherichia coli* (Fig. 1) (Heister et al. 2002). Transgene cassettes of interest can be inserted between the ITRs with classical cloning procedures.

1. Prepare VERO 2-2 cells for transfection.
   a. Maintain VERO 2-2 cells in DMEM/10% FBS containing 500 μg/ml G418. Propagate the culture twice per week by splitting 1 to 5 using 15 ml of fresh medium in a new 75-cm^2 tissue culture flask.
   b. One day before transfection, remove the culture medium. Wash cells twice with PBS.
   c. Add 3 ml of trypsin/EDTA. Incubate flasks for 10 minutes at 37°C to allow cells to detach.
   d. Count cells using a hemacytometer. Seed 1.2×10^6 cells per 60-mm-diameter tissue-culture plate in 3 ml of DMEM/10% FBS.
2. For each 60-mm-diameter tissue-culture plate, prepare two 15-ml conical tubes with 250 μl of Opti-MEM I.
   a. To one tube, add 0.5 μg of pHSV/AAV, 2 μg of fHSVΔpacΔ27ΔKn, and 0.2 μg of pEBHICP27. Mix and incubate for 5 minutes at room temperature.
      Replication-competent, packaging-defective HSV-1 cosmid DNA can be used as an alternative HSV-1 helper DNA (Fraefel 1997).
   b. To the other tube, add 16 μl of Lipofectamine. Vortex and incubate for 5 minutes at room temperature.
3. Combine the contents of the two tubes. Mix well and incubate for 30 minutes at room temperature.
4. Add 0.9 ml of Opti-MEM I to the tube containing the DNA-Lipofectamine transfection mixture.
5. Wash the VERO 2-2 cell cultures (Step 1) once with 2 ml of Opti-MEM I.
   The cells should be confluent at the time of transfection.
6. Aspirate all medium from the cells. Add the transfection mixture. Incubate in a humidified 37°C, 5% CO_2 incubator for 4 hours.
7. Aspirate the transfection mixture. Wash the cells three times with 2 ml of Opti-MEM I.
8. Add 3.5 ml of DMEM/6% FBS to the cells. Incubate in a humidified 37°C, 5% CO_2 incubator for 2–3 days.

Harvesting of Packaged Vectors

9. Scrape cells into the medium with a rubber policeman.
10. Transfer the suspension into a 15-ml conical tube. Place the tube into a beaker containing ice water.
11. Submerge the tip of the sonicator probe approximately 0.5 cm into the cell suspension. Sonicate for 20 seconds with 20% output energy (50 watts).
12. Remove cell debris by centrifugation at 1400*g* for 10 minutes at 4°C.
13. Filter the supernatant into a new 15-ml conical tube through a 0.45-μm syringe-tip filter attached to a 20-ml disposable syringe.
14. Remove a sample for titration (Step 19). Divide the remaining stock into 1-ml aliquots.
15. Freeze aliquots in a dry ice/ethanol bath. Store at –80°C.

    Alternatively, concentrate the stock before storage (see below). Vector stocks can be stored for at least 6 months at –80°C without a decrease in vector titers.

16. Concentrate the vector stocks as described below.
    a. Transfer the vector solution from Step 14 into a 30-ml centrifuge tube containing 25% sucrose in HBSS.
    b. Centrifuge in a Sorvall SS-34 rotor at 20,000 rpm for 2 hours at 16°C.
    c. Resuspend the pellet in approximately 300 μl of HBSS.
    d. Remove a sample for titration (Step 19). Divide the remaining stock into 30-μl aliquots.
    e. Freeze in a dry ice/ethanol bath. Store at –80°C.

Titration of Amplicon Stocks

To determine the concentration of infectious vector particles in an amplicon stock, cells are infected with vector stocks. Cells expressing the transgene are counted 24–48 hours postinfection to calculate the titer, expressed as transducing units (TU) per milliliter.

17. Plate VERO 76 cells at a density of 1.0×10^5 cells/well of a 24-well tissue-culture plate in 0.5 ml of DMEM/10% FBS. Incubate in a humidified 37°C, 5% CO_2 incubator overnight.
18. Aspirate the medium. Wash each well once with PBS.
19. Dilute 0.1, 1, and 5 μl of vector stock in 250 μl each in DMEM/2% FBS and apply to cells.
20. Incubate in a humidified, 37°C, 5% CO_2 incubator for 1–2 days.
21. Assay for presence of the transgene with an appropriate detection protocol (e.g., green fluorescence, X-gal staining, or immunocytochemical staining).
22. Count positive cells and calculate the vector titer by multiplying the number of transgene-positive cells by the dilution factor (TU/ml).

    Protocols to identify site-specific integration events and junctions between vector sequences and AAVS1 sequences in human cells infected with HSV/AAV hybrid vectors are described elsewhere (Heister et al. 2002; Wang et al. 2002).

ACKNOWLEDGMENTS

The authors are supported by the Julius-Muller Foundation, the Swiss National Science Foundation (Grant No. 3100A0-100195), and the Cancer League of Kanton Zurich.

REFERENCES

Fraefel C. 1997. Gene delivery using helper virus-free HSV-1 amplicon vectors. In *Current protocols in neuroscience* (ed. J.N. Crawley et al.), pp. 4.14.5–4.14.8. Wiley, New York.

Fraefel C., Song S., Lim F., Lang P., Yu L., Wang Y., Wild P., and Geller A.I. 1996. Helper virus-free transfer of herpes simplex virus type 1 plasmid vectors into neural cells. *J. Virol.* **10:** 7190–7197.

Heister T., Heid I., Ackermann M., and Fraefel C. 2002. Herpes simplex virus type 1/adeno-associated virus hybrid vectors mediate site-specific integration at the adeno-associated virus preintegration site, AAVS1, on human chromosome 19. *J. Virol.* **76:** 7163–7173.

Kotin R.M., Siniscalco, M.R., Samulski J., Zhu X.D., Hunter L., Laughlin C.A., McLaughlin S., Muzyczka N., Rocchi M., and Berns K.I. 1990. Site-specific integration by adeno-associated virus. *Proc. Natl. Acad. Sci.* **87:** 2211–2215.

Lim F., Hartley D., Starr P., Lang P., Song S., Yu L., Wang Y., and Geller A.I. 1996. Generation of high-titer defective HSV-1 vectors using an IE 2 deletion mutant and quantitative study of expression in cultured cortical cells. *Biotechniques* **3:** 460–469.

Linden R.M., Winocour E., and Berns K.I. 1996. The recombination signals for adeno-associated virus site-specific integration. *Proc. Natl. Acad. Sci.* **93:** 7966–7972.

Lusby E., Fife K.H., and Berns K.I. 1980. Nucleotide sequence of the inverted terminal repetition in adeno-associated virus DNA. *J. Virol.* **34:** 402–409.

Oehmig A., Fraefel C., Breakefield X.O., and Ackermann M. 2004. Herpes simplex virus type 1 amplicons and their hybrid virus partners, EBV, AAV, and retrovirus. *Curr. Gene Ther.* **4:** 385–408.

Saeki Y., Fraefel C., Ichikawa T., Breakefield X.O., and Chiocca E.A. 2001. Improved helper virus-free packaging system for HSV amplicon vectors using an ICP27-deleted, oversized HSV-1 DNA in a bacterial artificial chromosome. *Mol. Ther.* **3:** 591–601.

Saeki Y., Ichikawa T., Saeki A., Chiocca E.A., Tobler K., Ackermann M., Breakefield X.O., and Fraefel C. 1998. Herpes simplex virus type 1 DNA amplified as bacterial artificial chromosome in *Escherichia coli*: Rescue of replication-competent virus progeny and packaging of amplicon vectors. *Hum. Gene Ther.* **9:** 2787–2794.

Smith I.L., Hardwicke M.A., and Sandri-Goldin R.M. 1992. Evidence that the herpes simplex virus immediate early protein ICP27 acts post-transcriptionally during infection to regulate gene expression. *Virology* **186:** 74–86.

Srivastava A., Lusby E.A., and Berns K.I. 1983. Nucleotide sequence and organization of the adeno-associated virus 2 genome. *J. Virol.* **45:** 555–564.

Stavropoulos T.A. and Strathdee C.A. 1998. An enhanced packaging system for helper-dependent herpes simplex virus vectors. *J. Virol.* **9:** 7137–7143.

Wade-Martins R., Saeki Y., and Chiocca E.A. 2003. Infectious delivery of a 135-kb LDLR genomic locus leads to regulated complementation of low-density lipoprotein receptor deficiency in human cells. *Mol. Ther.* **7:** 604–612.

Wade-Martins R., Smith E.R., Tyminski E., Chiocca E.A., and Saeki Y. 2001. An infectious transfer and expression system for genomic DNA loci in human and mouse cells. *Nat. Biotechnol.* **19:** 1067–1070.

Wang Y., Camp S.M., Niwano M., Shen X., Bakowska J.C., Breakefield X.O., and Allen P.D. 2002. Herpes simplex virus type 1/adeno-associated virus *rep*$^+$ hybrid amplicon vector improves the stability of transgene expression in human cells by site-specific integration. *J. Virol.* **76:** 7150–7162.

24 Polyomaviruses: SV40

David S. Strayer, Christine Mitchell, Dawn A. Maier, and Carmen N. Nichols

Department of Pathology, Jefferson Medical College, Philadelphia, Pennsylvania 19107

ABSTRACT

Recombinant simian virus 40 (rSV40)-derived vectors are particularly useful for gene delivery to bone marrow progenitor cells and their differentiated derivatives, certain types of epithelial cells (e.g., hepatocytes), and central nervous system neurons and microglia. They integrate rapidly into cellular DNA to provide long-term gene expression in vitro and in vivo in both resting and dividing cells. We describe here techniques used to produce, purify, and quantitate these vectors. These procedures require only packaging cells (e.g., COS-7) and circular vector genome DNA. The only specialized equipment necessary is a quantitative real-time polymerase chain reaction (PCR) machine for titering. Amplification involves repeated infection of packaging cells with vector produced by transfection. Cotransfection is not required in any step. Viruses are purified by centrifugation using discontinuous sucrose or cesium chloride gradients and then titered by quantitative PCR. Resulting vectors are replication-incompetent and contain no detectable wild-type SV40 revertants. These approaches are simple, give reproducible results, and may be used to generate vectors that are deleted only for Tag, or for all SV40-coding sequences capable of carrying up to 5 kb of foreign DNA. Titers exceeding 10^{10} infectious units/ml may be readily achieved. These vectors are best applied to long-term expression of proteins normally encoded by mammalian cells or by viruses that infect mammalian cells, or of untranslated RNAs (e.g., RNA interference). The preparative approaches described facilitate application of these vectors and allow almost any laboratory to exploit their strengths for diverse gene delivery applications.

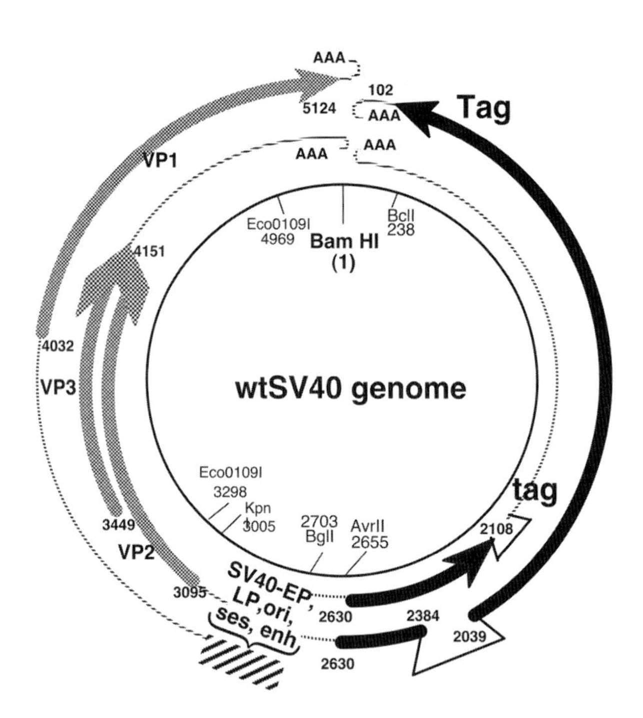

FIGURE 1. Transcriptional map of wild-type SV40. SV40 has a 5.25-kb double-stranded DNA genome. Key transcripts, both early and late, are shown. For our purposes, we have defined the unique SV40 BamHI site as the origin. Early transcription proceeds counterclockwise; late transcription progresses in a clockwise direction, from the opposite strand. Regulatory sequences include both early (EP) and late (LP) promoters, packaging sequences (ses), the origin of virus genome replication (ori), and enhancer sequences (enh). Polyadenylation sites for both early and late mRNAs are near the BamHI site. (Modified from Fried and Prives 1986; Cole and Conzen 2001.)

INTRODUCTION

SV40 is a polyomavirus with a 5.25-kb double-stranded circular DNA genome (Fig. 1), carrying three structural genes (VP1, VP2, VP3) and two main regulatory genes encoding large T antigen (Tag) and small T antigen (tag). Wild-type SV40 infects cells of many mammals via its receptor(s), variously reported as class I major histocompatibility complex (Breau et al. 1992; Stang et al. 1997) or GM1 ganglioside (Tsai et al. 2003). SV40 enters cells via lipid rafts, which deliver it to a microtubular transport system that carries the virion to the nucleus (Pelkmans et al. 2001, 2002). Without a cytoplasmic phase during cell entry, viral antigens are not processed until they are synthesized, following genome replication.

To make rSV40 gene delivery vectors, the Tag genes are excised. Tag is obligatory for viral DNA replication and late (i.e., structural) gene expression. Deletion of the gene encoding Tag means that the resulting viruses only replicate in cells that supply Tag in *trans*. In place of Tag, up to 3 kb of foreign DNA may be inserted, such as additional polymerase II (pol II) or pol III promoters and transgene-coding sequences (Fig. 2). The SV40-EP can be blocked by inserting polyadenylation (pA) signals. By using packaging cells that also supply SV40 capsid proteins, larger genomic deletions can be made to accommodate larger transgenes. A "minimal" rSV40 vector need only carry SV40 ori and packaging sequences (Fig. 2). However, it is usually convenient to retain SV40 pA and some untranslated sequences. The key characteristics and limitations of these vectors as gene-transfer vehicles are listed in Table 1.

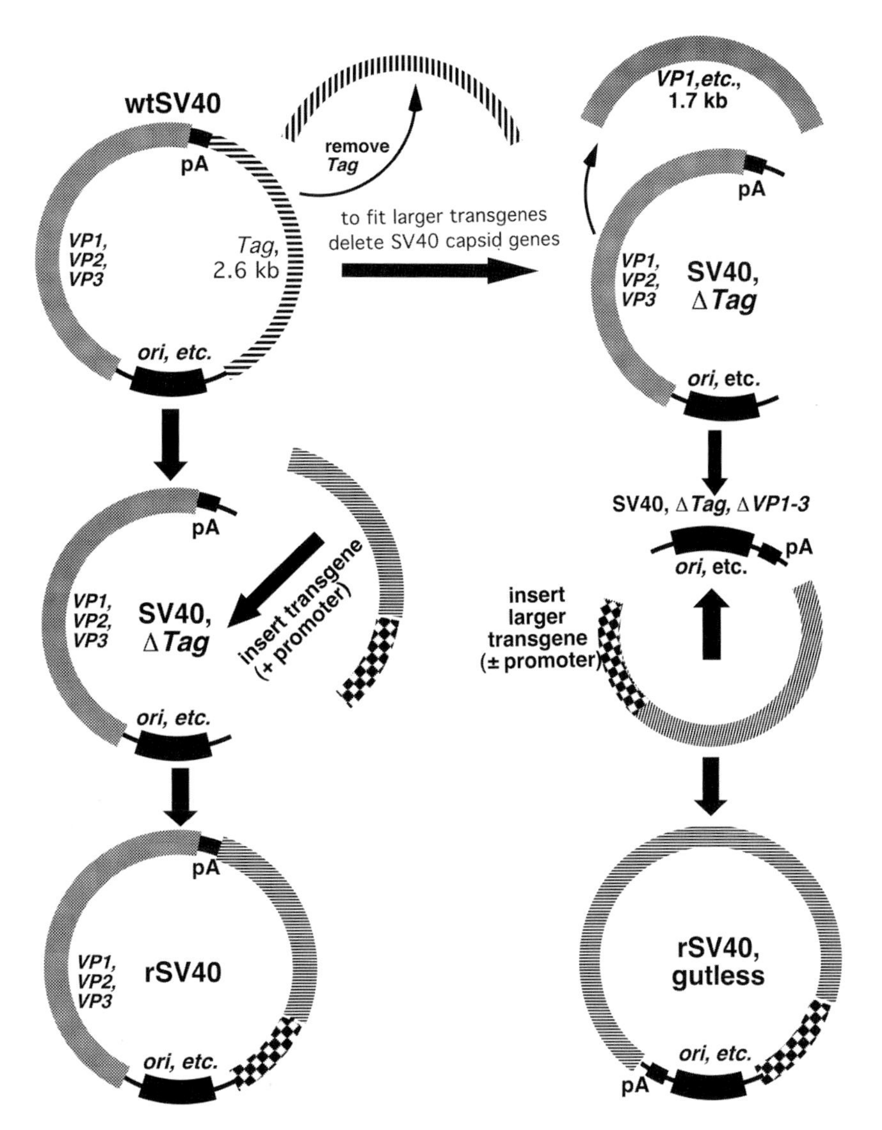

FIGURE 2. Making rSV40 vectors and gutless rSV40 vectors. Beginning with a cloned wild-type SV40 genome (*upper left*, carrier plasmid not shown), *Tag* and *tag* genes (2.6 kb) are excised, leaving the SV40 capsid genes and polyadenylation signal (pA), as well as the SV40 ori, encapsidation signal (ses), enhancer, and early promoter (which overlaps the ori, and so cannot be excised; ori, etc.). Transgenes (including an additional promoter such as CMV-IEP) of <3 kb in size can be cloned directly into this *Tag*-deficient genome (*middle, left*), resulting in an rSV40 containing the transgene, SV40 capsid genes, and the additional promoter. To accommodate a larger insert, SV40 capsid genes can also be excised (*upper right*) leaving ori, ses, promoter, and pA (*middle, right*). The larger transgene (with or without promoter, as desired) is then cloned into this "gutless" backbone, creating an rSV40 lacking all SV40-coding sequences.

Gene delivery efficiency and the level of transgene expression in transducing cells usually increase with increasing exposures to rSV40, both in vitro and in vivo. Increasing the multiplicity of infection for a single exposure does not improve delivery as effectively as does increasing the number of deliveries. Transduction efficiencies of >95% with rSV40s can be achieved with three sequential treatments, representing a cumulative multiplicity of infection of 10–16 in various organs and cell types, both in vitro and in vivo (Sauter et al. 2000; Strayer et al. 2002).

TABLE 1. Principal Traits of SV40-derived vectors

Attribute	Typical rSV40 characteristics
Titer	Typically, 10^{11} to 10^{12} infectious particles/ml.
Longevity of transgene expression	Once established, transgene expression persists indefinitely.
Gene delivery efficiency	In vitro: Usually >98%, without selection.
	In vivo: In discrete organs (e.g., liver), >90% of cells can be transduced.
Maximum insert size	Up to 5 kb.
Target cell tropisms	Almost all resting and dividing cells can be transduced.
Antigenicity imparted to target cells	None observed to date.
Neutralizing antibody	None observed to date.
"Gutless" vectors	Generated exactly as nongutless vectors. Yields are comparable to those of nongutless vectors.
Hardiness	Very robust; in lyophilized form, can be stored for long periods of time at 4°C and even at room temperature.
Promoters used	Constitutive pol II, pol III promoters; conditional pol III promoters.
Transgenes expressed	Almost all mammalian cDNAs; cDNAs of mammalian viruses.
Limitations	Nonvertebrate marker genes generally not expressed (e.g., GFP and β-galactosidase).
	Levels of protein expression are generally lower than seen with many other vector systems.

Protocol 1

Production of SV40-derived Vectors

A number of different approaches have been used to produce rSV40 vectors with variably different characteristics (Gething and Sambrook 1981; Asano et al. 1985; Strayer 1996, 1999; Strayer and Milano 1996; Rund et al. 1998; Strayer et al. 2001; Vera et al. 2004). Briefly, rSV40 genomes prepared by previously described methods (Strayer 1999; Strayer et al. 2002) are packaged in COS-7 cells (or other cell lines) that provide the deleted SV40 genes (usually Tag) in *trans*. Because native SV40 DNA is circular, vector genomes excised from the carrier plasmid must be recircularized before transfection into packaging cells. Vectors are purified by ultracentrifugation on a cesium chloride step gradient.

MATERIALS

CAUTION: See Appendix for appropriate handling of materials marked with <!>.

Reagents

COS-7 cells (ATCC)

Maintain cells in culture in DMEM-10 in 75-cm^2 flasks and passage every 3–4 days.

Cesium chloride gradient solutions (1.28 and 1.50 g/ml) <!>

Dissolve 30 and 40 g of solid CsCl in 70 and 54 ml of 0.02 M Tris-HCl (pH 7.4), respectively. Weigh 1-ml aliquots of each solution on an analytical balance to confirm densities. Filter-sterilize through a 0.22-µm filter.

Double detergent

5% deoxycholate (Sigma-Aldrich) <!>

10% Triton X-100 (Sigma-Aldrich) <!>

Dulbecco's modified Eagle's medium (DMEM; GIBCO-BRL)

Filter-sterilize through a 0.22-µm filter.

Ethanol (70% v/v) <!>

FuGENE-6 transfection reagent (Roche)

Growth medium (pH 7.4) (DMEM-10)

DMEM

10% normal calf serum (Cambrex)

1% penicillin/streptomycin (Cellgro) <!>

1% L-glutamine (Cellgro)

Filter-sterilize through a 0.22-µm filter.

Maintenance medium (pH 7.4) (DMEM-2)

DMEM

2% fetal bovine serum (Cellgro)

1% penicillin/streptomycin (Cellgro)

1% L-glutamine (Cellgro)

Filter-sterilize through a 0.22-µm filter.

10x Phosphate-buffered saline (PBS)

2 g of KCl <!>

80 g of NaCl
14.4 g of Na_2HPO_4 <!>
2.4 g of KH_2PO_4 <!>

Dilute to 1 liter in double-distilled (dd)H_2O. Before use, dilute to 1x PBS from the 10x stock with ddH_2O. Adjust pH to 7.4. Filter-sterilize through a 0.22-μm filter.

rSV40 DNA

Viral genome DNA containing the transgene of interest is excised from the carrier plasmid, gel-purified, and recircularized using standard techniques.

Trypsin/EDTA (Cellgro) <!>
Water, ddH_2O

Equipment

Cell scrapers (sterile, individually packaged)
Centrifuge tubes (plastic, clear, and opaque; Beckman)

Before use, sterilize centrifuge tubes by immersion in 70% ethanol for at least 15 minutes at room temperature.

CO_2 incubator (maintained at 37°C, 5–6% CO_2; NAPCO)
Cryovials (1.5 ml; Nalgene)
Dialysis buoys and cassettes (0.5–3.0 ml; Pierce)
Eppendorf tubes (1.5 ml)
Freezer (–80°C; Revco)
Laminar flow hood
Microscope (bright field)
Needles (18 gauge, sterile, individually wrapped)
Pipettes (sterile, individually wrapped, 2, 5, and 10 ml; Fisher)
Pipettors (10, 100, and 1000 ml; Ranin)
Syringes (3.0 and 5.0 ml, sterile)
Tissue-culture dishes (100 mm; Corning)
Tissue-culture flasks (75 cm^2; Corning)
Ultracentrifuge (Beckman) with SW 28.1 rotor with matching buckets

Sterilize the SW 28.1 rotor and buckets by autoclaving before use.

Vortex
Water bath sonicator

METHOD

1. One day before transfection, aspirate the DMEM-10 from two flasks of confluent COS-7 cells. Wash with 1 ml of PBS.

   The volumes used here are for 100-mm culture dishes. Scale accordingly for larger or smaller samples (Table 2).

TABLE 2. Transfection Conditions

Size of dish (mm)	μg of DNA*	μl of DMEM*	μl of lipid*
60	1–5	200	3–15
100	8–12	500	24–36
150	15–20	800	45–60

Asterisks indicate per dish.

2. Aspirate PBS. Add 1 ml of trypsin/EDTA.
3. Incubate for 1 minute at 37°C, or until cells begin to lift off from plastic.
   Monitor cells under a bright-field microscope.
4. Once cells have detached, add 10 ml of DMEM-10 to each flask. Pool the suspensions in a single flask.
5. Transfer the suspended cells into eight 100-mm tissue-culture dishes. Incubate overnight at 37°C with 5% CO_2.
   Cells should be 60–80% confluent at the time of transfection.
6. Aspirate media. Add 8 ml of DMEM-2 to each plate. Return to incubator.
7. Prepare transfection solution, using a separate Eppendorf tube for each plate.
   a. Add 500 μl of DMEM (serum-, antibiotic-, and glutamine-free) to each 1.5-ml tube.
   b. Add 30 μl of FuGENE to each tube. Pipette gently to mix.
      Add Fugene directly to the DMEM. Avoid contacting the sides of the tube.
   c. To each of six tubes, add 10 μg of rSV40 DNA. Pipette gently to mix.
      The seventh tube (without DNA) serves as a negative control. Add 10 μg of DNA expressing a control plasmid (e.g., one expressing enhanced green fluorescent protein) to the eighth tube as a positive control.
   d. Incubate for approximately 20 minutes at room temperature.
8. Remove the tissue-culture plates from the incubator. Add the lipid/DNA mixture, dropwise, to each dish. Swirl plates to distribute evenly.
9. Return the plates to the incubator for 72 hours at 37°C with 5% CO_2.
   After 48 hours, examine positive and negative controls for cytotoxicity and to determine transfection efficiency.
10. Seventy-two hours posttransfection, gently scrape cells into culture supernatant.
    Use a separate, individually wrapped, sterile cell scraper for each plate.
11. Pool cell lysates in 15- or 50-ml conical tubes (depending on total volume). Pipette cells gently to break up clumps.
12. Freeze at –80°C and then thaw to room temperature three times.
13. Sonicate lysate in a water bath sonicator.
    Lysate may be stored for months at –80°C.
14. To 10 ml of crude lysate, add 1 ml of double detergent. Vortex. Incubate for 15 minutes on ice.
15. Transfer lysate to centrifuge tubes. Place tubes into SW 28 buckets.
    Balance tubes with DMEM. Centrifuge all of the buckets even if only using some of them. Balance the empty buckets with water.
16. Centrifuge at 16,000*g* for 20 minutes at 4°C to remove cellular debris.
17. Transfer supernatants to 15-ml conical tubes. Discard pellets. Adjust volume to 13 ml with DMEM.
18. Add 2.5 ml of the 1.28-g/ml CsCl solution to the clear plastic tubes. Underlay with 1.5 ml of the 1.50-g/ml CsCl solution.
19. Carefully layer the 13 ml of supernatant above the upper CsCl solution. Avoid mixing layers.
20. Transfer tubes into buckets. Balance as described in Step 15. Centrifuge at 22,500 rpm for 3.5 hours at 4°C.

21. Use an 18-gauge needle to pierce the bottom of each tube.
22. Collect 0.5-ml fractions into Eppendorf tubes.

    The virus is most concentrated at the interface between the two CsCl solutions and should be in fractions 3–6. A change in flow rate should be apparent when this interface reaches the bottom of the tube.

23. Pool the fractions containing the virus, plus 1 fraction before and 1 fraction afterward.
24. Use Pierce 0.5–3-ml dialysis cassettes to dialyze the pooled fractions against PBS three times.

    Use 1 liter of PBS per 1 ml of sample being dialyzed for each round of dialysis.

Protocol 2

Titering Replication-defective rSV40s

Although wild-type SV40 is usually titered in monkey cells under agar overlays (Rosenberg et al. 1981), this method is not useful for rSV40 vectors; it requires weeks, it is cumbersome, and it does not always provide reproducible titers. Instead, replication-defective SV40 viruses can be titered by quantitative PCR (qPCR). qPCR measures the number of rSV40 genomes in purified viral stocks using primers specific for the rSV40, coupled with SYBR Green detection of PCR products. SYBR Green fluoresces strongly when bound to double-stranded DNA and the fluorescence increases proportionally to double-stranded DNA concentration. The fluorescent signal is measured during annealing and extension steps of qPCR. These measurements are displayed as real-time amplification plots (Fig. 3). The cycle threshold (C_t) is the first cycle at which the amplification fluorescence signal is significantly above background (Table 3). Higher levels of template present initially means that fewer cycles are needed to achieve fluorescence signals above background. Virus titers are determined by comparison to a standard curve of rSV40 plasmid DNA (Fig. 4).

qPCR can also be used to assess the purity of the sample. In melt (dissociation) curve analysis, fluorescence is monitored and PCR samples are subjected to gradual increases in temperature from 55°C to 95°C. As temperature increases, qPCR products in each tube will melt according to

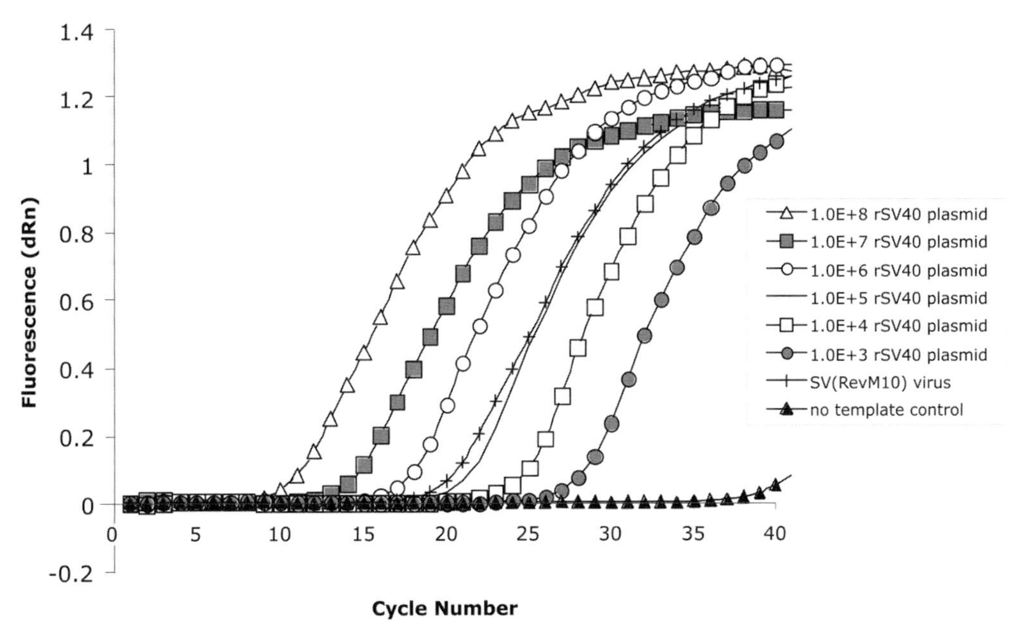

FIGURE 3. Amplification plots of standard curve dilutions. 1 x 10^8, 1 x 10^7, 1 x 10^6, 1 x 10^5, 1 x 10^4, and 1 x 10^3 input copies of rSV40 plasmid DNA; SV(RevM10) purified virus of unknown copy number; and "no template" control.

TABLE 3. Cycle Threshold Values from Amplification Plots in Figure 3

qPCR sample	Cycle threshold
1 x 10^8 rSV40 plasmid	10.4
1 x 10^7 rSV40 plasmid	13.95
1 x 10^6 rSV40 plasmid	17.22
1 x 10^5 rSV40 plasmid	20.79
1 x 10^4 rSV40 plasmid	24.08
1 x 10^3 rSV40 plasmid	27.79
SV(RevM10) virus	19.81
"No-template" control	no C_t

their size and composition. The temperature at which a qPCR product melts is seen as a peak, usually between 75°C and 90°C. A single peak denotes a single PCR product (Fig. 5). Secondary peaks or shoulders indicate the presence of a reaction product in addition to the product of interest, which may render the C_t value inaccurate.

COS-7 cells used to produce the viral stocks contain an SV40 genome integrated, so qPCR primers should amplify rSV40 vector sequences, not the integrated SV40 DNA. Primers used with SYBR Green should span 100–400 bp, be 15–30 nucleotides long, and have melting temperatures within 2° of each other. G/C content should be <50%, preferably 35-45%. Runs of the same base and G/C clamps at primers' 3′ ends should be avoided. A number of software programs are available online to assist in primer design (Primer3, http://frodo.wi.mit.edu/cgi-bin/primer3/primer3_www.cgi; Amplify, http://engels.genetics.wisc.edu/amplify/; BLAST, http://www.ncbi.nlm.nih.gov/BLAST/).

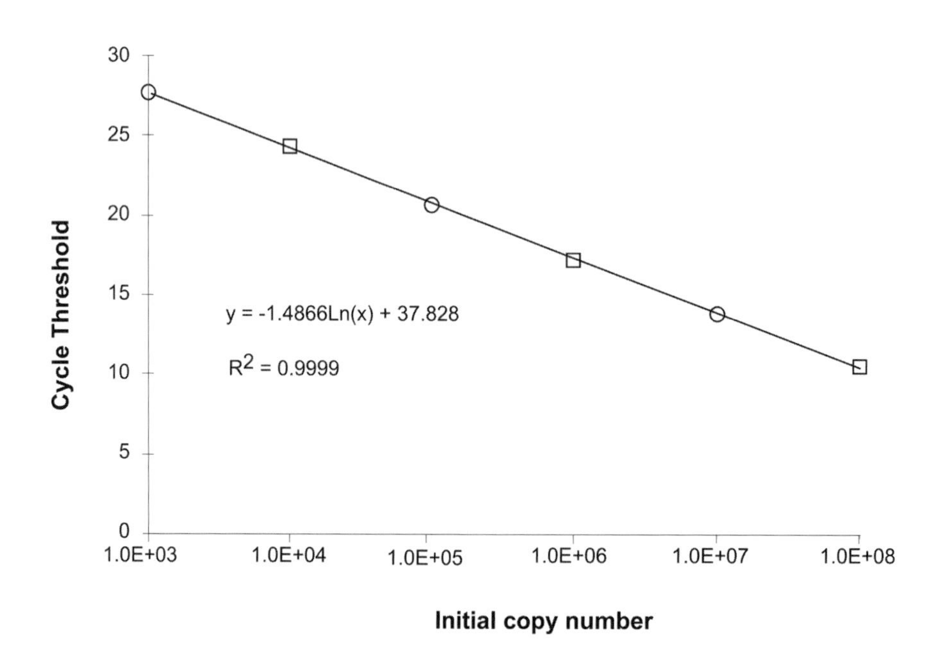

FIGURE 4. Standard curve of cycle threshold values. Cycle threshold values for the standard curve dilutions (from Table 3) plotted against the initial copy number. The input copy number in the SV(RevM10) reaction can be determined by substituting the cycle threshold value for y in the equation and solving for x. To determine the number of viral genomes/ml, multiply the input copy number by 5 (1:5 dilution in the RNase/DNase reaction) and then by 500 (2 µl of the RNase/DNase-treated sample used for qPCR).

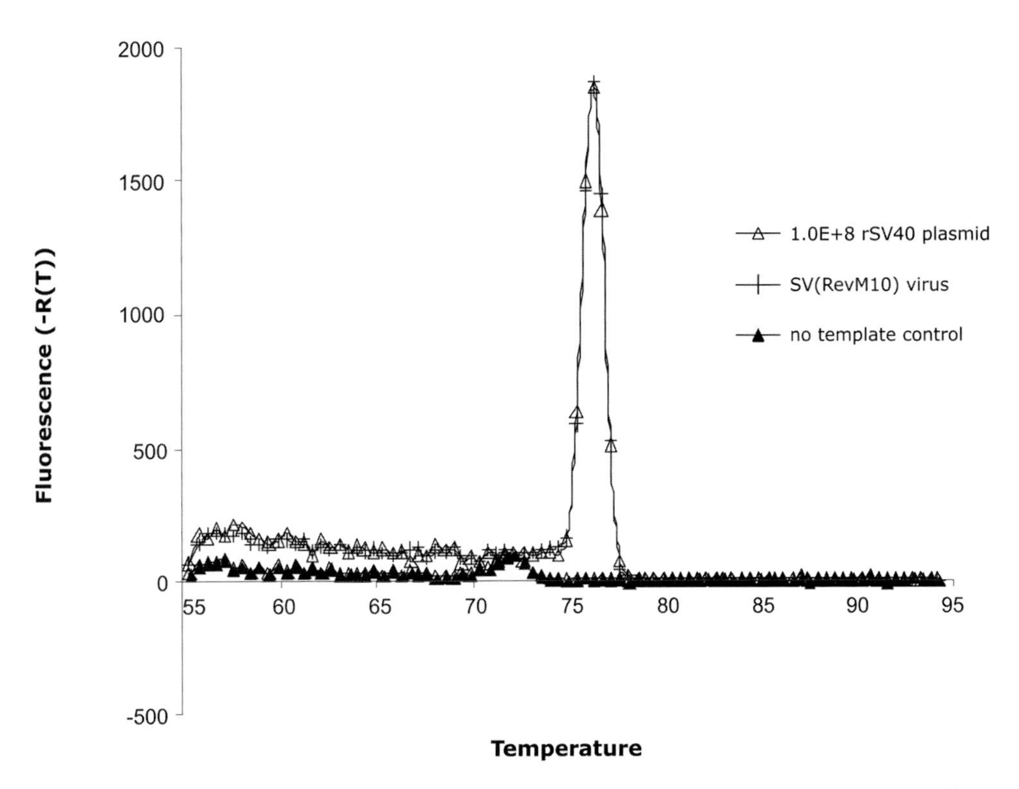

FIGURE 5. Dissociation (melt) curves. The dissociation curves for the standard curve dilution 1 x 10^8 and the SV(RevM10) reactions have identical melt peaks at 76.3°C. The "no-template control" reaction has a small peak at 72.1°C.

MATERIALS

CAUTION: See Appendix for appropriate handling of materials marked with <!>.

Reagents

Brilliant SYBR Green Master Mix (Stratagene)

After thawing this mix, store at 4°C. The SYBR Green and ROX components of the kit are light-sensitive and should be protected from light whenever possible.

DNA oligonucleotide primers (Integrated DNA Technologies)

Forward Primer: T7.1.fw; 5′ AAA CAG ATC AGA TCC AGA CAT GAT 3′

Reverse Primer: T7.277.rv; 5′ ATT CGC CCT ATA GTG AGT CGT ATT 3′

These primers anneal to the non-SV40 sequences of the vector to amplify a 277-bp product. T7.1.fw and T7.277.rv have melting temperatures of 63.99° and 64.48° in 100 mM salt, and a G/C content of 38.46% and 41.67%, respectively. Both are 24 nucleotides in length.

Phosphate-buffered saline (PBS)

Plasmid Maxi Kit (QIAGEN)

RNase (500 ng/μl; Roche Applied Sciences 1119915) <!>

RQ1 DNase (1 unit/μl; Promega)

Before using a new lot of DNase, determine the minimal concentration of enzyme needed to eliminate nonencapsidated DNA. Set up plasmid DNA reactions with 0–3 μl of DNase. Proceed with digestion, inactivation, and qPCR, including a standard curve, as described below. The concentration of DNase that reduces plasmid template by more than 2 logs compared to undigested control is used to treat viral samples.

RQ1 DNase reaction buffer
- 400 mM Tris (pH 8.0) <!>
- 100 mM $MgSO_4$ <!>
- 10 mM $CaCl_2$ <!>

RQ1 DNase stop solution
- 20 mM EGTA (pH 8.0)

Tris-HCl (1 M, pH 7.5; nuclease-free; Fisher)

Dilute to 10 mM Tris-HCl, pH 7.5 with nuclease-free water before use.

Water (PCR grade, nuclease-free; Fisher)

Equipment

Microcentrifuge tubes (nuclease-free; Fisher)
Minicentrifuge
Mx3000P qPCR Instrument (Stratagene) or other qPCR machine
qPCR optical tubes and caps (8x strip, nuclease-free; Stratagene)
Sharp precision barrier tips (nuclease-free; Denville Scientific)
Vortex

METHODS

Preparation of Standard Curve

1. Prepare plasmid DNA for standard curve using QIAGEN Plasmid Maxi Kit. Quantify by OD_{260}.

   The amount of DNA needed to make a stock solution is based on plasmid size; 1 μg of 1 kb of DNA is equal to 9.1×10^{11} molecules of double-stranded genomic DNA. Thus, 5×10^9 molecules of 1 kb DNA is equal to 5.49 ng.

2. Multiply 5.49 by the plasmid size to calculate the amount of DNA needed for 5×10^9 molecules of plasmid.
3. Dilute plasmid with 10 mM Tris-HCl (pH 7.5) to make a stock solution of 5.0×10^9 molecules/μl.
4. Quantify DNA by OD_{260} to check concentration of the stock and make any necessary corrections. Aliquot and store at –20°C.
5. Thaw the 5×10^9 molecules/μl stock. Vortex. Centrifuge briefly.
6. Prepare tenfold serial dilutions ranging from 5×10^8 to 5×10^2 molecules/μl.
   a. For each dilution, add 10 μl of the preceding higher concentration to 90 μl of 10 mM Tris-HCl (pH 7.5).
   b. Mix well by vortexing in several pulses of 3–5 seconds each.
   c. Pulse-centrifuge each dilution using the minicentrifuge.

      Prepare fresh dilutions for standard curves daily from the frozen stock solution. Diluted samples of DNA do not store well and tend to precipitate at –20°C.

DNase and RNase Digestion of Purified rSV40

7. Dialyze a 100-μl aliquot of purified rSV40 (from Protocol 1, Step 24) against an additional change of PBS to remove any additional components such as detergents that may inhibit qPCR. Use immediately or store at –80°C.

8. Thaw RQ1 DNase reaction buffer. Vortex to mix.

   Make sure salts are dissolved. If there is any white precipitate, incubate buffer at 37–50°C and vortex.

9. Dilute RNase 1:100 in 10 mM Tris-HCl (pH 7.5). Store on ice.

   Make a fresh dilution every day.

10. Thaw rSV40 stock on ice (~4°C).

11. Digest purified rSV40 with DNase/RNase.

    a. Mix 1 µl of RQ1 DNase reaction buffer and 1 µl of diluted RNase (5 ng/µl). Add a sufficient volume of 10 mM Tris-HCl such that the final volume of the reaction mixture is 100 µl.

    b. Add 20–34 µl of purified rSV40 viral stock. Vortex and centrifuge briefly. Transfer to ice.

    c. While on ice, add 1–3 µl of RQ1 DNase.

    The precise amount of DNase added is determined for each batch of DNase (see above).

12. Digest plasmid DNA with DNase/RNase.

    a. Mix 1 µl of RQ1 DNase reaction buffer and 1 µl of diluted RNase (5 ng/µl). Add a sufficient volume of 10 mM Tris-HCl such that the final volume of the reaction mixture is 100 µl.

    b. Add 10 µl of the 5×10^6 dilution of plasmid DNA and 20–34 µl of PBS. Vortex and centrifuge briefly. Transfer to ice.

    c. While on ice, add 1–3 µl of RQ1 DNase.

       The precise amount of DNase added is determined for each batch of DNase (see above).

13. Vortex both plasmid and viral samples briefly. Centrifuge briefly.

    Control reactions for plasmid and viral samples include all components except DNase.

14. Incubate samples for 20 minutes at room temperature.

15. Return tubes to ice. Add 1 µl of RQ1 DNase stop solution to each tube. Vortex and centrifuge briefly.

16. Incubate reactions for 10 minutes at 75°C to inactivate DNase. Centrifuge briefly to collect condensation.

    Inactivating DNase is critical to prevent digestion of the primers and template during qPCR.

17. Titer by qPCR immediately (see below) or store at –80°C for future use.

Titration by Quantitative Polymerase Chain Reaction (qPCR)

It is essential to avoid contamination of the qPCR reagents. Wear gloves at all times and use barrier tips for pipetting. If possible, aliquot reagents using a dedicated set of pipettors, in a clean area separate from where the template is added. Do not touch the edges of the reaction tubes or the insides of the tube lids.

18. Prepare a standard curve of tenfold serial dilutions of plasmid DNA from 5×10^7 to 5×10^2 molecules/µl (Steps 1–6).

19. Dilute ROX 1:500 in nuclease-free PCR-grade H_2O.

20. For each sample to be assayed, mix 12.5 µl of 2x master mix, 5 pmoles of Forward Primer, 5 pmoles of Reverse Primer, 0.375 µl of diluted ROX reference dye, and sufficient nuclease-free PCR-grade H_2O to adjust the final volume to 23 µl/sample.

    Prepare sufficient reaction mixture for each point of the standard curve, the unknown virus samples to be titrated, and "no added template" controls, in triplicate. Prepare additional reagent for at least one extra sample to allow for pipetting error.

21. Aliquot 23 μl of above reaction mix to the qPCR optical tubes. Take care not to introduce bubbles to the samples.
22. Vortex and briefly centrifuge plasmid and viral samples before adding to the qPCR assay tubes.
23. Add 2 μl of template (plasmid standard curve or viral unknown) to each tube. Titurate gently to mix. Avoid bubbles. Add 2 μl of 10 mM Tris-HCl to control tubes.
24. Cap tubes and centrifuge briefly. Load tubes into the qPCR machine.
25. Program the software of the qPCR machine with sample template and thermal profile (Strategene 2004). Perform PCR as follows:

    95°C for 10 minutes.
    40 cycles of 95°C for 30 seconds, 55°C for 60 seconds, and 72°C for 30 seconds
    Dissociation curve: Incremental temperature increase, 55–95°C, over 30 minutes

Analysis of qPCR Data

26. Plot the C_t values for each standard dilution against initial template quantity.

    This step can be performed automatically by the Mx3000P software or manually using graphing programs such as Excel.
27. Determine the initial copy number of the unknown samples being titered by comparison to the standard curve using the equation of the line, where y = the C_t value of the unknown (see Fig. 4).
28. Calculate the number of viral particles/ml using the following formula: Initial copy number x 5 x 500 = virus copies/ml.
29. Verify purity of sample by dissociation curve analysis. A single peak denotes a single PCR product.

    The purity and size of the PCR product can also be confirmed by agarose gel electrophoresis.

ACKNOWLEDGMENTS

This work reflects the most recent adaptations of established techniques used for more than 10 years in our laboratory. Many laboratory members and collaborators have helped develop new approaches that made much of our work possible and successful: Drs. Lokesh Agrawal, Omar Bagasra, Pierre Cordelier, Ling-Xun Duan, Harris Goldstein, Geetha Jayan, Maria Lamothe, Bianling Liu, Jean-Pierre Louboutin, Hayley McKee, Iwata Ozaki, Roger Pomerantz, Marlene Strayer, Danlan Wei, and Mark Zern. We are grateful to Dr. Janet S. Butel, Baylor College of Medicine, for generously giving us many of our original reagents and for her kind advice. Intellectual input and support of our program officers at the National Institutes of Health is gratefully acknowledged: Sandra Bridges, Scott Cairns, Kathy Kopniski, Diane Rausch, the late Nava Sarver, and Frosso Voulgaropoulou. This work was supported by NIH grants AI41399, AI48244, MH69122, MH70287, and RR13156.

REFERENCES

Asano M., Iwakura Y., and Kawade Y. 1985. SV40 vector with early gene replacement efficient in transducing exogenous DNA into mammalian cells. *Nucleic Acids Res.* **13:** 8573–8586.

Breau W.C., Atwood W.J., and Norkin L.C. 1992. Class I major histocompatibility protein share an essential component of the simian virus 40 receptor. *J. Virol.* **66:** 2037–2045.

Cole C.N. and Conzen S.D. 2001. *Polyomavirinae*: The viruses and their replication. In *Fields' virology* (ed. D.M. Knipe and P.M. Howley), pp. 2141–2174. Lippincott Williams & Wilkins, Philadelphia, Pennsylvania.

Fried M. and Prives C. 1986. The biology of simian virus 40 and polyomavirus. *Cancer Cells* **4:** 1–16.

Gething M.J. and Sambrook J. 1981. Cell-surface expression of influenza haemagglutinin from a cloned DNA copy of the RNA gene. *Nature* **293:** 620–625.

Pelkmans L., Kartenbeck J., and Helenius A. 2001. Caveolar endocytosis of simian virus 40 reveals a new two-step vesicular-transport pathway to the ER. *Nat. Cell Biol.* **3:** 473–483.

Pelkmans L., Puntener D., and Helenius A. 2002. Local actin polymerization and dynamin recruitment in SV40-induced internalization of caveolae. *Science* **296:** 535–539.

Rosenberg B.H., Deutsch J.F., and Ungers G.E. 1981. Growth and purification of SV40 virus for biochemical studies. *J. Virol. Methods* **3:** 167–176.

Rund D., Dagan M., Dalyot-Herman N., Kimchi-Sarfaty C., Schoenlein P.V., Gottesman M.M., and Oppenheim A. 1998. Efficient transduction of human hematopoietic cells with the human multidrug resistance gene 1 via SV40 pseudovirions. *Hum. Gene Ther.* **9:** 649–657.

Sauter B.V., Parashar B., Chowdhury N.R., Kadakol A., Ilan Y., Singh H., Milano J., Strayer D.S., and Chowdhury J.R. 2000. Gene transfer to the liver using a replication-deficient recombinant SV40 vector results in long-term amelioration of jaundice in Gunn rats. *Gastroenterology* **119:** 1348–1357.

Stang E., Kartenbeck J., and Parton R.G. 1997. Major histocompatibility complex class I molecules mediate association of SV40 with caveolae. *Mol. Biol. Cell* **8:** 47–57.

Stratagene 2004. *Introduction to quantitative PCR: Methods and application guide.* Stratagene, LaJolla, California.

Strayer D.S. 1996. SV40 as an effective gene transfer vector *in vivo. J. Biol. Chem.* **271:** 24741–24746.

———. 1999. Effective gene transfer using viral vectors based on SV40. *Methods Mol. Biol.* **133:** 61–74.

Strayer D.S. and Milano J. 1996. SV40 mediates stable gene transfer *in vivo. Gene Ther.* **3:** 581–587.

Strayer D.S., Lamothe M., Wei D., Milano J., and Kondo R. 2001. Generation of recombinant SV40 vectors for gene transfer. *Methods Mol. Biol.* **165:** 103–117.

Strayer D.S., Branco F., Zern M.A., Yam P., Calarota S.A., Nichols C.N., Zaia J.A., Rossi J., Li H., Parashar B., Ghosh S., and Chowdhury J.R. 2002. Durability of transgene expression and vector integration: Recombinant SV40-derived gene therapy vectors. *Mol. Ther.* **6:** 227–237.

Tsai B., Gilbert J.M., Stehle T., Lencer W., Benjamin T.L., and Rapoport T.A. 2003. Gangliosides are receptors for murine polyoma virus and SV40. *EMBO J.* **22:** 4346–4355.

Vera M., Prieto J., Strayer D.S., and Fortes P. 2004. Factors influencing the production of recombinant SV40 vectors. *Mol. Ther.* **10:** 780–791.

WWW RESOURCES

http://engels.genetics.wisc.edu/amplify/ (Amplify software).

http://frodo.wi.mit.edu/cgi-bin/primer3/primer3_www.cgi (Primer3 software).

http://www.ncbi.nlm.nih.gov/BLAST/ (Basic Local Alignment Search Tool [BLAST]).

25 SV40 In Vitro Packaging: A Pseudovirion Gene Delivery System

Chava Kimchi-Sarfaty* and Michael M. Gottesman[†]

*Center for Biologics Evaluation and Research, Food and Drug Administration; [†]Laboratory of Cell Biology, National Cancer Institute, National Institutes of Health, Bethesda, Maryland 20892

ABSTRACT

Nuclear extracts of *Spodoptera frugiperda* (*Sf9*) insect cells infected with baculoviruses encode the simian virus 40 (SV40) major coat protein (VP1). These extracts are able to package supercoiled plasmid DNA or RNA interference (RNAi) sequences in the presence of $MgCl_2$, $CaCl_2$, and ATP, thus forming SV40 pseudovirions in vitro. Such packaging has numerous advantages over other viral and nonviral delivery systems. Specifically, it provides a wide host range and high transduction efficiency. The only major disadvantage of this system is low expression per transduced cell observed in vitro, likely a result of DNA trapped in the cytoplasmic compartment of the cell that does not enter the cell nucleus.

INTRODUCTION

Empty SV40 capsids can disassemble by reducing disulfide bonds and reform in the presence of calcium ions around DNA to structure a pseudovirus (Colomar et al. 1993). Indeed, in the presence of salts and calcium ions, polyomavirus VP1 from bacteria (lacking VP2 or VP3 SV40 capsid proteins) assembles into viral capsids (Salunke et al. 1986). The SV40 capsid proteins VP1, VP2, VP3, and agno have the ability to package DNA carrying the SV40 origin of replication (Sandalon and Oppenheim 1997; Sandalon et al. 1997). Such fragments of host DNA encapsidated by polyomavirus protein coats can infect kidney cells from baby mice (Michel et al. 1967; Winocour 1968). Similarly, in a low ionic environment, supercoiled polyomavirus DNA can be encapsidated by polyomavirus-like particles and infect mouse and human embryonic cells (Qasba and Aposhian 1971).

The SV40 in vitro packaging delivery system is unique because it does not involve any SV40 sequences and also does not require packaging cell lines (Kimchi-Sarfaty et al. 2003). Other SV40 delivery methods require replacement of several of the wild-type viral genes by the transgene (Rund et al. 1998) This permits larger reporter genes than the normal capacity of 5.2 kb present in wild-type SV40. These systems also require a packaging cell line (usually COS-1 or COS-7 cells) that supplies the T antigen in *trans* and packages the virus (Rund et al. 1998). In contrast, the SV40 in vitro packaging delivery system can transport up to 17.7 kb of DNA with no SV40 sequence or packaging cell line. The advantages of the in vitro packaging delivery system over other viral and nonviral delivery systems are summarized in Table 1. The only major disadvantage of this system is low expression per transduced cell, observed in vitro, which is probably a result of DNA trapped in the cytoplasmic compartment of the cell that does not enter the cell nucleus (Kimchi-Sarfaty et al. 2004). The shape of in vitro pseudovirion particles is similar to that of SV40 wild-type viruses. However, the size of the particles might be affected by the size of the packaged DNA. For example, the pHaMDR1 plasmid (15.2 kb) forms particles 55 nm in diameter. These are, on average, 10 nm larger than wild-type SV40 (Fig. 1) (Kimchi-Sarfaty et al. 2003).

The pathway of the in vitro particles is known in human lymphoblastoid cells. Major histocompatibility complex (MHC) class I receptors are not necessary for their entry, although these often have an important role for both the SV40 wild-type and in vitro particle entry (Norkin 1999). Some of the packaged plasmids are observed in the Golgi 30 minutes after transduction and all eventually accumulate in the endoplasmic reticulum (ER). From the ER, the unpackaged plasmid DNA continues to the nucleus (Kimchi-Sarfaty et al. 2004). Expression of in-vitro-packaged DNA is transient (up to 3 weeks), but it can be maintained while selection is applied (Kimchi-Sarfaty et al. 2002; Kimchi-Sarfaty and Gottesman 2004). This pathway has also been used to facilitate small interfering RNA (siRNA) delivery to the cytoplasm. Using the same packaging protocol as for DNA, SV40 in vitro packaging of siRNA results in high delivery efficiency, as well as increased duration of expression compared to Lipofectamine-Plus (Kimchi-Sarfaty et al. 2005).

Chemoprotection can be achieved by delivering the multidrug transporter genes *MDR*1 (ABCB1) or *MXR* (ABCG2) and occurs at very high efficiency in human and mouse cells. Reporter genes such as green fluorescent protein (GFP) can also be delivered into different cell lines, including human lymphoblastoid cells and erythroleukemia cells (K562). Different promoters (such as SV40 or cytomegalovirus [CMV] with or without an intron) demonstrate very high efficiency with slight differences in expression (Fig. 2). However, introducing nuclear local-

TABLE 1. Advantages of the SV40 In Vitro Packaging System for Gene Delivery

Transduced cells	Very high ability to infect nondividing cells as well as dividing cells Wide variety of cell lines available in vitro (HeLa, Hek293, K562, lymphoblastoids, NIH-3T3) High transduction efficiency in vivo (tested in C57BL and nude mice, rats) High transduction efficiency for adherent cells or cells in suspension
Reporter gene	Reporter gene can be cloned into any plasmid DNA, under various promoters No SV40 sequences needed High capacity of the SV40 in vitro packaging system (up to 17.7 kb of the reporter gene plasmid DNA) No packaging cell line necessary siRNA can be packaged and delivered
Expression	Very high transduction efficiency (close to 100% of the cells express the reporter gene) In vitro system is short-term, with no likely integration into host DNA Expression of in vitro system is only maintained as long as selection is applied Very low (if any) immunogenicity No effect on cell viability Minimal to no cytotoxicity, both in vitro and in vivo

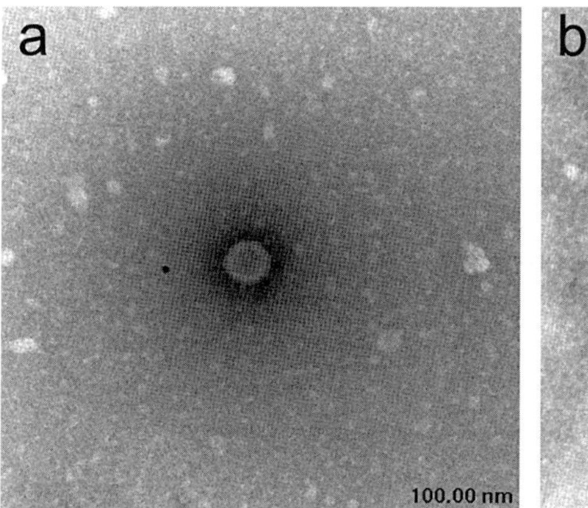

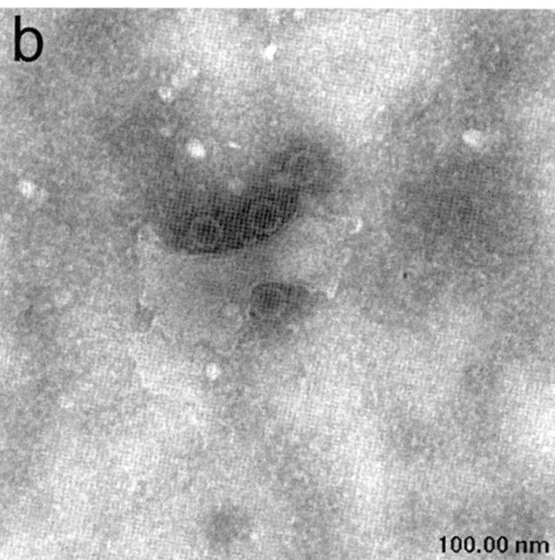

FIGURE 1. Electron micrographs of pseudovirion and wild-type SV40. Samples were adsorbed on a glow-discharged carbon-coated Formvarfilm grid, negatively stained and examined by electron microscopy. Images were captured with the Gatan Digital Image System of a Hitachi H7000 electron microscope operating at 75 kV. Magnification, 25,000x. (*a*) pHaMDR1 plasmid DNA (15.2 kb; the multidrug resistance gene is under the Harvey retroviral long terminal repeat promoter) was packaged in vitro using VP1-transduced-*Sf*9 nuclear extracts in the presence of $MgCl_2$, ATP, and $CaCl_2$. (*b*) Wild-type SV40 virions were prepared in COS-7 cells as described previously (Rund et al. 1998), stored at –20°C, thawed, and prepared for electron microscopy as described above. (Reprinted, with permission, from Kimchi-Sarfaty et al. 2003.)

ization signals from human immunodeficiency virus (HIV) or SV40 carried by the plasmid DNA does not change the efficiency or expression of a GFP reporter gene delivered by the in vitro particles (Kimchi-Sarfaty et al. 2002, 2003). In vivo, the GFP reporter gene is expressed in several tissues of the mouse after intraperitoneal injection (Kimchi-Sarfaty and Gottesman 2004). Similarly, when a toxin gene (*Pseudomonas* exotoxin; Kimchi-Sarfaty et al. 2006) and anti-angiogenic gene (pigment epithelium-derived factor) were delivered into adenocarcinomas in nude mice via SV40 in vitro packaging, tumor sizes were reduced significantly.

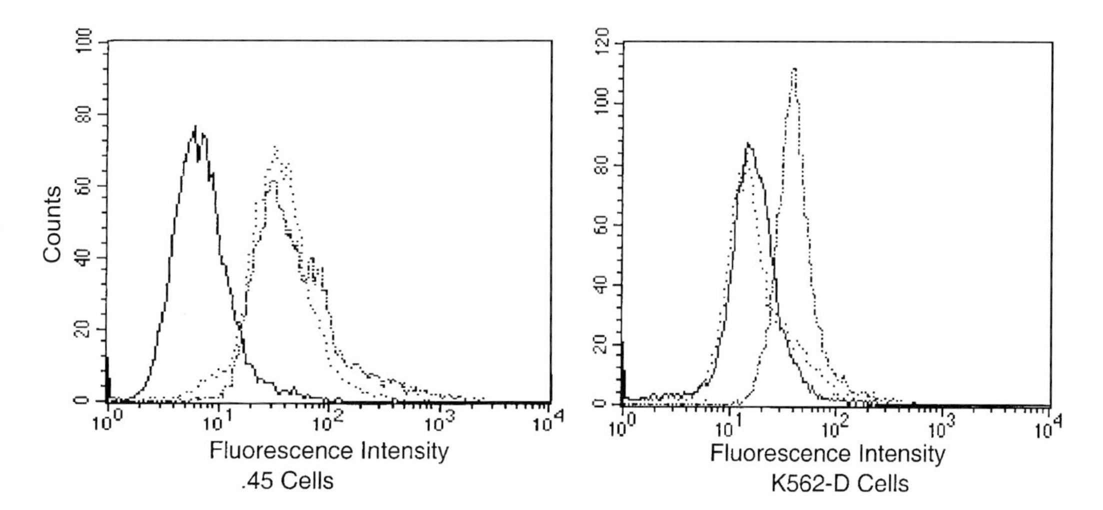

FIGURE 2. K562 human erythroleukemia cells and .45 human lymphoblastoid cells were incubated with the GFP reporter gene and packaged in vitro. Two different GFP constructs were used, employing either CMV or the SV40 promoter. (*Dashed line*) Cells infected with in-vitro-prepared CMV-EGFP-C1 vectors; (*dotted line*) cells infected with in-vitro-prepared SV40-EGFP-C1 vectors; (*solid line*) controls (empty capsids). (Reprinted, with permission, from Kimchi-Sarfaty et al. 2002.)

Protocol

SV40 In Vitro Delivery System

The protocol described here uses *S. frugiperda* (*Sf9*) insect cells that have been infected with baculoviruses encoding the SV40 major coat protein (VP1). Nuclear extracts (Schreiber et al. 1989; Sandalon and Oppenheim 2001) from these cells are then used to package supercoiled plasmid DNA or siRNA in the presence of $MgCl_2$, $CaCl_2$, and ATP to form SV40 pseudovirions in vitro (Fig. 3) (Sandalon et al. 1997; Kimchi-Sarfaty and Gottesman 2004; Kimchi-Sarfaty et al. 2005).

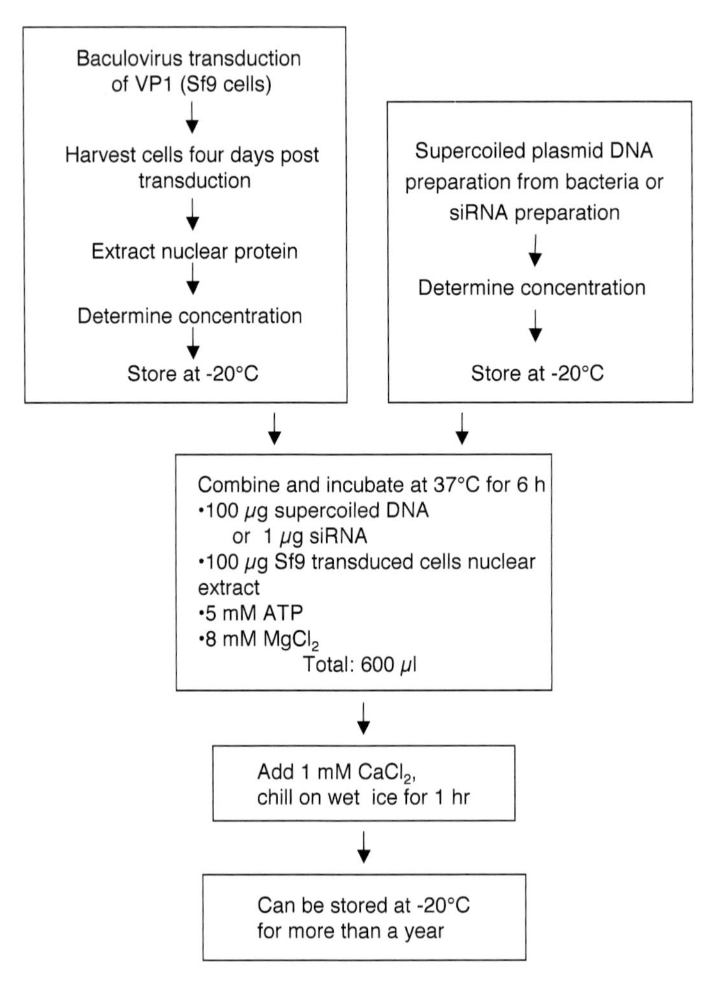

FIGURE 3. Flow chart illustrating steps involved in preparation of pseudovirions.

MATERIALS

CAUTION: See Appendix for appropriate handling of materials marked with <!>.

Reagents

ATP (100 mM; Roche)
Buffer I
  10 mM HEPES (pH 7.9)
  10 mM KCl <!>
  0.1 mM EDTA
  0.1 mM EGTA
    Sterilize by filtration through a 0.22-μm filter. Buffer I can be stored for 30 days at 4°C.
  Prior to use add:
  0.5 mM phenylmethylsulfonyl fluoride (PMSF) <!>
  1 mM dithiothreitol (DTT) <!>
    Dissolve 1 tablet of protease inhibitor (Roche 1873580) in 1 ml of PBS and dilute 1:50 in buffer I.
Buffer II
  20 mM HEPES (pH 7.9)
  0.4 M NaCl
  1 mM EDTA
  1 mM EGTA
    Sterilize by filtration through a 0.22-μm filter. Buffer II can be stored for 30 days at 4°C.
  Prior to use add:
  1 mM PMSF
  1 mM DTT
    Dissolve 1 tablet of protease inhibitor (Roche 1873580) in 1 ml of PBS and dilute 1:50 in buffer I.
$CaCl_2$ (10 mM) <!>
Cells
  Cell line of interest
    Maintain cell line to be transduced in plates or in suspension.
  Spodoptera frugiperda (*Sf9*) cells
Media
  Cell culture growth medium (Grace's insect medium containing lactalbumin and yeastolate)
  Complete growth medium
    Use a medium that is appropriate for the cell line of interest.
$MgCl_2$ (200 mM) <!>
Nonidet P-40 (NP-40; 10% v/v in PBS) <!>
Phosphate-buffered saline (PBS)
Plasmid DNA (up to 17.7 kb)
    To obtain the highest in vitro packaging efficiencies, purify the plasmid DNAs by equilibrium centrifugation in CsCl <!>–ethidium bromide <!> density gradients.
Small interfering RNA (siRNA)
VP1 baculovirus
    VP1 is cloned by introducing a StuI-BclI DNA fragment (SV40 coordinates 1463–2770) into the plasmid pVL1393 (PharMingen, San Diego, California) derived from *Autographa californica* nuclear polyhedrosis virus (AcMNPV) (Luckow and Summers 1988). This is then cleaved by the restriction endonucleases SmaI and BglII (Sandalon and Oppenheim 1997, 2001). Store aliquots of viral stock at –80°C.

Equipment

Incubator (humidified, 37°C, 5% CO_2)
Centrifuge tubes (40 ml, sterile; Sorvall)
Microfuge tubes (1.5 ml)
Orbital shaker
Spinner culture flasks (250 ml)
Tissue-culture dishes (60 mm)
Vortex with microfuge platform
Water bath (37°C)

METHODS

Transduction of *Sf*9 Cells

1. Infect a fresh culture of 5 x 10^8 *Sf*9 cells with the VP1 baculovirus in a 250-ml flask at a multiplicity of infection of ten.

   Scaling up to 500 ml is not recommended.

Nuclear Protein Extraction

Harvesting of the *Sf*9 cells should be done on day 4 or 5 postinfection. The level of VP1 expression and the function of the nuclear extracts in packaging should determine the harvest day.

2. Transfer the transduced cells from one 250-ml flask to precooled 40-ml tubes.
3. Prepare buffers I and II. Place them on ice. Cool microfuge tubes.
4. Wash the cells three times with cold PBS by centrifugation at 6000 rpm for 5 minutes at 4°C.
5. Remove the supernatant. Resuspend the pellet completely by vortexing.

   Do not freeze at this point.

6. Add 1 ml of buffer I. Resuspend the pellet gently by pipetting until it is evenly dispersed.
7. Gently add 28 ml of buffer I. Incubate for 15 minutes on ice.
8. Add 2.3 ml of 10% NP-40. Vortex for 10 seconds.
9. Centrifuge at 13,000 rpm in a refrigerated centrifuge (~4°C) for 3 minutes.
10. Remove supernatant with a pipette tip attached to a vacuum aspirator.

    The nuclear material is in the pellet.

11. Vortex the pellet. Add 5 ml of buffer II. Vortex well. Transfer to microfuge tubes.

    The total volume in each tube should not exceed 1 ml.

12. Place the tubes on a shaking platform attached to a vortex. Vortex for 15 minutes at 4°C.
13. Centrifuge the microfuge tubes at 13,000 rpm for 5 minutes at 4°C.
14. Aliquot the nuclear extract supernatant into prechilled tubes.
15. Determine the protein concentration of the nuclear extract.

    Note that the nuclear extract contains NP-40. Not all methods of determining protein concentration are compatible with NP-40 residues. Use for in vitro packaging only if the concentration exceeds 3 mg/ml.

16. Store nuclear extracts at –20°C to –80°C.

In Vitro Packaging

17. For each reaction of in vitro packaging mixture, add to a microfuge tube 24 µl of 200 mM $MgCl_2$, 30 µl of 100 mM ATP, and 100 µg of nuclear extract from VP1-transduced *Sf9* cells.
18. Add either 100 µg of plasmid DNA or 1 µg of siRNA.

    Control in vitro packaging reactions should be performed without nuclear extract, with nuclear extract from untransduced *Sf9* cells, or with a reporter DNA.
19. Add double-distilled H_2O to a final volume of 600 µl. Vortex well.
20. Incubate in a water bath for 6 hours at 37°C.
21. Add 66 µl of 10 mM $CaCl_2$ to each reaction tube. Vortex lightly.
22. Incubate for 1 hour on ice.
23. Store the in vitro packaging reaction tubes at –20°C until used.

Cell Transduction Using SV40 In Vitro Packaging

Variations on the original procedure can improve both the level of expression and the efficiency of transduction in several cell lines (Table 2).

24. About 16–24 hours before transduction, split a growing adherent culture of the cell line of interest by trypsinization.
25. Transfer the cells to 60-mm tissue-culture dishes at a density of 1×10^5 cells per dish.

    If cells are grown in suspension, place 1×10^5 cells per 60-mm dish in a total volume of 340 µl.
26. Add 5 ml of complete growth medium. Incubate at 37°C in a humidified incubator with an atmosphere of 5% CO_2.
27. Thaw in vitro packaging reaction tubes (from Step 23). Add the contents of one in vitro packaging reaction tube to each dish.

    Transduce cells with naked DNA as a control.
28. Place the dish on an orbital shaker located in the incubator. Shake for 2.5 hours.
29. Add 4 ml of fresh complete growth medium to each dish. Return to incubator.
30. Evaluate gene expression and function 24 hours or more posttransduction.

TABLE 2. Steps Added to the Original Procedure That Can Improve Expression and Efficiency of Transduction

Preparation of in vitro packaging reaction	Add 0.3 mM glutathione disulfide plus 3 mM reduced glutathione prior to Step 19.
	Digest 0.01 of in vitro packaging reaction with 0.1 unit of DNase I for 30 minutes on ice directly before transduction.
Cell transduction	Add 0.5 ng/ml phorbol 12-myristate 13-acetate to cells 5 days prior to transduction.
	Long-term expression can be achieved by applying selection, which should start 24–48 hours posttransduction using agents such as 600 µg/ml G418 for neomycin or 60 ng/ml colchicine for *MDR1*.
	Retransduce the cells 24 hours after the first transduction using the same transduction protocol.
	Apply two original reactions of in vitro vectors (from Step 14) concentrated to one reaction volume using a microconcentrator filtering system (Centricon centrifugal filter devices for volumes up to 2 ml, YM-100 m.w. membranes; Millipore).
	Add 10 ng/ml of the histone deacetylase inhibitor TSA to cells before transduction.

ACKNOWLEDGMENTS

We thank Kunio Nagashima (NCI-Frederick, SAIC Frederick) for taking the SV40 images in the electron microscope. We also express our thanks to Mr. George Leiman for editorial assistance and acknowledge productive collaborations and discussion with Dr. Ariella Oppenheim, Hadassah Medical Center, Jerusalem, Israel, on the SV40 in-vitro-packaged pseudovirions.

REFERENCES

Colomar M.C., Degoumois-Sahli C., and Beard P. 1993. Opening and refolding of simian virus 40 and in vitro packaging of foreign DNA. *J. Virol.* **67:** 2779–2786.

Kimchi-Sarfaty C. and Gottesman M.M. 2004. SV40 pseudovirions as highly efficient vectors for gene transfer and their potential application in cancer therapy. *Curr. Pharm. Biotechnol.* **5:** 451–458.

Kimchi-Sarfaty C., Arora M., Sandalon Z., Oppenheim A., and Gottesman M.M. 2003. High cloning capacity of in vitro packaged SV40 vectors with no SV40 virus sequences. *Hum. Gene Ther.* **14:** 167–177.

Kimchi-Sarfaty C., Ben-Nun-Shaul O., Rund D., Oppenheim A., and Gottesman M.M. 2002. In vitro-packaged SV40 pseudovirions as highly efficient vectors for gene transfer. *Hum. Gene Ther.* **13:** 299–310.

Kimchi-Sarfaty C., Brittain S., Garfield S., Caplen N.J., Tang Q., and Gottesman M.M. 2005. Efficient delivery of RNA interference effectors via in vitro-packaged SV40 pseudovirions. *Hum. Gene Ther.* **16:** 1110–1115.

Kimchi-Sarfaty C., Garfield S., Alexander N., Ali S., Cruz C., Chinnasamy D., and Gottesman M.M. 2004. The pathway of uptake of SV40 pseudovirions packaged in vitro: From MHC class I receptors to the nucleus. *Gene Ther. Mol. Biol.* **8:** 439–450.

Kimchi-Sarfaty C., Vieira W.D., Dodds D., Sherman A., Kreitman R.J., Shinar S., and Gottesman M.M. 2006. SV40 Pseudovirion gene delivery of a toxin to treat human adenocarcinomas in mice. *Cancer Gene Ther.* **3:** 648–657.

Michel M.R., Hirt B., and Weil R. 1967. Mouse cellular DNA enclosed in polyoma viral capsids (pseudovirions). *Proc. Natl. Acad. Sci.* **58:** 1381–1388.

Norkin L.C. 1999. Simian virus 40 infection via MHC class I molecules and caveolae. *Immunol. Rev.* **168:** 13–22.

Qasba P.K. and Aposhian H.V. 1971. DNA and gene therapy: Transfer of mouse DNA to human and mouse embryonic cells by polyoma pseudovirions. *Proc. Natl. Acad. Sci.* **68:** 2345–2349.

Rund D., Dagan M., Dalyot-Herman N., Kimchi-Sarfaty C., Schoenlein P.V., Gottesman M.M., and Oppenheim A. 1998. Efficient transduction of human hematopoietic cells with the human multidrug resistance gene 1 via SV40 pseudovirions. *Hum. Gene Ther.* **9:** 649–657.

Salunke D.M., Caspar D.L., and Garcea R.L. 1986. Self-assembly of purified polyomavirus capsid protein vp1. *Cell* **46:** 895–904.

Sandalon Z. and Oppenheim A. 1997. Self-assembly and protein-protein interactions between the SV40 capsid proteins produced in insect cells. *Virology* **237:** 414–421.

———. 2001. Production of SV40 proteins in insect cells and in vitro packaging of virions and pseudovirions. *Methods Mol. Biol.* **165:** 119–128.

Sandalon Z., Dalyot-Herman N., Oppenheim A.B., and Oppenheim A. 1997. In vitro assembly of SV40 virions and pseudovirions: Vector development for gene therapy. *Hum. Gene Ther.* **8:** 843–849.

Schreiber E., Matthias P., Muller M.M., and Schaffner W. 1989. Rapid detection of octamer binding proteins with "mini-extracts," prepared from a small number of cells. *Nucleic Acids Res.* **17:** 6419.

Winocour E. 1968. Further studies on the incorporation of cell DNA into polyoma-related particles. *Virology* **34:** 571–582.

26 Baculovirus-based Display and Gene Delivery Systems

Anna R. Mäkelä,* Wolfgang Ernst,† Reingard Grabherr,† and Christian Oker-Blom*

**NanoScience Center, Department of Biological and Environmental Science, FIN-40014 University of Jyväskylä, Finland; †University of Agriculture, Institute of Applied Microbiology, Muthgasse 18, A-1190 Vienna, Austria*

ABSTRACT

The baculovirus expression vector system has been used extensively for production of numerous proteins originating from both prokaryotic and eukaryotic sources. In addition to easy cloning techniques and abundant viral propagation, the system offers eukaryotic posttranslational modification machinery provided by the insect cell environment. The recently established eukaryotic molecular biology tool, the baculovirus display vector system (BDVS), allows combination of genotype with phenotype, enabling presentation of foreign peptides or even complex proteins on the baculoviral envelope or capsid. The strategy is of importance because it may be used to enhance viral binding and entry to mammalian cells as well as for production of antibodies against the displayed antigen. In addition, this technology should enable modifications of intracellular behavior, that is, trafficking of recombinant "nanoparticles," a highly relevant feature within the area of both targeted gene and protein delivery. In this chapter, the design and potential use of insect-derived baculoviral display vectors are discussed. The general procedures for generation of display libraries and characterization of these gene delivery vehicles are also described.

Continued

INTRODUCTION

Baculovirus Display Strategies

The prokaryotic phage display system has allowed a large variety of genes encoding different functions to be isolated from complex libraries. However, the limitations of prokaryotes, particularly with regard to folding requirements and posttranslational modifications of the displayed proteins, have led to the development of alternative eukaryotic display systems. During the last decade, baculovirus-based display strategies designed for presentation of foreign peptides or complex proteins on the viral surface have been established (Fig. 1) and evaluated for both in vitro and in vivo applications (Grabherr and Ernst 2001; Grabherr et al. 2001; Hüser et al. 2001; Hüser and Hofmann 2003; Oker-Blom et al. 2003). The tropism and transduction efficiency of *Autographa californica* multiple nucleopolyhedrovirus (*Ac*MNPV) have been altered by modifying the major envelope glycoprotein (gp64), by displaying foreign polypeptides as fusions to a heterologous membrane anchor or by viral pseudotyping (Oker-Blom et al. 2003). In addition, the major capsid protein of *Ac*MNPV (vp39) provides a compatible fusion partner for display of foreign proteins on the viral capsid (Kukkonen et al. 2003; Oker-Blom et al. 2003). These features, together with the ability to incorporate reporter genes such as green fluorescent protein (GFP), β-galactosidase, and/or luciferase under transcriptional regulation of mammalian promoters, have enabled monitoring of transduction efficiency in mammalian cells in vitro (Shoji et al. 1997; Yap et al. 1997; Condreay et al. 1999; Kost and Condreay 2002; Kost et al. 2005) and in tissues in vivo (Airenne et al. 2000; Lehtolainen et al. 2002; Hüser and Hofmann 2003; Tani et al. 2003; Li et al. 2004; Hoare et al. 2005). A number of systems to create recombinant baculoviral vectors, such as homologous recombination and site-specific transposition, have been described previously (Peakman et al. 1992; Airenne et al. 2003; Luckow et al. 1993). The present strategies for baculoviral display and the methodologies involved to engineer these potential gene delivery vehicles are outlined below.

Surface Display by Fusion to gp64

The major baculoviral envelope glycoprotein gp64 forms homotrimers on the surface of infected insect cells and on budded virions, forming typical peplomer structures (Oomens et al. 1995; Markovic et al. 1998). Analogous to established prokaryotic phage display systems, foreign proteins can be displayed on the surface of *Ac*MNPV after fusion to gp64 (Fig. 1A). First proofs of this principle were gained when several fusion variants of the gp64 gene incorporating glutathione-*S*-transferase were successfully expressed on the surface of infected cells and budded virions, and when the major surface glycoprotein (gp120) of human immunodeficiency virus type 1 (HIV-1) was incorporated into the amino terminus of gp64, illustrating functional ligand-binding activity (Boublik et al. 1995). In another study, the ectodomain of the envelope glycoprotein (gp41) of the same virus was fused to both native and truncated forms of gp64 (Grabherr et al. 1997). Similarly, the spike proteins E1 and E2 of rubella virus, as well as GFP, could be displayed on the surface of *Ac*MNPV (Mottershead et al. 1997).

These initial results led to further studies aimed at more specific targeting of the virus to mammalian cells. gp64-mediated display of single-chain antibody fragments (scFv) for the carcinoembryonic antigen and hapten 2-phenyloxazolone, the IgG-binding Z domains of protein A (Mottershead et al. 2000; Ojala et al. 2001), as well as avidin (Räty et al. 2004) in conjunction with "mammalianized" reporter genes, have clearly shown the potential of baculovirus with respect to

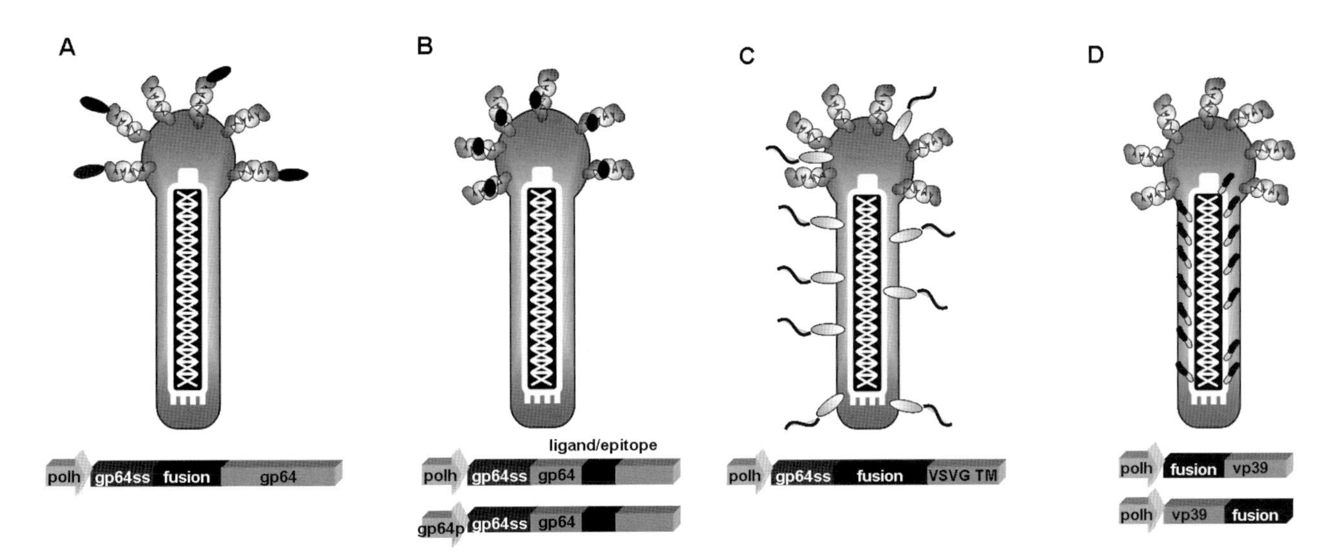

FIGURE 1. Schematic representation of different baculovirus display strategies. (*A*) The foreign peptides or proteins can be displayed as amino-terminal fusions to the second copy of *Ac*MNPV major envelope protein, gp64, or truncated versions thereof. (*B*) Peptide insertions can be placed within the native or additional coding sequence of gp64, and the modified protein is expressed under the gp64 promoter (where the native gp64-coding sequence has been omitted) or under the polh promoter, respectively. (*C*) The protein of interest can be fused to a heterologous membrane protein, such as VSV glycoprotein, or to its membrane-anchoring domain, enabling abundant and scattered presentation of the fusion construct around the viral particle. (*D*) In the baculovirus capsid display strategy, the fusion partner is placed either at the amino or carboxyl terminus of the major capsid protein, vp39, permitting efficient incorporation of the fusion moiety into the capsid structure. (gp64) Envelope glycoprotein of *Ac*MNPV; (gp64p) gp64 promoter; (gp64ss) gp64 signal sequence; (polh) polyhedrin promoter; (vp39) *Ac*MNPV capsid protein; (VSVG TM) membrane-anchoring domain of VSV-G protein.

targeted gene delivery. Significantly, the complement-regulatory protein, human decay-accelerating factor, was used to protect baculoviral display vectors from complement inactivation (Hüser et al. 2001). Permissive epitope insertions have also been achieved without altering or disturbing viral infectivity by directly modifying the coding sequence of the native gp64 (Fig. 1B) (Ernst et al. 1998, 2000; Spenger et al. 2002). The gp64-fusion protocol has also been used in antibody production (Lindley et al. 2000; Tami et al. 2000, 2004; Kaba et al. 2003), demonstrating that BDVS is a versatile tool applicable for the rapid production of monoclonal antibodies.

Surface Display via Alternative Membrane Anchors

Although foreign protein sequences have mainly been displayed on the viral surface by fusion to gp64, other viral glycoproteins have also been used. Both native and truncated forms of vesicular stomatitis virus G protein (VSV-G) have been exploited for pseudotyping baculoviruses and consequently for altering viral tropism (Fig. 1C). VSV-G enhances viral transduction of mammalian cells (Barsoum et al. 1997; Park et al. 2001; Pieroni et al. 2001; Tani et al. 2003; Kitagawa et al. 2005; Kaikkonen et al. 2006) and can replace the function of gp64 by restoring the ability of gp64-null viruses to assemble and produce infectious viral progeny (Mangor et al. 2001). Moreover, induction of humoral and cell-mediated immune responses has been achieved with a VSV-G-pseudotyped baculovirus expressing the hepatitis C viral glycoprotein, E2 (Facciabene et al. 2004). The transmembrane anchor of VSV-G has also been exploited for presentation of enhanced GFP (EGFP) (Chapple and Jones 2002) and the IgG-binding ZZ domains of protein A (Ojala et al. 2004) on the viral surface. To gain cancer cell-selective tropism of baculovirus, several tumor-homing peptides were displayed on the surface of baculovirus by fusion to the membrane anchoring domains of VSV-G (Makela et al. 2006). To increase specificity, the VSV-G fusion strategy was further modified by excluding the 21-amino-acid VSV-G ectodomain, known to mediate unspecific binding and transduction of the baculovirus vectors (Ojala et al. 2004; Kaikkonen et al. 2005). These vectors exhibited significantly improved binding and transgene delivery to both human breast carcinoma and hepatocarcinoma cells, highlighting the potential of targeted baculovirus vectors in cancer gene therapy. These strategies enable enhanced display due to the scattered surface distribution of both fusion partners, as opposed to gp64 fusions, which normally accumulate at the pole of the virion. In addition to the class I transmembrane VSV-G, the membrane-spanning region of a class II membrane protein (neuraminidase A of influenza virus) can serve as an amino-terminal anchor domain for efficient display of EGFP on the viral surface (Borg et al. 2004). BDVS can also be used for functional display of cellular membrane proteins on the surface of budded virions (Loisel et al. 1997; Masuda et al. 2003; Urano et al. 2003).

Capsid Display by Fusion to vp39

Presentation of a foreign protein moiety on the viral capsid was recently successfully conducted by fusing EGFP to both the amino and carboxyl termini of vp39 (Fig. 1D) (Kukkonen et al. 2003). This strategy should facilitate specific intracellular and nuclear targeting of the viral capsids, enabling more detailed studies on the mechanisms of baculoviral entry and nuclear import in mammalian cells. In fact, this study proposed that a block in transduction of mammalian cells is due to cytoplasmic trafficking or nuclear import instead of viral escape from the endosomes as previously suggested (Boublik et al. 1995; Boyce and Bucher 1996).

Generation of Display Libraries

As an alternative to homologous recombination, genes of interest can be inserted directly into the baculovirus genome (Kitts et al. 1990; Ernst et al. 1994; Lu and Miller 1996). Direct cloning leads to an increased number of recombinants. Direct insertion is the method of choice for generating gene expression libraries, where most possible variants must be generated in one step. Cloning efficiency (i.e., a high percentage of cleaved plasmid and insert, intact DNA termini, and pure material) is especially important when the main goal is a high number of progeny clones (10^5 vs. 10^3 by homologous recombination) (Ernst et al. 1998). Simultaneous cloning of a large number of genes serves to create a diverse library of recombinants. The most suitable clone can then be chosen according to selection criteria. Such a strategy can be used to generate a vast range of binding proteins such as antibody fragments. A specific binder is then selected by affinity screening, and the protein-encoding sequence is amplified by viral propagation. As an example of the use of baculovirus-infected *Sf*9 cells for surface display and library screening, a gp41-derived HIV-1 epitope was expressed in association with hemagglutinin of influenza A virus. This was displayed on the surface of baculovirus-infected insect cells as a library where each clone contained different amino acids adjacent to the epitope. This allowed for alteration of the structural environment such that the epitope was presented in the most accessible way, thereby increasing the binding capacity of the monoclonal antibody 2F5 (Ernst et al. 1998).

Protocol 1

Creation of Baculoviral Display Libraries

The baculovirus display vector system provides a number of advantages over prokaryotic systems. The baculoviruses permit larger gene insertions, are easily propagated, and can be grown to high titers. Furthermore, the eukaryotic system allows for posttranslational modifications, and surface modifications of the viral capsid enable specific targeting. Finally, such modified viruses are capable of transducing a wide variety of both dividing and nondividing mammalian cells. However, although the viruses do not replicate in mammalian cells, they are not entirely transcriptionally silent. They can also be highly antigenic when used in vivo, limiting their therapeutic use. The baculovirus vector *Ac*-omega used in the following protocol is a derivative of wild-type *Ac*MNPV containing a unique restriction site (SceI) downstream from the polyhedrin promoter, allowing linearization and direct ligation to the fragment to be inserted (Fig. 2) (Ernst et al. 1994).

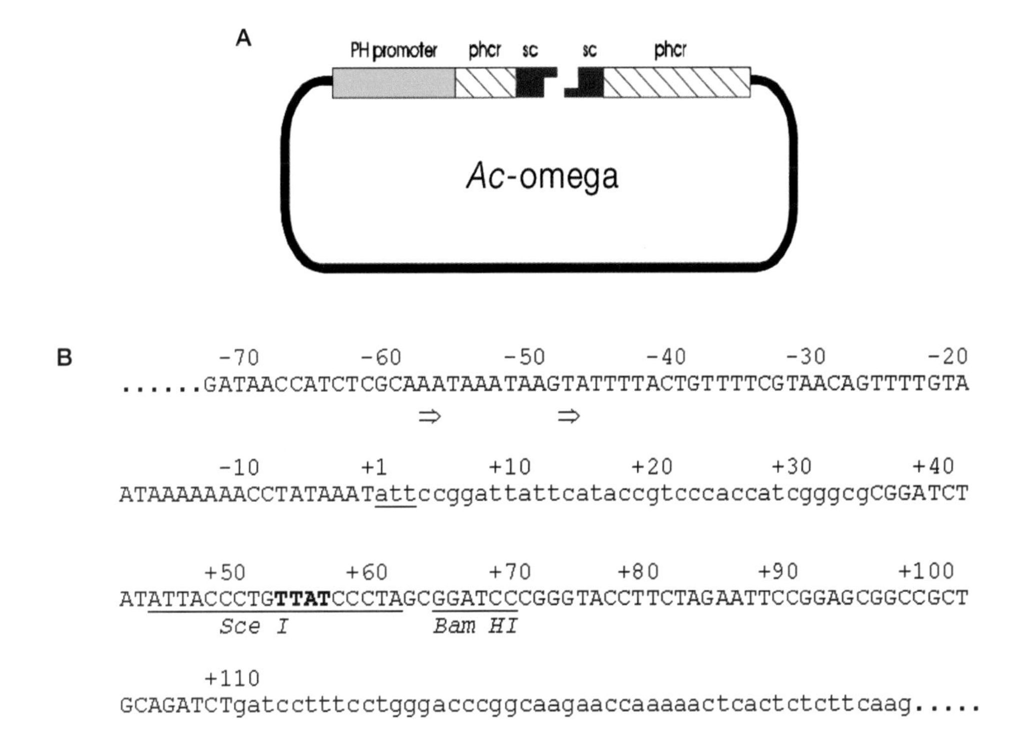

FIGURE 2. (*A*) A schematic diagram of *Ac*-omega containing the SceI single-cut site (sc) for direct gene insertion of DNA fragments treated with SceI meganuclease. The polyhedrin promoter (PH promoter), polyhedrin-coding residues (phcr), and multiple cloning site in *Ac*-omega are acquired from the transfer vector pVL1393. (*B*) The nucleotide sequence of the *Ac*MNPV derivative *Ac*-omega at the I-SceI recognition site including the polyhedrin promoter region and the 5′ and 3′ polyhedrin-coding residues. The arrow (position –57) indicates the 5′ end of mRNA. The arrow (position –48) marks the transcription start site. Natural polyhedrin ATG is mutated to ATT (+1, *underlined*). Lowercase letters from position +1 to +35 indicate 5′ polyhedrin-coding residues. The 18-bp I-SceI recognition site spans from nucleotide +45 to +62 (*underlined*). The I-SceI 3′ nonpalindromic overhang of four bases is also shown (*bold letters*). The target gene insertion site (only unidirectional) is located at position +58. Nucleotides +65 to +110 are derived from the multiple cloning site of transfer vector pVL1393. At position +111, the 3′ polyhedrin-coding region continues with nucleotide +172 of the natural polyhedrin gene (*lowercase letters*).

MATERIALS

CAUTION: See Appendix for appropriate handling of materials marked with <!>.

Reagents

Calf intestine alkaline phosphatase
DNA, *Ac*-omega baculoviral DNA and insert of interest
Ethanol (70%) <!>
IPL-41 (powdered medium, pH 6.2 +/– 0.1, osmolarity ~400 mOsmol)
Liposome formulation (Cellfectin; Invitrogen)
Phenol:chloroform <!>
Phosphate-buffered saline (PBS)
 2.7 mM KCl
 1.5 mM KH_2PO_4 <!>
 136.9 mM NaCl
 8.9 mM $Na_2HPO_4 \cdot 7H_2O$ <!>
  Store at 4°C.
I-SceI meganuclease (Boehringer-Mannheim)
*Sf*9 insect cells (ATCC, CRL-1711)
Spodoptera frugiperda (*Sf*9)

Equipment

Cell culture consumables for adherent cell cultures: T flasks, multiwell plates, Petri dishes
Cell incubator (27–28°C)
Disposables: single-use tubes, pipettes, tips, sterile filters
Other equipment: laminar flow hood, water bath, centrifuge for larger (>100 ml) volumes
Spinner flasks

METHOD

1. Digest 10 μg of *Ac*-omega viral DNA with 50 units of I-SceI meganuclease for 5 hours at 37°C in 100 μl of the manufacturer's recommended buffer.
2. Treat the linearized viral DNA with calf intestine alkaline phosphatase to remove 5′-phosphate groups.
3. Extract DNA with phenol:chloroform.
4. Precipitate DNA with ethanol.
5. Ligate insert of interest to 100–200 ng of purified, dephosphorylated viral *Ac*-omega DNA at a molar ratio of 1:40 as described by Ernst et al. (1994).
6. Incubate the ligation mixture overnight at 16°C.
7. Combine ligation mixture with 20 μl of Cellfectin (100 μl total volume). Incubate for 15 minutes at room temperature.
8. Transfect 2.5×10^6 *Sf*9 cells in culture with the above liposome formulation/ligation mixture. Incubate for 6 hours.
9. Add serum-containing medium. Incubate for 4–6 days at 27°C.
10. Perform plaque assay and analyze cells expressing the foreign protein by fluorescence-associated cell sorting (FACS; Ernst et al. 1998).

Protocol 2

Determination of Recombinant Display Viral Titer

The baculovirus titer can be determined by standard methods such as classical plaque assays or end-point dilution assays (Summers and Smith 1987; King and Possee 1992; O'Reilly et al. 1994). These are, however, often tedious and time-consuming. The procedure described here (modified from Volkman and Goldsmith 1982; BD Biosciences Clontech 2003) is fast and can be performed directly using marker genes such as GFP and β-galactosidase regulated by baculovirus-specific promoters or indirectly as an immunoassay with baculovirus-specific antibodies (e.g., anti-gp64).

MATERIALS

CAUTION: See Appendix for appropriate handling of materials marked with <!>.

Reagents

Antibodies
- Goat anti-mouse horseradish peroxidase (HRP)-conjugated secondary antibody
  - Dilute 1:400 in PBS/Tween before use.
- Mouse anti-gp64 antibody
  - Dilute in PBS/Tween before use.

Baculovirus
Blue horseradish peroxidase substrate
Ethanol <!>
Fixative (formyl-buffered acetone)
- 30% PBS
- 45% acetone <!>
- 25% formalin (diluted from 37% formaldehyde) <!>
  - Prepare a sufficient amount of fixative for 500 μl/slide fresh for each use.

HCl <!>
H_2O, deionized
Insect cell growth medium, serum-free (e.g., HyQ SFX [HyClone], Insect Xpress [Biowhittaker], or *Sf*-900 II SFM [Invitrogen])
Methylcellulose (0.6% w/v) in insect cell growth medium
- Autoclave methylcellulose, add growth medium, and stir overnight.

Mounting medium (e.g., Mowiol; Calbiochem)
Normal goat serum (NGS)
- Dilute 1:30 in PBS/Tween before use.

PBS/Tween (PBS containing 0.05% Tween 20)
Phosphate-buffered saline (PBS)
Spodoptera frugiperda (*Sf*9) cells (ATCC, CRL-1711)

Equipment

Microscope slides (12 well; e.g., Cel-line)

METHOD

1. Soak slides in a solution of 70% ethanol and 1% HCl overnight. Rinse ten times in tap water, followed by three rinses in deionized H_2O. Incubate overnight at 80°C.
2. Concentrate *Sf9* cells in log-phase to 1.5 x 10^6 cells/ml.
3. Apply 30 µl of *Sf9* cells to each well of the slides. Incubate in a humidified incubator for 30–60 minutes at 28°C to allow cells to attach.
4. Dilute baculoviral stock in medium.

   Final dilutions of 10^{-5}, 10^{-6}, and 10^{-7} are usually sufficient for titer determination.
5. Carefully remove the medium from the cells. Add 10 µl of virus inoculum to each well. Use four wells per dilution in duplicate (i.e., two slides per sample).
6. Incubate in a humidified incubator for 1 hour at 28°C.
7. Remove inoculum. Add 40 µl of methylcellulose to each well.
8. Incubate the slides in a sealed plastic bag containing a moist towel for 2–3 days at 28°C.
9. Add 40 µl of fixative to each well. Incubate for 5–10 minutes at room temperature.

   Do not overfix. It is not necessary to remove the methylcellulose.
10. Remove the fixative. Wash the wells three times with PBS/Tween, 5 minutes each wash.
11. Remove PBS/Tween. Add 40 µl of diluted NGS to each well. Incubate for 5 minutes at room temperature.
12. Remove the NGS. Add 15 µl of mouse anti-gp64 antibody to each well. Incubate for 30 minutes at room temperature.
13. Remove primary antibody. Wash twice with PBS/Tween, 5 minutes each wash.
14. Remove PBS/Tween. Add 15 µl of diluted HRP-conjugated goat anti-mouse IgG to each well. Incubate for 25 minutes at room temperature.
15. Remove secondary antibody. Wash three times with PBS/Tween, 5 minutes each wash.
16. Add 15 µl of peroxidase substrate to each well. Allow color development to occur for 1–3 hours.
17. Aspirate the substrate. Allow the slides to air-dry.
18. Mount dried samples.

    Dry slides can usually be stored for 1–2 weeks in the dark.
19. Use a light microscope to count the stained foci (a discrete cluster of 4–30 stained cells) in wells containing approximately 5–25 foci.
20. For each dilution, determine the average number of stained foci per well.
21. Multiply the average number of foci by the corresponding dilution factor and an inoculum normalization factor of 100 to convert the titer into pfu/ml:

    Virus titer (pfu/ml) = average number of foci per well x 1/dilution factor x 100.

Protocol 3

Immunofluorescence Analysis of Displayed Viral Proteins on Infected Insect Cells

After generating the display viral stock, it is important to confirm the presence and functionality of the displayed peptides or proteins on the viral particles before proceeding to further experiments. Accordingly, infected insect cells and budded virions can be analyzed by a variety of methods using appropriate antibodies, such as western blotting, enzyme-linked immunosorbent assay (ELISA) (Ojala et al. 2001, 2004), immunofluorescence analysis (IF), immunoelectron microscopy (IEM), fluorescence correlation spectroscopy (Toivola et al. 2002), and FACS (Ojala et al. 2001), as well as by atomic force microscopy. Two standard methods, IF and IEM, are described in detail in the following protocols.

MATERIALS

CAUTION: See Appendix for appropriate handling of materials marked with <!>.

Reagents

Antibodies
- primary antibody for the protein/peptide of interest
- secondary antibody, fluorophore-conjugated (e.g., Alexa Fluor Dyes; Invitrogen)

Baculovirus: recombinant display and wild type
Bovine serum albumin (BSA; 3%) in PBS
DABCO (25 mg/ml in Mowiol; Sigma-Aldrich)
Insect cell growth medium (serum-free)
Mowiol (Calbiochem)
Paraformaldehyde (4%) in PBS <!>
Phosphate-buffered saline (PBS)
Spodoptera frugiperda (*Sf*9; ATCC, CRL-1711)

Grow cells in suspension to approximately 2 x 10^6 cells/ml at the time of infection.

Triton X-100 <!>

Equipment

Cell incubator (e.g., orbital shaker for suspension cultures)
Coverslips
Microcentrifuge
Microscope slides
Plasticware and pipettes for cell culturing

METHOD

1. Infect approximately 10 ml of *Sf*9 cells with either recombinant or wild-type (control) viruses at a multiplicity of infection (moi) of 5–10. Incubate in an orbital shaker at 125 rpm for 24–48 hours at 28°C.
2. Harvest 0.5–1 ml (1×10^6 to 2×10^6 cells) from each infection. Centrifuge at 2000 rpm for 3 minutes at 4°C.
3. Aspirate the medium. Gently resuspend the cells with 0.5–1 ml of cold PBS.
4. Centrifuge as in Step 2. Aspirate the supernatant.
5. Optionally, the cells can be fixed and/or blocked and permeabilized.

   a. Incubate with 0.5–1 ml of 4% paraformaldehyde.

   b. Shake for 20 minutes at 4°C.

   c. Centrifuge as in Step 2. Aspirate the fixative.

      Alternatively, fixation can be performed following immunolabeling (after Step 10). If the displayed protein is already fluorescent, proceed directly to Step 11.

   d. Block and permeabilize the cells with 0.1% Triton X-100 in BSA for 10 minutes at room temperature.

      Permeabilization is not necessary if the antigenic epitope of the displayed protein recognized by the primary antibody is extracellular. In addition, some antibodies recognizing cytoplasmic domains of integral membrane proteins are able to penetrate the cell membrane without permeabilization.

6. Shake the cells with primary antibody diluted in BSA for 1 hour at 4°C.

   Volumes of approximately 200 μl are sufficient per 1×10^6 to 2×10^6 cells.

7. Shake three times with 0.5–1 ml of BSA, 15 minutes each wash at 4°C. Centrifuge the cells between the washes as in Step 2.
8. Add an appropriate fluorescent secondary antibody diluted in BSA. Incubate in the dark, with shaking, for 30 minutes at 4°C.

   Volumes of approximately 200 μl are sufficient per 1×10^6 to 2×10^6 cells.

9. Shake three times with 0.5–1 ml BSA, 15 minutes each wash at 4°C. Centrifuge the cells between the washes as in Step 2.
10. Aspirate the supernatant. Resuspend the cells in approximately 30–50 μl of DABCO.
11. Apply a few microliters of the cell suspension to microscope slides. Cover with coverslips. Allow to dry. Store in the dark at 4°C.

    Since the number of cells is reduced by 20–50% during the procedure, adjust the volume of DABCO accordingly for the final quantity of cells.

12. Analyze the cells with a conventional fluorescence microscope or by confocal microscopy.

Protocol 4

Immunoelectron Microscopy Analysis of Recombinant Display Viruses

This procedure is performed by using drops of staining reagents of appropriate volumes on Parafilm. Always float the grids on the drops (sample-side down) to avoid nonspecific labeling of the uncoated back of the grid. Never allow the grids to become wet on the back surface or dry out during the incubation procedure. The durations of the incubations below are recommended, but actual times, as well as antibody and protein-A-conjugated gold dilutions, must be empirically determined for each experiment.

MATERIALS

CAUTION: See Appendix for appropriate handling of materials marked with <!>.

Reagents

Antibodies
- primary antibody for the protein/peptide of interest
- secondary antibody (*optional*)

Baculovirus (sucrose-gradient-purified; Summers and Smith 1987; King and Possee 1992; O'Reilly et al. 1994)

Deionized distilled H_2O (ddH_2O)
> Sterilize by filtration through a 0.22-μm filter before use.

Fetal bovine serum (FBS; 10%) in PBS
> Sterilize by filtration through a 0.22-μm filter before use.

Fixative (*optional*): 0.1% glutaraldehyde/4% paraformaldehyde (in PBS) <!> or 2.5% glutaraldehyde (in PBS) <!>
> Sterilize by filtration through a 0.22-μm filter before use.

Glycine (0.15%) in PBS <!>
> Sterilize by filtration through a 0.22-μm filter before use.

Methylcellulose/uranyl acetate solution <!>
> Add 0.3 g of methylcellulose to 20 ml of boiling ddH_2O. Stir for 4–8 hours on ice (~4°C). Add 80 mg of uranyl acetate. Let stand for at least 24 hours at 4°C. Store at 4°C.

Phosphate-buffered saline (PBS, pH 7.4–7.5)
> Sterilize by filtration through a 0.22-μm filter. Store at 4°C.

Protein-A-conjugated nanogold particles (5–10-nm diameter)

Equipment

Carbon- and Formvar-coated metal grids
Filter paper
Filters (0.22 μm)
Parafilm
Syringes

METHOD

1. Prepare an appropriate dilution of the concentrated virus in PBS.

   a. To determine the optimal viral density on the grids, prepare fivefold or tenfold serial dilutions (e.g., nondiluted, 10^{-1}, 10^{-2}, 10^{-3}) of the concentrated baculoviral stock in PBS.

      Optionally, filter the dilutions through a 0.22-µm filter to disperse aggregates of viral particles.

   b. Aliquot 10 µl of each dilution to the grids. Allow particles to attach.

   c. Contrast and embed as described in Step 10. Allow to air-dry.

   d. Inspect grids with an electron microscope.

2. Attach 10 µl of optimal viral dilution (as determined in Step 1) to metal grids for 10–60 minutes at room temperature. The final viral density on the grids is dependent on the attachment time.

   As an optional step, fix the samples with fixative. Alternatively, fix the samples after Step 8.

3. Wash grids four times with 0.15% glycine within 10 minutes at room temperature.

4. Block nonspecific binding with 10% FBS for 10 minutes at room temperature. Remove excess buffer with filter paper.

5. Incubate with primary antibody diluted in 10% FBS for 15–60 minutes at room temperature.

   Drops of approximately 5–10 µl are sufficient per grid.

6. Wash five times with 10% FBS or 0.15% glycine within 15 minutes at room temperature. Remove excess buffer with filter paper.

   Optionally, incubate with bridging antibody diluted in 10% FBS for 15–60 minutes at room temperature. Wash as described in Step 7. Perform this step only if a primary antibody with weak binding capacity to protein A is used (e.g., sheep and goat IgG and some monoclonal antibodies) or when an enhancement of the gold signal is desired.

7. Incubate grids with protein-A-conjugated gold diluted in 10% FBS for 30–60 minutes at room temperature.

   Drops of approximately 10–20 µl are sufficient per grid.

8. Wash eight times with 10% FBS or 0.15% glycine within 20 minutes at room temperature.

9. Wash four times with ddH_2O within 8 minutes at room temperature.

   Optionally, negative-stain the grids with uranyl oxalate for 30 seconds at room temperature and allow to air-dry.

10. Incubate the samples with cold methylcellulose/uranyl acetate solution with two changes of solution of 10 seconds each on ice.

11. Transfer the grids to a third drop of cold methylcellulose/uranyl acetate solution. Incubate for 10 minutes on ice.

12. Remove the grids from the drops. Reduce the methylcellulose/uranyl acetate solution to a thin, even film with filter paper. Allow to air-dry for at least 10 minutes.

13. Examine the viral particles for nanogold particle distribution with a transmission electron microscope.

Protocol 5

Monitoring Baculovirus-mediated Efficiency of Gene Delivery

Luciferase molecules are nontoxic and emit light in direct proportion to their number in mammalian cells. This provides a sensitive and rapid assay for quantification of transgene expression without the need for illumination with an external excitation source.

MATERIALS

CAUTION: See Appendix for appropriate handling of materials marked with <!>.

Reagents

Cell culture growth medium (complete, serum-free, and/or selective)

D-Luciferin: 6 mM in 0.6 M sodium citrate buffer (pH 5.0) <!> (Sigma-Aldrich)

Store 2-ml aliquots at –20°C. Dilute 1:5 with H_2O before use. Once thawed, the solution can be stored for a few days at 4°C.

Mammalian cells (e.g., human liver cell line HepG2)

Phosphate-buffered saline (PBS)

Equipment

Cell culture dishes

Luminometer

Plates (96 well, white- or black-pigmented)

METHOD

1. Seed 1 x 10^6 HepG2 cells in a 60-mm Petri dish. Incubate the cells for 18–24 hours at 37°C in 5% CO_2.

   The initial number of cells needed for luminescence detection depends on the cell line, the recombinant baculovirus, and the detection sensitivity of the luminometer.

2. Dilute viral stock to 100 pfu/cell in 1 ml of serum-free medium.

   Unpurified viral stocks or sucrose-gradient-purified virus can be used. Note that determination of viral titers by conventional plaque assays defines the infectivity of baculovirus only with respect to insect cells. This may not reflect their situation with regard to mammalian cells. In such cases, it is recommended that the actual number of physical particles be determined by real-time polymerase chain reaction assays (Lo and Chao 2004).

3. Remove medium from the culture dish. Add virus to the cells. Incubate for 1 hour at 37°C in 5% CO_2.

4. Add 4–5 ml of complete growth medium containing 10% FBS. Incubate the samples for 24–48 hours at 37°C in 5% CO_2.

   The viral inoculum does not need to be removed. For alternative transduction protocols, see Chapter 29, this volume).

5. Remove the medium. Wash the cells once with PBS.
6. Detach the cells by trypsinization or by scraping and transfer to centrifuge tubes.
7. Centrifuge at 1500 rpm for 3 minutes. Resuspend the pellet in 100 μl of growth medium or PBS.
8. Transfer the cell suspension to a white- or black-pigmented 96-well plate. Add 100 μl of 1x luciferin solution (pH 5.0). Mix rapidly.

   Cells do not need to be lysed.
9. Measure the enzymatic activity of luciferase using a luminometer immediately after addition of the substrate.

ACKNOWLEDGMENTS

We are grateful to Professor Loy Volkman of the University of California at Berkeley for her valuable contribution to this chapter.

REFERENCES

Airenne K.J., Peltomaa E., Hytonen V.P., Laitinen O.H., and Yla-Herttuala S. 2003. Improved generation of recombinant baculovirus genomes in *Escherichia coli*. *Nucleic Acids Res.* **31:** e101.

Airenne K.J., Hiltunen M.O., Turunen M.P., Turunen A.M., Laitinen O.H., Kulomaa M.S., and Yla-Herttuala S. 2000. Baculovirus-mediated periadventitial gene transfer to rabbit carotid artery. *Gene Ther.* **7:** 1499–1504.

Barsoum J., Brown R., McKee M., and Boyce F.M. 1997. Efficient transduction of mammalian cells by a recombinant baculovirus having the vesicular stomatitis virus G glycoprotein. *Hum. Gene Ther.* **8:** 2011–2018.

BD Biosciences Clontech. 2003. *BacPAK™ baculovirus rapid titer kit user manual.* Clontech, Mountain View, California.

Borg J., Nevsten P., Wallenberg R., Stenstrom M., Cardell S., Falkenberg C., and Holm C. 2004. Amino-terminal anchored surface display in insect cells and budded baculovirus using the amino-terminal end of neuraminidase. *J. Biotechnol.* **114:** 21–30.

Boublik Y., Di Bonito P., and Jones I.M. 1995. Eukaryotic virus display: Engineering the major surface glycoprotein of the *Autographa californica* nuclear polyhedrosis virus (AcNPV) for the presentation of foreign proteins on the virus surface. *Biotechnology* **13:** 1079–1084.

Boyce F.M. and Bucher N.L. 1996. Baculovirus-mediated gene transfer into mammalian cells. *Proc. Natl. Acad. Sci.* **93:** 2348–2352.

Chapple S.D. and Jones I.M. 2002. Non-polar distribution of green fluorescent protein on the surface of *Autographa californica* nucleopolyhedrovirus using a heterologous membrane anchor. *J. Biotechnol.* **95:** 269–275.

Condreay J.P., Witherspoon S.M., Clay W.C., and Kost T.A. 1999. Transient and stable gene expression in mammalian cells transduced with a recombinant baculovirus vector. *Proc. Natl. Acad. Sci.* **96:** 127–132.

Ernst W.J., Grabherr R.M., and Katinger H.W. 1994. Direct cloning into the *Autographa californica* nuclear polyhedrosis virus for generation of recombinant baculoviruses. *Nucleic Acids Res.* **22:** 2855–2856.

Ernst W.J., Spenger A., Toellner L., Katinger H., and Grabherr R.M. 2000. Expanding baculovirus surface display. Modification of the native coat protein gp64 of *Autographa californica* NPV. *Eur. J. Biochem.* **267:** 4033–4039.

Ernst W., Grabherr R., Wegner D., Borth N., Grassauer A., and Katinger H. 1998. Baculovirus surface display: Construction and screening of a eukaryotic epitope library. *Nucleic Acids Res.* **26:** 1718–1723.

Facciabene A., Aurisicchio L., and La Monica N. 2004. Baculovirus vectors elicit antigen-specific immune responses in mice. *J. Virol.* **78:** 8663–8672.

Grabherr R. and Ernst W. 2001. The baculovirus expression system as a tool for generating diversity by viral surface display. *Comb. Chem. High Throughput Screen.* **4:** 185–192.

Grabherr R., Ernst W., Oker-Blom C., and Jones I. 2001. Developments in the use of baculoviruses for the surface display of complex eukaryotic proteins. *Trends Biotechnol.* **19:** 231–236.

Grabherr R., Ernst W., Doblhoff-Dier O., Sara M., and Katinger H. 1997. Expression of foreign proteins on the surface of *Autographa californica* nuclear polyhedrosis virus. *Biotechniques* **22:** 730–735.

Hoare J., Waddington S., Thomas H.C., Coutelle C., and McGarvey M.J. 2005. Complement inhibition rescued mice allowing observation of transgene expression following intraportal delivery of baculovirus in mice. *J. Gene Med.* **7:** 325–333.

Hüser A. and Hofmann C. 2003. Baculovirus vectors: Novel mammalian cell gene-delivery vehicles and their applications. *Am. J. Pharmacogenomics* **3:** 53–63.

Hüser A., Rudolph M., and Hofmann C. 2001. Incorporation of decay-accelerating factor into the baculovirus envelope generates complement-resistant gene transfer vectors. *Nat. Biotechnol.* **19:** 451–455.

Kaba S.A., Hemmes J.C., van Lent J.W., Vlak J.M., Nene V., Musoke A.J., and van Oers M.M. 2003. Baculovirus surface display of *Theileria parva* p67 antigen preserves the conformation of sporozoite-neutralizing epitopes. *Protein Eng.* **16:** 73–78.

Kaikkonen M.U., Räty J.K., Airenne K.J., Wirth T., Heikura T., and Yla-Herttuala S. 2006. Truncated vesicular stomatitis virus G protein improves baculovirus transduction efficiency in vitro and in vivo. *Gene Ther.* **13:** 304–312.

King L.A. and Possee R.D. 1992. *The baculovirus expression system*, 1st edition. Chapman & Hall, London, United Kingdom.

Kitagawa Y., Tani H., Limn C.K., Matsunaga T.M., Moriishi K., and Matsuura Y. 2005. Ligand-directed gene targeting to mammalian cells by pseudotype baculoviruses. *J. Virol.* **79:** 3639–3652.

Kitts P.A., Ayres M.D., and Possee R.D. 1990. Linearization of baculovirus DNA enhances the recovery of recombinant virus expression vectors. *Nucleic Acids. Res.* **18:** 5667–5672.

Kost T.A. and Condreay J.P. 2002. Recombinant baculoviruses as mammalian cell gene-delivery vectors. *Trends Biotechnol.* **20:** 173–180.

Kost T.A., Condreay J.P., and Jarvis D.L. 2005. Baculovirus as versatile vectors for protein expression in insect and mammalian cells. *Nat. Biotechnol.* **23:** 567–575.

Kukkonen S.P., Airenne K.J., Marjomaki V., Laitinen O.H., Lehtolainen P., Kankaanpaa P., Mahonen A.J., Räty J.K., Nordlund H.R., Oker-Blom C., Kulomaa M.S., and Yla-Herttuala S. 2003. Baculovirus capsid display: A novel tool for transduction imaging. *Mol. Ther.* **8:** 853–862.

Lehtolainen P., Tyynela K., Kannasto J., Airenne K.J., and Yla-Herttuala S. 2002. Baculoviruses exhibit restricted cell type specificity in rat brain: A comparison of baculovirus- and adenovirus-mediated intracerebral gene transfer in vivo. *Gene Ther.* **9:** 1693–1699.

Li Y., Wang X., Guo H., and Wang S. 2004. Axonal transport of recombinant baculovirus vectors. *Mol. Ther.* **10:** 1121–1129.

Lindley K.M., Su J.L., Hodges P.K., Wisely G.B., Bledsoe R.K., Condreay J.P., Winegar D.A., Hutchins J.T., and Kost T.A. 2000. Production of monoclonal antibodies using recombinant baculovirus displaying gp64-fusion proteins. *J. Immunol. Methods* **234:** 123–135.

Lo H.R. and Chao Y.C. 2004. Rapid titer determination of baculovirus by quantitative real-time polymerase chain reaction. *Biotechnol. Prog.* **20:** 354–360.

Loisel T.P., Ansanay H., St-Onge S., Gay B., Boulanger P., Strosberg A.D., Marullo S., and Bouvier M. 1997. Recovery of homogeneous and functional β_2-adrenergic receptors from extracellular baculovirus particles. *Nat. Biotechnol.* **15:** 1300–1304.

Lu A. and Miller L.K. 1996. Generation of recombinant baculoviruses by direct cloning. *Biotechniques* **21:** 63–68.

Luckow V.A., Lee S.C., Barry G.F., and Olins P.O. 1993. Efficient generation of infectious recombinant baculoviruses by site-specific transposon-mediated insertion of foreign genes into a baculovirus genome propagated in *Escherichia coli*. *J. Virol.* **67:** 4566–4579.

Makela A.R., Matilainen H., White D.J., Ruoslahti E., and Oker-Blom C. 2006. Enhanced baculovirus-mediated transduction of human cancer cells by tumor-homing peptides. *J. Virol.* **80:** 6603–6611.

Mangor J.T., Monsma S.A., Johnson M.C., and Blissard G.W. 2001. A GP64-null baculovirus pseudotyped with vesicular stomatitis virus G protein. *J. Virol.* **75:** 2544–2556.

Markovic I., Pulyaeva H., Sokoloff A., and Chernomordik L.V. 1998. Membrane fusion mediated by baculovirus gp64 involves assembly of stable gp64 trimers into multiprotein aggregates. *J. Cell Biol.* **143:** 1155–1166.

Masuda K., Itoh H., Sakihama T., Akiyama C., Takahashi K., Fukuda R., Yokomizo T., Shimizu T., Kodama T., and Hamakubo T. 2003. A combinatorial G protein-coupled receptor reconstitution system on budded baculovirus. Evidence for $G\alpha_i$ and $G\alpha_o$ coupling to a human leukotriene B_4 receptor. *J. Biol. Chem.* **278:** 24552–24562.

Mottershead D., van der Linden I., von Bonsdorff C.H., Keinanen K., and Oker-Blom C. 1997. Baculoviral display of the green fluorescent protein and rubella virus envelope proteins. *Biochem. Biophys. Res. Commun.* **238:** 717–722.

Mottershead D.G., Alfthan K., Ojala K., Takkinen K., and Oker-Blom C. 2000. Baculoviral display of functional scFv and synthetic IgG-binding domains. *Biochem. Biophys. Res. Commun.* **275:** 84–90.

Ojala K., Mottershead D.G., Suokko A., and Oker-Blom C. 2001. Specific binding of baculoviruses displaying gp64 fusion proteins to mammalian cells. *Biochem. Biophys. Res. Commun.* **284:** 777–784.

Ojala K., Koski J., Ernst W., Grabherr R., Jones I., and Oker-Blom C. 2004. Improved display of synthetic IgG-binding domains on the baculovirus surface. *Technol. Cancer Res. Treat.* **3:** 77–84.

Oker-Blom C., Airenne K.J., and Grabherr R. 2003. Baculovirus display strategies: Emerging tools for eukaryotic libraries and gene delivery. *Brief. Funct. Genomics Proteomics* **2:** 244–253.

Oomens A.G., Monsma S.A., and Blissard G.W. 1995. The baculovirus GP64 envelope fusion protein: Synthesis, oligomerization, and processing. *Virology* **209:** 592–603.

O'Reilly D.R., Miller L.K., and Luckow V.A. 1994. *Baculovirus expression vectors: A laboratory manual.* Oxford University Press, New York.

Park S.W., Lee H.K., Kim T.G., Yoon S.K., and Paik S.Y. 2001. Hepatocyte-specific gene expression by baculovirus pseudotyped with vesicular stomatitis virus envelope glycoprotein. *Biochem. Biophys. Res. Commun.* **289:** 444–450.

Peakman T.C., Harris R.A., and Gewert D.R. 1992. Highly efficient generation of recombinant baculoviruses by enzymatically mediated site-specific in vitro recombination. *Nucleic Acids. Res.* **20:** 495–500.

Pieroni L., Maione D., and La Monica N. 2001. In vivo gene transfer in mouse skeletal muscle mediated by baculovirus vectors. *Hum. Gene Ther.* **12:** 871–881.

Räty J.K., Airenne K.J., Marttila A.T., Marjomaki V., Hytonen V.P., Lehtolainen P., Laitinen O.H., Mahonen A.J., Kulomaa M.S., and Yla-Herttuala S. 2004. Enhanced gene delivery by avidin-displaying baculovirus. *Mol. Ther.* **9:** 282–291.

Shoji I., Aizaki H., Tani H., Ishii K., Chiba T., Saito I., Miyamura T., and Matsuura Y. 1997. Efficient gene transfer into various mammalian cells, including non-hepatic cells, by baculovirus vectors. *J. Gen. Virol.* **78:** 2657–2664.

Spenger A., Grabherr R., Tollner L., Katinger H., and Ernst W. 2002. Altering the surface properties of baculovirus *Autographa californica* NPV by insertional mutagenesis of the envelope protein gp64. *Eur. J. Biochem.* **269:** 4458–4467.

Summers M.D. and Smith G.E. 1987. *A manual of methods for baculovirus vectors and insect cell culture procedures*, 2nd edition. Texas Agricultural Experiment Station, College Station, Texas.

Tami C., Farber M., Palma E.L., and Taboga O. 2000. Presentation of antigenic sites from foot-and-mouth disease virus on the surface of baculovirus and in the membrane of infected cells. *Arch. Virol.* **145:** 1815–1828.

Tami C., Peralta A., Barbieri R., Berinstein A., Carrillo E., and Taboga O. 2004. Immunological properties of FMDV-gP64 fusion proteins expressed on *SF*9 cell and baculovirus surfaces. *Vaccine* **23:** 840–845.

Tani H., Limn C.K., Yap C.C., Onishi M., Nozaki M., Nishimune Y., Okahashi N., Kitagawa Y., Watanabe R., Mochizuki R., Moriishi K., and Matsuura Y. 2003. In vitro and in vivo gene delivery by recombinant baculoviruses. *J. Virol.* **77:** 9799–9808.

Toivola J., Ojala K., Michel P.O., Vuento M., and Oker-Blom C. 2002. Properties of baculovirus particles displaying GFP analyzed by fluorescence correlation spectroscopy. *Biol. Chem.* **383:** 1941–1946.

Urano Y., Yamaguchi M., Fukuda R., Masuda K., Takahashi K., Uchiyama Y., Iwanari H., Jiang S.Y., Naito M., Kodama T., and Hamakubo T. 2003. A novel method for viral display of ER membrane proteins on budded baculovirus. *Biochem. Biophys. Res. Commun.* **308:** 191–196.

Volkman L.E. and Goldsmith P.A. 1982. Generalized immunoassay for *Autographa californica* nuclear polyhedrosis virus infectivity in vitro. *Appl. Environ. Microbiol.* **44:** 227–233.

Yap C.C., Ishii K., Aoki Y., Aizaki H., Tani H., Shimizu H., Ueno Y., Miyamura T., and Matsuura Y. 1997. A hybrid baculovirus-T7 RNA polymerase system for recovery of an infectious virus from cDNA. *Virology* **231:** 192–200.

27 Safe, Simple, and High-capacity Gene Delivery into Insect and Vertebrate Cells by Recombinant Baculoviruses

Kari J. Airenne,* Olli H. Laitinen,*‡ Anssi J. Mähönen,*‡ and Seppo Ylä-Herttuala*†

*AI Virtanen Institute, Department of Biotechnology and Molecular Medicine and †Department of Medicine and Gene Therapy Unit, University of Kuopio, FI-70211 Kuopio, Finland; ‡Ark Therapeutics Oy, Neulaniementie 2 L9, FIN-70210 Kuopio, Finland

ABSTRACT

Baculoviruses have become standard tools to produce recombinant proteins in a eukaryotic milieu. Thousands of proteins have been successfully produced as intracellular or secreted proteins in insect cells. The ability of modified baculoviruses to transduce a wide range of both dividing and nondividing vertebrate cells has further expanded the utility of the baculoviral expression vector system (BEVS). This has also provided the basis for a universal expression process in which one-step cloning allows expression of desired genes in several different types of hosts. Recent findings demonstrated that baculoviruses are also able to mediate efficient gene delivery in mammals. Since, as nonmammalian viruses, they cannot replicate or express their own genes in vertebrate cells, baculoviruses make potentially useful gene therapy vectors. We have developed an improved transposition-based system (BVboost) for the generation of recombinant baculoviruses that bypasses the disadvantages of the original transposition-based generation of baculoviral genomes in *Escherichia coli* while remaining a simple, rapid, and convenient viral production technique. The method is based on the modified donor vector (pBVboost) and an improved selection scheme of the baculoviral genomes (bacmids) in *E. coli* with a mutated *sacB* gene. Recombinant bacmids can be generated at a frequency of approximately 10^7/µg of donor vector with a negligible background. The BVboost system also allows efficient setups for high-throughput screening and gene expression purposes.

INTRODUCTION

In the two decades since its introduction, the BEVS has become one of the most popular systems to produce recombinant proteins (Smith et al. 1983; O'Reilly et al. 1994; Kost et al. 2005). The finding that a modified baculovirus bearing a suitable promoter was able to efficiently transduce hepatocytes (Hofmann et al. 1995) stimulated further interest in BEVS as a potential gene delivery vector for cells other than those of insect origin. Indeed, a wide variety of vertebrate cells can be transduced by recombinant baculoviruses. Such viruses can also mediate efficient gene delivery into several tissues of mammals in vivo (Sandig et al. 1996; Condreay et al. 1999; Airenne et al. 2000; Haeseleer et al. 2001; Pieroni et al. 2001; Lehtolainen et al. 2002; Song et al. 2003; Leisy et al. 2003; Cheng et al. 2004; Hu 2005). The popularity of BEVS over other expression systems is based on several unique features such as safety, ability to accommodate large transgenes, and low cytotoxicity in vertebrate cells even at very high virus load (Kidd and Emery 1993; O'Reilly et al. 1994; Ghosh et al. 2002; Airenne et al. 2004; Kost et al. 2005).

Autographa californica Multicapsid Nucleopolyhedrovirus: The Workhorse of BEVS

Baculoviruses are a diverse group of viruses having a restricted host range often limited to specific invertebrate species, especially to insects (Gröner 1986). Baculoviruses are not known to infect vertebrate hosts. The double-stranded circular DNA genome (80–200 kbp) of baculoviruses (Summers and Anderson 1972; Burgess 1977) is condensed inside a flexible rod-shaped capsid, which is 25–50 nm in diameter and 200–320 nm in length (Harrap 1972) and which can expand to accommodate large DNA fragments (Fraser 1986). The most extensively studied NPV type I baculovirus is the *Autographa californica* multicapsid nucleopolyhedrovirus (*Ac*MNPV). Its genome (134 kbp) has been sequenced and predicted to contain 154 open reading frames (Ayres et al. 1994). The major glycoprotein associated with the *Ac*MNPV envelope is gp64. This protein has been successfully used to display numerous peptides and proteins on the baculoviral surface in order to study protein function as well as to enhance and target baculoviral transduction in vertebrate cells (Boublik et al. 1995; Ojala et al. 2001; Oker-Blom et al. 2003; Raty et al. 2004; Mäkelä et al., Chap. 28, this volume). *Ac*MNPV enters cells by endocytosis (Wang et al. 1997). The role of gp64 is essential for viral entry since it mediates pH-dependent escape of the *Ac*MNPV capsids from the endosomes (Blissard and Wenz 1992). The gp64 protein is also needed for viral attachment to the cell surface and efficient virion budding from the insect cells (Oomens and Blissard 1999). The cellular surface molecules for *Ac*MNPV attachment and entry are not known, but the large range of target cells suggests that the molecules are common cell surface components such as integrins, phospholipids, and heparan sulfate proteoglycans (Hynes 1992; Duisit et al. 1999; Tani et al. 2001). The most commonly used cell line for *Ac*MNPV propagation is *Spodoptera frugiperda* 9 (*Sf*9) cells (Vaughn et al. 1977).

To enable the expression of the recombinant protein in insect cells, the gene for the desired protein is usually placed under a strong polyhedrin promoter (*polh*) of *Ac*MNPV (O'Reilly et al. 1994). However, the use of the *polh* promoter may be restricted in some cases by the fact that it is activated very late in infection at a point when the host-cell machinery for posttranslational modifications is no longer working properly. Problems with the *polh* promoter have been encountered, especially with proteins whose biological activity depends on proper glycosylation (Kost et al. 2005). Other promoters, which will activate earlier than the *polh*, include the promoters for the *p10* gene (*p10*), the major capsid protein gene (*vp39*), the basic 6.9-kD protein gene (*cor*), and the viral *ie1* gene (*ie1*) (Miller 1993; O'Reilly et al. 1994). In vertebrate cells, a promoter active in the target cells such as cytomegalovirus (CMV), Rous sarcoma virus (RSV), or chicken β-actin promoter (CAG) must be used since the *polh* is inactive in these cells (Shoji et al. 1997; Airenne et al. 2004; Kost et al. 2005). Tissue-specific or regulated promoters can also be used (Park et al. 2001; Li et al. 2004, 2005; McCormick et al. 2004).

Construction of Recombinant Baculoviruses

The large size of the *Ac*MNPV genome limits direct cloning of desired genes into its genome. The homologous recombination procedure was therefore adapted to insert foreign genes into the baculoviral genome (Smith et al. 1983). In practice, the target gene is cloned into a transfer (donor) vector containing a baculoviral promoter flanked by baculoviral DNA derived from a nonessential locus such as the polyhedrin gene of *Ac*MNPV (the most common site for insertion of foreign DNA). The viral DNA and transfer plasmid are then cotransfected into insect cells where the recombination event takes place. The original homologous recombination protocol typically yielded only 0.1–1% of recombinant viruses. Several improvements have been made to this procedure such that a success rate of greater than 95% can currently be achieved (for review, see Davies 1994; Jones and Morikawa 1996).

The most popular approach for generating a recombinant baculovirus, however, uses a site-specific transposition with Tn*7* to insert foreign genes into bacmid DNA (baculoviral genome) propagated in *E. coli* cells. The recombinant bacmid-containing *E. coli* clones are selected by color (*lacZ*), and the bacmid purified from a single white colony is used to transfect insect cells (Luckow et al. 1993). This method is straightforward and rapid since it eliminates the need for time-consuming plaque purification. Pure recombinant viruses can be prepared within 7–10 days. However, the original system suffers from the poor selection scheme, and the identification of recombinant baculoviral genomes in *E. coli* is hampered by colony sectoring and multiple colony morphologies (Leusch et al. 1995). We have recently improved this method to bypass these problems by constructing a modified donor vector (pBVboost) bearing a mutated *sacB* gene and an *att*Tn*7* blocked *E. coli* strain DH10BacΔTn*7* (Airenne et al. 2003). Recombinant bacmids can be generated at a frequency of approximately 10^7/μg of donor vector with a negligible background. This easy to use and efficient BVboost system provides the basis for a high-throughput generation of recombinant baculoviruses as well as a more convenient way to produce single viruses (Fig. 1). A tetrapromoter variant of the pBVboost, pBVboostFG, enables an all-in-one strategy for gene expression in mammalian, bacterial, and insect cells (as well as in vivo) by one-step cloning of the desired gene(s), thus facilitating functional genomic approaches (Laitinen et al. 2005). This chapter provides instructions on how to prepare recombinant baculoviruses by the BVboost system in order to express the desired gene(s) in insect and/or vertebrate cells.

Advantages and Limits

Because the viral genome is prepared in *E. coli*, no plaque purification is needed to obtain pure viral clones. The selection scheme used in this method guarantees that practically all acquired baculoviral genomes are recombinant. These features also render the BVboost system attractive for high-throughput screening approaches. Baculovirus-mediated transduction of vertebrate cells is well tolerated, and no cytopathic effects have been reported even with a very high multiplicity of infection (moi 10,000) (Aoki et al. 1999). However, the level of transgene expression varies and is usually lower in cell types other than those of hepatic origin.

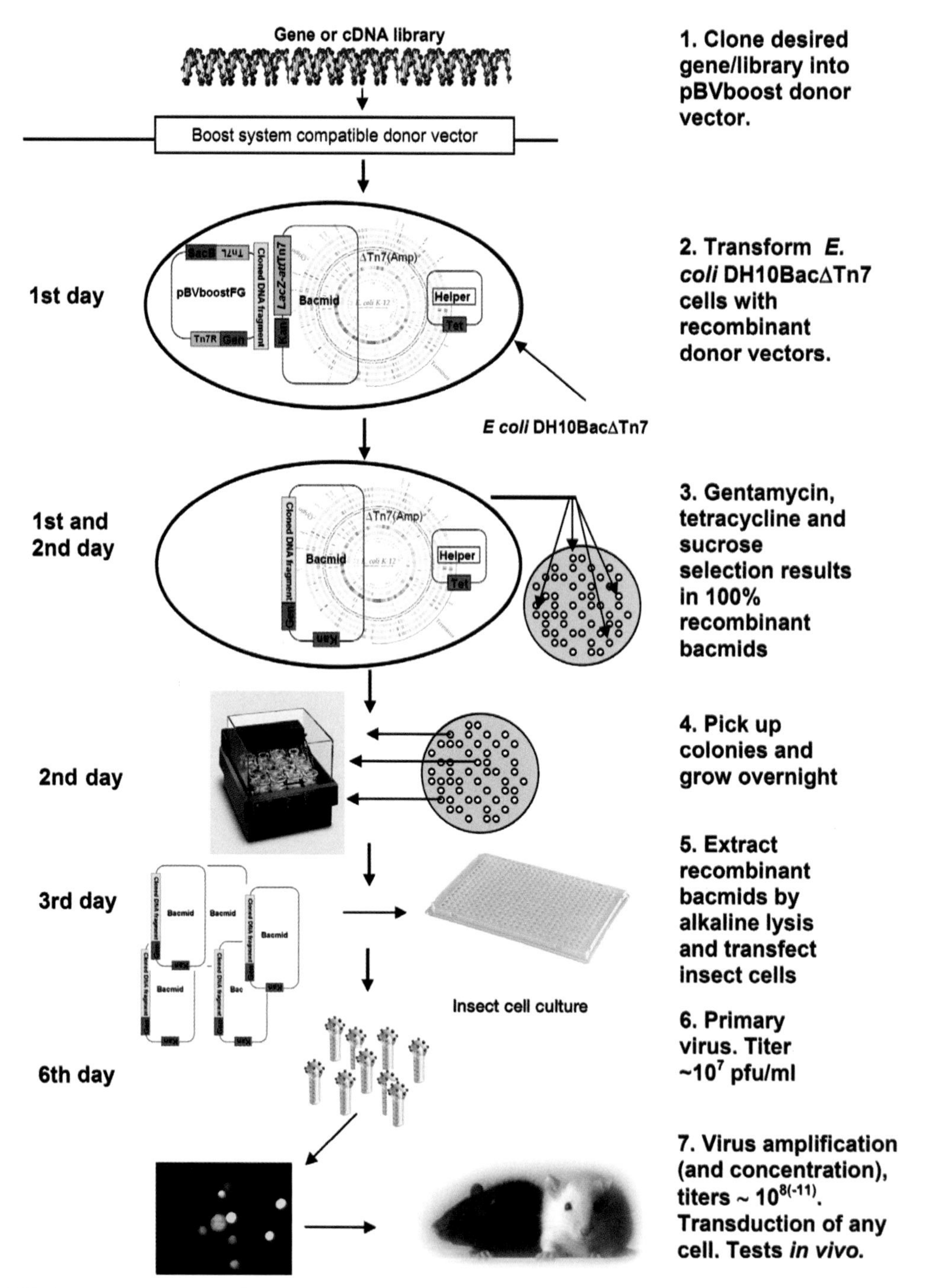

FIGURE 1. An overview of the pBVboost system to clone and generate recombinant baculoviruses (Airenne et al. 2003).

Protocol 1

Preparation of Recombinant Baculoviruses with the BVboost System

After cloning the desired gene/cDNA/library into a BVboost system-compatible donor plasmid (Fig. 2) (Airenne et al. 2003; Laitinen et al. 2005), the recombinant baculoviral genome is prepared simply by transforming electrocompetent DH10BacΔTn*7* *E. coli* cells with the donor (Fig. 1). Transfer from the donor vector into the baculoviral genome (bacmid) occurs via a Tn*7*-mediated site-specific transposition reaction in *E. coli* cells. The selection scheme guarantees that virtually all colonies are correct. The recombinant baculoviral genome is subsequently extracted from *E. coli* culture using a modified isolation procedure for large plasmids. To generate the recombinant viruses, insect cells are transfected with the isolated recombinant bacmid.

MATERIALS

CAUTION: See Appendix for appropriate handling of materials marked with <!>.

Reagents

Bacto-agar
Bluo-gal/X-gal (*optional*) <!>
Boost solution 1
  15 mM Tris-HCl (pH 8.0) <!>
  10 mM EDTA
  100 µg/ml RNase A <!>
    Filter-sterilize and store at 4°C.
Boost solution 2
  0.2 M NaOH <!>
  1% (w/v) SDS <!>
    Filter-sterilize and store at room temperature.
Boost solution 3 (3 M KCH_3COO at pH 5.5)
    Autoclave and store at room temperature.
Cell lines
  DH10BacΔTn*7* *E. coli* cells, electrocompetent (Airenne et al. 2003)
  *Sf*9 insect cells (ATCC CRL-1711)
Cellfectin (Invitrogen)
Donor plasmid (pBVboost or any of its derivatives; Fig. 2)
Ethanol (70%) <!>
Gentamicin
H_2O, sterile
Insect cell growth medium, serum-free (e.g., Insect Xpress [Biowhittaker] or Sf-900 II SFM [Invitrogen]).
Isopropanol <!>
Isopropyl-β-D-thiogalactopyranoside (IPTG) (*optional*) <!>

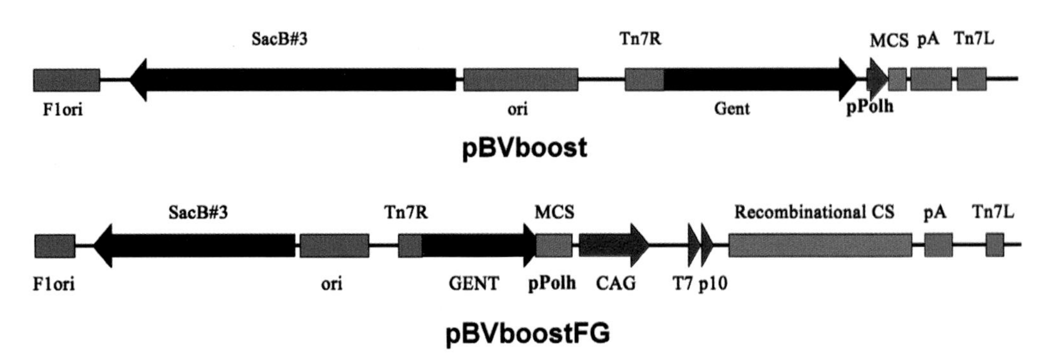

FIGURE 2. Maps of boost donor vectors pBVboost and pBVboostFG. (F1ori) Filamentous phage origin of replication; (SacB#3) mutant form of the levansucrase gene; ori, bacterial origin of replication; (Gent) gentamicin resistance gene; (Tn7R and Tn7L) left and right ends of bacterial transposon TnT7; (pPolh) baculovirus polyhedrin promoter; (MCS) multiple cloning site; (CAG) chicken β-actin promoter; (T7) bacteriophage T7 promoter; (p10) baculovirus p10 promoter; (pA) transcriptional terminator area; (recombinational CS) recombinational cloning site of bacteriophage λ.

LB medium
- 10 g of bacto-tryptone
- 5 g of bacto-yeast extract
- 10 g of NaCl
  - Add H_2O to 1 liter and adjust pH to 7.5 with 1 M NaOH.

S.O.C. medium
- 2 g of bacto-tryptone
- 0.5 g of yeast extract
- 1 ml of 1 M NaCl
- 0.25 ml of 1 M KCl <!>
- 1 ml of 2 M Mg stock solution (1 M $MgSO_4$, 1 M $MgCl_2$) <!>
  - Add H_2O to 1 liter.

Sucrose

TE buffer (pH 8.0)

Tetracycline <!>
- 10 mM Tris-HCl
- 1 mM EDTA

Equipment

Cell incubators for *E. coli* and insect cells

Centrifuges

Electroporator and cuvettes

Plasticware and pipettes for cell culturing

Sterile tubes

METHODS

Transposition into Baculoviral Genome in *E. coli*

1. Prepare LB_{stg}-agar plates.
   a. Dissolve 7.5 g of bacto-agar in 100 ml of 5x LB medium. Add water to 500 ml.
   b. Autoclave.

c. Equilibrate in a water bath to 53°C.

d. Add 10% sucrose, 10 μg/ml tetracycline, and 7 μg/ml gentamicin.

If transposition is to be confirmed by color selection, add 100 μg/ml Bluo-gal or 60 μg/ml X-gal and 100 mM IPTG (Steps 9 and 11).

e. Cast the plates.

For 15 plates, 100 ml of LB_{stg}-agar should be sufficient.

2. Thaw the electrocompetent DH10BacΔTn*7* *E. coli* cells on ice. Aliquot 40 μl of the thawed cells into a prechilled Eppendorf tube.
3. Add 1 ng of donor plasmid in 5 μl of TE to the cells. Mix gently.
4. Transfer the mixture into a prechilled cuvette. Electroporate according to the manufacturer's instructions.
5. Add 1 ml of room-temperature S.O.C. medium to the mixture.
6. Transfer the cells to a 2-ml Eppendorf tube. Incubate with shaking at 250 rpm for 4 hours at 37°C.
7. Serially dilute the cells using S.O.C. medium to 10^{-1}, 10^{-2}, and 10^{-3}.
8. Place 100 μl of undiluted sample and each serial dilution on the LB_{stg} plates. Spread evenly.
9. Incubate the plates for 24 hours at 37°C.

   The colonies thus generated contain recombinant baculoviral genomes. Transposition can be confirmed by color selection using the Bluo/X-gal- and IPTG-containing plates and extending the incubation time to 48 hours.

Isolation of Recombinant Bacmid (Baculoviral Genome) DNA

10. Pick colonies (from Step 9). In a 50-ml Nunc tube, inoculate 5 ml of LB medium supplemented with 10 μg/ml tetracycline and 7 μg/ml gentamicin with the selected colonies.
11. Incubate with shaking at 250 rpm for 16 hours at 37°C.

    To confirm suitable colony selection, streak them onto LB_{stg} plates containing Bluo-gal and IPTG and cultivate plates overnight at 37°C.
12. Transfer 1 ml of culture to a 1.5-ml tube. Centrifuge at 14,000*g* for 1 minute.
13. Remove the supernatant. Resuspend the cells in 300 μl of Boost solution 1 by gentle tituration.
14. Add 300 μl of Boost solution 2. Mix by inversion a few times. Incubate for 5 minutes at room temperature.
15. Slowly add 300 μl of Boost solution 3. Mix gently during addition. Incubate the mixture for 5 minutes on ice.

    A thick white precipitate of *E. coli* genomic DNA and proteins should form.
16. Centrifuge at 14,000*g* for 10 minutes.
17. Aliquot 0.8 ml of absolute isopropanol to new 2-ml tubes.
18. Gently transfer the supernatant to the tubes containing isopropanol. Avoid transferring the white precipitate. Mix gently by inverting the tubes several times. Incubate the tubes for 5 minutes on ice.
19. Centrifuge the sample at 14,000*g* for 15 minutes at 4°C.
20. Remove the supernatant. Add 1 ml of 70% ethanol to each tube. Wash the pellet by inverting the tube several times.
21. Centrifuge at 14,000*g* for 15 minutes at 16°C.

22. Remove the supernatant from the tube. Air-dry the pellet.

23. Dissolve the DNA in 40 μl of TE buffer by gently tapping the tube.

    The pellet can be left to dissolve overnight in a refrigerator. If not directly used for transfection, bacmid DNA can be stored at –20°C. Avoid repeated freeze-thaw cycles.

Transfection of Insect Cells with Recombinant Bacmid DNA to Generate Virus

24. Seed 1.5 x 10^6 *Sf*9 cells per each 35-mm well of a six-well plate in 2 ml of serum-free insect cell growth medium.

    Use cells in mid-log phase (1 x 10^6 to 2 x 10^6 cell/ml) with viability over 97%.

25. Allow cells to attach for 1 hour at 28°C.

26. Prepare the transfection solution.

    a. Dilute 5 μl of bacmid DNA (from Step 23) with insect cell growth medium to a final volume of 100 μl.

    b. Dilute 6 μl of thoroughly mixed Cellfectin transfection reagent with insect cell growth medium to a final volume of 100 μl.

    c. Combine the diluted bacmid DNA and transfection reagent. Mix gently. Incubate for 20–45 minutes at room temperature.

    d. For each transfection, add 0.8 ml of insect cell growth medium to the tubes containing the lipid-DNA complexes. Mix gently.

27. Remove media from cells. Overlay the 1 ml of diluted lipid-DNA complexes onto the cells. Incubate the cells for 5 hours at 28°C.

28. Remove the transfection mixture from the cells. Add 2 ml of insect cell growth medium. Incubate for 3–4 days at 28°C.

29. Transfer the supernatant from the cultured transfected cells to a sterile capped tube. Centrifuge at 500*g* for 5 minutes to pellet cell debris. Transfer the cleared virus-containing supernatant to a fresh tube.

    Primary viral stock can be stored in the dark for up to 6 months at 4°C, or for longer periods at –70°C. To maximize virus durability during storage, add fetal bovine serum to a final concentration of 2% to the primary viral stock.

30. Determine the viral titer by plaque assay or end point dilution procedures (O'Reilly et al. 1994).

Generation of Secondary Viral Stock

31. Infect a suspension culture of *Sf*9 cells (1 x 10^6 cells/ml in mid-log phase) with primary stock using an moi of 0.01–0.1.

    It is important to use a low moi in this step to avoid the accumulation of defective interfering particles. Routinely, 60 μl of primary stock leads to a suitable moi for preparing 30 ml of secondary stock.

32. Incubate the cells for 4 days at 28°C.

33. Centrifuge the culture at 1000*g* for 5 minutes. Transfer the supernatant to a fresh tube. Store protected from light at 4°C.

34. Determine the titer of the secondary viral stock.

    A good estimation for the secondary stock titer is $\geq 10^8$ pfu/ml.

Large-scale Viral Generation and Protein Production

35. Infect the desired volume of insect cells (1×10^6 cell/ml) with secondary stock using an moi of 0.01–0.1 (Step 36) or 5.0 (Step 37). Incubate the cells for 3–4 days at 28°C. Proceed to either Step 36 or Step 37.
36. Prepare the virus for gene-transfer experiments.
    a. Centrifuge the culture at 5000*g* for 20 minutes.
    b. Transfer the virus-containing supernatant to a sterile tube. Concentrate the virus (Airenne et al. 2000; Haeseleer et al. 2001).
    c. Determine the virus titer.
37. Recombinant protein production.
    a. Centrifuge the cells at 1000*g* for 20 minutes.
    b. Purify the expressed protein from medium or cells (or both), depending on the nature of the recombinant protein.

       The addition of a suitable affinity tag to the produced recombinant protein can significantly simply its purification (Arnau et al. 2006).

Protocol 2

Transduction of Vertebrate Cells with Recombinant Baculovirus

MATERIALS

CAUTION: See Appendix for appropriate handling of materials marked with <!>.

Reagents

Baculovirus (concentrated high-titer stock)
Cells (e.g., human liver cell line; HepG2), exponentially growing cultures
Cell culture growth medium (complete, serum-free, and selective)
Dulbecco's phosphate-buffered saline (PBS)
  2.7 mM KCl <!>
  1.5 mM KH_2PO_4 <!>
  136.9 mM NaCl
  8.9 mM $Na_2HPO_4 \cdot 7H_2O$ <!>
Sodium butyrate (1 M in PBS) (Sigma-Aldrich) (*optional*)
  Sterilize by filtration through a 0.2-μm filter.
Trypsin-versene (Cambrex) <!>

Equipment

Cell culture tubes
Centrifuge tubes (15 ml)
Pipettes (10 ml)
Tissue-culture dishes (six-well plate)

METHODS

This protocol is designed for cells grown in a six-well plate. If multiwell plates or dishes of different diameter are used, scale the cell density and reagent volumes accordingly.

Harvesting Exponentially Growing Cells

1. Remove the cell culture growth medium from the cells. Rinse once with 5–10 ml of sterile PBS.
2. Trypsinize the cells with approximately 1 ml of trypsin-versene per 90-mm culture dish.
3. Transfer the resulting cell suspension to a 15-ml centrifuge tube containing 4 ml of fresh growth medium. Centrifuge at 47–92*g* for 5–7 minutes at room temperature.
4. Remove the medium. Resuspend the cell pellet in 1–4 ml of fresh medium.
5. Determine cell density with a hemacytometer. Replate cells on six-well plates at a density of 1×10^5 to 2×10^5 cells.
6. Add 3 ml of growth medium per well. Incubate the cultures in a humidified incubator in an atmosphere of 5% CO_2 for 18–20 hours at 37°C.

   The exact density of cells on a plate depends on the size and growth rate of the cells and must be determined empirically. The best transduction efficiency will typically be achieved when cells are 50–75% confluent at the time of transduction.

Transduction

7. Replace the culture medium with 1 ml of prewarmed serum-free medium containing viruses at a desired moi. Coincubate the cells with the viruses for 2 hours in a humidified incubator in an atmosphere of 5% CO_2.

   Handle baculoviruses gently. Viruses do not stand for vigorous vortexing or pipetting.

8. Remove the virus-containing medium by aspiration. Add 3 ml of prewarmed serum-containing culture medium per well. Return plates to the incubator.

9. Examine the cells 24–72 hours after transduction.

TROUBLESHOOTING

A number of optional alterations to the incubation conditions can be used to enhance the efficiency of transduction of vertebrate cells with the baculovirus. Rather than using the concentrated viral stock in PBS (from Protocol 1, Step 36), the secondary high-titer viral stock (Protocol 1, Step 34) can be used diluted at least 1:1 in culture medium (Cheng et al. 2004; Hsu et al. 2004). Gently shaking the plates every 60 minutes during the incubation ensures even exposure of the cells to the virus (Step 7). Longer transduction times at lower temperatures (e.g., 25°C) combined with suitable transduction medium have also improved transduction results (Hsu et al. 2004). Transduction efficiency can also be improved by leaving the virus on the cells until they are examined. Additions to the growth medium such as microtubule depolymerizing agents (Salminen et al. 2005) or sodium butyrate to a final concentration of 2.5 to 5 mM (Condreay et al. 1999; Airenne et al. 2000) augment baculovirus-mediated gene delivery. Finally, due to the low cytotoxicity of baculoviruses, cells can be retransduced (i.e., superinfected) several times (Hu et al. 2003).

Transduction efficiency can also be enhanced by modifications made to the viruses during their construction. Recombinant baculoviruses having vesicular stomatitis virus glycoprotein (VSV-GED) or other envelope modifications can increase transduction efficacy (Barsoum et al. 1997; Ojala et al. 2001; Oker-Blom et al. 2003; Tani et al. 2003; Raty et al. 2004; Kitagawa et al. 2005; Kaikkonen et al. 2006). The choice of promoter can also make a difference. The chicken β-actin promoter has been shown to drive better transgene expression than cytomegalovirus in some cells (Shoji et al. 1997; Spenger et al. 2004). Cell-type-specific gene expression has also been achieved with tissue-specific promoters (Park et al. 2001; Li et al. 2004, 2005; McCormick et al. 2004).

REFERENCES

Airenne K.J., Mahonen A.J., Laitinen O.H., and Yla-Herttuala S. 2004. Baculovirus-mediated gene transfer: An evolving new concept. In *Gene and cell therapy*, 2nd edition (ed. N.S. Templeton), pp. 181–197. Marcel Dekker, New York.

Airenne K.J., Hiltunen M.O., Turunen M.P., Turunen A.M., Laitinen O.H., Kulomaa M.S., and Yla-Herttuala S. 2000. Baculovirus-mediated periadventitial gene transfer to rabbit carotid artery. *Gene Ther.* **7:** 1499–1504.

Aoki H., Sakoda Y., Jukuroki K., Takada A., Kida H., and Fukusho A. 1999. Induction of antibodies in mice by a recombinant baculovirus expressing pseudorabies virus glycoprotein B in mammalian cells. *Vet. Microbiol.* **68:** 197–207.

Arnau J., Lauritzen C., Petersen G.E., and Pedersen J. 2006. Current strategies for the use of affinity tags and tag removal for the purification of recombinant proteins. *Protein Expr. Purif.* **48:** 1–13.

Ayres M.D., Howard S.C., Kuzio J., Lopez-Ferber M., and Possee R.D. 1994. The complete DNA sequence of *Autographa californica* nuclear polyhedrosis virus. *Virology* **202:** 586–605.

Barsoum J., Brown R., McKee M., and Boyce F.M. 1997. Efficient transduction of mammalian cells by a recombinant baculovirus having the vesicular stomatitis virus G glycoprotein. *Hum. Gene Ther.* **8:** 2011–2018.

Blissard G.W. and Wenz J.R. 1992. Baculovirus gp64 envelope glycoprotein is sufficient to mediate pH-dependent membrane fusion. *J. Virol.* **66:** 6829–6835.

Boublik Y., Di Bonito P., and Jones I.M. 1995. Eukaryotic virus display: Engineering the major surface glycoprotein of the *Autographa californica* nuclear polyhedrosis virus (AcNPV) for the presentation of foreign proteins on the virus surface. *Biotechnology* **13:** 1079–1084.

Burgess S. 1977. Molecular weights of lepidopteran baculoviruses DNAs: Derivation by electron microscopy. *J. Gen. Virol.* **37:** 501–510.

Cheng T., Xu C.Y., Wang Y.B., Chen M., Wu T., Zhang J., and Xia N.S. 2004. A rapid and efficient method to express target genes in mammalian cells by baculovirus. *World J. Gastroenterol.* **10:** 1612–1618.

Condreay J.P., Witherspoon S.M., Clay W.C., and Kost T.A. 1999. Transient and stable gene expression in mammalian cells transduced with a recombinant baculovirus vector. *Proc. Natl. Acad. Sci.* **96:** 127–132.

Davies A.H. 1994. Current methods for manipulating baculoviruses. *Biotechnology* **12:** 47–50.

Duisit G., Saleun S., Douthe S., Barsoum J., Chadeuf G., and Moullier P. 1999. Baculovirus vector requires electrostatic interactions including heparan sulfate for efficient gene transfer in mammalian cells. *J. Gene Med.* **1:** 93–102.

Fraser M.J. 1986. Ultrastructural observations of virion maturation in *Autographa californica* nuclear polyhedrosis virus infected *Spodoptera frugiperda* cell cultures. *J. Ultrastruct. Mol. Struct. Res.* **95:** 189–195.

Ghosh S., Parvez M.K., Banerjee K., Sarin S.K., and Hasnain S.E. 2002. Baculovirus as mammalian cell expression vector for gene therapy: An emerging strategy. *Mol. Ther.* **6:** 5–11.

Gröner A. 1986. Specificity and safety of baculoviruses. In *The biology of baculoviruses* (ed. R.R. Granados and B.A. Federici), vol. 1, pp. 177–202. CRC Press, Boca Raton, Florida.

Haeseleer F., Imanishi Y., Saperstein D.A., and Palczewski K. 2001. Gene transfer mediated by recombinant baculovirus into mouse eye. *Invest. Ophthalmol. Vis. Sci.* **42:** 3294–3300.

Harrap K.A. 1972. The structure of nuclear polyhedrosis viruses. II. The virus particle. *Virology* **50:** 124–132.

Hofmann C., Sandig V., Jennings G., Rudolph M., Schlag P., and Strauss M. 1995. Efficient gene transfer into human hepatocytes by baculovirus vectors. *Proc. Natl. Acad. Sci.* **92:** 10099–10103.

Hsu C.S., Ho Y.C., Wang K.C., and Hu Y.C. 2004. Investigation of optimal transduction conditions for baculovirus-mediated gene delivery into mammalian cells. *Biotechnol. Bioeng.* **88:** 42–51.

Hu Y.C. 2005. Baculovirus as a highly efficient expression vector in insect and mammalian cells. *Acta Pharmacol. Sin.* **26:** 405–416.

Hu Y.C., Tsai C.T., Chang Y.J., and Huang J.H. 2003. Enhancement and prolongation of baculovirus-mediated expression in mammalian cells: focuses on strategic infection and feeding. *Biotechnol. Prog.* **19:** 373–379.

Hynes R.O. 1992. Integrins: Versatility, modulation, and signaling in cell adhesion. *Cell* **69:** 11–25.

Jones I. and Morikawa Y. 1996. Baculovirus vectors for expression in insect cells. *Curr. Opin. Biotechnol.* **7:** 512–516.

Kaikkonen M.U., Raty J.K., Airenne K.J., Wirth T., Heikura T., and Yla-Herttuala S. 2006. Truncated vesicular stomatitis virus G protein improves baculovirus transduction efficiency in vitro and in vivo. *Gene Ther.* **13:** 304–312.

Kidd I.M. and Emery V.C. 1993. The use of baculoviruses as expression vectors. *Appl. Biochem. Biotechnol.* **42:** 137–159.

Kitagawa Y., Tani H., Limn C.K., Matsunaga T.M., Moriishi K., and Matsuura Y. 2005. Ligand-directed gene targeting to mammalian cells by pseudotype baculoviruses. *J. Virol.* **79:** 3639–3652.

Kost T.A., Condreay J.P., and Jarvis D.L. 2005. Baculovirus as versatile vectors for protein expression in insect and mammalian cells. *Nat. Biotechnol.* **23:** 567–575.

Laitinen O.H., Airenne K.J., Hytonen V.P., Peltomaa E., Mahonen A.J., Wirth T., Lind M.M., Makela K.A., Toivanen P.I., Schenkwein D., Heikura T., Nordlund H.R., Kulomaa M.S., and Yla-Herttuala S. 2005. A multipurpose vector system for the screening of libraries in bacteria, insect and mammalian cells and expression in vivo. *Nucleic Acids Res.* **33:** e42.

Lehtolainen P., Tyynela K., Kannasto J., Airenne K.J., and Yla-Herttuala S. 2002. Baculoviruses exhibit restricted cell type specificity in rat brain: A comparison of baculovirus- and adenovirus-mediated intracerebral gene transfer in vivo. *Gene Ther.* **9:** 1693–1699.

Leisy D.J., Lewis T.D., Leong J.A., and Rohrmann G.F. 2003. Transduction of cultured fish cells with recombinant baculoviruses. *J. Gen. Virol.* **84:** 1173–1178.

Leusch M.S., Lee S.C., and Olins P.O. 1995. A novel host-vector system for direct selection of recombinant baculoviruses (bacmids) in *Escherichia coli*. *Gene* **160:** 191–194.

Li Y., Yang Y., and Wang S. 2005. Neuronal gene transfer by baculovirus-derived vectors accommodating a neurone-specific promoter. *Exp. Physiol* **90:** 39–44.

Li Y., Wang X., Guo H., and Wang S. 2004. Axonal transport of recombinant baculovirus vectors. *Mol. Ther.* **10:** 1121–1129.

Luckow V.A., Lee S.C., Barry G.F., and Olins P.O. 1993. Efficient generation of infectious recombinant baculoviruses by site-specific transposon-mediated insertion of foreign genes into a baculovirus genome propagated in *Escherichia coli*. *J. Virol.* **67:** 4566–4579.

McCormick C.J., Challinor L., Macdonald A., Rowlands D.J., and Harris M. 2004. Introduction of replication-competent hepatitis C virus transcripts using a tetracycline-regulable baculovirus delivery system. *J. Gen. Virol.* **85:** 429–439.

Miller L.K. 1993. Baculoviruses: High-level expression in insect cells. *Curr. Opin. Genet. Dev.* **3:** 97–101.

Ojala K., Mottershead D.G., Suokko A., and Oker-Blom C. 2001. Specific binding of baculoviruses displaying gp64 fusion proteins to mammalian cells. *Biochem. Biophys. Res. Commun.* **284:** 777–784.

Oker-Blom C., Airenne K.J., and Grabherr R. 2003. Baculovirus display strategies: Emerging tools for eukaryotic libraries and gene delivery. *Brief. Funct. Genomics Proteomics* **2:** 244–253.

Oomens A.G. and Blissard G.W. 1999. Requirement for GP64 to drive efficient budding of *Autographa californica* multicapsid nucleopolyhedrovirus. *Virology* **254:** 297–314.

O'Reilly D.R., Miller L.K., Luckow V.A. 1994. *Baculovirus expression vectors: A laboratory manual.* Oxford University Press, New York.

Park S.W., Lee H.K., Kim T.G., Yoon S.K., and Paik S.Y. 2001. Hepatocyte-specific gene expression by baculovirus pseudotyped with vesicular stomatitis virus envelope glycoprotein. *Biochem. Biophys. Res. Commun.* **289:** 444–450.

Pieroni L., Maione D., and La Monica N. 2001. In vivo gene transfer in mouse skeletal muscle mediated by baculovirus vectors. *Hum. Gene Ther.* **12:** 871–881.

Raty J.K., Airenne K.J., Marttila A.T., Marjomaki V., Hytonen V.P., Lehtolainen P., Laitinen O.H., Mahonen A.J., Kulomaa M.S., and Yla-Herttuala S. 2004. Enhanced gene delivery by avidin-displaying baculovirus. *Mol. Ther.* **9:** 282–291.

Salminen M., Airenne K.J., Rinnankoski R., Reimari J., Valilehto O., Rinne J., Suikkanen S., Kukkonen S., Yla-Herttuala S., Kulomaa M.S., and Vihinen-Ranta M. 2005. Improvement in nuclear entry and transgene expression of baculoviruses by disintegration of microtubules in human hepatocytes. *J. Virol.* **79:** 2720–2728.

Sandig V., Hofmann C., Steinert S., Jennings G., Schlag P., and Strauss M. 1996. Gene transfer into hepatocytes and human liver tissue by baculovirus vectors. *Hum. Gene Ther.* **7:** 1937–1945.

Shoji I., Aizaki H., Tani H., Ishii K., Chiba T., Saito I., Miyamura T., and Matsuura Y. 1997. Efficient gene transfer into various mammalian cells, including non-hepatic cells, by baculovirus vectors. *J. Gen. Virol.* **78:** 2657–2664.

Smith G.E., Summers M.D., and Fraser M.J. 1983. Production of human beta interferon in insect cells with a baculovirus expression vector. *Mol. Cell. Biol.* **3:** 2156–2165.

Song S.U., Shin S.H., Kim S.K., Choi G.S., Kim W.C., Lee M.H., Kim S.J., Kim I.H., Choi M.S., Hong Y.J., and Lee K.H. 2003. Effective transduction of osteogenic sarcoma cells by a baculovirus vector. *J. Gen. Virol.* **84:** 697–703.

Spenger A., Ernst W., Condreay J.P., Kost T.A., and Grabherr R. 2004. Influence of promoter choice and trichostatin A treatment on expression of baculovirus delivered genes in mammalian cells. *Protein Expr. Purif.* **38:** 17–23.

Summers M.D. and Anderson D.L. 1972. Granulosis virus deoxyribonucleic acid: A closed, double-stranded molecule. *J. Virol.* **9:** 710–713.

Tani H., Nishijima M., Ushijima H., Miyamura T., and Matsuura Y. 2001. Characterization of cell-surface determinants important for baculovirus infection. *Virology* **279:** 343–353.

Tani H., Limn C.K., Yap C.C., Onishi M., Nozaki M., Nishimune Y., Okahashi N., Kitagawa Y., Watanabe R., Mochizuki R., Moriishi K., and Matsuura Y. 2003. In vitro and in vivo gene delivery by recombinant baculoviruses. *J. Virol.* **77:** 9799–9808.

Vaughn J.L., Goodwin R.H., Tompkins G.J., and McCawley P. 1977. The establishment of two cell lines from the insect *Spodoptera frugiperda* (Lepidoptera; Noctuidae). *In Vitro* **13:** 213–217.

Wang P., Hammer D.A., and Granados R.R. 1997. Binding and fusion of *Autographa californica* nucleopolyhedrovirus to cultured insect cells. *J. Gen. Virol.* **78:** 3081–3089.

28 Alphaviruses: Semliki Forest Virus and Sindbis Virus as Gene Delivery Vectors

Kenneth Lundstrom

Flamel Technologies, 33 Avenue du Dr. Georges Lévy, 69693 Vénissieux, France

ABSTRACT

The alphaviruses Semliki Forest virus (SFV) and Sindbis virus (SIN) have been used frequently as expression vectors both in vitro and in vivo. In most cases, these systems consist of replication-deficient vectors that require a helper vector for packaging of recombinant particles. Recently, replication-proficient vectors have also been engineered. Alphaviral vectors can be used in various forms as nucleic-acid-based vectors (DNA and RNA) or infectious particles. High-titer viral production is achieved in less than 2 days and usually requires no further concentration or purification. The broad host range of alphaviruses facilitates studies in mammalian and nonmammalian cell lines, primary cells in culture, and in vivo. The strong preference for expression in neuronal cells has made alphaviruses particularly useful in neurobiological studies. Unfortunately, their strong cytotoxic effect on host cells, relatively short-term transient expression patterns, and the reasonably high cost of viral production remain drawbacks. However, novel mutant alphaviruses have demonstrated reduced cytotoxicity and prolonged expression. In particular, membrane proteins (which are generally difficult to express at high levels in recombinant systems) have generated high yields and facilitate applications in

Continued

structural biology. Alphaviruses have also been applied in vaccine development and gene therapy.

INTRODUCTION

Alphaviruses (Togaviridae) are represented by viruses with various degrees of pathogenicity to humans (Strauss and Strauss 1994). The SFV (Liljeström and Garoff 1991), SIN (Xiong et al. 1989), and Venezuelan equine encephalitis (VEE) virus (Davis et al. 1989) have all been developed into expression vectors (Fig. 1). Replication-deficient expression vectors (Fig. 1B) carry the alphaviral nonstructural protein genes (nsPs) and the gene of interest, whereas the structural proteins are provided from a helper vector in *trans* to generate replication-deficient particles. The replication complex (nsP1-4 proteins) generates approximately 200,000 RNA copies per cell (Strauss and Strauss 1994), and heterologous gene expression is driven by the subgenomic 26S promoter. Modifications of helper vectors make the generated recombinant SFV particles conditionally infectious (Berglund et al. 1993). Further safety improvements have been achieved by the split helper system, which eliminates homologous recombination (Smerdou and Liljeström 1999). In contrast, replication-competent vectors containing full-length genomes (Fig. 1C) have been engineered to provide continuous infection and production of viral progeny. In these vectors, a second subgenomic promoter is introduced downstream from the structural genes to drive expression of introduced foreign genes. For example, the replication-competent form of the nonpathogenic A7(74) strain demonstrated efficient transduction of pyramidal cells, interneurons, and glial cells in hippocampal slice cultures, as well as green fluorescent protein (GFP) expression in the central nervous system (CNS) after intraperitoneal injections (Vähä-Koskela et al. 2003). Finally, replacement of the SP6 RNA polymerase promoter with a eukaryotic polymerase-II-type promoter in both SFV (Berglund et al. 1996) and SIN (Dubensky et al. 1996) vectors generated DNA-based plasmid vectors (Fig. 1D) with no risk associated from infectious viral particles. Coinfection with a DNA-based helper vector allows viral production from DNA vectors (DiCiommo and Bremner 1998).

Alphaviruses are useful for many applications in molecular biology, neuroscience, drug discovery, structural biology, vaccine development, and gene therapy (Lundstrom et al. 2001; Lundstrom 2003c). Numerous G-protein-coupled receptors (GPCRs) and ligand-gated ion channels have been produced in large quantities for structural characterization using SFV vectors (Lundstrom 2003a). The mouse serotonin 5-HT_3 receptor was produced at 2–5 μg/ml in 11.5-liter bioreactors, purified to high homogeneity, and characterized by circular dichroism and single-particle imaging (Hovius et al. 1998). The human NK1 receptor was expressed in CHO cells in spinner flasks at 10 μg/ml, and receptor purification resulted in initial two-dimensional crystals (Lundstrom 2003b). Recently, a study on 101 GPCRs indicated that a large number of receptors could be expressed at structural biology-compatible levels (Hassaine et al. 2005).

The broad range of host cells has allowed gene delivery and expression studies from alphavirus vectors in many different types of cell lines and primary cells (Fig. 2; Table 1), especially in neurons (Olkkonen et al. 1993) and hippocampal slice cultures (Ehrengruber et al. 1999). Stereotactic injections of SIN and SFV into rodent brains showed local and transient expression of β-galactosidase

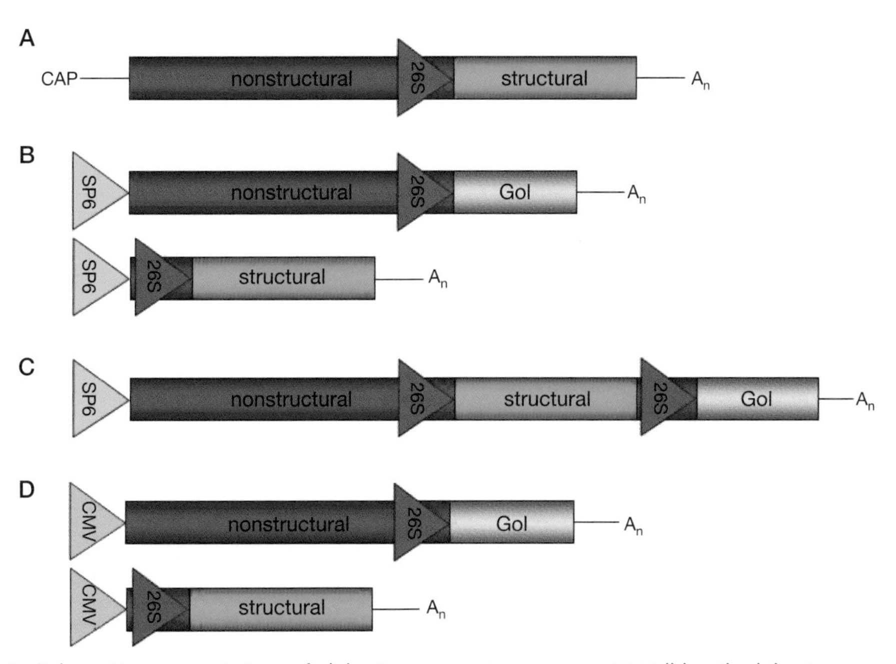

FIGURE 1. Schematic representations of alphavirus expression systems. (*A*) Full-length alphavirus genome. (*B*) Replication-deficient system with expression vector (*upper*) and helper vector (*lower*). (*C*) Replication-competent vector. (*D*) DNA-based system with expression vector (*upper*) and helper vector (*lower*). (GoI) Gene of interest; (CMV) cytomegalovirus promoter; (SP6) SP6 RNA polymerase promoter; (26S) alphavirus subgenomic promoter.

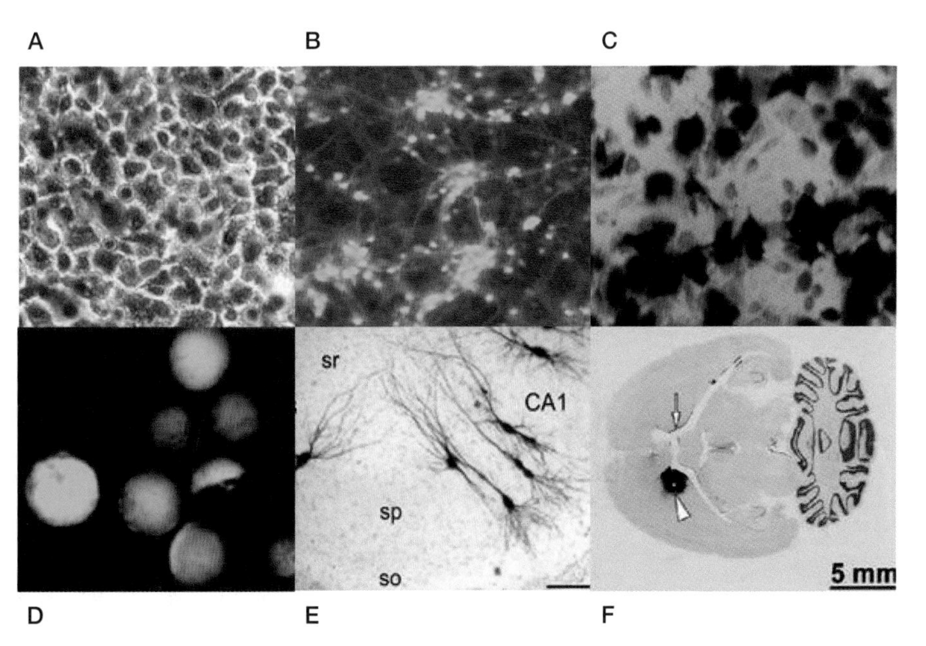

FIGURE 2. Examples of host range and expression patterns of SFV vectors. (*A*) BHK-21 cells infected with SFV-*lacZ*. (*B*) Primary rat hippocampal neurons infected with SFV-GFP. (*C*) Human prostate DU-145 cells infected with SFV-*lacZ*. (*D*) *Xenopus* oocytes infected with SFV-PD-GFP. (*E*) Rat hippocampal slice cultures infected with SFV-*lacZ*. (*F*) Stereotactic microinjection of SFV-*lacZ* into rat brain (1 day postinjection). (*E*, courtesy of Markus Ehrengruber; *F*, courtesy of Grayson J. Richards.)

TABLE 1. Host-cell Range of Alphaviral Vectors

Cell type	Transduction rate	Reference
Mammalian cell lines		
Acinar cells	low[a]	Arudchandran et al. (1999)
ALVA-31, human prostate tumor	high	Hardy et al. (2000)
BHK-21, baby hamster kidney	high	Liljeström and Garoff (1991)
CHO-K1, Chinese hamster ovary	high	Lundstrom et al. (1994)
C6, rat glioma	high	K. Lundstrom (unpubl.)
COS M6, green monkey kidney	high	Blasey et al. (1998)
COS-7, green monkey kidney	high	Liljeström and Garoff (1991)
C8166, human T lymphocyte	high	Hassaine et al. (2005)
DU-145, human prostate tumor	high	Hardy et al. (2000)
HeLa, human epithelial carcinoma	moderate	Liljeström and Garoff (1991)
HMC-1, human mast cell	low[a]	Arudchandran et al. (1999)
HOS, human osteosarcoma	high	Liljeström and Garoff (1991)
Hybridoma 179	moderate	Blasey et al. (1998)
JCA-1, human prostate tumor	high	Hardy et al. (2000)
Jurkat	low[a]	Arudchandran et al. (1999)
KU812, human basophil-like cell	low[a]	Arudchandran et al. (1999)
LnCAP, human prostate tumor	high	Hardy et al. (2000)
Mast cell, rat peritoneum	low[a]	Arudchandran et al. (1999)
MDCK, canine kidney	moderate	Olkkonen et al. (1993)
MME, mouse mammary epithelial	high	Olkkonen et al. (1993)
MOLT-4, lymphoblastic leukemia	low	Paul et al. (1993)
NIE 115, mouse neuroblastoma	moderate	Blasey et al. (1998)
OLF442, mouse olfactory epithelium	high	Monastyrskaia et al. (1999)
PC-3, human prostate tumor	high	Hardy et al. (2000)
PC12	moderate	K. Lundstrom (unpubl.)
PPC-1, human prostate tumor	high	Hardy et al. (2000)
Raji, human Burkitt lymphoma	moderate	Blasey et al. (1998)
RIN, renal	high	Lundstrom et al. (1997)
RPMI 8226, human myeloma	low	Blasey et al. (1998)
Swiss-3T3	low	Arudchandran et al. (1999)
TK6, human lymphoblast	high	Blasey et al. (1998)
TSU-PR1, human prostate tumor	moderate	Hardy et al. (2000)
V79, hamster lung fibroblast	high	Forsman et al. (2000)
WI-26 VA4, human lung fibroblast	high	Forsman et al. (2000)
Nonmammalian cell lines		
CEF, chicken embryo fibroblast	high	Xiong et al. (1989)
CHSE-214, salmon	high	K. Lundstrom (unpubl.)
C6/36 *Aedes albopictus* (insect)	high	Olson et al. (1994)
Grasshopper embryos	high	K. Lundstrom (unpubl.)
MOS-20, *Aedes aegypti* (insect)	high	Pettigrew and O'Neill (1999)
SL3, *Drosophila melanogaster* (insect)	moderate	K. Lundstrom (unpubl.)
Xenopus oocytes	high	K. Lundstrom (unpubl.)
Primary cells		
BAEC, bovine endothelial cells	low[a]	K. Lundstrom (unpubl.)
DRG, dorsal root ganglion	high	K. Lundstrom (unpubl.)
HPF, human primary fibroblast	high	Huckriede et al. (1996)
Myotubes, rat	high	K. Lundstrom (unpubl.)
NK cells	low[a]	Arudchandran et al. (1999)
NT2N, rat neurons	high	Cook et al. (1996)
PCN, primary rat cortical neurons	high	Olkkonen et al. (1993)
Prostate epithelial duct, biopsy	high	Hardy et al. (2000)
RBL-2H3, mast cell	low[a]	Arudchandran et al. (1999)
RHN, rat hippocampal neurons	high	Olkkonen et al. (1993)
Rat hippocampal slice cultures	high	Ehrengruber et al. (1999)
SCG, superior cervical ganglion	high	Ulmanen et al. (1997)

[a]Addition of polyethyleneglycol (PEG) is required.

in mouse caudatus putamen and nucleus accumbens septi (Altman-Hamamdzic et al. 1997) and rat amygdala and striatum (Lundstrom et al. 1999), respectively. Vaccine development using SFV, SIN, and VEE vectors in various formats (recombinant particles, naked RNA, and DNA plasmids) resulted in protection against challenges with lethal influenza (Schultz-Cherry et al. 2000) and Ebola viruses (Puschko et al. 2000) in animal models. Additionally, immunization has been carried out with tumor antigens to demonstrate both therapeutic and prophylactic efficacy in animal models (Colmenero et al. 1999).

Alphavirus vectors have also been applied in tumor therapy. Many human prostate tumor cell lines were efficiently transduced with SFV-*lacZ* particles, resulting in high β-galactosidase expression and SFV-induced apoptosis (Hardy et al. 2000). Tumor regression studies in mice with subcutaneously implanted H358a tumors led to substantial tumor regression (Murphy et al. 2000). In another study, SFV vectors carrying the therapeutic interleukin-12 (IL-12) subunits p40 and p35 were evaluated in a B16 tumor model, which demonstrated significant tumor regression and inhibition of tumor blood vessel formation (Asselin-Paturel et al. 1999). Additionally, intratumoral injections of recombinant SFV–IL-12 particles resulted in complete tumor regression in BALB/c mice implanted with K-BALB and CT26 xenografts (Chikkanna-Gowda et al. 2005). Recent findings from animal models suggest that systemically administered SIN–IL-12 particle vectors naturally target tumors (Tseng et al. 2004). In another approach, liposome-encapsulated SFV particles showed tumor targeting and protection against host immune system recognition in a pancreatic tumor model (Ren et al. 2003) and in a limited phase I clinical trial on kidney carcinoma and melanoma patients (Lundstrom and Boulikas 2003).

The inherent cytotoxicity to host cells has limited the application range of alphaviruses. However, several mutant vectors have been developed for both SFV (Lundstrom et al. 2001, 2003) and SIN (Agapov et al. 1998; Perri et al. 2000) to render them less cytotoxic. Two point mutations in the nsP2 gene of SFV generated a vector with higher transgene expression, less profound shutdown of host-cell protein synthesis, and substantially prolonged host-cell survival (Lundstrom et al. 2003). Many of the mutant alphavirus vectors have also demonstrated a temperature-sensitive phenotype that can be applied for expression studies only at the permissive temperature (Boorsma et al. 2000).

Protocol 1

Generation of Recombinant Alphaviral Vectors

The SFV- and SIN-based expression systems apply two vectors for recombinant particle production. Subcloning into the multiple cloning sites (MCS) of SFV and SIN expression vectors (Fig. 3) is done according to general procedures. However, it is advisable to initially clone polymerase chain reaction (PCR) fragments into other types of cloning vectors, such as the pCR4Blunt-TOPO vector (Invitrogen) because of the large size of the alphavirus vectors. The presence of inserts and their correct orientation can be analyzed by restriction endonuclease digestions. It is recommended that plasmid DNA Midipreps or Maxipreps be prepared for further procedures. Miniprep DNA can be subjected to in vitro transcription and successful viral production, although the quality and purity of DNA have crucial roles in obtaining high-titer recombinant alphaviral particles. In addition to the RNA-based vectors described below, DNA vectors with cytomegalovirus (CMV) or other RNA polymerase type II promoters can be used for direct plasmid DNA transfections. Cotransfection of SFV-based pSCA expression and pSCA helper vectors generates recombinant viral particles (DiCiommo and Bremner 1998).

MATERIALS

CAUTION: See Appendix for appropriate handling of materials marked with <!>.

Reagents

Agarose
Aprotinin (10 mg/ml) <!>
Baby hamster kidney cells (BHK-21)

Use adherent cultures of BHK-21 cells for generation of recombinant SFV and SIN particles and suspension cultures for large-scale protein production. Chinese hamster ovary cells (CHO-K1) and human embryonic kidney cells (HEK 293) can also be used for expression studies.

Cap analog ($m^7G[5']ppp[5']G$)
α-Chymotrypsin, 20 mg/ml
Complete BHK medium

This is Dulbecco's modified F-12 medium (GIBCO BRL) and Iscove's modified Dulbecco's medium (GIBCO BRL) (1:1) supplemented with 4 mM glutamate and 10% fetal calf serum (FCS).

Dithiothreitol (DTT; 50 mM) <!>
DMRIE-C (Invitrogen)
Ethanol (70% and 95% v/v) <!>
Opti-MEM I Reduced-Serum Medium (GIBCO BRL)
Phenol:chloroform:isoamyl alcohol (25:24:1, v/v/v) <!>
Plasmids
- pSINRep5 (Invitrogen)
- pSINrep504 (Frolov and Schlesinger 1994)
- SIN helper vector DH-BB
- SIN helper vector DH-BB(26S)5′SIN (Invitrogen)

All SIN and SIN helper vectors are linearized with XhoI.

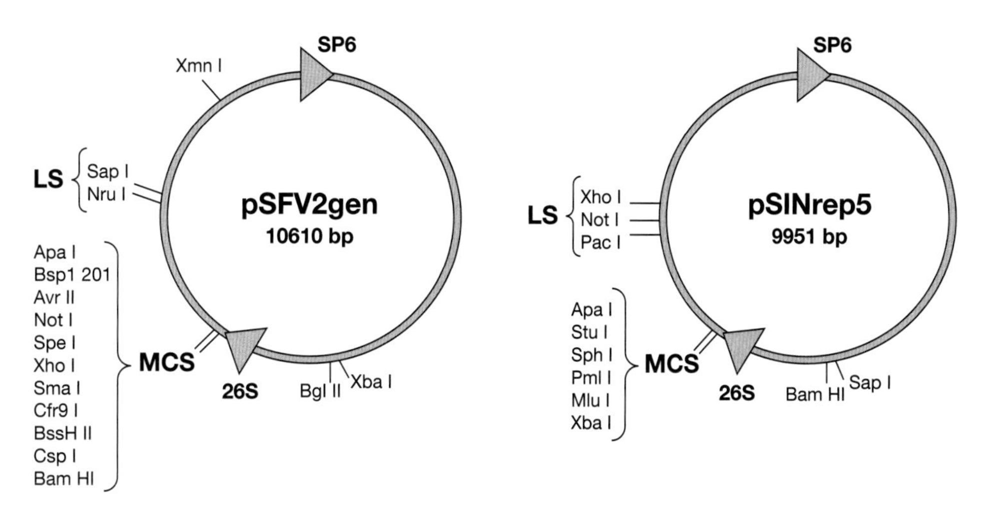

FIGURE 3. Plasmid maps of SFV and SIN expression vectors. (*A*) pSFV2gen and (*B*) pSINrep5 with indicated regions for MCS (multiple cloning sites) and LS (linearization sites). The LS cannot be used as cloning sites, as the region between the MCS and the LS contains the RNA replication and poly(A)$^+$ signals. The unique XmnI site at position 9929 in the SFV vector also cannot be used, as the RNA transcript becomes too long to function properly. The unique restriction sites BglII and XbaI (pSFV2gen) and BamHI and SapI (pSINrep5) can be used for determination of insert orientation.

pSFV1 (linearize with SpeI)
pSFV2gen (linearize with NruI)
pSFV-helper2 vector
pSFV-helperS2
pSFV-helperCS219A

Linearize all SFV helper vectors with SpeI.

Phosphate-buffered saline (PBS)
Restriction enzymes: NotI, NruI, PacI, SapI, SpeI, XhoI
RNase inhibitor (10–50 units/μl)
rNTP mix
10 mM rATP
10 mM rCTP
10 mM rUTP
5 mM rGTP
10x SP6 buffer
400 mM HEPES (pH 7.4)
60 mM magnesium acetate
20 mM spermidine <!>
15x SP6 buffer (commercially available)
SP6 RNA polymerase (10–20 units/μl)
Trypsin-EDTA
0.25% trypsin <!>
1 mM EDTA
Water, RNase-free

Equipment

Cell culture microwell plates (6, 12, and 24 well)
Electroporation cuvettes (0.2 and 0.4 cm)

TABLE 2. Electroporator Settings for BioRad Gene Pulser

	0.2-cm cuvette	0.4-cm cuvette
Capacitance extender	960 μF	960 μF
Voltage	1500 V	850 V
Capacitor	25 μF	25 μF
Resistance (pulse controller)	αΩ	disconnected
Expected time constant (tc)	0.8 sec	0.4 sec

Electroporator (BioRad Gene Pulser)

See Table 2 for settings. Consult manufacturer for appropriate conditions when using another electroporator.

Falcon tubes (15 and 50 ml)
Filters (sterile, 0.22 μm; Millipore)
Gel electrophoresis apparatus
Heating block
Incubator, 37°C, 5% CO_2
Microcentrifuge
Microcentrifuge tubes (1.5 ml)
Syringes (sterile, 1, 10, and 50 ml)
Tissue-culture flasks (T25, T75, and T175)
Water bath

METHODS

DNA Linearization

1. Linearize at least 5–10 μg of plasmid DNA (larger quantities can be stored at –20°C).
2. Purify linearized DNA by phenol:chloroform extraction followed by ethanol precipitation overnight at –20°C or 15 minutes at –80°C.
3. Centrifuge the ethanol precipitates at 18,000*g* for 15 minutes at 4°C.
4. Wash with 70% ethanol. Repeat centrifugation for 5 minutes.
5. Air-dry or lyophilize the DNA pellet.
6. Resuspend the pellet in RNase-free H_2O at a final concentration of 0.5 μg/μl.

   Confirm complete digestion by agarose gel electrophoresis.

In Vitro Transcription

The in vitro RNA transcription process is critical for obtaining high-titer viral stocks. It is recommended that fresh RNA preparations be made for electroporations, although RNA transcripts can be stored for weeks at –80°C. Under these conditions, approximately 20–50 μg of RNA should be obtained per transcription reaction. If large quantities of virus are needed, production can be scaled up by multiplying the volumes listed below and using aliquots for electroporations in parallel.

7. Prepare separate reaction mixtures (see Table 3) at room temperature for expression and helper vectors in sterile 1.5-ml microcentrifuge tubes. Add enzymes last.
8. Mix all reaction components and spin briefly in a microcentrifuge.
9. Incubate for 1 hour at 37°C (for SFV) or at 40°C (for SIN).
10. Remove 1–4-μl aliquots for RNA analysis on 0.8% agarose gels.

    High-quality RNA produces relatively thick bands with no smearing. The size of single-stranded RNA molecules cannot be compared directly to DNA markers as they migrate four

TABLE 3. Reaction Mixtures for SFV and SIN In Vitro Transcription

	SFV (μl)	SIN (μl)
Linearized plasmid DNA (2.5 μg)	5	5
10x SP6 buffer	5	–
15x SP6 buffer	–	10
10 mM Cap analog	5	5
50 mM DTT	5	5
rNTP mix	5	5
RNase-free H_2O	20	15
RNase inhibitor	1.5	1.5
SP6 RNA polymerase	3.5	3.5

times faster in gels. However, RNA from expression vectors has an approximate mobility of 8 kb (depending on insert size); helper RNA runs faster.

11. Use high-quality RNA immediately for transfection or store at –80°C for later use.

RNA Transfection

12. Transfect fresh or freshly thawed high-quality RNA into cells by electroporation or lipid-mediated transfection.

    If thawed RNA is used, reevaluate quality by gel electrophoresis before use.

    For electroporation of RNA

    BHK-21 cells are used preferentially for electroporation. Cells should not be passaged for more than 3 months, as they lose their viability. Cells prepared for electroporation should be plated less than 48 hours earlier and should not exceed 80% confluency.

    a. Plate 1:3 (overnight) or 1:5 (over two nights) sufficient quantities of BHK-21 cells in T175 flasks.

       Confirm cell growth and morphology prior to transfection.

    b. Wash cells once with PBS. Add 6 ml of trypsin-EDTA per T175 flask. Incubate for 5 minutes at 37°C.

    c. Resuspend cells to break up clumps. Add cell culture medium to 25 ml. Centrifuge at 800*g* for 5 minutes.

    d. Resuspend the cell pellet in a small volume (<5 ml) of PBS. Add PBS to 25 ml and recentrifuge at 800*g* for 5 minutes.

    e. Resuspend cells in approximately 2.5 ml of PBS per T175 flask (equivalent to 1×10^7 to 2×10^7 cells/ml).

       Use cells immediately for electroporation. Brief storage (<1 hour) on ice is acceptable.

    f. Transfer cell suspensions to electroporation cuvettes (0.4 ml in 0.2-cm cuvettes or 0.8 ml in 0.4-cm cuvettes).

    g. Add 20–45 μl of recombinant RNA and 20 μl of helper RNA (both prepared as described above, Steps 7–11) to the cell suspension.

    h. Place the cuvette in the electroporator holder. Apply two pulses.

    i. Dilute cells immediately 25-fold in cell culture medium. Transfer cells to cell culture flasks or plates.

    j. Incubate cells overnight at 37°C in an incubator with 5% CO_2.

       When using temperature-sensitive mutant alphavirus vectors, lower the culture temperature to 30°C or 33°C.

k. Harvest recombinant viral particles as described below (Steps 13–16).

For lipid-mediated transfection of RNA

As an alternative to electroporation, RNA can be transfected with lipofection reagents.

a. Plate 1.5×10^5 to 3×10^5 BHK-21 cells in 35-mm dishes or on six-well plates in 2 ml of medium.

   Culture the cells to approximately 80% confluency.

b. Wash cells with 2 ml of Opti-MEM I Reduced-Serum Medium.

c. To each of six 1.5-ml microcentrifuge tubes, add 1 ml of room-temperature Opti-MEM I reduced-serum medium.

d. Add 0, 3, 6, 9, 12, or 15 µl of DMRIE-C to the six tubes. Vortex briefly.

e. In a separate tube, mix 60 µl (~30 µg) of in-vitro-transcribed recombinant RNA and 30 µl (~12.5 µg) of helper RNA.

f. Add 15 µl of the RNA mixture to each of the six tubes containing the transfection mixture. Vortex briefly.

g. Add the lipid-RNA complexes immediately to the washed cells. Incubate for 4 hours at 37°C.

h. Replace the transfection medium with prewarmed (37°C) complete BHK medium.

i. Incubate cells overnight at 37°C in an incubator with 5% CO_2.

j. Harvest recombinant viral particles as described below (Steps 13–16).

Harvesting of Recombinant Viral Particles

Efficient production of recombinant alphaviral particles occurs within the first 24 hours. Generally, titers ranging from 10^8 to 10^9 infectious particles/ml are obtained. Extending the incubation time to 48 hours can result in slightly improved titers.

13. Examine virus-producing cells under a light microscope for contaminants and cytopathic effects.
14. Carefully aspirate and remove the medium from the BHK-21 cells.
15. Remove cell debris and possible contaminants by filter-sterilization through a 0.22-µm filter.
16. Store virus at –20°C (for weeks) or at –80°C (for years).

    Aliquot viral stocks to avoid repeated cycles of freezing/thawing, which can reduce the titers significantly.

Activation of Recombinant Particles

Viruses produced with conventional SFV and SIN helper vectors as well as with the split helper system (Smerdou and Liljeström 1999) generate directly infectious particles. However, use of the pSFV-Helper2 vector renders the produced SFV particles only conditionally infectious (Berglund et al. 1993). In this case, an additional activation procedure is required.

17. Add α-chymotrypsin to the viral stocks to a final concentration of 500 µg/ml. Incubate for 20 minutes at room temperature.
18. Add aprotinin to a final concentration of 250 µg/ml to stop the reaction.

    The activated SFV particles are now ready for infection.

Protocol 2

Purification Methods

Generally, purification or concentration of alphaviruses is not necessary. However, for instance, the medium derived from the BHK cells is toxic to primary neurons in culture. Including a purification step substantially improves the survival of the transduced neurons. Viral concentration and purification may also be advantageous for in vivo studies in animal models and are mandatory for any clinical applications.

MATERIALS

CAUTION: See Appendix for appropriate handling of materials marked with <!>.

Reagents

Absorption buffer
- 10 mM phosphate buffer (pH 7.5)
- 100 mM NaCl

Affinity chromatography resin (Matrex Cellufine sulfate; Millipore)
Elution buffer (1–2 M NaCl or KCl <!>)
Sucrose (20% and 50% w/v in H_2O)
TNE buffer
- 50 mM Tris-HCl (pH 7.4) <!>
- 100 mM NaCl
- 0.1 mM EDTA

Equipment

Centrifuge
Centriprep concentrators (Millipore)
Ultracentrifuge tubes
Ultracentrifuge with SW40 Ti, SW41 Ti, or SW28 rotor

METHODS

1. Prepare (and if necessary, activate) the viral stock as described (Protocol 1).
2. Purify and concentrate the viral stock.

For sucrose step-gradient ultracentrifugation

a. Pipette 1 ml of 50% sucrose into an ultracentrifuge tube.

b. Carefully overlay with 3 ml of 20% sucrose to establish a step-gradient.

c. Slowly add the viral stock (8 ml for SW41 Ti tubes or 9 ml for SW40 Ti tubes) on top of the sucrose gradient.

d. Centrifuge at 160,000*g* for 90 minutes at 4°C.

The virus will band close to the interface between the 20% and 50% sucrose layers.

e. Collect the virus by removing the medium fraction and the bottom 0.8 ml (50% sucrose).

Alternatively, viral stocks can be pelleted through a 20% sucrose cushion. Centrifuge in an SW28 rotor at 25,000 rpm for 2 hours at 4°C. Resuspend the pellet in 400 µl of TNE buffer.

For Centriprep concentration

Centriprep concentrators can be used for rapid and simple viral stock concentration.

a. Add the viral sample to the sample container of the Centriprep concentrator according to the manufacturer's instructions.

b. Centrifuge the assembled concentrator at the manufacturer's recommended speed until the fluid levels inside and outside the filtrate collector equilibrate.

c. Remove the Centriprep concentrator from the centrifuge. Snap off the air-tight seal cap. Decant the filtrate.

d. Replace the cap. Recentrifuge the concentrator.

e. Decant the filtrate. Remove the filtrate collector.

f. Collect the concentrated virus.

If further virus concentration is required, additional centrifugations are possible.

For affinity chromatography concentration

Affinity chromatography removes endotoxins and other contaminants as well as concentrating the virus approximately tenfold.

a. Pack the affinity chromatography column with Matrex Cellufine sulfate resin according to the manufacturer's instructions.

b. Equilibrate the column with several bed-volumes of the absorption buffer.

c. Load the viral sample at pH 7.5.

d. Wash the column with several bed-volumes of adsorption buffer to remove nonbinding contaminants.

e. Elute the virus with the elution buffer.

f. Collect several fractions. Determine fractions with peak viral concentration by test infections.

Protocol 3

Verification of Viral Titers

Prior to use in vitro or in vivo, it is essential to determine the titer of the generated alphaviral particles. As defective alphaviruses do not produce plaques, their titers cannot be determined by conventional methods. However, viral titers can be determined readily in cases where the recombinant viruses express reporter genes such as GFP or β-galactosidase, as well as indirectly by immunofluorescence methods. The potency of viral stocks can also be evaluated by light microscopic analysis. Alphavirus-infected cells show a dramatic decrease in growth and can be easily distinguished from noninfected control cells through their rounded morphology (Fig. 4).

MATERIALS

CAUTION: See Appendix for appropriate handling of materials marked <!>.

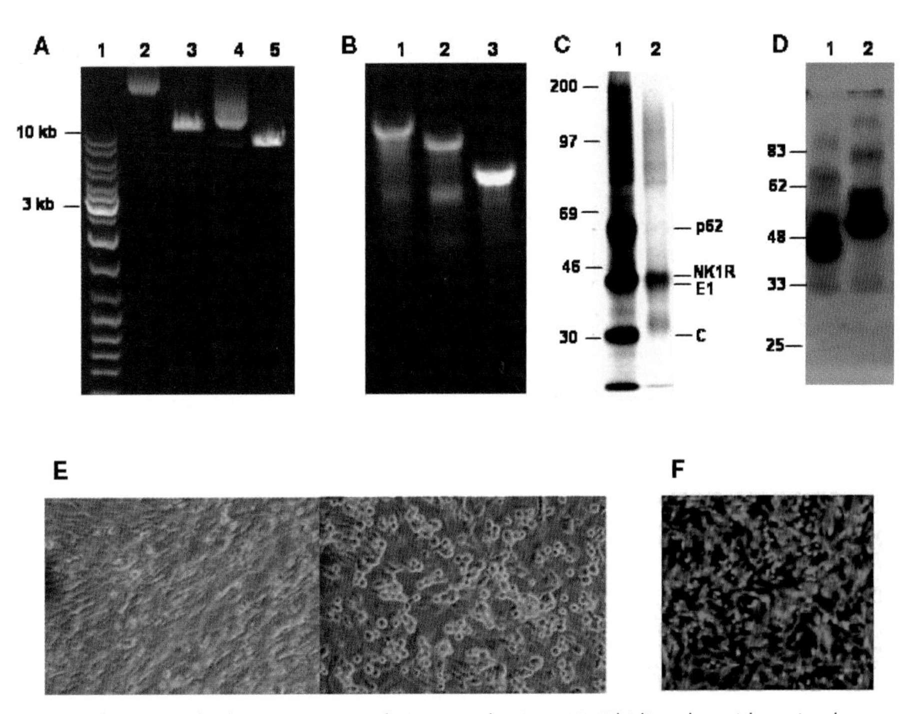

FIGURE 4. Quality control of various steps of virus production. (*A*) Ethidium-bromide-stained agarose gel of linearized expression and helper vectors. (*Lane 1*) 1-kb DNA ladder; (*lane 2*) pSFV-*lacZ*, undigested; (*lane 3*) SpeI-digested pSFV-*lacZ*; (*lane 4*) pSFV-Helper2, undigested; (*lane 5*) SpeI-digested pSFV-Helper2. (*B*) Ethidium-bromide-stained agarose gel of in-vitro-transcribed RNA. (*lane 1*) pSFV-CAP-NK1R (recombinant SFV with the neurokinin-1 receptor fused to the capsid sequence); (*lane 2*) pSFV1-GFP; (*lane 3*) pSFV-Helper2. (*C*) Metabolic labeling of electroporated (*lane 1*) and infected BHK-21 (*lane 2*) cells. (*D*) Western blot of infected BHK-21 cells (*lane 1*) pSFV2genB-5HT1BR; (*lane 2*) pSFV2genC-5HT1BR. (*E*) Morphological changes observed between infected cells (*right panel*) and noninfected control cells (*left panel*). (*F*) Fluorescence microscopy of SFV-GFP-infected cells.

Reagents

Antibodies
- primary antibody (target-antigen specific)
- secondary antibody, fluorescein-conjugated (primary-antibody specific)

BHK-21 cells

Depending on the study, other cells lines of interest can be used.

Blocking buffer: PBS containing 0.5% gelatin and 0.25% bovine serum albumin

Complete medium

As appropriate for the cell line used.

DABCO (1,4-diazabicyclo-[2.2.2]-octane) <!>

Methanol <!>

$MgCl_2$ (1 M) <!>

Store at room temperature.

Moviol 4–88

Phosphate-buffered saline (PBS)

Potassium ferricyanide (50 mM) <!>

Store at 4°C.

Potassium ferrocyanide (50 mM) <!>

Store at 4°C.

X-gal (2%) <!> in DMF <!> or DMSO <!>

Store at –20°C.

Virus stock

Recombinant alphaviruses expressing GFP, β-galactosidase, or the primary antigen of interest are prepared as described in Protocol 1.

Equipment

Cell culture microwell plates (6 or 12 well)

Coverslips

Fluorescence microscope

Glass slides

METHODS

Visualization of Gene Expression

1. Plate the cell line of interest in 6- or 12-well plates or directly on coverslips. Culture cells to a defined concentration.
2. Infect cells with serial dilutions (e.g., twofold dilutions in a range expected to give 20–50 positive cells per microscope field) of viral stock expressing the reporter gene or target antigen.
3. Culture cells overnight at 37°C.
4. Determine titers no later than 48 hours postinfection.

   At earlier time points, the signal may be suboptimal, especially for some mutant vectors. At later time points, the cytopathic effects result in increased cell detachment.

To determine titer by GFP detection

a. Examine wells or coverslips with a fluorescence microscope.

b. Count the number of GFP-positive cells (infectious particles).

To determine titer by X-gal staining

a. Prepare the X-gal staining solution fresh from individual stocks to final concentrations of 5 mM potassium ferricyanide, 5 mM potassium ferrocyanide, 2 mM $MgCl_2$, and 1 mg/ml X-gal in PBS.

b. Wash cells with PBS.

c. Fix in cold methanol for 5 minutes at –20°C.

d. Wash cells three times with PBS.

e. Stain cells in X-gal staining solution for at least 2 hours at 37°C or room temperature.

f. Count the X-gal-positive (i.e., blue) cells.

To determine titer by immunofluorescence

a. Rinse coverslips twice with PBS.

b. Fix in methanol for 6 minutes at –20°C.

c. Wash coverslips three times with PBS.

d. Incubate in blocking buffer for 30 minutes at room temperature to block nonspecific binding.

e. Replace the blocking buffer with primary antibody in blocking buffer. Incubate for 30 minutes at room temperature.

f. Wash coverslips three times with PBS. Incubate with secondary antibody for 30 minutes at room temperature.

g. Wash three times with PBS. Wash once with H_2O. Air-dry coverslips.

h. Mount coverslips on glass slides using 10 μl of Moviol 4-88 containing 2.5% DABCO.

i. Examine slides with a fluorescence microscope. Count the number of positive cells.

5. Multiply the number of positive cells by the dilution used. Express the titer as infectious particles/ml.

Protocol 4

Evaluation of Gene Expression in Cultured Cell Lines

Before any more detailed and large-scale expression is conducted, it is useful to rapidly evaluate the transgene expression from generated viral stocks. This approach also allows determination of the best expression conditions. Viral concentration, host cell line, and time of cell harvest (among other factors) have important roles in expression optimization. Expression levels and size of the gene product can be obtained relatively easily by western blotting. However, this requires available antibodies against the target protein or introduction of tags in the vector constructs. If neither of these criteria are available, transgene expression can be verified by metabolically labeling cells with [^{35}S]methionine.

MATERIALS

CAUTION: See Appendix for appropriate handling of materials marked with <!>.

Reagents

Acetic acid <!>
Amplify (Amersham) <!>
Antibodies
 primary antibody (target-antigen specific)
 secondary antibody (primary-antibody specific)
Cell line of interest: BHK-21, CHO-K1, HEK 293
Chase medium: Eagle's minimum essential medium (MEM) supplemented with 2 mM glutamine, 20 mM HEPES, and 150 μg/ml unlabeled methionine
ECL Chemiluminescence Kit (Amersham)
Lysis buffer
 50 mM Tris-HCl (pH 7.6) <!>
 150 mM NaCl
 2 mM EDTA
 1% (v/v) Nonidet P-40 (Sigma-Aldrich) <!>
Methanol <!>
[^{35}S]Methionine <!>
Milk
Phosphate-buffered saline (PBS)
Polyacrylamide <!>
Sodium dodecyl sulfate (SDS) <!>
Starvation medium: MEM (methionine-free) supplemented with 2 mM glutamine and 20 mM HEPES
Tris-buffered saline (TBS)
Tween-20
Virus stock
 Recombinant alphaviruses expressing the gene of interest are prepared as described in Protocol 1.

Equipment

Cell culture microwell plates (6, 12, or 24 well)
Hybond ECL nitrocellulose filter (Amersham)
Hyperfilm-MP
Radioactivity-intensifying screens
SDS-PAGE apparatus
Western blotting apparatus

METHODS

1. Infect appropriate cells (BHK-21, CHO-K1, HEK 293) cultured on 6-, 12-, or 24-well plates with serial dilutions of viral stocks.
2. Analyze gene expression.

   For western blotting

   a. Add 250 μl of lysis buffer per six-well plate at various times postinfection. Incubate for 10 minutes on ice.
   b. Detach the cells by resuspension.
   c. Load the samples and electrophorese on a 10–12% SDS-polyacrylamide gel under standard conditions.
   d. Electrotransfer the proteins from the gel to a nitrocellulose filter.
   e. Incubate the filter with TBS containing 5% milk and 0.1% Tween-20 for 30 minutes at 4°C.
   f. Incubate the filter with the primary antibody for 30 minutes at room temperature.
   g. Incubate the filter with the secondary antibody for 30 minutes at room temperature.
   h. Visualize specific bands with the ECL Chemiluminescence Kit according to the manufacturer's instructions.

   For metabolic labeling

   a. Remove medium from cells. Wash once with PBS.
   b. Add starvation medium. Incubate for 30 minutes at 37°C.
   c. Replace medium with 50–100 μCi/ml of [^{35}S]methionine in starvation medium.
   d. Incubate for 20 minutes at 37°C.
   e. Remove medium. Wash twice with PBS.
   f. Add chase medium. Incubate for 15 minutes to 3 hours.
   g. Remove chase medium. Wash once with PBS.
   h. Add 250 μl of lysis buffer per six-well plate. Incubate for 10 minutes on ice.
   i. Detach the cells by thorough resuspension.
   j. Load the samples and electrophorese on a 10–12% SDS-polyacrylamide gel under standard conditions.

k. Fix the gel in 10% acetic acid/30% methanol for 30 minutes at room temperature.

l. Replace the fixation solution with Amplify. Incubate for 30 minutes at room temperature.

m. Dry the gel. Apply radioactivity-intensifying screens and expose gel on Hyperfilm-MP for 2–24 hours (depending on signal intensity) at room temperature or at –80°C.

n. Develop film to visualize radioactive bands.

Infection of Primary Cells

Primary cells (and especially primary neurons) are sensitive to physical and chemical exposure. Avoid any extra handling of the cells. As the medium from BHK cells can present toxicity to neurons, purification and concentration procedures for the viral stocks are recommended.

1. Isolate primary rat hippocampal neurons from E17 stage embryos (e.g., from rat strain ROROspf120; BRL, Fullingsdorf, Switzerland).
2. Culture in DMEM (GIBCO BRL) supplemented with 10% horse serum.
3. Add appropriate (i.e., the lowest possible) virus concentration directly to the primary cells.
4. Avoid any unnecessary manipulation of cells.
5. Perform time-course studies to determine the optimal expression conditions.

Protocol 5

Delivery of Virus into Hippocampal Slice Tissue Culture

Gene delivery to hippocampal slice cultures requires special techniques and skills (Lundstrom and Ehrengruber 2003). As the organotypic slices are covered by a layer of glial cells that hinders the penetration of viral particles to the neurons, it is recommended that viral particles be injected manually into the extracellular space of the tissue (Ehrengruber et al. 1999).

MATERIALS

CAUTION: See Appendix for appropriate handling of materials marked with <!>.

Reagents

Cutting medium
- 10 mM $MgCl_2$ <!>
- 0.5 μM tetrodotoxin <!>

Rat hippocampal slice cultures

Organotypic slice cultures are prepared from rat hippocampus according to the roller-tube configuration (Gähwiler 1981).

Virus stock

Recombinant alphaviruses expressing the gene of interest are prepared as described in Protocol 1.

Equipment

Micromanipulator

Micropipettes

Prepare the micropipettes from glass capillaries on an electrode puller. Sterilize by autoclaving prior to use.

METHOD

1. Dilute viral stocks 10–1000-fold in cutting medium to allow identification of individual neurons.
2. Back-fill an autoclaved micropipette with 20–30 μl of diluted virus. Use a micromanipulator to inject into the pyramidal and granule cell layer of the tissue slice.
3. Move the micropipette into a neighboring site. Repeat injections 10–15 times per slice.
4. Incubate the slice in culture medium for 10–60 seconds to remove tetrodotoxin and any residual viral particles.
5. Incubate the slice cultures for 1–3 days for expression analysis.

Protocol 6

In Vivo Administration of Virus into Rodents

Alphaviruses have been used for stereotactic microinjections in rodents (Altman-Hamamdzic et al. 1997; Lundstrom et al. 1999). Administration was carried out without any purification or concentration of viral stocks. However, filter-sterilization is recommended to ensure that cell debris or other contaminants are not present.

MATERIALS

CAUTION: See Appendix for appropriate handling of materials marked with <!>.

Reagents

Animals: mouse or rat strain of choice
BHK-21 cells
Ketamine
Virus stock

Recombinant alphaviruses expressing the gene of interest are prepared as described in Protocol 1. Sterilize by filtration through a 0.22-μm filter before use.

Xylazine <!>

Equipment

Filters (sterile, 0.22 μm; Millipore)
Hamilton syringe (10 μl)
Injectors (stainless steel)
Microinfusion pump
Polyethylene tubing
Stereotactic equipment

METHOD

Conduct microinjections under general anesthesia with thermoregulatory control and oxygen supplementation.

1. Determine approximate titers of viral stocks in BHK-21 cells as described in Protocol 3.
2. Dilute virus appropriately for the experimental conditions.

   Use approximately 10^5 particles for infusion into striatum, amygdala, or cortex, or 10^7 particles for infusion into ventricles.
3. Anesthetize experimental subject by intraperitoneal injection of 200 mg of ketamine/10 mg xylazine per kilogram of body weight.
4. Secure subject in stereotactic device.
5. Apply virus using stainless steel injectors. Use a microinfusion pump connected via polyethylene tubing to a 10-μl Hamilton syringe.

6. Infuse viral solution at a rate of 0.5 µl/minute for 2 minutes.
7. Leave the needle in place for an additional 2 minutes.
8. Remove the needle slowly (i.e., over 1 minute).
9. Suture the wound. Keep animals warm after surgery for 3–4 hours.
10. Study general health (e.g., body weight, body temperature, and food intake) and behavioral effects (e.g., sensory and/or motor function, coordination, and exploratory behavior) relative to control animals.
11. At various time points postinjection, harvest tissue and determine transgene expression.

ACKNOWLEDGMENTS

Dr. Markus Ehrengruber (Brain Research Institute, Zurich, Switzerland) and Dr. Grayson J. Richards (F.Hoffmann-La Roche, Basel, Switzerland) are acknowledged for Figures 2E and 2F, respectively.

REFERENCES

Agapov E.V., Frolov I., Lindenbach B.D., Pragai B.M., Schlesinger S., and Rice C.M. 1998. Noncytopathic Sindbis virus RNA vectors for heterologous gene expression. *Proc. Natl. Acad. Sci.* **95:** 12989–12994.

Altman-Hamamdzic S., Groseclose C., Ma J.X., Hamamdzic D., Vrindavanam N.S., Middaugh L.D., Parratto N.P., and Sallee F.R. 1997. Expression of β-galactosidase in mouse brain: Utilization of a novel nonreplicative Sindbis virus vector as a neuronal gene delivery system. *Gene Ther.* **4:** 815–822.

Arudchandran R., Brown M.J., Song J.S., Wank S.A., Haleem-Smith H., and Rivera J. 1999. Polyethylene glycol-mediated infection of non-permissive mammalian cells with Semliki Forest virus: Application to signal transduction studies. *J. Immunol. Methods* **222:** 197–208.

Asselin-Paturel C., Lassau N., Guinebretiere J.M., Zhang J., Gay F., Bex F., Hallez S., Leclere J., Peronneau P., Mami-Chouaib F., and Chouaib S. 1999. Transfer of the murine interleukin-12 gene in vivo by a Semliki Forest virus vector induces B16 tumor regression through inhibition of tumor blood vessel formation monitored by Doppler ultrasonography. *Gene Ther.* **6:** 606–615.

Berglund P., Tubulekas I., and Liljeström P. 1996. Alphaviruses as vectors for gene delivery. *Trends Biotechnol.* **14:** 130–134.

Berglund P., Sjöberg M., Garoff H., Atkins G.J., Sheahan B.J., and Liljeström P. 1993. Semliki Forest virus expression system: Production of conditionally infectious recombinant particles. *Biotechnology* **11:** 916–920.

Blasey H.D., Brethon B., Hovius R., Lundström K., Rey L., Vogel H., Tairi A.-P., and Bernard A.R. 1998. Large scale transient 5-HT3 receptor production with the Semliki Forest Virus expression system. In *New developments and new applications in animal cell technology* (ed. O.W. Mertenet et al.), pp. 449–455. Kluwer Academic Publishers, Dordrecht.

Boorsma M., Nieba L., Koller D., Bachmann M.F., Bailey J.E., and Renner W.A. 2000. A temperature-regulated replicon-based DNA expression system. *Nat. Biotechnol.* **18:** 429–432.

Chikkanna-Gowda C.P., Sheahan B.J., Fleeton M.N., and Atkins G.J. 2005. Regression of mouse tumours and inhibition of metastases following administration of a Semliki Forest virus vector with enhanced expression of IL-12. *Gene Ther.* **12:** 1253–1263.

Colmenero P., Liljeström P., and Jondal M. 1999. Induction of P815 tumor immunity by recombinant Semliki Forest virus expressing the P1A gene. *Gene Ther.* **6:** 1728–1733.

Cook D.G., Sung J.C., Golde T.E., Felsenstein K.M., Wojczyk B.S., Tanzi R.E., Trojanowski J.Q., Lee V.M., and Doms R.W. 1996. Expression and analysis of presenilin 1 in a human neuronal system: Localization in cell bodies and dendrites. *Proc. Natl. Acad. Sci.* **93:** 9223–9228.

Davis N.L., Brown K.W., and Johnston R.E. 1989. In vitro synthesis of infectious Venezuelan equine encephalitis virus RNA from a cDNA clone: Analysis of a viable deletion mutant. *Virology* **171:** 189–204.

DiCiommo D.P. and Bremner R. 1998. Rapid, high level protein production using Semliki Forest virus vectors. *J. Biol. Chem.* **273:** 18060–18066.

Dubensky T.W., Jr., Driver D.A., Polo J.M., Belli B.A., Latham E.M., Ibanez C.E., Chada S., Brumm D., Banks T.A., Mento S.J., Jolly D.S., and Chang S.M.W. 1996. Sindbis virus DNA-based vectors: Utility for in vitro and in vivo gene transfer. *J. Virol.* **70:** 508–519.

Ehrengruber M.U., Lundstrom K., Schweitzer C., Heuss C., Schlesinger S., and Gähwiler B.H. 1999. Recombinant Semliki Forest virus and Sindbis virus efficiently infect neurons in hippocampal slice cultures. *Proc. Natl. Acad. Sci.* **96:** 7041–7046.

Forsman T., Lautala P., Lundstrom K., Monastyrskaia K., Ouzzine M., Burchell B., Taskinen J., and Ulmanen I. 2000. Production of human UDP-glucuronosyltransferases 1A6 and 1A9 using the Semliki Forest virus expression system. *Life Sci.* **67:** 2473–2484.

Frolov I. and Schlesinger S. 1994. Translation of Sindbis virus mRNA: Effects of sequences downstream of the initiating codon. *J. Virol.* **68:** 8111–8117.

Gähwiler B.H. 1981. Organotypic monolayer cultures of nervous tissue. *J. Neurosci. Methods* **4:** 329–342.

Hardy P.A., Mazzini M.J., Schweitzer C., Lundstrom K., and Glode L.M. 2000. Recombinant Semliki forest virus infects and kills human prostate cancer cell lines and prostatic duct epithelial cells ex vivo. *Int. J. Mol. Med.* **5:** 241–245.

Hassaine G., Wagner R., Kempf J., Cherouati N., Hassaine N., Prual C., André N., Reinhart C., Pattus F., and Lundstrom K. 2005. Semliki Forest virus vectors for overexpression of 101 G protein-coupled receptors in mammalian host cells. *Protein Expr. Purif.* **45:** 343–351.

Hovius R., Tairi A.-P., Blasey H., Bernard A., Lundstrom K., and Vogel H. 1998. Characterization of a mouse 5-HT3 receptor purified from mammalian cells. *J. Neurochem.* **70:** 824–834.

Huckriede A., Heikema A., Wilschut J., and Agsteribbe E. 1996. Transient expression of a mitochondrial precursor protein. A new approach to study mitochondrial protein import in cells of higher eukaryotes. *Eur. J. Biochem.* **237:** 288–294.

Liljeström P. and Garoff H. 1991. A new generation of animal cell expression vectors based on the Semliki Forest virus replicon. *Biotechnology* **9:** 1356–1360.

Lundstrom K. 2003a. Semliki Forest virus vectors for rapid and high-level expression of integral membrane proteins. *Biochim. Biophys. Acta* **1610:** 90–96.

———. 2003b. Semliki Forest virus vectors for large-scale protein production. *Methods Mol. Med.* **76:** 525–543.

———. 2003c. Alphavirus vectors for vaccine production and gene therapy. *Exp. Rev. Vaccines* **2:** 447–459.

Lundstrom K. and Boulikas T. 2003. Non-viral and viral vectors for gene therapy. *Technol. Cancer Res. Treat.* **2:** 471–485.

Lundstrom K. and Ehrengruber M.U. 2003. Semliki Forest virus (SFV) vectors in neurobiology and gene therapy. *Methods Mol. Med.* **76:** 503–523.

Lundstrom K., Pralong W., and Martinou J.C. 1997. Anti-apoptotic effect of Bcl-2 overexpression in RIN cells infected with Semliki Forest virus. *Apoptosis* **2:** 189–91.

Lundstrom K., Abenavoli A., Malgaroli A., and Ehrengruber M.U. 2003. Novel Semliki Forest virus vectors with reduced cytotoxicity and temperature sensitivity for long-term enhancement of transgene expression. *Mol. Ther.* **7:** 202–209.

Lundstrom K., Richards J.G., Pink J.R., and Jenck F. 1999. Efficient *in vivo* expression of a reporter gene in rat brain after injection of recombinant replication-deficient Semliki Forest virus. *Gene Ther. Mol. Biol.* **3:** 15–23.

Lundstrom K., Mills A., Buell G., Allet E., Adami N., and Liljeström P. 1994. High-level expression of the human neurokinin-1 receptor in mammalian cell lines using the Semliki Forest virus expression system. *Eur. J. Biochem.* **224:** 917–921.

Lundstrom K., Schweitzer C., Rotmann D., Hermann D., Schneider E.M., and Ehrengruber M.U. 2001. Semliki Forest virus vectors: Efficient vehicles for in vitro and in vivo gene delivery. *FEBS Lett.* **504:** 99–103.

Monastyrskaia K., Goepfert F., Hochstrasser R., Acuna G., Leighton J., Pink J.R., and Lundstrom K. 1999. Expression and intracellular localisation of odorant receptors in mammalian cell lines using Semliki Forest virus vectors. *J. Recept. Signal Transduct. Res.* **19:** 687–701.

Murphy A.M., Morris-Downes M.M., Sheahan B.J., and Atkins G.J. 2000. Inhibition of human lung carcinoma cell growth by apoptosis induction using Semliki Forest virus recombinant particles. *Gene Ther.* **7:** 1477–1482.

Olkkonen V.M., Liljeström P., Garoff H., Simons K., and Dotti C.G. 1993. Expression of heterologous proteins in cultured rat hippocampal neurons using the Semliki Forest virus vector. *J. Neurosci Res.* **35:** 445–451.

Olson K.E., Higgs S., Hahn C.S., Rice C.M., Carlson J.O., and Beaty B.J. 1994. The expression of chloramphenicol acetyltransferase in *Aedes albopictus* (C6/36) cells and *Aedes triseriatus* mosquitoes using a double subgenomic recombinant Sindbis virus. *Insect Biochem. Mol. Biol.* **24:** 39–48.

Paul N.L., Marsh M., McKeating J.A., Schulz T.F., Liljeström P., Garoff H., and Weiss R.A. 1993. Expression of HIV-1 envelope glycoproteins by Semliki Forest virus vectors. *AIDS Res. Hum. Retrovir.* **9:** 963–970.

Perri S., Driver D.A., Gardner J.P., Sherrill S., Belli B.A., Dubensky T.W., Jr., and Polo J.M. 2000. Replicon vectors derived from Sindbis virus and Semliki forest virus that establish persistent replication in host cells. *J. Virol.* **74:** 9802–9807.

Pettigrew M.M. and O'Neill S.L. 1999. Semliki Forest virus as an expression vector in insect cell lines. *Insect Mol. Biol.* **8:** 409–414.

Pushko P., Bray M., Ludwig G.V., Parker M., Schmaljohn A., Sanchez A., Jahrling P.B., and Smith J.F. 2000. Recombinant RNA replicons derived from attenuated Venezuelan equine encephalitis virus protect guinea pigs and mice from Ebola hemorrhagic fever virus. *Vaccine* **19:** 142–53.

Ren H., Boulikas T., Söling A., Warnke P.C., and Rainov N.G. 2003. Immunogene therapy of recurrent glioblastoma multiforme with a liposomally encapsulated replication-incompetent Semliki Forest virus vector carrying the human interleukin-12 gene—A phase I/II clinical protocol. *J. Neurooncol.* **64:** 147–154.

Schultz-Cherry S., Dybing J.K., Davis N.L., Williamson C., Suarez D.L., Johnston R., and Perdue M.L. 2000. Influenza virus (A/HK/156/97) hemagglutinin expressed by an alphavirus replicon system protects chickens against lethal infection with Hong Kong-origin H5N1 viruses. *Virology* **278:** 55–59.

Smerdou C. and Liljeström P. 1999. Two-helper RNA system for production of recombinant Semliki Forest virus particles. *J. Virol.* **73:** 1092–1098.

Strauss J.H. and Strauss E.G. 1994. The alphaviruses: Gene expression, replication, and evolution. *Microbiol. Rev.* **58:** 491–562.

Tseng J.C., Levin B., Hurtado A., Yee H., Perez de Castro I., Jimenez M., Shamamian P., Jin R., Novick R.P., Pellicer A., and Meruelo D. 2004. Systemic tumor targeting and killing by Sindbis viral vectors. *Nat. Biotechnol.* **22:** 70–77.

Ulmanen I., Peranen J., Tenhunen J., Tilgmann C., Karhunen T., Panula P., Bernasconi L., Aubry J.P., and Lundstrom K. 1997. Expression and intracellular localization of catechol O-methyltransferase in transfected mammalian cells. *Eur. J. Biochem.* **243:** 452–459.

Vähä-Koskela M.J.V., Tuittila M.T., Nygardas P.T., Nyman J.K.-E., Ehrengruber M.U., Renggli M., and Hinkkanen A.E. 2003. A novel neurotropic expression vector based on the avirulent strain A7(74) of Semliki Forest virus. *J. Neurovirol.* **9:** 1–15.

Xiong C., Levis R., Shen P., Schlesinger S., Rice C., and Huang H.V. 1989. Sindbis virus: An efficient broad host range vector for gene expression in animal cells. *Science* **243:** 1188–1191.

29 Gene Transfer into Mammalian Cells Using Targeted Filamentous Bacteriophage

Andrew Baird

La Jolla Institute for Molecular Medicine, San Diego, California 92121

ABSTRACT

Phage vectors, because of their genetic simplicity, are uniquely suited to methods that use directed evolution to genetically optimize vectors for therapeutic gene delivery. Moreover, because phage production is restricted to strain-specific bacteria, the hosts are equally amenable to genetic engineering, modification, and even genetic selections to optimize yield, genetic stability, manufacture, and cost. The choice of targeting ligand determines the specificity of targeted phage transduction. Genetic targeting is limited to proteins that can be efficiently expressed and biologically active following secretion into the periplasmic space of the bacteria and subsequent incorporation into the phage particle. Although it is possible to target phage with a fibroblast growth factor (FGF)2-pIII fusion protein (Kassner et al. 1999), the efficiency of FGF2 display is relatively low. The capacity of phage to display a chosen targeting ligand must be determined empirically and optimized. Alternatively, the targeting ligand can be selected after display in a phage library. The orientation of the reporter gene relative to the phage structural genes can affect vector transduction efficiency. Transduction efficiency of the MG4 phage vector (Kassner et al. 1999) in which the green fluorescent protein (GFP) cassette is in the antisense orientation relative to the phage sense strand is about threefold higher than the same vector containing the GFP cassette in the opposite orientation (MG3) (Larocca et al. 1999). We have also targeted phage particles for gene delivery using an avidin-biotin linkage. This allows selection of ligands without concern for their ability to be displayed genetically.

The phage particles are generally not toxic to mammalian cell lines. However, it is important that endotoxin be removed. DNase treatment is important to prevent nonspecific transfection of cells by any contaminating replicative form of the phage. Contamination by replicative-form phage DNA should be monitored before and after DNase treatment by evaluating the phage on an agarose gel. The timing, duration, and doses of genotoxic treatments must be optimized for each mammalian cell line used. Targeted phage will not produce infective phage particles after transfection of mammalian cells because bacterial promoters regulate all of the phage structural genes. Even if the phage proteins were expressed, the mechanism for phage packaging and the differences in the intracellular environment of mammalian cells versus bacteria make the probability of a productive infection negligible. Nevertheless, we recommend following the same biohazard safety precautions (BSL-2) for targeted phage as those used for nonreplication-competent adenoviral vectors.

INTRODUCTION

Phage-mediated gene transfer offers an alternative method of introducing genes into specific cell types because filamentous bacteriophage can be readily reengineered to transfer genes to mammalian cells by attaching a targeting ligand to the phage surface either noncovalently (Larocca et al. 1998) or genetically (Kassner et al. 1999; Larocca et al. 1999, 2001; Poul and Marks 1999; Mount et al. 2004). Successful gene transfer and subsequent protein expression is measured using a reporter gene such as GFP, neomycin phosphotransferase, or β-galactosidase. Theoretically, any gene with an appropriate mammalian transcriptional promoter and polyadenylation signal can be incorporated into a ligand-targeted phage vector. This combination of ligand retargeting and a mammalian expression cassette confers mammalian cell tropism. Once prepared, these modified phage act like nonproductive animal viral vectors that are propagated and manipulated genetically with all the conveniences of a phage vector.

Phage display is commonly used as a technique to identify peptides and proteins that can then be used as reagents themselves or used to retarget other particles (Kehoe and Kay 2005). By leaving the peptide moiety on the phage, a particle might be evolved and optimized for gene delivery. The molecular design of better phage is not new (Skiena 2001), and protein evolution in viral backgrounds is well described (Bamford et al. 2002; Baker et al. 2005; Briones and Bastolla 2005; Casjens 2005; Hambly and Suttle 2005). Molenaar et al. (2002), for example, showed that simply displaying peptides on phage could modify their natural pharmacokinetics when injected into mice. Indeed, the possibility of creating phage with altered pharmacokinetics when administered in vivo was first reported several years ago (Geier et al. 1973; Merril et al. 1996) and attributed to single-amino-acid changes in coat composition (Vitiello et al. 2005). Although their safety in humans (Bruttin and Brussow 2005) suggests that T4 phage may be an optimal starting point for directed evolution of a vector, we have focused on filamentous M13 phage (Larocca et al. 2002b), whereas other investigators have turned to similar approaches using mammalian viral systems (Spear et al. 2003; Perabo et al. 2006).

Gene transfer with phage, as with other mechanisms, is time- and dose-dependent. It is also specific for cell surface receptors. Transduced cells begin to appear at about 48 hours after the addition of phage, and the percentage of cells expressing reporter gene increases with time. Phage is internalized through interaction of the targeting ligand with its cognate receptors on the cell surface. Accordingly, ligand-targeted phage transduction is inhibited by competition with the free ligand or with a neutralizing antireceptor antibody (Kassner et al. 1999; Larocca et al. 1999). Little or no transduction occurs in the absence of a targeting ligand because phage particles have no native tropism for mammalian cells. In addition, transduction occurs with concentrations of phage as low as approximately 100 phage/cell and increases up to the highest concentration tested (~1 x 10^6 phage/cell). At these higher doses, internalization of the ligand-targeted phage is highly efficient, and phage protein is detectable in almost all cells. Transduction efficiencies of up to 40% have been described, using ligand- and antibody-targeted phage/phagemid (Larocca et al. 1998; Kassner et al. 1999; Larocca et al. 1999; Poul and Marks 1999; Burg et al. 2002).

Stable transformants of targeted cells can be isolated from phage-transduced cells using G418 drug selection, with sustained gene expression for several months. A kinetic analysis of the fate of particles reveals that the coat is metabolized within hours of internalization but that phage DNA can be recovered from cells long afterward. For this reason, it may be possible to optimize intracellular trafficking. Genotoxic treatments, which are thought to increase double-stranded DNA synthesis, improve transduction efficiency of single-stranded phage (Burg et al. 2002). Further improvements in transduction efficiency are possible by incorporating peptides into the phage coat protein that facilitate trafficking in the cell and by applying directed molecular evolution to genetically select improved phage from combinatorial libraries (Kassner et al. 1999; Larocca et al. 1999, 2001, 2002a,b).

Phage vectors are simple and convenient to produce in bacteria, can be specifically targeted to cells, and have the potential to be evolved genetically for specific applications. In addition, filamentous phage have an inherent capacity to package large DNA inserts because they are not limited in size by a preformed capsid but instead form their protein coat as they are extruded from bacteria. We have successfully transduced cells with phage vectors approaching 10 kb in length, including both the targeting ligand sequence and the mammalian expression cassette. For larger gene inserts that tend to be unstable in phage, we have recently engineered phagemid vectors that are much simpler and smaller (~6 kb). Phagemid vectors contain no phage DNA sequences except the origin of replication and are therefore prepared in bacteria by rescue with helper phage (Smith and Scott 1993; Kehoe and Kay 2005) or by reengineering the host itself. The simple genetics of filamentous phage vectors make them particularly adaptable for a wide variety of targeted gene-transfer applications (Larocca et al. 2001; Mount et al. 2004).

The flexible structure of the pIII coat protein is well suited for displaying a variety of biologically active peptide and protein sequences while retaining the structural integrity of the phage particle (Smith 1985; Smith and Scott 1993; Kehoe and Kay 2005). For example, biologically active hormones, cytokines, and growth factors have been displayed on phage (Bass et al. 1990; Gram et al. 1993; Saggio et al. 1995; Buchli et al. 1997; Merlin et al. 1997; Souriau et al. 1997; Vispo et al. 1997). To date, phage-mediated gene delivery has been performed with targeting ligands on pIII, but fusion to pVIII provides similarly targeted phage/phagemid (Petrenko and Smith 2000; Mount et al. 2004; Kehoe and Kay 2005). In fact, phage displaying multiple copies of a peptide on the major coat protein are rapidly internalized into mammalian cells (Ivanenkov et al. 1999). There are many examples of active pIII fusion proteins, but not all ligands are equally displayed. Differences in the ability of the fusion proteins to be secreted into the bacterial periplasmic space for packaging into the phage particle can significantly affect surface display. Thus, although many ligands are functional when displayed on phage, insufficient display is a limitation that should be considered when identifying new ligands that can target phage for gene delivery to cells. Alternatively, noncovalent display of the targeting ligand (Larocca et al. 1998) can be used when genetic display is not applicable.

Many types of ligands can be used for phage gene delivery, including those identified from phage libraries. For example, Poul and Marks (1999) have targeted M13 phage using an anti-Her2 single-chain antibody. Whereas ligand selection is often performed by panning phage on cells for binding and internalization (Barry et al. 1996; Pereira et al. 1997; Watters et al. 1997), we developed novel selection strategies called LIVE (ligand identification via expression) and SNAAP (selection of nucleic-acid-amplified phage) that directly select for ligands that bind and internalize into target cells for gene delivery (Kassner et al. 1999; Burg et al. 2004). Repeated rounds of phage transfection and recovery from transfected cells select gene targeting ligands. In these systems, a functional ligand can be enriched a millionfold after three to four rounds of selection.

Filamentous phage-mediated gene transfer, like that of adeno-associated virus (AAV), involves the introduction of single-stranded DNA that must be converted to double-stranded DNA for transgene expression. Accordingly, phage-mediated transduction can be increased by the same genotoxic treatments that enhance AAV efficiency (Yalkinoglu et al. 1988), such as camptothecin, hydroxyurea, heat shock, and UV irradiation. The degree of enhancement varies among cell lines, presumably because of individual differences in response to genotoxic stress. We obtained maximum enhancement of phage-mediated transduction (with minimal toxicity) on COS-1 and PC-3 cells using 5–10 μM camptothecin, 40 mM hydroxyurea, 50 J/m^2 UV irradiation, or 7-hour heat shock at 42.5°C.

Recombinant M13 phage vectors are adapted for gene transfer to mammalian cells by inserting a mammalian expression cassette into the intergenic region of the phage genome (Larocca et al. 1999). The modified phage vector contains the GFP gene expression cassette from pEGFP-N1 (Clontech) that encodes a mutagenized GFP (Cormack et al. 1996) which is optimized for visualization by fluorescent microscopy or fluorescence-activated cell sorting (FACS). It also contains the simian virus 40 (SV40) origin of replication from pEGFP-N1. Any phage vector that can be engineered for phage display (i.e., fUSE5 [Scott and Smith 1990], fAFF1 [Cwirla et al. 1990], and M13East [Giebel et al. 1995]) can be adapted for gene delivery in this manner, including phagemid vectors. When phagemids are rescued with helper phage, both wild-type pIII and the ligand-pIII fusion protein are incorporated into the phagemid particle (Kehoe and Kay 2005), resulting in monovalent display of the targeting ligand. Monovalent display, however, is sometimes less optimal because the number of ligands on the phage surface can significantly affect binding and internalization (Becerril et al. 1999). In this case, the system can be adapted for multivalent display by rescuing the phagemid with a gene-III-deleted helper phage as described by Raconjac et al. (1997).

Protocol

Gene Transfer into Mammalian Cells Using Targeted Filamentous Bacteriophage

The protocol presented here describes the use of targeted filamentous phage for gene delivery to mammalian cells. The final vector, although of low efficiency, is meant to serve as a starting point for a vector development platform that can use in vitro and in vivo techniques of combinatorial display to direct its evolution to high efficiency, high specificity, and eventually, safety in humans.

MATERIALS

CAUTION: See Appendix for appropriate handling of materials marked with <!>.

Reagents

β-mercaptoethanol <!>
Camptothecin <!> (10 mM) (Sigma-Aldrich C9911) in dimethyl sulfoxide <!> (DMSO)
 Store at –20°C.
Cell culture medium: RPMI 1640 + 10% fetal bovine serum (FBS), 0.1 mM nonessential amino acids, 1 mM sodium pyruvate, 2 mM L-glutamine, and 50 μg/ml gentamycin
DNase I (GIBCO 18068-015)
EDTA (0.5 M)
Fixative buffer (0.925% formaldehyde <!>, 0.02% sodium azide <!>, and 2% glucose in PBS at pH 7.4)
Glycerol (20%)
Host F′ bacterial strain (Stratagene 200249 XLI-Blue Competent Cells)
Hydroxyurea (1.0 M) <!> (Sigma-Aldrich H8627) in PBS
 Store at –20°C.
LB plates (1% tryptone, 0.5% yeast extract, 0.5% NaCl at pH 7.0, and 2% agar)
LB^+ plates (1% tryptone, 0.5% yeast extract, 0.5% NaCl at pH 7.0, 2% agar, and 60 μg/ml ampicillin <!>)
$MgCl_2$ (1 M) <!>
PC-3 cell line (NCI cell bank) and HT1229 (ATTC)
Phosphate-buffered saline (PBS)
PBS + AEBSF (4-[2-aminoethyl]-benzene sulfonyl fluoride; 0.2 mM) (Roche 1585916)
Replicative-form M13 phage DNA
SOC medium (GIBCO 15544-034)
Sodium chloride (NaCl; 1.5 M)/30% polyethylene glycol (PEG) 8000 <!>
Targeted phage particles containing reporter gene (GFP)
Triton X-114 (10%) <!>
Trypsin (0.25%) <!> (GIBCO 25200-056)
Ultraviolet irradiation <!> (Stratalinker UV cross-linker, Stratagene)
2x YT broth (1.6% tryptone, 1% yeast extract, 0.5% NaCl at pH 7.0, and 60 μg/ml ampicillin <!>)

Equipment

Centrifuge bottles
Cryovials
FACS/Flow cytometer with a fluorescein isothiocyanate (FITC) filter set
Incubator, preset to 37°C and 42°C
Microcentrifuge (Eppendorf)
Pipettes
Shaking platform
Syringe filters (0.45 μm)
Syringes
Tissue-culture dishes (12 well)
Tubes (sterile; Falcon 2059)
Water bath, preset to 55°C

METHODS

Transformation of Bacteria

Using standard molecular biology techniques, host bacteria are transformed with the replicative form of the recombinant phage vector that has been engineered to contain a mammalian expression cassette. In our studies, we have used cytomegalovirus (CMV) promoter and growth hormone (GH) polyadenylation DNAs obtained from commercial sources.

1. Thaw 100 μl of competent cells on ice.
2. Add 1.7 μl of β-mercaptoethanol to cells and incubate on ice for 10 minutes.
3. Mix 50 ng of replicative-form phage DNA with cells and incubate for 30 minutes on ice.
4. Heat-shock bacteria for 45 seconds by placing them at 42°C. Place bacteria on ice for 2 minutes.
5. Add 900 μl of SOC to cells and incubate with shaking at 250 rpm for 1 hour at 37°C.
6. Spread cells on LB^+ plates and incubate overnight at 37°C.

Preparation of Targeted Phage Particles

Using standard phage display techniques, the desired ligand is engineered onto gIII so that the final particle will display a ligand-pIII fusion capable of targeting the phage to the mammalian cell.

7. Inoculate phage-transformed bacteria into 2x YT + 60 μg/ml ampicillin. Grow bacteria with shaking at 300 rpm overnight at 37°C.
8. Centrifuge bacterial culture at 6000*g* for 10 minutes at 4°C.
9. Save the supernatant and add 1/5 volume of cold 1.5 M NaCl/30% PEG. Mix well and incubate for 2 hours on ice to precipitate phage.
10. Centrifuge at 15,000*g* for 30 minutes at 4°C to collect phage. Remove supernatant and all residual liquid.
11. Resuspend the phage pellet in PBS + 0.2 mM AEBSF and incubate for 30 minutes at 4°C.
12. Centrifuge at 20,000*g* for 20–30 minutes to remove debris.
13. Repeat Steps 9–11 if further concentration of phage is necessary.
14. Add $MgCl_2$ to 10 μM and 6 units of DNase I per milliliter of phage solution. Incubate for 20 minutes at room temperature and stop the reaction by adding 10 μl of 0.5 M EDTA per milliliter of phage solution.

15. Immediately add 1/5 volume of 1.5 M NaCl/30% PEG. Mix well and incubate for 2 hours on ice to precipitate phage.
16. Centrifuge at 15,000*g* for 30 minutes at 4°C to collect the phage. Remove supernatant and all residual liquid.
17. Resuspend the phage pellet in PBS + 0.2 mM AEBSF. Incubate for 5–15 minutes at 37°C and then for 30 minutes at 4°C.
18. Centrifuge at 20,000*g* for 20–30 minutes to remove debris.
19. Filter phage through a 0.45-μm filter, freeze in 20% glycerol, and store at –70°C.

Endotoxin Removal by Triton X-114 Phase Partitioning

It is critical to remove endotoxin before using the particles in any cell or animal studies.

20. Add 100 μl of 10% Triton X-114 per milliliter of sample and incubate for 30 minutes on ice with occasional vortexing. Incubate for 10 minutes at 37°C.
21. Centrifuge at 14,000 rpm in a microcentrifuge for 10 minutes at room temperature and save the aqueous (upper) phase.
22. Repeat phase partitioning (Steps 20 and 21) twice.

Quantification of Yields by Titering Phage for Plaque-forming Units

23. Prewarm LB plates at 37°C, melt top agar (LB), and place in a 55°C water bath.
24. Grow F′ bacteria (XLI-Blue) to OD_{600} = 0.5 and aliquot 300 μl of cells to Falcon 2059 tubes for each dilution to be tested.
25. Set up serial dilutions in PBS by starting with a 100-fold dilution (5 μl diluted in 500 μl of PBS) and repeating several times for desired dilution series. Add 100 μl from each dilution to bacterial cells.
26. Add 3 ml of top agar to each tube, briefly vortex, and pour on top of prewarmed LB plates. Allow top agar to harden and invert overnight at 37°C.
27. Count plaques and determine titer in pfu/ml by multiplying the number of plaques by the dilution and volume plated.

Transduction of Cultured Cells

28. Seed cells in 1 ml of culture medium in 12-well tissue-culture dishes. Incubate overnight at 37°C with 5% CO_2.

    The seeding density is determined by growth rate. After an overnight incubation, the cells should be 25% confluent in the 12-well dishes. We seed PC-3 cells at 2 x 10^4 cells/well.
29. Remove culture medium from the cells and replace with culture medium containing phage. A typical dose of targeted phage for highest transduction efficiency is 10^{11} pfu/ml. Incubate cells for 24–96 hours at 37°C with 5% CO_2. If genotoxic treatments are being used, incubate for 40 hours and proceed to Step 31.
30. Harvest cells for reporter gene analysis by removing the phage-containing culture medium and washing the cells with PBS. Remove the cells from the culture dishes by adding 150 μl of 0.25% trypsin and incubating for 2–3 minutes at 37°C. Once the cells have detached from the plate, add 350 μl of fixative buffer. Analyze the cells by flow cytometry with an FITC filter set.

Genotoxic Treatments

The phage are added to cultures and genotoxic treatments are started 40 hours later. Treatments should be optimized for each target cell line.

31. Prepare culture medium containing 10% FBS to use for heat shock and UV irradiation. Alternatively, add camptothecin to 1 μM and 100 μM or hydroxyurea to 10 mM and 100 mM.
32. Remove medium containing phage. Replace with one of the genotoxic media above or with culture medium containing 10% FBS for heat shock or UV irradiation.
33. Incubate directly for 7 hours at 37°C or 42°C (heat shock) or irradiate (10–100 J/m^2).
34. Wash three times in PBS.
35. Add fresh cell culture medium containing 10% FBS. Incubate an additional 24–48 hours at 37°C.
36. Wash three times with PBS and evaluate by FACS analysis or by direct observation of GFP.

ACKNOWLEDGMENTS

This work was supported in part by the National Institutes of Health.

REFERENCES

Baker M.L., Jiang W., Rixon F.J., and Chiu W. 2005. Common ancestry of herpesviruses and tailed DNA bacteriophages. *J. Virol.* **79:** 14967–14970.

Bamford D.H., Burnett R.M., and Stuart D.I. 2002. Evolution of viral structure. *Theor. Popul. Biol.* **61:** 461–470.

Barry M.A., Dower W.J., and Johnston S.A. 1996. Toward cell-targeting gene therapy vectors: Selection of cell-binding peptides from random peptide-presenting phage libraries. *Nat. Med.* **2:** 299–305.

Bass S., Greene R., and Wells J.A. 1990. Hormone phage: An enrichment method for variant proteins with altered binding properties. *Proteins* **8:** 309–314.

Becerril B., Poul M.A., and Marks J.D 1999. Toward selection of internalizing antibodies from phage libraries. *Biochem. Biophys. Res. Commun.* **255:** 386–393.

Briones C. and Bastolla U. 2005. Protein evolution in viral quasispecies under selective pressure: A thermodynamic and phylogenetic analysis. *Gene* **47:** 237–246.

Bruttin A. and Brussow H. 2005. Human volunteers receiving *Escherichia coli* phage T4 orally: A safety test of phage therapy. *Antimicrob. Agents Chemother.* **49:** 2874–2878.

Buchli P.J., Wu Z., and Ciardelli T.L. 1997. The functional display of interleukin-2 on filamentous phage. *Arch. Biochem. Biophys.* **339**: 79–84.

Burg M., Ravey E.P., Gonzales M., Amburn E., Faix P.H., Baird A., and Larocca D. 2004. Selection of internalizing ligand-display phage using rolling circle amplification for phage recovery. *DNA Cell Biol.* **23:** 457–462.

Burg M.A., Jensen-Pergakes K., Gonzalez A.M., Ravey P., Baird A., and Larocca D. 2002. Enhanced phagemid particle gene transfer in camptothecin-treated carcinoma cells. *Cancer Res.* **62:** 977–981.

Casjens S.R. 2005. Comparative genomics and evolution of the tailed-bacteriophages. *Curr. Opin. Microbiol.* **8:** 451–458.

Cormack B.P., Valdivia R.H., and Falkow S. 1996. FACS-optimized mutants of the green fluorescent protein (GFP). *Gene* **173:** 33–38.

Cwirla S.E., Peters E.A., Barrett R.W., and Dower W.J. 1990. Peptides on phage: A vast library of peptides for identifying ligands. *Proc. Natl. Acad. Sci.* **87:** 6378–6382.

Geier M.R., Trigg M.E., and Merril C.R. 1973. Fate of bacteriophage lambda in non-immune germ-free mice. *Nature* **246:** 221–223.

Giebel L.B., Cass R.T., Milligan D.L., Young D.C., Arze R., and Johnson C.R. 1995. Screening of cyclic peptide phage libraries identifies ligands that bind streptavidin with high affinities. *Biochemistry* **34:** 15430–15435.

Gram H., Strittmatter U., Lorenz M., Gluck D., and Zenke G. 1993. Phage display as a rapid gene expression system: Production of bioactive cytokine-phage and generation of neutralizing monoclonal antibodies. *J. Immunol. Methods* **161:** 169–176.

Hambly E. and Suttle C.A. 2005. The viriosphere, diversity, and genetic exchange within phage communities. *Curr. Opin. Microbiol.* **8:** 444–450.

Ivanenkov V.V., Felici F., and Menon A.G. 1999. Targeted delivery of multivalent phage display vectors into mammalian cells. *Biochim. Biophys. Acta* **1448:** 463–472.

Kassner P.D., Burg M.A., Baird A., and Larocca D. 1999. Genetic selection of phage engineered for receptor-mediated gene transfer to mammalian cells. *Biochem. Biophys. Res. Commun.* **264:** 921–928.

Kehoe J.W. and Kay B.K. 2005. Filamentous phage display in the new millennium. *Chem. Rev.* **105:** 4056–4072.

Larocca D., Jensen-Pergakes K., Burg M.A., and Baird A. 2001. Receptor-targeted gene delivery using multivalent phagemid particles. *Mol. Ther.* **3:** 476–484.

______. 2002a. Gene transfer using targeted filamentous bacteriophage. *Methods Mol. Biol.* **185:** 393–401.

Larocca D., Witte A., Johnson W., Pierce G.F., and Baird A. 1998. Targeting bacteriophage to mammalian cell surface receptors for gene delivery. *Hum. Gene Ther.* **9:** 2393–2399.

Larocca D., Burg M.A., Jensen-Pergakes K., Ravey E.P., Gonzalez A.M., and Baird A. 2002b. Evolving phage vectors for cell targeted gene delivery. *Curr. Pharm. Biotechnol.* **3:** 45–57.

Larocca D., Kassner P.D., Witte A., Ladner R.C., Pierce G.F., and Baird A. 1999. Gene transfer to mammalian cells using genetically targeted filamentous bacteriophage: *FASEB J.* **13:** 727–734.

Merlin S., Rowold E., Abegg A., Berglund C., Klover J., Staten N., McKearn J.P., and Lee S.C. 1997. Phage presentation and affinity selection of a deletion mutant of human interleukin-3. *Appl. Biochem. Biotechnol.* **67:** 199–214.

Merril C.R., Biswas B., Carlton R., Jensen N.C., Creed G.J., Zullo S., and Adhya S. 1996. Long-circulating bacteriophage as antibacterial agents. *Proc. Natl. Acad. Sci.* **93:** 3188–3192.

Molenaar T.J., Michon I., de Haas S.A., van Berkel T.J., Kuiper J., and Biessen E.A. 2002. Uptake and processing of modified bacteriophage M13 in mice: Implications for phage display. *Virology* **293:** 182–191.

Mount J.D., Samoylova T.I., Morrison N.E., Cox N.R., Baker H.J., and Petrenko V.A. 2004. Cell targeted phagemid rescued by preselected landscape phage. *Gene* **341:** 59–65.

Perabo L., Endell J., King S., Lux K., Goldnau D., Hallek M., and Buning H. 2006. Combinatorial engineering of a gene therapy vector: Directed evolution of adeno-associated virus. *J. Gene Med.* **8:** 155–162.

Pereira S., Maruyama H., Siegel D., Van Belle P., Elder D., Curtis P., and Herlyn D. 1997. A model system for detection and isolation of a tumor cell surface antigen using antibody phage display. *J. Immunol. Methods* **203:** 11–24.

Petrenko V.A. and Smith G.P. 2000. Phages from landscape libraries as substitute antibodies. *Protein Eng.* **13:** 589–592.

Poul M.A. and Marks J.D. 1999. Targeted gene delivery to mammalian cells by filamentous bacteriophage. *J. Mol. Biol.* **288:** 203–211.

Rakonjac J., Jovanovic G., and Model P. 1997. Filamentous phage infection-mediated gene expression: Construction and propagation of the gIII deletion mutant helper phage R408d3. *Gene* **198:** 99–103.

Saggio I., Gloaguen I., and Laufer R. 1995. Functional phage display of ciliary neurotrophic factor. *Gene* **152:** 35–39.

Scott J.K. and Smith G.P. 1990. Searching for peptide ligands with an epitope library. *Science* **249:** 386–390.

Skiena S.S. 2001. Designing better phages. *Bioinformatics* (suppl.) **1:** S253–261.

Smith G.P. 1985. Filamentous fusion phage: Novel expression vectors that display cloned antigens on the virion surface. *Science* **228:** 1315–1317.

Smith G.P. and Scott J.K. 1993. Libraries of peptides and proteins displayed on filamentous phage. *Methods Enzymol.* **217:** 228–257.

Souriau C., Fort P., Roux P., Hartley O., Lefranc M.P., and Weill M. 1997. A simple luciferase assay for signal transduction activity detection of epidermal growth factor displayed on phage. *Nucleic Acids Res.* **25:** 1585–1590.

Spear M.A., Schuback D., Miyata K., Grandi P., Sun F., Yoo L., Nguyen A., Brandt C.R., and Breakefield X.O. 2003. HSV-1 amplicon peptide display vector. *J. Virol. Methods* **107:** 71–79.

Vispo N.S., Callejo M., Ojalvo A.G., Santos A., Chinea G., Gavilondo J.V., and Arana M.J. 1997. Displaying human interleukin-2 on the surface of bacteriophage. *Immunotechnology* **3:** 185–193.

Vitiello C.L., Merril C.R., and Adhya S. 2005. An amino acid substitution in a capsid protein enhances phage survival in mouse circulatory system more than a 1000-fold. *Virus Res.* **114:** 101–103.

Watters J.M., Telleman P., and Junghans R.P. 1997. An optimized method for cell-based phage display panning. *Immunotechnology* **3:** 21–29.

Yalkinoglu A.O., Heilbronn R., Burkle A., Schlehofer J.R., and zur Hausen H. 1988. DNA amplification of adeno-associated virus as a response to cellular genotoxic stress. *Cancer Res.* **48:** 3123–3129.

30 Selection, Isolation, and Identification of Targeting Peptides for Ligand-directed Gene Delivery

Martin Trepel,* Wadih Arap,† and Renata Pasqualini†

*University of Freiburg Medical Center, Department of Hematology and Oncology and Institute for Molecular Medicine and Cell Research, D-79106 Freiburg, Germany; †The University of Texas M.D. Anderson Cancer Center, Houston, Texas 77030-4095

ABSTRACT

Parenchymal, stromal, and vascular endothelial cells within organs express differential surface receptors depending on their tissue localization and functional state in vivo. On the basis of this receptor diversity, random phage display peptide libraries can be selected to isolate peptide ligands that home to tissue-specific cell surface receptors. Following systemic administration of the library, homing bacteriophage clones can be recovered by harvesting target tissues, amplified by infection of a bacterial host, and validated. The recovered peptide ligands then serve to identify their corresponding receptors and to target agents to specific cell types. A functional vascular map of such tissue-specific and angiogenesis-related receptors has been increasingly extended and refined by screening strategies in isolated receptors, cell lines, experimental animal models, ex vivo in clinical samples, and directly in patients. Recently, a new class of ligand-directed hybrid adeno-associated virus (AAV) phage vectors has been designed for targeted gene delivery to accessible cell surface receptors after systemic administration. Ultimately, these advances may lead to clinical applications in human disease.

INTRODUCTION

Tissue-specific Vascular Receptor and Phage Display Library Selection

Tissues express specific patterns of cell surface proteins not only in their parenchymal and stromal components, but also in their vascular endothelium (for review, see Trepel et al. 2002; Hajitou et al. 2006a). Some of these cell surface proteins serve as growth factor receptors during normal development or pathological conditions. Moreover, tissue-specific endothelial cell surface proteins can serve as receptors for tissue-specific homing of circulating ligands or cells such as leukocytes. These tissue-specific endothelial receptors can be defined as functional vascular addresses by using systematic criteria (Marchiò et al. 2004). Such validated addresses can be targeted systemically through the circulation.

Ligand-directed targeting involves profiling of vascular addresses in normal organs as well as under pathological conditions. This can be readily accomplished by using random phage display peptide libraries. Such libraries represent large collections of phage particles displaying random peptides (typically ~10^9 unique sequences) (Smith and Scott 1993). Selection of phage libraries in vivo allows the recovery of displayed peptides that bind to tissue-specific vascular receptors and home preferentially to target organs. Briefly, libraries are injected intravenously, and the target tissue is surgically collected after a short circulation time. Phage clones displaying homing ligands present in the harvested tissue are rescued and reamplified by bacterial infection. Amplified mixtures are reinjected for further enrichment of clones displaying peptides with optimal homing capacity. After three or four rounds of selection, recovered phage clones are sequenced to identify the DNA corresponding to the inserts displayed, which should ideally share common peptide motifs (typically consisting of three to five residues). This technology has been used widely to identify tissue-specific ligand-receptor interactions in vivo. Different peptide motifs binding tissue-specific receptors have been recovered from several normal organs (Pasqualini and Ruoslahti 1996; Arap et al. 1998, 2002a,b; Rajotte et al. 1998; Porkka et al. 1999; Trepel et al. 2001; Essler and Ruoslahti 2002; Kolonin et al. 2002, 2004, 2006b; Laakkonen et al. 2002; Porkka et al. 2002). Likewise, targeting peptides recovered from known tissue-specific receptors or such receptors overexpressed on cells in vitro were used to study homing in vivo (Pasqualini et al. 1997; Burg et al. 1999; Arap et al. 2004; Marchiò et al. 2004; Zurita et al. 2004; Giordano et al. 2005).

An Emerging Human Vascular Ligand Receptor Map

The systematic mapping of receptors in human blood vessels (the so-called vascular ZIP codes) is required for development of clinically applicable targeted therapy. Recent studies showed that the tissue distribution of circulating peptides in vivo in humans is nonrandom and recovered a ligand interleukin-11 mimic peptide from the prostate (Arap et al. 2002b). Subsequently, the corresponding interleukin-11 receptor was validated as a morphologic and functional marker during human prostate cancer progression in a large panel of patient samples (Zurita et al. 2004). Recent refinements in methodology have enabled synchronous combinatorial selection of ligands from multiple organs in mice (Kolonin et al. 2006b) and have also adapted this strategy for use in patients (W. Arap et al., unpubl.). The challenges of accurate mapping are complicated further by studies in the mouse pancreas (Yao et al. 2005) which suggest that vascular heterogeneity in tissues may extend to functionally distinct regions within single organs.

Applications of Targeting Peptides

Receptor-targeting peptides can potentially serve in several applications. These include the identification of tissue-specific receptors accessible to circulating ligands (Rajotte and Ruoslahti 1999; Pasqualini et al. 2000; Giordano et al. 2001; Arap et al. 2002b; Kolonin et al. 2002, 2004; Christian et al. 2003; Yao et al. 2005), targeting therapeutic agents (Arap et al. 1998, 2002a, 2004; Ellerby et al. 1999; Koivunen et al. 1999b; Trepel et al. 2001; Curnis et al. 2002; Kolonin et al. 2004; Zurita et al. 2004) or diagnostic compounds (Chen et al. 2004; Kolonin et al. 2006a; Souza et al. 2006) to the tissue of interest, and targeting gene-transfer vectors to specific receptors (Reynolds et al. 1999; Nicklin et al. 2000, 2001; Trepel et al. 2000b; Grifman et al. 2001; White et al. 2001, 2004; Shi and Bartlett 2003). The latter application is the most pertinent to this chapter and hence is discussed in more detail.

Ligand-directed Hybrid Vectors for Targeted Transgene Delivery

Current gene therapy vectors present problems of unintended transduction of certain tissues, adverse immune reactions, and lack of efficient transduction of the cells of interest (Somia and Verma 2000; Trepel et al. 2000a; Thomas et al. 2003). Specific targeting of vectors offers a solution to these concerns. Peptides have been exploited for viral vector targeting by using bispecific molecular conjugates consisting of antivector antibodies and peptide ligands directed toward the target receptor (Trepel et al. 2000b), as well as insertion of specific peptide ligands into the vector capsid (Girod et al. 1999; Reynolds et al. 1999; Grifman et al. 2001; Nicklin et al. 2001; White et al. 2001, 2004; Loiler et al. 2003; Müller et al. 2003; Shi and Bartlett 2003; Work et al. 2004). The latter strategy has many advantages such as ease of handling, better stability in vitro and in vivo, maintenance of the small size of the vector particle, and avoidance of additional immunogenicity elicited by conjugates.

The incorporation of peptide ligand sequences isolated by phage display libraries into eukaryotic vectors is possible (Reynolds et al. 1999; Grifman et al. 2001; Loiler et al. 2003; Shi and Bartlett 2003; White et al. 2004), but the reported success rate is quite variable. Only about 20–30% of selected ligand peptides function as well in targeted phage particles as in modified common gene-transfer vector capsids such as adenovirus or AAV. One explanation is that the phage-derived peptides are selected only for cell or receptor binding but not for subsequent post-targeting cell entry required for gene transfer. In addition, the binding potential of a ligand peptide may change when incorporated within a viral envelope. Taking such limitations into account, random peptide-display libraries based on the gene therapy vector capsid itself were developed for AAVs (Müller et al. 2003; Perabo et al. 2003) and retroviruses (Bupp and Roth 2003; Khare et al. 2003a,b; Hartl et al. 2005). This approach allows the selection of peptide ligands specifically binding to a cell type of interest within the specific viral capsid protein context. Consequently, such selections yield vectors that specifically and efficiently transduce the cell types on which they have been selected.

These approaches—the use of phage-derived peptides for targeted gene delivery and the use of peptides in the structural protein context in which they have been selected—can now be combined. This is achieved by using the targeted phage particle itself for gene delivery. However, bacteriophage species have generally been considered unsuitable vehicles for mammalian cell transduction. On the one hand, phage particles have no tropism for mammalian cells (Barrow and Soothill 1997) that must be neutralized for retargeting, and in recent years they have even been used for transduction of eukaryotic cells (Larocca et al. 1999; Poul and Marks 1999; Piersanti et al. 2004). Still, inefficient transduction and immunogenicity have remained major obstacles for phage vectors to overcome for applications in eukaryotic cells. We have thus introduced the AAV phage (AAVP) vector system (Hajitou et al. 2006b). This is a new hybrid containing genetic *cis*-elements from AAV and from a single-stranded M13 bacteriophage derivative. Incorporation of AAV-inverted terminal repeats into the phage transgene cassette is associated with improved intracellular fate of the delivered transgene (Hajitou et al. 2006b). A targeted AAVP prototype has been established by targeting α_v integrins up-regulated in tumor vessels. AAVP can mediate tissue-specific ligand-directed transduction in vivo after systemic administration of the vector. AAVP-mediated transfer of reporter genes has also been used for molecular genetic imaging and suicide gene therapy strategies (Hajitou et al. 2006b). AAVP represents a new class of targeted prokaryotic/eukaryotic viral hybrid vectors that may serve in a wide range of applications in biomedical research. The method outlined here describes only the selection of tissue-targeted phage in vivo. Methods for modifying recovered phage into targeted AAVP are presented elsewhere (Hajitou et al. 2006b).

Protocol 1

Selection of Random Phage Display Peptide Libraries In Vivo

The following protocol is described for use in mice. However, it can be adapted for use in other species with modifications based on the dose of phage particles applied and phage recovery strategy. This procedure is confined to the screening of one organ at a time. However, an approach for simultaneous screening of multiple organs with phage display libraries in the mouse model has also been established (Kolonin et al. 2006b), following methodological principles similar to those outlined in this protocol.

MATERIALS

CAUTION: See Appendix for appropriate handling of materials marked with <!>.

Reagents

Anesthetic <!>

BALB/c mice (2 months old)

Despite being less cost-effective, nude mice are preferred to minimize fur-related bacterial cross-contamination. Most ligand receptors isolated thus far seem to be found across various mouse strains.

Dulbecco's modified Eagle's medium (DMEM) tissue-culture medium

K91*kan* bacteria

Any other F-pilus-positive *Escherichia coli* bacteria can be used for phage amplification.

Kanamycin <!>

Luria broth medium

NZY broth medium

Phage display random peptide libraries

The generation and production of phage display random peptide libraries have been described elsewhere (Smith and Scott 1993; Koivunen et al. 1999). Several libraries in different types of vectors are also commercially available. Suitable libraries comprise a diversity of $\geq 10^8$ unique phage peptide sequences. Usually, peptide lengths of ≤ 9 random residues are preferred as their selection can yield good affinity ligands. The potential diversity of larger library inserts is so high that such libraries may not be practically useful. Quality control of the phage library is crucial after each step of targeting in vivo. Amplified libraries (as opposed to primary libraries) may not always serve for this application if insertless or mutant phage clone frequencies are a concern.

Polyethylene glycol (PEG)/NaCl <!>

Dissolve 100 g of PEG (particle size 8000) and 110 g of NaCl in 450 ml of H_2O. Shake vigorously. Sterilize by autoclaving. Shake repeatedly while cooling.

Phosphate-buffered saline (PBS)

Protease inhibitor cocktail (Roche)

TB supplement

Dissolve 11.55 g of KH_2PO_4 and 105 g of K_2HPO_4 in 500 ml of H_2O. Sterilize by autoclaving. Dilute 1:10 in Terrific Broth prior to use.

Terrific Broth (TB) medium

Tetracycline <!>

Tris-buffered saline (TBS)

Equipment

Cannulae (butterfly IV 23-gauge blue)
Cell culture flasks
Glass tissue grinder
Luria Broth agar plates
Shaking incubator (37°C)
Surgical tools
Syringes

METHODS

Injection of Phage Display Library

1. Dilute the chosen phage library in DMEM or PBS to approximately 3.3×10^{10} transducing units (TU)/ml.
2. Administer 10^{10} TU of phage library intravenously into the tail vein of the subject animal.
   The volume should not exceed 300 µl.
3. Maintain animal viability for 3–5 minutes while the library circulates.

Transcardial Perfusion and Washing

Transcardial perfusion of the experimental animal prior to surgical harvesting of the target organ can decrease nonspecific background phage recovery. However, for certain organs (e.g., kidney), perfusion increases phage trapping and may worsen the outcome. In addition, perfusion is not recommended for the initial round of selection as excessive stringency can eliminate potential binding peptides present in small numbers. To decrease background phage recovery from the blood if perfusion is not performed, the mouse can be exsanguinated completely.

4. Ensure that the experimental animal is under very deep surgical anesthesia.
5. Dissect away the skin at the level of the diaphragm to expose the chest and abdominal walls.
6. Open the peritoneum just below the sternum.
   It is important not to damage the liver, heart, or any large blood vessels at this point.
7. Make a section underneath and along the coastal arch.
8. Dissect the lateral edges of the thorax through the diaphragm up to the axilla. Avoid damaging the lungs.
9. Fold the anterior chest wall cranially to expose the heart. Secure in position with a clamp.
10. Insert a cannula connected to a syringe containing approximately 5 ml of room-temperature DMEM in the left ventricle.
11. Make a small incision in the right atrium to provide a blood outlet.
12. Perfuse with low pressure to avoid vascular damage.
    Up to 50 ml can be perfused through the heart. The relative stringency is generally thought to be directly proportional to the volume used. During the first round of selection (if perfusion is performed), consider perfusion with lower volumes.
13. Surgically remove the target organ and at least one control organ (such as the lung or brain). Place on ice immediately to avoid bound phage internalization.
14. Weigh organs. Homogenize with a glass tissue grinder.

15. Add 1 ml of ice-cold DMEM supplemented with the protease inhibitor cocktail to the tissue homogenate. Vortex. Centrifuge at 3000 rpm for 4 minutes at 4°C. Remove the supernatant.
16. Repeat Step 15 (three washes total, or one to two washes for the first round of selection).
17. After the last centrifugation, remove the supernatant. Keep the pellet on ice until addition of bacteria.

Growth of K91*kan* Bacteria

18. Inoculate 5 ml of TB supplemented with 200 μg/ml kanamycin and 10% TB supplement with a streak from a K91*kan* agar plate.
19. Shake inoculum at regular speed for 2–4 hours at 37°C.

    It is recommended that the growth of the bacteria be started after injection of the library into the animal.
20. Dilute an aliquot of inoculum 1:10 with TB. Determine OD_{600}.

    When the OD_{600} is 0.16–0.20, decrease the shaker speed to regenerate sheared pili. Use the bacteria within 30 minutes.

Recovery of Phage from Cell Pellets

21. Add 1500 μl of the K91*kan* culture to the phage-cell pellet (from Step 17). Resuspend pellet gently but thoroughly. Incubate for 30 minutes at 37°C. Swirl or invert the sample at 10-minute intervals.
22. Transfer bacteria to a 500-ml cell culture flask. Add 100 ml of prewarmed NZY medium containing 0.2 μg/ml tetracycline. Incubate for 30 minutes at 37°C.
23. Remove 10- and 100-μl aliquots from the suspension. Plate on LB agar plates containing 40 μg/ml tetracycline to estimate the number of transducing phage units recovered from each tissue.
24. Adjust the tetracycline concentration in the remaining culture to 20 μg/ml. Grow overnight at 37°C in the shaker.

    Cultures should be grown for at least 12 hours, but no more than 16 hours.

Recovery of Phage from Bacterial Culture

25. Centrifuge the bacterial culture at 8000 rpm for 15 minutes. Decant the supernatant into a clean tube. Take care that the pellet is on the upper side of the tube when decanting.
26. Add 1.5 ml of PEG/NaCl per each 10 ml of supernatant. Shake well. Incubate for at least 1 hour on ice.

    Samples can be incubated with PEG/NaCl overnight at 4°C.
27. Centrifuge at 8000 rpm for 20 minutes at 4°C. Decant and discard the supernatant.
28. Insert the tube in the rotor such that the side of the tube containing the pellet is toward the outside of the rotor. Centrifuge the pellet again at 8000 rpm for 5 minutes to remove all PEG and to further concentrate the pellet.
29. Promptly and carefully remove the supernatant with a vacuum aspirator or a 200 μl-pipette.
30. Add TBS to the phage pellet (~200–400 μl, depending on the size of the pellet). Shake for 10 minutes.

    Avoid resuspending the phage pellet with the pipette as frothing can occur that might harm the phage or displayed peptides.

31. Transfer the solution to an Eppendorf tube. Centrifuge at 14,000 rpm for 10 minutes to remove any bacterial debris.
32. Carefully transfer the supernatant (phage solution) to a new labeled Eppendorf tube. Do not touch the pellet when transferring the supernatant.
33. Titer the recovered phage solution (see Protocol 2).
34. Repeat the phage selection (Steps 1–32) using 1×10^9 to 5×10^9 transducing units of the recovered phage for each successive round of selection.

    For most applications, three to four rounds are sufficient to yield tissue-specific phage.
35. Sequence colonies after the third round to identify the peptides recovered from the target tissue.

    The number of sequenced clones per selection round should be at least 30. For sequencing of the random phage insert, we use the primer 5′-GCAAGCTGATAAACCGATACAATT-3′. The recognition pattern for the insert in the phage genome is 5′-GCCGACGGGGCT–insert–GGGGCCGCTGGG-3′.

Protocol 2

Phage Titering

MATERIALS

CAUTION: See Appendix for appropriate handling of materials marked with <!>.

Reagents

K91*kan* bacteria
Kanamycin <!>
Terrific Broth (TB) medium
Tetracycline <!>

Equipment

Luria Broth agar plates
Shaking incubator (37°C)

METHOD

1. Grow K91*kan* bacteria as described in Protocol 1, Steps 18–20.
2. In separate Eppendorf tubes, prepare five serial dilutions of the phage stock (from Protocol 1, Step 32) corresponding to 10^{-5} to 10^{-9} μl of the initial stock.
3. Add 180 μl of the K91*kan* bacteria culture to each of the tubes.
4. Incubate phage with bacteria for 30 minutes at room temperature.
5. Plate 100 μl of each solution on LB plates containing tetracycline. Grow overnight in the incubator at 37°C.
6. Count the colonies on the plates.
7. Multiply the number of colonies by the dilution factor to determine the number of TU/μl of phage solution.

   Only plates with 20–600 colonies yield reliable results.

ACKNOWLEDGMENTS

This work was funded by grants from the National Institutes of Health (including the SPORE) and DOD (including the IMPACT) and by awards from the Gillson-Longenbaugh, the Keck Foundation, the Prostate Cancer Foundation (to R.P. and W.A.), and the Deutsche Forschungsgemeinschaft (to M.T.).

REFERENCES

Arap M.A., Lahdenranta J., Mintz P.J., Hajitou A., Sarkis A.S., Arap W., and Pasqualini R. 2004. Cell surface expression of the stress response chaperone GRP78 enables tumor targeting by circulating ligands. *Cancer Cell* **6:** 275–284.

Arap W., Pasqualini R., and Ruoslahti E. 1998. Cancer treatment by targeted drug delivery to tumor vasculature in a mouse model. *Science* **279:** 377–380.

Arap W., Haedicke W., Bernasconi M., Kain R., Rajotte D., Krajewski S., Ellerby H.M., Bredesen D.E., Pasqualini R., and Ruoslahti E. 2002a. Targeting the prostate for destruction through a vascular address. *Proc. Natl. Acad. Sci.* **99:** 1527–1531.

Arap W., Kolonin M.G., Trepel M., Lahdenranta J., Cardo-Vila M., Giordano R.J., Mintz P.J., Ardelt P.U., Yao V.J., Vidal C.I., et al. 2002b. Steps toward mapping the human vasculature by phage display. *Nat. Med.* **8:** 121–127.

Barrow P.A. and Soothill J.S. 1997. Bacteriophage therapy and prophylaxis: Rediscovery and renewed assessment of potential. *Trends Microbiol.* **5:** 268–271.

Bupp K. and Roth M.J. 2003. Targeting a retroviral vector in the absence of a known cell-targeting ligand. *Hum. Gene Ther.* **14:** 1557–1564.

Burg M.A., Pasqualini R., Arap W., Ruoslahti E., and Stallcup W.B. 1999. NG2 proteoglycan-binding peptides target tumor neovasculature. *Cancer Res.* **59:** 2869–2874.

Chen L., Zurita A.J., Ardelt P.U., Giordano R.J., Arap W., and Pasqualini R. 2004. Design and validation of a bifunctional ligand display system for receptor targeting. *Chem. Biol.* **11:** 1081–1091.

Christian S., Pilch J., Akerman M.E., Porkka K., Laakkonen P., and Ruoslahti E. 2003. Nucleolin expressed at the cell surface is a marker of endothelial cells in angiogenic blood vessels. *J. Cell Biol.* **163:** 871–878.

Curnis F., Arrigoni G., Sacchi A., Fischetti L., Arap W., Pasqualini R., and Corti A. 2002. Differential binding of drugs containing the NGR motif to CD13 isoforms in tumor vessels, epithelia, and myeloid cells. *Cancer Res.* **62:** 867–874.

Ellerby H.M., Arap W., Ellerby L.M., Kain R., Andrusiak, R., Rio G.D., Krajewski S., Lombardo C.R., Rao R., Ruoslahti E., Bredesen D.E., and Pasqualini R. 1999. Anti-cancer activity of targeted pro-apoptotic peptides. *Nat. Med.* **5:** 1032–1038.

Essler M. and Ruoslahti E. 2002. Molecular specialization of breast vasculature: A breast-homing phage-displayed peptide binds to aminopeptidase P in breast vasculature. *Proc. Natl. Acad. Sci.* **99:** 2252–2257.

Giordano R.J., Cardo-Vila M., Lahdenranta J., Pasqualini R., and Arap W. 2001. Biopanning and rapid analysis of selective interactive ligands. *Nat. Med.* **7:** 1249–1253.

Giordano R.J., Anobom C.D., Cardo-Vila M., Kalil J., Valente A.P., Pasqualini R., Almeida F.C., and Arap W. 2005. Structural basis for the interaction of a vascular endothelial growth factor mimic peptide motif and its corresponding receptors. *Chem. Biol.* **12:** 1075–1083.

Girod A., Ried M., Wobus C., Lahm H., Leike K., Kleinschmidt J., Deleage G., and Hallek M. 1999. Genetic capsid modifications allow efficient re-targeting of adeno-associated virus type 2. *Nat. Med.* **5:** 1052–1056.

Grifman M., Trepel M., Speece P., Gilbert L.B., Arap W., Pasqualini R., and Weitzman M.D. 2001. Incorporation of tumor-targeting peptides into recombinant adeno-associated virus capsids. *Mol. Ther.* **3:** 964–975.

Hajitou A., Pasqualini R., and Arap W. 2006a. Vascular targeting: Recent advances and therapeutic perspectives. *Trends Cardiovasc. Med.* **16:** 80–88.

Hajitou A., Trepel M., Lilley C.E., Soghomonyan S., Alauddin M.M., Marini F.C., III, Restel B.H., Ozawa M.G., Moya C.A., Rangel R., et al. 2006b. A hybrid vector for ligand-directed tumor targeting and molecular imaging. *Cell* **125:** 385–398.

Hartl I., Schneider R.M., Sun Y., Medvedovska J., Chadwick M.P., Russell S.J., Cichutek K., and Buchholz C.J. 2005. Library-based selection of retroviruses selectively spreading through matrix metalloprotease-positive cells. *Gene Ther.* **2:** 918–926.

Khare P.D., Russell S.J., and Federspiel M.J. 2003a. Avian leukosis virus is a versatile eukaryotic platform for polypeptide display. *Virology* **315:** 303–312.

Khare P.D., Rosales A.G., Bailey K.R., Russell S.J., and Federspiel M.J. 2003b. Epitope selection from an uncensored peptide library displayed on avian leukosis virus. *Virology* **315:** 313–321.

Koivunen E., Restel B.H., Rajotte D., Lahdenranta J., Hagedorn M., Arap W., and Pasqualini R. 1999a. Integrin-binding peptides derived from phage display libraries. *Methods Mol. Biol.* **129:** 3–17.

Koivunen E., Arap W., Valtanen H., Rainisalo A., Medina O.P., Heikkila P., Kantor C., Gahmberg C.G., Salo T., Konttinen Y.T., et al. 1999b. Tumor targeting with a selective gelatinase inhibitor. *Nat. Biotechnol.* **17:** 768–774.

Kolonin M.G., Pasqualini R., and Arap W. 2002. Teratogenicity induced by targeting a placental immunoglobulin transporter. *Proc. Natl. Acad. Sci.* **99:** 13055–13060.

Kolonin M.G., Saha P.K., Chan L., Pasqualini R., and Arap W. 2004. Reversal of obesity by targeted ablation of adipose tissue. *Nat. Med.* **10:** 625–632.

Kolonin M.G., Sun J., Do K.A., Vidal C.I., Ji Y., Baggerly K.A., Pasqualini R., and Arap W. 2006a. Synchronous selection of homing peptides for multiple tissues by in vivo phage display. *FASEB J.* **20:** 979–981.

Kolonin M.G., Bover L., Sun J., Zurita A.J., Do K.A., Lahdenranta J., Cardo-Vila M., Giordano R.J., Jaalouk D.E., Ozawa M.G., et al. 2006b. Ligand-directed surface profiling of human cancer cells with combinatorial peptide libraries. *Cancer Res.* **66:** 34–40.

Laakkonen P., Porkka K., Hoffman J.A., and Ruoslahti E. 2002. A tumor-homing peptide with a targeting specificity related to lymphatic vessels. *Nat. Med.* **8:** 751–755.

Larocca D., Kassner P.D., Witte A., Ladner R.C., Pierce G.F., and Baird A. 1999. Gene transfer to mammalian cells using genetically targeted filamentous bacteriophage. *FASEB J.* **13:** 727–734.

Loiler S.A., Conlon T.J., Song S., Tang Q., Warrington K.H., Agarwal A., Kapturczak M., Li C., Ricordi C., Atkinson M.A., et al. 2003. Targeting recombinant adeno-associated virus vectors to enhance gene transfer to pancreatic islets and liver. *Gene Ther.* **10:** 1551–1558.

Marchiò S., Lahdenranta J., Schlingemann R.O., Valdembri D., Wesseling P., Arap M.A., Hajitou A., Ozawa M.G., Trepel M., Giordano R.J., et al. 2004. Aminopeptidase A is a functional target in angiogenic blood vessels. *Cancer Cell* **5:** 151–162.

Müller O.J., Kaul F., Weitzman M.D., Pasqualini R., Arap W., Kleinschmidt J.A., and Trepel M. 2003. Random peptide libraries displayed on adeno-associated virus to select for targeted gene therapy vectors. *Nat. Biotechnol.* **21:** 1040–1046.

Nicklin S.A., White S.J., Watkins S.J., Hawkins R.E., and Baker A.H. 2000. Selective targeting of gene transfer to vascular endothelial cells by use of peptides isolated by phage display. *Circulation* **102:** 231–237.

Nicklin S.A., Von Seggern D.J., Work L.M., Pek D.C., Dominiczak A.F., Nemerow G.R., and Baker A.H. 2001. Ablating adenovirus type 5 fiber-CAR binding and HI loop insertion of the SIGYPLP peptide generate an endothelial cell-selective adenovirus. *Mol. Ther.* **4:** 534–542.

Pasqualini R. and Ruoslahti E. 1996. Organ targeting in vivo using phage display peptide libraries. *Nature* **380:** 364–366.

Pasqualini R., Koivunen E., and Ruoslahti E. 1997. α_v integrins as receptors for tumor targeting by circulating ligands. *Nat. Biotechnol.* **15:** 542–546.

Pasqualini R., Koivunen E., Kain R., Lahdenranta J., Sakamoto M., Stryhn A., Ashmun R.A., Shapiro L.H., Arap W., and Ruoslahti E. 2000. Aminopeptidase N is a receptor for tumor-homing peptides and a target for inhibiting angiogenesis. *Cancer Res.* **60:** 722–727.

Perabo L., Buning H., Kofler D.M., Ried M.U., Girod A., Wendtner C.M., Enssle R., and Hallek M. 2003. In vitro selection of viral vectors with modified tropism: The adeno-associated virus display. *Mol. Ther.* **8:** 151–157.

Piersanti S., Cherubini G., Martina Y., Salone B., Avitabile D., Grosso F., Cundari E., Di Zenzo G., and Saggio I. 2004. Mammalian cell transduction and internalization properties of λ phages displaying the full-length adenoviral penton base or its central domain. *J. Mol. Med.* **82:** 467–476.

Porkka K., Laakkonen P., Hoffman J.A., Bernasconi M., and Ruoslahti E. 2002. A fragment of the HMGN2 protein homes to the nuclei of tumor cells and tumor endothelial cells in vivo. *Proc. Natl. Acad. Sci.* **99:** 7444–7449.

Porkka K., Laakkonen P., Rajotte D., Hoffman J., and Ruoslahti E. 1999. Bone marrow homing peptides from phage display libraries. *Blood* **94:** 1107.

Poul M.A. and Marks J.D. 1999. Targeted gene delivery to mammalian cells by filamentous bacteriophage. *J. Mol. Biol.* **288:** 203–211.

Rajotte D. and Ruoslahti E. 1999. Membrane dipeptidase is the receptor for a lung-targeting peptide identified by in vivo phage display. *J. Biol. Chem.* **274:** 11593–11598.

Rajotte D., Arap W., Hagedorn M., Koivunen E., Pasqualini R., and Ruoslahti E. 1998. Molecular heterogeneity of the vascular endothelium revealed by in vivo phage display. *J. Clin. Invest.* **102:** 430–437.

Reynolds P., Dmitriev I., and Curiel D. 1999. Insertion of an RGD motif into the HI loop of adenovirus fiber protein alters the distribution of transgene expression of the systemically administered vector. *Gene Ther.* **6:** 1336–1339.

Shi W.F. and Bartlett J.S. 2003. RGD inclusion in VP3 provides adeno-associated virus type 2 (AAV2)-based vectors with a heparan sulfate-independent cell entry mechanism. *Mol. Ther.* **7:** 515–525.

Smith G.P. and Scott J.K. 1993. Libraries of peptides and proteins displayed on filamentous phage. *Methods Enzymol.* **217:** 228–257.

Somia N. and Verma I.M. 2000. Gene therapy: Trials and tribulations. *Nat. Rev. Genet.* **1:** 91–99.

Souza G.R., Christianson D.R., Staquicini F.I., Ozawa M.G., Snyder E.Y., Sidman R.L., Miller J.H., Arap W., and Pasqualini R. 2006. Networks of gold nanoparticles and bacteriophage as biological sensors and cell-targeting agents. *Proc. Natl. Acad. Sci.* **103:** 1215–1220.

Thomas C.E., Ehrhardt A., and Kay M.A. 2003. Progress and problems with the use of viral vectors for gene therapy. *Nat. Rev. Genet.* **4:** 346–358.

Trepel M., Arap W., and Pasqualini R. 2000a. Exploring vascular heterogeneity for gene therapy targeting. *Gene Ther.* **7:** 2059–2060.

———. 2001. Modulation of the immune response by systemic targeting of antigens to lymph nodes. *Cancer Res.* **61:** 8110–8112.

———. 2002. In vivo phage display and vascular heterogeneity: Implications for targeted medicine. *Curr. Opin. Chem. Biol.* **6:** 399–404.

Trepel M., Grifman M., Weitzman M.D., and Pasqualini R. 2000b. Molecular adaptors for vascular-targeted adenoviral gene delivery. *Hum. Gene Ther.* **11:** 1971–1981.

White S.J., Nicklin S.A., Sawamura T., and Baker A.H. 2001. Identification of peptides that target the endothelial cell-specific LOX-1 receptor. *Hypertension* **37:** 449–455.

White S.J., Nicklin S.A., Buning H., Brosnan M.J., Leike K., Papadakis E.D., Hallek M., and Baker A.H. 2004. Targeted gene delivery to vascular tissue in vivo by tropism-modified adeno-associated virus vectors. *Circulation* **109:** 513–519.

Work L.M., Nicklin S.A., Brain N.J.R., Dishart K.L., Von Seggern D.J., Hallek M., Buning H., and Baker A.H. 2004. Development of efficient viral vectors selective for vascular smooth muscle cells. *Mol. Ther.* **9:** 198–208.

Yao V.J., Ozawa M.G., Trepel M., Arap W., Mcdonald D.M., and Pasqualini R. 2005. Targeting pancreatic islets with phage display assisted by laser pressure catapult microdissection. *Am. J. Pathol.* **166:** 625–636.

Zurita A.J., Troncoso P., Cardo-Vila M., Logothetis C.J., Pasqualini R., and Arap W. 2004. Combinatorial screenings in patients: The interleukin-11 receptor α as a candidate target in the progression of human prostate cancer. *Cancer Res.* **64:** 435–439.

31 Rescue and Propagation of Tropism-Modified Measles Viruses

Takafumi Nakamura and Stephen J. Russell

The Molecular Medicine Program, Mayo Clinic College of Medicine, Rochester, Minnesota 55905

ABSTRACT

Reverse genetics techniques are available for rescue of infectious measles viruses (MVs) from cloned cDNA. This allows manipulation of the viral genome and has greatly contributed to the understanding of the molecular basis for MV replication, host-virus interactions, and viral pathogenicity. Such reverse genetics methods have also been applied to cancer virotherapy, because MV is emerging as a promising biological agent with potent oncolytic properties. Genetic engineering has been used to develop fully retargeted oncolytic MVs. However, ablation of the native receptor interactions is incompatible with viral growth on 293 or Vero rescue cells. To address this issue, the STAR (six-histidine tagging and retargeting) system uses a pseudoreceptor and a viral tag to allow rescue and propagation of fully retargeted viruses that no longer bind to native MV receptors. This protocol provides detailed instruction on the use of this system whereby new retargeted viruses can be rapidly generated in three simple steps. First, the cDNA encoding the targeting ligand is cloned into a chimeric H-protein shuttle vector that also encodes a carboxy-terminal six-histidine peptide tag. The chimeric H-protein-coding sequence is then cloned into a full-length MV cDNA. Finally, the retargeted virus is rescued from the cDNA and amplified on cells expressing the pseudoreceptor, a membrane-anchored anti-His tag antibody. Targeting is expected to further enhance the therapeutic index of oncolytic MVs, resulting in exclusive destruction of tumors with minimal collateral damage to normal tissues.

INTRODUCTION

MV, a member of the genus *Morbillivirus* in the Paramyxoviridae family, is an enveloped RNA virus that contains a single-strand, negative-sense, nonsegmented genome (Griffin 2001). The 15,894-kb MV genome comprises six nonoverlapping cistrons arranged 3′-N-P-M-F-H-L-5′ that code for a total of eight proteins. The genome is tightly associated with the nucleocapsid protein (N) in a helical ribonucleoprotein complex that serves as the template for transcription and replication and is packaged into progeny virions (Griffin 2001). The P cistron specifies three polypeptides: P, C, and V. The phospho (P) protein functions as a chaperone that interacts with and regulates the cellular localization of the N protein and assists in assembly of the ribonucleoprotein complex. The C and V polypeptides are nonstructural proteins that are translated from P mRNAs through the use of alternative reading frames (Griffin 2001). The viral envelope consists of the matrix (M) protein and two transmembrane glycoproteins, the fusion (F) and hemagglutinin (H) proteins. The M protein is also involved in viral budding (Griffin 2001). The H and F envelope glycoproteins mediate viral attachment and virus-cell membrane fusion during infection (Cathomen et al. 1998). Attachment to target cells is mediated by H and is followed by membrane fusion, mediated by F (Wild et al. 1994). The tropism of MV is determined by binding of the H protein to one of at least two possible cellular receptors, CD46 or SLAM. CD46 is a ubiquitous regulator of complement activation and is found in varying amounts on all human nucleated cells (Dorig et al. 1993; Naniche et al. 1993). SLAM is expressed only on activated T and B cells, dendritic cells, and macrophages (Tatsuo et al. 2000; Hsu et al. 2001; Minagawa et al. 2001). Finally, the L protein is the multifunctional catalytic subunit of the RNA-dependent RNA polymerase (Griffin 2001).

MV is emerging as a novel biological agent with a wide range of potential analytic and clinical uses. Attenuated measles has an excellent safety record and does not require high-level containment because most people have been infected or vaccinated against measles and have anti-measles antibodies. Recombinant MV can also be used for analysis of viral pathogenesis, as a vaccine platform for immunization against emerging viruses such as human immunodeficiency virus (HIV) and Ebola virus, and as a biotechnological tool for the isolation or optimization of novel polypeptide ligands. Perhaps most importantly, MV has potent oncolytic properties that may prove to be useful for cancer virotherapy. Replication-competent attenuated Edmonston B vaccine strains of MV (MV-Edm) have demonstrated potent antitumor activity against xenograft models of human multiple myeloma, ovarian cancer, lymphoma, and glioma (Grote et al. 2001; Peng et al. 2001, 2002; Phuong et al. 2003). The virus preferentially infects tumor cells and induces extensive, and ultimately lethal, cell-to-cell fusion via CD46, which is more highly expressed on tumor cells than normal cells (Anderson et al. 2004). This killing mechanism is distinct compared to most other oncolytic viruses. However, MV-Edm is not entirely tumor-specific and can infect a variety of cell types via its native receptors, CD46 and SLAM. Thus, targeting is required to further enhance the therapeutic index of oncolytic MVs, resulting in exclusive destruction of tumors with minimal collateral damage to normal tissues.

Reverse genetics techniques to generate infectious MVs from cloned cDNAs have been used extensively for analysis of clinical measles isolates and to generate genetically modified viruses, contributing substantially to the understanding of measles pathogenesis at the molecular, cellular, and whole-animal levels. The first reverse genetics system applied to MV used a full-length cDNA copy of the genome of the Edmonston B vaccine strain and the helper cell line, 293-3-46 (Radecke et al. 1995). This cell line stably expresses bacteriophage T7 RNA polymerase and measles N and P proteins. To rescue virus, the MV-Edm genomic cDNA clone was cotransfected into these cells with a plasmid coding for measles L protein, leading to generation of MV antigenomic RNA and its incorporation into a functional ribonucleoprotein complex.

The efficiency of this particular viral rescue system was improved simply by applying a heat shock treatment to the transfected 293-3-46 cells and then coculturing them with a monolayer of Vero cells (Parks et al. 1999).

The genome of attenuated MV can be easily expanded to incorporate additional transgenes. The expanded recombinant genomes are extremely stable during viral growth and do not easily rearrange or delete their foreign transgenes. This property can be exploited to efficiently retarget the virus by display of polypeptide ligands at the carboxy-terminal end of the H protein. In contrast to the other viruses for which this has been attempted, MV appears to have a remarkably flexible and adaptable entry mechanism that can utilize multiple alternative cellular receptors. At the first step of MV infection, the H protein mediates attachment to one of the viral receptors expressed on the cell surface of the target cell and signals the F protein to trigger cell fusion (von Messling et al. 2001). Engineering the H protein can therefore restrict and retarget cell fusion through various antibody-receptor interactions. The list of cell surface molecules through which MV entry can be redirected already includes carcinoembryonic antigen, CD20, CD38, epidermal growth factor (EGF) receptor, the VIII mutant of EGF receptor, insulin-like growth factor (IGF) receptor, a major histocompatibility complex (MHC) peptide complex, and $\alpha V\beta 3$ integrin. In chimeric H proteins displaying single-chain antibodies (scFvs) at their carboxyl termini, alanine substitutions at H-protein residues 481 and 533 (H_{AA}) efficiently ablated the interactions with native receptors CD46 and SLAM, providing an optimal platform for generation of fully retargeted H proteins (Nakamura et al. 2004). However, ablation of native receptor interactions is not compatible with viral rescue and growth on 293 or Vero cells using the rescue methods described above because the retargeted viruses no longer bind to the native measles receptors.

The rescue and propagation of retargeted MVs displaying a CD38 scFv– on Vero cell transfectants stably expressing human CD38 demonstrated the feasibility of retargeting (Hadac et al. 2004). Both the CD38-targeted viruses and the untargeted parental viruses grew efficiently on Vero-CD38 cells. In addition, fully CD38-retargeted viruses showed the expected host-range properties (CD46- and SLAM-blind, but efficiently entering cells through CD38). Subsequently, a more versatile method was developed to rescue and propagate retargeted viruses without having to go through the slow process of generating an antigen-specific viral rescue cell line for each new target. Vero-αHis cells expressing a membrane-anchored single-chain antibody that recognizes a six-histidine (H6) peptide were generated. Viruses incorporating the H6 peptide at the carboxyl terminus of their ablated, retargeted H proteins could be rescued and propagated on Vero-αHis cells under conditions where the interaction between H and its native receptor CD46 was absent (Fig. 1, following page). This STAR rescue system has been used to rescue retargeted MVs displaying many different cell-targeting polypeptide ligands (Nakamura et al. 2005).

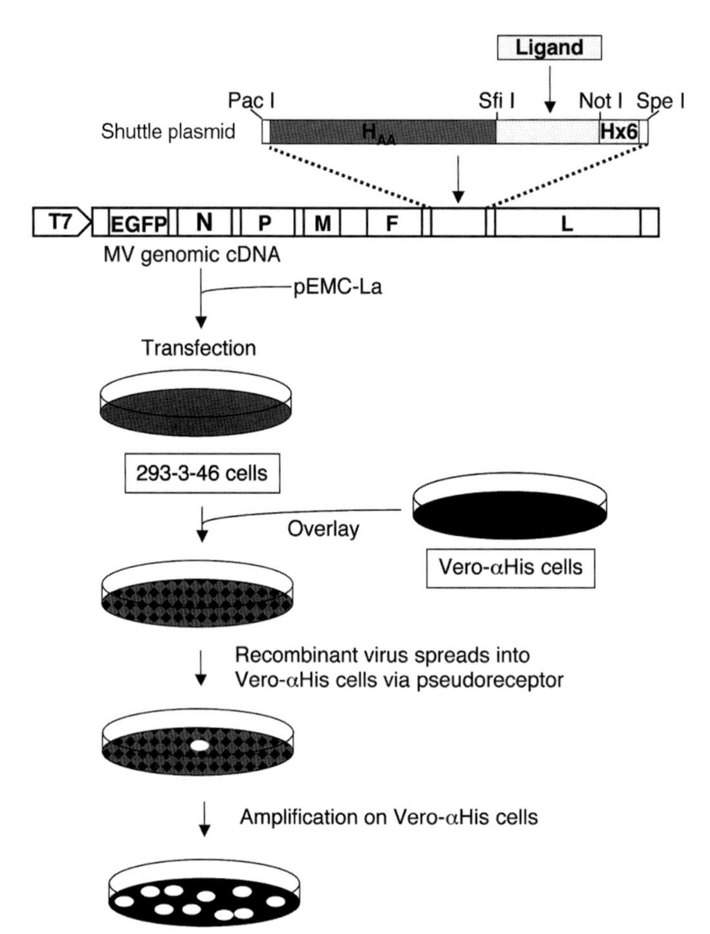

FIGURE 1. Flow diagram of STAR system. The pTNH6-Haa shuttle plasmid is constructed to encode the six-His tag (H X 6) and two alanine substitutions (H_{AA}) to ablate native binding. The cDNA for enhanced green fluorescent protein (GFP) is cloned upstream of the nucleocapsid gene N as an additional transcription unit. Enhanced GFP expression provides a convenient means to evaluate the specificity and efficiency of viral infection, both in cultured cells and in the tissues of euthanized mice. A different transgene can easily be inserted by using the MluI and AatII cloning sites flanking the enhanced GFP insert. The shuttle plasmid is cloned into the full-length MV genomic cDNA and cotransfected with the L-protein-expressing pEMC-La plasmid into 293-3-46 cells. After harvesting, the 293-3-46 cells are overlaid on a monolayer of Vero cells expressing a membrane-anchored anti-His tag antibody pseudoreceptor for rescue and amplification.

Protocol

Rescue and Propagation of Tropism-modified MV

The STAR system uses a pseudoreceptor and a viral tag to allow rescue and propagation of fully retargeted viruses that no longer bind to the native MV receptors (Nakamura et al. 2005). This enables development of fully retargeted MVs that can target a variety of tumors through a broad array of cell surface antigens and receptors. Using this system, new retargeted viruses can be rapidly generated in three simple steps: (1) cloning of the ligand cDNA into the chimeric H-protein shuttle vector, (2) generation of a full-length cDNA coding for the retargeted virus, and (3) rescue and propagation of the retargeted virus.

MATERIALS

CAUTION: See Appendix for appropriate handling of materials marked with <!>.

Reagents

Several manufacturers sell kits that provide 2x BBS, 2.5 M $CaCl_2$ and 2x HBS (e.g., ProFectin Mammalian Transfection System, Promega).

2x BBS

- 2.1 g of BES
- 3.3 g of NaCl
- 0.04 g of Na_2HPO_4 <!>

  Dissolve in 200 ml of deionized distilled H_2O (ddH_2O). Adjust pH to 6.95. Sterilize by filtration through a 0.22-μm filter. Store the filtered buffer in 5-ml aliquots at –20°C.

$CaCl_2$ (2.5 M) <!>

  Dissolve 7.4 g of $CaCl_2 \cdot 2H_2O$ in 20 ml of ddH_2O. Sterilize by filtration through a 0.22-μm filter. Store in 0.5-ml aliquots at –20°C.

Cells

- 293-3-46 cells

  Maintain these cells in complete growth medium containing 1.2 mg/ml G418.

- Vero-αHis cells

  These cells are grown in DMEM containing 5% FBS and 1 mg/ml G418.

Complete growth medium: Dulbecco's modified Eagle's medium (DMEM) supplemented with 10% fetal bovine serum (FBS)

G418 <!>

2x HBS (HEPES-buffered saline)

- 2 g of HEPES
- 3.3 g of NaCl
- 0.04 g of Na_2HPO_4

  Dissolve in 200 ml of ddH_2O. Adjust pH to 7.10 ± 0.05. Sterilize by filtration through a 0.22-μm filter. Store the filtered buffer in 5-ml aliquots at –20°C.

Opti-MEM Medium (Invitrogen)

Plasmids

- Shuttle plasmids: pTNH6-Haa (even) and pTNH6-Haa (odd)

  The above are unpublished vectors available on request from the authors.

p(+)MVeGFP (Duprex et al. 1999)
pEMC-La (Radecke et al. 1995)

To obtain the highest transfection efficiencies, purify all plasmid DNAs by equilibrium centrifugation in CsCl–ethidium bromide density gradients or commercial kits (e.g., QIAfilter Plasmid Midi Kit, QIAGEN).

Restriction enzymes: NotI, PacI, SpeI, SfiI
Trypsin <!>

Equipment

Cell scraper
Culture dishes (six well and 100 mm)
Incubator; humidified, 37°C, 5% CO_2
Parafilm
Pipette tips (sterile)
Pipettors
Water bath, 43.5°C

METHODS

Insertion of Targeting Ligand into MV Genome

1. Clone the cDNA for the polypeptide ligand of interest into SfiI and NotI sites of the shuttle vector. Care should be taken to maintain the reading frame between the H protein and the displayed ligand (see Fig. 1).

   Engineered MV genomes should be of hexameric length (rule of six) for efficient rescue. We therefore have developed two shuttle vectors with or without a 3-bp deletion in the noncoding region of the H mRNA. Depending on whether the ligand contains an odd or even number of amino acids, the pTNH6-Haa (odd) or pTNH6-Haa (even) plasmid should be used, respectively.

2. Insert a PacI/SpeI-digested fragment encoding the chimeric H protein into the corresponding sites of the full-length MV genomic cDNA clone p(+)MVeGFP.

Transfection for Viral Rescue

3. One day before transfection, harvest a 100-mm dish of confluent 293-3-46 cells by trypsinization. Suspend the cells in 12 ml of complete growth medium containing 1.2 mg/ml G418.
4. Transfer 2 ml of the suspended cells to each well of a six-well dish. Incubate the cultures for 20–24 hours at 37°C.

   The cells should be approximately 80% confluent at the time of transfection.

5. Three to six hours before transfection, replace the medium with 2 ml of fresh complete growth medium lacking G418.
6. Aliquot 250 μl of 2x HBS to a sterile tube.
7. In a second sterile tube, combine 5 μg of the recombinant full-length MV genomic cDNA clone with 5–50 ng of the L expression plasmid pEMC-La in a final volume of 225 μl in H_2O.
8. Add 25 μl of 2.5 M $CaCl_2$ to the DNA solution.
9. Slowly add the DNA solution (Step 8) dropwise to the 2x HBS (Step 6) while vortexing.
10. Incubate the combined solution for 30 minutes at room temperature. Vortex.

11. Immediately add the solution dropwise to the 293-3-46 cells in the six-well dish. Incubate the cells for 14–16 hours at 37°C.
12. Change the transfection medium for fresh complete growth medium without G418.
13. Wrap the six-well plate containing the transfected cells in Parafilm. Transfer to a water bath and incubate for 3 hours at 43.5°C to induce heat shock. Return plates to the 37°C incubator until harvesting for overlay.
14. One day before overlay (i.e., 48 hours after transfection of 293-3-46 cells), harvest exponentially growing Vero-αHis cells by trypsinization.
15. Transfer Vero cells to a 10-mm dish at a density of 1.5×10^6 to 2×10^6 cells/dish. Add 10 ml of complete growth medium containing 1 mg/ml G418. Incubate for 14–16 hours at 37°C.

    The cells should be approximately 70% confluent at the time of transfection.
16. Seventy-two hours after transfection, vigorously pipette the medium in the wells to detach the 293-3-46 cells and break the monolayer into small clumps. Harvest the dissociated cells along with 5 ml of medium.

    Do not use cell-dissociating agents.
17. Immediately overlay the 293-3-46 cell suspension onto the monolayer of Vero-αHis cells in the 100-mm dish. Incubate the culture for 2–7 days at 37°C.

    Several infectious centers should be visible as multinucleated syncytia, with enhanced green fluorescent protein (GFP) expression visible under fluorescence microscopy.

Propagation of Rescued Virus

18. One day before transfer of the rescued viral plaques, harvest exponentially growing Vero-αHis cells by trypsinization.
19. Transfer Vero cells to six-well dishes at a density of 2×10^5 cells/well. Add 2 ml of complete growth medium containing 1 mg/ml G418. Incubate the cultures for 14–16 hours at 37°C.

    The cells should be approximately 70% confluent at the time of transfer.
20. Immediately before transfer, aspirate the medium from the wells. Replace with 250 μl of Opti-MEM.
21. Aspirate the culture medium from the overlaid cells (from Step 17). Scratch and pick up each plaque using a P200 pipette with a sterile filtered tip.
22. Transfer plaques to the Vero-αHis cell monolayers growing in the six-well dish. Incubate the culture for 3 hours at 37°C to allow infection to occur.
23. Add 2 ml of complete growth medium without G418. Incubate the infected Vero cells for 2–5 days at 37°C.

    Observe the cells periodically during this period until more than 50% are incorporated in syncytia.
24. Aspirate the medium carefully. Add 1 ml of Opti-MEM. Scrape the cells into the medium.
25. Subject cells to two rounds of freeze-thawing
26. Centrifuge the suspensions at 2000 rpm for 10 minutes to pellet the cell debris. Aspirate and save the cleared cell lysate as a first seed stock.

    Store virus aliquots at –80°C.
27. Harvest exponentially growing Vero-αHis cells by trypsinization. Transfer cells to a 10-mm dish at a density of 1.5×10^6 to 2×10^6 cells/dish.
28. Add 10 ml of complete growth medium containing 1 mg/ml G418. Incubate the cultures for 14–16 hours at 37°C.

29. Infect the cells with 200–300 µl of the first seed stock (from Step 26) in 1 ml of Opti-MEM. Incubate for 3 hours at 37°C.
30. Add 9 ml of complete growth medium without G418. Incubate the infected cells for a further 2–3 days at 37°C.
31. When more than 80% of the cells are in syncytia, aspirate the medium carefully. Add 1 ml of Opti-MEM per dish.
32. Proceed as described in Steps 24–26 to produce second seed stock.
    Store virus aliquots at –80°C.

Titration of Viral Stocks

33. Plate Vero-αHis cells overnight at a density of 7000 cells/100 µl per well in 96-well plates.
34. Serially dilute the secondary viral seed stock (tenfold dilutions) in Opti-MEM from $1{:}10^1$ to $1{:}10^7$.
35. Add 50 µl of each viral dilution to eight wells in the 96-well plate. Incubate the plate for 6 days at 37°C.
36. Score each well as positive (+) or negative (–) for syncytia.
37. Determine the $TCID_{50}$ value using the Spearman and Kärber equation:

$$\log10\ (TCID_{50}/ml) = L + d(s - 0.5) + \log10\ (1/v)$$

where L is the negative log10 of the most concentrated viral dilution tested in which all wells are positive; d is the log10 of dilution factor; s is the sum of individual proportions (pi); pi is the calculated proportion of an individual dilution (i.e., number of positve wells/total number of wells/dilution); and v is the volume of inoculation (ml/well). For example, using the sample data presented in Table 1:

$$\begin{aligned}\log10\ (TCID_{50}/ml) &= 1 + \{1 \times [(8/8 + 7/8 + 2/8 + 0/8) - 0.5]\} + \log10\ (1/0.05)\\ &= 1 + (2.125 - 0.5) + 1.3\\ &= 3.925\\ TCID_{50}/ml &= 8.41 \times 10^3\end{aligned}$$

Large-scale Propagation of Virus

38. Harvest Vero-αHis cells by trypsinization.
39. Transfer cells to 10-mm dishes at a density of 1.5×10^6 to 2×10^6 cells/dish. Add 10 ml of complete growth medium containing 1 mg/ml G418. Incubate the cultures for 14–16 hours at 37°C.
40. Infect the cells using the secondary seed stock diluted in 1 ml of Opti-MEM to a multiplicity of infection of 0.02.
41. Proceed as described in Steps 29–32. Harvest the viral cell lysates when more than 80% of the cells are in syncytia.
    Store the virus aliquots at –80°C.

TABLE 1. Sample Data from Titration of Viral Stock

log10 of virus dilution	pi
–1	8/8
–2	7/8
–3	2/8
–4	0/8

42. Titrate viruses as described in Steps 33–37.

    Viral titers of 10^6 to 10^7 $TCID_{50}$/ml are ready for use in in vitro and in vivo experiments.

ACKNOWLEDGMENTS

This work is supported by National Institutes of Health grants CA100634 and HL66958 and the Harold W. Siebens Foundation.

REFERENCES

Anderson B.D., Nakamura T., Russell S.J., and Peng K.W. 2004. High CD46 receptor density determines preferential killing of tumor cells by oncolytic measles virus. *Cancer Res.* **64:** 4919–4926.

Cathomen T., Naim H.Y., and Cattaneo R. 1998. Measles viruses with altered envelope protein cytoplasmic tails gain cell fusion competence. *J. Virol.* **72:** 1224–1234.

Dorig R.E., Marcil A., Chopra A., and Richardson C.D. 1993. The human CD46 molecule is a receptor for measles virus (Edmonston strain). *Cell* **75:** 295–305.

Duprex W.P., McQuaid S., Hangartner L., Billeter M.A., and Rima B.K. 1999. Observation of measles virus cell-to-cell spread in astrocytoma cells by using a green fluorescent protein-expressing recombinant virus. *J. Virol.* **73:** 9568–9575.

Griffin D.E. 2001. Measles virus. In *Fields virology*, 4th edition (ed. D.M. Knipe and P.M. Howley), pp. 1402–1442. Lippincott Williams & Wilkins, Philadelphia, Pennsylvania.

Grote D., Russell S.J., Cornu T.I., Cattaneo R., Vile R., Poland G.A., and Fielding A.K. 2001. Live attenuated measles virus induces regression of human lymphoma xenografts in immunodeficient mice. *Blood* **97:** 3746–3754.

Hadac E.M., Peng K.W., Nakamura T., and Russell S.J. 2004. Reengineering paramyxovirus tropism. *Virology* **329:** 217–225.

Hsu E.C., Iorio C., Sarangi F., Khine A.A., and Richardson C.D. 2001. CDw150 (SLAM) is a receptor for a lymphotropic strain of measles virus and may account for the immunosuppressive properties of this virus. *Virology* **279:** 9–21.

Minagawa H., Tanaka K., Ono N., Tatsuo H., and Yanagi Y. 2001. Induction of the measles virus receptor SLAM (CD150) on monocytes. *J. Gen. Virol.* **82:** 2913–2917.

Nakamura T., Peng K.W., Harvey M., Greiner S., Lorimer I.A., James C.D., and Russell S.J. 2005. Rescue and propagation of fully retargeted oncolytic measles viruses. *Nat. Biotechnol.* **23:** 209–214.

Nakamura T., Peng K.W., Vongpunsawad S., Harvey M., Mizuguchi H., Hayakawa T., Cattaneo R., and Russell S.J. 2004. Antibody-targeted cell fusion. *Nat. Biotechnol.* **22:** 331–336.

Naniche D., Varior-Krishnan G., Cervoni F., Wild T.F., Rossi B., Rabourdin-Combe C., and Gerlier D. 1993. Human membrane cofactor protein (CD46) acts as a cellular receptor for measles virus. *J. Virol.* **67:** 6025–6032.

Parks C.L., Lerch R.A., Walpita P., Sidhu M.S., and Udem S.A. 1999. Enhanced measles virus cDNA rescue and gene expression after heat shock. *J. Virol.* **73:** 3560–3566.

Peng K.W., Ahmann G.J., Pham L., Greipp P.R., Cattaneo R., and Russell S.J. 2001. Systemic therapy of myeloma xenografts by an attenuated measles virus. *Blood* **98:** 2002–2007.

Peng K.W., TenEyck C.J., Galanis E., Kalli K.R., Hartmann L.C., and Russell S.J. 2002. Intraperitoneal therapy of ovarian cancer using an engineered measles virus. *Cancer Res.* **62:** 4656–4662.

Phuong L.K., Allen C., Peng K.W., Giannini C., Greiner S., TenEyck C.J., Mishra P.K., Macura S.I., Russell S.J., and Galanis E.C. 2003. Use of a vaccine strain of measles virus genetically engineered to produce carcinoembryonic antigen as a novel therapeutic agent against glioblastoma multiforme. *Cancer Res.* **63:** 2462–2469.

Radecke F., Spielhofer P., Schneider H., Kaelin K., Huber M., Dotsch C., Christiansen G., and Billeter M.A. 1995. Rescue of measles viruses from cloned DNA. *EMBO J.* **14:** 5773–5784.

Tatsuo H., Ono N., Tanaka K., and Yanagi Y. 2000. SLAM (CDw150) is a cellular receptor for measles virus. *Nature* **406:** 893–897.

von Messling V., Zimmer G., Herrler G., Haas L., and Cattaneo R. 2001. The hemagglutinin of canine distemper virus determines tropism and cytopathogenicity. *J. Virol.* **75:** 6418–6427.

Wild T.F., Fayolle J., Beauverger P., and Buckland R. 1994. Measles virus fusion: Role of the cysteine-rich region of the fusion glycoprotein. *J. Virol.* **68:** 7546–7548.

32 Picornavirus-based Expression Vectors

Steffen Mueller and Eckard Wimmer

Department of Molecular Genetics and Microbiology, Stony Brook University, Stony Brook, New York 11794

ABSTRACT

Picornarviruses comprise a large group of small, nonenveloped, positive-sense, single-stranded RNA viruses. The picornavirus life cycle is usually rapid and exclusively cytoplasmic, without integration into the host cell's genome or a nuclear phase. Due to their biology and genetic constraints, their utility for general gene delivery purposes is limited. However, they may prove useful as vaccine vectors. Furthermore, picornavirus-driven expression of various reporter genes or foreign RNA elements is of interest to the picornavirus molecular virologist.

INTRODUCTION

The genome of the picornaviruses, as exemplified by their most notable genera—enterovirus and rhinovirus, consists of one single-stranded, positive-sense RNA molecule of approximately 7500 nucleotides (Fig. 1A). On entry into the host cell's cytoplasm, the genomic RNA serves as an mRNA that encodes a single viral polyprotein of about 2200 amino acids. In a cascade of concerted autocatalytic cleavage events catalyzed by the virus-encoded proteinase moieties $2A^{pro}$, $3CD^{pro}$, and $3C^{pro}$, the polyprotein yields the set of functional viral proteins necessary for viral proliferation. Instead of the typical m^7G cap of eukaryotic mRNAs, the 5′ terminus of the genomic virion RNA is linked to a small, terminal, virus-encoded protein, VPg. The first approximately 100 nucleotides form a secondary structure crucial for genome replication, which in rhinovirus and enterovirus is termed the cloverleaf due to its characteristic three-stem loop arrangement. However, the majority of the relatively large 5′-nontranslated region (which varies in different picornaviral genera between 600 and 1100 nucleotides) is taken up by a *cis*-acting element that facilitates the assembly of the translation initiation complex. Because the assembly

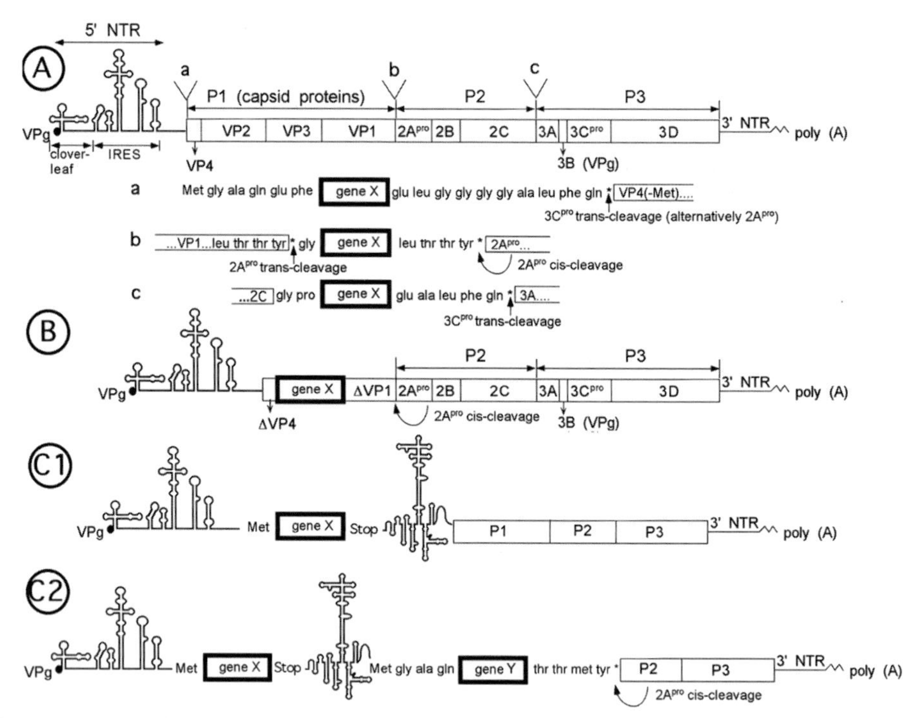

FIGURE 1. Overview of poliovirus vectors. (*A*) Polyprotein fusion vectors. The positions of three insertion sites (*a, b, c*) for in-frame coding sequences are indicated in the polioviral genome, with a detailed view of a representative insertion site given below. (*B*) Subgenomic replicon. An in-frame protein-coding sequence replaces part or all of the viral capsid sequence. (*C1*) Dicistronic virus. The native PV IRES drives expression of an exogenous protein, whereas a second encephalomyocarditis virus IRES drives translation of all PV-specific proteins. (*C2*) Dicistronic replicon. Analogous to *C1*, with most of the capsid-coding region deleted.

process does not occur at the 5′ terminus and is independent of a 5′ m^{7}G cap, this genetic element has been termed the internal ribosomal entry site (IRES). In fact, most picornaviruses very efficiently shut down the host cell's cap-dependent translation by virus-mediated proteolytic cleavage of the canonical translation initiation factor eIF4G, a component of the cap-binding complex eIF-4F. The 3′ end of the picornaviral genome contains a short, nontranslated region and a virus-encoded poly(A) tail, both of which are required for genome replication.

Here, we mainly focus on vector strategies that have been developed for poliovirus (PV). However, due to the close relationship among picornaviruses, these techniques can and have been adapted for many other picornaviral family members. For a more detailed review on the biology of poliovirus, see Mueller et al. (2005).

Expression of Foreign Sequences in a Polioviral Background

The life cycle of most picornaviruses is very rapid (as short as 6–7 hours), and due to the cytotoxic nature of several of their gene products (such as the proteinases 2A^{pro}, 3CD^{pro}, and 3C^{pro}) the typical infection results in a characteristic cytopathic effect (CPE) followed by host-cell lysis. Picornaviral vectors thus generally lend themselves only to transient expression. The advantage of this property is the temporally limited infection followed by clearance by the host immune system. Persistent infections are very rare. Picornaviruses generally infect their host through gastrointestinal or respiratory surfaces, thereby eliciting a strong mucosal immune response. In this regard, they may prove to be useful as recombinant vaccine vectors.

Picornaviruses are genetic minimalists. Their genomes are evolutionarily streamlined to such an extent that none of the gene products can be spared. Consequently, only relatively small exogenous sequences can be inserted and still remain within the maximal genome length constraints for encapsidation. Picornaviral replication is highly error-prone due to the low fidelity of the viral RNA polymerase 3D combined with a high frequency of homologous and illegitimate recombination events. This leads to rapid selection of smaller genomes by successive deletion of the insert, likely driven by the improved capability to replicate and (more importantly) encapsidate the resulting smaller genome. Generally, the larger the insertion, the quicker the selection of deletion variants will be. Although recombinant genomes carrying a foreign gene of up to 117% of the wild-type size have been reported to be viable (Alexander et al. 1994), only about 105% can realistically be stably maintained (Mueller et al. 1998). Genetic instability represents the major drawback of all picornavirus-based vectors (Mueller and Wimmer 2000; Crotty and Andino 2004). The advantages and disadvantages of the various types of picornavirus-based expression vectors are summarized in Table 1.

Polyprotein Fusion Vectors

Polyprotein fusion vectors are the most frequently utilized type of picornaviral vectors (Andino et al. 1994). An in-frame insertion of a foreign gene coding sequence into the open reading frame of the viral polyprotein can be placed such that synthesis of viral gene products is not affected. Three different insertion sites have been shown to be amenable for insertion: at the amino terminus upstream of the viral polyprotein (Fig.1Aa) (Andino et al. 1994; Mattion et al. 1994; Mueller and Wimmer 1998; Dobrikova et al. 2003), between the capsid sequence and $2A^{pro}$ (Fig. 1Ab) (Mattion et al. 1994; Mandl et al. 1998), and between 2C and 3A (Fig. 1Ac) (Paul et al. 1998). Of these, the insertion site between the P1 capsid region and $2A^{pro}$ appears to be the most favorable. The foreign protein is cleaved from the polyprotein by proteinase $2A^{pro}$ through addition of a $2A^{pro}$ cleavage-recognition site Leu-Thr-Thr-Tyr*Gly. Thus, the amino-terminal cleavage is between the final authentic amino acid of VP1 (Tyr) and an essential Gly at the amino terminus of the foreign protein. At the carboxyl terminus, *cis*-cleavage between Tyr and the authentic amino-terminal Gly of $2A^{pro}$ requires a minimal addition of Leu-Thr-Thr-Tyr (Fig. 1Ab). Although early versions of polyprotein fusion vectors used $3C^{pro}$ cleavage sites, it has become clear that $2A^{pro}$-mediated *trans*-cleavage is far more efficient (Tang et al. 1997; Zhao et al. 1999).

Subgenomic Replicons

During high-multiplicity infections, defective interfering (DI) particles arise. The genomes of these particles contain large in-frame deletions within the capsid-coding region. DI genomes retain all replicative functions and are encapsidated in *trans* by wild-type helper virus (for

TABLE 1. Major Properties of Various Picornavirus Expression Vectors

Vector type	Advantages	Disadvantages
Polyprotein fusion virus	1. Ease of propagation. 2. Good antigenic properties.	1. Small insert (1.1-kb maximum). 2. Genetically unstable.
Dicistronic virus	1. Ease of propagation. 2. Authentic protein expression (no fused amino acids).	1. Smallest insert (600-nucleotide maximum). 2. Genetically unstable.
Subgenomic replicon	1. No infectious virus (high safety). 2. Largest insert size (up to 3 kb). 3. Genetically more stable.	1. Encapsidation requires helper virus or capsid expression in *trans*. 2. Labor-intensive amplification. 3. Fusion peptides at amino and carboxyl termini of foreign protein.
Dicistronic replicon	1. No infectious virus (high safety). 2. Large insert size (up to 2.4 kb). 3. Authentic protein expression (no fused amino acids).	1. Encapsidation requires helper virus or capsid expression in *trans*. 2. Labor-intensive amplification.

review, see Wimmer et al. 1993). Consequently, it is possible to replace part or all of the capsid-coding region of poliovirus with an in-frame insertion of foreign coding sequences resulting in their expression (Fig. 1B) (Choi et al. 1991; Percy et al. 1992). Because (at least in enteroviruses) virtually the entire capsid-coding region can be replaced, the replicon strategy allows the largest possible insert size of all vector types (theoretically 3.5 kb, but practically 2.4 kb). In all replicon constructs described thus far, a minimum of the three amino-terminal amino acids of VP4 (GAQ) and four carboxy-terminal amino acids of VP1 (LTTY) remain fused to the foreign protein. Although the former could probably be eliminated, the latter are essential for 2A *cis*-cleavage to release the exogenous protein from the viral polyprotein and generate an authentic amino terminus for 2A. Other picornaviral genera (such as rhinoviruses) can contain an essential RNA structure termed the *cis*-acting replication element (cre), which cannot be deleted (McKnight et al. 1996). Like DI genomes, recombinant subgenomic replicons can be encapsidated by capsid precursors provided in *trans* by a wild-type helper virus. Alternatively, as shown for poliovirus, the capsid precursor P1 can be supplied by preinfection with a recombinant vaccinia virus (VV-P1) resulting in the production of infectious polioviral particles containing the replicon genome (Porter et al. 1993). Encapsidated replicons can be amplified by serial passaging in the presence of VV-P1. This method has the added benefit that the encapsidated replicons can easily be purified from the VV-P1 helper virus due to their different physical properties; PV particles have a different sedimentation coefficient and are resistant to strong detergents and low pH. In the case of a wild-type PV helper virus, separation of the encapsidated replicon would be impossible. Subgenomic replicons appear to be more genetically stable than other vectors described here, presumably because their genome size is equal to or smaller than that of the wild-type virus.

Dicistronic Viral Vectors

In dicistronic polioviral constructs, the viral polyprotein is translated in two parts independently from each other by insertion of a second IRES (Molla et al. 1992). This method has been adapted for the generation of dicistronic expression vectors where translation of a foreign gene is directed by the first IRES, whereas translation of the entire viral polyprotein is driven by a second IRES (Fig. 1C1) (Alexander et al. 1994). A beneficial feature of this type of construct is the ability to express authentic foreign proteins without requiring any amino- or carboxy-terminally fused peptides. Unfortunately, the maximum insert size is reduced even further, because much of the inserted sequence is taken up by the second IRES. The largest reported protein expressed in a dicistronic poliovirus is the gene for chloramphenicol (26 kD) with a total genome size of 117% of wild type. Insertion of the gene for firefly luciferase (67 kD) raised the genome size to 131%. This is well beyond the packaging limits and resulted in a lethal phenotype (Alexander et al. 1994).

Dicistronic Replicons

Although the dicistronic vectors discussed above produce viable viruses, the same technology can be adapted to produce dicistronic replicons by removing the viral capsid-coding sequences (Fig. 1C2). This frees up space for a larger insertion (2–2.5 kb) or even expression of two separate exogenous genes (Zhao et al. 2001) while maintaining a genome size within the constraints for encapsidation. Like their monocistronic counterparts, dicistronic replicons can be encapsidated in *trans* by VV-P1 helper virus (Johansen et al. 2000). The principle of dicistronic vectors was applied to construct the important dicistronic hepatitis C virus (HCV) replicons that enabled much of the current cell-based HCV research (Lohmann et al. 1999).

Protocol

Introducing Recombinant Picornaviral Genomes into Cells

Introduction of recombinant picornaviral genomes into the cell relies on the historic observation that the isolated virion RNA is infectious. This property extends to in-vitro-transcribed RNA as long as the authentic viral 5′ end is preserved. That said, up to two additional 5′-terminal guanine residues (remnants from the T7 RNA polymerase-based in vitro transcription), although reducing infectivity, can be tolerated. Additional nucleotides at the 3′ end are of far less consequence. Thus, any unique restriction site downstream from the poly(A) sequence (preferably as close as possible downstream) can be used to linearize the plasmid containing the viral genome before in vitro transcription.

MATERIALS

CAUTION: See Appendix for appropriate handling of materials marked with <!>.

Reagents

Agarose

Calf serum

DEAE-dextran (0.5 mg/ml; M_r = 500,000; Sigma-Aldrich) <!>

Add 20 mg of DEAE-dextran into a sterile Falcon tube containing 40 ml of HBSS. Shake vigorously for 5 minutes. Store for up to 1 month at 4°C.

DNase (RNase-free; Roche 10776785001)

Dulbecco's modified Eagle's medium (DMEM, serum-free; Invitrogen 11965-092)

Different cell types may require other media.

Ethanol (70% v/v and 100%) <!>

Hanks' balanced salt solution (HBSS; Invitrogen 14025-092)

HeLa R19 cells, exponentially growing

H_2O (ultrapure)

The use of ultrapure water (18.2 MΩ cm) in conjunction with dedicated chemicals used only for RNA work eliminates the need for autoclaving or DEPC treatment of aqueous solutions.

In vitro T7 RNA Polymerase Kit (Stratagene 600124, or equivalent)

Phenol:chloroform (50:50 v/v) <!>

Plasmid

Plasmids should be constructed such that the recombinant viral or replication genome expressing the gene of interest is inserted immediately downstream from a T7 RNA polymerase promoter.

Restriction enzyme, appropriate to linearize the plasmid of interest

TAE buffer

- 40 mM Tris <!>
- 1 mM EDTA
- 20 mM acetic acid <!>

Equipment

CentriSpin 20 columns (Princeton Separations, CS-201 or equivalent)
Eppendorf tubes
Freezer (–80°C)
Incubator (37°C, CO_2)
Microcentrifuge
Rocking platform
Tissue-culture dishes (35-mm diameter)

METHODS

Transcription of Recombinant RNA

A 50-µl transcription reaction should yield between 5 and 20 µg of RNA, sufficient for transfecting several 35-mm dishes. Anticipate using 2 µg of RNA per 35-mm dish. If more dishes are to be transfected, perform larger transcription reactions accordingly.

1. Linearize 1–2 µg of viral or replicon plasmid containing the gene of interest.
2. In-vitro-transcribe the linearized template using a commercial T7 RNA Polymerase Kit in a final reaction volume of 50 µl, according to the manufacturer's instructions.

   Optionally, the template DNA can be removed after transcription by adding 10 units of RNase-free DNase and incubating for 15 minutes at 37°C.
3. Determine RNA concentration.
4. Purify the RNA on a CentriSpin 20 column (or equivalent desalting spin column) according to the manufacturer's instructions.

   Alternatively, RNA may be purified by phenol:chloroform extraction and ethanol precipitation. If the concentration of the RNA transcript is high enough (>200 ng/µl), RNA purification can be omitted.
5. Assess the quality and integrity of the RNA on a TAE-buffered 1% agarose gel.
6. Use the RNA immediately for transfection or store at –80°C until needed.

Transfection of Cells with RNA

7. Twenty-four hours before RNA transfection, plate exponentially growing HeLa R19 cells in 35-mm dishes.

   At the time of transfection, cells should be 50–70% confluent.
8. For each dish to be transfected, mix 2 µg of in-vitro-transcribed RNA (from Step 6) with 300 µl of DEAE-dextran. Incubate for 30 minutes on ice.

   The volume of RNA added per 300 µl of DEAE-dextran should not exceed 10 µl. If the RNA concentration from the in vitro transcription reaction is higher than 200 ng/µl, it can be added directly without purification (omit Step 4).
9. Immediately before transfection, wash HeLa monolayers twice with cold HBSS.
10. Aspirate the HBSS from the dishes. Add 300 µl of DEAE-dextran/RNA complex to each dish. Place on a rocking platform for 30 minutes at room temperature.

    Include a control dish that receives the same treatment but without any RNA. Make sure that the cells do not dry out.
11. Add 2 ml of prewarmed serum-free DMEM to each dish. Incubate in a CO_2 incubator for 1 hour at 37°C.

12. Aspirate the medium. Replace with prewarmed DMEM containing 2% calf serum. Return to the CO_2 incubator at 37°C.
13. Examine plates periodically for virus-induced CPE (viral constructs) or perform expression assays for the inserted gene (replicon and viral constructs).

    Protein expression from subgenomic polioviral replicons peaks 6–12 hours posttransfection. Because no infectious particles are formed, there is no spread to other (nontransfected) cells. As transfected cells die, expression decreases rapidly. Cells transfected with viral constructs behave similarly during the first few hours posttransfection. However, infectious particles propagate, spread, and kill all cells within the dish. For wild-type polioviral RNA, complete CPE is usually observed within 24 hours posttransfection. Virus-based expression vectors are usually marked by delayed CPE (up to 4 days). Generally, the length of the delay in CPE as compared to wild-type virus often represents reduced fitness and genetic stability of the construct.
14. Freeze-thaw dishes three times to liberate virus from cells.
15. Transfer lysate to Eppendorf tubes. Centrifuge in a microcentrifuge at 14,000 rpm for 1 minute.
16. Store supernatants containing virus at –80°C.

    Virus can be amplified by passaging once or twice. Due to the general genetic instability of most constructs, it is advisable to use the earliest possible passages for any subsequent experiment and to frequently regenerate the virus by RNA transfection, rather than amplification by repeated passaging. Monitor genetic stability of the viral construct by reverse transcriptase–polymerase chain reaction using oligonucleotide primers flanking the inserted sequence (Mueller and Wimmer 1998).

REFERENCES

Alexander L., Lu H.H., and Wimmer E. 1994. Polioviruses containing picornavirus type 1 and/or type 2 internal ribosomal entry site elements: Genetic hybrids and the expression of a foreign gene. *Proc. Natl. Acad. Sci.* **91:** 1406–1410.

Andino R., Silvera D., Suggett S.D., Achacoso P.L., Miller C.J., Baltimore D., and Feinberg M.B. 1994. Engineering poliovirus as a vaccine vector for the expression of diverse antigens. *Science* **265:** 1448–1451.

Choi W.S., Pal-Gosh R.S., and Morrow C.D. 1991. Expression of human immunodeficiency virus type 1 (HIV-1) Gag, Pol, and Env proteins from chimeric HIV-1 poliovirus minireplicons. *J. Virol.* **65:** 2875–2883.

Crotty S. and Andino R. 2004. Poliovirus vaccine strains as mucosal vaccine vectors and their potential use to develop an AIDS vaccine. *Adv. Drug Deliv. Rev.* **56:** 835–852.

Dobrikova E.Y., Florez P., and Gromeier M. 2003. Structural determinants of insert retention of poliovirus expression vectors with recombinant IRES elements. *Virology* **311:** 241–253.

Johansen L.K. and Morrow C.D. 2000. Inherent instability of poliovirus genomes containing two internal ribosome entry site (IRES) elements supports a role for the IRES in encapsidation. *J. Virol.* **74:** 8335–8342.

Lohmann V., Korner F., Koch J., Herian U., Theilmann L., and Bartenschlager R. 1999. Replication of subgenomic hepatitis C virus RNAs in a hepatoma cell line. *Science* **285:** 110–113.

Mandl S., Sigal L.J., Rock K.L., and Andino R. 1998. Poliovirus vaccine vectors elicit antigen-specific cytotoxic T cells and protect mice against lethal challenge with malignant melanoma cells expressing a model antigen. *Proc. Natl. Acad. Sci.* **95:** 8216–8221.

Mattion N.M., Reilly P.A., DiMichele S.J., Crowley J.C., and Weeks-Levy C. 1994. Attenuated poliovirus strain as a live vector: Expression of regions of rotavirus outer capsid protein VP7 by using recombinant Sabin 3 viruses. *J. Virol.* **68:** 3925–3933.

McKnight K.L. and Lemon S.M. 1996. Capsid coding sequence is required for efficient replication of human rhinovirus 14 RNA. *J. Virol.* **70:** 1941–1952.

Molla A., Jang S.K., Paul A.V., Reuer Q., and Wimmer E. 1992. Cardioviral internal ribosomal entry site is functional in a genetically engineered dicistronic poliovirus. *Nature* **356:** 255–257.

Mueller S. and Wimmer E. 1998. Expression of foreign proteins by poliovirus polyprotein fusion: Analysis of genetic stability reveals rapid deletions and formation of cardioviruslike open reading frames. *J. Virol.* **72:** 20–31.

———. 2000. Picornaviruses as tools for antigen delivery. In *Viral vectors: Basic science and gene therapy* (ed. A. Cid-Arregui and A. García-Carrancá), pp. 543–562. Eaton Publishing, Natick, Massachusetts.

Mueller S., Wimmer E., and Cello J. 2005. Poliovirus and poliomyelitis: A tale of guts, brains, and an accidental event. *Virus Res.* **111:** 175–193.

Paul A.V., Mugavero J.A., Molla A., and Wimmer E. 1998. Internal ribosomal entry site scanning of the poliovirus polyprotein: Implications for proteolytic processing. *Virology* **250:** 241–253.

Percy N., Barclay W.S., Sullivan M., and Almond J.W. 1992. A poliovirus replicon containing the chloramphenicol acetyltransferase gene can be used to study the replication and encapsidation of poliovirus RNA. *J. Virol.* **66:** 5040–5046.

Porter D.C., Ansardi D.C., Choi W.S., and Morrow C.D. 1993.

Encapsidation of genetically engineered poliovirus minireplicons which express human immunodeficiency virus type 1 Gag and Pol proteins upon infection. *J. Virol.* **67:** 3712–3719.

Tang S., van Rij R., Silvera D., and Andino R. 1997. Toward a poliovirus-based simian immunodeficiency virus vaccine: Correlation between genetic stability and immunogenicity. *J. Virol.* **71:** 7841–7850.

Wimmer E., Hellen C.U., and Cao X. 1993. Genetics of poliovirus. *Annu. Rev. Genet.* **27:** 353–436.

Zhao W.D. and Wimmer E. 2001. Genetic analysis of a poliovirus/hepatitis C virus chimera: New structure for domain II of the internal ribosomal entry site of hepatitis C virus. *J. Virol.* **75:** 3719–3730.

Zhao W.D., Wimmer E., and Lahser F.C. 1999. Poliovirus/hepatitis C virus (internal ribosomal entry site-core) chimeric viruses: Improved growth properties through modification of a proteolytic cleavage site and requirement for core RNA sequences but not for core related polypeptides. *J. Virol.* **73:** 1546–1554.

33 Reverse Genetics of Influenza Viruses

Glenn A. Marsh and Peter Palese

Department of Microbiology, Mount Sinai School of Medicine, New York 10029

ABSTRACT

Influenza A virus is considered to be a good viral vector candidate because infection in humans and laboratory animals results in a strong, long-lasting cellular and humoral immune response. Therefore, influenza A viruses have been used for delivery of foreign antigens, with the purpose of eliciting immune responses against these antigens. The use of influenza virus as a vector requires the establishment of reverse genetics techniques, which allow for the direct genetic manipulation of the viral genome. This chapter describes methods utilized for the recovery of infectious influenza viruses from cDNA. These techniques have been used extensively in the development of vaccine candidates, for the incorporation and expression of foreign sequences, and in the study of influenza viral replication. An advantage in the use of influenza viruses as vectors for antigen delivery is the large pool of antigenically different viruses. This means that preexisting immunity to one strain does not prevent the use of an antigenically different strain.

INTRODUCTION

Influenza A virus, the prototypic member of the orthomyxovirus family, is an enveloped virus with a genome composed of eight single-stranded, negative-sense RNA segments that can encode 11 proteins. Transcription and replication of the viral genome occur in the nucleus of infected cells. The viral ribonucleoprotein (vRNP) complex, which is composed of the three virally encoded polymerase proteins PB1, PB2, and PA and the nucleoprotein NP, is thought to be responsible for many functions in the replication cycle of influenza viruses.

The generation of influenza viruses by reverse genetics is more complex than that of other negative-stranded viruses because eight different viral RNA segments with precise 5′ and 3′ ends must be generated. In addition, because viral RNA replication and transcription occur in the nucleus of infected cells, RNAs must be either transported to or generated in the nucleus. Initial experiments eventually leading to the genetic engineering of influenza viruses involved the reconstitution of functional RNP complexes in vitro (Honda et al. 1987; Szewczyk et al. 1988; Kobayashi et al. 1992). Using purified polymerase proteins, in vitro synthesis of short RNAs (Parvin et al. 1989; Kobayashi et al. 1992) and full-length vRNAs (Honda et al. 1990) demonstrated that the four influenza viral proteins PB1, PB2, PA, and NP were the minimal requirements for efficient RNA transcription.

Additional studies demonstrated the transfection of functional RNPs into cells. For this, virus-like RNA was generated in vitro from a plasmid containing the cDNA clone flanked by a T7 promoter and a hepatitis delta RNA-derived, self-splicing sequence. The RNA was mixed with purified virion NP and polymerase proteins and transfected to cells, before or after infection with a helper influenza virus to provide the remaining vRNPs (Luytjes et al. 1989; Enami et al. 1990; Enami and Palese 1991; Martin et al. 1992; Li et al. 1995). Rescue of infectious virus containing the cDNA-derived RNA required selection of the novel virus against the helper virus. Alternatively, cells were transfected with a plasmid construct containing a gene of interest that had been inserted within the 5′ and 3′ noncoding regions of the influenza genome segments. Negative-sense RNA transcription from the plasmid was directed by an RNA polymerase I promoter and terminator sequences. Cellular RNA polymerase I catalyzed the synthesis of vRNA from the plasmid construct and the viral polymerase proteins, and the additional genomic segments were supplied by infection with a helper virus (Neumann et al. 1994; Pleschka et al. 1996). Unfortunately, the need for such helper viruses proved to be a disadvantage in these early systems, because they had to be selected against to identify and purify the modified strain.

A decade after the initial influenza reverse genetics system had been described, influenza viruses were generated entirely from cloned cDNAs (Fodor et al. 1999; Neumann et al. 1999). In this system, cDNA from each of the eight genome segments was cloned in negative orientation between a truncated human RNA polymerase I promoter and the hepatitis delta virus ribozyme (Fodor et al. 1999). Transfection into Vero cells of eight vRNA-encoding plasmids along with four plasmids expressing the polymerase complex PB1, PB2, PA, and NP resulted in recovery of infectious virus (Fig. 1). Because helper virus was not required for the generation of recombinant virus, the cumbersome selection process proved to be unnecessary. Improvements to this system now include the transfection of cocultured 293T cells (a human cell line necessary, due to the human RNA polymerase I promoter) and Madin-Darby canine kidney (MDCK) cells.

Further improvements to these systems were reported, in which only eight plasmids were required (Hoffmann et al. 2000a,b). The plasmids contained cDNAs of genomic segments cloned in negative orientation with a human RNA polymerase I promoter at the 5′ end and the mouse RNA polymerase I terminator at the 3′ end. Downstream from the RNA polymerase I terminator was a cytomegalovirus (CMV) immediate-early promoter. A polyadenylation sequence was inserted at the other end, giving rise to a polymerase-II-driven mRNA transcript. The cellular RNA polymerase I was responsible for the transcription of the cDNA into vRNA. Expressed viral proteins and vRNAs then assembled and resulted in the budding of plasmid-only-derived infectious virus. Furthermore, by combining the eight vRNA-expressing plasmids into one plasmid, the efficiency of rescue in cells that are hard to transfect was improved (Neumann et al. 2005).

Influenza viruses have been prepared that express foreign genes, demonstrating their utility as a vector to deliver foreign antigens to the immune system. Several approaches have been successful for the expression of foreign antigens by influenza viruses (for review, see García-Sastre 2000). For example, the antigenic domains of the influenza hemagglutinin and neuraminidase glycoproteins have been replaced with epitopes from foreign proteins. The ectodomains of influenza viral surface glycoproteins were similarly replaced with those of foreign glycoproteins. Furthermore, existing viral genomic segments were modified to express influenza viral proteins fused to foreign proteins as polyproteins. These fusion proteins can be subsequently cleaved into two proteins.

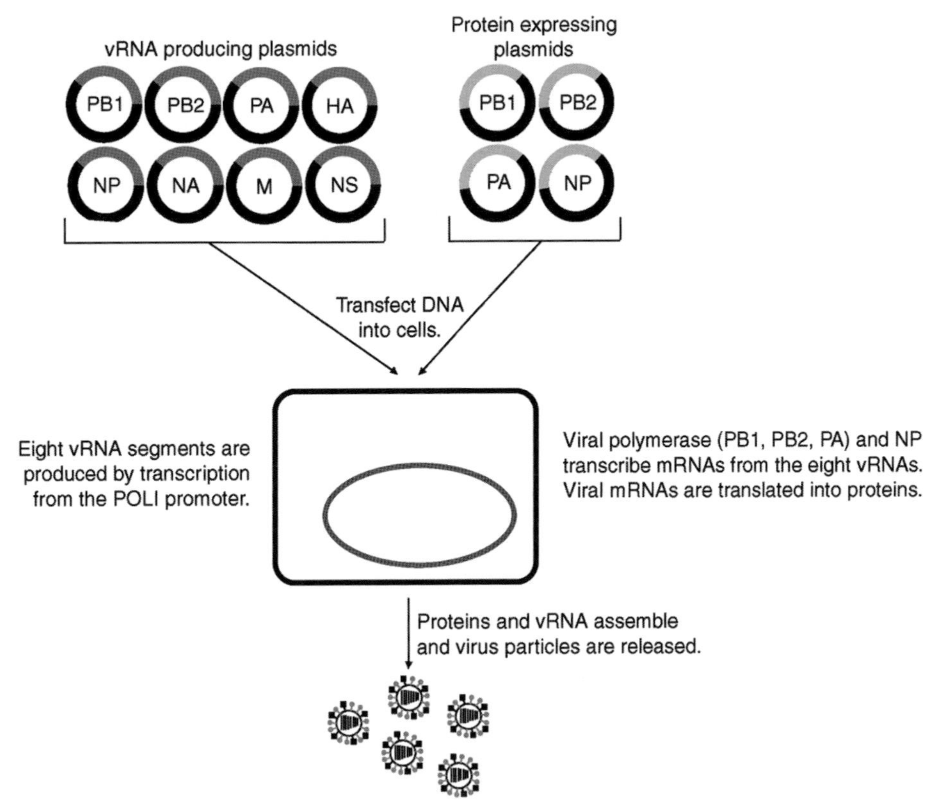

FIGURE 1. Schematic representation of the generation of influenza virus entirely from cDNA. Cells are cotransfected with plasmids encoding all eight segments of vRNA under the control of RNA polymerase I. Cellular RNA polymerase I synthesizes vRNAs, which are replicated and transcribed by the viral polymerase and NP proteins provided by the four protein-expressing plasmids.

Viruses could also be generated with foreign antigen encoded by a ninth RNA segment. Finally, influenza viruses can be used to express foreign antigens by using an internal ribosomal entry site (IRES) engineered into the viral mRNA.

Influenza viruses present a number of advantages over other viral systems for the expression of foreign proteins. Influenza viruses are nonintegrating and nononcogenic. Infection with influenza viruses elicits a strong, long-lasting immune response. In addition, because there is a large antigenic pool of viruses available, preexisting immunity to one strain does not prevent the use of an antigenically different strain. There are many different attenuated strains of influenza virus, and together with reverse genetics, it is possible to generate viral vectors with a good balance between attenuation and antigenicity. However, the capacity of influenza viruses to accommodate a foreign sequence is quite limited, and they require packaging signals on both the 5′ and 3′ ends of the vRNA. Indeed, the coding regions at both the 5′ and 3′ ends of the genomic RNA of the hemagglutinin (Watanabe et al. 2002), neuraminidase (Fujii et al. 2003), nonstructural (Fujii et al. 2005), and polymerase (Liang et al. 2005) segments aid in the packaging of vRNA segments into newly assembling virions.

Protocol

Generation of Influenza Virus by Reverse Genetics

MATERIALS

Reagents

Bovine serum albumin (BSA)
Cells
  239T cells (ATCC, CRL-11268)
  MDCK cells (ATCC, CCL-34)
    Cells should be grown to confluency in 75-cm^2 cell culture flasks before transfection.
Dulbecco's modified Eagle's medium (DMEM)
Fetal calf serum (FCS)
Lipofectamine 2000 Reagent (Invitrogen 11668-027)
Opti-MEM I Reduced-Serum medium (GIBCO 31985)
Phosphate-buffered saline (PBS)
Plasmid stocks (0.5 μg/μl), transfection quality
  protein expression constructs
    Prepare a separate construct for each of the following proteins: PA, PB1, PB2, and NP.
  RNA expression constructs
    Prepare a separate construct for each of the following eight influenza virus: vRNAs HA, M, NA, NP, NS, PA, PB1, and PB2.
TPCK-trypsin (Sigma-Aldrich T-8802)

Equipment

Cell culture dishes (35 mm^2)
Cell culture flasks (75 cm^2)
Incubator (37°C, 5% CO_2)
Tissue-culture plates (six well)

METHOD

1. Combine 0.5 μg each of the eight RNA expression plasmids and the four protein expression plasmids. Dilute to a final total volume of 100 μl with Opti-MEM medium at room temperature.
2. In a second tube, combine 6 μl of room-temperature Lipofectamine and 94 μl of Opti-MEM medium per transfection. Mix well. Incubate for 5 minutes at room temperature.
3. Add 100 μl of the transfection reagent master mix (Step 2) to the DNA mix (Step 1). Mix gently by tapping the tube. Incubate for 20 minutes at room temperature.
4. Trypsinize one 75-cm^2 flask each of confluent 293T and MDCK cells. Resuspend detached cells in 10 ml of room-temperature DMEM supplemented with 10% FCS (no antibiotics).

5. Centrifuge at approximately 300*g* for 5 minutes. Resuspend pelleted cells in 10 ml of DMEM supplemented with 10% FCS.
6. Mix 293T and MDCK cells in a 1:1 ratio. Aliquot 1 ml of mixed cells per 35-mm^2 dish.
7. Add DNA/Lipofectamine mix to cells in 35-mm^2 dishes. Rock the dishes back and forth to mix and distribute cells. Incubate for 6–24 hours at 37°C.
8. Replace media with Opti-MEM supplemented with 0.01% FCS, 0.3% BSA, and 1 μg/ml TPCK-trypsin. Return to incubator.
9. One day before virus passaging (i.e., about 24 hours posttransfection), seed six-well tissue-culture plates with MDCK cells. Incubate at 37°C.
10. Forty-eight hours posttransfection, collect the supernatant from the transfected mixed cells. Clarify supernatant by centrifugation at 7000*g*.
11. Remove media from MDCK cells in the six-well tissue-culture plates.

    MDCK cells should be 80–90% confluent at the time of infection (48 hours posttransfection).
12. Wash cells twice with PBS. Add 200 μl of the clarified supernatant (Step 10) to the cells. Incubate the plate with regular rocking for 1 hour at 37°C.
13. Add 2 ml of Opti-MEM supplemented with 0.01% FCS, 0.3% BSA, and 1 μg/ml TPCK-trypsin to the cells. Incubate at 37°C in a 5% CO_2 incubator.
14. Monitor cells infected with the transfection supernatant daily for the presence of CPE and HA activity in the culture supernatants.

REFERENCES

Enami M. and Palese P. 1991. High-efficiency formation of influenza virus transfectants. *J. Virol.* **65:** 2711–2713.

Enami M., Luytjes W., Krystal M., and Palese P. 1990. Introduction of site-specific mutations into the genome of influenza virus. *Proc. Natl. Acad. Sci.* **87:** 3802–3805.

Fodor E., Devenish L., Engelhardt O.G., Palese P., Brownlee G.G., and García-Sastre A. 1999. Rescue of influenza A virus from recombinant DNA. *J. Virol.* **73:** 9679–9682.

Fujii K., Fujii Y., Noda T., Muramoto Y., Watanabe T., Takada A., Goto H., Horimoto T., and Kawaoka Y. 2005. Importance of both the coding and the segment-specific noncoding regions of the influenza A virus NS segment for its efficient incorporation into virions. *J. Virol.* **79:** 3766–3774.

Fujii Y., Goto H., Watanabe T., Yoshida T., and Kawaoka Y. 2003. Selective incorporation of influenza virus RNA segments into virions. *Proc. Natl. Acad. Sci.* **100:** 2002–2007.

García-Sastre A. 2000. Transfectant influenza viruses as antigen delivery vectors. *Adv. Virus Res.* **55:** 579–597.

Hoffmann E., Neumann G., Hobom G., Webster R.G., and Kawaoka Y. 2000a. "Ambisense" approach for the generation of influenza A virus: vRNA and mRNA synthesis from one template. *Virology* **267:** 310–317.

Hoffmann E., Neumann G., Kawaoka Y., Hobom G., and Webster R.G. 2000b. A DNA transfection system for generation of influenza A virus from eight plasmids. *Proc. Natl. Acad. Sci.* **97:** 6180–6113.

Honda A., Ueda K., Nagata K., and Ishihama A. 1987. Identification of the RNA polymerase-binding site on genome RNA of influenza virus. *J. Biochem.* **102:** 1241–1249.

Honda A., Mukaigawa J., Yokoiyama A., Kato A., Ueda S., Nagata K., Krystal M., Nayak D.P., and Ishihama A. 1990. Purification and molecular structure of RNA polymerase from influenza virus A/PR8. *J. Biochem.* **107:** 624–628.

Kobayashi M., Tuchiya K., Nagata K., and Ishihama A. 1992. Reconstitution of influenza virus RNA polymerase from three subunits expressed using recombinant baculovirus system. *Virus Res.* **22:** 235–245.

Li S., Xu M., and Coelingh K. 1995. Electroporation of influenza virus ribonucleoprotein complexes for rescue of the nucleoprotein and matrix genes. *Virus Res.* **37:** 153–161.

Liang Y., Hong Y., and Parslow T.G. 2005. *cis*-Acting packaging signals in the influenza virus PB1, PB2, and PA genomic RNA segments. *J. Virol.* **79:** 10348–10355.

Luytjes W., Krystal M., Enami M., Parvin J.D., and Palese P. 1989. Amplification, expression and packaging of a foreign gene by influenza virus. *Cell.* **59:** 1107–1113.

Martin J., Albo C., Ortin J., Melero J.A., and Portela A. 1992. In vitro reconstitution of active influenza virus ribonucleoprotein complexes using viral proteins purified from insect cells. *J. Gen. Virol.* **73:** 1855–1859.

Neumann G., Zobel A., and Hobom G. 1994. RNA polymerase I-mediated expression of influenza viral RNA molecules. *Virology* **202:** 477–479.

Neumann G., Fujii K., Kino Y., and Kawaoka Y. 2005. An improved reverse genetics system for influenza A virus generation and its implications for vaccine production. *Proc. Natl. Acad. Sci.* **102:** 16825–16829.

Neumann G., Watanabe T., Ito H., Watanabe S., Goto H., Gao P., Hughes M., Perez D.R., Donis R., Hoffmann E., Hobom G., and Kawaoka Y. 1999. Generation of influenza A viruses entirely from cloned cDNAs. *Proc. Natl. Acad. Sci.* **96:** 9345–9350.

Parvin J.D., Palese P., Honda A., Ishihama A., and Krystal M. 1989. Promoter analysis of influenza virus RNA polymerase. *J. Virol.* **63:** 5142–5152.

Pleschka S., Jaskunas S.R., Engelhardt O.G., Zurcher T., Palese P., and García-Sastre A. 1996. A plasmid-based reverse genetics system for influenza A virus. *J. Virol.* **70:** 4188–4192.

Szewczyk B., Laver W.G., and Summers D.F. 1988. Purification, thioredoxin renaturation, and reconstituted activity of the three subunits of the influenza A virus RNA polymerase. *Proc. Natl. Acad. Sci.* **85:** 7907–7911.

Watanabe T., Watanabe S., Neumann G., Kida H., and Kawaoka Y. 2002. Immunogenicity and protective efficacy of replication-incompetent influenza virus-like particles. *J. Virol.* **76:** 767–773.

34 An Overview of Condensing and Noncondensing Polymeric Systems for Gene Delivery

Dinesh B. Shenoy and Mansoor M. Amiji

Department of Pharmaceutical Sciences, School of Pharmacy, Northestern University, Boston, Massachusetts 02115

INTRODUCTION

With the recent completion of the mapping of the human genome, genetic and molecular medicine holds significant promise for alleviation or cure for many diseases, which are currently untreatable or have poor prognoses (Collins and Mansoura 2001; Carroll 2003; Collins et al. 2003a,b). Gene therapy has been developed on two fronts: Initial efforts have been based on exploiting the cellular mechanisms existing within viruses by introducing therapeutic DNA into the virus genome (Carter and Samulski 2000; Walther and Stein 2000; Loser et al. 2002; Davidson and Breakefield 2003). With parallel advancement in polymer-based therapeutics, a majority of the current studies are based on synthetic nonviral vectors (Lollo et al. 2000; Fenske et al. 2001; Godbey and Mikos 2001; Schatzlein 2001; Liu and Huang 2002).

Polymeric gene delivery vectors are expected to have the following fundamental features: (1) an ability to interact specifically with the target tissue and cells and to deliver the genetic payload in a site-specific manner, (2) resistance to degradation by the metabolic or immune pathways and protection of the genetic material, (3) proven safety and minimal toxicity, and (4) an ability to express the therapeutic gene for a finite period of time in an appropriate regulated fashion.

Technically, the nonviral approach is relatively straightforward and does not provoke specific immune responses. However, delivery and targeting efficiency remain issues of concern compared to their viral counterparts.

The rationale of development of self-assembling synthetic vectors for DNA delivery is based on functional polymers. These polymers are designed to fulfill a number of biological functions, including encapsulating DNA within discrete particles and stabilizing it by protecting it with a hydrophilic polymer coating. The encapsulated DNA may be in a condensed or noncondensed form depending on the nature of the polymer and the technique used for formulating the vector system.

A polymeric gene delivery system must have three fundamental components for the product to be successful, (1) a gene that encodes a therapeutic protein, (2) a packing system to protect the gene from degradation during passage into a cell and to control production of the encoded protein, and (3) a mode of administration, such as by injection, through a catheter, or by aerosol. Once taken up by the target cells (typically by endocytosis), the carrier system must unpack the genetic material and facilitate its nuclear transport—the site of action for genetic expression. The whole process presents many barriers—at both tissue and cellular levels—and overcoming these hurdles is the principal objective for efficient polymer-based DNA therapeutics.

CONDENSING POLYMERIC SYSTEMS

DNA condensation has become an important area for research in diverse areas of science because it represents a process by which genetic information is packaged and protected within cells. Condensation is defined as a decrease in the volume occupied by the DNA molecule from a large domain occupied by a worm-like random coil, to a compact state in which the volume fractions of solvent and DNA are comparable (Bloomfield 1991). During the condensation process, extended DNA chains collapse into compact, orderly particles containing only one or a few molecules. The decrease in size of the DNA domain is striking, as is the characteristic toroidal morphology of the condensed particle.

Condensing agents generally work by either (1) decreasing repulsions between DNA segments (e.g., neutralizing phosphate charge and/or reorienting water dipoles near DNA surfaces by multivalent cations) or (2) making DNA-solvent interactions less favorable (e.g., by adding ethanol, which is a poorer solvent than water for DNA, or by adding another polymer, such as polyethylene glycol (PEG), which excludes volume). Multivalent cations may also cause localized bending or distortion of the DNA, thereby facilitating condensation (Bloomfield 1991).

The majority of condensing polymeric systems are based on electrostatic interactions between a negatively charged DNA molecule and a cationic polymer. In aqueous solutions, condensation normally requires cations of charge +3 or greater. DNA fragments shorter than about 400 base pairs will not condense into orderly, discrete particles. Thus, the net attractive interactions per base pair are likely very small—at least several hundred base pairs must interact, either intramolecularly or intermolecularly, to form a stably condensed particle. The supramolecular structures formed by interaction between DNA and a condensing polymer can be described as a hydrophobic core formed by DNA, the charge of which is compensated for by polycation blocks of the polymer, surrounded by a shell of hydrophilic polymer chains.

Depending on the agents used for condensing DNA, the complexes may be classified as polyplexes (complexation between DNA and the cationic polymer) (De Smedt et al. 2000; Hagstrom 2000; Wagner 2004) or lipoplexes (complexation between DNA and the cationic lipid) (Audouy and Hoekstra 2001; Ogris and Wagner 2002; Zhdanov et al. 2002; Pedroso de Lima et al. 2003; Tranchant et al. 2004). If the cationic polymer has a lipid entity attached to it, the resultant complex with DNA is called a lipopolyplex (Pampinella et al. 2000; Fenske et al. 2001; Tsai et al. 2002; Harvie et al. 2003). In all cases, the complexation ratio and the overall charge density are maintained such that a slightly positive charge prevails for the complexed DNA to facilitate cellular interactions and subsequent endocytosis. A comprehensive listing of cationic polymers and lipids (both natural and synthetic) is given in Table 1.

Typically, dilute solutions (micromolar units) of the DNA and the cationic polymer are mixed under gentle agitation. The polymer is maintained in excess, acts as a colloidal stabilizing agent, and imparts an overall positive charge to the system. Discrete particles of the complex in the nanometric size range are formed instantaneously (Fig. 1a). The polymer can be engineered to contain a hydrophilic chain, such as PEG, that acts as a sheath to the nanoparticles, or it may have a ligand block that will form the corona for the nanoparticle and acts as a homing (targeting) entity (Fig. 2). Such multifunctional block copolymers are increasingly popular for the development of DNA condensing systems that can be injected into the systemic circulation and find the target tissue and cells for efficient transfection.

The unpackaging of DNA from a condensing polymer presents a challenging problem from a design perspective because it requires the introduction of functionality that inherently opposes that required for condensing it. Release of DNA from self-assembled polycation/DNA complexes is a critical and poorly understood step in condensing polymer-based gene delivery systems. This process, however, creates opportunities for the design of new materials that balance these opposing criteria more effectively.

TABLE 1. Condensing Polymers Used in Gene Transfer

Polymer	References
Natural	
Poly-L-lysine	Read et al. (1999); Wang et al. (2001)
Chitosan	Borchard (2001); Liu et al. (2005)
Gelatin	Leong et al. (1998); Fukunaka et al. (2002)
Spermine/spermidine	Thomas et al. (1996)
Poly-L-ornithine	Brown et al. (2000)
Synthetic	
Polyethyleneimine	Forrest et al. (2003)
Polyethyleneimine-*g*-PEG-RGD	Kim et al. (2005)
Poly-β-amino esters	Anderson et al. (2005)
Cationic dendrimers	Bielinska et al. (1999); Vlasov et al. (2004); Zhang et al. (2005)
Cationic lipids	Lam et al. (2004); Janat-Amsbury et al. (2005)
Poly(L-lysine)-*g*-sulfonylurea	Kang et al. (2005)
Folate-polyethyleneimine-block-poly(L-lactide)[a]	Wang and Hsiue (2005)
Poly(dimethylaminoethyl methacrylate)-poly(butylmethacrylate)-PEG	Funhoff et al. (2005)
Poly(*N*-isopropyl acrylamide)	Zhang et al. (2004)
Acrylate and methacrylate polymers	Fonseca et al. (1999); Cortesi et al. (2004)
Polyphosphoramidates	Wang et al. (2004)
Polyaminoethyl propylene phosphate	Li et al. (2004)
Protamine	Dunne et al. (2003)
Polyphosphazenes	Luten et al. (2003)
Poly(amino ester)	Lim et al. (2002)
Galactose-PEG-polyethyleneimine[a]	Sagara and Kim (2002)
Folate-PEG-folate-graft-polyethyleneimine[a]	Benns et al. (2001)
Poly(ethyleneimine)-co-(N-[2-aminoethyl] ethyleneimine)-co-*N*-(*N*-cholesteryloxycarbonyl-[2-aminoethyl]ethyleneimine)	Han et al. (2001)
Poly[α-(4-aminobutyl)-L-glycolic acid]	Koh et al. (2000)
Cationic peptides	Fominaya et al. (2000)
1,4-dihydropyridine	Hyvonen et al. (2000)
Lactose-PEG-grafted poly-L-lysine	Choi et al. (1998)
Phosphatidylethanolamine	Hope et al. (1998)

[a]Hybrid multifunctional polymer

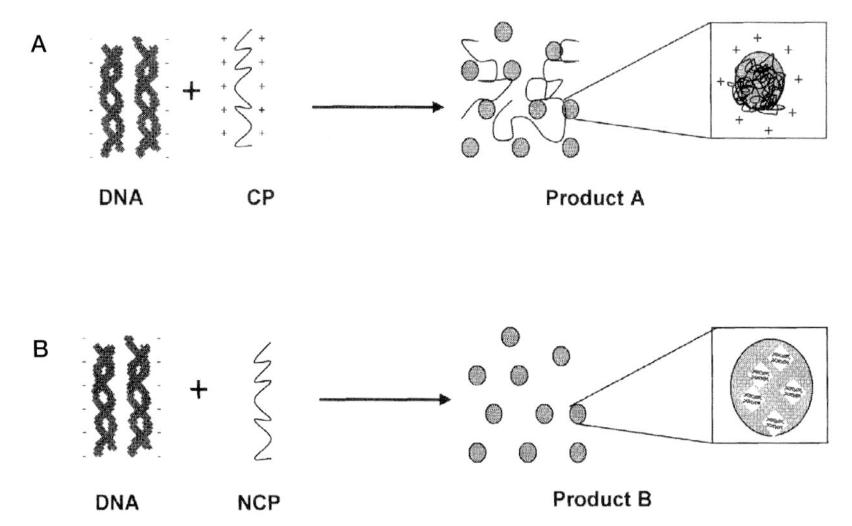

FIGURE 1. Schematic representation of gene delivery systems prepared from condensing polymers (CP) and noncondensing polymers (NCP). (*A*) Product A is principally a complex in the form of nanoparticles, stabilized by an excess of condensing polymer acting as a cationic colloidal stabilizer. (*B*) Product B represents a solid micro-nanoparticle, formed by the polymer that embeds the DNA within the micronanomatrix. The process of particle formation is generally assisted by a high-shear/pressure process.

NONCONDENSING POLYMERIC SYSTEMS

Whereas condensing polymers form self-assembled complexes with DNA to derive at a packed structure, noncondensing polymers depend on physical entrapment strategies to encase the DNA within a polymeric matrix (Lengsfeld et al. 2002; Otsuka et al. 2003; Ravi Kumar et al. 2004).

The design principle is based on precipitating the polymer in the presence of DNA in solution (Fig. 1B). The precipitation can be brought about by a reduction in solubility of the polymer either by changing the solvent or by adding another agent to bring about coacervation. Particle formation is generally assisted by high shear (e.g., with a Silverson high-speed homogenizer) or high-pressure (e.g., with a microfluidizer) for obtaining particles of the desired size. The resultant products are either micro-/nanocapsules (wherein the polymer forms a coat around a DNA-filled core) or micro-/nanoparticles (wherein the DNA is dispersed within the polymeric matrix that

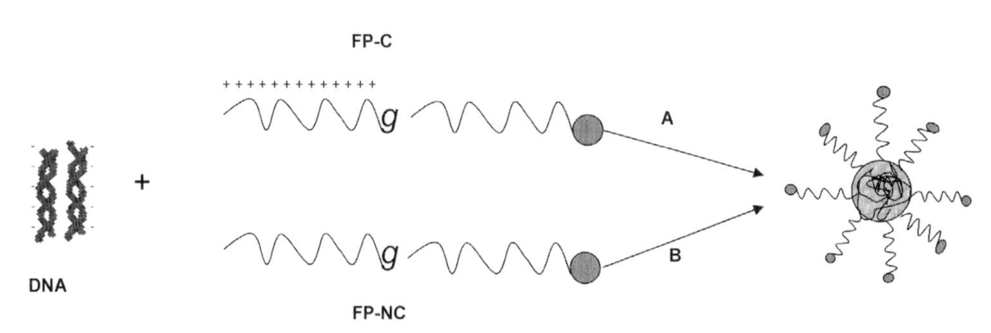

FIGURE 2. Schematic representation of the gene delivery system formed from functionalized condensing polymers (FP-C) and functionalized noncondensing polymers (FP-NC). Each polymer has a functional entity attached to a hydrophilic block (like PEG). The other block contributes to the formation of a particle core, either as a complex or as a matrix. Route A results essentially in the product described in Figure 1A, a complex in the form of nanoparticles, stabilized by the copolymeric block and having a functional corona. Route B results in a product similar to that in Figure 1B, with DNA immobilized within the micronanomatrix. The surface is stabilized by the copolymeric block and has a functional corona.

forms a solid monolithic system). In addition, it is necessary to add colloidal stabilizers to prevent particle aggregation. No interactions at the molecular level occur between DNA and the noncondensing polymer, usually resulting in reduced entrapment efficiency compared to condensing polymers. Through judicious selection of polymers, however, one can solve the encapsulation problem.

A striking advantage of these encapsulated systems is their ability to achieve controlled release of the DNA from the polymeric matrix system. The release of immobilized DNA from the micro-/nanomatrix is governed by the physicochemical properties of the polymer (molecular weight, solubility, hydrophilicity, pKa, etc.). If the polymer is water soluble (as is the case for many natural polymers), particle formation is brought about by coacervation induced by solvent or salt. In the case of water-insoluble polymers (the majority of synthetic polymers), particle formation is assisted by high-shear/pressure means.

The polymers may be tailored to have functional entities like a block of hydrophilic chains (e.g., PEG; Fig. 2) that can form a barrier on the particle surface to tackle premature clearance from systemic circulation by the reticuloendothelial system. In addition, one can attach targeting ligands to guide the particles within circulation to specific tissues and cells and promote efficient intracellular uptake.

HYBRID POLYMERS

Advances in functional polymer chemistry have resulted in biopolymers having several blocks and each may contribute a unique and desirable property to its DNA delivery systems. The polymer may be engineered to have a block that can form a complex with the DNA. Another block may contribute to formation of the matrix of the micro-/nanoparticle. A third block can make the par-

TABLE 2. Noncondensing Polymeric Systems Used in Gene Transfer

Polymer	References
Natural	
Alginate	Aggarwal et al. (1999)
Gelatin	Kaul and Amiji (2005); Zwiorek et al. (2005)
Synthetic	
Poly(lactide-co-glycolide)	Cohen et al. (2000); Laurencin et al. (2001); Eliaz and Szoka (2002); Dunne et al. (2003); Chun et al. (2004); Dawson et al. (2004); Garcia del Barrio et al. (2004); Khan et al. (2004); Kofron and Laurencin (2004); Oster et al. (2004)
Polyvinylpyrrolidone	Prokop et al. (2002)
Polycaprolactone	Shuai et al. (2005)
Poly(lactic acid)	Berton et al. (2001)
Poly(vinyl alcohol)	Paradossi et al. (2003); Oster et al. (2004)
Poly(D,L-lactide-co-4-hydroxy-L-proline)	Li and Huang (2004)
Poly(methyl methacrylate)	Caputo et al. (2004)
Poly(ethylene oxide)-poly(propylene oxide)-poly(ethylene oxide)	Liaw et al. (2001); Chang et al. (2004)
Cyclodextrin	Davis et al. (2004)
Maltodextrin	Huang et al. (2002)
Poly[α-(4-aminobutyl)-L-glycolic acid]	Lim et al. (2000)
Polyalkylcyanoacrylate	Fattal et al. (1998)

ticle stealth (less susceptible to recognition by the reticuloendothelial system due to the presence of a hydrophilized coating), and a fourth entity can impart target specificity (for hybrid polymers; see Tables 1 and 2).

FATE OF POLYMERIC GENE DELIVERY SYSTEMS IN CELLS

The majority of DNA delivery systems function at three general levels: DNA condensation /complexation or entrapment within a delivery system, endocytosis of the system, and nuclear targeting and entry.

First, the condensed form of DNA with a suitable cationic polymer or the entrapped form of DNA within a noncondensing polymer is taken up by cells, usually through endocytosis. Endocytosis is a multistep process involving cell-wall binding, internalization, formation of endosomes, fusion with lysosomes, and lysis. The extremely low pH and the enzymes within endosomes and lysosomes usually bring about degradation of DNA and the delivery system. Finally, DNA that has survived both endocytotic processing and cytoplasmic nucleases dissociates and is released from the delivery system, either before or after entering the nucleus. Nuclear entry is thought to occur through the pores (~10 nm in diameter) or during cell division. Nuclear uptake of DNA is considered to be the rate-limiting step in the overall cytosolic transport process. Once inside the nucleus, the transfection efficiency of delivered DNA is mostly dependent on the composition of the gene expression system.

The interactions between the carrier and the endosomal/cytosolic components are principally governed by the chemistry of the material that constitutes the nanocarrier. Hence, polymers must be designed so that DNA embedded within the nanomatrix remains stable during these intracellular interactions and to allow for efficient nuclear uptake.

APPLICATIONS

An explosion of research in the area of biopolymers for drug delivery applications has occurred during the last 20 years. At the same time, genetic therapies for diseases have suffered setbacks due to serious difficulties in safe delivery of the therapeutic genetic material to the desired areas of the body. For example, the death of a number of patients during clinical trials has resulted in discontinued use of retroviruses for the delivery of genetic materials. However, encouraging advancements in the field of nanotechnology and polymer engineering coupled with the successful completion of the Human Genome Project are reviving genetic therapy as a tool for curing diseases.

REFERENCES

Aggarwal N., HogenEsch H., Guo P., North A., Suckow M., and Mittal S.K. 1999. Biodegradable alginate microspheres as a delivery system for naked DNA. *Can. J. Vet. Res.* **63:** 148–152.

Anderson D.G., Akinc A., Hossain N., and Langer R. 2005. Structure/property studies of polymeric gene delivery using a library of poly(β-amino esters). *Mol. Ther.* **11:** 426–434.

Audouy S. and Hoekstra D. 2001. Cationic lipid-mediated transfection in vitro and in vivo (review). *Mol. Membr. Biol.* **18:** 129–143.

Benns J.M., Maheshwari A., Furgeson D.Y., Mahato R.I., and Kim S.W. 2001. Folate-PEG-folate-graft-polyethylenimine-based gene delivery. *J. Drug Target.* **9:** 123–139.

Berton M., Turelli P., Trono D., Stein C.A., Allemann E., and Gurny R. 2001. Inhibition of HIV-1 in cell culture by oligonucleotide-loaded nanoparticles. *Pharm. Res.* **18:** 1096–1101.

Bielinska A.U., Chen C., Johnson J., and Baker J.R., Jr. 1999. DNA complexing with polyamidoamine dendrimers: Implications for transfection. *Bioconjug. Chem.* **10:** 843–850.

Bloomfield V.A. 1991. Condensation of DNA by multivalent cations: Considerations on mechanism. *Biopolymers* **31:** 1471–1481.

Borchard G. 2001. Chitosans for gene delivery. *Adv. Drug Delivery Rev.* **52:** 145–150.

Brown M.D., Schatzlein A., Brownlie A., Jack V., Wang W., Tetley L., Gray A.I., and Uchegbu I.F. 2000. Preliminary characterization of novel amino acid based polymeric vesicles as gene and drug delivery agents. *Bioconjug. Chem.* **11:** 880–891.

Caputo A., Brocca-Cofano E., Castaldello A., De Michele R., Altavilla G., Marchisio M., Gavioli R., Rolen U., Chiarantini L., Cerasi A., Dominici S., Magnani M., Cafaro A., Sparnacci K., Laus M., Tondelli L., and Ensoli B. 2004. Novel biocompatible anionic polymeric microspheres for the delivery of the HIV-1 Tat protein for vaccine application. *Vaccine* **22:** 2910–2924.

Carroll S.B. 2003. Genetics and the making of *Homo sapiens. Nature* **422:** 849–857.

Carter P.J. and Samulski R.J. 2000. Adeno-associated viral vectors as gene delivery vehicles. *Int. J. Mol. Med.* **6:** 17–27.

Chang S.F., Chang H.Y., Tong Y.C., Chen S.H., Hsaio F.C., Lu S.C., and Liaw J. 2004. Nonionic polymeric micelles for oral gene delivery in vivo. *Hum. Gene Ther.* **15:** 481–493.

Choi Y.H., Liu F., Park J.S., and Kim S.W. 1998. Lactose-poly(ethylene glycol)-grafted poly-L-lysine as hepatoma cell-targeted gene carrier. *Bioconjug. Chem.* **9:** 708–718.

Chun K.W., Cho K.C., Kim S.H., Jeong J.H., and Park T.G. 2004. Controlled release of plasmid DNA from biodegradable scaffolds fabricated using a thermally induced phase-separation method. *J. Biomater. Sci. Polym. Ed.* **15:** 1341–1353.

Cohen H., Levy R.J., Gao J., Fishbein I., Kousaev V., Sosnowski S., Slomkowski S., and Golomb G. 2000. Sustained delivery and expression of DNA encapsulated in polymeric nanoparticles. *Gene Ther.* **7:** 1896–1905.

Collins F.S. and Mansoura M.K. 2001. The Human Genome Project. Revealing the shared inheritance of all humankind. *Cancer* **91:** 221–225.

Collins F.S., Morgan M., and Patrinos A. 2003a. The Human Genome Project: Lessons from large-scale biology. *Science* **300:** 286–290.

Collins F.S., Green E.D., Guttmacher A.E., and Guyer M.S. 2003b. A vision for the future of genomics research. *Nature* **422:** 835–847.

Cortesi R., Mischiati C., Borgatti M., Breda L., Romanelli A., Saviano M., Pedone C., Gambari R., and Nastruzzi C. 2004. Formulations for natural and peptide nucleic acids based on cationic polymeric submicron particles. *AAPS PharmSci.* **6:** E2.

Davidson B.L. and Breakefield X.O. 2003. Viral vectors for gene delivery to the nervous system. *Nat. Rev. Neurosci.* **4:** 353–364.

Davis M.E., Pun S.H., Bellocq N.C., Reineke T.M., Popielarski S.R., Mishra S., and Heidel J.D. 2004. Self-assembling nucleic acid delivery vehicles via linear, water-soluble, cyclodextrin-containing polymers. *Curr. Med. Chem.* **11:** 179–197.

Dawson M., Krauland E., Wirtz D., and Hanes J. 2004. Transport of polymeric nanoparticle gene carriers in gastric mucus. *Biotechnol. Prog.* **20:** 851–857.

De Smedt S.C., Demeester J., and Hennink W.E. 2000. Cationic polymer based gene delivery systems. *Pharm. Res.* **17:** 113–126.

Dunne M., Bibby D.C., Jones J.C., and Cudmore S. 2003. Encapsulation of protamine sulphate compacted DNA in polylactide and polylactide-co-glycolide microparticles. *J. Control. Release* **92:** 209–219.

Eliaz R.E. and Szoka F.C., Jr. 2002. Robust and prolonged gene expression from injectable polymeric implants. *Gene Ther.* **9:** 1230–1237.

Fattal E., Vauthier C., Aynie I., Nakada Y., Lambert G., Malvy C., and Couvreur P. 1998. Biodegradable polyalkylcyanoacrylate nanoparticles for the delivery of oligonucleotides. *J. Control. Release* **53:** 137–143.

Fenske D.B., MacLachlan I., and Cullis P.R. 2001. Long-circulating vectors for the systemic delivery of genes. *Curr. Opin. Mol. Ther.* **3:** 153–158.

Fominaya J., Gasset M., Garcia R., Roncal F., Albar J.P., and Bernad A. 2000. An optimized amphiphilic cationic peptide as an efficient non-viral gene delivery vector. *J. Gene Med.* **2:** 455–464.

Fonseca M.J., Storm G., Hennink W.E., Gerritsen W.R., and Haisma H.J. 1999. Cationic polymeric gene delivery of beta-glucuronidase for doxorubicin prodrug therapy. *J. Gene Med.* **1:** 407–414.

Forrest M.L., Koerber J.T., and Pack D.W. 2003. A degradable polyethylenimine derivative with low toxicity for highly efficient gene delivery. *Bioconjug. Chem.* **14:** 934–940.

Fukunaka Y., Iwanaga K., Morimoto K., Kakemi M., and Tabata Y. 2002. Controlled release of plasmid DNA from cationized gelatin hydrogels based on hydrogel degradation. *J. Control. Release* **80:** 333–343.

Funhoff A.M., Monge S., Teeuwen R., Koning G.A., Schuurmans-Nieuwenbroek N.M., Crommelin D.J., Haddleton D.M., Hennink W.E., and van Nostrum C.F. 2005. PEG shielded polymeric double-layered micelles for gene delivery. *J. Control. Release* **102:** 711–724.

Garcia del Barrio G., Hendry J., Renedo M.J., Irache J.M., and Novo F.J. 2004. In vivo sustained release of adenoviral vectors from poly(D,L-lactic-co-glycolic) acid microparticles prepared by TROMS. *J. Control. Release* **94:** 229–235.

Godbey W.T. and Mikos A.G. 2001. Recent progress in gene delivery using non-viral transfer complexes. *J. Control. Release* **72:** 115–125.

Hagstrom J.E. 2000. Self-assembling complexes for in vivo gene delivery. *Curr. Opin. Mol. Ther.* **2:** 143–149.

Han S., Mahato R.I., and Kim S.W. 2001. Water-soluble lipopolymer for gene delivery. *Bioconjug. Chem.* **12:** 337–345.

Harvie P., Dutzar B., Galbraith T., Cudmore S., O'Mahony D., Anklesaria P., and Paul R. 2003. Targeting of lipid-protamine-DNA (LPD) lipopolyplexes using RGD motifs. *J. Liposome Res.* **13:** 231–247.

Hope M.J., Mui B., Ansell S., and Ahkong Q.F. 1998. Cationic lipids, phosphatidylethanolamine and the intracellular delivery of polymeric, nucleic acid-based drugs (review). *Mol. Membr. Biol.* **15:** 1–14.

Huang C.Y., Ma S.S., Lee S., Radhakrishnan R., Braun C.S., Choosakoonkriang S., Wiethoff C.M., Lobo B.A., and Middaugh C.R. 2002. Enhancements in gene expression by the choice of plasmid DNA formulations containing neutral polymeric excipients. *J. Pharm. Sci.* **91:** 1371–1381.

Hyvonen Z., Plotniece A., Reine I., Chekavichus B., Duburs G., and Urtti A. 2000. Novel cationic amphiphilic 1,4-dihydropyridine derivatives for DNA delivery. *Biochim. Biophys. Acta* **1509:** 451–466.

Janat-Amsbury M.M., Yockman J.W., Lee M., Kern S., Furgeson D.Y., Bikram M., and Kim S.W. 2005. Local, non-viral IL-12 gene therapy using a water soluble lipopolymer as carrier system combined with systemic paclitaxel for cancer treatment. *J. Control. Release* **101:** 273–285.

Kang H.C., Kim S., Lee M., and Bae Y.H. 2005. Polymeric gene carrier for insulin secreting cells: Poly(L-lysine)-*g*-sulfonylurea for receptor mediated transfection. *J. Control. Release* **105:** 164–176.

Kaul G. and Amiji M. 2005. Tumor-targeted gene delivery using poly(ethylene glycol)-modified gelatin nanoparticles: In vitro and in vivo studies. *Pharm. Res.* **22:** 951–961.

Khan A., Benboubetra M., Sayyed P.Z., Ng K.W., Fox S., Beck G., Benter I.F., and Akhtar S. 2004. Sustained polymeric delivery of gene silencing antisense ODNs, siRNA, DNAzymes and ribozymes: In vitro and in vivo studies. *J. Drug Target.* **12:** 393–404.

Kim W.J., Yockman J.W., Lee M., Jeong J.H., Kim Y.H., and Kim S.W. 2005. Soluble *Flt-1* gene delivery using PEI-g-PEG-RGD conjugate for anti-angiogenesis. *J. Control. Release* **106:** 224–234.

Kofron M.D. and Laurencin C.T. 2004. Development of a calcium phosphate coprecipitate/poly(lactide-co-glycolide) DNA delivery system: Release kinetics and cellular transfection studies. *Biomaterials* **25:** 2637–2643.

Koh J.J., Ko K.S., Lee M., Han S., Park J.S., and Kim S.W. 2000. Degradable polymeric carrier for the delivery of IL-10 plasmid DNA to prevent autoimmune insulitis of NOD mice. *Gene Ther.* **7:** 2099–2104.

Lam J.K., Ma Y., Armes S.P., Lewis A.L., Baldwin T., and Stolnik S. 2004. Phosphorylcholine-polycation diblock copolymers as synthetic vectors for gene delivery. *J. Control. Release* **100:** 293–312.

Laurencin C.T., Attawia M.A., Lu L.Q., Borden M.D., Lu H.H., Gorum W.J., and Lieberman J.R. 2001. Poly(lactide-co-glycolide)/hydroxyapatite delivery of BMP-2-producing cells: A regional gene therapy approach to bone regeneration. *Biomaterials* **22:** 1271–1277.

Lengsfeld C.S., Manning M.C., and Randolph T.W. 2002. Encapsulating DNA within biodegradable polymeric microparticles. *Curr. Pharm. Biotechnol.* **3:** 227–235.

Leong K.W., Mao H.Q., Truong-Le V.L., Roy K., Walsh S.M., and August J.T. 1998. DNA-polycation nanospheres as non-viral gene delivery vehicles. *J. Control. Release* **53:** 183–193.

Li Y., Wang J., Lee C.G., Wang C.Y., Gao S.J., Tang G.P., Ma Y.X., Yu H., Mao H.Q., Leong K.W., and Wang S. 2004. CNS gene transfer mediated by a novel controlled release system based on DNA complexes of degradable polycation PPE-EA: A comparison with polyethylenimine/DNA complexes. *Gene Ther.* **11:** 109–114.

Li Z. and Huang L. 2004. Sustained delivery and expression of plasmid DNA based on biodegradable polyester, poly(D,L-lactide-co-4-hydroxy-L-proline). *J. Control. Release* **98:** 437–446.

Liaw J., Chang S.F., and Hsiao F.C. 2001. In vivo gene delivery into ocular tissues by eye drops of poly(ethylene oxide)-poly(propylene oxide)-poly(ethylene oxide) (PEO-PPO-PEO) polymeric micelles. *Gene Ther.* **8:** 999–1004.

Lim Y.B., Kim S.M., Suh H., and Park J.S. 2002. Biodegradable, endosome disruptive, and cationic network-type polymer as a highly efficient and nontoxic gene delivery carrier. *Bioconjug. Chem.* **13:** 952–957.

Lim Y.B., Han S.O., Kong H.U., Lee Y., Park J.S., Jeong B., and Kim S.W. 2000. Biodegradable polyester, poly[α-(4-aminobutyl)-L-glycolic acid], as a non-toxic gene carrier. *Pharm. Res.* **17:** 811–816.

Liu F. and Huang L. 2002. Development of non-viral vectors for systemic gene delivery. *J. Control. Release* **78:** 259–266.

Liu W., Sun S., Cao Z., Zhang X., Yao K., Lu W.W., and Luk K.D. 2005. An investigation on the physicochemical properties of chitosan/DNA polyelectrolyte complexes. *Biomaterials* **26:** 2705–2711.

Lollo C.P., Banaszczyk M.G., and Chiou H.C. 2000. Obstacles and advances in non-viral gene delivery. *Curr. Opin. Mol. Ther.* **2:** 136–142.

Loser P., Huser A., Hillgenberg M., Kumin D., Both G.W., and Hofmann C. 2002. Advances in the development of non-human viral DNA-vectors for gene delivery. *Curr. Gene Ther.* **2:** 161–171.

Luten J., van Steenis J.H., van Someren R., Kemmink J., Schuurmans-Nieuwenbroek N.M., Koning G.A., Crommelin D.J., van Nostrum C.F., and Hennink W.E. 2003. Water-soluble biodegradable cationic polyphosphazenes for gene delivery. *J. Control. Release* **89:** 483–497.

Ogris M. and Wagner E. 2002. Tumor-targeted gene transfer with DNA polyplexes. *Somat. Cell. Mol. Genet.* **27:** 85–95.

Oster C.G., Wittmar M., Unger F., Barbu-Tudoran L., Schaper A.K., and Kissel T. 2004. Design of amine-modified graft polyesters for effective gene delivery using DNA-loaded nanoparticles. *Pharm. Res.* **21:** 927–931.

Otsuka H., Nagasaki Y., and Kataoka K. 2003. PEGylated nanoparticles for biological and pharmaceutical applications. *Adv. Drug Delivery Rev.* **55:** 403–419.

Pampinella F., Pozzobon M., Zanetti E., Gamba P.G., McLachlan I., Cantini M., and Vitiello L. 2000. Gene transfer in skeletal muscle by systemic injection of DODAC lipopolyplexes. *Neurol. Sci.* **21:** S967–969.

Paradossi G., Cavalieri F., Chiessi E., Spagnoli C., and Cowman M.K. 2003. Poly(vinyl alcohol) as versatile biomaterial for potential biomedical applications. *J. Mater. Sci. Mater. Med.* **14:** 687–691.

Pedroso de Lima M.C., Neves S., Filipe A., Duzgunes N., and Simoes S. 2003. Cationic liposomes for gene delivery: From biophysics to biological applications. *Curr. Med. Chem.* **10:** 1221–1231.

Prokop A., Kozlov E., Moore W., and Davidson J.M. 2002. Maximizing the in vivo efficiency of gene transfer by means of nonviral polymeric gene delivery vehicles. *J. Pharm. Sci.* **91:** 67–76.

Ravi Kumar M., Hellermann G., Lockey R.F., and Mohapatra S.S. 2004. Nanoparticle-mediated gene delivery: State of the art. *Expert Opin. Biol. Ther.* **4:** 1213–1224.

Read M.L., Etrych T., Ulbrich K., and Seymour L.W. 1999. Characterisation of the binding interaction between poly(L-lysine) and DNA using the fluorescamine assay in the preparation of non-viral gene delivery vectors. *FEBS Lett.* **461:** 96–100.

Sagara K. and Kim S.W. 2002. A new synthesis of galactose-poly(ethylene glycol)-polyethylenimine for gene delivery to hepatocytes. *J. Control. Release* **79:** 271–281.

Schatzlein A.G. 2001. Non-viral vectors in cancer gene therapy: Principles and progress. *Anti-cancer Drugs* **12:** 275–304.

Shuai X., Merdan T., Unger F., and Kissel T. 2005. Supramolecular gene delivery vectors showing enhanced transgene expression and good biocompatibility. *Bioconjug. Chem.* **16:** 322–329.

Thomas T.J., Kulkarni G.D., Greenfield N.J., Shirahata A., and Thomas T. 1996. Structural specificity effects of trivalent polyamine analogues on the stabilization and conformational plasticity of triplex DNA. *Biochem. J.* **319:** 591–599.

Tranchant I., Thompson B., Nicolazzi C., Mignet N., and Scherman D. 2004. Physicochemical optimisation of plasmid delivery by cationic lipids. *J. Gene Med.* (suppl. 1) **6:** S24–35.

Tsai J.T., Furstoss K.J., Michnick T., Sloane D.L., and Paul R.W. 2002. Quantitative physical characterization of lipid-polycation-DNA lipopolyplexes. *Biotechnol. Appl. Biochem.* **36:** 13–20.

Vlasov G.P., Korol'kov V.I., Pankova G.A., Tarasenko I.I., Baranov A.N., Glazkov P.B., Kiselev A.V., Ostapenko O.V., Lesina E.A., and Baranov V.S. 2004. Dendrimers based lysine and their "starburst" polymeric derivatives: Prospects of use in compacting DNA and in vitro delivery of genetic constructs (in Russian). *Bioorg. Khim.* **30:** 15–24.

Wagner E. 2004. Strategies to improve DNA polyplexes for in vivo gene transfer: Will "artificial viruses" be the answer? *Pharm. Res.* **21:** 8–14.

Walther W. and Stein U. 2000. Viral vectors for gene transfer: A review of their use in the treatment of human diseases. *Drugs* **60:** 249–271.

Wang C.H. and Hsiue G.H. 2005. Polymer-DNA hybrid nanoparticles based on folate-polyethylenimine-block-poly(L-lactide). *Bioconjug. Chem.* **16:** 391–396.

Wang J., Gao S.J., Zhang P.C., Wang S., Mao H.Q., and Leong K.W. 2004. Polyphosphoramidate gene carriers: Effect of charge group on gene transfer efficiency. *Gene Ther.* **11:** 1001–1010.

Wang W., Tetley L., and Uchegbu I.F. 2001. The level of hydrophobic substitution and the molecular weight of amphiphilic poly-L-lysine-based polymers strongly affects their assembly into polymeric bilayer vesicles. *J. Colloid Interface. Sci.* **237:** 200–207.

Zhang J.T., Huang S.W., and Zhuo R.X. 2004. Temperature-sensitive polyamidoamine dendrimer/poly(*N*-isopropylacrylamide) hydrogels with improved responsive properties. *Macromol. Biosci.* **4:** 575–578.

Zhang X.Q., Wang X.L., Huang S.W., Zhuo R.X., Liu Z.L., Mao H.Q., and Leong K.W. 2005. In vitro gene delivery using polyamidoamine dendrimers with a trimesyl core. *Biomacromolecules* **6:** 341–350.

Zhdanov R.I., Podobed O.V., and Vlassov V.V. 2002. Cationic lipid-DNA complexes-lipoplexes-for gene transfer and therapy. *Bioelectrochemistry* **58:** 53–64.

Zwiorek K., Kloeckner J., Wagner E., and Coester C. 2005. Gelatin nanoparticles as a new and simple gene delivery system. *J. Pharm. Pharm. Sci.* **7:** 22–28.

35 Transfection of Hippocampal Neurons with Plasmid DNA Using Calcium Phosphate Coprecipitation

Bernhard Goetze and Michael Kiebler

Center for Brain Research, Medical University of Vienna, Department of Neuronal Cell Biology, 1090 Vienna, Austria

ABSTRACT

Many protocols have been described for the transfection of cells in culture using a DNA/calcium phosphate ($CaPO_4$) coprecipitate. Published for the first time more than 30 years ago (Graham and van der Eb 1973), this method is still widely used in many laboratories because of its clear advantages over other transfection methods. The chemicals needed are less costly than commercial transfection reagents, the method is easily adapted for transfection of established cell lines (both adherent and nonadherent cells) and primary cell cultures, and the creation of stably transfected cell lines is possible. Once established, the protocol is straightforward and easy to use. We have defined two critical parameters for reproducible and efficient transfection: the precise pH of the transfection medium and the incubation time of cells with the DNA/$CaPO_4$ coprecipitate. With the modifications described here, there is no need for additional cell culture equipment. We have found that this technique is most useful for high-efficiency transfections of primary hippocampal neurons and therefore focus on this particular application (Goetze et al. 2004).

INTRODUCTION

The basic principle of $CaPO_4$ transfection was first described in 1973 by Graham and van der Eb. Plasmid DNA and calcium phosphate form a DNA/$CaPO_4$ coprecipitate which, when added to cultured cells, is subsequently internalized by endocytosis. By unknown mechanisms, the DNA leaves the endocytic compartment and reaches the cytoplasm and then enters the nucleus, likely during breakdown of the nuclear envelope during cell division. Because neurons, which are postmitotic cells, can also be transfected by this method, it is likely that there exists a second unidentified route for the DNA to enter the nucleus. We have modified a protocol (described by Ishiura et al. in 1982) that has been successfully used for the transfer of phage particles into cell lines. In our method, BES-buffered phosphate saline (2x BBS) of precisely defined pH is dropped slowly onto a mixture of plasmid DNA and $CaCl_2$. This solution is added immediately to cells, and the DNA/$CaPO_4$ coprecipitate begins to form in the transfection medium; the size of the precipitate and its density increase with incubation time. The precipitate is washed off the neurons at some point between 45 minutes and 6 hours of incubation, depending on the goals of the experiment. As early as 3 hours after the start of incubation, the expression of a reporter plasmid can be monitored using a fluorescence marker (Goetze et al. 2004). The formation of the precipitate is critically dependent on the pH values of both the 2x BBS buffer and transfection medium. The HEPES-buffered transfection medium used in this protocol renders the medium independent of CO_2, in contrast to standard carbonate-buffered media. The pH of the medium is set to a fixed value of 7.4, the physiological pH of the neuronal growth medium used to culture the neurons (for a detailed cell culture protocol, see Goslin et al. 1998; Goetze et al. 2003). This constraint of pH allows for a controlled and linear formation of the precipitate, thereby increasing both transfection efficacy and cell survival during the transfection. Transfection conditions and efficiencies are optimized by using a series of 2x BBS buffer aliquots with pH values ranging from 6.80 to 7.30 in 0.05 pH steps to identify the optimal pH of 2x BBS individually for each plasmid. We have observed significant differences with respect to the rate of formation and size of the precipitate not only between different expression plasmids, but also between two preparations of the same plasmid. It is therefore necessary to optimize the conditions, including the pH of 2x BBS and the incubation time for each plasmid, to obtain reproducible transfection efficiencies and results. We have found that changes in the pH values of the transfection solutions are critically important for the size and formation of the precipitate (see Table 1) and that the size of the precipitate is an essential parameter for successful transfections of primary hippocampal neurons and can be fine-tuned with this protocol. We have optimized the protocol for the transfection of primary hippocampal neurons presented here; however, we have also used this approach to successfully transfect primary mixed glia cell populations, COS, BHK, and HeLa cells (B. Goetze and M. Kiebler, unpubl.).

TABLE 1. Relationship of pH to Precipitate Formation

Change in pH	Result
Increase in pH value of the 2x BBS/transfection medium	Faster formation and larger size of the DNA/$CaPO_4$ precipitate
Decrease in pH value of the 2x BBS/transfection medium	Slower formation and smaller size of the DNA/$CaPO_4$ precipitate

Protocol

Calcium-phosphate-mediated Transfection

The following protocol describes transfection of plasmid DNA into primary hippocampal neurons. To optimize the parameters of the method, vary the pH value of the transfection solutions as well as the incubation period after transfection. Typically, a set of three 2x BBS buffers, with different pH values (e.g., 6.90, 7.00, and 7.10) and different incubation times, can be used to establish near-optimal conditions for transfection with any given plasmid. The transfection efficacy can be determined 1 day after transfection or, if using a fluorescently tagged protein, simply compared by inspection under a fluorescence microscope.

The volumes of buffers and amounts of DNA used in this procedure are optimized for 2 ml of transfection medium and a total of 3 μg of plasmid DNA. It can be linearly scaled up or down according to the user's preference and goals of the experiment.

MATERIALS

Reagents

2x BES-buffered saline (BBS) (pH 6.85–7.20)
- 50 mM BES
- 1.5 mM Na_2HPO_4
- 280 mM $NaCl_2$

Adjust the pH of the buffer, selecting at least three values [see p. ???]; use fresh, thoroughly calibrated, and cleaned instruments. Use a working 2x BBS aliquot as a pH reference before preparing the next set of 2x BBS buffers. Sterilize the buffers by filtration and store at 4°C in sealed 50-ml glass bottles. The buffer may be stored for up to 12 months.

Cell line to be transfected (primary hippocampal neurons)

HBSS washing buffer (pH 7.3)
- 135 mM $NaCl_2$
- 20 mM HEPES
- 4 mM KCl
- 1 mM Na_2HPO_4
- 2 mM $CaCl_2$
- 1 mM $MgCl_2$
- 10 mM glucose

Adjust the pH to 7.3, sterilize by filtration, and store for up to 1 year at 4°C.

2.5 M $CaCl_2$

In our experience, repeated freezing and thawing of the 2x BBS and $CaCl_2$ solutions affect reproducibility of the transfections. It is therefore advisable to store all solutions and media at 4°C.

NMEM-B27 transfection medium (pH 7.4)

MEM (Invitrogen) is supplemented with 1 mM sodium pyruvate (Sigma-Aldrich), 15 mM HEPES (Invitrogen), 2 mM stable L-glutamine (PromoCell), 1x B27 supplement (Invitrogen), and 33 mM D-glucose (Sigma-Aldrich). Adjust the pH to 7.4, sterilize by filtration, and store in sealed 50-ml glass bottles for up to 2 months. A reference medium that is not used for transfections can be used for comparison to adjust the pH accurately and reproducibly by comparing the color of the phenol red.

Plasmid DNA

For the highest-quality plasmid preparation, purify the DNA using endotoxin-free Maxi-prep kits (e.g., QIAGEN endo-free Maxi Kit) or CsCl gradients. Assess the purity of DNA by measuring optical densities: An OD 260/280 of approximately 1.8 is preferred. Dissolve the dried plasmids in either H_2O or TE buffer. Before use, centrifuge the plasmids briefly at 4°C. Cotransfection may be performed with two plasmids. Usually, the cotransfection efficiency of two given plasmids mixed in a 1:1 ratio is well above 95% (Goetze et al. 2004). The use of three or more plasmids at a time has proven to be more difficult and must be carefully controled at the single-cell level with appropriate markers (B. Goetze and M. Kiebler, unpubl.).

Equipment

Fresh, thoroughly calibrated, and cleaned instruments for pH adjustment

METHOD

1. To prepare the cells for transfection, transfer them into 2 ml of transfection medium in a glass culture dish.
2. To prepare the coprecipitate, transfer the components listed below into a 1.5-ml centrifuge tube in the following order:

   6 µl of $CaCl_2$

   x µl of H_2O (mix well)

   3 µg of plasmid DNA (add the DNA very slowly while stirring with a pipette tip to thoroughly mix the components)

   The first three components should total 60 µl. Adjust the volume of H_2O accordingly.
3. Add 60 µl of 2x BBS dropwise to the calcium/DNA solution. Mix the solution well by tapping the tube five times after each addition of 3 drops. Do not vortex!
4. Add 120 µl of transfection mixture immediately to the cells in transfection medium. Incubate the cells at 36.5°C in a humidified incubator without a CO_2 supply.

   To increase the amount of precipitate resting on the cells, centrifuge the dish containing the cells (once the precipitate first becomes visible) in an 18-cm swing-out rotor at 1000 rpm for 2–6 minutes at room temperature. During subsequent incubation, inspect the cells every 10–15 minutes, because mortality increases rapidly with time. Typical indicators of mortality are the appearance of varicosities and loss of nuclear integrity. When the centrifugation step is included, the incubation time with the precipitate can be significantly shortened and efficacies increase up to 50% compared to those observed in noncentrifuged cells (B. Goetze et al., unpubl.).
5. After 45 minutes or up to 6 hours, the neurons should be evenly covered with a very fine precipitate that is barely visible with a 10x objective (Goetze et al. 2004).
6. Aspirate the transfection medium and replace it with prewarmed HBSS washing buffer. After 5 minutes, inspect the cells under a microscope. The precipitate should have dissolved completely. Do not wash the cells longer than 15 minutes.
7. Aspirate the HBSS washing buffer to remove the DNA/$CaPO_4$ precipitate and incubate the cells in normal culture medium.
8. Process the cells for analysis of expression of the transfected gene, using the appropriate assay (detection of a reporter, staining and microscopy, immunostaining, and immunoblotting, depending on the goals of the experiment).

TROUBLESHOOTING

Problem: Different preparations of 2x BBS buffer may give different transfection efficiencies.

Solution: This phenomenon is well known but not yet understood. If an individual batch of 2x BBS buffer does not yield optimal results, it is advisable to produce new buffers. In addition, shifts in the optimal pH value for a given plasmid have been observed after the production of a new set of 2x BBS buffers. This is less critical because the whole pH range typically shifts in parallel; therefore, the parameters for all plasmids in use tend to result in slightly more alkaline or acidic pH values. Sets of 2x BBS buffers not produced at the same time should therefore not be mixed.

Problem: When a given plasmid is expressed for several days in culture, overexpression artifacts are frequent.

Solution: This problem has been observed both when expression continues over several days and with "strongly expressing" plasmids incubated over shorter time periods. In some cases, precipitation and mislocalization of a fluorescently tagged protein have been observed after prolonged expression times (B. Goetze, unpubl.). This situation can be avoided by diluting the expression vector (in Step 3) with an "empty" vector of the same vector backbone (the parent vector that does not carry the insertion).

Problem: To adapt the protocol to other cell lines, some adjustments may need to be made to the procedure.

Solution: We have successfully transfected primary mixed glia cell populations, COS, BHK, and HeLa cells (B. Goetze et al., unpubl.). Generally, glia cells and established cell lines require larger precipitates than do neurons. Therefore, a longer washing period to remove the precipitate is needed. For further alternative information, see Lindell et al. (2004).

REFERENCES

Goetze B., Grunewald B., Baldassa S., and Kiebler M. 2004. Chemically controlled formation of a DNA/calcium phosphate coprecipitate: Application for transfection of mature hippocampal neurons. *J. Neurobiol.* **60:** 517–525.

Goetze B., Grunewald B., Kiebler M.A., and Macchi P. 2003. Coupling the iron-responsive element to GFP—An inducible system to study translation in a single living cell. *Sci. STKE* **2003:** PL12.

Goslin K., Asmussen H., and Banker G. 1998. Rat hippocampal neurons in low-density culture. In *Culturing nerve cells*, 2nd edition (ed. G. Banker and K. Goslin), pp. 339–370. MIT Press, Cambridge, Massachusetts.

Graham F.L. and van der Eb A.J. 1973. A new technique for the assay of infectivity of human adenovirus 5 DNA. *Virology* **52:** 456–467.

Ishiura M., Hirose S., Uchida T., Hamada Y., Suzuki Y., and Okada Y. 1982. Phage particle-mediated gene transfer to cultured mammalian cells. *Mol. Cell. Biol.* **2:** 607–616.

Lindell J., Girard P., Muller N., Jordan M., and Wurm F. 2004. Calfection: A novel gene transfer method for suspension cells. *Biochim. Biophys. Acta* **1676:** 155–161.

36 Gene Delivery to Skin Using Biolistics

William C. Heiser

Bio-Rad Laboratories, Life Science Group, Hercules, California 94547

ABSTRACT

Biolistics ("biological ballistics") or particle bombardment provides a rapid and simple procedure for delivering genes into cells (Klein et al. 1987; Yang et al. 1990). The technique has many advantages:

- Plasmids may be used for delivery—there is no need to construct complex biological vectors as for some viral delivery systems.
- Theoretically, DNA can be delivered to any cell type. DNA is carried through the cell wall, cell membrane, stratum corneum, or other protective structure, and delivery is independent of cell surface receptors or molecules.
- Genes, encoded by DNA or RNA, may be delivered to cells in vitro, ex vivo, or in vivo and are delivered to all cells concurrently.
- Delivery results in high local transformation efficiency.
- The technique is particularly suitable for inducing an effective immune response in animals. The epidermis, readily accessible, is a simple and effective target for vaccine studies. Delivery of DNA to the epidermis has been demonstrated to generate both antibody and cell-mediated immunity responses.

Despite these many advantages, the technique has some limitations. First, a special instrument is needed to deliver the DNA. Second, the region in which DNA is transferred is limited to the area of particle delivery. Finally, gene gun delivery usually elicits a humoral, rather than a cellular, immune response (Boyle and Robinson 2000).

INTRODUCTION

Particle bombardment provides a physical method for delivering nucleic acid into cells. Currently, Bio-Rad Laboratories—the only commercial source of instruments for this purpose—provides two instrument types: PDS-1000/He, the most suitable for gene delivery to microorganisms, plants, and in vitro to animal cells (McCabe et al. 1988; Heiser 1994), and the Helios Gene Gun (techniques are described here), which was specifically designed for gene delivery into mammalian cells in vivo, but it may also be used to deliver DNA ex vivo and in vitro. The following parameters should be optimized before starting a new study: the amount of DNA (μg) loaded per milligram of gold microcarriers, called the DNA-loading ratio (DLR); the amount of gold microcarriers per cartridge, called the microcarrier-loading quantity (MLQ); and the size of the gold microcarriers. Typical starting conditions for in vivo delivery to epidermal cells are 1 μg of DNA and 0.5 mg of either 1.0 or 1.6 μm of microcarriers, resulting in a DLR of 2 μg of DNA/mg microcarriers and a MLQ of 0.5 mg/cartridge. A lower MLQ can usually be used for in vitro delivery. To prepare the material to be delivered, a slurry of DNA and gold particles (microparticles or microcarriers) in a solution of spermidine is suspended in ethanol. The DNA-coated gold particles are delivered into and distributed evenly along the length of the tubing, which is subsequently cut into short sections (cartridges) to be used in the gene gun. Properly stored, the cartridges will last for at least 1 year with no significant decrease in activity, an important advantage when conducting a series of experiments over a period of time.

Protocol

Gene Delivery Using the Helios Gene Gun

The Helios Gene Gun uses a pulse of helium to launch the DNA-coated particles down the barrel of the gene gun, which serves to spread the gold particles onto the target cells as well as to reduce the blast effect of the helium shock wave as well as tissue damage. The laboratory of Shan Lu, at The University of Massachusetts Medical School, has provided many useful techniques for using the Helios Gene Gun to deliver DNA to skin (Wang et al. 2004).

MATERIALS

Reagents

Anesthetics (xylazine at 1 mg/ml and/or ketamine at 10 mg/ml)

Animal(s) targeted for gene delivery (usually ~20 g mice)

$CaCl_2$ (1.0 M)

Sterilize through a 0.22-µm filter. Store at room temperature.

Ethanol (100% dehydrated; Spectrum)

To avoid hydration, keep the bottle tightly capped when not in use.

Plasmid expressing the gene of interest

As control (if desired), include the parent plasmid lacking the gene of interest (the "empty vector"). Amplify plasmids according to standard methods (Sambrook and Russell 2001) and purify by available commercial methods (e.g., Quantum Prep or Aurum columns, Bio-Rad Laboratories). Suspend the plasmid at 0.5–2 mg/ml in TE.

Spermidine (50 mM, free base; Sigma-Aldrich)

Prepare a 1 M stock solution by dissolving 1.0 g of spermidine in 6.8 ml of sterile Milli-Q water and sterilizing through a 0.22-µm filter. Store aliquots of the stock solution at –70°C. Prepare a 50-mM solution by diluting the stock 1:20 with sterile Milli-Q water; store the working solution in single-use aliquots at –70°C.

Equipment

Analytical balance

Electric shaver (e.g., Oster clippers with a no. 40 surgical blade).

Helios Gene Gun System (Bio-Rad Laboratories)

The system includes the Helios Gene Gun, cartridge holders, barrel liners and O-rings, 9-volt battery, cartridge extractor tool, helium hose assembly, helium regulator, Tubing Prep Station, nitrogen hose, 10-cc syringes with fittings and silicone adapter tubing, Gold-Coat (Tefzel) tubing, 1.0- and 1.6-µm gold microcarriers, tubing cutter with razor blades, polyvinylpyrrolidone, desiccant pellets, and cartridge collection vials.

Helium gas (compressed, grade 4.5)

Nitrogen gas (compressed, grade 4.8) with regulator capable of adjustment to 1–3 psi

Peristaltic pump (optional)

Syringe (1 cc) with 25-gauge, one-half-inch needle

Ultrasonic water bath (e.g., Branson Model 1510)

METHODS

Preparation of DNA-coated Gold Particles

These steps describe preparing sufficient DNA-coated, gold microcarriers for one length of tubing (Steps 13–20), which generally will yield about 40–45 one-half-inch cartridges (Steps 21–23).

1. In a 1.5-ml centifuge tube, weigh 25 mg of gold microcarriers. When including control cartridges, weigh out a second 25-mg sample. Treat the samples in parallel, except for the addition of the plasmid sample (see Step 4).

   For control cartridges that may be used in experiments to check background levels of the gene of interest, prepare microcarriers either without DNA or with a plasmid lacking the gene of interest following this procedure through Step 12. Load these microcarriers into a second piece of Tefzel in Step 16.

2. Add 100 μl of 50 mM spermidine to each tube.
3. Vortex the gold and spermidine mixtures for a few seconds and then sonicate for 3–5 seconds in an ultrasonic water bath to completely disperse the microcarriers.
4. Add 50 μl of a 1-μg/μl plasmid stock to the gold solution to give a DNA-loading ratio of 2. Prepare control microcarriers as appropriate.
5. Mix the DNA and gold by vortexing at high speed for about 5 seconds.
6. While vortexing at moderate rate, carefully open the tube and slowly add 100 μl of the 1 M $CaCl_2$ solution dropwise to the mixture. Close the tube and vortex at high speed for about 5 seconds.
7. Allow the gold microcarriers to settle for 10 minutes at room temperature.
8. Centrifuge the microcarrier solution in a microcentrifuge for 15 seconds. Remove the supernatant and discard.
9. Vortex the tube briefly to resuspend the pellet in the remaining supernatant.
10. Add 1 ml of 100% ethanol to the tube. Centrifuge in a microcentrifuge for 5 seconds. Remove and discard the supernatant. Repeat this step twice.
11. Vortex the tube briefly to resuspend the pellet in the remaining supernatant. Transfer the suspension to a 15-ml disposable polypropylene centrifuge tube with screw cap. Rinse the gold from the centrifuge tube twice using 200 μl of fresh 100% ethanol and transfer it to the 15-ml centrifuge tube.
12. Add 2.6 ml of 100% ethanol to the centrifuge tube. Use this suspension immediately for tube preparation.

Loading the DNA Microcarrier Suspension into Tubing

13. Purge the Tefzel tubing with nitrogen to ensure that it is completely dry. Insert an uncut piece of Tefzel tubing into the opening on the right side of the Tubing Prep Station, through the tubing support cylinder, and into the O-ring on the far left side. Turn on the compressed nitrogen to 1–3 psi and then use the knob on the flowmeter to adjust the rate to 0.5 liters/minute for at least 15 minutes.
14. Insert a 10-ml syringe fitted with an 18-inch piece of silicone adapter tubing into the syringe sleeve and clamp it into the syringe holder on the base of the Tubing Prep Station. Remove the Tefzel tubing from the Tubing Prep Station. Use the knob on the flowmeter to turn off the flow of nitrogen.

Alternatively, if a peristaltic pump is to be used to remove the ethanol from the Tefzel tubing, calibrate the pump speed at 3.6–7.2 ml/minute using ethanol. Use the silicone adapter tubing to connect the pump tubing to the Tefzel tubing. The slower the speed of ethanol removal, the more ethanol removed from the Tefzel tubing.

15. Cut a 30-inch (75 cm) length of Tefzel tubing from one of the ends (include a length of tubing for each control). Use the tubing cutter to cut each end of the 30-inch tubing to prepare nondistorted ends. Insert one end of the Tefzel tubing into the end of adapter tubing fitted onto a 10-cc syringe. Check that the plunger of the syringe is pushed in completely before attaching the adapter tubing.
16. Vortex the microcarrier suspension; if necessary, sonicate briefly to achieve an even suspension. Invert the tube several times to resuspend the gold, immediately remove the cap, insert the free end of the Tefzel tubing, and, using the syringe on the other end, quickly draw the gold suspension approximately 2 inches (58 cm) into the Tefzel tubing. Do not vortex the microcarrier suspension while drawing it into the tubing. Do not try to remove all of the ethanol from the centrifuge tube.

    IMPORTANT: Avoid drawing any bubbles into the tubing.
17. Remove the Tefzel tubing from the centrifuge tube, hold it in a horizontal position, and continue drawing the suspension into the tubing for another 2–3 inches (6 cm). Immediately slide the loaded tube (with syringe attached) into the right side of the tubing for support cylinder on the Tubing Prep Station and then through the O-ring on the left side of the support cylinder.
18. Allow the microcarriers to settle for 3–5 minutes. Detach the Tefzel tubing from the adapter tubing and attach the Tefzel tubing to the adapter tubing on the 10-cc syringe clamped to the base of the Tubing Prep Station (Step 2). Use a syringe to remove the ethanol at a rate of 0.5–1 inch/second (this should require 30–60 seconds).

    If a peristaltic pump is used, attach the calibrated pump to the Tefzel tubing. It is not necessary to set up the 10-cc syringe in the syringe holder.
19. Detach the syringe adapter tubing from the Tefzel tubing. Immediately turn the Tefzel tubing in the Tubing Prep Station 180°. Wait 3–4 seconds and then turn on the Tubing Prep Station to start the tube rotating.
20. After rotating the tube for 20–30 seconds, open the valve on the nitrogen flowmeter to allow 0.35–0.4 liters/minute of nitrogen to dry the Tefzel tubing while it continues to rotate. Continue drying with rotation for 3–5 minutes and then turn off the motor on the Tubing Prep Station. Use the knob on the flowmeter to turn off the flow of nitrogen and then remove the coated tubing.

Preparing Cartridges for the Helios Gene Gun

21. Examine the Tefzel tubing to verify that the microcarriers are evenly distributed over the length of the tubing. Use a felt-tipped marking pen to mark the sections of the tubing in which the gold is sparsely or unevenly coated.

    This uneven distribution is usually limited to the outer 2 inches where the gold has settled, but may also include internal sections of the tubing. Ideally, the microcarriers should be spread uniformly over the entire inside surface of the Tefzel tubing; however, they may polarize to one side of the tubing while drying. As long as each 0.5-inch section of tubing visually contains the same amount of gold, the tubing can be used for cartridges.
22. Use scissors to cut off the unevenly coated tubing from one of the ends. Use the tubing cutter to cut the remaining tubing into 0.5-inch pieces.

23. Store the cartridges in a cartridge storage vial containing a desiccant pellet. Cap the vial tightly, label, wrap with Parafilm, and store at 4°C.

    When stored in this manner, cartridges are stable for at least 1 year.

Particle Delivery Using the Helios Gene Gun

The gene gun may be used to deliver DNA to any cell type in vivo; however, delivery to epithelial skin cells is by far easiest. With adjustment of the helium pressure for delivery, the majority of gold microcarriers may be delivered through the stratum corneum and into the epidermis without penetrating the dermal layer.

24. To activitate the Helios Gene Gun for gene delivery, install a battery and a barrel liner into the gene gun. Connect one end of the helium hose to the Helios Gene Gun and the other end of the base to the helium regulator. Open the helium tank main valve and adjust the helium regulator to the proper pressure.

    For gene delivery to mouse skin, 350 psi is usually optimum using 1.0- or 1.6-μm gold microcarriers.

25. Install an empty cartridge holder and fire one or two shots to pressurize the system. Remove the cartridge holder from the gene gun.
26. Load the cartridges prepared in Steps 21–23 into the cartridge holder. Insert the cartridge holder into the gene gun.

Preparation of Animals and Gene Delivery

27. Anesthetize the animals by injection with the appropriate agents, following the IACUC-approved protocol at your institution.

    We recommend injecting 200 μl of a mixture of xylazine (1 mg/ml) and ketamine (10 mg/ml) intraperitoneally into mice weighing 20 g.

28. Shave the target area of the animal, usually the chest or abdomen, and brush the fur away. The back of the ear is also a good target site for injection and does not need to be shaved.
29. Place the gene gun over the target site and fire the gene gun. Move the cartridge holder to the next position before firing the next cartridge.
30. After the appropriate period, use an appropriate assay to analyze for expression of the introduced gene.

TROUBLESHOOTING

Problem (Step 16): The solution has clumps and is not uniform.
Solution: Agglomerated particles will appear as pin-head-sized clusters of gold that fall to the bottom of the 15-ml tube very rapidly. Sonicate briefly to disrupt these clumps.

Problem (Steps 16 and 17): Distributing the gold solution evenly throughout the length of the tubing.
Solution: These steps must be done together and can be tricky. It is advisable to practice before using the DNA-coated gold. The objective is to prepare a suspension of gold particles in the 15-ml tube and to draw it into the Tefzel tubing as a **suspension** so that the same amount of gold is distributed throughout the entire length of the Tefzel tubing. It should be possible to draw nearly all of the gold into the tubing.

IN MEMORIAM

Bill Heiser passed away on Jan 19, 2006, after a long battle with cancer. He is greatly missed by his many friends and colleagues. The following was written by Theodore Friedmann, one of the editors of this volume.

When I asked Bill Heiser to prepare this chapter, he responded with great enthusiasm and apparent pleasure. He said that he eagerly looked forward to contributing to the book. His reaction was not the typical, more muted response that one usually gets when asking colleagues to write review chapters, but with his usual good grace and warmth he said that he'd like to take on this chore because for some time his illness had prevented him from having a more active hand in his research, which he very much missed. He was one of the first authors to submit their chapters, not only because of his usual conscientiousness, but also because he obviously understood that there was more than the usual editorial time limit to his writing. I am pleased to think that he derived some pleasure in preparing this chapter in the face of his illness.

REFERENCES

Boyle C.M. and Robinson H.L. 2000. Basic mechanisms of DNA-raised antibody responses to intramuscular and gene gun immunizations. *DNA Cell Biol.* **19:** 157–165.

Heiser W.C. 1994. Gene transfer into mammalian cells by particle bombardment. *Anal. Biochem.* **217:** 185–196.

Klein T., Wolf E., Wu R., and Sanford J. 1987. High velocity microprojectiles for delivering nucleic acids into living cells. *Nature* **327:** 70–73.

Sambrook J. and Russell D.W. 2001. *Molecular cloning: A laboratory manual*, 3rd edition. Cold Spring Harbor Laboratory Press, Cold Spring Harbor, New York.

Wang S., Joshi S., and Lu S. 2004. Delivery of DNA to skin by particle bombardment. *Methods Mol. Biol.* **245:** 185–196.

Yang N.-S., Burkholder J., Roberts B., Martinell B., and McCabe D. 1990. In vivo and in vitro gene transfer to mammalian somatic cells by particle bombardment. *Proc. Natl. Acad. Sci.* **87:** 9568–9572.

37 Optimizing Electrotransfection of Mammalian Cells In Vitro

Shulin Li

Department of Comparative Biomedical Sciences, School of Veterinary Medicine, Louisiana State University, Baton Rouge, Louisiana 70803

ABSTRACT

Electrotransfection can be used for studying gene function, promoter activity, and gene regulation; for generating transgenic or knockout embryo stem cells; and for transferring the tumor antigen-encoding RNA into dendritic cells (DC) for tumor vaccination, therapeutic genes into immune cells for treating tumors, and therapeutic genes into stem cells for treating different types of diseases (Table 1) (Li 2004). Electrotransfection is quick, easy, and inexpensive, compared to other transfection methods. The chief advantage is the ability to transfect virtually any cell line, regardless of morphology, size, passage, or cell type. The efficiency of transfection is greatly enhanced by the use of cell-specific electroporation buffers (Table 1). Electrotransfection also avoids the introduction of biological materials other than the targeted DNA or RNA molecules (an important issue for clinical applications). Furthermore, the process does not significantly change the micromolecular environment of cells as other transfection methods do; the introduction of other biological materials by chemical or viral methods causes a change in molecular environment and may affect results (Li et al. 2005). The supplies used for electrotransfection—electroporation cuvettes—may be reused, thus minimizing cost. The disadvantages are that electrotransfection is difficult to scale up for clinical applications, some of the electroporation buffers may compromise cell function in ways that are yet unknown, and the process may also cause cell death.

INTRODUCTION

Electroporation refers to the induction of cell membrane permeability by the external application of electric pulses to the cells, whereas electrotransfection is the introduction of plasmid DNA or

TABLE 1. Electroporation Buffers and Transfection Efficiency In Vitro

Electroporation buffer (suppliers)	Cell type	Cell number per milliliter	Genetic material	Transfection efficiency (%)	Source
Opti-MEM (Invitrogen)	DC	10–40 x 10^6	DNA RNA	45 89	Van Tendeloo et al. (2001)
Cyto Pulse Low-conductivity Media (Cyto Pulse Sciences)	DC	5 x 10^5	RNA	89–95	Michiels et al. (2005)
Opti-MEM (Invitrogen)	DC	4 x 10^6	RNA	70–80	Van Meirvenne et al. (2002)
OptiBuffer (Thermo Electron)	DC	50 x 10^6	RNA	100	Bonehill et al. (2003)
OptiBuffer (Thermo Electron)	T-cell MSC	50 x 10^6	RNA RNA	50 90	Smits et al. (2004)
Nucleofector Kits (Amaxa)	NK cells	10–100 x 10^6	DNA	50	Trompeter et al. (2003)
Nucleofector (Amaxa)	MSC	2 x 10^6	DNA	80	Haleem-Smith et al. (2005)
Opti-MEM (Invitrogen)	CD34 bone marrow cells	30 x 10^6	DNA	>30	Van Tendeloo et al. (2000)
MEM (Invitrogen)	primary neural cells	5 x 10^6	DNA	20	Mertz et al. (2002)
Hypo-osmolar/Iso-osmolar buffer (self-made)	CHO, B16F1	5 x 10^6	DNA	65–85	Cegovnik and Novakovic (2004)
Pulsing buffer	CHO7	80 x 10^6	DNA	80	Li et al. (2002)

RNA into the interior of cells via electric pulses. Electrotransfection, also known as electroinsertion or electrotransfer (Weaver 1993; Andre and Mir 2004; Golzio et al. 2004), was first described for mammalian cells in 1982 (Neumann et al. 1982). Wong and Neumann (1982) demonstrated expression of the thymidine kinase (*tk*) gene in mouse fibroblasts after electrotransfection of plasmid DNA. More than 400 cell lines have been tested with this technique, but the efficiency of transfection still varies (Cegovnik and Novakovic 2004). Most electroporation systems include a power supply and an electroporation chamber for holding electroporation cuvettes that contain a mixture of cells and DNA. Some electroporators are connected with different types of electrodes for applications both in vitro and in vivo. The most common electroporation systems are summarized in Table 2. For a detailed description of the types of electrodes, see Hofmann (1995).

The conventional model for DNA uptake by electroporation is the formation of transient electropores in the membrane, which allows large molecules such as DNA to diffuse from outside the cell membrane into the cytoplasm (Neumann et al. 1982; de Gennes 1999). The electropore may last from milliseconds to several minutes, depending on the electroporation field and the duration of electric pulses applied to the DNA and cell mixture (Klenchin et al. 1991; Gabriel and Teissie 1997). The higher the voltage and the longer the pulse duration, the larger the cell membrane electropore becomes. On the basis of this model, electrotransfection of DNA generally requires a longer pulse duration to generate relatively large pores for DNA uptake; however, small molecules require only a short pulse duration. An electric pulse duration of 1 msec is considered a threshold for inducing DNA uptake; but some reports demonstrate a successful electrotransfection using 100 μsec (Cegovnik and Novakovic 2004). The risk associated with increased electroporation field strength and long pulse duration is induction of nonreversible cell membrane leakage, causing cell rupture and cell death. This leakage usually occurs when the external electroporation field exceeds the

TABLE 2. Electroporator Systems

Manufacturer	Model	Web site address
Bio-Rad	Gene Pulser II	www.bio-rad.com
Thermo Electron	Celljectt Pro electroporator	www.thermo.com
BTX	ECM830	www.btxonline.com
Amaxa	Nucleofector device	www.amaxa.com
Tritech Research	Cloning gun (CG2)	www.tritechresearch.com
Cyto Pulse Sciences	PA-4000 Pulse generator	www.cytopulse.com
Eppendorf	Multiporator 4308	www.eppendorf.com
MaxCyte	Large-volume flow electroporation	www.maxcyte.com

threshold of transmembrane potential difference (250 mV) or when the transmembrane voltage reaches 0.5–1.5 V (Weaver 1993; Golzio et al. 2004). The transmembrane potential difference is dependent on membrane conductivities, pulsing buffer, cell size, cell shape, alignment of cell with electroporation field, and electroporation field intensity (Golzio et al. 2004). Transmembrane voltage is primarily dependent on the cell size and the angle between the applied electroporation field and the site on the cell membrane where electric voltage is measured (Weaver 1993).

Another model for DNA uptake by electroporation is electrophoresis. Because DNA is negatively charged, an electroporation field can induce migration from the external membrane of negatively charged DNA molecules, which then accumulate in the cytoplasm (Wolf et al. 1994). This model is supported by the fact that addition of DNA after electric pulses will not induce DNA uptake by the cells, even though the electropores in the cell membrane are not resealed. However, a recent detailed study using fluorescence-labeled DNA and video microscopic images indicates that DNA does not instantly translocate into the cytoplasm from the external cell membrane under electric pulses (Golzio et al. 2002). Instead, DNA first forms a complex or aggregate with the cell membrane within a millisecond after electroporation. Minutes after the electroporation, DNA is translocated into the cytoplasm and then slowly travels into the nucleus (Golzio et al. 2002); gene expression takes place only hours after transfection (Golzio et al. 2002). Thus, one of the limiting steps for electrotransfection seems to be translocating DNA from the cytoplasm into the nucleus. This idea is supported by the observation that use of Nucleofecter kits (Amaxa) greatly enhances electrotransfection efficiency because their electroporation buffers enhance DNA translocation into the nucleus. Similarly, the use of plasmid DNA containing nuclear localization signal sequences greatly enhances the level of gene expression in muscle fibers (Li et al. 2001).

Although cell membrane structures are very similar among different cell types, there is great variation in electrotransfection efficiency. One important variable that affects electrotransfection efficiency is the electroporation buffer (see Table 1): The use of an appropriate buffer can increase cell transfection efficiency 50–98%. The general rule is that a low-ionic-constant buffer will enhance cell transfection efficiency (Golzio et al. 2004); however, different cell lines require different buffer components (Cegovnik and Novakovic 2004). Amaxa has developed three buffers for transfecting 300 different types of cell lines at a high efficiency (Iversen et al. 2005). We have found that the key for high transfection efficiency is the nucleofector buffer, not the electroporation system; a comparable level of transfection efficiency can be achieved using Amaxa's buffer and other electroporation systems (S. Li, unpubl.).

Protocol

Electrotransfection of Mammalian Cells

The following protocol describes transfection of plasmid DNA into mammalian cell lines using electroporation. Parameters that affect the electrotransfection efficiency include the choice of electroporation buffer, plasmid DNA construct, and cell shape and size. In general, suspension cells are difficult to transfect, but adherent cells are relatively easy. Small-volume cells are difficult to transfect, whereas large-volume cells such as muscle cells are easy to transfect. The use of large plasmid DNA (>13 kb) seems to reduce transfection efficiency. Regardless of cell size and phenotype, a high concentration of cells in a small volume is required to increase transfection efficiency (Trompeter et al. 2003). A concentration of cells lower than 5×10^6 cells/ml will yield poor transfection efficiency (S. Li, unpubl.); therefore, a concentration of 1×10^7 per ml is recommended.

MATERIALS

Reagents

Cell culture medium (complete and serum-free)
Cell line to be transfected
Electroporation buffer

Commercially available electroporation buffers are listed in Table 1; store the buffer according to the manufacturer's instructions. Noncommercial electroporation buffer recipes are prepared as shown in Table 3; sterilize before use and store the buffer at room temperature.

Phosphate-buffered saline (PBS)
Plasmid DNA

Purify plasmid DNAs by column chromatography.

1x Trypsin-EDTA (0.05%)

Equipment

Electroporation cuvettes (BTX Instruments, Bio-Rad, Invitrogen, Amaxa, or Thermol Electron)

Cuvettes commonly used for mammalian cell electroporation are 1 cm wide with a 4-mm internal gap. Cuvettes with a 2-mm internal gap can be used with reduced pulse duration.

Electroporation system (see Table 2)

TABLE 3. Composition of Noncommericial ("Homemade") Electroporation Buffers

Name of EP Buffer	Composition of EP Buffer	Author
Hypo-osmolar EP buffer Iso-osmolar EP buffer	KCl (25 mM) KH_2PO_4 (0.3 mM) K_2HPO_4 (0.85 mM) Myo-inositol (90 mOsm/kg, pH 7.2)	Zimmermann (1996)
Pulsing buffer	KCl (125 mM) NaCl (15 mM) Glucose (3 mM) HEPES (25 mM) $MgCl_2$ (1.2 mM, pH 7.4)	Li et al. (2002)

EP=electroporation.

METHODS

Preparation of Cells for Electroporation

1. At 24–72 hours before electroporation, divide the cells and continue to grow them in an appropriate complete culture medium with serum.

   This procedure ensures that the cells are healthy and highly proliferative, which will increase tolerance to electroporation and enhance the transfection efficiency.

2. When the cultures reach mid- to late-logarithmic growth (or 60–90% confluence), harvest the cells.

   For adherent cells, detach from the plate with 1x trypsin-EDTA (prewarmed to 37°C) and wash once with PBS or selected washing buffer at room temperature.

3. Recover the cells by centrifugation at 250*g* for 5 minutes at 20°C. Resuspend them in a selected electroporation buffer in a volume of 5×10^5 to 5×10^6 cells per 100 μl.

4. Mix 2–10 μg of DNA with 100 μl of cells and transfer the mixture into an electroporation cuvette.

   Although sometimes recommended, it is not necessary to incubate the cell and DNA mixture before electroporation. The entire procedure should be conducted at room temperature.

Electroporation

5. Tap the cuvette with the mixture of DNA and cells (cells may precipitate when cuvettes sit for an extended time) and place the cuvette in the electroporation chamber.

6. Select the parameters appropriate for the electroporation device and the system under study and start the electroporation.

7. Immediately after electroporation, deliver 500–900 μl of complete culture medium into the cuvette. Use a disposable pipette to transfer the electrotransfected cells into an appropriate culture vessel, such as a 6-well plate. Add additional cell culture medium to the normal level for maintaining cell growth.

   For stable transfection, transfer only a portion of the electrotransfected cells (1/10–1/20 volume) from cuvettes into cell culture vessels.

8. Incubate the cells overnight under the appropriate growth conditions. Replace with fresh complete culture medium if there is obvious cell death in the cell culture.

Gene Expression Analysis

9. If the object of the study is stable transfection, go directly to Step 10. For transient expression, harvest the cells 24–72 hours after transfection. Determine gene expression as appropriate, using, e.g., flow cytometry, RNA hybridization, or immunological or enzymatic assays. If the transfected gene contains a fluorescence tag, examine it under a fluorescence microscope.

10. To isolate stable transfectants: After 48–72 hours of growth in complete medium, transfer the cells into the appropriate selection medium. Incubate further under appropriate conditions to allow resistant cells to grow.

EXAMPLE PROTOCOL AND PARAMETERS

1. Cells and concentration: SVEC (ATCC) at 106 cells/100 μliter cell culture medium.
2. Material delivered: 2 μg of pEGFP-N1 plasmid DNA (Clontech).
3. Electroporation buffer: DMEM cell culture medium.
4. Preparation:
   a. Detach cells in trypsin-EDTA and wash in serum-free DMEM medium once.
   b. Recover cells by centrifugation at 250g for 5 minutes at room temperature.
   c. Resuspend the cells in 100 μl of DMEM medium and mix with 2 μg of plasmid DNA.
   d. Transfer the mixture into an electroporation cuvette with a 4-mm internal gap.
   e. For electroporation: Place the cuvette in the electroporation chamber of the BTX ECM830 and provide one electric pulse with a 50-msec duration at 150 V/cm.

TROUBLESHOOTING

Problem (Step 3): Selecting an electroporation buffer.
Solution: If a high level of transfection efficiency is required, choose an appropriate cell-specific Amaxa buffer according to the manufacture's recommendation (see Table 1). These electroporation buffers are effective for cell lines that are difficult to transfect using other chemical methods and may increase the transfection efficiency up to 98%. Amaxa's electroporation buffers can be used in other electroporation systems; however, the electroporation parameters must be optimized. Note that the shelf life of reconstituted Amaxa buffers is only 3 months and the buffers are relatively expensive. If budget is restricted, we recommend testing the non-commercial pulsing buffer or hypo-osmolar/Iso-osmolar buffer (recipes given in Table 3) because these buffers are inexpensive, and a high level of transfection has been achieved in some cell lines (Table 1). For stable transfection, when the transfection efficiency is not a concern, we recommend using the cell culture medium for electroporation to reduce cost. For achieving a high level of RNA transfection, use Opti-MEM, OptiBuffer, or Cyto Pulse low-conductivity medium (Table 1).

Problem (Step 6): Selecting the proper electroporation parameters.
Solution: There are many types of electroporation systems for effectively transfecting cells in vitro (see Table 2), but it is important to note that the electroporation parameters cannot be translated from one system to another. For new users, the Amaxa electroporation system is recommended because this system has preset optimal electroporation programs for different cell lines and is easy to operate. For investigators interested in both in vitro and in vivo electrotransfer, BTX ECM830 is recommended because different electrodes can be adapted to this system for both in vitro and in vivo gene transfer. The standard electroporation parameters for cell transfection using BTX ECM830 are 150 V/cm, 50-msec duration, and one electric pulse. To maximize transfection efficiency, the user can change the duration of the pulse, the number of electric pulses, or pulse voltage. A button for adjusting pulse duration is not installed in some equipment; however, the equipment does have an electric capacitance button. The user can adjust the capacitance to change the pulse duration (pulse duration = capacitance x resistance [Weaver 1995]). Increased heat is generally not a concern for BTX ECM830, which releases square wavelength, but it may be a problem when multiple pulses or a long-pulse duration is selected. In these cases, we recommend chilling the cuvettes before electroporation (Reiss et al. 1986).

Problem (Step 7): Reusing cuvettes?

Solution: The use of new cuvettes for each transfection is recommended by most manufacturers, but this is not necessary. Cuvettes can be reused or recycled with the following treatment (S. Li, unpubl.):

1. Rinse the used cuvettes with tap water, followed by double-deionized H_2O and then 70% ethanol.
2. Place the cleaned cuvettes under a tissue culture hood and expose them to UV light for 1 hour.
3. Cap the cuvettes for the next electrotransfection. The number of times the cuvettes can be reused is unlimited.

ACKNOWLEDGMENTS

This work is supported by grants from NCI/NIH (RO1CA98928) and NIDCR/NIH (R21DE14682). This chapter was prepared with the assistance of Katie Watson and Ryan Craig.

REFERENCES

Andre F. and Mir L.M. 2004. DNA electrotransfer: Its principles and an updated review of its therapeutic applications. *Gene Ther.* (suppl. 1) **11:** S33–S42.

Bonehill A., Heirman C., Tuyaerts S., Michiels A., Zhang Y., van der Bruggen P., and Thielemans K. 2003. Efficient presentation of known HLA class II-restricted MAGE-A3 epitopes by dendritic cells electroporated with messenger RNA encoding an invariant chain with genetic exchange of class II-associated invariant chain peptide. *Cancer Res.* **63:** 5587–5594.

Cegovnik U. and Novakovic S. 2004. Setting optimal parameters for in vitro electrotransfection of B16F1, SA1, LPB, SCK, L929 and CHO cells using predefined exponentially decaying electric pulses. *Bioelectrochemistry* **62:** 73–82.

de Gennes P.G. 1999. Passive entry of a DNA molecule into a small pore. *Proc. Natl. Acad. Sci.* **96:** 7262–7264.

Gabriel B. and Teissie J. 1997. Direct observation in the millisecond time range of fluorescent molecule asymmetrical interaction with the electropermeabilized cell membrane. *Biophys. J.* **73:** 2630–2637.

Golzio M., Rols M.P., and Teissie J. 2004. In vitro and in vivo electric field-mediated permeabilization, gene transfer, and expression. *Methods* **33:** 126–135.

Golzio M., Teissie J., and Rols M.P. 2002. Direct visualization at the single-cell level of electrically mediated gene delivery. *Proc. Natl. Acad. Sci.* **99:** 1292–1297.

Haleem-Smith H., Derfoul A., Okafor C., Tuli R., Olsen D., Hall D.J., and Tuan R.S. 2005. Optimization of high-efficiency transfection of adult human mesenchymal stem cells in vitro. *Mol. Biotechnol.* **30:** 9–20.

Hofmann G.A. 1995. Instrumentation. In *Animal cell electroporation and electrofusion protocols* (ed. J.A. Nickoloff), pp. 41–60. Humana Press, Totowa, New Jersey.

Iversen N., Birkenes B., Torsdalen K., and Djurovic S. 2005. Electroporation by nucleofector is the best nonviral transfection technique in human endothelial and smooth muscle cells. *Genet. Vaccines Ther.* **3:** 2.

Klenchin V.A., Sukharev S.I., Serov S.M., Chernomordik L.V., and Chizmadzhev Yu A. 1991. Electrically induced DNA uptake by cells is a fast process involving DNA electrophoresis. *Biophys. J.* **60:** 804–811.

Li L.H., Shivakumar R., Feller S., Allen C., Weiss J.M., Dzekunov S., Singh V., Holaday J., Fratantoni J., and Liu L.N. 2002. Highly efficient, large volume flow electroporation. *Technol. Cancer Res. Treat.* **1:** 341–350.

Li S. 2004. Electroporation gene therapy: New developments in vivo and in vitro. *Curr. Gene Ther.* **4:** 309–316.

Li S., Wilkinson M., Xia X., David M., Xu L., Purkel-Sutton A., and Bhardwaj A. 2005. Induction of IFN-regulated factors and antitumoral surveillance by transfected placebo plasmid DNA. *Mol. Ther.* **11:** 112–119.

Li S., MacLaughlin F.C., Fewell J.G., Gondo M., Wang J., Nicol F., Dean DA., and Smith L.C. 2001. Muscle-specific enhancement of gene expression by incorporation of SV40 enhancer in the expression plasmid. *Gene Ther.* **8:** 494–497.

Mertz K.D., Weisheit G., Schilling K., and Luers G.H. 2002. Electroporation of primary neural cultures: A simple method for directed gene transfer in vitro (erratum *Histochem. Cell Biol.* [2003] **119:** 175). *Histochem. Cell Biol.* **118:** 501–506.

Michiels A., Tuyaerts S., Bonehill A., Corthals J., Breckpot K., Heirman C., Van Meirvenne S., Dullaers M., Allard S., Brasseur F., van der Bruggen P., and Thielemans K. 2005. Electroporation of immature and mature dendritic cells: Implications for dendritic cell-based vaccines. *Gene Ther.* **12:** 772–782.

Neumann E., Schaefer-Ridder M., Wang Y., and Hofschneider P.H. 1982. Gene transfer into mouse lyoma cells by electroporation in high electric fields. *EMBO J.* **1:** 841–845.

Reiss M., Jastreboff M.M., Bertino J.R., and Narayanan R. 1986. DNA-mediated gene transfer into epidermal cells using electroporation. *Biochem. Biophys. Res. Commun.* **137:** 244–249.

Smits E., Ponsaerts P., Lenjou M., Nijs G., Van Bockstaele D.R., Berneman Z.N., and Van Tendeloo V.F. 2004. RNA-based gene transfer for adult stem cells and T cells. *Leukemia* **18:** 1898–1902.

Trompeter H.I., Weinhold S., Thiel C., Wernet P., and Uhrberg M. 2003. Rapid and highly efficient gene transfer into natural killer cells by nucleofection. *J. Immunol. Methods* **274:** 245–256.

Van Meirvenne S., Straetman L., Heirman C., Dullaers M., De

Greef C., Van Tendeloo V., and Thielemans K. 2002. Efficient genetic modification of murine dendritic cells by electroporation with mRNA. *Cancer Gene Ther.* **9:** 787–797.

Van Tendeloo V.F., Ponsaerts P., Lardon F., Nijs G., Lenjou M., Van Broeckhoven C., Van Bockstaele D.R., and Berneman Z.N. 2001. Highly efficient gene delivery by mRNA electroporation in human hematopoietic cells: Superiority to lipofection and passive pulsing of mRNA and to electroporation of plasmid cDNA for tumor antigen loading of dendritic cells. *Blood* **98:** 49–56.

Van Tendeloo V.F., Willems R., Ponsaerts P., Lenjou M., Nijs G., Vanhove M., Muylaert P., Van Cauwelaert P., Van Broeckhoven C., Van Bockstaele D.R., and Berneman Z.N. 2000. High-level transgene expression in primary human T lymphocytes and adult bone marrow CD34+ cells via electroporation-mediated gene delivery. *Gene Ther.* **7:** 1431–1437.

Weaver J.C. 1993. Electroporation: A general phenomenon for manipulating cells and tissues. *J. Cell. Biochem.* **51:** 426-435.

———. 1995. Electroporation theory. Concepts and mechanisms. *Methods Mol. Biol.* **47:** 1–26.

Wolf H., Rols M.P., Boldt E., Neumann E., and Teissie J. 1994. Control by pulse parameters of electric field-mediated gene transfer in mammalian cells. *Biophys. J.* **66:** 524–531.

Wong T.K. and Neumann E. 1982. Electric field mediated gene transfer. *Biochem. Biophys. Res. Commun.* **107:** 584–587.

Zimmermann U. 1996. The effect of high intensity electric field pulses on eukaryotic cell membranes: Fundamentals and applications. In *Electromanipulation of cells* (ed. U. Zimmermann and G.A. Neil), pp. 1–106. CRC Press, Boca Raton, Florida.

38 Micro In Utero Electroporation for Efficient Gene Targeting in Mouse Embryos

Tomomi Shimogori

Critical Period Mechanisms Research Group, Institute of Physical and Chemical Research (RIKEN), Brain Science Institute, Saitama 351-0198, Japan

ABSTRACT

To understand the genetic mechanism of embryonic development, it is necessary to overexpress or misexpress target genes in specific areas and at specific times. Because of its easy access, the avian embryo has been widely used as a model system for the study of developmental events. The novel technique, in ovo electroporation, has been used successfully for introducing genes into chick embyos. In contrast, the inaccessibility of mammalian embryos, which are encased in the maternal uterus, has made in utero manipulations difficult or impossible for targeting specific areas at most developmental stages. To address this issue, we have developed a "micro in utero electroporation" technique in which DNA is introduced locally into the mouse embryo using fine tungsten and platinum wires placed into the ventricle. Although ectopic gene expression lasts only a few days, the temporal effect on brain maturation is testable from embryonic to postnatal stages. This procedure is simple and can be readily performed on most parts of the embryo. The method can be used with many different types of genes, as well as for simultaneously delivering multiple genes.

INTRODUCTION

Correct body patterning and wiring of neuronal networks in both vertebrates and invertebrate require spatiotemporal expression of specific genes during embryogenesis. Thus, the deliberate spatiotemporal manipulation of genes will be a powerful tool for understanding their true function during embryogenesis. For the study of developmental mechanisms in vertebrates, genetic manipulations are often performed in *Xenopus* and zebra fish, because of the accessibility and transparency of the embryos of these species. Understanding developmental mechanisms in complex organs, such as the brain, requires the use of mammalian experimental models. Mice are the most frequently used in this regard, because the function of specific genes can be studied in transgenic animals. However, spatiotemporal targeting of genes by electroporation has been considered difficult in mice due to the nontransparent nature and physical obstruction of the uterine wall.

The technique of in ovo electroporation in the chick has allowed the rapid and direct examination of the function of delivered genes (Muramatsu et al. 1997; Itasaki et al.1999; Nakamura et al. 2000). A logical step has been to adapt this approach to mouse embryos, thereby allowing rapid analysis of gene function, compared to the generation of knockout or transgenic lines, which takes much longer. Furthermore, in utero electroporation also provides the possibility of overexpressing or inhibiting target gene expression in a spatial and temporal manner. Several laboratories have pursued in utero electroporation in mouse embryos, with several reports of successful gene delivery (Saito and Nakatsuji 2001; Tabata and Nakajima 2001; Borrell et al. 2005). These reports show that genes can be delivered over a broad time period from embryonic (E) day 12.5 onward, using a tweezer-type electrode (such as model CUY650-P5 from NEPA Gene). The invention of needle-type electrodes has also enabled transfection of a spot area in chick in ovo electroporation (Momose et al. 1999), facilitating the analysis of spatial function of genes such as positional information of morphogens (Agarwalae et al. 2001). We have previously demonstrated the utility of micro in utero electroporation using needle-type electrodes for gene transfer into precisely targeted tissue of developing mouse embryonic brain (Fukuchi-Shimogori and Grove 2001). Because the insertion of electrodes can easily damage the uterus and cause low viability, the size of the electrodes must be precisely controlled. Visualization of intact embryos within the uterus is achieved by placing a fiber-optic light at a specific angle.

In this chapter, we describe how to make electrodes and perform a micro in utero electroporation that allows delivery of genes in early mouse embryos (as early as E10.5) in an area-specific manner. Note that this method of gene delivery is transient. If permanent gene expression is required, a Cre construct (encoding Cre-recombinase) and R26R reporter mouse line (Jackson Laboratories) may be used. To sequester specific gene function, it has been reported that electroporation of small interfering RNA (siRNA) constructs is efficient in chick embryo (Nakamura et al. 2004). The use of these constructs in in utero electroporation will become a powerful tool when combined with a genetic approach such as knockout and transgenic mice.

Protocol

Electroporation of Mouse Embryos In Utero

Delivery of genetic material into precisely targeted brain tissue of the developing mouse embryo is achieved using needle electrodes, placed on either side of the targeted area. A mixture of DNA and tracking dye microinjected into the tissue, and a pulse generator, is used to pass current pulses into the tissue. The procedure is visualized using fiber-optic light.

MATERIALS

Reagents

DNA to be delivered, solution (1 μg/μl)
Ethanol (50%)
Fast Green FCF protein staining reagent (Sigma-Aldrich)
Phosphate-buffered saline (PBS)
Pregnant mice (E10.5 or older)
Sodium pentobarbital

Equipment

Autoclip (9 mm, ROBOZ)
Electrodes

> To prepare the electrodes, tungsten wire and platinum wire (A-M Systems), encased in plastic tubes and connected to a metallic pin (WPI 5428), are sharpened with sandpaper (Fig. 1B). The electrodes are coated with a thin layer of nail polish for insulation. After the nail polish has dried, the pin is removed from the tip (~200 μm) with acetone-soaked cotton swabs.

Fiber-optic light (Leica)
Forceps (ROBOZ)
Glass capillary tubes (Stoelting Co.)

> These are pulled using a micropipette puller. The tip is pinched using forceps (Fig. 1A).

Micromanipulator (KD scientific)
Micropipette
Pulse generator (Model 2100, A-M Systems)
Scissors (ROBOZ)
Suture (3-0, Dexon II)

A.

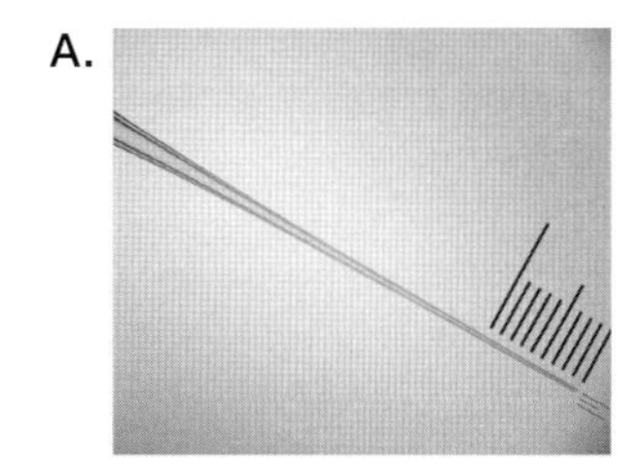

B.

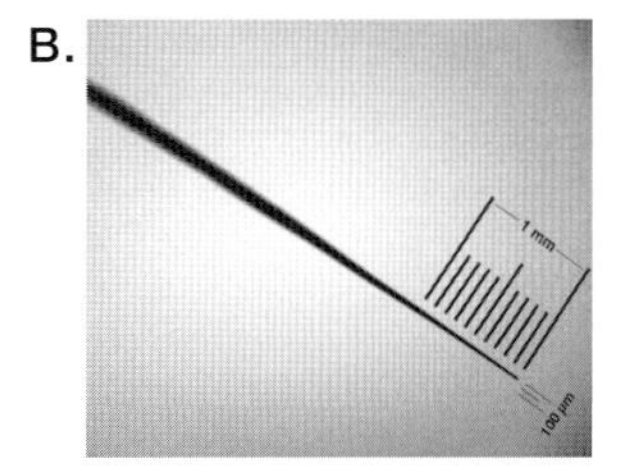

FIGURE 1. (*A*) Shape of glass capillary used for injection. The glass capillary tube is pulled using a micropipette puller. The tip is broken at about 20 μm external diameter. To prevent damage to the uterus, 1 mm from the tip should be no more than 50 μm. (*B*) Tungsten and platinum wires are sharpened with sand paper, and a thin coat of nail polish is applied.

METHOD

1. Anesthetize pregnant mice (E10.5 or older) intraperitoneally by injection of sodium pentobarbital (50 µg/g body weight). About 5–10 minutes after injection, use a razor blade to shave the fur over the abdomen and wash with 50% ethanol.
2. Use fine scissors to make an incision, 2.5 cm or less, in the abdominal cavity. Carefully remove all of the uterine horns with forceps, place them onto cotton gauze moistened with PBS, and then place them around the wound.

   It is important to keep the uterus moist at all times using PBS.
3. To visualize the embryos, use a fiber-optic light. Wet the tip of the fiber-optic cable and the uterus with warm PBS (37°C). Hold the fiber-optic light with the index and middle finger, and place the uterus between the fiber-optic light cable and thumb. Squeeze the uterus gently to push the embryos closer to the uterine wall (Fig. 2).

   The embryos may be rotated into position by rubbing the surface of the uterus with a cotton swab soaked in PBS.
4. Prepare a solution of 1 µl of DNA (1 µg/µl) with a concentration of 1% Fast Green protein dye.
5. Once the embryos are positioned properly, use a micropipette mounted in a micromanipulator to inject the DNA-Fast Green solution into the intracerebroventricle.
6. Insert a fine tungsten negative electrode and a platinum electrode into the uterus, so that the targeted area is placed between two electrodes (Fig. 2). Use a pulse generator to pass a series of three square-wave current pulses (7–10 V, 100 msec) three times at 1-second intervals.

   Note that shorter surgical times give better viability. A maximum surgery time of 30 minutes per pregnant dam is critical.

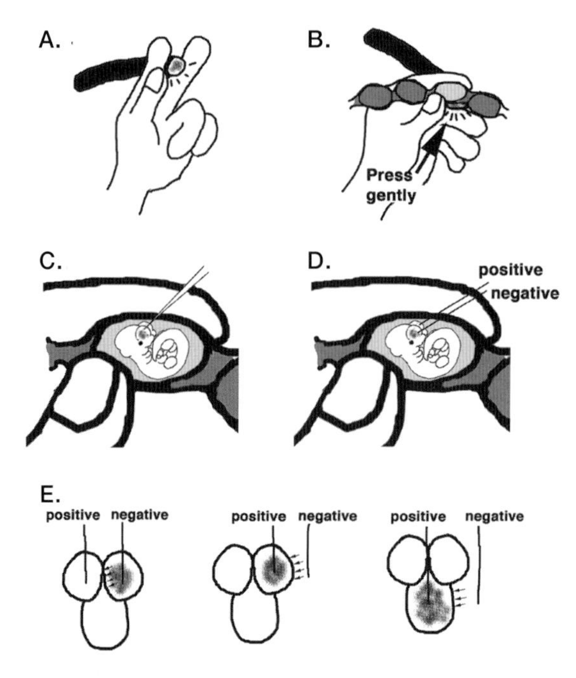

FIGURE 2. (*A*) Hold the fiber-optic cable between the index finger and middle finger. (*B*) Place the uterus horn between the tip of the light source and thumb. Press the uterus to position embryo close to the membrane. (*C*) Inject 1 µl of DNA solution into the ventricle. (*D*) Insert fine tungsten negative electrode and a platinum positive electrode into the left and right hemispheres, respectively, and then deliver a series of three square-wave current pulses. (*E*) Position two electrodes at different electroporation sites (*arrows*).

7. Place the uterine horn back into its original location with a small amount of PBS (~500 μl). Close the surgical incision in the uterine wall with sutures and close the skin with a 9-mm Autoclip. Keep the animal warm until it recovers from the anesthesia (~2 hours).

   The viability of embryos and efficiency of cells transfected vary with the age of the embryos and the location of the site of electroporation. In the case of E11.5 electroporation, 60% of the embryos generally survive, with 50% of surviving embryos showing effective electroporation. Conditions must be optimized depending on the stage and area of embryo undergoing electroporation.

ACKNOWLEDGMENTS

I thank Dr. N.P. Murphy for critically reading the manuscript.

REFERENCES

Agarwala S., Sanders T.A., and Ragsdale C.W. 2001. Sonic hedgehog control of size and shape in midbrain pattern formation. *Science* **291:** 2147–2150.

Borrell V., Yoshimura Y., and Callaway E.M. 2005. Targeted gene delivery to telencephalic inhibitory neurons by directional in utero electroporation. *J. Neurosci. Methods* **143:** 151–158.

Fukuchi-Shimogori T. and Grove E.A. 2001. Neocortex patterning by the secreted signaling molecule FGF8. *Science* **294:** 1071–1074.

Itasaki N., Bel-Vialar S., and Krumlauf R. 1999. "Shocking" developments in chick embryology: Electroporation and in ovo gene expression. *Nat. Cell Biol.* **1:** 203–207.

Momose T., Tonegawa A., Takeuchi J., Ogawa H., Umesono K., and Yasuda K. 1999. Efficient targeting of gene expression in chick embryos by microelectroporation. *Dev. Growth Differ.* **41:** 335–344.

Muramatsu T., Shibata O., Ryoki S., Ohmori Y., and Okumura J. 1997. Foreign gene expression in the mouse testis by localized in vivo gene transfer. *Biochem. Biophys. Res. Commun.* **233:** 45–49.

Nakamura H., Watanabe Y., and Funahashi J.A. 2000. Misexpression of genes in brain vesicles by in ovo electroporation. *Dev. Growth Differ.* **42:** 199–201.

Nakamura H., Katahira T., Sato T., Watanabe Y., and Funahashi J. 2004. Gain- and loss-of-function in chick embryos by electroporation. *Mech. Dev.* **121:** 1137–1143.

Saito T. and Nakatsuji N. 2001. Efficient gene transfer into the embryonic mouse brain using in vivo electroporation. *Dev. Biol.* **240:** 237–246.

Tabata H. and Nakajima K. 2001. Efficient in utero gene transfer system to the developing mouse brain using electroporation: Visualization of neuronal migration in the developing cortex. *Neuroscience* **103:** 865–872.

39 Lipoplex and LPD Nanoparticles for In Vivo Gene Delivery

Shyh-Dar Li,* Song Li,† and Leaf Huang*

*Division of Molecular Pharmaceutics, School of Pharmacy, University of North Carolina at Chapel Hill, North Carolina 27599; †Center for Pharmacogenetics, School of Pharmacy, University of Pittsburgh, Pennsylvania 15213

ABSTRACT

Research in gene therapy has focused on the development of suitable delivery systems. Although the majority of clinical trials have been based on the use of viral vectors, cationic liposomes are emerging as promising gene carriers as a result of their safety and versatility (Pedroso de Lima et al. 2001). Viral vectors may provoke mutagenesis and carcinogenesis, and repeated administration of a viral vector induces an immune response that abolishes the transgene expression (Liu and Huang 2002). In light of these limitations for viral vectors, cationic liposomes offer an attractive alternative. Despite the relatively low transfection efficiency of cationic liposome-based delivery systems compared to viral vectors, they remain a promising system for gene transfer because of their safety, versatility, ease of preparation and scale up, low immunogenicity, and high biocompatibility. Because of these features, this nonviral delivery system is widely used in both laboratory research and clinical trials.

INTRODUCTION

Lipoplex, also called cationic liposome–DNA complex, is formed via electrostatic interaction of anionic nucleic acids with cationic liposomes. The method is rapid and simple for the preparation of concentrated and homogeneous suspension of small unilamellar liposomes for laboratory-scale experiments. The procedure involves forming a thin film of dried lipids on the bottom of a glass tube, hydrating the lipid film in a desired aqueous solution, and passing the liposome suspension through polycarbonate filters of a desired pore size. Lipoplex is the most popular tool for gene transfer among nonviral vectors. However, cationic liposomes, especially those com-

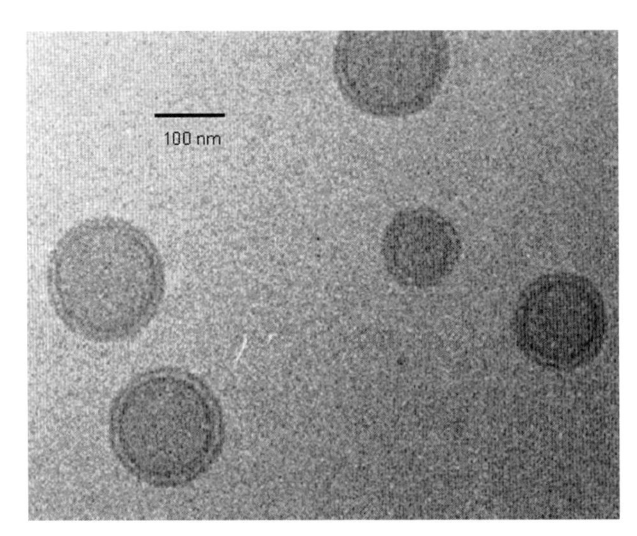

FIGURE 1. Cryoelectron microscopy picture of LPD.

posed of monovalent cationic lipids, cannot condense DNA efficiently, resulting in a very heterogeneous size distribution of the complex (from 300 nm to more than 2000 nm) (Pedroso de Lima et al. 2001). Considering that a gene delivery system should be small and **contain** condensed DNA, we developed a system for preparing liposome-polycation-DNA (LPD) nanoparticles (Gao and Huang 1996; Li and Huang 1997). LPD nanoparticles contain a highly condensed DNA core surrounded by lipid bilayers with an average size of approximately 100 nm (Fig. 1).

LPD nanoparticles are routinely used for systemic injection. Following systemic administration to mice, LPD effectively transfects all major organs, including heart, lung, liver, spleen, and kidney (Fig. 2). For example, 10 minutes after injection into the tail vein of a mouse, 30–40% and 35–45% of injected plasmid DNA is found in the lung and liver, respectively (Zhang et al. 2005). The highly positive charge of lipoplex and LPD causes them mainly to distribute to the lung, the first organ encountered after intravenous injection. The highest transgene activity is found predominantly in

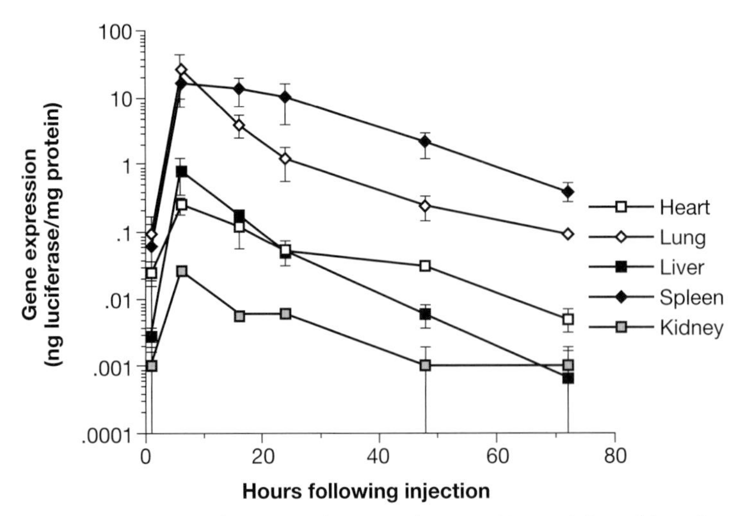

FIGURE 2. In vivo gene expression of LPD as a function of time. LPD was injected into the mice at 50-μg plasmid DNA (pCMVLacZ) per animal. At different times following the injection, mice were killed and major organs were assayed for luciferase gene expression. (Reprinted, with permission, from Li and Huang 1997.)

the pulmonary endothelial cells (Li and Huang 1997). Because the lung is the major organ for transgene expression, lipoplex is usually used to treat lung diseases, such as tumor metastases in lungs (Sakurai et al. 2003) and cystic fibrosis (Dass 2004). Furthermore, when delivered by LPD, retinoblastoma (Rb), a tumor suppressor gene, is effectively expressed in the lung of Rb (+/–) mice, which spontaneously develop primary tumors in the pituitary that later metastasize to the lung. With repeated injections, metastatic tumor cells undergo spontaneous apoptosis, and the frequency of lung tumor metastases is decreased (Nikitin et al. 1999).

This chapter describes the preparation, physical properties, and biological activities of these delivery systems.

Protocol

Preparation and Delivery of LPD Nanoparticles

Cationic liposomes are mixed with the DNA sample to form the lipoplex complex that is subsequently incorporated into the LPD nanoparticles. The nanoparticle complex is injected into mice, and expression of the transfected DNA is monitored with an appropriate assay.

MATERIALS

CAUTION: See Appendix for appropriate handling of materials marked with <!>.

Reagents

5x Dextrose solution (26% w/v) in sterile H_2O

Dissolve 13 g of dextrose in approximately 40 ml of sterile H_2O and then adjust to 50 ml with sterile H_2O. Filter the solution through a 0.2-µm membrane and store at room temperature for up to 1 year.

DOTAP (25-mg/ml stock solution) in chloroform <!> (Avanti Polar Lipids)

DOTAP (1,2-dioleoyl-3-trimethylammonium propane) is one of the most widely used cationic lipids for gene delivery. Some neutral colipids (e.g., DOPE [1,2-dioleoyl phosphatidylethanolamine] and cholesterol) can further improve gene transfection.

Cholesterol (20-mg/ml stock solution) in chloroform (Sigma-Aldrich)

Cholesterol can serve as a helper lipid for DOTAP to form more stable and efficient cationic liposomes compared to liposomes composed of DOTAP and DOPE when used in vivo (Li et al. 1998).

Plasmid DNA (1 mg/ml) in deionized H_2O

Plasmid DNA should be highly purified and endotoxin-free. We recommend diluting in H_2O or 5.2% dextrose solution, because salt interferes with the charge-charge interaction between DNA and cationic lipids and causes aggregate formation.

Protamine sulfate–USP (10 mg/ml, Elkins-Sinn)

Protamine sulfate (MW 4000–4250), an arginine-enriched cationic peptide, is applied in LPD formulation as a DNA condensation agent. One mole of protamine sulfate contributes 21 moles of positive charge and possesses high-compaction ability for DNA.

Sterile H_2O

Steam in an autoclave for 30 minutes at 121°C.

Equipment

Bath sonicator
Corex glass tube (30 ml)
LiposoFast extruder, with 1.0-ml syringes (Avestin)
Nitrogen (N_2) gas tank
Nuclepore (1.0, 0.4, and 0.1 µm) polycarbonate membrane filters (Corning, from VWR International)
Polyethylene conical tube (50 ml)
Vacuum desiccator

METHODS

Preparation of Cationic Liposomes

1. Rinse a 30-ml glass Corex tube three times with chloroform.

   There is no need to dry the residual chloroform.

2. Mix 0.8 ml of DOTAP stock solution and 0.55 ml of cholesterol stock solution in the glass tube.

   Currently, the cationic liposomes used in our laboratory consist of DOTAP and cholesterol at a molar ratio of 1:1. Lipoplex formed with 1 μg of plasmid DNA and cationic liposomes containing 36 nmoles of DOTAP and 36 nmoles of cholesterol has a charge ratio of 12:1 (+:–).

3. In a chemical fume hood, evaporate the chloroform to form a thin lipid film on the bottom of the glass tube by introducing a stream of N_2 gas down the side of the tube while rotating. To speed up the process, dip the tube in a warm water bath (35°C).

4. Place the glass tube in a vacuum desiccator for 2–3 hours for complete dryness. To prevent loss of lipid film under vacuum, cover the tube with aluminum foil and poke small holes in the foil with a needle.

5. Add 2 ml of sterile H_2O to the glass tube.

6. Suspend the lipids in solution by vortexing for 15 seconds at maximum speed until there are no lipid residuals or white precipitates in the tube.

   To facilitate the process, place the glass tube in a bath sonicator for several quick bursts (10–15 seconds).

7. Incubate the suspension either for 2–3 hours at room temperature or overnight at 4°C to allow for complete hydration for lipids.

   The longer the hydration, the easier the extrusion will be. Overnight hydration is recommended as a common practice.

8. Vortex the suspension and incubate it for 5–10 minutes in a 65°C water bath. If any lipid aggregates are observed, sonicate the suspension in a water bath sonicator until all aggregates disappear and then return it to the water bath.

9. Place two 1.0-μm polycarbonate filters in the extruder.

10. Heat the extruder to 65°C for 5 minutes.

    Heating the lipid suspension and extruder to 65°C maintains the lipids in a liquid state and allows for easier extrusion and improved lipid mixing.

11. Extrude the lipid suspension by passing the dispersion through the extruder five times. Return the extruder to the water bath.

    Following extrusion, the lipid suspension should be transformed from a cloudy suspension to a translucent one.

12. Repeat Steps 9–11 using 0.4- and 0.1-μm polycarbonate filters sequentially to obtain small unilamellar liposomes with a mean diameter of 100–200 nm.

    Cationic liposome suspensions are stable for several months when stored at 4°C.

Preparation of Lipoplex

13. Place 60 μl of 5x dextrose, 154 μl of sterile H_2O, and 86 μl of cationic liposomes in a 50-ml conical tube. Vortex at medium speed for 10 seconds.

    Lipoplex is prepared by gently mixing cationic liposomes and plasmid DNA. Volumes and quantities are given for 100 μg of plasmid DNA in a final volume of 600 μl. For larger doses, volumes and quantities should be proportionally scaled up.

14. Place 40 μl of 5x dextrose, 60 μl of sterile H_2O, and 100 μl of plasmid DNA in a 1.5-ml microcentrifuge tube. Gently tap tube to mix. **Do not vortex.**
15. While gently swirling the diluted liposomes suspension, slowly add the diluted DNA solution dropwise. Use a 1000-μm pipette tip to transfer the solution; 5–10 seconds is required to add 200 μm of DNA solution.
16. Incubate the complex for 10–15 minutes at room temperature before injection.

Preparation of LPD

17. Combine 60 μl of 5x dextrose, 148 μl of sterile H_2O, 86 μl of cationic liposomes, and 6 μl of protamine sulfate solution in a 50-ml conical tube and vortex at medium speed for 10 seconds.

    The optimal composition for LPD nanoparticles is 12 nmoles of DOTAP/12 nmoles cholesterol/0.6 μg of protamine sulfate/1 μg of plasmid DNA, which has a charge ratio of 4:1 (+:–) between DOTAP and DNA and a 1:1 charge ratio between protamine and DNA.

18. Add 40 μl of 5x dextrose, 60 μl of sterile H_2O, and 100 μl of plasmid DNA in a 1.5-ml centrifuge tube. Gently tap tube to mix. **Do not vortex.**
19. While gently swirling the diluted liposome-protamine suspension, slowly add the diluted DNA solution dropwise.

    Using a 1000-μl pipette tip, 5–10 seconds is required to add 200 μl of DNA solution. LPD nanoparticles are usually prepared immediately before use. However, the preparation can be stored as solution for at least 4 weeks at 4°C without losing activity. Alternatively, the preparation may be stored as a lyophilized powder for more than 1 year at room temperature and will be fully active after reconstitution (Li et al. 2000).

20. Incubate the complex for 10–15 minutes at room temperature and then inject the complex into the appropriate target site of the animal.
21. Use an appropriate assay to monitor expression of the DNA.

TROUBLESHOOTING

Problem (Step 15): The formation of white, string-like precipitates during mixing.
Solution: The formation of precipitates is commonly caused by salts present in the solution (TE, NaCl, etc.). The precipitates appear to be toxic to mice following intravenous injection. A proper mixing vessel, dropwise addition, and slow swirling of the mixture all help to prevent the formation of precipitate. If any precipitate forms, prepare the complex again. With practice, precipitate formation can be avoided. Remember **never** to vortex the DNA or the complex.

Problem (Step 20): Optimizing the transfection efficiency of the system for delivery into the lung.
Solution: The transfection efficiency of LPD and lipoplex are both affected by the overall charge ratio. Excess positive charge enhances the transfection efficiency in the lung. It is therefore practical to change the +:– charge ratio according to requirement. In the case of lipoplex, different amounts of liposomes can be complexed with DNA. However, in the case of LPD, a 1:1 charge ratio of protamine to DNA is an optimal choice for LPD formulation (Li and Huang 1997). We recommend keeping the ratio of protamine to DNA (1:1) as a constant and varying the amount of lipids. The charge ratio can be calculated according to how much charge each component carries; 1 nmole of DOTAP and protamine sulfate contributes 1 and 21 nmoles of positive charge, respectively. Cholesterol is neutral; 1 μg of DNA contains approximately 3.1 nmoles of negative charge.

Problem (Step 21): Problems of toxicity, even mortality, when attempting gene transfer using this method.

Solution: Systemic injection of lipoplex induces a rapid induction of proinflammatory cytokines in blood, such as TNF-α, IL-12, IL-6, and IFN-γ (Scheule et al. 1997). Mortality is frequently observed at high doses (>4 mg of plasmid DNA/kg body weight). Decreasing the dose can reduce mortality, but this also results in decreased transgene expression (Tousignant et al. 2000). It appears that unmethylated CpG sequences in plasmid DNA are responsible for the immunostimulatory response (Krieg et al. 1995). The toxicity of LPD is similar to that of lipoplex. Strategies to overcome this problem include modification of plasmid DNA to reduce the number of CpG motifs (Yew et al. 2000), use of a short DNA fragment amplified by polymerase chain reaction (PCR) instead of the intact plasmid (Hofman et al. 2001), use of a general immunosuppressant such as dexamethasone (Liu et al. 2004), and use of NF-κB decoy oligonucleotide (Tan et al. 1999).

ACKNOWLEDGMENTS

The work described here was supported by grants to L.H. from the National Institutes of Health (AI-48851, DK-54225, CA-74918, DK-44935, and AR-45925) and a grant to S.L. from the National Institutes of Health (HL-63080).

REFERENCES

Dass C.R. 2004. Lipoplex-mediated delivery of nucleic acids: Factors affecting in vivo transfection. *J. Mol. Med.* **82:** 579–591.

Gao X. and Huang L. 1996. Potentiation of cationic liposome-mediated gene delivery by polycations. *Biochemistry* **35:** 1027–1036.

Hofman C.R., Dileo J.P., Li Z., Li S., and Huang L. 2001. Efficient in vivo gene transfer by PCR amplified fragment with reduced inflammatory activity. *Gene Ther.* **8:** 71–74.

Krieg A.M., Yi A.K., Matson S., Waldschmidt T.J., Bishop G.A., Teasdale R., Koretzky G.A., and Klinman D.M. 1995. CpG motifs in bacterial DNA trigger direct B-cell activation. *Nature* **374:** 546–549.

Li B., Li S., Tan Y., Stolz D.B., Watkins S.C., Block L.H., and Huang L. 2000. Lyophilization of cationic lipid-protamine-DNA (LPD) complexes. *J. Pharm. Sci.* **89:** 355–364.

Li S. and Huang L. 1997. In vivo gene transfer via intravenous administration of cationic lipid-protamine-DNA (LPD) complexes. *Gene Ther.* **4:** 891–900.

Li S., Rizzo M.A., Bhattacharya S., and Huang L. 1998. Characterization of cationic lipid-protamine-DNA (LPD) complexes for intravenous gene delivery. *Gene Ther.* **5:** 930–937.

Liu F. and Huang L. 2002. Development of non-viral vectors for systemic gene delivery. *J. Control. Release* **78:** 259–266.

Liu F., Shollenberger L.M., and Huang L. 2004. Non-immunostimulatory nonviral vectors. *FASEB J.* **18:** 1779–1781.

Nikitin A.Y., Juárez-Párez M.I., Li S., Huang L., and Lee W.H. 1999. RB-mediated suppression of spontaneous multiple neuroendocrine neoplasia and lung metastases in $Rb^{+/-}$ mice. *Proc. Natl. Acad. Sci.* **96:** 3916–3921.

Pedroso de Lima M.C., Simões S., Pires P., Faneca H., and DüzgünesN. 2001. Cationic lipid-DNA complexes in gene delivery: From biophysics to biological applications. *Adv. Drug Delivery Rev.* **47:** 277–294.

Sakurai F., Terada T., Maruyama M., Watanabe Y., Yamashita F., Takakura Y., and Hashida M. 2003. Therapeutic effect of intravenous delivery of lipoplexes containing the interferon-β gene and poly I: Poly C in a murine lung metastasis model. *Cancer Gene Ther.* **10:** 661–668.

Scheule R.K., St George J.A., Bagley R.G., Marshall J., Kaplan J.M., Akita G.Y., Wang K.X., Lee E.R., Harris D.J., Jiang C., Yew N.S., Smith A.E., and Cheng S.H. 1997. Basis of pulmonary toxicity associated with cationic lipid-mediated gene transfer to the mammalian lung. *Hum. Gene Ther.* **8:** 689–707.

Tan Y., Li S., Pitt B.R., and Huang L. 1999. The inhibitory role of CpG immunostimulatory motifs in cationic lipid vector-mediated transgene expression in vivo. *Hum. Gene Ther.* **10:** 2153–2161.

Tousignant J.D., Gates A.L., Ingram L.A., Johnson C.L., Nietupski J.B., Cheng S.H., Eastman S.J., and Scheule R.K. 2000. Comprehensive analysis of the acute toxicities induced by systemic administration of cationic lipid:plasmid DNA complexes in mice. *Hum. Gene Ther.* **11:** 2493–2513.

Yew N.S., Zhao H., Wu I.H., Song A., Tousignant J.D., Przybylska M., and Cheng S.H. 2000. Reduced inflammatory response to plasmid DNA vectors by elimination and inhibition of immunostimulatory CpG motifs. *Mol. Ther.* **1:** 255–262.

Zhang J.S., Liu F., and Huang L. 2005. Implications of pharmacokinetic behavior of lipoplex for its inflammatory toxicity. *Adv. Drug Del. Rev.* **57:** 689–698.

40 Bioresponsive Targeted Charge Neutral Lipid Vesicles for Systemic Gene Delivery

Weijun Li and Francis C. Szoka, Jr.

Department of Biopharmaceutical Sciences and Pharmaceutical Chemistry, School of Pharmacy, University of California at San Francisco, California 94143-0446

ABSTRACT

This protocol describes a stepwise procedure to prepare nucleic acids encapsulated in a polyethylene glycol (PEG)-shielded nanolipoparticle (NLP) that contains a bioresponsive lipid and ligand for use in targeted delivery of DNA. The NLP differs from a traditional cationic lipid/DNA complex (lipoplex) in the method of preparation by ethanol dialysis, by incorporation of a PEG lipid in the bilayer that surrounds the DNA core, and in the small size of the particle diameter (<100 nm). The PEG lipid is used to stabilize the particles and extend the blood retention time in vivo. Incorporation of a low-pH-sensitive PEG lipid in the NLP enables release of the PEG chain in the low-pH environment of the endosome/lysosome compartment. Disulfide-containing cationic lipids can be included to adjust the NLP surface charge from positive to neutral or negative using disulfide exchange reactions. Finally, a ligand-lipid conjugate can be inserted into the NLP surface by a micelle transfer method to enhance the particle-cell recognition. This sequential assembly process provides several advantages for use of the NLP for systemic gene delivery: (1) The in vivo circulation time is extended due to the small particle size (diameter <100 nm), the PEG protection on the particle, and the neutral surface charge; (2) the low-pH-sensitive lipids enhance DNA unpacking and endosomal escape; and (3) ligands on the charge neutral or negative surface can target gene delivery to specific tissues or cells in vivo.

INTRODUCTION

Lipid-based colloidal nanoparticles have been extensively investigated as systemic gene delivery carriers in the past decade. Current efforts directed toward the development of gene-based drugs capable of treating acquired diseases, including cancer and inflammation, have required the design of gene vectors capable of accessing the distal sites of disease following systemic (intravenous) administration (Mahato and Kim 2002). Cationic liposomes, which are among the most efficient DNA transfection reagents in vitro, condense DNA into a cationic particle when the two components are mixed together. This cationic lipid–DNA complex (lipoplex) can protect DNA from enzymatic degradation and deliver DNA into cells by interacting with the negatively charged cell membrane. Unfortunately, the cationic surface also mediates strong interactions with plasma proteins upon injection into the blood as well as with the glycocalyx of many tissues (Duzgunes 2003). These interactions result in a rapid elimination of the lipoplex (half-life <5 minutes) from the blood stream by the reticuloendothelial system (RES) (Barron and Szoka 1999). The consequences are twofold: inflammatory adverse effects and decreased access to the target tissue.

To overcome cationic lipid-induced toxicity and extend the in vivo circulation time, DNA has been encapsulated into a PEG-shielded cationic liposomal bilayer by a detergent dialysis method (Wheeler et al. 1999; Zhang et al. 1999; Harvie et al. 2000; Choi et al. 2003; Li et al. 2005) or, more recently, by an ethanol dialysis method (Jeffs et al. 2005). The PEG-shielded cationic particles differ from lipoplexes by having a bilayer shell around the nucleic acid, smaller particle diameters (<100 nm), and better stability in vivo. These properties enable an extended circulation time of the particles (half-life: 1–10 hours) in vivo and increase the fraction of the injected particles that pass through the target site. Gene transfection activity mediated by the PEG-shielded particles in tumors is low but detectable and two orders of magnitude higher than reporter gene expression in other tissues (Fenske et al. 2001).

The PEG coat that enables long circulation interferes with particle uptake by cells and the subsequent escape of internalized DNA from the endosome into the cytoplasm. Thus, the transfection level in cell culture using the PEG-coated particles is orders of magnitude lower than that achieved with the lipoplex structures (Fenske et al. 2001). Removal of the PEG coat from the particle after it had bound to the target cell would expose the cationic surface and should lead to higher levels of gene transfer. We designed an acid-labile diorthoester conjugate between monomethyl-PEG and distearoylglycerol (PODS). This conjugate is stable when formulated into a cationic liposome at basic pH, but it is hydrolyzed in the lower-pH environment of the endosome. The acid-mediated hydrolysis would separate the PEG from the lipid anchor and uncoat the particle (Guo and Szoka 2001; Choi et al. 2003; Li et al. 2005). PODS was used to formulate a low-pH-sensitive, PEG-shielded, DNA-containing NLP based on the detergent dialysis method. This NLP exhibited better transfection activity than that observed for a pH-insensitive NLP (Li et al. 2005).

The surface properties of the pH-sensitive, PEG-shielded, cationic NLP can be further modified using a disulfide exchange protocol to create a particle with a charge neutral or negative surface (Huang et al. 2005). In this approach, a cationic lipid with a disulfide linkage between the cationic moiety and the lipid anchor that can be replaced by a disulfide exchange reaction with another thio compound is used to condense DNA into PEG-shielded NLP. Various thio-containing cations, linked to a variety of linkers (Guo and Szoka 2003), can be used for this purpose. In this protocol, we use cationic thiocholesterol derivatives (TCL) that are sensitive to reducing reagents under basic buffer conditions (Fig. 1). Reducing reagents such as cysteine or glutathione can be used to modify the particle surface and change the surface charge from positive to neutral or negative. The neutral surface increases the circulation time when the PEG NLP is injected into animals.

To further enhance the gene transfection activity of the DNA-containing particles in vivo, ligands for cell surface receptors can be attached onto the NLP to increase the particle uptake by cells that express the conjugate receptor. As an example, a nonspecific ligand, the TAT peptide, is used

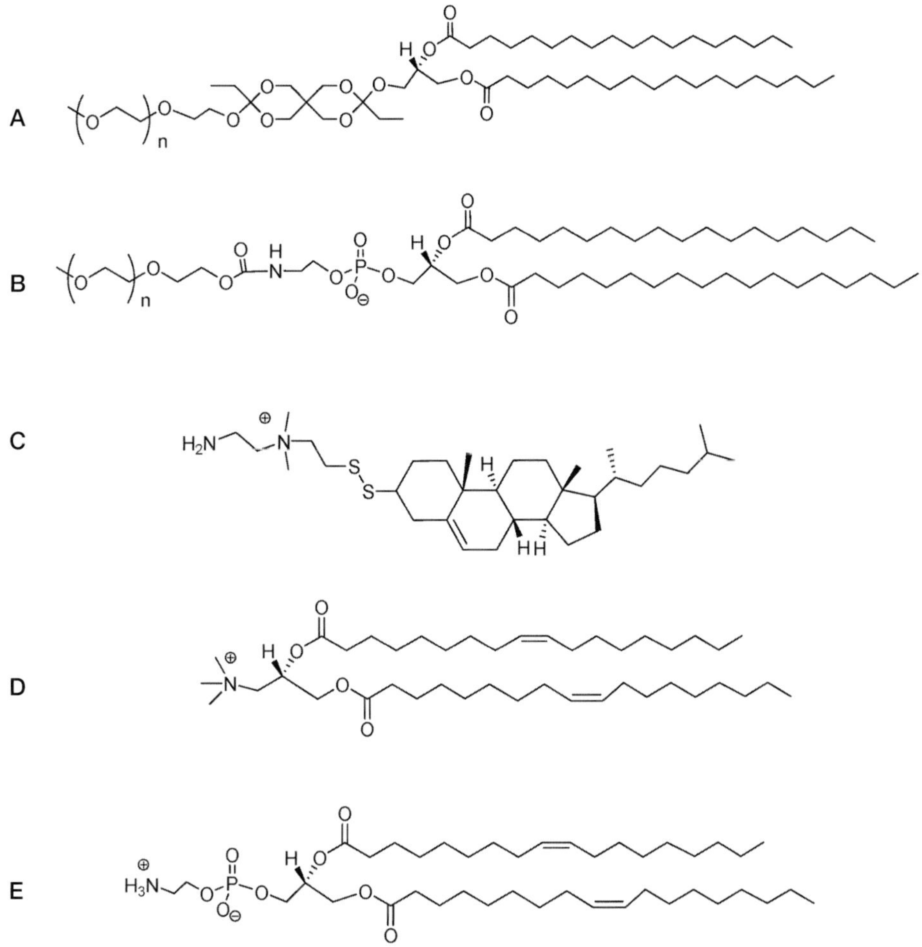

FIGURE 1. Structures of typical lipids used for the NLP formulation. These lipids are categorized into three types: (1) PEG lipids such as a low-pH-sensitive PEG lipid, PODS (*A*) or a pH-insensitive PEG lipid, PEG-DSPE (*B*); (2) cationic lipids such as a reduction-sensitive lipid, TCL (*C*) or a nonreduction-sensitive diacyl cationic lipid, DOTAP (*D*); and (3) helper lipids such as DOPE (*E*).

as a model ligand to illustrate how to insert the ligand onto the NLP surface with a "micelle transfer" method (Uster et al.1996; Huang et al. 2005).

In the following protocol, specific components are added in a sequential fashion to create the bioresponsive, targeted charge neutral lipid nanoparticle. First, DNA is encapsulated inside a cationic reducible lipid that contains a PEG lipid to create a small-diameter NLP with high DNA encapsulation efficiency, and nonencapsulated DNA is removed. Next, the cationic moiety is removed by a disulfide exchange reaction. Finally, the ligand attached to a lipid anchor is added as a micelle and transferred into the particle. The three steps of the particle assembly process are summarized below.

- ***Ethanol dialysis method to formulate the DNA-containing PEG-shielded NLP.*** It is possible to use either a detergent (octylglucoside) dialysis or an ethanol dialysis method for the particle formulation. The latter is recommended because it is more economical and easier to scale up. The PEG lipid, cationic lipid, and helper lipid are dissolved together in 50% (v/v) ethanolic buffer. DNA is dissolved in a separate aliquot of 50% (v/v) ethanolic buffer. The two solutions are mixed together, and upon mixing, the lipids and DNA form a lipid/DNA intermediate. The NLP is stabilized after the ethanol is removed from the mixture by continuous dialysis. The formulation method described in this protocol is also applicable for small interfering RNA (siRNA) or antisense oligonucleotide delivery.

- ***Surface modification of NLP with reducing reagent.*** The NLP surface charge can be changed from positive to neutral if the TCL lipid is used as the cationic lipid in the formulation step. A zwitterionic reducing reagent, cysteine, is then mixed with NLP at a molar ratio of 10:1 to the TCL lipid to exchange the TCL cationic headgroup to cysteine. An anionic reducing agent such as glutathione is used if a negative surface charge is desired. The excessive reducing agents are removed by dialysis.
- ***Micelle transfer method to insert ligand onto the NLP surface.*** The ligand molecules usually are attached to the lipid hydrophilic part through a PEG linker as a ligand-PEG-lipid conjugate. The chemical reaction for making the ligand-PEG-lipid conjugate are well documented elsewhere (Hermanson 1996). These ligand-PEG lipids usually form lipid micelles in aqueous buffer because of their large hydrophilic headgroup. The ligand-PEG-lipid micelles will insert themselves onto the surface of liposomes or liposome-like particles if they are incubated together (Uster et al. 1996). The insertion efficiency can be >90% if the ligand insertion is performed at elevated temperature (~55°C) for more than 1 hour.

It should be emphasized that particle size, stability, and transfection activity are dependent on the physicochemical properties of the specific lipids and/or ligands used in the formulation. Further optimization of this protocol by adjusting the lipid composition, concentration, and cationic lipid–DNA charge ratio (the molar ratio between the total amount of cationic lipid charge and DNA phosphate) may be necessary if lipids different from those listed in this protocol are used.

Protocol

Preparation and Delivery of Nanolipoparticles

Nanolipoparticles are prepared by encapsulating DNA into a PEG-shielded cationic liposomal bilayer using ethanol dialysis. The surface charge of the particle (the cationic headgroup) is changed in a disulfide exchange reaction from positive to neutral by treatment with a reducing agent or from positive to negative with an anionic reducing agent. A micelle carrying a desired ligand may be inserted into the surface of the particle to enhance uptake by specifically targeted cells.

MATERIALS

CAUTION: See Appendix for appropriate handling of materials marked with <!>.

Reagents

Cationic lipids (5 mg/ml)

The cationic lipid can be either a reduction-sensitive lipid such as TCL or a reduction-insensitive lipid such as 1,2-dioleoyl-3-trimethylammonium propane (DOTAP) (Fig. 1D). TCL lipid: (2-amino-ethyl)-dimethyl-(2-[3-cholesteryldithioethyl]) ammonium bromide hydrochloride (Fig. 1C) can be synthesized (Huang et al. 2005) and dissolved in methanol/chloroform (1:1, v/v) as a 5-mg/ml stock solution. DOTAP (Avanti Polar Lipids [Alabaster, Alabama]) can be diluted into pure chloroform as a 5-mg/ml stock solution.

Chloroform <!>

Ethanol

Helper lipid (10 mg/ml)

The fusogenic lipid, 1,2-dioleoyl-*sn*-glycero-3-phosphoethanolaimine (DOPE) (Fig. 1E), is widely used as a helper lipid to enhance gene transfection. DOPE (Avanti Polar Lipids) has a relatively small hydrophilic headgroup and is thought to induce membrane fusion and help the DNA escape from the endosome into cytosol in the DNA delivery process. DOPE can be diluted into pure chloroform as a 5-mg/ml stock solution.

Ligand-PEG lipid (0.05 mM)

The TAT peptide sequence is CGGGRKKRRQRRRGYG. The peptide is coupled to a maleimide-functional PEG lipid: 1,2-distearoyl-*sn*-glycero-3-phosphoethanolamine-*N*-maleimide(polyethylene glycol)2000 (Avanti Polar Lipids) through the amino-terminal cysteine residue as a TAT peptide-PEG-lipid conjugate. For more information on the chemical reaction, see Hermanson (1996). The TAT peptide-PEG-lipid conjugate can be dissolved in methanol:chloroform (1:1, v/v) as a 0.05 mM stock solution. Other ligand-PEG lipids can be used in place of the TAT-PEG lipid.

Mamalian cells (exponentially growing cultures)

or

Mice (female, 6–8 weeks old)

Methanol

PEG lipids (10 mg/ml)

The PEG lipid can be either a low-pH-sensitive lipid such as PODS (Fig. 1A) or a pH-insensitive lipid such as PEG-1,2-distearoyl-*sn*-glycero-3-phosphatidylethanolamine (PEG-DSPE) (Fig. 1B). PODS can be synthesized as described by Guo et al. (2004). PEG-DSPE is available from

Avanti Polar Lipids. The average molecular weight of the PEG chain in the PEG lipid is 2000. These PEG lipids can be dissolved in pure chloroform at 10 mg/ml.

Plasmid DNA (Advysis or other gene vector manufacturer)

If prepared in the laboratory, purify the plasmid DNA with one of the commercially available DNA purification kits.

siRNA or oligonucleotide (Invitrogen or other oligomer manufacturer)

10x Tris buffer (50 mM, pH 7.4 or 8.5)

The pH 7.4 buffer is used to prepare the NLP with the pH-insensitive PEG lipid, and the pH 8.5 buffer is used for the low-pH-sensitive PEG lipid (PODS). Dilute the 10x Tris buffer ten times as 1x Tris buffer before use. Reserve 5 ml of the 1x Tris buffer (filtered by a 0.22-µm pore size membrane) for suspending the lipid and dissolving the DNA.

10x Tris-buffered saline (TBS) (pH 7.4 or 8.5)

50 mM Tris

1.5 M NaCl

Adjust the pH to 7.4 or 8.5. The pH 7.4 buffer is used to prepare the NLP with pH-insensitive PEG lipid, and the pH 8.5 buffer is used for the low-pH-sensitive PEG lipid (PODS). Dilute to a 1x TBS buffer before use.

Equipment

Dialysis equipment (Pierce)

The dialysis equipment includes the Slide-A-Lyzer dialysis cassette with a cutoff molecular weight of 10,000 and a floating buoy.

Isotemperature water bath (55°C)

Rotorevaporator (Brinkmann Instruments)

Lipid-drying equipment is used to remove a large quantity of solvent from lipid mixtures and for depositing a film of lipid on a glass vessel. The lipid films are placed in a high-vacuum flask (300 ml) for at least 3 hours to remove any trace amount of organic solvent from the lipid film.

Sterile syringe (3 ml) and needle (18G1.5) (Becton Dickinson)

METHODS

Encapsulation of Nucleic Acids into NLP by the Ethanol Dialysis Method

1. Mix the PEG lipid, cationic lipid, and DOPE from stock solutions in a 16 x 100-mm glass tube with a screw cap (cleaned and sterilized before use) and dry by rotor evaporation under reduced pressure. Then place the tube containing the lipid film under high vacuum for at least 3 hours and preferably overnight. A typical sample is composed of 20 µmoles of total lipid with the molar ratio of PEG lipid:cationic lipid:helper lipid = 10:50:40 or 10:30:60.

   Glass pipettes are required for measuring and transferring lipids in organic solvents. After use, seal the lipid stock solutions immediately with Teflon tapes and store at –20°C. Add morpholine (molar ratio 20:1 to TCL) to the lipid mixture before adding PODS to prevent the possible degradation of PODS in the lipid-drying process if the sample contains both TCL and PODS.

2. Dissolve the dried lipids in 0.5 ml of pure ethanol for 30 minutes. Use a pipette to deliver 0.5 ml of 1x Tris buffer into the ethanolic lipid and immediately mix the ethanolic buffer on vortex at maximum speed for 2 minutes to generate a lipid monophase solution (Solution A). This solution may have a very slight opalescent appearance.

   To speed up the lipid dispersion into pure ethanol, heat the lipid mixture containing TCL to ~65°C in pure ethanol for 0.5 minute under argon protection and cool to room temperature for 5 minutes before mixing with the Tris buffer. It is normal if part of the lipids precipitate again in cooled ethanol.

3. Dissolve 1.100 mg of nucleic acid (plasmid DNA or other nucleotides) in 0.5 ml of 1x Tris buffer and 0.5 ml of ethanol (Solution B).

   Cap the tubes containing Solution A or B to minimize the ethanol evaporation in the sample preparation process.

4. In a typical preparation, use a 3-ml syringe with the 18G1.5 needle to inject the DNA solution (Solution B) into the lipid solution (Solution A) in 1 second. To emulsify, rapidly withdraw and expel the mixture in the syringe approximately ten times within 15 seconds. Vortex the mixture for 30 seconds at maximum speed.

5. After mechanical mixing, use the 3-ml syringe and 18G1.5 needle to transfer the lipid-DNA mixture into a Slide-A-Lyzer dialysis cassette. Dialyze the sample against 1 liter of 1x Tris buffer for 2 days at 4°C with four changes of the dialysis buffer.

   Immerse the dialysis cassette in the dialysis buffer for a 0.5-minute hydration period before use. Be cautious when filling the dialyzer to avoid needle contact with the dialysis membrane.

6. Change the dialysis buffer from 1 liter of 1x Tris buffer to 1 liter of 1x TBS buffer and dialyze the sample overnight at 4°C.

   The 1x TBS buffer containing 150 mM NaCl is more suitable for in vivo administration of the particles than the 1x Tris buffer with low ion strength.

7. Use the 3-ml syringe and 18G1.5 needle to withdraw the NLP sample from the Slide-A-Lyzer dialysis cassette. Transfer the sample into a screw-cap sterile tube and store the sample at 4°C.

Surface Modification of NLP by the Disulfide Exchange Reaction

8. Prepare the reducing reagents, e.g., cysteine, as a 0.1–0.2 M solution in pure H_2O or in 1x TBS buffer at pH 8.5 (for samples containing PODS) just before use.

9. Add the reducing reagent into the NLP solution (Step 7) with a molar ratio of 10:1 (amount of TCL in NLP). Vortex the mixture for 10 seconds, allow it to stand for 5 minutes at room temperature, and then transfer the sample into a dialysis cassette. Immediately dialyze the sample against 1 liter of 1x TBS at 4°C to remove the excess reducing reagent with one change of the dialysis buffer. Continue dialysis for a total of 1 day at 4°C.

Insert the Ligand-PEG-Lipid Conjugate onto the NLP Surface

10. Determine the amount of TAT peptide-PEG lipid that corresponds to 0.3 mol% of the total lipids in the NLP. Transfer that amount from the TAT peptide-PEG-lipid stock solution into a 16 x 100-mm sterilized screw-cap glass tube. Remove the organic solvent from the TAT lipid by rotary evaporation under reduced pressure and then place the tube under high vacuum overnight.

11. Hydrate the dried TAT-PEG-DPPE-lipid film directly with the NLP solution for 10 minutes at room temperature and then place the tube in a 55°C water bath for 1 hour. Make sure to flush the tube with argon or nitrogen and seal with the screw cap during these steps. During incubation, remove the tube and vortex the sample for 10 seconds every 10 minutes.

    This treatment results in the transfer of the TAT-PEG lipid from the glass surface into the NLP via a micelle intermediate. To assay transfection in vitro, perform Steps 12–13; to assay transfection in vivo, perform Steps 14 and 15.

Gene Transfection In Vitro

12. To assay gene transfer in vitro, grow mammalian cells, such as CV-1 cells, overnight in complete cell culture medium containing 10% fetal bovine serum in a 24-well plate at a concentration of 1×10^5 cells/well.

13. Change the medium, load the NLP encapsulating 2 μg of DNA into a well, and incubate the sample for 4–48 hours (the medium can be refreshed or not). Analyze the newly synthesized proteins using the appropriate activity assay.

    For siRNA or antisense oligonucleotide samples, assay the delivery efficiency by analyzing the gene knockdown effect by comparison with sample wells treated with NLP containing a control sequence.

Gene Transfection In Vivo

14. Inject a small volume of NLP samples (~200 μl) containing 10–50 μg of encapsulated plasmid DNA or other oligonucleotides into the tail vein of a mouse.
15. After a suitable period (e.g., 24 or 48 hours), sacrifice the animals and analyze the gene transfection activity in targeted tissues using the appropriate assay.

    For example, to assay transfection of the luciferase gene in different organs:

    i. Sacrifice the animal and remove and wash the organ (e.g., liver, spleen, lung, or heart) in cold phosphate-buffered saline.

    ii. Cut and transfer about 200 mg of tissue from the organ into a 1.5-ml centrifuge tube containing 1 g of Zircon Beads (diameter 2.4 mm; Biospec Products) and 1 ml of 1x Reporter Lysis Buffer (Promega).

    iii. Bead-beat the tissue and the beads on a Mini-Beadbeater (Biospec Products) for 30 seconds at 5000 beating rate/minute, to homogenize the sample.

    iv. Centrifuge the sample and combine 10 μl of supernatant with 100 μl of luciferase assay mix. Determine activity according to kit instructions.

TROUBLESHOOTING

Additional Materials and Instruments

PicoGreen dye (P7581; Molecular Probes, Inc.)
Photon correlation spectrometry (e.g., Malvern Zeta 3000 dynamic light scattering instrument, Malvern Instruments Ltd., Worcestershire, U.K.)
Fluorometer (excitation wavelength: 490 nm; emission wavelength: 515 nm; e.g., Quantech Fluorometer, Barnstead/Thermolyne, Dubuque, Iowa)

Problem: DNA encapsulation efficiency.
Solution: Analyze encapsulation efficiency by testing the ability of DNA to react with PicoGreen. When PicoGreen is added into DNA solutions, it intercalates into double-stranded DNA, emitting an intense fluorescence signal at 515 nm. DNA is considered to be encapsulated in the nanoparticle if it does not react with the PicoGreen dye in the absence of Triton X-100. The DNA encapsulation efficiency is calculated as $I\,(\%) = (F^\circ - F)/F^\circ \times 100$, where F° and F are the fluorescence values of nanoparticle and PicoGreen mixture in the presence or absence, respectively, of 0.4% Triton X-100 (Li et al. 2005).

Problem: Determining the particle size and surface charge of the NLP.
Solution: The average diameter and size distribution of NLP can be measured with photon correlation spectrometry on a Malvern Zeta 3000 dynamic light scattering instrument (Li et al. 2005). The zeta potential on the NLP surface can be measured with the electrophoretic mode on the same machine by diluting the particles 100-fold into a low-salt buffer (1 mM NaCl and 1 mM HEPES at pH 7.5).

Problem: The NLP sample is associated with a significant amount of free DNA.

Solution: The free DNA in the NLP sample can be removed by anionic exchange chromatography using an anionic exchange column with 1(diameter) x 3(height) cm of DEAE-Sepharose CL-6B media (Sigma-Aldrich). Apply 1.0 ml of NLP to the column and elute the sample as purified fractions using 10 ml of 150 mM NaCl, 5 mM Tris buffer (pH 7.4 or 8.5). The free plasmid DNA binding to the column can be eluted by applying 10 ml of 1 M NaCl in 5 M Tris buffer (pH 7.4 or 8.5). The UV absorbance of each fraction (0.5 ml) is measured at 260 nm (Li et al. 2005).

REFERENCES

Barron L.G. and Szoka F.C. 1999. The perplexing delivery mechanism of lipoplexes. In *Nonviral vectors for gene therapy* (ed. L. Huang et al.), pp. 229–266. Academic Press, San Diego, California.

Choi J.S., MacKay J.A., and Szoka F.C., Jr. 2003. Low-pH-sensitive PEG-stabilized plasmid-lipid nanoparticles: Preparation and characterization. *Bioconjug. Chem.* **14:** 420–429.

Duzgunes N., ed. 2003. Liposomes, Part C. *Methods Enzymol.*, vol. 373. Elsevier, New York.

Fenske D.B., MacLachlan I., and Cullis P.R. 2001. Long-circulating vectors for the systemic delivery of genes. *Curr. Opin. Mol. Ther.* **3:** 153–158.

Guo X. and Szoka F.C., Jr. 2001. Steric stabilization of fusogenic liposomes by a low-pH sensitive PEG-diortho ester-lipid conjugate. *Bioconjug. Chem.* **12:** 291–300.

———. 2003. Chemical approaches to triggerable lipid vesicles for drug and gene delivery. *Acc. Chem. Res.* **36:** 335–341.

Guo X., Huang Z., and Szoka F.C. 2004. Improved preparation of PEG-diortho ester-diacyl glycerol conjugates. *Methods Enzymol.* **387:** 147–152.

Harvie P., Wong F.M.P., and Bally M.B. 2000. Use of poly(ethylene glycol)-lipid conjugates to regulate the surface attributes and transfection activity of lipid-DNA particles. *J. Pharm. Sci.* **89:** 652–663.

Hermanson G.T. 1996. *Bioconjugate techniques.* Academic Press, San Diego, California.

Huang Z., Li W., MacKay J.A., and Szoka F.C., Jr. 2005. Thiocholesterol-based lipids for ordered assembly of bioresponsive gene carriers. *Mol. Ther.* **11:** 409–417.

Jeffs L.B., Palmer L.R., Ambegia E.G., Giesbrecht C., Ewanick S., and MacLachlan I. 2005. A scalable, extrusion-free method for efficient liposomal encapsulation of plasmid DNA. *Pharm. Res.* **22:** 362–372.

Li W., Huang Z., MacKay J.A., Grube S., and Szoka F.C. 2005. Low-pH-sensitive poly(ethylene glycol) (PEG)-stabilized plasmid nanolipoparticles: Effects of PEG chain length, lipid composition and assembly conditions on gene delivery. *J. Gene Med.* **7:** 67–79.

Mahato R.I. and Kim S.W., eds. 2002. *Pharmaceutical perspectives of nucleic acid-based therapeutics.* Taylor & Francis, London, United Kingdom.

Uster P.S., Allen T.M., Daniel B.E., Mendez C.J., Newman M.S., and Zhu G.Z. 1996. Insertion of poly(ethylene glycol) derivatized phospholipid into pre-formed liposomes results in prolonged in vivo circulation time. *FEBS Lett.* **386:** 243–246.

Wheeler J.J., Palmer L., Ossanlou M., MacLachlan I., Graham R.W., Zhang Y.P., Hope M.J., Scherrer P., and Cullis P.R. 1999. Stabilized plasmid-lipid particles: Construction and characterization. *Gene Ther.* **6:** 271–281.

Zhang Y.P., Sekirov L., Saravolac E.G., Wheeler J.J., Tardi P., Clow K., Leng E., Sun R., Cullis P.R., and Scherrer P. 1999. Stabilized plasmid-lipid particles for regional gene therapy: Formulation and transfection properties. *Gene Ther.* **6:** 1438–1447.

41 HVJ Liposomes and HVJ Envelope Vectors

Yasufumi Kaneda

Division of Gene Therapy Science, Graduate School of Medicine, Osaka University, Suita City, Osaka 565-0871, Japan

ABSTRACT

This chapter describes techniques for construction of fusion-mediated vectors based on inactivated HVJ (hemagglutinating virus of Japan; Sendai virus). HVJ liposomes are constructed by fusing liposomes containing DNA with inactivated HVJ. The HVJ envelope vector, a more simplified vector, incorporates DNA into inactivated HVJ particles without liposomes. Both vectors have many advantages. They can be used to introduce proteins, peptides, oligonucleotides (including antisense oligonucleotides, decoy oligonucleotides, and ribozymes), and short interfering RNA (siRNA), as well as plasmid DNA, into cultured cells in vitro and into organs in vivo. Fusion-mediated delivery avoids the degradation of therapeutic molecules before reaching the cytoplasm. Finally, repeated injection of the vector in vivo is not inhibited and even enhances the effects of the delivered molecules. These vectors have been used in many gene therapy experiments in animal models to address such problems as liver cirrhosis, hearing impairment, ischemic brain damage, peripheral arterial diseases, and cancers.

However, there are also problems with HVJ vectors. First, as is the case in many other delivery systems, gene expression is sometimes low. We found that this was mainly caused by epigenetic modification of a transgene in the nucleus (Yamano et al. 2000). With the histone deacetylase inhibitor, transgene expression in cancer masses increased by approximately 20 times over that seen without the inhibitor (Yamamoto et al. 2003). Second, because HVJ can fuse to almost all cells, HVJ vectors lose targeting ability. With injection, those vectors can be trapped in reticuloendothelial cells and accumulate in liver, spleen, and lung (Tsuboniwa et al. 2001). We have conjugated monoclonal antibody against Thy-1 antigen in glomerulus with liposomes and then fused the liposomes with inactivated HVJ to form HVJ immunoliposomes. With systemic injection of the liposomes, FITC oligonucleotides were delivered to rat glomerulus of both kidneys at approximately 90% efficiency (Tomita et al. 2002). Finally, HVJ is easily degraded in fresh serum, probably by complement lysis. But, by combining HVJ envelope vectors with cationized gelatin, we found that the vector was successfully stabilized in fresh serum from mice (Mima et al. 2005).

INTRODUCTION

One promising approach in vector development is the combination of viral and nonviral vectors to compensate for the limitations of one vector with the strength of another. This approach is likely to produce vectors with high efficiency and low toxicity. We have combined liposomes with UV-inactivated HVJ to construct liposomes with virus function (HVJ liposomes) (Nakanishi et al. 1985; Kaneda et al. 1989a,b; Kato et al. 1991). HVJ belongs to the mouse parainfluenza virus family and has two distinct envelope proteins for membrane fusion (Okada 1993). The structure of HVJ is shown in Figure 1.

The HN (hemagglutinating) protein binds to cell surface sialic acid receptors and degrades the receptor with its neuraminidase activity. Then, F (fusion protein) associates with lipid molecules such as cholesterol to induce cell fusion (Asano and Asano 1988). The HVJ liposomes (~400–500 nm in diameter) also contain these two proteins on the surface and fuse with the cell membrane to introduce DNA directly into the cytoplasm without degradation. The HVJ liposomes can encapsulate DNA smaller than 100 kb. RNA, oligodeoxynucleotides, and proteins can also be enclosed and delivered to cells (Dzau et al. 1996; Kaneda et al. 1999). We used FRET (fluorescence resonance energy transfer) to show that the degradation of oligodeoxynucleotides in the process of delivery to the cytoplasm is inhibited with HVJ liposomes, but not with simple cationic liposomes (Nakamura et al. 2001). One advantage of HVJ liposomes is the ability to perform repeated injections without inhibiting gene transfer. After repeated injections to rat liver, anti-HVJ antibody generated in the rat was shown not to be sufficient to neutralize HVJ liposomes (Hirano et al. 1998). Similarly, repetitive transfection in vivo was shown to be successful in cancer masses (Yamamoto et al. 2003), liver (Hirano et al. 1998), and skeletal muscle (Ueki et al. 1999).

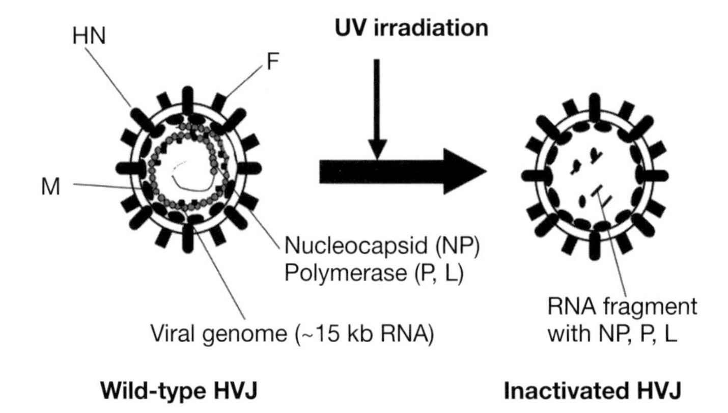

FIGURE 1. Structure of HVJ. Wild-type HVJ contains fusion proteins (HN and F), membrane-anchoring protein (M), nucleocapsid protein (NP), and polymerase (P and L) with approximately 15 kb of the viral RNA genome. Once wild-type HVJ infects cells, viral proteins and infectious viral particles are produced in the infected cells. With UV-light irradiation, viral RNA is degraded into very small pieces. Inactivated HVJ loses the ability either to produce viral proteins or to induce genome replication. Nevertheless, inactivated HVJ retains the robust fusion activity of wild HVJ.

To increase the efficiency of gene delivery by HVJ liposomes, we investigated the lipid components of liposomes. We concluded that the most efficient gene expression occurred with liposomes consisting of phosphatidylcholine, phosphatidylethanolamine, sphingomyelin, phosphatidylserine, and cholesterol at the molar ratio of 13.3:13.3:13.3:10 and 50, respectively (Saeki et al. 1997). The lipid components of the liposomes are very similar to those of the HIV envelope and mimic the red blood cell membrane (Chander and Schreier 1992). We called these liposomes HVJ-AVE (artificial viral envelope) liposomes. With HVJ-AVE liposomes, gene expression in heart, liver, and muscle was 5–10 times higher than that observed with various nonviral gene-transfer methods, such as conventional HVJ liposomes, cationic-lipid-mediated lipofection, and naked DNA injection (Kaneda et al. 1999).

Another improvement is the construction of cationic-type HVJ liposomes using cationic lipids. Of the cationic lipids, positively charged DC-cholesterol (3β-[*N*-(*N*′,*N*-dimethylaminoethane)carbamoyl] cholesterol) (Goyal and Huang 1995) has been shown to be the most efficient for gene transfer. For luciferase expression, HVJ-cationic DC liposomes were 100 times more efficient than were conventional HVJ-anionic liposomes (Saeki et al. 1997). HVJ-cationic liposomes were not appropriate for gene transfer to liver, kidney, heart, and muscle; but, compared to anionic-type HVJ-AVE liposomes, HVJ-cationic liposomes were much more effective for gene transfer to tumor masses or disseminated cancers (Mabuchi et al. 1997; Miyata et al. 2001; Otomo et al. 2001) in animal models.

A more direct and practical approach is the conversion of a fusigenic virion to a nonviral gene delivery particle. On the basis of this concept, we further developed the HVJ envelope vector (Kaneda et al. 2002) by incorporating exogenous DNA into inactivated HVJ particles by detergent treatment and centrifugation without liposomes. The resulting HVJ envelope vector introduced plasmid DNA efficiently and rapidly into both cultured cells in vitro and organs in vivo. Furthermore, proteins, synthetic oligodeoxynucleotides, and drugs have also been effectively introduced into cells using the HVJ envelope vector. The HVJ envelope vector has been approved for effective transfer of siRNA (Ito et al. 2003; Asada et al. 2004; Kuribara et al. 2004; Ohnuma et al. 2004) and is thus a promising tool for both ex vivo and in vivo gene therapy experiments. For example, gene transfer of hepatocyte growth factor into the cerebrospinal fluid in rats, using HVJ envelope vector, reduced, and in some cases prevented, hearing impairment and brain ischemia (Oshima et al. 2004; Shimamura et al. 2004). Transfer of the *lacZ* gene into the mouse uterus with the HVJ envelope vector yielded gene expression mainly in the glandular epithelium of the endometrium (Nakamura et al. 2003). Rad51 siRNA was successfully delivered to tumor masses by HVJ envelope vector, and these cancer cells had a greatly enhanced sensitivity to cisplatin (Ito et al. 2005). A pilot plan for human gene therapy has been established for commercial production of clinical-grade HVJ envelope vectors.

Protocol

Production of HVJ Liposomes and HVJ Envelope Vectors for Gene Delivery

We describe here methods for the preparation of HVJ liposomes and of HVJ envelope vectors and their use in delivery of plasmid DNA into various cells and tissues.

MATERIALS

CAUTION: See Appendix for appropriate handling of materials marked with <!>.

Reagents

BSS (buffered salt solution)
- 137 mM NaCl
- 5.4 mM KCl
- 10 mM Tris-HCl (pH 7.6)

Solubilize 8 g of NaCl, 0.4 g of KCl, and 1.21 g of Trizma base in distilled H_2O, adjust the pH to 7.6 with 1 M HCl, and bring the total volume to 1 liter with distilled H_2O. Sterilize by autoclaving and store at 4°C.

Bovine brain phosphatidylserine sodium salt (PS), chromatographically pure (Avanti Polar Lipids Inc.)

Chloroform (Sigma-Aldrich) <!>

Cholesterol (Sigma-Aldrich)

DC-cholesterol (Avanti Polar Lipids, Inc.)

DMSO (dimethylsulfoxide) (Sigma-Aldrich) <!>

DOPE (1,2-dioleoyl phosphatidylethanolamine) (Sigma-Aldrich)

HVJ seed from HVJ (Z strain, VR-105, American Type Culture Collection)

Store 100-μl aliquots of the best seed of HVJ in 10% DMSO.

Nitrogen gas <!>

Phosphate-buffered saline (PBS)

Phosphatidylcholine from egg yolk (PC) (Sigma-Aldrich)

Plasmid DNA

Purify plasmid DNA by column chromatography and dissolve the preparations in TE (4 mg/ml TE), BSS, or H_2O. Store the DNA at –20°C. The final concentration of DNA should be greater than 1 mg/ml.

Polypeptone solution
- 1% polypeptone (pancreatic digest of casein; Wako, Osaka, Japan)
- 0.2% NaCl (pH 7.2)

For 500 ml, solubilize 5 g of polypeptone and 1 g of NaCl in distilled H_2O, adjust the pH to 7.2 with 1 M NaOH, and bring to a total volume of 500 ml with distilled H_2O. Sterilize by autoclaving and store at 4°C.

Protamine sulfate (10 mg/ml in PBS)

Sphingomyelin (Sigma-Aldrich)

Sucrose gradient solutions

To prepare 30% and 60% (w/v) sucrose solutions, solubilize 150 g and 300 g of sucrose separately in BSS. Add BSS to a total volume of 500 ml, sterilize by autoclaving, and store at 4°C.

Tris-EDTA (TE)
10 mM Tris-Cl (pH 8.0)
1 mM EDTA
Triton X-100 (2% in TE)

Equipment

Cellulose acetate membrane filters (0.45 and 0.20 μm) (Nalgene)
Centrifuge (J2-HS) with JA-20 rotor and 35-ml tubes (Beckman Instruments) as well as a low-speed centrifuge (05PR-22, Hitachi, Tokyo, Japan)
Columns, endotoxin-free, for preparation of plasmid DNA (QIAGEN)
Conical tubes (50 ml) (Becton-Dickinson)
Glass tubes (24-mm caliber, 12 cm long) (Fujiston 24/40, Iwaki Glass Co. Ltd., Tokyo, Japan)
These tubes are custom made, but similar sterilized tubes resistant to chloroform are available. The new glass tubes were immersed in saturated KOH-ethanol solution for 24 hours, rinsed with distilled H_2O, and heated for 2 hours at 180°C before use.
Photometer (Spectrophotometer DU-68, Beckman Instruments)
Rotary evaporator (Type SR-650, Tokyo Rikakikai Inc., Tokyo, Japan)
Ultracentrifuge with RPS-40T rotor (55P-72, Hitachi)
Ultracentrifuge tubes (12 ml) (Hitachi)
Ultraviolet (UV) cross-linker (Spectrolinker XL-1000, Spectronics Co.)
Vacuum pump with a pressure gauge (Type Asp-13, Iwaki Glass Co. Ltd., Tokyo, Japan)

METHODS

Preparation of HVJ in Chick Eggs

1. Thaw the HVJ virus seed rapidly and dilute 1:1000 with polypeptone solution.
   Keep the diluted seed at 4°C before proceeding to the next step.
2. Observe embryonated eggs under illumination in a dark room at room temperature. Mark an injection point at about 0.5 mm above the chorioallantoic membrane. Disinfect the eggs with tincture of iodine and puncture at the marked point.
3. Inject 0.1 ml of the diluted seed into each egg using a 1-ml disposable syringe with a 26-gauge needle. Insert the needle vertically so as to stab the chorioallantoic membrane.
4. Inoculate the seed and cover the hole in the egg with melted paraffin. Incubate the eggs in sufficient moisture for 4 days at 36.5°C. Chill the eggs at 4°C for more than 6 hours before harvesting the virus.
5. To harvest the virus, partially remove the egg shell and transfer the chorioallantoic fluid into an autoclaved bottle using a 10-ml syringe with an 18-gauge needle. Keep the fluid at 4°C to avoid freezing.
   This step can be performed at room temperature. The virus is stable in the fluid for at least 3 months.

Purification of HVJ from Chorioallantoic Fluid

6. Transfer 200 ml of the chorioallantoic fluid into four 50-ml disposable conical tubes. Centrifuge at 3000 rpm (1000*g*) in a low-speed centrifuge for 10 minutes at 4°C.
7. Deliver aliquots of the supernatant into six tubes (Beckman JS-20) and centrifuge at 15,000 rpm (27,000*g*) for 30 minutes at 4°C.

8. Add about 5 ml of BSS to the pellet in each tube and store the materials overnight at 4°C.
9. Suspend the pellets gently, collect the suspension in two tubes, and centrifuge the tubes as described in Step 6.
10. Add about 5 ml of BSS to each pellet and incubate for more than 8 hours at 4°C.
11. Suspend the pellets gently, centrifuge at 3000 rpm (1000g) in a low-speed centrifuge for 15 minutes at 4°C. Transfer the supernatants to an aseptic tube and store at 4°C.
12. Dilute the supernatant tenfold and determine virus titer with a photometer by measuring the absorbance of at 540 nm. HVJ can be inactivated by UV irradiation (99 mjoules/cm^2) using a UV cross-linker.

    An optical density at 540 nm corresponds to 15,000 hemagglutinating units (HAU), which is well correlated with fusion activity. The supernatant (prepared as described) usually shows 20,000 to 30,000 HAU/ml. A virus solution aseptically prepared maintains fusion activity for 3 weeks. UV-inactivated HVJ can be stored in 10% DMSO for more than 6 months at –80°C or –170°C (in liquid nitrogen).

Preparation of Lipid Mixture

13. To prepare the lipid solution, dissolve dry reagents of DOPE (12.2 mg), sphingomyelin (11.5 mg), and cholesterol (23.8 mg) in 3870 μl of chloroform. Prepare a solution of PC chloroform (100 mg of PC in 1 ml of chloroform) and add 130 μl to the lipid solution.

    The 4000 μl of lipid solution is called a basal mixture for liposomes. The master mixture is ready to prepare anionic or cationic liposomes (as described below) or it can be stored at –20°C after infusion with nitrogen gas.

14. To prepare anionic lipid mixtures, add 10 mg of PS to the basal mixture. To prepare cationic lipid mixtures, add 6 mg of DC-cholestrol to the basal mixture.
15. Deliver aliquots of 0.5 ml of lipid solution into eight glass tubes; each will contain 8.8 mg of lipids. Keep the tubes on ice or at –20°C in nitrogen gas before evaporation.

    The lipid solution should be evaporated as soon as possible.

16. Connect the tube to a rotary evaporator; immerse the tip of the tube in a 40°C water bath. Evaporate the organic solvent in a rotary evaporator under vacuum for 5–10 minutes.

    Lipids appropriate for liposome preparation are those that stick to the inside of the tube in a thin layer. Lipids that accumulate at the bottom of the tube are inappropriate. Store the lipids at –20°C in nitrogen gas. The lipids can be used for 1 month.

Preparation and Delivery of HVJ Vectors

17. Prepare the HVJ vector for fusion-mediated gene transfer.

    For Preparation of HVJ liposomes Containing DNA (Figure 2)

    a. Deliver plasmid DNA (200 μg in 200 μl) into a lipid mixture in a glass tube (Step 16), agitate the tube intensely by vortexing for 30 seconds followed by incubating for 30 seconds at 37°C. Repeat this cycle eight times.

       With this method, plasmid DNA (up to 20 kb) can be enclosed at the rate of 10–30% in anionic liposomes or 50–60% in cationic liposomes.

    b. To prepare sized unilamellar liposomes, filter the liposome suspension through a 0.45-μm pore size cellulose acetate filter and then filter through a 0.2-μm filter.

       Sizing by an extruder with polycarbonate filters is better for preparing sized liposomes.

    c. Add 15,000 HAU of UV-inactivaed HVJ virus to the liposome suspension. Place the tube

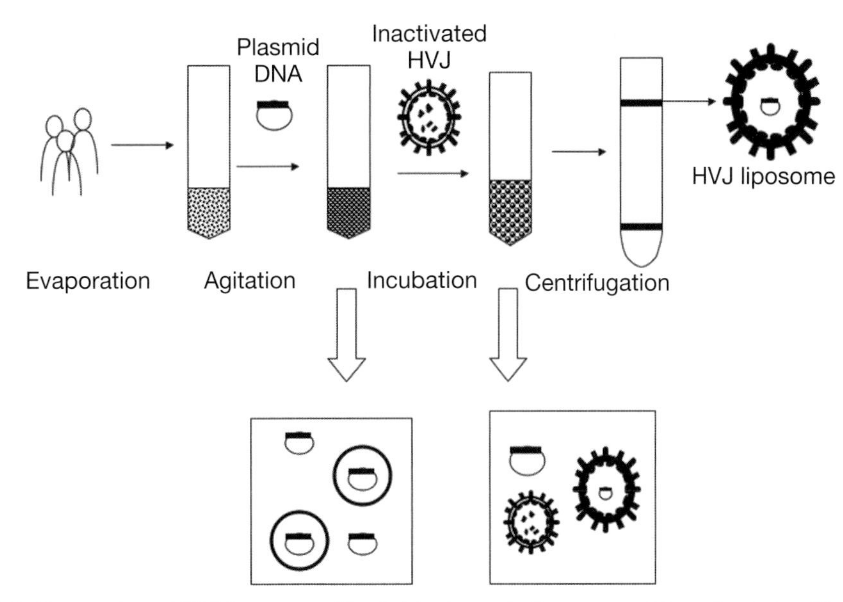

FIGURE 2. Preparation of HVJ liposomes. The lipid mixture is first prepared in a glass tube. Plasmid DNA is added to the mixture and DNA-loaded liposomes are prepared by vigorous agitation using a vortex mixer. Inactivated HVJ is then added to the liposome solution, and fusion between liposomes and HVJ is induced during incubation at 37°C. Sucrose density gradient centrifugation is performed to separate HVJ liposomes from free HVJ. HVJ liposomes are recovered from the top 30% sucrose layer, and free HVJ is precipitated between the 30% and 60% layer.

on ice for 5–10 minutes and then incubate for 1 hour at 37°C with shaking (120/min) in a water bath.

d. Deliver 1 ml of 60% and then 7 ml of 30% sucrose solution to a centrifuge tube and overlay the HVJ-liposome mixture. Fill the tube with BSS.

e. Separate the HVJ-liposome complexes from the free HVJ by sucrose gradient density centrifugation at 62,000*g* for 90 minutes at 4°C.

f. Remove the tubes from the centrifuge and carefully collect the conjugated liposomes only, just above the 30% layer.

   Free HVJ is concentrated at the layer between 30% and 60%. The final volume of the HVJ-liposome suspension should be approximately 1 ml. HVJ liposomes can be stored in 10% DMSO for more than 8 months at –80°C or –170°C (in liquid nitrogen), but freezing and thawing should be avoided.

For Preparation of HVJ Envelope Vector (Figure 3)

a. Suspend UV-inactivated HVJ (3 x 10^{10} particles, 10,000 HAU) in 40 μl of TE.

   If the inactivated HVJ suspension is more diluted, precipitate the HVJ particles by centrifugation at 10,000*g* for 5 minutes and then suspend them in an appropriate volume of TE to adjust the concentration.

b. Mix the virus suspension with plasmid DNA (200 μg in 50 μl) and 5 μl of Triton X-100. Centrifuge the mixture at 10,000*g* for 5 minutes at 4°C.

c. Remove the supernatant and suspend the vector in 300 μl of PBS.

   The vector can be stored for at least 1 week at 4°C.

18. Deliver the genetic material into cells or tissue, using the vector appropriate for the system under study.

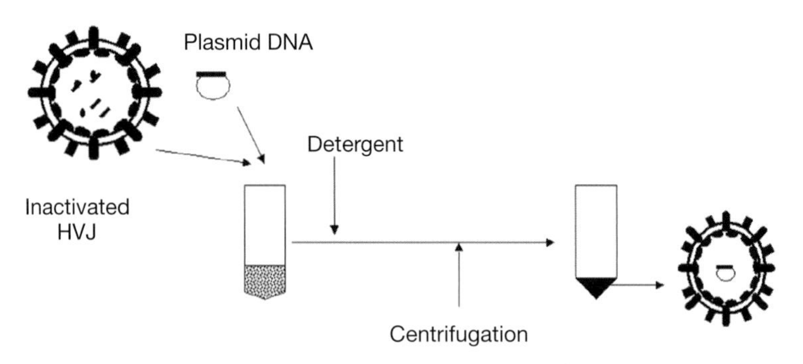

FIGURE 3. Preparation of HVJ envelope vector. Inactivated HVJ and plasmid DNA are mixed, and detergent (Triton X-100) is added. DNA is incorporated into the inactivated HVJ particle by centrifugation of the mixture to form the HVJ envelope vector. Unincorporated DNA can be removed by centrifugation.

For Gene Transfer into Cultured Cells Using Cationic HVJ-cationic Liposomes

a. Add 100 μl of a 1-ml HVJ-cationic liposome suspension to 2 x 10^6 cells in serum-containing culture medium (in a 6-well plate).

   HVJ-cationic liposomes should be used for in vitro gene transfer because they are approximately 100 times more efficient for gene transfer to cultured cells than are HVJ-anionic liposomes (Saeki et al. 1997).

b. Incubate the cells with the liposomes for 2 hours at 37°C.

c. Remove the medium, add fresh culture medium, and continue the culture.

For Gene Transfer to Animal Tissues Using HVJ-anionic Liposomes

Anionic liposomes are useful for gene transfer to liver, skeletal muscle, heart, lung, artery, brain, spleen, eye, and joint space of rodents, rabbits, dogs, lambs, and monkeys.

a. For gene transfer into rat liver, inject 2–3 ml of HVJ-anionic liposomes into the portal vein using a 5-ml syringe with a butterfly-shaped needle (Kaneda et al. 1989a,b) or directly inject the liposomes into the liver under the perisplanchnic membrane using a 5-ml syringe with a 27-gauge needle (Kato et al. 1991).

b. For gene transfer into rat kidney, inject 1 ml of anionic HVJ-liposome solution into the renal artery (Tomita et al. 1992; Isaka et al. 1993).

c. For gene transfer into rat carotid artery, fill a lumen of a segment of the artery with 0.5 ml of anionic HVJ-liposome complex for 20 minutes at room temperature using a cannula (Morishita et al. 1993).

d. For gene transfer to tumor masses or disseminated tumors, inject cationic HVJ liposomes (0.1–0.5 ml) directly into the tumor mass or tumor-infiltrating space (Mabuchi et al. 1997).

For Gene Transfer to Adherent Cells in Culture Using HVJ Envelope Vector

a. Mix 30 μl of vector suspension with 10 μl of protamine sulfate solution.

b. Add the mixture to cells cultured in a 6-well plate with 1 ml of normal medium containing fetal bovine serum.

c. Incubate cells with the vector for 10 minutes to 24 hours.

d. Change the medium and continue culturing.

For Gene Transfer to Floating Cells in Culture Using the HVJ Envelope Vector

a. Mix 30 μl of vector suspension with 10 μl of protamine sulfate solution.

b. Centrifuge the mixture of the vector and cells at 2,000 to 12,000 r.p.m. in a microcentrifuge for 10–30 minutes at 35°C.

c. Remove the supernatant, suspend the cells in 1–2 ml of fresh medium supplemented with serum, and transfer to a 6-well plate.

For Gene Transfer to Animal Tissues Using HVJ Envelope Vector

a. For in vivo transfection, inject HVJ envelope vector (3×10^9 to 9×10^9 particles) directly into animal tissues without protamine sulfate (Nakamura et al. 2003; Oshima et al. 2004; Shimamura et al. 2004; Ito et al. 2005).

The method is similar to that described for the use of HVJ liposomes.

Assessing Successful Gene Transfer

19. Monitor gene expression using the appropriate assay.

REFERENCES

Asada M., Ohmi K., Delia D., Enosawa S., Suzuki S., Yuo A., Suzuki H., and Mizutani S. 2004. Brap2 functions as a cytoplasmic retention protein for p21 during monocyte differentiation. *Mol. Cell. Biol.* **24:** 8236–8243.

Asano K. and Asano A. 1988. Binding of cholesterol and inhibitory peptide derivatives with the fusogenic hydrophobic sequence of F-glycoprotein of HVJ (Sendai virus): Possible implication in the fusion reaction. *Biochemistry* **27:** 1321–1329.

Chander R. and Schreier H. 1992. Artificial viral envelopes containing recombinant human immunodeficiency virus (HIV) gp160. *Life Sci.* **50:** 481–489.

Dzau V.J., Mann M., Morishita R., and Kaneda Y. 1996. Fusigenic viral liposome for gene therapy in cardiovascular diseases. *Proc. Natl. Acad. Sci.* **93:** 11421–11425.

Goyal K. and Huang L. 1995. Gene therapy using DC-Chol liposomes. *J. Liposome Res.* **5:** 49–59.

Hirano T., Fujimoto J., Ueki T., Yamamoto H., Iwasaki T., Morishita R., Sawa Y., Kaneda Y., Takahashi H., and Okada E. 1998. Persistent gene expression in rat liver in vivo by repetitive transfections using HVJ-liposome. *Gene Ther.* **5:** 459–464.

Isaka Y., Fujiwara Y., Ueda N., Kaneda Y., Kamada T., and Imai E. 1993. Glomerulosclerosis induced by in vivo transfection with TGF-β or PDGF gene into rat kidney. *J. Clin. Invest.* **92:** 2597–2601.

Ito M., Yamamoto S., Nimura K., Hiraoka K., Tamai K., and Kaneda Y. 2005. Rad51 siRNA delivered by HVJ envelope vector enhances the anticancer effect of cisplatin. *J. Gene Med.* **7:** 1044–1052.

Ito Y., Kawamata Y., Harada M., Kobayashi M., Fujii R., Fukusumi S., Ogi K., Hosoya M., Tanaka Y., Uejima H., Tanaka H., Maruyama M., Satoh R., Okubo S., Kizawa H., Komatsu H., Matsumura F., Noguchi Y., Shinohara T., Hinuma S., Fujisawa Y., and Fujino M. 2003. Free fatty acids regulate insulin secretion from pancreatic β cells through GPR40. *Nature* **422:** 173–176.

Kaneda Y., Iwai K., and Uchida T. 1989a. Increased expression of DNA cointroduced with nuclear protein in adult rat liver. *Science* **243:** 375–378.

———. 1989b. Introduction and expression of the human insulin gene in adult rat liver. *J. Biol. Chem.* **264:** 12126–12129.

Kaneda Y., Saeki Y., and Morishita R. 1999. Gene therapy using HVJ-liposomes: The best of both worlds? *Mol. Med. Today* **5:** 298–303.

Kaneda Y., Nakajima T., Nishikawa T., Yamamoto S., Ikegami H., Suzuki N., Nakamura H., Morishita R., and Kotani H. 2002. HVJ (hemagglutinating virus of Japan) envelope vector as a versatile gene delivery system. *Mol. Ther.* **6:** 219–226.

Kato K., Nakanishi M., Kaneda Y., Uchida T., and Okada Y. 1991. Expression of hepatitis B virus surface antigen in adult rat liver. *J. Biol. Chem.* **266:** 3361–3364.

Kuribara R., Honda H., Matsui H., Shinjyo T., Inukai T., Sugita K., Nakazawa S., Hirai H., Ozawa K., and Inaba T. 2004. Roles of Bim in apoptosis of normal and Bcr-Abl-expressing hematopoietic progenitors. *Mol. Cell. Biol.* **24:** 6172–6183.

Mabuchi E., Shimizu K., Miyao Y., Kaneda Y., Kishima H., Tamura M., Ikenaka K., and Hayakawa T. 1997. Gene delivery by HVJ-liposome in the experimental gene therapy of murine glioma. *Gene Ther.* **4:** 768–772.

Mima H., Tomoshige R., Kanamori T., Tabata Y., Yamamoto S., Ito S., Tamai K., and Kaneda Y. 2005. Biocompatible polymer enhances the in vitro and in vivo transfection efficiency of HVJ envelope vector. *J. Gene Med.* **7:** 888–897.

Miyata T., Yamamoto S., Sakamoto K., Morishita R., and Kaneda Y. 2001. Novel immunotherapy for peritoneal dissemination of murine colon cancer with macrophage inflammatory protein-1β mediated by a tumor-specific vector, HVJ-cationic liposomes. *Cancer Gene Ther.* **8:** 852–860.

Morishita R., Gibbons G.H., Kaneda Y., Ogihara T., and Dzau V.J. 1993. Novel and effective gene transfer technique for study of vascular renin angiotensin system. *J. Clin. Invest.* **91:** 2580–2585.

Nakamura H., Kimura T., Ikegami H., Ogita K., Kohyama S., Shimoya K., Tsujie T., Koyama M., Kaneda Y., and Murata Y. 2003. Highly efficient and minimally invasive in vivo gene transfer to the mouse uterus by hemagglutinating virus of Japan (HVJ) envelope vector. *Mol. Hum. Reprod.* **9:** 603–609.

Nakamura N., Hart D.A., Frank C.B., Marchuk L.L., Shrive N.G., Ota N., Taira T., Yoshikawa H., and Kaneda Y. 2001. Efficient transfer of intact oligonucleotides into the nucleus of ligament scar fibroblasts by HVJ-cationic liposomes is correlated with effective antisense gene inhibition. *J. Biochem.* **129:** 755–759.

Nakanishi M., Uchida T., Sugawa H., Ishiura M., and Okada Y. 1985. Efficient introduction of liposomes into cells using HVJ (Sendai virus). *Exp. Cell Res.* **159:** 399–409.

Ohnuma K., Yamochi T., Uchiyama M., Nishibashi K., Yoshikawa N., Shimizu N., Iwata S., Tanaka S., Dang N.H., and Morimoto C. 2004. CD26 up-regulates expression of CD86 on antigen-presenting cells by means of caveolin-1. *Proc. Natl. Acad. Sci.* **101:** 14186-14191.

Okada Y. 1993. Sendai-virus induced cell fusion. *Methods Enzymol.* **221:** 18–41.

Oshima K., Shimamura M., Mizuno S., Tamai K., Doi K., Morishita R., Nakamura T., Kubo T., and Kaneda Y. 2004. Intrathecal injection of HVJ-E containing HGF gene to cerebrospinal fluid can prevent and ameliorate hearing impairment in rats. *FASEB J.* **18:** 212–214.

Otomo T., Yamamoto S., Morishita R., and Kaneda Y. 2001. EBV replicon vector system enhances transgene expression in vivo: Applications to gene therapy for cancer. *J. Gene Med.* **3:** 345–352.

Saeki Y., Matsumoto N., Nakano Y., Mori M., Awai K., and Kaneda Y. 1997. Development and characterization of cationic liposomes conjugated with HVJ (Sendai virus): Reciprocal effect of cationic lipid for in vitro and in vivo gene transfer. *Hum. Gene Ther.* **8:** 2133–2141.

Shimamura M., Sato N., Oshima K., Aoki M., Kurinami H., Waguri S., Uchiyama Y., Ogihara T., Kaneda Y., and Morishita R. 2004. A novel therapeutic strategy to treat brain ischemia: Over-expression of hepatocyte growth factor gene reduced ischemic injury without cerebral edema in rat model. *Circulation* **109:** 424–431.

Tomita N., Higaki J., Morishita R., Kato K., Kaneda Y., and Ogihara T. 1992. Direct in vivo gene introduction into rat kidney. *Biochem. Biophys. Res. Commun.* **186:** 129–134.

Tomita N., Morishita R., Yamamoto K., Higaki J., Dzau V.J., Ogihara T., and Kaneda Y. 2002. Targeted gene therapy for rat glomerulonephritis using HVJ-immunoliposomes. *J. Gene Med.* **4:** 527–535.

Tsuboniwa N., Morishita R., Hirano T., Fujimoto J., Furukawa S., Kikumori M., Okuyama A., and Kaneda Y. 2001. Safety evaluation of HVJ-AVE liposomes in nonhuman primates. *Hum. Gene Ther.* **12:** 469–487.

Ueki T., Kaneda Y., Tsutsui H., Nakanishi K., Sawa Y., Morishita R., Matsumoto K., Nakamura T., Takahashi H., Okamoto E., and Fujimoto J. 1999. Hepatocyte growth factor gene therapy of liver cirrhosis in rats. *Nat. Med.* **5:** 226–230.

Yamamoto S., Yamano T., Tanaka M., Hoon D.S.B., Takao S., Morishita R., Aikou T., and Kaneda Y. 2003. A novel combination of suicide gene therapy and histone deacetylase inhibitor for treatment of malignant melanoma. *Cancer Gene Ther.* **10:** 179–186.

Yamano T., Ura K., Morishita R., Nakajima H., Monden M., and Kaneda Y. 2000. Amplification of transgene expression in vitro and in vivo using a novel inhibitor of histone deacetylase. *Mol. Ther.* **1:** 574–580.

42 Polylysine Copolymers for Gene Delivery

Sung Wan Kim

Department of Pharmaceutics and Pharmaceutical Chemistry, University of Utah, Salt Lake City, Utah 84112-5820

ABSTRACT

Polylysine and its copolymers have been extensively utilized as nonviral polymeric gene carriers. Whereas polylysine on its own is toxic to cells, polyethylene glycol covalently linked to polylysine has been shown to reduce toxicity and increase DNA transfection. The degradable polylysine analog polyaminobutyl glycolic acid was synthesized, and stearyl polylysine, demonstrating strongly hydrophobic interaction with low-density lipoprotein, was designed to deliver DNA. This "terplex" system demonstrated significant levels of transfection both in vitro and in vivo.

INTRODUCTION

In the early 1970s, poly-L-lysine (PLL), used to cover glass and plastic surfaces to facilitate cell adhesion, was shown to be able to condense DNA (Carroll 1972a). These early results derived from circular dichroism (CD), X-ray crystallography, birefringence, and enzymatic digestion studies of polynucleotide/polylysine complexes (Carroll 1972b; Laemmli 1975). PLL has an overall positive charge due to its many ε-amine groups (see structure shown in Fig. 1) and can compact or condense plasmid DNA into complexes (Pouton et al. 1998). The condensate shapes have been observed to be primarily toroidal (Wagner et al. 1991; Wolfert and Seymour 1996; Wolfert et al. 1996; Tang and Szoka 1997) or globular (Dunlap et al. 1997). Subsequent studies showed that these condensed DNA complexes could enter cells (Kawai and Nishizawa 1984), but the level of transfection was quite low and polylysine itself was

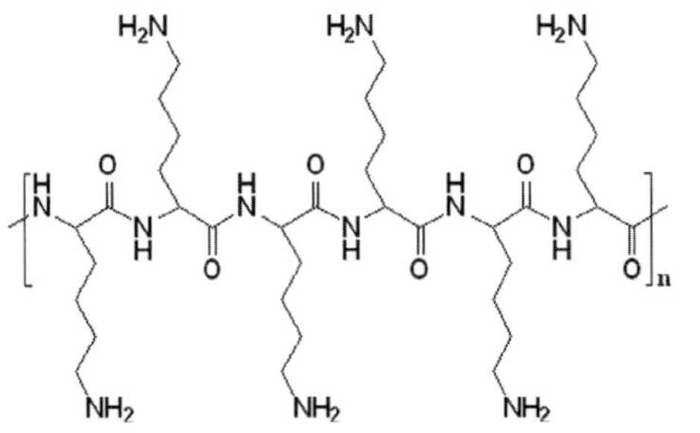

FIGURE 1. Structure of PLL.

toxic to the cells. Furthermore, PLL/plasmid DNA complexes sometimes precipitate out of solution depending on concentration, ionic strength, and type of solution (Choi et al. 1998a,b).

Efforts to decrease toxicity and increase transfection were attempted by either mixing PLL/DNA complexes with other adjuvants or chemically conjugating the adjuvants to PLL. For example, covalently linking polyethylene glycol to PLL was shown to decrease toxicity of the polymer, increase solubility of the complex, and increase transfection efficiency (Choi et al. 1998b; Toncheva et al. 1998). Further studies with PLL used targeting conjugates such as antibodies, carbohydrates, functional peptides, and other polymers. Table 1 summarizes the features of various polylysine-conjugated delivery systems, including the influence of molecular weight, transfection conditions, cell lines tested, and type of reporter gene used, as well as how adjuvants affect transfection.

The condensation of plasmid DNA by PLL occurs through the protonation of the ε-amino groups, followed by electrostatic interactions with the phosphate backbone. PLL on its own can efficiently condense plasmid DNA. But its use is limited because of the inability to effectively buffer the acidic environment at the late endosomal state, due to the high pKa (9.3–9.5) of the ε-amine. PLL modified with histidine residues (pKa 6.7–7.1), however, allows for proton buffering (Pack et al. 2000; Putnam et al. 2001; Bikram et al. 2003). PLL/plasmid DNA complexes are also prone to aggregation under physiological salt conditions. Poly[α-(4-aminobutyl)-L-glycolic acid] (PAGA), a more easily biodegradable analog of PLL, differs in having ester—instead of peptide—linkages in its backbone. PAGA is synthesized by converting the amino-terminal amine of CBZ ε-amine-protected lysine to a hydroxyl group via diazotization using sodium nitrite and sulfuric acid (Lim et al. 2000a,b). The resulting CBZ-L-oxylysine is polymerized via melting condensation at 150°C under vacuum, and the CBZ protecting groups are removed, revealing the primary amines. The PAGA/IL-12 (interleukin-12) complex has been used for prevention of insulitis for NOD mice (Koh et al. 2000).

PLL interacts with DNA to produce soluble complexes that exhibit a biphasic temperature profile: The lower temperature (T_m) corresponds to the melting of free DNA and the higher temperature (T_m') corresponds to the melting of the PLL/DNA complexes (Olins et al. 1967; Tsuboi et al. 1996). In addition, an increase in PLL concentration results in a decrease in the first melting temperature (less free DNA is present in the solution), but an increase in the second melting temperature, corresponding to an increase in complex formation. Thus, DNA in these complexes has increased stabilization against thermal denaturation, the complexes became less soluble, and the reaction between PLL and DNA is irreversible at low salt concentrations. Subsequent studies demonstrated that PLL almost exclusively selects for adenine-thymidine (A-T)-rich sequences (Shapiro et al. 1969). Two mechanisms for this selectivity are proposed: Lysine residues interact preferentially with an A-T pair at a specific site or a "group" property of the A-T-rich site shows selectivity. In addition, the interaction between PLL and DNA is irreversible at low salt concentrations but reversible at high salt concentrations. PLL was chosen as the typical model peptide because of its polycationic nature that interacts electrostatically with the negatively charged lipid bilayers under physiological conditions (Takahashi et al. 1991). In these experiments, the molecular weights of various PLL polymers were shown to induce aggregation and fusion of negatively

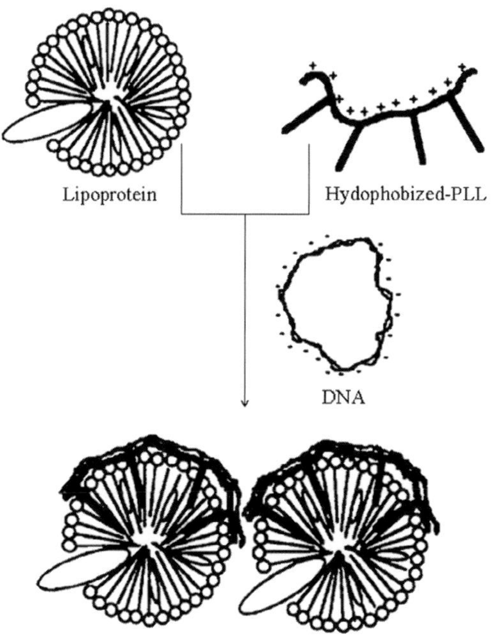

FIGURE 2. Formation of the terplex DNA delivery system. Equal amounts of LDL and H-PLL were mixed in phosphate-buffered saline at room temperature, to which varying amounts of plasmid DNA were added to form a stable terplex system (Kim et al. 1997).

charged unilamellar vesicles, accompanied by an increase in membrane permeability (Gad et al. 1982; Ohki and Daux 1986). In addition, a change in conformation of PLL from a random coil to a partially ordered configuration was observed (Hartsmann and Galla 1978; Walter et al. 1986). Subsequent experiments revealed that PLL assumes a β-sheet conformation on interaction with dipalmitoylphosphatidic acid (DPPA) lipid bilayer vesicles (Takahashi et al. 1991). Although PLL was shown to affect isolated vesicles, cumulative research findings together with the observation that PLL can fully condense DNA have encouraged studies to investigate whether PLL as carrier could effectively deliver DNA into cells with subsequent gene expression. We discuss here one system—terplex—that shows great promise for delivery of DNA into targeted cells.

The terplex system uses a balance of charge and hydrophobic interactions to stabilize the complex and has been proven to be an effective gene delivery vehicle in in vitro and in vivo myocardial infarction (Affleck et al. 2001). The separate entities of the complex are stearyl-PLL, low-density lipoprotein (LDL), and DNA (Fig. 2). The stearyl group has hydrophobic interactions with the lipoprotein that integrate into the LDL (Kim et al. 1997). The degree of the stearyl group conjugated to a PLL backbone was analyzed using ^{1}H-NMR (nuclear magnetic resonance). The peak area ratio of PLL ε-methylene groups at 3.0 ppm to the methyl group in the stearyl substitution at 8.0 ppm was used to determine the percent conjugation.

To determine the ability of the terplex system to sustain circulation of DNA in vivo, pharmacokinetic parameters were analyzed after systemic administration in mice. The terplex system showed enhanced circulation time and longer retention in the blood compared to naked DNA, and also a greater volume of distribution (Yu et al. 2001b). Reverse transcription–polymerase chain reaction (RT-PCR) was used to assess transgene expression in various organs after systemic administration, revealing higher levels of expression with the terplex system (Yu et al. 2001b).

These studies indicate that the terplex system is a promising delivery system for targeted gene delivery to LDL receptor-bearing cells. The capacity of the terplex system to condense DNA, neutralize charge, and limit toxic side effects has a role in its ability to transfect cells efficiently and derive from the separate components that interact to provide a multifunctional gene delivery vehicle (Yu et al. 2001a). The following protocols are based on our studies using the terplex system to deliver DNA in in vitro and in vivo systems.

TABLE 1. Polylysine Gene Delivery Systems

PLL-conjugate (% conjugation)	PLL (M.W.)	Transfection compared to just PLL	Visualization	Transfection adjuvants	Cell lines/in vivo	Special target/other	Reference
PLL-dextran (% conjugation not determined) graft	35 kD	No PLL transfection done for comparison	pNDA GFP (green fluorescent protein)		293 Sprague-Dawley rats		de Marco et al. (1998)
PLL-poly(diethyl-amino-ethyl-L-glutamine) block copolymer 33% PLL 67% D-L-glutamic acid	26 kD PLL M.W. of copolymer unknown	No PLL transfection done for comparison	pDNA β-gal	Chloroquine Serum-free	293		Dekie et al. (2000)
PLL-antibody (anti-CD4 antibody B-F5)	268 DP (degree of polymerization)	No transfection with just PLL in Jurkat cells; little transfection seen in K-562 cells	pDNA Luc (luciferase) GFP	PMA (phorbal myristate acetate) really helped transfection	Jurkat K562	Jurkat cell line antibody B-F5-specific; 95% of Jurkat cells express GFP when transfected	Puls and Minchin (1999)
PLL-TF (transferrin)	190 DP	2000-fold higher with adenovirus over that of PLL-Tf and even much better over that of PLL	pDNA Luc	Adenovirus particles 2% serum	HeLa WI-38 MRC-5 KB CFT1		Curiel et al. (1991)
PLL-TF (? %) PLL-Ab$_{CD3}$ (33% conjugation) PLL-streptavidin	300 DP	PLL better than PLL-Tf in H9 and Jurkat E6 in absence of chloroquine PLL-Tf, PLL-Ab	pDNA Luc β-gal	2–10% serum Chloroquine in some systems	Jurkat clone Eb-1 CCRF-CEM	Eb-1 cell line is CD3 antibody target	Buschle et al. (1995)
PLL-galactose (1% conjugation) graft	236 DP	PLL-gal much higher in hepatic cells, as well in the liver and spleen of rats	pDNA Luc β-gal	PBS	HepG2, Hu-H7, Sprague-Dawley rats	Hepatocytes (HepG2, Hu-H7) have asialoglycoprotein receptor	Perales et al. (1997)
PLL-lactose (68% conjugation)	30–50 kD	Transfection higher than PLL in hepatic cells	pDNA Luc	Chloroquine 1% serum	HepG2, HeLa 229	Hepatocyte (HepG2)	Midoux et al. (1993)
PLL-g-histidine (45% conjugation) graft	30–50 kD (190 DP)	Transfection higher than just PLL	pDNA Luc	Sometimes chloroquine Serum-free or 1–20% serum	16HBE, A549, HepG2, Rb-1, B16, HOS, MCF-7, COS, HeLa, 293T7	HepG2 and 293T7 show best transfection	Midoux and Monsigny (1999)
PLL-mannose (46% conjugation) PLL-glucose (46% conjugation)	30–50 kD (190 DP)	PLL-mannose transfects best; all carbohydrate-PLL better than just PLL	pDNA Luc CAT	Chloroquine 1% serum	Human blood monocyte-derived	Most Luc seen after 24 hours	Erbacher et al. (1996)
(PLL/pDNA complex)-PEG (PLL-TF/pDNA)-PEG	PLL (250 DP) PLL (19 DP) PEG (1,2,3,5,12 kD)	Only complement activation investigated	n.a.[a]	n.a.[a]	Sheep red blood cells (erthrocytes)	PLL/pDNA complex made in water and the complex reacted with NHS-PEG	Plank et al. (1996)

Polymer	Size	Transfection result	Reporter	Conditions	Cell lines	Comments	Reference
PLL-PEG-lactose (6,12,20,30% lac-PEG) PLL-PEG (25% conjugated) PLL-PEG-gal (23% conjugated)	PLL (25 kD, 120 DP) PEG (550 M.W.)	PLL-PEG-lac much better on HepG2 cells; lipofectin was control; PLL-PEG-lac better than lipofectin; transfection just as low in NIH-3T3 and A7R5 cell lines; PLL-PEG better than PLL	pDNA β-gal	10% serum Chloroquine Some conditions varied serum from 0% to 20%	HepG2 A7R5 NIH-3T3	Hepatocyte target (HepG2) for sugar; CD, atomic force microscopy of pDNA/polymer	Choi et al. (1998a, 1999)
PLL_{20k}-PEG_{500} (5%, 10% conjugated) $PLL_{11.4k}$-$dextran_{1k}$ (5% conjugated) PLL_{19k}-$pHPMA_{4.4k}$ (8% conjugated)	PLL (9.6, 22.4, 20 kD) PEG (5,12 kD)	PLL-PEG transfects better than PLL	pDNA β-gal	Polymer/pDNA Complexes mixed with INF7-SGSC EDP peptide Serum-free	HepG2 293		Toncheva et al. (1998)
PLL-stearyl (25% conjugated) mixed with LDL Complex trade name is Terplex	50 kD	Much better transfection than PLL	pDNA β-gal	Sometimes chloroquine Serum/serum-free	A7R5 CCD-32		Kim et al. (1998, 2000); Lentz and Kim (1998)
PLL-NLS (25-amino-acid nuclear localization sequence from SV40 large T antigen) cross-linked via SPDP	110 kD	PLL-NLS much better than PLL (twofold) when mixed with PLL-TF; use of a defective NLS mutant showed lower transfection than functional NLS, but higher than just PLL	pDNA β-gal	TF-PLL used in mixture with PLL-NLS Chloroquine 10% FCS	HTC rat hepatoma	Electron microscope done on complex; performed ELISA-based binding assay using α/β mouse importin to assess NLS accessibility in the PLL/pDNA conjugates	Chan and Jans (1999); Chan et al. (2000)
PLL-EGF (epidermal growth factor) conjugated via strepavidin-biotin bond PEI (DP 46,600)	DP 19, 36, 180	PLL(19)-EGF: 32% transfection PLL(36)-EGF: 38% transfection Not compared to just PLL	pDNA GFP (FACS analysis) Fluorescent microscopy PLL-EGF-Alexa 488 PDNA-Alexa 546	1% insulin-transferrin-selenium supplement 100 μM chloroquine	Swiss-3T3 NR6 fibroblasts transduced with EGF receptor	Cells have EGF-receptor-mediated endocytosis; up 50 10% of cell nuclei show fluorescence	Schaffer et al. (2000)
PLL-imidazole (73.5, 82.5, 86.5% conjugated)	M.W. 34,300	86.5% conjugate had best transfection; more imidazole groups meant better transfection	pDNA Luc	OPTI-MEM Media removed after 4 hours	HepG2 CRL1476 P388D1	PLL-imidazole less toxic than PLL; need particle sizes <150 nm	Putnam et al. (2001)
PLL-TF-HA2 HA2 is influenza endosomal disrupting peptide	DP:190 DP:290	EDP conjugates transfect much better than PLL; EDP not conjugated did not work	pDNA Luc β-gal	Chloroquine Some serum-free	K562 HeLa BNL Cl.2	Liposome leakage assay also done; chloroquine gets much higher transfection	Wagner et al. (1992)
PLL-insulin	90 and 120 kD	Not compared to PLL; transfection increased by receptor-mediated endocytosis (insulin receptors)	pDNA pSVRP-8 plasmid encodes SV40 large T antigen	Serum-free	PLC/PRF/5 hepatoma cells	SV40 expression detected by indirect immuno-fluorescence and FITC-labeled second antibody; cells express insulin-binding sites	Rosenkranz et al. (1992)

[a] not available

Protocol 1

In Vitro Transfection with Plasmid DNA

The following method may be used for optimizing relative components of the terplex system and transfection parameters for a given cell line.

MATERIALS

Reagents

A7R5 cells, a murine smooth-muscle cell line (American Type Cell Culture [ATCC])

Grow cells in Dulbecco's minimal essential medium (DMEM) supplemented with 10% fetal bovine serum at 37°C in a 5% CO_2 incubator.

Growth media and fetal bovine serum (FBS) (HyClone Laboratories)

o-nitrophenyl-β-D-galactopyranoside (ONPG)

Phosphate-buffered saline (PBS)

Plasmid DNA, pSV-β-galactosidase control vector (6821 bp, Promega)

Terplex complex

Prepare in formulations, varying the PLL or LDL.

Mix stearyl-PLL, synthesized by N-alkylation of PLL (M.W. 50,000 daltons) with stearyl bromide, in varying amounts with LDL in PBS at room temperature (for further details, see Kim et al. 1998b).

Equipment

Incubator, controlled at 5% CO_2 (e.g., Napco, Precision Scientific)

Spectrophotometer

METHOD

1. The day before transfection, seed A7R5 cells at a density of 2×10^4 cells/ml in 24-well plates and incubate at 37°C in a 5% CO_2 atmosphere.
2. Mix the various formulations of the DNA terplex system in PBS with a fixed amount of plasmid DNA: 1 μg of DNA per well of a 24-well plate for each transfection to be performed.

   It is best to prepare the terplex solution fresh for each use.
3. Transfect the cells by adding to each well 40 μl of the DNA mixture (as control) or the terplex system, including PLL, LDL, and plasmid DNA.
4. Incubate the cells for 48 hours at 37°C in a 5% CO_2 atmosphere, wash three times with PBS, and harvest for β-galactosidase activity assay.
5. Determine β-galactosidase activity in transfected cells by spectrophotometric detection at 420 nm using ONPG as a substrate.

Protocol 2

In Vitro Transfection with Oligonucleotide DNA

This protocol describes transfection of A7R5 cells or CCD-32 cells with the terplex system complexed with c-*myb* antisense oligonucleotide to analyze the antiproliferative effect of the transfected sequence.

MATERIALS

Reagents

A7R5 cells, a murine smooth-muscle cell line (American Type Cell Culture [ATCC])

Grow cells in DMEM supplemented with 10% FBS at 37°C in a 5% CO_2 incubator.

or

CCD-32 Lu, a human lung fibroblast cell line (ATCC)

Grow cells are grown in Eagle's minimum essential medium (EMEM) supplemented with 10% FBS, typically in 25-cm^2 polystyrene tissue culture flasks (T-25 flask, Falcon) until confluent. Trypsinize confluent cells to remove the cell monolayer and resuspend the cells for experiments or subculturing.

Growth media and fetal bovine serum (FBS, HyClone Laboratories)

MTT (3-[4,5-dimethylthiazol-2-yl]-2,5-diphenyltetrasodium bromide) (Sigma-Aldrich)

Oligonucleotides

c-*myb* antisense oligonucleotide (18-mer, 5′-GTG TCG GGG TCT CCG GGC-3′)

mismatched antisense oligonucleotide (18-mer, 5′-GTC TCC GGC TCA CCC GGG-3′)

The oligonucleotides are synthesized by phosphoramidite chemistry and purified by gel filtration at the Midland Certified Reagent Co.

Terplex complex

Prepare in formulations, varying the amount of oligonucleotide DNA (using c-*myb* antisense or antisense-mismatched control).

Mix stearyl-PLL, synthesized by N-alkylation of PLL (M.W. 50,000 daltons) with stearyl bromide, in fixed amounts with LDL in PBS with varying amounts of oligonucleotide at room temperature to create a stable terplex complex (for further details, see Kim et al. 1998b).

Equipment

Incubator, controlled at 5% CO_2 (e.g., Napco, Precision Scientific)

METHOD

1. The day before transfection, plate A7R5 cells or CCD-32 Lu cells in 96-well plates at a density of 5 x 10^4 cells/well and grow overnight at 37°C in a 5% CO_2 incubator.
2. Prepare the terplex formulations, varying the amounts of antisense oligonucleotide.
   It is best to prepare the terplex solution fresh for each use.
3. Transfect the cells by adding terplex complex to each well. Incubate the cells for 4 hours at 37°C in a 5% CO_2 incubator in the presence or absence of serum-containing media.
4. Determine the degree of cell proliferation using the MTT colorimetric assay (Mosmann 1983).

Protocol 3

In Vivo Animal Studies

This protocol gives an example of the use of tetraplex complexes to deliver plasmid DNA carrying the vascular endothelial growth factor VEGF-165 into rabbit myocardium, based on our experience with the procedure.

MATERIALS

Reagents

Buprenorphine
Isoflurane
Ketamine (50 mg/kg, intramuscularly)
Phosphate-buffered saline (PBS, pH 7.4)
Plasmid DNA, pCMV-VEGF

The plasmid carries the VEGF-165-coding region under the control of the cytomegalovirus (CMV) immediate-early promoter enhancer region and the chicken β-globulin intron. Plasmid DNA is purified by double CsCl gradient purification and confirmed spectrophotometrically at A260/A280 and by 0.8% agarose gel electrophoresis.

Rabbits, New Zealand White (2.5–3.5 kg) (Western Oregon)
Staining solutions
hematoxylin and eosin (H&E)
trichrome
Xylazine (8.8 mg/kg, intramuscularly)

Equipment

Angiocatheter
Electrocardiograph
Needles (30 gauge)
Rodent ventilator (Harvard Apparatus)
Standard planimetry software (NIH Image)
Suture (5-0 Prolene)
Ultrasound imaging system for echocardiography (General Electric Vivid Five or Hewlett-Packard 5500)

METHOD

1. Sedate New Zealand White rabbits, weighing 2.5–3.5 kg, with ketamine (50 mg/kg, intramuscularly) and xylazine (8.8 mg/kg, intramuscularly). Shave the rabbits and intubate them endotracheally.

   The number of animals prepared depends on the goal of the study. Assign the animals randomly to either a treatment group or a control group (see Step 6). In our studies, the operating surgeon was "blind" to the experimental group.

2. For each animal, prepare the left chest under sterile conditions, drape, and develop a surgical plane of anesthesia with isoflurane. Maintain the animal on a rodent ventilator with a respiratory rate of 60% and 40% FiO_2.
3. Perform a left thoracotomy through the third intercostals space, retract the lung, and open the pericardium widely.
4. Place a 5-0 Prolene suture around the first branch of the circumflex artery and perform a 10-second test occlusion to determine the area of the ventricle at risk.
5. Prepare the solutions to be injected. For each animal in the treatment group, prepare a 500-μl solution of 50 μg of LDL, 50 μg of stearyl-PLL, and 50 μg of pCMV-VEGF in PBS (pH 7.4). For each animal in the control group, prepare a 500-μl solution of 50 μg of LDL and 50 μg of stearyl-PLL in PBS (pH 7.4).

   It is best to prepare the terplex solutions fresh for each use.
6. Place a 30-gauge needle intramyocardially at the border zone surrounding the area at risk. Deliver the treatment solution (prepared in Step 5) in a series of four injections (125 μl each, total of 500 μl) into each of the treatment animals. In a similar manner, deliver the control solution in a series of foam injections (125 μl each, total of 500 μl) into each of the control animals.
7. Immediately after the injections, induce myocardial infarction as described by Affleck et al. (2001). Confirm ischemia by visual inspection of the myocardium and by electrocardiographic changes noted on continuous EKG monitoring.
8. After a period of observation, reinflate the lung with positive end expiratory pressure produced by temporary outflow occlusion. Evaculate residual air from the pleural space with an angiocatheter.
9. Reapproximate the ribs with a single "figure eight" stitch, close the muscular layers individually with a running absorbable stitch, and close the skin with a running 5-0 Prolene stitch.
10. Wean the rabbits and extubate them when they appear to be breathing spontaneously. Control postoperative pain with buprenorphine (0.05 mg/kg intramuscularly) every 12 hours for 48 hours.
11. Lightly sedate the rabbits with ketamine (20 mg/kg) and xylazine (5 mg/kg). Perform echocardiography immediately after surgery and on postoperative days 1 and 21.

    In our studies, an echocardiographer is blind to the treatment groups.
12. Collect images with either a Hewlett-Packard 5500 or General Electric Vivid Five using a 12-MHz probe.

    In our experiments, all images were obtained by a single experienced echocardiographer and reviewed by a second experienced echocardiographer, to confirm inter- and intraobserver reliability. Right parasternal long-axis views were used to obtain two-dimensional measurements of ejection fraction. The two-dimensional ejection fraction was derived from manually traced ventricular areas at end diastole and end systole. Three to five pairs of tracings are performed, and the reported value represents the mean. Right parasternal short-axis views were used with loops recorded at the base, midpapillary region, and apex. M-mode measurements were obtained at each level from which fractional shortening was derived, and the ejection fraction was calculated. The reported values of ejection fraction, fractional shortening, and left ventricular systolic and diastolic areas represent the mean of these values.
13. Sacrifice the animals on postoperative day 21 by exsanguinations after sedation with ketamine and inhaled isofluorane.
14. Perfuse the aortic root with 100 cm^3 of normal saline and fix the isolated hearts by perfusion with 100 mm Hg of buffered neutral formalin.

15. Cut short axis sections at 5-mm intervals from apex to base, embed whole-mount sections in paraffin, and stain with H&E and trichrome.
16. Measure the area of normal and infarcted left ventricular myocardium on trichrome-stained slides using standard planimetry software (NIH Image).

    The percentage of infarcted left ventricular myocardium is obtained by dividing the total area of infarcted myocardium (all sections) by the total area of all myocardium and multiplying by 100.

REFERENCES

Affleck D.G., Yu L., Bull D.A., Bailey S.H., and Kim S.W. 2001. Augmentation of myocardial transfection using TerplexDNA: A novel gene delivery system. *Gene Ther.* **8:** 349–353.

Bikram M., Ahn C.H., Chae S.Y., Lee M., Yockman J.W., and Kim S.W. 2003. Novel biodegradable PEG-PLL-His multiblock copolymers for nonviral carrier mediated gene therapy. *Mol. Ther.* **1:** S221.

Buschle M., Cotton M., Kirlappos H., Mechtler K., Schaffner G., Zauner W., Birnstiel M.L., and Wagner E. 1995. Receptor-mediated gene transfer into human T lymphocytes via binding of DNA/CD3 antibody particles to the CD3 T cell receptor complex. *Hum. Gene Ther.* **6:** 753–761.

Carroll D. 1972a. Complexes of polylysine with polyuridylic acid and other polynuceotides. *Biochemistry* **11:** 426–433.

———. 1972b. Optical properties of deoxyribonucleic acid-polylysine complexes. *Biochemistry* **11:** 421–426.

Chan C.K. and Jans D.A. 1999. Enhancement of polylysine-mediated transferrinfection by nuclear localization sequences: Polylysine does not function as a nuclear localization sequence. *Hum. Gene Ther.* **10:** 1695–1702.

Chan C.K., Senden T., and Jans D.A. 2000. Supramolecular structure and nuclear targeting efficiency determine the enhancement of transfection by modified polylysines. *Gene Ther.* **7:** 1690–1697.

Choi Y.H., Liu F., Park J., and Kim S.W. 1998a. Lactose-poly(ethylene glycol)-grafted poly-L-lysine as hepatoma cell-targeted gene carrier. *Bioconjug. Chem.* **9:** 708–718.

Choi Y.H., Liu F., Choi J.S., Kim S.W., and Park J.S. 1999. Characterizaion of a targeted gene carrier, lactose-polyethylene glycol-grafted poly-L-lysine and its complex with plasmid DNA. *Hum. Gene Ther.* **10:** 2657–2665.

Choi Y.H., Liu F., Kim J.S., Choi Y.K., Park J.S., and Kim S.W. 1998b. Polyethylene glycol-grafted poly-L-lysine as polymeric gene carrier. *J. Control. Release* **54:** 39–48.

Curiel D.T., Agarway S., Wagner E., and Cotton M. 1991. Adenovirus enhancement of transferrin-polylysine-mediated gene delivery. *Proc. Natl. Acad. Sci.* **88:** 8850–8854.

Dekie L., Toncheva V., Dubruel P., Schacht E.H., Barrett L., and Seymour L.W. 2000. Poly-L-glutamic acid derivatives as vectors for gene therapy. *J. Control. Release* **65:** 187–202.

de Marco G., Bogdanov A., Marecos E., Moore A., Simonova M., and Weissleder R. 1998. MR imaging of gene delivery to the central nervous system with an artificial vector (see comments). *Radiology* **208:** 65–71.

Dunlap D.D., Maggi A., Soria M.R., and Monaco L. 1997. Nanoscopic structure of DNA condensed for gene delivery. *Nucleic Acids Res.* **25:** 3095–3101.

Erbacher P., Bousser M.T., Raimond J., Monsigny M., Midoux P., and Roche A.C. 1996. Gene transfer by DNA/glycosylated polylysine complexes into human blood monocyte-derived macrophages. *Hum. Gene Ther.* **7:** 721–729.

Gad A.E., Silver B.L., and Eytan G.D. 1982. Polycation-induced fusion of negatively charged vesicles. *Biochim. Biophys. Acta* **690:** 124–132.

Hartmann W. and Galla H.J. 1978. Binding of polylysine to charged bilayer membranes: Molecular organization of a lipid peptide complex. *Biochim. Biophys. Acta* **509:** 474–490.

Kawai S. and Nishizawa M. 1984. New procedure for DNA transfection with polycation and dimethyl sulfoxide. *Mol. Cell. Biol.* **4:** 1172–1174.

Kim J.S., Maruyama A., Akaike T., and Kim S.W. 1998. Terplex DNA delivery system as a gene carrier. *Pharm. Res.* **15:** 116–121.

Kim J.-S., Maruyama A., Akaike T., and Kim S.W. 2000. Non-viral gene delivery strategies for enhanced cellular uptake. *Biomaterials and drug delivery toward new millenium* (ed. K.D. Park et al.), pp. 237–247. Han Rim Won Publishing, Seoul, South Korea.

Kim J.S., Maruyama A., Akaike T., and Kim S.W. 1997. In vitro gene expression on smooth muscle cells using a terplex delivery system. *J. Control. Release* **47:** 51–59.

Koh J.J., Ko K.S., Lee M., Han S., Park J.S., and Kim S.W. 2000. Degradable polymeric carrier for the delivery of IL-10 plasmid DNA to prevent autoimmune insulitis of NOD mice. *Gene Ther.* **7:** 2099–2104.

Laemmli U.K. 1975. Characterization of DNA condensates induced by poly(ethylene oxide) and polylysine. *Proc. Natl. Acad. Sci.* **72:** 4288–4292.

Lentz M. 2001. "Mechanistic study of gene transfection." Ph.D. thesis. University of Utah.

Lentz M.J. and Kim S.W. 1998. Terplex system for delivery of DNA and oligonucleotides. *Self-assembling complexes for gene delivery* (ed. A.V. Kabanov et al.), pp. 149–166. Wiley, New York.

Lim Y.B., Kim C.H., Kim K., Kim S.W., and Park J.S. 2000a. Development of a safe gene delivery system using biodegradable polymer, poly[α-(4-aminobutyl)-L-glycolic acid]. *J. Am. Chem. Soc.* **122:** 6524–6525.

Lim Y.B., Han S.O., Kong H.U., Lee Y., Park J.S., Jeong B., and Kim S.W. 2000b. Biodegradable polyester, poly[α-(4-aminobutyl)-L-glycolic acid], as a non-toxic gene carrier. *Pharm. Res.* **17:** 811–816.

Midoux P. and Monsigny M. 1999. Efficient gene transfer by histidylated polylysine/pDNA complexes. Bioconjug. Chem. **10:** 406–411.

Midoux P., Mendes C., Legrand A., Raimond J., Mayer R., Monsigny M., and Roche A.C. 1993. Specific gene transfer mediated by lactosylated poly-L-lysine into hepatoma cells. *Nucleic Acids Res.* **21:** 871–878.

Mosmann T. 1983. Rapid colorimetric assay for cellular growth and survival: Application to proliferation and cytotoxicity assays. *J. Immunol. Methods* **65:** 55–63.

Ohki S. and Duax J. 1986. Effects of cations and polyamines on the aggregation and fusion of phosphatidylserine membranes. *Biochim. Biophys. Acta* **861:** 177–186.

Olins D.E., Olins A.L., and Von Hippel P.H. 1967. Model nucleoprotein complexes: Studies on the interaction of cationic homopolypeptides with DNA. *J. Mol. Biol.* **24:** 157–176.

Pack D.W., Putnam D., and Langer R. 2000. Design of imidazole containing endosomolytic biopolymer for gene delivery. *Biotech. Bioeng.* **67:** 217–223.

Perales J.C., Grossmann G.A., Molas M., Liu G., Ferkol T., Harpst J., Oda H., and Hannson R.W. 1997. Biochemical and functional characterization of DNA complexes capable of targeting genes to hepatocytes via the asialoglycoprotein receptor. *J. Biol. Chem.* **272:** 7398–7407.

Plank C., Mechtler K., Szoka F.C., and Wagner E. 1996. Activation of the complement system by synthetic DNA complexes: A potential barrier for intravenous gene delivery. *Hum. Gene Ther.* **7:** 1437–1446.

Pouton C.W., Lucas P., Thomas B.J., Uduchi A.N., Milroy D.A, and Moss S.H. 1998. Polycation-DNA complexes for gene delivery: A comparison of the biopharmaceutical properties of cationic polypeptides and cationic lipids. *J. Control. Release* **53:** 289–299.

Puls R. and Minchin R. 1999. Gene transfer and expression of a non-viral polycation-based vector in $CD4^+$ cells. *Gene Ther.* **6:** 1774–1778.

Putnam D., Gentry C.A., Pack D.W., and Langer R. 2001. Polymer based gene delivery with low cytotoxicity by a unique balance of side chain termini. *Proc. Natl. Acad. Sci.* **98:** 1200–1205.

Rosenkranz A.A., Yachmenev S.V., Jans D.A., Serebryakova N.V., Murav'ev V.I., Peters R., and Sobolev A.S. 1992. Receptor-mediated endocytosis and nuclear transport of a transfecting DNA construct. *Exp. Cell Res.* **199:** 323–329.

Schaffer D.V., Fidelman N.A., Dan N., and Lauffenburger D.A. 2000. Vector unpacking as a potential barrier for receptor-mediated polyplex gene delivery. *Biotechnol. Bioeng.* **67:** 598–606.

Shapiro J.T., Leng M., and Felsenfeld G. 1969. Deoxyribonucleic acid-polylysine complexes: Structure and nucleotide specificity. *Biochemistry* **8:** 3119–3132.

Takahashi H., Matuoka S., Kato S., Ohki K., and Hatta I. 1991. Electrostatic interaction of poly(L-lysine) with dipalmitoylphosphatidic acid studied by X-ray diffraction. *Biochim. Biophys. Acta* **1069:** 229–234.

Tang M.X. and Szoka F.C. 1997. The influence of polymer structure on the interactions of cationic polymers with DNA and morphology of the resulting complexes. *Gene Ther.* **4:** 823–832.

Toncheva V., Wolfert M., Dash P., Oupicky D., Ulbrich K., Seymour L., and Schacht E., 1998. Novel vectors for gene delivery formed by self-assembly of DNA with poly(L-lysine) grafted with hydrophilic polymers. *Biochim. Biophys. Acta* **1380:** 354–368.

Tsuboi M., Matsuo K., and Ts'o P.O. 1996. Interaction of poly-L-lysine and nucleic acids. *J. Mol. Biol.* **15:** 256–267.

Wagner E., Cotton M., Foisner R., and Birnstier M.L. 1991. Transferrin-polycation-DNA complexes: The effect of polcations on the structure of the complex and DNA delivery to cells. *Proc. Natl. Acad. Sci.* **88:** 4255–4259.

Wagner E., Plank C., Zatloukai K., Cotton M., and Birnstiel M. 1992. Influenza virus hemagglutinin HA-2 N-terminal fusogenic peptides augment gene transfer by transferrin-polylysine-DNA complexes: Toward a synthetic virus-like gene-transfer vehicle. *Proc. Natl. Acad. Sci.* **89:** 7934–7938.

Walter A., Steer C.J., and Blumenthal R. 1986. Polylysine induces pH-dependent fusion of acidic phospholipids vesicles: A model for polycation-induced fusion. *Biochim. Biophys. Acta* **861:** 319–330.

Wolfert M.A. and Seymour L.W. 1996. Atomic force microscopic analysis of the influence of the molecular weight of poly(L)lysine on the size of polyelectrolyte complexes formed with DNA. *Gene Ther.* **3:** 269–273.

Wolfert M., Schacht E., Toncheva V., Ulbrich K., Nazarova O., and Seymour L. 1996. Characterization of vectors for gene therapy formed by self-assembly of DNA with synthetic block co-polymers. *Hum. Gene Ther.* **7:** 2123–2133.

Yu L., Nielson M., Han S.O., and Kim S.W. 2001a. TerplexDNA gene carrier system targeting artery wall cells. *J. Control. Release* **72:** 179–185.

Yu L., Suh H., Koh J.J., and Kim S.W. 2001b. Systemic administration of TerplexDNA system: Pharmacokinetics and gene expression. *Pharm. Res.* **18:** 1277–1283.

43 PEI Nanoparticles for Targeted Gene Delivery

Frank Alexis,* Jieming Zeng,* and Shu Wang*†

**Institute of Bioengineering and Nanotechnology, Singapore; †Department of Biological Sciences, National University of Singapore 138669*

ABSTRACT

Polycation/plasmid DNA nanoparticles are promising delivery systems for gene therapy, but their practical applications have been hampered by low gene-transfer efficiency. Incorporation of targeting peptides onto the particles is one way to improve cellular access and uptake in selected cell types. This chapter describes two methods for the preparation of functionalized nanoparticles based on polyethylenimine (PEI)/DNA complexes: (1) peptide conjugation to amine-reactive, cross-linker-modified PEI and formation of PEI peptide conjugate/DNA complexes and (2) the use of targeting peptides with a positively charged DNA-binding sequence to form noncovalent ternary complexes of PEI, the peptides, and plasmid DNA. The branched structure of PEI is an advantage for targeting peptide conjugation, allowing many copies of peptide per PEI molecule. One technical difficulty here is the choice of a cross-linker that can be dissolved and is stable in a PEI solution. Another is that PEI activation can sometimes lead to polymer cross-linking. The PEI/DNA nanoparticles can be used as vectors for cell-type-specific targeting, ensuring therapeutic efficacy in the cells of interest and limiting side effects caused by expression of exogenous genes in nontargeted cells. In addition to cell targeting, peptides with other functions can be incorporated into PEI/DNA nanoparticles for specific gene delivery purposes, such as endosomolysis or nuclear transport. Conjugation requires chemical reactions that may be harsh enough to inactivate sensitive moieties, such as proteins and peptides. For proteins or peptides difficult to purify or synthesize, the relatively large amounts of materials required for chemical conjugation may be prohibitively expensive, limiting scale up of vector preparation necessary for gene therapy applications. The use of ternary complexes circumvents the need for harsh and reagent-wasteful chemical processes and better preserves the biological activities of targeting moieties. However, the intrinsic competition between targeting peptides and PEI for DNA binding in this system requires optimization of the DNA/peptide/PEI ratio.

INTRODUCTION

Gene delivery using PEI involves condensation of negatively charged DNA by the cationic polymer into nanoparticles to facilitate DNA penetration through extra- and intracellular barriers. Recently, there has been significant effort directed to developing materials for gene delivery through receptor-mediated uptake, endosomal release, and nucleus targeting. Typically, functional moieties are chemically conjugated to cationic polymers that can be used as DNA vectors for targeted delivery into selected cells. The attachment of targeting moieties, such as transferrin, galactose, antibodies, and RGD peptides, to polymeric vectors has been achieved via the formation of either stable covalent bonds, thiol bonds, or enzymatic cleavable bonds (Hildebrandt et al. 2003; Kunath et al. 2003a,b; Kursa et al. 2003; Tang et al. 2003; Strehblow et al. 2005). Alternatively, functional moieties can be attached to PEI/DNA nanoparticles through noncovalent bonds. For example, chimeric peptides with a targeting domain and a DNA-binding sequence, such as SPKR repeats derived from the histone H1 or lysine repeats, were shown to interact electrostatically with DNA, followed by the addition of PEI to form ternary complexes (Rudolph et al. 2003; Ma et al. 2004; Zeng and Wang 2005).

This chapter describes the preparation of PEI/DNA nanoparticles for targeted gene delivery. This delivery strategy improves the efficiency of gene transfer by enhancing the entry of gene vectors into the desired cells and reducing uptake by nontarget cells. We describe here methods for (1) the conjugation of targeting peptides to PEIs, (2) the formation of DNA complexes using conjugated PEIs or nonconjugated PEIs together with targeting peptides, and (3) cell transfection using these complexes.

Protocol

Preparation and Transfection of PEI/DNA Nanoparticles

Peptide conjugation to PEI is a two-step reaction. PEI polymers are first modified by amine-reactive cross-linkers, and the activated PEI polymers then specifically react to a functional peptide group (e.g., the carboxyl or thiol group). Conjugation using heterobifunctional cross-linkers has the advantage of reducing the amount of reaction side products. In this protocol, we describe the use of SMCC, a heterobifunctional cross-linker, to form a stable bond between PEI and peptides containing thiol groups.

MATERIALS

Reagents

Dimethylsulfoxide (DMSO; Sigma-Aldrich)

DMEM (Dulbecco's modified Eagle's medium) cell culture medium with 10% fetal bovine serum (FBS)

Lithium chloride (Sigma-Aldrich)

Luciferase assay reagents (Promega)

Mammalian cells (exponentially growing)

Opti-MEM serum-free cell culture medium (Invitrogen)

PEI polymers (M.W. 600–1000 kD, Fluka; M.W. 750, 25, 2 kD, and 800 daltons, Sigma-Aldrich; M.W. 1.2, 10, or 70 kD, Polysciences)

Peptides, prepared using conventional solid-phase chemical synthesis method

Phosphate-buffered saline (PBS)

- 0.1 M sodium phosphate (pH 7.4)
- 0.15 M NaCl

Plasmid DNA encoding the luciferase reporter gene

5x Reporter lysis buffer (Promega), diluted to 1x in PBS

Succinimidyl-4-(*N*-maleimidomethyl)cyclohexane-1-carboxylate (SMCC; Pierce)

Ultrapure sterilized H_2O

Equipment

D_c protein assay kit (Bio-Rad)

Dialysis membranes (molecular-weight size-exclusion specification for purification of side reaction and excess products)

Freeze dryer

Luminometer

Magnetic stirrer and magnetic rod

Reaction vessel

Tissue-culture vessels

METHODS

Activation of PEI with a Cross-linker

1. Prepare a SMCC stock solution of 50 mM in DMSO.

   SMCC is moisture sensitive. Prepare the stock solution using high-quality anhydrous DMSO in a dry nitrogen atmosphere. When stored at 4°C, the SMCC solution remains stable for about 3 months. Perform Steps 1–10 in a chemical fume hood following chemical safe handling procedures.

2. Prepare a 10-mg/ml stock solution of PEI in DMSO. Add 2–5 mg of lithium chloride to increase the solubility of PEI.
3. Use a syringe to slowly add the SMCC solution into the PEI solution. Incubate the reaction for 2 hours at room temperature.

   The amount of SMCC solution added should be based on the desired molar ratio between SMCC and PEI.

4. Purify the modified PEI by dialysis against ultrapure water for 2 days, changing the water at least five times per day.
5. Collect the solution in a dialysis tube and freeze-dry the sample.

Conjugation of Peptide to Activated PEI

6. Prepare a peptide stock solution of 20–50 mM in PBS.
7. Prepare a 10-mg/ml stock solution of the activated PEI (Step 5) in PBS.
8. Slowly add the peptide solution to the activated PEI solution. Incubate the conjugation reaction for 24 hours at room temperature.

   The amount of the peptide added to the reaction should be based on the desired molar ratio in the final conjugate.

9. Purify the peptide-conjugated PEI by dialysis against ultrapure water for 2 days, changing the water at least 5 times a day.
10. Collect the solution in a dialysis tube and freeze dry the sample.

Preparation of PEI/DNA Complexes

11. Prepare the stock solutions.
    a. Prepare a 1-mg/ml plasmid DNA stock solution in ultrapure H_2O.
    b. Prepare a stock solution of PEI (Step 5) or peptide-conjugated PEI (Step 10) to contain 10 nmoles amino nitrogen per microliter in ultrapure H_2O (pH 7.2).
    c. For ternary complexes, prepare a 5-mg/ml stock solution of a targeting peptide linked with a DNA-binding sequence in ultrapure H_2O.

    Prepare Steps 11–22 under aseptic conditions using sterile reagents. Perform manipulation of the complexes in a horizontal flow hood at room temperature.

12. Prepare the working solutions.
    a. Dilute 1 µg of plasmid DNA (for transfection of cells seeded in a well of a 24-well plate) in 50 µl of Opti-MEM serum-free cell culture medium.
    b. Dilute the appropriate quantity of PEI or peptide-conjugated PEI in 50 µl of Opti-MEM serum-free cell culture medium.

c. For ternary complexes, dilute the appropriate quantity of the targeting peptide in 50 μl of Opti-MEM serum-free cell culture medium.

13. Add the peptide-conjugated PEI solution dropwise into the DNA solution while vortexing. *Or*, for ternary complexes, add the targeting peptide dropwise into the DNA solution while vortexing.

14. Incubate the mixture for 30 minutes at room temperature.

    The peptide-conjugated PEI/DNA complexes may now be used directly for transfection (Step 16).

15. To form ternary complexes, add the PEI solution dropwise into the targeting peptide/DNA complexes while vortexing. Incubate the ternary complexes for 30 minutes at room temperature.

    The ternary complexes may now be used directly for transfection (Step 16).

Transfection Assay

16. The day before transfection, harvest mammalian cells, grown in DMEM complete cell culture medium with 10% FBS, by trypsinization and replate them into 24-well plates at density of 50,000 cells/well. Incubate the cultures for 24 hours at 37°C in a humidified incubator with 5% CO_2.

17. Remove the medium from the wells and wash the cells twice with PBS (prewarmed to 37°C).

18. Add the following to the cells:

    100–150 μl/well of Opti-MEM serum-free cell culture medium

    100–150 μl/well of the gene transfection complex containing 1 μg of plasmid DNA (Step 14 or 15).

    Incubate the cells for 4 hours at 37°C in a humidified incubator with 5% CO_2.

19. Remove the transfection solution. Wash the cells twice with PBS (prewarmed to 37°C) and add 1 ml per well of DMEM complete cell culture medium with 10% FBS. Incubate the cells for 24–48 hours.

20. To assay for luciferase expression, lyse the cells by adding 100 μl per well of 1x reporter lysis buffer (dilute 5x stock solution with PBS).

21. To detect luciferase activity, add 20 μl of cell lysate to 100 μl of luciferase assay reagent. Measure luciferase activity with a luminometer.

22. To normalize the luciferase activity, determine the protein concentrations of the cell lysates using the D_c protein assay kit.

ACKNOWLEDGMENTS

This work was funded by the Institute of Bioengineering and Nanotechnology, Agency for Science, Technology, and Research (A* STAR), Singapore.

REFERENCES

Hildebrandt I.J., Ayer M., Wagner E., and Gambhir S.S. 2003. Optical imaging of transferrin targeted PEI/DNA complexes in living subjects. *Gene Ther.* **10:** 758–764.

Kunath K., Merdan T., Häberlein H., and Kissel T. 2003a. Integrin targeting using RGD-PEI conjugates for in vitro gene transfer. *J. Gene Med.* **5:** 588–599.

Kunath K., von Harpe A., Fischer D., and Kissel T. 2003b. Galactose-PEI-DNA complexes for targeted gene delivery: Degree of substitution affects complex size and transfection efficiency. *J. Control. Release* **88:** 159–172.

Kursa M., Walker G.F., Roessler V., Ogris M., Roedl W., Kircheis R., and Wagner E. 2003. Novel shielded transferring-polyethylene glycol-polyethylenimine/DNA complexes for systemic tumor targeted gene transfer. *Bioconjug. Chem.* **14:** 222–231.

Ma N., Wu S.S., Ma Y.X., Wang X., Zeng J., Tong G., Huang Y., and Wang S. 2004. Nerve growth factor receptor-mediated gene transfer. *Mol. Ther.* **9:** 270–281.

Rudolph C., Plank C., Lausier J., Schillinger U., Muller R.H., and Rosenecker J. 2003. Oligomers of the arginine-rich motif of the HIV-1 TAT protein are capable of transferring plasmid DNA into cells. *J. Biol. Chem.* **278:** 11411–11418.

Strehblow C., Schuster M., Moritz T., Kirch H.C., Opalka B., and Petri J.B. 2005. Monoclonal antibody-polyethylenimine conjugates targeting Her-2/*neu* or CD90 allow cell-type specific nonviral gene delivery. *J. Control. Release* **102:** 737–747.

Tang G.P., Zeng J.M., Gao S.J., Ma Y.X., Shi L., Li Y., Too H.P., and Wang S. 2003. Polyethylene glycol modified polyethylenimine for improved CNS gene transfer: Effects of PEGylation extent. *Biomaterials* **24:** 2351–2362.

Zeng J. and Wang S. 2005. Enhanced gene delivery to PC12 cells by a cationic polypeptide. *Biomaterials* **26:** 679–686.

44 Cyclodextrin-containing Polycations for Nucleic Acid Delivery

Jeremy D. Heidel

Calando Pharmaceuticals, Pasadena, California 91107

ABSTRACT

Numerous nonviral systems have been developed for the delivery of nucleic acids to cultured cells and to particular cell types in vivo. These systems vary with regard to their toxicity, immunogenicity, and ability to target particular cell surface receptors and/or cell types. A class of linear cationic polymers containing the sugar β-cyclodextrin has been shown to be effective at delivering a variety of nucleic acids in vivo, including plasmid DNA, DNAzymes, and short interfering RNAs (siRNAs). These polymer–nucleic acid complexes (polyplexes) can be further modified to incorporate a targeting ligand such as transferrin to induce preferential uptake of polyplexes by cells expressing high levels of the cognate receptor. Here, background information is presented on these materials as well as procedures for their use in vitro and in vivo.

INTRODUCTION

The promise of nucleic acids (NAs) as therapeutics remains exciting yet largely unrealized. This untapped potential does not owe to a lack of understanding of the relevant molecular mechanisms of action or an inability to design and prepare appropriate NAs. Rather, it results from the paucity of materials able to overcome the numerous challenges in delivery of these nucleic acids to the cells and subcellular locales required for their function. Many researchers have made considerable strides in development and understanding of vehicles for delivery of NA therapeutics. These vehicles fall into two broad categories: virus-based and nonviral. These nonviral vectors are typically composed of one or more polar lipids, synthetic polymers, or combinations of these and are cationic in nature such that they interact with NAs via electrostatics.

Two of the most common cationic polymers studied for NA delivery, poly(ethyleneimine) (PEI) and poly-L-lysine (PLL), are discussed in detail in Chapters 42, 43, and 51 of this volume. In the mid 1990s, Dr. Mark Davis at the California Institute of Technology developed synthetic polycations based on cyclical oligomers of glucose called cyclodextrins. Already known to be well tolerated at high doses as drug solubilizers, cyclodextrins are functionalized and incorporated within polymers to give cyclodextrin-containing polycations (CDPs). When added to a NA solution, they spontaneously condense the NAs into CDP-NA complexes that facilitate cellular uptake (Gonzalez et al. 1999). Since their initial characterization, dozens of variations have been prepared that systematically examined the effect of several distinct CDP properties on their function, including carbohydrate size and type, charge center, and spacing between charge centers (Hwang et al. 2001; Popielarski et al. 2003; Reineke and Davis 2003a,b). Furthermore, modification of CDP termini to incorporate imidazole moieties (im-CDP) (Fig. 1) reduced intracellular acidification of complex-containing endosomes and facilitated intracellular NA release (Hwang 2001; Davis et al. 2004; Kulkarni et al. 2005).

These im-CDP-NA complexes ("polyplexes") are suitable for transfecting cultured cells and achieving NA function (e.g., transgene expression). However, like polyplexes made with PEI or PLL, they require incorporation of an additional component to endow stability in physiological media required for in vivo application. This component is the neutral polymer poly(ethylene glycol) (PEG). The β-cyclodextrin (β-CD) moieties within im-CDP allow PEG to be incorporated into the CDP-NA formulations noncovalently. Having a cup-like three-dimensional structure with a relatively hydrophobic interior, β-CD can form noncovalent inclusion complexes with sufficiently small hydrophobic molecules. Adamantane (AD) is a small molecule known to form relatively strong inclusion complexes with β-CD in aqueous solutions, and AD conjugates to one end of linear PEG_{5000} to give an AD-PEG conjugate. When AD-PEG is combined with im-CDP and added to an NA solution, the resulting ternary complexes, although still containing condensed NA that is protected from serum nuclease degradation, no longer aggregate when incubated in physiological media (Pun and Davis 2002).

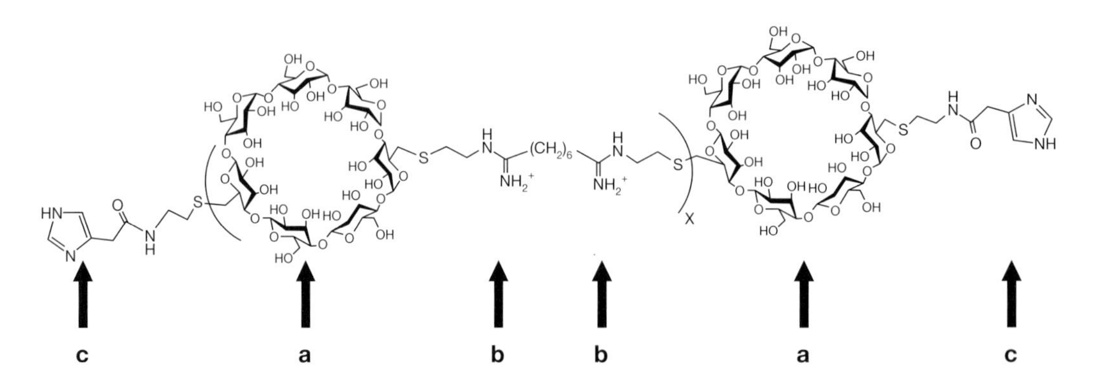

FIGURE 1. Structure and functions of imidazole-terminated CDP (im-CDP). β-cyclodextrin moieties (*a*) reduce toxicity and serve as sites for further modification. Amidine groups (*b*) provide positive charge needed to condense anionic nucleic acids. Terminal imidazole groups (*c*) buffer acidification of polyplex-containing endosomes and enhance intracellular nucleic acid release.

The final functionality incorporated within this NA delivery system is a targeting ligand to endow complexes with preferential uptake in the desired target cells or tissues. This is achieved by covalent attachment of a ligand such as a sugar or protein to the distal end (with respect to AD) of the AD-PEG conjugate to give an AD-PEG-Ligand molecule. Conjugates thus constructed containing lactose (Heidel 2005) and transferrin (Bellocq et al. 2003; Pun et al. 2004; Hu-Lieskovan et al. 2005) have successfully achieved uptake and/or NA function in vivo. Preparation of conjugates containing other possible types of ligands, including single-chain antibody fragments and full antibodies, is currently under study.

The use of CDP-NA polyplexes provides a number of advantages over traditional transfection methods. The condensation of NAs into complexes and modification of these polyplexes to be stable and targeted occur noncovalently; the fully formulated polyplexes form entirely by self-assembly (Fig. 2). Although the assembled polyplexes are too large to be cleared renally, each of the individual components (CDP, AD-PEG conjugates, nucleic acids) are small enough (~5 nm or below) for such clearance, reducing the possibility of deleterious effects of any uncomplexed material. Indeed, plasmid-containing polyplexes have been administered intravenously in mice at doses up to 550 mg/kg (40 mg/kg pDNA for 25-g mice) without any significant toxicity (Pun et al. 2004). In addition, unlike some lipid-based or virally derived vectors, polyplexes made with CDP have not exhibited any immunogenicity, even when containing siRNA possessing a putative immunostimulatory motif. This allows these polyplexes to be administered repeatedly; up to nine successive injections over 4 weeks have been performed with siRNA-containing polyplexes (Hu-Lieskovan et al. 2005). Finally, because only a single component (AD-PEG-Ligand) targets the polyplex to particular cell types, their uptake can be tuned simply by modifying the AD-PEG conjugate to contain the ligand of interest. The following protocols describe the preparation of im-CDP-NA complexes for both in vitro and in vivo applications.

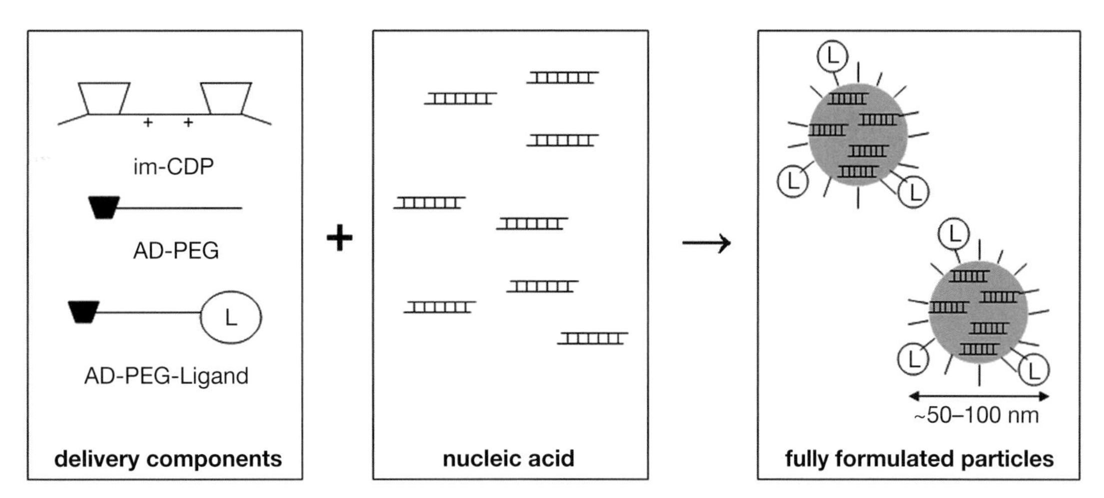

FIGURE 2. Self-assembly of fully formulated NA-containing particles. Addition of an aqueous solution of three synthetic components (im-CDP, AD-PEG, and AD-PEG-Ligand) to an aqueous NA solution results in the spontaneous formation of stable, targeted NA-containing particles by self-assembly.

Protocol 1

In Vitro Transfection Using CDPs

For in vitro transfection, salt stabilization and cell targeting are unnecessary, so polyplexes consist of CDP and nucleic acid only; no AD-PEG conjugates are included in these formulations. For typical plasmid DNA delivery experiments, up to 1 µg (5 µg for a six-well plate) of plasmid DNA is used per well. For typical siRNA delivery experiments, up to 20 nM siRNA (for each well of a 24-well plate, this equates to 53.2 ng of siRNA in a transfection volume of 200 µl) is used per well. For all in vitro delivery experiments, regardless of nucleic acid type or dose, CDP is mixed with the nucleic acid at a ratio of approximately 6:1 (w/w) or higher (corresponding to a charge ratio of 3/– +/– or higher).

MATERIALS

CAUTION: See Appendix for appropriate handling of materials marked with <!>.

Reagents

Complete growth medium (containing antibiotics and serum) as appropriate for the cell type of interest

Cyclodextrin-containing polycations (CDPs; 20 mg/ml)

Dissolve lyophilized solids in DNase-free, RNase-free H_2O. Stock solution should be colorless.

Nucleic Acids (80 µM)

DNAzymes (Eurogentec)

plasmid DNA (pDNA)

short interfering RNAs (siRNAs) (Dharmacon, Integrated DNA Technologies)

Amplify pDNA in DH5α bacteria (Invitrogen) and purify using Novagen's UltraMobius 1000 Kit. DNAzymes and siRNAs are prepared synthetically, purified by high-performance liquid chromatography (HPLC), and confirmed by gel electrophoresis. Dissolve nucleic acids in DNase-free, RNase-free H_2O or RNase-free potassium acetate buffer.

Opti-MEM I reduced-serum medium (Invitrogen)

Phosphate-buffered saline (PBS; sterile)

Trypsin <!>

Equipment

Cell culture plates (6 or 24 well)

Incubator, humidified (37°C, 5% CO_2)

METHOD

1. Twenty-four hours before transfection, harvest exponentially growing adherent cells by trypsinization. Transfer cells to a 6- or 24-well tissue-culture plate at a density of 5×10^4 cells/well (for 24-well plates) or 2.5×10^5 cells/well (for 6-well plates).
2. Add 1 ml (for 24-well plates) or 5 ml (for 6-well plates) of complete growth medium to each well. Incubate the plates for approximately 24 hours at 37°C in a humidified incubator with a 5% CO_2 atmosphere.

3. To prepare polyplex solutions, add 1 volume of CDP stock solution (typically 10 μl per well) to an equal volume of nucleic acid stock solution.

   For in vitro transfections, nucleic acid solutions more concentrated than 0.2 mg/ml can fail to condense completely upon addition of an equal volume of CDP solution. The resulting solutions are often hazy or milky and yield minimal gene expression, gene knockdown, etc.

4. Add 9 volumes of Opti-MEM to each polyplex volume (typically 180 μl of Opti-MEM per well of a 24-well plate).
5. Aspirate the complete growth medium from the well. Rinse gently with 0.5 ml of PBS.
6. Add 200 μl (for a 24-well plate) or 1000 μl (for a 6-well plate) per well of the polyplex/Opti-MEM solution. Return the plate to the incubator for 4 hours.
7. Aspirate the polyplex/Opti-MEM solution from the wells. Add 1 ml (for a 24-well plate) or 5 ml (for a 6-well plate) of complete growth medium.
8. At an appropriate time posttransfection for polyplex uptake, perform desired assays for gene expression, gene knockdown, etc.

Protocol 2

In Vivo Transfection Using CDPs

Salt stabilization and cell targeting are critical to the success of in vivo transfection using CDPs, so AD-PEG conjugates (both unmodified AD-PEG and an AD-PEG-Ligand conjugate) are included in these formulations. The amount of the AD-PEG-Ligand conjugate included depends on numerous factors, including its effect on polyplex stability (influenced by ligand size and charge) and the density of the cognate receptor on target cell type(s). Some targeting ligands may have extreme sizes or net charges that could present a challenge to their incorporation into these polyplex formulations. The typical dose of nucleic acid per subject depends on the nature of the nucleic acid and the type of experiment being performed. For typical intravenous delivery experiments in mice, up to 5 mg/kg (100 µg per 20-g mouse) pDNA is used per animal. Polyplexes containing DNAzyme and siRNA doses of up to 50 mg/kg (1 mg per 20-g mouse) and 5 mg/kg (100 µg per 20-g mouse), respectively, have been administered without adverse effects. A maximum tolerable dose has yet to be observed for these nucleic-acid-containing polyplex formulations.

MATERIALS

Reagents

Adamantane-poly(ethylene glycol) conjugates (50 mg/ml)
- AD-PEG, lactose-terminated (AD-PEG-Lac)
- AD-PEG, ligand-free (AD-PEG)
- AD-PEG, transferrin-conjugated (AD-PEG-Tf)

Dissolve lyophilized solids in DNase-free, RNase-free H_2O. Stock solutions should either be colorless (AD-PEG and AD-PEG-Lac) or red/orange (AD-PEG-Tf).

Cyclodextrin-containing polycations (CDPs; 20 mg/ml)

Dissolve lyophilized solids in DNase-free, RNase-free H_2O. Stock solution should be colorless.

Dextrose (D-glucose) (10% w/v in water [$D_{10}W$])

Dissolve dextrose in DNase-free, RNase-free H_2O. Sterilize by filtration through a 0.2-µm filter. For all in vivo delivery experiments, administer polyplexes in a final solution of 5% dextrose in H_2O.

Mice (20 g)

Nucleic acids (80 µM)
- DNAzymes (Eurogentec)
- plasmid DNA (pDNA)
- short interfering RNAs (siRNAs) (Dharmacon, Integrated DNA Technologies)

Amplify pDNA in DH5α bacteria (Invitrogen) and purify using Novagen's UltraMobius 1000 Kit. DNAzymes and siRNAs are prepared synthetically, purified by HPLC, and confirmed by gel electrophoresis. Dissolve nucleic acids in DNase-free, RNase-free H_2O or RNase-free potassium acetate buffer.

Equipment

Needles (27 gauge)

Syringes (1 cc)

METHOD

1. To prepare a mixture of synthetic components, combine CDP, AD-PEG, and AD-PEG-Ligand in H_2O at a final molar ratio of AD: β-CD of 1:1.
2. To prepare the polyplex solution, add 1 volume (~50 μl/20-g mouse) of synthetic component solution to an equal volume of nucleic acid stock solution.
3. Add 1 volume (~100 μl/20-g mouse) of sterile $D_{10}W$ to an equal volume of polyplex solution.
4. Load the final polyplex solution into a sterile 1-cc syringe affixed to a 27-gauge needle.
5. Prewarm the tail vein of the recipient mouse.
6. Inject the final polyplex solution into the tail vein over 2–3 seconds.

ACKNOWLEDGMENTS

I would like to acknowledge Dr. Mark Davis at Caltech, the principal inventor of these polycations and AD-PEG conjugates, in whose laboratory I learned how to prepare, characterize, and work with these materials. I also thank previous and current members of Dr. Davis' lab, including Suzie Pun, Nathalie Bellocq, Theresa Reineke, Swaroop Mishra, and Derek Bartlett, for their contributions to this body of work.

REFERENCES

Bellocq N.C., Pun S.H., Jensen G.S., and Davis M.E. 2003. Transferrin-containing, cyclodextrin polymer-based particles for tumor-targeted gene delivery. *Bioconjug. Chem.* **14:** 1122–1132.

Davis M.E., Pun S.H., Bellocq N.C., Reineke T.M., Popielarski S.R., Mishra S., and Heidel J.D. 2004. Self-assembling nucleic acid delivery vehicles via linear, water-soluble, cyclodextrin-containing polymers. *Curr. Med. Chem.* **11:** 1241–1253.

Gonzalez H., Hwang S.J., and Davis M.E. 1999. New class of polymers for the delivery of macromolecular therapeutics. *Bioconjug. Chem.* **10:** 1068–1074.

Heidel J.D. 2005. "Targeted, systemic non-viral delivery of small interfering RNA in vivo." Ph.D. thesis. California Institute of Technology, Pasadena.

Hu-Lieskovan S., Heidel J.D., Bartlett D.W., Davis M.E., and Triche T.J. 2005. Sequence-specific knockdown of EWS-FLI1 by targeted, nonviral delivery of small interfering RNA inhibits tumor growth in a murine model of metastatic Ewing's sarcoma. *Cancer Res.* **65:** 8984–8992.

Hwang S.J. 2001. "Rational design of a new class of cyclodextrin-containing polymers for gene delivery." Ph.D. thesis. California Institute of Technology, Pasadena.

Hwang S.J., Bellocq N.C., and Davis M.E. 2001. Effects of structure of β-cyclodextrin-containing polymers on gene delivery. *Bioconjug. Chem.* **12:** 280–290.

Kulkarni R.P., Mishra S., Fraser S.E., and Davis M.E. 2005. Single cell kinetics of intracellular, nonviral, nucleic acid delivery vehicle acidification and trafficking. *Bioconjug. Chem.* **16:** 986–994.

Popielarski S.R., Mishra S., and Davis M.E. 2003. Structural effects of carbohydrate-containing polycations on gene delivery. 3. Cyclodextrin type and functionalization. *Bioconjug. Chem.* **14:** 672–678.

Pun S.H. and Davis M.E. 2002. Development of a nonviral gene delivery vehicle for systemic application. *Bioconjug. Chem.* **13:** 630–639.

Pun S.H., Tack F., Bellocq N.C., Cheng J., Grubbs B.H., Jensen G.S., Davis M.E., Brewster M., Janicot M., Janssens B., Floren W., and Bakker A. 2004. Targeted delivery of RNA-cleaving DNA enzyme (DNAzyme) to tumor tissue by transferrin-modified, cyclodextrin-based particles. *Cancer Biol. Ther.* **3:** 641–650.

Reineke T.M. and Davis M.E. 2003a. Structural effects of carbohydrate-containing polycations on gene delivery. 1. Carbohydrate size and its distance from charge centers. *Bioconjug. Chem.* **14:** 247–254.

Reineke T.M. and Davis M.E. 2003b. Structural effects of carbohydrate-containing polycations on gene delivery. 2. Charge center type. *Bioconjug. Chem.* **14:** 255–261.

45 Bionanocapsules Using the Hepatitis B Virus Envelope L Protein

Tadanori Yamada,*† Joohee Jung,* Masaharu Seno,†‡ Akihiko Kondo,†§ Masakazu Ueda,†¶ Katsuyuki Tanizawa,*† and Shun'ichi Kuroda*†

**Department of Structural Molecular Biology, Osaka University, Ibaraki, 567-0047, Japan; †Beacle Inc., Okayama 701-1221, Japan; ‡Graduate School of Natural Science and Technology, Okayama University, Okayama 700-8530, Japan; §Faculty of Engineering, Kobe University, Kobe, Hyogo 657-8501, Japan; ¶Keio University, School of Medicine, Tokyo 160-8582, Japan*

ABSTRACT

Hepatitis B virus (HBV) envelope L proteins, when synthesized in yeast cells, form a hollow bionanocapsule (BNC) in which genes (including large plasmids up to 40 kbp), small interfering RNA (siRNA), drugs, and proteins can be enclosed by electroporation. BNCs made from L proteins have several advantages as a delivery system: Because they display a human liver-specific receptor (the pre-S region of the L protein) on their surface, BNCs can efficiently and specifically deliver their contents to human liver-derived cells and tissues ex vivo (in cell culture) and in vivo (in a mouse xenograft model). Retargeting can be achieved simply by substituting other biorecognition molecules such as antibodies, ligands, receptors, and homing peptides for the pre-S region. In addition, BNCs have already been proven to be essentially safe for use in humans during their development as an immunogen of hepatitis B vaccine.

BNCs do have several limitations. First, electroporation of BNCs is not always efficient. We recently have found, however, that liposome-containing materials fuse efficiently with BNCs (J. Jung et al., in prep.). Second, wild-type BNCs can infect only human and chimpanzee liver. To facilitate in vivo assays, we are now establishing transgenic rats expressing HBV receptor in their liver. Finally, although BALB/c mice did not exhibit an anti-BNC antibody response after intravenous administration of BNCs (three times, 2-week intervals, 10 μg, without adjuvant), low-immunogenic and low-antigenic BNCs should be developed for long-term and repetitive administration.

INTRODUCTION

Viral vectors, including adenoviruses, adeno-associated viruses, and retroviruses, have been used widely for gene therapy and medical science, because they display more efficient gene transfer than most nonviral methods. A major drawback of current viral vectors, however, is the lack of specificity in targeting. This targeting problem can result in unexpected side effects such as the production of neutralizing antibodies, inflammation, germ-line gene transfer, and oncogenesis. Additionally, the mass production of viral vectors for clinical use is time-consuming and labor-intensive, and it poses liability considerations for most manufacturers. These disadvantages prompted us to develop a gene-transfer method that does not involve viral genomes, is easily adapted to mass production, and maintains high transfection efficiency and target specificity.

HBV is a human liver-specific virus that includes the S, M, and L envelope (Env) proteins. Around 1990, HBV Env proteins were produced in yeast cells as BNCs, and the S and M particles were used as immunogens in hepatitis B vaccines that were proven to be safe for humans (Kuroda et al. 1989, 1991). The L particles, which are much more efficiently synthesized in yeast cells than S and M particles, have an average diameter of 100 nm with no HBV genome inside (Kuroda et al. 1992; Yamada et al. 2001). The amino-terminal region of the L protein, specifically amino acid residues 3–77 (HBV [ayw]) or 14–88 (HBV [adr]), is called the pre-S region. It functions as a human liver-specific receptor and is indispensable for HBV infectivity. We have taken advantage of its presence on the surface of L particle BNCs to deliver both genes and drugs with high ex vivo and in vivo transfection efficiencies and precise targeting to human liver cells (Fig. 1) (Yamada et al. 2003).

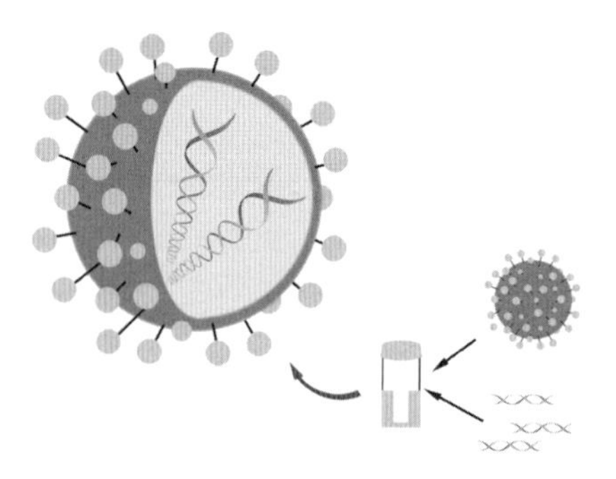

FIGURE 1. Introduction of nucleic acids into BNCs by electroporation.

Protocol

Electroporation and Use of L Particles as Bionanocapsules

MATERIALS

Reagents

Materials to be incorporated into the BNCs

Possible materials include 20 μg of mammalian expression plasmid for green fluorescent protein (e.g., pEGFP; Clontech), 20 μg of FITC-labeled dextran (m.w. 3,000–10,000; Molecular Probes), 20 μg of FITC-labeled polystyrene beads (100- or 40-nm diameter; FluoSpheres, Molecular Probes), 20 μg of purifed GFP, or 1 mM Calcein (Sigma-Aldrich).

Negative control cells (either nonhuman or nonliver; e.g., WiDr, A431)

Phosphate-buffered saline (PBS) (for recipe, see Sambrook and Russell 2001)

Purified BNC solution

Synthesize BNCs up to approximately 40% (w/w) of total soluble protein in *Saccharomyces cerevisiae* strain AH22R$^-$, which harbors the expression plasmid for HBV Env L protein (pGLDLIIP39-RcT) and has been described by Kuroda et al. (1992). Purify the BNCs from crude extract using ultracentrifugation as described by Yamada et al. (2001). Filter the solution (~200–300 μg of protein/ml in PBS) with sterile 0.45-μm pore discs and store at 4°C. Repeated freeze-thaw cycles will destroy the capsule structure of the BNCs.

Target cells: Human liver-derived cells (e.g., HepG2, HuH-7, NuE)

Xenograft mouse model: BALB/c nude mice (nu/nu, 10–11 weeks old, male) bearing tumors (NuE and WiDr, ~1 cm in diameter)

Equipment

Electroporation cuvettes (4-mm gap)

Electroporation system (Gene Pulser II, Bio-Rad)

Fluorescence microscope

METHODS

Loading the BNCs

1. Mix the material to be incorporated into the BNCs (see Reagents) with purified BNCs (100 μg of protein) in 500 μl of PBS.
2. Transfer the mixture to an electroporation cuvette with a 4-mm gap and electroporate with a Gene Pulser II electroporation system. Settings are typically 220 volts and 950 μF for about 20 msec.

   It is not necessary to remove unloaded materials from the BNC preparation after electroporation.

Ex Vivo Experiments with BNCs

3. Add the mixture of electroporated BNCs directly to a culture of approximately 5×10^4 human liver-derived cells in the presence of 10% FBS. For DNA-loaded BNCs, add <100 ng of BNCs that contain 20 ng of plasmid. Observe GFP-derived fluorescence with a fluorescence microscope 3 days after transfection. For BNCs loaded with non-DNA materials, use <10 μg of electroporated BNCs and observe fluorescence on day 1 after transfection.

In Vivo Experiments with BNCs

4. Inject the electroporated BNCs (<100 μg in <100 μl) containing plasmid or non-DNA materials intravenously into the tail vein of a xenografted mouse. Observe fluorescence of tumors and tissues with a fluorescence microscope 3–7 days after injection.

REFERENCES

Kuroda S., Fujisawa Y., Iino S., Akahane Y., and Suzuki H. 1991. Induction of protection level of anti-pre-S2 antibodies in humans immunized with a novel hepatitis B vaccine consisting of M (pre-S2 + S) protein particles (a third generation vaccine). *Vaccine* **9:** 163–169.

Kuroda S., Itoh Y., Miyazaki T., Otaka-Imai S., and Fujisawa Y. 1989. Efficient expression of genetically engineered hepatitis B virus surface antigen P31 proteins in yeast. *Gene* **78:** 297–308.

Kuroda S., Otaka S., Miyazaki T., Nakao M., and Fujisawa Y. 1992. Hepatitis B virus envelope L protein particles: Synthesis and assembly in *Saccharomyces cerevisiae*, purification and characterization. *J. Biol. Chem.* **267:** 1953–1961.

Sambrook J. and Russell D. 2001. *Molecular cloning: A laboratory manual*, 3rd edition. Cold Spring Harbor Laboratory Press, Cold Spring Harbor, New York.

Yamada T., Iwabuki H., Kanno T., Tanaka H., Kawai T., Fukuda H., Kondo A., Seno M., Tanizawa K., and Kuroda S. 2001. Physicochemical and immunological characterization of hepatitis B virus envelope particles exclusively consisting of the entire L (pre-S1 + pre-S2 + S) protein. *Vaccine* **19:** 3154–3163.

Yamada T., Iwasaki Y., Tada H., Iwabuki H., Chuah M.K.L., VandenDriessche T., Fukuda H., Kondo A., Ueda M., Seno M., Tanizawa K., and Kuroda S. 2003. Nanoparticles for the delivery of genes and drugs to human hepatocytes. *Nat. Biotechnol.* **21:** 885–890.

46 Formulations of Solid Lipid Nanoparticles for Transfection of Mammalian Cells In Vitro

Carsten Rudolph and Joseph Rosenecker

Department of Pediatrics, Ludwig-Maximilians University, 80337 Munich, Germany

ABSTRACT

Solid lipid nanoparticles (SLNs) offer several technological advantages over standard DNA carriers such as cationic lipids or cationic polymers. However, in the absence of endosomolytic agents such as chloroquine, gene-transfer efficiency mediated by SLN-derived gene vectors consisting of optimized lipid composition remains lower compared to those achieved with standard transfection agents. This protocol describes the incorporation of a dimeric human immunodeficiency virus type-1 (HIV-1) TAT peptide into SLN gene vectors to increase gene-transfer efficiency. This results in higher transfection rates than for standard transfection agents in vitro, but with lower toxicity.

INTRODUCTION

SLNs were invented at the beginning of the 1990s and were produced either by high-pressure homogenization or by microemulsion techniques. From the point of view of production and regulatory aspects, high-pressure homogenization is considered as the method of choice. SLNs consist of a solid matrix. SLNs are comparable with parenteral emulsions, except that in SLN formulations, the liquid lipid (oil) is replaced by solid lipid. Cationic SLNs condense DNA into nanometric colloidal particles capable of transfecting mammalian cells in vitro (Olbrich et al. 2001; Tabatt et al. 2004). In these studies, cationic SLNs were produced by hot homogenization using either Compritol ATO 888 (a mixture of mono-, di-, and triglycerides of behenic acid) or paraffin as matrix lipid, a mixture of Tween-80 and Span 85 as tenside and either EQ1 (*N,N*-di-(β-steaorylethyl)-*N,N*-dimethylammonium chloride) or cetylpyridinium chloride as the charge carrier. More detailed analyses of cationic lipid and matrix lipid composition of SLNs revealed that plasmid DNA binding, cytotoxicity, and transfection efficiency were dependent on the structure of the cationic lipid and the matrix lipid. Whereas SLNs made from two-tailed cationic lipids were well-tolerated in cell culture, SLNs made from one-tailed cationic detergents were highly toxic. These studies revealed that optimal SLN formulations for gene transfer were made from 1,2-dioleoyl-*sn*-glycero-3-trimethylammoniumpropane (DOTAP) as the cationic lipid and cetylpalmitate as the matrix lipid.

Compared with standard DNA carriers such as cationic lipids or cationic polymers, SLNs offer several technological advantages. They are relatively easy to produce without the use of organic solvents (Mehnert and Mader 2001), and large-scale production is possible with qualified production lines (Müller et al. 2000b; Dingler and Gohla 2002). They exhibit good stability during long-term storage (Freitas and Müller 1999) and are amenable to both steam sterilization (Schwarz et al. 1994) and lyophilization (Schwarz and Mehnert 1997). Perhaps most importantly, as substances that are generally recognized as safe (Müller et al. 2000a), they are less toxic (Olbrich et al. 2001) than cationic polymers such as polyethylenimine (PEI) (Bragonzi et al. 2000; Gebhart and Kabanov 2001). However, in the absence of endosomolytic agents such as chloroquine, gene-transfer efficiency mediated by SLN-derived gene vectors (even when the lipid composition is optimized) remains lower than those observed with standard transfection agents such as PEI 25 kD (Olbrich et al. 2001; Tabatt et al. 2004).

Recent studies have shown that precompaction of DNA with oligomers of the HIV-1 TAT peptide for the formulation of gene vector complexes led to an increase of up to two orders of magnitude in gene-transfer efficiency (Rudolph et al. 2003). The dimeric TAT peptide was found to be most efficient. This effect was related to the unique features of the HIV-1 TAT peptide which represents a protein transduction domain (Frankel and Pabo 1988; Fawell et al. 1994) and a nuclear localization sequence (Truant and Cullen 1999). This protocol describes the formulation of ternary gene vector complexes consisting of DNA precompacted with a dimeric TAT peptide (TAT_2), which is then completed by the addition of a cationic SLN gene carrier. When the plasmid DNA is first complexed with the TAT_2 peptide under the given conditions (corresponding to a charge ratio of ±1) the resulting intermediate complexes have a zeta potential of approximately –20 mV. The zeta potentials are measured electrophoretically (ZetaPALS/Zeta Potential Analyzer; Brookhaven Instruments Corporation, Austria). The following settings were used: 10 subrun measurements/sample; viscosity for H_2O 0.89 cP; beam mode F(Ka) = 1.50 (Smoluchowsky); temperature 25°C. These preformed negatively charged TAT_2-plasmid DNA (pDNA) gene vector complexes allow the binding of positively charged SLNs to their surface through electrostatic interaction. The formulation method apparently results in a shell-like ternary complex with the TAT_2 peptide bound to the plasmid DNA in the core of the complex and a layer of SLN at the periphery of the complex. Therefore, our experience is that the pH of the solutions is important to result in negatively charged TAT_2-pDNA complexes. Such gene vectors increase gene-transfer efficiency, resulting in higher transfection rates than for standard transfection agents in vitro, but with lower toxicity.

Protocol

Formation of SLN-Gene Vector Complexes for In Vitro Transfection

The ternary SLN-gene vector complexes usually result in transfection levels equal to or higher than those observed with gene vector complexes formulated with branched PEI 25 kD. One significant advantage of using this method is the low cytotoxicity of the SLN gene vectors. The application of the gene-transfer technique is limited to relatively low pDNA concentrations of the resulting complexes (10 μg/ml). At higher concentrations, the particles tend to aggregate and precipitate. Therefore, their use for in vivo application, which generally requires high pDNA concentrations, is limited. A schematic drawing of the ternary gene vector formulation is shown in Figure 1.

MATERIALS

CAUTION: See Appendix for appropriate handling of materials marked with <!>.

Reagents

Cell line to be transfected
Cell growth medium (serum-free), as appropriate for cell line of interest
Fetal calf serum (FCS)
Gentamycin (Invitrogen 15710–049)
HBS (HEPES-buffered saline)
  150 mM NaCl
  10 mM HEPES (pH 7.4)
Penicillin/streptomycin (GIBCO 15140–122) <!>
Plasmid of interest
Solid lipid nanoparticles (SLNs)
  Cetylpalmitate (Henkel, Düsseldorf, Germany)
  DOTAP (Sigma-Aldrich)
  Span 85 (ICI Surfactants, Eversberg, Belgium)
  Tween-80 (ICI Surfactants, Eversberg, Belgium)

SLNs are produced by hot high-pressure homogenization as described previously (Müller et al. 2000b). Briefly, the solid lipids are heated to approximately 10°C above their melting points.

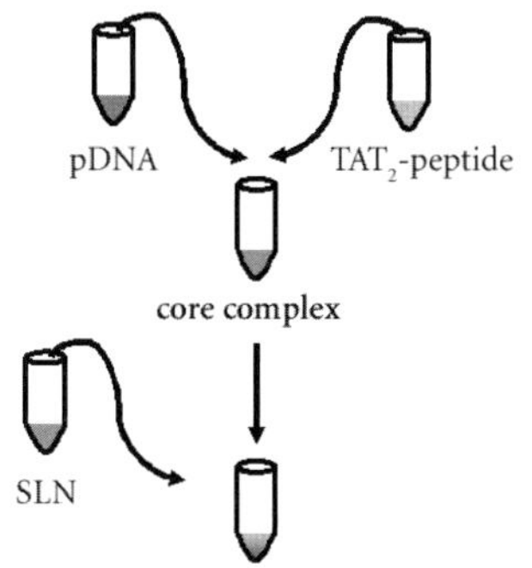

FIGURE 1. Formulation of ternary SLN-gene vectors. The desired plasmid DNA is added to the TAT_2 peptide solution, mixed, and incubated for 10 minutes at room temperature before the solution containing the SLNs is added.

The surfactants Tween-80 and Span 85 are mixed in a 7:3 ratio. The surfactant mix is combined (2% w/w) with 1% (w/w) cationic DOTAP in a hot aqueous solution. Molten matrix lipid (cetylpalmitate, 4% w/w) and the hot aqueous solution of surfactants and DOTAP are combined into a preemulsion by stirring with a high-speed stirrer for 1 minute. Batches of SLNs containing the DOTAP are homogenized at 85°C at a pressure of 480 bar and four homogenization cycles using a high-pressure homogenizer.

TAT_2 peptide

The TAT_2 peptide comprises the following sequence, $C(YGRKKRRQRRRG)_2$, containing the arginine-rich motif of the HIV-1 TAT protein. The TAT_2 peptide can be synthesized on an automatic synthesizer according to a standard Fmoc protocol, purified by reversed-phase HPLC and analyzed by mass spectroscopy. The free sulfhydryl groups are modified by dithiodipyridin reaction (Plank et al. 1999). This modification is important to avoid disulfide formation of the TAT_2 peptide in aqueous solution.

Equipment

Automatic synthesizer (431 A; Applied Biosystems)
Cell culture plates (24 well)
High-pressure homogenizer (EmulsiFlex-B3; Avestin Inc., Ottawa, Canada)
High-speed stirrer (Ultra Turrax T25; Jahnke & Kunkel, Germany)
Incubator (37°C, 5% CO_2)
Reversed-phase high-performance liquid chromatography (HPLC)

METHODS

Preparation of SLN-Gene Vector Complexes

1. For each well of a 24-well plate, prepare the following solutions.
   a. Dilute 1 μg of pDNA in HBS to a final volume of 50 μl.
   b. Dilute 0.65 μg of TAT_2 peptide in HBS to a final volume of 50 μl.
   c. Dilute 2.5 μg of SLN in HBS to a final volume of 50 μl.
2. To prepare the core complex, add the plasmid DNA solution to the TAT_2 solution. Mix vigorously by pipetting up and down 10 times. Incubate for 10 minutes at room temperature.
3. To prepare the ternary complex, add the SLN solution to the core complex. Mix by pipetting up and down 10 times. Incubate for 10 minutes at room temperature.

In Vitro Transfection

4. One day prior to transfection, seed the cells to be transfected in a 24-well plate in cell growth medium supplemented with FCS.
   Cell should be 60–70% confluent at the time of transfection.
5. Aspirate the complete cell growth medium from each well. Replace with 850 μl of serum-free medium before transfection.
6. Add 150 μl of SLN-gene vector ternary complex solution (from Step 3) to the cells. Incubate for 4 hours at 37°C in a 5% CO_2 atmosphere.
7. Aspirate the transfection medium. Replace with cell growth medium supplemented with 10% FCS and 0.1% (v/v) penicillin/streptomycin and 0.5% (v/v) gentamycin.
8. Measure the gene-transfer efficiency at the desired timepoint after transfection.

ACKNOWLEDGMENTS

The authors would like to acknowledge Dr. Kerstin Tabatt and Prof. Dr. R.H. Müller for providing the SLN formulations and the Deutsche Forschungsgemeinschaft Ro994/2–1 for funding of the work.

REFERENCES

Bragonzi A., Dina G., Villa A., Calori G., Biffi A., Bordignon C., Assael B.M., and Conese M. 2000. Biodistribution and transgene expression with nonviral cationic vector/DNA complexes in the lungs. *Gene Ther.* **7:** 1753–1760.

Dingler A. and Gohla S. 2002. Production of solid lipid nanoparticles (SLN): Scaling up feasibilities. *J. Microencapsul.* **19:** 11–16.

Fawell S., Seery J., Daikh Y., Moore C., Chen L.L., Pepinsky B., and Barsoum J. 1994. Tat-mediated delivery of heterologous proteins into cells. *Proc. Natl. Acad. Sci.* **91:** 664–668.

Frankel A.D. and Pabo C.O. 1988. Cellular uptake of the tat protein from human immunodeficiency virus. *Cell* **55:** 1189–1193.

Freitas C. and Müller R.H. 1999. Correlation between long-term stability of solid lipid nanoparticles (SLN) and crystallinity of the lipid phase. *Eur. J. Pharm. Biopharm.* **47:** 125–132.

Gebhart C.L. and Kabanov A.V. 2001. Evaluation of polyplexes as gene transfer agents. *J. Control. Release* **73:** 401–416.

Mehnert W. and Mader K. 2001. Solid lipid nanoparticles: Production, characterization and applications. *Adv. Drug Delivery Rev.* **47:** 165–196.

Müller R.H., Mader K., and Gohla S. 2000a. Solid lipid nanoparticles (SLN) for controlled drug delivery—A review of the state of the art. *Eur. J. Pharm. Biopharm.* **50:** 161–177.

Müller R.H., Dingler A., Schneppe T., and Gohla S. 2000b. Large scale production of solid lipid nanoparticles (SLN™) and nanodispersions (DissoCubes™). In *Handbook of pharmaceutical controlled release technology* (ed. D. Wise), pp. 359–376. Marcel Dekker, New York.

Olbrich C., Bakowsky U., Lehr C.M., Müller R.H., and Kneuer C. 2001. Cationic solid-lipid nanoparticles can efficiently bind and transfect plasmid DNA. *J. Control. Release* **77:** 345–355.

Plank C., Tang M.X., Wolfe A.R., and Szoka F.C., Jr. 1999. Branched cationic peptides for gene delivery: Role of type and number of cationic residues in formation and in vitro activity of DNA polyplexes. *Hum. Gene Ther.* **10:** 319–332.

Rudolph C., Plank C., Lausier J., Schillinger U., Müller R.H., and Rosenecker J. 2003. Oligomers of the arginine-rich motif of the HIV-1 TAT protein are capable of transferring plasmid DNA into cells. *J. Biol. Chem.* **8:** 11411–11418.

Schwarz C. and Mehnert W. 1997. Freeze-drying of drug-free and drug-loaded solid lipid nanoparticles (Sln). *Int. J. Pharm.* **157:** 171–179.

Schwarz C., Mehnert W., Lucks J.S., and Müller R.H. 1994. Solid lipid nanoparticles (Sln) for controlled drug delivery. I. Production, characterization and sterilization. *J. Control. Release* **30:** 83–96.

Tabatt K., Sameti M., Olbrich C., Müller R.H., and Lehr C.M. 2004. Effect of cationic lipid and matrix lipid composition on solid lipid nanoparticle-mediated gene transfer. *Eur. J. Pharm. Biopharm.* **57:** 155–162.

Truant R. and Cullen B.R. 1999. The arginine-rich domains present in human immunodeficiency virus type 1 Tat and Rev function as direct importin β-dependent nuclear localization signals. *Mol. Cell. Biol.* **19:** 1210–1217.

47 PEGylated Poly-L-lysine DNA Nanoparticles

Pamela B. Davis* and Tomasz H. Kowalczyk†

*Department of Pediatrics, Case Western Reserve University School of Medicine, Cleveland, Ohio 44106; †Copernicus Therapeutics, Inc., Cleveland, Ohio 44016

ABSTRACT

PEGylated poly-L-lysine DNA nanoparticles are soluble and stable in saline and tissue fluids, transfect nondividing cells (Liu et al. 2003), display minimal toxicity (Ziady et al. 2003b), and are effective in vivo (Ziady et al. 2003a) and in humans (Konstan et al. 2004). They are easy to prepare in a reliable and reproducible fashion. These properties represent a substantial advance for nonviral gene transfer.

INTRODUCTION

DNA nanoparticles composed of plasmid DNA compacted with polylysine conjugated with polyethylene glycol (PEG) are easy to prepare in a reproducible fashion, because they self-assemble from purified components. Compaction protects the plasmid DNA from degradation. The nanoparticles are stable for years at 4°C, for months at room temperature, and for days at body temperature. They can be lyophilized in the presence of a cryoprotectant or aerosolized with the proper equipment and reconstituted without loss of activity.

PEGylated DNA nanoparticles, without any targeting moieties, transfect the airway epithelium, the retina, and neural tissue in vivo. Cells derived from these tissues and grown in primary culture are also susceptible to transfection by these particles, although most immortalized cell lines are not, which limits experimentation in vitro. The route of entry into cells is not established, but once in the cells, the nanoparticles can enter the nucleus of nondividing cells and promote transgene expression. This property appears to depend on the minor diameter of the particles being less than that of the nuclear pore (Liu et al. 2003). In addition to their favorable physical properties, stability, and ability to transfect nondividing cells, DNA nanoparticles demonstrate remarkably low toxicity in vivo (Ziady et al. 2003b). Large doses of nanoparticles are needed to compensate for their relative inefficiency as gene-transfer reagents, so a lack of toxicity is especially critical for their successful use in gene therapy.

Protocol

Preparation and Analysis of PEGylated Poly-L-lysine DNA Nanoparticles

This protocol describes the conjugation of methoxy-PEG-maleimide with the peptide CK_{30} and the compaction of DNA with the resultant PEGylated polylysine. It also describes the analyses used to check the morphology and colloidal stability of the nanoparticles. These assays should be performed each time the nanoparticles are prepared, because although the compaction procedure is very reproducible, variations in product quality do sometimes occur (i.e., the particles are unstable or have an unacceptable morphology). Variations seem to happen most often when the source of plasmid or method of plasmid production is changed.

MATERIALS

CAUTION: See Appendix for appropriate handling of materials marked with <!>.

Reagents

CK_{30}

This peptide, which contains an amino-terminal cysteine and 30 lysine residues, is custom-synthesized as a trifluoroacetate (TFA) salt by UCB Bioproducts using solid-phase methods and analyzed by high-performance liquid chromatography (HPLC), mass spectrometry, and quantitative amino acid analysis. HPLC routinely demonstrates >95% purity (CK_{28}–CK_{31} content) and <1% of peptide dimer. When the lyophilized peptide is stored under argon at –20°C for up to 1 year, it is >90% in monomeric form based on a fast protein liquid chromatography Resource S profile and a quantitative 4,4′-dithiodipyridine assay.

Dextrose (5%) or NaCl (0.9%) (see Step 8)

Dimethylsulfoxide (DMSO) <!>

Methoxy-PEG-maleimide (mPEG-MAL-10k) (10 kD)

mPEG-MAL-10k is available from Nektar Therapeutics (formerly Shearwater Polymers). Its molecular mass (+/– 1 kD) and polydispersity are confirmed by gel filtration. More than 80% of the PEG molecules should contain functional maleimide groups, as assayed by ^{1}H-NMR (nuclear magnetic resonance), and impurities detected by gel filtration (PEG dimer) should constitute <10% of the weight. Stabilization of compacted DNA can also be achieved with 5- or 20-kD PEG.

Phosphate-buffered saline (PBS) (0.1 M, pH 7.2)/EDTA (5 mM) (for PBS recipe, see Sambrook and Russell 2001)

Plasmid DNA (0.2 mg/ml in H_2O)

Grow plasmids in bacteria and purify by double equilibrium centrifugation in CsCl–ethidium bromide gradients after alkaline lysis. Precipitate DNA twice using 0.1 volume of 3 M sodium acetate and 2.5 volumes of ethanol and resuspend in H_2O. Determine the concentration by absorption spectroscopy. Treat the DNA preparation twice with RNase A <!> and RNase T1 and resuspend to a final concentration of 1.5–2 mg/ml. Ensure that no substantial contamination with bacterial genomic DNA or RNA is present in the plasmid preparations. Plasmid preparations should contain <30% open circular or linear DNA.

Sterile H_2O for injection

This is available commercially from Baxter. It is used to prepare solutions of DNA and PEGylated peptide.

Trifluoroacetic acid (0.1%) or 50 mM ammonium acetate (see Step 5) <!>

Equipment

Agarose gel and electrophoresis apparatus
Dialysis equipment
Sephadex G15 column
Spectrophotometer
Transmission electron microscope

METHODS

Preparation of PEGylated Polylysine (CK_{30}PEG10k)

1. Perform a 4,4′-dithiodipyridine assay (Grassetti and Murray 1967) to ensure that the CK_{30} contains the expected concentration of reactive SH groups.

   The expected value is calculated based on peptide weight, purity, and H_2O content. The CK_{30} is acceptable starting material if the measured value is at least 80% of the expected value.

2. Prepare a solution containing 60 μmoles CK_{30} (TFA salt) in 15 ml of 0.1 M PBS (pH 7.2)/5 μM EDTA.

3. Dissolve 66 μmoles of the mPEG-MAL-10k in 15 ml of DMSO. Add dropwise to the CK_{30} solution for approximately 5 minutes while vortexing at room temperature. Continue mixing the reaction for about 1 hour.

   The reaction contains 10% molar excess of mPEG-MAL-10k (based on maleimide reactivity). It is strongly exothermic. At pH 7, the reaction of maleimide with sulfhydryls is 1000 times faster than its reaction with amines (Hermanson 1996).

4. Perform a 4,4′-dithiodipyridine assay to verify that the conjugation reaction is complete. The fraction of CK_{30} that is PEG-substituted should be near 100%.

   The PEGylated CK_{30} should be used only if the concentration of reactive sulfhydryl groups has been reduced by at least 90%.

5. Fractionate the reaction mixture on a Sephadex G15 column equilibrated either with 0.1% trifluoroacetic acid or 50 mM ammonium acetate. Determine which fractions contain peptide by measuring absorbance at 220 nm. Pool and lyophilize those fractions.

   The lyophilized PEGylated polylysine can be stored for at least 2 years at –20°C.

Preparation of Nanoparticles

6. Resuspend the CK_{30}PEG10k in H_2O to a concentration of 7.1 mg/ml (TFA salt) or 6.4 mg/ml (acetate salt).

7. Add 0.9 ml of DNA (0.2 mg/ml in H_2O) in 100-μl aliquots to 0.1 ml of the resuspended CK_{30}PEG10k for approximately 2 minutes at room temperature while vortexing.

   The DNA concentration in the final solution is 0.18 mg/ml, and the end point ratio of positive to negative charges (NH_3^+/PO_4^-) is 2:1.

8. Dialyze the compacted DNA sample in either 5% dextrose or 0.9% NaCl to remove free CK_{30}PEG10k and unreacted PEG and store at 4°C.

Analysis of Nanoparticles

9. Visualize the particles by transmission electron microscopy. Compacted DNA must meet the following specifications: Particles must be nonaggregated, electron-dense, and oval (when

TFA is the counterion) or rod-like in shape (when in acetate form). The size of the particles must be consistent with the size of the compacted plasmid.

10. Analyze the particles on an agarose gel. No free or degraded DNA should be detected, and the particles themselves should remain in the well or migrate slightly upward. Particles treated with either 75% mouse serum (for 2 hours at 37°C) followed by trypsin digestion (2.5% for 40 minutes at room temperature) to uncomplex the DNA or incubated with DNase should demonstrate that >95% of compacted DNA remains intact, although nicking of the supercoiled DNA is expected.
11. Measure the colloidal stability of CK_{30}PEG10k-compacted DNA in physiologic saline by sedimentation of the DNA: Centrifuge the solution of nanoparticles at 3400*g* for 1 minute at room temperature. The ratio of the absorbance at 260 nm (A_{260}) of the supernatant to the A_{260} of the starting material should be 1 ± 10%.
12. Determine the turbidity of the DNA solution (Ziady et al. 2003a). The slope of the line obtained by plotting the log of apparent absorbance versus log of the wavelength (330–415 nm) should be in the range of –3.5 to –4.5.

REFERENCES

Grassetti D.R. and Murray J.F., Jr. 1967. Determination of sulfhydryl groups with 2.2′- or 4,4′-dithiodipyridine. *Arch. Biochem. Biophys.* **119:** 41–49.

Hermanson G.T. 1996. *Bioconjugate techniques*. Academic Press, San Diego, California, p. 148.

Konstan M.W., Davis P.B., Wagener J.S., Hilliard K.A., Stern R.C., Milgram L.J., Kowalczyk T.H., Hyatt S.L., Fink T.L., Gedeon C.R., Oette S.M., Payne J.M., Muhammad O., Ziady A.G., Moen R.C., and Cooper M.J. 2004. Compacted DNA nanoparticles administered to the nasal mucosa of cystic fibrosis subjects are safe and demonstrate partial to complete cystic fibrosis transmembrane regulator reconstitution. *Hum. Gene Ther.* **15:** 1255–1269.

Liu G., Li D., Pasumarthy M.K., Kowalczyk T.H., Gedeon C.R., Hyatt S.L., Payne J.M., Miller T.J., Brunovskis P., Fink T.L., Muhammad O., Moen R.C., Hanson R.W., and Cooper M.J. 2003. Nanoparticles of compacted DNA transfect postmitotic cells. *J. Biol. Chem.* **278:** 32578–32586.

Sambrook J. and Russell D. 2001. *Molecular cloning: A laboratory manual*, 3rd edition. Cold Spring Harbor Laboratory Press, Cold Spring Harbor, New York.

Ziady A.G., Gedeon C.R., Miller T., Quan W., Payne J.M., Hyatt S.L., Fink T.L., Muhammad O., Oette S., Kowalczyk T., Pasumarthy M.K., Moen R.C., Cooper M.J., and Davis P.B. 2003a. Transfection of airway epithelium by stable PEGylated poly-L-lysine DNA nanoparticles in vivo. *Mol. Ther.* **8:** 936–947.

Ziady A.G., Gedeon C.R., Muhammad O., Stillwell V., Oette S.M., Fink T.L., Quan W., Kowalczyk T.H., Hyatt S.L., Payne J., Peischl A., Seng J.E., Moen R.C., Cooper M.J., and Davis P.B. 2003b. Minimal toxicity of stabilized compacted DNA nanoparticles in the murine lung. *Mol. Ther.* **8:** 948–956.

48 Water-soluble Lipopolymers and Lipopeptides for Nucleic Acid Delivery

Ram I. Mahato,*† Zhaoyang Ye,* and Sung Wan Kim‡

*Departments of *Pharmaceutical Sciences and †Biomedical Engineering, University of Tennessee Health Science Center, Memphis, Tennessee 38163; ‡Department of Pharmaceutical Sciences and Pharmaceutical Chemistry, University of Utah, Salt Lake City, Utah 84112*

ABSTRACT

Water-soluble lipopolymers and lipopeptides are nonviral gene-transfer reagents that combine the advantages of lipids, which increase permeability of DNA through cell membranes, with the DNA-condensing and enhanced endosomal release properties of polycations. The lipopolymers are formed by conjugating cholesteryl chloroformate with the primary or secondary amines of 1800-dalton branched polyethyleneimine (PEI), and the lipopeptides are composed of a human protamine-derived peptide that has been incubated with a reaction of *O*-(*N*-succinimidyl)-*N*,*N*,*N*′*N*′,-tetramethyluronium tetrafluoroborate (TSTU) with lithocholic acid in thepresence of excess diisopropyl ethylamine (DIPEA). The use of the 1800-dalton PEI avoids the cytotoxicity problems associated with higher-molecular-mass PEI.

INTRODUCTION

Cationic liposomes and polymers form complexes with plasmid DNA via electrostatic interaction. As candidates in the search for effective, nonviral gene delivery systems, they efficiently transfect cells in culture but diffuse poorly within tissue and can be highly toxic to cells. In addition, usual methods for preparing liposomes can produce variable products and involve the use of hazardous organic solvents such as chloroform. Another potential group of gene-transfer reagents is pep-

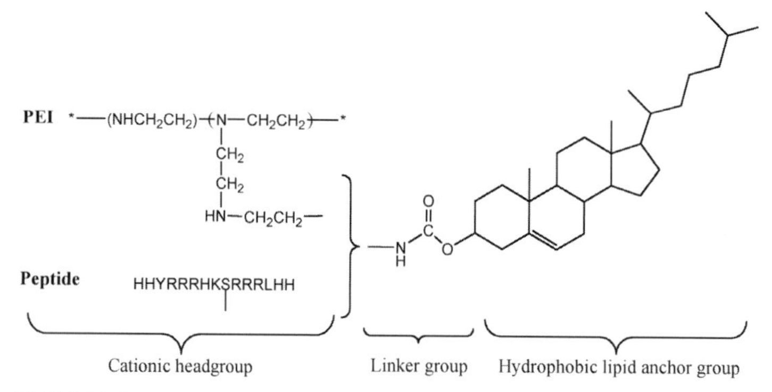

FIGURE 1. Basic components of water-soluble lipopolymers/lipopeptides.

tides. Peptides that possess DNA condensing, endosomolytic, or nuclear localization signal properties have been used either alone or in combination with liposomes and polymers and have also had some success (Mahato et al. 2005).

Gene carriers that combine the concepts of water solubility, amphiphilicity, lipid-mediated membrane interactions, and endosomal buffering would be an exciting option for nucleic acid delivery. To this end, we have developed water-soluble lipopolymers/lipopeptides by conjugating cholesterol onto PEI of 1800 daltons or peptides derived from protamine sulfate (Han et al. 2001; Mahato et al. 2001; Wang et al. 2002; Mahato et al. 2004, 2005). They have three components (see Fig. 1): a headgroup, a linker, and a lipid anchor. Low-molecular-mass PEI (PEI ≤1800 daltons is nontoxic) or peptides are used as a hydrophilic headgroup for DNA condensation, endosomal release of plasmid DNA in the cytoplasm, and traffic to the nucleus. Cholesterol is used as the hydrophobic lipid anchor, which can form a stable micellar complex with the hydrophilic headgroup in an aqueous environment and may shield pDNA particles from erythrocytes and plasma proteins. The biodegradable linker between the headgroup and the lipid anchor group is often used to minimize their cytotoxicity. The lipopolymer/pDNA and lipopeptide/pDNA complexes efficiently transfect cells in vitro and after intratumoral injection into tumor-bearing mice (see Fig. 2) with little toxicity.

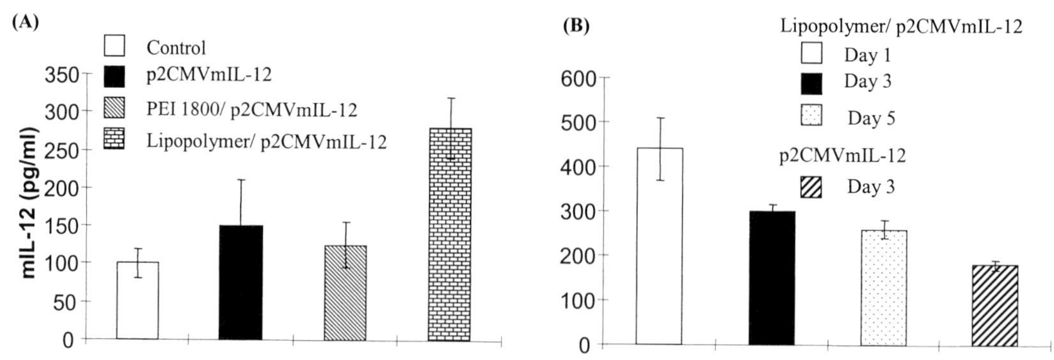

FIGURE 2. IL-12 gene expression after intratumoral injection of lipopolymer-p2CMVmIL-12 complexes into CT-26 tumor-bearing BALB/c mice. *(A)* Effect of gene carriers; *(B)* persistence of mIL-12 gene expression.

Protocol

Formation and Use of Water-soluble Lipopolymer-pDNA and Lipopeptide-pDNA Complexes

The following protocol describes the synthesis and characterization of water-soluble lipopolymer/lipopeptide–DNA complexes and their use in transfection of cultured cells and injection into tumor-bearing mice. The therapeutic gene used in this case encodes the murine interleukin 12 (IL-12) subunits p35 and p40, each under the transcriptional control of a separate cytomegalovirus (CMV) promoter, and the target cells are C-26 colon carcinoma cells. The reporter plasmid contains the luciferase gene driven by a CMV promoter. Lipopolymer/lipopeptide–DNA complexes should be freshly prepared before each nucleic acid delivery experiment.

MATERIALS

CAUTION: See Appendix for appropriate handling of materials marked with <!>.

Reagents

Important: All solvents are HPLC grade.

Branched polyethyleneimine (PEI; 1800 daltons; Polysciences) <!>
CT-26 colon adenocarcinoma cells
Cholesteryl chloroformate <!>
Diethyl ether <!>
Diisopropylethylamine solution (DIPEA; 1 M in DMF) <!>
Dimethylformamide (DMF) <!>
Ethidum bromide (0.5 μg/ml in 1x TBE) <!>
Fetal bovine serum (FBS)
Glucose (5%) for injection into mice
Hydrochloric acid (HCl; 0.1 N) <!>
Lithocholic acid
Methanol <!>
Methylene chloride <!>
Mice (5-week-old BALB/c; Simonsen Laboratories)
Phosphate-buffered saline (PBS; tissue culture grade)
Reporter gene: Plasmid DNA encoding luciferase driven by CMV promoter (pCMV-Luc)
RPMI 1640 tissue-culture medium
Therapeutic gene: p2CMVmIL-12, encoding murine IL-2 subunits p35 and p40, each under the transcriptional control of a separate CMV promoter
Triethylamine <!>
Trifluoroacetic acid (TFA; 95% and 5%) <!>
Tris–boric acid–EDTA buffer (TBE; available commercially or for recipe, see Sambrook and Russell 2001)
O-(*N*-succinimidyl)-*N,N,N',N'*-tetramethyluronium tetrafluoroborate (TSTU) solution (1 M in DMF)

Equipment

Agarose gel electrophoresis equipment
Bicinchoninic acid (BCA) Protein Assay Kit (Pierce)
Enzyme-linked immunosorbent assay (ELISA) for mIL-12 p70 (Pharmingen)
High-performance liquid chromatography (HPLC) equipment, including a C18 column (Vydac)
Luminometer (Dynex Technologies)
Matrix-assisted laser desorption–time-of-flight (MALDI-TOF) mass spectrometry equipment (e.g., Perspective Voyager-DE STR, Applied Biosystems)
Nuclear magnetic resonance (NMR) equipment (Varian)
Peptide synthesis equipment (e.g., Applied Biosystems 433A peptide synthesizer and related equipment)
Standard tissue-culture equipment and supplies
Zeta potential analyzer (ZetaPALS, Brookhaven Instruments)

METHODS

Synthesis and Characterization of Water-soluble Lipopolymer

1. Stir 3 g of PEI (1800 daltons) in a mixture of 10 ml of anhydrous methylene chloride and 100 μl of triethylamine for 30 minutes on ice.
2. Dissolve 1 g of cholesteryl chloroformate in 5 ml of ice-cold methylene chloride and slowly add it to the PEI mixture from Step 1. Stir for 12 hours on ice. Dry the resulting product, dissolve in 50 ml of 0.1 N HCl, and filter.
3. Extract the aqueous solution with 100 ml of methylene chloride and filter. Concentrate the product, precipitate with a large excess of acetone, and dry. Wash with methanol and diethyl ether.
4. To verify the identity of the product, determine its molecular mass using MALDI-TOF mass spectrometry with *trans*-4-hydroxy-3-methoxycinnamic acid as a matrix. Determine the structure of the peptide using ^{1}H-NMR. The water-soluble lipopolymer can be stored for 6 months at –20°C.

Synthesis and Characterization of Water-soluble Lipopeptides

5. Synthesize the following peptide: His-His-Tyr-Arg-Arg-Arg-His-Cys-Ser-Arg-Arg-Arg-Leu-His-His. This sequence corresponds to amino acid residues 51–63 of human protamine, with the replacement of cysteine-57 with a lysine and the addition of amino-terminal and carboxy-terminal histidines. Synthesize the peptide with standard protocols for 9-fluorenylmethoxycarbonyl (Fmoc)-based solid-phase synthesis. Use these protecting groups: 2,2,4,6,7-pentamethyldihydrobenzofurane-5-sulfonyl (Pfp) for arginine side chains, trityl (Trt) for histidine side chains, and 1-(4,4-dimethyl-2,6-dioxocylcohexylidene) 3-methylbutyl (Dde) for lysine side chains. Use tertiary-butoxycarbonyl (t-Boc) for all other amino acids.
6. Treat the peptide (still attached to the resin) briefly with 5% TFA to remove the Dde protective group from the ε-amine of lysine side chains. Wash the peptide multiple times with 100% methanol. This procedure should yield approximately 50 μmoles of peptide for further reactions.
7. Gradually add 100 μl of TSTU solution (1 M in DMF) to 100 μl of lithocholic acid solution (1 M in DMF) in the presence of threefold molar excess of DIPEA (1 M in DMF). Incubate with mild shaking for 2 hours at room temperature.

8. Add the resin with the washed peptide from Step 6 to the TSTU/lithocholic acid/DIPEA mixture. Incubate with shaking for 12 hours at room temperature.
9. Wash the reaction mixture with DMF. Cleave the peptide from the resin by incubating with 95% TFA for 90 minutes at room temperature. Separate the supernatant from the resin by centrifugation at 1000*g*. Purify the steroidal peptide using reverse-phase HPLC on a C18 column. Confirm the composition of the peptide by amino acid analysis. Determine the molecular mass using MALDI-TOF mass spectrometry. Determine the concentration of the peptide by measuring the absorbance of tyrosine at 274.5 nm ($\varepsilon_{274.5}$ = 1400 $M^{-1}cm^{-1}$).

   The water-soluble lipopeptides can be stored for 6 months at –20°C.

Preparation of Lipopolymer-pDNA and Lipopeptide-pDNA Complexes

10. Prepare lipopolymer-pDNA or lipopeptide-pDNA complexes by mixing pDNA and lipopolymer or lipopeptide at various N:P ratios ranging from 1 to 25 in 5% glucose. Incubate for 15–20 minutes at room temperature.

    N:P is the ratio of nitrogen atoms of water-soluble lipopolymer or lipopeptide to phosphate of plasmid DNA.
11. Analyze the samples by electrophoresis through an agarose gel in 1x TBE. Stain the gel with 0.5 μg/ml ethidium bromide and examine on a UV illuminator.
12. Measure the mean particle size and zeta potential of lipopolymer-pDNA or lipopeptide-pDNA complexes using phase analysis light scattering (ZetaPALS).

In Vitro Transfection and Luciferase Activity Assays

13. Grow and maintain CT-26 colon adenocarcinoma cell lines in RPMI 1640 medium supplemented with 10% FBS in a humidified incubator (5% CO_2) at 37°C.
14. Seed CT-26 cells in six-well plates at 3 x 10^5 cells per well in RPMI 1640 containing 10% FBS. Transfect cells that have reached 70% confluency with freshly prepared lipopolymer-pDNA or lipopeptide-pDNA complexes at a dose of 2.5 μg/well in the absence of serum.
15. Incubate the cells in the presence of complexes for 6 hours in the CO_2 incubator at 37°C. Replace the medium with 2 ml of complete medium and incubate for an additional 36 hours at 37°C.
16. Wash the cells with PBS. Lyse the cells and determine the total protein content using a BCA Protein Assay Kit.
17. Measure luciferase activity using a luminometer and report in terms of relative light units (RLU)/mg total protein.

Intratumoral Gene Delivery

18. To generate tumors, inject five-week-old BALB/c mice subcutaneously in the left flank with 100 ml of a suspension containing 1 x 10^6 CT-26 cells. Begin tumor treatment 10–15 days later, when the tumor size reaches approximately 100–120 mm^3.
19. Inject 50 μl of freshly prepared lipopolymer-p2CMVmIL-12 or lipopeptide-p2CMVmIL-12 complexes into the tumor-bearing mice at a dose of 25 μg of pDNA per mouse. Inject naked p2CMVmIL-12 and 5% glucose into the tumor as controls.
20. After a single intratumoral injection, monitor the mice for tumor growth. Report tumor progression in terms of tumor volume over a period of 48 days.
21. Harvest the tumors 48 hours after injection. Chop them into small pieces and re-culture for 24 hours at 37°C. Analyze the culture supernatants for mIL-12 p70 by ELISA.

ACKNOWLEDGMENTS

We thank Dong-an Wang, Sang-oh Han, and Anurag Maheshwari for performing these experiments.

REFERENCES

Han S.O., Mahato R.I., and Kim S.W. 2001. Water-soluble lipopolymer for gene delivery. *Bioconjug. Chem.* **12:** 337–345.

Mahato R.I., Han S.O., and Furgeson D.Y. 2004. Cationic lipopolymer as biocompatible gene delivery agent. U.S. Patent no. 6,696,038 B1.

Mahato R.I., Maheshwari A., and Kim S.W. 2005. Soluble steroidal peptides for nucleic acid delivery. U.S. Patent no. 6,875,611 B2.

Mahato R.I., Lee M., Han S.O., Maheshwari A., and Kim S.W. 2001. Intratumoral delivery of p2CMVmIL-12 using water-soluble lipopolymers. *Mol. Ther.* **4:** 130–138.

Sambrook J. and Russell D. 2001. *Molecular cloning: A laboratory manual,* 3rd edition. Cold Spring Harbor Laboratory Press, Cold Spring Harbor, New York.

Wang D.A., Narang A.S., Kotb M., Gaber A.O., Miller D.D., Kim S.W., and Mahato R.I. 2002. Novel branched poly(ethylenimine)-cholesterol water-soluble lipopolymers for gene delivery. *Biomacromolecules* **3:** 1197–1207.

49 Cationic Polysaccharides for DNA Delivery

Ira Yudovin-Farber, Hagit Eliyahu, and Abraham J. Domb

Department of Medicinal Chemistry and Natural Products, School of Pharmacy, Faculty of Medicine, The Hebrew University of Jerusalem, Jerusalem 91120, Israel

ABSTRACT

Polycations are effective nonviral carriers for gene delivery systems. These carriers vary in molecular weight, polymer structure, polymer:DNA ratio, molecular architecture, and the ability to introduce target-specific moieties. Polycations are capable of complexing various plasmids and transfecting them into different cells to produce a high yield of a desired protein. Cationic polysaccharides are attractive candidates for gene delivery. They are natural or seminatural, nontoxic, biodegradable, and biocompatible materials that can be modified for improved physicochemical properties. Cationic polysaccharides are synthesized by conjugation of various oligoamines to oxidized polysaccharides via reductive amination. These conjugates have been rigorously tested for gene delivery in cultured cells and in animals. From more than 300 polysaccharide-oligoamine derivatives tested, only dextran-spermine (D-SPM) was found to be highly effective in gene transfection, both in vitro and in vivo.

INTRODUCTION

Nucleic acid delivery has many applications in basic science, biotechnology, agriculture, and medicine. One of the main applications is DNA or RNA delivery for gene therapy purposes. Gene therapy, an approach for treatment or prevention of diseases associated with defective gene expression, involves the insertion of a therapeutic gene into cells, followed by expression and production of the required proteins. This approach enables the replacement of damaged genes or inhibition of expression of undesirable genes. After two decades of research, there are two major methods for delivery of genes. The first method, considered the dominant approach, utilizes viral vectors and is generally an efficient tool of transfection. Attempts, however, to resolve drawbacks of viral vectors (e.g., high risk of mutagenicity, immunogenicity, low production yield, and limited gene size) led to the development of alternative methods that make use of nonviral delivery systems. Nonviral gene carriers, which are more chemically defined than viral vectors, are primarily composed of polymers or lipids that undergo dissolution or formation of liposomes, micelles, or similar structures in aqueous solutions.

Polycations are a leading class of nonviral gene delivery "self-assembled" systems, which form spontaneous complexes with negatively charged nucleic acids. Among the various polycations used in gene delivery and transfection, cationic polysaccharides are considered to be attractive candidates. They are natural, water soluble, biodegradable, and biocompatible materials that can be modified simply for improved physicochemical properties. Cationic polysaccharides commonly used in gene delivery and transfection include chitosan and its derivatives, and diethylaminoethyl-dextran (DEAE-dextran). The low toxicity of chitosan and the availability of primary amines along the polysaccharide allow various modifications, including quaternization of amine groups, ligand attachment, and conjugation with hydrophilic polymers and endosomolytic peptides.

In initial screening experiments, more than 300 transfection conjugates based on polysaccharide-oligoamine combinations were synthesized (Azzam et al. 2002b) in our laboratory. The variables in the synthesis procedures included the use of (1) several polysaccharide backbones (e.g., linear dextran or pullulan and branched arabinogalactan of various molecular weights and oxidation degrees), (2) several oligoamines (e.g., 2–6 amino groups conjugated in various mole ratios to saccharide units), and (3) several reaction conditions (e.g., variations of pH and temperature). All polymers were synthesized using the reductive amination method, with the following stages: (1) oxidation of the desired polysaccharide using potassium periodate (KIO_4) to obtain polyaldehydes, (2) conjugation of the desired oligoamine to obtain imine conjugates, and (3) reduction using sodium borohydride to obtain stable amine conjugates (Fig. 1). These conjugates were then rigorously tested for transfection in cultured cells.

To our surprise, only D-SPM, a water-soluble polycation prepared from highly oxidized dextran (i.e., at a 1:1 IO_4^-/saccharide mole ratio) and spermine as the oligoamine, was active in transfection. It was highly effective, both in vitro (Azzam et al. 2003) and in vivo (Hosseinkhani et al. 2004). Using the intramuscular and intranasal routes of administration, Eliyahu et al. (2006a) showed that transfection occurred primarily in the bronchial epithelial cells, pneumocytes, and bronchial alveoli of the lungs, in the fibrocytes, and in the hepatocytes. Polyplexes based on D-SPM showed systemic transfection on local administration (H. Eliyahu et al., in prep.) with good

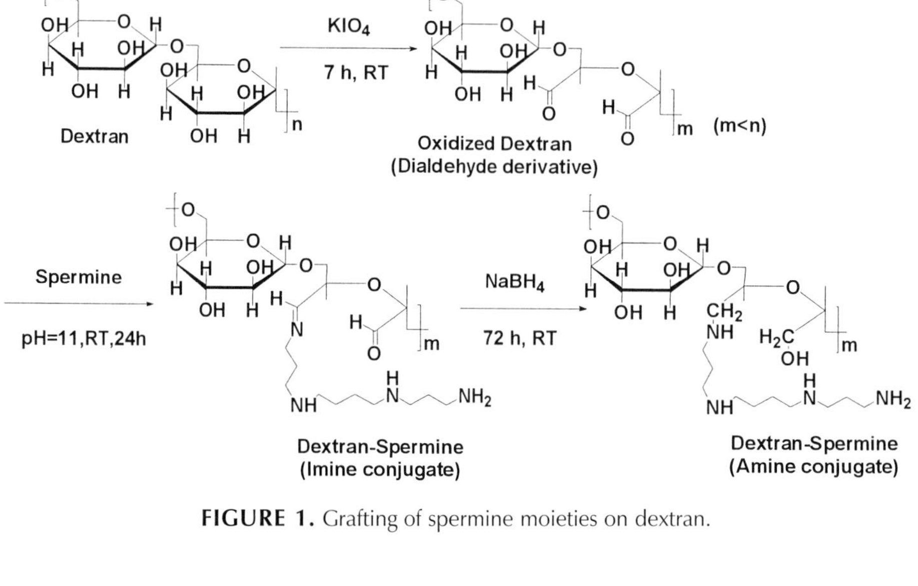

FIGURE 1. Grafting of spermine moieties on dextran.

tolerability and low toxicity (Eliyahu et al. 2006a). Transfection efficacy was highly dependent on the charge ratio. Studying the relationship among chemical structure, physical parameters, and transfection of cells, we determined that the crossed-linked SPM, as well as the large amount of unprotonated secondary amines found in D-SPM but not in other conjugates tested, are of great importance for the polymer's transfection activity. Similar to D-SPM, linear polyethyleneimine (L-PEI) contains a large fraction of secondary amines, and branched polyethyleneimine (B-PEI) contains a large fraction of secondary and tertiary amino groups. The large fraction of secondary and tertiary amines enables these polymers to act as a "proton sponge" in the endosome, enhancing endosomal escape. The transfection efficiency in serum-free medium of PEIs, which are considered the gold standard for nonviral transfection, was similar to that of D-SPM (Eliyahu et al. 2005; H. Eliyahu et al., in prep.).

Although reductive amination reactions between polyaldehydes and oligoamine are considered to be nonreproducible because of random branching, branching was reduced by reaction in a highly diluted system and dropwise addition of the polyaldehyde solution to the oligoamine. The availability of a large number of primary amine groups along the polycation permits chemical modifications that lead to altered physicochemical properties. Such modifications include attachment of polyethylene glycol (PEG) to serve as steric stabilizer (Hosseinkhani et al. 2004), partial hydrophobization (e.g., by an oleyl moiety) (Azzam et al. 2004), attachments of fluorescence markers for trafficking and mechanistic studies, targeting molecules, and peptides such as nuclear localizing sequences (NLSs). The transfection efficiency of cationic complexes is usually decreased significantly when transfection is performed in serum-rich medium. To improve the transfection in the presence of serum, PEGylated derivatives of D-SPM were used in vitro and in vivo. Intravenous and intramuscular administration of the polyplexes, based on PEGylated–D-SPM derivatives and pSV-LacZ, resulted in higher gene expression in the liver in comparison with the non-PEGylated–D-SPM–pSV-LacZ complex (Table 1) (Hosseinkhani et al. 2004).

Another modification of D-SPM involved partial hydrophobization by an oleyl moiety (ODS). When the water-soluble D-SPM-based polyplexes were used, transfection decreased with increasing serum concentration in cell culture in a concentration-dependent manner, reaching 95% inhibition at 50% serum in the cell growth medium. In contrast, polyplexes based on ODS were good transfection agents in serum-rich medium (50% serum). The efficient transfection obtained in zebra fish and mice demonstrated the potential of ODS to serve as an efficient nonviral vector for in vivo transfection as well (Eliyahu et al. 2005).

TABLE 1. Gene Expression at Different Organs 2 Days after Intravenous Injection of 5% PEGylated–Dextran-Spermine–pSV-LacZ Complex and Other Agents

	Specific β-gal activity (mU/mg protein)			
Organ	PBS[a]	PEGylated–D-SPM[b]	Free plasmid DNA[c]	PEGylated–D-SPM–pSV-LacZ complex[d]
Blood	6.22 ± 0.82[e]	6.42 ± 3.18	6.24 ± 0.73	7.12 ± 1.56
Heart	7.23 ± 2.23	4.45 ± 2.40	5.54 ± 3.23	8.23 ± 1.10
Lung	6.94 ± 3.12	8.25 ± 0.75	9.02 ± 0.52	7.29 ± 0.14
Liver	7.45 ± 1.03	8.25 ± 2.55	8.17 ± 1.23	**28.45 ± 3.38**
Spleen	5.87 ± 1.14	4.45 ± 3.46	7.30 ± 1.05	9.13 ± 1.45
Kidney	6.98 ± 1.08	6.45 ± 2.34	7.34 ± 1.34	**17.30 ± 1.15**
Gastrointestinal tract	7.14 ± 4.20	5.32 ± 0.34	6.23 ± 5.32	9.10 ± 1.02
Carcass	8.72 ± 2.14	9.04 ± 1.01	8.06 ± 1.17	9.26 ± 1.08
Excretion	7.94 ± 0.15	8.16 ± 0.23	6.24 ± 3.90	5.45 ± 3.34

[a]Phosphate-buffered saline control (200 μl per single injection).

[b]PEGylated–dextran-spermine-based conjugate (250 μg in 200 μl of PBS per mouse).

[c]Free plasmid DNA (50 μg in 200 μl of PBS per mouse).

[d]5% PEGylated–dextran-spermine–pSV-LacZ at a weight-mixing ratio of 5 (polycation/DNA) and 50 μg/mouse of the plasmid in 200 μl of PBS.

[e]Mean ± S.D.

Protocol

Synthesis of Cationic Polysaccharides and Use for In Vitro Transfection

This protocol describes the synthesis of cationic polysaccharides and their use for DNA transfection in vitro.

MATERIALS

CAUTION: See Appendix for appropriate handling of materials marked with <!>.

All solvents and reagents are of analytical grade.

Reagents

BCA Assay Kit (Pierce)
β-gal ELISA (enzyme-linked immunosorbent assay) Kit (Roche) or β-gal Assay Kit (Invitrogen)
Calcium phosphate reagent (Sigma-Aldrich)
Cells to be transfected, e.g.,
  Murine C3H10T1/2 progenitor
  human embryonic kidney (HEK293)
  Chinese hamster ovary (CHO)
  human cervix carcinoma (HeLa)
  murine fibroblasts (NIH3T3)
  EPC
  COS-7
Dextran (average molecular mass of 40 kD) (Sigma-Aldrich)
Fetal calf serum (FCS) (Beit Haemek, Israel)
Glutamine (4 mM) (Beit Haemek, Israel)
HEPES-buffered saline (HBS; 150 mM NaCl, 20 mM HEPES at pH 7.1)
  Filter-sterilize through a 0.2-μm membrane or autoclave and store at 4°C.
Human growth hormone (hGH) ELISA kit (Roche)
Hydroxylamine hydrochloride <!>
Lipid formulations
  DOTAP/Chol 1/1 (Avanti Polar Lipids Inc., Alabama)
  Transfast (Promega)
  FuGENE 6 (Roche)
Luciferase Reporter Gene Assay (Roche) or Luciferase Assay Kit (Promega)
Media
  Complete medium: DMEM supplemented with 10% (w/w) FCS, 0.2 mM L-glutamine, 1 mg/ml penicillin, and 100 units/ml streptomycin <!>
  Dulbecco's modified Eagle's medium (DMEM)
Nitrogen atmosphere (for storing the conjugate)
Penicillin (Beit Haemek, Israel)
Plasmids
  Use pEGFP-C1 plasmid under the control of the CMV (cytomegalovirus) promoter, pLNCluc plasmid containing the firefly luciferase gene, pLacZ under the control of the SV40 or CMV pro-

moter, and pCMVhGH for quantitative and qualitative assays. Purify plasmids using the Nucleobond AX 500 column (Machery-Nagel, Duren, Germany) or the QIAGEN Plasmid Mega Kit (Hilden, Germany). The concentration of DNA can be quantified by determination of organic phosphorus, which represents DNA negative charges (Eliyahu et al. 2005).

Potassium periodate (KIO_4) (Sigma-Aldrich) <!>
Sodium borohydride ($NaBH_4$) (Sigma-Aldrich) <!>
Spermine (Sigma-Aldrich) <!>
Streptomycin (Beit Haemek, Israel) <!>

Equipment

Cellulose dialysis tubing (3500 cut-off)
Dishes (6 well)
Dowex-1 (acetate form) anion exchange column
Fluorescence microscope (Model Axiovert 35, Zeiss, Jena, Germany)
Lyophilizer
Spectrometers, NMR (nuclear magnetic resonance) and infrared

METHODS

Synthesis of Cationic Polysaccharides: Oxidation of Dextran

1. Dissolve dextran (10 g, 62.5 mmoles of glucose units) in 200 ml of double-deionized H_2O.
2. Add potassium periodate at a 1:1 mole ratio (glucose/IO_4^-) and stir the mixture in the dark for 6–8 hours at room temperature.
3. Purify the resulting polyaldehyde derivative from iodate (IO_3^-) and unreacted periodate (IO_4^-) by Dowex-1 (acetate form) anion exchange chromatography, followed by extensive dialysis against double-deionized H_2O (3500 cut-off cellulose tubing) for 3 days at 4°C.
4. Freeze-dry the purified polyaldehyde derivative. This will produce a white powder.

   The average yield is 70%. Fourier transform-infrared (FT-IR) (KBr) = 1724 cm^{-1} (C = O).
5. Determine the aldehyde content by titration with hydroxylamine hydrochloride (Zhao and Heindel 1991).

Oligoamine Conjugation

6. Dissolve oxidized dextran (1 g, 0.75–6.56 mmoles of aldehyde groups) in 100 ml of double-deionized H_2O. Add dialdehyde solution slowly over several hours to a basic solution containing 1.5 equimolar amount of spermine dissolved in 50 ml of borate buffer (0.1 M, pH 11).
7. Stir the mixture for 24 hours at room temperature. Add $NaBH_4$ (1 g, 4 equimolar) to reduce the imine bonds to amines. Continue stirring for 48 hours under the same conditions.
8. Repeat the reduction with an additional portion of $NaBH_4$ (1 g, 4 equimolar) under the same conditions for 24 hours.
9. Pour the resulting light-yellow solution into a dialysis tube (3500 cut-off cellulose tubing) and dialyze it against double-deionized H_2O for 3 days at 4°C.
10. Freeze-dry the dialysate and store under nitrogen atmosphere. The average yield is 50% (w/w). The degree of conjugation can be determined by nuclear magnetic resonance (NMR) analysis of the compound. (m) Multiplet, (H) hydrogen.

^{1}H-NMR (D_2O):

1.645 ppm (m, 4H, Dextran-$NH(CH_2)_3NHCH_2\underline{CH_2}\underline{CH_2}CH_2NH(CH_2)_3NH_2$)

1.804 ppm (m, 4H, Dextran-$NHCH_2\underline{CH_2}CH_2NH(CH_2)_4$ $NHCH_2\underline{CH_2}CH_2NH_2$)

2.815 ppm (m, 12H, Dextran-$NH\underline{CH_2}CH_2\underline{CH_2}NH\underline{CH_2}CH_2CH_2\underline{CH_2}NH\underline{CH_2}CH_2\underline{CH_2}NH_2$)

3.30–4.45 ppm (m, glucose hydrogens)

5.01 ppm (m, 1H, anomeric hydrogen)

FT-IR (KBr)

1468 cm^{-1} (-CH_2- aliphatic), 1653 cm^{-1} (-NH_2, primary amine)

2935 cm^{-1} (C-C, aliphatic) and 3297 cm^{-1} (-NH, -OH groups)

11. Determine the primary amine content by the TNBS method (Azzam et al. 2002a).

In Vitro Transfection Procedure (Fig. 2)

12. Twenty-four hours before transfection, seed 1.3 x 10^5 cells to be transfected in complete medium in six-well dishes and incubate them overnight at 37°C in a humidified atmosphere of 5% CO_2.

    The cells should be 70–80% confluent at the time of transfection.

13. Replace complete medium with 1 ml of DMEM.
14. In every transfection experiment, use polyplexes of several polymer-to-DNA w/w ratios (e.g., 2.5, 5, 7.5, 10, 12.5, and 15 μg of polymer to 1 μg of DNA). Dilute polycation/DNA complex mixtures to a final volume of 200 μl with DMEM or 20 mM HBS and allow to stand for 15–30 minutes at room temperature.
15. Add either polyplexes or lipoplexes (control) to the dish.

    Run the controls at the same time as testing the polyplexes. Cell transfection with the calcium phosphate reagent can be performed according to a well-documented protocol (Wigler et al. 1978, 1979). Transfection using DOTAP/Chol 1/1, Transfast, and FuGENE 6–based lipid formulations can be performed according to manufacturers' protocols.

16. After 4 hours, replace DMEM with 1 ml of complete medium.
17. Quantify the hGH produced in the supernatant 24 hours after transfection using the hGH ELISA Kit.

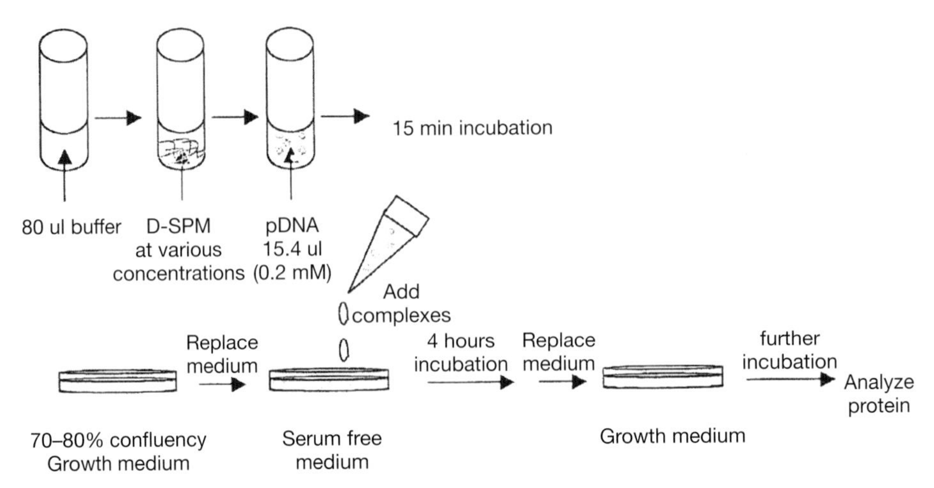

FIGURE 2. Transfection procedure.

18. Quantify β-galactosidase activity in the lysate 48 hours after transfection using a β-gal ELISA Kit or the β-gal Assay Kit.
19. Quantify luciferase activity 48 hours after transfection using the Luciferase Reporter Gene Assay or Luciferase Assay Kit.
20. Estimate the cells transfected with D-SPM–GFP (green fluorescent protein) polyplexes by using fluorescence microscopy 48 hours after transfection. Calculate the yield of transfection (% transfection) by dividing the number of fluorescent cells in a field of a particular well by the total number of cells in the same field.

    In some cases, the degree of gene expression can be normalized to total protein content using the standard BCA Assay Kit.

ACKNOWLEDGMENTS

This study was supported in part by the AFIRST, French-Israeli Cooperation, the U.S.–Israeli BiNational Fund, and the Polygene.

REFERENCES

Azzam T., Eliyahu H., Makovitzki A., and Domb A.J. 2003. Dextran-spermine conjugate: An efficient vector for gene delivery. *Macromol. Symp.* **195:** 247–261.

Azzam T., Eliyahu H., Makovitzki A., Linial M., and Domb A.J. 2004. Hydrophobized dextran-spermine conjugate as potential vector for in vitro gene transfection. *J. Control. Release* **96:** 309–323.

Azzam T., Eliyahu H., Shapira L., Linial M., Barenholz Y., and Domb A.J. 2002a. Polysaccharide-oligoamine based conjugates for gene delivery. *J. Med. Chem.* **45:** 1817–1824.

Azzam T., Raskin A., Makovitzki A., Brem H., Vierling P., Lineal M., and Domb A.J. 2002b. Cationic polysaccharides for gene delivery. *Macromolecules* **35:** 9947–9953.

Eliyahu H., Azzam T., Barenholz Y., and Domb A.J. 2006a. Dextran-spermine-based polyplexes—Evaluation of transgene expression and of local and systemic toxicity in mice. *Biomaterials* **27:** 1636–1645.

Eliyahu H., Siani S., Azzam T., Domb A.J., and Barenholz Y. 2006b. Relationships between chemical composition, physical properties and transfection efficiency of polysaccharide-spermine conjugates. *Biomaterials* **27:** 1646–1655.

Eliyahu H., Makovitzki A., Azzam T., Zlotkin A., Joseph A., Gazit D., Barenholz Y., and Domb A.J. 2005. Novel dextran-spermine conjugates as transfecting agents: Comparing water-soluble and micellar polymers. *Gene Ther.* **12:** 494–503.

Hosseinkhani H., Azzam T., Tabata Y., and Domb A.J. 2004. Dextran-spermine polycation: An efficient nonviral vector for in vitro and in vivo gene transfection. *Gene Ther.* **11:** 194–203.

Wigler M., Pellicer A., Silverstein S., and Axel R. 1978. Biochemical transfer of single-copy eukaryotic genes using total cellular DNA as donor. *Cell* **14:** 725–731.

Wigler M., Pellicer A., Silverstein S., Axel R., Urlaub G., and Chasin L. 1979. DNA-mediated transfer of the adenine phosphoribosyltransferase locus into mammalian cells. *Proc. Natl. Acad. Sci.* **76:** 1373–1376.

Zhao H. and Heindel N.D. 1991. Determination of degree of substitution of formyl groups in polyaldehyde dextran by the hydroxylamine hydrochloride method. *Pharm. Res.* **8:** 400–402.

50 Sustained Release of Plasmid DNAs Encoding Platelet-derived Growth Factor and Hyaluronan Synthase 2 from Cross-linked Hyaluronan Matrices and Films

Weiliam Chen

Department of Biomedical Engineering, State University of New York, Stony Brook, New York 11794-2580

ABSTRACT

Natural carbohydrate is a class of underexplored polymers for gene delivery. The noninflammatory and nonimmunogenic properties of hyaluronan (HA) are of particular importance when clinically relevant situations are considered. Moreover, the presence of hyaluronidase in vivo enables any vehicle fabricated from HA to be degraded by enzyme-mediated erosion. When DNA is entrapped in a cross-linked HA vehicle, HA-DNA fragments are released upon digestion by hyaluronidase. These fragments could serve dual roles as microcarriers of DNA and its protective mechanism. This chapter describes two HA-based water-insoluble systems (matrix and film) designed specifically for future clinical applications (Kim et al. 2003, 2005). pDNA encoding platelet-derived growth factor (PDGF) is coupled to the matrices (Kim et al. 2003) that could be implanted into chronic wounds to accelerate their healing. This system also provides an HA-enriched environment that is known to be critical for wound repair (Bronson et al. 1987). pDNA encoding hyaluronan synthase 2 (HAS2) is coupled to the film (Kim et al. 2005) that could initially serve as a physical barrier and subsequently a pDNA reservoir for sustaining HAS2 transfection. This would lead to continual HA production for preventing post-surgical adhesion. The main advantages of these HA-based systems are that they are fully biocompatible and their degradation is almost completely controlled by enzymatic erosion of hyaluronidase. The DNA-HA complex fragments formed by degradation could act as microcarriers and thus protect the DNA. Finally, the vehicles serve relevant clinical functions. The major limitation of these vehicles is that their transfection efficiency is not expected to be high in an in vivo environment.

INTRODUCTION

HA is a truly biocompatible natural polymer. It is a highly conserved, biocompatible, and biodegradable glycosaminoglycan composed of D-glucuronic acid alternating with *N*-acetyl-D-glucosamine (Toole 1990; Laurent and Fraser 1992; Goa and Benfield 1994). Other than its critical role in wound repair (Bronson et al. 1987), HA is known to have great lubricating ability and has been administered exogenously for preventing postsurgical adhesion. However, HA in its native state is highly soluble and can be eliminated rapidly in vivo, thereby limiting its usefulness in wound repair and preventing postsurgical adhesion. We have designed both cross-linked HA matrix and film. The former could serve as a temporary artificial wound bed and double as a delivery vehicle for controlled release of pDNA encoding PDGF for sustaining transfection leading to accelerated wound repair. The latter could initially serve as a physical barrier and subsequently, a pDNA reservoir for sustaining HAS2 transfection. This would lead to continual HA production for prolonged prevention of postsurgical adhesion.

Protocol

Preparation of HA-DNA Matrices and Films

This protocol describes preparation of HA-DNA matrices and films and assays for verification of their bioactivities.

MATERIALS

CAUTION: See Appendix for appropriate handling of materials marked with <!>.

Reagents

Adipic dihydrazide (ADH; Sigma-Aldrich)
Anti-Flag antibody (Sigma-Aldrich)
COS-1 cells (ATCC, Rockville, Maryland)
Dimethylformamide (DMF) (Sigma-Aldrich) <!>
Dulbecco's modified Eagle's medium (DMEM) (Cellgro, Heindon, Virginia)
Ethyl-3-[3-dimethyl amino]propyl carbodiimide (EDCI) (Sigma-Aldrich) <!>
Fetal bovine serum (FBS) (Cellgro)
FGM-2 BulletKits medium (Clonetics, Walkersville, Maryland)
FuGENE transfection reagent (Roche Diagnostics)
HA Quantitative Kit (Corgenix, Westminster, Colorado)
Hoescht 33342 fluorescent dye (ex:485 nm, em:530 nm; Molecular Probes) <!>
Hyaluronan (HA) (Kraeber GmbH & Co., Waldhofstr, Germany) stock solution (10 mg/ml, i.e., 1%)
Hyaluronidase (Sigma-Aldrich)
Hydrochloric acid (HCl; 1 N) <!>
Isopropanol (80%, 90%, and 100%; Fischer Chemical)
MC285 rabbit anti-mouse HAS2 monoclonal antibody (from Dr. John McDonald, University of Utah) and peroxidase-labeled goat anti-rabbit secondary antibody
Normal neonatal human dermal fibroblasts (NNHDF; Clonetics)
Phosphate-buffered saline (PBS; Sigma-Aldrich)
Plasmid DNA (pDNA, various concentrations) in sterile H_2O
pFlag-CMV-5a mammalian expression vector
This vector should have the cDNAs encoding PDGF or HAS2 cloned into it.
Quantikine ELISA (enzyme-linked immunosorbent assay) Kit for PDGF (R&D Systems)

Equipment

Aluminum weighing pan (autoclaved, 4-cm diameter) (VWR Scientific)
Cytofluorometer (Cytofluor-II, Biosearch, Bedford, Massachusetts)
Incubator, preset to 37°C (VWR Scientific)
Lyophilizer (Freezemobile 12EL, Gardiner, New York)
Orbital shaker (LabQuake shaker L-1237, Lab Industries, Berkeley, California)

METHODS

Preparation of HA-DNA Matrices or Films

1. Gently mix 1 ml of appropriate pDNA solution with 20 ml of a 1% HA solution.
2. Deposit 8 ml of the HA/pDNA mixture in each aluminum pan. Then proceed as follows.

For HA-DNA matrices

a. Snap freeze and lyophilize the HA/pDNA mixture to form solid HA-DNA matrices.

b. Prepare 100 ml of a reagent mixture by mixing 100 mg of ADH in a solution of 90% DMF/10% H_2O. Incubate the HA-DNA matrices for 30 minutes in the reagent mixture.

c. Dissolve 120 mg of EDCI in the reagent mixture and adjust it to pH 5 with HCl (1 N) to initiate the cross-linking reaction.

d. Decant the reagent solution and replace it with 100 ml of 90% isopropanol, followed by several additional extractions with 90% isopropanol and then pure isopropanol.

e. Dry the HA-DNA matrices by aspiration. Dried matrices are depicted in Figure 1a (Kim et al. 2003).

For HA-DNA films

a. Air-dry the HA/pDNA mixture to form an HA-DNA film.

b. Dissolve 10 mg of ADH in 100 ml of 80% isopropanol. Incubate the dried film in this solution for 30 minutes, for partial rehydration.

c. Add 1 ml of EDCI solution (12 mg/ml in H_2O) to the reagent mixture and adjust it to pH 5 with HCl (1 N) to initiate the cross-linking reaction.

d. Decant the reagent solution and replace it with 100 ml of 90% isopropanol, followed by several additional extractions with 90% isopropanol and then pure isopropanol.

e. Air-dry the HA-DNA film. A dried film is depicted in Figure 1b (Kim et al. 2003).

Verification of Bioactivities

3. Prepare a hyaluronidase solution of 10 units/ml in PBS. Digest the matrices or films with this solution and collect the releasate samples at different times (Kim et al. 2005). Then proceed as follows.

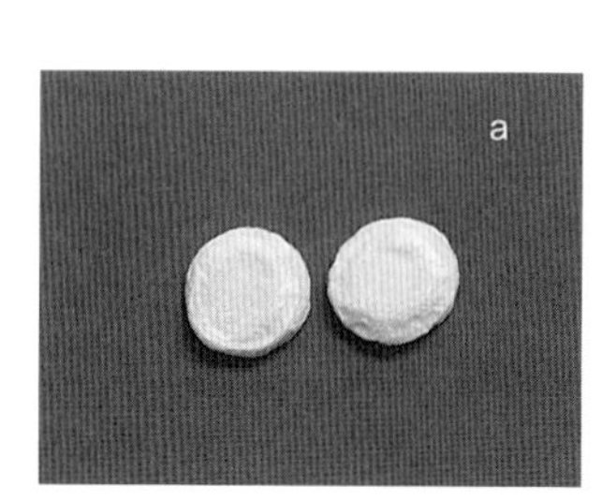

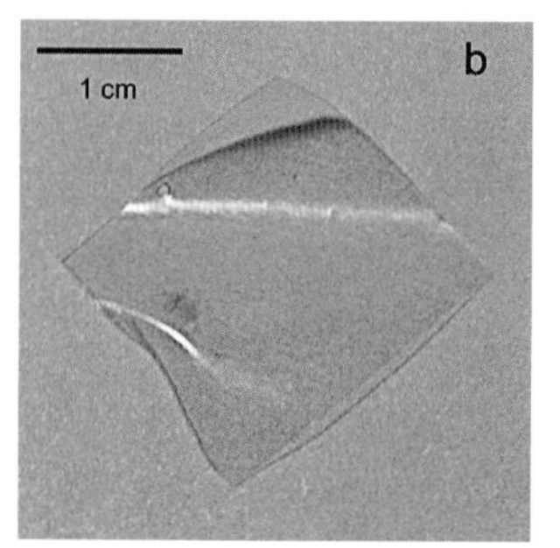

FIGURE 1. (*a*) DNA-HA matrices; (*b*) DNA-HA film. (*a*, Reprinted, with permission, from Kim et al. 2003; *b*, reprinted, with permission, from Kim et al. 2005 [© Elsevier].)

For HA-DNA matrices

a. Grow COS-1 cells in FGM-2 BulletKits medium (with 0.5% FBS but without the growth supplements). Transfect the cells with the releasate samples using a FuGENE-mediated protocol (Kim et al. 2003).

b. Recover the medium after 3 days of incubation and dilute 1:2 with the modified FGM-2 BulletKits medium.

c. Add the diluted medium to NNHDF cells at a density of 5×10^4 cells/cm^2. Harvest and freeze cells daily for 5 days.

d. Assess the extent of cell proliferation of all samples using Hoescht 33342 dye and a cytofluorometer (Kim et al. 2003).

e. Corroborate the exogenous origin of PDGF by immunostaining for Flag (Kim et al. 2003). See Figure 2 for typical results.

For HA-DNA films

a. Grow COS-1 cells in DMEM. Transfect the cells with the releasate samples using a FuGENE-mediated protocol (Kim et al. 2005).

b. Quantify the levels of HA in the media conditioned by the transfected cells using an HA Quantitative Kit. See Table 1 for some sample results (Kim et al. 2005).

c. Detect expression of HAS2 using an MC285 rabbit anti-HAS2 antibody (with a peroxidase-labeled goat anti-rabbit secondary antibody).

d. Corroborate the exogenous origin of HAS2 expression by immunostaining for Flag (Kim et al. 2005).

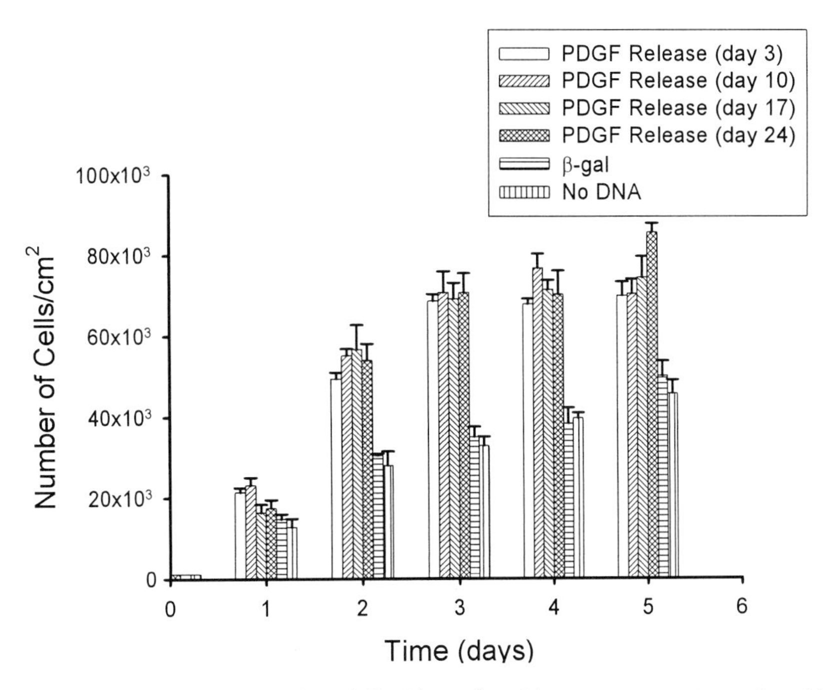

FIGURE 2. Induction of human neonatal dermal fibroblast cell proliferation using PDGF produced by cells transfected with releasate from DNA-HA matrices. (Reprinted, with permission, from Kim et al. 2003 [© Elsevier].)

TABLE 1. Synthesis of HA by Cells Transfected with Releasates From DNA-HA Films

DNA release sample 1	DNA release sample 2	+ Control	– Control
114.84 ng/ml	92.74 ng/ml	96.42 ng/ml	35.89 ng/ml

Modified, with permission, from Kim et al. (2005).

ACKNOWLEDGMENTS

These studies were supported by the National Institutes of Health (R43 AG17778 and R43 HL66876).

REFERENCES

Bronson R.E., Bertolami C.N., and Siebert E.P. 1987. Modulation of fibroblast growth and glycosaminoglycan synthesis by interleukin-1. *Collagen Relat. Res.* **7:** 323–332.

Goa K.L. and Benfield P. 1994. Hyaluronic acid. A review of its pharmachology and use as a surgical aid in ophthalmology, and its therapeutic potential in joint disease and wound healing. *Drugs* **47:** 536–566.

Kim A., Checkla D., Dehazya P., and Chen W. 2003. Characterization of hyaluronan-DNA matrices for sustained gene transfer. *J. Control. Release* **90:** 81–95.

Kim A., Yellen P., Yun Y.H., and Chen W. 2005. Delivery of a vector encoding mouse hyaluronan synthase 2 gene transfer via a crosslinked hyaluronan film. *Biomaterials* **26:** 1586–1593.

Laurent T.C. and Fraser J.R. 1992. Hyaluronan. *FASEB J.* **6:** 2397–2404.

Toole B.P. 1990. Hyaluronan and its binding proteins, the hyaladherins. *Curr. Opin. Cell Biol.* **2:** 839–844.

51 Linear Polyethylenimine: Synthesis and Transfection Procedures for In Vitro and In Vivo

Manfred Ogris and Ernst Wagner

Pharmaceutical Biology-Biotechnology, Department of Pharmacy, Ludwig-Maximilians-Universität, D-81377 Munich, Germany

ABSTRACT

Linear polyethylenimine (LPEI) is an efficient transfection reagent for a broad range of cell lines and primary cells and is also useful for local or systemic gene delivery in vivo. In contrast to many other nonviral transfection methods, LPEI is less dependent on mitosis (Brunner et al. 2002) and also transfects postmitotic cells. Transfections can be performed with a simple protocol. The presence of serum (up to 10%) during transfection does not markedly decrease transfection efficiency, but it significantly reduces toxicity, especially to primary cells. This is a major advantage compared to most lipidic transfection systems. Toxicity can also be kept to a minimum using low, optimized amounts of the transfection reagent. However, LPEI can be quite toxic when applied at elevated doses in vitro or in vivo. As for most other lipid- or polymer-based transfection systems, the performance of LPEI with cells in suspension is rather low. Nevertheless, if LPEI polyplexes containing cell-binding ligands (e.g., transferrin) are used, performance can be considerably increased (Kircheis et al. 2001). LPEI is a fully synthetic product that can be easily synthesized from rather inexpensive precursors and is suitable for large-scale transfections.

INTRODUCTION

LPEI has been described as one of the most potent transfection reagents in vitro and in vivo (Ferrari et al. 1997; Goula et al. 1998a,b). When complexed with nucleic acids, LPEI exhibits a biophysical behavior different from that of branched PEI (BPEI), resulting in formation of aggregates on the cell surface of initially small polyplexes (Wightman et al. 2001). Within the cytoplasm,

LPEI releases DNA more easily than BPEI (Itaka et al. 2004). These effects, added to the endosomal buffering of PEI (Sonawane et al. 2003), are mainly responsible for its high transfection efficiency. After adding polyplexes, plates can be centrifuged for 5 minutes at 280*g* using a swing-out rotor, which can further increase transfection efficiency (Boussif et al. 1996).

Depending on the salt content of the buffer used for complex formation, LPEI forms either small particles 50–100 nm in diameter in the absence of salt or large aggregates in 150 mM NaCl (Goula et al. 1998b; Wightman et al. 2001). For in vitro transfections, aggregated polyplexes usually give the best transfection results, whereas for in vivo applications, small particles generated in glucose are preferred (Goula et al. 1998a,b; Wightman et al. 2001). Because LPEI can be quite toxic when applied at elevated doses in vitro or in vivo (Chollet et al. 2002; Boeckle et al. 2004), titration to find the optimal DNA concentration is advisable. We recommend using an N:P (nitrogen:phosphate) ratio of 6, although an N:P ratio range of 4–10 is potentially suitable for transfection. If toxicity is observed, either the N/P ratio can be lowered or the incubation time of polyplexes with cells can be reduced to 1 hour. The synthesis of LPEI described below should be performed in a chemistry laboratory. Alternatively, the commercial product (available from PolyPlus, Illkirch, France [JetPEI], www.polyplus-transfection.com) can be used.

Protocol

Synthesis of Linear Polyethylenimine and Use in Transfection

This protocol describes the synthesis of LPEI from a precursor polymer and the generation of DNA/LPEI polyplexes. Transfection protocols for cells adherent to tissue culture dishes or grown in suspension are given, as well as in vivo applications.

MATERIALS

CAUTION: See Appendix for appropriate hadling of materials marked with <!>.

Reagents

Copper(II) sulfate (CuII; analytical grade) <!>
Culture medium appropriate for target cells
HEPES-buffered glucose (HBG; 5% glucose w/v, 20 mM HEPES at pH 7.1)
Filter-sterilize using a 0.2-μm membrane and store at 4°C.
HEPES-buffered saline (HBS; 150 mM NaCl, 20 mM HEPES at pH 7.1)
Filter-sterilize using a 0.2-μm membrane or autoclave and store at 4°C.
Hydrochloric acid (30%, analytical grade) <!>
Mice, for in vivo experiment
Phosphorus pentoxide <!>
Plasmid DNA of choice
Poly(2-ethyl-2-oxazolin) (50 kD; Sigma-Aldrich 37,284-6)
Sodium acetate (analytical grade) <!>
Sodium hydroxide (50% w/v, analytical grade) <!>
Sulfuric acid (30%, analytical grade) <!>
Target cells of choice

Equipment

Beaker
Büchner funnel
Contact thermometer
Filter paper
Filters (0.2 μm)
Ice bath
Multiwell plates
Oil bath
pH paper
Phase separator
Reflux condenser
Syringe (1-ml insulin, 33-gauge needle)
UV spectrophotometer
Water bath, 37°C, for in vivo experiment

METHODS

Synthesis and Purification of LPEI

Appropriate safety measures must be applied for this synthesis! LPEI is synthesized by acid catalyzed hydrolysis of poly(2-ethyl-2-oxazolin). Complete deacylation is necessary for optimal transfection performance of the product (Thomas et al. 2005).

1. Dissolve 45 g of poly(2-ethyl-2-oxazolin) in 200 ml of 30% sulfuric acid.
2. Heat the reaction mixture and stir for 6 days. Remove propionic acid released during the reaction by azeotropic distillation.
3. Keep the volume of the mixture constant by dropwise addition of distilled H_2O.
4. When no acidic vapors are detectable with pH paper (after 5–6 days at 106°C), transfer the solution to a beaker in an ice bath and stop the reaction by adding sodium hydoxide in small portions under stirring. **Use caution: This is an exothermic reaction!**
5. Cool the solution further until a precipitate forms.

   Slow precipitation will increase the yield.
6. Wash the precipitate with distilled H_2O on a Büchner funnel covered with filter paper until the wash water is neutral.
7. Dry the product (LPEI) over phosphorus pentoxide for 1 week and store it thereafter dry at room temperature.
8. Resuspend 100 mg of LPEI in 1 ml of distilled H_2O. Add hydrochloric acid dropwise to pH 7.1 and adjust to 2 ml with distilled H_2O.
9. Sterilize the LPEI solution through 0.2-μm filters. Check the concentration after filtration with the copper complex assay (described in next section).
10. Store the stock solution obtained (50 mg/ml LPEI) either directly or after further dilution to a working concentration of 1 mg/ml either at 4°C or, for extended periods of time, frozen at –80°C.

Quantification of LPEI

LPEI can be quantified by a copper complex assay as described by Ungaro et al. (2003). LPEI forms a dark blue cuprammonium complex after addition of CuII ions.

11. Prepare 200 ml of a 0.1 M Na-acetate buffer (pH 5.4).
12. Dissolve 23 mg of $CuSO_4$ in 100 ml of 0.1 M Na-acetate (pH 5.4) and stir until the solution is clear. This is the Cu(II) solution.
13. Prepare a standard curve (five concentrations, in duplicate) with 1–10 μg of LPEI dissolved in 100 μl of distilled H_2O plus blank (distilled H_2O only).
14. Dilute different amounts of the LPEI solution to be analyzed in a total volume of 100 μl.
15. Add 100 μl of Cu(II) solution to the samples. Mix and incubate for 5 minutes at ambient temperature.
16. Measure the absorption of the solution at 285 nm on a UV spectrophotometer against the blank and calculate the LPEI concentration using the standard curve.

Polyplex Preparation

17. Prepare the polyplexes.

    To calculate the N:P ratio: An average of 330 daltons is assumed per phosphate per nucleotide and 43 daltons for nitrogen in LPEI. For example, to generate polyplexes at N:P 6 containing 1 μg of DNA, 43 x (6/330) = 0.78 μg of PEI is used. A 1-mg/ml LPEI solution is 23.3 mM for N.

For in vitro transfection

a. Dilute plasmid DNA and LPEI separately in HBS in equal volumes to 40 μg/ml (DNA) and 31 μg/ml (LPEI, in the case of N:P 6), respectively.

b. Transfer the LPEI solution to the DNA solution and rapidly mix by pipetting up and down five to ten times.

c. Allow the polyplexes to stand for 20 minutes at ambient temperature before transfection.

Polyplexes generated in salt-containing buffer aggregate with time (Wightman et al. 2001).

For in vivo transfection

a. Dilute plasmid DNA and LPEI separately in HBG in equal volumes to 400 μg/ml (DNA) and 310 μg/ml (LPEI, in the case of N:P 6), respectively.

Plasmid and buffers used must be free of endotoxin.

b. Transfer the LPEI solution to the DNA solution and rapidly mix by pipetting up and down five to ten times.

c. Allow the polyplexes to stand for 20 minutes at ambient temperature before injection.

Transfection

18. Carry out transfection

For transfection of adherent cells

a. Seed cells in 48-well plates 1 or 2 days before transfection.

Cells should have a confluency of 50–70%.

b. Aspirate medium from the cells. Replace the medium with prewarmed fresh medium (100–200 μl/cm^2 growth area).

c. Add polyplexes to the cells (final concentration of 0.5–5 μg/ml based on DNA). Gently swirl by slow hand rotation and then incubate in a cell incubator.

d. After 4 hours, replace the transfection solution with fresh culture medium.

For transfection of cells in suspension

a. Centrifuge logarithmically growing cells and resuspend in fresh culture medium at a concentration of 3×10^5 cells/ml.

b. Add 0.5 ml of cell suspension to each well of a 48-well plate. Add polyplex solution containing 1–5 μg of DNA, gently swirl by slow hand rotation, and then incubate in a cell incubator.

c. After 4 hours of incubation, centrifuge the plates for 5 minutes at 280*g*, carefully remove approximately 400 μl of the supernatant with a pipette, and add 1 ml of fresh medium.

Maximal transgene expression is observed between 24 and 48 hours after transfection, depending on the cell line and the transgene used.

For in vivo transfection

a. For systemic application in mice, immobilize the animal and immerse the tail for 1 minute in a 37°C water bath to allow the veins to dilate.

b. Draw the polyplex solution into a 1-ml insulin syringe (33-gauge needle) and remove any air bubbles. Inject the polyplex solution (200–250 ml) intravenously within approximately 5 seconds.

c. When injecting polyplexes into tissues (e.g., tumor), keep the needle in place for at least 1 minute to avoid leaking of the solution from the injection canal.

REFERENCES

Boeckle S., von Gersdorff K., van der Piepen S., Culmsee C., Wagner E., and Ogris M. 2004. Purification of polyethylenimine polyplexes highlights the role of free polycations in gene transfer. *J. Gene Med.* **6:** 1102–1111.

Boussif O., Zanta M.A., and Behr J.P. 1996. Optimized galenics improve *in vitro* gene transfer with cationic molecules up to 1000-fold. *Gene Ther.* **3:** 1074–1080.

Brunner S., Furtbauer E., Sauer T., Kursa M., and Wagner E. 2002. Overcoming the nuclear barrier: Cell cycle independent nonviral gene transfer with linear polyethylenimine or electroporation. *Mol. Ther.* **5:** 80–86.

Chollet P., Favrot M.C., Hurbin A., and Coll J.L. 2002. Side effects of a systemic injection of linear polyethylenimine-DNA complexes. *J. Gene Med.* **4:** 84–91.

Ferrari S., Moro E., Pettenazzo A., Behr J.P., Zacchello F., and Scarpa M. 1997. ExGen 500 is an efficient vector for gene delivery to lung epithelial cells *in vitro* and *in vivo*. *Gene Ther.* **4:** 1100–1106.

Goula D., Benoist D., Mantero S., Merlo G., Levi G., and Demeneix B.A. 1998a. Polyethylenimine-based intravenous delivery of transgenes to mouse lung. *Gene. Ther.* **5:** 1291–1295.

Goula D., Remy J.S., Erbacher P., Wasowicz M., Levi G., Abdallah B., and Demeneix B.A. 1998b. Size, diffusibility and transfection performance of linear PEI/DNA complexes in the mouse central nervous system. *Gene Ther.* **5:** 712–717.

Itaka K., Harada A., Yamasaki Y., Nakamura K., Kawaguchi H., and Kataoka K. 2004. In situ single cell observation by fluorescence resonance energy transfer reveals fast intra-cytoplasmic delivery and easy release of plasmid DNA complexed with linear polyethylenimine. *J. Gene Med.* **6:** 76–84.

Kircheis R., Wightman L., Schreiber A., Robitza B., Rossler V., Kursa M., and Wagner E. 2001. Polyethylenimine/DNA complexes shielded by transferrin target gene expression to tumors after systemic application. *Gene Ther.* **8:** 28–40.

Sonawane N.D., Szoka F.C., Jr., and Verkman A.S. 2003. Chloride accumulation and swelling in endosomes enhances DNA transfer by polyamine-DNA polyplexes. *J. Biol. Chem.* **278:** 44826–44831.

Thomas M., Lu J.J., Ge Q., Zhang C., Chen J., and Klibanov A.M. 2005. Full deacylation of polyethylenimine dramatically boosts its gene delivery efficiency and specificity to mouse lung. *Proc. Natl. Acad. Sci.* **102:** 5679–5684.

Ungaro F., De Rosa G., Miro A., and Quaglia F. 2003. Spectrophotometric determination of polyethylenimine in the presence of an oligonucleotide for the characterization of controlled release formulations. *J. Pharm. Biomed. Anal.* **31:** 143–149.

Wightman L., Kircheis R., Rossler V., Carotta S., Ruzicka R., Kursa M., and Wagner E. 2001. Different behavior of branched and linear polyethylenimine for gene delivery *in vitro* and *in vivo*. *J. Gene Med.* **3:** 362–372.

52 Protein Nanospheres for Gene Delivery: Preparation and In Vitro Transfection Studies with Gelatin Nanoparticles

Sushma Kommareddy and Mansoor M. Amiji

Department of Pharmaceutical Sciences, School of Pharmacy, Northeastern University, Boston, Massachusetts 02115

ABSTRACT

Nanoparticles have been widely used to overcome the barriers for drug delivery. Nanoparticles prepared from natural polymers have a significant advantage over those prepared from synthetic polymers. This chapter describes the protocols for preparation of nanoparticles from the biopolymer gelatin and in vitro cell culture studies using them. Presented here are the steps involved in preparation of nanoparticles by desolvation using ethanol and the precautions to be taken during the encapsulation of the payload. Trafficking of these nanoparticles in the cells can be best carried out by encapsulation of electron-dense material (e.g., gold) in the nanoparticles and visualizing them in the cell by transmission electron microscopy (TEM). Preparation and encapsulation of the gold nanoparticles within the protein nanospheres, cell culture, and sample preparation for TEM are described. Also provided are cell culture techniques for carrying out qualitative and quantitative analysis of transfection using the DNA encapsulated nanoparticles.

Continued

INTRODUCTION

Colloidal systems, including nanoparticles, have revolutionized the application of delivery systems by providing a mechanism for drug or gene transport to the desired physiological site of action with a suitable release profile. Polymers used for the preparation of nanoparticles may be synthetic or natural in origin (Douglas et al. 1987; Speiser 1991; Goa and Benfield 1994; Mehvar 2000; Suh and Matthew 2000; Cascone et al. 2001; Djagny et al. 2001; Moghimi et al. 2001; Soppimath et al. 2001; Brigger et al. 2002; Chuang et al. 2002; Lavasanifar et al. 2002; Barratt 2003; Panyam and Labhasetwar 2003). An ideal polymer should be biocompatible, biodegradable with minimum toxicity, sterile and nonpyrogenic, and must have a high capacity to accommodate drugs and genes and protect them from degradation.

Owing to their biocompatibility and biodegradability, polymers of natural origin are preferred over synthetic ones. Natural macromolecules such as proteins and polysaccharides offer a wide range of choices of biodegradable carriers for preparation of particles in the nano-size range. Proteins such as albumin, gelatin, zein, and poly(amino acids) are the most widely used (Courts 1954; Flory and Weaver 1960; Goa and Benfield 1994; Farrugia and Groves 1999a,b; Djagny et al. 2001; Chuang et al. 2002; Kaul and Amiji 2002, 2004a,b, 2005; Lavasanifar et al. 2002; Kaul et al. 2003). Their nonantigenic nature, ease of preparation, and the possibility of covalent modification through amino or carboxyl groups of the proteinaceous matrix have led to a wide variety of applications for protein-based nanoparticles. Advances in biotechnology have meant that nanoparticles with well-defined physicochemical properties can be synthesized reproducibly from gelatin and other proteins.

Gelatin is one of the most versatile proteins, being widely used in food products and pharmaceuticals. It is a hydrocolloid obtained by partial hydrolysis of type I collagen extracted from bovine or porcine skin and bones (Courts 1954; Flory and Weaver 1960; Farrugia and Groves 1999b). The large number of pendant functional groups throughout the polymeric chain can be chemically derivatized to induce novel functions. The reactive amine groups of gelatin are functionalized using polyethylene glycol (PEG), which forms a dense, hydrophilic cloud surrounding the particle, thus preventing the opsonization of particles by the reticuloendothelial system (RES) of the liver and spleen and hence improving the circulation time of the particles in blood (Kaul and Amiji 2004a, 2005). Thiolation of gelatin produces nanoparticles that are stable during blood circulation, releasing their payload in a reducing environment of high glutathione concentration (Kommareddy and Amiji 2005). Other modifications include covalent conjugates of gelatin with copolymers and cationic complexes (Leong et al. 1998; Kim and Byun 1999; Kaul and Amiji 2002, 2004a,b, 2005; Ma et al. 2002; Kaul et al. 2003; Kushibiki et al. 2004; Michaelis et al. 2004).

Preparation of nanoparticles from a preformed polymer is relatively straightforward and highly reproducible. However, molecular-weight heterogeneity of proteins makes it a challenge to produce stable nanoparticles. Marty et al. (1978) first observed that the complex phase behavior of gelatin was affected by temperature and other experimental conditions. Farrugia et al. have investigated the effect of temperature, pH, and ethanol concentration on the molecular-weight distribution profile of gelatin in solution in order to develop a robust method for the preparation of gelatin nanoparticles (Farrugia and Groves 1999b). Many research groups have successfully prepared nanoparticles from gelatin and its derivatives, and these have been evaluated for a range of biomedical applications (Oppheneium 1986; Leo et al. 1997; Truong-Le et al. 1998; Cascone et al. 2002; Gupta et al. 2004; Vandervoort and Ludwig 2004).

Desolvation is the most popular technique to prepare gelatin nanoparticles, followed by techniques such as two-step desolvation using acetone or coacervation using sodium sulfite

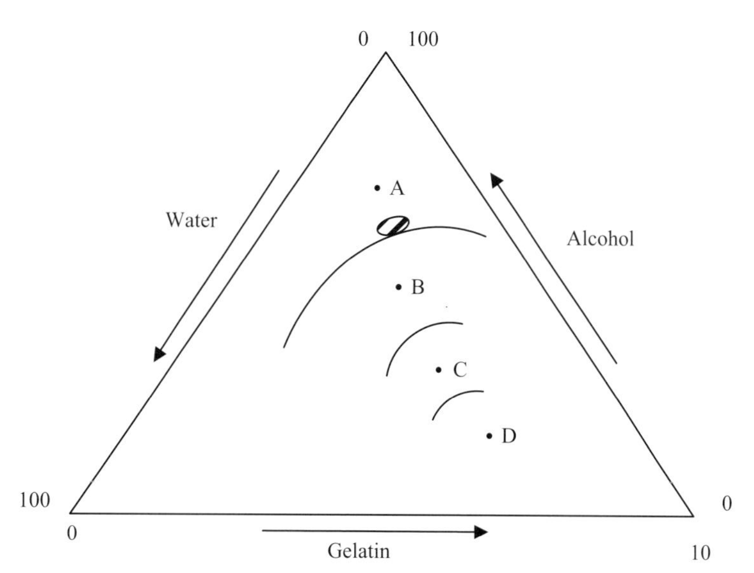

FIGURE 1. Phase diagram for the preparation of gelatin nanoparticles using a three-component system of gelatin-ethanol-water with ethanol and water expressed as % v/v and gelatin expressed as % w/v. *(A)* Region of rolling up of protein molecules; *(B)* coacervate region; *(C)* two-phase region; *(D)* flocculate region.

(Leong et al. 1998; Truong-Le et al. 1998, 1999; Coester et al. 2000; Brzoska et al. 2004; Michaelis et al. 2004). In desolvation, solvents of different polarity and hydrogen bonding, when added to solution of proteins, cause rolling up or controlled precipitation of proteins due to the displacement of water molecules from the surface of proteins (Weber et al. 2000). Desolvation using ethanol is a relatively simple one-step process wherein the nanoparticles are formed by rolling-up of gelatin molecules before the gelatin solution even enters the coacervate region which is followed by the two-phase region and flocculate region (for details, see Fig. 1) (Kreuter 1978; Marty et al. 1978). As a result, stable nanoparticles can be achieved with a size range of 100–200 nm. The size range and yield are affected by pH and stirring conditions (Kreuter 1994; Farrugia and Groves 1999b). The particles are further hardened by cross-linking with aldehydes such as formaldehyde or glutaraldehyde. The excess of glyoxal can be quenched by using sodium metabisulfite or any other simple amino acid such as glycine. Our laboratory has standardized procedures for the preparation of nanoparticles from gelatin and its functionalized derivatives using the solvent displacement process (Kaul and Amiji 2002; Kaul et al. 2003). The scanning electron microscopy (SEM) image of the freeze-dried, empty gelatin nanoparticles prepared by this technique can be seen in Figure 2.

On the basis of the method used for their formation, nanoparticles can be classified into either nanospheres (or nanoparticles), which are monolithic systems having a solid matrix or

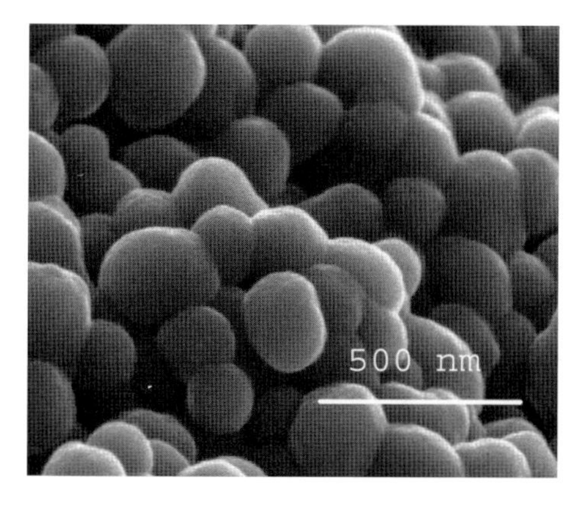

FIGURE 2. Scanning electron microscopy image of gelatin nanoparticles prepared by desolvation using ethanol.

nanocapsules, which have a hollow interior that is filled with a compound of interest and is surrounded by a polymeric shell. Gelatin nanoparticles are usually loaded with plasmid DNA or other hydrophilic macromolecules (Leo et al. 1997, 1999; Leong et al. 1998; Truong-Le et al. 1998, 1999; Coester et al. 2000; Kaul and Amiji 2002, 2004a,b, 2005; Kaul et al. 2003; Brzoska et al. 2004; Gupta et al. 2004; Zwiorek et al. 2004, 2005; Zillies and Coester 2005). Drugs, and especially plasmid DNA, may also be adsorbed to the surface of the nanoparticles.

After cellular uptake, DNA-containing nanoparticles travel through the endosomal/lysosomal vesicular transport pathway until they reach the perinuclear region of the cell. Efficient transfection is dependent on the ability of DNA to remain stable during vesicular transport and enter the nucleus. Failure of plasmid DNA to cross the nuclear membrane is the major hurdle for improving transfection efficiency with nonviral vectors. To characterize the trafficking pattern, fluorescent payloads and gold nanoparticles are used as markers. Gold nanoparticles are used because of their high electron density (which allows for easier visualization with electron microscopy) and their ability to complex with proteins (Kaul and Amiji 2004b). The cells are grown in vitro on Aclar sheets and incubated with the nanoparticles (Fig. 3). The particles that are taken up can be visualized in different parts of the cell under a TEM following sample preparation (Frens 1973; Horisberger 1979).

DNA-loaded gelatin nanoparticles are used in transfection studies and other gene delivery applications. The qualitative analysis of these studies is carried out by fluorescence microscopy and quantitative analysis by fluorescence-activated cell sorting (FACS). The plasmid DNA is released from the nanoparticles due to the acidic environment and the enzymes in the endolysosome. The principle involved in the transfection studies is represented in Figure 4. Calcium phosphate, diethylaminoethyl (DEAE) dextran, and cationic liposomes are some of the reagents that are used as transfection agents. Gorman et al. (1982) have introduced the concept of the reporter gene with a sensitive assay system for the gene product using the bacterial chloramphenicol acetyltransferase (CAT) gene and associated CAT assay system. Several reporter gene systems have been developed, including luciferase, β-galactosidase, alkaline phosphatase, and green fluorescent protein (GFP).

In our laboratory, we have carried out cell-trafficking studies using gelatin and PEGylated gelatin nanoparticles encapsulated with gold particles and observed the localization of the nanoparticles in the perinuclear region of the cells. Studies have also been conducted using plasmid DNA encapsulated gelatin and PEGylated gelatin nanoparticles. Upon transfection, most of the administered nanoparticles were internalized by NIH-3T3 fibroblast cells within the first 6 hours of incubation. GFP expression was observed as early as 12 hours after incubation of nanoparticles and remained stable for up to 96 hours (Kaul and Amiji 2004b). Other investigators have also shown that nanoparticles prepared from gelatin and its functionalized derivatives can be used for gene and drug delivery, both in vitro and in vivo (Leo et al. 1997, 1999; Leong et al. 1998; Truong-Le et al. 1998, 1999; Kaul and Amiji 2002, 2004a,b, 2005; Kaul et al. 2003; Zwiorek et al. 2004, 2005; Zillies and Coester 2005).

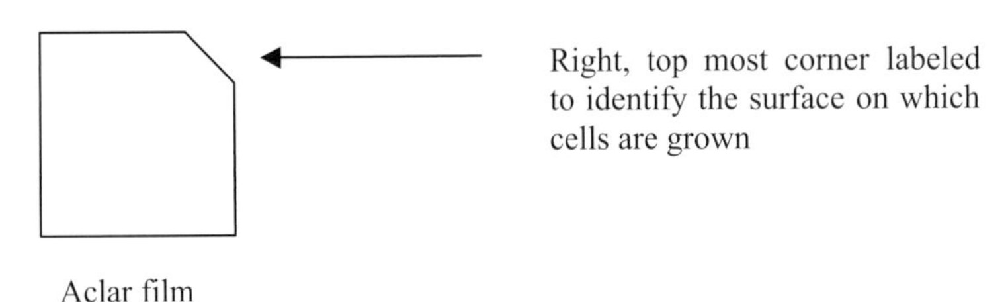

FIGURE 3. Preparation of Aclar film for cell trafficking of nanoparticles.

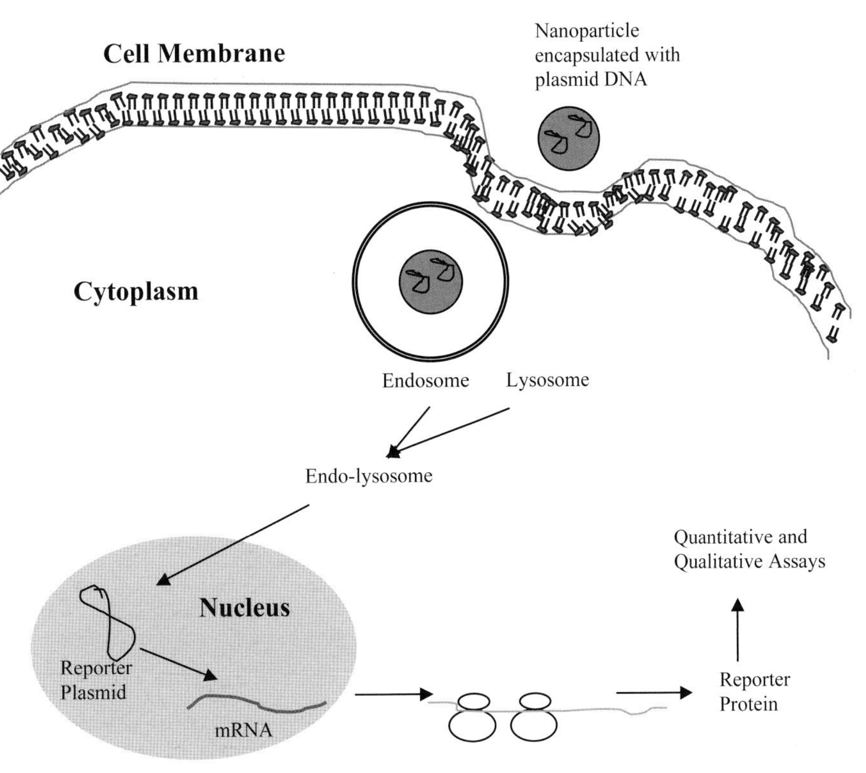

FIGURE 4. Principle involved in qualitative and quantitative analysis of transfection in cells using nanoparticles.

Protocol 1

Preparation and Loading of Gelatin Nanoparticles

This protocol describes the synthesis of nanoparticles using gelatin and its derivatives, encapsulation of DNA within the nanoparticles, and estimation of encapsulation efficiency.

MATERIALS

CAUTION: See Appendix for appropriate handling of materials marked with <!>.

Reagents

Ethanol (200 proof)
Gelatin (Type B, Bloom strength 225)
Glycine (0.1 M) <!>
Glyoxal (40% glutaraldehyde) <!>
Phosphate-buffered saline (PBS)

To prepare 1 liter of PBS, dissolve 8 g of sodium chloride (NaCl), 0.2 g of potassium chloride (KCl) <!>, 1.44 g of sodium phosphate (dibasic) (Na_2HPO_4) <!>, and 0.24 g of potassium dihydrogen phosphate (KH_2PO_4) <!> (monobasic) in 900 ml of H_2O. Adjust the pH to 7.4 and make up the volume to 1 liter.

PicoGreen <!> or another fluorescent dye that binds to double-stranded DNA
Protease
Sodium hydroxide <!> (NaOH; 0.1 N)

Equipment

Lyophilizer
Ultracentrifuge
Water bath, preset to 37°C

METHODS

Preparation of Nanoparticles Using Gelatin

1. Prepare a 1% (w/v) solution of gelatin by adding 200 mg of gelatin to 20 ml of H_2O. Dissolve by placing it in a water bath at 37°C.
2. Adjust the pH of the gelatin solution to 7.0 using 0.1 N NaOH.

   For DNA encapsulation, proceed to Step 9.
3. Add approximately 65 ml of ethanol with vigorous stirring, until a cloudy solution is obtained.

   Temperature and pH must be maintained for maximum yield of the nanoparticles. A cloudy solution without any particles visible to the naked eye indicates the formation of nanoparticles.
4. Dilute 0.1 ml of 40% (w/v) glutaraldehyde to 1 ml. Add the glyoxal to the nanoparticulate solution slowly over a period of 10 minutes with continuous stirring.

   This will cross-link the particles but prevent aggregation and cross-linking of two or more particles together.
5. Add 10 ml of 0.1 M glycine to quench the excess aldehyde groups of glyoxal.

Separation and Purification

6. Collect the nanoparticles by centrifugation in an ultracentrifuge at 16,000 rpm for 30 minutes.
7. Resuspend the pellet in H_2O and centrifuge repeatedly to remove any traces of ethanol and unreacted glycine.
8. After repeated washings, resuspend the nanoparticulate pellets in 1 or 2 ml of H_2O and freeze-dry to obtain a free-flowing product.

Incorporation of the Payload in Gelatin Nanoparticles

9. Incorporate the payload in the gelatin nanoparticles.

   For DNA encapsulation

   a. Dissolve the plasmid DNA to be encapsulated in the gelatin solution. The concentration of DNA in the gelatin solution can vary from 0.1% (w/w) to 1% (w/w).

      To avoid precipitation, add the DNA to the gelatin solution after the adjustment of pH to ensure that the DNA is physically encapsulated rather than electrostatically complexed or adsorbed to the surface.

   b. Prepare and purify the nanoparticles from the solution containing DNA by following Steps 3–8 above.

   For adsorption of DNA on to the surface of nanoparticles

   a. Prepare empty nanoparticles and then incubate with drug/DNA solution. A positive surface charge on gelatin nanoparticles can be obtained by adjusting the pH of the suspending buffer.

   b. Separate the nanoparticles from the unbound DNA by centrifugation as described previously.

Determination of DNA Encapsulation/Surface Adsorption

10. Plot a standard curve of the DNA solution using an appropriate analytical method. Use PicoGreen or another fluorescent dye that binds to double-stranded DNA.
11. Separate the DNA-loaded nanoparticles by centrifugation.
12. Wash the nanoparticles repeatedly to remove any surface-bound DNA.
13. Collect the supernatant and the washings. Estimate the free DNA remaining.
14. Subtract the total amount of the DNA present in the aqueous extract from that in the original reaction medium. Calculate the entrapment efficiency from the ratio of the drug entrapped to that added.

$$\text{Efficiency (\%)} = \frac{\text{Drug Entrapped}}{\text{Drug Added}} \times 100$$

15. Alternatively, measure DNA encapsulation or adsorption directly by digesting the DNA-containing gelatin nanoparticles in PBS (pH 7.4) containing 0.2 mg/ml protease. After about 30 minutes of incubation, when all of the gelatin nanoparticles are digested, a clear solution remains. Determine the amount of DNA by fluorescence measurements using PicoGreen or a similar dye.

Protocol 2

Intracellular Trafficking Studies

This protocol describes the preparation of gold particles, their encapsulation in protein nanoparticles, culture conditions, and sample preparation for TEM.

MATERIALS

CAUTION: See Appendix for appropriate handling of materials marked with <!>.

Reagents

Cells to be treated with gold nanoparticles
Distilled, deionized H_2O (ddH_2O)
Ethanol (200 proof) (30%, 50%, 75%, 85%, 90%, and 100%)
Gelatin (Type B, Bloom strength 225)
Glutaraldehyde (2.5% in 0.1 M sodium cacodylate buffer) <!>
Glycine (0.1 M) <!>
Glyoxal (40% glutaraldehyde) <!>
Gold chloride (4.34% w/v) <!>
Media: growth medium for cells and serum-free medium
Osmium tetroxide (1% [w/v] in 0.1 M sodium cacodylate buffer) <!>
Sodium cacodylate buffer (0.1 M) <!>
Sodium hydroxide (NaOH; 0.1 N) <!>
Spurr's resin <!>
Trisodium citrate (1% w/v, freshly prepared) <!>

Equipment

Aclar film
Aluminum plate
Incubator preset to 37°C
Lyophilizer
Oven preset to 60°C
Reflux condenser
Round-bottomed flask (250 ml)
Tissue-culture plates (six well)
Transmission electron microscope
Ultracentrifuge
Ultramicrotome with a diamond knife
Water bath, preset to 37°C

METHODS

Preparation of Gold Nanoparticles

1. Pour approximately 99 ml of ddH_2O into a round-bottomed flask (250 ml) fitted to a reflux condenser that is hooked to an external water supply.
2. Add 230 µl of 4.34% (w/v) gold chloride solution to the reaction vessel and bring to a boil while stirring continuously.
3. Add approximately 2.5 ml of freshly prepared 1% (w/v) trisodium citrate to the boiling reaction mixture.
4. Heat the resulting solution for an additional 10 minutes, until the gray solution turns to dark purple, burgundy, and then deep red.
5. Heat the reaction vessel for 2–3 minutes, until a stable red-orange color is obtained.
6. Cool the solution to room temperature and store at 4°C.
7. Separate the gold nanoparticles by centrifugation at 32,000 rpm for 1 hour.

   The nanoparticles will have a size range of 10–15 nm.

Encapsulation of the Gold Particles in Protein Nanospheres

8. Prepare a 1% (w/v) solution of gelatin by adding 200 mg of gelatin to 20 ml of H_2O. Dissolve by placing it in a water bath at 37°C.
9. Mix the gold nanoparticles (the pellet obtained in Protocol 2) in aqueous solution with the gelatin solution at 37°C.
10. Adjust the pH of the resulting solution to 7.0 using 0.1 N NaOH. Next, follow Steps 3–5 of Protocol 1 for preparation of nanoparticles. To encapsulate the gold particles in gelatin nanoparticles, disperse them in gelatin solution and prepare them as described in Protocol 1.
11. Centrifuge the gelatin nanoparticles encapsulated with gold particles and freeze-dry to obtain a free-flowing product following Steps 6–8 of Protocol 1.

Cell Culture

12. Cut Aclar sheets into squares that will fit into the wells of six-well tissue-culture plates.

    Mark the top surface of the Aclar sheets by cutting them in one of the corners, in order to avoid confusion about the surface on which the cells are grown (as seen in Fig. 3).
13. Seed cells in appropriate growth medium onto the Aclar sheets and grow to semiconfluence at 37°C.

    Cells can be seeded at a density of 1×10^6 cells per well; based on their doubling time, the cells would grow to semiconfluence within 24–48 hours.
14. Suspend the gelatin nanoparticles encapsulated with gold colloids in serum-free medium and incubate with the cells at 37°C.

    To monitor the uptake of nanoparticles, vary the time of incubation from 1 to 6 hours. Multiple samples can be taken at different points within this time interval and monitored using TEM.
15. Remove the nanoparticle suspension and replace with culture medium.
16. Sample the cells periodically. Prepare for TEM as described in the next section.

Sample Preparation for TEM

17. Carry out primary fixation of the samples for 1 hour with 2 ml of 2.5% (w/v) glutaraldehyde in 0.1 M sodium cacodylate buffer.
18. Perform three 10-minute washes with 0.1 M sodium cacodylate buffer at 4°C.
19. Postfix the cells using 1% (w/v) osmium tetroxide in 0.1 M sodium cacodylate buffer for 1 hour at room temperature. Follow this with three 10-minute washes with 0.1 M sodium cacodylate buffer.
20. Dehydrate the cells by incubating them in increasing concentrations of ethanol (30%, 50%, 75%, 85%, and 95%) for 10 minutes each.
21. Carry out the final dehydration in 100% ethanol for 1 hour at room temperature.
22. Infiltrate the samples by incubating in a 1:1 solution of ethanol and Spurr's resin for 1.5 hours, followed by infiltration in 100% Spurr's resin for 2 hours.
23. Place the Aclar film with the dehydrated and infiltrated cells at the bottom of an aluminum plate with Spurr's resin and polymerize for 24 hours in a hot-air oven at 60°C.
24. Section the samples in Spurr's resin using a diamond knife on a ultramicrotome.
25. Stain the sections in osmium tetroxide and view under TEM.

The steps involved in sample preparation for TEM are summarized in Table 1.

TABLE 1. Summary of Sample Preparation for TEM

Process	Chemical treatment	Temperature	Time involved
Primary fixation	Cells on Aclar film are fixed with 2.5% glutaraldehyde in 0.1 M sodium cacodylate buffer	room temperature	1 hour
Washings	Three times with 0.1 M sodium cacodylate buffer	4°C	10 minutes
Secondary fixation	1% (w/v) osmium tetroxide	room temperature	1 hour
Washings	Three times with 0.1 M sodium cacodylate buffer	4°C	10 minutes
Dehydration	30% ethanol 50% ethanol 75% ethanol 85% ethanol 95% ethanol absolute ethanol	room temperature	10 minutes 10 minutes 10 minutes 10 minutes 10 minutes 1 hour
Infiltration of resin	1:1 mixture of ethanol and Spurr's resin	room temperature	1.5 hours
Embedding	100% Spurr's resin	room temperature	2 hours
Curing	Aclar film is placed at the bottom of an aluminum plate in Spurr's resin	60°C	24 hours

Protocol 3

Cell Transfection and Analysis

This protocol describes cell transfection using DNA-loaded gelatin nanoparticles, followed by qualitative analysis with fluorescence microscopy and quantitative analysis by FACS.

MATERIALS

CAUTION: See Appendix for appropriate handling of materials marked with <!>.

Reagents

Cells in culture
Isopropyl alcohol (70%) <!>
Media: growth medium appropriate for cells and serum-free medium appropriate for cells
Mounting medium (biomeda)
Phosphate-buffered saline (PBS; sterile)
Reporter plasmid DNA (e.g., enhanced GFP-N1, CMV-β)
Trypan blue <!>
1x Trypsin-EDTA (0.05% trypsin and 0.53 mM EDTA)

Equipment

Fluorescence-activated cell sorter (FACS)
Fluorescence microscope (e.g., Olympus BX61)
Glass coverslips for fluorescence microscopy
Glass slides
Incubator preset to 37°C (5% CO_2 atmosphere)
Neubauer slide (slide with counting chambers)
Tissue-culture plates (six well)

METHODS

Preparation of Cells and Transfection

1. Soak glass coverslips for a few minutes in 70% alcohol. Use forceps to hold the coverslips and pass them through the flame of a spirit lamp.
2. Place the disinfected glass coverslips at the bottom of six-well tissue-culture plates.
3. Trypsinize the cells in culture using a 1x trypsin-EDTA solution. Add medium containing serum to inactivate the trypsin-EDTA.
4. Centrifuge the trypsinized cells in a benchtop centrifuge and suspend in fresh medium.
5. Add approximately 50 µl of the cell suspension to 200 µl of trypan blue to stain the dead cells. Count the number of living cells in the suspension on a neubauer slide and calculate the number of cells/ml of cell suspension according to the instructions provided for the neubauer slide.

6. Seed the six-well plates containing glass coverslips with 10^5 cells/well in growth medium.
7. Grow the cells to semiconfluence in an incubator at 37°C in 5% CO_2 atmosphere.
8. Disperse the DNA-loaded or fluorescently labeled gelatin nanoparticles in serum-free medium and add to cells such that each well is incubated with nanoparticles containing 20 µg of DNA.
9. Incubate the cells with the DNA-containing nanoparticles for 6 hours. At the end of the transfection period, siphon off the serum-free media containing the nanoparticles, wash with sterile PBS, and replace with growth medium for the cells.

Analysis of the Cells

10. Analyze the cells by one of the following methods.

For fluorescence microscopy

a. At different time intervals after transfection (allowing at least 6 hours for the expression of the transfected reporter plasmid), remove the medium from each well and wash the coverslips three times with sterile PBS.
b. Mount the coverslips onto clean glass slides containing fluorescence-free mounting medium.
c. Observe the uptake and distribution of the nanoparticles in the cells using a fluorescence microscope.

Figure 5 represents the six-well culture plates with sterilized coverslips. The fluorescence images of NIH-3T3 cells transfected with gelatin nanoparticles encapsulated with plasmid DNA (enhanced GFP-N1) are shown in Figure 6.

For cell culture for FACS analysis

Follow Steps 3–9 for cell preparation and transfection, using six-well plates without coverslips, and then analyze as follows.

a. Allow sufficient time for the cells to synthesize the protein encoded by the reporter plasmid.
b. Depending on the type of plasmid DNA used, analyze the cells to quantify the transfection efficiency. If an enhanced GFP reporter plasmid is used, perform FACS analysis for GFP protein expression. If luciferase is used, quantify luminescence. For β-galactosidase, use visible chromogen development techniques.

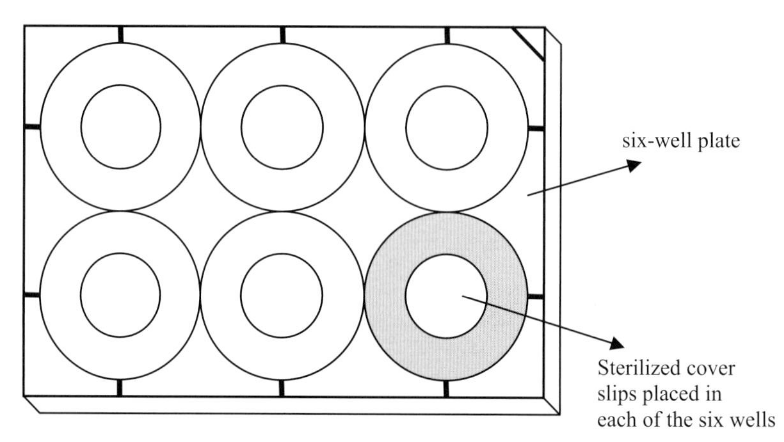

FIGURE 5. Diagrammatic representation of six-well culture plates with sterilized coverslips for fluorescence microscopy studies.

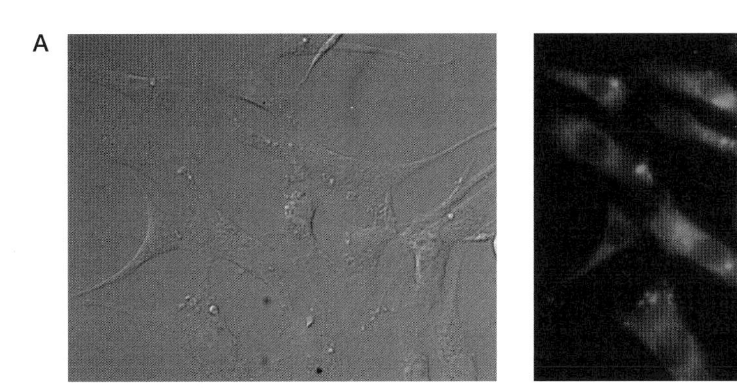

FIGURE 6. Differential interference contrast (DIC) (*A*) and fluorescence images (*B*) of NIH-3T3 cells transfected with gelatin nanoparticles encapsulated with pEGFPN1 plasmid DNA, observed 24 hours after transfection with an Olympus BX61 fluorescence microscope.

REFERENCES

Barratt G. 2003. Colloidal drug carriers: Achievements and perspectives. *Cell. Mol. Life Sci.* **60:** 21–37.

Brigger I., Dubernet C., and Couvreur P. 2002. Nanoparticles in cancer therapy and diagnosis. *Adv. Drug Delivery Rev.* **54:** 631–651.

Brzoska M., Langer K., Coester C., Loitsch S., Wagner T.O., and Mallinckrodt C. 2004. Incorporation of biodegradable nanoparticles into human airway epithelium cells—In vitro study of the suitability as a vehicle for drug or gene delivery in pulmonary diseases. *Biochem. Biophys. Res. Commun.* **318:** 562–570.

Cascone M.G., Lazzeri L., Carmignani C., and Zhu Z. 2002. Gelatin nanoparticles produced by a simple W/O emulsion as delivery system for methotrexate. *J. Mater. Sci. Mater. Med.* **13:** 523–526.

Cascone M.G., Barbani N., Cristallini C., Giusti P., Ciardelli G., and Lazzeri L. 2001. Bioartificial polymeric materials based on polysaccharides. *J. Biomater. Sci. Polym. Ed.* **12:** 267–281.

Chuang V.T., Kragh-Hansen U., and Otagiri M. 2002. Pharmaceutical strategies utilizing recombinant human serum albumin. *Pharm. Res.* **19:** 569–577.

Coester C.J., Langer K., van Briesen H., and Kreuter J. 2000. Gelatin nanoparticles by two step desolvation—A new preparation method, surface modifications and cell uptake. *J. Microencapsul.* **17:** 187–193.

Courts A. 1954. The N-terminal amino acid residues of gelatin 2. Thermal degradation. *Biochem. J.* **58:** 74–79.

Djagny V.B., Wang Z., and Xu S. 2001. Gelatin: A valuable protein for food and pharmaceutical industries: Review. *Crit. Rev. Food Sci. Nutr.* **41:** 481–492.

Douglas S.J., Davis S.S., and Illum L. 1987. Nanoparticles in drug delivery. *Crit. Rev. Ther. Drug Carrier Syst.* **3:** 233–261.

Farrugia C.A. and Groves M.J. 1999a. The activity of unloaded gelatin nanoparticles on murine melanoma B16-F0 growth in vivo. *Anticancer Res.* **19:** 1027–1031.

———. 1999b. Gelatin behaviour in dilute aqueous solution: Designing a nanoparticulate formulation. *J. Pharm. Pharmacol.* **51:** 643–649.

Flory P.J. and Weaver E.S. 1960. Helix–coil transitions in dilute aqueous collagen solutions. *J. Am. Chem. Soc.* **82:** 4518–4525.

Frens G. 1973. Controlled nucleation for the regulation of the particle size in monodisperse gold solution. *Nat. Phys. Sci.* **241:** 20–22.

Goa K.L. and Benfield P. 1994. Hyaluronic acid. A review of its pharmacology and use as a surgical aid in ophthalmology, and its therapeutic potential in joint disease and wound healing. *Drugs* **47:** 536–566.

Gorman C.M., Moffat L.F., and Howard H.F. 1982. Recombinant genomes which express chloramphenicol acetyltransferase in mammalian cells. *Mol. Cell. Biol.* **2:** 1044–1051.

Gupta A.K., Gupta M., Yarwood S.J., and Curtis A.S. 2004. Effect of cellular uptake of gelatin nanoparticles on adhesion, morphology and cytoskeleton organisation of human fibroblasts. *J. Control. Release* **95:** 197–207.

Horisberger M. 1979. Evaluation of colloidal gold as a cytochemical marker for transmission and scanning electron microscopy. *Biol. Cell* **36:** 253–258.

Kaul G. and Amiji M. 2002. Long-circulating poly(ethylene glycol)-modified gelatin nanoparticles for intracellular delivery. *Pharm. Res.* **19:** 1062–1068.

———. 2004a. Biodistribution and targeting potential of poly(ethylene glycol)-modified gelatin nanoparticles in subcutaneous murine tumor model. *J. Drug Target.* **12:** 585–591.

———. 2004b. Cellular interactions and in vitro DNA transfection studies with poly(ethylene glycol)-modified gelatin nanoparticles. *J. Pharm. Sci.* **94:** 184–198.

———. 2005. Tumor-targeted gene delivery using poly(ethylene glycol)-modified gelatin nanoparticles: In vitro and in vivo studies. *Pharm. Res.* **22:** 951–961.

Kaul G., Lee-Parsons C., and Amiji M. 2003. Poly(ethylene glycol)-modified gelatin nanoparticles for intracellular delivery. *Pharm. Eng.* **23:** 1–5.

Kim K.J. and Byun Y. 1999. Preparation and characterizations of self-assembled PEGylated gelatin nanoparticles. *Biotechnol. Bioprocess Eng.* **4:** 210–214.

Kommareddy S. and Amiji M. 2005. Preparation and evaluation of thiol-modified gelatin nanoparticles for intracellular DNA delivery in response to glutathione. *Bioconjug. Chem.* **16:** 1423–1432.

Kreuter J. 1978. Nanoparticles and nanocapsules: New dosage forms in the nanometer size range. *Pharm. Acta Helv.* **53:** 33–39.

———. 1994. Drug targeting with nanoparticles. *Eur. J. Drug Metab. Pharmacokinet.* **19:** 253–256.

Kushibiki T., Matsuoka H., and Tabata Y. 2004. Synthesis and physical characterization of poly(ethylene glycol)-gelatin conjugates. *Biomacromolecules* **5:** 202–208.

Lavasanifar A., Samuel J., and Kwon G.S. 2002. Poly(ethylene oxide)-block-poly(L-amino acid) micelles for drug delivery. *Adv. Drug Delivery Rev.* **54:** 169–190.

Leo E., Cameroni R., and Forni F. 1999. Dynamic dialysis for the drug release evaluation from doxorubicin-gelatin nanoparticle conjugates. *Int. J. Pharm.* **180:** 23–30.

Leo E., Arletti R., Forni F., and Cameroni R. 1997. General and cardiac toxicity of doxorubicin-loaded gelatin nanoparticles. *Farmaco* **52:** 385–388.

Leong K.W., Mao H.Q., Truong-Le V.L., Roy K., Walsh S.M., and August J.T. 1998. DNA-polycation nanospheres as non-viral gene delivery vehicles. *J. Control. Release* **53:** 183–193.

Ma J., Cao H., Li Y., and Li Y. 2002. Synthesis and characterization of poly(DL-lactide)-grafted gelatins as bioabsorbable amphiphillic polymers. *J. Biomater. Sci. Polym. Ed.* **13:** 67–80.

Marty J.J., Oppenheim R.C., and Speiser P. 1978. Nanoparticles—A new colloidal drug-delivery system. *Pharm. Acta Helv.* **53:** 17–23.

Mehvar R. 2000. Dextrans for targeted and sustained delivery of therapeutic and imaging agents. *J. Control. Release* **69:** 1–25.

Michaelis M., Langer K., Arnold S., Doerr H.W., Kreuter J., and Cinatl J., Jr. 2004. Pharmacological activity of DTPA linked to protein-based drug carrier systems. *Biochem. Biophys. Res. Commun.* **323:** 1236–1240.

Moghimi S.M., Hunter A.C., and Murray J.C. 2001. Long-circulating and target-specific nanoparticles: Theory to practice. *Pharmacol. Rev.* **53:** 283–318.

Oppheneium R.C. 1986. Nanoparticulate drug delivery systems based on gelatin and albumin. In *Polymeric nanoparticles and microspheres* (ed. P. Guiot and P. Couvreur), pp. 1–25. CRC Press, Boca Raton, Florida.

Panyam J. and Labhasetwar V. 2003. Biodegradable nanoparticles for drug and gene delivery to cells and tissue. *Adv. Drug Delivery Rev.* **55:** 329–347.

Soppimath K.S., Aminabhavi T.M., Kulkarni A.R., and Rudzinski W.E. 2001. Biodegradable polymeric nanoparticles as drug delivery devices. *J. Control. Release* **70:** 1–20.

Speiser P.P. 1991. Nanoparticles and liposomes: A state of the art. *Methods Find. Exp. Clin. Pharmacol.* **13:** 337–342.

Suh J.K. and Matthew H.W. 2000. Application of chitosan-based polysaccharide biomaterials in cartilage tissue engineering: A review. *Biomaterials* **21:** 2589–2598.

Truong-Le V.L., August J.T., and Leong K.W. 1998. Controlled gene delivery by DNA-gelatin nanospheres. *Hum. Gene Ther.* **9:** 1709–1717.

Truong-Le V.L., Walsh S.M., Schweibert E., Mao H.Q., Guggino W.B., August J.T., and Leong K.W. 1999. Gene transfer by DNA-gelatin nanospheres. *Arch. Biochem. Biophys.* **361:** 47–56.

Vandervoort J. and Ludwig A. 2004. Preparation and evaluation of drug-loaded gelatin nanoparticles for topical ophthalmic use. *Eur. J. Pharm. Biopharm.* **57:** 251–261.

Weber C., Coester C., Kreuter J., and Langer K. 2000. Desolvation process and surface characterisation of protein nanoparticles. *Int. J. Pharm.* **194:** 91–102.

Zillies J. and Coester C. 2005. Evaluating gelatin based nanoparticles as a carrier system for double stranded oligonucleotides. *J. Pharm. Pharm. Sci.* **7:** 17–21.

Zwiorek K., Kloeckner J., Wagner E., and Coester C. 2004. In vitro gene transfection with surface modified gelatin nanoparticles. In *Proceedings of the International Conference on MEMS and NANO and Smart Systems*, August, 2004, pp. 60–63. IEEE Computer Society, Banff, Alberta, Canada.

———. 2005. Gelatin nanoparticles as a new and simple gene delivery system. *J. Pharm. Pharm. Sci.* **7:** 22–28.

53 Vesicular Stomatitis Virus-G Conjugate

Atsushi Miyanohara

UCSD Program in Human Gene Therapy, University of California School San Diego of Medicine, La Jolla, California 92093-0692

ABSTRACT

The fusiogenic envelope G glycoprotein of the vesicular stomatitis virus (VSV-G) that has been used to pseudotype retrovirus and lentivirus vectors can be used alone as an efficient vehicle for gene transfer. VSV-G protein is secreted into the culture medium as sendimentable vesicles from cells transfected with a VSV-G expression plasmid in the absence of other viral components. The VSV-G vesicles in the conditioned medium can be partially purified by pelleting through sucrose cushion ultracentrifugation. Protein–DNA complexes are formed by mixing the VSV-G vesicles with naked plasmid DNA. Such complexes demonstrate markedly enhanced transfection efficiency when added to the culture medium of recipient cells. The cell tropism of VSV-G-DNA complex-mediated gene transfer resembles that of VSV-G-pseudotyped retrovirus and lentivirus vectors, and the complex is therefore particularly useful for transfection of cells that are refractory to other methods. Still, some cells are refractory to VSV-G-mediated transfection. It should also be noted that overdose of VSV-G can be quite toxic to the recipient cells. The primitive complexes formed by mixing a viral fusiogenic envelope protein with naked DNA may represent a step toward fusing useful features of viral and nonviral vectors for safer and more efficient gene transfer.

INTRODUCTION

The VSV-G protein is a useful reagent for the study of many aspects of cell biology as well as virology. It is also well known that the VSV-G protein can replace the envelope proteins of a number of other viruses to produce pseudotyped virus particles with new host ranges and cell tropisms (Zavada 1982). VSV-G has therefore been used as an efficient surrogate envelope protein to produce pseudotyped retrovirus and lentivirus vectors (Emi et al. 1991; Burns et al. 1993; Yee et al. 1994b; Naldini et al. 1996; Poeschla et al. 1996). Because of the stability and pantropic characteristics of the VSV-G protein, VSV-G-pseudotyped vectors can be produced to very high titers and can mediate efficient gene transfer into many cell types, including nonmammalian cells, that are refractory to other methods of gene transfer (Burns et al. 1993). We have demonstrated that VSV-G protein is produced and efficiently released into culture medium as sedimentable vesicles from cells transfected with a plasmid expressing VSV-G in the absence of other viral components. The VSV-G vesicles partially purified from such conditioned medium can be introduced into the membrane of spikeless, immature, noninfectious murine leukemia virus (MLV)- and human immunodeficiency virus type 1 (HIV1)-based virus-like particles (VLPs) under cell-free conditions to assemble infectious virus particles in vitro (Abe et al. 1998b; Sharma et al. 2000). We have also demonstrated that VSV-G can be physically incorporated into lipofectin–DNA complexes to form fusiogenic VSV-G liposomes (VSV-G virosomes) that demonstrate an increased transfection efficiency that is not subject to serum-mediated inhibition (Abe et al. 1998a). We have further found that VSV-G vesicles can associate with naked DNA in the absence of other viral components or other fusiogenic factors and that such VSV-G-DNA complexes demonstrate markedly enhanced transfection efficiency in a variety of recipient cells (Okimoto et al. 2001).

Whatever is the mechanism for the DNA–VSV-G-mediated gene transfer, it may be more reminiscent of virus transduction than physical transfection. This is suggested by the greatly increased susceptibility of several cell lines that are relatively resistant to other transfection methods but highly susceptible to VSV-G-pseudotyped retrovirus and lentivirus vectors. The experiments were intended to develop hybrid DNA-based virus-like vectors that can be assembled fully in vitro under cell-free conditions from viral components. The primitive form of DNA–VSV-G complexes can be further modified to add more advantageous features such as cell targeting, efficient nuclear transport, and site-specific integration.

Protocol

Preparation of VSV-G Conjugate and Use in Gene Transfer

This protocol describes simple methods for preparation of VSV-G and for gene transfer with DNA–VSV-G complexes.

MATERIALS

Reagents

Anti-VSV-G monoclonal antibody P5D4 (Sigma-Aldrich)
BCA Protein Assay Kit (Pierce, Rockford, Illinois)
Culture medium for HEK293 cells and recipient cells
HEK293 cells (ATCC)
Phosphate-buffered saline (PBS)
Plasmids
  Reporter gene expression plasmid (e.g., EGFP-N1, BD Biosciences, Palo Alto, California)
  VSV-G expression plasmid pHCMV-G (Yee et al. 1994a,b)
Polybrene (4-mg/ml stock)
Reagents for
  SDS-PAGE and silver staining
  western blot analysis
Recipient cells
Sucrose (20%)/PBS

Equipment

Centrifuge with SW 28 rotor and tubes (Beckman)
Equipment for
  SDS-PAGE and silver staining
  western blot analysis
Filters (0.45 μm)
Microcentrifuge tubes
Plate (sixwell)
Tissue-culture dish (10 cm)
Water bath preset to 37°C

METHODS

Production of VSV-G

Any transfection method can be used to introduce the VSV-G expression plasmid into the cells to produce VSV-G protein. We routinely use HEK293 cells because of their extreme high efficiency of transfection by any method including the $CaPO_4$-DNA coprecipitation method (Yee et al. 1994a,b).

1. Twenty-four hours before transfection, seed the HEK293 cells in a 10-cm culture dish in appropriate medium at a cell density of approximately 70% and grow the cells overnight.

   The cell density should be 90–100% by the time of transfection.

2. Transfect the HEK293 cells with 10–20 µg of pHCMV-G plasmid by the standard $CaPO_4$-DNA coprecipitation method.

   Because HEK293 cells are sensitive to the toxicity of the $CaPO_4$ precipitate, transfection time should be shorter than 8 hours.

3. The following day, replace the medium with fresh medium.

   Formation of a weak syncytium caused by the fusiogenic activity of the VSV-G may be visible but should not be too apparent at this time point. Continue to incubate overnight.

4. The syncytium should now be more apparent. Harvest the medium, filter through a 0.45-µm filter, and store at –70°C. Add fresh medium to the cells and continue to incubate overnight.

5. Many cells may now be floating. Harvest the medium and filter again as in Step 4. Store at –70°C or proceed to the purification.

Partial Purification of the VSV-G Vesicles in the Conditioned Medium (Simple Method)

6. Partially purify the VSV-G vesicles in the conditioned medium as follows:

   a. Thaw the medium harvested on the second and third days posttransfection in a 37°C water bath and mix. Pour the medium into the centrifuge tube of the Beckman SW 28 rotor and centrifuge at 25,000 rpm for 2 hours at 4°C. Discard the supernatant and resuspend the pellet in 5–10 ml of PBS. For larger volumes (>200 ml) of conditioned medium, centrifuge at 6000–7000 rpm overnight with a large-capacity rotor to collect the VSV-G vesicles.

   b. To remove insoluble debris, centrifuge at low speed (e.g., 3000 rpm for 10 min at 4°C). Place the supernatant onto a 20% sucrose/PBS cushion (5 ml) in a tube for the SW 28 rotor, overlay with PBS to fill the tube, and centrifuge at 25,000 rpm for 6 hours to overnight. After centrifugation, remove the supernatant completely and resuspend the pellet in 100–200 µl of PBS.

   c. Measure the protein concentration of the VSV-G preparation with the BCA Protein Assay Kit. Estimate the degree of purification by SDS-PAGE followed by silver staining and western blot analysis, using the anti-VSV-G monoclonal antibody P5D4.

Transfection with VSV-G (in a Six-well Plate)

7. Twenty-four hours before transfection, seed the recipient cells in appropriate culture medium in a six-well plate and grow them overnight.

8. Reduce the medium in the wells to about 1 ml/well and add polybrene to the medium (4 µg/ml). In a microcentrifuge tube, mix 2–5 µg of plasmid DNA (e.g., pEGFP-N1) and 10 µl of VSV-G solution containing approximately 1 µg of protein. Add the VSV-G-DNA mixture to the recipient cells immediately.

   Alternatively, add the plasmid DNA directly to the well containing 1 ml of medium, incubate for about 15 minutes, and then add the polybrene and VSV-G to the cell medium.

9. After addition of the mixture, mix well and incubate in a regular cell culture incubator for 2–6 hours and then add 1 ml/well fresh medium and continue the incubation. If there is much debris, wash the cells and add 2 ml of fresh medium.

10. Assess the cells for transfection efficiency at day 1–3 by measuring the transgene expression.

ACKNOWLEDGMENTS

This work was supported in part by National Institutes of Health grant HL66941.

REFERENCES

Abe A., Miyanohara A., and Friedmann T. 1998a. Enhanced gene transfer with fusogenic liposomes containing vesicular stomatitis virus G glycoprotein. *J. Virol.* **72:** 6159–6163.

Abe A., Chen S.-T., Miyanohara A., and Friedmann T. 1998b. In vitro cell-free conversion of noninfectious Moloney retrovirus particles to an infectious form by the addition of the vesicular stomatitis virus surrogate envelope G protein. *J. Virol.* **72:** 6356–6361.

Burns J.C., Friedmann T., Driever W., Burrascano M., and Yee J.K. 1993. Vesicular stomatitis virus G glycoprotein pseudotyped retroviral vectors: Concentration to very high titer and efficient gene transfer into mammalian and non-mammalian cells. *Proc. Natl. Acad. Sci.* **90:** 8033–8037.

Emi N., Friedmann T., and Yee J.K. 1991. Pseudotype formation of murine leukemia virus with the G protein of vesicular stomatits virus. *J. Virol.* **65:** 1202–1207.

Naldini L., Blomer U., Gallay P., Ory D., Mulligan R., Gage F., Verma I.M., and Trono D. 1996. In vivo gene delivery and stable transduction of nondividing cells by a lentiviral vector. *Science* **272:** 263–267.

Okimoto T., Friedmann T., and Miyanohara A. 2001. VSV-G envelope glycoprotein forms complexes with plasmid DNA and MLV retrovirus-like particles in cell-free conditions and enhances DNA transfection. *Mol. Ther.* **4:** 232–238.

Poeschla E., Corbeau P., and Wong-Staal F. 1996. Development of HIV1 vectors for anti-HIV gene therapy. *Proc. Natl. Acad. Sci.* **93:** 11395–11399.

Sharma S., Miyanohara A., and Friedmann T. 2000. Separable mechanisms of attachment and cell uptake during retrovirus infection. *J. Virol.* **74:** 10790–10795.

Yee J.K., Friedmann T., and Burns J.C. 1994a. Generation of high-titer, pantropic retroviral vectors with very broad host range. *Methods Cell Biol.* **43:** 99–112.

Yee J.K., Miyanohara A., LaPorte P., Bouic K., Burns J.C., and Friedmann T. 1994b. A general method for the generation of high-titer, pantropic retroviral vectors: Highly efficient infection of primary hepatocytes. *Proc. Natl. Acad. Sci.* **91:** 9564–9568.

Zavada J. 1982. The pseudotype paradox. *J. Gen. Virol.* **63:** 15–24.

54 High-throughput Methods for Screening Polymeric Transfection Reagents

Gregory T. Zugates, Daniel G. Anderson, and Robert Langer

Department of Chemical Engineering, Massachusetts Institute of Technology, Cambridge, Massachusetts 02139

ABSTRACT

In our efforts to identify cationic degradable materials for gene delivery, we have developed high-throughput methods to assess polymer-mediated transfection. One strength of this technique is its modular design with 96-well plates that allows many transfection parameters to be varied and optimized in parallel. By varying material properties and transfection conditions, the effects of polymer structure, polymer:DNA ratio, DNA concentration, cell type, and cell density can be investigated. Hundreds of polymers can be tested in quadruplicate in a single day with a minimal amount of reagents used. The technique can easily be automated to efficiently and reproducibly test large material libraries. One limitation is that many plate types, solutions, and equipment must be stocked and sterilized. In addition, because all polymers are processed simultaneously in very small volumes, it is difficult to validate each step for each polymer to ensure solution uniformity and adequate polymer–DNA complexation. Despite these drawbacks, this high-throughput screening method has already been used successfully in the development of efficient polycation vectors. It may be a useful platform for the future discovery of gene delivery systems.

INTRODUCTION

The clinical success of gene therapy requires the development of a safe and efficient delivery system for DNA. Although viral vectors have proved to be highly effective, concerns over their safety and immunogenicity have prompted the exploration of nonviral alternatives (Nishikawa and Huang 2001; Check 2003). One class of nonviral vectors with particular promise is cationic polymers (De Smedt et al. 2000; Merdan et al. 2002; Wagner 2004), which can associate electrostatically with plasmid DNA to form nanocomplexes. In some cases, this is sufficient for cellular uptake and transfection.

Many cationic polymers have been explored for their ability to condense DNA and mediate cellular transfection. These include polypeptides (e.g., poly-L-lysine [PLL] and polyhistidine), polysaccharides (e.g., chitosan and diethylaminoethyl-dextran), polyethylenimines (PEI) and various derivatives, and both linear and dendritic poly(amido amine)s (PAMAMs) (Merdan et al. 2002; Pack et al. 2005). In addition, hydrolyzable polycations have been synthesized to provide a mechanism for DNA unpackaging and polymer elimination. Examples of erodible polymers developed for gene delivery include polyphosphoesters, poly(4-hydroxy-L-proline ester), poly(α-[4-aminobutyl]-L-glycolic acid), and the poly(β-amino ester)s (Lim et al. 1999, 2000; Lynn and Langer 2000; Zhao et al. 2003). Many materials developed for gene therapy have tertiary amines built into the polymer framework. These groups can buffer against endosomal acidification and prevent the acid-induced degradation of DNA. Furthermore, the osmotic pressure gradient that results is hypothesized to induce endosomal rupture and the release of polymer–DNA complexes into the cytosol (Boussif et al. 1995). Although much progress has been made, the precise mechanisms by which polymers facilitate gene delivery remain unclear. This missing link between chemical structure and cellular function has provided strong motivation for the continued development of new polycationic materials.

Central to the synthesis of polycationic transfection reagents is a robust and reliable method to screen for efficacy. There are numerous parameters that must be controlled and optimized, such as polymer–DNA-binding conditions (mixing method, pH, ionic strength, incubation time, concentrations, ratios), transfection media (type, serum content), DNA dose and incubation time on the cells, cell specificity, and assay conditions. Given the wide parameter space and increasingly large libraries of materials, high-throughput screening methods are necessary to assess transfection efficiencies (Regelin et al. 2001; Anderson et al. 2003). These screens can be performed manually or by automated systems and generally use microtiter plates of cells.

In our efforts to synthesize large libraries of poly(β-amino ester)s for gene delivery, we have developed a high-throughput, 96-well plate method to test their in vitro ability to transfect cells (Akinc et al. 2003b; Anderson et al. 2003, 2005). A schematic diagram of the process is shown in Figure 1. The transfection process involves four separate transfer steps over five 96-well plates. The first plate on the left contains aqueous polymer solutions, with the 12 wells in the first row containing ten different polymers to be tested, a positive control (e.g., PEI) and a blank (i.e., no polymer). These concentrated solutions are diluted in the second 96-well plate (Step 1). Each row in this plate contains a separate dilution so that multiple polymer:DNA ratios can be easily tested. Only two ratios are shown, but more dilutions can be added, depending on the number of polymer:DNA ratios desired. At this point, four aliquots of each diluted polymer solution are added to four rows of a plate containing DNA (Step 2). This allows polymer–DNA complexes to form at each ratio and in quadruplicate. The complexes are then diluted in media to obtain the desired DNA concentration (Step 3) and finally transferred to the cells (Step 4). This procedure allows for the testing of 11 polymers (including the postive control) at two polymer–DNA ratios in quadruplicate. The following discussion emphasizes important considerations in testing hydrolyzable pH-responsive polycations such as the poly(β-amino ester)s.

Cells are seeded in 96-well plates one day in advance of transfection to allow the cells sufficient time to adhere to the plate surface. Opaque white plates are preferred for luminescence assays because they minimize signal bleed between wells when measuring the expressed protein levels. The optimal cell density may be different for each polymer and cell type (Gebhart and Kabanov 2001). It is rather easy to scale up the procedure to test a range of cell densities if desired.

The high-throughput transfection begins by making aqueous polymer and DNA solutions. For many hydrophilic polymers such as PLL, PEI, and the PAMAMs, solutions can easily be prepared over a wide pH range by dissolving the polymer in an appropriate buffer. However, some cationic polymers are almost completely insoluble at physiologic pH and higher, but they can dissolve to a modest extent in acidic buffers. Aqueous stock solutions of these polymers can be made

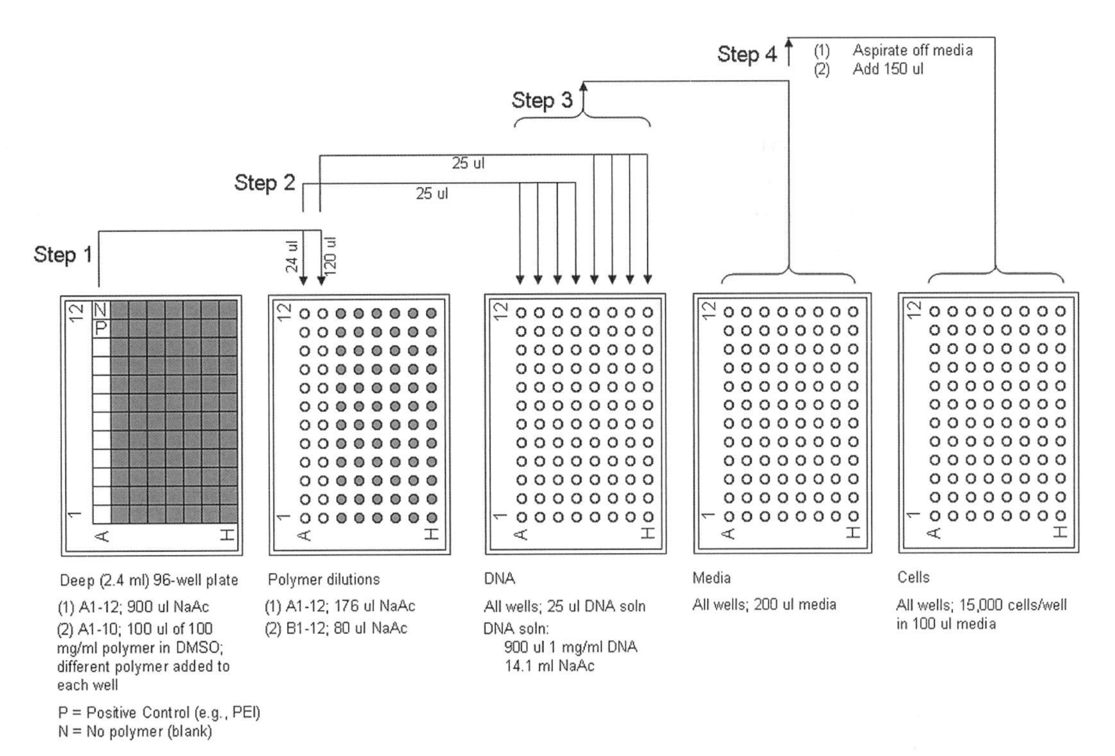

FIGURE 1. Layout of 96-well plates and steps involved in a high-throughput transfection. All values shown for each solution and transfer step are given in microliters per well. In addition, all four transfer steps are performed with a 12-channel pipettor. Shaded wells indicate rows in the plate that are not used. All solutions should be made and plated as indicated below each plate before any of the steps above the plates are performed. The first plate contains sodium acetate buffer (*1*) and is used to dissolve the polymers (*2*), with each well containing its own polymer. The second plate contains only buffer and is used to dilute the polymer to the appropriate concentration (Step 1). The two dilutions shown correspond to a 20:1 and 100:1 polymer:DNA weight ratio. Each solution is made on a 200 μl per well scale so that there is enough to test each polymer in quadruplicate. Four aliquots of each ratio are mixed with four rows of DNA to generate the fourfold replication (Step 2). The polymer–DNA complexes are then diluted into media (Step 3) and added to the cells (Step 4).

by dissolving the polymer at a high concentration in a water-miscible solvent of low toxicity and diluting into an acidic aqueous buffer, such as acetate or another non-amine-based buffer. The low pH is necessary to solubilize most poly(β-amino ester)s and to ensure that the tertiary amines in the backbone are in a protonated state to facilitate DNA complexation, but does not significantly affect DNA integrity, especially in the short term of only a few hours (Walter et al. 1999). The ionic strength is selected to solubilize the polymer and account for its buffering capacity without noticeably decreasing the pH of the medium. From the stock solution, dilutions are made so that different polymer:DNA ratios can be tested (Fig. 1, Step 1). This parameter has a critical effect on transfection efficiency (Akinc et al. 2003a).

A working DNA solution is made by diluting a concentrated stock solution into the acidic buffer. The DNA concentration is an important variable that affects the physical properties of polymer–DNA complexes and their subsequent transfection. Although more plasmid can lead to increased cellular uptake and protein expression, toxicity can occur due to the increased amounts of cationic polymer. Therefore, the dose of DNA must be controlled to permit measurable protein expression while limiting cytotoxicity. The plasmid used is pCMV-Luc, where CMV is the cytomegalovirus promoter and Luc stands for the luciferase reporter protein. This expression system is used because (1) it is very sensitive, (2) the relative light unit (RLU) output scales linearly with luciferase concentration over several orders of magnitude, and (3) it can be analyzed conveniently in a 96-well plate format.

Using a multichannel pipettor or robotic fluid-handling system, the diluted polymer is added to the DNA to form polymer–DNA nanoparticles (Fig. 1, Step 2). The solution is left undisturbed for 5 minutes to allow the formation of polymer–DNA nanoparticles. Positively charged complexes with effective diameters of 70–200 nm are produced, depending on the polymer and polymer:DNA ratio (Anderson et al. 2005). Variables that affect the size, charge, and morphology of the complexes include the polymer structure (i.e., chemical functionalities and molecular weight), polymer and DNA concentrations (absolute and relative), mixing kinetics, association time, temperature, and solvent properties (i.e., pH and ionic strength) (Reschel et al. 2002; Neu et al. 2005).

Separate plates are used to dilute polymer–DNA complexes before they are added to the cells, which allows for the desired concentration and pH adjustments. This step is generally performed separate from the cells to prevent detachment and lysis that could result from vigorous pipetting. Polycation efficiency tends to be reduced as the serum concentration is increased, due to the interactions of negatively charged serum proteins with the positively charged polymer–DNA complexes (Forrest et al. 2004).

After an aliquot of the complexes is diluted into the transfecting medium (Fig. 1, Step 3), the nanoparticle suspension is added to the cells (Fig. 1, Step 4). The polymer–DNA incubation time can affect the transfection results, especially for very early times. We selected 1 hour because longer incubations did not result in much higher protein expression. After the complexes are removed, the cells are returned to the incubator in fresh medium.

The protein expression levels are quantified 3 days after the transfection. This duration is optimal, but high protein levels can be measured after just 1 day. Luciferase protein can be detected using Bright Glo Assay Kits (Promega) with luminescence quantified in a plate reader using a 1-second count time per well. The RLUs are compared to a luciferase standard curve to give the mass of expressed luciferase per well. Although this assay is very sensitive, it is inherently unstable with a half-life of approximately 30 minutes. Therefore, measurements are recorded very shortly after addition of substrate, and the luminescence integration time is minimized so that the entire plate is read within 2 minutes.

Protocol

High-throughput Methods for Screening Polymeric Transfection Reagents

This protocol describes a high-throughput 96-well plate method to screen polymeric transfection reagents for their ability to transfect cells in vitro. This method can easily be extended to screen hundreds of polymers in a 4-day experiment.

MATERIALS

CAUTION: See Appendix for appropriate handling of materials marked with <!>.

Reagents

Bright-Glo Kits (Promega)
COS-7 cells, an African green monkey kidney fibroblast-like cell line suitable for transfection by vectors requiring expression of SV40 T antigen (ATCC: CRL-1651)
Dimethyl sulfoxide (DMSO, >99.7% sterile-filtered) (Sigma-Aldrich) <!>
Dulbecco's modified Eagle's medium (DMEM; 500 ml containing 50 ml of fetal bovine serum [FBS] and 5 ml of penicillin-streptomycin <!>) (Invitrogen 21063-029, 10082-147, 25200-056)
Firefly luciferase protein (Promega)
pCMV-Luc DNA (1 mg/ml in H_2O, stored at –20°C) (Elim Biopharmaceuticals)
PEI, 25-kD branched (Sigma-Aldrich) <!>
Sodium acetate solution (3 M, pH 5.2, 0.2-μm filtered) (Sigma-Aldrich) <!>
Trypsin (Invitrogen 25200-056)

Equipment

Aspirator wand (12 channel; V&P Scientific), sterilized by autoclaving
Centrifuge tube (15 ml, sterile)
Half area plate, 96 well (Corning Costar 3695)
  Sterilize by UV treatment in laminar flow cell culture hood for at least 1 hour.
Hemocytometer (VWR)
Incubator (37°C, 5% CO_2; Forma Scientific)
Microcentrifuge tubes (1.5 ml, sterile; Eppendorf, VWR)
Micropipettors (Eppendorf)
  5–50-μl 12 channel
  50–300-μl 12 channel
  100–1000-μl single channel
  20–200-μl single channel
Orbital shaker (optional)
Pipette tips, sterilized by autoclaving for 30 minutes (18 psi, 120°C)/30-minute drying time
  1–200-μl (four) boxes (U.S. Scientific)
  100–1000 μl (U.S. Scientific)
Pipetting reservoirs, sterile (VWR)
Plates (96 well)
  Multiwell flat-bottom plates with lids, (two; BD Falcon, sterile, BD 353072)
  Multiwell plates, 96 well (2.4 ml), clear V-bottom, Greiner Polypropylene (Sigma-Aldrich M1561)

Tissue-culture-treated, white 96-well plate (sterile; Corning Costar 3917)
Plate Luminometer (Perkin Elmer)
Timer (VWR)
Tissue-culture filter unit, 500 ml, 0.2-μm cellulose acetate (sterile; Nalgene)
Water bath, room temperature

METHOD

All work should be conducted in a laminar flow cell culture hood using sterilized reagents and equipment. It is assumed that the reader is familiar with basic, sterile cell culture techniques required to grow, passage, and plate cells.

1. One day before transfection, seed COS-7 cells in a white, 96-well tissue-culture-treated plate at 1.5×10^4 cells/well (Fig. 1, plate 5).
2. Use sterilized microcentrifuge tubes to dissolve 20 mg of each polymer in 200 μl of DMSO to prepare 100-mg/ml polymer solutions. For the positive control, dissolve 10 mg of PEI in 1 ml of DMSO to prepare a 10-mg/ml PEI solution.

   This lower concentration is selected to optimize the polymer:DNA ratio for PEI.
3. Prepare 500 ml of 25 mM sodium acetate buffer (pH 5.2) by diluting 4.2 ml of 3 M sodium acetate into 495.8 ml of distilled H_2O. Sterilize by vacuum filtration through a Nalgene tissue-culture filter unit.
4. Transfer 30 ml of 25 mM sodium acetate buffer to a pipetting reservoir. Use a multichannel pipettor to add 176 μl per well to row A and 80 μl to row B of a BD Falcon 96-well plate (Fig. 1, plate 2). Following the protocol described here, these volumes will ultimately give 20:1 and 100:1 polymer:DNA ratios, respectively. Also, add 900 μl/well to row A of a Greiner deep 96-well plate (Fig. 1, plate 1).
5. Thaw the pCMV-Luc DNA stock solution in a room-temperature water bath. In a 15-ml sterile centrifuge tube, dilute 600 μl of the 1-mg/ml DNA stock into 9.4 ml of 25 mM sodium acetate buffer. Transfer the diluted DNA solution to a pipetting reservoir and add 25 μl per well to a 96-well half area plate (Fig. 1, plate 3).
6. Warm 25 ml of medium in a 37°C water bath for at least 15 minutes. Transfer the medium to a pipetting reservoir and add 200 μl per well to a BD Falcon 96-well plate (Fig. 1, plate 4).
7. For each 100-mg/ml polymer/DMSO solution and the positive control, add 100 μl to a well in row A of the Greiner deep 96-well plate containing 900 μl of sodium acetate buffer (Fig. 1, plate 1). Vigorously pipette the solution several times to ensure polymer dissolution and homogeneity.

   All subsequent additions and transfers between plates are done using 12-channel pipettors.
8. To prepare polymer dilutions (Fig. 1, Step 1):
   a. Add 24 μl of polymer solution from row A of the Greiner deep 96-well plate to row A of the polymer dilution plate containing 176 μl of sodium acetate buffer.
   b. Add 120 μl of polymer solution from row A of the Greiner deep 96-well plate to row B of the polymer dilution plate containing 80 μl of sodium acetate buffer.
   c. Vigorously pipette the solution several times to ensure that the solution is well mixed.
9. To prepare polymer–DNA complexes (Fig. 1, Step 2):
   a. Add 25 μl of polymer from row A of the polymer dilution plate to 25 μl of DNA in rows A, B, C, and D of the DNA plate.

b. Add 25 μl of polymer from row B of the polymer dilution plate to 25 μl of DNA in rows E, F, G, and H of the DNA plate.

c. Set the timer for 5 minutes. For each addition, vigorously pipette the solution several times to ensure adequate mixing and promote polymer–DNA association. Make sure to change pipette tips between each addition to eliminate the possibility of polymer–DNA contamination between wells. It is important to be consistent with the mixing technique to get the most reproducible results.

10. At the end of 5 minutes, add 30 μl of polymer–DNA complexes from row A of the DNA plate to row A of the media plate containing 200 μl of media (Fig. 1, Step 3). Repeat for the remaining rows. For each addition, pipette the solution several times to ensure adequate mixing. Make sure to change pipette tips between each addition to eliminate the possibility of polymer–DNA contamination between wells.

11. Add polymer–DNA complexes to the cells (Fig. 1, Step 4):

    a. For each row of the cell plate, remove the media over the cells using a 12-channel aspiration wand. Gently slide the tips of the wand down the side of the wells until the bottom is reached or no more media is aspirated and then quickly remove.

    b. Add 150 μl of polymer–DNA complexes from row A of the media plate to row A of the cells.

    c. Repeat for the remaining rows. Change pipette tips between each addition. For each addition to the cell plate, the pipette tips should be angled and pressed against the side of the well as opposed to being placed directly over top the cells. This method minimizes the force and turbulence of the fluid, which can dislodge the cells.

    d. Return cells to the incubator and set the timer for 1 hour.

12. After the 1-hour incubation, aspirate the polymer–DNA solution from the cells and add 105 μl of fresh medium to each well. Return the cells to the incubator.

13. Perform the luciferase protein assay. This assay is generally performed 1–5 days posttransfection, with 3 days being optimal for COS-7 cells.

    a. Thaw the Bright Glo Luciferase Assay Kit by placing the two vials in a room-temperature water bath for at least 1 hour.

    b. Add the buffer to the substrate vial. Recap, invert a few times to mix, and then dispense into a reservoir.

    c. Use a multichannel pipettor to add 100 μl of the solution to each well of the cell plate. After the last addition, set the timer for 2 minutes. Gently tap the plate or place it on an orbital shaker to promote mixing during this time.

    d. After 2 minutes, measure the luminescence in a plate reader with an integration time of 1 second per well. Compare to a standard curve to calculate the mass of luciferase protein per well.

    For more details and considerations, see the supplier's technical manual.

REFERENCES

Akinc A., Anderson D.G., Lynn D.M., and Langer R. 2003a. Synthesis of poly(β-amino ester)s optimized for highly effective gene delivery. *Bioconjug. Chem.* **14:** 979–988.

Akinc A., Lynn D.M., Anderson D.G., and Langer R. 2003b. Parallel synthesis and biophysical characterization of a degradable polymer library for gene delivery. *J. Am. Chem. Soc.* **125:** 5316–5323.

Anderson D.G., Lynn D.M., and Langer R. 2003. Semi-automated synthesis and screening of a large library of degradable cationic polymers for gene delivery. *Angew. Chem. Int. Ed. Engl.* **42:** 3153–3158.

Anderson D.G., Akinc A., Hossain N., and Langer R. 2005.

Structure/property studies of polymeric gene delivery using a library of poly(β-amino esters). *Mol. Ther.* **11:** 426–434.

Boussif O., Lezoualc'h F., Zanta M.A., Mergny M.D., Scherman D., Demeneix B., and Behr J.P. 1995. A versatile vector for gene and oligonucleotide transfer into cells in culture and in vivo: Polyethylenimine. *Proc. Natl. Acad. Sci.* **92:** 7297–7301.

Check E. 2003. Harmful potential of viral vectors fuels doubts over gene therapy. *Nature* **423:** 573–574.

De Smedt S.C., Demeester J., and Hennink W.E. 2000. Cationic polymer based gene delivery systems. *Pharm. Res.* **17:** 113–126.

Fischer D., Li Y.X., Ahlemeyer B., Krieglstein J., and Kissel T. 2003. In vitro cytotoxicity testing of polycations: Influence of polymer structure on cell viability and hemolysis. *Biomaterials* **24:** 1121–1131.

Forrest M.L., Meister G.E., Koerber J.T., and Pack D.W. 2004. Partial acetylation of polyethylenimine enhances in vitro gene delivery. *Pharm. Res.* **21:** 365–371.

Gebhart C.L. and Kabanov A.V. 2001. Evaluation of polyplexes as gene transfer agents. *J. Control. Release* **73:** 401–416.

Gosselin M.A., Guo W.J., and Lee R.J. 2001. Efficient gene transfer using reversibly cross-linked low molecular weight polyethylenimine. *Bioconjug. Chem.* **12:** 989–994.

Kircheis R., Wightman L., and Wagner E. 2001. Design and gene delivery activity of modified polyethylenimines. *Adv. Drug Delivery Rev.* **53:** 341–358.

Lim Y.B., Choi Y.H., and Park J.S. 1999. A self-destroying polycationic polymer: Biodegradable poly(4-hydroxy-L-proline ester). *J. Am. Chem. Soc.* **121:** 5633–5639.

Lim Y.B., Han S.O., Kong H.U., Lee Y., Park J.S., Jeong B., and Kim S.W. 2000. Biodegradable polyester, poly[α-(4-aminobutyl)-L-glycolic acid], as a non-toxic gene carrier. *Pharm. Res.* **17:** 811–816.

Lynn D.M. and Langer R. 2000. Degradable poly(β-amino esters): Synthesis, characterization, and self-assembly with plasmid DNA. *J. Am. Chem. Soc.* **122:** 10761–10768.

Merdan T., Kopecek J., and Kissel T. 2002. Prospects for cationic polymers in gene and oligonucleotide therapy against cancer. *Adv. Drug Delivery Rev.* **54:** 715–758.

Neu M., Fischer D., and Kissel T. 2005. Recent advances in rational gene transfer vector design based on poly(ethylene imine) and its derivatives. *J. Gene Med.* **7:** 992–1009.

Nishikawa M. and Huang L. 2001. Nonviral vectors in the new millennium: Delivery barriers in gene transfer. *Hum. Gene Ther.* **12:** 861–870.

Pack D.W., Hoffman A.S., Pun S., and Stayton P.S. 2005. Design and development of polymers for gene delivery. *Nat. Rev. Drug Discov.* **4:** 581–593.

Regelin A.E., Fernholz E., Krug H.F., and Massing U. 2001. High throughput screening method for identification of new lipofection reagents. *J. Biomol. Screen.* **6:** 245–254.

Reschel T., Konak C., Oupicky D., Seymour L.W., and Ulbrich K. 2002. Physical properties and in vitro transfection efficiency of gene delivery vectors based on complexes of DNA with synthetic polycations. *J. Control. Release* **81:** 201–217.

Wagner E. 2004. Strategies to improve DNA polyplexes for in vivo gene transfer: Will "artificial viruses" be the answer? *Pharm. Res.* **21:** 8–14.

Walter E., Moelling K., Pavlovic J., and Merkle H.P. 1999. Microencapsulation of DNA using poly(DL-lactide-co-glycolide): Stability issues and release characteristics. *J. Control. Release* **61:** 361–374.

Zhao Z., Wang J., Mao H.Q., and Leong K.W. 2003. Polyphosphoesters in drug and gene delivery. *Adv. Drug Delivery Rev.* **55:** 483–499.

55 Poly(Lactic Acid) and Poly(Ethylene Oxide) Nanoparticles as Carriers for Gene Delivery

Noémi S. Csaba, Alejandro Sánchez, and Maria Jose Alonso

Department of Pharmaceutical Technology, School of Pharmacy, University of Santiago de Compostela, 15782, Santiago, Spain

ABSTRACT

New nanoparticulate blend compositions based on poly(D,L lactic-co-glycolic acid) (PLGA) and polyoxyethylene (PEO) derivatives have been designed as transmucosal gene carriers. These nanosystems benefit from the inherent biodegradability and low toxicity of their components and the mild conditions required for their preparation. In addition, specific advantages of these nanoparticles for in vivo gene delivery are (1) their adequate DNA loading capacity, (2) their ability to control the release of the encapsulated DNA for extended periods of time while preserving its delicate conformational structure as well as its biological activity, and (3) their capacity to overcome the nasal mucosa barrier and transport the associated model DNA vaccine, leading to a significant systemic immune response against the encoded protein.

INTRODUCTION

Among the strategies explored so far to achieve efficient gene transfer, the design of biodegradable nanoparticles based on a combination of poly(L-lactic acid) (PLA) and PLGA with PEO derivatives (in the form of copolymers or blends) represents a promising alternative for in vivo applications. Indeed, these new biodegradable nanoparticle formulations have already shown potential as transmucosal carriers of bioactive proteins and antigens (Sánchez et. al 2003; Vila et al. 2004b). Considering these promising results, we have recently evaluated the potential of these nanocarriers for transmucosal gene delivery. The development of these nanostructures, which are composed of either PLA-PEG copolymers or blends of physically entangled PLGA:poloxamer or PLGA:poloxamine (blend nanoparticles), was intended to address three key issues: (1) the stabilization of the DNA molecules in the nanoparticle matrix, (2) the controlled release of the encapsulated genetic material for extended periods of time, and (3) the facilitation of the effective transport of the genetic material across mucosal surfaces. These nanosystems have a size range of 100–400 nm and a negative surface charge (between –45 and –25 mV). These parameters can be modulated by adjusting the composition and architectural organization (copolymer PLA-PEG vs. physical blend of PLA/PLGA and PEO derivative, type of PEO derivative) (Pérez et al. 2001; Csaba et. al. 2004). The DNA encapsulation efficiency of these nanocarriers is high and dependent on their composition, varying between 30% and 90%. Moreover, these systems exhibit the ability to control the release of the encapsulated plasmid DNA for variable periods of time, from 1 week up to several weeks. More specifically, nanocarriers based on physical blends of PLGA and PEO derivatives released the encapsulated plasmid for 1–2 weeks, whereas those made of PLA-PEG copolymers gave a more extended release (Pérez et al. 2001; Csaba et. al 2005). This different controlled release behavior indicates the possibility of appropriate carrier selection, depending on application requirements.

The major limitation of these nanosystems is their low "in vitro" transfection efficiency, as compared to that of the reagents commercialized for cell transfection applications. An additional limitation could be related to the fabrication conditions, which imply the need to use organic solvents for the encapsulation of the genetic material into the polymeric nanomatrix. This limitation comes from the fact that PLA, PLGA, PLA-PEG, and some PEO derivatives are relatively hydrophobic and hence cannot be dissolved in aqueous media. Therefore, these nanosystems should not be considered as cell transfection vehicles.

The main advantages of these nanoparticles consisting of PLA and PEO and derivatives are their biodegradability and low toxicity. Consequently, these nanocarriers should be seen as potential candidates for in vivo gene delivery. An additional advantage of these systems is related to their

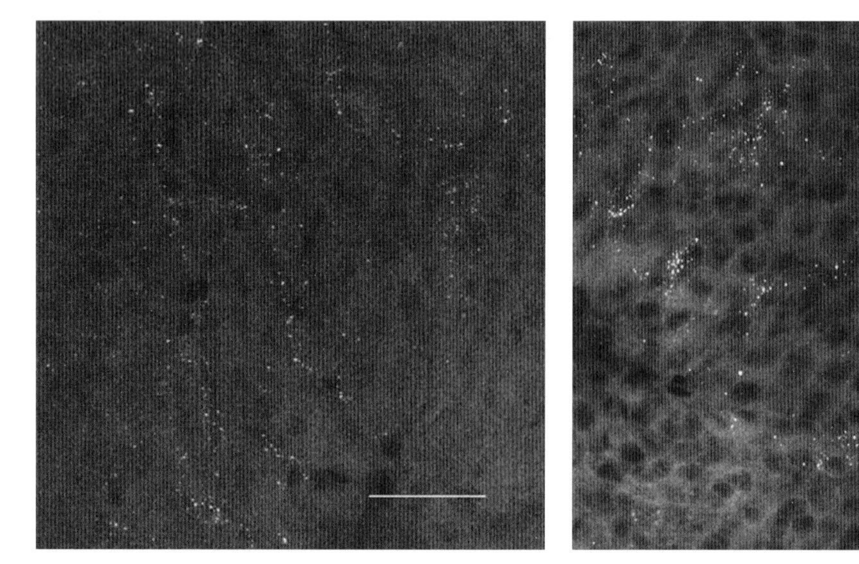

FIGURE 1. Confocal laser-scanning micrographs of (*A*) surface and (*B*) accumulated internal cross sections of nasal mucosa excised following administration of rhodamine-loaded nanoparticles prepared from PLA-PEG copolymer. Bar, 25 μm. (Reprinted, with permission, from Vila et al. 2004a [© Elsevier].)

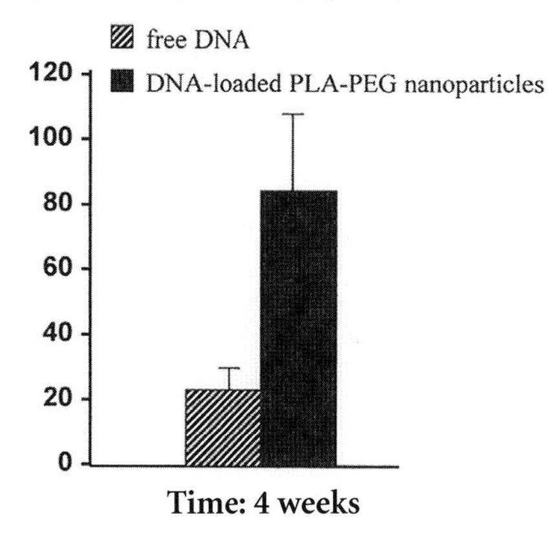

FIGURE 2. Specific serum IgG antibody levels after intranasal administration of DNA, free and encapsulated in PLA-PEG nanoparticles. (Reprinted, with permission, from Vila et al. 2002 [© Wiley].)

ability to preserve the stability and, as already mentioned, to control the release of the encapsulated DNA for extended periods of time. Thus, they have potential for long-term gene expression. Finally, these nanocarriers have the capacity to overcome not only cellular barriers, but also mucosal barriers. For example, using confocal microscopy, we were able to visualize the transport of fluorescent nanoparticles across the nasal mucosa and conclude that the presence of PEO has a critical role in their stability and further transport across the nasal mucosa (Fig. 1) (Tobío et al. 2000; Vila et al 2004a; Csaba et al. 2005). Furthermore, using pCMV–β-gal encoding β-galactosidase as a model plasmid, we could verify that both types of nanocarriers, based on block copolymer (PLA-PEG) or physical blends (PLGA:poloxamer), were able to transport the associated plasmid DNA across the nasal mucosa. Indeed, as shown in Figure 2, a single dose of these DNA-loaded PLA-PEG nanoparticles administered intranasally to mice led to a significant systemic immune response to the encoded protein (Vila et. al. 2002), this response being significantly higher than that corresponding to the naked DNA. A very similar pattern was recently observed for the blend nanocarriers consisting of PLGA and poloxamer. Overall, these results highlight the positive behavior of these nanoparticles as transmucosal gene delivery carriers and their potential application for mucosal delivery of DNA vaccines.

Protocol

Preparation of PLA and PEO Nanoparticles as Carriers for Gene Delivery

This protocol describes the preparation and characterization of DNA-loaded nanoparticles composed of PLA and PEO and the use of these particles as transmucosal gene delivery carriers.

MATERIALS

CAUTION: See Appendix for appropriate handling of materials marked with <!>.

Reagents

Agarose gel (1%)
Ethanol
Ethyl acetate <!>
Methylene chloride <!>
Phosphotungstic-acid-staining solution (2%) <!>
PicoGreen reagent <!> (Molecular Probes, Invitrogen)
Plasmid: pCMV–β-gal vector (Elim Biopharmaceuticals)
Poloxamers with different hydrophilic-lipophilic balance (HLB) values and chain lengths: Pluronic F68 and Pluronic L121 (BASF, Germany)
Poloxamines with different HLB values and chain lengths: Tetronic 904 and Tetronic 908 (BASCOM, Belgium)
Poly(L-lactic acid)–poly(ethylene glycol) (PLA-PEG) copolymer (Alkermes, Ohio)
Poly(D,L-lactic-co-glycolic acid) (PLGA; Boehringer Ingelheim)
Polyvinyl alcohol (PVA) <!> aqueous solutions (1% and 0.3% w/v)
Reagents for ELISA (enzyme-linked immunosorbent assay)

Equipment

Equipment for agarose gel electrophoresis and ELISA
Sonicator (e.g., Sonifier 250, Branson, Barcelona, Spain)
Spectrofluorometer
Transmission electron microscope
Vacuum evaporator
Zetasizer 3000 HS (Malvern Instruments)

METHODS

Nanoparticle Preparation

1. Prepare nanoparticles of desired composition.

   For PLGA: polyoxyethylene blend nanoparticles

   Prepare according to the solvent displacement technique.

a. Dissolve 500 μg of plasmid DNA in a small aqueous phase (200 μl).

b. Dissolve 50 mg of PLGA and 50 mg of appropriate PEO derivative (poloxamer or poloxamine) in 2 ml of the organic phase (methylene chloride).

c. Mix the aqueous phase from Step a with the organic phase from Step b by vortexing, thus forming a water-in-oil emulsion.

d. Incorporate 25 ml of a polar solvent (i.e., ethanol) into the emulsion to cause polymer precipitation in the form of nanoparticles.

e. Add 25 ml of H_2O, and allow the organic solvents to evaporate. Collect the nanoparticles in the form of an aqueous suspension.

For PLA-PEG nanoparticles

Prepare either by the emulsion-solvent displacement technique described above or by the double emulsion technique.

a. Dissolve 200 μg of DNA in a small aqueous phase volume (200 μl).

b. Dissolve 40 mg of PLA-PEG in 1.5 ml of a 1:1 mixture of ethyl acetate:methylene chloride.

c. Mix the aqueous phase from Step a and the organic phase from Step b by sonication for 5 seconds (output = 10 units from Branson Sonifier 250).

d. Add this emulsion to 2 ml of an aqueous PVA solution (1% w/v) and sonicate again for 5 seconds (output = 10) to form a double (water-in-oil-in-water) emulsion.

e. Pour the resulting emulsion into 50 ml of a 0.3% (w/v) PVA aqueous solution under moderate magnetic stirring to allow the solidification of the nanodroplets.

f. Eliminate the organic solvents by evaporation under vacuum.

Nanoparticle Isolation

2. Isolate the nanoparticles by centrifugation cycles of 10,000*g* for 1 hour at 15°C.

   These conditions allow sedimentation of a maximum amount of nanoparticles and prevent the sedimentation of the nonencapsulated plasmid DNA.

3. Resuspend the nanoparticles in H_2O by hand shaking or vortexing.

Characterization of DNA-loaded Nanoparticles

4. Determine the size of the nanoparticles using photon correlation spectroscopy (Zetasizer HS 3000).
5. Obtain ζ potential distribution values using laser-Doppler anemometry (Zetasizer HS 3000).
6. Observe nanoparticle size and morphology by transmission electron microscopy (TEM), using 2% phosphotungstic-acid-staining solution (Harris et al. 1999).
7. Determine the DNA association efficiency by spectrofluorometry using the PicoGreen reagent according to the manufacturer's instructions (Molecular Probes, Invitrogen).
8. Assess the structural integrity of plasmid DNA released or extracted from the nanoparticles using standard 1% agarose gel electrophoresis.

In Vivo Gene Expression

In vivo gene expression can be studied in mice by measuring the immune response generated against the encoded protein.

9. Introduce a 10-μl drop of the appropriate nanoparticle suspension into each nasal cavity of a mouse (6–9 mice) at 15-minute intervals (a total amount of 25 μg of CMV–β-gal is administered by 3 droplets per nasal cavity).
10. Take blood samples from the tail vein of the animals at the time points of interest. Centrifuge to collect serum at 3000*g* for 5 minutes at 4°C and store at –20°C.
11. Determine IgG and IgA antibody levels using ELISA.

ACKNOWLEDGMENTS

This work was financed by the Xunta de Galicia. The authors are grateful to Rafael Romero for his work in the in vivo studies.

REFERENCES

Csaba N., Gonzalez L., Sanchez A., and Alonso M.J. 2004. Design and characterisation of new nanoparticulate polymer blends for drug delivery. *J. Biomater. Sci. Polym. Ed.* **15:** 1137–1151.

Csaba N., Sanchez A., and Alonso M.J. 2005. PLGA-poloxamer and PLGA-poloxamine blend nanoparticles: New carriers for gene delivery. *Biomacromolecules* **6:** 271–278.

———. 2006. PLGA:poloxamer and PLGA:poloxamine blend nanostructures for transmucosal delivery of genetic vaccines: In vitro and in vivo evaluation of uptake and transfection efficiency. *J. Control. Release* (in press).

Harris J.R., Roos C., Djalali R., Rheingans O., Maskos M., and Schmidt M. 1999. Application of the negative staining technique to both aqueous and organic solvent solutions of polymer particles. *Micron* **30:** 289–298.

Pérez C., Sanchez A., Putnam D., Ting D., Langer R., and Alonso M.J. 2001. Poly(lactic acid)-poly(ethylene glycol) nanoparticles as new carriers for the delivery of plasmid DNA. *J. Control. Release* **75:** 211–224.

Sánchez A., Tobio M., Gonzalez L., Fabra A., and Alonso M.J. 2003. Biodegradable micro- and nanoparticles as long-term delivery vehicles for interferon-alpha. *Eur. J. Pharm. Sci.* **18:** 221–229.

Tobío M., Sanchez A., Vila A., Soriano I., Evora C., Vila-Jato J.L., and Alonso M.J. 2000. The role of PEG on the stability in digestive fluids and in vivo fate of PEG-PLA nanoparticles following oral administration. *Colloid. Surf. B Biointerfaces* **18:** 315–323.

Vila A., Gill H., McCallion O., and Alonso M.J. 2004a. Transport of PLA-PEG particles across the nasal mucosa: Effect of particle size and PEG coating density. *J. Control. Release* **98:** 231–244.

Vila A., Sanchez A., Perez C., and Alonso M.J. 2002. PLA-PEG nanospheres: New carriers for transmucosal delivery of proteins and plasmid DNA. *Polym. Adv. Technol.* **13:** 851–858.

Vila A., Sanchez A., Evora C., Soriano I., Vila-Jato J.L., and Alonso M.J. 2004b. PEG-PLA nanoparticles as carriers for nasal vaccine delivery. *J. Aerosol Med.* **17:** 174–185.

56 Biodegradable Nanoparticles

Jaspreet K. Vasir* and Vinod Labhasetwar*†

*Department of Pharmaceutical Sciences, College of Pharmacy, †Department of Biochemistry and Molecular Biology, Nebraska Medical Center, Omaha, Nebraska 68198-6025

ABSTRACT

Biodegradable nanoparticles (NPs) are colloidal particles, typically 100 nm in diameter, and are formulated using biodegradable polymers such as poly(DL-lactide-coglycolide) (PLGA) or polylactide (PLA). These NPs are taken up by cells by an endocytic process and have been shown in our studies to escape the endolysosomal compartment rapidly, thus protecting NPs and the encapsulated DNA from the degradative environment present in endolysosomes (Panyam et al. 2002). Encapsulated plasmid DNA entrapped in NPs is thus protected from degradation by both extra- and intracellular nucleases. It is released slowly, sustaining gene delivery and gene expression. Unlike other nonviral gene delivery systems, NPs thus constitute a sustained gene expression vector.

INTRODUCTION

NPs are colloidal particles, typically 100 nm in diameter, with a gene of interest encapsulated inside the polymeric matrix (Panyam and Labhasetwar 2003; Vasir and Labhasetwar 2006). They are formulated using FDA-approved biodegradable and biocompatible polymers: PLGA or PLA. These polymers are available in different molecular weights and copolymer compositions having different degradation rates. A polymer composition can be selected to achieve the desired release rates and duration of gene expression. PVA is a commonly used emulsifier in the formulation of PLGA NPs because the particles formed are relatively uniform and small in size. They can also be redispersed easily in aqueous media because of surface-associated polyvinyl alcohol (PVA) (Sahoo et al. 2002). Furthermore, the type (molecular weight and degree of hydrolysis) of PVA used for emulsification influences NP surface properties, DNA loading, and particle size, and thus gene transfection. On the basis of the results of our previous studies, PVA with a molecular weight (m.w.) of 30,000–70,000 and 87–89% hydrolyzed was found to be optimal for the NP formulation used for gene transfection (Prabha and Labhasetwar 2004a).

These NPs are negatively charged at physiological pH, do not undergo aggregation in the presence of serum in media, and are nontoxic when tested at up to 1000 μg/ml concentration in different cell lines. The encapsulated DNA is released slowly inside the cells due to degradation of the polymeric matrix of NPs, thus resulting in sustained gene delivery and hence gene transfection (Prabha and Labhasetwar 2004b). Bovine serum albumin (BSA) is used along with DNA to facilitate the release of encapsulated DNA from NPs. Specifically, acetylated BSA is used, since it is nuclease-free and thus will help prevent DNA from degrading. Since the DNA is released slowly, NP-mediated gene transfection is lower but sustained, as opposed to the higher but transient gene expression observed with lipid- or polymer-based complexes where most of the delivered DNA is available quickly for transfection. Sustained gene expression is advantageous, especially when the half-life of the expressed protein is very low or chronic gene delivery is required for therapeutic efficacy (Prabha and Labhasetwar 2004b).

Protocol

Preparation of Biodegradable Nanoparticles and Use in Transfection

This protocol describes a method for nanoencapsulation of DNA and the subsequent use of NPs for transfection.

MATERIALS

CAUTION: See Appendix for appropriate handling of materials marked with <!>.

Reagents

Acetylated BSA (Ac-BSA; Sigma-Aldrich)
Cell culture lysis reagent (CCLR; Promega) or Triton X-100 (1.0% w/v) solution <!>
Cells to be transfected
Chloroform <!>
Fetal bovine serum (FBS)
H_2O (sterile)
Isopropyl alcohol <!>
Luciferase assay substrate (if using luciferase as a marker gene) (Promega)
Medium for cells
  Complete growth medium (RPMI 1640 or other medium, depending on the cell line) containing 10% (v/v) FBS
  RPMI 1640 or any appropriate serum-free medium, depending on the cell line
MicroBCA Kit (Pierce)
Phosphate-buffered saline (PBS, pH 7.4)
Plasmid DNA (10 mg/ml)
PLGA polymer (50:50 polylactide to glycolide ratio, Inherent viscosity 1.32 dl/g in hexafluoroisopropanol; Birmingham Polymers Inc., Alabama)
Polyvinyl alcohol (PVA) (m.w. 30,000–70,000; 87–89% hydrolyzed; Sigma-Aldrich) <!>
Tris-EDTA buffer (pH 8.0, autoclaved) <!>

Equipment

Centrifuge tube (50 ml)
Equipment for lyophilization
Filter (0.22 μm; Millipore)
Glass vial (5 ml)
Ice bath
Incubator preset to 37°C
Microcentrifuge tubes
Plate (24 well)
Shaker-incubator, preset to 37°C
Sonicator, microtip (staged-type) probe (XL 2015 sonicator ultrasonic processor; Misonix Inc.)
UV spectrophotometer
Vacuum desiccator

METHODS

NPs containing plasmid DNA are formulated by an emulsification solvent evaporation technique. Typically, a batch of NPs with 30–90 mg of polymer is prepared. The following procedure applies to a 30-mg batch.

Preparation of Solutions

1. Prepare polymer solution: Dissolve 30 mg of PLGA in 1 ml of chloroform in a 5-ml glass vial with magnetic stirring.

   It takes 3–4 hours to dissolve the polymer completely.

2. Prepare plasmid DNA solution: Mix 1 mg of plasmid DNA solution (10 mg/ml) and 200 μl of Tris-EDTA buffer. Add 2 mg of nuclease-free Ac-BSA without vortexing the tube. Gentle intermittent tapping of the tube helps dissolve the BSA.

   It takes approximately 3 hours to dissolve the BSA completely. Alternatively, the DNA solution can be kept overnight with BSA at 4°C to allow the BSA to dissolve completely.

3. Prepare 2.0% (w/v) PVA solution: Sprinkle 0.2 g of PVA (m.w. 30,000–70,000) slowly over 10 ml of cold Tris-EDTA buffer while stirring on a magnetic stirrer. It takes about 30 minutes to dissolve the PVA. Centrifuge the PVA solution at 1000 rpm for 10 minutes at 4°C. Filter through a sterile 0.22-μm filter to remove any undissolved PVA from the solution. Add 10 μl of chloroform to saturate the PVA solution.

Nanoencapsulation of DNA

4. Add the plasmid DNA solution (as prepared above) to the polymer solution in two aliquots of 100-μl each, vortexing for 1 minute after each addition. This forms a water-in-oil emulsion.

5. Sonicate the emulsion using a microtip (staged-type) probe sonicator set at 55-watt energy output for 2 minutes over an ice bath.

   This process reduces the droplet size of the emulsion (primary emulsion). Clean and rinse the sonicator probe with isopropyl alcohol before using it for emulsification. Place the probe approximately in the middle of the emulsion and avoid contact with the wall of the tube during sonication.

6. Add this primary emulsion in two portions to 6 ml of PVA solution in a 50-ml centrifuge tube, vortexing for 1 minute after each addition. This forms a water-in-oil-in-water double emulsion. Sonicate this emulsion for 5 minutes as in Step 5.

7. Stir the resulting emulsion overnight (~18 hours) in the same tube at room temperature inside a chemical hood with gentle magnetic stirring to allow chloroform to evaporate. Avoid excessive turbulence.

8. To ensure complete removal of chloroform, stir the resulting suspension of NPs in a vacuum desiccator placed on a magnetic stir plate for an additional hour.

9. Recover NPs by ultracentrifugation at 110,000*g* for 20 minutes at 4°C. Remove and collect the supernatant.

10. Resuspend the pellet in 5 ml of Tris-EDTA buffer, first by flushing the buffer over the pellet repeatedly until the complete pellet is suspended and then by sonicating the suspension for 30 seconds over an ice bath as described in Step 5.

11. Repeat Steps 9 and 10 to remove unencapsulated DNA and PVA.

12. Use the supernatant in Step 9 and the washing in Step 11 to determine the amount of DNA that is not entrapped, by measuring the absorbance at 260 nm in a UV spectrophotometer. Use the supernatant and washing from the NPs, but without DNA, to zero the instrument.

13. Resuspend the pellet in about 1–2 ml of sterile H_2O and sonicate the suspension for 1 minute. Centrifuge the NP suspension at 1000 rpm for 10 minutes at 4°C to remove any large aggregates of NPs if present.

14. Collect the supernatant in preweighed sterile microcentrifuge tube(s) (make aliquots if necessary), freeze for 45 minutes at –80°C, and then lyophilize for 2 days in a vial/container to obtain a solid dry powder. Store the lyophilized NPs at 4°C.

    The difference in the weight of the tube in the absence and presence of NPs gives the yield of NPs.

Transfection Protocol

15. Seed cells to be transfected in a 24-well plate at a density of 3.5×10^4 cells/well/ml in complete growth medium (containing 10% v/v serum). Keep $n = 6$ wells for each sample of NPs. Seed cells at the same density for a control lane, to which no NPs will be added.

16. Allow the cells to attach and grow in the plate for 24 hours.

17. Use sterile conditions to prepare a stock suspension of DNA-loaded NPs (4 mg in 0.5 ml of RPMI 1640 or in any appropriate serum-free medium depending on the cell line). Sonicate in a water bath sonicator for 10 minutes.

18. Dilute the NP suspension to 9 ml with RPMI 1640 (or suitable medium used for cell growth) containing 10% FBS.

19. Aspirate the medium from the wells and add 1 ml of NP suspension to each well.

20. Incubate the plate for 24 hours at 37°C in 5% CO_2.

21. Aspirate the NP suspension from the wells and replace with fresh medium. Replace medium on alternate days thereafter, with no further addition of NPs.

22. Prepare cell lysates at appropriate time intervals to use for measurement of gene expression levels:

    a. Wash the cells twice with cold PBS (pH 7.4).

    b. Add 0.1 ml of 1x cell culture lysis reagent (CCLR) or 1.0% (w/v) Triton X-100 solution per well.

    c. Shake the plates in a shaker-incubator for 30 minutes at 37°C to allow for cell lysis.

23. Determine the gene expression levels in the cell lysates. If luciferase is used as a marker gene, measure the expression levels using chemiluminescence intensity with a luciferase assay substrate. Determine the protein concentration using the microBCA kit. Normalize the levels of luciferase gene expression per milligram of cell protein.

REFERENCES

Panyam J. and Labhasetwar V. 2003. Biodegradable nanoparticles for drug and gene delivery to cells and tissue. *Adv. Drug Delivery Rev.* **55:** 329–347.

Panyam J., Zhou W.Z., Prabha S., Sahoo S.K., and Labhasetwar V. 2002. Rapid endo-lysosomal escape of poly(DL-lactide-co-glycolide) nanoparticles: Implications for drug and gene delivery. *FASEB J.* **16:** 1217–1226.

Prabha S. and Labhasetwar V. 2004a. Nanoparticle-mediated wild-type p53 gene delivery results in sustained antiproliferative activity in breast cancer cells. *Mol. Pharm.* **1:** 211–219.

———. 2004b. Critical determinants in PLGA/PLA nanoparticle-mediated gene expression. *Pharm. Res.* **21:** 354–364.

Sahoo S.K., Panyam J., Prabha S., and Labhasetwar V. 2002. Residual polyvinyl alcohol associated with poly (D,L-lactide-co-glycolide) nanoparticles affects their physical properties and cellular uptake. *J. Control. Release* **82:** 105–114.

Vasir J.K. and Labhasetwar V. 2006 Polymeric nanoparticles for gene delivery. *Expert. Opin. Drug Deliv.* **3:** 325–344.

57 Transposon-mediated Delivery of Small Interfering RNA: *Sleeping Beauty* Transposon

Bradley S. Fletcher

Department of Pharmacology and Therapeutics, University of Florida, College of Medicine, Gainesville, Florida 32610-0267 and Medical Research Service, Department of Veteran Affairs Medical Center, Gainesville, Florida 32610

ABSTRACT

RNA interference (RNAi) has become a powerful tool for genetic manipulation of mammalian cells. Stable RNAi delivery systems use viral or nonviral vectors to facilitate expression. Virus-based systems are certainly efficient, but they require technical expertise in both viral production and transduction. Nonviral approaches are generally simpler, but they rely on random integration into the target genome, which occurs with low frequency. Additional steps are therefore required to isolate clones with adequate gene knockdown. Short-hairpin-shaped RNAs (shRNAs) generated from transcription units within the cell can function as active small interfering RNAs (siRNAs) following processing and lead to gene knockdown. Here we describe a nonviral plasmid-based approach to deliver polymerase-III (pol III)-driven shRNA cassettes with high efficiency using the *Sleeping Beauty* (SB) transposon. The advantages of this system include the ease of gene delivery with standard plasmid-based transfection reagents, the increased efficiency of genomic integration due to transposition, and the likelihood of multiple integration events (about six per genome), thereby improving hairpin expression. No expertise in viral packaging or transduction is required, because only transfection of plasmid DNA is necessary. The disadvantages include the fact that certain cell types may be difficult to transiently transfect and that SB-mediated transposition is less active in certain cell types. Overall, transposon-mediated delivery of siRNA is a simple approach for the generation of targeted gene knockdown in mammalian cell lines.

INTRODUCTION

RNAi is an evolutionarily conserved process that silences gene expression through double-stranded RNA species in a sequence-specific manner (McManus and Sharp 2002). siRNAs can promote sequence-specific degradation and/or translational repression of target RNA by activation of the RNA-induced silencing complex (RISC) (Tang 2005). Traditionally, silencing in mammalian cells had been achieved by transfection of synthetically derived siRNA duplexes, resulting in transient gene suppression of the target sequence (Elbashir et al. 2001). As the technology was advanced, inhibitory shRNAs could be produced by transcription from RNA pol-III-driven promoters, such as H1 or U6 (Brummelkamp et al. 2002; Paddison et al. 2002; Paul et al. 2002) or cytomegalovirus (CMV)-enhanced pol III promoters (Xia et al. 2003). Following transcription, the shRNAs are processed by the enzyme Dicer into active siRNA (Dykxhoorn et al. 2003). This approach allows for the continuous production of siRNA within cells using a DNA template and offers increased options for delivery of the pol-III-driven transcriptional units. Recent advancements in the field of RNAi have led to a better understanding of the normal functions and regulation of the RISC complexes (Tang 2005). Insights from structural analyses are leading to more reliable tools for target site selection, thereby increasing the efficacy of gene repression (Schwarz et al. 2003; Reynolds et al. 2004; Boese et al. 2005; Heale et al. 2005).

A number of different viral vectors, as well as plasmid DNAs, have been utilized to deliver shRNA to mammalian cells. Here, the use of the *Tc1/mariner* DNA transposon SB as a tool to deliver shRNA encoding transcriptional units is discussed. The SB transposon system uses a "cut-and-paste" mechanism to insert the transposon into random TA dinucleotides within the target genome. Our laboratory has exploited the SB system to generate a transposon-based RNAi delivery vector for the creation of knockdown cell lines (Heggestad et al. 2004). The transposon vector generated in this work and an example of shRNA processing are shown in Figure 1. A similar approach has been successful in down-regulating the huntingtin gene in cell culture (Chen et al. 2005), and a system based on the *Frog Prince* (FP) transposon has also been described recently (Kaufman et al. 2005). Although these approaches have only been applied to cell culture systems, they have potential for application in animal models.

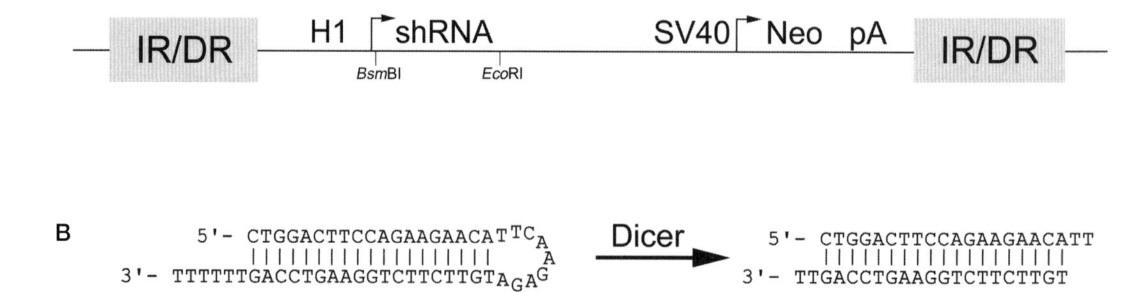

FIGURE 1. Schematic representation of the plasmid pMaleficent and produced shRNA. (*A*) The transposon is bounded by IR/DR elements and contains the H1 promoter driving the shRNA and the SV40 promoter driving neomycin resistance. (*B*) An example of an shRNA directed against lamin A is shown before and after processing by Dicer.

Protocol

Transposon siRNA Delivery

This protocol describes the use of the *Sleeping Beauty* (SB) transposon system to deliver shRNAs, which can then be processed and used for gene knockdown.

MATERIALS

CAUTION: See Appendix for appropriate handling of materials marked with <!>.

Reagents

Agarose gel (0.9%), prepared by standard methods
BGH reverse primer (5′-TAGAAGGCACAGTCGAGG-3′) (Invitrogen)
LipofectAMINE Plus (Invitrogen)
NEBuffer 3 (New England BioLabs)
Oligonucleotides encoding shRNAs to target the gene of interest

High-performance liquid chromatography (HPLC)-purified oligonucleotides with 5′ phosphate groups attached are recommended (Operon Biotechnologies, Huntsville, Alabama).

QIAEX II Gel Extraction Kit (QIAGEN)
QIAGEN Maxi Prep
Restriction enzymes: BsmBI and EcoRI (New England BioLabs)
SB transposon plasmid containing a pol III promoter to drive shRNA expression

pMaleficent (Heggestad et al. 2004) or an alternative is the FP system, pFP/Neo-H1 (Kaufman et al. 2005).

SB transposase expression plasmid: pCMV-SB10 (Ivics et al. 1997)

Other hyperactive transposase mutants can be used, such as pCMV-HSB17 (Baus et al. 2005) or pCMV-HSB4 (Yant et al. 2004). If using FP, the transposase plasmid is pFV-FP (Kaufman et al. 2005).

Selection media containing G418 (0.5–1 mg/ml) <!>
T4 DNA ligase (New England BioLabs)
Target cell line and methods for transient transfection

Equipment

Boiling water bath
Tissues-culture plates (10 cm)
Water baths at 16°C, 37°C, and 55°C

METHODS

Synthesis of Oligonucleotides

1. Design oligonucleotide pairs that target the gene of interest.

   Several research articles can be used to facilitate target site selection (Schwarz et al. 2003; Reynolds et al. 2004; Boese et al. 2005; Heale et al. 2005). For online help, information can be obtained from OligoEngine (http://www.cerebre.com/index.html) or Ambion (http://www.ambion.com/techlib/tb/tb_506.html). Synthesize oligonucleotides corresponding to at least three different target sites, as well as a control oligonucleotide pair.

2. For pMaleficent, design oligonucleotides to have overhangs compatible with ligation into the BsmBI and EcoRI (5′ and 3′ ends, respectively) restriction sites and have the appropriate loop structure and termination signal (Heggestad et al. 2004). For the FP vector (pFP/Neo-H1), the overhangs are BglII and HindIII.
3. Anneal the oligonucleotide pairs (50 pmoles of each primer in a total of 50 μl of distilled H_2O) by boiling in H_2O for 2 minutes followed by slowly cooling the boiled samples to room temperature (over about 20 minutes).

Ligation into Transposon

4. Cut the transposon plasmid pMaleficent (2 μg) with approximately 20 units of EcoRI in NEBuffer 3 for 1 hour at 37°C (reaction volume 20 μl).
5. Add 10 units of the enzyme BsmBI to the same reaction tube and incubate for an additional 2 hours at 55°C.
6. Separate the restriction digestion on a 0.9% agarose gel and purify the excised band of about 6 kb using the QIAEX II Gel Extraction Kit.
7. Ligate the gel-purified vector (1 μl or ~50 ng) to the annealed oligonucleotides (5 μl) using T4 DNA ligase in a 20-μl reaction. Incubate overnight at 16°C.
8. Following bacterial transformation of the ligation reaction (~3 μl), evaluate ampicillin-resistant colonies for inserts and sequence clones using the BGH reverse primer (5′-TAGAAG GCACAGTCGAGG-3′).
9. Once a correct clone is identified, perform a large-quantity plasmid prep (QIAGEN Maxi Prep).

Transfection into Cells

10. Deliver high-quality plasmid DNA to the target cells using a transfection reagent such as LipofectAMINE Plus, or other approaches that yield high transfection efficiencies.
11. A 4:1 mass-based ratio of pMaleficent to transposase-expressing plasmid is suggested.
12. Two days after transfection, split and count the cells. Plate a range of 5×10^3 to 1×10^5 cells in 10-cm plates in selection medium containing G418 (0.5–1 mg/ml) for a total of 14 days. Replace medium every 7 days.

    The number of cells plated depends on the transfection efficiency and the activity of SB transposition within the cells.
13. Following selection, pool resistant colonies together or isolate and clone single resistant colonies. Freeze cell lines and then analyze for knockdown of the gene of interest by western blot or other methods.

REFERENCES

Baus J., Liu L., Heggestad A.D., Sanz S., and Fletcher B.S. 2005. Hyperactive transposase mutants of the *Sleeping Beauty* transposon. *Mol. Ther.* **12:** 1148–1156.

Boese Q., Leake D., Reynolds A., Read S., Scaringe S.A., Marshall W.S., and Khvorova A. 2005. Mechanistic insights aid computational short interfering RNA design. *Methods Enzymol.* **392:** 73–96.

Brummelkamp T.R., Bernards R., and Agami R. 2002. A system of stable expression of short interfering RNAs in mammalian cells. *Science* **296:** 550–552.

Chen Z.J., Kren B.T., Wong P.Y., Low W.C., and Steer C.J. 2005. *Sleeping Beauty*-mediated down-regulation of huntingtin expression by RNA interference. *Biochem. Biophys. Res. Commun.* **329:** 646–652.

Dykxhoorn D.M., Novina C.D., and Sharp P.A. 2003. Killing the messenger: Short RNAs that silence gene expression. *Nat. Rev. Mol. Cell Biol.* **4:** 457–467.

Elbashir S.M., Harborth J., Lendeckel W., Yalcin A., Weber K., and Tuschl T. 2001. Duplexes of 21-nucleotide RNAs mediate RNA interference in cultured mammalian cells. *Nature* **411:** 494–429.

Heale B.S., Soifer H.S., Bowers C., and Rossi J.J. 2005. siRNA target site secondary structure predictions using local stable substructures. *Nucleic Acids Res.* **33:** e30.

Heggestad A.D., Notterpek L., and Fletcher B.S. 2004. Transposon-based RNAi delivery system for generating knockdown cell lines. *Biochem. Biophys. Res. Commun.* **316:** 643–650.

Ivics Z., Hackett P.B., Plasterk R.H., and Izsvák Z. 1997. Molecular reconstruction of *Sleeping Beauty*, a *Tc1*-like transposon from fish, and its transposition into human cells. *Cell* **91:** 501–510.

Kaufman C.D., Izsvák Z., Katzer A., and Ivics Z. 2005. *Frog Prince* transposon-based RNAi vectors mediate efficient gene knockdown in human cells. *J. RNAi Gene Silenc.* **1:** 97–104.

McManus M.T. and Sharp P.A. 2002. Gene silencing in mammals by small interfering RNAs. *Nat. Rev. Genet.* **3:** 737–747.

Paddison P.J., Caudy A.A., and Hannon G.J. 2002. Stable suppression of gene expression by RNAi in mammalian cells. *Proc. Natl. Acad. Sci.* **99:** 1443–1448.

Paul C.P., Good P.D., Winer I., and Engelke D.R. 2002. Effective expression of small interfering RNA in human cells. *Nat. Biotechnol.* **29:** 505–508.

Reynolds A., Leake D., Boese Q., Scaringe S., Marshall W.S., and Khvorova A. 2004. Rational siRNA design for RNA interference. *Nat. Biotechnol.* **22:** 326–330.

Schwarz D.S., Hutvagner G., Du T., Xu Z., Aronin N., and Zamore P.D. 2003. Asymmetry in the assembly of the RNAi enzyme complex. *Cell* **115:** 199–208.

Tang G. 2005. siRNA and miRNA: An insight into RISCs. *Trends Biochem. Sci.* **30:** 106–114.

Xia X.G., Zhou H., Ding H., Affar el B., Shi Y., and Xu Z. 2003. An enhanced U6 promoter for synthesis of short hairpin RNA. *Nucleic Acids Res.* **31:** e100.

Yant S.R., Park J., Huang Y., Mikkelsen J.G., and Kay M.A. 2004. Mutational analysis of the N-terminal DNA-binding domain of sleeping beauty transposase: Critical residues for DNA binding and hyperactivity in mammalian cells. *Mol. Cell. Biol.* **24:** 9239–9247.

58 Efficient DNA Delivery into Mammalian Cells by Displaying the TAT Transduction Domain on Bacteriophage λ

Jehangir Wadia, Akiko Eguchi, and Steven F. Dowdy

Howard Hughes Medical Institute and Department of Cellular and Molecular Medicine, UCSD School of Medicine, La Jolla, California 92093-0686

ABSTRACT

The ability to modulate cellular function by the transfer and expression of novel genes or by affecting the levels of endogenous proteins by genetic means has been of tremendous benefit in studying cellular functions and offers great promise in the treatment of a variety of diseases. Consequently, the development of novel and efficient nonviral DNA delivery systems is an important goal. Recently, small cationic peptides, termed peptide/protein transduction domains (PTDs), have been identified that effectively deliver a wide variety of cargoes, including DNA, into all cells. Expression of the human immunodeficiency virus type 1 (HIV-1) TAT PTD on the surface of bacteriophage λ results in the efficient delivery of plasmid DNA into a variety of cells in a concentration-dependent manner without cytotoxicity.

INTRODUCTION

Peptide/Protein Transduction Domains

PTDs are a novel class of small cationic peptide motifs that are able to transduce rapidly into almost 100% of all cells both in vitro and in vivo in a receptor-independent, concentration-dependent manner (Dietz and Bahr 2004; Zorko and Langel 2005). Fusion of these cationic PTD peptides to a wide variety of cargoes, including proteins, oligonucleotides, small-molecule drugs, and 40-nm iron particles, is sufficient to direct their intracellular accumulation with no apparent toxicity (for review, see Snyder and Dowdy 2005; Wadia and Dowdy 2005).

One of the most characterized PTDs is from HIV-1 TAT protein. TAT protein is a viral regulatory factor involved in *trans*-activation of the HIV long terminal repeat and therefore has a critical role in viral replication (Sodroski et al. 1985). In 1988, two groups independently found that extracellular TAT protein could enter cells and *trans*-activate the viral promoter within cells in culture (Frankel and Pabo 1988; Green and Loewenstein 1988). In an effort to improve the cellular uptake and activity of conjugated proteins, Vives et. al. (1997) further characterized shorter domains of the TAT protein that were sufficient for cell internalization. Starting with a peptide encompassing residues 37–60 of TAT that included both the basic region and the putative α-helical domain, a series of truncations at either the carboxyl or amino terminus were constructed. In this way, the minimal transduction domain was found to consist of amino acids 47–57 (YGRKKRRQRRR). Since the initial discovery of TAT transduction, novel transduction domains have been identified within several other proteins including Antennapedia (Antp) (Perez et al. 1992; Fujimoto et al. 2000; Thoren et al. 2000) and by synthetic peptoid carriers such as homopolymers of arginine (poly-Arg) (Rothbard et al. 2002; Uemura et al. 2002; Wender et al. 2002). Although there does not appear to be any homology between the primary and secondary structures of these protein transduction domains, the rate of cellular uptake has been found to correlate strongly with the number of basic residues present. This observation suggests the presence of a common internalization mechanism that is likely to be dependent on an interaction between the charged side groups of the basic residues and lipid phosphates and negatively charged proteoglycans on the cell surface (Wender et al. 2000; Futaki et al. 2001).

Although different PTDs show similar characteristics for cellular uptake, they vary somewhat in their efficacy for transporting protein cargo into cells. To date, TAT-fusion proteins have shown markedly better cellular uptake than similar fusions using the peptide sequence from Antp, although recently devised peptide sequences such as the retro-inverso form of TAT or poly-Arg appear to increase cellular uptake severalfold (Wender et al. 2000; Futaki et al. 2001). For example, the Antennapedia protein transduction domain can transduce into cells when associated with chemically synthesized peptides; however, the efficiency dramatically decreases with the incorporation of larger proteins (Kato et al. 1998; Chen et al. 1999). Taken together, these observations suggest that a threshold basic charge is required to deliver larger cargo. Consistent with this notion, TAT and poly-Arg have been used extensively and successfully in preclinical mouse models of cancer and stroke (Wadia et al. 2004; Snyder and Dowdy 2005).

Mechanism of Cellular Uptake

Early observations regarding the nature of TAT-mediated cellular uptake relied on visualization and were strongly influenced by the cell surface binding of these domains. Subsequently, it was discovered that following fixation, extracellular bound TAT proteins were redistributed within the cytoplasm and nucleus of the cell, resulting in apparent internalization (Lundberg and Johansson 2002; Vives et al. 2003). To determine the mechanism of TAT-mediated internalization while avoiding these potential pitfalls, Wadia et. al. (2004) used a phenotypic assay for cellular uptake based on the transduction of TAT-Cre recombinase. In this system, exogenous TAT-Cre protein must enter the cell, translocate to the nucleus, and excise the transcriptional termination sequence (STOP) DNA segment in live cells in a nontoxic fashion before scoring positive for enhanced GFP (green fluorescent protein) expression (Snyder and Dowdy 2005; Wadia and Dowdy 2005). Treatment of reporter T cells with TAT-Cre for as little as 15 minutes was sufficient to induce half-maximal recombination of the cell population, confirming that the cellular uptake was a rapid process. Moreover, internalization and recombination were found to be negatively affected by coincubation with sulfated proteoglycans, indicating that interaction with negatively charged cell surface constituents is a necessary event before internalization.

Endocytosis is an essential cellular process for the uptake of a wide variety of extracellular factors that occurs through functionally distinct mechanisms (Conner and Schmid 2003). Early visu-

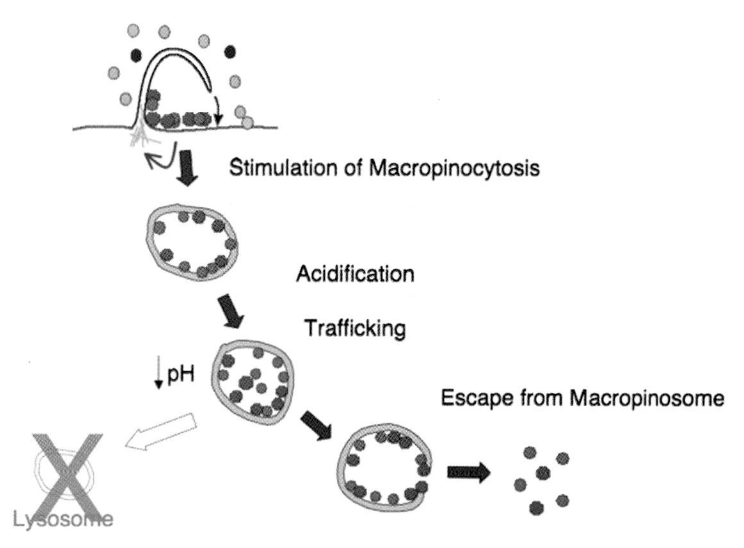

FIGURE 1. Mechanism of TAT-mediated protein transduction. Cationic PTDs stimulate macropinocytosis, a specialized form of fluid-phase endocytosis. Macropinosomes undergo pH drop, but do not traffic to lysosomes. The rate-limiting step for delivery of cargo into the cell is escape from macropinosome.

alization-only studies suggested that uptake of full-length TAT protein and recombinant TAT-fusion proteins occurs by endocytosis (Silhol et al. 2002; Console et al. 2003; Fittipaldi et al. 2003; Lundberg et al. 2003). Similarly, fluorescently labeled TAT-Cre colocalized with endosomal markers in live cells. However, by treating cells with a variety of cholesterol-depleting agents, endocytotic inhibitors, and dominant negatives, it was determined that internalization of TAT-fusion proteins and peptides occurs by a specific endocytosis pathway, lipid-dependent macropinocytosis (Fig. 1) (Wadia et al. 2004). Moreover, binding of the TAT protein transduction domain to the cell surface was sufficient to stimulate macropinocytotic uptake. Although macropinosomes are thought to be inherently leaky vesicles compared to other types of endosomes (Norbury et al. 1995; Meier et al. 2002), the majority of full-length TAT-fusion proteins remained trapped within macropinosome compartments up to 24 hours following treatment, suggesting that the release into the cytoplasm remains a rate-limiting step.

Delivery of DNA by TAT Peptide Displaying Bacteriophage λ

Bacteriophage-based peptide display systems rely on the fact that coat or capsid proteins of bacteriophages can be expressed as chimeras with foreign peptides without loss of function (Maruyama et al. 1994). Encapsulation of DNA coding for chimeric surface proteins into phage particles requires only the expression of a short DNA segment on the phage genomic DNA for efficient packaging. In this way, a variety of peptides can be screened for biological interactions, while the genetic information encoding each variant resides within the phage particle. Peptide display systems have been established using filamentous, T7, and λ bacteriophages. Among these, λ offers several advantages for TAT-mediated DNA delivery. First, λ particles can express foreign peptides on their surfaces more abundantly than other types of phages as chimeras either with the D-head protein (420 copies/particle) or with the major tail V protein (200 copies/particle). Second, λ encapsulates large duplex DNA fragments up to 50 kb in size into a tightly packed proteinaceous shell that is completely protected from destructive environmental nucleases. Third, loading of DNA into λ can be performed entirely in a test tube. Finally, λ can be easily adapted to large-scale production, and because it is physically stable, it can be purified easily under extremely stringent conditions.

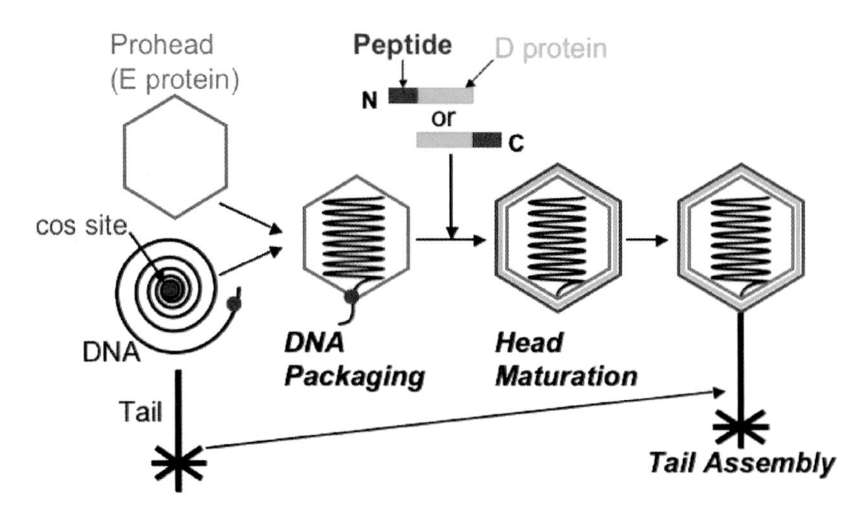

FIGURE 2. Process for preparing λ particles displaying foreign peptides. First, DNA is packaged into bacteriophage λ prohead consisting of E protein by recognition of the COS site. Next, chimeric D protein with foreign peptide assembles onto the prohead structure. Finally, tail and head tightly bind each other.

To construct the bacteriophage λ delivery vehicle, the TAT transduction domain was fused to the amino terminus of the phage D protein to permit expression on the phage surface (Fig. 2). Phage particles carrying the TAT-D protein (TAT-phage) were prepared by simultaneously inducing the replication of the lysogenic phage genome and production of TAT-D protein in bacteria. Phage particles were recovered by chloroform treatment and equilibrium centrifugation with cesium chloride. To measure cellular uptake and DNA gene expression, recombinant TAT-phage particles containing the luciferase gene were incubated with cells for 6 hours, and luciferase activity was measured after 48 hours. Naked DNA or wild-type phage did not result in any luciferase activity in the cell lines tested. In contrast, TAT phage was able to induce significant reporter gene activity in the cell lines tested (Table 1). Because the TAT peptide has been shown to bind to the cell surface, the enhanced delivery of phage may be due to increased absorption of the phage particle to the plasma membrane. However, phage particles expressing high (RGD phage) and low (VN-phage) affinity cell-surface-binding ligands failed to induce any luciferase activity, indicating that the TAT

TABLE 1. Induction of Marker Gene Expression by Various Recombinant Phage

	Luciferase activity (RLU/well)					
Phage	COS-1	VA13/2RA	293	NIH-3T3	HeLa	A431
Tat	33,400 ± 13,000	5590 ± 1785.6	1820 ± 1000	1660 ± 642	282.2 ± 97.7	56.7 ± 10
NLS	<10	<10	<10	<10	<10	<10
VN	<10	n.d.	n.d.	<10	<10	n.d.
RGD	<10	<10	n.d.	<10	<10	n.d.
Wild type	<10	<10	<10	<10	<10	<10
Naked DNA	<10	<10	<10	<10	<10	<10

Cells (2.5×10^4) were incubated with 500 μl of medium containing 2.5×10^8 pfu (plaque-forming units) of various recombinant phage or 10 ng (2.5×10^8 copies) of purified phage genomic DNA for 6 hours at 37°C. Cells were then washed twice with medium and cultured for 48 hours. Luciferase activity was determined as described in Methods. (RLU) Relative light units; (n.d.) not determined.

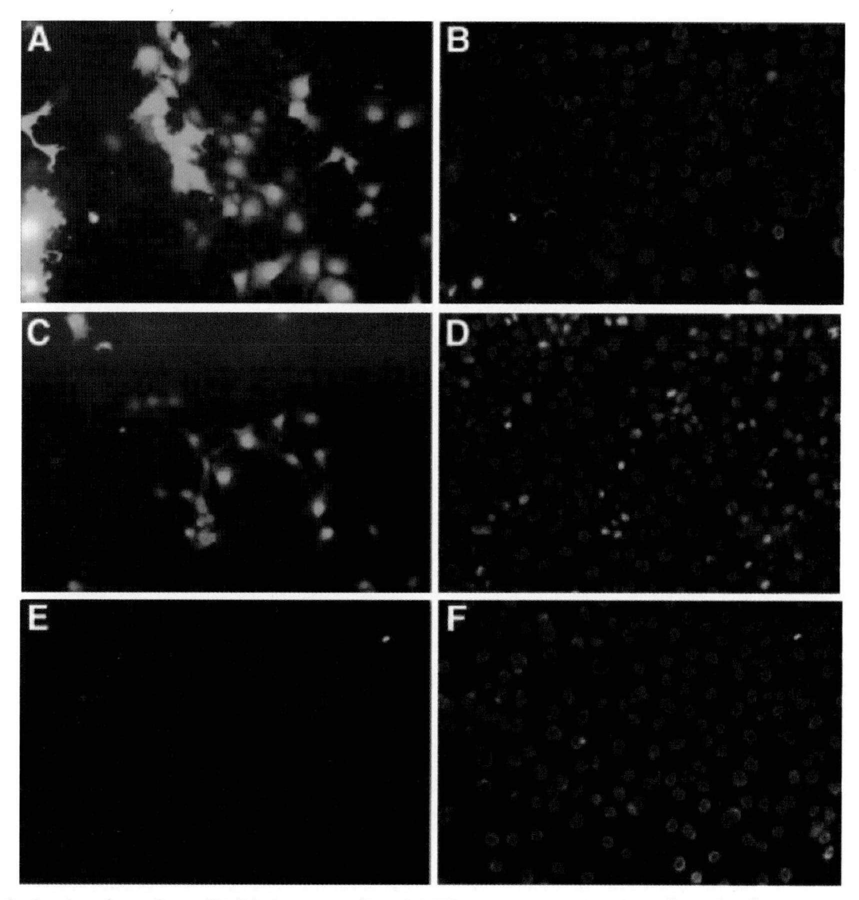

FIGURE 3. In situ detection of TAT-phage-mediated GFP gene expression in cultured cells. COS-1 cells (2.5 x 10^4) were incubated for 6 hours at 37°C, with 500 μl of medium containing 2.5 x 10^{10} pfu of TAT phage (GFP) (*A,B*), 2.5 x 10^9 pfu of TAT phage (GFP) (*C,D*), or 2.5 x 10^{10} pfu of wild-type phage (GFP) (*E,F*). After 48 hours, the cells were counterstained with DAPI and examined with fluorescence microscopy, using a GFPA cube (*A,C,E*) or a WU cube (*B,D,F*), as described in Methods.

peptide facilitated phage uptake and cytoplasmic release. The efficacy of TAT-phage-mediated gene delivery could be reproduced in situ by observing cellular GFP fluorescence following exposure of cells to TAT phage encapsulating the GFP gene (Eguchi et al. 2001). GFP expression in COS-1 cells increased with the dose of TAT phage. Cells incubated with 10^6 and 10^5 particles/cell resulted in strong GFP expression in 30% and 12% of cells, respectively, whereas incubation with wild-type phage at a dose of 10^6 did not result in any detectable fluorescence (Fig. 3).

Similar to other nonviral gene delivery systems, effective delivery of TAT phage requires incubation with cells of 1 hour or more, with gene expression increasing in proportion to incubation time (up to 6 hours) and dose. Moreover, gene delivery by TAT phage has several advantages over other systems: (1) TAT phage delivery is not affected by the presence of serum, (2) the phage DNA is encapsulated and protected from nucleases, and (3) TAT phage is nontoxic to cells at doses up to 10^6 particles/cell for 20 hours (Nakanishi et al. 2003). In summary, the expression of the TAT peptide transduction domain sequence on the surface of bacteriophage λ is sufficient for the intracellular delivery of phage particles and subsequent gene expression in a variety of cells without cytotoxicity.

Protocol

DNA Delivery into Mammalian Cells Using Bacteriophage λ Displaying the TAT Transduction Domain

This protocol describes the preparation of recombinant λ particles with a TAT peptide transduction domain sequence on their surface and their use in delivery of plasmid DNA into a variety of cells

MATERIALS

CAUTION: See Appendix for appropriate handling of materials marked with <!>.

Reagents

Cell lines: COS-1, 293 (both obtained from the RIKEN Cell Bank, Wako, Saitama, Japan), A431, NIH-3T3, WI38/VA13/2RA (obtained from the Health Science Research Resource Bank, Tokyo, Japan), and HeLa

Cesium chloride <!>

Chloroform <!>

DAPI (4′,6-diamidine-2-phenylindole HCl) <!>

DNase I (Sigma-Aldrich)

H-SM buffer (10 mM HEPES-NaOH, 10 mM $MgSO_4$ <!>, and 100 mM NaCl at pH 7.5)

LE392 cells

Luciferase Assay System (Promega)

Lysogenic *Escherichia coli*, λ D1180-GFP, or D1180 luciferase

Media

- Dulbecco's modified minimum essential medium (DMEM), supplemented with 10% fetal calf serum (FCS) for COS-1, 293, A431, and NIH-3T3 cells
- Eagle's minimum essential medium (MEM), supplemented with 10% FCS for WI38/VA13/2RA and HeLa cells
- Luria-Bertani (LB) medium (10 g/liter tryptone [Becton Dickinson], 5 g/liter yeast extract [Becton Dickinson], and 10 g/liter NaCl/10 mM $MgSO_4$ <!>/100 μg/ml ampicillin <!>)

Phage genomic DNA (10 ng, purified; 2.5×10^8 copies)

Plasmids: pTrc-D, pTrc-TAT-D, pTrc-RGD-D, pTrc-VN-D, and pTrc-NLS-D, constructed from pTrcHisA (Invitrogen)

Phage

- Recombinant phage of various types (2.5×10^8 pfu)
- TAT phage (2.5×10^{10} or 2.5×10^9 pfu)
- Wild-type phage (2.5×10^{10} pfu)

Equipment

Dialysis tubing

Fluorescence microscope

GFPA cube and WU cube (Olympus Optical Co. Ltd., Tokyo, Japan)

Glass chamber slides
Incubator preset to 32°C, 37°C, 38°C, and 45°C
Plates (24 well)

METHODS

Preparation of Recombinant λ Particles

1. Culture lysogenic *E. coli* transformed with plasmid in LB/10 mM $MgSO_4$/100 μg/ml ampicillin at 32°C until OD_{600} equals 0.3.
2. Culture the bacteria for 15 minutes at 45°C and then for 180 minutes at 38°C.
3. Recover the phage particles from the bacteria with chloroform (final 10%) and DNase I (final 10 μg/ml) treatment.
4. Purify the phage particles by two rounds of cesium chloride equilibrium centrifugation, followed by dialysis against H-SM buffer.
5. Check the titer by plaque assay, using LE392 cells as host, at 37°C.

Induction of Marker Gene Expression by Various Recombinant Phage

6. Seed the cells to be transfected at 2.5×10^4 cells/well in 24-well plates and culture for 12 hours.
7. Wash the cells once with medium and incubate them with 500 μl of medium containing 2.5×10^8 pfu of various recombinant phage or 10 ng (2.5×10^8 copies) of purified phage genomic DNA for 6 hours at 37°C.
8. Wash the cells twice with medium and then culture for 48 hours.
9. Harvest the cells and quantify luciferase activity with the Luciferase Assay System. Express results as average relative light units (RLUs) with standard deviations.

In Situ Detection of TAT-phage-mediated GFP Gene Expression in Cultured Cells

10. Seed COS-1 cells at $2.5 \times 10_4$ cells/well in glass chamber slides and culture for 12 hours.
11. Wash the cells once with medium and incubate with 500 μl of medium containing 2.5×10^{10} or 2.5×10^9 pfu of TAT phage or 2.5×10^{10} pfu of wild-type phage for 6 hours at 37°C.
12. Wash the cells twice with medium and then culture for 48 hours.
13. Harvest the cells and detect the GFP with fluorescence microscopy, using a GFPA cube. Localize the cell nucleus with fluorescence microscopy using DAPI and a WU cube.

REFERENCES

Chen Y.N., Sharma S.K., Ramsey T.M., Jiang L., Martin M.S., Baker K., Adams P.D., Bair K.W., and Kaelin W.G., Jr. 1999. Selective killing of transformed cells by cyclin/cyclin-dependent kinase 2 antagonists. *Proc. Natl. Acad. Sci.* **96:** 4325–4329.

Conner S.D. and Schmid S.L. 2003. Regulated portals of entry into the cell. *Nature* **422:** 37-44.

Console S., Marty C., Garcia-Echeverria C., Schwendener R., and Ballmer-Hofer K. 2003. Antennapedia and HIV transactivator of transcription (TAT) "protein transduction domains" promote endocytosis of high molecular weight cargo upon binding to cell surface glycosaminoglycans. *J. Biol. Chem.* **278:** 35109–35114.

Dietz G.P. and Bahr M. 2004. Delivery of bioactive molecules into the cell: The Trojan horse approach. *Mol. Cell. Neurosci.* **27:** 85–131.

Eguchi A., Akuta T., Okuyama H., Senda T., Yokoi H., Inokuchi H., Fujita S., Hayakawa T., Takeda K., Hasegawa M., and Nakanishi M. 2001. Protein transduction domain of HIV-1 TAT protein promotes efficient delivery of DNA into mammalian cells. *J. Biol. Chem.* **276:** 26204–26210.

Fittipaldi A., Ferrari A., Zoppe M., Arcangeli C., Pellegrini V., Beltram F., and Giacca M. 2003. Cell membrane lipid rafts mediate caveolar endocytosis of HIV-1 TAT fusion proteins. *J. Biol. Chem.* **278:** 34141–34149.

Frankel A. and Pabo C. 1988. Cellular uptake of the TAT protein from human immunodeficiency virus. *Cell* **55:** 1189–1193.

Fujimoto K., Hosotani R., Miyamoto Y., Doi R., Koshiba T., Otaka A., Fujii N., Beauchamp R.D., and Imamura M. 2000. Inhibition of pRb phosphorylation and cell cycle progression by an *antennapedia*-p16^{INK4A} fusion peptide in pancreatic cancer cells. *Cancer Lett.* **159:** 151–158.

Futaki S., Suzuki T., Ohashi W., Yagami T., Tanaka S., Ueda K., and Sugiura Y. 2001. Arginine-rich peptides. An abundant source of membrane-permeable peptides having potential as carriers for intracellular protein delivery. *J. Biol. Chem.* **276:** 5836–5840.

Green M. and Loewenstein P. 1988. Autonomous functional domains of chemically synthesized human immunodeficiency virus TAT *trans*-activator protein. *Cell* **55:** 1179–1188.

Kato D., Miyazawa K., Ruas M., Starborg M., Wada I., Oka T., Sakai T., Peters G., and Hara E. 1998. Features of replicative senescence induced by direct addition of antennapedia-p16^{INK4A} fusion protein to human diploid fibroblasts. *FEBS Lett.* **427:** 203–208.

Lundberg M. and Johansson M. 2002. Positively charged DNA-binding proteins cause apparent cell membrane translocation. *Biochem. Biophys. Res. Commun.* **291:** 367–371.

Lundberg M., Wikstrom S., and Johansson M. 2003. Cell surface adherence and endocytosis of protein transduction domains. *Mol. Ther.* **8:** 143–150.

Maruyama I.N., Maruyama H.I., and Brenner S. 1994. Lambda foo: A lambda phage vector for the expression of foreign proteins. *Proc. Natl. Acad. Sci.* **91:** 8273–8277.

Meier O., Boucke K., Hammer S.V., Keller S., Stidwill R.P., Hemmi S., and Greber U.F. 2002. Adenovirus triggers macropinocytosis and endosomal leakage together with its clathrin-mediated uptake. *J. Cell Biol.* **158:** 1119–1131.

Nakanishi M., Eguchi A., Akuta T., Nagoshi E., Fujita S., Okabe J., Senda T., and Hasegawa M. 2003. Basic peptides as functional components of non-viral gene transfer vehicles. *Curr. Protein Pept. Sci.* **4:** 141–150.

Norbury C.C., Hewlett L.J., Prescott A.R., Shastri N., and Watts C. 1995. Class I MHC presenTATion of exogenous soluble antigen via macropinocytosis in bone marrow macrophages. *Immunity* **3:** 783–791.

Perez F., Joliot A., Bloch-Gallego E., Zahraoui A., Triller A., and Prochiantz A. 1992. Antennapedia homeobox as a signal for the cellular internalization and nuclear addressing of a small exogenous peptide. *J. Cell Sci.* **102:** 717–722.

Rothbard J.B., Kreider E., VanDeusen C.L., Wright L., Wylie B.L., and Wender P.A. 2002. Arginine-rich molecular transporters for drug delivery: Role of backbone spacing in cellular uptake. *J. Med. Chem.* **45:** 3612–3618.

Silhol M., Tyagi M., Giacca M., Lebleu B., and Vives E. 2002. Different mechanisms for cellular internalization of the HIV-1 TAT-derived cell penetrating peptide and recombinant proteins fused to TAT. *Eur. J. Biochem.* **269:** 494–501.

Snyder E.L. and Dowdy S.F. 2005. Recent advances in the use of protein transduction domains for the delivery of peptides, proteins and nucleic acids in vivo. *Expert Opin. Drug Deliv.* **2:** 43–51.

Sodroski J., Patarca R., Rosen C., Wong-Staal F., and Haseltine W. 1985. Location of the *trans*-activating region on the genome of human T-cell lymphotrophic virus type III. *Science* **229:** 74–77.

Thoren P.E., Persson D., Karlsson M., and Norden B. 2000. The Antennapedia peptide penetratin translocates across lipid bilayers—The first direct observation. *FEBS Lett.* **482:** 265–268.

Uemura S., Rothbard J.B., Matsushita H., Tsao P.S., Fathman C.G., and Cooke J.P. 2002. Short polymers of arginine rapidly translocate into vascular cells. *Circ. J.* **66:** 1155–1160.

Vives E., Brodin P., and Lebleu B. 1997. A truncated HIV-1 TAT protein basic domain rapidly translocates through the plasma membrane and accumulates in the cell nucleus. *J. Biol. Chem.* **272:** 16010–16017.

Vives E., Richard J.P., Rispal C., and Lebleu B. 2003. TAT peptide internalization: Seeking the mechanism of entry. *Curr. Protein Pept. Sci.* **4:** 125–132.

Wadia J.S and Dowdy S.F. 2005. Transmembrane delivery of protein and peptide drugs by TAT-mediated transduction in the treatment of cancer. *Adv. Drug Delivery Rev.* **57:** 579–596.

Wadia J.S., Stan R.V., and Dowdy S.F. 2004. Transducible TAT-HA fusogenic peptide enhances escape of TAT-fusion proteins after lipid raft macropinocytosis. *Nat. Med.* **10:** 310–315.

Wender P.A., Rothbard J.B., Jessop T.C., Kreider E.L., and Wylie B.L. 2002. Oligocarbamate molecular transporters: Design, synthesis, and biological evaluation of a new class of transporters for drug delivery. *J. Am. Chem. Soc.* **124:** 13382–13383.

Wender P.A., Mitchell D.J., Pattabiraman K., Pelkey E.T., Steinman L., and Rothbard J.B. 2000. The design, synthesis, and evaluation of molecules that enable or enhance cellular uptake: Peptoid molecular transporters. *Proc. Natl. Acad. Sci.* **97:** 13003–13008.

Zorko M. and Langel U. 2005. Cell-penetrating peptides: Mechanism and kinetics of cargo delivery. *Adv. Drug Delivery Rev.* **57:** 529–545.

59 Cell-penetrating Peptide-mediated Delivery of Peptide Nucleic Acid Oligomers

Pontus Lundberg,* Kalle Kilk,† and Ülo Langel*

*Department of Neurochemistry, University of Stockholm, S-106 91 Stockholm, Sweden; †Department of Biochemistry, Faculty of Medicine, Tartu University, National and European Centre of Excellence of Molecular and Clinical Medicine, Ravila 19, 51014 Tartu, Estonia

ABSTRACT

Techniques for knockdown of specific genes and nonviral DNA delivery are important for studying protein functions, as well as a means for future therapeutics. The problem with many techniques is low bioavailability and the toxicity, especially in in vivo applications. The discovery that some cationic peptides are able to translocate across the plasma membrane, and more importantly, to carry a cargo many times their own size, might improve the bioavailability of hydrophilic macromolecules. These cationic peptides, often referred to as cell-penetrating peptides, have been used to transport a wide variety of cargos, including gene-regulating oligonucleotides and analogs, both in vitro and in vivo. One oligonucleotide analog that shows great potential in protein knockdown is peptide nucleic acid (PNA). It is a DNA-mimicking molecule, with the advantages that it binds more strongly to DNA and RNA than DNA itself does and is more stable than its DNA counterpart. Furthermore, it is also easily synthesized and modified using standard peptide synthesis protocols.

INTRODUCTION

PNAs are synthetic molecules combining the properties of peptides and nucleic acids (Fig. 1) (Nielsen et al. 1991, 1994), where the phosphorylated sugar moieties of nucleotides are replaced by an aminoethylglycine (peptide) backbone, thus allowing the synthesis of PNA using the *t*-Boc or Fmoc solid-phase peptide synthesis strategy (Awasthi and Nielsen 2002). The increased flexi-

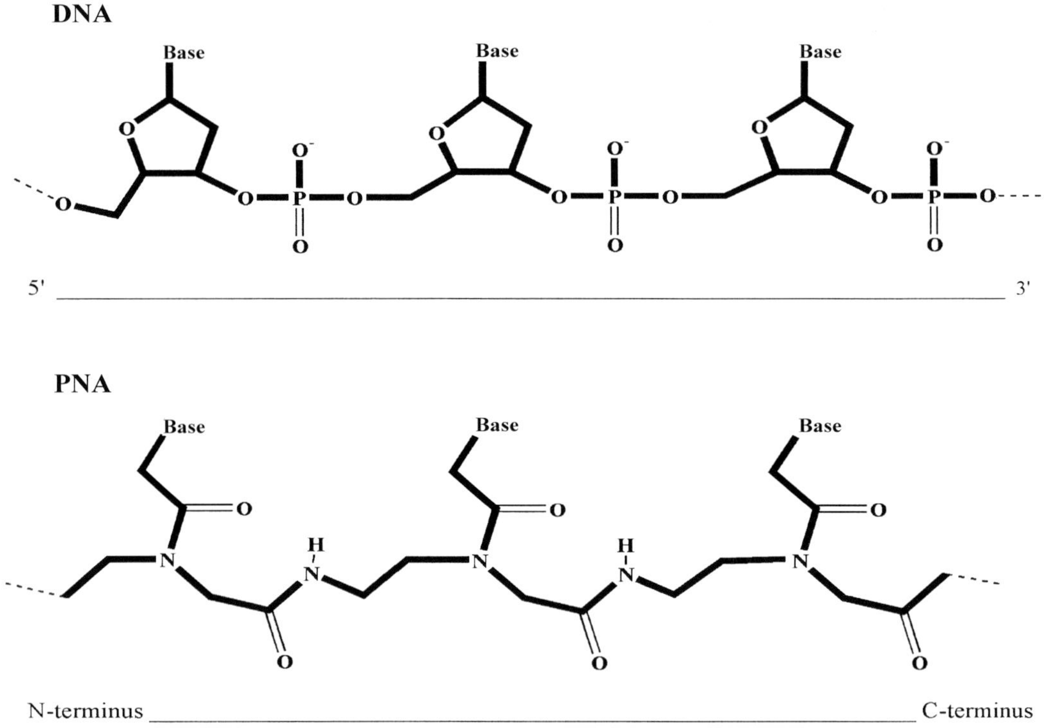

FIGURE 1. Comparison between DNA and PNA structures, where base indicates any of the four nucleobases.

bility and lack of negative charges allow PNA to bind more tightly to complementary DNA or mRNA, as compared to ordinary oligonucleotides (Egholm et al. 1993; Giesen et al. 1998), and also render PNA stable to nuclease degradation. All of the above-mentioned properties make PNA an attractive tool in antisense studies (protein knockdown by application of mRNA targeting oligonucleotides). The limitation, thus far, is its bioavailability. There are some examples where cell-penetrating peptides (CPPs, also sometimes referred to as protein transduction domains) (Lundberg and Langel 2003) have been conjugated to PNA, thereby increasing the cellular uptake of PNA. In 1998, Pooga et al. (1998) reported in vitro and in vivo effects of an antisense PNA conjugated via a disulfide bridge to a CPP, and, at approximately the same time, a nonreducible CPP-PNA conjugate was reported to have a biological effect (Aldrian-Herrada et al. 1998). The following year, a nuclear localization signal (NLS)-PNA conjugate was shown to increase both oligonucleotide and plasmid delivery, where the PNA sequence was complementary to a part of the DNA. This permitted them to hybridize, thus allowing delivery of the DNA (Brandén et al. 1999). A similar method was shown to be efficient, in which a decoy DNA with an overhang complementary to the PNA sequence was delivered. Binding to NF-κB occurred, with subsequent blocking of the interleukin-1β-activation induced by NF-κB (Fisher et al. 2004). The same approach has also been used to target the Myc protein (El-Andaloussi et al. 2005). Different strategies for formation and use of the CPP-PNA conjugates are summarized below (Fig. 2; Table 1). Due to space limitation, only the protocol for disulfide conjugation with CPP is described here, because this is a complicated but crucial step in many CPP-PNA applications. For further reading concerning PNA delivery and other nucleotide analogs, see recent reviews from the groups of P. Nielsen and M. Gait (Koppelhus and Nielsen 2003; Zatsepin et al. 2005).

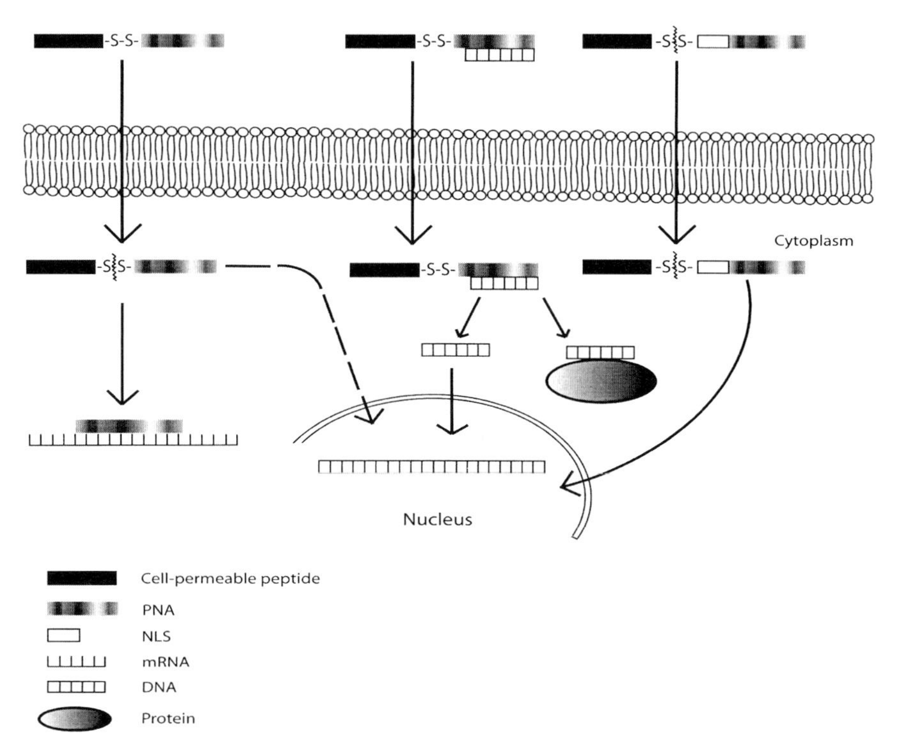

FIGURE 2. Schematic picture illustrating PNA delivery using a CPP as a vector.

TABLE 1. Different Methods to Improve PNA Delivery by Peptide Conjugation

Conjugation strategy	Cell type	Target	Reference
CPP-S-S-PNA	Bowes melanoma cells	galanin receptor 1	Pooga et al. (1998)
CPP-PNA	primary rat neurons	pre-pro-oxytocin	Aldrian-Herrada et al. (1998)
NLS-PNA	COS-7, 3T3, and HeLa	PNA used for DNA delivery	Brandén et al. (1999)
CPP-S-S-NLS-PNA	human prostate cancer cells	–	Braun et al. (2002)
CPP-S-S-PNA-Decoy	Rinm5F	NF-κB through DNA decoy delivery	Fisher et al. (2004) El-Andaloussi et al. (2005)

For protocols, see the references.

Protocol

Conjugation of Peptide and PNA followed by Delivery of Conjugate into Cells

This protocol presents a synthesis strategy for covalently linking a cell-penetrating peptide to PNA using a disulfide bridge. This can improve existing protocols for protein knockdown as well as DNA delivery, due to its relatively high efficacy and low toxicity.

MATERIALS

CAUTION: See Appendix for appropriate handling of materials marked with <!>.

Reagents

Sodium acetate buffer (0.1 M, pH 5.5)
Cell line expressing the protein of interest grown in a 24-well plate
CPP-S-S-PNA solution in sterile H_2O (0.1 mM), prepared in Steps 1–9
CPP-S-S-scrambled control PNA (0.1 mM), prepared in Steps 1–9
Cysteine-containing PNA and peptide, one of which must have a 3-nitro-2-pyridinesulphenyl (Npys)-derivatized cysteine
Dimethylsulfoxide (DMSO, at least 99.9% pure) <!>
Dimethylformamide (DMF, at least 99.8% pure) <!>
Eluent A for reverse phase–high-performance liquid chromatography (HPLC): 99.9% H_2O (HPLC grade) + 0.1% TFA <!>
Eluent B for reverse phase–HPLC: 99.9% acetonitrile (HPLC grade) <!> + 0.1% TFA <!>
Growth media, serum-free and complete, suitable for selected cell line
Trifluoroacetic acid (TFA) <!> (or 6 M guanidinium hydrochloride <!> or 7 M urea <!> if TFA is not available)

Equipment

C-18 column for reverse phase–HPLC (e.g., Supelco discovery 25 cm x 10 mm x 5 μM)
Equipment for lyophilization and for quantification of the protein of interest (e.g., western blot and radioligand binding assay)
Incubator, preset to 37°C
Mass spectrometer
Microcentrifuge tubes (two 1.5 ml)
Plate (24 well)
Spectrophotometer
Water bath, preset to 55°C

METHODS

Conjugation of Peptide and PNA via a Disulfide Bridge

1. Weigh 0.5–2 mg of peptide and 1 molar equivalent of PNA in separate microcentrifuge tubes.

2. Dissolve the PNA in 200 µl of deoxygenized DMSO. If solubility is a problem, warm the tube for 5 minutes to 55°C and then continue with the suspension even if the PNA is not dissolved.
3. Dissolve the peptide in 100 µl of 0.01 M acetate buffer (pH 5.5).
4. Add 200 µl of DMF (recommended) or DMSO to either of the solutions.
5. Mix the two solutions and vortex thoroughly. If the PNA remains insoluble, centrifuge at 10,000*g* (maximum speed on a tabletop centrifuge) for 2 minutes and pipette 5–30 µl of TFA into the pellet (depending on the pellet size). Wait 20 seconds and vortex again. If TFA is not available, use 6 M guanidinium chloride or 7 M urea instead.

   Alternatively, to increase the PNA solubility, two lysines can be attached carboxy- and amino-terminally. This enables disulfide bridge formation to be performed in a 0.01 M acetate buffer and acetonitrile (50/50).
6. Shake or stir at least 2 hours at room temperature or leave overnight. Avoid direct contact with light.
7. Separate the reaction products by semipreparative reverse phase–HPLC:
   a. The gradient depends on peptide hydrophobicity. For hydrophobic peptides, use isocratic 20% eluent B for 5 minutes, followed by a gradient increase of eluent B to 100% in 40 minutes.

      Peptides consisting mainly of arginines and lysines have HPLC profiles very similar to those of PNA oligomers, making efficient separation difficult. A gradient starting from 0% of eluent B and increasing to 80% B within 60 minutes may be attempted.
   b. Detection wavelengths are 218 nm, the absorbance maximum for peptide bond, and 260 nm, which indicates PNA nucleobases. Use 260 nm for a single-wavelength detector.
8. Collect peaks absorbing both wavelengths in comparable scale.
9. Lyophilize fraction(s) and store in a dark place at –20°C. Avoid repeated freezing-thawing of dissolved conjugate. Apply mass spectrometry analysis of conjugate to verify the correct product.

Delivery of CPP-S-S-PNA Conjugate into Cells for Antisense Application

10. At least 1 day before the experiment, seed the cell line expressing the protein of interest such that the cells reach 40–60% confluency on the day of the experiment. Six wells are required (two parallels for antisense, two for control PNA, and two with no treatment).
11. Replace the culture medium with fresh serum-free medium.
12. Prepare a fresh 0.1 mM solution of CPP-S-S-PNA conjugate or thaw an aliquot of premade conjugate (stored at –20°C).

    A serial dilution from 0.1 mM down to 5 µM can be made to evaluate the antisense effect at lower concentrations. This requires more cells and conjugate than listed in this protocol.
13. Heat CPP-S-S-PNA solution(s) to 55°C for 1 minute.
14. Add 10 µl of 0.1 mM CPP-S-S-PNA solution per 1 ml of medium in the cell culture wells to reach a final concentration of 1 µM of the CPP-S-S-PNA conjugate.
15. Incubate for 3 hours at 37°C and then change to complete growth medium.
16. Incubate the cells at 37°C for at least double the time of the target protein's half-life.
17. Harvest the cells and quantify the protein with an appropriate assay.

REFERENCES

Aldrian-Herrada G., Desarmenien M.G., Orcel H., Boissin-Agasse L., Mery J., Brugidou J., and Rabie A. 1998. A peptide nucleic acid (PNA) is more rapidly internalized in cultured neurons when coupled to a retro-inverso delivery peptide. The antisense activity depresses the target mRNA and protein in magnocellular oxytocin neurons. *Nucleic Acids Res.* **26:** 4910–4916.

Awasthi S.K. and Neilsen P.E. 2002. Parallel synthesis of PNA-peptide conjugate libraries. *Comb. Chem. High Throughput Screen* **5:** 253–259.

Brandén L.J., Mohamed A.J., and Smith C.I. 1999. A peptide nucleic acid-nuclear localization signal fusion that mediates nuclear transport of DNA. *Nat. Biotechnol.* **17:** 784–787.

Braun K., Peschke P., Pipkorn R., Lampel S., Wachsmuth M., Waldeck W., Friedrich E., and Debus J. 2002. A biological transporter for the delivery of peptide nucleic acids (PNAs) to the nuclear compartment of living cells. *J. Mol. Biol.* **318:** 237–243.

Egholm M., Buchardt O., Christensen L., Behrens C., Freier S.M., Driver D.A., Berg R.H., Kim S.K., Norden B., and Nielsen P.E. 1993. PNA hybridizes to complementary oligonucleotides obeying the Watson-Crick hydrogen-bonding rules. *Nature* **365:** 566–568.

El-Andaloussi S., Johanssen H., Magnusdottir A., Järver P., Lundberg P., and Langel Ü. 2005. TP10, a delivery vector for decoy oligonucleotides targeting the Myc protein. *J. Control Release* **110:** 189–201.

Fisher L., Soomets U., Cortes Toro V., Chilton L., Jiang Y., Langel Ü., and Iverfeldt K. 2004. Cellular delivery of a double-stranded oligonucleotide NFκB decoy by hybridization to complementary PNA linked to a cell-penetrating peptide. *Gene Ther.* **11:** 1264–1272.

Giesen U., Kleider W., Berding C., Geiger A., Orum H., and Nielsen P.E. 1998. A formula for thermal stability (Tm) prediction of PNA/DNA duplexes. *Nucleic Acids Res.* **26:** 5004–5006.

Koppelhus U. and Nielsen P.E. 2003. Cellular delivery of peptide nucleic acid (PNA). *Adv. Drug Delivery Rev.* **55:** 267–280.

Lundberg P. and Langel Ü. 2003. A brief introduction to cell-penetrating peptides. *J. Mol. Recognit.* **16:** 227–233.

Nielsen P.E., Egholm M., and Buchardt O. 1994. Peptide nucleic acid (PNA). A DNA mimic with a peptide backbone. *Bioconjug. Chem.* **5:** 3–7.

Nielsen P.E., Egholm M., Berg R.H., and Buchardt O. 1991. Sequence-selective recognition of DNA by strand displacement with a thymine-substituted polyamide. *Science* **254:** 1497–1500.

Pooga M., Soomets U., Hällbrink M., Valkna A., Saar K., Rezaei K., Kahl U., Hao J.X., Xu X.J., Wiesenfeld-Hallin Z., Hökfelt T., Bartfai T., and Langel Ü. 1998 Cell penetrating PNA constructs regulate galanin receptor levels and modify pain transmission in vivo. *Nat. Biotechnol.* **16:** 857–861.

Zatsepin T.S., Turner J.J., Oretskaya T.S., and Gait M.J. 2005. Conjugates of oligonucleotides and analogues with cell penetrating peptides as gene silencing agents. *Curr. Pharm. Des.* **11:** 3639–3654.

60 Conditional Mutagenesis of the Genome Using Site-specific DNA Recombination

Kazuaki Ohtsubo and Jamey D. Marth

Department of Cellular and Molecular Medicine and the Howard Hughes Medical Institute University of California, San Diego, La Jolla, California 92093

ABSTRACT

Altering the genome of intact cells and organisms by site-specific DNA recombination has become an important gene-transfer methodology. DNA modifications produced by gene transfer and homologous recombination are typically static once integrated among target cell chromosomes. In contrast, the inclusion of exogenous recombinase target sequences within transferred DNA segments allows subsequent modifications to previously altered genomic structure that increase the utility of gene transfer and enhance experimental design. Creating tissue- and cell-type-specific genetic lesions in animal models, indelibly marking progenitors for cell fate mapping, inducing large-scale chromosomal rearrangements, and complementing gene defects in studies of phenotypic maintenance and reversion are all possible by directing recombinase expression using gene transfer among experimentally modified genomes. Moreover, this approach is effective in providing controlled data establishing genotype-phenotype relationships and allows for the excision of introduced marker genes that can affect neighboring chromatin structure and function. Although early work involved the yeast Flp recombinase, most studies in mammalian systems have used the Cre recombinase derived from bacteriophage P1. Both enzymes are members of the integrase family of recombinases but bind to distinct DNA target signals. Cre recombinase operates on the 34-bp *loxP* sequence and like Flp performs conservative recombination involving DNA segments positioned among these target sites. We describe here procedures for introducing *loxP* sites among vectors and the mammalian genome using a gene-targeting protocol and subsequently altering transferred alleles bearing these recombinase recognition sites.

INTRODUCTION

The modification of intact genomes among cells and organisms can be accomplished with a variety of DNA vectors and gene-transfer strategies. Such approaches have yielded fundamental insights linking gene function with biological mechanisms governing ontogeny, physiology, and disease. Early gene-transfer techniques were limited to the DNA structure designed into each vector. This resulted in irreversible modifications to the genome that precluded access to effective control genomes and further limited experimental potential. In contrast, the more recently acquired technology to excise or otherwise mutate transferred DNA sequences among populations of cells is now an essential tool in state-of-the-art gene-transfer and experimental design. This is accomplished by the use of site-specific DNA recombinase enzymes and their substrates by various gene-transfer methodologies. The now many published studies that have applied conditional mutagenesis have greatly expanded knowledge of genotype-phenotype relationships in development, physiology, and disease etiology. Moreover, the ability to generate precisely controlled data by this approach, by linking the recombined DNA to the phenotype, permits unambiguous assignments of genetic and physiologic relationships, which often fail to exist otherwise and are not often considered in small- or large-scale approaches to gene mutagenesis in vivo.

Numerous enzymes exist with DNA recombination activity. These can be found in the genomes of organisms separated by millions of years of evolution, including bacteriophages and mammals. Enzymes that function by conservative DNA recombination include the integrase family of site-specific DNA recombinases. Members of this family include the Flp recombinase of the yeast 2-μm circle and the commonly employed Cre recombinase of bacteriophage P1 (Abremski and Hoess 1984; Andrews et al. 1985). The basic mechanisms of action are the same for Flp and Cre and are exemplified by the Cre recombinase (Hoess and Abremski 1990). No cofactors have been identified as necessary. Cre recombination depends on a high-energy phosphotyrosine linkage intermediate without the need for ATP or DNA topoisomerase activity and can recombine isolated DNA in vitro. As these enzymes perform conservative DNA recombination, no base pairs are added or deleted in DNA strand cleavage, exchange, or resolution.

Recombination begins by binding to two separate DNA target sequences (*loxP* sites for Cre and *frt* sites for Flp). These are 34-bp sequences composed of two 13-bp repeats separated by an 8-bp spacer sequence. The asymmetric 8-bp spacer provides directionality to the *loxP* target sites. Sequence identity in this spacer region is important for recombination to proceed via the Holliday junction intermediate. DNA positioned between two direct repeats of *loxP* is excised. The excised DNA may again recombine with target sequences in the presence of Cre, in the reverse reaction; however, this appears to be infrequent within intact cells as excised DNA can be rapidly eliminated by degradation (Fig. 1A). Inverted repeats of *loxP* create inversion of the intervening DNA sequence, which yield various frequencies of nonrecombined and inverted *loxP*-flanked DNA (Fig. 1B).

Both Cre and Flp recombinases have been used in heterogenic genomes to effect DNA recombination following gene transfer. Studies using Flp in vitro demonstrated significant and experimentally useful recombinase activity within the genome of both intact *Drosophila* and mammalian cell lines (Golic and Lindquist 1989; O'Gorman et al. 1991). Cre recombinase had been shown to function in gene-transfer studies of mammalian cell types in vitro (Sauer and

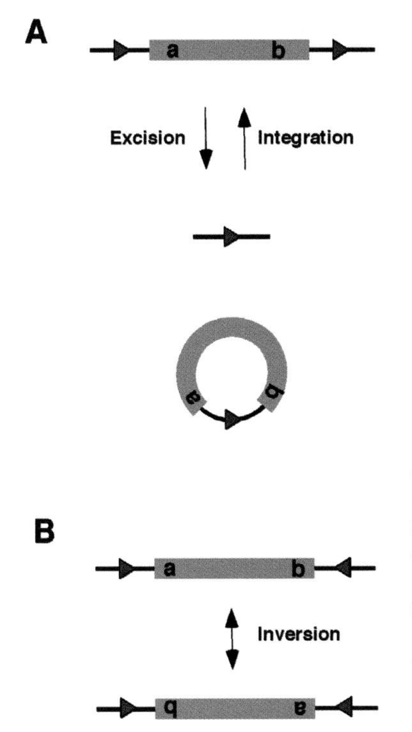

FIGURE 1. Cre-*loxP* recombination. (*A*) DNA between direct repeats of *loxP* sites (*red*) can be excised. This results in a residual *loxP* site remaining in the target locus and an excised circular *loxP*-bearing DNA structure. In the presence of sufficient Cre, excision will dominate among intact cells due to degradation of the circularized extrachromosomal product. (*B*) Reverse recombination reaction, resulting in site-specific DNA integration can also be achieved in the presence of inverted *loxP* repeats. This reaction can continue, and upward of 50% of *loxP*-flanked DNA may be inverted among recombined genomes. In both *A* and *B,* the outcome depends on *loxP* orientations among the DNA substrates employed and Cre expression levels.

Henderson 1988). However, the Cre gene sequence was not expected to be expressed as effectively as a transgene in transgenic mice since the DNA sequence is of prokaryotic origin. Nor was it clear whether primary and multiple diploid cell types could support Cre recombination. It was therefore remarkable and unexpected when transgenic mice were found to support bacteriophage Cre gene expression and recombination activity in multiple cell types in vivo (Lakso et al. 1992; Orban et al. 1992). From a study designed to assess transgenic Cre recombinase efficiency, chromosomal site dependence of action, length influences of intervening chromosomal DNA sequence, and heritability of recombined DNA structure among daughter cells that have shut down Cre transgene expression in development, it was proposed that Cre-*loxP* recombination could provide an in vivo approach to create tissue-specific gene inactivation models, enhance cell fate determination, assess dependence on transgene activity in the maintenance of cellular phenotypes and disease such as cancer, and perhaps promote heterologous chromosomal recombination (Orban et al. 1992). Subsequent placement of *loxP* sites in defined mammalian genetic loci confirmed that Cre-*loxP* recombination was indeed an effective novel approach for more refined investigations of mammalian gene function (Gu et al. 1994; Hennet et al. 1995; Ramirez-Solis et al. 1995; Smith et al. 1995; Marth 1996).

Among the useful features of site-specific DNA recombination is the ability to remove marker genes, which are typically included for screening and selection strategies in gene transfer. That chromosomally integrated marker gene cistrons such as those containing neomycin phosphotransferase can alter the activity of neighboring and endogenous genetic loci remains an important consideration in experimental design (Fiering et al. 1993). The ability to target DNA vector integration to chromatin bearing a single *loxP* site has provided the potential to obtain more reproducible gene expression (Fukushige and Sauer 1992). Many features in the application of conditional mutagenesis by site-specific DNA recombination have been developed, including the use of inducible promoters to control Cre expression and the production of conditional hypomorphic alleles (Kühn et al. 1995; Logie and Stewart 1995; Metzger et al. 1995; Meyers et al. 1998). Moreover, the derivation of mice that faithfully express β-galactosidase in any tissue or cell type upon Cre recombination has provided a more accurate way to assess endogenous promoter activity (Mao et al. 1999; Soriano 1999). In addition, the ability to investigate the function of essential genes residing on the X chromosome among intact mouse models, including the reproducible generation of viable female heterozygotes, has been possible thus far only by conditional mutagenesis using Cre-*loxP* recombination (Shafi et al. 2000; O'Donnell et al. 2004). The germ-line expression of Cre in parental gametes has further enabled the highest level of experimental control by linking phenotypes observed with the deletion of *loxP*-flanked DNA, thereby abolishing the need for multiple lines of mice produced by distinct embryonic stem (ES) cell clones (Shafi et al. 2000; and see below).

The efficiency of recombination has an important role in whether the experimental design is applicable to conditional mutagenesis by Cre, Flp, or other recombinase systems. Cre recombinase efficacy is much higher in vivo than that measured by expression of Flp. Mutants of Flp bearing increased stability at 37°C have increased recombinase efficacy per unit of Flp (Bucholz et al. 1998). At present, translating this into high-efficiency recombination among multiple mammalian cell types in vivo remains to be achieved. It is also unclear whether the native *frt* sequence configuration derived from yeast reduces the efficiency of Flp recombination among mammalian cell types in vivo, as an adjacent third *frt* site is present, although it is not required for recombination in vitro (Senecoff et al. 1985). The often high efficiency of native Cre recombination can be improved by including eukaryotic nuclear localization and translational initiation consensus signals. Versions of Cre lacking these modifications or containing one of the two have been generated and cannot be identified accurately among the many vectors bearing Cre that are now in circulation. Knowing which Cre is being employed can be relevant for studies that seek variation in the efficiency of Cre recombination among target cell populations.

In producing chimeric cell populations, the Flp recombinase may also be a good choice. Flp has been used successfully for production of hypomorphic alleles and in combination with Cre for multiple-tiered recombination strategies (Meyers et al. 1998; Soukharev et al. 1999; Nagy 2000). In gauging the efficiency of Cre recombination in vivo, most studies have found significant variation among targeted cell populations and even among genetically similar littermates. The use of anti-Cre antibodies and markers of DNA recombination can yield discordancies suggesting variable access of Cre to the nucleus. Other factors that affect recombination among disparate *loxP* target sites likely include neighboring chromatin structure as well as base-pair distances between *loxP* sites. Virtually all studies have reported less than quantitative recombination, with some fraction of the target cell population typically lacking evidence of Cre activity by the retention of *loxP*-flanked DNA. Depending on the experimental rationale, this can be either advantageous or disadvantageous. The advantage arises in the formation of chimeric cell populations that provide experimental systems for uncovering nearest-neighbor effects in mechanisms of cell function. The disadvantage is obvious if the recombination frequency among the targeted cell population is too low to induce a measurable physiologic phenotype. Production and characterization of Cre transgenic mice have included many lines with well-characterized and experimentally effective recombination efficiencies among distinct cell types. These live-animal genomes permit the deduction of otherwise undecipherable genotype-phenotype relationships. It is likely that such key reagents, as well as additional *loxP*-flanked genomic targets, will continue to be developed (Nagy 2000).

In the absence of *loxP* sites, Cre expression can be mutagenic and deleterious to cell viability in some contexts (Schmidt et al. 2000; de Alboran et al. 2001). The levels of Cre expression may need to be unusually high for this to occur. Studies using multiple independent transgenic Cre lines bearing different expression patterns and levels in vivo have not thus far observed phenotypes due to Cre expression itself, and no illegitimate recombination involving introduced *loxP* sites has been detected. However, such genomic recombination may in some cases involve the presence of pseudo-*loxP* sites that are homologous but nonidentical to wild-type *loxP* sites (Sauer 1992). Such events are likely to be rare and phenotypically silent among targeted somatic cell populations. This reflects the improbability that even a single wild-type 34-bp *loxP* site exists among mammalian genomes. Nevertheless, the inclusion of appropriate control genomes in experimental design including the Cre transgene itself is essential in formulating conclusions.

Methods of introducing *loxP* sites and the Cre recombinase into cells and genomes vary with the experimental design and are not limited to the use of transgenic mice. The transfer of *loxP* sequences into vectors and into the genome can be achieved by basic gene-transfer protocols involving technologies covered in other chapters in this volume. Although Cre recombines large segments of *loxP*-flanked DNA, it generally seems that smaller is better when a high efficiency of recombination is sought. *loxP*-flanked DNA sequences of approximately 10 kb among more than two dozen distinct loci have been observed to recombine at high frequency. The effect of *loxP*-flanked DNA structure (which is dependent on chromosomal location, cell type, and the configuration of vector sequences) inhibits predictability. Gene-transfer vectors that have previously included Cre expression involve recombinant adenoviral, lentiviral, and retroviral systems, as well as various plasmid-based vectors. The choice among gene-transfer systems is typically determined by the target cell population chosen and then considering the best means of achieving gene transfer in the relevant cell types and intact organisms. Protocols for the use of Cre recombinase-bearing vectors are basically identical to those that employ various other eukaryotic cDNAs in gene-transfer studies.

Incorporating Cre-*loxP* recombination approaches into gene-transfer experimental design is relatively straightforward. Difficulties are occasionally encountered with steps in the protocols, particularly involving *loxP* site verification and multiple allele acquisition following Cre gene transfer in vitro. Specific experimental techniques are detailed below, beginning with a typical protocol for obtaining vector insertion by homologous recombination in ES cells. For those investigators who wish to acquire characterized Cre-based expression systems, such reagents are well-documented among the extensive scientific and academic literature and are typically made freely available to the research community through the laboratory of origin. In addition, Jackson Laboratories maintains an increasing stock of Cre transgenic lines developed initially by many separate laboratories, as well as mice bearing various *loxP*-flanked target alleles. For further information, see http://www.mshri.on.ca/nagy/ and http://www.jax.org/resources/mouse_resources.html/.

Protocol 1

Genome Modification by Inclusion of *loxP* Transgene Sequences

The vector to insert *loxP* sites into the genome is constructed using, for example, the pflox vector to introduce three *loxP* sites into the targeted transgenic linear genome sequence by homologous recombination (Fig. 2A). A typical gene-targeting approach to create a conditional null allele is presented here, where the initial placement of *loxP* sites is nondeleterious to allele function. This can be modified to include knockins of point mutations, with such mutations flanked by *loxP* sites that can then be recombined by Cre expression. The choice of sequence or regulatory element to modify is dependent on the experimental design. Consideration must be given to the possible production of truncated and altered gene sequence products, or otherwise aberrantly functioning alleles.

The pflox vector has two juxtaposed selection marker genes, neomycin phosphotransferase (*neo*) and herpes simplex virus thymidine kinase (*tk*). The former is used for positive selection for the presence of the integrated parental type-3 allele (3 *loxP* sites) *F[tkneo]* (Fig. 2B). The latter is used for subsequent negative selection against recombinant alleles retaining the *loxP*-flanked *neo-tk* selectable marker array. The *tk* gene can be removed and Cre recombinants that lack the *neo* gene can be screened for by other means, such as polymerase chain reaction (PCR). The *tk* gene must be removed prior to its presence in the germ line of male mice, otherwise spermatogenic failure occurs. The *loxP*-flanked target gene sequence lacking the marker genes is designated as *F* (floxed) or a type-2 allele (2 *loxP* sites) and becomes the precise control genotype (Fig. 2B). Cre recombination among the most distal *loxP* sites results in a type-1 or deleted (Δ) allele (1 *loxP* site remains), which further lacks the experimentally targeted *loxP*-flanked sequence (Fig. 2B). One of two possible orientations in the placement of short and long arms of genomic DNA sequence is depicted. The use of isogenic DNA with linear sequence homology ranging from 4 to 12 kb indeed provides the highest frequencies of homologous recombination (te Riele et al. 1992). When using PCR to detect homologous recombinants, a short arm of approximately 1–2 kb is best to achieve a robust sensitivity in the assay. In this regard, it is helpful to produce a control vector bearing the partially integrated structure; this allows for PCR testing of the oligonucleotides chosen (Fig. 2C). Subsequent analyses of *loxP* presence and position following gene-transfer and Cre recombination are facilitated using the *loxP* probe sequence from the ploxP^2 plasmid as a probe for genomic analyses (Fig. 2D).

MATERIALS

CAUTION: See Appendix for appropriate handling of materials marked with <!>

Reagents

Agarose, electrophoresis grade
2x Cell-freezing medium (40:40:20 DMEM/FBS/DMSO v/v/v)
50x Denhardt's solution
  1 g of Ficoll (Amersham Biosciences 17-0300-50)
  1 g of polyvinylpyrrolidone (Sigma-Aldrich P 0930) <!>
  1 g of bovine serum albumin (Calbiochem 126609)
    Dissolve in 100 ml of autoclaved distilled H_2O.

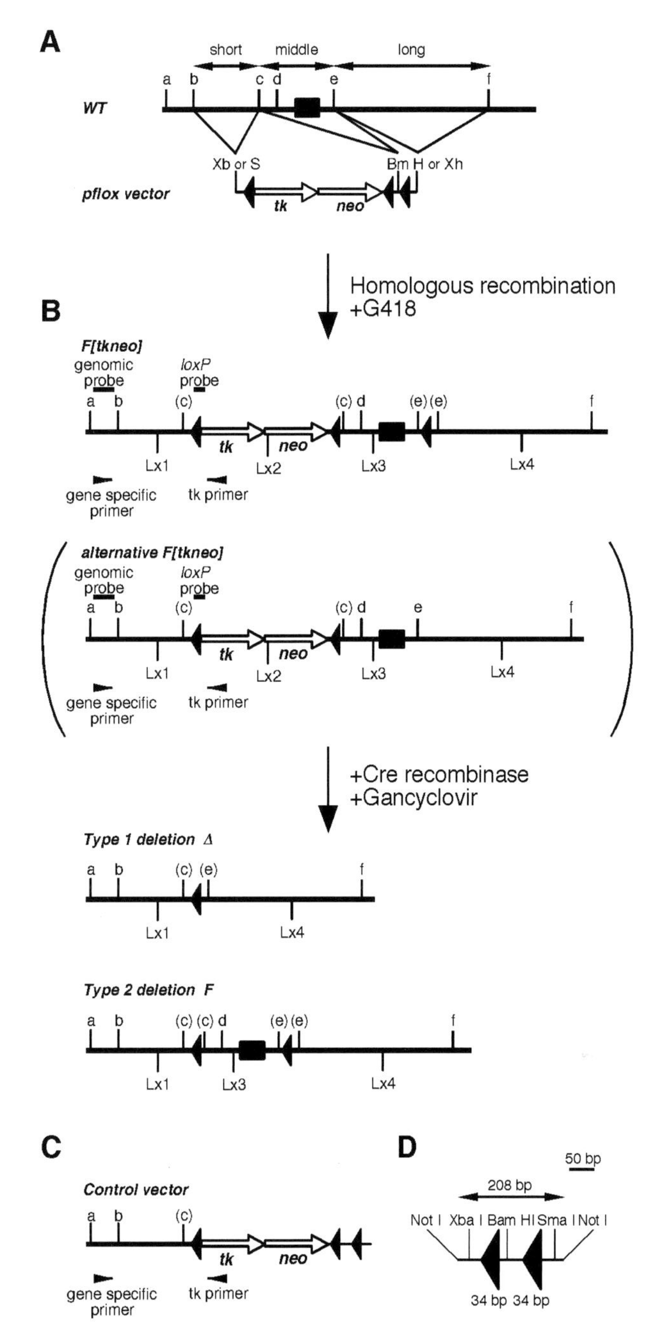

FIGURE 2. Modifying genomes by the inclusion of *loxP* sequences and Cre recombination. (*A*) The targeting vector is constructed by the insertion of DNA fragments (*short, middle,* and *long*) contiguous with the genome into the pflox vector. The middle fragment contains the genetic sequence/element chosen for deletion in the experimental strategy. (*black arrowheads*) *loxP* sites. Restriction enzyme sites present in the pflox vector are indicated: (Bm) BamHI; (H) HindIII; (S) SalI; (Xb) XbaI; (Xh) XhoI; (*a–f*) unique restriction enzyme sites in genomic DNA. (*B*) Vector integration into the genome by homologous recombination produces the *F[tkneo]* allele in the ES cell. The *F[tkneo]* allele can be converted into type-1 (Δ) or type-2 (*F*) alleles by Cre recombinase activity. An alternative allelic structure is possible following crossover during recombination within the middle arm of homology (within *large parenthesis*). This would generate a PCR-positive indication of homologous recombination, depending on primer positioning, but result in the absence of the distal *loxP* site that flanks the DNA sequence selected for future elimination by Cre recombinase. The frequency of this basically useless internal crossover is almost always low, but may be influenced by the chromatin context and size of the *loxP*-flanked DNA fragment with respect to the size of the adjacent short-arm genomic DNA sequence. (*Small arrowheads*) Positions of PCR primers used during screening for homologous recombinant ES cell clones are denoted. (*Small parentheses*) Restriction enzyme sites destroyed during construction of the targeting vector. Southern blot analysis is accomplished using a genomic probe residing adjacent to sequences transferred by the vector. Examples are provided of restriction enzyme site positions useful for allele analysis by the genomic probe (*a–d*) and the *loxP* probe (Lx1–4). (*C*) Structure of a sample control vector used to develop sensitive specific PCR screening for the *F[tkneo]* allele. (*D*) Structure of the isolated (i.e., NotI digest) *loxP* probe from ploxP[2].

Dextran sulfate (10%; Sigma-Aldrich D7037)

Dimethylsulfoxide (DMSO; Sigma-Aldrich D2650) <!>

DNA target clone for gene transfer

> This is typically an isogenic genomic fragment of approximately 20 kb.

Dulbecco's modified Eagle's medium (DMEM; GIBCO/Invitrogen 11960–044)

Embryonic stem (ES) cells

> Various mouse ES cell lines are available from academic and commercial sources. Some, such as the R1 line (Nagy et al. 1993), do not require a supportive "feeder-cell" underlay and can be maintained on gelatinized plates in the presence of leukemia inhibitory factor.

ES cell culture medium

- 500 ml of DMEM
- 15% FBS (94 ml stock)
- 0.1 mM nonessential amino acid (6.2 ml stock)
- 1 mM sodium pyruvate (6.2 ml stock)
- 55 μM 2-mercaptoethanol (620 μl stock)
- 100 units/ml:100 units/ml:2 mM penicillin:streptomycin:L-glutamine (6.2 ml stock) <!>
- 1000 units/ml LIF (ESGRO) (62 μl stock)

Ethanol (70% and 100%) <!>

Fetal bovine serum (FBS; Hyclone SH30070.03)

G418 (200 mg/ml; GIBCO/Invitrogen 11811-031) <!>

Gelatin (0.1%)

> Dissolve 0.5 g of gelatin (Type A, from porcine skin; Sigma-Aldrich G2500) in 500 ml of tissue-culture-grade H_2O. Sterilize by autoclaving.

Hybridization buffer

- 35% formamide <!>
- 4x SSC
- 25 mM NaH_2PO_4
- 10% dextran sulfate
- 4x Denhardt's
- 100 mg/ml salmon sperm DNA
- 0.15x SET

Hybriwash buffer

- 5 ml of 20x SSC
- 1 g of SDS <!>
- 1 liter of H_2O

Leukemia inhibitory factor (LIF; 10^7 units/ml) (ESGRO, Chemicon ESG1107)

Lysis buffer

- 10 μM Tris-HCl (pH 7.5) <!>
- 10 μM EDTA
- 10 μM NaCl
- 0.5% sarcosyl
- 1 mg/ml proteinase K <!>

MEM nonessential amino acids (10 mM; GIBCO/Invitrogen 11140-050)

2-Mercaptoethanol (stock is 55 mM; GIBCO/Invitrogen 21985-023) <!>

Mineral oil (Sigma-Aldrich M8410)

NaCl-ethanol <!>

> Dissolve 3 ml of 5 M NaCl in 100 ml of 100% ethanol.

NaH_2PO_4 (1 M) <!>

> Dissolve 1.7 g of sodium phosphate dibasic (Sigma-Aldrich S9390) and 10.56 g of monobasic (J.T. Baker 4011-01) in 100 ml of H_2O.

Oligonucleotides
- gene-specific primer(s) (must be chosen from target allele of interest)
- TK-specific primer (5′-TTCGAATTCGCCAATGACAAGACGCTG-3′)

PCR kit (Takara Ex Taq polymerase; TAKARA, TAKRR001)

Penicillin:streptomycin:L-glutamine (10,000 units/ml, 10,000 units/ml, and 200 mM, respectively) (GIBCO/Invitrogen 10378-016) <!>

Phosphate-buffered saline (PBS; GIBCO/Invitrogen 14190-144)

Plasmid DNA
- pflox vector (variants may be substituted) (Chui et al. 1997)
- ploxP[2] (for preparing the *loxP*-bearing probe by NotI digestion) (Chui et al. 1997)

  Other plasmids may be constructed if convenient to include genomic fragments of the surrounding and target DNA sequence. These are used to probe allele structure before and after homologous recombination and after Cre-mediated recombination.

Prehybridization buffer
- 6x SSC
- 20 mM NaH_2PO_4
- 0.4% SDS
- 5x Denhardt's
- 100 μg/ml salmon sperm DNA

Proteinase K (1 mg/ml; Sigma-Aldrich P6556)

Salmon sperm DNA (Sigma-Aldrich D1626)

1x SET
- 10 mM Tris-HCl (pH 7.5)
- 5 mM EDTA
- 1% SDS

Sodium pyruvate (100 mM; GIBCO/Invitrogen 11360-070)

20x SSC

  Dissolve 175.3 g of NaCl and 88.2 g of sodium citrate <!> in 1 liter of H_2O. Adjust pH to 7.0.

TE buffer
- 10 mM Tris-HCl (pH 7.5)
- 1 mM EDTA

Trypsin-EDTA (0.05%; GIBCO/Invitrogen 25300-054) <!>

Equipment

Electroporation cuvette (Gene pulser cuvette, 0.4-cm electrode gap; Bio-Rad 165-2088 or equivalent)

Electroporator (Gene Pulser II and Capacitance Extender; Bio-Rad or equivalent)

Hemocytometer

Hybri oven (Model 400 H.I.; Robbins Scientific)

Incubator (37°C, 5% CO_2)

Microscope (inverted)

Nylon membrane (Hybond N^+; Amersham RPN303B)

PCR machine (GeneAmp PCR System 2700, Applied Biosystems or equivalent)

PCR tubes (200 μl; Applied Biosystems N8010540)

Pipettors (P20 and P200 Pipetman)

Plate-sealing sheet (Seal Plate; Excel scientific STR-SEAL-PLT)

Styrofoam box (sized to fit 96-well plates)

Tissue-culture dishes (gelatinized)

  Gelatinize tissue-culture dishes (10 cm, NUNCLON Δ Surface; NUNC 172958) with 3 ml of gelatin solution for 1 hour at room temperature and then remove the excess solution.

Tissue-culture plates (96 well, flat bottom, gelatinized)

Gelatinize 96-well plates (flat bottom, NUNCLON Δ Surface; NUNC 167008) with 50 μl of gelatin solution for 1 hour at room temperature and then aspirate the excess solution.

Tissue-culture plates (96-well plates, U bottom) (NUNCLON Δ Surface; NUNC 163320)

Water baths preset at 37°C and 55°C

X-ray film (BioMax MS; Kodak 8294985)

METHODS

Preparation and Verification of DNA

A control vector can be constructed in parallel that extends the boundary of the *F[tkneo]* targeting vector to include the chromosomal oligonucleotide primer to be used in PCR detection of homologous recombination of the targeting vector. Use PCR to test for amplification sensitivity and in background of wild-type ES cell genomic DNA.

1. Digest the genomic/transgene clone of the gene of interest with appropriate restriction endonucleases.
2. Insert DNA fragments into the restriction enzyme sites of pflox vector using enzymatically produced blunt ends as necessary.

   Use XbaI and/or SalI sites for short-arm insertion, the BamHI site for middle-arm insertion, and the HindIII and/or XhoI sites for long-arm insertion (Fig. 2A).
3. Linearize the targeting vector (or isolate the targeting construct away from the plasmid vector backbone) with an appropriate restriction endonuclease (e.g., NotI).
4. Confirm DNA quantity and quality by agarose gel electrophoresis.
5. Store the isolated and linearized DNA targeting vector in nuclease-free TE buffer in 70% ethanol until use.

ES Cell Culture and Gene Transfer by Electroporation

6. Thaw and culture R1 or similar ES cells on gelatinized tissue-culture dishes in ES cell culture medium.

   Change medium every day until cells are 70% confluent.
7. Change the culture medium 4 hours before electroporation.
8. Wash cells with PBS. Add 1 ml of trypsin-EDTA. Incubate for 5 minutes at 37°C.
9. Agitate gently to detach cells from the plate. Add 9 ml of ES culture medium. Pipette gently to resuspend cells and break up clumps.
10. Transfer the cell suspension to a 50-ml tube. Centrifuge at 190*g* for 5 minutes.
11. Aspirate the supernatant. Resuspend the cells with 30 ml of ice-cold sterile PBS.
12. Count the cells using a hemocytometer.
13. Centrifuge at 190*g* for 5 minutes. Aspirate the supernatant. Resuspend the cells with ice-cold sterile PBS to adjust the cell concentration to 10^7 cells per 800 μl.
14. Transfer the cell suspension to a sterile electroporation cuvette. Mix with 10 μg of the linearized targeting vector. Incubate for 10 minutes on ice.
15. Set the electroporator to the following conditions: voltage, 240 V and capacitance, 500 μF. Place the cuvette in the holder with electrodes facing the connectors. Deliver the electrical pulse.
16. Incubate cuvette for 10 minutes on ice.

17. Plate cells at varying dilutions (0.5 x 10^6 to 2 x 10^6 to cells/dish). Incubate for 24 hours.
18. To start positive selection, change medium to ES cell culture medium supplemented with 200 μg/ml (active concentration) G418. Continue incubation.

    Change culture media every 1–2 days, until G418-resistant colonies grow enough to isolate (~2 mm). This usually takes 7–10 days.
19. Prepare gelatinized flat-bottom 96-well plates containing 150 μl of G418-supplemented ES cell culture medium. Prepare U-bottom 96-well plates containing 70 μl of trypsin/ EDTA.
20. Replace the culture medium in the dish containing colonies with PBS. Place the dish on the stage of the inverted microscope.
21. Use a P20 Pipetman set to 10 μl to pick up the drug-resistant colonies by suction. Transfer to a trypsin-containing 96-well plate.
22. Prepare cells for additional culture, frozen storage, and/or DNA preparation for PCR.

    To avoid the cell-freezing step, PCR screening must be completed on fewer cells and prior to ES cell confluence in culture of the master plates.

For freezing ES cell clones in master plates

a. Gently suspend cells using a P200 Pipetman. Transfer 50 μl of the cell suspension to one 96-well plate (master plate for cell culture and/or freezing). Transfer the remaining (~30 μl) in the plate containing the trypsin-EDTA to a gelatinized 96-well flat-bottom plate (the replica plate for DNA preparation, see below) and add 150 μl of culture medium. Return both master and replica plates to culture incubator.
b. Culture cells in G418-supplemented ES medium for 3–5 days.

   Most of the clones in the master plates will be ready to freeze (~70% confluent). Continue ES cell culture of replica plates until confluent for highest yield of DNA for PCR analysis.
c. Wash wells twice with sterile PBS.
d. Add 50 μl of trypsin-EDTA. Incubate for 5 minutes at 37°C.
e. Add 50 μl of ice-cold 2x freezing medium. Gently suspend cells by pipetting.
f. Add 100 μl of ice-cold mineral oil.
g Seal plates with Parafilm. Transfer in a styrofoam box to a –80°C freezer.

For preparing DNA from replica plates

a. Wash cultured replica plates with PBS. Add 50 μl of lysis buffer.
b. Seal each plate with a plate-sealing sheet. Incubate for 2–3 hours at 55°C.
c. Add 150 μl of NaCl-ethanol. Incubate for 30 minutes at room temperature.
d. Decant the supernatant by tilting the plate slowly or centrifuge plate and lightly aspirate.
e. Wash wells three times with 200 μl of 70% ethanol. Allow the DNA to dry on the bench.
f. Add 30 μl of TE buffer to each well. Seal each plate with a plate-sealing sheet. Incubate overnight at 37°C prior for use as a DNA preparation.

For quick screening by PCR without freezing master plates

With either approach, combine and label DNA samples for initial PCR to represent rows of 8 or 12 wells and when PCR sensitivity in detecting the control vector is sufficiently high.

a. Use a P200 Pipetman to gently resuspend cells placed in the 96-well replica plate bearing trypsin-EDTA.

b. Transfer the suspension to a sterile 15-ml conical tube containing at least 5 ml of ES cell culture medium.

c. Centrifuge the tube to pellet the cells. Aspirate medium.

d. Wash cells with 500 μl of PBS. Transfer to a 1.5-ml Eppendorf tube and centrifuge to pellet the cells. Aspirate the PBS.

e. Resuspend the cells in 200 μl of autoclaved distilled H_2O. Place on dry ice immediately.

f. Boil tubes for 5 minutes. Centrifuge briefly to remove moisture from the tube walls.

g. Add 50 μl of proteinase K. Incubate at least 5 hours (or overnight) at 55°C.

h. Boil for 5 minutes to heat-inactivate the proteinase K.

23. Subject 20 μl of this preparation to PCR screening.

24. Expand ES cells representing PCR-positive clones.

25. Analyze clones of interest for the *loxP* sequence by genomic Southern blotting.

    For *loxP* genomic Southern blotting, use one or more DNA probes outside of the vector sequence used for gene transfer, as well as the *loxP* probe (isolated as a NotI fragment; Fig. 2D) to detect the presence of all *loxP* sites. Alternatively, include a restriction enzyme site during gene vector production that can distinguish the alleles.

Protocol 2

Cre Recombinase Gene Transfer In Vitro and Detection of *loxP*-dependent Recombination

Correctly targeted ES cell clones bearing the *F[tkneo]* allele (as established in Protocol 1) are typically subjected to in vitro Cre gene transfer to generate ES cell subclones bearing either type-1 (Δ) or type-2 (*F*) alleles. Expression of the Cre recombinase would normally of course also lead to the production of a distinct type-2 allele (2 *loxP* sites) bearing a *loxP*-flanked *tkneo* marker array. These are, however, eliminated by negative selection using ganciclovir. Apparently, the level and persistence of Cre enzyme activity transferred determine the frequency of type-1 and type-2 allele structures among ganciclovir-resistant ES cell clones. A problem sometimes encountered is the appearance of apparently wild-type ES cells lacking *loxP* sites (which are inherently ganciclovir-resistant). These wild-type ES cells may result from incomplete selection of cells that were not transduced by the vector following the initial gene transfer and selection steps and should only be present, if at all, in a small number of the total clones isolated. The selection of ES subclones bearing the type-2 allele and lacking *tkneo* are best for generating parental cell and mouse lines for subsequent Cre recombination among daughter cells and offspring. This effectively controls against the presence of spontaneous mutations within the ES cell genome as it allows linkage of the deletion of *loxP*-flanked DNA (the genotype) with the phenotype, avoiding the need for multiple independent parental ES cell clones and mouse lines. This procedure employs type-2 ES cells to generate chimeric mice that are then crossed to germ-line Cre-expressing mice, such as ZP3-Cre transgenic mates (Shafi et al. 2000; and see below). The additional time needed to breed the mice (~2–3 months) is typically less troublesome than the cost and effort of maintaining multiple clone-derived lines of mice. Another problem sometimes encountered is the lack of type-2 alleles in the presence of type-1 alleles following Cre electroporation and ganciclovir selection. This can be mitigated by reducing the amount of Cre plasmid used in the electroporation to a level 10–100-fold lower than that used in the initial targeting gene transfer. In addition, less-effective Cre variants that lack nuclear localization signals and eukaryotic translational consensus motifs also work well in this regard.

MATERIALS

Reagents and equipment required for Cre recombinase gene transfer and detection of the *loxP*-dependent recombination are the same as those described in Protocol 1, with the following additions:

Reagents

Ganciclovir (2 mM; InvivoGen, sud-gcv) <!>
Mice: ZP3-Cre transgenic (Shafi et al. 2000; available from Jackson Laboratories)
Oligonucleotides
  Cre forward primer (5′-CTGCATTACCGGTCGATGCA-3′)
  Cre reverse primer (5′-ACGTCCACCGGCATCAACGT-3′)
Plasmid DNA: pCre vector (variants may be substituted)

This vector includes a modified Cre sequence bearing both a nuclear localization signal and a Kozak translational consensus element. A less-robust Cre expression vector may be useful when desiring the production of type-2 alleles following electroporation.

METHODS

Cre Gene Transfer In Vitro to Establish Targeted ES Cells Bearing Type-2 Alleles

1. Culture parental ES cells bearing the *F[tkneo]* allele (from Protocol 1) in G418-supplemented ES cell culture medium on gelatinized dishes until 70% confluent.
2. Prepare cultures for electroporation as described in Protocol 1, Steps 7–13.
3. Electroporate 1 x 10^7 cells with 0.2–2 μg of the supercoiled form of the Cre expression vector as described in Protocol 1, Steps 14–16.

   The supercoiled structure reduces the likelihood of Cre expression vector integration into the ES cell genome. The low level of plasmid used increases the frequency of type-2 alleles generated.
4. Plate cells on ten gelatinized dishes at various dilutions (2 x 10^4 to 2 x 10^5) in ES cell culture medium. Culture cells for 3 days.
5. To begin negative selection, change medium to ES cell culture medium supplemented with 2 μM ganciclovir. Continue incubation. Change the medium every day for 4–5 days.
6. Pick and process clones as described in Protocol 1, Steps 19–24.

   Culture cells in ES cell culture media lacking G418 and ganciclovir.
7. Analyze DNA samples prepared from replica plates by genomic and *loxP* Southern blot hybridization.

   Expected genomic fragment sizes using both probes separately should be observed as planned during targeting vector construction. In genomic probe assays, the type-2 allele often appears with a mobility similar or identical to that of the wild-type allele. The *loxP* probe can distinguish these two allelic structures.

Conditional and Systemic Gene Mutagenesis In Vivo by Cre-*loxP* Recombination

During cell culture and selection, some ES cell clones fail to form germ-line chimeric mice. Although karyotyping and in vitro differentiation assays are helpful in eliminating the most aneuploid and defective of isolated ES cell clones, the lack of observable aneuploidy and the presence of robust in vitro differentiation are still insufficient for predicting success. This is where multiple ES cell clones become useful, and only one germ-line transmission event with a type-2 allele is needed to establish a well-controlled genomic modification of the mouse germ line. Moreover, animals bearing type-2 alleles are useful in achieving gene mutagenesis with Cre-expressing viral vectors and in the establishment of ex vivo cell systems for in vitro experimentation.

8. Produce chimeric mice from the desired ES cell clones.

   Typically two to three independent type-2 mutant alleles should be included to increase the chance of obtaining a germ-line chimeric animal.
9. Once germ-line transmission of the ES cell genome to offspring is obtained, cross type-2 heterozygotes with mice bearing germ-line Cre expression, such as ZP3-Cre transgenic mice, to achieve transmission of a germ-line type-1 allele.

   At very high efficiency in vivo, type-2 alleles are exposed to Cre expression during oogenesis and are recombined to become type-1 alleles among offspring.
10. Breed mice bearing type-1 or type-2 alleles with mouse-strain-specific backgrounds. Alternatively, cross Cre-*loxP* mice to produce homozygous type-1 and type-2 germ lines (Fig. 3A).

    Control mice bearing the type-2 allele can be bred to Cre transgenic mice bearing defined somatic cell expression profiles, yielding tissue- and cell-type-specific mutagenesis by type-1 allele formation in an otherwise type-2 somatic and germ-line genetic background (Fig. 3B) (Marth 1996).

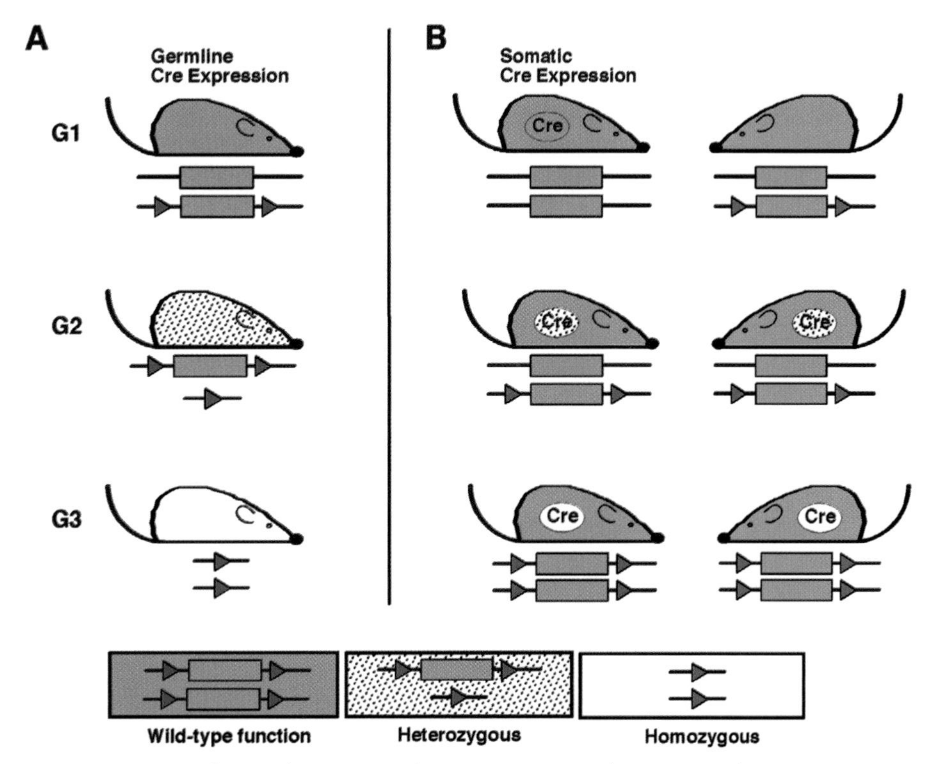

FIGURE 3. In vivo manipulation of genomic and transgene structure by Cre recombinase expression. (*A*) Mice bearing *loxP*-flanked germ-line alleles (allelic structures indicated below mice) are bred with transgenic mates bearing gamete Cre expression (oocytes or spermatids) in generation 1 (G1) to achieve systemic heterozygous (G2) and homozygous (G3) mutations in subsequent generations. (*B*) Somatic cell Cre expression provides for tissue- and cell-type-specific recombination that can be controlled further by inducible promoter systems. In both *A* and *B*, the presence in parallel of mice bearing the ancestral homozygous *loxP*-flanked loci provides a control genome allowing phenotypes to be assigned to the deletion of *loxP*-flanked DNA.

ACKNOWLEDGMENTS

The authors acknowledge support from the National Institutes of Health (DK4247) and the Howard Hughes Medical Institute (J.D.M.).

REFERENCES

Abremski K. and Hoess R. 1984. Bacteriophage P1 site-specific recombination. Purification and properties of the Cre recombinase protein. *J. Biol. Chem.* **259:** 1509–1514.

Andrews B.J., Proteau G.A., Beatty L.G., and Sadowski P.D. 1985. The FLP recombinase of the 2 micron circle DNA of yeast: Interaction with its target sequences. *Cell* **40:** 795–803.

Bucholz F., Angrand P.O., and Stewart A.F. 1998. Improved properties of FLP recombinase evolved by cycling mutagenesis. *Nat. Biotechnol.* **16:** 657–662.

Chui D., Oh-Eda M., Liao Y.-F., Panneerselvam K., Lal A., Marek K.W., Freeze H.H., Moremen K.W., Fukuda M.N., and Marth J.D. 1997. Alpha-mannosidase-II deficiency results in dyserythropoiesis and unveils an alternate pathway in oligosaccharide biosynthesis. *Cell* **90:** 157–167.

de Alboran I.M., O'Hagan R.C., Gartner F., Malynn B., Davidson L., Rickert R., Rajewsky K., DePinho R.A., and Alt F.W. 2001. Analysis of C-MYC function in normal cells via conditional gene-targeted mutation. *Immunity* **14:** 45–55.

Fiering S.C., Kim C., Epner E., and Groudine M. 1993. An "in-out" strategy using gene targeting and FLP recombinase for the functional dissection of complex DNA regulatory elements: Analysis of the β-globin locus control region. *Proc. Natl. Acad. Sci.* **90:** 8469–8473.

Fukushige S. and Sauer B. 1992. Genomic targeting with a positive-selection *lox* integration vector allows highly reproducible gene expression in mammalian cells. *Proc. Natl. Acad. Sci.* **89:** 7905–7909.

Golic K.G. and Lindquist S. 1989. The FLP recombinase of yeast catalyzes site-specific recombination in the *Drosophila* genome. *Cell* **59:** 499–509.

Gu H., Marth J.D., Orban P.C., Mossmann H., and Rajewsky K. 1994. Deletion of a DNA polymerase β gene segment in T cells using cell type-specific gene targeting. *Science* **265:** 103–106.

Hennet T., Hagen F.K., Tabak L.A., and Marth J.D. 1995. T cell-specific deletion of a polypeptide *N*-acetylgalactosaminyl-transferase gene by site-directed recombination. *Proc. Natl. Acad. Sci.* **92:** 12070–12074.

Hoess R.H. and Abremski K. 1990. The Cre-lox recombination system. In *Nucleic acids and molecular biology* (ed. F. Eckstein and D.M. Lilley), vol. 4, pp. 99–109. Springer-Verlag, Berlin and Heidelberg, Germany.

Kühn R., Schwenk F., Aguet M., and Rajewsky K. 1995. Inducible gene targeting in mice. *Science* **269:** 1427–1429.

Lakso M., Sauer B., Mosinger B., Jr., Lee E.J., Manning R.W., Yu S.H., Mulder K.L., and Westphal H. 1992. Targeted oncogene activation by site-specific recombination in transgenic mice. *Proc. Natl. Acad. Sci.* **89:** 6232–6236.

Logie C. and Stewart A.F. 1995. Ligand-regulated site-specific recombination. *Proc. Natl. Acad. Sci.* **92:** 5940–5944.

Mao X., Fujiwara Y., and Orkin S.H. 1999. Improved reporter strain for monitoring Cre recombinase-mediated DNA excisions in mice. *Proc. Natl. Acad. Sci.* **96:** 5037–5042.

Marth J.D. 1996. Recent advances in gene mutagenesis by site-directed recombination. *J. Clin. Invest.* **97:** 1999–2002.

Metzger D., Clifford J., Chiba H., and Chambon P. 1995. Conditional site-specific recombination in mammalian cells using a ligand-dependent chimeric Cre recombinase. *Proc. Natl. Acad. Sci.* **92:** 6991–6995.

Meyers E.N., Lewandowski M., and Martin G.R. 1998. An *Fgf8* mutant allelic series generated by Cre- and Flp-mediated recombination. *Nat. Genet.* **18:** 136–141.

Nagy A. 2000. Cre recombinase: The universal reagent for genome tailoring. *Genesis* **26:** 99–109.

Nagy A., Rossant J., Nagy R., Abramow-Newerly W., and Roder J.C. 1993. Derivation of completely cell culture-derived mice from early passage embryonic stem cells. *Proc. Natl. Acad. Sci.* **90:** 8424–8428.

O'Donnell N., Zachara N.E., Hart G.W., and Marth J.D. 2004. *Ogt*-dependent X-chromosome-linked protein glycosylation is a requisite modification in somatic cell function and embryo viability. *Mol. Cell. Biol.* **24:** 1680–1690.

O'Gorman S., Fox D.T., and Wahl G.M. 1991. Recombinase-mediated gene activation and site-specific integration in mammalian cells. *Science* **251:** 1351–1355.

Orban P.C., Chui D., and Marth J.D. 1992. Tissue- and site-specific DNA recombination in transgenic mice. *Proc. Natl. Acad. Sci.* **89:** 6861–6865.

Ramirez-Solis R., Liu P., and Bradley A. 1995. Chromosome engineering in mice. *Nature* **378:** 720–724.

Sauer B. 1992. Identification of cryptic *lox* sites in the yeast genome by selection for Cre-mediated chromosomal translocations that confer multiple drug resistance. *J. Mol. Biol.* **223:** 911–928.

Sauer B. and Henderson N. 1988. Site-specific DNA recombination in mammalian cells by the Cre recombinase of bacteriophage P1. *Proc. Natl. Acad. Sci.* **85:** 5166–5170.

Schmidt E.E., Taylor D.S., Prigge J.R., Barnett S., and Capecchi M.R. 2000. Illegitimate Cre-dependent chromosome rearrangement in transgenic mouse spermatids. *Proc. Natl. Acad. Sci.* **97:** 13702–13707.

Senecoff J., Bruckner R.C., and Cox M.M. 1985. The FLP recombinase of the yeast 2-μm plasmid: Characterization of its recombination site. *Proc. Natl. Acad. Sci.* **82:** 7270–7274.

Shafi R., Iyer S.P.N., Ellies L.G., O'Donnell N., Marek K.W., Chui D., Hart G.W., and Marth J.D. 2000. The O-GlcNAc transferase gene resides on the X chromosome and is essential for embryonic stem cell viability and mouse ontogeny. *Proc. Natl. Acad. Sci.* **97:** 5735–5739.

Smith A.J., De Sousa M.A., Kwabi-Addo B., Heppell-Parton A., Impey H., and Rabbitts P. 1995. A site-directed chromosomal translocation induced in ES cells by Cre-*loxP* recombination. *Nat. Genet.* **9:** 376–385.

Soriano P. 1999. Generalized *lacZ* expression with the ROSA26 Cre reporter strain. *Nat. Genet.* **21:** 70–71.

Soukharev S., Miller J.L., and Sauer B. 1999. Segmental genomic replacement in embryonic stem cells by double *lox* targeting. *Nucleic Acids Res.* **27:** e21.

te Riele H., Maandag E.R., and Berns A. 1992. Highly efficient gene targeting in embryonic stems cells through homologous recombination with isogenic DNA constructs. *Proc. Natl. Acad. Sci.* **89:** 5128–5132.

WWW RESOURCES

http://www.mshri.on.ca/nagy/

http://www.jax.org/resources/mouse_resources.html

61 Expression and Validation of Ribozyme and Short Hairpin RNA in Mammalian Cells

Mohammed Amarzguioui

The Biotechnology Centre of Oslo, University of Oslo, Gaustadalleen 21, 0349 Oslo, Norway

ABSTRACT

Recent years have witnessed an explosion in the application of RNA interference (RNAi) for functional genomics research. RNAi is the most recent and most successful member of a diverse class of gene expression modulators, all having in common an antisense-based mechanism of action. There are two principal methods for delivery of antisense effectors into cells: exogenous delivery of presynthesized molecules and endogenous expression from an appropriate vector construct. Down-regulation of gene expression by most delivery-based methods is transient in nature and may not be suitable for studies of proteins with long half-lives, and poor or variable transfection efficiencies represent additional potential problems. In contrast, vector-based approaches allow the generation of stable transfectants (Brummelkamp et al. 2002; Paddison et al. 2002; Perry et al. 2005; Yeo et al. 2005), thus improving experimental reproducibility and, in case of transient experiments, the possibility to sort or enrich for or track transfected cells by inclusion of a reporter gene in the expression vector. Expression from stable transfectants can also be made inducible (van de Wetering et al. 2003; Bowden and Riegel 2004; Gupta et al. 2004; Tiscornia et al. 2004), allowing temporal control of target gene knockdown. Furthermore, the use of viral vectors for delivery of the expression constructs provides easy and efficient stable expression even in cells that are typically difficult to transfect (Shen et al. 2003; Tiscornia et al. 2003; Morris and Rossi 2004; Li et al. 2005). All of these properties make the vector-based approaches highly suitable for antisense-based functional genomics applications.

INTRODUCTION

Antisense-mediated inhibition of gene expression has intrigued researchers since the first demonstration of proof-of-principle in the late 1970s (Zamecnik and Stephenson 1978). The general premise of the various incarnations of this technology is that introduction of a stretch of nucleic acids with perfect complementarity to a gene can inhibit its expression by disrupting the flow of information from gene to protein. Two classes of antisense effectors that are compatible with vector-based delivery systems are discussed here: (1) hammerhead ribozymes and (2) short hairpin RNAs (shRNAs). For a review of these and related approaches, see Scherer and Rossi (2003).

Ribozymes are short (~50 nucleotides) naturally occurring catalytic RNA motifs that in nature are associated with self-cleavage (Forster and Symons 1987), but, in the laboratory, these sequences can also be designed to cleave nonnatural targets in *trans* (Haseloff and Gerlach 1989). The catalytic strand contains most of the conserved sequences, and the only substrate sequence requirement for the hammerhead ribozyme is NU↓H at the cleavage site (↓indicates site of cleavage) (Zoumadakis and Tabler 1995). Their cleavage properties give ribozymes an advantage over basic antisense transcripts since they do not rely on a blocking mechanism of action and can in theory achieve multiple turnover. The efficacies of both basic antisense and catalytic RNAs are strongly dependent on target mRNA secondary structure and the accessibility of the target for hybridization. Various experimental and computational techniques for determination of hybridization accessible sites have been described previously (Amarzguioui et al. 2000; Scherr et al. 2000; Sohail et al. 2001; Ding and Lawrence 2003).

The most recent addition to the antisense field is RNAi (Fire et al. 1998; Hannon 2002). In the initiation stage of the RNAi pathway, the RNase-III-like enzyme Dicer (Bernstein et al. 2001) processes long double-stranded RNA (dsRNA) to 21–22 mer duplexes having 3′ overhangs 2 nucleotides in length. The processing products, short interfering RNA (siRNA), are utilized by the RNA-induced silencing complex (RISC) to recognize and guide cleavage of homologous mRNA (Elbashir et al. 2001; Hammond et al. 2001; Liu et al. 2004). RNAi can also be induced by endogenous expression of shRNAs in which two strands of an siRNA are closed with a short loop in one end of the duplex (Brummelkamp et al. 2002; Paddison et al. 2002). Alternatively, the two strands of the siRNA can be transcribed separately (N.S., Lee et al. 2002). shRNAs are modeled on endogenously expressed microRNAs (miRNAs) that normally mediate RNAi through a translational inhibition mechanism (He and Hannon 2004). The sequence of the RNAi effectors appears to be the single most important determinant of functionality (Khvorova et al. 2003; Schwarz et al. 2003), with target secondary structure having at most a moderate role. This realization has resulted in the identification of design rules and procedures that substantially increase the success rate of rationally designed siRNA/shRNA (Amarzguioui and Prydz 2004; Reynolds et al. 2004; Huesken et al. 2005). Recent reports indicate improved efficacy of siRNA and shRNA of longer than standard stem structures (Kim et al. 2005; Rose et al. 2005; Siolas et al. 2005). For an overview of these results and their implications, as well as other design considerations, see Amarzguioui et al. (2006).

Expression vectors generally contain the effector transcription unit under a promoter suitable for high-level expression, transcription termination sequences if appropriate, and optionally additional sequences to facilitate site-specific integration and/or selection of transfected or transduced clones in cases of stable expression. Three distinct expression strategies are typically employed: (1) polymerase-II-driven expression cassettes, (2) tRNA-based expression cassettes, and (3) expression from self-contained RNA polymerase III promoters.

Expression from polymerase II promoters typically results in generation of heterogeneous transcripts with extraneous vector-derived sequences that may interfere with the functionality of the embedded effector molecule. *cis*-inhibitory effects may be reduced by incorporating ribozyme target sequences upstream and downstream from the effector, allowing release of the latter from the larger transcript through *cis*-cleavage (Xing et al. 1995; Perry et al. 2005). shRNAs expressed within the context of a complex primary transcript are typically inactive (Krol et al. 2004), probably due to lack of proper processing. Functional processing may be achieved through the use of an miRNA expression background, in which the region of the primary miRNA transcript from which the mature miRNA is generated is replaced with the siRNA sequences the investigator wants to express (Zeng et al. 2003; Zeng and Cullen 2005). The resulting chimeric transcript is efficiently processed through the natural pathway for miRNA processing, involving the nucleases Drosha and Dicer (Y., Lee et al. 2002; Y., Lee et al. 2003).

An alternative strategy for expression of short antisense molecules within a larger transcript while minimizing inhibition by the extraneous sequences utilizes tRNA as the expression background. Because the tRNA structure is known, effector cassettes can be inserted at carefully selected sites to minimize any inhibitory interactions with the extraneous tRNA-derived sequences (Amarzguioui and Prydz 1998; Kato et al. 2001; Sano and Taira 2004). This approach also appears to be compatible with expression of functional shRNA (Oshima et al. 2003).

The most robust and widely applicable strategy for expression of antisense effectors involves transcription from self-contained RNA polymerase III promoters, such as U6 (Paddison et al. 2002), H1 (Brummelkamp et al. 2002), and 7SK (Czauderna et al. 2003), to generate transcripts having clearly defined start and termination sites. In this system, transcription is initiated at a precise position outside of the promoter sequence and is terminated as the polymerase encounters a stretch of five to six thymidines immediately after the cassette encoding the desired transcript. The expression cassettes are generated by either polymerase chain reaction (PCR) or oligo cassette-based strategies. This chapter describes a version of the cassette-based strategy. For more detailed description of the other alternatives, see Amarzguioui et al. (2005).

The method described in this chapter involves the annealing of two complementary oligos, generating a double-stranded oligo cassette with appropriate overhangs for directional cloning immediately downstream from the promoter. To ensure that transcription starts at the correct position and produces a transcript with no extraneous 5′ sequences, targeted mutations are introduced at the end of the promoters to generate a unique cloning site. Targeted mutation of the H1 promoter to introduce a BglII site has been shown to be compatible with shRNA-mediated silencing of expression (van de Wetering et al. 2003). The U6 promoter has likewise been modified to introduce a BglII site to enable cloning of the same oligo cassette into multiple promoter backgrounds (M. Amarzguioui et al., unpubl.). The relevant U6 promoter modifications are illustrated in Figure 1. The modified U6 promoter was amplified by PCR, and a BclI+NheI PCR fragment

A.

```
Comparison of wildtype and modified (U6M) U6 promoter:

  U6          gatttcttggctttatatatcttgtggaaaggacgaaacaccG
              ||||||||||||||||||||||||||||||||||||||.|.|.
  U6M         gatttcttggctttatatatcttgtggaaaggacgaaagatct
                                                    BglII
```

B.

```
Vector fragment (pcDNA3-U6M):

gacgaaagatctgctagcgtttaaacttaagcttggtaccgagctcggatccactagtcc
----:----|----:----|----:----|----:----|----:----|----:----|
ctgctttctagacgatcgcaaatttgaattcgaaccatggctcgagcctaggtgatcagg
       BglII                        KpnI
```

FIGURE 1. Excerpt of promoter and vector sequences. (*A*) Comparison of wild-type and modified U6 promoters. Nucleotide in bold underlined uppercase letter in wild-type sequence represents transcription start site. (*B*) Excerpt of expression vector with unique sites for oligo cassette cloning indicated.

was cloned directionally into the BglII+NheI sites of the expression plasmid pcDNA3.1+ (Invitrogen), generating the cloning vector pcDNA3-U6M. The PCR-based cloning introduces additional BamHI and NotI sites upstream of the promoter in order to facilitate restriction analysis and subcloning. Design of oligonucleotide cassettes for hammerhead ribozyme and shRNA expression is illustrated in Figure 2.

Other versions of this expression system that rely on vector backbones which introduce additional functionalities have been described. For example, the U6 promoter and tetracycline-responsive versions of this promoter (Czauderna et al. 2003) and the H1 promoter (van de Wetering et al. 2003) have been cloned into the expression plasmid pcDNA5/FRT/TO (Invitrogen), generating the respective expression vectors pFRT-U6M, pFRT-U6tet, and pFRT-H1tet (M. Amarzguioui et al., unpubl.). All of the vectors are compatible with the cloning methodology described below. The vector backbone contains an Flp recombinase recognition sequence, allowing site-specific integration in cells containing such sequences, upon cotransfection with Flp recombinase. In cells lines expressing the tetracycline repressor, the tet-responsive versions allow doxycycline-inducible expression. The two functionalities can be combined in cell lines with the appropriate genetic background (Flp-In T-Rex 293 cells, Invitrogen). Recombinase-mediated stable expression and tetracycline-inducible expression are described in more detail elsewhere in this book.

```
Target sequence:
    5' GGCAGACUGAGUCGUGAUAUCUACA 3'

Ribozyme sequence:
    5' UAUCACcugaugagcgcguaagcgcgaaACUCAGUCUGCC 3'

ShRNA sequence:
    5' GGCAGACUGAGUCGUGAUAUCUACAccugacccaUGUAGAUAUCACGACUCAGUCUGCC 3'


Design of ribozyme oligo cassette:

A) 5' GATCTATCACctgatgagcgcgtaagcgcgaaACTCAGTCTGCCTTTTTGTAC  3'
B) 3'     ATAGTGgactactcgcgcattcgcgcttTGAGTCAGACGGAAAAA      5'

Design of shRNA oligo cassette:

A) 5' GATCGGCAGACTGAGTCGTGATATCTACAcctgacccaTGTAGATATCACGACTCAGTCTGCCTTTTTGTAC 3'
B) 3'     CCGTCTGACTCAGCACTATAGATGTggactgggtACATCTATAGTGCTGAGTCAGACGGAAAAA     5'
```

FIGURE 2. Design of oligo cassettes for a hammerhead ribozyme and shRNA targeting the same sequence. The ribozyme binds to the 19-nucleotide target sequence with 6+12-nucleotide hybridizing arms (*underlined positions*). The components of the ribozyme-encoding oligo cassette include (in 5′ to 3′ order) a 4-nucleotide BglII overhang, a 6-nucleotide stem I hybridizing arm, catalytic core, and stem-loop II sequences (*lowercase letters*), 12-nucleotide stem III hybridizing arm, terminator (TTTTT), and a 4-nucleotide KpnI compatible overhang. The shRNA is designed to generate an siRNA targeting the same sequence as the hammerhead. The stem region of the shRNA is extended to 25 bp in order to improve processing and efficacy (Rose et al. 2005; Siolas et al. 2005). In addition to the same overhangs and terminator sequences at equivalent positions, the shRNA cassette consists of a 25-nucleotide sense strand (*uppercase*, the 19-nucleotide bold part of which represents the duplex region of the expected processing product), a 9-nucleotide loop (*lowercase*), and a 25-nucleotide antisense strand encoding region.

Protocol

Construction and Expression of Oligonucleotide-based Antisense Cassettes

This protocol describes simple and robust methods for the construction and cloning of expression constructs that can be used to deliver antisense effectors, exemplified by the hammerhead ribozyme and shRNA, into cultured cell lines. The protocol also describes the construction of reporter vectors to be used for target validation. Due to the variable efficacy of the antisense effectors, it is advisable to design multiple constructs targeting different sites. Once the different constructs have been generated, their relative efficacy can be readily determined through reporter cotransfection experiments in which a stretch of cDNA encompassing all target sites is cloned directionally into the 3′ UTR (untranslated region) of a reporter (psiCheck2, Promega). Successful cleavage of the target site results in degradation of reporter mRNA, with concomitant decrease in translated product, which is detected by a luminescence-based assay system.

MATERIALS

Reagents

Adherent mammalian (preferably human) cell lines (e.g., HEK293 and HCT116)

Agar plates with LB medium and ampicillin (50 μg/ml)

Dissolve 15 g of Bacto-agar in 1.0 liter of LB medium and sterilize by autoclaving. Cool to 50°C in a temperature-controlled water bath and add 0.50 ml of 100 mg/ml ampicillin (dissolved in water).

ATP (GE Healthcare Life Sciences)

Calf intestinal alkaline phosphatase (New England BioLabs)

Column-based Miniprep Kit (various suppliers)

Complete and basal cell culture growth medium

10x DNA ligase buffer (500 mM Tris-HCl (pH 7.5), 100 mM $MgCl_2$, 100 mM DTT, 10 mM ATP, and 250 μg/ml bovine serum albumin [optional])

Dual Luciferase Reporter Assay System (Promega E1910 or E1960)

Expression vectors

- pcDNA3-U6M, pFRT-U6M, pFRT-U6tet, pFRT-H1tet (available from author)
- pcDNA3.1 (Invitrogen V790-20; used to generate pcDNA3-U6M)
- pcDNA5/FRT/TO (Invitrogen K6500-01; used to generate pFRT-X vectors)
- psiCheck2 (Promega C8021)

Gel Extraction Kit (various suppliers)

Heat shock competent bacteria (any of the commonly used strains, including DH5α and DH10B)

Lipofectamine 2000 (Invitrogen)

LB (Luria-Bertani) medium

Weigh in 10 g of tryptone, 5 g of yeast extract, and 10 g of NaCl into 900 ml of deionized H_2O. Adjust the pH to 7.0 with 5 N NaOH, and adjust volume to 1000 ml. Autoclave to sterilize.

1x NEB2 buffer (50 mM NaCl, 10 mM Tris-HCl (pH 7.9), 1 mM DTT, and 10 mM $MgCl_2$)

Plasmid DNA (column-purified or equivalent quality)

10x PNK buffer (700 mM Tris-HCl (pH 7.6), 100 mM $MgCl_2$, and 50 mM DTT)

Polynucleotide kinase (New England BioLabs)

Restriction enzymes (New England BioLabs)

BglII (R0144), KpnI (R0142), XhoI (R0146), NotI (R0189), BamHI (R0136)

Sequencing primers

U6-For	5′ ccaaggtcgggcaggaag 3′	(T_m = 60.9°C)
BGH-Rev	5′ aggggcaaacaacagatggc 3′	(T_m = 62.1°C)
RLuc-For	5′ gaggacgctccagatgaaatg 3′	(T_m = 59.5°C)
TK-Rev	5′ tgagagtgtttcgttccttccc 3′	(T_m = 60.8°C)

T4 DNA ligase (New England BioLabs)

Target-specific oligonucleotides

Oligonucleotides should be synthesized at lowest possible scale and purified by desalting only (PCR-quality purification). Dissolve in 10 mM Tris pH 7.5 buffer, quantify by UV spectroscopy (If provided quantity is known, just resuspend in the appropriate volume of buffer), and adjust concentration of stock solution to 100 μM.

1 M Tris buffer (pH 7.5):

Weigh in 6.05 g of Tris base, dissolve in 30 ml of deionized H_2O, adjust pH with 5 M HCl, and adjust final volume to 50 ml with H_2O.

Equipment

Bacterial culture plates
Bacterial culture shaking incubator
Luminometer (injector-equipped plate reader recommended)
Tissue-culture dishes (100 mm) or flasks (T-75)
Tissue-culture plates (24 well)

METHODS

Preparation of Oligo Cassette

1. Prepare a separate 20 μl 10 μM phosphorylation reaction for each oligonucleotide.

100 μM stock oligo	2.0 μl
10x PNK buffer	2.0 μl
10 mM ATP	2.0 μl
PNK	5 μl
H_2O	13.5 μl

   Incubate for 1 hour at 37°C.

2. Combine 10 μl of each complementary phosphorylated oligonucleotide. Incubate in a heating block for 10 minutes at 75°C to inactivate the enzyme and anneal by subsequent slow cooling (remove the metal block from heating source and allow the temperature to drop gradually to 30–40°C).

   The final concentration of oligo duplex cassette is 5 μM.

3. Dilute an aliquot of double-stranded oligo cassette 25-fold, to a final concentration of 200 nM (9 ng/μl for a typical 70-bp oligo cassette).

Preparation of Vector Fragment

4. Digest 3–4 µg of the appropriate vector (pcDNA3-U6M, pFRT-U6M, pFRT-U6tet, or pFRT-H1tet) with 30–40 units each of BglII and KpnI in 50 µl of 1x NEB2 buffer for 2–4 hours at 37°C.
5. Add 1 µl of 10 units/µl calf intestinal alkaline phosphatase and incubate for a further 1 hour.

   Dephosphorylation reduces background in case of incomplete digestion by one of the enzymes in the previous step.
6. Separate the digested fragments on a preparative 1.0% agarose gel. Use a clean scalpel to excise linearized vector fragment. Set aside a small amount of the digestion reaction (1–2 µl) for use as a control in Step 8.
7. Extract the linearized vector from the gel using any of the various available gel-extraction kits, following the manufacturer's recommended protocol, and elute the purified DNA in 30–40 µl of the supplied elution buffer or 10 mM Tris buffer (pH 7.5).
8. Estimate the vector concentration by electrophoresis of a 1-µl sample of eluted DNA through an analytical agarose gel. Include aliquots of the restriction digest reaction set aside in Step 6 as a concentration marker.

Ligation of Linearized Vector and Oligo Cassette Insert

9. Set up a reaction to ligate the linearized vector and oligo cassettes.

vector (typically 1–2 µl)	50 ng
200 nM oligo cassette	1.0 µl
10x DNA ligase buffer	0.5 µl
T4 DNA ligase	0.5 µl
H_2O	to 5.0 µl

   Incubate for 30 minutes at room temperature.

   Include a negative ligation control with no insert, by preparing a master mix containing all components except the insert.

Transformation of Competent Cells

10. Mix 3 µl of the ligation reaction with 30–50 µl of heat shock competent bacteria and incubate for 30 minutes on ice. Store the remainder of the ligation reaction in the freezer in case it is needed for retransformation.
11. Incubate the transformation mixture for 35–40 seconds at 42°C. Snap-cool for 2 minutes on ice and then add 700 µl of LB medium and incubate on a shaker for 45–60 minutes at 37°C.
12. Plate half of the mixture onto agar plates containing 50 µg/ml ampicillin and incubate overnight at 37°C.

Analysis of Transformants

13. Inspect the agar plates the day after transformation; the number of colonies in the plates for transformations with insert should be much higher than in the control plate. Use a sterile pipette tip to pick two to three colonies from each construct and inoculate into 5 ml of LB medium supplemented with 50–100 µg/ml ampicillin. Culture in a shaking incubator overnight (16–18 hours) at 37°C.

14. Isolate miniprep DNA from transformants by any standard methodology (a column-based approach is recommended), following the manufacturer's recommended protocol. Elute DNA in 50 µl of the supplied elution buffer or 10 mM Tris (pH 7.5).

    The purity of column-isolated DNA is generally suitable for cotransfection experiments for initial evaluation of construct efficacy (Steps 20–28).

15. Perform analytical restriction digestions on miniprep DNA (0.3–0.5 µg in 10 µl of reaction volume) with either BamHI or NotI to verify the presence of insert. Include the cloning vector as negative control in these analyses.

    Insertion of a typical shRNA cassette increases the size of the released fragment by 40–50 bp relative to cloning plasmid (327- and 364-bp fragments for BamHI and NotI restrictions, respectively). A plasmid map of a representative shRNA expression construct is shown in Figure 3A. The enzymes indicated are suitable for restriction analysis and subcloning.

16. Sequence positive clones identified in Step 15 using U6-For and/or BGH-Rev primers.

Construction of Dual Luciferase Reporter to Evaluate Effector Construct

17. Amplify a region of target gene cDNA that encompasses all target sites as well as at least 100 nucleotides each upstream of the 5′-most target sites and downstream from the 3′-most site.

    Including additional target-specific sequences outside of the minimal target region reduces the chances that altered secondary structure of the target may influence results. Tag the forward and reverse primers for amplification with recognition sequences for XhoI and NotI, respectively, for directional cloning into the same sites in psiCheck2. Make sure to include an additional 4–6 nucleotides outside of the enzyme recognition sequences in order to improve digestion efficiency of the PCR product.

18. To clone the target cDNA, follow Steps 4–15, with the following modifications: In Step 9 (ligation), the insert should be in two- to threefold molar excess with respect to the linearized vector. In Step 15 (analysis of clones), analyze the transformant DNA by digestion with XhoI+NotI to release the insert.

19. Sequence positive clones, identified by restriction endonuclease digestion, using RLuc-For and TK-Rev primers.

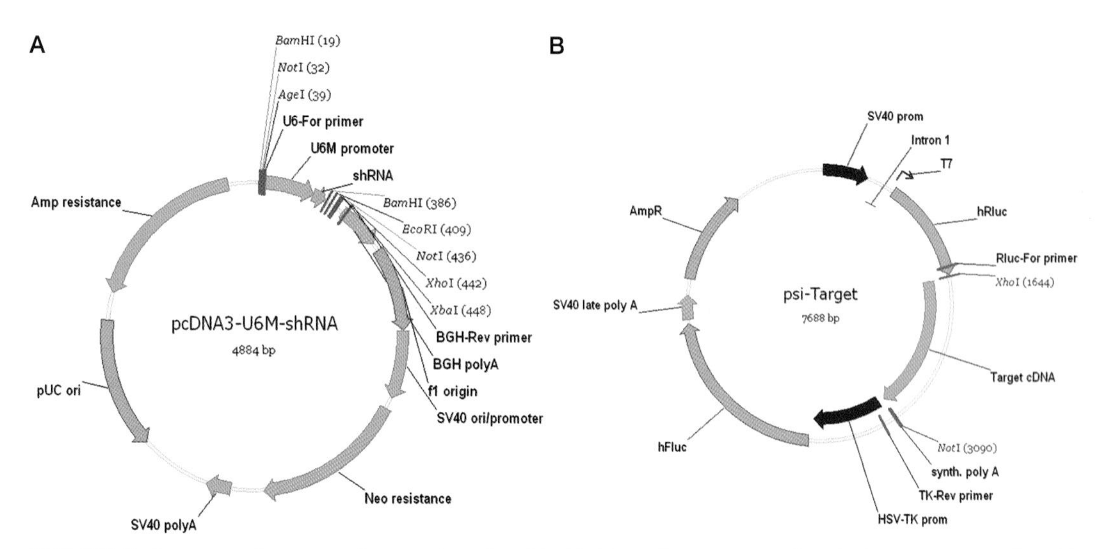

FIGURE 3. Representative maps of expression plasmids. (*A*) Map of a representative shRNA expression plasmid. (*B*) Map of a dual luciferase reporter vector in which a 1.4-kb cDNA fragment of the target has been cloned into the XhoI+NotI sites of psiCheck2.

Cotransfections with Dual Luciferase Reporter and Effector Constructs

20. At 16–24 hours before transfection, plate any commonly used cell line (HEK293, HCT116, etc.) in 24-well plates (0.5 ml/well) at a density such that the cells are 60–80% confluent at the time of transfection.
21. For triplicate samples, prepare a batch dilution of Lip/med (Lipofectamine 2000/medium) sufficient for all (*N*) samples according to the following formula:

    Lip/med: 50 x *N* µl of basal medium + 1.5 x *N* µl of Lipofectamine 2000

    Incubate for 5–10 minutes at room temperature.
22. Complex the reporter and effector DNAs (including an irrelevant effector as control) for transfections in triplicate wells (conditions: 100 ng of reporter, 100 ng of effector, 0.5 µl of Lipofectamine 2000 per well) according to the following protocol:

100 ng/µl reporter	3.0 µl
100 ng/µl effector construct	3.0 µl
basal medium	50 µl
Mix well with DNA.	
Lip/med (see Step 20)	50 µl

    Mix well with medium-diluted DNA and incubate the complexes for 30 minutes at room temperature (see Step 21).
23. During incubation of the complexes, remove the medium from the cells and add 300 µl of fresh full medium to each well. Alternatively, only remove 200 µl from each well.
24. Dilute the complexes with 500 µl of full medium and add 200 µl of diluted complex per well for each triplicate sample.
25. Incubate the complexes with cells (no medium replacement necessary) until the time of harvest, 24 hours posttransfection.
26. Remove the medium and lyse the cells in 150 µl of passive lysis buffer (supplied with the Dual Luciferase Reporter Assay System Kit) per well. Incubate under gentle shaking for 15 minutes at room temperature.
27. Measure reporter expression on a luminometer, preferably equipped with reagent injectors, using the Dual Luciferase Reporter Assay System Kit, essentially as described by the manufacturer. For conservation of reagent, use 50 µl of each reagent per 10 µl of lysate sample.
28. Normalize target-specific Renilla luciferase expression to the internal control Firefly luciferase expression for all replicates and samples. Set the average expression ratio for control samples (cells transfected with irrelevant control effector) to 100% and calculate relative expression levels for other samples and replicates accordingly. Perform at least two independent experiments and select two constructs, demonstrating the most efficient expression, for further study.

REFERENCES

Amarzguioui M. and Prydz H. 1998. Hammerhead ribozyme design and application. *Cell Mol. Life Sci.* **54:** 1175–1202.

———. 2004. An algorithm for selection of functional siRNA sequences. *Biochem. Biophys. Res. Commun.* **316:** 1050–1058.

Amarzguioui M., Rossi J.J., and Kim D. 2005. Approaches for chemically synthesized siRNA and vector-mediated RNAi. *FEBS Lett.* **579:** 5974–5981.

Amarzguioui M., Brede G., Babaie E., Grotli M., Sproat B., and Prydz H. 2000. Secondary structure prediction and in vitro accessibility of mRNA as tools in the selection of target sites for ribozymes. *Nucleic Acids Res.* **28:** 4113–4124.

Amarzguioui M., Lundberg P., Cantin E., Hagstrom J., Behlke M.A., and Rossi J.J. 2006. Rational design and in vitro and in vivo delivery of Dicer substrate siRNA. *Nat. Protocols* **1:** 508–517.

Bernstein E., Caudy A.A., Hammond S.M., and Hannon G.J. 2001. Role for a bidentate ribonuclease in the initiation step of RNA interference. *Nature* **409:** 363–366.

Bowden E.T. and Riegel A.T. 2004. Tetracycline-regulated expression of hammerhead ribozymes in vivo. *Methods Mol. Biol.* **252:** 179–194.

Brummelkamp T.R., Bernards R., and Agami R. 2002. A system for stable expression of short interfering RNAs in mammalian cells. *Science* **296:** 550–553.

Czauderna F., Santel A., Hinz M., Fechtner M., Durieux B., Fisch G., Leenders F., Arnold W., Giese K., Klippel A., and Kaufmann J. 2003. Inducible shRNA expression for application in a prostate cancer mouse model. *Nucleic Acids Res.* **31:** e127.

Ding Y. and Lawrence C.E. 2003. A statistical sampling algorithm for RNA secondary structure prediction. *Nucleic Acids Res.* **31:** 7280–7301.

Elbashir S.M., Harborth J., Lendeckel W., Yalcin A., Weber K., and Tuschl T. 2001. Duplexes of 21-nucleotide RNAs mediate RNA interference in cultured mammalian cells. *Nature* **411:** 494–498.

Fire A., Xu S., Montgomery M.K., Kostas S.A., Driver S.E., and Mello C.C. 1998. Potent and specific genetic interference by double-stranded RNA in *Caenorhabditis elegans*. *Nature* **391:** 806–811.

Forster A.C. and Symons R.H. 1987. Self-cleavage of plus and minus RNAs of a virusoid and a structural model for the active sites. *Cell* **49:** 211–220.

Gupta S., Schoer R.A., Egan J.E., Hannon G.J., and Mittal V. 2004. Inducible, reversible, and stable RNA interference in mammalian cells. *Proc. Natl. Acad. Sci.* **101:** 1927–1932.

Hammond S.M., Boettcher S., Caudy A.A., Kobayashi R., and Hannon G.J. 2001. Argonaute2, a link between genetic and biochemical analyses of RNAi. *Science* **293:** 1146–1150.

Hannon G.J. 2002. RNA interference. *Nature* **418:** 244–251.

Haseloff J. and Gerlach W.L. 1989. Sequences required for self-catalysed cleavage of the satellite RNA of tobacco ringspot virus. *Gene* **82:** 43–52.

He L. and Hannon G.J. 2004. MicroRNAs: Small RNAs with a big role in gene regulation. *Nat. Rev. Genet.* **5:** 522–531.

Huesken D., Lange J., Mickanin C., Weiler J., Asselbergs F., Warner J., Meloon B., Engel S., Rosenberg A., Cohen D., et al. 2005. Design of a genome-wide siRNA library using an artificial neural network. *Nat. Biotechnol.* **23:** 995–1001.

Kato Y., Kuwabara T., Warashina M., Toda H., and Taira K. 2001. Relationships between the activities in vitro and in vivo of various kinds of ribozyme and their intracellular localization in mammalian cells. *J. Biol. Chem.* **276:** 15378–15385.

Khvorova A., Reynolds A., and Jayasena S.D. 2003. Functional siRNAs and miRNAs exhibit strand bias. *Cell* **115:** 209–216.

Kim D.H., Behlke M.A., Rose S.D., Chang M.S., Choi S., and Rossi J.J. 2005. Synthetic dsRNA Dicer substrates enhance RNAi potency and efficacy. *Nat. Biotechnol.* **23:** 222–226.

Krol J., Sobczak K., Wilczynska U., Drath M., Jasinska A., Kaczynska D., and Krzyzosiak W.J. 2004. Structural features of microRNA (miRNA) precursors and their relevance to miRNA biogenesis and small interfering RNA/short hairpin RNA design. *J. Biol. Chem.* **279:** 42230–42239.

Lee N.S., Dohjima T., Bauer G., Li H., Li M.J., Ehsani A., Salvaterra P., and Rossi J. 2002. Expression of small interfering RNAs targeted against HIV-1 *rev* transcripts in human cells. *Nat. Biotechnol.* **20:** 500–505.

Lee Y., Jeon K., Lee J.T., Kim S., and Kim V.N. 2002. MicroRNA maturation: Stepwise processing and subcellular localization. *EMBO J.* **21:** 4663–4670.

Lee Y., Ahn C., Han J., Choi H., Kim J., Yim J., Lee J., Provost P., Radmark O., Kim S., and Kim V.N. 2003. The nuclear RNase III Drosha initiates microRNA processing. *Nature* **425:** 415–419.

Li M.J., Kim J., Li S., Zaia J., Yee J.K., Anderson J., Akkina R., and Rossi J.J. 2005. Long-term inhibition of HIV-1 infection in primary hematopoietic cells by lentiviral vector delivery of a triple combination of anti-HIV shRNA, anti-CCR5 ribozyme, and a nucleolar-localizing TAR decoy. *Mol. Ther.* **12:** 900–909.

Liu J., Carmell M.A., Rivas F.V., Marsden C.G., Thomson J.M., Song J.J., Hammond S.M., Joshua-Tor L., and Hannon G.J. 2004. Argonaute2 is the catalytic engine of mammalian RNAi. *Science* **305:** 1437–1441.

Morris K.V. and Rossi J.J. 2004. Anti-HIV-1 gene expressing lentiviral vectors as an adjunctive therapy for HIV-1 infection. *Curr. HIV Res.* **2:** 185–191.

Oshima K., Kawasaki H., Soda Y., Tani K., Asano S., and Taira K. 2003. Maxizymes and small hairpin-type RNAs that are driven by a tRNA promoter specifically cleave a chimeric gene associated with leukemia in vitro and in vivo. *Cancer Res.* **63:** 6809–6814.

Paddison P.J., Caudy A.A., Bernstein E., Hannon G.J., and Conklin D.S. 2002. Short hairpin RNAs (shRNAs) induce sequence-specific silencing in mammalian cells. *Genes Dev.* **16:** 948–958.

Perry W.L., III, Shepard R.L., Sampath J., Yaden B., Chin W.W., Iversen P.W., Jin S., Lesoon A., O'Brien K.A., Peek V.L., et al. 2005. Human splicing factor SPF45 (*RBM17*) confers broad multidrug resistance to anticancer drugs when overexpressed—A phenotype partially reversed by selective estrogen receptor modulators. *Cancer Res.* **65:** 6593–6600.

Reynolds A., Leake D., Boese Q., Scaringe S., Marshall W.S., and Khvorova A. 2004. Rational siRNA design for RNA interference. *Nat. Biotechnol.* **22:** 326–330.

Rose S.D., Kim D.H., Amarzguioui M., Heidel J.D., Collingwood M.A., Davis M.E., Rossi J.J., and Behlke M.A. 2005. Functional polarity is introduced by Dicer processing of short substrate RNAs. *Nucleic Acids Res.* **33:** 4140–4156.

Sano M. and Taira K. 2004. Ribozyme expression systems. *Methods Mol. Biol.* **252:** 195–207.

Scherer L.J. and Rossi J.J. 2003. Approaches for the sequence-specific knockdown of mRNA. *Nat. Biotechnol.* **21:** 1457–1465.

Scherr M., Rossi J.J., Sczakiel G., and Patzel V. 2000. RNA accessibility prediction: A theoretical approach is consistent with experimental studies in cell extracts. *Nucleic Acids Res.* **28:** 2455–2461.

Schwarz D.S., Hutvagner G., Du T., Xu Z., Aronin N., and Zamore P.D. 2003. Asymmetry in the assembly of the RNAi enzyme complex. *Cell* **115:** 199–208.

Shen C., Buck A.K., Liu X., Winkler M., and Reske S.N. 2003. Gene silencing by adenovirus-delivered siRNA. *FEBS Lett.* **539:** 111–114.

Siolas D., Lerner C., Burchard J., Ge W., Linsley P.S., Paddison P.J., Hannon G.J., and Cleary M.A. 2005. Synthetic shRNAs as potent RNAi triggers. *Nat. Biotechnol.* **23:** 227–231.

Sohail M., Hochegger H., Klotzbucher A., Guellec R.L., Hunt T., and Southern E.M. 2001. Antisense oligonucleotides selected by hybridisation to scanning arrays are effective reagents in vivo. *Nucleic Acids Res.* **29:** 2041–2051.

Tiscornia G., Singer O., Ikawa M., and Verma I.M. 2003. A general method for gene knockdown in mice by using lentiviral vectors expressing small interfering RNA. *Proc. Natl. Acad. Sci.* **100:** 1844–1848.

Tiscornia G., Tergaonkar V., Galimi F., and Verma I.M. 2004. CRE recombinase-inducible RNA interference mediated by lentiviral vectors. *Proc. Natl. Acad. Sci.* **101:** 7347–7351.

van de Wetering M., Oving I., Muncan V., Pon Fong M.T., Brantjes H., van Leenen D., Holstege F.C., Brummelkamp T.R., Agami R., and Clevers H. 2003. Specific inhibition of gene expression using a stably integrated, inducible small-interfering-RNA vector. *EMBO Rep.* **4:** 609–615.

Xing Z., Mahadeviah S., and Whitton J.L. 1995. Antiviral activity of RNA molecules containing self-releasing ribozymes targeted to lymphocytic choriomeningitis virus. *Antisense Res. Dev.* **5:** 203–212.

Yeo M., Rha S.Y., Jeung H.C., Hu S.X., Yang S.H., Kim Y.S., An S.W., and Chung H.C. 2005. Attenuation of telomerase activity by hammerhead ribozyme targeting human telomerase RNA induces growth retardation and apoptosis in human breast tumor cells. *Int. J. Cancer* **114:** 484–489.

Zamecnik P.C. and Stephenson M.L. 1978. Inhibition of Rous sarcoma virus replication and cell transformation by a specific oligodeoxynucleotide. *Proc. Natl. Acad. Sci.* **75:** 280–284.

Zeng Y. and Cullen B.R. 2005. Efficient processing of primary microRNA hairpins by Drosha requires flanking nonstructured RNA sequences. *J. Biol. Chem.* **280:** 27595–27603.

Zeng Y., Yi R., and Cullen B.R. 2003. MicroRNAs and small interfering RNAs can inhibit mRNA expression by similar mechanisms. *Proc. Natl. Acad. Sci.* **100:** 9779–9784.

Zoumadakis M. and Tabler M. 1995. Comparative analysis of cleavage rates after systematic permutation of the NUX↓ consensus target motif for hammerhead ribozymes. *Nucleic Acids Res.* **23:** 1192–1196.

62 Mifepristone-inducible Gene Regulatory System

Kurt Schillinger,* Xiangcang Ye,† Sophia Tsai,* and Bert W. O'Malley*

*Department of Molecular and Cellular Biology, Baylor College of Medicine, Houston, Texas 77030; †Department of Molecular Pathology, University of Texas M.D. Anderson Cancer Center, Houston, Texas 77030

ABSTRACT

The Mifepristone-inducible gene regulatory system (MIGRS) enables the control of target gene expression through administration of the antiprogestin Mifepristone (Mfp). The system utilizes a chimeric *trans*-activator protein containing a truncated form of the human progesterone receptor ligand-binding domain (hPR-LBD), the DNA-binding domain of the yeast Gal4 transcription factor, and a *trans*-activation domain. In the presence of Mfp, but not endogenous progesterone, this *trans*-activator binds to upstream activation sequences placed within the promoter of a target gene and initiates the transcription of the target gene.

The major limitation of the MIGRS is basal transgene expression derived from ligand-independent interaction of the GeneSwitch protein with the Mfp-responsive promoter of target genes. Recent modifications have reduced basal expression and thus have significantly increased the fold-induction of target gene expression. However, the very low levels of target gene expression observed even with the latest-generation GeneSwitch protein must be considered by investigators working with highly cytotoxic target genes.

The MIGRS possesses important advantages. It works well with tissue-specific promoters, allowing both spatial and temporal regulation of target gene expression. It relies on Mfp, which has been studied in humans for 25 years and has been shown to be well-tolerated by patients. Mfp has been approved for experimental use in humans for induction of pregnancy termination (see Escudero et al. 2005) at doses more than 100-fold higher than those currently required for induction of transgene expression in transgenic or helper-dependent adenoviral (HD-Ad)-infected mice harboring the MIGRS. The lipophilic nature of Mfp allows it to penetrate the blood-brain barrier, making the MIGRS ideal for control of target gene expression following vector delivery to the brain or in transgenic animals where GeneSwitch protein expression must be restricted to the central nervous system (see Burcin et al. 1999). These properties make it an attractive candidate for a variety of in vivo applications, including future gene therapy in humans.

INTRODUCTION

The centerpiece of the MIGRS is a chimeric *trans*-activator protein, called the GeneSwitch protein, composed of the DNA-binding domain from the yeast Gal4 transcription factor, a truncated form of the hPR-LBD, and a *trans*-activation domain derived either from the herpes simplex virus (HSV) VP16 protein or from the p65 subunit of human NF-κB. In the absence of the progesterone antagonist Mfp, this protein is thought to exist in a predominantly monomeric state, complexed with heat shock proteins and other molecular chaperones in the cytoplasm. Binding of Mfp to the truncated progesterone LBD of the GeneSwitch protein causes a conformational change that results in nuclear translocation and homodimerization. In the nucleus, the activated GeneSwitch homodimer binds through its Gal4 DNA-binding domains to Gal4 upstream activation sequences (UAS) placed within the promoter of target genes. Placing these yeast-specific enhancer sequences in the promoter of target genes ensures that target gene transcription can be activated by only the Mfp-bound GeneSwitch protein, as no mammalian transcription factors have been identified that bind the UAS. After binding to the UAS, the transcriptional activation domain of the GeneSwitch protein promotes preinitiation complex formation at an adjacent TATA box and induces transcription and expression of a downstream target gene (Fig. 1). The MIGRS may sometimes produce low levels of target gene expression in the absence of Mfp. This

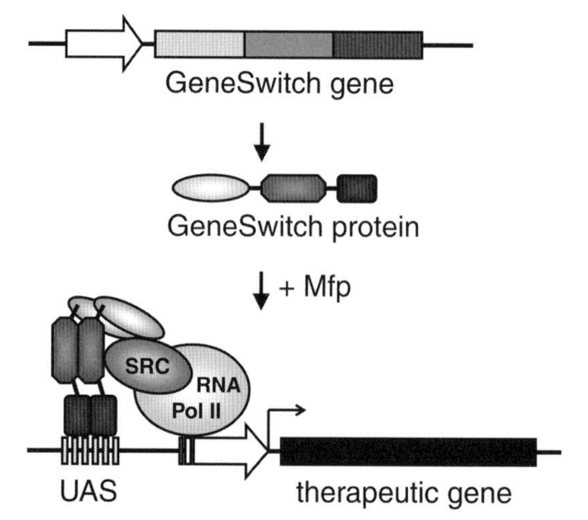

FIGURE 1. Structure and function of the Mifepristone-inducible gene regulatory system (MIGRS). In the presence of Mfp, the chimeric GeneSwitch protein dimerizes, interacts with upstream activation sequences (UAS) in the promoter of the target gene, and initiates transcription of the target gene.

ligand-independent transcription activation, termed "basal expression," is thought to be caused by weak interaction of the GeneSwitch protein with the UAS in a ligand-independent manner and is a feature of all inducible systems. Modifications in the GeneSwitch protein (described below and summarized in Fig. 2) have focused on increasing the ratio of ligand-dependent expression to basal expression, a measurement described as "fold-induction." This important parameter is used to evaluate the effectiveness of the regulation delivered by gene regulatory systems.

The key component of the chimeric GeneSwitch protein is the carboxy-terminally truncated hPR-LBD. As initially observed by Vegeto et al. (1992), removal of the carboxy-terminal 42 amino acids from the LBD of the hPR is sufficient to prevent binding of endogenous progesterone, but to allow the progesterone antagonist Mfp to act as an agonist for this mutant hPR. Wang et al. (1994) then combined the 42-amino-acid truncated hPR-LBD with the DNA-binding domain (residues 1–94) from the yeast Gal4 transcription factor and the *trans*-activation domain of the HSV VP16 protein (residues 411–487) to create the first-generation GeneSwitch protein GLVP. In vitro, GLVP demonstrated low basal expression, but also low levels of Mfp-inducible expression, resulting in 7–50-fold induction of reporter gene expression, depending on the Mfp-responsive promoter used to flank the reporter gene (see Wang et al. 1994).

The second-generation GeneSwitch protein $GL_{914}VP_{c'}$ arose from two important observations by Wang et al. First, it was found that lengthening the truncated hPR-LBD domain to include amino acids 880–914 of the hPR increased Mfp-induced activation of target gene expression in GLVP. Second, moving the VP16 *trans*-activation domain to the carboxyl terminus of GLVP significantly increased the potency of this chimeric *trans*-activator in the presence of Mfp (see Wang

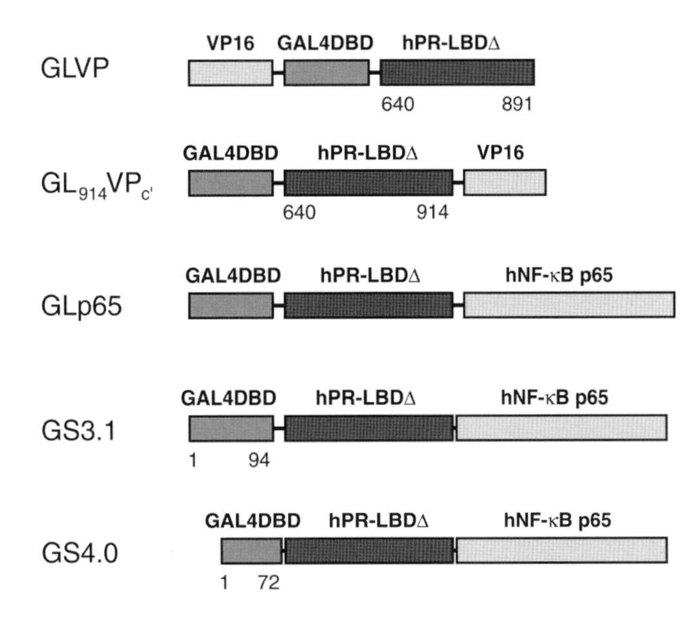

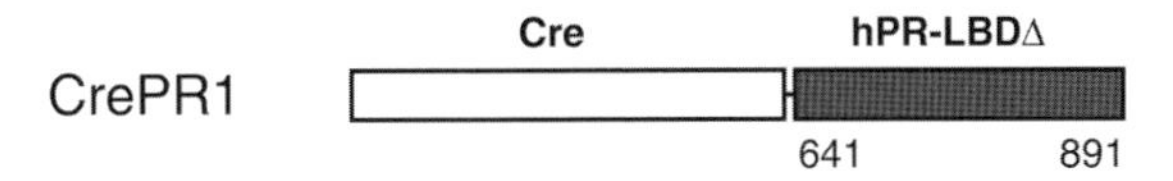

FIGURE 2. Current versions of the Mifepristone-inducible gene regulatory system chimeric *trans*-activator (GeneSwitch protein).

et al. 1997b). The GeneSwitch protein $GL_{914}VP_{c'}$, which contained both of these modifications, was shown to be eight to ten times more potent than GLVP at inducing reporter gene expression. It activated target gene expression at doses of Mfp approximately tenfold lower than those required for GLVP. However, $GL_{914}VP_{c'}$ also demonstrated elevated levels of basal expression, making the fold-induction comparable to that of GLVP (see Wang et al. 1997b).

Significant improvements in both basal and Mfp-induced expression were achieved with the third-generation GeneSwitch protein, GLp65. Burcin et al. (1999) replaced the viral VP16 *trans*-activation domain of $GL_{914}VP_{c'}$ with one from the p65 subunit of human NF-κB (residues 283–551). As a result, GLp65 demonstrated significantly reduced basal expression and comparable levels of Mfp-induced reporter gene expression when compared to $GL_{914}VP_{c'}$. Consequently, fold-induction with GLp65 was markedly improved over $GL_{914}VP_{c'}$ (see Burcin et al. 1999). Replacement of the highly immunogenic and potentially cytotoxic VP16 domain with one from a human protein resulted in a chimeric GeneSwitch protein that was 86% human in its sequence.

Further modifications have produced the fourth- and fifth-generation GeneSwitch proteins, GLp65.1 and GS4.0, respectively. GLp65.1 differs from GLp65 only in that it contains a three-amino-acid linker, rather than a seven-amino-acid linker, between the hPR-LBD and p65 domains. Functionally, GLp65.1 is very similar to GLp65 (see Abruzzese et al. 1999). GS4.0, in contrast, differs significantly from GLp65. Nineteen amino acids have been removed from the carboxy-terminal portion of the Gal4 DNA-binding domain and the link between the Gal4 DNA-binding domain and the hPR-LBD has been reduced from seven to two amino acids. These modifications have reduced the ability of sequences contained within the Gal4 DNA-binding domain to promote ligand-independent protein dimerization, further reducing basal expression when compared with GLp65 (see Nordstrom 2002). However, the utility of this vector has not been thoroughly tested in vivo.

Recent modifications of the MIGRS have resulted in the development of a spatiotemporally restricted gene targeting system allowing recombination of a target gene in a specific tissue and in response to Mfp administration. This strategy utilizes a unique fusion protein, termed CrePR1, which is composed of the Mfp-responsive, truncated PR-LBD initially described by Vegeto et al. fused to bacterial Cre recombinase (see Cao et al. 2001). In the absence of Mfp, CrePR1 is retained in the cytoplasm, complexed to heat shock proteins and other molecular chaperones. Upon administration of Mfp, however, CrePR1 translocates to the nucleus and mediates site-specific recombination at *loxP*-flanked target sites. As a result, tissue-specific, site-specific, and Mfp-inducible recombination can be achieved (see Cao et al. 2001). Of course, because use of CrePR1 relies on genomic recombination, rather than induction of target gene transcription, this system differs significantly from systems utilizing the GeneSwitch protein, in that Mfp administration results in permanent alterations in gene expression for a given cell type expressing CrePR1. In addition, if CrePR1 is expressed in a stem cell population, these alterations in gene expression have the potential to be permanent for the life of the animal.

General Design Considerations

The way in which the MIGRS is used is determined by the desired characteristics of target gene expression. Experiments requiring high levels of target gene expression with little concern for basal expression differ markedly from those requiring tight control of target gene expression. The key elements that can be modified include the generation of GeneSwitch protein used, the promoter driving the GeneSwitch protein, and the promoter driving target gene expression.

Later-generation GeneSwitch proteins allow reasonably good maximum inducible levels of transgene expression with very low basal levels of expression. The properties of Mfp-inducible target gene expression are also controlled by the way in which the GeneSwitch protein itself is expressed. Strong constitutive promoters such as the cytomegalovirus (CMV) or Rous sarcoma virus long terminal repeat (RSV-LTR) give generally high maximum inducible levels of expression but also increase basal expression. Tissue-specific promoters may decrease both basal and maximum inducible expression and may prolong inducibility in vivo when compared with strong constitutive promoters (see Ye et al. 2002). Finally, an autoregulatable promoter has been designed for the GeneSwitch protein, consisting of four copies of the UAS in front of a minimal HSV thymidine kinase promoter. In the absence of Mfp, low levels of the GeneSwitch protein are produced. Mfp administration simultaneously stimulates GeneSwitch protein expression via a positive feedback control circuit and induces target gene expression. The result is extremely low basal expression and maximum target gene expression that is generally lower than that seen with other promoters (see Abruzzese et al. 2000).

The promoter that flanks the target gene is also important. In general, the TATA box from the E1B gene of serotype-5 adenovirus (E1B TATA) is more stringent than the HSV thymidine kinase minimal promoter when linked to target gene expression. The minimal thymidine kinase promoter drives target gene expression to higher levels than the E1B TATA, but it is also associated with higher levels of basal expression. Similarly, promoters containing more copies of Gal4 UAS (up to six) have allowed greater maximum inducible target gene expression but have also produced higher levels of basal expression (see Ye et al. 2002).

Since its inception, the MIGRS has undergone numerous modifications aimed at reducing target gene expression in the absence of Mifepristone while maximizing target gene expression upon administration of Mifepristone. As a result, the latest-generation MIGRS chimeric *trans*-activators have been shown through numerous experiments, both in vitro and in vivo, to produce tightly controlled, highly inducible transgene expression.

Protocol 1

Use of the MIGRS in Cultured Cells

We strongly recommend in vitro testing of inducible vectors (plasmid or viral) prior to application in vivo. This may be accomplished using standard cell culture techniques with attention to restrictions imposed by tissue-specific expression of the GeneSwitch protein and tropisms of viral vectors. As with use of plasmids and transient transfection, conditions for in vitro testing with viral vectors are largely determined through empirical means and are dictated by cell type, viral vector, and promoter usage. The MIGRS has been used extensively in a variety of in vitro systems including cell lines such as HeLa, CV1, HepG2, and primary chicken myoblast. For a summary of important examples and references for use of the MIGRS with transient transfection, see Table 1.

MATERIALS

CAUTION: See Appendix for appropriate handling of materials marked with <!>.

Reagents

Cell culture medium appropriate for target cells
Dimethylsulfoxide <!> (DMSO; Sigma-Aldrich D8418)
Ethanol (80%)
Hank's balanced salt solution (HBSS)
Mifepristone <!> (Mfp, RU-486; Sigma-Aldrich M8046)

Store Mfp powder in a dark bottle at about 4°C. To prevent exposure to condensation, allow Mfp to warm to room temperature in a closed container prior to weighing. Protect all solutions containing Mfp from exposure to light at all times. Prepare a 10^{-2} M Mfp stock solution by dissolving Mfp powder in DMSO. Store solution for up to 6 months at –20°C. For cell culture experiments, prepare a working solution of Mfp by diluting the Mfp stock solution to 10^{-5} M with 80% ethanol. Store solution for up to 6 months at –20°C. Use this working solution to generate a 10^{-8} M or 10^{-7} M concentration of Mfp in cell culture medium.

TABLE 1. Summary of MIGRS Use in Various Cell Culture Systems

Regulator	Regulator promoter	Target construct	Mfp	Induction[a]	Cells	Reference
GLVP	CMV	4xUAS-tk-CAT	10^{-7} M	10–15-fold	CV-1	Wang et al. (1994)
	CMV	4xUAS-TATA-CAT	10^{-7} M	50-fold	CV-1	Wang et al. (1994)
	RSV	4xUAS-TATA-hGH	10^{-8} M	24-fold	HepG2	Wang et al. (1997b)
$GL_{914}VP_{c'}$	TTRB	4xUAS-TATA-luc	10^{-8} M	20-fold	HeLa	Burcin et al. (1999)
GLp65	TTRB	4xUAS-TATA-luc	10^{-8} M	n.a.	HeLa	Burcin et al. (1999)
GLp65.1	CMV	3xUAS-TATA-SEAP	10^{-8} M	44-fold	HeLa	Abruzzese et al. (1999)
	CMV	6xUAS-TATA-SEAP	10^{-8} M	100-fold	HeLa	Abruzzese et al. (1999)
	CMV	6xUAS-TATA-SEAP	10^{-8} M	10–133-fold	COS-1, HepG2, NIH-3T3	Abruzzese et al. (2000)
	autoinducible	6xUAS-TATA-SEAP	10^{-8} M	221–6845-fold	COS-1, HepG2, NIH-3T3	Abruzzese et al. (2000)
	autoinducible	6xUAS-TATA-VEGF	10^{-8} M	>2500-fold	293	Abruzzese et al. (2000)
	autoinducible	6xUAS-TATA-Epo	10^{-8} M	766-fold	COS-1	Abruzzese et al. (2000)
GS4.0	α-actin	6xUAS-TATA-GHRH	10^{-8} M	n.a.	primary chick myoblasts	Draghia-Akli et al. (2002)

[a]n.a. indicates not available.

Minimal essential medium (MEM) supplemented with glutamine, antibiotics, and 10% fetal calf serum (FCS)
Stripped fetal calf serum (SFCS)
Target cells
Transfection reagent of choice

Equipment

CO_2 cell culture
Incubator, preset to 37°C

METHOD

1. Deliver the vector into the cells

   For delivery of plasmid vectors

   a. Carry out the general procedure for transfection according to the manufacturer's protocol accompanying the transfection reagents.

      When the MIGRS and target genes are introduced to cells on separate plasmids, adjustments in the ratio of these plasmids can be used to influence basal and maximum inducible levels of target gene expression. For the GLVP GeneSwitch protein, the molar ratio of GLVP plasmid:target gene plasmid should range from 1:10 to 1:2.5. For the GLp65 GeneSwitch protein, however, plasmid ratios of 1:1 give excellent results due to the lower basal activity of GLp65 (see Ye et al. 2002).

      Strong constitutive promoters may result in higher basal target gene expression and will require even lower GeneSwitch:target gene plasmid ratios. Ultimately, the molar ratio of plasmids should be determined empirically to achieve desired levels of basal and maximum inducible target gene expression.

   b. Immediately after transfection, wash the cells with HBSS and incubate in Mfp-free cell culture medium for at least 3 hours.

   c. Replace this medium with a prewarmed solution containing 10^{-8} M Mfp in cell culture medium with 5–10% SFCS.

   For delivery of recombinant viral vectors

   a. Infect the cells using conditions determined by the nature of the viral vector and the cell type.

      HSV vectors, first-generation adenoviral vectors, and HD-Ad vectors have been tested in Vero, HeLa, and HepG2 cells, respectively.

   b. After infection, remove excess vector by washing cells with HBSS.

   c. Transfer the cells to standard cell culture medium and allow them to recover for 3–12 hours.

   d. Subsequent to recovery, change the standard medium to cell culture medium containing 5–10% SFCS and 10^{-7} M Mfp.

2. Incubate the cells in Mfp-containing medium for 24–72 hours, depending on the cell type.
3. Harvest the cells and analyze for expression of the target gene (see Ye et al. 2002).

SPECIFIC EXAMPLES

A single HSV vector carrying both GLVP and LacZ target gene expression cassettes was used to transduce Vero cells at a multiplicity of infection (moi) of 0.1. X-gal staining was used to evaluate LacZ expression after 24 hours of incubation in a solution containing 10^{-7} M Mfp in MEM supplemented with glutamine, antibiotics, and 10% FCS (see Oligino et al. 1998). In an alternative strategy, two first-generation adenoviral vectors, one encoding GLVP and one encoding chloramphenicol acetyltransferase (CAT) as a target gene, were used to coinfect HeLa cells at 10 pfu/cell. Cells were incubated for 16 hours in cell culture medium containing concentrations of Mfp ranging from 10^{-7} M to 2×10^{5} M. Cells were then harvested and CAT expression was quantified (see Molin et al. 1998). Finally, a single HD-Ad vector has been described that carried both GLp65 and hGH target gene expression cassettes; 5×10^{7} infectious units (IU) of this vector were used to transduce 6×10^{5} HepG2 cells. Virus was then removed, and cells were allowed to recover for 3 hours in cell culture medium supplemented with 5% SFCS. Medium was changed to one containing 5% SFCS and 10^{-8} M Mfp; 24 hours later, hGH expression was measured in the cell culture medium (see Burcin et al. 1999).

Protocol 2

In Vivo Use of the MIGRS with Plasmid Injection and/or Electroporation

The properties of Mfp-inducible target gene expression in vivo, including basal expression, maximum inducible levels, and kinetics of target gene expression, are dependent on many factors. These include the method of gene delivery, the tissue transduced, the promoter used to drive GeneSwitch protein expression, and the half-life of the protein expressed from the target gene. The actual conditions under which the MIGRS produces a desired result must be determined empirically. Plasmids encoding GLp65, GLp65.1, and GS4.0, in combination with plasmids encoding target genes, have been used successfully for direct muscle injection and electroporation in mice. Analogous to the use of the MIGRS in transient transfection assays, optimization of the in vivo performance of this system can be accomplished through adjustments in the ratio of GeneSwitch protein plasmid to target gene plasmid, the choice of promoters used to drive GeneSwitch protein expression, and the quantity of DNA injected into the muscle. The following protocol serves as a starting point for application in vivo.

MATERIALS

CAUTION: See Appendix for appropriate handling of materials marked with <!>.

Reagents

Mice

Mifepristone <!> (Mfp, RU-486; Sigma-Aldrich M8046)

For intraperitoneal injection and gastroesophageal lavage, dilute the 10^{-2} M stock solution as prepared in Protocol 1 to 2×10^{-4} M in pure sesame oil. Store this solution at 4°C overnight to allow the solution to become heterogeneous. Prior to use in animals, place the solution in a dark location and allow to warm to room temperature for approximately 1 hour. This solution may be stored for up to 3 weeks at 4°C with no decrease in Mfp activity. Consistent, sustained delivery of Mfp to animals may be achieved by subcutaneous implantation of timed-release, biodegradable pellets containing Mfp. Best results are obtained using the proprietary technology of Innovative Research of America (Sarasota, Florida). To determine the quantity of Mfp that must be incorporated into a single pellet, multiply the desired daily dose of Mfp (μg/kg) by the average weight (kg) of the animals. Then, multiply by the number of days over which Mfp release is desired.

Plasmid of choice

Plasmid preparation kit (large scale)

Polyvinylpyrrolidine <!> or saline

Sesame oil (Sigma-Aldrich S3547)

Equipment

Electroporation device with two parallel plate caliper electrodes (*optional*, see Step 3)

METHOD

1. Purify plasmids using commercially available large-scale (250–500 μg) plasmid preparation kits according to the manufacturer's instructions.
2. Suspend the purified plasmid in either saline or polyvinylpyrrolidine at a concentration of 1 mg/ml.
3. Inject 25–50 μg of DNA directly into each tibialis cranialis or gastrocnemius muscle of anaesthetized mice.

   The efficiency of muscle transfection can be enhanced by following this injection with electroporation using two parallel plate caliper electrodes and square wave electroporation. Typical voltage settings for the electroporation are 375 V/cm delivered twice in two 25-μsec-long square wave pulses.
4. Following injection and/or electroporation, allow the mice to recover for approximately 14 days from the minor inflammatory response that accompanies plasmid introduction.
5. Administer Mfp to the mice at a dose of 250–330 μg/kg using the 2×10^{-4} M sesame oil Mfp solution and either intraperitoneal injection or esophageal lavage.

   Alternatively, implant biodegradable, timed-release pellets containing Mfp to obtain sustained induction of target gene expression.
6. After single-dose administration of Mfp, analyze target gene expression at a timepoint determined by the plasma or cytoplasmic half-life of the target gene product.

   If the half-life is not known, perform initial experiments to determine empirically the kinetics of target gene expression subsequent to Mfp administration.

SPECIFIC EXAMPLES

After injection and electroporation of a 1:1 (by weight) mixture of two plasmids (one encoding GLp65.1 driven by the CMV promoter and the second encoding SEAP in the context of three Gal4 UAS sites and a consensus TATA box) into the hind limb muscles (75 μg/muscle) of BALB/c mice, Abruzzese et al. (1999) demonstrated an average 19-fold induction of SEAP expression with administration of 330 μg/kg Mfp. Peak serum levels of SEAP were observed 2–3 days after a single dose of Mfp. Abruzzese et al. (2000) used an autoinducible promoter to drive GLp65.1 expression and a target plasmid containing SEAP in the context of six Gal4 UAS sites and a consensus TATA box. Basal SEAP expression was significantly reduced, and a 214-fold increase in SEAP expression was observed with comparable maximum SEAP expression levels after a single dose of Mfp at 330 μg/kg. Draghia-Akli et al. (2002) used injection and electroporation to introduce 10 μg of a 1:10 ratio of two plasmids (one encoding GS4.0 driven by a muscle-specific skeletal α-actin promoter, and the second encoding a mutated porcine growth-hormone-releasing hormone [GHRH] in the context of six copies of Gal4 UAS and a TATA box) into the left tibialis anterior muscle of subacute combined immunodeficiency (SCID) mice; 21 days later, 250 μg/kg Mfp was administered intraperitoneally for 3 consecutive days. On the fourth day, serum insulin-like growth factor 1 (IGF-1) levels were measured and showed consistent 1.1-fold to 1.7-fold increases in response to Mfp-induced GHRH expression. This Mfp-inducible expression was maintained over a period of 149 days (see Draghia-Akli et al. 2002). These different approaches to the use of the MIGRS with plasmid injection and electroporation demonstrate the versatility and consistency of the system in vivo.

Protocol 3

Virus-mediated Gene Transfer

The MIGRS has been studied in vivo in conjunction with gene delivery by both recombinant HSV vectors and helper-dependent or "gutless" adenoviral vectors. The use of viral vectors differs markedly from the use of naked plasmid injection and electroporation. Most significantly, the GeneSwitch and target protein cassettes have been simultaneously cloned into single vectors. In addition, to facilitate liver transduction, HD-Ad vectors have been delivered systemically, through tail-vein injection. The use of tissue-specific promoters to drive expression of the GeneSwitch protein is advisable. Such promoters, because they are often less powerful than strong constitutive viral promoters, ensure that GeneSwitch protein overexpression, and consequently elevated basal levels of target gene protein expression, do not occur with the high cell-transduction efficiencies of viral vectors (see Ye et al. 2002). In addition, when viral vectors are systemically administered, target gene expression can be effectively restricted to a desired tissue even though multiple different tissues may be transduced.

MATERIALS

CAUTION: See Appendix for appropriate handling of materials marked with <!>.

Reagents

Mifepristone <!> (Mfp, RU-486; Sigma-Aldrich M8046)

For intraperitoneal injection and gastroesophageal lavage, dilute the 10^{-2} M stock solution as prepared in Protocol 1 to 2 x 10^{-4} M in pure sesame oil. Store this solution overnight at 4°C to allow the solution to become heterogeneous. Prior to use in animals, place the solution in a dark location and allow to warm to room temperature for approximately 1 hour. This solution may be stored for up to 3 weeks at 4°C with no decrease in Mfp activity.

Rodents

Sesame oil (Sigma-Aldrich S3547)

Equipment

Stereotactic apparatus or needles and syringes for tail-vein injection (see Step 1)

METHOD

1. Introduce viral vectors encoding the MIGRS and target gene cassettes into rodents by either stereotactic injection or systemic tail-vein injection (see Oligino et al. 1998; Burcin et al. 1999; Schillinger et al. 2005).

   In both instances, introduction of large quantities of virus is associated with higher basal levels of target gene expression. The factors that determine basal target gene expression and fold-induction include the ratio of IUs to particles for a viral preparation, synthetic properties of the tissue being transduced, and the promoter used to drive GeneSwitch protein expression. Optimization of transduction conditions is empirical, and it is advisable to perform a preliminary dosing experiment with all viral preparations.

2. Allow the animals approximately 1–2 weeks to recover from the local inflammatory response that accompanies viral vector delivery.
3. Administer Mfp in sesame oil (2×10^{-4} M) by either intraperitoneal injection or esophageal lavage.
4. Analyze target gene expression at an appropriate timepoint.

   This timepoint is determined by the plasma or cytosolic half-life of the target protein, or empirically if the half-life is not known.

SPECIFIC EXAMPLES

The earliest use of the MIGRS in conjunction with viral vectors was performed by Oligino et al. using an HSV vector called GVHLZ. This was a double-recombinant vector encoding both GLVP driven by a CMV promoter and the bacterial *lacZ* gene in the context of five copies of the Gal4 UAS sequence and the adenovirus E1B TATA; 2×10^6 pfu of GVHLZ was injected stereotactically into the hippocampus of 200–250-g Sprague-Dawley rats. Mfp (25 mg/kg) was administered intraperitoneally 12 and 36 hours after infection with GVHLZ. At 48 hours after infection, the animals were sacrificed and β-galactosidase (β-gal) expression was assessed in the brain. Extracts from the brain of animals treated with Mfp showed 150-fold greater levels of β-gal activity than controls. In animals receiving GVHLZ and no Mfp, β-gal activity was barely elevated above background levels, indicating reasonably tight control of β-gal expression with the MIGRS (see Oligino et al. 1998).

The MIGRS also has been used with HD-Ad or "gutless" adenoviral vectors. Burcin et al. (1999) described the HD-Ad vector hGH-GLp65, which encoded GLp65 driven by a liver-specific transthyretin promoter and hGH in the context of four copies of the Gal4 UAS and an E1B TATA; 1×10^9 infectious particles of hGH-GLp65 were injected into the tail vein of 20–25-g C57BL/6J mice. Two weeks after injection, a single dose of 250 µg/kg of Mfp was administered intraperitoneally, and the kinetics of hGH induction were investigated. hGH was detectable as early as 3 hours after administration of Mfp, reached peak levels at 12 hours, and was undetectable by 192 hours. Basal levels of hGH were detectable in the serum of mice 8 days after infection with hGH-GLp65, and 50,000-fold induction of hGH expression was possible with administration of 250 µg/kg Mfp 14 days after infection. This induction could be repeated up to 5 times over a 12-week period with comparable results. Finally, prolonged induction of hGH expression in mice infected with hGH-GLp65 was demonstrated over a 4-week period after the subcutaneous implantation of biodegradable pellets containing 360 µg of Mfp (see Burcin et al. 1999).

Protocol 4

Transgenic Models

The MIGRS has been very useful in the study of gene function through transgenic mouse models. The strategy used for application of the system involves the generation of two separate transgenic lines, using the method described in Protocol 3.

MATERIALS

CAUTION: See Appendix for appropriate handling of materials marked with <!>.

Reagents

Mifepristone <!> (Mfp, RU-486; Sigma-Aldrich M8046)

For intraperitoneal injection and gastroesophageal lavage, dilute the 10^{-2} M stock solution as prepared in Protocol 1 to 2×10^{-4} M in pure sesame oil. Store this solution overnight at 4°C to allow the solution to become heterogeneous. Before use in animals, place the solution in a dark location and allow to warm to room temperature for approximately 1 hour. This solution may be stored for up to 3 weeks at 4°C with no decrease in Mfp activity. Consistent, sustained delivery of Mfp to animals may be achieved by subcutaneous implantation of timed-release, biodegradable pellets containing Mfp. Best results are obtained using the proprietary technology of Innovative Research of America (Sarasota, Florida). To determine the quantity of Mfp that must be incorporated into a single pellet, multiply the desired daily dose of Mfp (μg/kg) by the average weight (kg) of the animals. Then, multiply by the number of days over which Mfp release is desired.

Mice

Sesame oil (Sigma-Aldrich S3547)

Equipment

Equipment for northern blot or quantitative reverse-transcriptase PCR (see Step 4)

METHOD

1. To create the first transgenic founder line, introduce a vector for which expression of the GeneSwitch protein is driven by a tissue-specific promoter, ensuring eventual tissue-specific expression of the target gene.
2. To create the second transgenic founder line, introduce a vector for which the target gene is placed in the context of a minimal promoter, such as HSV thymidine kinase or adenovirus E1B TATA, preceded by multiple copies of Gal4 UAS.
3. To activate target gene expression (as with MIGRS in gene therapy models in Protocol 3), administer 330 μg/kg of Mfp dissolved in sesame oil (2×10^{-4} M) through intraperitoneal injection or esophageal lavage.

   Biodegradable timed-release pellets containing Mfp may be implanted subcutaneously to enable consistent, prolonged target gene expression.

4. Characterize GeneSwitch protein expression for each transgenic founder line using northern blot or quantitative reverse transcriptase–polymerase chain reaction (PCR) analysis.

   Typically, high levels of GeneSwitch protein expression will result in high levels of maximally induced target gene expression but will also result in relatively higher basal levels of expression.

5. Once appropriate expression in the founder lines is confirmed, cross the separate transgenic founder lines to generate bitransgenic mice simultaneously possessing tissue-specific GeneSwitch protein expression and, as a result, tissue-specific, Mfp-inducible target gene expression.

   Analogous to the characteristics of transgenic induction in gene therapy models, it has been shown that basal and induced target gene expression in transgenic models is determined by the levels of GeneSwitch protein expressed in a given tissue. Levels of GeneSwitch protein, in turn, are determined by the promoter used to express the GeneSwitch protein, as well as the chromosomal location and number of copies of GeneSwitch protein expression cassettes incorporated into a given transgenic founder line.

Specific Examples: Creation of Bitransgenic Mice

In initial studies of the MIGRS in transgenic mice, Wang et al. (1997a) generated bitransgenic mice simultaneously harboring a cassette driving expression of the GLVP GeneSwitch protein from a liver-specific transthyretin promoter as well as a cassette encoding the hGH gene in the context of a minimal HSV thymidine kinase promoter and four copies of the Gal4 UAS. In the absence of Mfp, minimal levels of hGH were detectable in the serum of these bitransgenic mice. After a single intraperitoneal injection of Mfp, a 5800–33,000-fold increase in serum hGH levels was observed. Bitransgenic mice receiving an intraperitoneal injection of Mfp every 2 days showed a 30–37% increase in weight over 10 days (see Wang et al. 1997a). The liver-specific GLVP-expressing mice of Wang et al. were also crossed with mice harboring a target gene cassette encoding the Inhibin-A gene in the context of four copies of Gal4 UAS and a minimal HSV thymidine kinase promoter by Pierson et al. (2000). Resultant male bitransgenic mice demonstrated physiologic (436 pg/ml) and supraphysiologic (1.2 ng/ml) levels of Inhibin-A expression with intraperitoneal administration of Mfp at doses of 250 μg/kg and 500 μg/kg, respectively. Furthermore, breeding these bitransgenic mice into an Inhibin-α null genetic background and then activating Inhibin-A expression with the implantation of timed-release Mfp pellets (6 μg/day) at 3 weeks of age prevented the development of sex-cord tumors associated with the Inhibin-α null phenotype (see Pierson et al. 2000).

The GLp65 GeneSwitch protein also has been used to enable Mfp-inducible target gene expression in bitransgenic animals. Zhao et al. (2001) generated a GeneSwitch founder line possessing a GLp65 expression cassette driven by a lung-specific human surfactant protein-C promoter. These founder mice were crossed to mice possessing a cassette encoding fibroblast growth factor-3 (FGF-3) in the context of an elastase promoter and four copies of Gal4 UAS. Intraperitoneal doses of Mfp ranging from 66 to 200 μg/kg for 30 days resulted in a dose-dependent expression of FGF-3 in the lungs of the resultant bitransgenic mice. Implantation of 60-day timed-release 500-μg Mfp pellets in bitransgenic mice also resulted in consistent elevations in lung FGF-3 expression, with subsequent macrophage infiltration and type II pneumocyte proliferation in the adult lung (see Zhao et al. 2001). Mfp-inducible FGF-3 expression using the GLp65 GeneSwitch protein has also been investigated in the mammary gland. Ngan et al. (2002) generated a founder line expressing GLp65 from the mammary-gland-specific mouse mammary tumor virus (MMTV) promoter and crossed these mice with the UAS-FGF-3 founder line used by Zhao et al. (2001). A dose-dependent increase in FGF-3 expression in the mammary glands of resulting bitransgenic mice could be induced by implanting timed-release Mfp pellets ranging in dose from 150 to 450 μg/kg. Mfp-inducible ectopic expression of FGF-3 in female bitransgenic mice at early puberty resulted in abnormal mammary gland development (see Ngan et al. 2002). Finally, cardiac-specific expression of GLp65 has been described by Lutucuta et al. (2004) in mice bearing a GLp65 cassette driven by a murine α-myosin heavy-chain promoter. Crossing these founders to mice harboring expression cassettes encoding cardiac troponin T-Q92 (cTnT-Q92) in the context of four copies of Gal4 UAS and an E1B TATA box resulted in bitransgenic mice demonstrating Mfp-inducible, cardiac-specific cTnT-Q92 expression. Expression of cTnT-Q92 was detected consistently after daily administration of Mfp at 1000 μg/kg/day for at least 16 days. Subcutaneous implantation of Mfp pellets releasing 1000 μg/kg/day also successfully induced cTnT-Q92 expression in these mice for approximately 70 days. Short-term induction of cTnT-Q92 was associated with enhanced systolic function in bitransgenic mice, whereas long-term cTnT-Q92 expression resulted in significant cardiac interstitial fibrosis. Both of these changes were reversible upon withdrawal of Mfp (see Lutucuta et al. 2004).

Specific Examples: Use of the Cre-PR1 System in Transgenic Animals

The Cre-PR1-inducible gene recombination system has been studied almost exclusively in transgenic animal models. Using tissue-specific expression of Cre-PR1 in combination with Mfp administration, gene recombination can be targeted in a specific spatiotemporal pattern. Cao et al. (2001) generated a mouse model of epidermolysis bullosa simplex (EBS-DM) using bitransgenic mice expressing Cre-PR1 from an epithelial cell-specific promoter and containing a mutant keratin-14 (K14) expression cassette that could be transcriptionally activated after Cre-mediated excision of an interfering *neo* gene. Topical application of Mfp to the front legs and paws of these bitransgenic mice for 2–7 days resulted in the formation of blisters filled with fluid, histologically consistent with EBS-DM (see Cao et al. 2001). A similar strategy was employed by Arin et al. (2001) to establish an inducible model of epidermolytic hyperkeratosis (EHK). In the bitransgenic mice of this model, Mfp-inducible Cre-PR1-mediated recombination activated expression of a mutant form of keratin-10, resulting in phenotypic lesions characteristic of EHK upon topical Mfp administration (see Arin et al. 2001). Finally, Mfp-inducible gene excision models have also been generated for the postnatal heart and adult brain of mice. Minamino et al. (2001) used the cardiac-specific αMHC (major histocompatibility complex) promoter to express Cre-PR1 in the ROSA-Cre background. Intraperitoneal administration of 250 μg of Mfp for 5 days to 3-week-old mice led to markedly increased expression of *lacZ* in the heart (see Minamino et al. 2001). Kellendonk et al. (1999) and Kitayama et al. (2001) generated similar bitransgenic reporter strains using CamKIIα, Thy-1, and GluRγ2 brain-specific gene promoters to express Cre-PR1 in the ROSA-Cre background.

REFERENCES

Abruzzese R.V., Godin D., Burcin M., Mehta V., French M., Li Y., O'Malley B.W., and Nordstrom J.L. 1999. Ligand-dependent regulation of plasmid-based transgene expression in vivo. *Hum. Gene Ther.* **10:** 1499–1507.

Abruzzese R.V., Godin D., Mehta V., Perrard J.L., French M., Nelson W., Howell G., Coleman M., O'Malley B.W., and Nordstrom J.L. 2000. Ligand-dependent regulation of vascular endothelial growth factor and erythropoietin expression by a plasmid-based autoinducible geneswitch system. *Mol. Ther.* **2:** 276–287.

Arin M.J., Longley M.A., Wang X.J., and Roop D.R. 2001. Focal activation of a mutant allele defines the role of stem cells in mosaic skin disorders. *J. Cell Biol.* **152:** 645–649.

Burcin M.M., Schiedner G., Kochanek S., Tsai S.Y., and O'Malley B.W. 1999. Adenovirus-mediated regulable target gene expression in vivo. *Proc. Natl. Acad. Sci.* **96:** 355–360.

Cao T., Longley M.A., Wang X.J., and Roop D.R. 2001. An inducible mouse model for epidermolysis bullosa simplex: Implications for gene therapy. *J. Cell Biol.* **152:** 651–656.

Draghia-Akli R., Malone P.B., Hill L.A., Ellis K.M., Schwartz R.J., and Nordstrom J.L. 2002. Enhanced animal growth via ligand-regulated GHRH myogenic-injectable vectors. *FASEB J.* **16:** 426–428.

Escudero E.L., Boerrigter P.J., Bennink H.J., Epifanio R., Horcajadas J.A., Olivennes F., Pellicer A., and Simon C. 2005. Mifepristone is an effective oral alternative for the prevention of premature luteinizing hormone surges and/or premature luteinization in women undergoing controlled ovarian hyperstimulation for in vitro fertilization. *J. Clin. Endocrinol. Metab.* **90:** 2081–2088.

Kellendonk C., Tronche F., Casanova E., Anlag K., Opherk C., and Schutz G. 1999. Inducible site-specific recombination in the brain. *J. Mol. Biol.* **285:** 175–182.

Kitayama K., Abe M., Kakizaki T., Honma D., Natsume R., Fukaya M., Watanabe M., Miyazaki J., Mishina M., and Sakimura K. 2001. Purkinje cell-specific and inducible gene recombination system generated from C57BL/6 mouse ES cells. *Biochem. Biophys. Res. Commun.* **281:** 1134–1140.

Lutucuta S., Tsybouleva N., Ishiyama M., Defreitas G., Wei L., Carabello B., and Marian A.J. 2004. Induction and reversal of cardiac phenotype of human hypertrophic cardiomyopathy mutation cardiac troponin T-Q92 in switch on-switch off bigenic mice. *J. Am. Coll. Cardiol.* **44:** 2221–2230.

Minamino T., Gaussin V., DeMayo F.J., and Schneider M.D. 2001. Inducible gene targeting in postnatal myocardium by cardiac-specific expression of a hormone-activated cre fusion protein. *Circ. Res.* **88:** 587–592.

Molin M., Shoshan M.C., Ohman-Forslund K., Linder S., and Akusjarvi G. 1998. Two novel adenovirus vector systems permitting regulated protein expression in gene transfer experiments. *J. Virol.* **72:** 8358–8361.

Ngan E.S., Ma Z.Q., Chua S.S., DeMayo F.J., and Tsai S.Y. 2002. Inducible expression of FGF-3 in mouse mammary gland. *Proc. Natl. Acad. Sci.* **99:** 11187–11192.

Nordstrom J.L. 2002. Antiprogestin-controllable transgene regulation in vivo. *Curr. Opin. Biotechnol.* **13:** 453–458.

Oligino T., Poliani P.L., Wang Y., Tsai S.Y., O'Malley B.W., Fink D.J., and Glorioso J.C. 1998. Drug inducible transgene

expression in brain using a herpes simplex virus vector. *Gene Ther.* **5:** 491–496.
Pierson T.M., Wang Y., DeMayo F.J., Matzuk M.M., Tsai S.Y., and Omalley B.W. 2000. Regulable expression of inhibin a in wild-type and inhibin α null mice. *Mol Endocrinol.* **14:** 1075–1085.
Schillinger K.J., Tsai S.Y., Taffet G.E., Reddy A.K., Marian A.J., Entman M.L., Oka K., Chan L., and O'Malley B.W. 2005. Regulatable atrial natriuretic peptide gene therapy for hypertension. *Proc. Natl. Acad. Sci.* **102:** 13789–13794.
Vegeto E., Allan G.F., Schrader W.T., Tsai M.J., McDonnell D.P., and O'Malley B.W. 1992. The mechanism of RU486 antagonism is dependent on the conformation of the carboxy-terminal tail of the human progesterone receptor. *Cell.* **69:** 703–713.
Wang Y., DeMayo F.J., Tsai S.Y., and O'Malley B.W. 1997a. Ligand-inducible and liver-specific target gene expression in transgenic mice. *Nat. Biotechnol.* **15:** 239–243.
Wang Y., O'Malley B.W., Jr., Tsai S.Y., and O'Malley B.W. 1994. A regulatory system for use in gene transfer. *Proc. Natl. Acad. Sci.* **91:** 8180–8184.
Wang Y., Xu J., Pierson T., O'Malley B.W., and Tsai S.Y. 1997b. Positive and negative regulation of gene expression in eukaryotic cells with an inducible transcriptional regulator. *Gene Ther.* **4:** 432–441.
Ye X., Schillinger K., Burcin M.M., Tsai S.Y., and O'Malley B.W. 2002. Ligand-inducible transgene regulation for gene therapy. *Methods Enzymol.* **346:** 551–561.
Zhao B., Chua S.S., Burcin M.M., Reynolds S.D., Stripp B.R., Edwards R.A., Finegold M.J., Tsai S.Y., and DeMayo F.J. 2001. Phenotypic consequences of lung-specific inducible expression of FGF-3. *Proc. Natl. Acad. Sci.* **98:** 5898–5903.

63 Dimerizer-mediated Regulation of Gene Expression

Victor M. Rivera, Lori Berk, and Tim Clackson

ARIAD Gene Therapeutics, Inc., Cambridge, Massachusetts 02139

ABSTRACT

Several systems have been developed that allow transcription of a target gene to be chemically controlled, usually by an allosteric modulator of transcription factor activity. An alternative is to use chemical inducers of dimerization, or “dimerizers,” to reconstitute active transcription factors from inactive fusion proteins. The most widely used system employs the natural product rapamycin, or a biologically inert analog, as the dimerizing drug. A key feature of this system is the tightness of regulation, with basal expression usually undetectable and induced expression levels comparable to constitutive promoters. In our experiments, the use of the minimal interleukin-2 (IL-2) promoter is an important determinant of this; substitution of a minimal simian virus 40 (SV40) or cytomegalovirus (CMV) promoter results in significantly higher levels of basal expression. The key factor dictating the successful use of the system is achieving high expression levels of the activation domain fusion protein. In the context of clinical gene therapies, the system has the advantage of being built exclusively from human proteins, potentially minimizing immunogenicity in the clinical setting. The dimerizer system has been successfully incorporated into diverse vector backgrounds and has been used to achieve long-term regulated gene expression in vitro and in vivo. This chapter provides guidance in designing constructs and experiments to achieve dimerizer-regulated expression of a target gene both in vitro and in vivo.

INTRODUCTION

Activation of gene expression in eukaryotes is controlled by the induced binding of transcription factor proteins to target genes. Transcription factors are bifunctional proteins that recognize specific DNA sequences near target genes and then recruit the transcriptional machinery of the cell to activate transcription. The two domains responsible for these activities, the DNA-binding domain and the transcriptional activation domain, are functionally separable and can reconstitute a sequence-specific transcriptional activator even when expressed as individual proteins and brought together via a noncovalent interaction.

This modular architecture has been exploited to develop a general method for controlling gene transcription using chemical "dimerizers" to induce the interaction of engineered proteins (Ho et al. 1996; Rivera et al. 1996). A dimerizer is a cell-permeant organic molecule with two separate motifs that each bind with high affinity to a specific protein module (Spencer et al. 1993). By fusing such modules to a DNA-binding domain and an activation domain, the reconstitution of a functional transcription factor, and therefore the expression of a target gene, can be made absolutely dependent on the presence of dimerizer (Fig. 1) (for review, see Pollock and Clackson 2002).

In principle, the expression of any cloned gene can be brought under dimerizer control by equipping the gene with upstream sequences that are recognized by the engineered DNA-binding domain. Following introduction of the modified gene into cells that also express the engineered transcription factor proteins, addition of dimerizer will lead to dose-dependent activation of target gene expression. Because the transcription factor fusion proteins have no affinity for one another in the absence of dimerizer, regulation is characterized by an extremely low, usually undetectable, basal level of gene expression (Rivera et al. 1996; Pollock et al. 2000). In addition, the highly potent activation domains incorporated into the system typically lead to high maximal levels of induced gene expression, equivalent to those obtained with strong constitutive promoters/enhancers (Pollock et al. 2000).

Although several dimerizer systems have been characterized, the most widely adopted is based on rapamycin and its analogs. Rapamycin is a natural product that functions as a heterodimerizer and has the advantages of oral bioavailability, favorable pharmacology, and chemical accessibility to allow ready preparation of analogs. The rapamycin system has been used to regulate gene expression in vitro in a variety of cell lines transfected transiently or stably and has been successfully incorporated into most in vivo vector contexts, including adenovirus, adeno-associated virus (AAV), oncoretrovirus, lentivirus, herpes simplex virus, and naked DNA (for review, see Pollock and Clackson 2002). Use in experimental animals has been particularly well-characterized with adenovirus and AAV vectors; e.g., tightly controlled erythropoietin (Epo) production in response

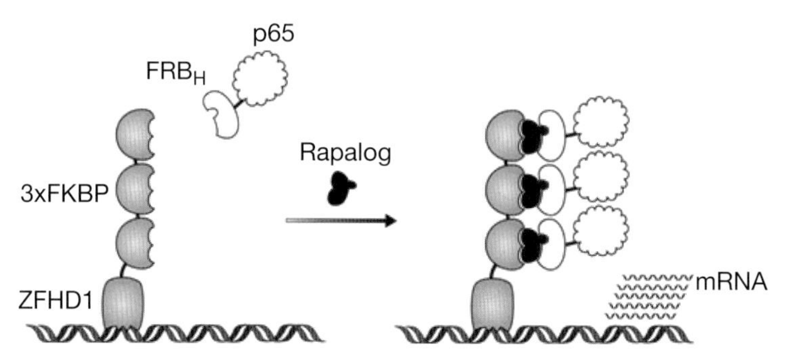

FIGURE 1. Scheme for dimerizer-induced gene expression using a rapamycin analog (rapalog). For an explanation of abbreviations, see Introduction.

to rapamycin has been demonstrated in nonhuman primates for more than 6 years following a single administration of AAV vectors (Rivera et al. 2005). For a complete list of publications describing the use of dimerizers to regulate transcription, see http://www.ariad.com/regulationkits/reg_ref3.html.

Components of the Rapamycin-inducible System

Rapamycin functions by binding with high affinity to the cytoplasmic protein FKBP12 (FKBP, for FK506-binding protein) and then to a large Ser-Thr protein kinase called FRAP (FKBP-rapamycin-associated protein, also known as mTOR [mammalian target of rapamycin]), thereby acting as a heterodimerizer to join the two proteins together (Choi et al. 1996). To control transcription of a target gene, a DNA-binding domain is fused to one or more FKBP domains, and a transcriptional activation domain is fused to a 93-amino-acid portion of FRAP, termed FRB, that is sufficient for binding the FKBP-rapamycin complex (Chen et al. 1995). Only in the presence of rapamycin are the two fusion proteins dimerized and therefore capable of activating the transcription of a gene equipped with binding sites for the DNA-binding domain (Rivera et al. 1996).

In some cases, the use of rapamycin may be compromised by the antiproliferative/immunosuppressive effects that result from inhibiting endogenous FRAP kinase activity. To overcome this limitation, the system has been engineered to function with inert analogs of rapamycin, termed rapalogs. These compounds have been chemically modified so that they no longer bind tightly to wild-type endogenous FRAP, greatly reducing immunosuppressive activity. They can, however, bind to a modified FRAP that contains a single, designed, amino acid change (T2098L). Incorporation of this mutation into the FRB domain fused to the activation domain allows a rapalog to be used to heterodimerize the engineered transcription factor fusion proteins specifically without interfering with the activity of endogenous FRAP (Pollock et al. 2000, 2002; for review, see Pollock and Clackson 2002; Rivera et al. 2005). This redesigned system also retains the ability to respond to rapamycin itself.

AP21967 has reduced antiproliferative activity compared to rapamycin (Fig. 3) and is the most recommended rapalog for most in vitro and in vivo applications. However, rapamycin can often provide higher induced expression levels than the rapalog AP21967 and thus can be used when antiproliferative/immunosuppressive activity is not an issue. In vivo, dosing rapamycin intermittently (as opposed to daily) dramatically reduces its immunosuppressive activity. The rapalog AP23102 has no residual antiproliferative activity and can be used in vitro when such a profile is required. It has not been tested in vivo. Rapamycin is able to cross the blood-brain barrier in mice. It is not known whether this is the case for AP21967 and AP23102. Regarding delivery of dimerizer, rapamycin and rapalogs can also be delivered orally to experimental animals (Magari et al. 1997; Rivera et al. 2005). However, the required doses are higher, with bioavailability approximately 15%. Rapamycin and rapalogs have low aqueous solubilities. Delivery to animals in drinking water has not been reported.

The rapamycin system is built using only human proteins to minimize the potential for immunogenicity in clinical applications. The DNA-binding domain is ZFHD1, a composite human DNA-binding domain with novel DNA recognition specificity (Pomerantz et al. 1995). ZFHD1 is composed of two zinc finger domains from the human transcription factor Zif268, joined to a homeodomain derived from the human transcription factor Oct-1. ZFHD1 binds with high affinity and specificity to a unique composite DNA-binding sequence, but not to Zif268- or Oct-1-binding sites. Typically, multiple copies of the ZFHD1-binding site are included upstream of target genes to obtain robust gene activation. The activation domain comprises the carboxy-terminal 191 amino acids from the p65 subunit of human NF-κB (Schmitz and Baeuerle 1991), which substantially outperforms the commonly used herpes virus VP16 activation domain (Rivera et al. 1996; Natesan et al. 1999).

Expected Results for In Vitro Experiments

Figure 2A shows the effects on reporter gene expression of adding increasing amounts of AP21967, AP23102, or rapamycin to HT1080 fibrosarcoma cells stably transfected with vectors expressing a secreted alkaline phosphatase target gene and the regulated transcription factors. In the absence of dimerizer, target gene expression is undetectable. Maximal induction typically occurs at 10–30 nM, with AP21967 generally requiring approximately threefold higher concentrations than AP23102 or rapamycin to achieve equivalent levels of induction. In general, induction should be maximal by 24–48 hours after addition of dimerizer. The decay kinetics of expression will depend on many factors, including the half-lives of the transcriptional complex, the mRNA, and the target protein. Significantly elevated gene expression usually persists for many days. When long-term induction is desired, we recommend replenishing dimerizer with each change of medium (e.g., every 3 days).

Expected Results for In Vivo Experiments

Figure 2B shows levels of human growth hormone (hGH) measured in the serum of mice treated with varying doses of AP21967. The mice had been transduced, by hydrodynamic injection of DNA into the tail vein, with vectors expressing an hGH target gene and the regulated transcription factors. A 3-mg/kg dose of AP21967 induced hGH to levels 10,000-fold over background. hGH levels were measured in serum 24 hours after administration of AP21967. In our experience, a 3- to 10-mg/kg dose of AP21967 delivered intraperitoneally to mice gives a maximal response, with an approximately 30-fold lower dose giving little or no response. The kinetics of induction and decay will depend on many factors, but in general expression should be maximal by about 24–48 hours after administration of dimerizer. Decay kinetics will be dictated by the half-lives of the dimerizer, transcription complex, mRNA, and protein, whichever is longest (Clackson 2000). Usually, a single dose of dimerizer will provide expression for many days.

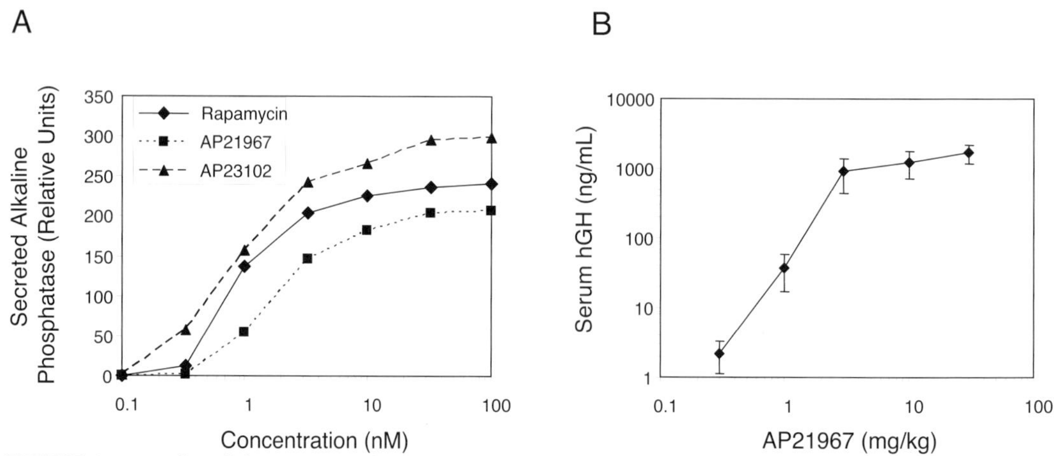

FIGURE 2. Examples of dimerizer-regulated gene expression. (*A*) In vitro and (*B*) in vivo studies as described in the text.

Protocol 1

Dimerizer-mediated Regulation of Gene Expression In Vitro

This protocol describes the preparation of vectors for rapamycin- or rapalog-inducible gene expression, followed by induction of gene expression in vitro.

MATERIALS

CAUTION: See Appendix for appropriate handling of materials marked with <!>.

Reagents

The plasmid constructs and rapalog dimerizers described in this chapter are available from ARIAD as the ARGENT Regulated Transcription Plasmid Kit (www.ariad.com/regulationkits).

Culture medium appropriate for target cells

Dimerizers (chemical structures of the dimerizers described here are shown in Fig. 3)

Rapalogs AP21967 and AP23102 (ARIAD)

Rapamycin (Sigma-Aldrich R0395)

Compounds are typically obtained as dry powders and should be stored in a tightly sealed container in the dark at –20°C. For in vitro use, the compounds should be dissolved in absolute ethanol as described in Methods.

Ethanol (e.g., AAPER alcohol, 200 proof), ice-cold

Plasmid constructs (maps of plasmid constructs commonly used to create vectors for dimerizer-inducible expression are shown in Fig. 4)

Transcription factor plasmid: pC_4N_2-R_HS/ZF3

This plasmid vector drives expression of the regulated transcription factors (Fig. 4, *top*) and contains a CMV enhancer/promoter (C) driving expression of the following fusion proteins from a bicistronic transcript:

- An activation domain fusion (R_HS) which consists of the FRB fragment of human FRAP (R_H), fused to an activation domain derived from the p65 subunit of human NF-κB (S; amino acids 361 to 551). The FRB domain consists of amino acids 2021–2113 of FRAP. The T2098L mutation is incorporated to allow binding of rapalogs.
- A DBD fusion (ZF3) which consists of the ZFHD1 DBD (Z) and three tandemly repeated copies of human FKBP12 (F3).

Both fusion proteins contain an amino-terminal nuclear localization signal (N_2, from the human c-*myc* gene). The two coding regions are separated by an internal ribosome entry sequence (IRES) derived from the encephalomyocarditis virus to allow translation of the second cistron (ZF3) and are followed by a 3′-untranslated region (UTR) from the rabbit β-globin gene that includes its final intron and poly(A) signal.

The optimal configuration of the transcription factor fusions is one in which ZFHD1 is fused to three tandemly reiterated FKBP12 domains and the p65 activation domain to a single FRB domain. This configuration theoretically allows recruitment of up to three activation domains per DNA-binding site (see Fig. 1).

HO, MeO, OH, MeO, OMe, HO, N, O

Rapamycin

MW 914.18 Da

32, OH, 26, 16, H, N

AP21967

MW 1017.4 Da

300x reduced mTOR inhibition

28, HN, 16

AP23102

MW 1019.3 Da

>10000x reduced mTOR inhibition

FIGURE 3. Chemical structures of rapamycin and rapalogs used for inducing gene expression. The three numbered positions on AP21967 and two on AP23102 indicate those positions that are altered with respect to rapamycin. Also shown are molecular weights (MW) and the reduced antiproliferative activities of the rapalogs on mammalian cells, as compared to rapamycin.

pC_4N_2-R_HS/ZF3 (7811 bp)

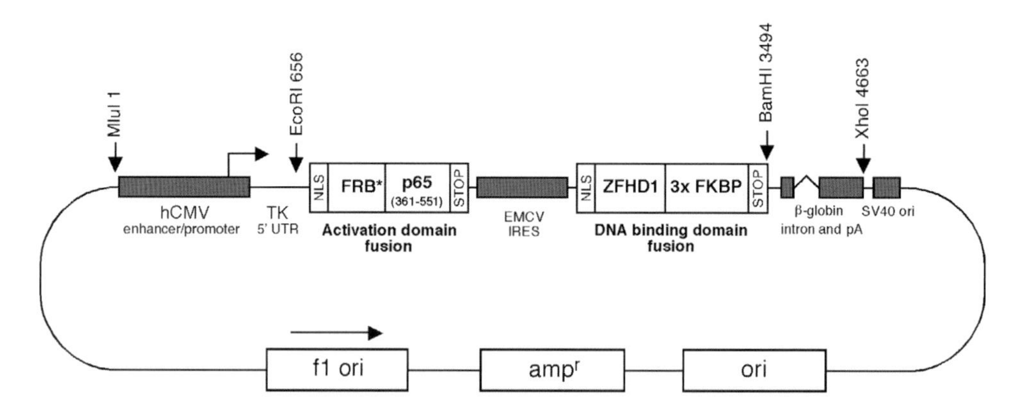

pZ_{12}I-PL-2 (3528 bp)

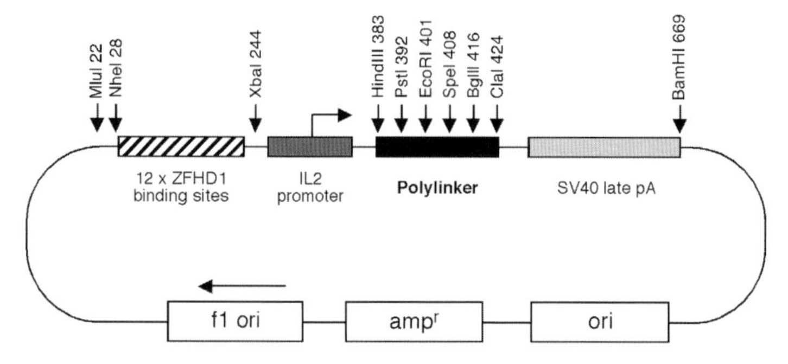

FIGURE 4. Maps of constructs for dimerizer-inducible gene expression. (*Top*) Transcription factor expression plasmid pC_4N_2-R_HS/ZF3. (*Bottom*) Target gene plasmid pZ_{12}I-PL-2. Abbreviations are defined in the Materials section of the text.

Target gene plasmid: $pZ_{12}I$-PL-2

The coding region of the gene to be regulated is inserted into this target gene vector (Fig. 4, *bottom*). The vector contains 12 ZFHD1-binding sites and a minimal human IL-2 gene promoter ($Z_{12}I$), upstream of a polylinker (PL) followed by a 3′ UTR containing a polyadenylation signal from the SV40 late gene (–2). The polylinker consists of the following restriction sites: HindIII, PstI, EcoRI, SpeI, BglII, or ClaI.

Control target gene plasmid: $pZ_{12}I$-hGH-2

This positive control vector contains the hGH cDNA inserted into $pZ_{12}I$-PL-2 as a 759-bp HindIII-EcoRI fragment.

Target cells

Equipment

Microcentrifuge tube or glass vial
Vortex mixer

METHODS

Preparation of Vectors

Vectors are derived from the starting constructs listed in Materials, using standard cloning techniques.

1. Construct the target gene vector by inserting the coding region of the gene to be regulated into the polylinker of $pZ_{12}I$-PL-2. This will place its expression under control of the dimerizer-regulated transcription factors.

   Alternatively, remove the $Z_{12}I$ control region from $pZ_{12}I$-PL-2 (e.g., using 5′ MluI or NheI sites and 3′ HindIII, PstI, EcoRI, SpeI, BglII, or ClaI sites) and insert it upstream of the coding region of the gene of interest located in a different vector. This vector must also provide a suitable 3′ UTR.

2. Optimize the transcription factor vector, if necessary.

   In certain situations, it may be necessary to replace the CMV enhancer that drives expression of the regulated transcription factors with an alternate enhancer, e.g., one that is more active in the target cell population. To do so, replace the MluI-EcoRI fragment of pC_4N_2-R_HS/ZF3 with the desired enhancer and 5′ UTR sequence.

3. Subclone the vector components, if necessary.

   For certain applications, it is desirable to subclone the regulated transcription factor and target gene cassettes into alternate vector backbones, e.g., AAV, adenoviral, lentiviral, and retroviral vectors or a vector that contains a selectable marker. This can be accomplished by transferring the entire transcription factor expression cassette (e.g., using MluI-XhoI sites) or just the coding region (e.g., EcoRI-BamHI), if appropriate, into another vector. The entire target gene cassette can be transferred using flanking 5′ MluI or NheI and 3′ BamHI sites.

   The transcription factor and target gene cassettes can also be combined into a single vector. For example, vectors in which the target gene cassette is cloned directly downstream from the transcription factor cassette, in the same orientation, have been constructed successfully and retain the tight level of regulation achieved by cotransduction of separate vectors (Rivera et al. 2005).

Introduction of Vectors into Target Cells

4. Introduce the transcription factor and target gene vectors into the target cells of interest using the appropriate technique.

For transient transfection

a. As a starting point, introduce the transcription factor and target gene plasmids in a 2:1 ratio.

b. If the expression level of the transcription factors is too high, transcription of a transiently transfected target gene can be reduced or "squelched" (Natesan et al. 1997).

It may be necessary to further optimize plasmid ratios for a given situation and to limit the amount of transcription factor plasmid introduced by adding carrier DNA.

For stable transfection

When making stable cells lines (a situation in which squelching generally does not apply; see Natesan et al. 1997), the level of induction of target gene expression should correlate with the level of expression of the activation domain fusion. The enhancer that drives expression of the transcription factor fusion must be optimal for the cells being transduced. The CMV enhancer is generally suitable for this purpose, but in certain cell types, alternate enhancers might be more appropriate and may need to be cloned into the transcription factor vector (see Step 2, above). To generate stable cell lines containing both the transcription factor and target gene plasmids, we recommend transfecting them into cells sequentially.

a. Stably integrate the transcription factor vector plasmid.

b. Screen individual clones by transiently transfecting the target gene of interest or an easily assayed target gene (e.g., $pZ_{12}I$-hGH-2).

c. Select the clone with lowest background and highest rapamycin- or AP21967-dependent induction.

d. Stably integrate the target gene plasmid.

e. Screen individual clones for the lowest background and highest levels of dimerizer-dependent target gene expression.

The transcription factor and target gene plasmids must be either cotransfected with vectors containing distinct selectable markers or subcloned into vectors that contain selectable markers (see Step 3 above).

For infection with viral vector

Cells can be transduced with appropriate viral vectors, such as retroviral vectors (Pollock et al. 2000). Retroviral versions of the transcription factor and target gene vectors are available from ARIAD (www.ariad.com/regulationkits).

Dimerizer-mediated Gene Expression in Target Cells

5. Prepare dimerizer. The example given is AP21967. Rapamycin (914.18 daltons) or AP23102 (1019.3 daltons) can be prepared and stored similarly.

a. To prepare lyophilized AP21967 (molecular mass 1017.4 daltons) for in vitro use, add ice-cold absolute ethanol to the lyophilized material to make a 1 mM solution (e.g., dissolve 250 μg of AP21967 in 246 μl of ethanol). Seal and vortex periodically over a period of a few minutes to dissolve the compound. Keep on ice while dissolving to minimize evaporation.

b. Store the dimerizer stock solution in the dark at –20°C in a glass vial or a microcentrifuge tube. Store further dilutions in ethanol in the same way. Stocks stored in this way should be stable for at least one year.

At the bench, keep solutions in ethanol on ice and open for as short a time as possible, to prevent evaporation and consequent changes in concentration.

For most applications, AP21967 is the recommended dimerizer, as it has reduced anti-proliferative activity compared to rapamycin (Fig. 3).

c. Prepare culture medium containing the desired concentration of dimerizer by adding dimerizer directly from ethanol stocks or by diluting serially in medium just before use.

In the latter case, the highest concentration should not exceed 5 μM, to ensure complete solubility in the (aqueous) medium. In either case, the final concentration of ethanol in the medium added to mammalian cells should be kept below 0.5% (a 200-fold dilution of a 100% ethanol solution) to prevent detrimental effects of the solvent on the cells.

6. Replace medium on target cells with medium containing dimerizer.

   A single administration of dimerizer is typically sufficient to induce expression for 3 days. For longer-term experiments, prepare fresh solution of medium containing dimerizer each time cells are fed (e.g., every 3 days).

7. Determine dose-response and kinetics of expression. In initial experiments, test AP21967 across a broad range of concentrations (e.g., 0.01 to 1000 nM) to provide a complete dose-response profile (see Troubleshooting).

TROUBLESHOOTING

Problem: Induced expression in vitro is weak or undetectable.

Solution: For transient transfections, it may be necessary to allow time following transfection for the transcription factor proteins to accumulate to adequate levels (e.g., 24 hours). Use the inducible hGH plasmid ($pZ_{12}I$-hGH-2) as a control to test the transcription factor vector and the transfection procedure. hGH can be quantitated using a kit from Roche (1 585 878). Check the expression of the transcription factor fusion proteins by western blot. Anti-p65 antibodies (Santa Cruz Biotechnology, sc-372) can be used to detect the R_HS activation domain fusion protein (~48 kD). Adequate levels of this protein are the most critical to effective induction. If R_HS expression levels are low, try increasing the amount of pC_4N_2-R_HS/ZF3 vector introduced into cells. (Levels of the ZF3 DBD fusion protein are usually undetectably low, as its expression is driven off the IRES sequence.)

Problem: Basal (uninduced) expression is high.

Solution: Reduce the concentration of vectors introduced into cells. For stable transfections, assay individual clones to identify those showing tight regulation of expression.

Protocol 2

Dimerizer-mediated Regulation of Gene Expression In Vivo

This protocol describes how to achieve rapamycin- or rapalog-inducible gene expression in vivo.

MATERIALS

CAUTION: See Appendix for appropriate handling of materials marked with <!>.

Reagents

Dimerizers (chemical structures of the dimerizers described here are shown in Fig. 3)
  Rapalogs AP21967 and AP23102 (ARIAD)
  Rapamycin (Sigma-Aldrich R0395)
    Compounds are typically obtained as dry powders and should be stored in a tightly sealed container in the dark at –20°C.
N,N-Dimethylacetamide <!> (DMA; e.g., Sigma-Aldrich D5511)
Mice
Plasmid constructs (see Protocol 1)
Polyethylene glycol 400 <!> (PEG-400; e.g., J.T. Baker U216–07)
Polyoxyethylene-sorbitan monooleate (Tween-80; e.g., Sigma-Aldrich P1754)

Equipment

Glass vial
Vortex mixer

METHOD

1. Use appropriate techniques to introduce transcription factor and target gene vectors into experimental animals. For example, mice can be coinjected intramuscularly with separate AAV vectors (Rivera et al. 1999, 2005).
2. Prepare dimerizer for intraperitoneal injection into mice. The example given is AP21967. Rapamycin can be formulated, stored, and administered similarly.
   a. For a 10-mg/kg dose, solubilize AP21967 in DMA to a concentration of 10 mg/ml. Mix or vortex gently for a few minutes until dissolved completely. This stock solution can be stored for up to 2 months at –20°C.
      To deliver a dose lower than 10 mg/kg, begin with a more dilute stock of AP21967 in DMA. For most applications, AP21967 is the recommended dimerizer, as it has reduced antiproliferative activity compared to rapamycin (Fig. 3).
   b. Prepare a stock of PEG-400 and Tween-80 in a 9:1 ratio.
      This stock can be prepared beforehand and stored for up to 1 month at room temperature.
   c. Shortly before use, mix equal parts of AP21967/DMA and PEG/Tween to create a 5-mg/ml dosing solution. Vortex to mix fully.

3. Deliver the appropriate volume of the 5-mg/ml AP21967 dosing solution in a dose volume of 2 ml/kg intraperitoneally, e.g., to deliver a 10-mg/kg dose to a 20-g mouse, administer 40 μl intraperitoneally.

   This formulation is only suitable for intraperitoneal delivery. Since DMA can be toxic to mice, we recommend performing practice injections using vehicle only (on nontransduced animals) prior to initiation of studies. Vehicle-dosed animals should always be included as controls.

4. Determine dose response and kinetics of expression. In initial experiments, test a range of doses of AP21967 (e.g., from 0.3 to 10 mg/kg) to find the most effective for the specific application.

   In studies to date, AP21967 was nontoxic to mice at intraperitoneal doses up to 30 mg/kg. Initial studies should use a single dose of dimerizer to determine induction and decay kinetics (see Troubleshooting).

TROUBLESHOOTING

Problem: Induced expression is weak or undetectable.

Solution: Always check vectors for ability to provide inducible expression in vitro before using them for in vivo studies. Try to increase expression levels of the transcription factor fusion proteins by increasing the dose of pC_4N_2-R_HS/ZF3 vector. Carefully check the formulation of dimerizer to ensure that adequate drug levels are reaching the target cells. Use the inducible hGH plasmid (pZ_{12}I-hGH-2) as a control (as described in Troubleshooting in Protocol 1).

Problem: Basal (uninduced) expression is high.

Solution: Reduce the dose of each vector.

REFERENCES

Chen J., Zheng X.F., Brown E.J., and Schreiber S.L. 1995. Identification of an 11-kDa FKBP12-rapamycin-binding domain within the 289-kDa FKBP12-rapamycin-associated protein and characterization of a critical serine residue. *Proc. Natl. Acad. Sci.* **92:** 4947–4951.

Choi J., Chen J., Schreiber S.L., and Clardy J. 1996. Structure of the FKBP12-rapamycin complex interacting with the binding domain of human FRAP. *Science* **273:** 239–242.

Clackson T. Regulated gene expression systems. 2000. *Gene Ther.* **7:** 120–125.

Ho S.N., Biggar S.R., Spencer D.M., Schreiber S.L., and Crabtree G.R. 1996. Dimeric ligands define a role for transcriptional activation domains in reinitiation. *Nature* **382:** 822–826.

Magari S.R., Rivera V.M., Iuliucci J.D., Gilman M., and Cerasoli F., Jr. 1997. Pharmacologic control of a humanized gene therapy system implanted into nude mice. *J. Clin. Invest.* **100:** 2865–2872.

Natesan S., Rivera V.M., Molinari E., and Gilman M. 1997. Transcriptional squelching re-examined. *Nature* **390:** 349–350.

Natesan S., Molinari E., Rivera V.M., Rickles R., and Gilman M. 1999. A general strategy to enhance the potency of chimeric transcription factors. *Proc. Natl. Acad. Sci.* **96:** 13898–13903.

Pollock R. and Clackson T. 2002. Dimerizer-regulated gene expression. *Curr. Opin. Biotechnol.* **13:** 459–467.

Pollock R., Giel M., Linher K., and Clackson T. 2002. Regulation of endogenous gene expression with a small-molecule dimerizer. *Nat. Biotechnol.* **20:** 729–733.

Pollock R., Issner R., Zoller K., Natesan S., Rivera V.M., and Clackson T. 2000. Delivery of a stringent dimerizer-regulated gene expression system in a single retroviral vector. *Proc. Natl. Acad. Sci.* **97:** 13221–13226.

Pomerantz J.L., Sharp P.A., and Pabo C.O. 1995. Structure-based design of transcription factors. *Science* **267:** 93–96.

Rivera V.M., Ye X., Courage N.L., Sachar J., Cerasoli F., Jr., Wilson J.M., and Gilman M. 1999. Long-term regulated expression of growth hormone in mice after intramuscular gene transfer. *Proc. Natl. Acad. Sci.* **96:** 8657–8662.

Rivera V.M., Gao G.P., Grant R.L., Schnell M.A., Zoltick P.W., Rozamus L.W., Clackson T., and Wilson J.M. 2005. Long-term pharmacologically regulated expression of erythropoietin in primates following AAV-mediated gene transfer. *Blood* **105:** 1424–1430.

Rivera V.M., Clackson T., Natesan S., Pollock R., Amara J.F., Keenan T., Magari S.R., Phillips T., Courage N.L., Cerasoli F., Jr., Holt D.A., and Gilman M. 1996. A humanized system for pharmacologic control of gene expression. *Nat. Med.* **2:** 1028–1032.

Schmitz M.L. and Baeuerle P.A. 1991. The p65 subunit is responsible for the strong transcription activating potential of NF-κB. *EMBO J.* **10:** 3805–3817.

Spencer D.M., Wandless T.J., Schreiber S.L., and Crabtree G.R. 1993. Controlling signal transduction with synthetic ligands. *Science* **262:** 1019–1024.

64 RheoSwitch System: A Highly Sensitive Ecdysone Receptor-based Gene Regulation System Induced by Synthetic Small-molecule Ligands

Prasanna Kumar and Anand Katakam

RheoGene, Inc., Norristown, Pennsylvania 19403

ABSTRACT

The RheoSwitch system is a highly efficient ecdysone receptor-based gene regulatory system that is activated by synthetic small-molecule ligands. It consists of two chimeric proteins derived from the ecdysone receptor and RXR in a two-hybrid format that heterodimerizes and activates transcription from a responsive promoter in the presence of the ligand. The RheoSwitch system functions in animal cells, yeast, and plants.

Advantages include an excellent safety profile, lack of pleiotropic effects in mammalian cells, and the high sensitivity and bioavailability of the diacylhydrazine ligands. The ligands are pharmacologically inert. Since the aqueous solubility of the ligands is low, the ligands are given orally or intraperitoneally for in vivo applications. There is dose-dependent induction over a wide range of ligand concentrations. The system has significant flexibility in that it can be formatted into multivector or single-vector versions. Incorporation can occur into adenoviral, adeno-associated viral (AAV), retroviral, and lentiviral vectors. A repertoire of inducible promoters incorporated into different versions of the system provides desired levels of regulation of transgene expression. These properties make the RheoSwitch system ideal for cell and gene therapy applications, basic research on cellular pathways and in transgenic animals carrying precisely regulatable transgenes. The RheoSwitch system in various viral vectors has been developed only recently and extensive in vivo evaluations of the vectors are under way.

INTRODUCTION

Early versions of the gene regulation systems for mammalian cells utilized endogenous promoters regulated by factors such as heat shock or certain metal ions, but those systems suffered from pleiotropic effects on cellular functions. The next generation of regulatory systems used heterologous or highly engineered proteins aimed at reducing the pleiotropic effects and regulating a promoter that drives the gene of interest in response to a ligand. One of these systems is based on the ecdysone receptor (EcR), an insect nuclear receptor that can bind to the molting hormone 20-hydroxy-ecdysone and activate transcription of several genes involved in molting (Palli et al. 2005). *Drosophila* EcR was initially used to develop a gene regulatory system for mammalian cells that is activated by ecdysteroids (Christopherson et al. 1992). Subsequently, an improved system using *Drosophila* EcR and human RXR-α was developed (No et al. 1996). Another regulatory system is based on the *Bombyx mori* EcR that heterodimerizes with endogenous RXR (Suhr et al. 1998).

The RheoSwitch system, developed by RheoGene, Inc., uses a mutagenized EcR from *Choristoneura fumiferana* (Palli et al. 2003). It consists of two nuclear receptor components in a "two-hybrid" format and is activated by a synthetic small-molecule ligand. The ligand-binding component is a fusion protein between the DEF domains of EcR and the DNA-binding domain (DBD) of yeast Gal4 (Gal4-EcR). The other receptor component is a fusion protein between the EF domains of a chimeric RXR derived from human and locust sequences and the acidic activation domain of VP16 (VP16-RXR). The two receptor fusion proteins are expressed under the control of constitutive promoters. Binding of the ligand to the Gal4-EcR enhances its heterodimerization with the VP16-RXR and also activates the heterodimer to function as an efficient transcription factor that can induce gene expression from a responsive (inducible) promoter that contains binding sites for Gal4 (Fig. 1).

The inducible promoter consists of five copies of a Gal4-binding site inserted upstream of a minimal promoter derived from natural or synthetic sequences. Various inducible promoters have been incorporated into different versions of the RheoSwitch system. They vary from promoters that tightly control the basal expression levels and induce moderate levels of expression of genes that require stringent regulation to promoters that are slightly leaky in the absence of ligand, but induce very high levels of expression for applications where the absolute amount of gene expression is important.

Ligands

The ligands used for activation of the RheoSwitch system are derivatives of the diacylhydrazine class of compounds (Dhadialla et al. 1998). Some, such as RSL1 (see Fig. 2), have been evaluated for their effect on gene expression profiles by microarray analysis of 39,500 human gene

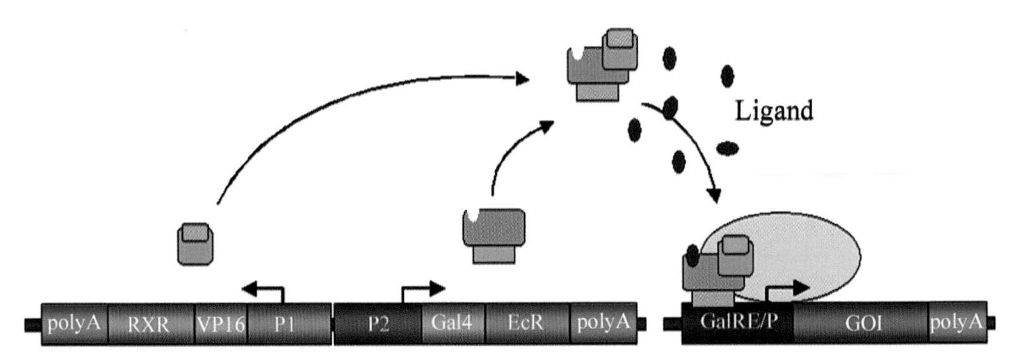

FIGURE 1. RheoSwitch system in dual-vector format: The two receptor components are expressed under constitutive promoters (P1 and P2). Ligand binding to EcR enhances heterodimerization as well as activation of the receptors to induce transcription of the gene of interest (GOI) from the inducible promoter (GalRE/P).

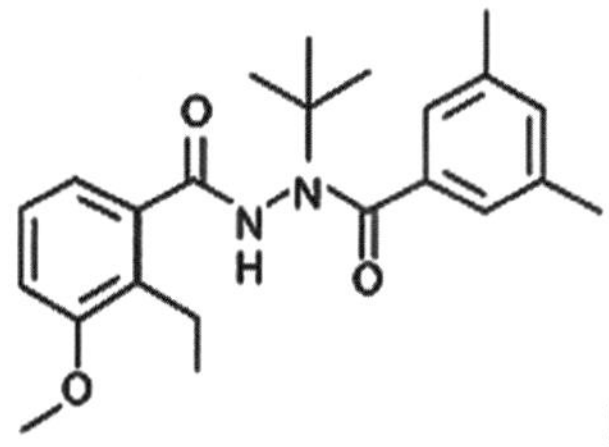

FIGURE 2. Structure of ligand RSL1.

sequences. There is almost no change in the profiles of the endogenous genes. Toxicological studies on selected diacylhydrazine ligands show no detectable toxicity in rodents. Studies on one of the diacylhydrazine ligands in rodents have demonstrated that the ligand can cross the blood brain barrier and activate gene expression in the brain.

EcR Mutants and the RheoPlex System

The EcR ligand-binding domain has been subjected to extensive mutagenesis to identify mutations that exhibit enhanced ligand sensitivity and selectivity (Kumar et al. 2002, 2004). A current standard version of the RheoSwitch system contains mutations at two positions (V390I and Y410E). This receptor is activated by picomolar concentrations of ligands such as RSL1. Mutants have also been developed that are activated only by diacylhydrazine ligands and are completely unresponsive to ecdysteroids.

In the RheoPlex system, multiple genes can be regulated independently within the same cell or organism. The first version of this system uses two EcR mutants: Mutant 1 responds only to ligand 1 and mutant 2 responds only to ligand 2. The ligand-binding domains of EcR mutants 1 and 2 are fused to two different DBDs, and these fusion proteins use the same heterodimer partner VP16-RXR. In response to their cognate ligands, EcR mutants 1 and 2 activate transcription from two separate inducible promoters containing the appropriate binding sites for the corresponding DBDs with no cross-activation. Figure 3 shows independent activation of two genes by two different ligands in a cotransfection experiment followed by detection of the genes by immunofluorescence.

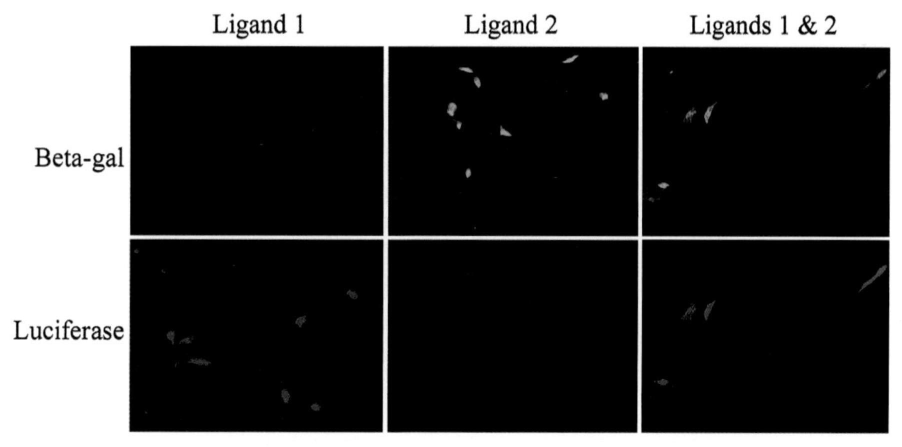

FIGURE 3. Independent regulation of firefly luciferase and *lacZ* under the RheoPlex system using two ligands. The panels under each ligand treatment represent the cells from the same field double-stained by immunofluorescence for β-galactosidase (FITC) and firefly luciferase (TRITC) and photographed under the corresponding filters. Ligand 1 induces only luciferase and not β-galactosidase, ligand 2 induces only β-galactosidase and not luciferase, ligands 1and 2 together induce both the proteins.

Vector Formats

The RheoSwitch system has been assembled into dual-vector and single-vector systems. Figure 1 shows the dual-vector system. The receptor vector expresses the two receptor components from constitutive promoters. The inducible expression vector contains the inducible promoter and the gene of interest. Single-vector systems can be used in direct plasmid transfections or incorporated into viral vectors. The single-vector and dual-vector systems for plasmid transfection contain selection markers such as hygromycin and neomycin for creating stable cell lines harboring the integrated plasmid(s). This system is being evaluated for the creation of transgenic mice carrying transgenes regulated by the ligand. In the vectors designed for viral delivery, the two receptor components are expressed from a single constitutive promoter as a bicistronic message in order to reduce the total size of the components. The RheoSwitch system has been incorporated into adenoviral, AAV, retroviral, and lentiviral vectors. Adenoviral, retroviral, and lentiviral versions are in the two-hybrid format and can accommodate coding sequences up to approximately 3 kb of the gene of interest. The AAV version is configured in a two-hybrid format for expression of relatively smaller genes (<900 bp). In addition, a single-receptor system has been developed that consists of a single fusion protein of a VP16 activation domain, Gal4-DBD, and the DEF domains of the EcR that binds the ligand and activates transcription from the inducible promoter. The single-receptor system can accommodate transgene sequences that are less than 2 kb in the AAV vector. The viral vector versions have been developed recently and have been evaluated in vitro and in vivo. Additional in vivo evaluations are under way.

Using tissue-specific constitutive promoters for the expression of the receptor components, the RheoSwitch system can be restricted to specific cell types. In addition to the VP16 activation domain, other activation domains of mammalian origin have been used in different versions of the RheoSwitch system (Karzenowski et al. 2005). Table 1 lists the mammalian cell lines and primary cells in which the system has been used.

TABLE 1. Mammalian Cell Lines and Primary Cells in which the RheoSwitch System Has Been Successfully Used

Cell line	Description
NIH-3T3	mouse embryonic fibroblast
B16-F1	mouse skin melanoma cells
C17.2	mouse neural stem cells
DC2.4	mouse dendritic cells
CHO	Chinese hamster ovary
PC-12	rat adrenal pheochromocytoma
CV-1	monkey kidney fibroblast
A549	human lung carcinoma
HCT 116	human colon carcinoma
HeLa	human cervix carcinoma
HepG2	human liver carcinoma
MCF-7	human mammary gland carcinoma
HEK-293	human embryonic kidney
HT1080	human fibrosarcoma
LAPC4	human prostate cancer
Primary Cells	
Mouse embryonic stem cells	
Rat bone marrow stromal cells	
Human and mouse T cells	
Human and mouse hematopoietic stem cells	
Human mesenchymal stem cells	
Rat aortic smooth-muscle cells	

The Inducible Expression Vector

The inducible expression vector of the RheoSwitch system contains a multiple cloning site just downstream from the inducible promoter. The gene of interest is inserted at any of the convenient restriction sites using standard molecular cloning methods (Sambrook and Russell 2001). The minimal 5′ UTR (untranslated region) required for efficient translation is provided in the vector between the transcription start site and the multiple cloning site. However, a consensus Kozak sequence at the 5′ end of the coding sequence for the gene of interest (GCCACC**ATG**G, where the bold letters ATG code for the initiator methionine) needs to be provided where possible. The most important bases in the sequence are a purine at position –3 and a G at position +4 relative to the initiation site.

In Vivo Applications of the RheoSwitch System

The RheoSwitch system can be delivered in vivo by viral or nonviral vectors. For viral delivery, RheoSwitch vectors can be transferred as a single cassette into commercial or proprietary adenoviral, AAV, retroviral, or lentiviral vectors. Briefly, the desired number of purified viral particles is injected either intravenously or directly into the target tissue. The ligand dissolved in dimethylsulfoxide (DMSO) is administered by intraperitoneal injection. The optimal time for ligand administration after vector delivery depends on the viral vector used. For mice, an initial dosage of 50 mg/kg body weight at a concentration of 20 mg/ml DMSO is tested. (For a weight of about 20 g, a dose of 1 mg of ligand in 50 μl of DMSO is appropriate.) For continuous gene expression, the ligand can be administered in the animal feed. Gene expression is evaluated at desired time points. For AAV vectors, a delay of 3–4 weeks is needed for conversion of the genome from single-stranded DNA to double-stranded DNA. Nonviral methods of in vivo delivery, such as in vivo electroporation, are described in Chapter 38 of this volume.

Contact Information and Marketing Partners

New England BioLabs (NEB) is marketing a dual-vector version of the RheoSwitch system in which the inducible promoter provides extremely tight regulation of the transgene. The receptor vector is named pNEBR-R1 and the inducible expression vector for inserting the gene of interest is named pNEBR-X1. Inducible vectors for higher expression levels of the gene of interest and several cell lines containing genomic integration of the receptor components are also available from NEB. For more details on products available from NEB visit www.neb.com and for special research needs, partnerships, collaborations, and licensing for therapeutic development using the technology, visit www.rheogene.com.

Protocol

Gene Regulation In Vitro Induced by a Synthetic Small-molecule Ligand, RSL1

This protocol describes transfection of cells in culture using the RheoSwitch system, in which gene expression is regulated by a synthetic small molecule ligand, RSL1.

MATERIALS

CAUTION: See Appendix for appropriate handling of materials marked with <!>.

Reagents

Ampicillin (100 mg/ml aqueous solution) <!>
Bovine serum (Invitrogen)
Cells to be transfected
Dimethylsulfoxide (DMSO; Sigma-Aldrich) <!>
Ligand: RSL1 (5 mM in DMSO; stable at room temperature, m.w. 382.5) (New England BioLabs)
Lipofectamine 2000 (Invitrogen)
Luciferase assay system (Bright-Glo; Promega)
Nucleic acids
- Inducible expression vector: pNEBR-X1 (New England BioLabs)
- Normalization plasmid: pRL-CMV (*optional*, see Step 2a) (Promega)
- Receptor vector: pNEBR-R1 (New England BioLabs)

  Bacteria (XL1-Blue or other) harboring the plasmids are grown in Luria-Bertani (LB) medium containing 100 μg/ml ampicillin. The plasmids used for transfection are prepared endotoxin-free. Column purifications (e.g., Endo-free Maxikit from QIAGEN) or CsCl–gradient methods are suitable.

Opti-MEM I (Invitrogen) or serum-free medium
Passive lysis buffer (Promega)
TRIZOL reagent (Invitrogen) or RNeasy Kit (QIAGEN)

Equipment

Cell culture plates (48 well in this example) and 96-well opaque assay plates (VWR or Fisher Scientific)
Microcentrifuge tubes (1.7 ml)
Microplate luminometer (Dynex Technologies, Inc.) for luminescence assays
Platform shaker

METHODS

Transfection

The RheoSwitch system is compatible with various transfection reagents, including Lipofectamine 2000 (Invitrogen), FuGENE 6 (Roche Applied Science), GenePORTER 2 (Gene Therapy Systems), SuperFect (QIAGEN), and calcium phosphate coprecipitates. Electroporation and nucleofection

(Amaxa Biosystems) can also be used. Transfection can be performed in many plate formats (96-well, 48-well, or 24-well plates) and on a larger scale. Here, Lipofectamine 2000 is used for transfection of cells in a 48-well plate with pNEBR-R1 as the receptor vector and pNEBR-X1 containing firefly luciferase as the inducible expression vector. The procedure may be modified according to the well format used.

1. If using adherent cells, seed cells on the day prior to transfection in 250 µl of medium without antibiotics such that the cells are 80–90% confluent at the time of transfection. If using cells in suspension, seed 2×10^5 to 4×10^5 cells/well in 250 µl of medium without antibiotics just before starting the transfection procedure.

   The use of Lipofectamine 2000 in media containing antibiotics causes cytotoxicity.

2. Prepare each transfection mix for six wells as described below. Perform replicates of each transfection for statistical analysis of the final results.

   a. Dilute the RheoSwitch plasmids by adding 500 ng of the receptor vector (pNEBR-R1) and 1.5 µg of the inducible expression vector (pNEBR-X1) to 160 µl of Opti-MEM I (or serum-free medium) in a 1.7-ml microcentrifuge tube.

      If desired, a control reporter gene such as *Renilla* luciferase or lacZ driven by a constitutive promoter may be included for normalization of transfection efficiency and the assay between wells. To do so, add 50 ng of the normalization plasmid to the microcentrifuge tube.

   b. Add 6 µl of Lipofectamine 2000 to 160 µl of Opti-MEM I (or serum-free medium) in another 1.7-ml microcentrifuge tube.

   c. After 5 minutes at room temperature, combine the diluted DNA and diluted Lipofectamine 2000, mix gently, and incubate for 20 minutes.

3. Dispense 50 µl of the transfection mix to each well (total six wells).

4. Return the plate to the incubator.

   For optimal transfection efficiency, vary the amount of DNA and the DNA to Lipofectamine 2000 ratio to determine the most suitable amounts for the cell type used.

Induction of Gene Expression by Ligand

Test various ligand concentrations as follows to determine the optimal amount of ligand required for the desired level of gene induction.

5. Dilute the 5 mM stock solution of RSL1 (in DMSO) to different working stock solutions in DMSO. Suggested stock concentrations are 0.8, 4, 20, 100, and 500 µM.

6. Dilute each of the working stock solutions by 1000-fold in the cell culture medium (to create concentrations of 0.8, 4, 20, 100, and 500 nM RSL1) and mix well. Prepare only the required volume of the medium containing the ligand.

7. At 4–6 hours after transfection, remove medium from the transfected wells and add 250 µl of the medium containing different concentrations of the ligand.

8. Return the plate to the incubator.

   The presence of the transfection mix does not affect the ability of the ligand to induce gene expression, and, in most cases, it is not necessary to remove medium containing the transfection mix from the cells. This is especially useful when a large number of transfection assays are performed. In such instances, the above protocol can be modified as follows.

9. Prepare a series of media containing twice the desired concentration of ligand (e.g., 0, 1.6 nM, 8 nM, 40 nM, 200 nM, and 1 µM RSL1 ligand).

10. Follow Steps 1 and 2 of the transfection protocol above.

11. While incubating the transfection mix for the formation of DNA-Lipofectamine 2000 complexes, remove the medium from the wells and add 75 µl of fresh medium.
12. Dispense 50 µl of the transfection mix to each well and rock the plate gently (total 125 µl of medium per well).
13. Add 125 µl of the medium prepared in Step 9, containing twice the desired concentration of the ligand.

    The ligand will be diluted to half the original concentration. There is no need to incubate the cells between Steps 12 and 13.

14. Rock the plate gently and return it to the incubator.

Assay for the Induction of Gene Expression

The induction of gene expression can be detected within a few hours after ligand treatment at the mRNA level or with a highly sensitive method such as chemiluminescence or green fluorescent protein (GFP) at the protein level. Expression will be sufficient in 24 hours and maximal between 48 and 72 hours after ligand treatment. It is not necessary to change the medium or to add more ligand (unless the medium is changed).

15. To detect gene induction at the mRNA level, isolate total RNA from the cells and then perform a northern blot, or reverse-transcribe RNA and then perform real-time quantitative polymerase chain reaction (PCR). The plate format used for transfection should be chosen depending on the RNA isolation method used. To isolate RNA, use TRIZOL reagent or RNeasy Kit.
16. Protein expression can be detected and/or quantified in intact cells, cell extracts, or culture medium, depending on whether the protein is intracellular or secreted. Detection methods include intrinsic protein-fluorescence (e.g., GFP and its fusion proteins where the induction can be visualized in intact cells), chemiluminescence, western blotting, ELISA (enzyme-linked immunosorbent assay), and other functional assays. Prepare cell extracts by removing the medium from the wells and adding lysis buffer compatible with the assays to be performed on the lysate. For the luciferase expression vector used in the transfection protocol, assay induction as follows:

    a. Lyse the cells by adding 100 µl of 1X passive lysis buffer to the wells and rocking for 15 minutes on a platform shaker.

    b. Transfer 20 µl of the lysate to a white opaque 96-well plate.

    c. Dispense 50 µl of the luciferase substrate to the wells and then read the plate in the microplate luminometer.

    A sample assay result for firefly luciferase reporter is shown in Figure 4.

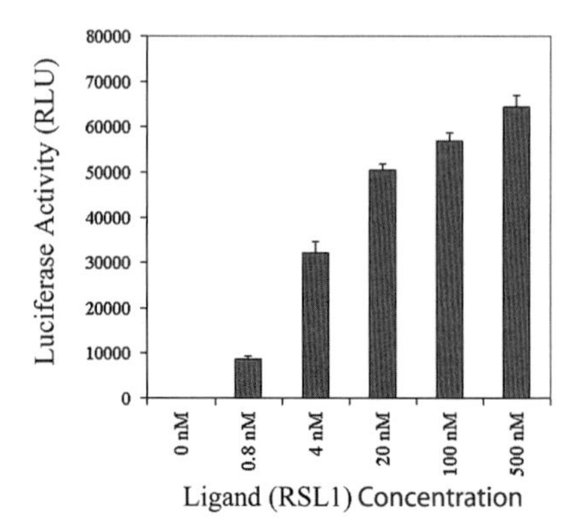

FIGURE 4. Induction of firefly luciferase in NIH-3T3 cells in a 48-well plate by different concentrations of ligand RSL1 after 48 hours. (RLU) Relative light units.

ACKNOWLEDGMENTS

The authors profusely thank Drs. Dean E. Cress, Malla Padidam, Subba Reddy Palli, David W. Potter, and Camille J. Tornetta for their suggestions and editing during the preparation of this chapter.

REFERENCES

Christopherson K.S., Mark M.R., Bajaj V., and Godowski P.J. 1992. Ecdysteroid-dependent regulation of genes in mammalian cells by a *Drosophila* ecdysone receptor and chimeric transactivators. *Proc. Natl. Acad. Sci.* **89:** 6314–6318.

Dhadialla T.S., Carlson G.R., and Le D.P. 1998. New insecticides with ecdysteroidal and juvenile hormone activity. *Annu. Rev. Entomol.* **43:** 545–569.

Karzenowski D., Potter D.W., and Padidam M. 2005. Inducible control of transgene expression with ecdysone receptor: Gene switches with high sensitivity, robust expression, and reduced size. *Biotechniques* **39:** 191–200.

Kumar M.B., Fujimoto T., Potter D.W., Deng Q., and Palli S.R. 2002. A single point mutation in ecdysone receptor leads to increased ligand specificity: Implications for gene switch applications. *Proc. Natl. Acad. Sci.* **99:** 14710–14715.

Kumar M.B., Potter D.W., Hormann R.E., Edwards A., Tice C.M., Smith H.C., Dipietro M.A., Polley M., Lawless M., Wolohan P.R., Kethidi D.R., and Palli S.R. 2004. Highly flexible ligand binding pocket of ecdysone receptor: A single amino acid change leads to discrimination between two groups of nonsteroidal ecdysone agonists. *J. Biol. Chem.* **279:** 27211–27218.

No D., Yao T.P., and Evans R.M. 1996. Ecdysone-inducible gene expression in mammalian cells and transgenic mice. *Proc. Natl. Acad. Sci.* **93:** 3346–3351.

Palli S.R., Hormann R.E., Schlattner U., and Lezzi M. 2005. Ecdysteroid receptors and their applications in agriculture and medicine. *Vitam. Horm.* **73:** 59–99.

Palli S.R., Kapitskaya M.Z., Kumar M.B., and Cress D.E. 2003. Improved ecdysone receptor-based inducible gene regulation system. *Eur. J. Biochem.* **270:** 1308–1315.

Sambrook J. and Russell D.W. 2001. *Molecular cloning: A laboratory manual*, 3rd edition. Cold Spring Harbor Laboratory Press, Cold Spring Harbor, New York.

Suhr S.T., Gil E.B., Senut M.C., and Gage F.H. 1998. High level transactivation by a modified *Bombyx* ecdysone receptor in mammalian cells without exogenous retinoid X receptor. *Proc. Natl. Acad. Sci.* **95:** 7999–8004.

65 Site-specific Integration with Phage ϕC31 Integrase

R. Tyler Hillman and Michele P. Calos

Department of Genetics, Stanford University School of Medicine, Stanford, California 94305-5120

ABSTRACT

Few nonviral techniques exist for efficient and stable eukaryotic gene transfer and fewer still are broadly useful in both cell culture and whole-organism applications. ϕC31 integrase, a site-specific bacteriophage recombinase, is able to catalyze chromosomal transgene insertion under a diverse range of experimental and therapeutic conditions. The enzyme recognizes and catalyzes unidirectional recombination between attachment motifs found in phage and bacterial genomes (*attP* and *attB* sites, respectively). Use of ϕC31 integrase for gene transfer requires that an *attB* sequence be cloned into a transgene-bearing plasmid. When this modified plasmid is introduced into cells alongside integrase-expressing plasmid, ϕC31 integrase is able to catalyze insertion of the transgene plasmid into one of a limited pool of sites in the target genome that exhibit sequence similarity to wild-type *attP*. Its advantages thus include its relative site specifity and use in a wide variety of applications, as well as its relative simplicity. Integration catalyzed by the enzyme is efficient due to its ability to function in the eukaryotic environment and the absence of any appreciable back reaction. Efficient delivery of ϕC31 integrase and *attB* donor plasmid to the tissue or cells of interest remains the most challenging aspect of the system. Unlike viral methods of genome manipulation, use of ϕC31 integrase almost always requires an additional method of stimulating cellular DNA uptake. However, the relative simplicity of the plasmid-based system means that nearly any proven method of introducing exogenous DNA into cells can be used with ϕC31 integrase. Indeed, most classes of transfection technology have at some time been shown to be suitable for this application.

INTRODUCTION

The ability to manipulate eukaryotic genomes has expanded simultaneously in several distinct yet interrelated directions. Most methods available for stable transgene integration are useful in one or more of the following three broadly defined applications: manipulation of cells in culture, construction of transgenic organisms, and genetic therapies. The Cre and FLP recombinases are prototypical of techniques used in the first two applications. Use of these enzymes requires that target sequences be artificially placed in the genome before manipulation, limiting their applicability in the de novo creation of transgenic cells (Sauer and Henderson 1988, 1990; O'Gorman et al. 1991). In addition, efficiency of the integration reaction catalyzed by Cre and FLP is hindered by its bidirectionality. Viral gene delivery vectors, especially those of the lentivirus subfamily, have recently become popular for the stable modification of cells in culture. Lentiviral vectors are able to infect even noncycling cells with high efficiency and integrate their payloads into unmodified chromosomes. However, lentiviral integration is highly promiscuous and tends to favor sites proximal to actively expressing genes (Schroder et al. 2002; Mitchell et al. 2004). Little control can be exercised over lentiviral integration multiplicity and specificity. Similarly, the *Sleeping Beauty* transposon also mediates random integration (Vigdal et al. 2002). Both *Sleeping Beauty* and the viral systems have size limits in transgene capacity as well.

ϕC31 integrase is a member of the serine recombinase protein family that has been developed as a system for genetic manipulation (Groth et al. 2000; Thyagarajan et al. 2001; Groth and Calos 2004). Phage integrases such as ϕC31 recognize bacterial chromosome attachment sequences (*attB* sites) and phage genome attachment sequences (*attP* sites) of 30–40 bp and catalyze a recombination reaction between the two (Fig. 1a). The process of phage integration typically cleaves the *attB* and *attP* sites and joins them to each other, generating two hybrid sequences (*attL* and *attR*) that flank the integrated phage genome. The usefulness of the enzyme for stable gene transfer is accentuated by the relatively specific nature of its site preference in contrast to the

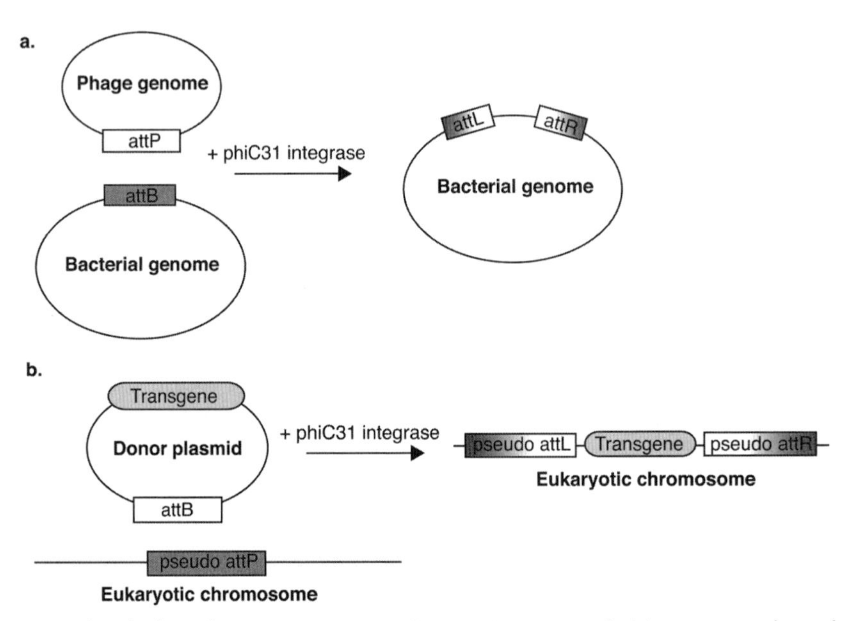

FIGURE 1. Normal and adapted ϕC31 integrase reactions. (*a*) In nature, ϕC31 integrase catalyzes the host integration of the phage genome via recombination between *attP* and *attB* sites. (*b*) We have adapted ϕC31 integrase for the purpose of eukaryotic genomic integration. In the presence of enzyme, *attB*-bearing donor plasmid will undergo chromosomal integration in an efficient and site-specific manner.

aforementioned technologies. ϕC31 integrase recognizes certain endogenous sequences in eukaryotic chromosomes as *attP* sites and can integrate into them an *attB*-bearing transgene plasmid (Fig. 1b) (Thyagarajan et al. 2001). A number of these pseudo *attP* sites have been identified in both human and mouse cells, and published primer pairs exist to screen clones for the most common integration events (Thyagarajan et al. 2001; Olivares et al. 2002; Chalberg et al. 2006). Such pseudo *attP* sites have been found in every mammalian genome examined to date, but they remain sufficiently rare on a genomic scale to give integration pronounced site specificity. The forward reaction catalyzed by ϕC31 integrase proceeds without additional cofactors in mammalian cells, whereas the reverse excision reaction does not (Thorpe and Smith 1998). This property lends the integration reaction a considerable degree of efficiency.

One mode in which ϕC31 integrase has been used for genetic manipulation involves the cloning of a plasmid vector containing both the transgene of interest and an *attB* sequence. In the presence of integrase, such a construct will undergo efficient unidirectional integration at a chromosomal pseudo-*attP* site (Thyagarajan et al. 2001; Thyagarajan and Calos 2005). The ϕC31 integrase system has also achieved considerable success in an analogous application as a tool for gene therapy and the construction of transgenic organisms. However, such uses reside outside the scope of this chapter (Olivares et al. 2002; Allen and Weeks 2005; Chalberg et al. 2005). The enzyme can also catalyze efficient integration, deletion, and cassette exchange at preintegrated *att* sites (Thomason et al. 2001; Thyagarajan et al. 2001; Belteki et al. 2003; Groth et al. 2004).

This chapter describes detailed procedures for using ϕC31 integrase to achieve efficient stable transfection of unmodified mammalian cells in culture. Special emphasis is given to the use of drug selection for the purpose of isolating and characterizing individual transgenic clones. Plasmid delivery methods vary widely by application and tissue type, but the fundamental principles governing use of the ϕC31 integrase system do not. As such, the methods detailed herein can easily be generalized to other situations and cell types.

The ϕC31 integrase system can be used in conjunction with nearly any DNA transfection method. For cell culture, such methods include lipophilic reagents (FuGENE6, Roche; Lipofectamine 2000, Invitrogen), standard millisecond pulse electroporation (ECM 830, BTX Molecular Delivery Systems), and Nucleofector technology (Amaxa). The choice of delivery method depends on cell type and experimental context, but all should be considered relatively interchangeable with regard to ϕC31 integrase usage. Before using the ϕC31 integrase system for stable transfection, an *attB* sequence must be added to the plasmid bearing the transgene of interest. This can be accomplished using standard cloning techniques, restriction enzymes, and buffers.

A PCR primer panel has been developed to screen cells for the presence of specific ϕC31 integrase-mediated integration events. Although more than 100 ϕC31 integration sites have been found in human cell lines, the three sites detected by the primers in Table 1 appear to be preferred sites that may be used by the enzyme in many cell types (Chalberg et al. 2006). These primer pairs and the expected polymerase chain reaction (PCR) product sizes are detailed in Table 1. There is some evidence that the integration site found on chromosome 19 may be the most preferred site and so could be probed first in any screen using these primers. Integration sites are partially palindromic, allowing integration to occur in either orientation. Such events are arbitrarily designated forward and reverse in Table 1, with primer sequences provided to detect either orientation.

TABLE 1. PCR Primers for Detecting ϕC31 Integration Sites in Human Cells

Vector primers	Name	Sequence (5′ > 3′)
First round (*attL*)	*attBF2*	ATGTAGGTCACGGTCTCGAAGC
First round (*attR*)	*ChoSeqR*	TCCCGTGCTCACCGTGACCAC
Second round (*attL*)	*attBF3*	CGAAGCCGCGGTGCG
Second round (*attR*)	*attBR2*	ACTACCGCCACCTCGAC

		First round		Second round		Product size (bp) with indicated vector primer	
Genomic site	Orientation	Name	Sequence (5′ > 3′)	Name	Sequence (5′ > 3′)	*attBF3*	*attBR2*
Xq22.1	F	Xq22-A1	CTGGAGTACAGTGGCGCGATCT	Xq22-A2	CTGCCTCAGCCTCCCGAGTAGC	506	
	R	Xq22-932+	CACCCAGGCCATTAACAACTC	Xq22-738+	CCCCAATTTTTCTGTTTAAGAT	766	
8p22	F	psA-R	AAGTCTTCTGGCTATACAGG	psLL1	TTGCATGGCCTCATTTCCGTC	323	362
	R	psA-F2	TATACCAGACCAGTAGAAAG	psA-F	ATTTGTAGAACTATTATGGG	183	222
19q13.31	F	chr19-B1	CCAAGCGTCATCAGAAGTCCAACGAC	chr19-B2	ATAGTCCCAGCGACAGTGAGCAATTC	331	370
	R	chr19-A1	TTGTGTGATTCTGCTGCCTTACACCA	chr19-A2	GCAGGGTAGCAGATGGAACACTTAGC	409	448

Protocol

Site-specific Integration with Bacteriophage ϕC31 Integrase

This protocol describes the use of ϕC31 integrase in mammalian cell culture for the creation of clonal lines exhibiting robust and stable expression of an experimental transgene.

MATERIALS

CAUTION: See Appendix for appropriate handling of materials marked with <!>.

Reagents

Agarose gel and loading buffer
Cells: Immortalized mammalian cell line or primary cells
DNA ligase and appropriate ligation buffer
DNeasy Tissue Kit (QIAGEN) or equivalent for genomic DNA extraction
Electrocompetent DH10B *Escherichia coli*
Fetal bovine serum (FBS)
FuGENE6 transfection reagent (Roche) or equivalent
G418 <!> or other appropriate drug for selection
Geneticin sulfate (GIBCO) or other appropriate drug for mammalian selection
Media
- Dulbecco's modified Eagle's medium (DMEM) with high glucose
- Luria-Bertani (LB) agar plates containing appropriate selection antibiotic
- LB medium containing appropriate selection antibiotic
- Opti-MEM I medium (Invitrogen)
- Supplemented DMEM: DMEM with high glucose to which 9% FBS and 1% penicillin/streptomycin <!> have been added

Plasmids
- Integrase expression plasmid, pCMVInt (Groth et al. 2000), and appropriate control plasmid, pCMVmInt (Olivares et al. 2002).
  - pCMVmInt is identical to pCMVInt except for a mutation at the catalytic serine that renders the ϕC31 integrase enzyme it produces inactive.
- Plasmid containing transgene of interest in a species- and cell-type-appropriate expression cassette.
  - If drug selection is to be performed, include an appropriate mammalian antibiotic resistance cassette.
- Plasmid containing either the minimal (34 bp) or full-length (~300 bp) *attB* sequence (Groth et al. 2000)
- pEGFP-C1 reporter plasmid

Qiafilter Plasmid Maxi Kit (QIAGEN) or equivalent
QIAquick Gel Extraction Kit (QIAGEN) or equivalent
PCR oligonucleotides, as described in Table 1
Phosphate-buffered saline (PBS)
Restriction enzymes and buffers appropriate for the experiment
Trypsin (0.05%) <!>/EDTA (1 mM) (Invitrogen or equivalent)

Equipment

Centrifuge (tabletop)
Electroporator (Bio-Rad) or other means of bacterial transformation
Flow cytometer
Hemacytometer
Incubator, preset to 37°C
Microcentrifuge tubes
Tissue-culture plates (six well and 100 mm)
Tubes (12 x 75 Falcon)

METHOD

Transfection

1. To make the donor plasmid that carries the *attB* site and the transgene of interest, clone the ϕC31 integrase *attB* sequence into a vector expressing the transgene. Note that integration efficiency is not significantly dependent on *attB* position or orientation. This plasmid construction step requires appropriate enzymes and reagents, such as LB agar plates containing selection antibiotic and a QIAquick Gel Extraction Kit. Transform the donor vector into electrocompetent DH10B *E. coli* cells and inoculate an appropriate culture volume of LB media. It is possible to obtain DNA of sufficient purity for subsequent steps by using a Qiafilter Plasmid Maxi Kit.
2. Twenty-four hours before transfection, harvest adherent mammalian cells with 0.05% trypsin/1 mM EDTA. Centrifuge them in a tabletop centrifuge at 1200 rpm for 10 minutes and resuspend the pellet in a suitable volume of PBS. Perform a cell count using a hemacytometer. Plate cells in four wells of a six-well plate at a 30% seeding density and incubate at an atmosphere appropriate to the cell type overnight at 37°C.
3. Aliquot 100 µl of Opti-MEM I into each of four microcentrifuge tubes. To each tube, add 3 µl of room-temperature FuGENE6, mix gently, and then incubate for 5 minutes at room temperature.
4. Add DNA to these tubes in the following manner:

   Tube 1: no DNA
   Tube 2: 1 µg of pEGFP-C1 reporter plasmid
   Tube 3: 900 ng of pCMVInt + 100 ng of transgene plasmid
   Tube 4: 900 ng of pCMVmInt + 100 ng of transgene plasmid

   Gently mix tube contents and incubate for 25 minutes at room temperature.
5. Aspirate medium from cells and replace with 2 ml of supplemented DMEM per well. Then add 100 µl of each respective lipid–DNA complex in Opti-MEM I dropwise to the appropriate well. Incubate overnight at 37°C.
6. Twenty-four hours following transfection, assay transfection efficiency using flow cytometry as follows:

   a. Harvest cells from "no DNA" and "1 µg of pEGFP-C1" wells with 1 ml of 0.05% trypsin/1 mM EDTA.

   b. Collect cells by centrifuging in a tabletop centrifuge at 1200 rpm for 10 minutes.

c. Resuspend pellet in 200 μl of PBS and transfer to 12 x 75 Falcon tubes for flow cytometric analysis of GFP (green fluorescent protein) expression.

In general, a transfection efficiency greater than 15% is sufficient. For stably transfected pools, proceed to the alternative protocol that follows.

7. Trypsinize and replate each of the two remaining wells on three 100-mm culture dishes in supplemented DMEM. For each well, seed one 100-mm plate with 1/3 of the original well volume, another with 1/10, and the last with 1/30.

   This plating gradient lessens the impact of transfection efficiency on the ability to discern distinct drug-resistant colonies following a period of selection.

8. Twenty-four hours after replating, aspirate medium and replace with supplemented DMEM to which 1.25 mg/ml G418 (w/v), or other appropriate drug concentration, has been added.

9. Replace selection medium every 4 days, taking care not to dislodge nascent colonies.

   Following 14 days of drug selection, distinct colonies should be readily identifiable.

ALTERNATIVE PROTOCOL

Stably Transfected Pools

Following Step 6 above, proceed with this protocol.

1. Trypsinize cells from the two wells remaining after flow cytometry. Collect cells by centrifuging at 1200 rpm in a tabletop centrifuge for 10 minutes. Resuspend the cell pellet in supplemented DMEM.
2. Plate the entire volume of each well into separate 100-mm cell culture plates. Allow cells to grow to confluency and passage as usual.
3. After several passages, measure transgene expression using an appropriate assay. The cells initially transfected with ϕC31 integrase and transgene *attB* plasmid should be at least several-fold enriched for transgene expression when compared to the control plate (Thyagarajan and Calos 2005).

SUPPLEMENTAL PROTOCOL

Integration Site PCR Screen

1. In preparation for an integration PCR panel screen, prepare genomic DNA from pools of transfected cells or a clonal line using a DNeasy Tissue Kit or equivalent according to the manufacturer's protocol.
2. Perform the integration screen using the primers from Table 1 with a standard nested PCR protocol. In brief, loci are first amplified from the genomic DNA template using the *attBF2* and *ChoSeqR* first-round primers paired in turn with each respective site-specific primer. The amplified regions from the first round are then used as template for a second PCR with either the *attBF3* or *attBR2* primers and the appropriate second-round primer from Table 1.
3. Visualize PCR products from the second round using agarose gel electrophoresis.

   The presence of a band of the size indicated in Table 1 will establish the location of transgene plasmid integration.

ACKNOWLEDGMENTS

R.T.H. is a Howard Hughes Medical Institute Medical Student Research Training Fellow. This work was supported by National Institutes of Health grants DK58187 and HL68112 to M.P.C.

REFERENCES

Allen B.G. and Weeks D.L. 2005. Transgenic *Xenopus laevis* embryos can be generated using ϕC31 integrase. *Nat. Methods* **2:** 975–979.

Belteki G., Gertsenstein M., Ow D.W., and Nagy A. 2003. Site-specific cassette exchange and germline transmission with mouse ES cells expressing ϕC31 integrase. *Nat. Biotechnol.* **21:** 321–324.

Chalberg T.W., Genise H.L., Vollrath D., and Calos M.P. 2005. ϕC31 integrase confers genomic integration and long-term transgene expression in rat retina. *Invest. Ophthalmol. Vis. Sci.* **46:** 2140–2146.

Chalberg T.W., Portlock J.L., Olivares E.C., Thyagarajan B., Kirby P.J., Hillman R.T., Hoelters J., and Calos M.P. 2006. Integration specificity of phage ϕC31 integrase in the human genome. *J. Mol. Biol.* **357:** 28–48.

Groth A.C. and Calos M.P. 2004. Phage integrases: Biology and applications. *J. Mol. Biol.* **335:** 667–678.

Groth A.C., Fish M., Nusse R., and Calos M.P. 2004. Construction of transgenic *Drosophila* by using the site-specific integrase from phage ϕC31. *Genetics* **166:** 1775–1782.

Groth A.C., Olivares E.C., Thyagarajan B., and Calos M.P. 2000. A phage integrase directs efficient site-specific integration in human cells. *Proc. Natl. Acad. Sci.* **7:** 5995–6000.

Mitchell R.S., Beitzel B.F., Schroder A.R., Shinn P., Chen H., Berry C.C., Ecker J.R., and Bushman F.D. 2004. Retroviral DNA integration: ASLV, HIV, and MLV show distinct target site preferences. *PLoS Biol.* **2:** E234.

O'Gorman S., Fox D.T., and Wahl G.M. 1991. Recombinase-mediated gene activation and site-specific integration in mammalian cells. *Science* **251:** 1351–1355.

Olivares E.C., Hollis R.P., Chalberg T.W., Meuse L., Kay M.A., and Calos M.P. 2002. Site-specific genomic integration produces therapeutic Factor IX levels in mice. *Nat. Biotechnol.* **20:** 1124–1128.

Sauer B. and Henderson N. 1988. Site-specific DNA recombination in mammalian cells by the Cre recombinase of bacteriophage P1. *Proc. Natl. Acad. Sci.* **85:** 5166–5170.

———. 1990. Targeted insertion of exogenous DNA into the eukaryotic genome by the Cre recombinase. *New Biol.* **2:** 441–449.

Schroder A.R., Shinn P., Chen H., Berry C., Ecker J.R., and Bushman F. 2002. HIV-1 integration in the human genome favors active genes and local hotspots. *Cell* **110:** 521–529.

Thomason L.C., Calendar R., and Ow D.W. 2001. Gene insertion and replacement in *Schizosaccharomyces pombe* mediated by the *Streptomyces* bacteriophage ϕC31 site-specific recombination system. *Mol. Genet. Genomics* **265:** 1031–1038.

Thorpe H.M. and Smith M.C. 1998. In vitro site-specific integration of bacteriophage DNA catalyzed by a recombinase of the resolvase/invertase family. *Proc. Natl. Acad. Sci.* **95:** 5505–5510.

Thyagarajan B. and Calos M.P. 2005. Site-specific integration for high-level protein production in mammalian cells. *Methods Mol. Biol.* **308:** 99–106.

Thyagarajan B., Olivares E.C., Hollis R.P., Ginsburg D.S., and Calos M.P. 2001. Site-specific genomic integration in mammalian cells mediated by phage ϕC31 integrase. *Mol. Cell. Biol.* **21:** 3926–3934.

Vigdal T.J., Kaufman C.D., Izsvak Z., Voytas D.F., and Ivics Z. 2002. Common physical properties of DNA affecting target site selection of *Sleeping Beauty* and other Tc1/*mariner* transposable elements. *J. Mol. Biol.* **323:** 441–452.

66 Creating Zinc Finger Nucleases to Manipulate the Genome in a Site-specific Manner Using a Modular-assembly Approach

Matthew Porteus

Department of Pediatrics and Biochemistry, University of Texas Southwestern Medical Center, Dallas, Texas 75390

ABSTRACT

Homologous recombination is the most precise way to manipulate the genome. As a tool, it has been used extensively in bacteria, yeast, murine embryonic stem (ES) cells, and a few other specialized cell lines, but it has not been available to investigators in other systems, such as mammalian somatic cell genetics. Recently, work has shown that the creation of a gene-specific DNA double-strand break can stimulate homologous recombination by several thousandfold in mammalian somatic cells. These double-strand breaks can now be created in mammalian genomes by zinc finger nucleases (ZFNs). ZFNs are artificial proteins in which a zinc finger DNA-binding domain is fused to a non-specific nuclease domain. This chapter describes how to identify potential targets for ZFN cutting, make ZFNs to cut this target site, and test whether the newly designed ZFNs are active in a mammalian cell-culture-based system. It also focuses on how to assemble ZFNs to recognize target sequences of the form 5′GNNGNNGNN-3′. Further study is needed to determine whether efficient and reliable assembly of ZFNs is possible to recognize targets not of this form. In the next several years, it is likely that there will be improvements in making the individual finger modules. High-throughput methods that combine both selection and assembly may be developed, broadening the number of sites that can be targeted, optimizing the overall structure of the ZFNs, and increasing the possible experimental and therapeutic uses for ZFNs.

INTRODUCTION

In this chapter, gene targeting is defined as the transfer of genetic information from an introduced fragment of DNA to the genome by homologous recombination. It represents the most precise way to manipulate the genome and can be used to introduce changes both small (such as single-nucleotide changes) and large (such as the introduction of several thousand base pairs of new DNA) into the genetic material of a cell. Gene targeting has been a powerful experimental tool for scientists in the study of bacteria, yeast, murine ES cells (Capecchi 1989), and certain other specialized vertebrate cell lines such as the chicken DT40 cell line (Buerstedde and Takeda 1991). Moreover, because of its precision, gene targeting would be an ideal way to perform gene therapy. However, the broad experimental use of gene targeting in vertebrate cells or its use for therapeutic purposes has been precluded by its low spontaneous rate. In HEK-293 cells, for example, the spontaneous rate of gene targeting is approximately 10^{-6} (or one event per every million cells transfected) (Porteus and Baltimore 2003).

A critical finding in the development of homologous recombination as a tool was the discovery that a DNA double-strand break in the genomic target gene could stimulate targeting at that locus by several thousandfold (Jasin 1996). These experiments were generally performed by introducing the recognition site for the I-SceI (Sce) homing endonuclease (a DNA endonuclease with an 18-bp recognition site) into a reporter gene and then introducing the modified reporter as a transgene into a cell of interest. The rate of gene targeting is then measured by the correction of the mutated reporter gene after the introduction of the I-SceI nuclease and a donor plasmid to serve as a template for homologous recombination. An obvious limitation to this strategy is that no human gene has a natural recognition site for the I-SceI nuclease.

The next critical finding was the discovery that ZFNs, previously known as chimeric nucleases, could stimulate gene targeting in human somatic cells (Porteus and Baltimore 2003). ZFNs are artificial proteins that fuse a zinc finger DNA-binding domain to a nonspecific nuclease domain (Kim et al. 1996). So far, the only nuclease domain examined in a ZFN is the one derived from the FokI type-IIS restriction endonuclease. The ability of ZFNs to stimulate gene targeting was first demonstrated using model ZFNs in which the zinc finger domain was derived from a zinc finger protein with a known binding site. Subsequently, ZFNs were made de novo to recognize novel target sequences from natural genes (Alwin et al. 2005; Porteus 2006). These nucleases could stimulate targeting several thousandfold using a transgenic green fluorescent protein (GFP) reporter assay. More significantly, Urnov et al. (2005) demonstrated that ZFNs could stimulate targeting at the endogenous IL2RG locus in up to 20% of cells. These authors designed the IL2RG ZFNs by assembling zinc finger domains from a proprietary archive and then subsequently optimizing the zinc fingers by using highly specialized knowledge of zinc finger–DNA binding. This archive and specialized knowledge, however, are not generally available to investigators.

Construction of New Zinc Finger Proteins by Modular Assembly

An alternative "modular-assembly" approach is to assemble a zinc finger protein using previously described individual zinc fingers. This approach depends on two crucial features: (1) the modular nature of zinc finger binding to DNA and (2) the published data sets of individual zinc fingers and their cognate DNA-binding sites. The modular nature of zinc finger binding was revealed by the crystal structure of the zinc finger domain (Pavletich and Pabo 1991; Pabo et al.

2001). An individual zinc finger consists of 30 amino acids arranged in a ββα structure that is stabilized by chelating a single zinc ion. The α-helix of the zinc finger lies in the major groove. The binding of DNA is mediated by amino acid residues –1 to 6 of the α-helix with respect to the beginning of the helix. A zinc finger protein consists of a series of individual zinc fingers. Thus, a three-finger protein is one that has three individual zinc finger domains. The modular nature of zinc finger binding has two aspects: (1) Each individual finger seems to bind a nonoverlapping triplet independently of its neighboring finger, and (2) each finger makes contacts with each base of the triplet with a single amino acid. From these two modular aspects, two predictions were made: (1) By altering the amino acid contact residues of an individual finger, one could create a new finger that binds to a different 3-bp sequence, and (2) by shuffling different individual zinc fingers, one could create a new zinc finger protein with a new target site specificity. For a three-finger protein, for example, that would mean creating a protein with a novel 9-bp binding site. The testing, successes, and limitations of the modular model of zinc finger binding are well-described elsewhere (Wolfe et al. 2000; Pabo et al. 2001). A variety of individual fingers have been published that bind to unique triplet-binding sites, including triplets of the form 5′-GNN-3′ (Segal et al. 1999; Dreier et al. 2000; Liu et al. 2002), 5′-ANN-3′, and 5′-CNN-3′ (Dreier et al. 2001, 2005).

Using the modular nature of binding and the published data sets, a number of different artificial transcription factors have been made (Segal et al. 2003). Almost all of them have been designed to recognize target sequences rich in 5′-GNN-3′ target site triplets. The reason for this bias may be because the published individual zinc fingers that recognize GNN triplets are of higher quality than those that bind to non-GNN triplets. Alternatively, fingers that bind GNN triplets may be more modular in their binding and depend less on the context of the binding of the neighboring domains. Careful studies of zinc finger DNA binding have shown that binding is not completely modular and that binding of a given finger depends on its neighbors. Thus, there is "context" dependence. An important caveat to the modular-assembly approach to designing new zinc finger proteins is that this context dependence is ignored.

Identification of a ZFN Target Site

I will first define some terms. To cut DNA, the FokI nuclease domain must dimerize. Dimerization occurs when two ZFNs bind to their cognate binding sites in the correct orientation. Each ZFN binds to its own 9-bp recognition site. A single ZFN-binding site will be called a "ZFN half-site" because a single site is not sufficient for cleavage. A pair of ZFN half-sites oriented in inverse orientation (necessary for nuclease dimerization) is called a "ZFN full-site." The spacer between the two ZFN half-sites should be 5 or 6 bp. There is evidence from the mammalian system and others that longer spacers are less effective. Studies of shorter spacers and systematic comparison of different spacer lengths in mammalian cells have not been done. If the spacer between two half-sites is 6 bp, it is called a "ZFN full-site(6)"and if the spacer is 5 bp, it is called a "ZFN full-site(5)".

The first step to designing a pair of ZFNs is to identify a ZFN full-site (with either a 5-bp or 6-bp spacer) in the target gene of interest. Because the published zinc finger domains that recognize 5′GNN-3′ seem to work best in the modular-assembly, I limited my search to half-site targets of the form 5′GNNGNNGNN-3′. Alwin et al. (2005) were able to make an active ZFN that recognized a target site 5′-AGG GAT AAC-3′. Future work will determine how often ZFNs can be made using the modular-assembly approach to recognize target sequences that do not contain all 5′GNN-3′ triplets (Alwin et al. 2005). I used the program DNAsis to find a sequence either of the form 5′-NNCNNCNNCnnnnnnGNNGNNGNN-3′(ZFN full-site[6]) or of the form 5′-NNCNNCNNCnnnnnGNNGNNGNN-3′ (ZFN full-site[5]). In the GFP-coding region, there are three ZFN full-sites(6) and two ZFN full-sites(5). Other search programs can be used to identify possible ZFN full-sites. Porteus assembled ZFNs to target the most 5′ ZFN full-site(6) (Porteus 2006).

After identification of a potential ZFN full-site, the next step is to design the zinc finger proteins to recognize each ZFN half-site. There are two major published data sets for zinc finger domains that recognize GNN triplets (Segal et al. 1999; Liu et al. 2002). I have used the data set from Liu et al. (2002), but have not formally compared it with the Segal et al. (1999) data set to determine which may be better. Porteus (2006) and Alwin et al. (2005) both attempted to make ZFNs to recognize two ZFN half-sites in the GFP gene. Porteus was successful in making active ZFNs to both half-sites using the data set of Liu et al. (2002), whereas Alwin et al. (2005) were only successful at one of the two using the data set of Segal et al. (1999). Further experiments will determine which is the better data set to use. Using the Liu et al. (2002) data set, we have been successful, albeit with varying degrees of efficiency, in ten out of ten attempts at targeting ZFN half-sites of the form 5′-GNNGNNGNN-3′ as measured by the GFP gene-targeting reporter system described below.

One of the ZFN full-site(6) targets in the GFP gene is the following: 5′-ACCATCTTC ttcaag GACGACGGC-3′. Therefore, one ZFN must be designed to recognize the ZFN half-site(1) 5′-GAAGATGGT-3′ and the other to recognize the ZFN half-site(2) 5′-GACGACGGC-3′. Zinc fingers bind to DNA in an "antiparallel" fashion such that finger 1 (the most amino-terminal finger) binds to the most 3′ triplet. Two recent reviews show a schematic representation of this antiparallel binding (Durai et al. 2005; Porteus and Carroll 2005). Thus, finger 1 of GFP1-ZFN must bind to 5′-GGT-3′, finger 2 to 5′-GAT-3′, and finger 3 to 5′-GAA-3′. The amino acid composition of the recognition α-helix is then deduced by using Figure 2 of Liu et al. (2002). For GFP1-ZFN, this results in the α-helix of finger 1 being QSSHLTR, finger 2 being TSGNLVR, and finger 3 being QSGNLAR (using the standard one-letter amino acid abbreviations). Table 1 of Porteus (2006) describes the amino acid content of the α-helix with the desired DNA target binding site for five other ZFNs including GFP2-ZFN. Table 2 of Liu et al. (2002) shows that the order of the fingers is reversed from the order they occur in the designed ZFN. For example, since finger 1 of GFP1-ZFN is designed to recognize the triplet 5′GGT-3′, which is the 3′-most triplet of the overall binding site, the amino acids from "Finger 3" in Table 2 of Liu et al. (2002) are used because that is the 3′-most triplet in their table.

Using PCR to Assemble a New Three-finger Protein

An overlapping polymerase chain reaction (PCR) strategy is used to assemble the new zinc finger proteins (schematized in Fig. 1 and previously used by others [Segal et al. 2003]). Each finger is amplified independently using a general primer at the 5′ end and a finger-specific primer at the 3′ end. The Zif268 backbone is used as a template for each PCR as there is enough heterogeneity in the backbone to allow assembly of the three fingers in the correct order in the final step. In some artificial ZFN constructs, such as QQR-ZFN, the nucleotide sequence surrounding the recognition helix of each finger is identical, which makes it very difficult to assemble the fingers in the correct orientation by overlap PCR in the next step. Each finger fragment is amplified so that it has a 15-bp overlap with its neighboring finger. The individual fingers are then assembled using an overlap PCR strategy (see Fig. 1). The PCR product is digested with BamHI and SpeI and cloned into pBluescript that has been digested with BamHI/SpeI. The three-finger cassette is then sequenced to verify that no errors were created in the PCR process. We used a high-fidelity polymerase initially, but subsequently found that standard *Taq* was accurate enough and did not create undesired point mutations. The sequence of oligonucleotides used in this process is shown in Table 1.

Testing a New Three-finger ZFN

There are a number of ways to test the activity of a newly designed ZFN. Protein can be made via in vitro transcription/translation or from some other expression system, and the ability to cut a DNA template in vitro can be tested. In the original design of ZFN, Chandrasegaran and colleagues

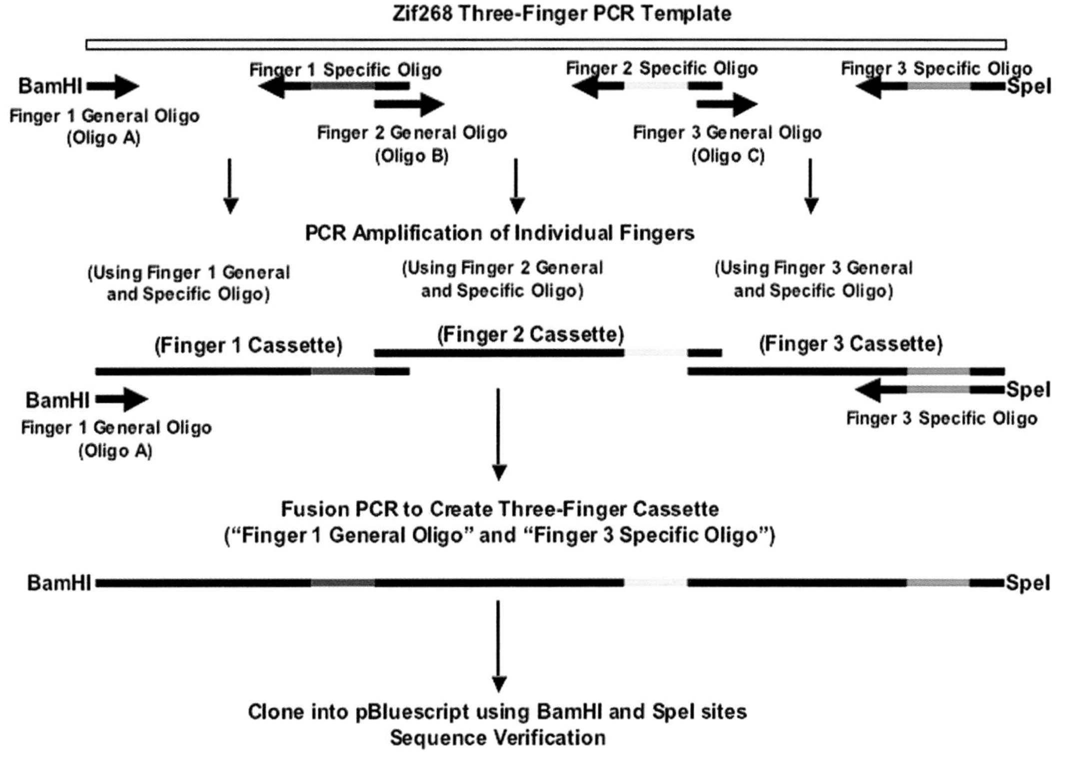

FIGURE 1. Schematic overview of assembly of new zinc finger protein. The experimental details of the procedure are outlined in the text. The colors represent the unique amino acids of the zinc finger α-helix that mediate specific DNA binding for each finger.

TABLE 1. Oligonucleotides to Assemble New Three-finger Protein

Finger 1 General Oligo (Oligo A): 5′-CAGTGGCGGCCGCTCTAGAAC-3′

Finger 2 General Oligo (Oligo B): 5′-CATATCCGCATCCATACC-3′

Finger 3 General Oligo (Oligo C): 5′-CACATCCGCACCCACACA-3′

Finger 1 Specific Oligo: 5′ GTA TGG ATG CGG ATA TG **antisense for amino acid codons –1 to 6** AGA AAA GCG GCG ATC GC 3′

e.g., GFP1-ZFN Finger 1 (target triplet GGT): 5′ GTA TGG ATG CGG ATA TG **CCT CGT GAG GTG AGA AGA CTG** AGA AAA GCG GCG ATC GC 3′

Finger 2 Specific Oligo: 5′ GTG TGG GTG CGG ATG TG **antisense for amino acid codons –1 to 6** ACT GAA GTT ACG CAT GC 3′

e.g., GFP1-ZFN Finger 2 (target triplet GAT): 5′ GTG TGG GTG CGG ATG TG **GCG GAC AAG GTT GCC ACT AGT** ACT GAA GTT ACG CAT GC 3′

Finger 3 Specific Oligo: 5′ TTG ACT AGT TG GTC CTT CTG TCT TAA ATG GAT TTT GGT ATG **antisense for amino acid codons –1 to 6** GGC AAA CTT CCT CCC 3′

e.g., GFP1-ZFN Finger 3 (target triplet GAA): 5′ TTG ACT AGT TG GTC CTT CTG TCT TAA ATG GAT TTT GGT ATG **GCG GGC AAG GTT ACC CGA CTG** GGC AAA CTT CCT CCC 3′

Pick codons such that there are no BamHI (GGATCC) or SpeI (ACTAGT) sites in oligonucleotides.

This table lists the sequences of the oligonucleotides used in the PCR assembly approach described in the text and shown in Figure 1. The finger-specific oligonucleotides are antisense in orientation. To make these oligonucleotides, one must first deduce the correct amino acids for the recognition helix using the appropriate data set. These are then reverse-translated into the codons for each amino acid, avoiding codons that are poorly used in mammalian cells. Finally, the reverse complement of that 21-nucleotide sequence is made and substituted in at the location labeled "antisense for amino acid codons –1 to 6."

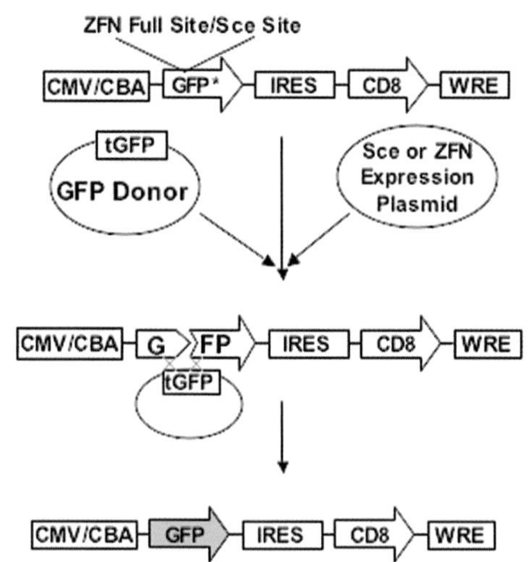

FIGURE 2. Schematic of GFP gene-targeting reporter system. To measure gene targeting, a cell line is created with a single copy of the GFP reporter gene. The GFP reporter gene contains recognition sites for the I-SceI endonuclease, the GFP-ZFNs, and a full-target site created from the Zif268 half-site and the target site for the newly assembled ZFN. Nuclease expression constructs are then transfected into the reporter line along with a repair donor plasmid using standard techniques. Three days after transfection, the cells are analyzed by flow cytometry for the number of GFP-positive cells. (IRES) Internal ribosomal entry site; (WRE) woodchuck posttranscriptional regulatory element; (CMV/CBA) hybrid cytomegalovirus-chicken β-actin enhancer promoter.

evaluated their proteins by in vitro cutting using bacterially expressed and purified proteins (Kim et al. 1996; Smith et al. 1999, 2000). Alwin et al. (2005) described a cell-based transcription assay to test the new zinc finger portion of the ZFN for binding to its cognate binding site. In our lab, we use a GFP-based reporter system. This system has been described elsewhere and is schematized in Figure 2 (Porteus and Baltimore 2003; Porteus and Carroll 2005).

Protocol 1

Creating ZFNs Using a Modular-assembly Approach

This protocol first describes how to use PCR to assemble a new three-finger protein and then how to create a new ZFN.

MATERIALS

CAUTION: See Appendix for appropriate handling of materials marked with <!>.

Reagents

Agarose gel (2.5%)
Calf intestinal alkaline phosphatase
dNTP (10 mM each)
Escherichia coli for transformation
$MgCl_2$ (50 mM) <!>
Plasmids: pBluescript (pBS-ZF) and pcDNA6 (Invitrogen)
PCR Fragment Purification Kit
10x Polymerase buffer
Restriction endonucleases: ApaI, BamHI, and SpeI
T7 primer
Taq polymerase
Zif-ZFN DNA Template

Equipment

Heating block or water bath at 37°C
Thermal cycler programmed with desired amplification protocol

METHODS

PCR Protocol for Amplification of Individual Fingers

1. Mix the following reagents:

Zif-ZFN DNA template	100 ng
10x polymerase buffer	5 µl
$MgCl_2$ (50 mM)	2 µl
dNTP (10 mM each)	1 µl
finger 1 (or 2 or 3) specific oligo	30 pmoles
finger 1 (or 2 or 3) general oligo	30 pmoles
Taq polymerase	2 units
H_2O to final volume of 50 µl	

2. Amplify using the parameters below:

Cycle number	Denaturation	Annealing	Polymerization
First cycle	5 min at 94°C		
15 cycles	30 sec at 94°C	45 sec at 50°C	60 sec at 72°C
Last cycle			5 min at 72°C

Hold at 4°C.

3. Purify individual finger fragments on a 2.5% agarose gel into 30 μl of buffer.

   size of finger 1 fragment = 161 bp

   size of finger 2 fragment = 100 bp

   size of finger 3 fragment = 124 bp

PCR Assembly of Three-finger Protein

4. Mix the following reagents:

finger 1 fragment	10 μl
finger 2 fragment	10 μl
finger 3 fragment	10 μl
10x polymerase buffer	5 μl
$MgCl_2$ (50 mM)	2 μl
dNTP (10 mM each)	1 μl
finger 1 general oligo	30 pmoles
finger 3 specific oligo	30 pmoles
Taq polymerase	2 units
H_2O to final volume of 50 μl	

   Cycling parameters: See above PCR Protocol for Amplification of Individual Fingers.

5. Purify the PCR product using a PCR Fragment Purification Kit.
6. Digest the PCR product with BamHI and SpeI.
7. Purify the digested fragment on a 2.5% agarose gel. Expected size of fragment = ~312 bp.
8. Clone the digested/purified PCR fragment into BamHI/SpeI-digested pBluescript (pBS-ZF).
9. Sequence to confirm the new three-finger protein.

Creating a ZFN from the New Three-finger Protein

10. Digest pcDNA6 with BamHI/ApaI (2 hours at room temperature followed by 2 hours at 37°C). Treat with calf intestinal alkaline phosphatase and isolate the 5.1-kb fragment by gel purification. This is the vector fragment.
11. Digest GFP1-ZFN with SpeI/ApaI (2 hours at room temperature followed by 2 hours at 37°C) and isolate the approximately 591-bp fragment by gel purification. This is the FokI nuclease domain (Fn) fragment.
12. Digest the pBS-ZF clone with BamHI/SpeI and isolate the approximately 312-bp fragment by gel purification. This is the three-finger protein fragment.

13. Ligate the three fragments together using a molar ratio of vector:Fn fragment:ZF fragment of 1:4:4 and standard procedures.
14. Transform into *E. coli* using standard procedures.
15. Identify positive clones by a BamHI/ApaI digest.

    Correct clones should have a single 5.1-kb band and an approximately 909-bp band.
16. Sequence with T7 primer to confirm that the zinc finger is intact and that the junction between zinc finger protein and nuclease domain is in-frame.

    This cloning strategy will create a ZFN that has five amino acids between the terminal histidine of the zinc finger and the first residue of the nuclease domain. In the published literature, this has been called an L0 ZFN, but in the future, it is likely to be called an L5 ZFN.

Protocol 2

Testing the New Three-finger ZFN Using a GFP Reporter System

This protocol describes how to test the ZFN using a cell-based GFP reporter assay.

MATERIALS

CAUTION: See Appendix for appropriate handling of materials marked with <!>.

Reagents

E. coli for transformation
FACS buffer
  Phosphate-buffered saline (PBS)
  2% calf serum
  1 mM EDTA
  0.1% sodium azide <!>
G418 <!>
HEK-293 cells
MACS buffer
  PBS
  2% calf serum
  1 mM EDTA
Media
  Full Dulbecco's modified Eagle's medium (DMEM): DMEM/10% bovine growth serum (Hyclone)/2 mM L-glutamine
  Serum-free DMEM
Plasmids
  GFP expression plasmid
  GFP repair donor (RS2700 plasmid)
  I-SceI expression plasmid
  Parental reporter plasmid (pPC17)
  Zif-ZFN expression plasmid
Restriction endonucleases: BamHI, EcoRI, HindIII, SpeI, and XhoI
Target-site oligonucleotides (see Step 2)

Equipment

Electroporator (BTX ECM399)
Flow cytometer
Magnetic bead-sorting system (Miltenyi, Auburn, California)
Tissue-culture plates (10 cm and 24 well)

METHODS

Construction of GFP Reporter Plasmid

1. Digest parental reporter plasmid (pPC17) with XhoI and HindIII and isolate the approximately 9-kb plasmid by gel purification.
2. Order target-site oligonucleotides in which one half-site is the binding site for Zif-ZFN (a ZFN with known activity) and the other half-site is the new target site. Design the pair of oligonucleotides so that, when annealed, they create XhoI and HindIII compatible overhangs. Place an EcoRI site 5′ to the ZFN full-site in order to provide a new restriction enzyme site to determine which clones have acquired the oligonucleotide insert. For example, to test HGBZF1-ZFN which has a target site of 5′-GAGGTTGCT-3′, we ordered oligonucleotides with the following sequence:

   A: 5′-AGCT GAATTC CGCCCACGC ggatcc GAGGTTGCT-3′

   B: 5′-TCGA AGCAACCTC ggatcc GCGTGGGCG GAATTC-3′
3. Anneal oligonucleotides A and B using standard procedures and ligate into the purified reporter plasmid vector using molar ratio of vector: annealed oligonucleotide of 1:100.
4. Transform the ligation into *E. coli* using standard procedures.
5. Analyze colonies by an EcoRI digest.

   Clones that contain the oligonucleotide will have fragments of 4400, 2100, and 1700 and a doublet at 400 bp.
6. Sequence clones with appropriate fragment sizes using primer A220A (5′-ACCGGCAAGCT-GCCCGTGCCCTGG-3′) to confirm single-copy insertion of the oligonucleotide into the vector.
7. Midiprep-sequence confirmed clones to prepare DNA of quality and quantity to create a reporter cell line using standard procedures.

Construction of GFP Reporter Cell Line

We create reporter cell lines in HEK-293 cells because they are easy to grow and easy to transfect. The following describes our method for creating 293 GFP gene-targeting cell lines using electroporation.

8. Grow HEK-293 cells (not HEK-293-T cells which are already resistant to G418) to mid-log phase in a 10-cm plate in full DMEM.
9. Trypsinize the cells and wash once in 10 ml of serum-free DMEM. Resuspend at 10^6 cells/ml in serum-free DMEM.
10. Remove 400 µl of cells and mix with 10 µg of super-coiled reporter plasmid.

    Using these ratios, we find that cell lines contain single-copy integrants.
11. Incubate the cells for 5 minutes on ice and then electroporate in a BTX ECM399 electroporator at 150 volts.
12. Allow the cells to recover for 5 minutes on ice and then add them to 10 ml of full DMEM in a 10-cm plate.
13. The following day, add G418 to a final concentration of 500 µg/ml to begin selection.
14. Grow cells in this medium and gently change the medium every 3–4 days to avoid disrupting any colonies that are forming.

15. After 2 weeks of G418 selection, individual colonies are clearly evident. Generate a monoclonal reporter cell line by picking individual colonies and analyzing each colony for surface expression of the human CD8α (CD8) gene. Identify those clones with high, consistent surface CD8 expression and use as a reporter.

    For simple testing of ZFNs, a polyclonal population is adequate. The following steps outline how to generate a polyclonal population.

16. At 2 weeks, trypsinize the plate briefly to remove the colonies but not to cut CD8 on the cell surface.
17. Wash the trypsinized cells in 10 ml of MACS buffer and then resuspend in 100 μl of MACS buffer.
18. Purify $CD8^+$ cells with CD8 microbeads using the Miltenyi magnetic bead-sorting system.

    We have found that about 30–50% of the G418-resistant colonies are $CD8^+$. This $CD8^+$ G418-resistant polyclonal population (called "ZFN Reporter Line") is usually adequate for quantitative testing of the newly designed ZFN.

Testing New ZFN Using a New GFP Reporter Cell Line by Calcium Phosphate Transfection

19. The day prior to transfection, plate the ZFN reporter line at 10^5 cells/well in a 24-well tissue-culture plate in full DMEM. Plate 12 wells.
20. Just prior to transfection, replace the medium with 500 μl of full DMEM.
21. Transfect each of the wells with the following plasmids by the calcium phosphate procedure.

    Well 1: 200 ng of GFP expression plasmid + 200 ng of I-SceI expression plasmid

    Wells 2–4: 200 ng of GFP repair donor (RS2700 plasmid) + 200 ng of I-SceI expression plasmid

    Well 5: 200 ng of GFP expression plasmid + 100 ng of GFP-ZFN1 + 100 ng of GFP-ZFN2

    Wells 6–8: 200 ng of GFP repair donor (RS2700 plasmid) + 100 ng of GFP-ZFN1 + 100 ng of GFP-ZFN2

    Well 9: 200 ng of GFP expression plasmid +100 ng of Zif-ZFN expression plasmid + 100 ng of new ZFN expression plasmid

    Wells 10–12: 200 ng of GFP repair donor (RS2700 plasmid) +100 ng of Zif-ZFN expression plasmid + 100 ng of new ZFN expression plasmid

22. Incubate the calcium phosphate/DNA precipitate on the cells for 8–16 hours.
23. Remove the medium and precipitate and replace with 1 ml of full DMEM.
24. About 66–78 hours after transfection, trypsinize the cells briefly and wash in 1 ml of FACS buffer. Resuspend in 400 μl of FACS buffer.
25. Analyze the cells by flow cytometry for GFP expression.
26. Determine the rate of gene targeting by normalizing the rate of GFP-positive cells in the experimental condition (the average of wells 10–12) to the transfection efficiency (determined from well 9). Compare the activity of the new ZFN to other ZFNs by normalizing the efficiency of targeting using the ZFN to the efficiency of targeting using I-SceI (determined by wells 1–4) or to the efficiency of targeting using the GFP-ZFNs (determined by wells 5–8).

    In this system, the background rate of targeting without a double-strand break is one in a million.

TROUBLESHOOTING

Problem: There are multiple potential ZFN full-sites to choose from.

Solution: With data sets of zinc fingers that recognize all 5′-GNN-3′, 5′-ANN-3′, and 5′-CNN-3′ triplets, it is likely that multiple potential full ZFN sites can be identified within a gene of interest. Although algorithms are being generated to identify which sites might be better than others, those algorithms are in the prototype stage of development, have not been systematically verified and tested, and remain unpublished. Thus, there is currently no good mechanism to pick one site over another, other than to pick sites that are rich in 5′-GNN-3′ triplets.

Problem: How close to the desired target site should the full ZFN site be?

Solution: Work from the Jasin and Nickoloff laboratories has shown that the frequency of targeting decreases with distance from the site of the double-strand break. Targeting adjacent to the site of the double-strand break was 100% but decreased to approximately 20% at 100 bp from the site of the break and to 1–5% at 1 kb from the site of the break (Elliott et al. 1998). We have shown that insertions of 1000 kb can occur up to 500 kb from the site of the ZFN-induced double-strand break (Porteus 2006). Thus, the ZFN cleavage site should be chosen as close to the desired target as possible.

ACKNOWLEDGMENTS

M.H.P. thanks the Burroughs-Wellcome fund and the University of Texas Southwestern Medical Center for their support of the research in his laboratory. The work in M.H.P.'s lab is also supported by National Institutes of Health grant K08 HL70268.

REFERENCES

Alwin S., Gere M.B., Guhl E., Effertz K., Barbas C.F., III, Segal D.J., Weitzman M.D., and Cathomen T. 2005. Custom zinc-finger nucleases for use in human cells. *Mol. Ther.* **12:** 610–617.

Buerstedde J.M. and Takeda S. 1991. Increased ratio of targeted to random integration after transfection of chicken B cell lines. *Cell* **67:** 179–188.

Capecchi M. 1989. Altering the genome by homolgous recombination. *Science* **244:** 1288–1292.

Dreier B., Segal D.J., and Barbas C.F., III. 2000. Insights into the molecular recognition of the 5′-GNN-3′ family of DNA sequences by zinc finger domains. *J. Mol. Biol.* **303:** 489–502.

Dreier B., Beerli R.R., Segal D.J., Flippin J.D., and Barbas C.F., III. 2001. Development of zinc finger domains for recognition of the 5′-ANN-3′ family of DNA sequences and their use in the construction of artificial transcription factors. *J. Biol. Chem.* **276:** 29466–29478.

Dreier B., Fuller R.P., Segal D.J., Lund C.V., Blancafort P., Huber A., Koksch B., and Barbas C.F., III. 2005. Development of zinc finger domains for recognition of the 5′-CNN-3′ family DNA sequences and their use in the construction of artificial transcription factors. *J. Biol. Chem.* **280:** 35588– 35597.

Durai S., Mani M., Kandavelou K., Wu J., Porteus M.H., and Chandrasegaran S. 2005. Zinc finger nucleases: Custom-designed molecular scissors for genome engineering of plant and mammalian cells. *Nucleic Acids Res.* **33:** 5978–5990.

Elliott B., Richardson C., Winderbaum J., Nickoloff J.A., and Jasin M. 1998. Gene conversion tracts from double-strand break repair in mammalian cells. *Mol. Cell. Biol.* **18:** 93–101.

Jasin M. 1996. Genetic manipulation of genomes with rare-cutting endonucleases. *Trends Genet.* **12:** 224–228.

Kim Y.G., Cha J., and Chandrasegaran S. 1996. Hybrid restriction enzymes: zinc finger fusions to *Fok* I cleavage domain. *Proc. Natl. Acad. Sci.* **93:** 1156–1160.

Liu Q., Xia Z., Zhong X., and Case C.C. 2002. Validated zinc finger protein designs for all 16 GNN DNA triplet targets. *J. Biol. Chem.* **277:** 3850–3856.

Pabo C.O., Peisach E., and Grant R.A. 2001. Design and selection of novel Cys_2His_2 zinc finger proteins. *Annu. Rev. Biochem.* **70:** 313–340.

Pavletich N.P. and Pabo C.O. 1991. Zinc finger-DNA recognition: Crystal structure of a Zif268-DNA complex at 2.1 Å. *Science* **252:** 809–817.

Porteus M.H. 2006. Mammalian gene targeting with designed zinc finger nucleases. *Mol Ther.* **13:** 438–446.

Porteus M.H. and Baltimore D. 2003. Chimeric nucleases stimulate gene targeting in human cells. *Science* **300:** 763.

Porteus M.H. and Carroll D. 2005. Gene targeting using zinc finger nucleases. *Nat. Biotechnol.* **23:** 967–973.

Segal D.J., Dreier B., Beerli R.R., and Barbas C.F., III. 1999. Toward controlling gene expression at will: Selection and design of zinc finger domains recognizing each of the 5′-GNN-3′ DNA target sequences. *Proc. Natl. Acad. Sci.* **96:** 2758–2763.

Segal D.J., Beerli R.R., Blancafort P., Dreier B., Effertz K., Huber A., Koksch B., Lund C.V., Magnenat L., Valente D., and

Barbas C.F., III. 2003. Evaluation of a modular strategy for the construction of novel polydactyl zinc finger DNA-binding proteins. *Biochemistry* **42:** 2137–2148.

Smith J., Berg J.M., and Chandrasegaran S. 1999. A detailed study of the substrate specificity of a chimeric restriction enzyme. *Nucleic Acids Res.* **27:** 674–681.

Smith J., Bibikova M., Whitby F.G., Reddy A.R., Chandrasegaran S., and Carroll D. 2000. Requirements for double-strand cleavage by chimeric restriction enzymes with zinc finger DNA-recognition domains. *Nucleic Acids Res.* **28:** 3361–3369.

Urnov F.D., Miller J.C., Lee Y.L., Beausejour C.M., Rock J.M., Augustus S., Jamieson A.C., Porteus M.H., Gregory P.D., and Holmes M. C. 2005. Highly efficient endogenous human gene correction using designed zinc-finger nucleases. *Nature* **435:** 646–651.

Wolfe S.A., Nekludova L., and Pabo C.O. 2000. DNA recognition by Cys_2His_2 zinc finger proteins. *Annu. Rev. Biophys. Biomol. Struct.* **29:** 183–212.

67 Assembly of De Novo Bacterial Artificial Chromosome–based Human Artificial Chromosomes

Joydeep Basu and Huntington F. Willard

Institute for Genome Sciences and Policy, Duke University, Durham, North Carolina 27708

ABSTRACT

The effective transfer of genes requires efficient delivery and long-term maintenance of exogenous DNA such that physiologically relevant and appropriately regulated expression of the therapeutic gene is maintained in the host target cell. The optimal gene delivery platform should possess an unlimited payload capacity and should not integrate into the host genome. Instead, it should be maintained in the host cell as an autonomously replicating element and should be able to demonstrate long-term, regulated gene expression. Although a number of potential approaches are being explored, these requirements may be achieved through the generation of human artificial chromosomes (HACs) formed by the de novo assembly of individual, cloned, functional chromosomal units (Basu and Willard 2005).

Continued

INTRODUCTION

To ensure proper segregation in mitosis, HACs must include a centromere. The centromere is the *cis*-acting chromosomal locus responsible for directing formation of the kinetochore, the protein-DNA complex that mediates attachments to and movements of the chromosome along the spindle apparatus (Cleveland et al. 2003; Amor et al. 2004). In humans, the DNA component of the centromere consists of α-satellite DNA, a repetitive element based on the hierarchical organization of an approximately 171-bp monomeric unit, itself tandemly concatemerized over several thousand kilobases at the centromere of all normal human chromosomes (Rudd and Willard 2004). Definitive evidence for the causal role of α-satellite DNA in centromere activity was provided by the development of methodologies to construct and manipulate synthetic α-satellite arrays shown to be capable of forming centromeres de novo, as first described by Harrington et al. (1997).

Although the term "human artificial chromosome" has been used loosely in the literature to refer to a variety of engineered microchromosome platforms (for review, see Basu and Willard 2005), the term is most rigorously used to describe an entity formed de novo from individual, structurally defined, chromosomal components, as first demonstrated more than 20 years ago for the construction of yeast artificial chromosomes (YACs) in *Saccharomyces cerevisiae* (Murray and Szostak 1983). Analagous to YAC vectors, HACs have clearly defined centromeric, replication origin and, where relevant, telomeric elements, all of which are cloned into an appropriate vector backbone, usually a bacterial artificial chromosome (BAC) vector, although YAC and PAC (P1-based artificial chromosome) vectors have also been used. BAC-based artificial chromosomes may be either linear or circular (Ebersole et al. 2000). They minimally require only a mammalian selectable marker and a cloned α-satellite array, which may be of natural or synthetic origin, to nucleate de novo centromere formation. The construction of these unimolecular BAC-based vectors has been greatly facilitated by the development of recombinogenic techniques for the manipulation of BACs containing multiple large (>100 kb), repetitive arrays as described in Basu et al. (2005a,b) and Kotzamannis et al. (2005).

The critical development facilitating the establishment of HAC technology was the demonstration that cloned α-satellite DNA can form de novo centromeres (for review, see Basu and Willard 2005). Both synthetic and genomic arrays of α-satellite DNA from a number of human chromosomes have now been shown to be competent to drive formation of de novo centromeres, albeit at somewhat variable frequencies (Harrington et al. 1997; Ikeno et al. 1998; Grimes et al. 2002; Basu et al. 2005b; Kaname et al. 2005). Sequences other than α-satellite, however, do not appear to be competent for de novo centromere formation. The molecular or genomic characteristics of α-satellite DNA responsible for this difference are unknown. Recent efforts to optimize HAC formation have focused on improving de novo centromere formation by manipulating the density and distribution of CENP-B box elements within α-satellite DNA. The CENP-B box is the biochemically defined motif "PyTTCGTTGGAAPuCGGGA," which is minimally responsible for mediating binding of the constitutive centromeric protein CENP-B to human α-satellite DNA. Experiments with synthetic α-satellite arrays in which the density and distribution of CENP-B box elements have been systematically varied suggest strongly a causal link between the density of CENP-B boxes and de novo centromere seeding efficiency (Ohzeki et al. 2002; Basu et al. 2005b). Furthermore, it has been shown that synthetic arrays containing additional CENP-B boxes above and beyond that found in nature may form de novo centromeres even more efficiently (Basu et al. 2005b).

From the biotechnological perspective, the ultimate aim of constructing HACs is to deliver a genomic copy of a human gene of therapeutic interest and to express it in relevant recipient cells in a properly regulated manner. Several proof-of-principle studies have now demonstrated

TABLE 1. Strategies for the Construction of De Novo Human Artificial Chromosomes

	Composition	Vector	α-satellite	Genomic DNA	Methodology
1	Bimolecular	BAC	synthetic-17	100–200 kb (defined)	ligation of centromere BAC with genomic BAC (Basu et al. 2005a)
2	Unimolecular	BAC	natural-21	156 kb HPRT	multistep recombinogenic assembly in *E. coli* (Kotzamannis et al. 2005)
3	Unimolecular	BAC	synthetic-17	100–200 kb (defined)	transposition of α-satellite array into genomic BAC target (Basu et al. 2005b)
4	Unimolecular	BAC	natural-17	160 kb HPRT	Cre/lox-mediated assembly in *E. coli* (Mejia and Larin 2000; Mejia et al. 2001)
5	Unimolecular	BAC	synthetic-17	none	assembled by traditional cloning (Grimes et al. 2002; Rudd et al. 2003)
6	Mixed species	PAC	natural-21	140 kb HPRT	PACs mixed and transfected directly into HT1080 (Grimes et al. 2001)
7	Mixed species	BAC	natural-21	180 kb GCH1	BACs mixed and transfected directly into HT1080 (Ikeno et al. 2002)
8	Unimolecular	YAC	natural-21	none	recombinogenic assembly in yeast (Ikeno et al. 1998)
9	Mixed species	BAC	synthetic-17	random-sheared	nonhomologous recombination of individual components transfected directly into HT1080 (Harrington et al. 1997)

expression of a human gene from artificial chromosomes, in particular, expression of HPRT from an approximately 150-kb fragment containing the human *HPRT* gene in a circular BAC construct, together with α-satellite sequence that was either present in the same BAC vector (Mejia et al. 2001) or cotransfected as a separate BAC species (Grimes et al. 2001). The cotransfection approach has also been used to generate artificial chromosomes expressing the *GCH I* locus (Ikeno et al. 2002). In addition, our laboratory has recently reported expression of the β-globin gene from custom-built HACs incorporating the entire 200-kb β-globin genomic locus (Basu et al. 2005a). We have also constructed artificial chromosomes that carry 159- and 208-kb genomic fragments containing the human growth hormone and polycystic kidney disease I loci, respectively (Basu et al. 2005a).

When considering the potential application of current strategies for HAC formation in gene therapy, it is helpful to evaluate the different alternative approaches currently being developed, as well as their caveats and limitations. A side-by-side comparison of the different vector systems for assembling de novo HACs is shown in Table 1. The principal technical difficulties that remain to be overcome before de novo HACs can be considered as practical gene delivery vectors are summarized below.

- *Size and composition of matter:* Thus far, de novo HACs are typically 2–5 Mb in size and are formed by the uncontrolled concatemerization of the input DNA species.
- *Cell-line specificity:* Thus far, de novo HACs have been created only in the fibrosarcoma cell line HT1080 and other immortalized cell lines (293, HeLa).
- *Species specificity:* Cloned α-satellite DNA has only been shown to nucleate de novo centromere formation in human cell lines. Assembly of mouse artificial chromosomes will require construction and evaluation of synthetic mouse minor satellite arrays in a de novo centromere formation assay.

Protocol 1

Transfection of α-satellite BACs into HT1080 Cells

Creation of a de novo HAC minimally requires a cloned α-satellite array of either natural or synthetic origin in a BAC or YAC vector that contains a mammalian selectable marker (conferring resistance to, e.g., neomycin, puromycin, or blasticidin). This protocol describes methods for transfecting mammalian cells with BAC vectors containing large (>50 kb) α-satellite arrays.

MATERIALS

CAUTION: See Appendix for appropriate handling of materials marked with <!>.

Reagents

BAC DNA

Prepare endotoxin-free BAC DNA using a QIAGEN-tip 500 column. BAC DNAs are typically resuspended to concentrations of 0.5–1.0 µg/µl in endotoxin-free H_2O.

Fetal bovine serum (FBS; Hyclone)

FuGENE-6 (Roche)

Commercially available DNA delivery systems, including transfection reagents and electroporation devices, are typically evaluated based on the efficiency of delivery of small (<5 kb) reporter plasmids. Few, if any, data are available on the effectiveness of these methods for the delivery of high-molecular-weight (>100 kb) DNA. In our experience, platforms optimized for smaller DNA constructs do not necessarily translate well to the delivery of large BACs. We have examined most of the well-known commercially available transfection reagents; in our hands, the FuGENE-6 reagent (Roche) delivers large BACs most efficiently and is also straightforward to use, with minimal toxicity. We have also worked with Lipofectamine 2000 (Invitrogen), but it is more cumbersome, especially when performing multiple transfections.

G418 (50 mg/ml; Mediatech) <!>

HT1080 fibrosarcoma cell line

Culture medium for HT1080 is 10% FBS in αMEM, supplemented with 50 µg/ml of gentamycin (GIBCO).

Minimal essential medium, Alpha (αMEM), supplemented with L-glutamine (e.g., Cellgro or GIBCO)

Puromycin (10 mg/ml; Invivogen) <!>

Equipment

Cloning rings

Standard tissue-culture equipment, including 100-mm dishes and T-25 flasks

METHOD

Transfection of α-satellite BACs into HT1080 Cells

1. The day before transfection, seed 100-mm tissue-culture dishes with enough HT1080 cells to achieve 50–75% confluency on the following day. Seed at least two dishes for each DNA species to be transfected.
2. For each dish of cells, dilute 4–6 μg of DNA in 100 μl of serum-free αMEM in a sterile 1.5-ml microfuge tube. Slowly add 15–20 μl of FuGENE-6 reagent, mixing gently by flicking the side of the tube with a finger. Check the solution carefully for the presence of a white precipitate. If a precipitate is observed, try to disperse it by slow pipetting with a wide-bore tip.

   In our hands, a 3:2 ratio of FuGENE-6:DNA (volume:mass) gives best results for transfection of large BACs, but this ratio should be optimized in each laboratory. Be gentle! Although supercoiled, high-molecular-weight BACs are surprisingly resistant to shearing damage, linearized high-molecular-weight DNA is very susceptible.
3. Incubate the FuGENE/DNA mixture for 30–45 minutes at room temperature, with occasional gentle flicking to mix. Confirm that no precipitates have formed.
4. With a wide-bore pipette tip, gently add the FuGENE/DNA mixture to the HT1080 cells, rocking the plates continuously to ensure even distribution.
5. Incubate the plates overnight at 37°C.
6. The following day, aspirate the medium from the plates and replace it with the appropriate selective medium (e.g., medium containing, for example, either 3 μg/ml puromycin or 600 μg/ml G418 as appropriate).

   G418 is also called neomycin or geneticin. Do not confuse it with the antibiotic gentamycin!
7. After colonies appear, pick at least 25–30 colonies for further analysis. Use cloning rings to transfer individual clones into T-25 flasks for further analysis. During this period, add fresh selective medium every 3–5 days.

   It may take 10–14 days for visible colonies to appear. When transferring individual clones, seed a second T-25 or T-75 flask to provide material for freeze-down and long-term storage.

Protocol 2

Identification of Candidate De Novo HACs by FISH

Protocols 2 and 3 describe a two-step cytogenetic approach for the identification of HACs in the clonal cell lines isolated from Protocol 1. In the first step (Protocol 2), fluorescence in situ hybridization (FISH) is used with α-satellite probes to identify putative HACs as autonomous elements containing the α-satellite array present in the BAC vector. In the second step (Protocol 3), de novo centromere formation is confirmed by using immuno-FISH to show BAC-vector-specific hybridization and the presence of functional centromeres as evidenced by the assembly of key kinetochore-specific antigens such as the centromere-specific histone H3 variant CENP-A (Amor et al. 2004).

For the sake of brevity, the protocol assumes that a synthetic or naturally derived chromosome-17-based α-satellite array is being used to provide centromere function. It is well established that chromosome-17-specific α-satellite DNA can nucleate de novo centromere formation effectively without the detectable incorporation of host sequence elements (Harrington et al. 1997; Rudd et al. 2003; Basu et al. 2005a,b). Centromere function can also be provided by α-satellite arrays originating from other chromosomes (Kaname et al. 2005). In these cases, a FISH probe complementary to the selected α-satellite array should be substituted in place of the Vysis CEP-17 (or equivalent noncommercial) probe discussed below.

MATERIALS

CAUTION: See Appendix for appropriate handling of materials marked with <!>.

Reagents

Acetic acid <!>
CEP-17 α-satellite probe (Vysis) or equivalent noncommercial probe
Colcemid solution (10 μg/ml; KaryoMAX, GIBCO) <!>
Ethanol (70%, 80%, and 100%; store at –20°C) <!>
Formamide (Chemicon International) <!>
H_2O, Milli-Q-treated (ultrapure)
HT1080 fibrosarcoma cells, transfected with BAC DNA and grown as individual clones in T-25 flasks (see Protocol 1)
Hybrisol VII (Qbiogene)
KCl (0.075 M) <!>

Dissolve 5.59 g of KCl in 1 liter of double-distilled H_2O (ddH_2O) to form a hypotonic solution. Filter-sterilize through a 0.22-μm filter. Warm to 37°C before use.

Methanol (100%) <!>
Methanol/acetic acid solution (3:1 v/v)
Mounting medium (antifade) with DAPI <!> (Vectashield, Vector Laboratories)
Phosphate-buffered saline (PBS)
20x SSC

Dissolve 175.3 g of NaCl and 88.2 g of sodium citrate <!> in 800 ml of H_2O. Adjust pH to 7.0 with 14 N HCl <!>. Adjust volume to 1 liter with H_2O. Dispense into aliquots and sterilize by autoclaving. Dilute before use.

Trypsin (0.05% solution) <!>

Equipment

Benchtop microscope equipped for phase-contrast optics
Clinical centrifuge
Coplin jars for cytology
Coverslips (no. 1, various sizes; Fisherfinest, Fisher Scientific)
Fluorescence microscope
Humidifier
Microscope slides (Fisher Scientific 12-550-43)
Pasteur pipettes (9 inches, glass; Fisher 13-678-20D)
Standard tissue-culture equipment, including 15-ml conical tubes and T-25 flasks
Water bath, preset to 37°C

METHODS

Preparation of the Clonal Cell Lines for Chromosome Spreads

1. Begin with individual HT1080 clones that have been seeded in T-25 flasks (see Protocol 1). When the cells are nearly confluent, add 16.5 μl of colcemid solution to each flask. Incubate for 40–60 minutes at 37°C. Monitor the flasks for the presence of rounded, mitotic cells.
2. Warm 50 ml of 0.075 M KCl in a water bath preset to 37°C. Label a 15-ml conical tube for each cell line.
3. Transfer the cell supernatants to the conical tubes. Wash the flasks with PBS and aspirate the excess liquid.
4. Add 1 ml of trypsin to each T-25 flask. Incubate at 37°C until the cells have detached. Return the supernatants to the flasks, pipette vigorously to resuspend the cells, and transfer the cell suspensions to the conical tubes.
5. Centrifuge the tubes in a clinical centrifuge at 1500–2000 rpm for 5–10 minutes at room temperature.
6. Aspirate the liquid from the cell pellets. Add 10 ml of 0.075 M KCl to each cell pellet while shaking. Incubate in the tissue-culture hood for 12 minutes at room temperature. Centrifuge as in Step 5.

   In our hands, 12 minutes is the ideal time for swelling of HT1080. Some optimization may be required.
7. Aspirate the excess liquid from the pellets. To avoid cell clumping, resuspend the pellets by flicking the tubes.
8. Fix the cells by adding dropwise 1 ml of a 3:1 (v/v) solution of methanol:acetic acid. Add an additional 4 ml of methanol:acetic acid, for a total of 5 ml. Invert the tubes.
9. Centrifuge as in Step 5. Store the pellets in the methanol:acetic acid solution at –20°C until ready to prepare the chromosome spreads.

Preparation of Chromosome Spreads for FISH

The preparation of high-quality chromosome spreads for FISH is notoriously susceptible to local environmental conditions and to the experience of the individual investigator. Many aspects of this procedure are more "art" than anything else, and different laboratories may adopt somewhat different approaches. The following protocol should be thought of as an initial guideline; fine-tuning and optimization are required for each local laboratory environment.

Do the chromosome "drops" in a small, enclosed room where the local humidity may be controlled. For example, turn on two or three water baths and a humidifier in the room several hours before performing the dropping procedure.

10. Remove the fixed cells from storage at –20°C (see Step 9). Pellet the cells by centrifugation at 1000–2000 rpm in a clinical centrifuge for 5–10 minutes at room temperature. Aspirate the excess fixative and resuspend the cell pellet by flicking.
11. Make fresh fixative (30 ml of methanol + 10 ml of acetic acid) and place it in the dropping room. Place several new microscope slides in a Coplin jar filled with 100% methanol.
12. Use a glass Pasteur pipette to add enough fresh fixative to the resuspended cell pellet such that the solution is slightly cloudy yet clear enough to see the pipette.
13. Wipe a microscope slide dry and blow/huff vigorously on the slide at least three times to moisten; condensation should be visible on the dropping surface of the slide. Hold the slide at about waist level and the glass pipette with the cells at shoulder level. Make sure that the cells are properly resuspended (absolutely no clumps!). Let 1–2 drops of the cell suspension fall onto the slide.
14. Blow three more times vigorously onto the slide's surface to moisten. Hold the slide over the steam of a humidifier for about 10 seconds. Blow again three times to further moisten the dropping surface. Set the slide upright to dry. Wipe the back of the slide with a lint-free tissue if required.

    Do not hold the slide too close to the humidifier. The objective is to keep the slide moist, but not dripping wet.

15. When the slide is dry, examine the quality of the spreads using a phase-contrast microscope and a 20x objective.

    Chromosomes should be crisp and uniformly dark, well spread, yet still unambiguously together as a group clearly originating from the lysis of a single cell. If chromosomes are not spread well, try adding acetic acid directly into the fix and repeat. Alternatively, try dropping from a greater height, use a 1:1 ratio of methanol:acetic acid, or increase or decrease the local humidity.

16. Add 10 ml of new fixative to any remaining cells and centrifuge in a clinical centrifuge at 1000–2000 rpm for 5–10 minutes. Store the cell pellets in fixative at –20°C until needed.
17. Dry the slide more completely by leaving it on the bench or in a vacuum chamber overnight. Store the slide in a desiccator for 1–14 days before proceeding to hybridization.

Hybridization of Chromosome Spreads with α-satellite Probes

18. Incubate dropped, dried slides in 2x SSC for 30–90 minutes in a water bath preset to 37°C.
19. Carry cold 70% ethanol solution (stored at –20°C) in a Coplin jar to the water bath. Dehydrate the chromosome spreads by performing the following series of ethanol washes at –20°C. Keep the Coplin jars in the bottom of the freezer during the washes.
    a. Soak in 70% ethanol for 2 minutes.
    b. Soak in 80% ethanol for 2 minutes.
    c. Soak in 100% ethanol for 2 minutes.
    d. Remove the slides from the ethanol wash one at a time. Blot the bottom surface of each slide dry with a lint-free tissue.
    e. Allow the slides to dry for about 10 minutes on the bench.

20. Denature the samples by soaking the slides in 70% formamide/2x SSC (pH 7.0) for 2 minutes at 72°C. Repeat the ethanol wash series in Step 19. Carry the cold 70% ethanol in a Coplin jar to the 72°C water bath. Allow the slides to dry on the bench for about 10 minutes to overnight.
21. For each slide to be probed, add 0.5 μl of Vysis CEP-17 probe (or equivalent noncommercial probe) to 20 μl of Hybrisol VII. Denature by incubating for 5 minutes at 72°C and then place on ice.
22. Add the denatured probe directly to each slide and cover with a coverslip. *Do NOT seal with rubber cement or nail polish.* Incubate overnight at 37°C in a humid chamber (e.g., a plastic box with wet paper towels; use two 10-ml pipettes to create a rack for the slides to sit on).
23. The following day, pass the slides through the following series of washes. The coverslips should slide off during the first wash.
    a. Wash slides in 65% formamide/2x SSC (pH 7.0) for 8 minutes at 42°C (high-stringency wash for higher-order α-satellite).
    b. Repeat the wash with fresh 65% formamide/2x SSC (pH 7.0).
    c. Wash slides in 2x SSC for 8 minutes at 37°C.
    d. Rinse the slides in a Coplin jar filled with Milli-Q-treated H_2O. Shake off the excess H_2O and wipe the back of the slides dry with a lint-free tissue.
24. Add 20–40 μl DAPI/antifade mounting medium (Vectashield) to each damp slide. Use a P20 pipette tip to gently lower a no. 1 22 x 50 coverslip onto each slide. Gently blot excess mounting medium by pressing the surface of the coverslip with a lint-free tissue. Seal the edges of the coverslip with nail polish. Keep the slides in a slide folder in the dark for at least 2 hours at room temperature before examining under a fluorescence microscope. Save slides at 4°C for long-term storage.

Protocol 3

Confirmation of De Novo Centromere Formation on HACs

After clonal cell lines containing candidate artificial chromosomes have been identified, the following protocol is used to further confirm that the candidate extrachromosomal elements contain functional centromeres and BAC-vector-specific sequences. This protocol is based on a combined FISH/immunocytochemistry strategy that uses a BAC-vector-specific FISH probe and CENP-A-specific antibodies to provide evidence of a functional kinetochore. Examples of de novo HACs identified and characterized by these techniques are shown in Figure 1.

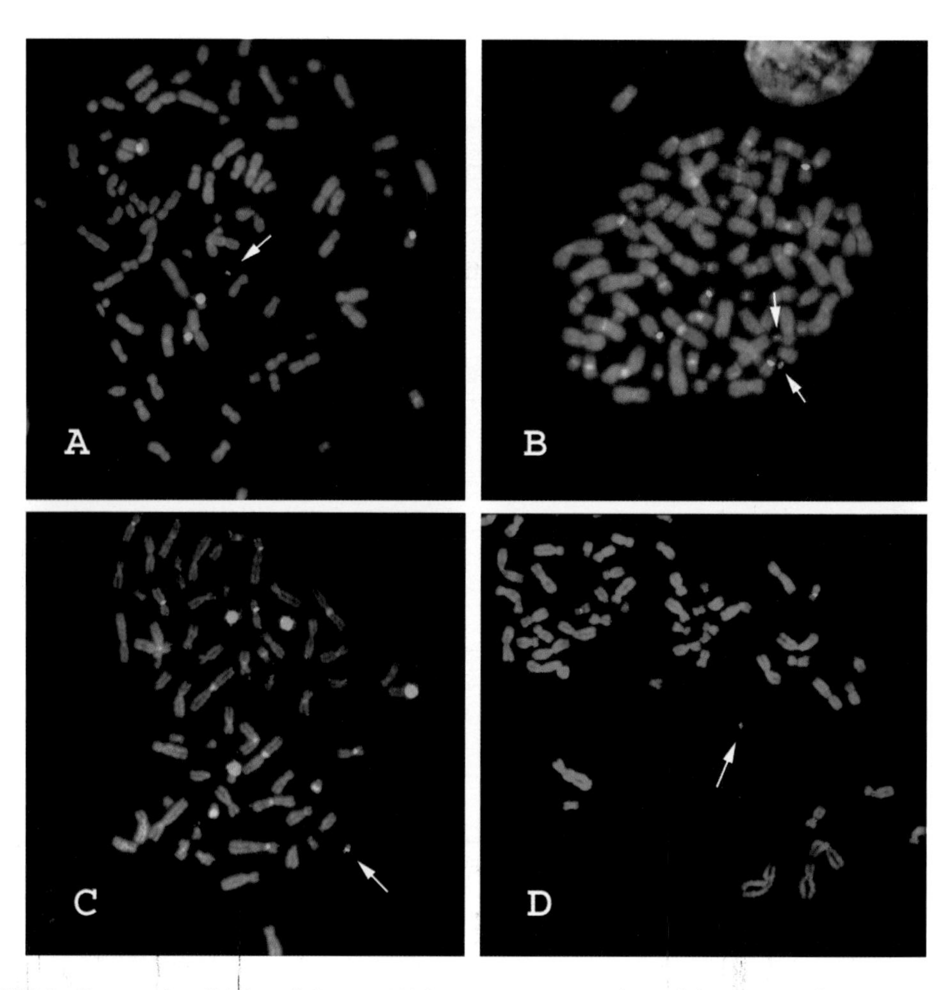

FIGURE 1. Cytogenetic validation of de novo HAC vectors containing the β-globin genomic locus. (*A*) FISH with D17Z1 α-satellite (Vysis CEP-17) (*green*) and BAC-vector (*red*). (*B*) FISH with D17Z1 α-satellite (*green*) and β-globin genomic locus (*red*). (*C*) FISH with D17Z1 α-satellite (*green*) and telomere DNA (*red*). (*D*) Immuno-FISH with D17Z1 α-satellite (*green*) and anti-CENP-C (*red*). In all cases, DNA is in *blue* (DAPI). Arrows point to de novo HACs. Reproduced from Basu et al. (2005b), under terms of Open Access agreement.

MATERIALS

CAUTION: See Appendix for appropriate handling of materials marked with <!>.

Reagents

Colcemid solution (10 µg/ml; KaryoMAX, GIBCO) <!>
Ethanol (70%, 80%, and 100%; store at –20°C) <!>
FISH probe (see Protocol 4)
Formaldehyde (4%)/PBS/0.2% Triton X-100
  48 ml of PBS
  2 ml of 37% formaldehyde <!>
  100 µl of Triton X-100 <!>
Formamide (Chemicon International) <!>
HT1080 fibrosarcoma cells, transfected with BAC DNA and grown as individual clones in T-25 flasks (see Protocol 1)
KCl (0.075 M) <!>

Dissolve 5.59 g of KCl in 1 liter of ddH_2O to form a hypotonic solution. Filter-sterilize through a 0.22-µm filter. Warm to 37°C before use.

Methanol <!>/acetic acid <!> solution (3:1 v/v)
Mounting medium with DAPI <!> (Vectashield, Vector Laboratories)
Phosphate-buffered saline (PBS)
PBS/0.2% Triton X-100
Primary antibody (anti-CENP-A mouse monoclonal antibody; Stressgen Biotechnologies)
Secondary antibody (e.g., anti-mouse IgG/rhodamine or fluorescein; Jackson Immunoresearch)
20x SSC

Dissolve 175.3 g of NaCl and 88.2 g of sodium citrate <!> in 800 ml of H_2O. Adjust pH to 7.0 with 14 N HCl <!>. Adjust volume to 1 liter with H_2O. Dispense into aliquots and sterilize by autoclaving. Dilute before use.

Equipment

Benchtop microscope equipped for phase-contrast optics
Clinical centrifuge
Fluorescence microscope
Shandon Cytospin 4 cytocentrifuge (Thermo Electron)
Shandon Double Cytofunnels (Thermo Electron 5991039)
Shandon Double Cytoslide, coated (Thermo Electron 5991055)
Standard tissue-culture equipment, including 15-ml conical tubes and T-25 flasks

METHOD

1. Grow the selected clonal cell lines in T-25 flasks. Once the flasks are 80–90% confluent, add 25 µl of Colcemid solution to each flask. Incubate for approximately 45 minutes at 37°C. Use a microscope to monitor the flasks for the presence of rounded, mitotic cells.
2. After the Colcemid incubation is complete, transfer most of the medium from each flask to 15-ml conical tubes. Bang the T-25 flasks hard on the benchtop 15–20 times. Use the medium in the conical tubes to rinse off the newly detached mitotic cells. Return the medium (with cells) to the 15-ml conical tubes.

3. Centrifuge the cells in a clinical centrifuge at 2000 rpm for 10 minutes at 4°C. Aspirate the medium and gently flick the tubes to resuspend the pellets in the drops that remain.
4. Add 1 ml dropwise of prewarmed (37°C) 0.075 M KCl to each tube of cells with agitation. Add an additional 4 ml of prewarmed 0.075 M KCl to each tube. Place the remaining 0.075 M KCl on ice and incubate the cells for 12 minutes at 37°C.
5. Label and assemble the Shandon Double Cytofunnels. Use corresponding coated microscope slides designed for the Cytospin, with two spots on each slide for cell deposition.
6. Centrifuge the cells in a clinical centrifuge at 1000 rpm for 5 minutes at room temperature. Decant the 0.075 M KCl. Resuspend the cells in 3 ml of ice-cold 0.075 M KCl. Pipette vigorously with a transfer pipette to ensure complete resuspension.
7. Add 250–300 μl of cell suspension to each Cytofunnel. Centrifuge at 2000 rpm for 10 minutes at room temperature.
8. Remove the slides from the Cytospin and allow them to dry for about 10 seconds. Do not allow them to dry to completion. Incubate the slides in PBS/0.2% Triton X-100 for 6 minutes at room temperature.
9. Check for spreads with a phase-contrast microscope and a 20x objective.

   Chromosomes will be very light, with very little contrast, and therefore somewhat difficult to see. If a few spreads can be seen, however, it is safe to assume that numerous additional spreads are present. If the spreads are too dense, dilute the cells with more iced 0.075 KCl and repeat the Cytospin step.
10. Fix the cells in 4% formaldehyde/PBS/0.2% Triton X-100 for 10 minutes at room temperature.
11. Wash the slides in PBS for 5 minutes and immediately incubate with the primary antibody (anti-CENP-A mouse monoclonal, diluted in PBS). Place a coverslip on each slide (do not seal) and incubate in a humid chamber overnight at 37°C.

    Test a range of antibody dilutions from 1/50 to 1/500.
12. Wash the slides twice for 5 minutes each in PBS.
13. Add 100 μl of a 1:200 dilution of secondary antibody in PBS (e.g., anti-mouse IgG conjugated to FITC/TRITC) to each slide. Cover with an unsealed coverslip and incubate in a humid chamber for 1 hour at 37°C.
14. Wash the slides twice for 5 minutes each in PBS.
15. Fix the samples again by soaking in 4% formaldehyde/PBS/0.2% Triton X-100 for 10 minutes at room temperature (see Step 10).
16. Wash the slides twice for 5 minutes each in H_2O.
17. Fix the samples by soaking in methanol:acetic acid (3:1 v/v) for 15 minutes at room temperature. Allow the slides to dry in the dark for 5 minutes at room temperature.
18. Denature the samples by soaking in 70% formamide/2x SSC for 14 minutes at 85°C.
19. Dehydrate the samples by performing the following series of ethanol washes at –20°C.
    a. Soak in 70% ethanol for 2 minutes.
    b. Soak in 80% ethanol for 2 minutes.
    c. Soak in 100% ethanol for 2 minutes.
    d. Allow the slides to dry in the dark at room temperature.
20. When the 80% ethanol dehydration is started in Step 19, denature the FISH probe by incubating for 5 minutes at 85°C (see Protocol 4). Store the probe at 37°C until use.

21. Apply 7–10 µl of probe to each circle of cells and cover with a 22 x 22-mm-square coverslip (do not seal). Incubate overnight at 37°C in a humidified chamber (e.g., a plastic box with wet paper towels; use two 10-ml pipettes to create a rack for the slides to sit on).
22. Wash the slides twice for 10 minutes each in 50% formamide/2x SSC at 42°C and then wash in 2x SSC for 10 minutes at 42°C.
23. Dry the back of each slide. Add three 8-µl drops of Vectashield per slide, and cover with a no. 1 24 x 50 coverslip. Seal the edges of the coverslip with nail polish, and store the slides at 4°C until ready to examine under a fluorescence microscope.

Protocol 4

Preparation of the FISH Probe

This procedure labels approximately 1 μg of BAC DNA. This is enough for ten FISH experiments (one target area equal to 22 x 22 mm).

MATERIALS

CAUTION: See Appendix for appropriate handling of materials marked with <!>.

Reagents

BAC DNA template

Prepare a 0.2 μg/μl to 1 μg/μl solution of gel-purified BAC vector backbone DNA in Tris-EDTA (10 mM Tris, 1 mM EDTA, pH 8.5) buffer.

COT-1 DNA (Invitrogen)

dNTP mix (0.1 mM)

Mix together 10 μl each of 0.3 mM dATP, 0.3 mM dCTP, and 0.3 mM dGTP.

dTTP (0.1 mM)

Add 10 μl of 0.3 mM dTTP to 20 μl of nuclease-free H_2O.

dUTP (0.2 mM; SpectrumGreen, SpectrumOrange, or SpectrumRed; Vysis)

Add 10 μl of 1 mM dUTP to 40 μl of nuclease-free H_2O.

Ethanol (100%) <!>

H_2O (nuclease-free)

Human placental DNA (Sigma-Aldrich)

Hybrisol VII (Qbiogene)

Nick Translation Kit (Vysis)

Sodium acetate (3 M, pH 5.2)

Equipment

Standard equipment for agarose gel electrophoresis, including size markers

METHODS

Labeling the Probe

1. Cool a microcentrifuge tube by placing it on ice.
2. Add the following components to the tube in the order listed. Briefly centrifuge and vortex the tube before adding the enzyme.

(17.5-x) µl of nuclease-free H_2O
x µl of extracted DNA (1 µg)
2.5 µl of dUTP (0.2 mM SpectrumGreen, SpectrumOrange, or SpectrumRed)
5 µl of dTTP (0.1 mM)
10 µl of dNTP mix
5 µl of 10x nick translation buffer
10 µl of nick translation enzyme

50 µl Total volume

3. Briefly centrifuge and vortex the tube. Incubate for 8–16 hours at 15°C.
4. Stop the reaction by heating in a 70°C water bath for 10 minutes. Chill on ice.

   As the amount of enzyme and incubation time are increased, the size distribution shifts to progressively smaller probe fragments. To produce smaller probe fragments, use the following conditions (listed in order of decreasing fragment size): 5 µl of enzyme mix + 8-hour incubation, 5 µl of enzyme mix + 16-hour incubation, 10 µl of enzyme mix + 8-hour incubation, and 10 µl of enzyme mix + 16-hour incubation. Adjust the amount of nuclease-free H_2O to keep the total reaction volume at 50 µl.
5. Check the size of the probe by electrophoresis of a 10-µl aliquot on a 2% agarose gel. Most of the probe should run with the 300-bp marker.

Precipitating the Probe

6. Pipette 5 µl (~100 ng of probe) of the nick translation reaction mixture into a microfuge tube. Add 1 µg of COT-1 DNA, 2 µg of human placental DNA, and 4 µl of purified H_2O to the tube. Add 1.2 µl (0.1 volume) of 3 M sodium acetate (pH 5.2) and then add 30 µl (2.5 volumes) of 100% ethanol to precipitate the DNA. Vortex briefly and place on dry ice for 15 minutes.
7. Pellet the DNA in a microfuge at 12,000 rpm for 30 minutes at 4°C.
8. Remove the supernatant and dry the pellet for 10–15 minutes under a vacuum at room temperature.

   The precipitated probe can be stored as a pellet indefinitely at –20°C.
9. Resuspend the pellet in 10 µl of Hybrisol VII. Before use, denature the probe by heating it for 5 minutes in a water bath preset to 85°C (see Protocol 3, Step 20).

   The resuspended probe can be stored at 4°C for less than 2 weeks or at –20°C indefinitely.

REFERENCES

Amor D.J., Kalitsis P., Sumer H., and Choo K.H. 2004. Building the centromere: From foundation proteins to 3D organization. *Trends Cell Biol.* **14:** 359–368.

Basu J. and Willard H.F. 2005. Artificial and engineered chromosomes: Non-integrating vectors for gene therapy. *Trends Mol. Med.* **11:** 251–258.

Basu J., Compitello G., Stromberg G., Willard H.F., and Van Bokkelen G. 2005a. Efficient assembly of de novo human artificial chromosomes from large genomic loci. *BMC Biotechnol.* **5:** 21.

Basu J., Stromberg G., Compitello G., Willard H.F., and Van Bokkelen G. 2005b. Rapid creation of BAC-based human artificial chromosome vectors by transposition with synthetic alpha-satellite arrays. *Nucleic Acids Res.* **33:** 587–596.

Cleveland D.W., Mao Y., and Sullivan K.F. 2003. Centromeres and kinetochores: From epigenetics to mitotic checkpoint signaling. *Cell* **112:** 407–421.

Ebersole T.A, Ross A., Clark E., McGill N., Schindelhauer D., Cooke H., and Grimes B. 2000. Mammalian artificial chromosome formation from circular alphoid input DNA does not require telomere repeats. *Hum. Mol. Genet.* **9:** 1623– 1631.

Grimes B.R., Rhoades A.A., and Willard H.F. 2002. α-satellite DNA and vector composition influence rates of human artificial chromosome formation. *Mol. Ther.* **5:** 798–805.

Grimes B.R., Schindelhauer D., McGill N.I., Ross A., Ebersole T.A., and Cooke H.J. 2001. Stable gene expression from a mammalian artificial chromosome. *EMBO Rep.* **2:** 910–914.

Harrington J.J., Van Bokkelen G., Mays R.W., Gustashaw K., and Willard H.F. 1997. Formation of de novo centromeres and construction of first-generation human artificial microchromosomes. *Nat. Genet.* **15:** 345–355.

Ikeno M., Inagaki H., Nagata K., Morita M., Ichinose H., and Okazaki T. 2002. Generation of human artificial chromosomes expressing naturally controlled guanosine triphosphate cyclohydrolase I gene. *Genes Cells* **7:** 1021–1032.

Ikeno M., Grimes B., Okazaki T., Nakano M., Saitoh K., Hoshino H., McGill N.I., Cooke H., and Masumoto H. 1998. Construction of YAC-based mammalian artificial chromosomes. *Nat. Biotechnol.* **16:** 431–439.

Kaname T., McGuigan A., Georghiou A., Yurov Y., Osoegawa K., De Jong P.J., Ioannou P., and Huxley C. 2005. Alphoid DNA from different chromosomes forms de novo minichromosomes with high efficiency. *Chrom. Res.* **13:** 411–422.

Kotzamannis G., Cheung W., Abdulrazzak H., Perez-Luz S., Howe S., Cooke H., and Huxley C. 2005. Construction of human artificial chromosome vectors by recombineering. *Gene* **351:** 29–38.

Mejia J.E. and Larin Z. 2000. The assembly of large BACs by in vivo recombination. *Genomics* **70:** 165–170.

Mejia J.E., Willmott A., Levy E., Earnshaw W.C., and Larin Z. 2001. Functional complementation of a genetic deficiency with human artificial chromosomes. *Am. J. Hum. Genet.* **69:** 315–326.

Murray A.W. and Szostak J.W. 1983. Construction of artificial chromosomes in yeast. *Nature* **305**: 189–193.

Ohzeki J., Nakano M., Okada T., and Masumoto H. 2002. CENP-B box is required for de novo centromere chromatin assembly on human alphoid DNA. *J. Cell Biol.* **159:** 765–775.

Rudd M.K. and Willard H.F. 2004. Analysis of the centromeric regions of the human genome assembly. *Trends Genet.* **20:** 529–533.

Rudd M.K., Mays R.W., Schwartz S., and Willard H.F. 2003. Human artificial chromosomes with alpha-satellite-based de novo centromeres show increased frequency of nondisjunction and anaphase lag. *Mol. Cell. Biol.* **23**: 7689–7697.

68 Delivery of Naked DNA Using Hydrodynamic Injection Techniques

David L. Lewis,* Mark Noble,* Julia Hegge,* and Jon Wolff†

*Mirus Bio Corporation, Madison, Wisconsin 53719; †Department of Pediatrics, Waisman Center, University of Wisconsin, Madison, Wisconsin 53705

ABSTRACT

Among the various viral and nonviral approaches to deliver nucleic acids and express genes in animals, the use of naked plasmid DNA has several advantages. First, plasmid DNA can be manipulated using standard recombinant DNA techniques, circumventing the need for specialized skills or equipment. Second, it is relatively inexpensive to produce, even in large quantities. Third, because it is nonimmunogenic under most circumstances, it can be administered multiple times without induction of an antibody response (Jiao et al. 1992). Contrary to common belief, long-term foreign gene expression from naked plasmid DNA is possible in dividing or slowly dividing cells even without chromosomal integration (Wolff et al. 1992; Miao et al. 2000; Herweijer et al. 2001; Zhang et al. 2004). Although delivery of nucleic acid by nonviral means has been historically inefficient, expression levels from the intravascular delivery of naked DNA are approaching what can be achieved from viral vectors.

INTRODUCTION

Delivery of Naked Nucleic Acids Using the Vasculature

The observation that naked nucleic acids could transfect cells in animals was first made after direct injection of mRNA or plasmid DNA into skeletal muscle (Wolff et al. 1990). It was subsequently found that skeletal muscle was not unique in this regard, as foreign gene expression could also be detected after direct injection of heart, thyroid, skin, and liver tissues (Acsadi et al. 1991; Kitsis and Leinwand 1992; Hickman et al. 1994; Sikes et al. 1994; Hengge et al. 1996; Yang and Huang 1996; Li et al. 1997). In all cases, the number of cells transfected was limited, and expression was mostly localized to the site of injection.

An advance in nucleic acid delivery using intravascular injection occurred when it was discovered that rapid injection of naked plasmid DNA in a relatively large volume into the portal vein of mice caused substantial foreign gene expression in hepatocytes (Budker et al. 1996). Expression levels in the liver were orders of magnitude higher than had been previously observed using direct injection (Hickman et al. 1994). In subsequent studies, it was shown that rapid injection of large volumes into the inferior vena cava or bile duct in mice and rats and the bile duct in dogs also resulted in high levels of reporter gene expression in the liver (Zhang et al. 1997).

Perhaps the most commonly used method of hydrodynamic injection involves injection via the tail vein in mice. When this method is used to deliver plasmid DNA, the highest levels of reporter gene expression are found in the liver, where 10–40% of hepatocytes are transfected. Expression can also be detected in other organs including the kidney, spleen, lung, and heart, although expression in these organs is typically two to three orders of magnitude lower than in the liver (Liu et al. 1999; Zhang et al. 1999). Some transient liver toxicity is associated with this procedure. Tail-vein injection can also be accomplished in rats (Maruyama et al. 2002).

Efficient delivery of plasmid DNA to skeletal muscle has also been accomplished by hydrodynamic injection of plasmid DNA into the iliac artery of rats and rhesus monkeys after clamping blood vessels leading into and out of the limb (Budker et al. 1998; Zhang et al. 2001). More recently, intravenous routes of injection have also proven highly effective for delivery of plasmid DNA and short interfering RNA (siRNA) (Hagstrom et al. 2004; Liang et al. 2004). Injection via the venous route is simpler to perform than injection via the artery because surgery is not required. This also allows multiple injections to be performed over time to further increase the number of cells transfected.

Applications of Hydrodynamic Injection

The administration of naked nucleic acids into animals is increasingly being used as a research tool to elucidate mechanisms of gene expression and the role of genes and their cognate proteins in the pathogenesis of disease in animal models (Herweijer et al. 2001; Hodges and Scheule 2003). For example, hydrodynamic tail-vein injection can be used to deliver plasmids to express potential therapeutic proteins such as growth factors, cytokines, clotting factors, and antibodies in small animal models, thereby allowing assessment of its therapeutic utility and determination of its mode of action. Hydrodynamic tail-vein injection of DNA or RNA sequences has created mouse models for acute hepatitis-B virus infection and human hepatitis δ virus replication (Chang et al. 2001; Yang et al. 2002).

Hydrodynamic delivery of transgenes to muscle also has a wide range of applications. For example, plasmids encoding dystrophin have been delivered in an attempt to alleviate symptoms of muscular dystrophy in animal models (Zhang et al. 2004). Expression of proteins designed to be secreted in the bloodstream has also been accomplished. Although protein secretion is not as efficient from muscle as from other organs, namely, the liver, proteins that elicit biological effects at low serum concentrations such as erythropoietin can be effectively produced in muscle (Hagstrom et al. 2004).

Finally, hydrodynamic delivery of plasmid DNA can be used for genetic immunization to provide a source of polyclonal antibodies, or the splenocytes can be used for the generation of monoclonal-antibody-producing hybridomas (Bates et al. 2006). This chapter describes hydrodynamic injection techniques that allow delivery of plasmid DNA to the liver or skeletal muscle of mice and rats.

Protocol 1

Delivery of Plasmid DNA to Liver in Mice via Hydrodynamic Tail-vein Injection

The tail-vein procedure is best performed using an appropriate restraining device, without anesthetizing mice, as the use of anesthetics sometimes results in morbidity. Modifications for plasmid DNA delivery to the liver of rats are listed in the alternative protocol that follows.

MATERIALS

CAUTION: See Appendix for appropriate handling of materials marked with <!>.

Reagents

Ringer's solution

Prepare Ringer's solution (0.85% NaCl, 0.03% KCl <!>, 0.03% $CaCl_2$ <!>) as a 10x stock in deionized H_2O and filter sterilize. Prepare the working stock at a 1x concentration in deionized sterile H_2O and store at room temperature.

Mice

We use the ICR or C57BL/6 strains of mice for the bulk of our studies. Other strains including BALB/c, ddY, and transgenics have also been used.

Plasmid DNA

For best results, prepare endotoxin-free plasmid DNA. Companies such as QIAGEN manufacture plasmid DNA preparation kits for this purpose. Treat plasmid preparations containing endotoxin with commercial endotoxin removal kits. Avoid Tris-based resuspension buffers in the final preparation, as these buffers are toxic. The choice of enhancer, promoter, and other elements controlling transcription can have an impact on both the amount and duration of transgene expression. For example, plasmids containing a viral enhancer (e.g., the cytomegalovirus [CMV] early enhancer/promoter) will give very high expression initially in liver but are shut down in the liver in less than 24 hours. Plasmids containing elements that allow long-term expression in the liver have been described previously (Miao et al. 2000; Wooddell et al. 2005).

Equipment

Conical tubes (plastic, 50 ml; see Step 1)
Disposable syringes (3 ml)
Heat lamp
Syringe needles (27 gauge)

METHOD

1. Restrain the mouse during the procedure in a 50-ml plastic conical tube that has a 3–5-mm hole cut in the bottom to facilitate the animal's breathing. Place the mouse head first in the tube and thread the tail through a small slit cut in the cap of the tube.

   Alternatively, commercially available tail tubes for mice and rats can be used for restraint.

2. Dilate tail vessels before injection by warming the mouse under a heat lamp for 10 minutes. Place the heat lamp 2–3 feet away and observe the mice closely to ensure that they do not overheat.

   This procedure facilitates tail-vein visualization and ensures optimal injections. It is more difficult to visualize the tail vein in black mice. In these cases, we have found that spraying 70% isopropanol on the tail increases the contrast between the tail vein and skin.

3. Add plasmid DNA (generally 5–100 μg, endotoxin-free) to sterile, room-temperature, 1x Ringer's solution. The total volume of solution injected should be approximately equal to 10% of the animal's body weight (i.e., 2 ml for a 20-g mouse).

   The use of lower volumes results in suboptimal delivery efficiencies. Do not use normal saline solution as an injection solution, because it may lead to postinjection complications. The amount of nucleic acid injected can be scaled up or down while keeping the total injection volume constant.

4. Fit a 3-ml syringe with a 27-gauge, 0.5-inch long syringe needle. Place the syringe needle into the dilated tail vein, preferably midway in the tail or near the distal end.

   It is best to insert nearly the full length of the needle into the vein to prevent accidental release while injecting.

5. Gently inject a small volume of the plasmid DNA solution to ensure that the needle is properly placed in the vein. Once a clear injection pathway is established, dispense the complete volume into the tail vein in 5–7 seconds.

   Maximum delivery is attained by quick injection of the entire contents of the syringe at a constant speed. Immediately after the procedure, mice may display a short period of inactivity and labored breathing, but these effects do not typically last more than 15–20 minutes.

Alternative Protocol 1

Hydrodynamic Tail-vein Injection in Rats

The protocol used for hydrodynamic tail-vein injection in rats is similar to that used for mice, with the following modifications.

ADDITIONAL MATERIALS

CAUTION: See Appendix for appropriate handling of materials marked with <!>.

Reagents

Isoflurane <!>
Rats
Warm water (see Step 4)

Equipment

Butterfly infusion set (21 gauge x 3/4 inch syringe needle)
Disposable syringes (60 ml)
Dissecting microscope
Harvard Apparatus PHD 2000 syringe pump
Surgvet/Anesco Isotec 4 flowmeter

METHOD

1. Dilute plasmid DNA (50–500 μg, endotoxin-free) in a volume of sterile, room-temperature, 1x Ringer's solution equal to 10% of the rat's body weight (i.e., 12 ml for a 120-g rat).

   Use proportionally less volume for rats weighing more than 200 g. Inject no more than 18 ml.

2. Place the injection solution in a 60-ml syringe that is attached to a Butterfly infusion set (21-gauge winged needle).
3. Anesthetize rats with 1–2% isoflurane throughout the procedure using a Surgvet/Anesco Isotec 4 flowmeter set to deliver approximately 0.4 liter/min.
4. Submerse the tail in warm water to dilate the tail vein.

   This makes the vein easier to visualize through the skin and facilitates insertion of the injection needle.

5. With the aid of a dissecting microscope, insert the 21-gauge winged syringe needle end of the Butterfly infusion into the tail vein with the bevel side down. Place the syringe in a programmable Harvard Apparatus PHD 2000 syringe pump.
6. Before delivery, inject a minimal amount of fluid by tapping on the syringe head to ensure that the needle is properly inserted into the tail vein. Then inject the solution at the pump's maximum speed (100 ml/min).

   If a syringe pump is not available, the fluid can be delivered manually, but manual delivery may result is less than optimal delivery efficiency.

Protocol 2

Delivery of Plasmid DNA to Limb Muscle in Mice via Hydrodynamic Saphenous Vein Injection

This procedure is used for delivery of plasmid DNA to skeletal muscle in mice. Modifications for plasmid DNA delivery to rats are listed in the alternative protocol that follows.

MATERIALS

CAUTION: See Appendix for appropriate handling of materials marked with <!>.

Reagents

Isoflurane <!>

Isotonic saline

Prepare isotonic saline (0.9% NaCl) in deionized H_2O. Filter-sterilize and store at room temperature.

Ketoprofen <!> or acetaminophen <!> (see Step 9)

Mice

We use the ICR or C57BL/6 strains of mice for the bulk of our studies. Other strains including BALB/c, ddY, and transgenics have also been used.

Plasmid DNA

Prepare plasmid DNA as described for hydrodynamic tail-vein injection (Protocol 1). Unlike in liver, long-term expression from muscle can be obtained using plasmids containing viral or endogenous enhancer/promoter combinations. Maximal expression in muscle, however, is not attained until 4–7 days after delivery.

Equipment

Disposable syringes (10 ml)

Dissecting microscope (optional)

Forceps (several sizes)

Gel-foam

Harvard Apparatus PHD 2000 syringe pump

Medical tape

Needle driver

Razor

Small retractors

Surgvet/Anesco Isotec 4 flowmeter

Suture (4-0 Vicryl)

Syringe needles (30 gauge)

Polish the needles using a Drummel tool to decrease the outer diameter.

Tourniquet and hemostat

A simple but effective tourniquet for mice and rats can be prepared by cutting the tip off a finger of a latex glove and then cutting 1 cm from the end (see Step 2).

METHOD

1. Anesthetize mice with 1–2% isoflurane to a surgical plane depth using a Surgvet/Anesco Isotec 4 flowmeter set at approximately 0.4 liter/min.
2. Shave the leg of the animal to be injected and use a tourniquet to restrict blood flow to and from the limb. Thread the leg through the tourniquet until the tourniquet is around the proximal aspect of the limb (just proximal to or partially over the quadriceps muscle group). Tighten the tourniquet by twisting it and clamp the twist with a hemostat. Tape the limbs to the surgical table to maintain dorsal recumbency.

   The position of the tourniquet will roughly correspond to the limit of nucleic acid delivery area or target area.
3. Make a small midline incision on the medial side of the leg near the ankle to expose the distal great saphenous vein.
4. Dilute plasmid DNA (50–300 μg, endotoxin-free) in 1.0 ml of sterile, room-temperature, isotonic saline.

   This volume is optimal for a 25-g mouse but does not vary significantly with the size of the mouse.
5. Place the injection solution in a 3-ml syringe and attach PE-10 catheter tubing. Attach a polished syringe needle (30-gauge) to the catheter tubing and insert it into the great saphenous vein using a needle driver.

   A dissecting microscope can be used to visualize needle insertion more easily. During injection, hold the needle in place by using a cotton swab and applying light pressure. It is important, however, to allow the needle to move with the limb during injection so as not to puncture the vein.
6. Deliver the entire volume of the nucleic acid solution at a rate of 8 ml/min using a programmable Harvard Apparatus PHD 2000 syringe pump. Leave the tourniquet in place during the injection and for 2 minutes after injection. Then remove the catheter and tourniquet. Control bleeding from the saphenous vein with gel-foam or gentle pressure.
7. Close the incision with a 4-0 Vicryl suture.
8. Monitor the animals until they have fully recovered from anesthesia. Use heat lamps to prevent loss of body heat and facilitate recovery.
9. Administer 5 mg of ketoprofen per gram of body weight by subcutaneous injection once a day for the first 2 days after the surgical procedure.

   Alternatively, use acetaminophen (100 mg/kg/day, in drinking water) for the first 2 days after the surgical procedure.

Alternative Protocol 2

Intravenous Injection into the Saphenous Vein of Rats

The protocol used for intravenous injection into the saphenous vein of rats is identical to that used for mice, with the following minor modifications.

ADDITIONAL MATERIALS

Equipment

Butterfly infusion set (25 gauge x 3/4 inch syringe needle)
Disposable syringes (10 ml)

METHOD

1. Dilute plasmid DNA (50–750 μg, endotoxin-free) in a volume of sterile, room-temperature, isotonic saline equal to 0.03 times the rat's weight (i.e., 3.6 ml for a 120-g rat).
2. Place the injection solution in a 10-ml syringe and attach it to a Butterfly infusion set (25-gauge winged needle).
3. Insert the syringe needle into the saphenous vein with the bevel side down.
4. Deliver the entire volume of the nucleic acid solution at a rate of 10 ml/min with a programmable Harvard Apparatus PHD 2000.

REFERENCES

Acsadi G., Jiao S.S., Jani A., Duke D., Williams P., Chong W., and Wolff J.A. 1991. Direct gene transfer and expression into rat heart in vivo. *New Biol.* **3:** 71–81.

Bates M., Zhang G., Sebestyen M.G., Neal Z.C., Wolff J.A., and Herwiejer H. 2006. Genetic immunization for antibody generation in research animals by intravenous delivery of plasmid DNA. *BioTechniques* **40:** 199–208.

Budker V., Zhang G., Knechtle S., and Wolff J.A. 1996. Naked DNA delivered intraportally expresses efficiently in hepatocytes. *Gene Ther.* **3:** 593–598.

Budker V., Zhang G., Danko I., Williams P., and Wolff J. 1998. The efficient expression of intravascularly delivered DNA in rat muscle. *Gene Ther.* **5:** 272–276.

Chang J., Sigal L.J., Lerro A., and Taylor J. 2001. Replication of the human hepatitis delta virus genome is initiated in mouse hepatocytes following intravenous injection of naked DNA or RNA sequences. *J. Virol.* **75:** 3469–3473.

Hagstrom J.E., Hegge J., Zhang G., Noble M., Budker V., Lewis D.L., Herweijer H., and Wolff J.A. 2004. A facile nonviral method for delivering genes and siRNAs to skeletal muscle of mammalian limbs. *Mol. Ther.* **10:** 386–398.

Hengge U.R., Walker P.S., and Vogel J.C. 1996. Expression of naked DNA in human, pig, and mouse skin. *J. Clin. Invest.* **97:** 2911–2916.

Herweijer H., Zhang G., Subbotin V.M., Budker V., Williams P., and Wolff J.A. 2001. Time course of gene expression after plasmid DNA gene transfer to the liver. *J. Gene Med.* **3:** 280–291.

Hickman M.A., Malone R.W., Lehmann-Bruinsma K., Sih T.R., Knoell D., Szoka F.C., Walzem R., Carlson D.M., and Powell J.S. 1994. Gene expression following direct injection of DNA into liver. *Hum. Gene Ther.* **5:** 1477–1483.

Hodges B.L. and Scheule R.K. 2003. Hydrodynamic delivery of DNA. *Expert Opin. Biol. Ther.* **3:** 911–918.

Jiao S., Williams P., Berg R.K., Hodgeman B.A., Liu L., Repetto G., and Wolff J.A. 1992. Direct gene transfer into nonhuman primate myofibers in vivo. *Hum. Gene Ther.* **3:** 21–33.

Kitsis R.N. and Leinwand L.A. 1992. Discordance between gene regulation in vitro and in vivo. *Gene Expr.* **2:** 313–318.

Li K., Welikson R.E., Vikstrom K.L., and Leinwand L.A. 1997. Direct gene transfer into the mouse heart. *J. Mol. Cell. Cardiol.* **29:** 1499–1504.

Liang K.W., Nishikawa M., Liu F., Sun B., Ye Q., and Huang L. 2004. Restoration of dystrophin expression in mdx mice by intravascular injection of naked DNA containing full-length dystrophin cDNA. *Gene Ther.* **11:** 901–908.

Liu F., Song Y., and Liu D. 1999. Hydrodynamics-based transfection in animals by systemic administration of plasmid DNA. *Gene Ther.* **6:** 1258–1266.

Maruyama H., Higuchi N., Nishikawa Y., Kameda S., Iino N., Kazama J.J., Takahashi N., Sugawa M., Hanawa H., Tada N., Miyazaki J., and Gejyo F. 2002. High-level expression of naked DNA delivered to rat liver via tail vein injection. *J. Gene Med.* **4:** 333–341.

Miao C.H., Ohashi K., Patijn G.A., Meuse L., Ye X., Thompson A.R., and Kay M.A. 2000. Inclusion of the hepatic locus control region, an intron, and untranslated region increases and stabilizes hepatic factor IX gene expression in vivo but not in vitro. *Mol. Ther.* **1:** 522–532.

Sikes M.L., O'Malley B.W., Jr., Finegold M.J., and Ledley F.D. 1994. In vivo gene transfer into rabbit thyroid follicular cells by direct DNA injection. *Hum. Gene Ther.* **5:** 837–844.

Wolff J.A., Ludtke J.J., Acsadi G., Williams P., and Jani A. 1992. Long-term persistence of plasmid DNA and foreign gene expression in mouse muscle. *Hum. Mol. Genet.* **1:** 363–369.

Wolff J.A., Malone R., Williams P., Chong W., Acsadi G., Jani A., and Felgner P. 1990. Direct gene transfer into mouse muscle in vivo. *Science* **247:** 1465–1468.

Wooddell C.I., Van Hout C.V., Reppen T., Lewis D.L., and Herweijer H. 2005. Long-term RNA interference from optimized siRNA expression constructs in adult mice. *Biochem. Biophys. Res. Commun.* **334:** 117–127.

Yang J.P. and Huang L. 1996. Direct gene transfer to mouse melanoma by intratumor injection of free DNA. *Gene Ther.* **3:** 542–548.

Yang P.L., Althage A., Chung J., and Chisari F.V. 2002. Hydrodynamic injection of viral DNA: A mouse model of acute hepatitis B virus infection. *Proc. Natl. Acad. Sci.* **99:** 13825–13830.

Zhang G., Budker V., and Wolff J.A. 1999. High levels of foreign gene expression in hepatocytes after tail vein injections of naked plasmid DNA. *Hum. Gene Ther.* **10:** 1735–1737.

Zhang G., Budker V., Williams P., Subbotin V., and Wolff J.A. 2001. Efficient expression of naked DNA delivered intraarterially to limb muscles of nonhuman primates. *Hum. Gene Ther.* **12:** 427–438.

Zhang G., Vargo D., Budker V., Armstrong N., Knechtle S., and Wolff J.A. 1997. Expression of naked plasmid DNA injected into the afferent and efferent vessels of rodent and dog livers. *Hum. Gene Ther.* **8:** 1763–1772.

Zhang G., Ludtke J.J., Thioudellet C., Kleinpeter P., Antoniou M., Herweijer H., Braun S., and Wolff J.A. 2004. Intraarterial delivery of naked plasmid DNA expressing full-length mouse dystrophin in the mdx mouse model of duchenne muscular dystrophy. *Hum. Gene Ther.* **15:** 770–782.

69 Nonviral Gene Transfer across the Blood-brain Barrier with Trojan Horse Liposomes

William M. Pardridge

Department of Medicine, University of California, Warren Hall 13-164, Los Angeles, California 90024

ABSTRACT

Nonviral plasmid DNA is delivered to the brain via a transvascular route across the blood-brain barrier (BBB) following an intravenous administration of the DNA encapsulated in the interior of Trojan horse liposomes (THLs), also called PEGylated immunoliposomes (PILs). The surface of the 100-nm liposomes is covered with several thousand strands of polymer, such as 2000-dalton polyethylene glycol (PEG), or PEG 2000. The tips of 1–2% of the PEG strands are conjugated with a targeting monoclonal antibody (MAb) that acts as a molecular Trojan horse (MTH). The MTH binds to a receptor on the BBB and brain cell membrane, such as the transferrin receptor or the insulin receptor, and this binding triggers receptor-mediated transcytosis of the THL across the BBB in vivo, and receptor-mediated endocytosis into brain cells beyond the BBB. The persistence of transgene expression in the brain is inversely related to the rate of degradation of the episomal plasmid DNA. THL technology enables adult transgenics in 24 hours, i.e, an exogenous gene is widely expressed in the majority of cells in brain, or other organs, within 1 day of a single intravenous administration without viruses. Applications of the THLs include (1) tissue-specific gene expression with tissue-specific promoters, (2) complete normalization of striatal tyrosine hydroxylase in experimental Parkinson's disease following intravenous tyrosine hydroxylase gene therapy, (3) 100% increase in survival time of mice with brain tumors following weekly intravenous antisense gene therapy using THLs, and (4) 90% increase in survival time with weekly intravenous RNA interference (RNAi) gene therapy in mice with intracranial brain tumors.

INTRODUCTION

Nonviral forms of gene transfer have generally employed polyplexes of a cationic polymer and anionic DNA. These polyplexes, which form small stable structures in water or media of low ionic strength, form micron-size aggregates in physiological saline (Plank et al. 1999; Simberg et al.

2003). This aggregation underlies the high potency of the structures in cell culture and the limited efficacy in animals following intravenous injection. The microparticles trigger uptake by cells in culture via phagocytosis, which produces expression of the transgene in tissue culture (Matsui et al. 1997; Niidome et al. 1997). However, in vivo, the aggregates are rapidly sequestered within the first vascular bed encountered after an intravenous injection, which is the pulmonary circulation. They embolize in the pulmonary capillaries, which is why gene expression in the lung is log orders of magnitude greater than in peripheral tissues (Hong et al. 1997; Song et al. 1997). The intravenous administration of cationic polyplexes does not result in any gene expression in the brain.

THLs and cationic liposomes have markedly different molecular formulations. Cationic liposomes form sandwich-like structures with anionic DNA, and the DNA is exposed to any surrounding nuclease (Simberg et al. 2001). In contrast, in THLs, a single supercoiled plasmid DNA is encapsulated in the interior of a 100-nm liposome, which contains a small amount of cationic lipid, but has a net anionic lipid charge (Pardridge 2003). This encapsulation renders the plasmid DNA insensitive to the ubiquitous endo- and exonucleases present in vivo (Shi and Pardridge 2000).

Plasmid DNA must be engineered to incorporate the appropriate promoter and 3′-untranslated region (UTR). Some promoters will enable transgene expression in cell culture following transfection with cationic lipids. Cationic polyplexes are injurious to cells with a narrow therapeutic index and may activate transcription pathways so that weak promoters permit transcription of the transgene in cell culture (Kofler et al. 1998). Such a promoter may not drive transgene expression in vivo following delivery with THLs. Moreover, certain 3′ UTRs may contain a sequence that enables translation repression, thus blocking transgene expression in vivo (Luikenhuis et al. 2004).

Liposomes are rapidly cleared by the reticuloendothelial system in vivo owing to rapid absorption of serum proteins at their surface. However, this protein absorption is minimized by conjugating several thousand strands of polymer, e.g., PEG 2000, to the surface of the liposome (Papahadjopoulos et al. 1991). A PEGylated liposome with encapsulated DNA would allow minimal gene transfer across cell membranes in vivo. The PEGylated lipsome is poorly recognized by any cell-membrane-internalizing mechanism, other than phagocytic uptake mechanisms in liver and spleen. A PEGylated liposome is converted to a THL when the tips of 1–2% of the polymeric strands are conjugated with a receptor-specific MTH (Fig. 1A). The MTH binds a cell membrane receptor, triggering internalization. The degree of internalization is a function of the receptor specificity. When an IgG isotype control antibody replaces the receptor-specific targeting MAb on

A

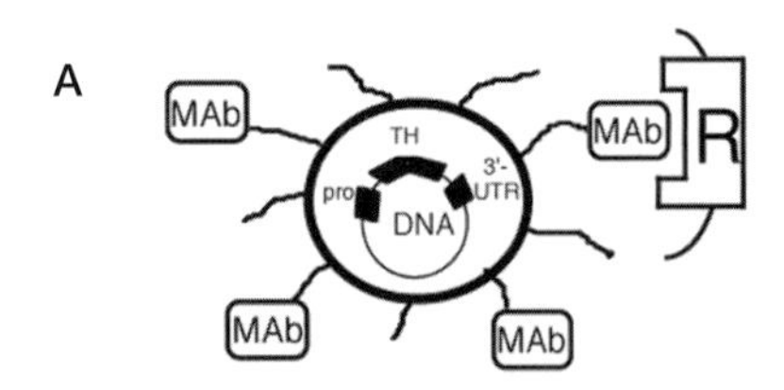

B

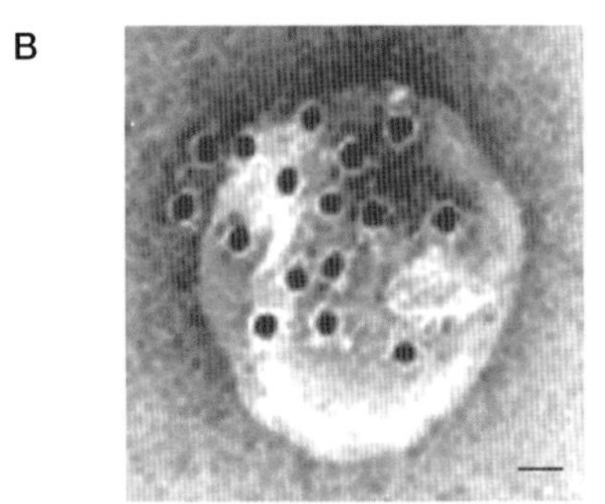

FIGURE 1. (*A*) Diagram of a supercoiled expression plasmid DNA encapsulated in an 85-nm PEGylated immunoliposome targeted to a cell membrane receptor (R) with a receptor-specific, endocytosing monoclonal antibody (MAb). Tissue-specific expression of the plasmid is regulated by the promoter (pro), which is inserted 5′ of the gene. TH is tyrosine hydroxylase. (*B*) Transmission electron microscopy of a PIL. The MAb molecule tethered to the tips of the 2000-dalton polyethylene glycol (PEG) is bound by a conjugate of 10-nm gold and a secondary antibody. The position of the gold particles shows the relationship of the PEG extended MAb and the liposome. Magnification, 20 nm. (Reprinted, with permission, from Zhang et al. 2003b.)

the THL, gene expression in vivo is not observed (Shi et al. 2001). An electron micrograph of a THL is shown in Figure 1B. In this study, the THL was mixed with an anti-mouse antibody conjugated to 10-nm gold. The position of the 10-nm gold particles, which are about the same size as a MAb, shows the relationship of the MAb and the surface of the liposome; the PEG 2000 bridges the liposome surface and the MAb.

The receptor-specific MAbs that have been used to deliver plasmid DNA to adult rats, mice, and Rhesus monkeys are species-specific (Pardridge 2001). Gene delivery to the mouse uses the 8D3 rat MAb to the mouse transferrin receptor, which is not active in rats, whereas delivery to rats uses the OX26 murine MAb to the rat transferrin receptor. These antibodies are not active in primates or humans. Gene delivery to Old World primates, but not to New World primates, uses the 83-14 murine MAb to the human insulin receptor (HIR). This MAb cannot be used in humans, owing to immune reactions, but a genetically engineered form whose activity is identical to that of the original can be used in humans. Since 1–3 mg of purified MAb is used in each THL production run, the hybridoma secreting the MAb must be obtained so that adequate amounts of MAb can be produced with either ascites or large volumes of serum-free conditioned medium. It is likely that MAbs to receptors other than the BBB transferrin or insulin receptor could be used, but these have not been validated. The MAb to the receptor must be an "endocytosing antibody," as not all receptor MAbs enable endocytosis into the cell following surface membrane binding to the receptor.

Tissue-specific Expression of Reporter Genes in Brain In Vivo

The THL is the only formulation, viral or nonviral, that enables expression of a transgene throughout the brain following a single intravenous administration. This is illustrated in Figure 2 for mice, rats, and Rhesus monkeys. A β-galactosidase expression plasmid was encapsulated in THLs targeted with the 8D3 MAb and injected intravenously in adult mice; 48 hours later, the brain was removed and β-galactosidase histochemistry was performed on frozen sections. As shown in Figure 2A, the transgene was expressed in all parts of the brain, including the brain stem (Zhu et al. 2004), and in all parts of the retinal pigmented epithelium, but not in the photoreceptor cells (Fig. 2B). Gene expression was not observed in the photoreceptor layer, because there are few transferrin receptors in the outer nuclear layer of the photoreceptor cell (Zhu et al. 2002). Insulin receptor is expressed in the outer nuclear layer, and when the transgene was targeted with the HIR MAb, gene expression in the photoreceptor cells was observed (Zhang et al. 2003c).

A β-galactosidase expression plasmid was encapsulated in THLs targeted with the 83-14 MAb (Zhang et al. 2003a), and 48 hours after an intravenous injection in the adult Rhesus monkey, there was global expression of the transgene in the brain (Fig. 2C) and also widespread expression in liver and spleen, but not in heart, skeletal muscle, or fat (Fig. 2D). The THL must traverse two barriers prior to entry into the nucleus: barrier 1 is the vascular endothelial barrier (the BBB in brain) and barrier 2 is the plasma membrane of the parenchymal cells of the target organ. The insulin receptor is present on both barriers 1 and 2 in brain and eye, but only on barrier 2 in heart, skeletal muscle, and fat, thus explaining the distribution of gene expression. Insulin receptor is not expressed on barrier 1 in liver or spleen, but the THLs freely cross the porous vascular barrier in these organs, which are perfused by a sinusoidal capillary bed. In contrast, heart, skeletal muscle, and fat are perfused by a capillary bed with a continuous endothelial barrier. New targeting MAbs that enable transport through the vascular endothelial barrier in peripheral tissues must be discovered before THLs can be used to target genes to these tissues.

The delivery of genes to the brain with THLs that target the BBB transferrin or insulin receptor can result in ectopic expression of the transgene in peripheral tissues due to the expression of the transferrin or insulin receptor in those organs. However, if the transgene is incorporated in an expression plasmid under the influence of a brain-specific promoter, such as the 5′-flanking

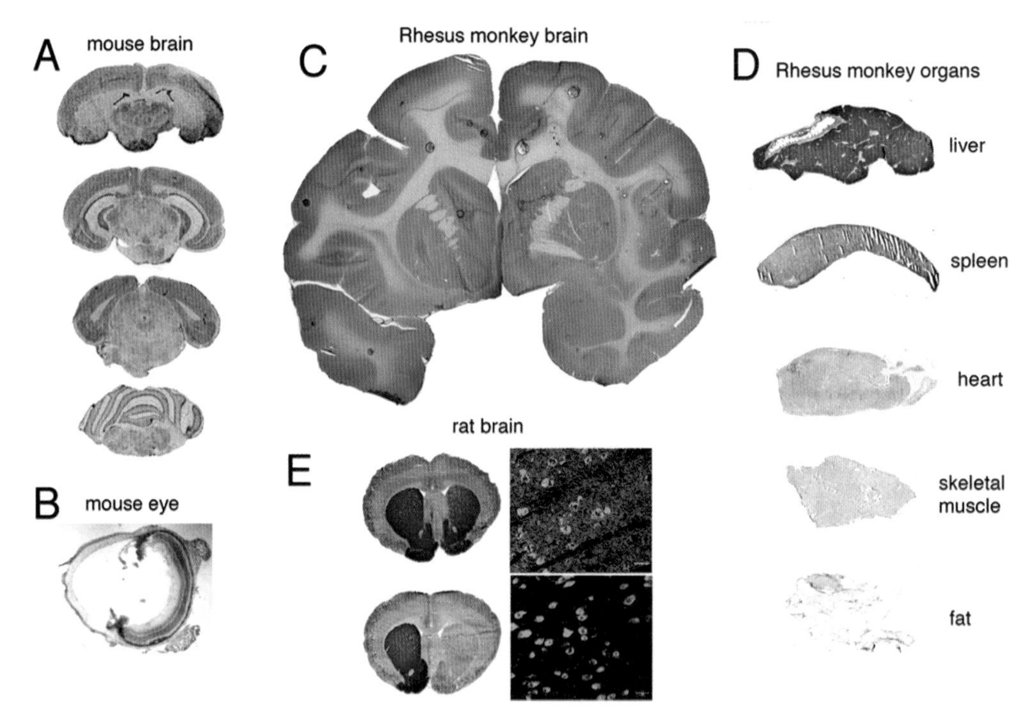

FIGURE 2. (*A,B*) β-galactosidase histochemistry of brain (*A*) and eye (*B*) of the adult mouse removed 48 hours after the intravenous injection of a β-galactosidase expression plasmid encapsulated in PILs targeted with the 8D3 rat MAb to the mouse transferrin receptor. (*C,D*) β-galactosidase histochemistry of brain (C) and peripheral organs (*D*) of the adult Rhesus monkey removed 48 hours after the intravenous injection of a β-galactosidase expression plasmid encapsulated in PILs targeted with the 83-14 murine MAb to the HIR. There is no gene expression in heart, skeletal muscle, or fat. (*E*) Tyrosine hydroxylase immunocytochemistry (*left panels*) or confocal microscopy (*right panels*) of rat brain removed 72 hours after the intravenous injection of a tyrosine hydroxylase expression plasmid encapsulated in PILs targeted with either the OX26 mouse MAb to the rat transferrin receptor (*top panels*) or a mouse IgG2a isotype control antibody (*bottom panels*). Three weeks prior to gene administration, the rats were injected with 6-hydroxydopamine into the medial forebrain bundle on the right side, which caused a complete loss of immunoreactive tyrosine hydroxylase in the striatum ipsilateral to the neurotoxin lesion. Gene therapy with PILs targeted with the transferrin receptor MAb causes a complete normalization of striatal tyrosine hydroxylase (*top panels*). There is no restoration of striatal tyrosine hydroxylase if the tyrosine hydroxylase expression plasmid is encapsulated in PILs targeted with an isotype control antibody that does not target a BBB receptor (*bottom panels*). None of the sections in panels *A, C,* or *D* are counterstained. (*A,* Reprinted, with permission, from Zhu et al. 2004; *B,* reprinted, with permission, from Zhu et al. 2002 [© Association for Research in Vision and Ophthalmology]; *C and D,* reprinted, with permission, from Zhang et al. 2003a [© Elsevier]; *E,* reprinted, with permission, from Zhang et al. 2004a.)

sequence of the glial fibrillary acidic protein (GFAP) gene, then ectopic expression is eliminated in the mouse (Shi et al. 2001). Similarly, transgene expression in the brain or peripheral tissues is not observed following the intravenous injection of THLs carrying a gene under the influence of the 5′-flanking sequence of the rhodopsin gene, although gene expression in ocular structures is observed in the adult Rhesus monkey (Zhang et al. 2003c). The combination of tissue-specific gene promoters and THLs enables highly localized expression of transgenes.

Intravenous Gene Therapy with THLs

THLs were produced that carried an expression plasmid encoding a 700-nucleotide antisense RNA against the human epidermal growth factor receptor (EGFR) (Zhang et al. 2002a). Mice with intracranial human brain tumors were treated with weekly intravenous injections of THLs that were doubly targeted with two MAbs. The rat 8D3 MAb to the mouse transferrin receptor delivers the THL across the mouse BBB of the brain tumor, and the murine 83-14 MAb to the HIR delivers the THL across the brain tumor cell plasma membrane. Weekly intravenous injections of the THLs produced a 100% increase in survival time (Zhang et al. 2002a).

Mice with intracranial brain tumors were also treated with THLs targeted with the 8D3 and 83-14 MAbs that encapsulated an expression plasmid encoding a short hairpin RNA (shRNA) that knocks down the EGFR via RNAi (Zhang et al. 2004b). Weekly intravenous RNAi gene therapy produced a 90% increase in survival time in adult subacute combined immunodeficiency (SCID) mice with intracranial human brain tumors. This is the first demonstration that the survival time in experimental tumors can be prolonged with intravenous RNAi gene therapy.

In adult rats with experimental Parkinson's disease, THLs carrying a tyrosine hydroxylase expression plasmid produced a 100% normalization of striatal tyrosine hydroxylase, as shown in Figure 2E (top panels). If the THL carrying the tyrosine hydroxylase expression plasmid was targeted with a mouse IgG2a isotype control antibody, there was no restoration of striatal tyrosine hydroxylase (Fig. 2E, bottom panels) (Zhang et al. 2004a).

THLs could be used to deliver plasmids containing the inverted terminal repeats that allow for permanent yet random integration of the host genome. However, this is not desirable owing to the insertional mutagenesis that follows random integration of the host genome. The use of plasmid DNA that functions episomally allows for reversible gene therapy, and episomal gene therapy can be given at repeat intervals, similar to other medicines.

Protocol

Preparation of THLs

This protocol describes the preparation of PEGylated immunoliposomes, also known as THLs, which can then be used for gene transfer in vitro or in vivo.

MATERIALS

CAUTION: See Appendix for appropriate handling of materials marked with <!>.

Reagents

[α-^{32}P]dNTP <!> for nick translation
Chloroform <!>
EDTA
Ellman's reagent <!> (5,5′-dithiobisnitrobenzoic acid, DTNB)
HEPES (0.05 M, pH 7.0)
Lipids (e.g., Avanti Polar Lipids)
  DDAB (dimethyldioctadecylammonium bromide) <!> (631 daltons)
  DSPE-PEG 2000 <!> (distearoylphosphatidylethanolamine-PEG 2000) (2748 daltons)
  DSPE-PEG 2000-maleimide <!> (2955 daltons)
  POPC (1-palmitoyl-2-oleoyl-*sn*-glycerol-3-phosphocholine) (760 daltons)
Liquid nitrogen or a dry ice/ethanol bath <!>
Nitrogen gas
NSP (^{3}H-*N*-succinimidyl propionate <!>) or [^{125}I]iodine <!> and chloramine T <!> or Iodogen <!>
Pancreatic endonuclease I <!> and exonuclease III <!>
Plasmid DNA
  Maxipreparations of plasmid DNA generally start with 200 μg of plasmid DNA.
Receptor-specific MAbs (see Introduction for details)
Sepharose CL-4B column (1.5 x 20 cm)
Sodium borate buffer (0.15 M, pH 8.5)/EDTA (0.1 mM)
Tricholoroacetic acid (TCA) <!>

Equipment

Bath sonicator
Extruder, hand-held (e.g., Liposofast hand-held extruder, Avestin) and stacked polycarbonate filters, two each of pore size of 400 nm and 100 nm (50 nm may also be needed; see Step 4)
Gel-filtration chromatography apparatus
Glass tubes (12 x 75 mm)
Millipore Millex-GV filter (0.22 μm)
Particle size analyzer
Rocker platform
Rotary evaporator (*optional,* see Step 4)
Scintillation counter
Spectrophotometer
Vortex mixer

METHOD

1. Radiolabel the MAb.

   The number of MAb molecules conjugated to the THL must be determined as part of the quality control assessment of each THL production run.

   a. Label the ε-amino group of external lysine moieties on the MAb with NSP.

      Alternatively, radiolabel with [^{125}I]iodine and chloramine T or Iodogen. Discard the ^{125}I-labeled MAb after 1 month. The ^{3}H-labeled MAb can be used for 6–12 months if stored at −20°C.

   b. Measure the TCA precipitability of the radiolabeled MAb at the beginning of each THL production run.

   c. Purify any MAb that has a TCA precipitation of less than 95% by gel-filtration chromatography.

2. Measure the amount of DNA encapsulated in the interior of the THL as part of the quality control for each THL production run by radiolabeling an aliquot of the plasmid DNA with ^{32}P using nick translation. Discard the ^{32}P-labeled plasmid DNA after 10 days of storage at 4°C.

3. Conjugate a thiol group of the MAb to the maleimide moiety of the DSPE-PEG 2000-maleimide to form a stable thio-ether linkage. Thiolate the MAb as described below:

   a. To a 12 x 75-mm glass tube, add

      3.0 mg of MAb (20 nmoles)

      2 μCi of ^{3}H-labeled MAb

      ~200 μl of 0.15 M sodium borate buffer (pH 8.5)/0.1 mM EDTA

   b. Add 1200 nmoles of Traut's reagent.

      This is a 60:1 molar ratio relative to the MAb, which is used if the targeting MAb is a mouse IgG2a isotype (e.g., OX26 or 83–14 MAb). If the targeting MAb is a rat IgG (e.g., 8D3 MAb), a 40:1 molar ratio is used. The intent is to add 1–1.5 thiol groups per MAb.

   c. Confirm the number of thiol groups added per MAb with Ellman's reagent and a spectrophotometric assay.

      If the number of thiol groups added per MAb is greater than 1.5, intermolecular MAb cross-linking may occur. If the number of thiol groups added per MAb is less than 1.0, then the efficiency of conjugation of the MAb to liposome will be reduced.

4. For liposome production and extrusion, proceed as follows:

   a. To a 12 x 75-mm glass tube containing 1 ml of chloroform, add

      18.6 μmoles of POPC

      0.6 μmole of DDAB

      0.6 μmole of DSPE-PEG 2000

      0.2 μmole of DSPE-PEG 2000-maleimide

   b. Completely evaporate the lipid under a stream of nitrogen gas while vortexing continuously to produce a thin lipid layer film.

      Alternatively, use a rotary evaporator. This thin lipid layer should be formed carefully.

c. Hydrate the lipids by adding 300 μl of 0.05 M HEPES (pH 7.0) so that the lipids are approximately 100 mM.

The hydration and subsequent vortexing of the lipids should be done carefully and without interruption.

d. Place the hydrated lipid preparation in a bath sonicator for 2 minutes and then vortex at maximum speed for 1 minute. Determine the diameter of the liposome with a Particle Size Analyzer that employs quasi-elastic light-scattering measurements.

Following hydration, vortexing, and bath sonication, the size of vesicles should be 500–1000 nm. Monitor the size of the liposome throughout the procedure.

e. Add the plasmid DNA (100–250 μg) with approximately 1 μCi of ^{32}P-labeled plasmid DNA to the hydrated lipids.

f. Freeze the mixture in either liquid nitrogen or a dry ice/ethanol bath for 5 minutes and then thaw the lipids for 10 minutes at 37°C. Repeat the cycle five to seven times.

The plasmid DNA is encapsulated in large liposomes with the repeat freeze/thaw cycles. Check the sample for excess air bubbles, which should be discharged, as their presence can impair the subsequent extrusion process.

g. Add HEPES buffer (0.05 M, pH 7.0) to a final volume of about 500 μl or 40 mM lipid.

h. To reduce the size of the liposomes, pass the lipid/DNA mixture through two stacked polycarbonate filters with a pore size of 400 nm in a hand-held extruder. Repeat five times. Disassemble the extruder and place two stacked polycarbonate filters with a pore size of 100 nm in it. Pass the lipid/DNA mixture five times through the 100-nm-pore-size filters.

The optimal diameter of the liposomes at this point should be 80–120 nm. It may be necessary to pass the mixture through two filters with a 50-nm pore size in order to achieve this.

5. Perform nuclease digestion as described below:

a. Incubate the liposomes with pancreatic endonuclease I and exonuclease III for 60 minutes at 37°C.

If the extrusion process was done correctly, approximately 40–60% of the DNA is encapsulated in 100-nm liposomes and the remainder is exteriorized. This exteriorized DNA can be toxic in vivo and must be removed. Nuclease digestion is recommended for this purpose (Monnard et al. 1997).

b. Stop the reaction by adding EDTA to a final concentration of 20 mM.

The removal of external DNA with a strong anion-exchange column such as DEAE is not recommended, because the column does not removed anionic DNA that is electrostatically attached to the cationic lipid residues on the surface of the liposome. The use of a DEAE column will produce a THL preparation with significant exteriorized DNA that can elicit toxic and inflammatory reactions in vivo (Norman et al. 2000).

6. Perform ovenight conjugation and gel-filtration column chromatography as described below:

a. Mix the nuclease-treated PEGylated liposomes with the thiolated MAb.

b. Cap the solution, gas with nitrogen, and rock slowly overnight at room temperature.

c. Apply the preparation to a 1.5 x 20-cm column of Sepharose CL-4B equilibrated with 0.05 M HEPES (pH 7.0). Elute the column at room temperature with the HEPES buffer at 1 ml/min and collect 1-ml fractions (Huwyler et al. 1996).

d. Count an aliquot of each fraction for both ^{3}H and ^{32}P radioactivity. For ^{3}H/^{32}P double-isotope liquid scintillation counting, count the ^{3}H in a window of 0–16 keV and the ^{32}P in a window of 16–1700 keV.

e. Determine the percent of DNA encapsulation from the amount of ^{32}P radioactivity in the first or liposome peak off the CL-4B column.

f. Calculate the number of MAb molecules conjugated per individual THL from the ^{3}H radioactivity in the same peak.

The THL peak off the column migrates at about fraction 10, a 10-ml elution volume. The unconjugated MAb elution volume is approximately 25 ml, and the nuclease-digested nucleic acid elutes at 30–35 ml. There is no spillover of ^{3}H into the ^{32}P channel, but there is an approximately 2% spillover of ^{32}P into the ^{3}H channel. This spillover should be accounted for in the calculations of ^{3}H and ^{32}P radioactivity.

7. Before using the THLs, sterilize them by passage through a 0.22-μm Millipore Millex-GV filter, which does not alter their structural integrity (Zhang et al. 2002b). Apply the THLs to cells in culture or inject into animals within 24 hours of production, preferably on the same day as elution from the CL-4B column. THLs may be stored overnight at 4°C.

ACKNOWLEDGMENTS

This work was supported by a grant from the U.S. Department of Defense, and by National Institutes of Health grant R01-NS-53540.

REFERENCES

Hong K., Zheng W., Baker A., and Papahadjopoulos D. 1997. Stabilization of cationic liposome-plasmid DNA complexes by polyamines and poly(ethylene glycol)-phospholipid conjugates for efficient in vivo gene delivery. *FEBS Lett.* **400:** 233–237.

Huwyler J., Wu D., and Pardridge W.M. 1996. Brain drug delivery of small molecules using immunoliposomes. *Proc. Natl. Acad. Sci.* **93:** 14164–14169.

Kofler P., Wiesenhofer B., Rehrl C., Baier G., Stockhammer G., and Humpel C. 1998. Liposome-mediated gene transfer into established CNS cell lines, primary glial cells, and in vivo. *Cell Transplant.* **7:** 175–185.

Luikenhuis S., Giacometti E., Beard C.F., and Jaenisch R. 2004. Expression of MeCP2 in postmitotic neurons rescues Rett syndrome in mice. *Proc. Natl. Acad. Sci.* **101:** 6033–6038.

Matsui H., Johnson L.G., Randell S.H., and Boucher R.C. 1997. Loss of binding and entry of liposome-DNA complexes decreases transfection efficiency in differentiated airway epithelial cells. *J. Biol. Chem.* **272:** 1117–1126.

Monnard P.A., Oberholzer T., and Luisi P. 1997. Entrapment of nucleic acids in liposomes. *Biochim. Biophys. Acta* **1329:** 39–50.

Niidome T., Ohmori N., Ichinose A., Wada A., Mihara H., Hirayama T., and Aoyagi H. 1997. Binding of cationic α-helical peptides to plasmid DNA and their gene transfer abilities into cells. *J. Biol. Chem.* **272:** 15307–15312.

Norman J., Denham W., Denham D., Yang J., Carter G., Abouhamze A., Tannahill C.L., MacKay S.L.D., and Moldawer L.L. 2000. Liposome-mediated, nonviral gene transfer induces a systemic inflammatory response which can exacerbate pre-existing inflammation. *Gene Ther.* **7:** 1425–1430.

Papahadjopoulos D., Allen T.M., Gabizon A., Mayhew E., Matthay K., Huang S.K., Lee K.D., Woodle M.C., Lasic D.D., Redemann C., and Martin F.J. 1991. Sterically stabilized liposomes: Improvements in pharmacokinetics and antitumor therapeutic efficacy. *Proc. Natl. Acad. Sci.* **88:** 11460–11464.

Pardridge W.M. 2001. *Brain drug targeting: The future of brain drug delivery.* Cambridge University Press, Cambridge, United Kingdom.

———. 2003. Gene targeting in vivo with pegylated immunoliposomes. *Methods Enzymol.* **373:** 507–528.

Plank C., Tang M.X., Wolfe A.R., and Szoka F.C., Jr. 1999. Branched cationic peptides for gene delivery: Role of type and number of cationic residues in formation and in vitro activity of DNA polyplexes. *Hum. Gene Ther.* **10:** 319–332.

Simberg D., Weisman S., Talmon Y., Faerman A., Shoshani T., and Barenholz Y. 2003. The role of organ vascularization and lipoplex-serum initial contact in intravenous murine lipofection. *J. Biol. Chem.* **278:** 39858–39865.

Simberg D., Danino D., Talmon Y., Minsky A., Ferrari M.E., Wheeler C.J., and Barenholz Y. 2001. Phase behavior, DNA ordering, and size instability of cationic lipoplexes. *J. Biol. Chem.* **276:** 47453–47459.

Shi N. and Pardridge W.M. 2000. Non-invasive gene targeting to the brain. *Proc. Natl. Acad. Sci.* **97:** 7567–7572.

Shi N., Zhang Y., Boado R.J., Zhu C., and Pardridge W.M. 2001. Brain-specific expression of an exogenous gene after i.v. administration. *Proc. Natl. Acad. Sci.* **98:** 12754–12759.

Song Y.K., Liu F., Chu S., and Liu D. 1997. Characterization of cationic liposome-mediated gene transfer in vivo by intravenous administration. *Hum. Gene Ther.* **8:** 1585–1594.

Zhang Y., Schlachetzki F., and Pardridge W.M. 2003a. Global nonviral gene transfer to the primate brain following intravenous administration. *Mol Ther.* **7:** 11–18.

Zhang Y., Zhu C., and Pardridge W.M. 2002a. Antisense gene therapy of brain cancer with an artificial virus gene delivery system. *Mol Ther.* **6:** 67–72.

Zhang Y., Lee, H.J., Boado R.J., and Pardridge W.M. 2002b. Receptor-mediated delivery of an antisense gene to human brain cancer cells. *J. Gene Med.* **4:** 183–194.

Zhang Y., Calon F., Zhu C., Boado R.J., and Pardridge W.M. 2003b. Intravenous non-viral gene therapy causes normalization of striatal tyrosine hydroxylase and reversal of motor impairment in experimental Parkinsonism. *Hum. Gene Ther.* **14:** 1–12.

Zhang Y., Schlachetzki F., Li J.Y., Boado R.J., and Pardridge W.M. 2003c. Organ-specific gene expression in the Rhesus monkey eye following intravenous non-viral gene transfer. *Mol. Vis.* **9:** 465–472.

Zhang Y., Schlachetzki F., Zhang Y.F., Boado R.J., and Pardridge W.M. 2004a. Normalization of striatal tyrosine hydroxylase and reversal of motor impairment in experimental Parkinsonism with intravenous non-viral gene therapy and a brain-specific promoter. *Hum. Gene Ther.* **15:** 339–350.

Zhang Y., Bryant J., Zhang Y.F., Charles A., Boado R.J., and Pardridge W.M. 2004b. Intravenous RNAi gene therapy targeting the human EGF receptor prolongs survival in intracranial brain cancer. *Clin. Cancer Res.* **10:** 3667–3677.

Zhu C., Zhang Y, and Pardridge W.M. 2002. Widespread expression of an exogenous gene in the eye after intravenous administration. *Invest. Opthalmol. Vis. Sci.* **43:** 3075–3080.

Zhu C., Zhang Y., Zhang Y.F., Li J.Y., Boado R.J., and Pardridge W.M. 2004. Organ specific expression of lacZ gene controlled by the opsin promoter after intravenous gene administration in adult mice. *J. Gene Med.* **6:** 906–912.

70 Sonoporation: An Efficient Technique for the Introduction of Genes into Chick Embryos

Sho Ohta,* Ogino Yukiko,* Kentro Suzuki,* Mika Kamimura,* Katsuro Tachibana,[†] and Gen Yamada*

**Center for Animal Resources and Development (CARD) and Graduate School of Medical and Pharmaceutical Sciences, Kumamoto University, Honjo 2-2-1, Kumamoto 860-0811, Japan;* [†]*Department of Anatomy, Fukuoka University School of Medicine, 7-45-1 Nanakuma, Jonan-ku, Fukuoka 814-0180, Japan*

ABSTRACT

Sonoporation is a new method for the introduction of genes into tissues such as those of chick embryos. It enables gene incorporation into target tissue by the combination of ultrasound exposure and subsequent rupture of microbubbles. Sonoporation has several advantages: (1) It is simple, involving two main steps—preparation of a microbubble-DNA mixture, followed by injection and ultrasound treatment of the target tissue; (2) it allows highly efficient gene incorporation into mesenchymal tissues; (3) it does not induce significant tissue damage—most of the sonoporated chick embryos survived without showing significant embryonic abnormalities or cell death; and (4) it can be used to introduce genes into several chick embryonic tissues (e.g., branchial arch and lateral plate mesoderm). To demonstrate the utility of sonoporation, an expression vector containing the *Sonic hedgehog* (*Shh*) gene was incorporated into the developing chick limb bud, leading to additional digit formation in the *Shh*-expressing embryos. Sonoporation may not be appropriate for the introduction of genes into tissues with cavitated or open structures. Sonoporation into the neural tube and limb ectoderm often resulted in dispersed gene expression due to diffusion of the injected microbubbles before ultrasound treatment. Despite these problems, sonoporation is a useful method for the analysis of gene function(s) in vivo and brings a new area of gene transfer to medical, molecular, and developmental biology.

INTRODUCTION

Gene transfer is useful for analyzing gene functions that underlie vertebrate embryogenesis (Yamada et al. 1997; Ohta et al. 2003). Retrovirus- or adenovirus-mediated gene-transfer methods enable efficient manipulation of transgenes (Morgan and Fekete 1996). However, these methods have problems with respect to temporal and spatial regulation of gene expression. Electroporation has been developed to introduce genes in a regulated manner (Nakamura and Funahashi 2001). This technique has some disadvantages, including tissue damage. To circumvent such problems, we have developed a novel gene-transfer technique for developing embryos called sonoporation, which is based on completely different principles.

Sonoporation is an efficient gene incorporation method using microbubbles and ultrasound together. It was originally developed as a drug-delivery system for fields such as gene therapy (Tachibana and Tachibana 1995, 2001). It has been applied to various embryonic chick tissues (Table 1). Here, we explain the procedure of sonoporation applied to an embryonic chick organ (an example of a limb bud is shown at Hamburger and Hamilton [HH] stage 20–21) (Fig. 1). The procedure involves preparation of the DNA-microbubble mixture, injection into target tissue(s), and exposure to ultrasound. Three parameters of ultrasound are important: intensity (W/cm^2); duty cycle (%), which gives the rate of pulse wave irradiation; and exposure time (seconds). In our experiments, settings of 2.0 W/cm^2, duty cycle 20%, and 60 seconds resulted in highly efficient gene incorporation into chick limb bud mesenchymal cells. To demonstrate the utility of sonoporation, an expression vector containing the *Shh* gene was introduced into the developing chick limb bud. Additional digit formation was observed in such *Shh*-expressing embryos (Fig. 2c–e). Genes can also be sonoporated into mouse embryonic explants and some adult mouse organs (S. Ohta, unpublished data).

Sonoporation may not be appropriate for incorporation into tissues with cavitated or open structures, where diffusion of microbubbles can lead to dispersed gene expression. Using a higher concentration of DNA or mixing DNA with a viscous solvent, which increases the viscosity of the DNA-microbubble mixture, might help prevent the diffusion of the injected mixture. Some new microbubbles with positive electric charge have also been developed. Use of a smaller and finer ultrasound emission probe with a 0.5-mm diameter rather than 3-mm diameter might enable more restricted embryonic gene transfer. Despite some drawbacks, sonoporation achieves efficient gene incorporation into various tissues without significant tissue damage (Ohta et al. 2003).

TABLE 1. Parameters of Gene Sonoporation into Various Chick Tissues

Injection site and stages (HH stages)	DNA solution (μl): microbubble (μl)	Ultrasound irradiation time (seconds)
Branchial arch (HH stages 20–21)	20:10	60
Lateral plate mesoderm (HH stages 15–16)	20:20	60
Neural tube (HH stages 10–11)	20:20	10

Parameters for gene transfer into various chick tissues are summarized. Genes introduced into branchial arch, lateral plate mesoderm, and neural tube were treated with an ultrasound intensity of 2.0 W/cm^2 and duty cycle of 20%. The concentration of DNA solution was 1.0–2.0 μg/μl.

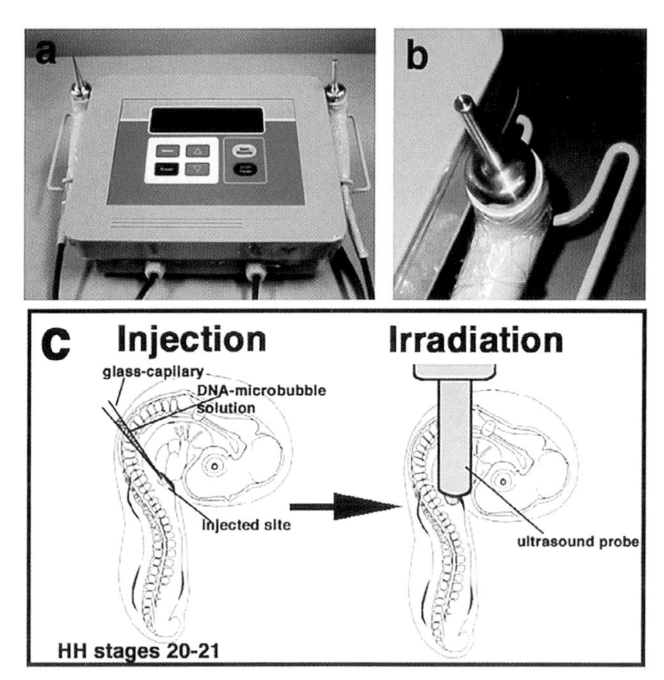

FIGURE 1. (*a*) Sonitron 2000N; (*b*) Ultrasound probe. (*c*) DNA-microbubble mixture was injected into chick limb bud at HH stages 20–21. For efficient gene transfer, it is necessary to keep the DNA-microbubble mixture in the target tissue. Immediately after the injection, ultrasound irradiation was performed with a 3-mm diameter ultrasound probe.

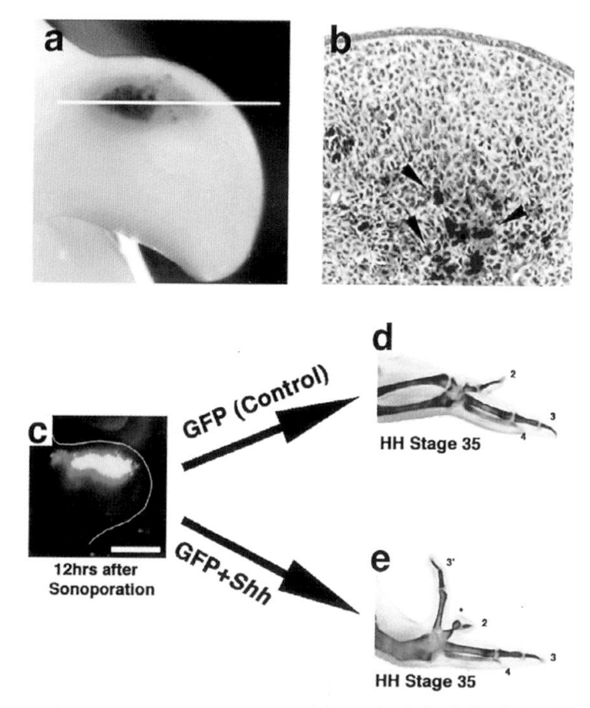

FIGURE 2. (*a*) LacZ expression vector was sonoporated into chick limb bud. LacZ expression was detected in anterior limb bud region after 12 hours. (*b*) In the transverse section of the limb bud treated with expression vector, the LacZ-positive cells were detected in the limb bud mesenchyme. (*c, d, e*) Additional digit formation as the result of ectopic *Shh* expression in the anterior limb bud. To visualize the region of gene expression following sonoporation, pCAGGS-c*Shh* and pCAGGS-GFP were cointroduced into the anterior limb bud region. After 12 hours, GFP (green fluorescent protein) expression was detected in the anterior limb bud region. Additional digit formation was observed in the ectopic *Shh*-expressing embryos.

Protocol

Sonoporation for Gene Transfer into Embryos

This protocol describes the use of sonoporation to introduce vectors into embryonic chick limb bud.

MATERIALS

CAUTION: See Appendix for appropriate handling of materials marked with <!>.

Reagents

Alcohol (70%)
Expression vector, e.g., pEGFP (Clontech; 632319) with a DNA concentration of 1.0–2.0 µg/µl
Dissolve the DNA in TE solution (pH 8.0).
Fertilized chick eggs
Hank's solution (10 mM HEPES, 140 mM NaCl, 27 mM KCl <!>, 0.1% glucose, 0.34 mM Na_2HPO_4 <!>, 0.5 mM $CaCl_2$ <!>, 0.5 mM $MgCl_2$ <!>, at pH 7.4)
OPTISON: microbubble reagent (Amersham Health 2707-03)
Store refrigerated at 2–8°C.
Sodium chloride (NaCl; 0.9%) or phosphate-buffered saline (PBS)
TE solution (10 mM Tris-HCl <!>, 1 mM EDTA at pH 8.0)

Equipment

Black ink (Pelikan)
Forceps (watchmaker's, straight no. 5)
Glass microneedles prepared from a glass capillary tube (G-1; Narishige)
Incubator preset to 38.5°C
Micropipettor
Microscope
Petri dish (35 mm)
Scissors (straight, fine)
Sonitron 1000N or Sonitron 2000N (RichMar, Inola, Oklahoma or Nepagene, Japan)
Syringe (5 ml) and 18-gauge needle for drawing up egg thin albumen
Syringe (1 ml) and 26-gauge needle for injecting ink
Tube (0.5 ml)
Vinyl tape

METHODS

Preparation of the DNA-Microbubble Mixture

1. Prepare 5–10 µl of stock DNA solution (e.g., pEGFP 1.0–2.0 µg/µl) in a 0.5-ml tube.
2. Add 10–15 µl of 0.9% NaCl or PBS to the tube (DNA final concentration 0.25–1.0 µg/µl).

3. Gently shake the vial of OPTISON until the solution becomes homogeneous. Add 10–20 μl of the microbubble solution to the tube containing the DNA. Mix very well and keep on ice.

   The DNA-microbubble mixture is suitable for ten rounds of injections. Use it within 2 hours of preparation.

Sonoporation

For example, use for gene transfer for chick limb bud HH stages 20–21.

4. Incubate fertilized chick eggs at 38.5°C until HH stages 20–21.
5. Sterilize the egg shells with 70% alcohol. Make a small hole at the blunt end of the eggs. Draw 4–5 ml of thin albumen into a 5-ml syringe fitted with an 18-gauge needle.
6. Cut the shell with straight fine scissors to open a window for the operation.
7. Dilute black ink about tenfold with Hank's solution. To visualize chick embryos, inject 0.05–0.1 ml of the diluted black ink under the blastoderm using a 1-ml syringe fitted with a 26-gauge needle.
8. Use fine watchmaker's forceps to cut the vitelline membrane adjacent to the chick limb bud. Apply a few drops of Hank's solution, if required.
9. Mix the DNA-microbubble mixture very well with a micropipettor. Spot 2–3 μl of the DNA-microbubble mixture on a 35-mm Petri dish. Draw up 0.25–0.5 μl of the spotted DNA-microbubble mixture into a glass needle. Make sure that microbubbles are included in the DNA-microbubble mixture.
10. Inject the DNA-microbubble mixture into limb bud mesenchyme (see Fig. 1c). Since the DNA-microbubble mixture tends to leak from the injected site, adjust the method of injection depending on the target organ. After injection, confirm the presence of microbubbles in the injected region by looking for a white color under bright-field microscopy (white color by reflected light).
11. Expose the injected limb bud region to ultrasound by touching the probe (3-mm diameter) of the sonicator to the injected region gently. Use settings of 2.0 W/cm^2, duty cycle 20%, and 60 seconds (see Fig. 1a,b). The white color in the injected region disappears after ultrasound irradiation.
12. After ultrasound exposure, cover the window with vinyl tape and incubate the egg.
13. Use LacZ (Fig. 2a,b) or green fluorescent protein (GFP) (Fig. 2c–e) to visualize the region of the chick embryo where the sonoporated DNA is expressed.

ACKNOWLEDGMENTS

We thank Drs. Olivier Pourquie, Richard R. Behringer, Andrew McMahon, Juan-Carlos Izpisua Belmonte, Urlich Ruther, Matthew Scott, and Cliff Tabin for reagents and/or suggestions, and Shiho Kitagawa for her assistance. Contacts with companies included Richmar (Mr. Jim Whittaker [jwhittaker@richmarweb.com] and Mr. Ken Coffey [coffeyk@aol.com]; Web site [www.richmarweb.com]) and Nepagene (Mr. Yasuhiko Hayakawa [y-hayaka@nepagene.jp]; Web site [www.nepagene.jp]). This research was supported in part by a grant-in-aid for 21st Century COE Research from Ministry of Education, Culture, Sports, Science and Technology and by a grant-in-aid for Scientific Research on Priority Areas (A), General Promotion of Cancer Research in Japan.

REFERENCES

Morgan B.A. and Fekete D.M. 1996. Manipulating gene expression with replication-competent retroviruses. *Methods Cell Biol.* **51:** 185–218.

Nakamura H. and Funahashi J. 2001. Introduction of DNA into chick embryos by in ovo electroporation. *Methods* **24:** 43–48.

Ohta S., Suzuki K., Tachibana K., and Yamada G. 2003. Microbubble-enhanced sonoporation: Efficient gene transduction technique for chick embryos. *Genesis* **37:** 91–101.

Tachibana K. and Tachibana S. 1995. Albumin microbubble echo-contrast material as an enhancer for ultrasound accelerated thrombolysis. *Circulation* **92:** 1148–1150.

———. 2001. The use of ultrasound for drug delivery. *Echocardiography* **18:** 323–328.

Yamada G., Nakamura S., Haraguchi R., Sakai M., Terashi T., Sakisaka S., Toyoda T., Ogino Y., Hatanaka H., and Kaneda Y. 1997. An efficient liposome-mediated gene transfer into the branchial arch, neural tube and the heart of chick embryos; a strategy to elucidate the organogenesis. *Cell. Mol. Biol.* **43:** 1165–1169.

71 Genetic Manipulation of Mammalian Cells by Microinjection

David W. Rose

Department of Medicine and Moores Cancer Center, University of California, San Diego, La Jolla, California 92093-0673

ABSTRACT

Of the many methods available for the introduction of DNA and other biological materials into cells, microinjection of individual mammalian cells is one of the more technically difficult to perform. The technique is also reliant upon comparatively expensive equipment and demands the experimental skills of an experienced operator. There are, however, very good reasons to use this approach. It allows the introduction of molecules into a defined population of cells at a defined concentration, and the timing of the experiment can be controlled stringently, minimizing problems associated with overexpression. Perhaps the most powerful aspect of microinjection is the ability to introduce several types of reagents into cells simultaneously, including DNA constructs, a labeled dextran to mark the injected cells, antibodies, short interfering RNAs (siRNAs), and peptides (Lavinsky et al. 1998; Rose et al. 1998; Jepsen et al. 2000; Zhu et al. 2006). No other techniques available provide these capabilities.

Delivery by microinjection can be used for any type of cell that is adherent in culture, including primary cells. Because siRNA can be easily and rapidly generated for any target gene, it is relatively simple to assess many effects of the knockout of any gene in any type of adherent cell in a matter of a few days. In addition, delivery by microinjection assures that every cell receives the siRNA at a relatively equal concentration. Many of the disadvantages of the approach relate to the challenging aspects of the technique itself: It is not trivial to learn, and it is an undertaking that requires attention to detail. It also suffers the technical disadvantage that a limited number of cells are involved, which often does not permit the subsequent analysis of effects upon some biochemical parameters. The major limitation of the approach is the small amount of material obtained, but as the sensitivity of analysis techniques increases, so does the usefulness of this method.

INTRODUCTION

A variety of transgenic animals have been generated using microinjection as a means of introduction into zygotes of plasmid constructs (Gordon et al. 1980; Gordon 1993). Microinjection as a means of reducing the expression or activity of an intracellular target protein has also been used fairly extensively, initially through the use of microinjected antibodies (Bar-Sagi 1995; Lamb and Fernandez 1997; Torchia et al. 1997). This technique was later advanced by addition to the antibody mixture of plasmid constructs that were used to rescue the activity of the affected intracellular protein (Li et al. 1998). The introduction of mutations into these DNA constructs prior to reintroduction into cells made it possible to use the rescue approach to explore structure-function relationships and to relate protein domains to biological activity in living cells (McInerney et al. 1998; Li et al. 2003). Another important advance came with the realization that microinjection of siRNA provided a complement to the use of antibodies in microinjection experiments (Perissi et al. 2004). This approach is in fact advantageous in many ways to the use of antibodies, because the knockdown of a protein target with siRNA does not affect the activity of other proteins that might be in a complex with the target, as is the case for antibodies. It is also possible to easily design and synthesize siRNA against virtually any target. Many of the basic aspects of microinjection, particularly of antibodies and DNA constructs, have been described in detail elsewhere (Rose et al. 1999; Ikeda 2004). We emphasize here recent advances in the use of microinjection in combination with siRNA technology.

Protocol

Microinjection of siRNA into Mammalian Cells

In these experiments, several hundred cells out of a much larger population are injected and identified by the presence of the fluorescent dextran. Rescue experiments using expression constructs for the target gene can also be conducted as long as they are not recognized by the siRNA. We routinely use constructs from another species to avoid this issue, but it is also possible to introduce silent mutations into the target gene to prevent recognition. A description of injection of a small population of cells is given in the box on page 721.

MATERIALS

CAUTION: See Appendix for appropriate handling of materials marked with <!>.

Reagents

Cell line to be injected (adherent line used here as example)

Dextran, fluorescently labeled (Molecular Probes/Invitrogen)

> We recommend the use of dextrans that are chemically modified with amino groups that render them aldehyde-fixable. It is also advisable to use dextran with a molecular mass above 70 kD to prevent movement through the nuclear membrane.

Oligonucleotide siRNAs

> The siRNAs to target the gene of interest and to create controls can be designed using online software available through several distributors. Stock solutions are prepared from annealed oligos at 20 μM that are stored at –20°C.

siRNA sample buffer

5 mM Sodium phosphate (pH 7.2)

100 mM KCl <!>

Equipment

Glass coverslips (acid-washed)

Epifluorescence microscope for analysis of the results of most experiments

Inverted light microscope of good quality with several objectives

Microinjection system, semiautomated, including controls for injection pressure and timing as well as a micromanipulator and joystick.

> Microinjection equipment is available from several manufacturers and has been described in detail elsewhere (Rose et al. 1999; see Ikeda 2004 for a detailed description; King 2004). Microinjection needles are commercially available; we have had good success making our own on a commercial needle puller.

METHOD

1. Prepare working solutions of the siRNA oligonucleotides (specific and control oligos) at a concentration of 50 nM in siRNA sample buffer. Dissolve the fluorescently labeled dextran at 5 mg/ml in siRNA sample buffer.

2. To prepare the siRNAs for microinjection, dilute the working solution of siRNA oligonucleotides tenfold into the labeled dextran solution.

   This picomolar concentration is very effective and is low enough to avoid off-target nonspecific effects.

3. Seed adherent cells onto acid-washed glass coverslips and grow to the appropriate density, typically about 50% confluent.

   At this point, the cells can be deprived of serum if needed to render them quiescent.

4. Carry out nuclear microinjection using standard procedures with samples prepared from specific or control siRNAs (see, e.g., King 2004 and Fig. 1).

   DNA, peptides, and other biomolecules can be included as appropriate for the system under study (Zhu et al. 2006).

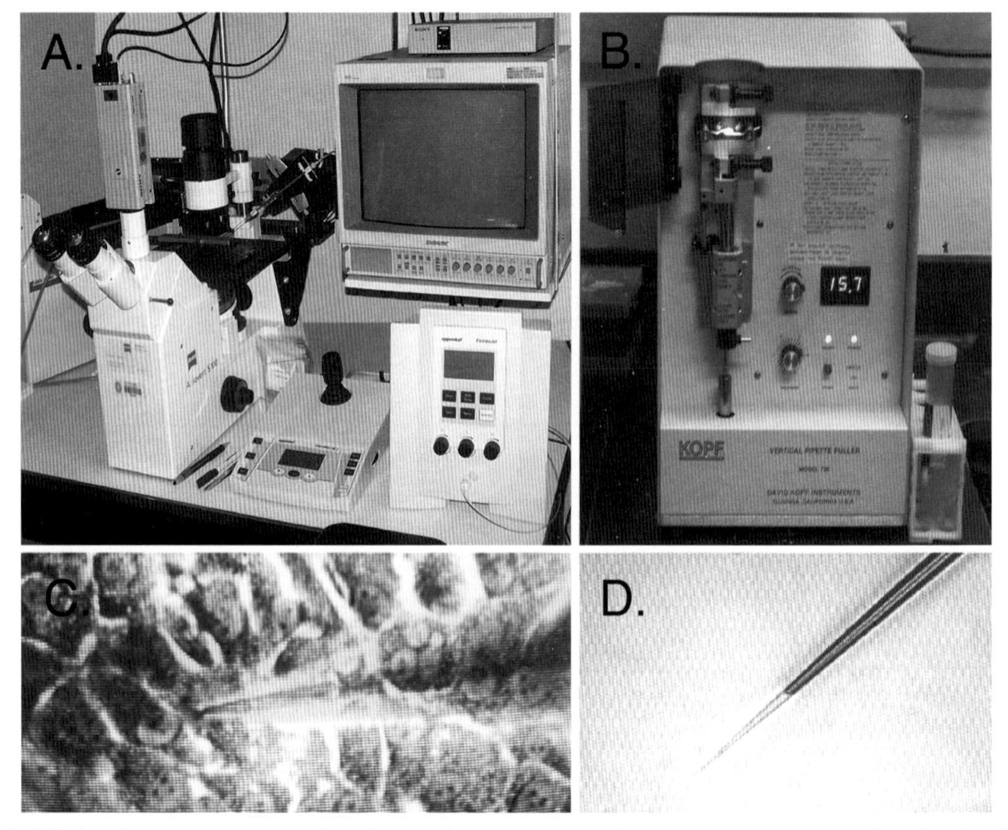

FIGURE 1. Microinjection setup. (*A*) Typical semiautomated microinjection apparatus, consisting of a vibration isolation air table, inverted microscope, micromanipulator, video setup, and microinjector. Motors mounted on the stage of the microscope are operated by the joystick, allowing fine movement of the needle, attached by a stylus. Positive pressure is controlled by the microinjector during and between injections. (*B*) Vertical needle puller. Two identical needles are created simultaneously from a fine capillary tube. (*C*) Nuclear injection process visualized on video monitor. (*D*) 320x view of an ideal, fine-tipped needle appropriate for nuclear microinjection of DNA and antibodies.

MICROINJECTION OF EVERY CELL IN A SMALL POPULATION

Another practical extension of siRNA involves the microinjection of every cell in a small population, resulting in what amounts to a functional knockout of any desired target within 48 hours or less. Using RT-PCR or quantitative PCR (qPCR), it is possible to demonstrate that effective ablation of the target mRNA has occurred using the RNA extracted from fewer than 100 cells. It is also possible to use the same approach with selected primers to quantitate the downstream effects of siRNA injection upon the expression of any other gene. This can be done easily using qPCR, and, when an amplification step is included, the RNA can also be analyzed using low-density gene expression arrays (Perissi et al. 2004; Zhu et al. 2006).

1. Seed the cells in a very small volume (typically 1 µl or less) in the center of a glass coverslip.
2. Allow the cells to adhere for about 20 minutes and then add additional medium.

   With the appropriate dilution, it is feasible to select any number of cells desired in a very small area in which the injection of every cell is possible.
3. Carry out microinjection and incubate the cells to allow knockdown.
4. Harvest RNA from the cells and analyze as appropriate.

5. Following microinjection, incubate the cells for a period of time sufficient to allow knockdown of the targeted protein.

   The siRNA effect is both specific and rapid. In most cases, target mRNA becomes undetectable by reverse transcript–polymerase chain reaction (RT-PCR) in less than 6 hours after injection. This time period of course varies, depending on the half-life of the target, but most regulatory proteins are relatively short-lived.
6. Analyze the cells as appropriate for the system under study.

   One of the most important aspects of this approach is the analysis used to evaluate the effects of microinjection. The phenotype generated by the siRNA must be scrutinized using an objective endpoint, preferably one that allows each individual injected cell to be scored as positive or negative. Excellent examples include bromodeoxyuridine incorporation (Li et al. 2002), immunocytochemical staining for an appropriate marker (Klappacher et al. 2002), or the use of a co-injected promoter/reporter construct (Perissi et al. 2004; Zhu et al. 2006).

ACKNOWLEDGMENTS

We thank Claudia Santiago and Kelly Ou for administrative assistance, and members of the Rose laboratory, particularly Paul F. Kotol, Nathaniel dela Paz, and Radin Aur for their contributions to the development of novel experimental approaches.

REFERENCES

Bar-Sagi D. 1995 Mammalian cell microinjection assay. *Methods Enzymol.* **255:** 436–442.

Gordon J.W. 1993. Production of transgenic mice. *Methods Enzymol.* **25:** 747–771.

Gordon J.W., Scangos G.A., Plotkin D.J., Barbosa J.A., and Ruddle F.H. 1980. Genetic transformation of mouse embryos by microinjection of purified DNA. *Proc. Natl. Acad. Sci.* **77:** 7380–7384.

Ikeda S.R. 2004. Expression of G-protein signaling components in adult mammalian neurons by microinjection. *Methods Mol Biol.* **259:** 167–181.

Jepsen K., Hermanson O., Onami T.M., Gleiberman A.S., Lunyak V., McEvilly R.J., Kurokawa R., Kumar V., Liu F., Seto E., et al. 2000. Combinatorial roles of the nuclear receptor corepressor in transcription and development. *Cell.* **102:** 753–763.

King R. 2004. Gene delivery to mammalian cells by microinjection. *Methods Mol. Biol.* **45:** 167–174.

Klappacher G.W., Lunyak V., Sykes D., Sawka-Verhelle D., Sage J., Brard G., Ngo S.D., Gangardharan D., Jacks T., Kamps M.P., et al. 2002. Identification of an induced Ets repressor complex reveals a combinatorial mechanism directing growth arrest during terminal macrophage differentiation. *Cell* **109:** 169–180.

Lamb N.J.C. and Fernandez A. 1997. Microinjection of antibodies into mammalian cells. *Methods Enzymol.* **283:** 72–83.

Lavinsky R.M., Jepsen K., Heinzel T., Torchia J., Mullen T.-M., Schiff R., Del-Rio A.L., Ricote M., Ngo S., Gemsch J., et al. 1998. Diverse signaling pathways modulate nuclear receptor recruitment of N-CoR and SMRT complexes. *Proc. Natl. Acad. Sci.* **95:** 2920–2925.

Li J., Tang H., Mullen T.-M., Westberg C., Rose D.W., and Wong-Staal F. 1998. A role for RNA Helicase A in post-transcriptional regulation of HIV. *Proc. Natl. Acad. Sci.* **96**: 709–714.

Li X., Perissi V., Liu F., Rose D.W., and Rosenfeld M.G. 2002. Tissue-specific regulation of retinal and pituitary precursor cell proliferation. *Science* **297:** 1180–1183.

Li X., Oghi K.A., Zhang J., Krones A., Bush K.T., Glass C.K., Nigam S.K., Aggarwal A.K., Maas R., Rose D.W., and Rosenfeld M.G. 2003. Eya protein phosphatase activity regulates Six1-Dach-Eya transcriptional effects in mammalian organogenesis. *Nature* **426:** 247–254.

McInerney E., Rose D.W., Flynn S.E., Westin S., Mullen T.-M., Krones A., Inostroza J., Nolte R.P., Assa-Munte N., Milburn M., Glass C.K., and Rosenfeld M.G. 1998. Coactivator LXXLL recognition motifs confer specificity in nuclear receptor transcriptional activation. *Genes Dev.* **12:** 3357–3368.

Perissi V., Aggarwal A., Glass C.K., Rose D.W., and Rosenfeld M.G. 2004. A corepressor/coactivator exchange complex required for transcriptional activation by nuclear receptors and other regulated transcription factors. *Cell* **116:** 511–526.

Rose D.W., Mullen T.-M., Rosenfeld M.G., and Glass C.K. 1999. Functional characterization of coactivators using mammalian cell microinjection. In *Nuclear receptors: A practical approach* (ed. D. Picard), pp. 119–135. Oxford University Press, New York.

Rose D.W., Xiao S., Pillay T.S., Kolch W., and Olefsky J.M. 1998. Prolonged versus transient roles for early cell cycle signaling components. *Oncogene* **17:** 889–899.

Torchia J., Rose D.W., Inostroza J., Kamei Y., Glass C.K., and Rosenfeld M.G. 1997. The transcriptional coactivator p/CIP binds CBP and mediates nuclear receptor function. *Nature* **387:** 677–685.

Zhu P., Baek S.-H., Bourk E.M., Ohgi K.A., Garcia-Bassets I., Sanjo H., Akira S., Kotol P.F., Glass C.K., Rosenfeld M.G., and Rose D.W. 2006. Macrophage/cancer cell interactions mediate hormone resistance through a conserved nuclear receptor derepression pathway. *Cell* **124:** 615–629.

72 Magnetofection

Christian Plank* and Joseph Rosenecker†

*Institute of Experimental Oncology, Technische Universität München, 81675 Munich, Germany;
†Department of Pediatrics, Ludwig-Maximilians-University, 80337 Munich, Germany

ABSTRACT

Magnetofection is defined as nucleic acid delivery guided and mediated by magnetic force acting on associations of magnetic particles and nucleic acids or nucleic acid vectors. Vectors are bound to magnetic, usually iron oxide, nanoparticles, in most cases by noncovalent bonds. Magnetic force accumulates and/or holds magnetic vectors in a target tissue against hydrodynamic forces. In cell culture, magnetic vectors are magnetically sedimented on the target cells within minutes. Thus, the diffusion barrier to nucleic acid delivery is overcome, the full vector dose comes in contact with the target cells, and introduction of genetic material is synchronized. Nucleic acid delivery is greatly accelerated and its efficiency with many, if not most, vector types is improved. Magnetofection is applicable to small and large nucleic acids. Other advantages include low-dose requirements, the possibility of confining nucleic acid introduction to a localized area (magnetic targeting), and the amenability to high-throughput automation. Due to the favorable dose-response profile and the rapid kinetics, vector-related toxicity can be kept low (Krotz et al. 2003).

Magnetofection does have limitations. In some cases, similar efficiencies can be achieved with simple centrifugation procedures (Huth et al. 2004), although Magnetofection is much more convenient. Magnetofection works better for adherent cells than for suspension cells. The real limitations of the method are encountered in vivo. No suitable magnetic field technology is available to accumulate magnetic nanoparticles against high blood flow rates. Limitations of standard vectors, such as opsonization and immune activation, also exist for Magnetofection. However, it can be expected that partial solutions to the magnetic field technology and to improved magnetic vector formulation will be available in the near future.

INTRODUCTION

The probability of vector success (functional delivery of a nucleic acid to the desired subcellular localization) is the product of probabilities of overcoming the individual barriers to delivery. If the probability of vector-target cell contact is low to start with, the efficacy of the overall delivery process will be low as well, independent of vector type. Luo and Saltzman (2000) have pointed out that DNA transfection efficiency is limited by a simple physical barrier: low DNA concentration at the cell surface. Vector-cell contact, at least in cell culture, is driven by diffusion, an exceedingly slow process. Below the saturation of uptake processes, any measure that increases vector concentrations at target cell surfaces at a given vector dose should increase nucleic acid introduction efficiencies. This can be achieved by physical force, including gravitational and centrifugal force and hydrodynamic force, and by electric and magnetic fields. Magnetofection exploits magnetic force for vector accumulation and/or retention at target sites. The procedure sediments the full vector dose on the target cells. Therefore, the vector dose can usually be decreased significantly compared to standard transfections (up to one tenth of the usual dose).

The mechanism of Magnetofection is probably the same as that for standard transfection and transduction concerning vector uptake into cells (Huth et al. 2004). Magnetic iron oxide particles are cointernalized with vectors into cells and are biodegradable over long time periods. Magnetofection is applicable to small and large nucleic acids, which can be synthetic, including small interfering RNA (siRNA), or of biological origin (plasmid DNA and larger constructs), naked or in vector formulation (lipoplexes, polyplexes) (Scherer et al. 2002). It is also applicable to viral vectors, having been demonstrated for adenovirus (Pandori et al. 2002; Scherer et al. 2002; Mok et al. 2005), adeno-associated virus (AAV) (Mah et al. 2002), retrovirus (Hughes et al. 2001; Scherer et al. 2002), lentivirus (Haim et al. 2005) baculovirus (Raty et al. 2004), and measles virus (Kadota

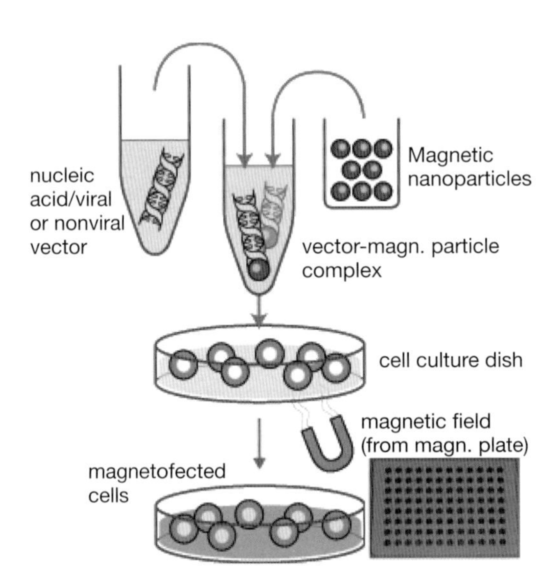

FIGURE 1. Principle of Magnetofection. In its simplest form, Magnetofection consists of mixing a naked nucleic acid or a vector of interest with polycation- or polyanion-coated magnetic nanoparticles. The components will self-assemble by electrostatic interaction and/or by salt-induced aggregation. Alternatively, biological interactions such as biotin-streptavidin or chemical coupling with biofunctional linkers are an option. The magnetic fields of strong permanent magnets positioned on the lower side of cell culture dishes will sediment the full vector dose on the cell culture and thus assist in nucleic acid introduction. For this purpose, magnetic plates for formats of various cell culture dishes have been constructed (Scherer et al. 2002).

et al. 2005). Depending on viral surface characteristics, the association between virus and magnetic particles proceeds by electrostatic interactions.

Streptavidin-coated magnetic particles have also been used. The association of vectors with these particles requires that at least one vector component (nucleic acid and/or polycation or cationic lipid) or the viral vector itself be biotinylated. Procedures for various viruses that have been associated with streptavidin magnetic particles are available (Hughes et al. 2001; Mah et al. 2002; Pandori et al. 2002; Raty et al. 2004; Mok et al. 2005). Magnet pieces, which can be disk-shaped or rectangular, are attached to the bottom side of cell culture dishes using adhesive tape. Nucleic acid delivery within a cell culture dish will then be confined to an area defined by the shape of the magnet. Larger pieces of permanent magnets or strong electromagnets producing large field gradients may be required for Magnetofection in vivo. However, to date, no magnet technology is commercially available that is especially designed for in vivo magnetic drug targeting. The interested reader may contact the authors of this chapter for further advice.

Most of the viral protocols cited here were developed independently. Most authors do not use the "Magnetofection" nomenclature for their procedures. Here, the term Magnetofection is used for simplicity to cover all forms of magnetically guided nucleic acid delivery. Several reviews have been written on the subject (Plank et al. 2003a,b; Schillinger et al. 2005). Figure 1 summarizes the protocol.

Protocol

Use of Magnetic Nanoparticles for Nucleic Acid Delivery

This protocol describes the use of magnetic nanoparticles for delivery of nucleic acid to target cells, using either nonviral or viral vectors.

MATERIALS

CAUTION: See Appendix for appropriate handling of materials marked with <!>.

Reagents

Cells

Target cells are plated as for standard nucleic acid delivery, preferably the day prior to Magnetofection. For nonviral Magnetofection, the cells should be 60–80% confluent. For suggested cell numbers in various plate formats, see Table 1. The cells should be covered with a volume of medium equaling the "Final Transfection Volume" minus the "Volume of Vector Preparation" in Table 1.

Dulbecco's modified Eagle's medium (DMEM) or other appropriate culture medium

Magnetic nanoparticles

Iron oxide nanoparticles with polycation or polyanion surface coatings (containing amino, carboxylic acid, or phosphate groups) or with a streptavidin coating can be obtained from OZ Biosciences (Marseille, France; www.ozbiosciences.com), chemicell (Berlin, Germany; www.chemicell.com), New England BioLabs (www.neb.com), Bangs Laboratories (Fishers; www.bangslabs.com), Miltenyi Biotech (Bergisch-Gladbach, Germany; www.miltenyibiotech.com), Dynal (www.dynalbiotech.com), and Promega (www.promega.com). The diameter should not exceed 1 μm and should preferably be less than 500 nm. Magnetic nanoparticles designed and optimized for Magnetofection are commercially available under the names CombiMAG, PolyMAG, and silenceMAG from OZ Biosciences or chemicell.

Nucleic acid vectors

Any known vector can be used. If streptavidin-coated magnetic particles are used, vectors must be biotinylated. For instructions, see protocols of providers of biotinylation reagents such as

TABLE 1. Cell Numbers, DNA/Nucleic Acid Doses and Transfection Volumes Suggested for Nonviral Magnetofection

Tissue-culture dish	Cell number	DNA quantity (μg)	Volume of vector preparation (ml)	Final transfection volume (ml)
96 well	0.5–2 x 10^4	0.01–1	0.05	0.2
24 well	0.5–1 x 10^5	0.1–10	0.1–0.2	0.5–1
12 well	1–2 x 10^5	0.2–20	0.15–0.3	1.5
6 well	2–4 x 10^5	0.5–35	0.2–0.6	2–3
60-mm dish	5–10 x 10^5	1–100	0.5–1.6	5–8
90- to 100-mm dish	10–20 x 10^5	1–60	0.5–2	10–20
T-25 flask	5–10 x 10^5	0.2–20	0.5–2	5–10
T-75 flask	20–50 x 10^5	1–60	0.5–2	15–20

Transfection volume = total transfection volume = volume of vector preparation plus volume of medium covering the cells. As a rule of thumb, nonviral vectors should be prepared in a manner such that the final nucleic acid concentration in the vector preparation does not exceed 100 μg/ml, and preferably does not exceed 50 μg/ml.

Sigma-Aldrich or Pierce. With polycation or polyanion-coated magnetic particles such as CombiMAG, no vector biotinylation is required. Viral vectors can be biotinylated and associated with streptavidin-coated magnetic particles, or native vectors can be incubated with magnetic particles (preferably polycation-coated, although polyanion-coated particles have been used with lentiviruses [Haim et al. 2005]).

Phosphate-buffered saline (PBS), HEPES-buffered saline (HBS), 0.9% sodium chloride (NaCl), or serum-free cell culture medium for vector preparation

Transfection reagent of choice

Equipment

Incubator preset to 37°C

Microcentrifuge tube

Permanent magnets

Rare earth permanent magnets (such as neodymium-iron-boron magnets) are most suitable. Manufacturers include IBS Magnet (Berlin, Germany), Dexter Magnetic Technologies (Chicago, Illinois), Bisbell Magnetic Products Ltd. (Burton-on-Trent, UK), or amf magnetics (Mascot, Australia). Magnetic plates designed and optimized for Magnetofection are commercially available from OZ Biosciences (Marseille, France; www.ozbiosciences.com) or chemicell (Berlin, Germany; www.chemicell.com).

Suitable pieces of permanent magnets are disk-shaped and are 0.5, 1, 1.5, 2 cm or more in diameter and 0.5 cm in height, or rectangular pieces of 2 x 1 x 0.5 cm with remanences around one Tesla.

CAUTION: RARE-EARTH PERMANENT MAGNETS PRODUCE STRONG MAGNETIC FIELDS AND MUST BE HANDLED WITH CARE. THESE MAGNETS CAN DELETE MAGNETIC STORAGE DEVICES AND MAY INTERFERE WITH PROPER OPERATION OF ELECTRONIC OR ELECTROMECHANIC DEVICES SUCH AS CARDIAC PACEMAKERS.

Tissue-culture dishes or flasks of choice (see Table 1)

METHOD

The ratios of magnetic particles to gene vectors must be optimized for each particle and vector type. For detailed optimized protocols, see www.ozbiosciences.com.

1. Prepare a 1 mg/ml dilution of magnetic nanoparticles in deionized H_2O.

   Most suppliers ship magnetic nanoparticles as aqueous suspensions of several mg/ml. Polycation- and polyanion-coated magnetic particles will start to aggregate in salt-containing solution.

2. Calculate the amount of magnetic nanoparticles required. For nonviral vectors, use 0.2–4 μg of surface-coated magnetic nanoparticles per microgram of nucleic acid to be delivered; 15 μg of polyethylenimine-coated magnetic particles are sufficient to bind 10 billion adenoviral particles. For suggested nucleic acid doses in various plate formats, see Table 1 for nonviral Magnetofection and Table 2 for viral Magnetofection. Add the required amount of magnetic particles to a microcentrifuge tube.

3. Preparation of nonviral or viral vector

   For nonviral vector

   a. Prepare the vector according to the instructions of the supplier of the transfection reagent. It is essential to prepare the vector in a salt-containing solution, preferably in serum- and supplement-free cell culture medium such as DMEM, 0.9% NaCl, PBS, or HBS.

   b. Mix the nucleic acid and transfection reagent and immediately transfer this preparation to the magnetic particles. Mix well using a pipette.

TABLE 2. Cell Numbers, Magnetic Particle Doses and Transfection Volumes Suggested for Viral Magnetofection

Tissue-culture dish	Cell number	Magnetic particles (μg per well)	Volume of vector preparation (ml)	Final transfection volume (ml)
96 well	0.5–2 x 10^4	0.1–1	0.05	0.2
24 well	0.5–1 x 10^5	1–10	0.1–0.2	0.5–1
12 well	1–2 x 10^5	2–20	0.15–0.3	1.5
6 well	2–4 x 10^5	4–40	0.2–0.6	2–3
60-mm dish	5–10 x 10^5	10–50	0.5–1.6	5–8
90–100-mm dish	10–20 x 10^5	30–150	0.5–2	10–20
T-25 flask	5–10 x 10^5	10–50	0.5–2	5–10
T-75 flask	20–50 x 10^5	30–150	0.5–2	15–20

Transfection volume = total transfection volume = volume of vector preparation plus volume of medium covering the cells.

For viral vector

a. If required, dilute the aliquot of viral preparation with serum-free cell culture medium or other salt-containing buffer. Alternatively, directly use an aliquot of culture supernatant from a producer cell line.

b. Add the viral dilution to the magnetic particles and mix using a pipette.

IMPORTANT: Do not use polycationic additives such as Polybrene as is often used for standard transduction protocols.

4. Incubate for 15–30 minutes at room temperature. For the viral protocol, incubate on ice. During this time, magnetic particles and vectors associate.
5. Add the magnetic vector preparation to the target cells. Place the cell culture dish on a commercially available magnetic plate or attach suitable pieces of permanent magnets to the bottom side of the cell culture dish using adhesive tape.
6. Incubate in a cell culture incubator or at room temperature for 15–30 minutes and then remove the magnet.
7. *Optional:* Change medium or supplement with more medium and continue cultivation under standard conditions until the transfection experiment is analyzed.

Magnetofection can be carried out using streptavidin-coated magnetic particles and biotinylated vectors according to the above protocol. A potential limitation with nonviral vectors is that usually transfection reagents such as cationic lipids or polycations are used in excess over the nucleic acid component. These excess reagents may not be directly associated with the nucleic acid and may block binding sites on streptavidin-coated magnetic particles.

REFERENCES

Haim H., Steiner I., and Panet A. 2005. Synchronized infection of cell cultures by magnetically controlled virus. *J. Virol.* **79:** 622–625.

Hughes C., Galea-Lauri J., Farzaneh F., and Darling D. 2001. Streptavidin paramagnetic particles provide a choice of three affinity-based capture and magnetic concentration strategies for retroviral vectors. *Mol. Ther.* **3:** 623–630.

Huth S., Lausier J., Gersting S.W., Rudolph C., Plank C., Welsch U., and Rosenecker J. 2004. Insights into the mechanism of magnetofection using PEI-based magnetofectins for gene transfer. *J. Gene Med.* **6:** 923–936.

Kadota S.I., Kanayama T., Miyajima N., Takeuchi K., and Nagata K. 2005. Enhancing of measles virus infection by magnetofection. *J. Virol. Methods* **128:** 61–66.

Krotz F., Wit C., Sohn H.Y., Zahler S., Gloe T., Pohl U., and Plank C. 2003. Magnetofection—A highly efficient tool for antisense oligonucleotide delivery in vitro and in vivo. *Mol. Ther.* **7:** 700–710.

Luo D. and Saltzman W.M. 2000. Enhancement of transfection by physical concentration of DNA at the cell surface. *Nat. Biotechnol.* **18:** 893–895.

Mah C., Fraites T.J.J., Zolotukhin I., Song S., Flotte T.R., Dobson J., Batich C., and Byrne B.J. 2002. Improved method of recombinant AAV2 delivery for systemic targeted gene therapy. *Mol. Ther.* **6:** 106–112.

Mok H., Palmer D.J., Ng P., and Barry M.A. 2005. Evaluation of polyethylene glycol modification of first-generation and helper-dependent adenoviral vectors to reduce innate immune responses. *Mol. Ther.* **11:** 66–79.

Pandori M.W., Hobson D.A., and Sano T. 2002. Adenovirus-microbead conjugates possess enhanced infectivity: A new strategy to localized gene delivery. 2002. *Virology* **299:** 204–212.

Plank C., Anton M., Rudolph C., Rosenecker J., and Krotz F. 2003a. Enhancing and targeting nucleic acid delivery by magnetic force. *Expert Opin. Biol. Ther.* **3:** 745–758.

Plank C., Schillinger U., Scherer F., Bergemann C., Remy J.S., Krotz F., Anton M., Lausier J., and Rosenecker J. 2003b. The magnetofection method: Using magnetic force to enhance gene delivery. *Biol. Chem.* **384:** 737–747.

Raty J.K., Airenne K.J., Marttila A.T., Marjomaki V., Hytonen V.P., Lehtolainen P., Laitinen O.H., Mahonen A.J., Kulomaa M.S., and Yla-Herttuala S. 2004. Enhanced gene delivery by avidin-displaying baculovirus. *Mol. Ther.* **9:** 282–291.

Scherer F., Anton M., Schillinger U., Henke J., Bergemann C., Kruger A., Gansbacher B., and Plank C. 2002. Magnetofection: Enhancing and targeting gene delivery by magnetic force in vitro and in vivo. *Gene Ther.* **9:** 102–109.

Schillinger U., Brill T., Rudolph C., Huth S., Gersting S., Krotz F., Hirschberger J., Bergemann C., and Plank C. 2005. Advances in Magnetofection—Magnetically guided nucleic acid delivery. *J. Magn. Magn. Mat.* **293**: 501–508.

73 Photochemical Internalization for Light-directed Gene Delivery

Anette Bonsted,* Anders Høgset,† Ernst Wagner,‡ and Kristian Berg*

*Department of Radiation Biology, Institute for Cancer Research, Rikshospitalet-Radiumhospitalet HF, Montebello, N-0310 Oslo, Norway; †PCI Biotech AS, Hoffsvn 48, N-0310 Oslo, Norway; ‡Pharmaceutical Biology-Biotechnology, Department of Pharmacy, Ludwig-Maximilians-Universitat, D-81377 Munich, Germany

ABSTRACT

Photochemical internalization (PCI) enables efficient transfection by light-directed disruption of endocytic vesicles within cells (Berg et al. 1999). The technology is based on photosensitizers that localize in the membranes of endocytic vesicles and, following activation by light, induce rupture of these membranes. Endocytosed compounds (e.g., nucleic acids) are released into the cell cytosol where they may act on their target or further translocate to the nucleus (Fig. 1).

PCI substantially improves the delivery of nonviral vectors, such as plasmid polyplexes and peptide nucleic acids (Høgset et al. 2000; Folini et al. 2003). PCI also enhances gene transfer from adenovirus and adeno-associated virus vectors (Høgset et al. 2002; Bonsted et al. 2005). The technology can be used to transfect or transduce cells in vitro and in vivo when 100% survival of the treated cell population is not required. The effect of the photochemical treatment depends on exposure to light, establishing PCI as a highly specific treatment modality. Moreover, dual targeting can be achieved by combining the photochemical treatment with gene vectors targeted to specific cell surface receptors (Kloeckner et al. 2004). Performing PCI with gene vector formulations that are not efficient on their own may reduce the risk of side effects due to inadvertent expression of transgenes in nontarget cells. Furthermore, the enhanced efficiency obtained with PCI could minimize vector-related toxicity and immunogenicity because of the lower amounts of vector needed. Photochemical treatment with adenovirus can be applied to slowly dividing cells or nondividing cells. The cytotoxicity induced by the photochemical treatment may be a drawback for some gene therapy applications, but it has been successfully exploited in photodynamic therapy (PDT) for cancer therapy. It is possible to reduce the cytotoxicity by reducing the light dose. However, this would also decrease the effect of the photochemical treatment on transfection/transduction. Thus, at present, PCI may be best suited for treatments where tissue removal is the main goal.

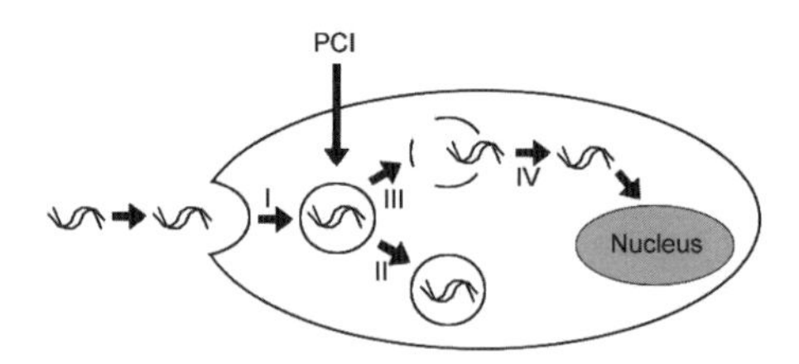

FIGURE 1. PCI-mediated delivery of nucleic acids. The nucleic acids are endocytosed and localize in endocytic vesicles (I). If not translocated, the nucleic acids are degraded inside the endocytic vesicles (II). PCI permeabilizes the endocytic membranes (III), leading to the release of the nucleic acids into the cytosol (IV).

INTRODUCTION

The efficiency of transfection by PCI is influenced by the photosensitizer, the light dose, and the gene vector applied, as described here. Photosensitizers are compounds that, upon absorption of light at specific wavelengths, induce chemical or physical alterations in other chemical entities. The photochemical reactions induced by the photosensitizers utilized in PCI proceed mainly via formation of singlet oxygen ($^1O_2$), a highly reactive form of oxygen generated after interaction between the excited photosensitizer and ground-state molecular oxygen (O_2). Singlet oxygen has a short lifetime (<0.04 µsec) and a short range of action (10–20 nm) in cells (Moan and Berg 1991). Accordingly, only structures very close to the photosensitizer will be affected following light exposure, whereas distant molecules will be left unaffected.

To date, aluminum phthalocyanine with two sulfonate groups on adjacent phenyl rings ($AlPcS_{2a}$) and mesotetraphenylporphine with two sulfonate groups on adjacent phenyl rings ($TPPS_{2a}$) have been the most efficient photosensitizers for enhancing gene delivery (Fig. 2A) (Prasmickaite et al. 2001). These amphiphilic photosensitizers localize in the membranes of the endocytic vesicles (Berg and Moan 1994). The light source may be any source emitting light of wavelengths absorbed by the photosensitizer. Typically, a red light source (peak wavelength at 670 nm) is used for the excitation of $AlPcS_{2a}$ and a blue light source (peak wavelength at 420

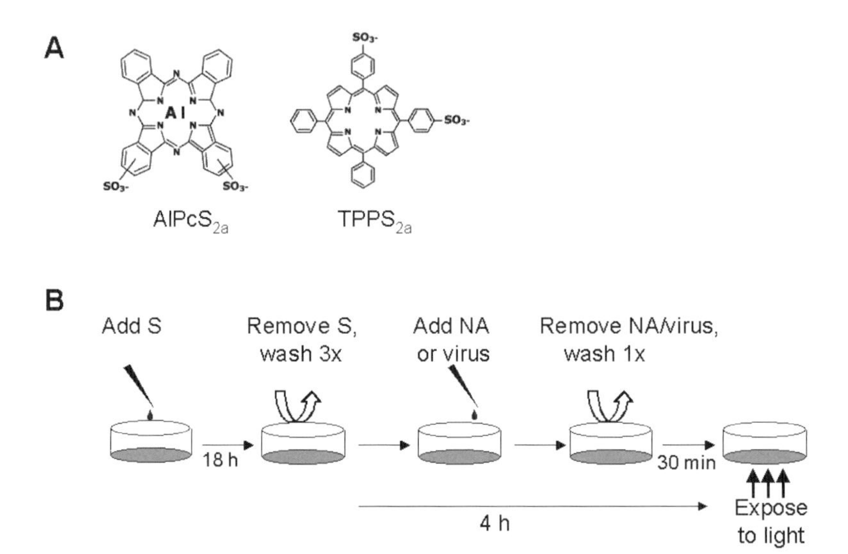

FIGURE 2. (*A*) Chemical structures of photosensitizers used for PCI; (*B*) PCI experimental scheme. The photosensitizer (S) is added to the cultured cells for 18 hours. The cells are washed, incubated with the desired nucleic acids or recombinant virus, and exposed to light at an appropriate wavelength.

nm) for $TPPS_{2a}$. Photosensitizers that localize to cellular structures other than the endocytic vesicles are not efficient for inducing the PCI effect (Berg et al. 1999; Prasmickaite et al. 2001).

PCI enhances gene delivery from several gene vector constructs of both nonviral and viral origin (Table 1). Of note, the relative effect of PCI is most pronounced at suboptimal doses of DNA-

TABLE 1. In Vitro and In Vivo Gene Therapy Studies Performed with PCI

Vector	Photosensitizer	Cell line	Reference
		(a) *In vitro experiments*	
Untargeted nonviral vectors			
poly-L-lysine (with reporter gene EGFP)	$TPPS_{2a}$ $A1PcS_{2a}$	HCT116, THX; HCT116, THX; THX; CME-1, SW 982; U87MG, GaMg	Høgset et al. (2000) Prasmickaite et al. (2000) Prasmickaite et al. (2001) Dietze et al. (2003) Bonsted et al. (2005)
PEI (with reporter genes EGFP and luciferase, therapeutic gene HSV-thymidine kinase)	$TPPS_{2a}$ $A1PcS_{2a}$	THX, HCT116; U87MG, HUH7, A431, HepG2	Prasmickaite et al. (2000) Prasmickaite et al. (2004) Kloeckner et al. (2004)
Cationic lipids (with reporter gene EGFP)	$A1PcS_{2a}$	HCT116, THX, BL2-G-E6 HCT116, THX; HCT116, THX;	Berg et al. (1999) Prasmickaite et al. (2000) Prasmickaite et al. (2001)
Peptide nucleic acids (PNA)	$TPPS_{2a}$ $A1PcS_{2a}$	DU145	Folini et al. (2003)
Targeted nonviral vectors			
EGFR-targeted PEI (with reporter genes EGFP and luciferase, therapeutic gene HSV-thymidine kinase)	$TPPS_{2a}$ $A1PcS_{2a}$	HUH7, HepG2, A431	Kloeckner et al. (2004)
Glucosylated PEI (with *p53* gene)	$TPPS_{2a}$	FaDu, PANC3	Ndoye et al. (2004a,b)
PEI (with PTEN gene)	$TPPS_{2a}$	Ishikawa	Maurice-Duelli et al. (2004)
Untargeted viral vectors			
AdCMV-*lacZ*	$AlPcS_{2a}$ $TPPS_{2a}$	WiDr, HCT116; CME-1, SW 982; WiDr, FEMXIII, HeLa, HuFib; U87MG, GaMg; WiDr, A549, FEMXIII, HeLa, HuFib, RaJi; THX*, OHS*, SaOS*, FLS*	Høgset et al. (2002) Dietze et al. (2003) Bonsted et al. (2004) Bonsted et al. (2005) Engesæter et al. (2005)
AdCMV-EGFP	$TPPS_{2a}$	WiDr*, HCT116*, THX*	
AdCMV-*luc*	$TPPS_{2a}$	WiDr*	
AdCMV-TRAIL	$TPPS_{2a}$	WiDr, HCT116, OHS*, SaOS*	B.Ø. Engesæter et al. (in prep.)
Targeted viral vectors			
AdRGD-GFP	$TPPS_{2a}$	WiDr	Engesæter et al. (2005)
EGFR-targeted AdCMV-*lacZ*	$TPPS_{2a}$	DU145, RD	Bonsted et al. (2005)
		(b) *In vivo experiments*	
Plasmid (with *p53* gene)	$AlPcS_{2a}$	human head and neck cancer implants	Ndoye et al. (2006)

Asterisks indicate unpublished experiments.

(EGFR) Epidermal growth factor receptor; (AdRGD-GFP) adenovirus containing an RGD peptide epitope in the HI loop of the fiber knob and expressing green fluorescent protein controlled by the CMV promoter.

vector complexes (e.g., plasmid DNA-polyethylenimine polyplexes at nitrogen to phosphate [N:P] ratios of ≤4 and recombinant adenovirus vector doses at multiplicities of infection of ≤20). Usually, the cells are first preloaded with the sensitizer and treated with a gene vector construct, followed by light exposure (Høgset et al. 2000; Prasmickaite et al. 2000). This procedure is explained in detail below. Alternatively, the cells may be treated with the photosensitizer and exposed to light before the gene vector is added, as long as the gene vector is added within 3–6 hours after light exposure (Prasmickaite et al. 2002).

The photochemical dose to be used for induction of photochemical gene delivery must be determined individually for every cell line, photosensitizer, and light source. The survival of cells preloaded with the photosensitizer as a function of the light dose should be determined. Typically, 5–20 μg/ml $AlPcS_{2a}$ or 0.2–1 μg/ml $TPPS_{2a}$ is used. The light dose killing about 50% of the cells is recommended as a starting point for obtaining photochemically enhanced gene delivery. Cell survival can be measured by one of the common cell survival tests such as the MTT (3-[4,5-dimethylthiazol-2-yl]-2,5-diphenyltetrasodium bromide) test, clonogenic analysis, protein synthesis, or any other test established in individual laboratories. All indirect cell survival tests should be performed 24–48 hours after exposure of the cells to light, because at least 24 hours is required for the cytotoxic effect of the photochemical treatment to be fully revealed.

Protocol

Photochemically Enhanced Gene Delivery

This protocol describes the use of photosensitizers to enhance introduction of DNA into cells exposed to light.

MATERIALS

CAUTION: See Appendix for appropriate handling of materials marked with <!>.

Reagents

Cells to be transfected

Cell culture medium appropriate for cells to be transfected

Nucleic acids or recombinant viruses

> Freshly dilute the nucleic acids (*optional*, DNA polyplexes) in cell culture medium before the application to the cells or freshly dilute the recombinant adenovirus, adeno-associated virus, or possibly other viruses that are endocytosed to the desired multiplicity of infection, preferably in PBS or cell culture medium without serum.
>
> The amount of DNA or viral vectors to use may depend on the cell line and the vector. We recommend using DNA concentrations of 0.2–5 μg/ml or viral doses of 5 infectious particles per cell as a starting point.

Phosphate-buffered saline (PBS)

Photosensitizer: $AlPcS_{2a}$ or $TPPS_{2a}$ (Porphyrin Products, Logan, Utah)

> For the stock solution of the photosensitizer, dissolve 5 mg of $AlPcS_{2a}$ or 2 mg of $TPPS_{2a}$ in 0.2 ml of 0.1 M NaOH. Dilute with sterile PBS to a final volume of 1 ml. Sterilize by filtration and store at –20°C in small aliquots for up to 6 months.
>
> Protect the photosensitizer solutions from light to avoid photoinduced damage. Also avoid long-term storage or repeated freezing/thawing, which may cause the photosensitizers to aggregate and reduce their efficacy. Treat the photosensitizer solution in a sonicator bath for a few seconds if complete solubilization is difficult.

Sodium hydroxide (NaOH; 0.1 M) <!>

Equipment

Incubator (CO_2), preset to 37°C

Light source for excitation of $AlPcS_{2a}$ at 670 nm or $TPPS_{2a}$ at 420 nm: LumiSource (PCI Biotech AS, Oslo, Norway)

Tissue-culture dishes

METHOD

For a simplified experimental scheme, see Figure 2B.

1. Seed cell culture dishes with cells to be transfected in appropriate culture medium and allow the cells to attach to the substratum.

There is not a certain size of cell culture dish that works best. The choice of culture dish is determined by the method of analysis. For example, if the investigator wishes to study transgene expression in cells by flow cytometry analysis, 75,000 cells per well in a 12-well plate (Nunc, Roskilde, Denmark) would be sufficient.

Leave cells 6 hours before adding the photosensitizer.

2. Prepare the working solution of the photosensitizer in cell culture medium just before applying it to the cells. Use 5–20 μg/ml $AlPcS_{2a}$ or 0.2–1 μg/ml $TPPS_{2a}$, as described in the Introduction.

   The optimal photosensitizer concentration is dependent on the light source in use, that is, fluence rate (light intensity) and wavelength overlap with the absorption spectrum of the photosensitizer.

   From this point on, perform all steps in subdued light to avoid uncontrollable activation of the photosensitizer and to protect the cells from photochemical damage. For in vitro studies, it may be sufficient to turn off the light at the sterile bench during the procedure.

3. Remove the cell culture medium and add cell culture medium containing the photosensitizer. Incubate for 16–18 hours at 37°C in a CO_2 incubator.

4. Remove the medium containing photosensitizer and wash the cells three times with photosensitizer-free cell culture medium. Chase for 4 hours at 37°C. During the chase period, add the nucleic acids (*optional*, DNA polyplexes) or recombinant virus vector solutions to the cells at the desired time point. The suggested incubation time is 3.5 hours for nonviral gene vectors and 30 minutes for recombinant viral vectors.

   For drug delivery by PCI, incubate the cells with the photosensitizer and the macromolecule at the same time (e.g., for 18 hours before removal of the photosensitizer). However, for nucleic acid delivery, incubation with the photosensitizer before the nucleic acids is recommended.

5. Remove the vector and wash the cells once. Incubate for 30 minutes at 37°C in a CO_2 incubator.

6. Expose the cells to light, using the light dose empirically determined as described in the Introduction.

   To reduce damage to the plasma membrane, incubate the cells in photosensitizer-free medium for 1–4 hours before exposure to light.

7. Grow the cells in the dark for another 24–48 hours and analyze for transgene expression.

REFERENCES

Berg K. and Moan J. 1994. Lysosomes as photochemical targets. *Int. J. Cancer* **59:** 814–822.

Berg K., Selbo P.K., Prasmickaite L., Tjelle T.E., Sandvig K., Moan J., Gaudernack G., Fodstad Ø., Kjølsrud S., Anholt H., Rodal G.H., Rodal S.K., and Høgset A. 1999. Photochemical internalization: A novel technology for delivery of macromolecules into cytosol. *Cancer Res.* **59:** 1180–1183.

Bonsted A., Høgset A., Hoover F., and Berg K. 2005. Photochemical enhancement of gene delivery to glioblastoma cells is dependent on the vector applied. *Anticancer Res.* **25:** 291–298.

Bonsted A., Engesæter B.Ø., Høgset A., Mælandsmo G.M., Prasmickaite L., Kaalhus O., and Berg K. 2004. Transgene expression is increased by photochemically mediated transduction of polycation-complexed adenoviruses. *Gene Ther.* **11:** 152–160.

Bonsted A., Engesæter B.Ø., Høgset A., Mælandsmo G.M., Prasmickaite L., D'Oliveira C., Hennink W.E., van Steenis J.H., and Berg K. 2006. Photochemically enhanced transduction of polymer-complexed adenovirus targeted to the epidermal growth factor receptor. *J. Gene Med.* **8:** 286–297.

Dietze A., Bonsted A., Høgset A., and Berg K. 2003. Photochemical internalization enhances the cytotoxic effect of the protein toxin gelonin and transgene expression in sarcoma cells. *Photochem. Photobiol.* **78:** 283–289.

Engesæter B.Ø., Bonsted A., Berg K., Høgset A., Engebråten O., Fodstad Ø., Curiel D.T., and Mælandsmo G.M. 2005. PCI-enhanced adenoviral transduction employs the known uptake mechanism of adenoviral particles. *Cancer Gene Ther.* **12:** 439–448.

Folini M., Berg K., Millo E., Villa R., Prasmickaite L., Daidone M.G., Benatti U., and Zaffaroni N. 2003. Photochemical internalization of a peptide nucleic acid targeting the catalytic subunit of human telomerase. *Cancer Res.* **63:** 3490–3494.

Høgset A., Prasmickaite L., Tjelle T.E., and Berg K. 2000. Photochemical transfection: A new technology for light-induced, site-directed gene delivery. *Hum. Gene Ther.* **11:** 869–880.

Høgset A., Engesæter B.Ø., Prasmickaite L., Berg K., Fodstad Ø., and Mælandsmo G.M. 2002. Light-induced adenovirus gene transfer, an efficient and specific gene delivery technology for cancer gene therapy. *Cancer Gene Ther.* **9:** 365–371.

Kloeckner J., Prasmickaite L., Høgset A., Berg K., and Wagner E. 2004. Photochemically enhanced gene delivery of EGF receptor-targeted DNA polyplexes. *J. Drug Target.* **12:** 205–213.

Maurice-Duelli A., Ndoye A., Bouali S., Leroux A., and Merlin J.L. 2004. Enhanced cell growth inhibition following PTEN nonviral gene transfer using polyethylenimine and photochemical internalization in endometrial cancer cells. *Technol. Cancer Res. Treat.* **3:** 459–465.

Moan J. and Berg K. 1991. The photodegradation of porphyrins in cells can be used to estimate the lifetime of singlet oxygen. *Photochem. Photobiol.* **53:** 549–553.

Ndoye A., Merlin J.L., Leroux A., Dolivet G., Erbacher P., Behr J.P., Berg K., and Guillemin F. 2004. Enhanced gene transfer and cell death following p53 gene transfer using photochemical internalisation of glucosylated PEI-DNA complexes. *J. Gene Med.* **6:** 884–894.

Ndoye A., Bouali S., Dolivet G., Leroux A., Erbacher P., Behr J.P., Berg K., Guillemin F., and Merlin J.L. 2004. Sustained gene transfer and enhanced cell death following glucosylated-PEI-mediated p53 gene transfer with photochemical internalisation in p53-mutated head and neck carcinoma cells. *Int. J. Oncol.* **25:** 1575–1581.

Ndoye A., Dolivet G., Høgset A., Leroux A., Fifre A., Erbacher P., Berg K., Behr J.P., Guillemin F., and Merlin J.L. 2006. Eradication of p53-mutated head and neck squamous cell carcinoma xenografts using nonviral gene therapy and photochemical internalization. *Mol. Ther.* **13:** 1156–1162.

Prasmickaite L., Høgset A., and Berg K. 2001. Evaluation of different photosensitizers for use in photochemical gene transfection. *Photochem. Photobiol.* **73:** 388–395.

Prasmickaite L., Høgset A., Tjelle T.E., Olsen V.M., and Berg K. 2000. The role of endosomes in gene transfection mediated by photochemical internalisation. *J. Gene Med.* **2:** 477–488.

Prasmickaite L., Høgset A., Olsen V.M., Kaalhus O., Mikalsen S.O., and Berg K. 2004. Photochemically enhanced gene transfection increases the cytotoxicity of the herpes simplex virus thymidine kinase gene combined with ganciclovir. *Cancer Gene Ther.* **11:** 514–523.

Prasmickaite L., Høgset A., Selbo P.K., Engesæter B.Ø., Hellum M., and Berg K. 2002. Photochemical disruption of endocytic vesicles before delivery of drugs: A new strategy for cancer therapy. *Br. J. Cancer* **86:** 652–657.

74 Pronuclear Microinjection in Mice

Walter Tsark

Transgenic/Knockout Mouse Core Facility, City of Hope Beckman Research Institute, Duarte, California 91010

ABSTRACT

The purpose of this chapter is to describe the techniques used to produce transgenic mice by pronuclear microinjection in order to facilitate comparison with other methods of producing transgenic animals presented in this volume. Pronuclear microinjection is a well-established technique involving the injection of DNA into the haploid pronuclei of a fertilized egg. Transgenic mice have been produced with simple promoter/cDNA constructs and large genomic constructs such as cosmids, yeast artificial chromosomes (YACs), and bacterial artificial chromosomes (BACs). The major elements of transgenic mouse production are micro-tool fabrication, mouse embryo production and culture, pronuclear microinjection, and surgical embryo transfer. This chapter provides instruction on these major elements of transgenic mouse production with the exception of surgical embryo transfer. This subject is covered in detail in Mann 1993, Hogan et al. 1994, and Nagy et al. 2003.

Continued

INTRODUCTION

The specific term for an animal created for the purpose of expressing an exogenous DNA sequence integrated into its genome is "transgenic" (Gordon and Ruddle 1981).

The thousands of citations produced by the keyword search "transgenic mouse and not knockout" of the PubMed database (www.pubmed.gov) provide an approximation of the scientific value of studying transgenes in mice. Currently, the two most common techniques used to produce transgenic mice are pronuclear microinjection and germ-line chimera production with transfected mouse embryonic stem cells (see, e.g., Gossler et al. 1986). In pronuclear microinjection, the transgene sequence is injected into the haploid pronuclei of a fertilized one-cell mouse embryo with a fine glass microneedle. The essential methodology of pronuclear microinjection has not changed significantly from the original description (Gordon et al. 1980). Subsequent technical refinements have resulted in a routine efficiency of transgenesis in mice of approximately 20% and 60% of the mice derived from microinjected embryos (Mann and McMahon 1993). Pronuclear microinjection has been used successfully to produce transgenic animals in several species other than mice and rats, but the efficiency is somewhat lower (Hammer et al. 1985). One of the advantages of pronuclear microinjection over transfection of embryonic stem cells is that each experiment generates several founder transgenic animals directly. In addition, the technique has been applied successfully to a very wide range of construct sizes: small RNAs, plasmid vectors, and large genomic constructs (cosmids, BAC, and YAC vectors).

The disadvantages of pronuclear microinjection include the expense for equipment, and the fact that gaining proficiency with pronuclear microinjection requires substantial training. In addition, the efficiency of transgenic mouse production by pronuclear microinjection varies among different strains of laboratory mice (Mann and McMahon 1993; Auerbach et al. 2003). Several inbred, F1, and F2 hybrid strains are used routinely for transgenic mouse production (e.g., C57BL/6J, FVB, and [C57BL/6J x CBA/J]F1 or F2 hybrids). Unfortunately, not all inbred strains can be used with the same rates of success. One notable example of this is the 129S strain, used commonly for gene-targeting experiments.

Mann and McMahon (1993) described the isolation of DNA for microinjection by zonal rate centrifugation through a linear sucrose gradient and demonstrated that this method produces DNA of consistently high purity optimal for efficient production of transgenic mice. A drawback of this technique is that small transgene inserts may be difficult to separate from a vector backbone of similar size (Sambrook et al. 1998). This problem can be addressed by reducing the vector fragments in size using a restriction enzyme that cuts within the vector sequence only. Transgenic mice have been produced successfully with larger fragments isolated from cosmid vectors that were purified by sucrose gradient centrifugation (see, e.g., Clarke et al. 1998). Large genomic constructs such as YACs and BACs are prepared using methods to reduce DNA shearing (Hiemisch et al. 1998; Nagy et al. 2003). These large genomic constructs do not necessarily need to be linearized prior to pronuclear microinjection.

Transgenes introduced into the pronucleus tend to integrate in tandem arrays at a single location in the genome rather than as single copies. Integration of the transgene into the genome within a region of heterochromatin may influence the expression of the transgene (Martin and Whitelaw 1996; Milot et al. 1996), and elements within the promoter are sometimes silenced by methylation (Lettmann et al. 1991; Chevalier-Mariette et al. 2003). Flanking the transgene construct with chromatin insulator elements can help to reduce the effects of incorporation into heterochromatin (McKnight et al. 1992).

The success and efficiency of producing transgenic mice depend on a combination of several important factors: the skill of the microinjectionist, purity of the DNA used for microinjection, embryo culture conditions, the success of surgical transfer of the microinjected embryos into the reproductive tract of pseudopregnant recipient females, and the characteristics of the mouse strain used. Techniques for production of transgenic mice by pronuclear microinjection were first summarized in 1983 (Gordon and Ruddle 1983). A number of other authors have provided their own improvements on the original technique in subsequent years (Gordon 1993; Mann and McMahon 1993; Hogan et al. 1994; Polites and Pinkert 1994; Si-Hoe and Murphy 1999; Nagy et al. 2003).

Protocol 1

Preparation of Microinjection Chambers and Micro-tools

This protocol describes the preparation of microinjection chambers and micro-tools. Puliv-type microinjection chambers made from glass microscope slides (Latham and Solter 1993) are easy and inexpensive to make and can be used with differential interference contrast (DIC) optics or Hoffman Modulation Contrast optics (Fig. 1). Both microinjection needles (Femtotips II, Eppendorf) and holding pipettes (Vacutips, Eppendorf) are available from commercial sources, but fabricating tools "in-house" is probably still the norm for most laboratories. The basic set of micromanipulation equipment can be used for other common micromanipulation techniques (injection of embryonic stem cells into blastocyst embryos, subzonal injection of lentiviral vectors).

MATERIALS

CAUTION: See Appendix for appropriate handling of materials marked with <!>.

Reagents

Ethanol (95% and 70%)
Hydrochloric acid (1%) <!>
SafetyCoat Nontoxic Coating (JTBaker 4017-01) or Sigmacote <!> (Sigma-Aldrich)

Equipment

Blunt 4-inch stainless-steel needle (Custom order: Hamilton Co., Reno, Nevada) or a blunt 18-gauge hypodermic needle fitted with a piece of 1/32 internal diameter (ID) Tygon tubing (Saint-Gobain Performance Plastics AJC40001) to fill holding pipettes from the back end
Diamond or carbide pencils
Glass slides (25 x 75 mm, precleaned) compatible with DIC optics
High-vacuum grease (Dow Corning 1597418)

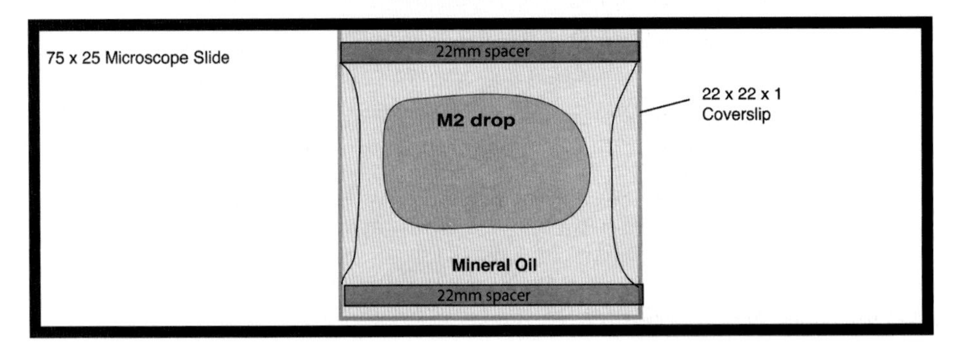

FIGURE 1. Microinjection chamber prepared for use. A 300–400-µl drop of M2 medium is sandwiched between the slide and coverslip and surrounded by light mineral oil.

Micro-bunsen burner

The burner is modified to produce a very small flame by replacing the burner tube with a blunt 18-gauge needle.

Microcapillary tubes (100 μl)

Cut the tubes into 22-mm pieces with diamond or carbide pencils (prepare two pieces for each chamber).

Microforge (DeFonbrune-type or Narishige-type) fitted with a binocular compound microscope; 10x oculars with a reticule micrometer in one eyepiece, objectives 4x, 10x, and 20x (long working distance)

P-87/P-97 Flaming-Brown horizontal micropipette puller (Sutter Instruments, Novato California) equipped with a box-type platinum filament (2.5 x 2.5 mm or 3 x 3 mm)

Plastic bulb pipette (disposable)

Plastic slide box for dust-free storage

Silicone adhesive (GE Silicone II), high-vacuum grease, or wax to attach the microcapillary spacers to the slide

Thick-walled microcapillary (1.0-mm outer diameter [OD] x 0.58 mm ID) for holding pipettes (World Precision Instruments, Sutter Instruments)

Thin-walled single-barrel microcapillary with filament (1.0 mm OD x 0.75 mm ID) for microinjection needles (World Precision Instruments, Sarasota, Florida; Sutter Instruments)

Syringe used to apply positive pressure to the needle during siliconization (*optional,* see Step 13)

A blunt 18-gauge hypodermic needle fitted with a 6–8-cm length of Tygon tubing (1/32 ID x 3/32 OD x 1/32 wall) (Saint-Gobain Performance Plastics AJC40001) is attached to a 20-cc disposable syringe with a Luer lock tip.

METHODS

Preparation of Microinjection Chamber

1. Use a small amount of silicone adhesive (GE Silicone II) to glue a 22-mm piece of 100-μl capillary tube to a 25 x 75-mm precleaned glass slide parallel to the long side and 2 mm from the edge.
2. Glue a second 22-mm piece of capillary tubing to the other side of the slide in the same relative position as the first piece and align the ends of the spacers. Allow the adhesive to dry in a running tissue-culture hood overnight.
3. Clean the slides thoroughly with 1% hydrochloric acid, several washes in distilled H_2O, and a final rinse in 95% ethanol.
4. Use a disposable plastic bulb pipette to coat the area of the slide between the spacers with a siliconizing solution and then rinse several times with a strong stream of distilled H_2O.

   SafetyCoat Nontoxic Coating is preferable, but Sigmacote can be used.
5. Allow the microinjection slides to dry in a tissue-culture hood and store in a plastic slide box for dust-free storage.

Micro-tool Fabrication: Pulling Embryo-holding Pipettes

6. Heat a 5-mm section of the thick-walled microcapillary (1.0 mm OD x 0.58 mm ID) at a point two-thirds from one end. Pull a 30-mm section of 100–150 μm.

7. Nick the pulled thinned section with a carbide-tipped pencil at approximately 10–15 mm from the shoulder of the longer side of the pipette. Pull the long and short sections apart to produce a break.
8. Inspect the face of the pipette tip on the microforge to ensure that it is symmetrical with no chips or cracks or the finished holding pipette will not hold the embryos in place correctly during microinjection. If necessary, break the pipette on the microforge with the "thick" filament to produce a flat symmetrical face.
9. Use the thick filament of the microforge to smooth and narrow the opening at the tip to 15–20 µm, and, if necessary, introduce a 5° bend at the shoulder of the pipette to allow the holding pipette to enter the microinjection drop parallel to the floor of the chamber.
10. Polish the large end of the holding pipette with the flame of the microbunsen burner, but take care not to close the opening.

Micro-tool Fabrication: Pulling Microinjection Needles

11. To produce microinjection needles on the Sutter Instruments P-97 pipette puller, start with the values suggested by the manufacturer: Heat = Ramp value; Pull = 50; Velocity = 80; Delay = 90; Pressure = 200. Adjust the parameters to produce a needle with 7–9-mm overall taper that narrows rapidly from about 20 µm to a fine point over the last 0.5 mm.
12. Gently "chip" the point of the needle against the glass bead of a microforge to create a sharp tip with an opening of between 0.2 and 0.5 µm (Mann and McMahon 1993).
13. Optional step: Siliconizing the microinjection needle prior to "chipping" the point helps to reduce adhesion of cellular material to the needle and can increase the number of eggs that can be injected with a single needle. Perform as follows:
    a. Attach the end of the needle to a 20-cc syringe fitted with a 10-cm length of 1/32-inch ID Tygon tubing.
    b. Apply positive pressure to the syringe and then dip the tip of the needle into a reservoir of SafetyCoat Nontoxic coating diluted 1:5 with 95% ethanol twice for 5 seconds.
    c. Immediately rinse the tip of the needle by dipping it into 95% ethanol, next into 70% ethanol, and then into 0.22-µm filter-sterilized distilled H_2O.
    d. Allow the needles to dry and then chip the point on the microforge as described in Step 12.

Angle Adjustments to the Holding Pipette and Microinjection Needle

14. No further modification of microinjection needles or holding pipettes is required after finishing the tips when using the Leica-type micromanipulators. When using the rail-mounted Narishige-type micromanipulators, it is necessary to create a 30–35° bend in both the holding pipette and microinjection needle to allow the pipette tips to enter into the microinjection drop parallel to the floor of the chamber.

Protocol 2

Preparation of DNA for Microinjection

This protocol describes the preparation of DNA for microinjection using zonal rate centrifugation through a linear sucrose gradient (Sambrook et al. 1989; Mann and McMahon 1993).

MATERIALS

CAUTION: See Appendix for appropriate handling of materials marked with <!>.

Reagents

Agarose gel
Ammonium acetate (5 M) <!>
DNA microinjection buffer (10 mM Tris-HCl <!> at pH 7.5 and 0.1 mM EDTA meeting all USP Monograph requirements for water for injection [WFI; Invitrogen])
  Filter-sterilize through a 0.22-µm surfactant-free cellulose acetate filter (Millipore).
Ethanol (100% and 95%)
Phenol: chloroform <!>
Plasmid vector
Restriction enzymes appropriate for digesting plasmid vector
Sucrose solutions
  10% (w/v) and 40% (w/v) in 1 M NaCl, 10 mM Tris-HCl <!> at pH 8.0, and 1 mM EDTA prepared with WFI H_2O. Filter-sterilize through a 0.22-µm surfactant-free cellulose acetate filter (Millipore).
TE buffer (10 mM Tris-Cl <!> at pH 8.0 and 1 mM EDTA)

Equipment

Centricon-100 microconcentrator (Amersham)
Gradient former (e.g., SG30 gradient former; Amersham)
Low-flow peristaltic pump (0.03–8.2 ml/min) (VWR Scientific)
Microcentrifuge tubes (1.7 ml, sterile)
  If necessary, rinse with 0.22-µm filter-sterilized endotoxin-free distilled H_2O to remove any dust particles.
Needle (25 gauge)
Rubber stopper
  This stopper should have an 18-gauge hypodermic needle pushed through the center from top to bottom and should fit tightly into the top of the ultracentrifuge.
Ultracentrifuge tubes (Beckman Ultra-Clear 344060)
Ultracentrifuge and a swinging-bucket ultracentrifuge rotor (e.g., SW 40 Ti; Beckman-Coulter)

METHOD

1. Digest 50 μg of plasmid vector with restriction enzymes to release the transgene insert from the plasmid vector backbone. Cosmid DNA can be linearized with restriction enzymes before purification.
2. Extract with phenol:chloroform and then precipitate the digested plasmid with 1 volume of 5 M ammonium acetate and 2.5 volumes of 100% ethanol. Centrifuge at 13,000*g* for ≥30 minutes and then resuspend in 50 μl of TE buffer.
3. Pour two linear 10–40% sucrose gradients of approximately 14 ml total volume in the ultracentrifuge tubes following the gradient-former manufacturer's instructions.
4. Load the digested DNA onto one of the gradients. Use the other gradient for separation of another fragment or as balance.
5. Balance the gradients precisely with microinjection buffer. Centrifuge plasmid digests at 35,000 rpm for 14 hours at 4°C and cosmids at 35,000 rpm for 24 hours at 4°C.
6. Set up 80 sterile dust-free 1.7-ml microcentrifuge tubes in racks while the ultracentrifuge is slowing. Remove approximately 2 ml from the top of the gradient containing the DNA and insert the stopper/needle into the top of the ultracentrifuge tube.
7. Place a finger over the opening in the stopper and then push a 25-gauge needle into the bottom of the ultracentrifuge tube. The tip of the 25-gauge needle should protrude ≤5 mm into the gradient. Control the rate of outflow with the needle in the stopper and collect about 0.2-ml fractions in each collection tube.
8. Analyze a 5-μl aliquot of every fourth fraction on an agarose gel to identify the first and last fractions containing only the transgene insert. Repeat the gel analysis with all of the fractions at the boundary between vector + insert and insert-only.
9. Combine all of the fractions containing only the transgene insert. Perform five to seven consecutive buffer changes with microinjection buffer using a Centricon-100 microconcentrator. The yield will be approximately 25–50% of the starting amount of DNA for plasmids and 10–25% for cosmids.
10. Quantitate the concentration of DNA accurately by spectrophotometry and analyze a small sample on an agarose gel compared to the appropriate size standards.
11. Store the concentrated DNA at 4°C and dilute the DNA stock to between 2 and 3 μg/ml with microinjection buffer. Prepare several 100-μl aliquots of DNA for microinjection in sterile dust-free microcentrifuge tubes and store for up to 1 week at 4°C. The stock DNA can be frozen at –20°C, but take care to resuspend the DNA fully before preparing dilutions for microinjection.

Protocol 3

Superovulation and Embryo Collection

This protocol describes the procedures for superovulation of embryo donor female mice and for collection of embryos. Mice are maintained on a 14-hour light/10-hour dark cycle (e.g., lights on at 6 am, lights off at 8 pm). The age of the females superovulated and dosages of hormones are strain-dependent, but several inbred and hybrid strains (C57BL/6J, FVB, F1[B6 x CBA]) are superovulated at 3–5 weeks of age.

MATERIALS

CAUTION: See Appendix for appropriate handling of materials marked with <!>.

Reagents

Bovine testicular hyaluronidase type IV-S (10 mg/ml; Sigma-Aldrich) in M2 (40x stock).

Store aliquots at –20°C. Dilute to 250 µg/ml with M2 before use.

CO_2 for euthanizing embryo donor mice (*optional*, see Step 4b)

Embryo donor mice

Human chorionic gonadotropin (HCG) (Sigma-Aldrich)

Dilute to 500 IU/ml in sterile 0.9% saline. Store sterile 0.1-ml aliquots at –80°C for up to 6 months.

Light mineral oil (Sigma-Aldrich, batch tested for toxicity or embryo-tested)

Keep 30–40 ml of mineral oil in a loosely capped bottle in the incubator and use this "equilibrated" mineral oil to cover drops of CZB (–) medium.

Media

CZB medium: Bicarbonate-buffered embryo culture medium without glucose, used for all one-cell embryos regardless of strain (Chatot et al. 1989). Prepare stock solutions using salts with sterile endotoxin-free distilled H_2O (see Nagy et al. 2003).

M2 medium: HEPES-buffered embryo handling medium (Quinn et al. 1982). Prepare from stock solutions with sterile endotoxin-free distilled H_2O (refer to Nagy et al. 2003) or purchase (Sigma-Aldrich or Specialty Media/Chemicon, Temecula, California). Other mouse embryo culture media (KSOM-AA, M16) can be purchased (Specialty Media/Chemicon).

Pregnant mare's serum gonadotropin <!> (PMSG) (e.g., Sigma-Aldrich or the National Hormone Peptide Program [Dr. A.F. Parlow, Harbor UCLA Medical Center, Los Angeles, California]).

Prepare a stock solution of 500 IU/ml in sterile 0.9% saline. Store sterile 0.1-ml aliquots at –80°C for up to 6 months.

Saline (0.9%, sterile) for diluting hormone stocks before injection

Equipment

Incubator, preset to 37°C (humidified, 5% CO_2)

Mouth pipette for embryo handling

Attach a mouth piece for a 100-µl microcapillary to one end of a 16–18 inch length of latex tubing (1/8 x 3/16 x 1/32 inch). Cut off the tip of a P-1000 aerosol barrier pipette tip to 2–3 mm and insert it into the opposite end of the latex tubing. The large end of a Pasteur pipette fashioned into an embryo handling pipette fits into the large end of the aerosol barrier tip. The mouth pieces are available in bulk (HPI Hospital Products, Apopka, Florida).

Petri dishes (3 cm)
Stereo-zoom dissecting microscope with trans-illuminated base for moving embryos between media droplets (e.g., Leica MZ7.5)
Watchmaker's forceps (Dumont no. 5) (two pairs) and fine dissecting scissors (one pair) (World Precision Instruments; Roboz Surgical Instrument Co., Inc., Gaithersburg, Maryland)

METHOD

Superovulation of Embryo Donor Females

1. Females from strains of mice used commonly in pronuclear microinjection experiments are superovulated at 3–5 weeks of age before the onset of an endogenous estrus cycle. Some strains of mice respond to hormonal stimulation more efficiently after the onset of the endogenous estrus cycle.
2. The standard dose of PMSG is 5 IU/mouse and is delivered by intraperitoneal injection between 6 and 10 hours after the beginning of the light cycle (e.g., 12 p.m. to 4 p.m. for lights on at 6 a.m.). Increasing or decreasing the standard dosage by 2.5 IU/mouse may work better for some strains of mice.
3. Ovulation is stimulated by intraperitoneal injection of HCG at a dosage of 5 IU/mouse 46–48 hours after injection of PMSG. The dosage of HCG usually matches the dosage of PMSG delivered earlier. The embryo donor females are mated to fertile male mice immediately following injection with HCG. The yield of normal fertilized zygotes from inbred strains (e.g., C57BL/6J) may improve when the females are not paired with the fertile males until 4–6 hours after injection of HCG. The detection of a copulatory plug in the vaginal canal of the superovulated female is strong evidence that mating has occurred.
4. The recovery and maintenance of fertilized one-cell embryos are carried out as described below.
   a. Prepare four 3-cm Petri dishes each containing five rows of five 10-μl drops of CZB embryo culture medium, label ("Wash1," "Pre," "Wash2," and "Injected"), and cover with mineral oil equilibrated in a 5% CO_2/95% air cell culture incubator. Allow the dishes to equilibrate in the tissue-culture incubator for ≥30 minutes before introducing embryos.
   b. Euthanize embryo donor mice with inhaled CO_2, or by cervical dislocation, approximately 10–12 hours after the middle of the dark cycle.
   c. Dissect both of the oviducts from each mouse, taking care not to tear the swollen ampulla containing the one-cell ova surrounded by the cumulus mass.
   d. Rinse the oviducts well in M2 HEPES-buffered embryo handling medium.
   e. Prepare a 0.5-ml drop of embryo-handling medium containing approximately 250 μg/ml of type IV-S bovine testicular hyaluronidase.
   f. Release the cumulus mass from each oviduct into the drop of M2 medium containing the hyaluronidase by tearing the ampulla using watchmaker's forceps.
   g. Stir the cumulus masses in the drop containing hyaluronidase until the cumulus cells disperse. Collect the eggs and rinse immediately in M2 medium without hyaluronidase. Work quickly.
   h. Rinse groups of 40–60 embryos through six drops in the "Wash1" dish and then place into a drop of medium in the "Pre" culture dish. Transfer the embryos to the CO_2 incubator. The pronuclei should become clearly visible between 10 and 14 hours after the middle of the dark cycle, and microinjection should commence as soon as possible after the pronuclei are detected.

Protocol 4

Microinjection

This protocol describes the procedures for setting up a microinjection drop and performing pronuclear microinjection.

MATERIALS

CAUTION: See Appendix for appropriate handling of materials marked with <!>.

Reagents

DNA solution for microinjection (prepared in Protocol 2)
Eggs to be injected (prepared in Protocol 3)
Fluorinert (Sigma-Aldrich)

Fluorinert is used for filling the embryo-holding pipette when using oil-filled micrometer syringes.

High-vacuum grease (Dow Corning 1597418)
M2 medium
Petri dishes containing CZB medium drops ("Wash2" dish,"Injected" dish prepared in Protocol 3, Step 4a)

Equipment

Absorbent paper
Cover glass (22 x 22 x 1 mm)
Embryo holding pipette (prepared in Protocol 1)
Inverted microscope (high quality) with 4x bright field, 10x bright field, and 30–40x DIC or Hoffman-Modulation contrast optics for microinjection (e.g., Leica Microsystems, Wexlar, Delaware; Nikon USA, Melville, New York; Olympus America Inc., Melville, New York; Carl Zeiss Imaging Inc., Thornwood, New York)

A 26 X 76 x 1.5-mm slot machined into the center of the microscope stage insert holds the microinjection chamber in precise alignment. The lid of a 60-mm or 100-mm plastic Petri dish can be used with Hoffman Modulation Contrast optics.

Manual micrometer syringes to control the suction of the embryo holding pipettes (e.g., Eppendorf, Narishige)
Microinjection chamber (prepared in Protocol 1)
Microinjection needle (prepared in Protocol 1)
Micromanipulator pair (left and right)

Mechanical micromanipulators (Leitz) require a base plate with mounting rails, and hydraulic micromanipulators (Narishige International, New York) require mounting rails that attach to the microscope.

Pasteur pipette (sterile)
Pipette holders/Instrument sleeves (left and right)
Pneumatic antivibration table (Kinetic Systems Inc., Boston, Massachusetts)
Pneumatic microinjection apparatus (e.g., Femtojet Eppendorf AG, Hamburg)

Syringe used to remove bubbles from the microinjection needle

A blunt 18-gauge hypodermic needle fitted with a 6–8-cm length of Tygon tubing (1/32 ID x 3/32 OD x 1/32 wall) (Saint-Gobain Performance Plastics AJC40001) attached to a 20-cc disposable syringe with a Luer lock tip

UV spectrophotometer

METHODS

Setting Up the Microinjection Drop

1. If the 2 μg/ml DNA solution prepared in Protocol 2 was frozen previously, mix it thoroughly to ensure that the DNA is dissolved completely. Check the concentration of the solution with a UV spectrophotometer (A_{260} nm).
2. Centrifuge the DNA solution at 14,000*g* for 30 minutes at 4°C and then place the solution on ice.
3. Coat the top of both spacers of the microinjection chamber (prepared in Protocol 1) with high-vacuum grease and pipette 300–400 μl of M2 medium into an oval drop between the spacers. Place a coverslip onto the spacers and push down gently to seal. Fill the space around the drop with light mineral oil (see Fig. 1).
4. Fill the embryo holding pipette with Fluorinert, insert the pipette into the instrument sleeve attached to the micrometer syringe, and place the instrument sleeve into the micromanipulator.
5. Adjust the angle of the holding pipette, so that the tip is parallel to the surface of the microinjection chamber.
6. Cover the opening in the microscope stage with a piece of absorbent paper to prevent fluid from the holding pipette from dripping onto the microscope objectives.
7. Pull a sterile Pasteur pipette to 10–12 cm in length with a diameter of about 0.15 mm and draw up the DNA solution from the top layer only to avoid particles. Insert the loading pipette into the shoulder of the microinjection needle and dispense 2–5 μl of DNA solution.
8. The DNA solution will not flow from the microinjection needle unless all bubbles are removed. This process can be accelerated significantly using the following technique.
   a. Fill the 20-cc syringe fitted with the Tygon tubing adaptor with air and insert the large end of the needle into the adaptor.
   b. Bend the tubing so that the microinjection needle points down and apply positive pressure while tapping the needle gently to dislodge any bubbles.
9. Insert the microinjection needle into the instrument sleeve and load the instrument sleeve into the micromanipulator. The instrument sleeves should be aligned with the holding clamp of the micromanipulator, so that the micro-tools are centered and exactly parallel to the long axis of the injection chamber. Adjust the tilt of the instrument sleeve so that the microinjection needle is nearly parallel to the surface of the microinjection chamber.
10. Adjust the injection pressure (Pi) of the pneumatic microinjector to approximately 70–100 hPa and the holding pressure (Ph) to approximately 50% of Pi.
11. Advance the instrument sleeve in the micromanipulator clamp until the tip of the needle is about 2 mm from the edge of the microinjection chamber and then tighten the holding clamp.
12. Wash the eggs from the CZB microdrops in the 37°C/CO_2 incubator through several drops of M2 medium under light mineral oil.

13. Remove the microinjection chamber from the microscope stage and deposit the eggs in a vertical line in the center of the drop of M2, starting at the midpoint of the drop and extending toward the top.
14. Under low magnification, advance the holding pipette and microinjection needle into the drop until they are visible in the same field of view as the embryos. Keep the holding pipette and microinjection needle separated to avoid damaging the tip of the needle.
15. Switch to the 10x objective and adjust the height of the holding pipette and microinjection needle until they are in the same focal plane as the embryos. Repeat the adjustment under the highest magnification.
16. The tip of the holding pipette should almost touch the surface of the injection chamber and should be in the same focal plane as the embryos. Adjust the angle and tilt of the holding pipette, so that the holding pipette moves without visible vibration.

Pronuclear Microinjection

17. Pick up an egg with the holding pipette and inspect it carefully under high magnification. There should be only two pronuclei visible in a fertilized egg and one or more nucleoli visible within the pronucleus. The presence of a single pronucleus indicates that the egg is not fertilized, and eggs with >2 pronuclei are not capable of normal development.
18. Focus on the tip of the microinjection needle and look for a very small eddy of fluid at the tip when the "inject" function of the pneumatic injector is activated. If no flow is visible, then it may be necessary to open the tip slightly by pushing it very gently into the holding pipette. If triggering the "inject" function produces a large stream of fluid from the tip, the needle will cause excessive numbers of eggs to lyse. Replace the needle before proceeding if the flow of DNA solution is absent or excessive.
19. Adjust the position of the egg so that the larger pronucleus and the tip of the holding pipette are both in sharp focus.

    The target pronucleus should be positioned near the center of the egg or the egg plasma membrane may prove difficult to penetrate. Ideally, the target pronucleus will be positioned so that the microinjection needle will not hit any of the nucleoli after if penetrates the pronuclear membrane.
20. Make sure that the embryo is held firmly by the holding pipette and then position the microinjection needle until the tip is directly below the target pronucleus (6 o'clock). Adjust the height of the needle (*Z* axis) until the tip of the microinjection needle and the pronuclear membrane are in exactly the same focal plane.

    The pronuclear membrane will appear as a fine circular line, and the nucleoli are often also in sharp focus. The tip of the microinjection needle may be difficult to see clearly, so adjust the height of the needle until the upper and lower edges are sharp and merge at the tip of the needle (Fig. 2).

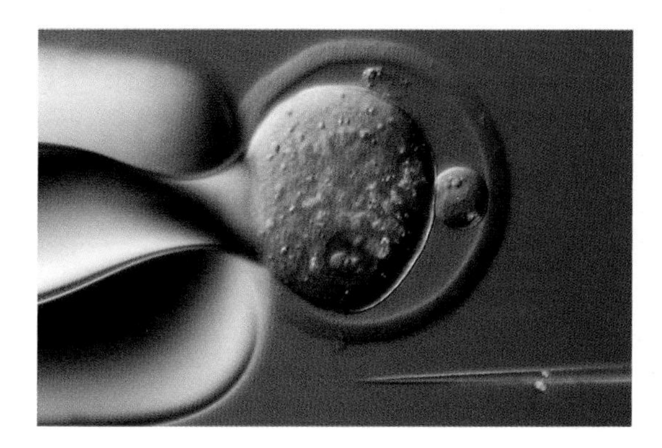

FIGURE 2. Correct positioning of the pronucleus relative to the holding pipette and parfocal alignment of the target pronucleus with the microinjection needle.

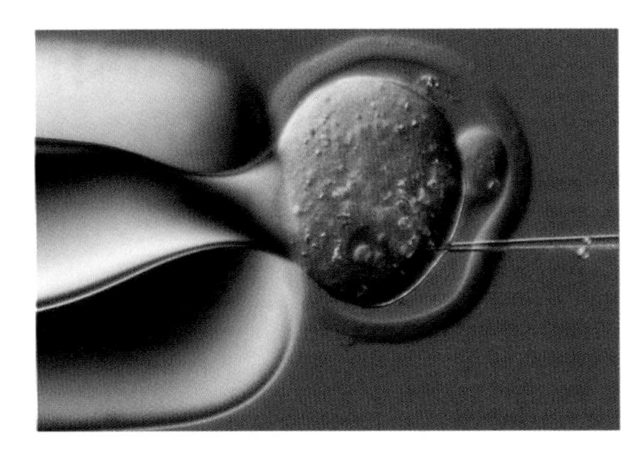

FIGURE 3. Penetration of the pronucleus by the microinjection needle. Note that the egg plasma membrane "relaxes" when punctured by the microinjection needle and that the microinjection needle avoids contact with the nucleoli as it enters the pronucleus.

21. Position the needle opposite the pronucleus and aim the tip at an area of the pronucleus devoid of nucleoli. Push the needle through the zona pellucida and egg plasma membrane into the pronucleus with a single smooth motion. The surface of the egg will indent, and then return to a spherical shape once the needle punctures the zona pellucida and egg plasma membrane (Fig. 3). Continue to push the needle tip into, and often through, the pronucleus, and trigger the injection. If the needle pierces the pronuclear membrane, then the pronucleus will swell with the inflow of DNA solution (Fig. 4). Continue injecting until the volume of the pronucleus increases by 30–50%. Withdraw the needle from the egg very rapidly following successful microinjection of the pronucleus and purge the needle several times with the "clear" function.
22. If the needle fails to pierce the egg plasma membrane or pronuclear membrane, a small "bubble" will appear at the tip of the needle. If this occurs, withdraw the needle from the egg. Bring the pronuclear membrane and needle into focus and repeat the microinjection procedure.
23. Following microinjection, wash the injected eggs through five to six drops of CO_2-equilibrated CZB medium ("Wash2" dish), and culture them for about 60 minutes ("Injected" dish) to assess the survival rate. (These dishes were prepared in Protocol 3, Step 4a.)

    Postmicroinjection survival of the eggs depends on individual skill and the strain of mouse used. Typically, more than 90% of the eggs should survive the microinjection process.

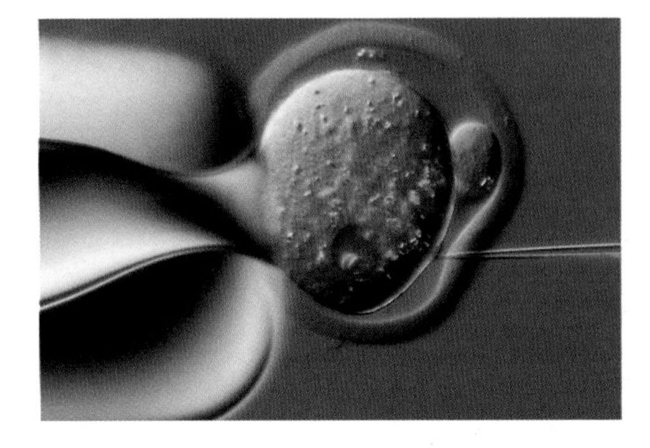

FIGURE 4. Successful pronuclear microinjection. Note the moderate swelling of the target pronucleus following microinjection of the DNA solution.

Protocol 5

Embryo Transfer

This protocol describes the preparation of pseudopregnant recipient mice and the transfer of embryos to them. Descriptions of the procedure for producing vasectomized male mice can be found in several of the volumes cited previously (Hogan et al. 1994; Nagy et al. 2003). The surgical techniques required to transfer one-cell eggs and preimplantation embryos (two-cell to blastocyst) to the oviducts of pseudopregnant recipient females have been described in considerable detail elsewhere (Mann 1993; Hogan et al. 1994; Nagy et al. 2003). A general description is included here.

MATERIALS

CAUTION: See Appendix for appropriate handling of materials marked with <!>.

Reagents

Anaesthetic of choice <!>

Eggs/embryos that have been microinjected (as in Protocol 4)

Ideally, the eggs surviving the microinjection procedure are transferred to the oviducts of a 0.5-dpc (days postcoitus) pseudopregnant recipient female following a short incubation. If no recipient females are available, culture the eggs in vitro in CZB medium to more advanced preimplantation stages.

Recipient female mice

Use hybrid (e.g., F1 or F2 C57BL/6J X CBACa/J) or outbred female mice (e.g., Swiss Webster) of 6–9 weeks of age.

Vasectomized male mice

Equipment

Glass transfer pipette (120–150 μm)

Instruments for exposing reproductive tract of pseudopregnant female mice

Stereo-zoom dissecting microscope with fiber optic spot illumination for surgical embryo transfer (e.g., Leica MZ7.5)

Surgical staple (9 mm)

METHODS

Preparation of Pseudopregnant Female Mice

1. Mate recipient female to vasectomized male mice on the evening before the embryo transfer surgery is scheduled.
2. Check females for the presence of a copulatory plug in the vaginal canal before 12 o'clock noon on the next day.

   A female mouse found with a copulatory plug is termed 0.5 dpc, as mating is assumed to have occurred at the middle of the dark cycle.

Transfer of Embryos to Pseudopregnant Recipient Females

3. Anesthetize a 0.5-dpc pseudopregnant recipient female mouse. Prepare a sterile surgical field and exteriorize the reproductive tract on one side through a mid-dorsal laparotomy.
4. Load the eggs/embryos into a 120–150-μm glass transfer pipette. Perforate or tear the ovarian bursa carefully to minimize bleeding and introduce the eggs/embryos into the ostium of the oviduct.
5. Return the reproductive tract to the peritoneal cavity and repeat the procedure on the contralateral side if necessary.
6. Close the incision in the skin with a 9-mm surgical staple.

   With practice, between 20% and 40% of the microinjected embryos should produce live pups, so transferring about 30 embryos to each recipient female will result in a "normal"-sized litter.

ACKNOWLEDGMENTS

Sincere thanks to Jeffery R. Mann, now at the University of Melbourne in Australia, who provided me with the opportunity to learn the art and science of embryology from a true master.

REFERENCES

Auerbach A.B., Norinsky R., Ho W., Losos K., Guo Q., Chatterjee S., and Joyner A.L. 2003. Strain-dependent differences in the efficiency of transgenic mouse production. *Transgenic Res.* **12:** 59–69.

Chatot C.L., Ziomek C.A., Bavister B.D., Lewis J.L., and Torres I. 1989. An improved culture medium supports development of random-bred 1-cell mouse embryos in vitro. *J. Reprod. Fertil.* **86:** 679–688.

Chevalier-Mariette C., Henry I., Montfort L., Capgras S., Forlani S., Muschler J., and Nicolas J.F. 2003. CpG content affects gene silencing in mice: Evidence from novel transgenes. *Genome Biol.* **4:** R53.

Clarke P., Mann J., Simpson J.F., Rickard-Dickson K., and Primus F.J. 1998. Mice transgenic for human carcinoembryonic antigen as a model for immunotherapy. *Cancer Res.* **58:** 1469–1477.

Gordon J.W. 1993. Production of transgenic mice. *Methods Enzymol.* **225:** 747–771.

Gordon J.W. and Ruddle F.H. 1981 Integration and stable germ-line transmission of genes injected into mouse pronuclei. *Science* **214:** 1244–1246.

———. 1983. Gene transfer into mouse embryos: Production of transgenic mice by pronuclear injection. *Methods Enzymol.* **101:** 411–433.

Gordon J.W., Scangos G.A., Plotkin D.J., Barbosa J.A., and Ruddle F.H. 1980. Genetic transformation of mouse embryos by microinjection of purified DNA. *Proc. Natl. Acad. Sci.* **77:** 7380–7384.

Gossler A., Doetschman T., Korn R., Serfling E., and Kemler R. 1986. Transgenesis by means of blastocyst-derived embryonic stem cell lines. *Proc. Natl. Acad. Sci.* **83:** 9065–9069.

Hammer R.E., Pursel V.G., Rexroad C.E., Wall R.J., Bolt D.J., Ebert K.M., Palmiter R.D., and Brinster R.L. 1985. Production of transgenic rabbits, sheep and pigs by microinjection. *Nature* **315:** 680–683.

Hogan B., Beddington R., and Costantini F. 1994. *Manipulating the mouse embryo,* 2nd edition. Cold Spring Harbor Laboratory Press, Cold Spring Harbor, New York.

Lettmann C., Schmitz B., and Doerfler W. 1991. Persistence or loss of preimposed methylation patterns and de novo methylation of foreign DNA integrated in transgenic mice. *Nucleic Acids Res.***19:** 7131–7137.

Mann J.R.M 1993. Guide to techniques in mouse development. *Methods Enzymol.* **225:** 782–793.

Mann J.R.M. and McMahon A.P. 1993. Guide to techniques in mouse development. *Methods Enzymol.* **225:** 771–781.

Martin D.I. and Whitelaw E. 1996. The vagaries of variegating transgenes. *Bioessays* **18:** 919–923.

McKnight R.A., Shamay A., Sankaran L., Wall R.J., and Hennighausen L. 1992. Matrix-attachment regions can impart position-independent regulation of a tissue-specific gene in transgenic mice. *Proc. Natl. Acad. Sci.* **89:** 6943–6947.

Milot E., Strouboulis J., Trimborn T., Wijgerde M., de Boer E., Langeveld A., Tan-Un K., Vergeer W., Yannoutsos N., Grosveld F., and Fraser P. 1996. Heterochromatin effects on the frequency and duration of LCR-mediated gene transcription. *Cell* **87:** 105–114.

Nagy A., Gertsenstein M., Vintersten K., and Behringer R. 2003. *Manipulating the mouse embryo,* 3rd edition. Cold Spring Harbor Laboratory Press, Cold Spring Harbor, New York.

Polites H.G. and Pinkert C.A. 1994. Transgenic animal production using DNA microinjection. In *Transgenic animal technology: A laboratory handbook* (ed. C.A. Pinkert), pp. 15–68. Academic Press, San Diego, California.

Quinn P., Barros C., and Whittingham D.G. 1982. Preservation of hamster oocytes to assay the fertilizing capacity of human spermatozoa. *J. Reprod. Fertil.* **66:** 161–168.

Sambrook J., Fritsch E.F., and Maniatis T. 1989. Molecular Cloning: A laboratory manual, 2nd edition, pp. 285–287. Cold Spring Harbor Laboratory Press, Cold Spring Harbor, New York.

Shashikant C.S., Bieberich C.J., Belting H.-G., Wang J.H., Borbely M.A., and Ruddle F.H. 1995. Regulation of HOXC-8 during mouse embryonic development: Identification and characterization of critical elements involved in early neural tube expression. *Development* **121:** 4339–4347.

Si-Hoe S.L. and Murphy D. 1999. Production of transgenic rodents by the microinjection of cloned DNA into fertilized one-celled eggs. *Methods Mol. Biol.* **97:** 61–100.

75 Knockdown Transgenic Mice Generated by Silencing Lentiviral Vectors

Oded Singer, Gustavo Tiscornia, and Inder M. Verma

The Salk Institute for Biological Studies, Laboratory of Genetics, La Jolla, California 92037

ABSTRACT

This chapter describes the use of lentiviral vectors to deliver genes and small interfering RNA (siRNA)-expressing cassettes into preimplantation embryos. This technique allows for the rapid development of transgenic animals and transgenic knockdown animals.

INTRODUCTION

For the purpose of generating transgenic and knockout animals, preimplantation embryos must be harvested and manipulated in vitro. Although the generation of transgenic animals has been performed by pronuclear injection of DNA into single-cell embryos, the generation of mouse knockouts is time-consuming and laborious. An embryonic stem (ES) knockout line must be generated, characterized, and injected into a blastocyst in order to obtain a chimeric founder that can be subsequently bred to homozygosity. Taking advantage of the unique ability of lentiviral vectors to generate transgenic animals (Lois et al. 2002; Pfeifer et al. 2002), lentiviruses expressing short hairpin RNAs (shRNAs) from polymerase III (pol III) promoters such as H1 (Tiscornia et al. 2003) and mU6 (Rubinson et al. 2003) can be used to generate transgenic knockdown mice. Two methods are available for delivering lentiviral particles to the embryo to obtain lentiviral transgenesis: zona pellucida removal and subzonal injection. For a detailed description of harvesting and manipulating mouse embryos, see Hogan et al. (1994). The zona pellucida removal protocol is able to achieve up to 100% transgenesis and does not require the use of a micromanipulator. However, zona removal is toxic to the embryos and results in low survival of embryos. The subzonal injection protocol is able to achieve up to 100% transgenesis (Fig. 1) and is not toxic for the embryos; therefore, survival of embryos is much higher. Nevertheless, the use of micromanipulators requires some practice.

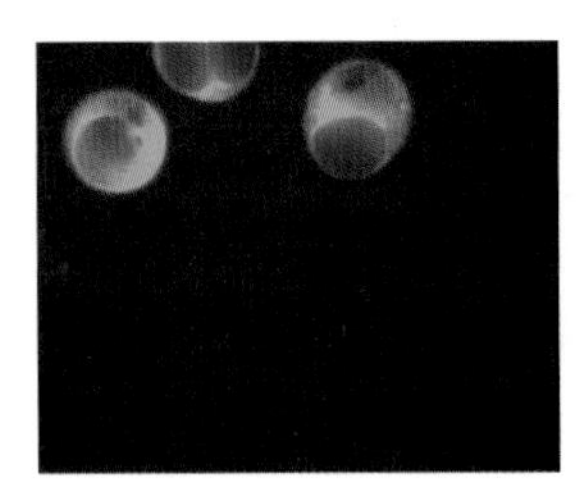
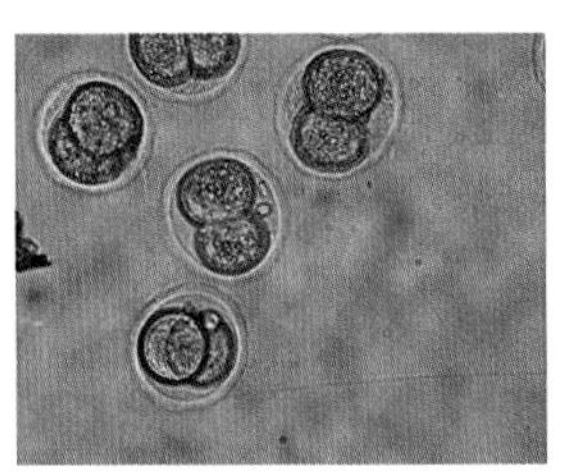
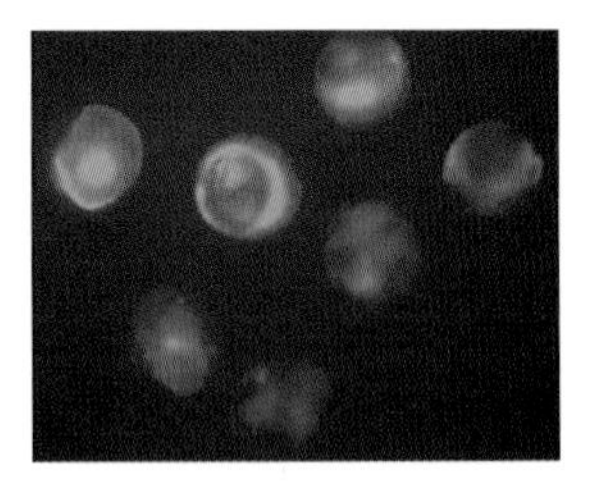

FIGURE 1. Single-cell embryos injected with lentiviral vector expressing green fluorescent protein (GFP) driven by a CAG promoter and cultured in medium. (*Left*) GFP-positive embryos were viewed and photographed in fluorescent and visible field (24 hr postinjection). (*Right*) GFP-positive embryos were viewed and photographed again in fluorescent field only (72 hr postinjection).

Protocol

Zona Pellucida Removal and Subzonal Injection Methods

This protocol describes two methods to deliver genes and siRNA-expressing cassettes into preimplantation mouse embryos using lentiviral vectors.

MATERIALS

CAUTION: See Appendix for appropriate handling of materials marked with <!>.

Reagents

Acidic Tyrode solution

Mix the following in 80 ml of H_2O: 0.8 g of NaCl, 0.02 g of KCl <!>, 0.024 g of $CaCl_2$•$2H_2O$ <!>, 0.01 g of $MgCl_2$•$6H_2O$ <!>, 0.1 g of glucose, and 0.4 g of polyvinylpyrrolidone (PVP) <!>. Adjust the pH to 2.5 and final volume to 100 ml.

Gonadotropin from pregnant mare serum (PMS; 25 IU/ml in saline) (Sigma-Aldrich G 4527) <!>
Human chorionic gonadotropin (hCG; 25 IU/ml in saline) (Sigma-Aldrich C 8554)
Hyaluronidase solution in M2 (Specialty Media MR-051-F)
Media
- M2 (Specialty Media MR-015-D)
- M16 (Specialty Media MR-010-D)

Mice
- B6D2F1 females (6–8 weeks; Haarlan)
- Pseudopregnant females, timed (2.5 days postcoitum [dpc]; 0.5 dpc is optional for subzonal injection)

Mineral oil (Fischer 0121-1)
Virus to be used for transgenesis (10^9 viral particles/ml)

Equipment

Coverslips

To prepare pipettes, pull borosilicate glass capillary by softening in Bunsen burner flame, until the outside diameter of the drawn-out region of the pipette tip is approximately 80–120 µm. Use the 10-µl calibrated micropipettes for this step (VWR 53432-728).

Prepare by pulling borosilicate glass capillary (soften in Bunsen burner flame) until the internal diameter of tip is approximately 200 µm. Use the 50-µl micropipettes for this step (VWR 53432-783).

Incubator, preset to 37°C; 5% CO_2
Inverted microscope
Microcentrifuge
Micromanipulating equipment
- Cell-tram oil pump (Eppendorf) to operate the holding needle
- Manual micromanipulator pair (Leitz ACS01) for holding and injection needles
- Micropipettes (ICSI, sterile) (MIC-cust-0, ID 4–5 µm, OD 5–6 µm, bevel length 50° [8–9

µm] with a spike Humagen, Virginia)
Screw-actuated syringe (SAS11/2-E [ZMS]) to operate the injection needle
Needles (30 gauge)
Pasteur pipette
Stereoscope
Surgical instruments: blunt curved forceps (2), fine-point forceps (2), small scissors

METHOD

Preparation of Mice from which Embryos will be Harvested

1. Inject 8–12 B6D2F1 females (age 6–8 weeks, Haarlan) intraperitoneally with 5 units of PMS.
2. Exactly 48 hours later, inject with 5 units of hCG. Transfer females to male cages (one female per one male).
3. Perform either zona pellucida removal or subzonal injection.

For zona pellucida removal

a. Forty-eight hours after hCG injection, surgically collect oviducts under sterile conditions and transfer to a 50-µl droplet of M2 under mineral oil. Insert a blunt 30-gauge needle in the infundibulum and flush out embryos (2–4-cell stage) with M2 medium. Wash embryos in several successive M2 droplets by pipetting up and down three to four times. Repeat the procedure in M16 medium. Work quickly and incubate embryos at 37°C until the next step.

b. To remove the zona pellucida, transfer embryos to a droplet of acidic Tyrode solution, pipette up and down, and then transfer to a second acidic Tyrode solution droplet. Incubate at room temperature until the zona pellucida dissolves (30 sec to 1 min). Wash embryos twice in 50-µl droplets of M2 under mineral oil using a Pasteur pipette. Repeat the washing procedure three times in M16 droplets under mineral oil.

c. For the transduction of virus particles, use a viral titer of 10^9 viral particles/ml. Dilute virus 100-fold in M16. Transfer embryos to a 30-µl droplet of diluted virus, pipette up and down, and then transfer individual embryos to a single 10-µl droplet of diluted virus under mineral oil. Incubate embryos for 48 hours at 37°C in 5% CO_2.

d. After 48 hours, embryos should have developed into blastocysts. Use a transfer needle to transfer developed blastocysts into the uterus of a 2.5-dpc pseudopregnant female (Hogan et al. 1994).

For subzonal injection

a. Twenty-four hours after hCG injection, surgically collect oviducts under sterile conditions and transfer to a 50-µl droplet of M2 under mineral oil. Use a pair of fine forceps to gently release the embryos (single-cell stage) from the swollen ampulla (upper portion of the oviduct) and transfer embryos to hyaluronidase solution in M2 medium to enzymatically digest the cumulus cells surrounding the embryos. Transfer the embryos to fresh M2 medium to wash off the hyaluronidase solution and place in 50-µl M16 droplets under mineral oil.

b. Thaw a viral preparation aliquot (10^9 viral particles/ml) and centrifuge briefly (5 sec on high speed) to prevent clogging of cellular debris at the tip of the micropipette.

c. Prepare the micromanipulator. After assembling the injection needle, spot 5 µl of viral suspension on a coverslip and use negative flow to fill the needle. If the needle is clogged, use short positive flow to release the debris. Transfer single-cell embryos to a droplet of M16 covered with mineral oil.

Microinjection can be performed using "homemade" micropipettes (Lois et al. 2002) or custom-made sterile ICSI micropipettes.

d. Use an inverted microscope to perform the injection. Load the micropipette containing the virus onto the micromanipulator and lower the tip of the pipette into the mineral oil until the tip is visible within the optical field. While holding the embryo with the holding needle, turn on the positive flow and gently push the micropipette tip through the zona layer into the perivitelline space without harming the embryo cell membrane. Leave the tip in the perivitelline space for 5–10 seconds. Some swelling of the perivitelline space indicates positive flow. Incubate embryos in M16 under mineral oil in a 37°C, 5% CO_2 incubator until implantation.

e. Implant embryos immediately in the oviduct of 0.5-dpc timed pseudopregnant females or culture the embryos for 2–3 days until blastocyst formation and implant the blastocysts in the uterus of 2.5-dpc timed pseudopregnant females.

REFERENCES

Hogan B., Beddington R., Costantini F., and Lacy E., eds. 1994. *Manipulating the mouse embryo: A laboratory manual*, 2nd edition. Cold Spring Harbor Laboratory Press, Cold Spring Harbor, New York.

Lois C., Hong E.J., Pease S., Brown E.J., and Baltimore D. 2002. Germline transmission and tissue-specific expression of transgenes delivered by lentiviral vectors. *Science* **295:** 868–872.

Pfeifer A., Ikawa M., Dayn Y., and Verma I.M. 2002. Transgenesis by lentiviral vectors: Lack of gene silencing in mammalian embryonic stem cells and preimplantation embryos. *Proc. Natl. Acad. Sci.* **99:** 2140–2145.

Rubinson D.A., Dillon C.P., Kwiatkowski A.V., Sievers C., Yang L., Kopinja J., Rooney D.L., Ihrig M.M., McManus M.T., Gertler F.B., Scott M.L., and Van Parijs L. 2003. A lentivirus-based system to functionally silence genes in primary mammalian cells, stem cells and transgenic mice by RNA interference. *Nat. Genet.* **33:** 401–406.

Tiscornia G., Singer O., Ikawa M., and Verma I.M. 2003. A general method for gene knockdown in mice by using lentiviral vectors expressing small interfering RNA. *Proc. Natl. Acad. Sci.* **100:** 1844–1848.

Appendix: Cautions

GENERAL CAUTIONS

Please note that the Cautions Appendix in this manual is not exhaustive. Readers should always consult individual manufacturers and other resources for current and specific product information. Chemicals and other materials discussed in text sections are not identified by the icon (!) used to indicate hazardous materials in the protocols. However, they may be hazardous to the user without special handling. Please consult your local safety office or the manufacturer's safety guidelines for further information. The following general cautions should always be observed.

- **Become completely familiar** with the properties of substances used before beginning the procedure.
- **The absence of a warning** does not necessarily mean that the material is safe, because information may not always be complete or available.
- **If exposed to toxic substances**, contact your local safety office immediately for instructions.
- **Use proper disposal procedures** for all chemical, biological, and radioactive waste.
- **For specific guidelines on appropriate gloves**, consult your local safety office.
- **Handle concentrated acids and bases** with great care. Wear goggles and appropriate gloves. A face shield should be worn when handling large quantities.

  Do not mix strong acids with organic solvents because they may react. Sulfuric acid and nitric acid especially may react highly exothermically and cause fires and explosions.

  Do not mix strong bases with halogenated solvent because they may form reactive carbenes that can lead to explosions.
- **Handle and store pressurized gas containers** with caution as they may contain flammable, toxic, or corrosive gases; asphyxiants; or oxidizers. For proper procedures, consult the Material Safety Data Sheet that must be provided by your vendor.
- **Never pipette solutions using mouth suction.** This method is not sterile and can be dangerous. Always use a pipette aid or bulb.
- **Keep halogenated and nonhalogenated solvents separately** (e.g., mixing chloroform and acetone can cause unexpected reactions in the presence of bases). Halogenated solvents are organic solvents such as chloroform, dichloromethane, trichlorotrifluoroethane, and dichloroethane. Some nonhalogenated solvents are pentane, heptane, ethanol, methanol, benzene, toluene, *N,N*-dimethylformamide (DMF), dimethyl sulfoxide (DMSO), and acetonitrile.
- **Laser radiation**, visible or invisible, can cause severe damage to the eyes and skin. Take proper precautions to prevent exposure to direct and reflected beams. Always follow manufacturer's safety guidelines and consult your local safety office. See flash lamps caution below for more detailed information.
- **Flash lamps**, due to their light intensity, can be harmful to the eyes. They also may explode on occasion. Wear appropriate eye protection and follow the manufacturer's guidelines.
- **Photographic fixatives and developers** also contain chemicals that can be harmful. Handle them with care and follow manufacturer's directions.
- **Power supplies and electrophoresis equipment** pose serious fire hazards and electrical shock hazards if not used properly.
- **Microwave ovens and autoclaves** in the lab require certain precautions. Accidents have occurred involving their use (e.g., when melting agar or bacto-agar stored in bottles or sterilizing). If the screw top is not completely removed and insufficient space is

available for the steam to vent, the bottles can explode and cause severe injury when the containers are removed from the microwave or autoclave. Always completely remove bottle caps before microwaving or autoclaving. An alternative method for routine agarose gels that do not require sterile agar is to weigh out the agar and place the solution in a flask.

- **Ultrasonicators** use high-frequency sound waves (16–100 kHz) for cell disruption and other purposes. This "ultrasound," conducted through air, does not pose a direct hazard to humans, but the associated high volumes of audible sound can cause a variety of effects including headache, nausea, and tinnitus. Direct contact of the body with high-intensity ultrasound (not medical imaging equipment) should be avoided. Use appropriate ear protection and display signs on the door(s) of laboratories in which the units are used.
- **Use extreme caution when handling cutting devices** such as microtome blades, scalpels, razor blades, or needles. Microtome blades are extremely sharp! Use care when sectioning. If unfamiliar with their use, have someone demonstrate proper procedures. For proper disposal, use the "sharps" disposal container in your lab. Discard used needles unshielded, with the syringe still attached. This prevents injuries (and possible infections) when manipulating used needles because many accidents occur while trying to replace the needle shield. Injuries may also be caused by broken Pasteur pipettes, coverslips, or slides.

GENERAL PROPERTIES OF COMMON CHEMICALS

The hazardous materials list can be summarized in the following categories.

- Inorganic acids such as hydrochloric, sulfuric, nitric, or phosphoric are colorless liquids with stinging vapors. Avoid spills on skin or clothing. Spills should be diluted with large amounts of water. The concentrated forms of these acids can destroy paper, textiles, and skin as well as cause serious injury to the eyes.
- Inorganic bases such as sodium hydroxide are white solids that dissolve in water and under heat development. Concentrated solutions will slowly dissolve skin and even fingernails.
- Salts of heavy metals are usually colored powdered solids that dissolve in water. Many are potent enzyme inhibitors and therefore toxic to humans and to the environment (e.g., fish and algae).
- Most organic solvents are flammable volatile liquids. Avoid breathing the vapors, which can cause nausea or dizziness. Also avoid skin contact.
- Other organic compounds, including organosulphur compounds such as mercaptoethanol or organic amines, can have very unpleasant odors. Others are highly reactive and should be handled with appropriate care.
- If improperly handled, dyes and their solutions can stain not only your sample, but also your skin and clothing. Some of them are also mutagenic (e.g., ethidium bromide), carcinogenic, and toxic.
- Nearly all names ending with "ase" (e.g., catalase, β-glucuronidase, or Zymolase) refer to enzymes. There are also other enzymes with nonsystematic names such as pepsin. Many are provided by manufacturers in preparations containing buffering substances, etc. Be aware of the individual properties of materials contained in these substances.
- Toxic compounds are often used to manipulate cells. They can be dangerous and should be handled appropriately.
- Be aware that the toxicological properties of several of the compounds listed have not been thoroughly studied with respect to. Handle each chemical with the appropriate respect. Although the toxic effects of a compound can be quantified (e.g., LD50 values), this is not possible for carcinogens or mutagens for which one single exposure can have an effect. Also realize that dangers related to a given compound may also depend on its physical state (fine powder vs. large crystals/diethylether vs. glycerol/dry ice vs. carbon dioxide under pressure in a gas bomb). Anticipate under which circumstances during an experiment exposure is most likely to occur and how best to protect yourself and your environment.

HAZARDOUS MATERIALS

In general, proprietary materials are not listed here. Kits and other commercial items as well as most anesthetics, dyes, fixatives, and stains are also not included. Anesthetics also require special care. Follow the manufacturer's safety guidelines that accompany these products.

Acetaminophen is harmful by inhalation, ingestion, or skin absorption. Wear appropriate gloves and safety goggles and always use in a chemical fume hood. Do not breathe the dust.

Acetic acid (glacial) is highly corrosive and must be handled with great care. It may be a carcinogen. Liquid and mist cause severe burns to all body tissues. It may be harmful by inhalation, ingestion, or skin absorption. Wear appropriate gloves and goggles and use in a chemical fume hood. Keep away from heat, sparks, and open flame.

Acetonitrile (methyl cyanide) is very volatile and extremely flammable. It is an irritant and a chemical asphyxiant that can exert its effects by inhalation, ingestion, or skin absorption. Treat cases of severe exposure as cyanide poisoning. Wear appropriate gloves and safety glasses and use only in a chemical fume hood. Keep away from heat, sparks, and open flame.

Ammonium acetate, NH_4Ac, $H_3CCOONH_4$, may be harmful by inhalation, ingestion, or skin absorption. Wear appropriate gloves and safety glasses and use in a chemical fume hood.

Ammonium sulfate, $(NH_4)_2SO_4$, may be harmful by inhalation, ingestion, or skin absorption. Wear appropriate gloves and safety glasses.

Ampicillin may be harmful by inhalation, ingestion, or skin absorption. Wear appropriate gloves and safety glasses and use in a chemical fume hood.

Anesthetics, *follow* **manufacturer's safety guidelines**

Bisbenzimide may be harmful by inhalation, ingestion, or skin absorption. Wear appropriate gloves and safety glasses and use in a chemical fume hood. Do not breathe the dust.

Bleach (Sodium hypochlorite), NaOCl, is poisonous, can be explosive, and may react with organic solvents. It may be fatal by inhalation and is also harmful by ingestion and destructive to the skin. Wear appropriate gloves and safety glasses and use in a chemical fume hood to minimize exposure and odor.

Boric acid, H_3BO_3, may be harmful by inhalation, ingestion, or skin absorption. Wear appropriate gloves and goggles.

5-Bromo-4-chloro-3-indolyl-β-D-galactopyranoside (BCIG; X-gal) is toxic to the eyes and skin and may be harmful by inhalation, ingestion, or skin absorption. Wear appropriate gloves and safety goggles.

Bromophenol blue may be harmful by inhalation, ingestion, or skin absorption. Wear appropriate gloves and safety glasses and use in a chemical fume hood.

$CaCl_2$, *see* **Calcium chloride**

Calcium chloride, $CaCl_2$, is hygroscopic and may cause cardiac disturbances. It may be harmful by inhalation, ingestion, or skin absorption. Do not breathe the dust. Wear appropriate gloves and safety goggles.

Carbon dioxide, CO_2, in all forms may be fatal by inhalation, ingestion, or skin absorption. In high concentrations, it can paralyze the respiratory center and cause suffocation. Use only in well-ventilated areas. In the form of dry ice, contact with carbon dioxide can also cause frostbite. Do not place large quantities of dry ice in enclosed areas such as cold rooms. Wear appropriate gloves and safety goggles.

Cesium chloride, CsCl, may be harmful by inhalation, ingestion, or skin absorption. Wear appropriate gloves and safety glasses.

CH_3CH_2OH, *see* **Ethanol**

$CHCl_3$, *see* **Chloroform**

$C_7H_7FO_2S$, *see* **Phenylmethylsulfonyl fluoride**

Chloramine is corrosive and causes burns. It may be harmful by inhalation, ingestion, or skin absorption. Wear appropriate gloves and safety glasses and use in a chemical fume hood. Do not breathe the dust.

Chloramphenicol is a potential carcinogen and may be harmful by inhalation, ingestion, or skin absorption. Wear appropriate gloves and safety glasses and use in a chemical fume hood.

Chloroform, $CHCl_3$, is irritating to the skin, eyes, mucous membranes, and respiratory tract. It is a carcinogen and may damage the liver and kidneys. It is also volatile. Avoid breathing the vapors. Wear appropriate gloves and safety glasses and always use in a chemical fume hood.

Chloroquine may be harmful by inhalation, ingestion, or skin absorption. Prolonged exposure can lead to permanent eye damage. Wear appropriate gloves and safety goggles.

Cholesteryl chloroformate may be harmful by inhalation, ingestion, or skin absorption. Wear appropriate gloves and safety glasses and use in a chemical fume hood.

Citric acid is an irritant and may be harmful by inhalation, ingestion, or skin absorption. It poses a risk of serious damage to the eyes. Wear appropriate gloves and safety goggles. Do not breathe the dust.

CO_2, *see* **Carbon dioxide**

Cobalt nitrate, $Co(NO_3)_2 \cdot 6H_2O$, is a strong oxidizer and may be harmful by inhalation, ingestion, or skin absorption. Wear appropriate gloves and safety goggles.

Colcemid is highly toxic and may cause organ failure or death if inhaled or swallowed. It may cause reproductive or fetal effects. It is harmful by inhalation, ingestion, and skin absorption. Wear appropriate gloves and safety goggles and use only in a chemical fume hood. Do not breathe the dust.

$Co(NO_3)_2 \cdot 6H_2O$, *see* **Cobalt nitrate**

Crystal Violet can cause severe burns. It may be harmful by inhalation, ingestion, and skin absorption. Wear appropriate gloves and safety goggles and use in a chemical fume hood, Do not breathe the dust.

CsCl, *see* **Cesium chloride**

DAPI, *see* **4′,6-Diamidine-2-phenylindole dihydrochloride**

DCM, *see* **Dichloromethane**

DDAB, *see* **Dimethyldioctadecylammonium bromide**

DEAE, *see* **Diethylaminoethanol**

Deoxycholate (DOC) may be harmful by inhalation, ingestion, or skin absorption. Do not breathe the dust. Wear appropriate gloves and safety glasses.

4′,6-Diamidine-2-phenylindole dihydrochloride (DAPI) is a possible carcinogen. It may be harmful by inhalation, ingestion, or skin absorption. It may also cause irritation. Avoid breathing the dust and vapors. Wear appropriate gloves and safety glasses and use in a chemical fume hood.

Dichloromethane (DCM), CH_2Cl_2 (also known as **Methylene chloride**) is toxic if inhaled, ingested, or absorbed through the skin. It is also an irritant and is suspected to be a carcinogen. Wear appropriate gloves and safety glasses and use in a chemical fume hood. Do not breathe the vapors.

DIEA, *see* **Diisopropylethylamine**

Diethylaminoethanol (DEAE) may be harmful by inhalation, ingestion, or skin absorption. Wear appropriate gloves and safety glasses and use in a chemical fume hood.

Diethyl ether, Et_2O or $(C_2H_5)_2O$, is extremely volatile and flammable. It is irritating to the eyes, mucous membranes, and skin. It is also a CNS depressant with anesthetic effects. It may be harmful by inhalation, ingestion, or skin absorption. Avoid breathing the vapors. Wear appropriate gloves and safety glasses and always use in a chemical fume hood. Explosive peroxides can form during storage or on exposure to air or direct sunlight. Keep away from heat, sparks, and open flame.

Diethylene ether, *see* **1,4-Dioxane**

Diisopropylethylamine (DIEA, DIPEA) is extremely destructive to the mucous membranes, upper respiratory tract, skin, and eyes. It may be harmful by ingestion or skin absorption. Inhalation may be fatal. Wear appropriate gloves and safety glasses and always use in a chemical fume hood. Keep away from heat sparks, and open flame.

Dimethyldioctadecylammonium bromide (DDAB) is an irritant and may be harmful by inhalation, ingestion, or skin absorption. Wear appropriate gloves and safety glasses. Do not breathe the dust.

***N,N*-Dimethylacetamide** (DMA) is toxic and may cause harm to the unborn child. It may be harmful by inhalation, ingestion, or skin absorption. Wear appropriate gloves and safety glasses and use in a chemical fume hood. Do not breathe the mist or vapor. It is also flammable. Keep away from heat, sparks, and open flame.

***N,N*-Dimethylformamide** (DMF), $HCON(CH_3)_2$, is a possible carcinogen and is irritating to the eyes, skin, and mucous membranes. It can exert its toxic effects through inhalation, ingestion, or skin absorption.

Chronic inhalation can cause liver and kidney damage. Wear appropriate gloves and safety glasses and use in a chemical fume hood.

Dimethylsulfoxide (DMSO) may be harmful by inhalation or skin absorption. Wear appropriate gloves and safety glasses and use in a chemical fume hood. DMSO is also combustible. Store in a tightly closed container. Keep away from heat, sparks, and open flame.

Dithiothreitol (DTT) is a strong reducing agent that emits a foul odor. It may be harmful by inhalation, ingestion, or skin absorption. When working with the solid form or highly concentrated stocks, wear appropriate gloves and safety glasses and use in a chemical fume hood.

1,4-Dioxane is highly flammable in both liquid and vapor forms. It is a possible carcinogen and is highly toxic by inhalation, ingestion, or skin absorption. Do not breathe the vapor. Wear appropriate gloves and safety glasses. Keep away from heat, sparks, and open flame.

DIPEA, *see* **Diisopropylethylamine**

5,5′-Dithiobis(nitrobenzoic acid) (DTNB) is an irritant and may be harmful by inhalation, ingestion, or skin absorption. Wear appropriate gloves and safety glasses. Do not breathe the dust.

DMF, *see* ***N,N*-dimethylformamide**

DMSO, *see* **Dimethyl sulfoxide**

DOC, *see* **Deoxycholate**

Dry ice, *see* **Carbon dioxide**

DTNB, *see* **5,5′-Dithiobis-(2-nitrobenzoic acid)**

DTT, *see* **Dithiothreitol**

Dyes, *follow* **manufacturer's safety guidelines**

EDC, *see* ***N*-ethyl-*N*′-(dimethylaminopropyl)-carbodiimide**

EDCI (Ethyl-3-[3-dimethylamino] propyl carbodiimide), *see* ***N*-Ethyl-*N*′-(dimethylaminopropyl)-carbodiimide**

Ellman's reagent, *see* **5,5′-Dithiobis(2-nitrobenzoic acid) (DTNB)**

Ethanol (EtOH), CH_3CH_2OH, may be harmful by inhalation, ingestion, or skin absorption. Wear appropriate gloves and safety glasses.

Ethidium bromide is a powerful mutagen and is toxic. Consult the local institutional safety officer for specific handling and disposal procedures. Avoid breathing the dust. Wear appropriate gloves when working with solutions that contain this dye.

Ethyl acetate may be fatal by ingestion and harmful by inhalation or skin absorption. Wear appropriate gloves and safety goggles. Do not breathe the dust. Use in a well-ventilated area.

1-Ethyl-3-[3-dimethylaminopropyl] carbodiimide (EDC), *see* ***N*-Ethyl-*N*′-(dimethylaminopropyl)-carbodiimide**

***N*-Ethyl-*N*′-(dimethylaminopropyl)-carbodiimide (EDC)** is irritating to the mucous membranes and upper respiratory tract. It may be harmful by inhalation, ingestion, or skin absorption. Wear appropriate gloves and safety glasses. Handle with care.

Ethyl-3-[3-dimethylamino] propyl carbodiimide (EDCI), *see* ***N*-Ethyl-*N*′-(dimethylaminopropyl)-carbodiimide**

EtOH, *see* **Ethanol**

Et_2O or $(C_2H_5)_2O$, *see* **Diethyl ether**

Exonuclease III may be harmful by inhalation, ingestion, or skin absorption. Wear appropriate gloves and safety goggles.

Formaldehyde, HCHO, is highly toxic and volatile. It is also a possible carcinogen. It is readily absorbed through the skin and is irritating or destructive to the skin, eyes, mucous membranes, and upper respiratory tract. Avoid breathing the vapors. Wear appropriate gloves and safety glasses and always use in a chemical fume hood. Keep away from heat, sparks, and open flame.

Formamide is teratogenic. The vapor is irritating to the eyes, skin, mucous membranes, and upper respiratory tract. It may be harmful by inhalation, ingestion, or skin absorption. Wear appropriate gloves and safety glasses and always use a chemical fume hood when working with concentrated solutions of formamide. Keep working solutions covered as much as possible.

G418 (an aminoglycosidic antibiotic) is toxic and may cause harm to the unborn child. It may be harmful by inhalation, ingestion, or skin absorption. Wear appropriate gloves and safety goggles and use in a chemical fume hood. Do not breathe the dust.

Glacial acetic acid, *see* **Acetic acid (glacial)**

Glutaraldehyde is toxic. It is readily absorbed through the skin and is irritating or destructive to the skin, eyes, mucous membranes, and upper respiratory tract. Wear appropriate gloves and safety glasses and always use in a chemical fume hood.

Glycine may be harmful by inhalation, ingestion, or skin absorption. Wear gloves and safety glasses. Avoid breathing the dust.

Gold chloride solution is an irritant and may be harmful by inhalation, ingestion, or skin absorption. Wear gloves and safety glasses. Avoid breathing the dust.

Guanidine hydrochloride is irritating to the mucous membranes, upper respiratory tract, skin, and eyes. It may be harmful by inhalation, ingestion, or skin absorption. Wear appropriate gloves and safety glasses. Avoid breathing the dust.

Guanidinium hydrochloride, *see* **Guanidine hydrochloride**

^{3}H, *see* **Radioactive substances**

H_2SO_4, *see* **Sulfuric acid**

H_3BO_3, *see* **Boric acid**

$H_3CCOONH_4$, see **Ammonium acetate**

HCl, *see* **Hydrochloric acid**

HCHO, *see* **Formaldehyde**

H_3COH, *see* **Methanol**

$HCON(CH_3)_2$, *see* ***N,N*-Dimethylformamide**

$HOCH_2CH_2SH$, *see* **β-Mercaptoethanol**

Hoescht No. 33342, *see* **Bisbenzimide**

Hydrochloric acid, HCl, is volatile and may be fatal if inhaled, ingested, or absorbed through the skin. It is extremely destructive to mucous membranes, upper respiratory tract, eyes, and skin. Wear appropriate gloves and safety glasses and use with great care in a chemical fume hood. Wear goggles when handling large quantities.

Hygromycin B is highly toxic and may be fatal if inhaled, ingested, or absorbed through the skin. Wear appropriate gloves and safety goggles and use only in a chemical fume hood. Do not breathe the dust.

^{125}I, *see* **Radioactive substances**

IAA, *see* **Isoamyl alcohol**

Iodogen may be harmful by inhalation, ingestion, or skin absorption. Wear appropriate gloves and safety goggles. Do not breathe the dust.

IPTG, *see* **Isopropyl-β-D-thiogalactopyranoside**

Isoamyl alcohol (IAA) may be harmful by inhalation, ingestion, or skin absorption and presents a risk of serious damage to the eyes. Wear appropriate gloves and safety goggles. Keep away from heat, sparks, and open flame.

Isofluorane (Isoflurane) is an irritant and may be harmful by inhalation, ingestion, or skin absorption. Chronic exposure may be harmful. Wear appropriate gloves and safety glasses.

Isopropanol is flammable and irritating. It may be harmful by inhalation, ingestion, or skin absorption. Wear appropriate gloves and safety glasses. Do not breathe the vapor. Keep away from heat, sparks, and open flame.

Isopropyl-β-D-thiogalactopyranoside (IPTG) may be harmful by inhalation, ingestion, or skin absorption. Wear appropriate gloves and safety glasses.

Kanamycin may be harmful by inhalation, ingestion, or skin absorption. Wear appropriate gloves and safety glasses. Use only in a well-ventilated area.

KCl, *see* **Potassium chloride**

Ketoprofen is toxic. It is harmful by inhalation, ingestion, or skin absorption. Wear appropriate gloves and safety goggles and always use in a chemical fume hood.

$K_3Fe(CN)_6$, *see* **Potassium ferricyanide**

$K_4Fe(CN)_6 \cdot 3H_2O$, *see* **Potassium ferrocyanide**

K_2HPO_4/KH_2PO_4, *see* **Potassium phosphate**

Liquid nitrogen (LN_2) can cause severe damage due to extreme temperature. Handle frozen samples with extreme caution. Do not breathe the vapors. Seepage of liquid nitrogen into frozen vials can result in an exploding tube upon removal from liquid nitrogen. Use vials with O-rings when possible. Wear cryo-mitts and a face mask. No not allow the liquid nitrogen to spill onto your clothes. Do not breathe the vapors.

LN_2, *see* **Liquid nitrogen**

Magnesium chloride, $MgCl_2$, may be harmful by inhalation, ingestion, or skin absorption. Wear appropriate gloves and safety glasses and use in a chemical fume hood.

Magnesium sulfate, $MgSO_4$, may be harmful by inhalation, ingestion, or skin absorption. Wear appropriate gloves and safety glasses and use in a chemical fume hood.

Maleimide is extremely harmful and may be fatal by inhalation, ingestion, or skin absorption. Do not breathe the dust. Wear appropriate gloves and safety goggles and use in a chemical fume hood.

MeOH or H_3COH, *see* **Methanol**

β-Mercaptoethanol (2-Mercaptoethanol), $HOCH_2CH_2$-SH, may be fatal if inhaled or absorbed through the skin and is harmful if ingested. High concentrations are extremely destructive to the mucous membranes, upper respiratory tract, skin, and eyes. β-Mercaptoethanol has a very foul odor. Wear appropriate gloves and safety glasses and always use in a chemical fume hood.

Methanol, MeOH or H_3COH, is toxic and can cause blindness. It may be harmful by inhalation, ingestion, or skin absorption. Adequate ventilation is necessary to limit exposure to vapors. Avoid inhaling these vapors. Wear appropriate gloves and safety goggles and use only in a chemical fume hood.

Methyl cyanide, *see* **Acetonitrile**

Methylene chloride, *see* **Dichloromethane**

$MgCl_2$, *see* **Magnesium chloride**

$MgSO_4$, *see* **Magnesium sulfate**

Mifepristone is toxic and may cause harm to the unborn child. It may impair fertility and damage the female reproductive system. Wear appropriate gloves and safety goggles. Do not breathe the dust.

$NaBH_4$, *see* **Sodium borohydride**

Na_2HPO_4, *see* **Sodium hydrogen phosphate**

NaH_2PO_4, *see* **Sodium dihydrogen phosphate**

NaN_3, *see* **Sodium azide**

NaOAc, *see* **Sodium acetate**

NaOCl, *see* **Bleach**

NaOH, *see* **Sodium hydroxide**

NH_4Ac, *see* **Ammonium acetate**

$(NH_4)_2SO_4$, *see* **Ammonium sulfate**

Nitrogen (gaseous or liquid) may be harmful by inhalation, ingestion, or skin absorption. Wear appropriate gloves and safety glasses. Consult your local safety office for proper precautions.

OsO_4, *see* **Osmium tetroxide**

Osmium tetroxide (osmic acid), OsO_4, is highly toxic if inhaled, ingested, or absorbed through the skin. Vapors can react with corneal tissues and cause blindness. There is a possible risk of irreversible effects. Wear appropriate gloves and safety goggles and always use in a chemical fume hood. Do not breathe the vapors.

^{32}P, *see* **Radioactive substances**

Pancreatic endonuclease I may be harmful by inhalation, ingestion, or skin absorption. Wear appropriate gloves and safety goggles.

Paraformaldehyde is highly toxic and may be fatal. It may be a carcinogen. It is readily absorbed through the skin and is extremely destructive to the skin, eyes, mucous membranes, and upper respiratory tract. Avoid breathing the dust or vapor. Wear appropriate gloves and safety glasses and use in a chemical fume hood. Keep away from heat, sparks, and open flame.

PEG, *see* **Polyethyleneglycol**

PEI, *see* **Polyethylenimine**

Penicillin G (Procaine salt) may cause allergic respiratory and skin reactions and may be harmful by inhalation, ingestion, or skin absorption. Wear appropriate gloves. Do not breathe the dust.

Phenazine methosulfate (PMS) is an irritant and may be harmful by inhalation, ingestion, or skin absorption. Wear appropriate gloves and safety glasses.

Phenol is extremely toxic, highly corrosive, and can cause severe burns. It may be harmful by inhalation, ingestion, or skin absorption. Wear appropriate gloves, goggles, protective clothing, and always use in a chemical fume hood. Rinse any areas of skin that come in contact with phenol with a large volume of water and wash with soap and water; do not use ethanol!

Phenylmethylsulfonyl fluoride (PMSF), $C_7H_7FO_2S$, is a highly toxic cholinesterase inhibitor. It is extremely destructive to the mucous membranes of the respiratory tract, eyes, and skin. It may be fatal by inhalation, ingestion, or skin absorption. Wear appropriate gloves and safety glasses and always use in a chemical fume hood. In case of contact, immediately flush eyes or skin with copious amounts of water and discard contaminated clothing.

Phosphotungstic acid can cause severe irritation and may be fatal. It is harmful by inhalation, ingestion, or skin absorption. Wear appropriate gloves and safety goggles. Do not breathe the dust.

PicoGreen contains **Dimethylsulfoxide** (**DMSO**); see **Dimethylsulfoxide** (**DMSO**)

PMS, *see* **Phenazine methosulfate**

PMSF, *see* **Phenylmethylsulfonyl fluoride**

Polyethyleneglycol (**PEG**) may be harmful by inhalation, ingestion, or skin absorption. Wear appropriate gloves and safety glasses. Do not breathe the vapor.

Polyethylenimine (**PEI**) may cause eye and skin burns. It may be harmful by inhalation, ingestion, or skin absorption. Wear appropriate gloves and safety goggles.

Polyvinyl alcohol may be harmful by inhalation, ingestion, or skin absorption. Wear appropriate gloves and safety glasses.

Potassium chloride, **KCl**, may be harmful by inhalation, ingestion, or skin absorption. Wear appropriate gloves and safety glasses.

Potassium ferricyanide, $K_3Fe(CN)_6$, may be fatal by inhalation, ingestion, or skin absorption. Wear appropriate gloves and safety glasses and always use with extreme care in a chemical fume hood. Keep away from strong acids.

Potassium ferrocyanide, $K_4Fe(CN)_6 \cdot 3H_2O$, may be fatal by inhalation, ingestion, or skin absorption. Wear appropriate gloves and safety glasses and always use with extreme care in a chemical fume hood. Keep away from strong acids.

Potassium phosphate, K_2HPO_4/KH_2PO_4, may be harmful by inhalation, ingestion, or skin absorption. Wear appropriate gloves and safety glasses. Do not breathe the dust. *$K_2HPO_4 \cdot 3H_2O$ is dibasic and KH_2PO_4 is monobasic.*

Proteinase K is an irritant and may be harmful by inhalation, ingestion, or skin absorption. Wear appropriate gloves and safety glasses.

Puromycin is toxic and may be carcinogenic. It may be harmful by inhalation, ingestion, or skin absorption. Wear appropriate gloves and safety glasses.

PVP, *see* **Polyvinylpyrrolidone**

Radioactive substances: When planning an experiment that involves the use of radioactivity, include the physicochemical properties of the isotope (half-life, emission type, and energy), the chemical form of the radioactivity, its radioactive concentration (specific activity), total amount, and its chemical concentration. Order and use only as much as really needed. Always wear appropriate gloves, lab coat, and safety goggles when handling radioactive material. **X rays** and **gamma rays** are electromagnetic waves of very short wavelengths either generated by technical devices or emitted by radioactive materials. They may be emitted isotropically from the source or may be focused into a beam. Their potential dangers depend on the time period of exposure, the intensity experienced, and the wavelengths used. Be aware that appropriate shielding is usually of lead or other similar material. The thickness of the shielding is determined by the energy(s) of the X rays or gamma rays. Consult the local safety office for further guidance in the appropriate use and disposal of radioactive materials. Always monitor thoroughly after using radioisotopes. A convenient calculator to perform routine radioactivity calculations can be found at http://graphpad.com/quickcalcs/index.cfm/.

RNase A is an irritant and may be harmful by inhalation, ingestion, or skin absorption. Wear appropriate gloves and safety glasses. Do not breathe the dust.

SDS, *see* **Sodium dodecyl sulfate**

Sigmacote, *see* **Silane**

Silane is extremely flammable and corrosive. It may be harmful by inhalation, ingestion, or skin absorption. Keep away from heat, sparks, and open flame. The vapor is irritating to the eyes, skin, mucous membranes, and upper respiratory tract. Wear appropriate gloves and safety goggles and always use in a chemical fume hood.

Sodium acetate (**NaOAc**), *see* **Acetic acid**

Sodium azide, NaN_3, is highly poisonous. It blocks the cytochrome electron transport system. Solutions containing sodium azide should be clearly marked. It may be harmful by inhalation, ingestion, or skin absorption. Wear appropriate gloves and safety goggles and handle with great care in a chemical fume hood. Sodium azide is an oxidizing agent and should not be stored near flammable chemicals.

Sodium borohydride, $NaBH_4$, is corrosive and causes burns. It may be harmful by inhalation, ingestion, or skin absorption. Wear appropriate gloves and safety goggles and use in a chemical fume hood.

Sodium cacodylate may be carcinogenic and contains arsenic. It is highly toxic and may be fatal by inhalation, ingestion, or skin absorption. It also may cause harm to the unborn child. Effects of contact or inhalation may be delayed. Do not breathe the dust. Wear appropriate gloves and safety goggles and use only in a chemical fume hood. *See also* **Cacodylate.**

Sodium citrate, *see* **Citric acid**

Sodium deoxycholate is irritating to mucous membranes and the respiratory tract and may be harmful by inhalation, ingestion, or skin absorption. Wear appropriate gloves and safety glasses when handling the powder. Do not breathe the dust.

Sodium deoxycholic acid, *see* **Sodium deoxycholate**

Sodium dihydrogen phosphate, NaH_2PO_4 **(sodium phosphate, monobasic),** may be harmful by inhalation, ingestion, or skin absorption. Wear appropriate gloves and safety glasses and use in a chemical fume hood.

odium dodecyl sulfate (SDS) is toxic, an irritant, and poses a risk of severe damage to the eyes. It may be harmful by inhalation, ingestion, or skin absorption. Wear appropriate gloves and safety goggles. Do not breathe the dust.

Sodium hydrogen phosphate, Na_2HPO_4 **(sodium phosphate, dibasic),** may be harmful by inhalation, ingestion, or skin absorption. Wear appropriate gloves and safety glasses and use in a chemical fume hood.

Sodium hydroxide, NaOH, and **solutions containing NaOH** are highly toxic and caustic and should be handled with great care. Wear appropriate gloves and a face mask. All other concentrated bases should be handled in a similar manner.

Sodium hypochlorite, NaOCl, *see* **Bleach**

Spurr's resin contains chemicals that are carcinogens and are toxic. It may be harmful by inhalation, ingestion, or skin absorption. Wear gloves and safety glasses and use in a chemical fume hood.

Streptomycin is toxic and a suspected carcinogen and mutagen. It may cause allergic reactions. It may be harmful by inhalation, ingestion, or skin absorption. Wear appropriate gloves and safety glasses.

Sulfuric acid, H2SO4, is highly toxic and extremely destructive to tissue of the mucous membranes and upper respiratory tract, eyes, and skin. It causes burns, and contact with other materials (e.g., paper) may cause fire. Wear appropriate gloves, safety glasses, and lab coat and use in a chemical fume hood.

Tetracycline may be harmful by inhalation, ingestion, or skin absorption. Wear appropriate gloves and safety glasses and use in a chemical fume hood.

TFA, *see* **Trifluoroacetic acid**

Triethylamine is highly toxic and flammable. It is extremely corrosive to the mucous membranes, upper respiratory tract, eyes, and skin. It may be harmful by inhalation, ingestion, or skin absorption. Wear appropriate gloves and safety glasses and use in a chemical fume hood. Keep away from heat, sparks, and open flame.

Trifluoroacetic acid (TFA) (concentrated) may be harmful by inhalation, ingestion, or skin absorption. Concentrated acids must be handled with great care. Decomposition causes toxic fumes. Wear appropriate gloves and a face mask and use in a chemical fume hood.

Tris may be harmful by inhalation, ingestion, or skin absorption. Wear appropriate gloves and safety glasses.

Trisodium citrate, *see* **Citric acid**

Triton X-100 causes severe eye irritation and burns. It may be harmful by inhalation, ingestion, or skin absorption. Wear appropriate gloves and safety goggles. Do not breathe the vapor.

Trypsin may cause an allergic respiratory reaction. It may be harmful by inhalation, ingestion, or skin absorption. Do not breathe the dust. Wear appropriate gloves and safety goggles. Use with adequate ventilation.

Urea may be harmful by inhalation, ingestion, or skin absorption. Wear appropriate gloves and safety glasses.

X-gal may be toxic to the eyes and skin. Observe general cautions when handling the powder. Note that stock solutions of X-gal are prepared in DMF, an organic solvent. For details, see ***N,N*-dimethylformamide (DMF).** *See also* **5-Bromo-4-chloro-3-indolyl-β-D-galactopyranoside (BCIG).**

Index

A

C

D

H

I

J

K

L

M

Q

R

S

T

U

V

W

X

Y

Z